Twort's Water Supply

Twort's Water Supply

Seventh Edition

Malcolm J. Brandt
BSc, FICE, FCIWEM, MIWater

K. Michael Johnson
BSc, FICE, FCIWEM

Andrew J. Elphinston
BSc, MIChemE, MCIWEM

Don D. Ratnayaka
BSc, DIC, MSc, FIChemE, FCIWEM

AMSTERDAM • BOSTON • HEIDELBERG • LONDON
NEW YORK • OXFORD • PARIS • SAN DIEGO
SAN FRANCISCO • SINGAPORE • SYDNEY • TOKYO

Butterworth-Heinemann is an imprint of Elsevier

Butterworth-Heinemann is an imprint of Elsevier
The Boulevard, Langford Lane, Kidlington, Oxford OX5 1GB, United Kingdom
50 Hampshire Street, 5th Floor, Cambridge, MA 02139, United States

Notices
Knowledge and best practice in this field are constantly changing. As new research and experience broaden our understanding, changes in research methods, professional practices, or medical treatment may become necessary.

Practitioners and researchers must always rely on their own experience and knowledge in evaluating and using any information, methods, compounds, or experiments described herein. In using such information or methods they should be mindful of their own safety and the safety of others, including parties for whom they have a professional responsibility.

To the fullest extent of the law, neither the Publisher nor the authors, contributors, or editors, assume any liability for any injury and/or damage to persons or property as a matter of products liability, negligence or otherwise, or from any use or operation of any methods, products, instructions, or ideas contained in the material herein.

British Library Cataloguing-in-Publication Data
A catalogue record for this book is available from the British Library

Library of Congress Cataloging-in-Publication Data
A catalog record for this book is available from the Library of Congress

ISBN: 978-0-08-100025-0

For Information on all Butterworth-Heinemann publications
visit our website at https://www.elsevier.com

Publisher: Joe Hayton
Acquisition Editor: Ken McCombs
Editorial Project Manager: Peter Jardim
Production Project Manager: Kiruthika Govindaraju
Designer: Greg Harris

Typeset by MPS Limited, Chennai, India

Cover images:

Black Esk Dam
Dumphries & Galloway, Scotland.
Raising of the Black Esk earthfill dam to increase storage by 40%, works designed by Black & Veatch. The work, completed in 2013, raises the overflow level by 2.5 m and involved the innovative application of precast concrete sections in "piano-key" form to raise the crest of the existing bellmouth spillway and to accommodate the peak design flood outflow of 183 m^3/s, with a flood surcharge of 0.97 m.

Photographs by courtesy of Scottish Water.

Contents

Authors' Biographic Details

The four authors, with combined experience of over 160 years, working for Binnie & Partners and then Black & Veatch, have written this text with the practicing water supply engineer in mind. However, the book will also be essential reading for students of water engineering.

Malcolm J. Brandt
Global Practice Technology Leader
Civil engineer specializing in water supply planning and distribution system management; responsibilities include research in the United Kingdom and overseas, as well as project execution in the United Kingdom, Europe, Asia, Middle East, and North and South America.

K. Michael Johnson
Technical Director
Civil engineer for design of large and small works, such as dams, spillways, intakes, pipelines, tunnels, pumping stations, conventional water treatment, reverse osmosis, and water storage in the United Kingdom, South and Central America, Middle East, and Far East.

Andrew J. Elphinston
Technical Director
Chemical engineer specializing in potable water treatment, including both conventional and advanced treatment processes, with experience in the United Kingdom, North and Southern Africa, and Far East.

Don D. Ratnayaka
Global Practice Technology Leader
Chemical engineer with extensive specialist knowledge of water treatment techniques; responsible for all aspects of water treatment process design for projects in the United Kingdom, Europe, Africa, Asian subcontinent, Middle East, Far East, and Australia.

Foreword

Twort's Water Supply (and its predecessor—*Water Supply*) has been the first reference for many water engineers in the United Kingdom and internationally for over 50 years. In an age when a huge amount of information is accessible on the internet, some of unknown provenance, it is invaluable to the engineer engaged in water supply, and to students of the subject, to have a concise source of reliable guidance to hand. With the help of specialist contributors from the United Kingdom and the United States, the authors have managed to convey a large amount of information in a single volume. They have ensured, by means of extensive cross references, that the attention of the reader is drawn to all significant related matters, across all the disciplines involved in engineering new water supply facilities or in getting the most out of existing ones.

In the last decade the strains of rapid industrialization, rising population, and conflict are making themselves increasingly felt on basic infrastructure in many parts of the world, including water supplies and sanitation facilities. For huge numbers of people the provision and maintenance of even a basic supply of safe water seems to be an endless struggle. Meanwhile, the focus of attention globally, but particularly in post-industrialized nations, has moved to the quality of water in river basins, to more effective use of resources and assets, and to the treatment of the increasing number of pollutants of concern.

This Seventh Edition has been thoroughly updated to reflect current issues. It brings in two new chapters to allow extensive new material to be included on waste and energy efficiency and on water intakes, and for those needing further information, the references have been updated and increased in number. It is confidently expected that this new edition, like its predecessors, will prove to be an essential reference for those involved in the education and practice of water supply systems.

Nigel J.D. Graham MA (Cantab), MSc, PhD, ScD (Cantab), FICE, FIChemE, FCIWEM
Department of Civil and Environmental Engineering,
Imperial College, London
October 2015

Preface

Since the first edition of the book was published in 1963, *Water Supply* has provided the practicing water engineer with a practical treatise on all aspects of the water supply in one book. The readership of more recent editions has broadened to include students of engineering and it is now used as a standard text in many universities as well as a reference manual for water industry professionals worldwide.

The Sixth Edition recognized the valuable contribution of Alan Twort, the lead author and editor of the first five editions, by including his name in the title *Twort's Water Supply.* We are pleased to continue to honor Alan's valuable contribution to water engineering in the same way in this edition.

The Seventh Edition benefits from the extensive experience of nearly 30 UK and US specialists from across the water industry. The content has been expanded, revised, and updated to reflect the significant changes in the industry including international regulation, recognition of performance with respect to the customer, conservation, and the efficient use of resources and technology. Practice in the United States, the United Kingdom, Europe, and worldwide is included. In addition, the index has been considerably enlarged to improve accessibility.

The book is structured to reflect the process of water supply from understanding demand for water, statutory, regulatory, water quality, financing, and economic aspects of the provision of wholesome water to the consumer, and then through the water cycle from abstraction through treatment to storage and distribution. The content addresses hydraulic and system design, pipelines, valves, pumps and other MEICA plant, dams, intakes and waste treatment, and energy demand reduction.

The text gives updated standards for drinking water and describes in detail the significance of the large number of chemicals and organisms, which now present, in raw waters, a potential hazard to human health. Conventional and specialized treatment processes, disinfection, and waste treatment are covered in five separate chapters. A new chapter brings together the storage and dosing of treatment chemicals and sampling. Developing technologies, such as membrane filtration and advanced treatment methods for micropollutants, are also dealt with. Desalination by reverse osmosis and thermal processes is covered. A new chapter on intakes has allowed major expansion on this subject and coverage of pipelines has been increased.

The authors are grateful to the many contributors and reviewers who have aided the production of this Seventh Edition and to Black & Veatch for permitting use of their work.

Malcolm J. Brandt
K. Michael Johnson
Andrew J. Elphinston
Don D. Ratnayaka

Abbreviations for Organizations

ADB	Asian Development Bank, Manila, Philippines
API	American Petroleum Institute, Washington, DC, USA
ASCE	American Society of Civil Engineers, Reston, VA, USA
ASME	American Society of Mechanical Engineers, New York, USA
AWWA	American Waterworks Association, Denver, USA
AwwaRF	American Waterworks Association Research Foundation, Denver, USA
BDS	British Dam Society, London, UK
BGS	British Geological Survey, Nottingham, UK
BHRA	British Hydromechanics Group, Cranfield, UK
BRE	Building Research Establishment, Watford, UK
BSI	British Standards Institution, London, UK
BW	British Water, London, UK
CIRIA	Construction Industry Research & Information Association, London, UK
CIWEM	Chartered Institution of Water & Environmental Management, London, UK
CRC	Chemical Rubber Company, FL, USA
CEH	Centre for Ecology & Hydrology (previously IoH), Wallingford, UK
CEU	Council of the European Union, Brussels, Belgium
Defra	Department for Environment, Food and Rural Affairs, London, UK
DETR	Department of Environment, Transport and Regions, London, UK
DFID	Department for International Development, London, UK
DHSS	Department of Health and Social Services, London, UK
DoE	Department of the Environment, London, UK
DWI	Drinking Water Inspectorate, London, UK
EA	The Environment Agency, Bristol, UK
EC	European Community (now European Union)
EEA	European Environment Agency, Copenhagen, Denmark
FAO	Food and Agriculture Organization, Rome, Italy
FWR	Foundation for Water Research, Marlow, UK
HMSO	Her Majesty's Stationery Office (now TSO), London, UK
HSE	Health and Safety Executive, London, UK
IAHS	International Association of Hydrological Sciences, Wallingford, UK
IAM	Institute of Asset Management, Bristol, UK
IChemE	Institution of Chemical Engineers, Rugby, UK
ICOLD	International Committee of Large Dams, Paris, France
IEC	International Electrotechnical Commission, Geneva, Switzerland
IET	Institute of Engineering Technology, London, UK
IoH	Institute of Hydrology, Wallingford, UK
IMechE	Institute of Mechanical Engineers, London, UK
ISO	International Organisation for Standardisation, Geneva, Switzerland
IUVA	International Ultraviolet Association, Ontario, Canada

IWA/IWSA	International Water Association, The Hague, Netherlands
IWEM/IWES	Institution of Water Engineers (& Scientists), London, UK
NEWA	New England Waterworks Association, Massachusetts, USA
NERC	Natural Environmental Research Council, Swindon, UK
NFPA	National Fire Protection Agency, Massachusetts, USA
NFPA	National Fire Protection Association, Ascot, UK
Ofwat/WSRA	Water Services Regulation Authority, Birmingham, UK (Term Ofwat still used)
PWC	Portsmouth Water Company, Portsmouth, UK
SEPA	Scottish Environmental Protection Agency, Stirling, Scotland, UK
STW	Severn Trent Water, Coventry, UK
SWTE	Society for Water Treatment and Examination, London, UK
TSO	The Stationery Office (previously HMSO), London, UK
TWU	Thames Water Utilities, Reading, UK
UKWIR	UK Water Industry Research Ltd., London, UK
UN	United Nations, New York, USA
UNESCO	UN Educational, Scientific and Cultural Organisation, Paris, France
UNICEF	UN Children's Fund, New York, USA
USACE	US Army Corps of Engineers, Washington, DC, USA
USBR	United States Bureau of Reclamation, Washington, DC, USA
US EPA	US Environmental Protection Agency, USA
USGS	United States Geological Survey, Washington, USA
WEDC	Water Education Development Centre, Loughborough, UK
WHO	World Health Organization, Geneva, Switzerland
WMO	World Meteorological Organization, Geneva, Switzerland
WRc	Water Research Centre, Medmenham, UK
WSA	Water Services Association, London, UK
WSRT—Aqua	Water Supply: Research Technology—Aqua, IWA, The Hague, Netherlands

Contributing Authors, Reviewers and Advisors

CONTRIBUTING AUTHORS

Black & Veatch

Ian Barker BSc, FIET	Chief Engineer (Electrical)
Dipankar Basak BEng, MCIWEM, MICorr	Chief Pipeline Engineer, (now with MWH)
Nicholas Burns BS, MS	Ozone Technology Leader (Ozone)
Peter B. Clark MA, MSc, FICE	Consultant (Hydraulics)
Scott Freeman BS, BA	Membrane Technology Leader (Membrane processes)
John Hall BSc, MCIWEM	Consultant (Hydrology, Surface Supplies and Floods)
Tim A. Holmes BSc	Chief Instrumentation and Control Engineer (I&C)
Robert Hulsey BS, MS	Global Practice Technology Leader - Water (Advanced Oxidation Processes and UV)
Roger A. Middleton MPhil, FIMechE	Consultant (Sustainability and Renewable Energy)
John L. Petrie BSc, MSc, FGS	Consultant (Hydrogeologist, Groundwater Supplies)
Hari R. Shah BSc, MICE, MASCE	Consultant (Water Intakes)
Bryan Townsend BS, MS	UV Technology Leader (Advanced Oxidation Processes and UV)
David J. Winzer BSc, MSc, MICE	Team Leader, Pipelines and Outfalls

Others

Simon Cole BSc, Dphil, MCIWEM,	Microbiology & Public Health Manager Wessex Water (Microbiology)
Sharon Evans MSc, MRSC, MBA	Head of Water Quality, Dwr Cymru Welsh Water (Water Quality Issues and Standards)
Vikki Williams BSc, MCIWEM, MIAM	Principal Consultant – Energy and Utilities, PA Consulting Group

REVIEWERS AND ADVISORS

John Ackers BSc, FICE	Black & Veatch
Michelle Ashford BEng, MCIWEM, MIWater	Bristol Water
Terry Heard BSc, MRSC, MCIWEM	Consultant
Stephen Murray BSc, Eng, MICE	Black & Veatch
James Ostrowski BSc, MIChemE, MCIWEM	Black & Veatch
Alvin J. Smith BSc, MSc	Consultant
Claire R. Stacey FRSC, MIWO	Consultant
Sarah Styan BA, MA	Black & Veatch
H. Jane Walbancke BSc, MSc, PhD, FGS, MICE	Black & Veatch

ATTRIBUTION OF ILLUSTRATIONS

The names of the organisations acknowledged in connection with illustrations are those current at the time the illustrations were produced. Where an organisation has since changed its name or has been acquired by another, the new name is given below:

Sir Alexander Binnie, Son & Deacon	Black & Veatch
Binnie & Partners	Black & Veatch
Binnie Black & Veatch	Black & Veatch
Paterson Candy Ltd	Black & Veatch

CHAPTER

The Demand for Potable Water

1

1.1 CATEGORIES OF CONSUMPTION

The demand for potable water is made up of authorized consumption by domestic and non-domestic consumers and water losses. Domestic consumers use water within the household; for drinking, personal hygiene, cooking and cleaning, and outside the dwelling: for cleaning patios, irrigating gardens, filling ponds and swimming pools, and washing cars. Non-domestic consumption comprises industrial, commercial, institutional and agricultural demand legitimately drawn from potable water mains. This category also includes legitimate public use for irrigating public parks and green areas, street cleaning, flushing water mains and sewers and for firefighting.

Water is delivered to premises via service connections, the size of which depends on the demand from the premises. The majority of service connections feed single premises, but some supply a group of adjacent premises or a private development, such as an industrial estate, commercial complex or group of dwellings accessed from a private road.

Commercial and industrial supplies are generally fitted with an operational revenue meter because they represent a major source of income to a water utility. Small shops and offices occupied only in the daytime are also generally metered even though their consumption is small. In many countries where the national government, state or city owns the utility, large quantities of potable water used for watering public parks and green areas and within government offices, military establishments and other institutional buildings are often not metered nor accounted for as revenue.

Domestic revenue meters are widespread although not universal. For example in the USA, mainland Europe and Australia meter penetration is effectively 100%; in England and Wales in 2012 nearly 40% of domestic supplies were metered with individual meter penetration ranging between 10% and 75% for the 22 water companies; while in Scotland and Northern Ireland domestic consumers are not currently metered. However, the underlying issue related to the effectiveness of domestic metering is one of operability not its coverage.

The world population is estimated to be in excess of 7 billion and international organizations estimate that over 1.4 billion people live in extreme poverty; nearly 900 million people live in

Twort's Water Supply. DOI: http://dx.doi.org/10.1016/B978-0-08-100025-0.00001-6

slum dwellings and over 700 million do not have access to an improved drinking water supply (WB, 2013; WHO, 2014; UNESCAP, 2013). Accordingly, not all consumers are supplied through a domestic service connection. In many cities, more typically in Asia and Africa, supplies are not 24-hour, do not reach 100% of the urban population or the pressure is so low that many consumers receive an intermittent supply. The Asian Development Bank (ADB) survey of 2005 (ADB, 2007a) showed that nine of 40 Asian water utilities surveyed did not have a 24-hour supply and that only eight of the utilities supplied 100% of the population in their service area, a connection being either in-house or a communal yard tap or standpipe.

A survey by the ADB published in 1997 (ADB, 1997) reported that of 27 cities in Asia with populations over 1 million, 15 were fully metered but six metered less than 7% of their service connections (Kolkata 0%, Karachi 1%). Further surveys by the ADB in 2005 and 2006 (ADB, 2007a, b) also highlighted that of 44 water utilities in Asia only 14 provided 100% individual service connections; in the remainder communal connections ranged between 2% and 70% of the population served.

Consumers who do not have direct access to a drinking water supply rely on collecting water from standpipes and public taps located in the street, on water tanker deliveries or on collecting water from an uncontrolled alternative source. Unmetered standpipe supplies to urban slums and rural communities are usually given free.

Water losses comprise the leakage and wastage from the distribution network.

1.2 LEVELS OF TOTAL CONSUMPTION

The usual measure of total consumption is the amount supplied from sources per head of population. However, in many cases the population served is not known accurately. In large cities there may be thousands of commuters coming in daily from outside; in holiday areas the population may double for part of the year. Other factors having a major influence on consumption figures are:

- sufficiency of supplies and pressures to meet the demand, 24-hour or intermittent;
- population served by standpipes and tankers;
- population with waterborne sanitation;
- efficiency in metering and billing and in controlling leakage and wastage;
- proportion of supply going to relatively few large industrial consumers;
- the climate.

The supply situation described in Section 1.1 and the factors listed above mean that comparing average total consumption between utilities is not informative; e.g. high consumption could result from large industrial demand, whereas low consumption could be due to a shortage of resources. However, the general range of total supplies per capita is:

- from 500 to 900 lcd (litres per capita per day) in the big industrial cities;
- from 200 to 500 lcd for many major cities and urban areas throughout the world;
- from 75 to 150 lcd in areas where supplies are restricted, where there are many street standpipes or where much of the population has private wells.

1.3 DOMESTIC DEMAND

As for total consumption data, domestic consumption figures reported by various countries are not necessarily comparable, mainly because there is no assurance that the figures quoted are produced on the same basis. However, data for 84 cities are presented in Table 1.1. Average domestic plus small trade consumption reported for 20 European countries for 2010 and 2012 (IWA, 2014) centred about 130–160 lcd in a range of 61–360 lcd, with Marseille in France having the highest reported consumption and three other listed cities recording over 200 lcd. For 12 countries in Asia consumption was generally between 200 and 250 lcd in a range of 40–427 lcd, the highest consumption being in Darwin, Australia and the lowest in Hetauda, Nepal; 11 cities reported consumption over 250 lcd. In the USA the typical in-house consumption (excluding cooling water) is 180–230 lcd, but the total household and small trade consumption can be significantly higher as the table suggests.

In-house domestic consumption is influenced by many factors including the class of dwelling, number of people in the household, changes in household income, ablution habits, culture, religion, differences in climate including seasonal variations, and the number and capacity of water fittings installed. The effect of such factors is illustrated in Table 1.2 by the range of metered and unmetered household consumptions for the water-only companies in England and Wales; the reported higher domestic consumption reflecting the fact that 77% of the population that they serve resides in the more affluent southeastern parts of England. The lower figures for metered consumption are not representative because metering is optional for most householders in the UK and those choosing to have a metered supply tend to be those expecting to pay less for their supply because of their low consumption, e.g. single occupants or elderly retired people. This is reflected in the household size, the 'occupancy ratio'.

When estimating consumption for a whole area of a distribution system, it is the average occupancy which is important. In England and Wales average occupancy according to census figures (Census, 2012) showed a decline from 2.70 in 1981 to 2.36 in 2011 and is expected to continue to fall slowly. The 2001 figures varied from 2.10–2.20 in retirement areas to 2.30–2.60 in the more dense populated urban areas. In the USA mean household occupancy reported by the US Census Bureau (US Census, 2012) declined from 2.63 people in 1990 to 2.34 in 2010. In many countries in Africa, the Indian sub-continent and in Southeast Asia, etc. the average occupancy is five or six people per household. Table 1.3 shows the influence of occupancy ratio on household demand in selected parts of England; smaller households have proportionately larger per capita usage.

Water usage varies both in quantity and timing during the week, the pattern for working days being relatively consistent with morning and evening peaks coinciding with leaving for and returning from work and school. However, at weekends demand tends to increase with diurnal peaks often later than on working days and of greater magnitude. Cultural and religious characteristics of a supply area and religious days and public holidays can also influence weekly and seasonal demand patterns. For example during Ramadan, the Muslim holy month of fasting, domestic per capita demand increases and the diurnal pattern switches from day to night with peaks coinciding with sunrise and sunset.

Table 1.1 Domestic and small business per capita consumption for countries and cities with reference year (litres per capita per day; lcd)

Argentina, Buenos Aires, 2010[1]	261	Gabon, 2010[1]	143	New Zealand, Whangarei 2012[1]	179
Australia, Darwin, 2010[1]	425	Hungary, Budapest, 2012[1]	150	Norway, Oslo, 2012[1]	160
Australia, Perth, 2010[1]	287	Hungary, Miskolc, 2012[1]	80	Poland, Wroclaw, 2012[1]	116
Australia, Sydney, 2010[1]	200	India, Jamshedpur, 2007[3]	302	Poland, Bialystok, 2012[1]	91
Australia, Melbourne, 2010[1]	148	India, Mumbai, 2007[3]	191	Portugal, Lisbon, 2012[1]	173
Austria, Vienna, 2012[1]	143	India, Bhopal, 2007[3]	72	Portugal, Porto, 2012[1]	70
Bangladesh, Dhaka, 2007[2]	140	Indonesia, Bandar Aceh, 2007[2]	90	Romania, Suceava, 2012[1]	125
Belgium, Antwerp, 2012[1]	148	Iran, Tehran, 2012[1]	260	Romania, Satu Mare, 2012[1]	80
Belgium, Brussels, 2012[1]	95	Iran, Esfahan, 2012[1]	162	Singapore 2010[1]	154
Brazil, Brasilia Federal, 2012[1]	187	Italy, Rome, 2012[1]	214	Slovakia, Bratislava, 2010[1]	98
Brazil, San Paulo, 2012[1]	176	Italy, Bologna, 2012[1]	147	South Korea, Incheon, 2012[1]	296
Bulgaria, Sofia, 2012[1]	134	Japan, Yokohama, 2012[1]	288	South Korea, Seoul, 2012[1]	286
Bulgaria, Razgrad, 2012[1]	77	Japan, Tokyo, 2012[1]	225	South Korea, Ulsan, 2012[1]	252
Canada, Richmond 2010[1]	240	Japan, Sapporo 2012[1]	202	Spain, Madrid, 2003[1]	124
Canada, Winnipeg 2012[1]	172	Lithuania, Vilnius, 2012[1]	80	Spain, Barcelona, 2012[1]	105
China, Shenzhen, 2012[1]	319	Lithuania, Siauliai, 2012[1]	61	Sri Lanka, Colombo, 2007[2]	120
China, Gwangzhou, 2012[1]	252	Malawi, Blantyre, 2010[1]	75	Sweden, Linköping, 2012[1]	240
China, Hong Kong, 2012[1]	209	Mauritius, 2012[1]	173	Sweden, Stockholm, 2012[1]	198
China, Shanghai, 2012[1]	200	Mexico, Monterrey, 2012[1]	301	Switzerland, Basel 2012[1]	207
China, Macao, 2012[1]	194	Mexico, Guadalajara, 2012[1]	268	Switzerland, Zurich 2012[1]	153
China, Beijing, 2012[1]	150	Mexico, San Luis Potosi, 2012[1]	182	UK, London, 2012[1]	163
Chinese Tiawan, Taipei, 2012[1]	340	Morocco, Marrakech, 2010[1]	130	UK, Manchester, 2012[1]	134
Cyprus, Larnaca, 20121	153	Morocco, Casablanca 2010[1]	95	UK, Birmingham, 2012[1]	126
Denmark, Copenhagen, 2012[1]	104	Nepal, Kathmandu, 2007[2]	90	USA, Miami, 2012[1]	526
Finland, Helsinki, 2012[1]	184	Nepal, Hetauda, 2007[2]	40	USA, New York, 2012[1]	469
France, Marseille, 2012[1]	360	Netherlands, Amsterdam, 2012[1]	130	USA, Denver, 2012[1]	322
France, Paris, 2012[1]	120	New Zealand, Tauranga 2012[1]	270	Viet Nam, Ho Chi Min, 2007[2]	150
France, Lille, 2012[1]	98	New Zealand, Wellington 2012[1]	230	Viet Nam, Song Cau, 2007[2]	135

Sources of information: [1]IWA, 2014; [2]ADB, 2008; [3]ADB, 2007b.

Table 1.2 Domestic consumption in England & Wales, 2009/10. Reported average consumption excluding supply pipe leakage (lcd) and range

	10 Regional water and sewerage companies	12 Water-only companies
Population in households	44.83 million	10.19 million
Unmetered households	152 (130–171)	168 (127–183)
Metered households	125 (110–146)	141 (113–156)
(Percent households metered)	(37.0%)	(37.6%)
(Percent population metered)	(32.6%)	(32.7%)
(Occupancy ratio unmetered households)	(2.6)	(2.7)
(Occupancy ratio metered households)	(2.1)	(2.1)
Mean per capita consumption	143.2	158.7
All companies mean per capita consumption	146.1	
All companies average occupancy ratio	2.4	

Sources of information: Ofwat, 2010.

Table 1.3 Domestic consumption per capita (pcc) by household size (lcd)

Number in household		1	2	3	4	5	6
Portsmouth Water Company[1]	lcd	252	183	140	130	108	101
% of company average pcc	%	156	113	87	80	67	63
Thames Water Utilities[2]	lcd	215	184	159	141	135	123
% of company average pcc	%	131	112	97	86	82	75
Severn Trent Water[3]	lcd	189	153	122	117	109 (5+)	
% of company average pcc	%	139	113	90	86	80	
CCWater – average water use[4]	lcd	181	151	124	113	100	91
% of 'average water use' pcc	%	121	100	83	75	66	61

Sources of information: [1]PWC, 2005; [2]TWU, 2007; [3]STW, 2007; [4]CCWater, 2014.

Components of Domestic In-House Consumption

In-house water usage is for drinking, personal hygiene, WC flushing, showers, baths and hand basins, cooking, cleaning and laundry, including washing machines and dishwashers. Water consumption in developed countries has increased as a consequence of trends to install dishwashers and modern bathrooms including fitting high-capacity power showers. Conversely during the last decade there has been increased awareness in some sectors of society of the need to conserve water, resulting in a worldwide trend for water utilities and environmental agencies to promote water conservation and encourage the householder to install lower flow, smaller capacity and more efficient fittings.

Table 1.4 presents in-house domestic consumption in litres per person per day from detailed studies. Again the figures are not strictly comparable because they are obtained by different monitoring methods and small sample sets. In the 'diary' method for monitoring household consumption, residents record each day the number of times each fitting is used over a period. The frequencies of use for each type of fitting are multiplied by the average consumption for the fitting, determined from separate measurements, to derive the household consumption. In the 'data logging' method, the flow records from pulsed output consumer meters recording at frequent time intervals are analysed; the different uses being identified by the flow pattern. In a few instances, utilities have installed meters at every point of use in a property. The table shows that, with the exception of a few anomalous total in-house figures, per capita consumption is relatively consistent across the reported figures, the higher figures tending to represent older housing stock before modernization.

Table 1.4 Domestic household consumption by usage excluding supply pipe leakage (lcd)

Location	Internal total[a]	WC flush	Personal hygiene	Kitchen uses[b]	Laundry[c]
Australia, Perth – single residence, 2003[1]	155	33	51	29	42
Australia, Perth – multiple residence, 2003[1]	166	28	55	40	43
Australia, Sydney[2]	184	45	66	24	49
Singapore, 2004[3]	162	26	64	36	31
Thailand, Bangkok[2]	169	27	68	29	45
United Kingdom, Thames, metered h/hold, 2006[4]	140	39	71	10	21
United Kingdom, Thames, unmetered h/hold, 2006[4]	153	46	77	11	21
United Kingdom, Severn Trent, metered h/hold, 2007[5]	121	41	40	21	19
United Kingdom, Severn Trent, unmetered h/hold, 2007[5]	132	45	47	23	18
United States, Seattle, no conservation, 2000[6]	240	71	48	65	56
United States, Seattle, high efficiency, 2000[6]	151	30	43	43	35
United States, Study, no conservation, 1997[7]	220	69	51	43	56
United States, Study, with conservation, 1997[7]	164	39	42	42	40
United States, Westminster, pre 1977 houses, 1995[7]	235	70	53	56	56
United States, Westminster, post 1988 houses, 1995[7]	179	53	53	23	49

Sources of information: [1]Water Corp, 2003; [2]White, 2004; [3]PUB, 2005; [4]TWU, 2007; [5]STW, 2007; [6]Aquacraft, 2000; [7]AwwaRF, 1999.
Notes: [a]Adjusted for net of premise leakage and excludes rounding of component figures.
[b]Includes consumption and dishwasher use.
[c]Includes washing machine use.

Table 1.4 also shows that toilet flushing represents the single largest use of water. In the UK, water bylaws introduced in 1994 resulted in cistern capacity being reduced from 9 litres (13 litres in Scotland) to 7.5 litres. The capacity was further reduced for new installations in 2001 to 6 litres. More recently dual flushing toilets with 6/3 litre capacities have become available. The water utilities are also promoting displacement devices, e.g. the 'Hippo the Water Saver', a stiff plastic bag immersed in the cistern that further reduces the flush capacity. The full impact of introducing reduced capacity cisterns on flushing demands will only be realized over about 20 years, the average life of a cistern, as older ones are replaced with the newer specification tanks. There have been similar trends in reducing flow rates and capacities of other in-house appliances including flow restrictors, efficient spray devices on taps and showers and smaller capacity washing machines. Reduced capacity fittings are also being installed in systems and locations where water resources are limited, including Australia, Singapore and in many cities in North America under the US Energy Policy Act 1992. Table 1.5 compares the capacity of fittings in different countries. However, water-efficient fittings are only effective where they are maintained. If an appliance malfunctions or a dripping or a leaking fitting is not maintained, as can be common with flap valve cisterns, the water losses from the defective fittings can represent a significant proportion of the household demand.

Water used in air conditioners and humidifiers is not included in the US figures in Table 1.3. In some parts of western USA, where the climate is hot and arid, evaporative or 'desert' coolers are used; a fan draws air through a vertical porous pad of cellulose fibre, down which water is trickled. Recirculating units consume about 12–15 l/h. Coolers that bleed off part of the surplus water to reduce deposits on the porous pad can use up to 40 l/h. The impact of cooling units on domestic consumption depends on the percentage of dwellings equipped with them and the duration of the hot season. Estimates of the additional quantity, above the annual average daily domestic consumption, attributable to cooling units vary between about 90 lcd, in areas where the summer climate is exceptionally hot and dry, e.g. Arizona where the average temperature in July is 40°C with 29% daytime humidity, and about 5 lcd for coastal areas. In hot and wet climates, as in the tropics, evaporative cooling is less effective and electrical air conditioners are used instead.

Outdoor Domestic Use for Garden Irrigation and Bathing Pools

Outdoor water uses can represent a significant proportion of domestic consumption. In hotter climates in the USA up to 60% of the average annual day demand can be used for irrigation, the demand being sustained throughout most of the year. In Australia 35% of the average annual day demand is for outdoor uses including irrigation and yard cleaning. In Europe it represents about 2% of the total demand (Ofwat, 2007).

In the UK garden watering can increase daily consumption by up to 50% during a prolonged dry period but the total amount used in a year depends on whether the summer is 'dry' or 'wet'. In the north of UK prolonged dry periods are rare. In the drier southeastern part of England, garden watering has been estimated to account for about 5–10 lcd (3–6%) of the total household demand during recent years of prolonged low summer rainfall. However, dry summer periods in the UK are often of relatively short duration and the time-lag between the start of a dry period and the build-up of garden watering demand means that the peak of the latter is short-lived. The demand for garden irrigation is more significant for the seasonal peak than the average annual per capita consumption.

Table 1.5 Typical water appliance usages

	United States	United Kingdom	Singapore[a]			Australia[5,6]
			1 Tick	***2 Tick***	***3 Tick***[a]	
WC toilets – flush size (l/flush)	6[1]	6, but cisterns installed before 1 July 1999 can be replaced by same volume[2] Flushing or pressure flushing cistern[2]	>4.0–4.5	>3.5–4.0	3.5 or less	Efficient = average 3.6 per 5 flush cycles (6/3 capacity) Old = 12
– Low flush			>2.5–3.0	>2.5–3.0	2.5 or less	
– Flush type	Flap-valve[1]	7.5 l/unit/h + fitted with demand control[3]				On demand flush control
– Urinals		1.5 with pressure flushing valve[3]				Most efficient = 1.5
Showerhead flow (l/min)	11–30[1] (Low flow = 7)	Power shower = 12 + [3]	>7.0–9.0	>5.0–7.0	5.0 or less	A = 9–12
		'Water Saver' heads = 4–9[3]				AA = 6.8–9
		Ultra-low flow = 1.5 (non-domestic)[3]				AAA = <6.8
Taps (faucets): (l/min)	10–26[1]					For all taps:
Bath		12–16 (19 mm)[2]				A = aerator & positive seal
Washbasin		6–9 (13 mm)[2]	4.0–6.0	2.0–4.0	2.0 or less	AA = as A with flow control
Kitchen		6–12 (13 mm)[2]	6.0–8.0	4.0–6.0	4.0 or less	AAA = as AA with self-closing
All taps			Self-closing delayed action[4,b]			
Constant flow regulators		Advice[3]	Yes[4,c]			Advice
Washing machines: (5 kg load)		<50 l/cycle[3]	Full load only[4,c]			Efficient = 45 l/cycle Inefficient = 180 l/cycle
Dishwasher: (12 place setting load)		16 l/cycle[3]	Full load only[4,c]			Efficient = 12 l/cycle Inefficient = 36 l/cycle

Sources of information: [1]USA Energy Policy Act 1992; [2]UK Water Supply (Water Fittings) Regulations 1999; [3]UK Environment Agency, Water Resources, CWB Fact cards, 2006 website; [4]PUB, 2014 (Tick water conservation rating system); [5]Water Services Association, Buyers Guide to Saving Water, 2006 website; [6]Water Smart Guidelines, Master Plumbers and Mechanical Services Association of Australia, 2006 website (Water conservation Rating system A to AAAA).

Notes: [a]Mandatory Water Efficiency Labelling Scheme (Tick Rating).
[b]Mandatory (non-domestic premises).
[c]Water saving advice.

Water used to fill and top up swimming pools depends on the incidence of such pools in an area. Coupled with car washing and miscellaneous other outdoor uses, estimated outdoor consumption excluding garden irrigation represents about 3–13% of the total average residential consumption, or 5–20 lcd.

In response to the 2007–08 prolonged drought in Australia, some municipalities imposed restrictions on the use of water for filling swimming pools, washing down outside surfaces and cars, and for irrigating gardens and lawns. Restrictions included banning the use of hosepipes, sprinklers and irrigation devices and limiting garden watering to alternate days using a handbucket. Less stringent restrictions on outdoor water use were imposed by a number of water companies in southeast England during the 2006 drought. Although specific restrictions may be lifted at the end of a drought, it is likely that governments and utilities will increasingly use restrictions on outdoor water use to support initiatives to reduce domestic demand. Indeed, with the increasing awareness of the scarcity of fresh water and need for water conservation, schemes for rainwater harvesting and recycling household grey water are being promoted by government agencies and water utilities. If the campaigns are successful, they will have a significant impact on reducing the demand for mains water for outside use and irrigation demand.

1.4 STANDPIPE DEMAND

Various researchers have suggested that a minimum quantity of 50 lcd should be available to consumers using standpipe supplies. However, there is no international minimum standard beyond:

- Water service level to promote good health (WHO, 2003), where;
    - Basic access: up to 20 lcd, collected up to 1 km from household with up to 30 min collection time. Service level represents high level of health concern.
    - Intermediate access: about 50 lcd, delivered through yard tap or to within 100 metres of dwelling. Service level represents low level of health concern.
    - Optimal access: above 100 lcd, delivered through multiple taps within the dwelling. Service level represents very low level of health concern.
- Key indicators in meeting a minimum standard (Sphere, 2011) for humanitarian crises and disaster relief are;
    - Average use for drinking, cooking and personal hygiene: at least 15 lcd;
    - Maximum distance from household to nearest water supply point: 500 metres;
    - Queuing time at a water source: no more than 30 min.

Standpipe consumption is influenced by the distance over which consumers must fetch water, the usages permitted from the standpipe, the degree of control exercised over use at the standpipe, and the daily hours of supply. The range of uses can be as follows:

1. Water taken away for drinking and cooking only.
2. Additionally for household cleaning, clothes washing, etc.
3. Bathing and laundering at the standpipe.
4. Additionally for watering animals at or near the standpipe.
5. In all cases: spillage, wastage and cleansing vessels at the standpipe.

The quantity that an individual can carry and use is directly related to the travel time and carrying distance. Where the supply is for drinking and cooking only, the basic minimum requirement for direct consumption is about 8 lcd (WHO, 2003) but spillage and wastage at the tap increase the minimum consumption to 10 lcd. Consumption for items (**1**) and (**2**) combined is about 15–20 lcd but wastage and spillage raise this to 25 lcd, a suggested minimum design figure where the water is to be carried over more than a few hundred metres. Where there is little control exercised over consumers' usage of water at a standpipe and where bathing and clothes washing takes place near it, then at least 50 lcd needs to be provided. If purpose-built bathing and laundering facilities are provided consumption rises to 65 lcd. In India 50 lcd is the usual standpipe design allowance, in Indonesia the quantity is 15–20 lcd where the water is sold from standpipes.

In rural Egypt where uncontrolled all-purpose usage tended to occur, including watering of animals, surveys indicated consumptions of between 45 and 70 lcd (Binnie, 1979).

In low-income communities in the lesser developed countries it is often the practice that one householder has an external 'yard tap' which he permits his neighbours to use, usually charging them for such use. These are, in effect, 'private standpipes' and, because carrying distances are short, the consumption can be up to about 90 lcd based on numbers reliant on the yard tap for water.

A public standpipe should supply at a rate sufficient to fill consumers' receptacles in a reasonably short time, otherwise consumers may damage the standpipe in an attempt to get a better flow. The design of the standpipe should permit a typical vessel to be stood on the ground below the tap so that water is not wasted; the tap, typically 19 mm (3/4 in) size, should provide a good flow under low pressure conditions so that it is not vandalized and should be constructed of readily available materials to reduce the temptation for theft. The hours of supply need to be adequate morning and evening for the number of people using the facility and proper drainage should be provided at the standpipe to take away spillage.

1.5 SUGGESTED DOMESTIC DESIGN ALLOWANCES

Domestic in-house consumption for average middle class properties having a kitchen, a bath facility and waterborne sanitation falls into a fairly narrow range of 120–160 lcd, irrespective of climate or country. A system that uses a class of dwelling unit type (flat or house, etc.) and value (size, age, etc.), such as the ACORN geo-demographic classification system (www.caci.co.uk), is the most practicable basis for estimating domestic consumption where more accurate data are not available. From visual identification of the predominant housing stock in an area and knowledge of the average occupancy, the demand for the area can be estimated with reasonable accuracy.

Suggested design allowances of domestic per capita consumptions by dwelling type are given in Table 1.6.

Table 1.6 In-house domestic water and standpipe demand – suggested design allowances (lcd)

Type of property and income group[a]	UK and Europe[b]	Hotter climates
1. Villas, large detached houses, luxury apartments, wealthy villages and suburbs; four plus bedrooms, two or three WCs, bath and shower, kitchen and utility room (High-income groups – 25%)	190 *180*	230–250
2. Detached and suburban houses, large flats, rented properties (professionals and students sharing); three or four bedrooms, one or two WCs, bath and shower and kitchen (Upper middle-income groups – 10%)	165 *150*	200–230
3. Flats, semi-detached and terrace houses, retirement flats; two or three bedrooms, one or two WCs, bath or shower and kitchen (Average middle-income groups – 25%)	150 *130*	180–200
4. Semi-detached and terrace houses and flats, two or three bedrooms, one WC, bathroom and kitchen (Lower middle-income groups – 15%)	140 *120*	
Developing countries: Tenement blocks:		
• Block centrally metered		160 +
• Individual household metered		130
5. Small houses, cottages and flats, sustainable housing with separate kitchen and bathroom, high occupancy (Low-income groups – 25%)	90–110 *75–90*	
Developing countries: Tenement blocks, high-density occupation with one shower, one Asian toilet, one or two taps:		
• Building centrally metered or free		130 +
• Individually metered		90
Developing countries: Lowest income groups:		
Low grade tenement blocks with one or two room dwellings and high density occupation:		
• Communal washrooms (unmetered)		110
• One tap and one Asian toilet per household: block metered		90
• One tap dwellings with shared toilet or none; dwellings with intermittent supplies		50–55
Developing countries: Standpipe supplies		
• Urban areas with no control		70 +
• Rural areas under village control		45
• Rural with washing and laundering facilities at the standpipe		65
• Absolute minimum for drinking, cooking and spillage allowance		25

Notes: [a]Income category and percentage of UK population are indicative only. Within an area of similar housing stock there will be a range of household sizes and incomes.
[b]Figures in italics are for metered consumption.
Figures exclude lawn and garden irrigation, bathing pool use and use of evaporative coolers; but include an allowance for unavoidable consumer wastage.

1.6 NON-DOMESTIC DEMAND

Non-domestic demands comprise:

Industrial: Factories, industries, power stations, docks, etc.
Commercial: Shops, offices, restaurants, hotels, railway stations, airports, small trades, workshops, etc.
Institutional: Hospitals, schools, universities, government offices, military establishments, etc.
Agricultural: Use for crops, livestock, horticulture, greenhouses, dairies and farmsteads.

Industrial demand for water can be divided into four categories:

1. ***Power station demand*** – mostly for cooling but increasingly for flue gas treatment. Cooling water is usually abstracted direct from rivers or estuaries and returned to the same with little loss. However, evaporative cooling is employed, sometimes using reclaimed water. Supplies are not normally taken from the public supply except for some supplies to water cooled air conditioning systems for commercial and office buildings. Power station usage may be in conflict with other water usage in some locations (Murrant, 2015).
2. ***Major industrial demand*** – consumption greater than 1000 m^3/d, e.g. for paper making, chemical manufacturing, production of iron and steel and oil refining. Large capacity water supplies tend to be obtained either from private sources or a 'raw water' supply provided by the water utility. The raw water is distributed through a public non-potable network or a dedicated pipeline to the industry and may receive disinfection treatment to reduce the health risk to people who could come into contact with it. The user would normally treat the water to the quality required for his processes including additional treatment and 'polishing' where the supply is derived from a potable supply. Non-potable supplies are always reported separately from the 'public' water supply in statistics.
3. ***Large industrial demand*** – factories using 100–500 m^3/d for uses such as food processing, vegetable washing, drinks bottling and chemical products. These demands are often met from the public supply. Generally the supply receives additional treatment on-site to meet process requirements.
4. ***Medium to small industrial demand*** – factories and all kinds of small manufacturers using less than 50 m^3/d, the great majority taking their water from the public supply.

All industrial premises provide a potable supply for their staff for hygiene and catering. Generally this 'domestic use' supply is obtained from the public system, but occasionally it may be supplied from the treated water used in the industrial processes.

Estimating industrial demand can be complex. The same industry in a different environment can use significantly different quantities of water per processed unit. For example, the specific water use of industrial production of raw steel from delivered ore in eight European countries is reported to range from 0.6 to 600 litres/kg (average of 90 litres/kg); for paper produced from dry pulp, the range is 15–500 litres/kg of product (average of 140 litres/kg) (EEA, 1999). These broad ranges demonstrate how variations in production process, water use, water efficiency, water recycling and possibly tariff structure can all influence the specific usage by an industry. There may also be differences in how each industry or country reports the statistics.

Existing industrial demand can be best estimated by measuring the daily and weekly demand for the specific consumer at the point of supply using the consumption survey approach outlined in Section 1.18. Ideally diurnal and seasonal variations should also be measured to ensure that the range of demand on the system is understood especially where the industry takes water as required and has seasonal production variations, e.g. seasonal food processing. Industries without on-site storage can impose onerous operational performance characteristics on a network such as surge and short-term high flows; these may adversely affect the system and other consumers. Often it is found that about 90% of the total demand in a large industrial area is accounted for by only 10–15% of the industrial consumers. A consumption survey can therefore be selectively targeted to monitor the major users. However, it is important to check the accuracy of the meters used to ensure that the demand is being measured accurately and that usage lost through meter error is not being incorrectly reported as transmission mains losses (Section 1.8).

For new industries, the preferred approach is to adopt the forecast demand for water proposed by the developer, by industrial area or individual development site. Where no other information is available, industrial demand can be derived using typical usage by plot area and by industry category, e.g. car production, chemical industries, electronics, food production, general industrial and service industries. Only a few service industries, e.g. drinks bottling, laundries, ice and concrete block manufacturing, use large quantities of process water. Many light industries, such as those involved in printing, timber products and garment making, use water only for staff hygiene and catering. Typical ranges of industrial usage are given in Table 1.7; however, as discussed above, they should be used with caution.

Table 1.7 Commercial, industrial and institutional water consumption allowances

Usage	Consumption allowance
Industrial areas	50–100 m^3/ha. Product specific demand should be assessed where practical
Light industry that includes food and drinks processing	1–1.5 m^3/d per employee
Light industry that excludes large water consumption	0.25–0.5 m^3/d per employee
Light industry and warehouse; cleaning and sanitation use only	50 l/day per employee
For small trades, small lock-up shops and offices in urban areas	Up to 25 lcd (applied as a per capita allowance to the whole urban population)
Offices[a]	50–75 l/day per employee
Department stores[a]	75–135 l/day per employee
Restaurants, bars	75–120 l/day per customer/table seating
Hospitals	350–500 l/day per bed
Hotels	250–400 l/day per bed; up to 750 l/day per bed for luxury hotels in hot climates
Schools[a]	25 l/day per pupil and staff for small schools; rising to 75 l/day per pupil and staff in large schools

Note: [a]Applies to the days when those establishments are open.

Commercial and institutional consumption comprises the demand from shops, offices, schools, restaurants, hotels, hospitals, small workshops and similar activities common in urban areas. In England the overall average commercial and institutional demand is equivalent to about 25 lcd over the whole population served. This includes domestic use by people living in non-domestic premises or in attached living quarters. In the USA commercial and institutional demand can be significantly higher because of the high consumption for water cooled air conditioning systems and for outdoor irrigation. For example in office buildings, the indoor use for employees for personal hygiene and catering can represent 35–40% of the total building demand. Similarly for schools the internal use may represent only 20% of the total (AwwaRF, 2000). Typical allowances made for demand in certain types of commercial and institutional premises in the UK are also given in Table 1.7.

Most water for agriculture, including crop irrigation, horticulture and greenhouses, is taken direct from rivers or boreholes because it does not need to be treated. The principal use of the public supply is for the watering of animals via cattle troughs, for cleaning down premises and for milk bottling. Table 1.8 gives estimates of such consumption.

Table 1.8 Agricultural water demands – suggested allowances

Livestock	Dairy: farming	40–100 l/day per animal in milk[a]
	Dairy: process cleaning	20–50 l/day per animal[a]
	Beef cattle	25–80 l/day per animal for drinking[a]
	Beef abattoir use	Average 1500 litres per animal
	Horses	30–55 l/day per animal[a]
	Sheep	5–15 l/day per animal for drinking[a]
	Pigs	10–20 l/day per animal for drinking[a]
	Poultry (eggs)	20–40 l/day per 100 birds[a]
	Poultry (meat)	15–25 l/day per 100 birds[a]
	Poultry meat processing	15–20 l/bird
Crop irrigation		130 m^3/ha per week during growing season[b]
Glass house crop production		20–30 m^3/d per hectare (or more) in growing season 10–15 m^3/d per hectare in winter

Notes: [a]Livestock consumption depends on season, age of animal and production stage. Consumption can increase by 20–30% in extreme heat conditions.
[b]Demand depends on crop and rainfall during growing season.

1.7 PUBLIC AND MISCELLANEOUS USE OF WATER

The quantities used to water and maintain parks, green areas, ornamental ponds, fountains and gardens attached to public buildings have to be assessed for each particular case in relation to the area to be watered and the demand from the type of cover planted, e.g. the area of grass or flower beds, types of plants and shrubs, etc. The estimate should include potentially high seasonal variation especially in hot dry climates. Often the quantity of water used for public watering is only limited by the available supply. However, in some cities, raw water or 'grey water' is used for these purposes. Other miscellaneous uses include supplies to government-owned properties, street cleaning, flushing water mains and sewers, and for firefighting. Where supplies to public buildings, such as government offices, museums, universities and military establishments, are not paid for the demand can be substantial compared with the usage in equivalent private sector buildings. Water used for street cleaning, flushing, firefighting and system maintenance activities can be assessed from the records of the time, duration and equipment used.

In the UK supplies to public parks and buildings and to government and local authority offices is generally metered and should thus form part of the metered consumption. Unbilled and unmeasured legitimate water usage would be for firefighting and for routine maintenance activities, such as testing fire hydrants, sewer cleansing and flushing dead ends of mains. Temporary connections for building sites, which used not to be recorded, are now metered and the consumption is billed. The total of the miscellaneous unbilled and unmeasured demand is estimated by utilities in England and Wales to be about 1.7% of the total input into distribution, equivalent to about 4.5 lcd for the total population.

1.8 WATER LOSSES

Water losses comprise the leakage and wastage from the distribution network; these and other components of non-legitimate use are categorized as:

- ***Apparent losses***: source and revenue meter errors, unauthorized or unrecorded consumption, and
- ***Real losses***: leakage from transmission and distribution mains and service pipes upstream of consumers' meters, from valves, hydrants and washouts, and service reservoir leakage and overflows.

Comparing water losses has in the past been complicated by inconsistent use and different interpretations of terms such as 'unaccounted-for water', 'non-revenue water', 'legitimate usage' and 'losses'. Therefore, from the late 1990s the International Water Association (IWA) has been promoting an internationally consistent set of terms and definitions for the components of water losses within the water balance. Figure 1.1 illustrates the IWA terminology and relationship between the terms that are being used increasingly worldwide. Reducing water losses, leakage and wastage is a water utility's high priority when managing the water supply and demand balance. Water losses, a component of '*non-revenue water*' (NRW), are made up of *apparent* and *real* losses and '*unbilled authorized consumption*'. Unbilled authorized consumption, essentially unbilled metered and unmetered consumption, can be managed effectively either by installing permanent

<table>
<tr><td rowspan="9">System input volume</td><td rowspan="4">Authorized consumption</td><td rowspan="2">Billed authorized consumption</td><td>Billed metered consumption (including water exported)</td><td rowspan="2">Revenue water</td></tr>
<tr><td>Billed unmetered consumption</td></tr>
<tr><td rowspan="2">Unbilled authorized consumption</td><td>Unbilled metered consumption</td><td rowspan="7">Non revenue water (NRW)</td></tr>
<tr><td>Unbilled unmetered consumption (e.g. fire demand)</td></tr>
<tr><td rowspan="5">Water losses</td><td rowspan="2">Apparent losses</td><td>Unauthorized consumption (e.g. illegal connections, meter tampering or bypassing)</td></tr>
<tr><td>Customer metering inaccuracies</td></tr>
<tr><td rowspan="3">Real losses</td><td>Leakage from transmission and distribution mains</td></tr>
<tr><td>Leakage and overflows at utility's storage facilities</td></tr>
<tr><td>Leakage from service connections up to the customer's meter</td></tr>
</table>

FIGURE 1.1

IWA standard international water balance and terminology.

meters to measure consumption or by monitoring the supplies regularly to assess demand and identify changes in demand patterns.

Apparent losses represent unauthorized consumption that is not measured or billed to the consumer, e.g. illegal connections, meter tampering or bypassing and meter inaccuracies. Apparent losses include consumer meter inaccuracy and errors in the meter reading and billing processes that are key to identifying and eliminating unauthorized consumption. Metering inaccuracies, the largest proportion of apparent losses, can be minimized by maintaining the meters (inspection, recalibration and replacement) and by managing the billing procedures to minimize errors in data entry and billing. Managing unauthorized consumption is complicated because of the difficulties in quantifying the illegal usage and locating the connections. Consequently unauthorized consumption tends to be included in the legitimate per capita consumption figures.

Real losses generally represent the majority of NRW. Real losses comprise leakage, overflow and wastage from trunk mains, distribution pipework, storage facilities and service connections between the distribution pipework and consumers' premises. Leaks occur from pipes, pipe joints and fittings; valves, hydrants and washouts; and from service pipes upstream of consumers' meters or boundary stopcocks. The ferrule connections of service pipes to mains are often a major cause of distribution leakage. Hence distribution losses are influenced both by the length of mains serving consumers and the number of service pipe connections per kilometre.

Water losses cannot be measured directly but have to be estimated by measuring the total input into a system and deducting the amount supplied for legitimate consumption. The estimate includes an allowance for leakage from supply pipes and plumbing systems where the consumer's supply is not metered. Water utilities worldwide report a range of figures for NRW and losses because the figures are influenced by a variety of factors such as the age and condition of the pipes, supply pressures, efficiency of leak and waste prevention measures, how the unmetered demand is estimated and the methodology used for compiling the statistics. Reported leakage and NRW figures from utilities worldwide can range from 5% to 10% of the distribution input (the quantity of potable water supplied) for well managed systems, up to 40% and 60% or more for systems in poor condition, where there is a history of long term under investment in network maintenance and rehabilitation. Systems with intermittent supplies also exhibit high leakage rates. Table 1.9 relates levels of losses to a range of infrastructure characteristics.

Table 1.9 Typical figures of non-revenue water

Percentage of total supply	Typical circumstances applying
5–15%	Small systems with little leakage; residential parts of large systems with little leakage
16–20%	Usual lowest reported for whole cities, often associated with active leak control strategy
20–25%	Achievable in large systems with active leakage and waste control methods and good system monitoring and network data
25–35%	Reported for large systems comprising old mains and service pipes in moderate to poor condition, lower meter coverage and poor data
35–55% and greater	Systems with many old mains and service pipes in poor condition; inefficient metering, lack of attention to leaks and consumer wastage and limited financial resources

The wide range of figures reflects the variety of methods used to estimate water losses as well as the range of actual losses themselves. Apart from metering errors there is always some unmeasured consumption that has to be estimated. High figures in excess of 30% may be partly due to leakage from pipes and partly due to lack of valid consumption data. Losses in a new or extensively renewed system should be low, say 5–10%, but low reported figures could also result from liberal estimates of unmetered consumption, or by excluding trunk mains losses or meter inaccuracy from the calculation.

Expressing water losses as a percentage of the distribution input may be appropriate within a utility because the data used and methodology of the calculation is understood and can be applied consistently. However, it is not suitable for comparing different organizations with different physical and operational characteristics and different per capita and non-domestic demands as shown in the following example. Company A reports a total per capita supply (total authorized consumption and water losses) of 300 lcd and water losses at 20% of the total demand; the losses represent 60 lcd. Company B reports an overall per capita supply of 500 lcd and losses of 15%; the equivalent losses are 75 lcd. The calculation illustrates that using percentages as a performance indicator between utilities understates the real losses of utilities with high usage and misrepresents the losses for utilities with lower unit usage, but higher reported percentage losses. Quoting percentage losses can also disguise other system differences such as utilities that deliver a large proportion of their supply through a few connections to industry compared with utilities with no large industrial supplies. Furthermore, percentage losses can decline as consumption rises but not because the losses have actually been reduced.

In recognizing the need for consistent reporting, the IWA Task Force on Water Losses developed the Infrastructure Leakage Index, *ILI* for reporting and comparing real water losses:

$$ILI = \frac{CARL}{UARL}$$

where *CARL* is the current annual real loss derived from the annual volume of real losses expressed either as litres/day, litres per connection per day or litres per kilometre of main per day for the hours in the day when the system is pressurized. *UARL* is the system-specific unavoidable annual real losses, the technically achievable lowest real water loss based on pipe burst frequency, duration and flow rates, and system pressures for well-run systems in good condition. *UARL* is made up of: Background (Unavoidable) Losses + Reported Bursts + Unreported Bursts, or:

$$UARL\ (in\ l/day) = (18L_m + 0.8N_c + 25L_p)P$$

where:

L_m = Total length of mains in km
N_c = Number of service connections
L_p = Total length of underground supply pipe in km
P = Average zone operating pressure in metres

The equation can be reconfigured to calculate UARL in different units such as litres per kilometre of main per day per metre pressure or in gallons and miles. The UARL coefficients given in Table 1.10 were derived from international data for minimum background loss rates and typical burst flow rates and frequencies (Lambert, 1999). The calculation assumes that there is a linear relationship between leakage and pressure and it can be modified to take account of intermittent periods of supply.

Table 1.10 UARL (Unavoidable Annual Real Loss) coefficients

	Per day/metre of pressure	Background losses	Reported bursts	Unreported bursts	UARL total
Distribution mains	l/km	9.6	5.8	2.6	18.0
Service pipe to property boundary	l/connection	0.60	0.04	0.16	0.80
Service pipe to property boundary to meter	l/km	16.0	1.9[a]	7.1[a]	25.0

Source of information: Lambert, 1999.
Note: [a]Assumes 15 m average length of underground service pipe within property boundary.

For systems with 24-hour supplies and supply pressure above about 20 metres, it is suggested that ILIs up to two represent networks where water losses are being managed efficiently and further reductions would need to be assessed carefully in relation to the cost of achieving additional savings. ILIs between two and eight represent networks where water losses could be reduced, the higher the index the greater the potential for savings. ILIs over eight represent systems with unacceptably high leakage and where leakage reduction programmes should be implemented as a high priority. Equivalent breakpoints for developing countries, networks with intermittent supplies and low supply pressures are less than 4, 4–16, and over 16, respectively (Liemberger, 2005).

Table 1.11 gives guidance on unavoidable real losses for a range of average operating pressures and connection densities for 24-hour pressurized systems.

Table 1.11 Unavoidable real losses (meter at boundary) (l/service connection/day – l/conn/d) and suggested performance indicator for developed and developing countries

Connection density	Average operating pressure in metres				
Number per km of main	**20 m**	**40 m**	**60 m**	**80 m**	**100 m**
20	34	68	112	146	170
40	25	50	75	100	125
60	22	44	66	88	110
80	21	41	62	82	103
100	20	39	59	78	98

Source of information: IWA, 2000.

In the UK the water regulator WSRA, also known as Ofwat, uses two performance indicators, '*litres/property/day*' – l/prop/d and '*m^3/kilometre/day* – m^3/km/d' to assess and compare total leakage. However, these measures also need to be viewed with caution when making international performance comparisons. Issues that need recognizing include: a service connection may supply a single property or multiple dwelling units but the reported leakage is only on the service pipe; utilities supplying rural areas have long mains serving few connections per km; and utilities supplying urban areas can have a high number of connections per km of main. Table 1.12 presents the 2009/10 leakage statistics for England and Wales and also comparative figures for the UK and selected international countries for 2004/05.

Some of the difference between the leakage figures of the UK regional (water and sewerage) and water-only companies may stem from differences of approach in estimating losses, or because the larger regional companies have a larger scale of problems to deal with. However, physical factors also contribute to the difference. The regional companies supply the largest urban areas in the country, which tend to have older systems than those of the water-only companies; some include coal-mining areas where ground settlement has disturbed mains and several have to supply hilly areas requiring high distribution pressures.

The water from leaks is not actually 'wasted' from the hydrological cycle. Much of it percolates underground and recharges aquifers. Hydrologists often take account of leakage when assessing groundwater flows (Section 4.5).

Table 1.12 International comparisons of leakage performance measures

	l/prop/d	m^3/km/d
England & Wales leakage statistics 2009/10 (Ofwat, 2010)		
10 Regional water and sewerage companies	138.4 (90–186)	9.9 (5.5–21.3)
12 Water-only companies	108.6 (68–132)	8.4 (5.5–12.6)
Industry average	135.1	9.7
Comparative international leakage statistics 2004/05 (Ofwat, 2007)		
England & Wales (23 utilities)	150 (70–248)	11 (6–29)
Australia (8 utilities)	83 (44–115)	5 (3–9)
Netherlands (5 utilities)	17 (4–35)	1 (1–2)
Portugal	126 (62–268)	8 (1–31)
Scandinavia (4 utilities)	80 (33–114)	15 (5–21)
USA (9 utilities)	314 (114–657)	14 (6–26)

1.9 REAL LOSSES (LEAKAGE) FROM 24-HOUR SUPPLY SYSTEMS

The level of leakage from a system depends on the success of the water utility's loss reduction programme, the age and condition of the system and the system operating pressure. Policies for loss reduction depend on the financial and manpower resources a water utility can allocate to leakage control, both for 'one-off' exercises to reduce current levels of leakage to a satisfactory level and for continued application of leakage control to maintain that level.

The age of a distribution system is a major factor influencing real losses. High losses quoted by several UK water utilities are primarily due to the advanced age of many of their mains and service pipes. Thames Water, with the highest leakage rate in England and Wales, reports the average age of its mains in London to be over 100 years, with a third being over 150 years old compared with 60–70 years for other UK companies. Many European cities report average ages of between 40 and 50 years.

Table 1.13 summarizes international burst rates by material type. Table 1.14 presents some typical 'background' leakage levels, which are estimated to occur on the UK and international water distribution networks together with indicative leak flow rates.

Leaks on distribution systems break out continuously so that the total leakage from a system for a period is the aggregate sum of each leak-rate multiplied by the time it runs before repair. Hence the frequency with which all parts of a system can be tested for leaks influences the level of leakage experienced. Obviously there is a practical limitation to the cycle time for retesting for leaks; therefore, some level of leakage is unavoidable. There is also a need to determine the economic level of resources that should be put into leak detection and repair. This is discussed in Section 16.8.

Table 1.13 International comparison of burst frequencies

Pipe material	CI	DI	Steel	PE	PVC	AC	All mains
Burst rate per 100 km of main per year							
United Kingdom; all companies	21.1	4.2	11.1	3.1	8.7	9.4	27.2
United Kingdom; 11 companies – range	12–30	3–7	5–23	1–11	4–19	6–31	12–31
Australia – 3 cities	22.3	1.6	9.8	–	–	8.5	34.3
Australia – range	13–25	–	–	–	–	7–54	2–72
Canada	39.0	9.7	–	–	1.2	7.3	36.7
West Germany	19.0	2.0	–	10.3	6.0	6.0	–
East Germany	41.0	–	–	74.0	14.0	34.0	–
% material of total length of mains							
United Kingdom; all companies	64	10	4	8	1	13	
United Kingdom; 11 companies – range	43–83	2–16	0–6	2–17	4–25	0–20	
Australia – 3 cities	68	10	1	0	1	20	
Australia – range	31–78	0–18	0–2	0	0–2	1–69	
Canada	51	23	0	0	11	15	

Source of information: UKWIR, 2001.

Table 1.14 Background night flow losses and UK industry average leak flow rates

Estimated background night flow losses at 50 m pressure				
Infrastructure element		**Good**	**Average**	**Poor**
Trunk mains[1]	l/km/hr	100	200	400
Distribution mains[2]	l/km/hr	20	40	60
Communication pipe[2] (see Fig. 16.1)	l/conn/hr	1.5	3.0	4.5
Service pipe (UGSP)[2] (see Fig. 16.1)				
Either:average length 15 m	l/conn/hr	0.25	0.50	0.75
or:	l/km/hr	16.7	33.3	50.0
In-house plumbing losses[2]	l/property/hr	0.25	0.50	0.75

Average flow rate by type of leak and typical range based on natural rate of rise calculation (m^3/hr)			
	Average[3]	**Reported[2]**	**Detected[2]**
Mains leak	3.0	0.5–1.6	0.2–1.1
Mains fittings (valves, hydrants)	0.15	0.1–0.2	0.1–0.2
Service pipe – communication pipe	0.4	0.3–1.1	0.2–0.6
Service pipe – supply pipe	0.4	0.3–0.9	0.2–0.5
Communication and supply pipe fittings	0.1		

Source of information: [1]WRc, 1994, [2]UKWIR, 2011,[3]UKWIR, 2006b.

Leakage from service reservoirs can be found by direct static testing. Acceptance figures for a new concrete service reservoir would normally be a drop in water level of 1/500th of its depth (up to 5 m deep) over a 7-day test period, equivalent to about 0.03% per day of the average daily supply with diurnal turnover, which is a negligible amount. Overflow discharge pipelines should be so designed that any overflow can be detected and therefore stopped.

The data used to assess NRW and water losses and hence whether action is necessary to reduce leakage need to be validated. Meter error resulting from the under-recording of consumers' revenue meters can represent a significant percentage of NRW. Low flow consumer demands, particularly at night, are frequently below a revenue meter's stalling speed and hence go unrecorded. Over a period of time they can, in total, represent an appreciable demand on the system. The methods described in Section 1.19 can be used to assess the accuracy of consumers' revenue meters and the magnitude of their likely under-recording. It is essential to monitor and assess all unbilled legitimate uses regularly.

1.10 SUPPLY PIPE LEAKAGE AND CONSUMER WASTAGE

The service pipe connects the main in the street to the consumer's premises. In the UK it is considered in two parts; the '*communication pipe*', maintained by the utility, which goes up to the boundary stopcock and meter where fitted, and the '*supply pipe*', maintained by the customer, which completes the connection onto the property plumbing network. Leaks on communication pipes are accounted for in the utilities' reported distribution losses.

In the UK many of the service pipes over 50 years old are made of galvanized iron or lead. In the installation of 50 000 meters on household supplies in the Isle of Wight during the National Metering Trials, 1989–92, it was reported that 8000 service pipes were either repaired or replaced in part or in total, most defects being found on the customer's supply pipe downstream of the boundary stopcock. This represents one in six of such pipes being found faulty (Smith, 1992). For 2009–10, utilities in England and Wales reported that the leakage estimated from unmetered consumer supply pipes averaged about 39 l/prop/day (range 14–65 l/prop/day), but on properties where meters were installed at the property boundary, supply pipe leakage was reported to average 17 l/prop/day (range 2–27 l/prop/day) (Ofwat, 2010).

Within premises there can be leakage and wastage, termed 'plumbing losses'. Where supplies are plentiful or water is cheap, or where waste conservation measures are slack, consumer wastage can be 50 lcd or more. Where water bylaws or local regulations require, among other things, that all WC and storage cisterns are float valve controlled and the cistern overflow pipe must discharge outside the premises or be visible within the sanitary fitting, overflows can be easily noticed or heard by waste inspectors and leakage technicians checking premises at night; the nuisance created by the overspill at the premises may sometimes motivate the occupier to take remedial action. Some wastage caused by dripping taps is, however, unavoidable. Block metering of multiple occupancy buildings tends to result in high consumer wastage where one meter measures the supply to all the households and a single landlord pays the water charges and recovers the cost of wastage through the rents charged. Thus individual householders are not aware that they are paying for their own and others wastage. Table 1.6 shows that 25–40% extra domestic consumption needs to be allowed for in-block metered premises.

1.11 MINIMUM NIGHT FLOW AS INDICATOR OF LEAKAGE AND WASTAGE

The minimum night flow (MNF) to a section of the distribution system can act as an indicator of distribution leakage and consumer wastage. MNF is the measured flow into a controlled area of the network, e.g. a District Meter Area (DMA), during the period of minimum demand. The MNF typically occurs during the night between 01:00 and 04:00 hours but the characteristics of the area can have a significant impact on when the MNF occurs, e.g. DMAs containing a high density of bars and nightclubs in tourist areas. There is always some legitimate demand for water at night, which has to be deducted from the minimum recorded flow. Although MNF tests do not measure quantities of water lost, they are a good indicator of the condition of a system; Table 1.15 shows an interpretation of results obtained. However, the test is impracticable if the supply is intermittent or houses have large storages that fill at night. In the UK domestic house storage tanks are relatively small and therefore are usually full before an MNF test takes place. It is important, therefore, to understand the characteristics of consumer demand and plumbing as well as how the distribution system performs when analysing night flow records.

Table 1.15 Figures for minimum night flow (MNF) per connection (l/hr)

MNF per connection	Interpretation
1.7	Average minimum night use per household (WRc, 1994) (or 0.6 l/hr × number of people in household)
5	About the lowest found in practice in parts of systems in good condition
7	A frequent 'target level' for distribution districts, indicating good control over leakage and wastage
9	Experienced on large systems where there is a fair amount of nocturnal demand and/or some distribution leakage and consumer wastage
11	Indicative of substantial night demand and/or considerable distribution leakage and/or consumer wastage

The lowest night flow in small residential areas comprises distribution system and supply pipe leakage, legitimate night domestic demand plus unavoidable consumer wastage from dripping taps. However, in larger test areas legitimate use is likely to include non-domestic night-time consumption at premises such as hospitals, nursing homes, police and fire stations, railway stations, airports, clubs, as well as industrial demand from units working night shifts. This non-domestic consumption needs to be monitored or measured during the period of the flow test. Large users may be continuously monitored. The consumption of metered consumers can be assessed by reading their meters at the start and end of a night test or, if not practical because of the number of meters to be read or manpower resource constraints, instead their typical rates of night consumption can be measured before the test takes place and deducted from the area night flow measurement.

1.12 VARIATIONS IN DEMAND (PEAKING FACTORS)

Consumers draw water as and when they require it and consequently the demand for water varies both diurnally and seasonally. Demand over a 24-hour period varies from the MNF to a peak hour demand, although the rate of draw may be reduced where the supply passes through in-house storage. Demand also varies: by the day of the week, weekend demand patterns being different to weekday usage; and seasonally, depending on temperature variations; and for regional reasons, e.g. city demands can be observed to fall during holiday seasons when at the same time demand in tourist areas rises significantly to cater for the influx of visitors.

The variations in system diurnal and seasonal demand are usually derived as ratios (factors) or percentages of the annual Average Day Demand (ADD). Peaking factors are influenced by the size of the area being monitored. The demand from an individual consumer can have a significant impact on the flow profile of a small supply area such as a rural community or an area with a single large industrial unit, e.g. a seasonal food processing factory. However, the overall peaking factor reduces with the number and mix of consumers and the impact of an abnormal user can be less apparent.

Maximum and Minimum Hourly Rate of Consumption

The diurnal demand depends on the size of the population and the type of commercial and industrial usage in the area served. In the UK the domestic peak period for residential areas is typically between 07:00 and 09:00 hours reflecting households getting up and preparing for work and school; peak hour factors vary between about 2.25 and 1.75 depending on the size of the area. The peak period can start earlier and be shorter in communities with high proportions of workers who commute and in rural areas; peaking factors being up to 2.5 or even 3.0 for small areas. For mixed residential and industrial areas, the peak is typically at mid-day with factors typically between about 1.5 and 1.75. Similar peak factor ranges are observed worldwide but cultural, religious and lifestyle differences can greatly influence the amplitude, duration and time of the peak.

Diurnal factors also vary between weekday and weekends. Traditionally domestic peak hour factors were influenced by weekday household water consumption. Increasingly flow profiles are demonstrating changing water use patterns to higher domestic consumption at weekends, the peak hour occurring later in the day, between 09:00 and 11:00 in the morning or in the evening and the higher demand period lasting for longer in the day. There are also indications that the peak day demand can vary between a Saturday and Sunday in areas that otherwise appear to have similar characteristics.

Garden watering demand can create an evening peaking factor of 3.0 or more. In the USA with its hot summers, where peak hourly factors for in-house demand are reported to be between 3.0 in eastern states and up to 5.0 for western states, sprinkler demand can increase the factor to 6.0.

Seasonal Variations

The Maximum or Peak Day Demand is the maximum daily demand reported in a year. The quantity is significant for understanding the potential absolute demand on a system and the centres of the high demands. However, it is not as important as the demand during a maximum week in the year, expressed as the Average Day Peak Week (ADPW) demand. The ADPW is typically only a few percent below the peak day demand but it is important because the daily overdraw above the average for 7 consecutive days cannot usually be met from the amount of service reservoir storage provided. Therefore, utilities typically manage their resources and storage to maintain an overall weekly resource balance. This means that the maximum output of the source works must be at least equal to the average daily demand for the 7 days of the peak week. Typical system ratios of the ADPW to the annual ADD are given in Table 1.16. In temperate climates a design figure of 140% is often adopted.

Table 1.16 Range of peak system demands as percentage of Average Day Demand

Location, etc.	ADPW/ADD
UK. Seaside and holiday resorts	130–150%
UK. Residential towns, rural areas	120–130%
UK. Industrial towns	115–125%
UK. Peak due to garden watering in prolonged hot dry weather	150–170%
USA. Typical peak domestic demands – in-house only	
– Western states	180–190%
– Eastern states	130–140%
USA. Typical peak domestic demands due to lawn sprinkling	
– Western states	220–340%
– Eastern states	200–300%
Worldwide. Cities with hot dry summers	135–145%
Worldwide. Cities in equable climates	125–135%
Worldwide. Cities with substantial industrial demand	110–125%

1.13 GROWTH TRENDS OF CONSUMPTION AND FORECASTING FUTURE DEMAND

The change in consumption of potable water in England and Wales since 1970 is shown in Figure 1.2. Over the period 1970–2007 total per capita consumption rose from 277 lcd to a peak of 331 lcd in 1995–96 and thereafter has reduced over the last 20 years to about 265 lcd. The figures are the aggregate of the individual water company estimates. The reduction is primarily due to increased work on leakage reduction but the decline in industrial demand and measures to restrain domestic consumption have contributed. Variations in the summer climate, which affect household and garden water use, may also have had an impact.

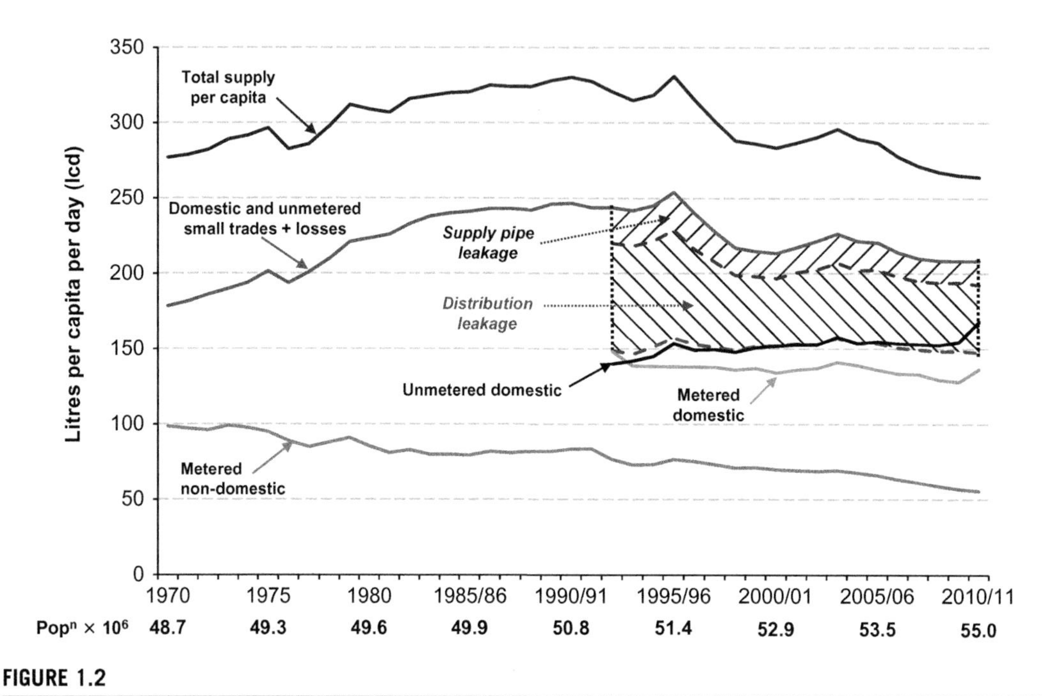

FIGURE 1.2

Change in per capita consumption, England and Wales, 1970–2011.

Long-term forecasting of water demand presents problems. Figure 1.2 shows how a period of sharp economic decline in 1992 reduced domestic consumption immediately and trade consumption the following year. Such incidents interrupt previous trends of increase that, in the case of domestic consumption in developed countries, tend to be asymptotic to some future maximum demand per capita. The ultimate maximum unconstrained level of domestic consumption depends on household wealth and housing and standards of water fittings installed, coupled with the policy of the water supplier. However, where resources are scarce additional supplies are difficult or expensive to procure; or where construction of a new water supply scheme would meet with strong environmental opposition, measures to restrain the rise of domestic demand may need to be adopted.

1.14 WATER CONSERVATION AND DEMAND MANAGEMENT

Water scarcity and the need to conserve resources are recognized worldwide as challenges for the near future. Water conservation and environmentally sustainable use of water will be increasingly implemented to manage the water balance, particularly where water resources are limited and the areas are subjected to droughts. Much emphasis is being placed in Australia, the USA and in the drier parts of south-east England on the adoption of demand constraint measures. In England emphasis is being placed on extending the metering of domestic supplies, installing water saving fittings and on increasing measures to reduce leakage and wastage. In the USA, as required by the

Safe Drinking Water Act 1996, US EPA published draft guidelines in 1998 to water suppliers for 'conservation planning', i.e. measures to induce economy in the use of water. The guidelines propose three sequential levels of approach. The first comprises universal metering, loss control (i.e. leakage and wastage reduction) and public education. The second and third levels include such measures as water audits, pressure management, re-use and recycling, and integrated resource management. However, emphasis on demand constraint does not necessarily apply in all countries, where some utilities may be reluctant to curb the demand from metered industrial consumers and from metered households occupied by higher income groups; the payments made by them comprise a major part of the utility's income and are needed to cross-fund supplies given free through standpipes or at below cost to low-income groups. The conditions and position adopted by each utility vary. Where more plentiful supplies exist there may be less incentive to conserve water.

In many countries restricted hours of supply have to be adopted in order to prevent consumption and losses exceeding available supplies. Metering is adopted for the same purpose but it must be efficient to be effective. Intermittent supplies bring many problems. Consumers store water when the supply is on, but throw away the unused balance when the supply next comes on, believing the new supply is 'fresher'. Consumers may leave taps open so as not to miss when the supply comes on again, allowing storage vessels to overflow. Intermittent supplies make leak detection and prevention of consumer wastage very difficult. Typically the hours of supply have to be reduced to at most 4 hours in the morning and 4 hours in the evening, and frequently less, to gain control of consumption. To some extent intermittent supplies are self-defeating: more consumer wastage and more distribution leakage occur because of the difficulty of maintaining the system in a good state; if mains become emptied contaminated groundwater may enter the pipes and endanger the health of consumers. The situation is often exacerbated by loss of income due to difficulties with metering and income collection. Nevertheless, many utilities worldwide have to adopt intermittent supplies.

Water conservation measures that can be effective on 24-hour supplies include:

- imposing temporary bans on the use of water for washing vehicles, on refilling swimming pools and ponds, and on the use of hosepipes and sprinkler equipment for watering gardens during drought or shortage of supplies. During the 2006 hosepipe bans in southern England, seasonal reductions of up to 15% were reported;
- good publicity can achieve a temporary and short-term reduction in demand, perhaps as much as 10%;
- metering domestic supplies can curb excessive consumption, especially water used for lawns and gardens, provided the tariff structure imposes a financial penalty when consumption exceeds a reasonable amount;
- promoting the use of low water use fittings (Table 1.5) can make significant reductions as illustrated in Table 1.4;
- using pressure management control to reduce leakage and to extend the life of the pipes, improves the reliability of network control valves which can operate within their designed range, thereby reducing wastage through the valve malfunctioning;
- keeping operational pressures at the minimum necessary to maintain levels of service reduces water taken unnecessarily;
- flow limiters and throttles on service pipes can curb consumption but are not always effective. If set too low, consumers leave taps open to fill containers which overspill. They can also be bypassed in an attempt to get a better supply.

Commercial and institutional demand can be constrained by metering all consumers: both large and small shops, offices and other business premises. Wastage through plumbing fittings from these non-domestic consumers is frequently high because no one working in the premises is responsible for paying the water charges and the premises are unoccupied outside working hours. Many cases have been reported of night and weekend flows to unoccupied premises being nearly as high as daytime flows when staff are present, especially in government offices in some countries. Manufacturers are also often unaware of the potential financial savings they can achieve by adopting water conservation measures, the lower usage reducing both their water purchase and effluent discharge costs.

Table 1.17 summarizes the opportunities and constraints of water saving technologies, metering and tariffs and leakage reduction, the three primary measures for managing consumption.

Table 1.17 Water conservation measures

Component	Opportunities	Constraints
Water saving technologies	• Low and ultra-low water use efficient appliances • Standards and regulations will achieve greatest savings	• Policing installation and retrofit • Does not stop excessive use • Time to promote and develop • Unlikely to achieve 100% acceptance
Metering and tariffs	• Reduces demand, short and long term • Consumers responsible for their impact on environment • Charge real cost of usage	• To be effective, need both • Ability to pay and need to maintain supplies to low-income and vulnerable consumers • Cost of meter installation and long-term maintenance
Leakage control	• Proactive control can reduce losses down to economic level of leakage (ELL) • Short-term local solution	• Reduction to ELL is one-off win, thereafter marginal impact on resources • Savings depend on maintaining ELL/leakage reduction • Is ELL optimum level?

Estimating future demands in developed countries can be approached by combining the forecast trends in population and per capita consumption growth with implementing water conservation measures within the forecast period. The trends provide an upper limit to the projections. The conservation measures will produce a range of achievable forecasts provided there is the political and social will to conserve water and adopt a realistic tariff for water and provided the utility manages, and is seen to be managing, system water losses and leakage.

1.15 THE QUESTION OF METERING DOMESTIC SUPPLIES IN THE UK

Reducing demand for fresh water is an environmental benefit to all. The aim of metering is to restrain rises in domestic consumption, especially in areas where additional sources of water are difficult or expensive to procure. With rising demand for water, increasing difficulty in developing new supplies, and periodic droughts, the question of whether universal metering of the 65% unmetered domestic supplies in England and Wales should be adopted has been debated repeatedly by a cross section of stakeholders. The questions centre mainly on cost, fairness to the consumer and what reduction of consumption could be achieved.

The savings in consumption achieved by metering are difficult to assess. The National Metering trials gave erratic results, the best information coming from the Isle of Wight where 50 000 meters installed resulted in a 5–9% reduction (WSA, 1993). More recently a report for UK Water Industry Research (UKWIR) gave potential savings of between 10% and 15% based on trials and compulsory programmes (UKWIR, 2006a). It is probably safe to assume that metering can achieve at least 7.5% reduction, equivalent to about 12 lcd on current domestic consumption in England and Wales of 150–160 lcd. Greater benefit may be achieved through the use of seasonal tariffs to curb both seasonal demand and the peak day and hour ratios, both of which cause problems to utilities. A reduction of about 30 l/property/day in supply pipe leakage could also eventually be expected where meters are installed outside (Section 1.10). This is a useful benefit, but whether universal domestic metering is justified by the cost is a complex matter. Water companies in England and Wales install domestic meters free when requested by the householder, on change of property ownership and to all new newly built houses.

The National Metering Trials, 1989–92, also found that only about 30% of properties could be metered inside and for 5% of properties a water meter could not be installed due to plumbing difficulties or high cost. In many older housing areas up to 20 properties can be fed by one common supply pipe: one utility was said to have 650 000 properties fed through joint service pipes (Roberts, 1986). Common supply pipes are often laid at the rear of terraced properties and occasionally the supply pipe is laid through the roof space of a terrace of properties. Multi-occupancy buildings, such as blocks of flats, are also mostly fed by one metered riser supply pipe; inserting meters on short off-takes to individual dwelling units could cause unacceptable disruption of kitchen or bathroom fitments.

The average water standing charge for household customers, covering meter provision, meter reading and extra billing costs, can represent a substantial proportion of a metered consumer's bill. Any offset saving of cost due to the reduced production of water achieved by metering is unlikely to be significant unless the reduction in consumption is sufficient to avoid the need to develop a new source of supply. However, while companies continue to have the right to use the rateable value of houses as a basis for unmeasured charges for water and sewerage, voluntary domestic metering is likely to increase only in those areas where rateable values are high and household occupancy is low. Hence substantial extension of domestic metering is likely to occur only in conjunction with a process of designating water scarce areas.

The question of fairness of metering to householders also causes debate. The large family will obviously pay more than the small family; but if a tariff is adopted which provides a large family with a sufficient basic water allowance at low cost, then a single occupant can take several times his basic need at the same low cost. The average occupancy ratio of metered households in England and Wales at 2.1 is significantly lower than for unmetered dwellings at 2.7 (Table 1.2). At present, charging for unmetered supplies according to property value seems reasonably fair, since there is a relationship between property value and consumption per capita. However, the metering of domestic supplies is so widely practised in many countries that on the face of it, it is difficult to maintain that domestic metering creates an unacceptable injustice. When taking into account tariff subsidies, the argument becomes less straightforward and utility specific.

1.16 EFFECT OF PRICE ON WATER DEMAND

In their Environmental assessment report No. 1 titled *Sustainable Water Use in Europe* (EEA, 1999), the European Environment Agency concluded that although water pricing is difficult to use as a demand management measure, increasing water tariffs is a useful tool to make users more responsible for their water use when applied in conjunction with other water conservation initiatives. They cite a case in Hungary where demand fell 50% over an 11-year period during which the price of water increased from HUF 2 to HUF 120 per m^3. The report concluded that the dramatic price rise contributed to the reduction in demand.

It is generally true that the demand for a product or service reduces with increasing price to the customer and vice versa and there are formulae, guidelines and rules of thumb for predicting the impact of a price change on demand. Where the product or service is subsidized, the impact of cost changes becomes more difficult to predict because it is dependent on whether the demand is suppressed or not, the level of subsidy in place and the affluence of the community. However, there is limited documented evidence of case studies demonstrating the relationship between demand and water tariffs.

Price elasticity of demand is the term used to measure the percentage change in demand resulting from a percentage change in price. The relationship is represented by:

$$\text{Price elasticity of demand } e = \frac{\Delta Q/Q}{\Delta P/P}$$

where Q is the demand at price P per unit of consumption. Since price increases tend to cause a reduction in demand, e is negative. When $e = -1.0$ increases in P cause proportional decreases in Q. A value of e between 0 and -1 indicates a high degree of price inelasticity, e.g. at $e = -0.2$ a 25% price increase of P would cause Q to decrease by 5%. Where e is greater than -1 price is considered elastic and a given change in price will result in a greater change in demand.

Many of the documented studies of demand analyses conclude that water tariff changes fall within the price inelastic range: that their impact on demand had been proportionately less than the price increases. However, the conclusions from a study of eight systems, six systems in Europe and two towns in Australia also concluded that there was more elasticity in 'Increasing Block Tariffs' for residential usage (Metaxas, 2005).

Values of e are difficult to determine accurately because environmental or economic conditions before and after price rise are often not the same. In addition, the elasticity value must be influenced by the size of the price rise, a large one-off price rise having a greater influence than smaller rises annually. A time lag also occurs between a rise in price and any observable effect on demand. Consequently e values quoted show a wide variation for apparently similar situations. The generally reported range for price elasticity of demand is between -0.1 and -0.3 for average annual in-house domestic demand. In the Hungarian example discussed above the price elasticity of demand in each year remained in the inelastic range averaging about -0.2 (range -0.01 to -0.5).

Despite the difficulty of getting an accurate measure of e, the elasticity of demand is important to a utility. Industrial and trade consumers and the higher income householders are often a major source of income to many water utilities overseas. Hence the elasticity of their demand is an important factor, because a price rise may cause them to reduce their take, thereby not producing the anticipated proportionate increase of income.

US EPA in their Water Conservation Plan Guidelines (US EPA, 1998) suggests the following benchmark figures for the impact of price rises on demand:

- 10% increase in residential prices reduces domestic demand by 2–4% ($e = -0.2$ to -0.4);
- 10% increase in non-domestic prices reduces non-domestic demand by 5–8% ($e = -0.5$ to -0.8);
- Increasing block tariff rates reduces demand by 5%.

1.17 ASSESSING FUTURE DEMAND IN DEVELOPING COUNTRIES

In developing countries consumption is often limited to the amount of water available, with the consequence that there can be much unsatisfied demand. Estimating the long-term potential demand therefore involves a different approach. Although population growth forecasts may be available, they can be unreliable because of unexpected changes in birth rates. In addition a major proportion of the population growth in many cities in developing countries is often caused by migration from rural to urban areas. Hence estimating future water demand of such a city involves the following steps:

1. Plot the population trend for the past 10–20 years and assess the likely proportion due to migration and that due to natural increase of the existing population.
2. Divide the supply area into different classes of housing and assess in which classes of housing the main rises of population have occurred and are likely to occur in the future.
3. Assess typical rates of domestic per capita consumption by class of housing.
4. Seek likely figures for future migration and natural increase and allocate these to appropriate classes of housing.
5. Investigate future proposals for new housing development (master plans, development plans and proposals).

This approach can be used to build up forecasts of future demand. However, the estimates may need to be adjusted to ensure that they represent a realistic continuation of historic consumption trends. Generally speaking, consumption forecasts for more than 10 years ahead tend to be unreliable.

Growth of commercial and institutional demands is often estimated as a per capital allowance on the population growth because these activities tend to relate to the size of the population served. Increases in manufacturing and industrial demand are generally dominated by the needs of a relatively few major industries whose development plans should be ascertained. However, the sum of their estimated individual future demands should be adjusted down to allow for the probability that not all the developments are likely to go ahead during the period of forecast.

1.18 CONSUMPTION SURVEYS

Although the total supply available for an area can seem to be adequate when expressed as the amount available per head of population, this statistic may conceal the fact that, due to excessive wastage and leakage in some areas and by some consumers, there is unsatisfied demand in other parts of the system. Symptoms of such problems include: apparent high demand or water losses; consumers in some areas not receiving an adequate supply or pressure; metering and billing practices appear to be inefficient. All these situations occur worldwide in utilities constrained by lack of money and technical resources, resulting in poor system performance and service to the consumer. The water balance equation for a suppressed demand is:

$$\text{Available supply} = \text{legitimate potential demand} + \text{water losses} - \text{unsatisfied demand}$$

In order to understand the demand in an area, it may be necessary to carry out a consumption survey. This provides information to analyse the performance of the system in terms of delivering an adequate supply to meet the demand for water. The survey addresses how much, when and for what purposes water is being used. The steps involved in a consumption survey are:

1. Using consumer complaints and system monitoring data, identify areas in the distribution network where pressures are below a (service) level sufficient to deliver an adequate supply to consumers both at peak demand times and during the whole of the daytime.
2. Check the accuracy of source meters supplying the area. This may involve installing a temporary check meter adjacent to the meter under investigation or monitoring the source meter while discharging a measured quantity into a tank or through a controlled fire hydrant or washout point where the flow rate is also monitored.
3. Check consumer revenue meter readings against observed check-readings over a period and compare with the billing records:
   - Quantify the typical number of revenue meters found stopped or faulty at any one time and investigate what billings are made when meters are found stopped.
   - Determine the average age of revenue meters and how frequently they are tested, recalibrated and repaired; establish the performance characteristics of meters, of different manufacturers, installed in the network.
   - Test some meters of typical size and age for accuracy. Where more than 15% of meters are found stopped at any one time or where many meters are over 10 years old and not regularly tested and repaired, a substantial amount of under-recording must be suspected.

4. Assess typical domestic consumption per capita by classifying dwellings into five or six classes and test-metering 30–35 dwellings typical of each class (Section 1.19).
5. Test-meter a few typical standpipes and estimate the population reliant on each. Estimate the typical standpipe consumption per capita.
6. Examine the revenue meters on all large trade and industrial supplies. Compare observed readings against billing system data and investigate inconsistencies. Check the accuracy of meters found in poor condition.
   - Identify the largest potential trade consumers and check their billing records to see whether usage appears reasonable in relation to the size of the supply pipe, the hours of take, and the amount of water likely to be used for their activities.
   - Check for major consumers missed from the billing system; those with two feeds to the premises only one of which is metered; or for which the billing is estimated from a previous lower usage activity (common where there has been a change of owner or activity).
7. Monitor and estimate or meter the amount of water supplied to legitimate unmetered consumers such as government or municipal offices, public parks and gardens.

The information obtained during the survey then needs to be compiled to estimate the probable total potential demand on the system. On a map of the distribution system identify supply districts and mark areas of the different classes of housing. Using appropriate population densities per hectare (or census data) and measuring the area of each class of housing within a district, estimate the total domestic demand per district by using the appropriate consumption per capita (Step **4**).

For each district any standpipe consumption, commercial, industrial and other non-domestic demands should be added, apportioned according to the characteristics of the district. Major trade consumers should be individually assigned. An allowance for a reasonable quantity of unavoidable distribution leakage should be added to the total domestic and non-domestic demand. This gives the total average daily demand on the whole system, broken down into sufficiently small supply districts for the demand in each district to be distributed to 'nodal points' of the mains layout, i.e. to key junctions on the mains. These 'nodal demands' can be used for the basis of a hydraulic analysis of flows in the distribution system, as described in Section 15.15. The whole exercise will determine the adequacy of the distribution system to meet the demands.

All meters measuring the output of sources have also to be checked for accuracy; unless regularly serviced they will almost certainly be in error. Consumer meter inaccuracy is categorized as an apparent water loss in the water balance (Fig. 1.1).

1.19 TEST-METERING IN-HOUSE DOMESTIC CONSUMPTION

Sample sets of domestic properties are monitored when individual supplies are not metered or where records of metered domestic consumption are not sufficiently reliable to assess an average domestic consumption per person. The sample set should be representative of the consumers in the area and should include about 35 properties from each of a maximum of five or six household categories, individually or collectively metered. A sample size of at least 30 properties is required to provide a representative estimate of the mean per capita consumption for a given household category. By monitoring 35 properties per category, about 30 valid sets of results should be obtained allowing for equipment failures and data lost from individual properties during the survey. Larger sample sets are difficult to monitor because of the need to ensure that, during the period of the survey, all meters continue to work accurately, the service pipes remain leak free and to keep a check on the number of people in each household. The consumer categories are usually related to housing type and their number is restricted because of the difficulties of distinguishing between ranges of consumers with any certainty. Furthermore, since the intention is to monitor small sets of specifically identified properties and household categories, the sample set should not be chosen randomly but specifically identified in order to avoid introducing a bias of the sample mean towards the higher or lower end of the range within the class. The test period should be 2–4 weeks, avoiding holiday times and, if possible, extremes of weather. Meter readings and occupancy rates should be recorded weekly.

Such tests generally show a wide scatter of results. The mean per capita consumption can be substantially influenced by a few households where the consumption seems extraordinarily high. However, provided high consumption readings are not due to meter reading error, the figures should be included because domestic consumption is so variable. The limitation of the exercise is that the sample size is too small to evaluate the incidence of such high consumption which may be, e.g one in 20, so that a sample size of 30 may contain no such high consumers, or one or two. However, if five classes of housing are adopted there will be at least five separate samples from which to judge, roughly, the frequency of such exceptional consumption. The mean domestic per capita consumption is derived from the total consumption in the 30 or so households tested in a given category, divided by the total occupancy during the test period.

Where individual properties cannot be metered, the flow through a main supplying a number of properties of the same class can be monitored instead. This approach will not reveal individual high or low household consumption and will include leakage from the supply main and service pipes downstream of the metering point. However, it is a useful supplementary method of assessing mean consumption if reasonably leak-free conditions can be assured.

Statistical analysis techniques can be used to define the size of the sample sets. However, because the per capita consumption varies so greatly, the statistical standard deviation of samples is seldom below 30 lcd and often higher. Samples sets would need to be very large to get a useful degree of accuracy in the mean, e.g. with a standard deviation of 30 lcd, a sample size of 865 is needed for 95% probability that the population mean lies within ±2 lcd of the sample mean. To test-meter such a large sample is impracticable. If, alternatively, existing billing records are used for the analysis, the mean will include metering and billing errors, consumer wastage and leakage, and it will be difficult to know accurately the population in residence during the billing period. Furthermore, relying on analysis of existing billing records is inadvisable because an important purpose of test-metering is to check the validity of billings.

In the UK, some utilities have been running long-term monitoring surveys usually with the agreement of the householder, often employees of the organization, in order to establish long-term trends in consumption. However, the difficulty for the utilities is maintaining a long-term stable sample set, monitoring for changes in household characteristics and continuing the survey when properties change ownership.

REFERENCES

ADB (1997). *Second Water Utilities Data Book, Asian and Pacific Region, 1997*. ADB.

ADB (2007a). *Data Book of Southeast Asian Water Utilities 2005*. ADB.

ADB (2007b). *2007 Benchmarking Data Book of Water Utilities in India*. ADB.

ADB (2008). *Asian Sanitation Data Book 2008*. ADB.

Aquacraft (2000). *Seattle Home Water Conservation Study*. Aquacraft, Boulder, Colorado, USA.

AwwaRF (1999). *Resident End Uses of Water*. AwwaRF.

AwwaRF (2000). *Commercial and Institutional End Use of Water*. AwwaRF.

Binnie & Partners (1979). *Report on Egyptian Provincial Water Supplies*. Binnie-Taylor, Egypt.

CCWater (2014). *Save Water and Money, Average Water Use*. Consumer Council for Water, ccwater.org.uk/savewaterandmoney/averagewateruse/ (accessed on 13 November 2014).

Census UK (2012). *2011 Census, Population and Household Estimates for England and Wales – Unrounded Figures for the Data Published 16 July 2012*. Office of National Statistics.

EEA (1999). *Environmental Assessment Report No 1 "Sustainable Water Use in Europe"*. EEA.

IWA (2000). *The Blue Pages, Losses from Water Supply Systems and Recommended Performance Measures*. IWA Publishing.

IWA (2014). *International Statistics for Water Services*. IWA Publishing.

Lambert, A. O., Brown, T. G., Takizawa, M. and Weimer, D. (1999). A review of performance indicators for real losses from water supply systems. *Journal of Aquatic* **48**, 227–237, IWSA (now IWA) Publishing.

Liemberger, R. and Mckenzie, R. (2005). Accuracy limitations of the ILI – is it an appropriate indicator for developing countries. *Proc. IWA Leakage Conference, September 2005, Halifax, Nova Scotia*. IWA Publishing.

Metaxas, S. and Charalambous, E. (2005). Residential price elasticity of demand for water. *Water Science & Technology: Water Supply* **5**(6), 183–188, IWA Publishing.

Murrant, D., Quinn, A. and Chapman, L. (2015). The water-energy nexus: future water resources availability and its implications on UK thermal power generation. *Water and Environmental Journal* **29**(2015), 307–319, CIWEM.

Ofwat (2007). *International Comparison of Water and Sewerage Services, 2007 Report Covering Period 2004–05*. Ofwat.

Ofwat (2010). *Service and Delivery – Performance of the Water Companies in England and Wales 2009–10 Report, Supporting Information*. Ofwat.

PUB (2005). *Towards Environmental Sustainability*. Public Utilities Board, Singapore, Chapter 2.

PUB (2014). *Handbook on Application for Water Supply, 2014*. 6th Edn. Public Utilities Board, Singapore.

PWC (2005). *Annual Report 2003/04*. Portsmouth Water Company, UK.

Roberts, K. F. (1986). Requirements for trial programme. *Proc. WRc Seminar on Water Metering, April. WRc, Swindon, UK*.

Smith, A. (1992). The reduction of leakage following water metering on the Isle of Wight. *Journal of CIWEM* p. 516.

Sphere (2011). *Humanitarian Charter and Minimum Standards in Humanitarian Response, 2011 Edn*. The Sphere Project, third reprint 2013.

STW (2007). Personal communication. Severn Trent Water, UK.

TWU (2007). Personal communication. Thames Water Utilities Limited, UK.

UKWIR (2001). *Understanding Burst Rate Patterns of Water Pipes*. UKWIR.

UKWIR (2006a). *Critical Review of Relevant Research Concerning the Effects of Charging and Collection Methods on Water Demand, Different Customer Groups and Debt*. UKWIR.

UKWIR (2006b). *Natural Rate of Rise in Leakage*. UKWIR.

UKWIR (2011). *Managing Leakage 2011*. UKWIR.

US Census (2012). *2010 Census of Population and Housing Unit Counts*. US Census Bureau.

US EPA (1998). *Water Conservation Plan Guidelines, 1998*. US EPA.

UNESCAP (2013). *Statistics Yearbook for Asia and the Pacific, 2013*. United Nations Economic and Social Commission for Asia and the Pacific.

Water Corp (2003). *Domestic Water Use Study in Perth Western Australia, 1998–2001*. Water Corporation, Australia.

WB (2013). *World Bank Indicators 2013*. World Bank.

WHO (2003). *Domestic Water Quality, Service Level and Health*. WHO.

WHO (2014). *Progress on Drinking Water and Sanitation, 2014 Update*. WHO.

White, S., Milne, G. and Reidy, C. (2004). End use analysis, issues and lessons. *Water Science & Technology: Water Supply* **4**(3), 57–65, IWA Publishing.

WRc (1994). *Managing Leakage*. WRc, Swindon, UK.

WSA (1993). *Water Metering Trials: Final Report*. WSA, London, UK.

CHAPTER

Water Supply Regulation, Protection, Organization and Financing

2

DEVELOPMENT, REGULATION AND PROTECTION

2.1 CONTROL OF PUBLIC WATER SUPPLIES

All governments must exercise some control over public water supply to ensure that everyone, irrespective of income, receives an adequate supply of water that is fit for human consumption. Regulations will usually include service level provisions that aim to ensure that access to water for basic needs is available to the whole population. The quantities of water to be provided are discussed in Chapter 1; service levels are discussed in Section 16.2. The requirement for water to be fit for human consumption embraces physical, chemical, bacteriological and aesthetic qualities (Chapter 7). Aesthetic criteria are used so that other unsafe supplies are not used for drinking and cooking.

Controls affecting water supplies are usually set out in legal enactments passed by national or state governments or imposed by law making organizations representing groups of countries, such as the European Union. Controls in developed regions may be very sophisticated and wide ranging but those in poor regions may need to be restricted to the essential elements for minimum levels of service and provision of a *'safe'* supply.

To ensure a safe supply many governments require their water utilities to comply with the water quality standards recommended by the World Health Organization (WHO) as a minimum, including the implementation of water safety plans to assess and manage risk throughout the whole water supply cycle; from catchment to consumer (Section 7.82 et seq.). Some countries set extensive national water quality standards, which are legally enforceable, and systems of inspection may be used to ensure they are achieved. However, compliance with such standards is expensive; consequently countries with low per capita incomes may need to limit national standards to those that require the water to be free of bacteriological contamination and substances obviously injurious to health.

Since all persons need wholesome water, it may have to be supplied to low-income households at a charge below the cost of production, or free of charge via standpipes set up in the street.

Twort's Water Supply. DOI: http://dx.doi.org/10.1016/B978-0-08-100025-0.00002-8

2.2 CONTROL OF ABSTRACTIONS

Governments also have to control abstractions of water from natural sources and may need to prioritize use for different purposes, such as domestic, commercial, industry, energy and agriculture. In a world of limited resources abstractions must make best use of the available water and do as little harm to the environment as possible. Far sighted governments will want to ensure that abstractions are sustainable and may need to limit 'mining' of fossil water.

Irrigation and impounding for hydropower can have a major impact on stream flows, sediment transport and water quality. Where catchments cross frontiers, numerous international (usually bi-lateral) agreements on impounding and abstraction have been made. Examples include: the 1892, 1956 and 1970 treaties affecting the regulation of the river Rhine; the 1978 Great Lakes Water Quality Agreement; the 1944 Treaty on the Utilization of the River Colorado; and the 1926 USSR–Persia Frontier Water Convention (Teclaf, 1981).

It is clear that water resources have long been a potential source of dispute between neighbouring countries. With ever-increasing demand due to population growth, pressure on shared resources is bound to increase. In 1992 the UN adopted the *Convention on the Protection and Use of Trans-Boundary Watercourses* (UN, 1992). The convention came into force in 1996 after nearly 25 years of work by the UN International Law Commission (Eckstein, 2005), and there have been two further amendments in 2003 and 2012. The purpose of the convention is to provide a framework for international agreements. It is aimed principally at surface watercourses but also covers linked aquifers that discharge to a terminus in common with the associated watercourse. In this respect the Convention recognizes the 'drainage basin' concept enshrined in the 'Helsinki Rules' (ILA, 1966). However, certain types of aquifer are not covered. Examples are the Nubian Sandstone aquifer beneath Chad, Egypt, Libya and Sudan and the 'Mountain Aquifer', recharged in largely Palestinian territories and becoming 'confined' as it flows under Israel.

2.3 PUBLIC WATER SUPPLIES IN THE USA

Public water supplies in the United States are characterized by numerous small supply systems as shown in Table 2.1. The majority of small systems are privately owned; the larger systems are predominantly municipal and publicly owned. The share of public ownership of all systems increased from 20% in 1830 to 50% in 1880 and 70% in 1924 (Melosi, 2000). The physical isolation of many small communities in a large country is partly a cause of this fragmentation, but an additional factor (Okun, 1995; MacDonald, 1997) is that developers have often preferred to site new residential communities outside the limits of major urban areas, where land is cheaper and property taxes lower; individual small water supply systems have thus been established instead of connecting the new communities to the nearest existing system. Economic circumstances in the last decade have reduced the rate of urban build in the USA. Consequently there is increased focus on extending the life of ageing infrastructure and on alternative means of financing, including private sector participation (B&V, 2016).

Table 2.1 Number of water utilities in the USA in mid-2013

US EPA designation	Population served	No. of water utilities		Total served (million)	
Very small	25–500	121 695	(81.1%)	13.879	(4.4%)
Small	501–3300	18 843	(12.6%)	24.791	(7.8%)
Medium	3301–10 000	5151	(3.4%)	29.949	(9.4%)
Large	10 001–100 000	3870	(2.5%)	110.302	(34.7%)
Very large	Over 100 000	427	(0.3%)	139.204	(43.8%)
Total		149 986		318.125	

Source of information: US EPA, 2013.

The US Environmental Protection Agency

US EPA sets drinking water quality standards under the Safe Drinking Water Act (SDWA) (US EPA, 1996). It can issue Regulations stipulating:

- an MCLG (maximum contaminant level goal) which is not mandatory;
- an MCL (maximum contaminant level) which is mandatory;
- a 'Treatment Rule' which is mandatory and which sets out the type(s) of water treatment to be adopted where an MCL is not appropriate or sufficient (Section 7.65) and for minimizing the incidence of disinfection by-products.

An 'MCLG' is defined under the 1996 Act as – 'a level at which no known or anticipated adverse effect on human health occurs and that allows for an adequate margin of safety'. An 'MCL' is defined as – 'a level as close to the MCLG as feasible'. 'Feasible' is defined as – 'practicable according to current treatment technology, provided no adverse effect is caused on other treatment processes used to meet other water quality standards' (Pontius, 1997). MCL values under the 1996 Act are shown in Table 7.1.

Under the previous Safe Drinking Water Acts of 1974 and 1986, US EPA had issued Regulations covering some 85 organic and inorganic substances, and 16 others covering radionuclides and microbial levels. US EPA had not been required to assess either the cost of implementing a regulation or the health benefits it would achieve. This gave rise to widespread criticism that some costs had been imposed on public water suppliers that could not be justified by the incidence of such contaminants or the numbers of persons at risk.

Under the 1996 Act US EPA is now required to support any proposed new Regulation by publishing a report on:

- the risk the contaminant presents to human health;
- the estimated occurrence of the contaminant in public supplies;
- the population groups and numbers of persons estimated to be affected by the contaminant;
- the benefits of reduced risks to health the proposed Regulation should achieve;
- the estimated costs to water utilities of implementing the Regulation;
- estimated changes to costs and benefits of incremental changes to the MCL value proposed;
- the range of uncertainties applying to the above evaluations.

US EPA has to publish and keep updated every 5 years a list of contaminants likely to require regulation, choosing at least five of them to Regulate every 5 years and giving priority to those posing the greatest health risk. A period of 18–27 months for public comment and US EPA's consultation with certain authorities must be allowed before a proposed Regulation containing an MCL is promulgated, i.e. formally put into operation.

To follow these new procedures US EPA has first to collect country-wide data revealing the incidence of various types of contaminants in public water supplies and the effectiveness of current treatment processes in reducing such contaminants to required levels. All systems supplying a population of 100 000 or more were required to submit detailed reports on the results of 18–24 months comprehensive sampling programmes, with lesser sampling programmes being required for smaller systems serving upwards of 10 000 population. To help small water systems serving 10 000 population or less, the 1996 Act required US EPA to provide a list of treatment techniques to help them achieve the required quality standards. US EPA must also provide small systems with 'variances' which permit the adoption of treatment processes that will achieve nearest compliance with an MCL, taking into consideration a system's resources and the quality of its source water. Government funds made available to States are to be used to assist in training waterworks operators to standards set by US EPA. The funds are also intended to help these small systems obtain technical assistance with water quality compliance problems and source protection.

The radical changes made by the 1996 SDWA achieve a more realistic approach to setting of water quality standards, and make it more practicable for utilities to comply with them. However, the performance of the many small utilities continues to present a weakness, unless they can be provided with sufficient day-to-day technical and laboratory assistance to ensure they achieve compliance with the many sophisticated drinking water standards in force.

2.4 PUBLIC WATER SUPPLIES IN MAINLAND EUROPE

The manner of development of public water supplies across Europe was influenced by the interplay between differing administrative and legal systems (Newman, 1996) and by differing cultures. Centralizing influences include monarchies, Napoleon and communism. Decentralizing influences include federal governments and the strength of municipalities who were instrumental in water supply development in most European countries. In France this interplay led to the operation, by a few large companies, of a large number of supplies owned by municipalities. This contrasts with a process, more widespread elsewhere, of transfer of private commercial water supply systems to public ownership in order to better control quality and meet the needs of all citizens. Private participation in public water supplies in Europe has a long history from early commercial enterprises, to long-term concessions in several large cities, such as Barcelona and to privatization in the UK and Eastern Europe.

With a few exceptions, public water supplies to large cities in Europe commenced between 1850 and 1900, in some cases using private companies. Initial drivers included firefighting, industrial demand and public health. By about 1950 most remaining supplies in private ownership had been taken over by municipalities. In Eastern Europe water supplies were generally nationalized between 1945 and 1950 but were re-privatized or transferred to municipal ownership around 1995.

Privatization of some city water supplies occurred generally between 1990 and 2005 but some cases of transfer back to municipal ownership have occurred, e.g. Grenoble in 2000 (Juuti, 2005).

All members of the European Economic Community, 28 states in 2016, have to comply with the same European legislation (EC directives). However, the national institutions responsible for different aspects of water supply and environmental legislation vary widely.

The European Water Framework Directive

The Water Framework Directive (2000/60/EC) requires Member States of the European Union (EU) to control, protect and improve the water resources of the State. Member States have to define appropriate river basin districts within their country boundaries, and set up 'a competent authority' for each. For river basins that cross-national boundaries, international river basin districts must be set up. In England and Wales, Regulations SI 2003/3242, which came into force on 2 January 2004, requires the Environment Agency (EA) and the National Assembly for Wales to undertake the duties imposed by the EC Directive. In Scotland implementation of the Directive is the responsibility of the Scottish Environment Protection Agency (SEPA), and in Northern Ireland the responsibility of the Northern Irish government.

For each river basin district the EC Directive required base information to be identified and reported to the EC Commission. All forms of water have to be assessed – rivers, lakes, 'artificial surface waters', such as reservoirs, underground water, together with basin-related estuarial waters (termed 'transient waters') and coastal waters one nautical mile out to sea.

A programme of measures ('*River Basin Management Plan*' – RBMP) had to be drawn up for each river basin district in order to achieve the Directive's environmental objectives by 22 December 2015. The RBMPs had to be operational by 2012. The objectives are to achieve 'Good status' for surface waters and groundwater within each basin, as defined in Tables set out in Annex V to the Directive. For surface waters the biological quality, hydromorphological quality and physico-chemical quality (inclusive of synthetic and non-synthetic pollutants) are stipulated. For groundwater, 'Good status' comprises achieving a balance between abstraction and recharge and preventing or limiting the ingress of pollutants.

The provisions of the Directive directly affecting water and sewerage suppliers are:

- adoption of water-pricing policies which provide incentives for users to use water resources efficiently;
- basing water charges on 'the principle of recovery of the costs of water, including environmental and resource costs';
- setting charges for wastewater disposal 'in accordance with the polluter pays principle';
- conducting an economic analysis to decide the most effective combination of water abstraction and wastewater disposal measures that should be adopted.

Compliance with the requirements of the Directive may pose substantial administrative cost on member states, especially those that have not hitherto managed water resources by river basin, or do not already have a national organization responsible for the water environment. In the UK the use of water resources has been planned on a river basin basis since 1973; consequently considerable information was already available to meet the Directive requirements, although not to the extensive detail required by the Directive.

2.5 PUBLIC WATER SUPPLIES IN THE ENGLAND AND WALES

In the UK up to about 60 years ago the building of water abstraction works for public water supply had be to sanctioned by a Private Act passed by Parliament, so called since it gave powers to a particular (i.e. 'private') body. This procedure enabled interested parties to lodge objections. A parliamentary committee would hear the arguments and approve or modify the proposed works. This procedure was later modified to enable the government to obtain an 'Order' authorizing the works after achieving agreement between all the parties involved.

This unplanned approach and the demand for more waterworks led to the setting up of 26 River Authorities in England and Wales under the 1963 Water Resources Act. The River Authorities had control over use of water by licensing all water abstractions in each river catchment (except those for domestic use in a single dwelling). A National Water Resources Board allowed development of inter-basin transfer schemes, such as those implemented in the North-west and East of England. In 1973 10 new multi-purpose regional Water Authorities were set up to cover England and Wales. Each Water Authority took over all the water supply undertakings in its area (except specially licensed water companies), plus all the sewerage works belonging to the local authorities, and all the duties of the previous River Authorities. The authorities had powers to licence all water abstractions and sewage and wastewater discharges in their basins and therefore, were able to tackle growing river pollution. However, with this decentralization the ability to plan strategic use of water for national benefit was reduced.

The Privatization of Water in England and Wales and Ofwat

The success of the Water Authorities created in 1973 was limited by Government restrictions on borrowing by the public sector, imposed to contain inflation. This made it difficult to continue to improve many of the sewerage systems they had taken over and to meet more rigorous public water supply quality requirements set by a 1980 European Commission directive. The solution adopted was to privatize the Water Authorities by changing them to private companies. This would move their large outstanding debts and future borrowings out of the public sector. The debts would become the debts of private companies.

The privatization process overcame three difficulties. A new body, the National Rivers Authority, was set up in 1989 to take over public duties that posed a potential conflict of interest – licensing of water abstractions and wastewater discharges, the control of pollution, inland navigation, fishing and amenity protection, flood protection, land drainage and coastal defences. A Director General of Water Services (DG), with a department known as the Office of Water Services (Ofwat), was appointed to control charges to consumers and to set service levels. On the other hand, investment in the industry, where organic growth prospects were limited (99% of householders already had a water supply and 96% were connected to the sewerage system (WCA, 1992)), was made more attractive by allowing formation of 'Holding Companies' that would own the Water Service Companies and be able to undertake other sorts of commercial activity. The result was a successful stock market flotation of shares in the new Holding Companies.

After privatization under the UK Water Act 1989 four 'Consolidation Acts' were passed to tidy up the legislation and avoid the need to refer to previous Acts. The four Acts were:

- The Water Industry Act 1991.
- The Water Resources Act 1991.
- The Statutory Water Companies Act 1991.
- The Land Drainage Act 1991.

Under Section 2.10 of the Water Act 2003 Ofwat was renamed the Water Services Regulation Authority (WSRA). However, in the UK and overseas it continues to be referred to and known as Ofwat.

The DG has legal duties to protect consumers and powers to control charges to consumers, and could require the improvement of water and sewerage services given to consumers. The charges were to be set for 5-year periods and charges from one period to the next could not be increased by more than 'the current rate of inflation $\pm K$', known as the '*K-factor*'. It was expected that rising 'real' costs would be partially or wholly offset by efficiency gains, which were to be assessed by comparison between companies, using information provided by the companies in periodic reviews.

Leading into the price review for the regulatory period 2015–20 Ofwat moved to the concept of delivering benefit for the customer; '*outcomes*' rather than the historic focus on 'outputs' or 'inputs'. The intent of the new approach is to ensure companies are accountable for delivering outcomes while being incentivized to deliver through the price review process. '*Willingness to pay*' surveys were used to offer the consumer alternative enhanced or reduced levels of service with consequential increased costs or reduced performance. The survey conclusions were used in the price review to support the proposed investment programme. Performance is being assessed using a suite of 'Measures of Success' for each company, each Measure of Success defining a specific performance metric. Combinations of measures relate to a specific outcome.

Experience with Water Privatization in the UK. Released from government constraints on borrowing, the new Water Service Companies were able to raise finance and progress with many needed improvement works. However, the Holding Companies had mixed success with other commercial ventures (Isack, 1995; CN, 1994; NCE, 1994). After a period of experimentation they settled with their core activities, plus a few related activities, such as solid waste disposal or acting as consultant to overseas water undertakings.

In a drive for profit in a restricted regime, some Holding Companies transferred their design staffs and laboratories to separate subsidiary companies, using these and equipment supply companies they had purchased to supply goods and services they needed. The DG decided such supplies had to be put to open competitive bidding from others and that a subsidiary could not be favoured in the bidding process.

With the *K-factor* derived from information provided by the companies, it was inevitable that such information was coloured to produce results favourable to the companies. The DG had to audit these submissions using a team of inspectors, so as to control the excessive profit that might result. In some cases of serious distortion, the companies agreed to give rebates to customers; in other cases the DG imposed penalties such as requiring additional expenditure to reduce leakage.

Overall, privatization has worked, but it has been difficult for the regulator to respond to a continual stream of developments, including take-over of some Holding Companies by large foreign firms whose other subsidiaries provide goods and services commonly used in the water industry.

Under commercial, regulatory and political pressures the industry continues to change. For example, the Water Act 2014 has introduced:

- Retail competition into the Water Sector by 2017 with the purpose of developing a seamless retail water market for non-household customers across England and Scotland. This will allow non-household customers to choose their water supplier for retail activities (meter reading, billing and account management). The Act precludes water undertakers from holding a water supply licence; where they wish to do so, they must set up a separate retail company and apply for a supply licence. There is a provision in the Act for retail exit for those water suppliers who do not wish to continue to supply non-household customers.
- Upstream competition by 2019. This will allow new entrants to supply non-household customers with water and wastewater services. Further provisions will allow self-supply licences and permit new entrants to connect their distribution systems with a water company's distribution system.

Drinking Water Inspectorate (DWI). Formed in 1990 after the privatization of the English and Welsh water industries, DWI plays a key role in England, under the Department for Environment, Food and Rural Affairs (Defra), and in Wales ensuring that consumers receive a quality of water that complies strictly and continuously with the Water Supply (Water Quality) Regulations 2010 (see Chapter 7). Similar bodies regulate water quality in Scotland and Northern Ireland. The role of the DWI is to monitor public water suppliers' performance, to give guidance on best practice in the use of reliable quality monitoring procedures and to advise where new or improved techniques need to be adopted.

Private water supplies, supplying water to a single property or several properties through a piped network are covered by the Private Water Supplies Regulations 2010 and are the responsibility of Local Authorities in England and Wales. DWI provides technical advisory support to the authorities.

Environmental Protection and Pollution Control

The Environment Agency (UK). In the early 1990s it became increasingly evident there was a need to protect the whole environment from all forms of pollution. Therefore, the Environment Act 1995 set up the Environment Agency (EA) for England and Wales to take over the duties of the National Rivers Authority but with much wider powers. The Agency's three main functions in relation to the use of water resources were:

- licensing all water abstractions and discharges of wastewater;
- preventing, controlling and reducing the pollution of all waters;
- undertaking flood protection and coastal defence measures.

Other duties were promoting the environmental and recreational benefits of water, fisheries protection, land drainage, controlling inland navigation and complying with special protection measures required by the Government in any area designated as a 'water protection zone'. These provisions applied to all water in rivers, lakes and underground and also estuarial and coastal waters 3 miles out to sea.

To cater for a wider diversity of abstraction licence situations than had been foreseen in the 1995 Act, the Water Act 2003 was passed. Part 1 of this extended the water abstraction licensing powers of the EA. The most important of the changes affecting water and sewerage services were (figures in brackets denote the Sections of the 2003 Act referred to):

- The EA has to issue separate licences for abstraction and for impounding *(3) (12)*, and can require alteration of impounding works if they cause environmental damage *(4)*.
- It can vary a licence for impounding in a way that requires the impounding works to be modified *(22)*.
- It can transfer an abstraction licence from one water utility to another, the latter compensating the former *(26)*.
- After consultation with the WSRA, the EA can propose that a bulk supply agreement between water utilities be entered into where necessary for the proper use of water resources *(31)*.
- Water undertakers have a duty to conserve water resources *(82)* and must produce water resource management plans *(62)* and drought plans *(63)*. They must fluoridate water supplied if the appropriate health authority requests it *(58)*.
- The WSRA must protect the interests of water and sewerage customers by promoting effective competition for supply of services required by the Water Service Companies *(39)*, and statutory water utilities must disclose if they link the remuneration of the directors to standards of performance *(50)*.

A further addition to the Agency's responsibilities was made in 2003 by The Water Environment (Water Framework Directive) Regulations (SI 2003/3242). Under these, the EA was appointed the 'competent authority' to carry out the requirements of the EC Water Framework Directive 2000/60 (Section 2.4).

HM Inspectorate of Pollution (HMIP) was first set up under the Control of Pollution Act 1974 and its scope was widened under the Environmental Protection Act 1990. It now forms part of the EA and deals more especially with controlling the discharge of industrial wastes (solid, liquid and gaseous), including trade wastes discharged to public sewers. Several EC directives set limits for the discharge of certain dangerous substances to the aquatic environment. In response, Defra maintains a 'Red-List' of substances needing priority control with the objective of reducing discharge loads to the absolute minimum. Except for the metals mercury and cadmium, all the Red-List substances are organic compounds, mostly pesticides and herbicides; although some are particular wastes from certain types of industry. Two different quality standards have to be applied to the discharge of wastes to surface waters – the more stringent governs:

- an 'environmental quality standard' (EQS), or
- 'best available technology not entailing excessive cost' (BATNEEC) – sometimes alternatively denoted as a 'uniform emission standard' (UES).

An EQS is set by estimating the effect the discharge of a substance has on the environment; this involves research and monitoring of such discharges by the EA, followed by the setting of an appropriate EQS by Defra. For the BATNEEC or UES standards, HMIP approaches an industrial polluter who is producing or discharging a Red-List substance to assess what process should be applied to render it suitable for discharge and to set limits for the amount discharged. Neither the relative volumes of the discharge and the receiving water nor the existence of other discharges of similar substances are taken into account, since receiving waters have no acceptable assimilative capacity for Red-List substances. HMIP has to deal also with radioactive wastes under the Radioactive Substances Act 1993.

Agricultural Pollution forms another area of pollution control. 'Point source' pollution can occur from slurries, silage effluents, yard washings and vegetable processing wastes. These all have high BOD (biological oxygen demand), silage effluents exceptionally so, and the slurries and yard washings have high ammonia content as well. Guidance to farmers and grants for measures taken are provided by Defra in the publication *Protecting our Water, Soil and Air, A Code of Good Agricultural Practice for Farmers, Growers and Land Managers*, 2009. The legal requirements on farmers are set out in the *Control of Pollution (Silage, Slurry and Agricultural Fuel Oil) Regulations 2010* (Statutory Instrument No. 2010/639). Many treatment methods have been tried on farm wastes but the simplest of them is to store slurries and sludge drainage for 9 to 12 months in open ponds, after which they can be sprinkled evenly over grassland in the right weather conditions, avoiding any direct run-off to a watercourse (Barker, 1991).

'***Diffuse Pollution***' (not arising at a point) of the environment occurs from the use of pesticides (including herbicides) and therefore control has to be applied to the usage. Metaldehyde pesticides are currently attracting considerable attention from the English and Welsh water companies and the DWI. However, not all such pollution comes from farming. Increasing detection of the presence of the herbicides atrazine and simazine in groundwater was attributed mainly to their use by road and rail authorities.

Dealing with the many types of pesticides and herbicides used is a complex matter, as illustrated by the fact that Defra lists about 400 approved compounds. Some compounds have been banned, including DDT, aldrin, dieldrin and chlordane. An added complication is that farmers rotate the use of different pesticides from year to year to avoid build-up of resistance; this increases the difficulty of monitoring for such contaminants. For the 2015–20 investment period, some water companies have agreed to work with farmers to reduce or eliminate the use of some pesticides including compensating farmers who use more expensive less harmful chemicals.

Defra designates 'nitrate sensitive areas' within which farmers are assisted to adopt practices designed to limit, and possibly reduce, the amount of nitrates found in water. EC Directive 91/676 made a similar policy mandatory on member states, stipulating the need to identify 'vulnerable zones' where underground or surface waters used for drinking would contain more than 50 mg/l nitrate if no protective action were taken. Among the practices required by the EC Directive are measures to limit the application of nitrogenous fertilizers and manures and to prohibit the application of nitrogenous fertilizers during certain periods of the year.

Solid Waste Disposal sites are, in the first place, the responsibility of the County Council Waste Regulation Authorities. However, before licensing a site a County Authority must obtain agreement of the EA and of any water undertaking likely to be affected. Compliance with EC Directive 80/68, *Protection of Groundwater against Pollution Caused by Certain Dangerous Substances*, is required. This prohibits certain substances from waste disposal landfill sites entering groundwater and limits the amount allowable for others; it has led to use of membrane seals under landfill.

The substances prohibited are: cadmium, cyanides, mercury; mineral oils and hydrocarbons; organohalogen, organophosphorous and organotin compounds; and those possessing carcinogenic, mutagenic or teratogenic properties.

Substances that must be limited (although no limit criteria are stated) include: the inorganic substances normally limited in drinking water supplies e.g. arsenic, chromium, etc. (Table 7.1(A) of the Directive) plus biocides, taste or odour producing substances and toxic compounds of silica.

Contaminated Land has to be identified and registered by the local authority who must inform the owner thereof and the EA. The local authority can require the owner to carry out remediation of the land or can undertake remediation itself, charging the cost to the owner. If, however, the EA decides the land is a '*special site*', it takes over responsibility for enforcing or undertaking the necessary remediation measures. A special site as defined by Defra is one which would or might cause serious harm, or serious pollution of 'controlled waters', as defined in the Water Resources Act 1991, Section 104.

2.6 PUBLIC WATER SUPPLIES IN SCOTLAND AND NORTHERN IRELAND

Since 2002 in Scotland and 2007 in Northern Ireland a statutory corporation owned by the national government has been responsible of supplying water within each country, each corporation being managed by three regulators.

Scottish Water (SW) supplies water to a population of over 5.1 million, operating 256 water treatment works, and 290 water supply zones, covering an area of 79 800 km^2. SW is regulated by:

- Water Industry Commission for Scotland (WICS), the economic regulator;
- Drinking Water Quality Regulator for Scotland (DWQR); created under the Water Industry (Scotland) Act 2002, the Regulator ensures the quality of drinking water through a process of inspections and monitoring in compliance of the requirements of the Water Supply (Water Quality) Regulation (Scotland) 2001. The Regulator is supported by a small team within the Drinking Water Quality Division of the Scottish Government; and
- Scottish Environment Protection Agency (SEPA) was set up to take over the work of the previous 10 River Purification Authorities (RPAs) in controlling abstractions, authorizing effluent discharges and monitoring their quality.

SW has the duty to promote the conservation and effective use of water resources, a duty previously held by the Secretary of State for Scotland. Although, unlike the previous publicly owned water undertakings in England and Wales before privatization, SW was given powers to seek private (i.e. commercial) finance for capital investment projects termed Private Finance Initiatives (PFIs). The device has not been used for the provision of water supply services, only for wastewater assets.

From 1 April 2008 the provision of water and waste water services was split into wholesale and retail, with SW being responsible for the provision of water to all domestic consumers and bulk supplies to 'Licensed Providers' (LPs) who sell water to non-domestic customers; the retail market. A Central Marketing Agency (CMA), was also set up to administer the market and to establish, maintain and administer the standards expected of market participants, and the processes for calculating charges between retailers, reconciling retail and wholesale charges and for non-domestic customers switching between LPs.

Northern Ireland Water (NIW) supplies water to about 1.8 million people, operating 25 water treatment works and 50 water supply zones, covering an area of 13 840 km^2.

The utility is regulated by:

- Utility Regulator is the economic regulator;
- Drinking Water Inspectorate, a unit within the Northern Ireland Environment Agency, is responsible for ensuring the quality of drinking water through a process of inspections and monitoring in compliance of the requirements of the Water Supply (Water Quality) Regulations (Northern Ireland) 2007;
- Department of the Environment (NI), as set out in the Water (Northern Ireland) Act 1972, is responsible for the aquatic environment including water abstraction and impounding through the Water Abstraction and Impounding (Licensing) Regulations (Northern Ireland), 2006.

ORGANIZATION

2.7 ORGANIZATION OF A WATER UTILITY

The organizational structures of water utilities are diverse; depending on:

- the size of the utility;
- its breadth of responsibilities which could include wastewater disposal, highway and storm drainage, flood defence and regulation of discharges by others;
- its financial structure;
- how it procures its services;
- how 'vertically integrated' it is: bulk supplier, distribution, wholesaler, retailer.

However it is now rare for water utilities to be responsible for flood defence from rivers and the sea and for them to regulate discharges direct to watercourses or the sea by others; generally these duties remain a part of central government. The description below is limited to those water utilities that are responsible for water supply to, and possibly wastewater disposal from, both domestic and non-domestic customers.

There is wide diversity of water utilities in Europe (Section 2.4). In the UK most companies remain vertically integrated: that is they are responsible for the supply of all water related services from water resource development to customer billings; however, this is likely to change from 2017 when wholesale and retail separation takes place for England and Wales (Section 2.5). A similar model to that described in Section 2.6 for Scotland is currently being proposed but with the additional opportunity for competition in upstream resource development.

In the past water undertakers carried out most duties in-house. However, increasingly, many duties are outsourced. It is now common to see engineering design and construction together with scientific services such as water quality sampling being outsourced. In an extreme case all basic duties could be outsourced with the water utility retaining only a small core of senior managers and regulatory personnel in-house.

However a water utility procures its services, the same functions, covering the following main areas, are needed:

- ***The Board*** is responsible for policy and corporate governance. The Board will be assisted by an audit committee that undertakes internal audits to ensure that correct corporate governance is being followed.
- ***The Chief Executive and Executive Directors*** are responsible for implementing the Board's policy. They are therefore responsible for the day-to-day running of the utility.
- ***Operations*** are responsible for the technical operations, likely to be split into a number of departments such as water resources, water treatment, water distribution, leakage control, emergency planning and response and asset maintenance. Where relevant there may be equivalent wastewater responsibilities.
- ***Customer Services*** generally comprise two parts: customer billing and customer contacts. The former prepares and issues bills, deals with customer queries and manages debt recovery. The latter responds to all non-billing related customer contacts. These are generally of a technical nature such as interruptions to supply, water quality and flooding from sewers. Customer services work closely with operations to rectify technical deficiencies that impact the customer. A major part of customer services is the call centre, a vital part of any water utility as it is where it interfaces most directly with the customer. Where water supply connections are metered, meter reading is the responsibility either of customer services or of the billing section. In the separate water market as due in the UK under the Water Act 2014 (Section 2.5), the wholesaler will retain responsibility for domestic customers, water quality and other performance measures and the retailer will be responsible for the non-domestic market.
- ***Strategic, Asset Management and Capital Investment Planning***. The water industry is capital intensive with large programmes of capital investment. The water utilities' assets have to be managed through the asset life cycle of:

  Appraise → Design → Procure → Construct/install/demolish and replace → Operate and maintain → Appraise

  Apart from the operation of the asset, all parts of the life cycle are the responsibility of the capital investment department. Many of the detailed aspects of the department can be outsourced but key functions, such as overall programme management, strategic planning and client project management, are likely to be retained in-house.
- ***Scientific Services*** comprise water and wastewater sampling and laboratory testing. This is a key function since the maintenance of water and wastewater quality standards are central to the responsibilities of any water undertaking.

These main functional areas will be assisted by a number of support departments, such as Finance and Accounting; Human Resources; Payroll; Regulation; Health and Safety.

2.8 STAFFING LEVELS

Water utility staffing levels expressed as a number per property served depend on many of the considerations covered in Section 2.7. Staffing levels are influenced by numerous factors including the size and characteristics of the company supply area, the level of service provided to the consumer, staff salaries, work culture, organizational structure and the extent of outsourcing, control and optimization. Typical staffing levels for water utilities in the United Kingdom range between 0.7 and 1.6 employees per 1000 connections. These levels represent the situation where the majority of the capital programme is outsourced to external designers and contractors but all other functions remain in-house. International staffing levels range between 1.5 and about 3.0 employees per 1000 connections for utilities with similar characteristics to UK companies. International utilities that utilize less sophisticated technology, deliver a lower level of service or intermittent supplies, or where labour costs are low, tend to operate with higher staffing levels, typically between 2 and 20 employees per 1000 connections.

PROJECT APPRAISAL AND FINANCING OF CAPITAL WORKS

2.9 APPRAISAL REQUIREMENTS

For a water supply project involving capital works to proceed:

- the need for it has to be demonstrated; e.g. demand, quality and levels of service, public health and efficiency gains;
- where government regulations or financing institution rules require, the project is justifiable on economic and environmental grounds and that the selected option is the most favourable;
- any loans required for financing can be repaid.

The extent of study required depends on the owner's status (public or private), sources of finance proposed and the legislative framework applying. Aspects of some common project 'drivers' are discussed elsewhere, e.g. demand (Chapter 1); water quality (Chapter 7); service levels (Section 16.2); energy efficiency (Section 13.2).

Economic and financial appraisals are discussed in the following sections. The reader will find useful background in *An Introduction to Engineering Economics* (ICE, 1969) and is referred to literature published by agencies such as the UN and World Bank. The ADB *Guidelines for the Economic Analysis of Water Supply Projects* (ADB, 1998) and the *Handbook* on the same subject (ADB, 1999) cover the principles and methods involved, as well as specific ADB requirements. Some of the key terms used in economic and financial appraisal are explained in Table 2.2.

Many of the variables used in project appraisal cannot be foreseen accurately; especially the less easily quantifiable environmental and social costs and benefits. For this reason studies usually include analyses of sensitivity of the conclusions to possible differences (e.g. in price) and of risks (e.g. to construction period or customers' ability to pay). The sensitivity analyses should be taken into account in any decision making.

Table 2.2 Project appraisal terms

Term	Meaning	Explanation/use
(A)IEC	(Average) Incremental Economic Cost	Method of comparing options as alternative to NPV or IRR. It provides information on long-run marginal costs (LRMC) and is useful in tariff determination
CBA	Cost–benefit analysis: • Financial • Economic	Comparison of costs and benefits using constant: • financial (actual) prices • economic (true or real) prices after (1) removing distorting effects such as subsidy, shortage, artificial exchange rates and (2) allowing for opportunity costs
EDR	Equalizing Discount Rate	For comparison of two options by finding the discount rate at which preference for one changes to the other. Requires more calculation than NPV
IRR	Internal Rate of Return	The discount rate that results in zero NPV. It should be greater than the minimum return required on capital (OCC). Requires more calculation than NPV. EIRR is the Economic IRR and FIRR is the Financial IRR (shows profitability)
LCA	Least cost analysis	Selection of option with lowest IEC. Can be determined using NPV, IRR, EDR or payback period methods
NEB	Net Economic Benefit	Economic benefit for each year (for use in discounting)
NFB	Net Financial Benefit	Financial benefit (revenues less costs) for each year (for use in discounting)
NPV	Net Present Value	The value today of a future cash flow. Determined by discounting future inflows (revenue) and outflows (construction, and operational and maintenance costs). A simple method but it provides no information about the unit cost of water. ENPV is the Economic NPV
OC(C)	Opportunity Cost (of Capital)	Value of commodity (such as capital or water) if used for other purposes than the project
PP	Payback period	Length of time from initial capital outlay (e.g. for construction) to when revenue equals that outlay
r	Discount rate	The return which capital could obtain on open market investment
STPR	Social Time Preference Rate	Rate for discounting given in Green Book (HM Treasury, 2003)
SP	Shadow price	Price that reflects scarcity or real value of a resource

2.10 COMPARISON OF PROPOSED CAPITAL PROJECTS

Increasingly infrastructure investment plans are derived from the optimization of a set of individual initiatives required to maintain and extend the service and thereby achieve the required outcome for consumers. The optimization process is driven by a set of criteria, including minimum acceptable service, compliance, cost and risk. Comparisons of capital schemes should therefore include both Capital and Operational expenditure ('Capex' and 'Opex'), termed Total expenditure ('Totex'), for all costs of all scheme components.

Discounting to Compare Present Values of Projects. If two or more different projects are possible to meet an expected demand for more water, they will almost certainly differ in cost and probably also in the timing of the capital outlay required. They may also differ in the balance of capital to operating cost. One way of comparing them is by 'discounting' each scheme's costs to the present, to obtain the '*total present value*' cost of each project. This allows the timing of expenditure to be taken into account.

The present value, P, of £X due to be paid in n years' time is taken as $P = £X/(1 + r)^n$, where r is the discount rate expressed as a decimal. This is the inverse of the compound interest calculation, i.e. P invested at $r\%$ compound interest accumulates to £X in n years' time. To compare projects (or options), the compared project must meet the same objectives to the same extent. An example is meeting demand increases over the same period – usually 20 or 25 years. The capital expenditure and works renewal costs, plus the running costs, for each year are estimated and discounted to give their equivalent present value cost and the total is summed for the chosen period.

The UK Government 'Green Book' (HM Treasury, 2003) adopts the concept of the '*Social Time Preference Rate*' (STPR) and recommends a rate of 3.5%. This is a 'real' rate applicable to analyses based on current costs without allowance for inflation. The STPR is made up of two components – time preference and an allowance for real future growth in per capita consumption. Both are related to society's perception of the excess of the value of present consumption over future consumption. Sensitivity checks using values either side of the recommended 3.5% may need to be carried out. For very long periods (over 30 years) the Green Book suggests that discount rates reducing with time are appropriate. Outside UK the advice of the relevant government or funding agency should be taken on discount rates to be used. The present discount rates contrast with those adopted in the 1980s at a time of high interest rates (and inflation).

The higher the discount rate, the smaller is the '*present value*' of a future cost. Hence high discount rates tend to favour schemes which can be built in stages, or which have a lower capital cost despite having higher running costs.

Inflation of Prices. The treatment of inflation should be consistent. Present prices can be used with a 'real' discount rate (excluding inflation) or costs that are escalated for inflation may be taken with a discount rate that includes inflation. Nevertheless, inflation is not usually taken into account when discounting, on the assumption that it affects all prices proportionately. History, however, shows otherwise: as standards of living have increased, labour costs have inflated more than material prices due to increased machine production. Fuel oil prices doubled in 1973 and then fell back, then rose again later, the cycle repeating itself in the last decade. Hence, possible differences of inflation should be allowed for in sensitivity analyses (Section 2.9).

Shadow Pricing of costs is sometimes adopted when market prices do not represent 'true' costs. Thus, if unemployment is high, the shadow price for labour is the wage paid (including on-costs) less the cost to society of that person when unemployed (e.g. unemployment benefits). Taxes are excluded from shadow prices because they represent only a transfer of money from one section of society to another. Where the unofficial exchange rate implies that offshore goods and services cost more in local currency than apparent from the official exchange rate, the shadow price for purchase of offshore goods may need to be based on the unofficial rate because this represents the real cost of offshore purchases. Such shadow pricing has the benefit of favouring a scheme that uses more local labour or inshore goods than another does, an advantage to the country in which the project is to be built. However, carried to extremes, shadow pricing is complex, because each item price has to be broken down into its component parts e.g. labour, fuel, plant, materials, etc. and also broken down into offshore and onshore elements, with taxes taken off. Consequently shadow pricing tends to be adopted only when a funding agency requires it, possibly for labour only.

Short- and Long-Term Marginal Costing. When a utility has spare capacity at its source works, it can meet additional demand at little extra cost because over 90% of a water company's annual expenditure is fixed, '*short-term marginal cost*'. Debt repayments are fixed and the operational costs involved in the day-to-day running of a water system are virtually fixed because less than 10% of the cost varies with the amount of water delivered. However, if new major works have to be built, the cost of the supply rises sharply because the loan repayments on the new works have to be met immediately, whereas the initial rise in consumption can be small and the increase is gradual. To avoid this sudden rise in charges, '*long-term marginal costing*' can be used, under which account is taken of the cost of the next stage (or stages) of capital expenditure required. This means that, before the new source works are actually required, consumers pay more than current costs; therefore money is set aside to fund partially the cost of the new works; price rises are thus smoothed out. The early increase in charges may induce metered consumers to curb their demand, which itself may postpone the date when the new works are required.

How much current consumers should be asked to pay for works that benefit future consumers is debatable; but often current consumers benefit from works still in use but long since paid for by previous consumers. Arguments about the monetary value to be put on paid-off assets still in use are only relevant in the relatively rare case when a water utility is put up for sale. However, more importantly, a utility should not become so heavily burdened by long-term loans, that it finds itself in financial difficulty if unexpected heavy financial commitments should arise. This situation occurred in England, where many new schemes had been built in the 1960s and 1970s to meet swiftly rising demand for water. As a consequence, the water authorities were already burdened with long-term debt at a time when they were required to meet new standards of water quality as well as for effluent discharges. Hence, when the UK Government decided to privatize the water authorities (Section 2.5), it wrote off £5.2 billion of their long-term debt (Kinnersley, 1994) in order to make them financially attractive.

The ***Internal Rate of Return*** (IRR) is another way of comparing projects (Table 2.2). In this case the 'value' of 'the outputs' has to be assessed. For water supply projects it is difficult to establish an appropriate value for the supply. Tariff charges can be used but it may be more appropriate to attach a value to the service that takes account of secondary benefits such as health, environment, comfort, recreation, trade, industry and food production. One option is to relate the 'value' of domestic supply to a proportion of average family household income. The value of water used for trade or industry generally might be judged from the elasticity of the demand for it, if that can be found out (Section 1.16). Often, however, the funding or loan agency will give guidance on what value to adopt.

Using the 'value' (in 'real' money terms) of, say, a cubic metre of water sold, one can calculate the present value of all the water supplied over the planning horizon. Then by trial and error one can find what discount rate makes the present value in real terms of the output equal to the present value of all the expenditure needed to construct and maintain the project. That rate is '*the return on capital*' or the IRR. The usefulness of the procedure to a funding agency or government is that it puts projects on a comparable basis indicating which projects give '*the best value for money*' from a funding point of view.

Cost–Benefit Analysis (CBA). A full social cost–benefit analysis tries to evaluate all the '*benefits*' and '*disbenefits*' which accrue from a project over a period, in order to compare options. When applied to a major water project, such as an impounding reservoir, it is difficult to put any clear 'value' to many of the disbenefits, such as the loss in perpetuity of the valley land covered by the reservoir. An '*Environmental Impact Assessment*' (EIA), now required for most major water projects, is a more informative approach because it sets out all the results of building a project whether such results can be evaluated or not. Whatever the method of comparing the social and environmental impacts of different courses of action, it is important that both benefits and disbenefits are taken into account.

'Whole Life Cost–Benefit Analyses' (WLCBA) generally compare a number of options to derive a preferred scheme solution. The selected scheme will exhibit a relatively high level of confidence for the early years where the basis of the development is founded on more reliable data. Predictions and hence options for later in the period will be less certain and therefore the scheme solution will need to have inbuilt alternative options and opportunities to manage divergence from the most likely sequence of interventions. Therefore the WLCBA needs to include the concepts of '*Best Case*' versus '*Least Worse Case*' options and '*Regret Cost*' within the analysis.

If the development plan, and hence proposed collection of assets, is set at the start of the development, there is a high risk that the installed assets may not have the optimum capacity and layout for the ultimate final development. The WLCBA should therefore include a regret cost. Predictions and hence options for later in the development period will be less certain. Therefore, the development plan should have inbuilt alternatives and opportunities to manage divergence from the most likely sequence of interventions. The regret cost represents the cost of not having the right solution. It is valued as the cost of overpaying for a larger asset that is built early to satisfy an ultimate demand which is not realized plus the cost of corrective interventions to deliver the required output in the right location. An example is a treatment works and associated pipework constructed to meet a long-term development forecast but with an immediate need; however, the longer term planning assumptions do not materialize and development takes place elsewhere. Hence there is a cost of the oversized assets as built and the additional cost of building the facilities in the alternative location. The regret cost is included in the WLCBA as part of scheme evaluation. The purpose of the approach is to minimize regret costs over multiple scenarios. In essence regret costs are an additional factor in a multi-criteria decision analysis.

In practice an optimized investment plan represents the 'least worse' set of interventions that would deliver the required outcomes. Over a short-investment planning period there is greater opportunity to manage change. However longer term planning decisions that are implemented early represent a significantly greater risk that they are not the ultimate optimum output.

The best case option may not be the least cost option, but will address the investment requirements in the shorter term taking into account all the necessary technical and softer opportunities. At face value and using traditional CBA optimization tools the best case solution would not necessarily stand up to financial scrutiny. However, by introducing regret cost into the analysis, best case options become more viable because they present the opportunity to limit the need for redundant assets longer term when the plan does not develop in line with the least worse interventions.

2.11 COMMENTS ON THE USE OF DISCOUNTING

From the end of the 20th century project evaluation has extended into a wide ranging assessment covering a number of social and environmental aspects as well as economic comparisons. The weight given to economic studies has reduced. One reason is the difficulty in accurately forecasting future requirements and costs and the balance between future costs of different types. It is possible to carry out sensitivity checks to test the impact of a number of possible variables such as discount rate, energy costs and labour costs. However, the resulting economic analysis remains only part of the evaluation process (Sections 2.10 and 2.12). It can help decision making but should not be the basis of it.

Economic comparisons of long-term developments have been used in the past to plan sequences of investment to meet demands projected over a long period. However, the result depends on the timing of the analysis. Due to the effect of discounting future costs, the earlier projects in the sequence with the least NPV will tend to be those requiring less capital expenditure and more operating costs. If medium term expenditure plans are based on periodic updates of the economic analyses, high capital cost projects with lower operating cost are likely never to get built. This is not necessarily compatible with a desire to conserve resources, particularly energy.

2.12 SUSTAINABILITY, RESILIENCE AND ENGINEERING CHOICES

The definition of sustainable development adopted by the UK water industry, in common with UK Government, is 'development which satisfies people's "basic" needs and provides a better quality of life without compromising the quality of life of future generations' (WSI, 2007). Resilience is defined as 'the ability of assets, networks and systems to anticipate absorb adapt to and/or rapidly recover from a disruptive event' (HM Cabinet Office, 2011). It should, therefore, be clear that, as for many other systems (transport, hospitals, emergency services, education and law and order), absence of spare capacity or redundancy in a water supply system in normal circumstances renders it incapable of dealing with a disruptive event of any magnitude without trauma to the consumer or wider public.

Implementation of the Water Framework Directive in Europe has influenced the way water is made use of in the environment and has already directly affected water supplies. However, the way water supply developments are carried out must include efficient use of non-renewable resources (including fossil fuels), waste minimization and recycling; these aspects are discussed in Chapter 13.

The choices made in implementing water supply developments increasingly will be influenced by sustainability, security of supply, consequences of failure to supply and resilience.

Balancing the effects of use of different kinds of resource against different forms of consequence and against cost is a matter for widely different interpretation. Making such choices in a way all can agree will require consensus on the approach and on the weight to be applied to each factor.

2.13 FINANCING OF CAPITAL WORKS

A water utility often borrows money to finance construction of major new capital works because the cost of such works is too large to meet from current income; however, it can be borne reasonably by future consumers who will benefit from the works. Before agreeing to make a loan a financial institution will need to be confident that it will be repaid. The most straightforward way to demonstrate capacity to repay a loan is by use of cash flow analysis. This will form the basis of a financial plan and will show annual expenditures, including construction and operation and maintenance costs, and revenues. It will also include inflows from loan payments and outflows on repayment. If funds are to be raised by sale of 'equity' (share capital) receipts from the sale and future dividends would also need to be included.

Loan repayments are typically 10–15 years for plant and machinery; 20–30 years for buildings; and 50–60 years for major civil works such as dams and land. Routine capital works expenditure on extensions to pipelines can normally be met from current income. Three methods of loan repayment are possible: Capital plus interest, Annuity and Sinking fund. If a loan of £X is repayable over n years at r% interest rate (where r is expressed as a decimal) the annual amounts payable under the three different methods are as follows:

Method	Annual payment
Capital plus interest	$X/n + r$ times the loan balance outstanding at the start of the year
Annuity	$Xr(1 + r)^n/[(1 + r)^n - 1]$
Sinking fund	$Xr/[(1 + r)^n - 1]$ to sinking fund + Xr interest on loan.

The annuity repayment method is the most usual. The capital plus interest method results in reducing yearly payments. The second and third are equivalent if the interest received on the sinking fund is the same as that charged on the loan; the third tends to be adopted only if the interest rate on the sinking fund investment is more than the interest charged on the loan.

Public water authorities may be able to borrow from government, from an international lending agency, or may be authorized to borrow from the money market. Private companies usually borrow from the money market. In theory they can also raise more money by offering further shares for sale on the market if their full share capital has not been issued; but this would rarely be adopted for new works. Of course if an undertaking has an excess of income over expenditure, part of the excess can be set aside to build up a fund to help finance new capital works. If a variety of capital projects has to be funded, short-term borrowing from current funds or from the bank may be adopted to fund the capital outlay, until the time is considered appropriate for raising a single long-term loan. Much depends on interest rates prevailing and whether these are expected to move up or down in the future.

2.14 DEPRECIATION AND ASSET MANAGEMENT PLANNING

'Depreciation' is an accountancy term for the practice of writing down the initial cost of an asset annually. In company accounts the amount an asset is depreciated is debited against income, with the result that an equivalent amount of money (in cash or securities) is set aside and allocated to a depreciation fund, which would be available later to meet the cost of renewing the asset when it becomes worn out. Depreciation has been defined as representing – 'the consumption of assets by current users thereof'. The amount of the depreciation depends on accountancy practice; it needs to be sufficient to write off the whole cost of the asset before it has to be replaced. However, adjustment may be needed to allow for higher cost of replacement due to inflation or 'betterment' for new standards and improved efficiency.

The technique of depreciation is particularly suited to manufacturing companies. Application to a water utility may not produce the desired result with respect to some large, long-life assets, such as dams and large pipelines. These may continue in use and to present a liability long after they cease to have any 'book' value. Nevertheless, it is prudent to build up a 'contingency fund' to meet future repair or renewal costs.

Of more value in the financial planning of a water utility, is the 'Asset Management Plan'. This requires assessing the condition of each asset and estimating its remaining life and replacement cost. However, each of these estimates is prone to errors, which may be compounded by difficulty in forecasting future interest rates. All assets should be treated individually; as a result the exercise is a major task for an asset rich entity such as a water utility. Usually a long-term plan for asset renewal is developed with a horizon of say 20–30 years and a 5-year rolling programme of renewals is produced in more detail with estimates of the year-by-year expenditure involved.

2.15 PRIVATE SECTOR PARTICIPATION IN WATER SUPPLY

Since the later 1980s there has been increasing interest in using the private (i.e. commercial) sector to provide some or all of the functions of a water supply company. Whilst the public sector needs always to retain ultimate control of water supply, the advantages of private sector participation can be:

- commercial funds for capital and improvement works may be easier to obtain than government or state funds (at least without adding to government debt);
- specialized technical and managerial skills not generally available within a small organization can be brought in to benefit a water undertaking;
- improved efficiencies of water services can be obtained by setting appropriate targets as contractual obligations on a private company.

Private companies have always been used by publicly owned water utilities to provide such services as the design and construction of new works, mains repair gangs, analysis of water samples and the like. A wider variety of contractual arrangements for private sector inputs has evolved, the two main types being: '*Design and Build*' (or '*Engineer and Construct*') contracts, typically for new treatment works, and '*Operations*' contracts for undertaking operational activities and provide technical resources, e.g. leak detection teams.

The '*Build, Own, Operate and Transfer*' (BOOT) contract was developed for the design and construction of new works. A private company, backed by a financial institution bank, or using a loan provided by its government or a funding agency, finances and builds a new treatment works, operates the new assets for a term of years using revenues to cover operating costs and to repay any loans and, at the end of the contract, transfers the assets to the public water utility. The operation period is typically 20–25 years. Construction costs may be repaid by the water utility to a defined schedule and payment for the operation of the works is on a time-related or works output basis.

A '*Build, Operate and Transfer*' (BOT) contract is similar but excludes financing. It is usually used where the contractor is paid in stages for design and construction of some works, after which he proceeds to operate the works for a given period on some agreed basis of payment. The contractor is responsible for all repairs and maintenance of the works during the operation period, at the end of which ownership is transferred to the water authority. Appropriate incentives can be included. For example, if the contract includes a fixed sum for maintenance, the contractor will endeavour to build to achieve minimum maintenance.

2.16 PRIVATE SECTOR OPERATION AGREEMENTS

Operations contracts involve a private company taking over and managing some or all of the operational activities associated with a set of assets for a defined period. In practice virtually all operations can be outsourced should the utility so desire. The principal types of operational agreements are as follows.

Contracting Out. A private company takes responsibility for a fixed duration for a task such as meter reading; or billing and revenue collection; or the operation of a set of assets such as a water treatment plant. The responsibility of the private company is limited to its contracted task.

Management Contracts. A private company contracts to provide a service to improve certain water supply operations for a fixed fee, or a fee partly or wholly based on achieving some specified performance target, such as: reduction of leakage or betterment of service levels to consumers. In practice management contracts have had mixed success globally. The more successful contracts are those where the contractor is given the greatest freedom to control and manage the assets, thereby deliver or outperform the performance goals and benefit from performance based payment terms. Contracts which restrict a contractor's actions over the operation of an asset or where an uncoordinated third party is also working on the same assets at the same time generally do not deliver the desired outcome either for the asset owner or the contractor.

Leasing (or Affermage). Under leasing a private company takes over all, or some of a water utility's operations for a fee, or a share of the income collected from consumers. Specific targets are normally set. Leasing does not usually involve the company (i.e. lessee) in financing any new capital works. The lessee is principally responsible for the management and technical aspects of running the works including their routine maintenance. If the lessee's payment comprises a share of the utility's income, he has opportunities for increasing his profit by improving the efficiency of the water utility. The duration of leasing agreements is normally 10–15 years.

In many French cities, mainly for the operation of treatment works, leasing (termed Affermage) has been widely used for many years for fixed periods. One of the reasons for this is that, whereas in the UK there are only 12 major water and sewerage companies with 14 smaller water-only companies, France has over 29 000 separate water utilities (MoE, 2008) and Germany has 6210 (GWS, 2011). The use of large companies allows a degree of aggregation where many small undertakings exist.

Concession. Under a concessionary agreement a private company takes over full responsibility for a whole system, including planning and funding all capital works needed for rehabilitation and expansion of the system, taking on most risks in return for receiving all, or part of the total income generated from water sales and other charges. This motivates the private company to use its expertize to develop the system in the most efficient manner in order to maximize its income. The duration of such agreements is normally 20–30 years.

The drawing up and settlement of any long-term contract is a complex matter on which it is essential to take experienced advice. It may take 2 years or more to produce bid documents, evaluate bids and negotiate a contract. Bidding costs for contractors are high; few bids may be received or bids may only be received for the less onerous and less risky contracts. Regulator and supervisory bodies will need to be set up to monitor performance and adjust tariffs.

With concessions it is difficult to achieve a balance between income and expenditure through the period and to ensure that investment decisions are not distorted by the time remaining under the concession. It is also difficult to assess the risks, misjudgement of which can lead to an unfair financial advantage or penalty for one party or the other, thus leading to disputes. A number of disputes have arisen over the information provided by the utility about the condition of assets (particularly buried assets) at the time of bidding.

2.17 CHARGING FOR PUBLIC WATER SUPPLIES

In some countries water supplies are given free or below cost as a social service to those on very low incomes. Most standpipe supplies in low-income countries are given free. The consequent financial cost has to be recovered either by the water supplier charging other consumers more; '*cross subsidy*', or by the state or government meeting the deficit; '*subsidy*'. In the 1940s to 1960s in the UK the extension of water mains to rural areas had to be funded by central government or local government through a series of Rural Water Supply Acts because the cost of the long water mains required to supply rural dwellers did not give an economic return for the water undertakings and was too expensive for rural dwellers to finance.

Where domestic supplies are unmetered, as is common practice in the UK, a charge related to the value of a householder's dwelling is applied to domestic consumers; this can result in a socially equitable structure since most low-income householders occupy low-value property. Where domestic supplies are metered, a basic quantity can be allowed for each household (per charge period) at a charge sufficiently low for the low-income householders to afford. Neither arrangement is perfect since there are some low-income householders who occupy large properties because of a large family size and, with metering; if the lowest rate of charge covers basic needs for a large family then it gives a small family a liberal supply at the cheapest rate.

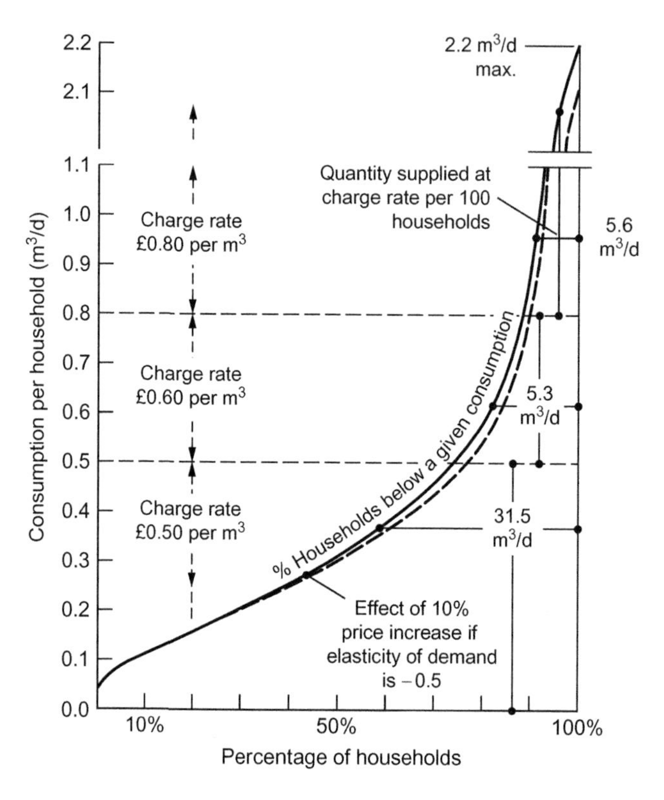

FIGURE 2.1

Cumulative frequency graph of consumption per household.

With charges fixed according to property value, it is relatively simple for the water utility to set them at a level that will provide the required income because the number of properties in each valuation range is known. However, if all supplies are metered, the setting of an appropriate tariff to produce a given income has to take the elasticity of the demand into account (Section 1.16). This is particularly important because the margin of a utility's annual income over expenditure to pay loan charges and to provide a profit may be heavily dependent on the larger consumptions taken by the wealthier domestic consumers. A graph of the percentage of households taking less than a given amount needs to be produced, as shown in Figure 2.1. The area under this graph represents the quantity of water taken under each charge band. Assuming the elasticity (e) of demand for the lowest band is -0.3, and 31.5 m^3/d is taken per 100 households within that band when the price is £1.00/m^3; then the amount taken (V) when the price is raised by 10% to £1.10/m^3 is given by:

$$V = 31.5 \times [(1.00)/(1.10)]^{0.3} = 30.6\ \text{m}^3/\text{d per 100 households}$$

With this value of e, a 10% rise in the rate of charge results in a 6.9% increase of income for that band. Proceeding thus one can find the approximate total revised income to be expected from domestic properties, and a similar approach can be used for finding the income to be expected from a charge increase on metered trade and industrial demand. Comments on appropriate values for price elasticity are given in Section 1.16.

2.18 COMPARISON OF CHARGES FOR WATER AND OTHER DATA

Table 2.3 summarizes the results of a worldwide tariff survey for 2014/15 published annually in Global Water Intelligence. The data demonstrate the ranges in tariffs both globally and within a region. However, it is worth noting that the data sets include linear, rising and decreasing tariffs, that local currencies were converted to US$ at 2015 mid-year exchange rates and that individual tariffs may be influenced by operational, management and social structures and strategies of the organization and may include incentives and subsidies.

Table 2.3 Tariffs for water in 370 cities of the world for 2015[a]

Region	Number in sample	Number making no direct charge	Average tariff (US$/m^3)[b]	Tariff range (US$/m^3)
Western Europe – North	50	1	2.29	0.80–3.76
Western Europe – South	18		1.12	0.23–2.50
Eastern Europe	40		0.87	0.10–1.80
Middle East	14		0.88	0.03–2.48
Central Asia	22	1	0.54	0.06–1.82
South Asia	23		0.11	0.03–0.29
China/Mongolia	26		0.40	0.25–0.68
Far East	21		0.85	0.22–1.38
South East Asia	15		0.33	0.06–1.21
Australia and New Zealand	6		2.52	1.03–3.45
North Africa	8		0.44	0.10–0.80
Sub Saharan Africa	30		0.62	0.14–1.63
Latin America	32		0.71	0.18–1.20
Caribbean	9		1.19	0.01–3.64
US and Canada	56		1.56	0.43–2.72

Source of information: GWI, 2015 (2015_GWI_Global_Water_Tariff_Survey.xlsx).
Notes: [a]Tariffs do not include connection or indirect charges.
[b]Tariffs are not weighted for volume of total supply in region.

Tariffs in the US are significantly lower than in comparable developed countries. Black & Veatch's 2016 Strategic Directions: Water Industry Report (B&V, 2016) identifies a looming funding gap in US water utilities and the need to raise awareness of the value of water. The report also indicates that there is increasing interest in the USA in private sector participation in water project funding.

In 2013–14 over 45% of households in England and Wales were connected to a metered supply, with the proportion ranging between suppliers from 23% to 76%. Meter penetration is forecast to reach 90% by about 2050. The average household bills for water 2013/14 were as follows:

	Unmetered £ per annum		**Metered £ per annum**	
England and Wales[1]	Median	Range	Median	Range
Water Services Co's. (10 No.)	210	165–347	167	131–210
Water-only Co's. (12 No.)	200	98–255	153	88–200
Scotland[2]	187	125–374	Not applicable	

Sources of information: [1]Ofwat. *Average household bills information 2013–14.*
[2]Scottish Water website; Council Tax Band range; Median is Band D.

Tariffs for metered domestic supplies in England and Wales averaged about £1.30/m^3 (range £0.65–£2.05) plus a standing charge in most cases of between £23 and £96 per annum.

REFERENCES

ADB (1998). *Guidelines for the Economic Analysis of Water Supply Projects.* ADB.

ADB (1999). *Handbook for the Economic Analysis of Water Supply Projects.* ADB.

Barker, P. (1991). Agricultural pollution control and abatement in the upper Thames region. *Journal IWEM,* June, pp. 318–325.

B&V (2016). *2016 Strategic Directions: Water Industry Report*, Black & Veatch Insights Group.

CN (1994). *Construction News.* 1 December.

Eckstein, G. A. (2005). Hydrological perspective of the status of groundwater resources in the UN watercourse convention. *Columbia Journal of Environmental Law,* June.

GWI (2015). *GWI 2015 Tariff Survey.* Global Water Intelligence, Oxford (2015_GWI_Global_Water_Tariff_Survey.xlsx).

GWS (2011). *Profile of the German Water Sector, 2011.* Gass und Wasser mbH, Bonn.

HM Cabinet Office (2011). *Keeping the Countryside Running: Natural Hazards and Infrastructure.* London.

HM Treasury (2003). *Appraisal and Evaluation in Central Government.* Treasury Guidelines, London (Note: Originally published in 2003 with additional pages published in 2011).

ICE (1969). *An Introduction to Engineering Economics.* ICE.

ILA (1966). Helsinki rules on the uses of waters of international rivers and comments. *Report of the Fifty-Second Conference 484, Article II.* International Law Associates.

Isack, F. (1995). Money down the drain. *Construction News,* 31 August 1995, pp. 16–17.

Juuti, P. S. and Katko, T. S. (2005). *Water, Time and European Cities. Water Time Project*. EC Contract EVK4-2005-0095.

Kinnersley, D. (1994). *Coming Clean: The Politics of Water and the Environment*. Penguin Books.

MacDonald, J. A., et al., (1997). Improving service to small communities. *Journal AWWA*, January, pp. 58–64.

Melosi, M. V. (2000). *The Sanitary City. Urban Infrastructure in America From Colonial Times to the Present*. John Hopkins University Press.

MoE (2008). *Public Water Supply and Sanitation in France (with data for 2004)*. Ministry of Ecology, Energy, Sustainable Development and the Sea, France.

NCE (1994). *New Civil Engineer*. 1 December 1994 and 9 February 1995.

Newman, P. and Thornley, A. (1996). *Urban Planning in Europe. International Competition*. National Systems and Planning.

Okun, D. A. (1995). Addressing the problems of small water systems. *IWSA Conf. Proceedings 1995*, Special Subject 9, pp. 439–442.

Pontius, F. W. (1997). Future directions in water quality regulations. *Journal AWWA*, March, pp. 40–48.

Teclaf, L. A. and Utton, A. E. (1981). *International Goundwater Law*. Oceans Publishers Inc.

UN (1992). *Convention on the Protection and Use of Trans-Boundary Watercourses and International Lakes*. UN.

US EPA (1996). *Safe Drinking Water Act (SDWA) 1996*. US EPA.

US EPA (2013). *Public Water System Inventory Data*. US EPA.

WSI (2007). *Water Services Infrastructure Guide: A Planning Framework*. Water Services Infrastructure Group.

CHAPTER

Hydrology and Surface Supplies 3

PART I HYDROLOGICAL CONSIDERATIONS

3.1 INTRODUCTION

When a new water supply is required, it is necessary to assess whether the desired extra water is available and how its development might impact on existing water users and on water quality. In developed countries, reference should be made to existing river basin or water resource plans or, in Europe, to river basin management plans under the European Water Framework Directive (Section 2.4). A detailed impact assessment is likely to be required in support of any application for permission to abstract. In developing countries, a new assessment of the water resources may be needed to consider all possible sources and define a suitable development plan. The types of development are as follows:

Surface water sources – (1) direct supply from an impounding reservoir or lake, supplemented, if necessary, by gravity feed from an adjacent catchment or pumped inflow from another source; (2) abstraction from a river or canal, supplemented if necessary by releases from a storage reservoir; (3) collection of rainfall–run-off from the roofs of buildings or bare catchments and feeding to storage tanks.

Groundwater sources – (1) springs, wells and boreholes; (2) adits and collecting galleries driven underground; (3) riverside wells or sub-surface extraction wells sunk in the bed of a river course or 'wadi'; (4) artificial recharge of aquifers.

Water reclamation schemes – (1) desalination of brackish water or seawater; (2) re-use of acceptable wastewater discharges by appropriate treatment; (3) de-mineralization or other treatment of a minewater, including blending with a freshwater supply.

Other types of scheme – (1) integrated (conjunctive) use of surface water, groundwater or water reclamation schemes; (2) transfer of a bulk supply from another supplier or river basin.

Twort's Water Supply. DOI: http://dx.doi.org/10.1016/B978-0-08-100025-0.00003-X

Studies of the hydrology of a catchment should focus on dry years or critical drought periods but the overall available resource is defined by a catchment water balance. This quantifies for a given period of one or more water years:

- rainfall on catchment less losses from evaporation, transpiration and natural surface run-offs (i.e. after correcting for abstractions and discharges);
- resultant percolation into aquifers that results in outflow from the catchment below ground rather than as surface run-off;
- *plus* inflow *less* outflow across underground water boundaries;
- *minus* abstractions from wells and boreholes;
- *plus or minus* change in soil moisture content, underground water storage and water in transition zone;
- balance remaining = water unaccounted for.

If all inflows and outflows have been accounted for a balance of zero will be obtained. This will not be true where there are over-year storage effects (e.g. in reservoirs or aquifers) or where there are significant errors in the components in the water balance. However, once a reasonably satisfactory balance has been achieved there can be confidence that the hydrology of the catchment has been properly appraised.

3.2 CATCHMENT AREAS

A precise knowledge of the catchment area draining to a surface water source is an essential prerequisite for any hydrological assessment of that source. Where this is not known, it must be determined, usually from an examination of contours on a topographic map.

The catchment boundary is located by defining the direction of downhill flow at right angles to the contour lines. Particular care must be taken in areas of low relief and where watercourses, marked on maps, appear to cross the catchment divide. If not the top pound of a canal, these watercourses may well be a contour leat or catchwater designed to augment the flow into a reservoir or to bring water to a mill or mine from an adjacent catchment. Field visits to key sections of the topographic divide may be required to resolve uncertainties, identify apparent cross-boundary channels and check whether they are active or derelict, thereby carrying no flow into or out of the catchment.

Large scale maps showing topography in sufficient detail can be hard to obtain in some parts of the world. For very large basins, satellite imagery now widely available on the Internet may prove adequate. Alternatively, 1:500 000 scale maps produced by the USA Air Force in their 'World Tactical Pilotage' series (obtainable in the UK from the Director of Military Survey, Ministry of Defence or major map sellers) are recommended. Large inaccessible basins can be examined from satellite photographs, libraries of which are now available internationally. However, the best sources are those giving digital descriptions of terrain and rivers on CD-ROM; the most detailed global coverage being from ESRI, Redlands, California, the major American geographic information software company.

Catchment areas can be determined quickly from geographic information systems (GISs) and by the use of digital terrain models. These include computer programs that will define a catchment boundary from any gridded data set. One such program, available from the Centre for Ecology and Hydrology (CEH), permits catchment area and numerous other parameters to be computed for any catchment in the UK (NERC, 2009). However, care must be taken on small catchments ($<10\ km^2$) and in areas of low relief where the stream drainage network is poorly defined. Manual adjustments to computed values may be required in some cases.

The catchment contributing to a groundwater source may not be readily defined. In many cases, it corresponds closely to the surface water catchment so that water table contours reflect surface topography, albeit with a reduced vertical range. However, there are cases where surface and groundwater catchments differ markedly, making precise catchment water balance calculations difficult or impossible. Groundwater catchment divides can migrate seasonally, especially in response to pumped abstractions. Monitoring networks need to have very high densities of observation wells in order to track these variations, but this is rarely practicable except under research funding.

3.3 DATA COLLECTION

Good hydrological data forms the basis on which water schemes must be planned and designed. The collection, archiving and dissemination of hydrological data are expensive. It is usually a government funded activity. This funding can be vulnerable when governments try to reduce public spending as the impact is not felt immediately and the benefits of data collection are not always obvious. However, it has been demonstrated that the availability of reliable hydrological data results in benefits an order of magnitude larger than the costs of collection (Simpson, 1987). The need for more data is particularly important where climatic or environmental conditions are changing and water supplies are critical. In Nigeria, for example, it was reported in 1998 that failure of a number of large scale resource developments to achieve their forecasted output was partly due to the absence of accurate data on rainfall and river flows (Tor, 1998). If large capital sums are not to be wasted, the basic hydrological data on which water schemes are designed must be adequate and reliable. Where good hydrological data are available, very large savings can be made.

To assess a potential surface water source the prime need is for long streamflow records. Similarly, for groundwater sources, long records of changing aquifer water levels are required. Streamflow records need to be naturalized (Section 3.9) to eliminate the effects of any artificial influences. For this, details are required of all abstractions, effluent returns and reservoir storage changes in the catchment. Similarly, in order to interpret changes in groundwater levels correctly, details should be kept of pumped output from wells together with pumping and rest water levels.

Before embarking on detailed analyses, it is essential to check the validity of all basic data used. This may involve visits to monitoring sites to check their condition and probable degree of accuracy. If relevant data do not exist, temporary gauges or permanent systems must be set up to acquire it. Once data is obtained it should be carefully filed or archived so that it is permanently available for subsequent analyses such as updating yield estimates when catchment or other conditions change.

3.4 STREAMFLOW MEASUREMENT

River or stream flow records taken in the vicinity of an existing or proposed intake or dam site are an invaluable aid to the assessment of the potential yield of a source. The longer the period of record the more reliable any yield estimates are likely to be but even a very short record is often a significant improvement over estimates derived from generalized regional relationships.

Flows are often obtained by measuring the 'stage level' of a river, i.e. the elevation at some location of the water surface above an arbitrary zero datum. Continuous or regular measurements of stage are then converted to discharge by means of a rating curve. Increasingly, use is being made of methods that measure flow velocity directly, such as ultrasonic and electromagnetic devices.

The simplest way to measure stage level is by a permanent staff gauge, set so that its zero is well below the lowest possible flow. Although such gauges are simple and inexpensive, they must be read frequently when the water level is changing rapidly to define the shape of the streamflow hydrograph. It is preferable to construct a stilling well to house a data logger or chart recorder connected to a shaft encoder, pressure transducer or float (Plate 2(a)). Electronic data loggers have marked advantages for data processing but the chart record gives an important visual check of existing conditions and may still have a role to play in areas where a power supply cannot be provided or for providing a backup record.

A rating curve for a site can be obtained in a variety of ways of which the two most common are by means of velocity–area methods using a current meter (USGS, 1968, 1969) or by means of weirs or flumes which are mostly permanent structures (Plate 2(b)).

Current meter measurements are most often used when large flows have to be measured and the available fall is small. They are also often desirable for smaller rivers with sediment laden flows. Current meter gauging stations are relatively easy to set up since they often require little modification of the existing channel. Since each potential gauging site is unique, each requires a careful pre-assessment of the width and depth of the channel, likely flood velocities and alternative ways of current meter measurement. Wherever possible, current meter gauging stations should be located at accessible sites in straight uniform channel reaches with relatively smooth banks, no obstructions and a stable bed.

Current meter measurements can be made by wading with a current meter attached to a graduated wading rod to measure depth (Plate 2(c)). Wading can often be carried out safely in water depths up to about 1 m flowing at less than about 1 m/s. However, the force exerted by flowing water increases with the square of the velocity and even 0.5 m depth can be enough to sweep a person off their feet, depending on the footing. Otherwise, current meter observations should be carried out from above the water. Where the current is strong the current meter should be ballasted with a weight; in extreme cases, this may be so heavy that a crane is necessary. Access may be from the deck of a single span bridge, a specially constructed cableway or from a boat. An alternative faster and safer technique for the operator is the acoustic Doppler current profiler (ADCP) which uses ultrasonic technology to measure integrated flow velocity across the river. The ADCP can be mounted on a boat or dragged across the surface of a river from the bank on a small raft.

At current meter stations, the cross section is divided into vertical sections in which velocity measurements are taken at set depths. The mean velocity in each section is determined and applied to the area it represents. The river discharge is computed from the sum of the individual section values. The number and spacing of the vertical sections should be such that no section accounts for more than 10% of the river flow. Velocity measurement points are normally located by means of a tagged tape or wire stretched across the river or from graduations painted on the deck of a bridge. Detailed observations for many different types of river have shown that the mean velocity of a vertical section can be closely approximated from the mean of two observations made at 0.2 and 0.8 of the depth. If the water depth in the vertical section is less than about 0.6 m, or if time is limited, then one observation at 0.6 of the depth will approximate to the average over the whole depth.

The Water Resources Division of the United States Geological Survey has produced some excellent publications on current meter gauging techniques (USGS, 1968, 1969); while standard practice in the United Kingdom is contained in BS EN 748:2007, BS 1100:1998 and BS ISO 1438:2008.

Despite the relatively low capital cost of velocity–area stations, the need for sufficient current meter measurements over a wide range of river levels and for a repeat of the process each time a major flood is thought to have shifted the river bed profile, make heavy demands on staff time. These factors and the difficulty of access and gauging when the river is in flood make completing a rating curve take a long time. The alternative of a standard gauging structure, therefore, is often more attractive to the engineer, particularly for catchments of less than 500 km^2.

Dilution gauging and ultrasonic gauging (BS EN 6416:2005) can also be used to help define the stage/discharge relationship at sites where conditions are difficult for current meter measurements and to check an existing rating curve. The choice of gauging method depends on channel and streamflow characteristics, staff time availability and cost.

Dilution gauging (USGS, 1985) is a flow measurement technique that is particularly well suited to small turbulent streams with rocky beds where the shallow depths and high velocities are unsuitable for accurate current meter gauging. The approach can also be used to calibrate non-standard gauging structures. With dilution gauging, the discharge is measured by adding a chemical solution of known concentration to the flow and measuring the dilution of the solution some distance downstream where the chemical is completely mixed with the stream flow. Sodium dichromate is the most commonly used chemical although dyes such as Rhodomine B have the advantage that they can be easily detected at very low concentrations. With the commonly used 'gulp injection' method a known volume of chemical is added to the stream flow as quickly as possible in a single 'gulp' and downstream samples are used to construct a graph of concentration against time. It follows that if a known volume of chemical V of concentration C_1 is added to a streamflow and the varying downstream concentration C_2 is measured regularly then

$$VC_1 = Q\int_{t_1}^{t_2} C_2 dt$$

A graph of C_2 against time is drawn and the area below it, between t_1 when the chemical just starts to be detected in the stream and t_2 when it ceases to be detectable, is measured. This gives the integral on the right, hence Q can be found.

Ultrasonic gauging uses the transmission of sound pulses to measure the mean velocity at a prescribed depth across a river channel. Two sets of transmitters/receivers are usually located on either bank of a rectangular channel, offset at an angle of about 45° to the direction of flow.

They send ultrasonic pulses through the water, mean water velocity at pulse level being a function of the difference in pulse travel times in upstream and downstream directions. In essence, ultrasonic flow measurement is a velocity–area method requiring a survey of the channel cross section at the gauging site. If used in conjunction with a water level recorder, ultrasonic measurements can provide a complete record of stream flows.

Sharp edged plate weirs, of rectangular or vee-shape depending upon the sensitivity required, are generally only suitable for spring flows or for debris-free small streams. The need to keep the weir nappe aerated at all stages limits their use but they are frequently used for low flow surveys as they can be rapidly placed in small channels (BS ISO 1438:2008).

Of all the weirs that designers have tried, perhaps the most successful has been that of the ***Crump weir*** (Section 14.14 and Herschy, 1977). It has a simple and efficient flow characteristic while operating up to the total head at which tailwater depth reaches 75% of that upstream (relative to crest elevation). Using the crest tapping designed by Crump it is possible to go further and attain reasonable results up to 90% submergence, but this versatility is marred by a tendency for the crest tappings to block in floods carrying sediment. A minimum head of 600 mm on the weir is necessary for accuracy, but with compounding of the weir crest this can be achieved. The flat-vee variant is another option.

The accuracy of streamflow data obtained from weirs can often be poorer than generally realized. Laboratory rating conditions rarely appear in real rivers; deterioration of the weir crest and siltation upstream, resulting in non-standard approach conditions, can cause errors in excess of 10%. Drowning out of the weir at high flows can result in gross over-estimation of flood discharges.

Critical depth flumes are appropriate on smaller catchments (area under about 100 km^2) which have a wide flow variation and where sensitive results are required. Such flumes force flow to attain critical depth whatever the upstream head (Section 14.15 and Ackers, 1978). A unique upstream head–discharge relationship is then created as for a rectangular throat.

Calibration of an existing sluice structure may be achieved using formulae derived from laboratory model tests but these should be checked by current meter measurements wherever possible. Sometimes long, relatively homogeneous, sluice keeper's records exist (Sargent, 1992), and the calibration of such sites can then produce records of flow of several decades in length for a modest cost.

3.5 RAINFALL MEASUREMENT

Precipitation is measured with a rain gauge, the majority of which are little more than standard cylindrical vessels so designed that rainfall is stored within them and does not evaporate before it can be measured (WMO, 2008). In an effort to ensure that consistent measurements of the precipitation reaching the ground are obtained, observers are recommended to use standard instruments, which are set up in a uniform manner in representative locations. Many national meteorological institutions provide pamphlets designed to ensure good standard observation practice and the World Meteorological Organization (WMO) plays an effective coordinating role.

The standard daily rain gauge in the UK is the Meteorological Office Mark II instrument. It consists of a 127 mm diameter copper cylinder with a chamfered rim made of brass. Precipitation that falls on the rain gauge orifice drains through a funnel into a removable container from which the rain may be poured into a graduated glass measuring cylinder. Monthly storage gauges are designed to measure the rainfall in remoter areas and are invaluable on the higher parts of reservoir catchments. The Seathwaite gauge is a monthly storage gauge developed for use in the Lake District in North West England.

Ideally, rainfall should be measured at ground level but this gives rise to problems due to rain splashing into the gauge. The higher the rim is placed, the more some rain will be blown away from the gauge orifice and goes unrecorded. All standard storage gauges in the United Kingdom are set into the ground with their rims level and 300 mm above the ground surface, which should be covered by short grass or gravel to prevent any rain splash. Many international gauges are set with their rims 1 m high; these can be expected to read 3% lower than the standard British gauge.

In the United Kingdom, daily storage gauges are inspected each day at 09:00 hours and any rainfall collected is attributed to the previous day's date. If the inner container of a rain gauge overflows as the result of exceptional rainfall, or possibly because of irregular emptying, it is important that the surplus water held in the outer casing should also be recorded. Monthly storage gauges are usually inspected on the first day of each month to measure the previous month's rainfall total. Corrections may need to be made to the measurements taken at any gauges visited later than the standard time during spells of wet weather.

Problems occur in snow prone areas. Small quantities of sleet or snow which fall into a rain gauge will usually melt to yield their water equivalent, but if the snow remains in the collecting funnel it must be melted to combine with any liquid in the gauge. If there is deep fresh snow lying on the ground at the time of measurement, possibly burying the gauge, a core of the snow should be taken on level ground and melted to find the equivalent rainfall. Countries with snow cover throughout the winter months require regular snow course surveys (Hudleston, 1933) to measure precipitation.

Continuously recording rain gauges are invaluable for flood studies. The original type gives a daily chart recording of the accumulated contents of a rain-filled container, which empties by a tilting siphon principle each time 5 mm has collected. A more recent development is the tilting bucket gauge linked to a logger, which runs for at least 1 month. Each time the bucket tilts to discharge 2 mm the event is recorded in a computer compatible form. A daily or monthly storage gauge is often installed on the same site as a recording gauge to serve as a check gauge and to avoid the possibility of loss of data due to instrument failure. Plate 2(d) shows an automatic rain gauge with telemetry for remote monitoring.

Particular care must be taken when siting a new rain gauge from which the resulting records are to be published. The gauge should be placed on level ground, ideally in a sheltered location with no ground falling away steeply on the windward side. Obstructions such as trees and buildings, which affect local wind flow, should be a distance away from the gauge of at least twice their height above it. In particularly exposed locations, such as moorlands, it used to be standard British practice to install a turf wall (Hudleston, 1933) around the gauge but it has proved difficult to sustain the level of turf maintenance that is needed.

Rain gauges provide a spot sample of the rain falling over a catchment area. The number of gauges required to give a reliable estimate of catchment rainfall increases where rainfall gradients are marked. A minimum density of 1 per 25 km^2 should be the target, bearing in mind that significant thunderstorm systems may be only about 20 km^2 in size. In hilly country, where orographic effects may lead to large and consistent rainfall variations in short distances, it may be necessary to adopt the high densities suggested in Table 3.1 for the first few years. Thereafter high densities are only required where control accuracy necessitates it.

Table 3.1 Rain gauges required in a hill area (IWE, 1937)

Catchment area (km^2)	4	20	80	160
Number of gauges	6	10	20	30

In large areas of the tropics, there is great variation in rainfall from place to place on any one day but only a relatively small variation in annual totals. In such areas the rain gauge densities of Table 3.1 will be excessive and it is better to concentrate on obtaining homogeneous records of long duration at a few reliable sites.

Measurement of Catchment Rainfall

There are several methods for computing catchment precipitation from rain gauge measurements. The more frequently used techniques include simple numerical procedures (averaging or interpolation), interpolation from isohyetal maps or Thiessen polygons and trend surfaces.

The simplest objective method of calculating the average monthly or annual catchment rainfall is to sum the corresponding measurements at all gauges within or close to the catchment boundaries and to divide the total by the number of gauges. The *arithmetic mean* provides a reliable estimate provided the whole catchment is of similar topography and the rain gauge stations are fairly evenly distributed. If accurate values of area rainfall are obtained first from a large number of rainfall stations, by one or other of the more time-consuming methods described below, the mean of the corresponding measurements from a smaller number of stations may provide equally acceptable results. In the Thames basin, for example, it was found that the annual catchment rainfall for the 9980 km^2 area derived from the arithmetic mean of 24 well distributed representative gauges was within $\pm 2\%$ of the value computed by a more elaborate method using measurements from 225 stations.

The *isohyetal method* is generally considered to be the most accurate method of computing catchment rainfall although a good understanding of the rainfall of a region is needed to ensure its reliability. The monthly or annual rainfall total recorded by each gauge within or close to the catchment boundaries is plotted on a contour base map. Isohyetal lines (lines joining points of equal rainfall) are then drawn on the map taking into account the likely effects of topography on the rainfall distribution. The total precipitation over the catchment for the period considered is obtained by measuring the areas between isohyets by GIS techniques or by planimeter. The mean catchment rainfall is calculated by summing the products of the areas between each pair of isohyets and the corresponding mean rainfall between them, and then dividing by the total catchment area.

The most popular method of weighting gauge readings objectively by area has been that of Thiessen (*Thiessen polygons*). An area around each gauge is obtained by drawing a bisecting perpendicular to the lines joining gauges, as shown in Figure 3.1. The portion of each resulting polygon lying within the catchment boundary is measured and the rainfall upon each is assumed to equal the gauge reading. The total precipitation is the weighted average of these values. One drawback is that, if the gauges are altered in number or location, major alterations to the polygonal

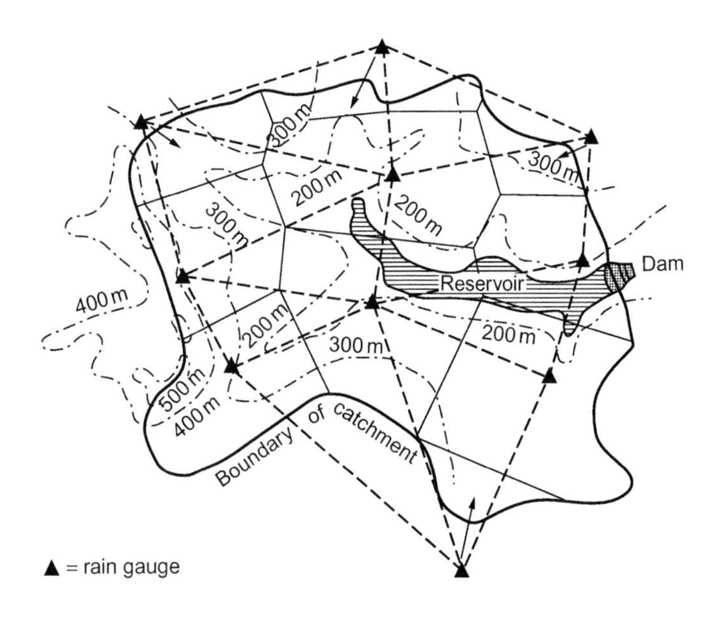

FIGURE 3.1

Thiessen's method of estimating general rainfall over an area.

pattern ensue. To maintain homogeneity it is better to estimate any missing individual gauge values. The gauges must also be reasonably evenly distributed if the results are to lie within a few per cent of the isohyetal method. The approach is not particularly good for mountainous areas because no account is taken of the effects of altitude on rainfall when deriving the Thiessen coefficients for individual polygons.

In mountainous areas, where there may be few stations, the main difficulty is to allow for the influence of topography. One widely used approach for such areas is to develop a multivariate regression model using parameters such as elevation, orientation, exposure or distance from the sea, and then to use a *numerical interpolation procedure* such as Kriging (Creutin, 1982) to smooth out residual discrepancies from the regression correlation.

Progress has been made in rainfall estimation by both weather radar (Lau, 2006) and satellite although the establishment and operation of such networks is the domain of national meteorological organizations rather than individual projects. The strength of both approaches lies in the spatial view they afford with the former being particularly useful for flood forecasting. Radar images for low altitude scans are calibrated to ground measurements of actual rainfall and rainfall values assigned to each pixel of the image, normally at 5 or 15 minute intervals. Pixels cover 1 km squares close to the radar increasing to 2 or 5 km squares further away. GIS techniques can be used to compute catchment rainfall for each time step when digital catchment boundaries are applied to the gridded data. Satellite estimates are far more approximate, being related to cloud top temperature and only indirectly to actual rainfall amount. For large basins in the tropics it is now possible to obtain public domain estimates of $0.5^\circ \times 0.5^\circ$ grid satellite 'monthly rain estimates' from the Climate Analysis Center, Washington, DC.

3.6 EVAPORATION AND TRANSPIRATION MEASUREMENT

Evaporation is a key part of the hydrological cycle in that, globally, about 75% of total annual precipitation is returned to the atmosphere by the processes of evaporation and transpiration. Water evaporates to the air from any open water surface or film of water on soil, vegetation or impervious surfaces such as roads and roofs. The rate of evaporation varies with the colour and reflective properties of the surface (the albedo) and with climatic factors. Energy from solar radiation is the main factor for which air temperature is often used as a proxy measurement. However, wind speed, the relative humidity of the air and the temperature of the water are also important.

Transpiration is the water used by plants. A small part of this water is retained in the plant tissue but most passes through the roots to the stem or trunk and is transpired into the atmosphere through the leaves. As it is almost impossible under field conditions to differentiate between evaporation and transpiration when the ground is covered with vegetation, the amounts of water used by both processes are usually combined and referred to as '*evapotranspiration*'.

Evapotranspiration losses vary with the same meteorological factors as evaporation, but depend additionally on the incidence of the precipitation, the characteristics and stage of development of the vegetation, and the properties of the soil.

Evaporation can be measured directly or estimated indirectly (WMO, 1966), although both have their difficulties. The US Class A evaporation pan is one of the most commonly used instruments for measuring evaporation directly. It is constructed of galvanized iron or monel metal, 1.21 m in diameter and 255 mm deep, and is set on a standard wooden framework 100 mm above ground level, thus allowing air to circulate all round it. As a result, measured evaporation is higher than that of a natural water surface and a reduction factor must be applied. This is generally taken to be 0.7, but it can vary between 0.35, in areas of low humidity, very strong wind and surrounded by bare soil and 0.85 where high humidity and light winds prevail (Doorenbos, 1976). The British Symons sunken tank is 1.83 m square and 610 mm deep, with the rim 75 mm above ground level. It is more nearly a model of reservoir evaporation but suffers from inconsistent results if it is not in tight contact with the surrounding ground. The heat storage of a small tank is correspondingly small, whereas a large lake takes time to warm up or cool down. As a consequence, tank results do not quite match the evaporation of a nearby lake in regions with strong seasonal temperature variations. Peak lake evaporation rates in the Kempton Park experiment (Lapworth, 1965) occurred up to a month after peak tank measurements; this was explained by heat storage theory. Annual open water evaporation ranges from 700 mm in Northern Europe, through 1500 mm in much of the tropics, to more than 2500 mm in hot arid zones.

Attempts have been made to measure evapotranspiration experimentally by means of percolation gauges (Rodda, 1976) and, on a larger scale, by the use of lysimeters. However, no standard way of measuring it routinely has yet been devised. Indirect techniques therefore have to be relied on.

Penman's formula (Penman, 1963) for predicting water surface evaporation is recognized as the most accurate of these indirect techniques. It is based on physical principles, but involves the use of data which may not always be available, e.g. measurements of radiation (or sunshine duration), wind run at 2 m above ground, vapour pressure and air temperature, all of which should be taken at the same site. At altitudes above about 1000 m, McCulloch's fuller version of Penman's equation

(McCulloch, 1965) should be used as it makes express allowance for the corresponding pressure drop and adjusts the radiation term for latitude.

Penman also showed how simple coefficients could be applied to his open water evaporation figures to obtain the evapotranspiration rate from a grassed surface. The use of the latter became standardized as being the evapotranspiration (designated 'ET_o') '*from green grass surface cover 80–150 mm high, actively growing*'. Crop coefficients k_c could then be applied to give the evapotranspiration of various types of crops at various stages of development and meteorological conditions, e.g. $ET_{(crop)} = k_c.ET_o$.

Thornthwaite's formula for evapotranspiration from short vegetative cover, widely used in the USA, is empirical and simpler than Penman's, being dependent on sunshine hours and mean monthly temperature. Thornthwaite's method has been widely used, but it is strictly valid only for climates similar to that of eastern USA where the method was developed. The method tends to give evaporation estimates higher than those produced by the Penman formula, particularly during the summer months.

More recent guidance to assessing the irrigation requirements of growing crops are provided in the FAO's *Irrigation and Drainage Paper 49, CROPWAT: A Computer Program for Irrigation Planning and Management*, 1993 (Smith, 1993) and *Paper 56, Crop Evapotranspiration – Guidelines for Computing Crop Water Requirements*, 1998 (Allen, 1998). They compute evapotranspiration ET_o values according to the Penman–Monteith method, to which quoted crop coefficients, k_c are then applied. An earlier FAO publication, *Irrigation and Drainage Paper 24', Crop Water Requirements*, 1976, was based on the Blaney–Criddle formula, which uses temperature and length of daylight hours to give ET_o values, which are then modified by values given for a range of humidities and wind speeds.

Evapotranspiration formulae assume no shortage of water to meet crop growth and potential evaporation. In dry periods when moisture availability becomes limiting, actual evapotranspiration is less than the computed potential. A good explanation of all the foregoing indirect methods of estimating evapotranspiration is given by Wilson (1990).

In the UK, the Meteorological Office (Thompson, 1981) has used an adaptation of the Penman–Monteith equation in their MORECS, and more recently MOSES, evaporation estimation systems. Up-to-date values of actual and potential evaporation are computed and published regularly, as well as precipitation and soil moisture deficits for a network of 40 km × 40 km grid squares covering the whole country. Similar systems are likely to exist in other countries and practitioners will in most cases find it preferable to use published data rather than derive their own evaporation estimates from climate data.

3.7 SOIL MOISTURE MEASUREMENT

Agriculturalists have always been interested in studying soil moisture. More recently, the potential impact of soil moisture content on run-off has been understood by a wider audience. The temporary storage of rainfall in the soil and aquifer layers of a catchment can be significant in the overall catchment water balance. Soil moisture variations occur predominantly in the first metre below the surface. The water content in very sandy soil may vary from 3% to 10% from the driest (*wilting point*) condition to the wettest drained state (*field capacity*), or from 20% to 40% in a clay soil. Thus, the maximum range of water storage in 1 m of soil may be as much as 200 mm; markedly higher values apply to peat. Additional water may be held under waterlogged conditions whenever the drainage rate is lower than the rainfall intensity.

Knowledge of the types and distributions of soils over the area of interest is an essential prerequisite for the selection of sampling points at which to measure soil moisture. For England and Wales, this information can be obtained from maps published by the Soil Survey and Land Research Centre at Silsoe, while for Scotland soil maps can be obtained from the Macaulay Land Use Institute, Aberdeen. In other countries, similar sources should be consulted but it may be necessary to engage a soil scientist to carry out soil surveys in the catchments where detailed hydrological measurements are required.

A common method of measuring soil moisture is by gravimetric determination, whereby a soil sample of known volume is removed from the ground with a soil auger. It is weighed, dried in a special oven and then reweighed. The method is accurate provided care is taken with the measurements and the method is often used to calibrate other techniques. However, the method is time-consuming and requires laboratory facilities. It is also a destructive process and has obvious limitations where regular sampling is required. A soil capacity probe (Dean, 1987) can be used to provide direct field measurements.

3.8 CATCHMENT LOSSES

A significant proportion of rainfall is lost by immediate evaporation or by the later transpiration of growing vegetation. In some cases, there will also be deep infiltration that enters aquifers and eventually emerges in the sea without reappearing on the surface. Catchment losses are best estimated from a water balance conducted over a number of years on the catchment concerned or from one with similar rainfall, geology and land use.

A typical loss rate in England would be 450 mm per annum, with figures significantly above 500 mm per annum generally only occurring where afforestation predominates in a high rainfall area. Table 3.2 gives some idea of the variation in loss in different regions of the world. Although it might be thought that losses would be higher in years hotter than average, this is often more than offset by the concurrent dryness of the weather, which leads to a deficit of moisture in the soil and suppression of transpiration by growing vegetation. As soil dries out and approaches wilting point evapotranspiration rates can drop to only about one tenth of the potential. Some plants, such as pine trees, are more successful than others in control of their water use in drought conditions.

Losses do not always decrease at cooler altitudes because advected wind energy and higher radiation may offset the lower temperatures. Detailed measurements at experimental catchments by CEH have shown that losses increase with the density and height of natural vegetation growth and crops, particularly in mature coniferous forests. However, predicting forest annual loss compared with that from grazed pasture is still not easy. It has been shown that the forest loss is due to intercepted raindrops being evaporated back into the atmosphere at rates up to five times normal transpiration values for short grass. This is because water laid out in thin films on vegetation can take up available heat in the atmosphere more readily; the sight of a forest steaming gently in a short

Table 3.2 Typical catchment losses in various parts of the world

Country	Location	Catchment cover	Annual rainfall (mm)	Annual loss (mm)	Marked seasonal variation
Nigeria	Ibadan	Rain forest	2500	2350	No
Malaysia	Johor basin	Forest/oil palm	2320	1240	No
Sri Lanka	Kirindi Oya	Mixed forest	1650	1230	Yes
Hong Kong	Islands	Grass	2100	1050	No
Zaire	Fimi	Rain forest	1700	1040	No
Thailand	Chau Phraya	Forest/rice paddy	1130	1000	No
Japan	Ota	Conifer forest	1615	890	Yes
Australia	Perth	Mixed grass/forest	875	760	Yes
S. Africa	Transvaal	Mixed grass/forest	870	760	Yes
Kenya	Tana	Forest/savannah	1100	730	Yes
India	Bombay	Rain forest	2550	700	Yes
Zimbabwe	Low Velt	Mixed grass/forest	655	560	Yes
Lesotho	Maseru	Grassland	600	530	Yes
Holland	Castricum	Low vegetation	830	450	Yes
Algeria	Hamman Grouz	Scrub	420	400	Yes
Russia	Moscow	Agricultural	525–600	375	Yes
Iraq	Adheim basin	Scrub/grassland	420	350	Yes
S. Korea	Had basin	Forest	1180	320	Yes
Oman	Oman	Rock	160	130	Yes
Iran	Khatunabad	Bare ground	150–550	50–200	Yes
Britain	South England	Pasture/arable	600–900	450–530	Yes
	Midlands	Pasture/arable	650–850	440	Yes
	Central Wales	Moorland/forest	1500–2300	480–530	Yes
	Pennies	Moorland/forest	1150–1800	410–460	Yes
	NE England	Moorland/pasture	700–1250	380	Yes
	S. Scotland	Moorland/pasture	600–1800	360–410	Yes
	N. Scotland	Moorland/pasture	1250–2500	330–380	Yes

spell of sunshine between showers is not uncommon. To quantify the extra loss to be expected requires an idea of the depth of water the canopy of vegetation can hold during a shower and the frequency of showers (Calder, 1992). One point to note is that whilst rain is being evaporated off the outside of leaves, water is not transpired through them. As a result, perhaps only 90% of forest interception losses are in addition to the catchment losses that would prevail anyway.

A review of 94 catchment experiments in Africa, Asia, Australia and North America concluded that conifer forests could be expected to reduce catchment water yield, on average, by 40 mm per 10% forest cover (Bosch, 1982). More recent studies (Kirby, 1991), at Plynlimon in the UK, found that the magnitude of the reduction in water yield was 29 mm per 10% forest cover, which equates to a 15% reduction in the water yield for a completely forested upland catchment. It should be noted, however, that while heavy afforestation of catchments reduces their overall water yield, the results of the Plynlimon study and a study of catchment data from 40 European agencies (Gustard, 1989) found no statistically significant relationship between the proportion of forest cover and measures of low flow. These findings support the view that low flows in upland headwater catchments are primarily influenced by drainage from minor sources of groundwater, which remain largely unaffected by forest interception losses.

3.9 STREAMFLOW NATURALIZATION

In most river basins, very few flow regimes can be considered entirely natural, i.e. free of artificial influences such as abstractions, discharges or storage effects from impounding reservoirs. It is therefore not usually sufficient to evaluate a potential resource directly from the 'as-measured' stream flow records. The need to *naturalize* the available flow records should be considered at the start of any major water resource assessment. This involves considering the extent of artificial influences within a catchment, the scale of their impact on flows and the purpose for which the flow sequences are required.

Flow naturalization does not normally correct for anthropogenic effects such as urbanization or land use changes. Except on very small catchments, these effects are not usually significant. The main types of abstraction or artificial losses from a river system are public water supply abstractions; irrigation and other abstractions for agriculture; power station cooling water abstractions; and industrial (non-cooling) water abstractions. The more common gains are from sewage and industrial effluent returns; irrigation return flows; and inter-basin transfers.

The need for flow naturalization arises for three main reasons:

- artificial influences have most impact on low flows which are the focus of water resource assessments;
- the long-term variability of flow records is characterized by comparison with a suitable statistical distribution; natural flow records tend to conform to such distributions, those affected by artificial influences may not;
- options for water resource development are often assessed by means of a computer simulation model; model inflow sequences need to be free of any historic artificial influences embedded in the observed flow record so as not to bias results or lead to the double-counting of influences.

In the UK, all time series of daily river flow data held on the National River Flow Archive have been graded according to the degree of artificial influence at low flows. Elsewhere, review of abstraction licences/permits, where they exist, can often give an initial indication of the scale of abstractions, even though actual abstracted quantities will almost certainly be less than the licensed total.

The type of abstraction is also relevant. Power stations are often licensed to abstract very large quantities of river water for cooling, although most of this water is returned to the river a short distance downstream. The net effect on river flow is therefore small except immediately downstream of the abstraction. Similarly, large groundwater abstractions may be from parts of the aquifer not in hydraulic connection with the river and therefore have little or no effect on river flows. Some influences are seasonal so winter abstractions, in temperate climates e.g. may have no effect on summer low flows, although they could affect the refilling of storage.

The purpose of flow naturalization should be clearly born in mind and the parameters that are critical to the study in question should be assessed, e.g. water levels, minimum flows or flow volume over a critical period. It may be considered, for example, that if gauged flows are within 10% of natural flows over the critical period, then naturalization is not necessary.

If naturalization is required, the two most commonly used methods are naturalization by decomposition and rainfall–run-off modelling. Naturalization by decomposition involves breaking the observed flow record down into its component parts. If there are no reservoirs within the catchment then the natural flow is deduced by quantifying all the artificial components in the observed flow record at the appropriate time step (daily or monthly) and using the following equation:

Natural flow = Gauged river flow + sum of all upstream abstractions − sum of all upstream discharges and return flows to river.

This process is critically dependent on the availability of good quality, complete data for both the observed stream flows and for artificial influences (abstractions and return flows) upstream of the location of interest. In catchments where there are many small abstractions that individually have little or no impact on river flows, these may be filtered so that the smallest are either ignored or combined. It may be adequate to consider only the main public water supply abstractions or sewage treatment works discharges. These are usually the largest artificial influences and are often monitored so that specific time series data are available.

Factors may be applied to certain types of abstraction to account for the proportion of abstracted water that is returned to the river. If time series data are not available, a seasonal profile may be applied to the mean annual abstraction where quantities vary over the year. Factors may also be applied to groundwater abstractions according to their degree of impact on river flows. Impacts may be time-lagged depending on the length of the flow path between river and groundwater table. The results of regional groundwater models may help decide how best to allow for the effect of groundwater abstractions in the naturalization process. A detailed discussion of these and other aspects is contained in EA Guidelines (EA, 2001) and in Hall (1994).

For catchments that contain a reservoir, allowance needs to be made for the storage and attenuation of flows and for additional evaporation losses from the reservoir surface, which can be significant in warm semi-arid environments. In order to unravel storage effects, time series data are required for changes in reservoir level and/or storage, draw-off to supply and spills/releases to the channel downstream. If data on reservoir spills are not available, the naturalization can still be carried out on dry season flows, which are usually the most critical from a water resources point of view.

If the available data are inadequate for a decomposition approach, it may be possible to generate natural flow sequences by means of rainfall–run-off modelling. Nevertheless, this approach still requires data to represent rainfall and evaporation at an appropriate time scale over the catchment in question. Ideally, a brief period of observed (natural) river flows is also required with which to calibrate the model, although if this is not available, some spot gaugings at low flow may suffice. Often, the available data permits naturalization by decomposition for the recent historic period with rainfall–run-off modelling required to extend flow sequences back to earlier years.

3.10 LONG-TERM AVERAGE CATCHMENT RUN-OFF

Wherever possible, average catchment run-off should be calculated from a long streamflow record, which, if subject to artificial influences, has been naturalized as described in Section 3.9. Where streamflow records are either short-term or non-existent, long-term average run-off can be estimated by:

- correlating the brief records available for the study catchment with those of a long record station in a catchment with similar characteristics;
- deducting loss estimates from catchment rainfall figures;
- use of a rainfall/run-off model.

Correlations between long and short record stations are best carried out using monthly data. The use of daily figures is more time-consuming and often produces a large scatter while annual values provide too few points. An initial mathematical 'best fit' relationship between the stations may be derived by computer but a graphical plot should always be produced as a check on the mathematical fit. A manual adjustment should be made if, for example, the mathematically computed relationship is unduly influenced by the values for a few high flow months, or if the relationship implies unreal intercept values.

An estimate of average run-off for a catchment for which no streamflow records exist can be obtained by deducting a value for average annual catchment losses from the average annual catchment rainfall for a suitable standard period (e.g. 1961–90). The latter can be obtained from an isohyetal map of the region or from a GIS-based gridded database such as the FEH CD-ROM (NERC, 2009) (Section 3.2). If possible, the value for average annual catchment losses should be based on the typical average annual loss value obtained from similar gauged catchments in the region. However, if no such data exist, estimates of actual evapotranspiration can provide a reasonable measure of catchment losses as long as there are no significant abstractions or groundwater outflows from the catchment. In the UK, estimates of actual evapotranspiration for a network of grid squares covering the whole country have been published by the Meteorological Office from its MORECS (Thompson, 1981; Hough, 1997) and more recently, MOSES calculation systems. If actual evaporation data are not available, values can be estimated from catchment average annual potential evapotranspiration (Smith, 1975) multiplied by the adjustment factors listed in Table 3.3.

Table 3.3 Adjustment factor for estimating actual evaporation in the United Kingdom (Gustard, 1992)

Standard average annual rainfall (mm)	500	600	700	800	900	1000	>1100
Adjustment factor	0.88	0.90	0.92	0.92	0.94	0.96	1.00

For most regions, rainfall stations are more numerous and have longer records than streamflow measurement stations. Rainfall/run-off models such as HYSIM (Manley, 1978) or HEC-HMS (USACE, 2000, 2010) are therefore commonly used to extend short-term stream flow records. Typically, a chosen model is calibrated by adjusting the model parameters so as to produce the best possible match between predicted and measured flows. The calibrated model is then used to extend the short-term stream flow data to cover the longer period of rainfall records. One of the usual checks on the synthetic stream flow data generated by rainfall/run-off models is whether they accurately reproduce the long-term mean run-off estimated by other means, particularly over a standard period.

3.11 MINIMUM RAINFALLS

Experience of past recorded droughts and low rainfall is an important factor in assessing probable future conditions that may be encountered. In the variable climate of the UK a spell of 73 days without any rain at all was recorded at Mile End in London from 4 March 1893 (Holford, 1977). At the other extreme, in desert climates many years may have no rainfall at all. At Calama, which is in the Atacama desert of northern Chile, it is believed that virtually no rain fell for 400 years until a sudden storm in 1972.

The most notable droughts in England and Wales were the following:

- 1920–21 Southeast England. Annual rainfall lowest in over 100 years; spring sources hard hit as the autumn rainfall was insufficient to prevent flow recession, which began in a dry spring and continued until January 1922 in many parts;
- 1933–34 Wales and mid-England. Two dry summers with a remarkably dry winter intervening;
- 1943–44 Southern England. A similar pattern to 1933–34; very low flows experienced in spring-fed rivers because preceding years were also dryish;
- 1947–49 Exceptionally low summer rainfall and high temperatures affected sources reliant on river flows or with little storage;
- 1959 Similar to 1949;
- 1975–76 Many low flow records broken because of low summer rainfall;
- 1988–92 South and East England. A succession of dry winters taxed groundwater supplies; run-off deficits in parts of East England were the largest for 150 years;
- 1995–96 Pennines. Two dry summers with a very dry winter intervening;
- 2004–06 Southeast England. Two consecutive dry winters combined with elevated temperatures; steep declines in reservoir storage and groundwater levels;
- 2010–12 Central Eastern and Southern England. Two consecutive dry winters combined with dry spring 2011 resulting in low reservoir levels and very low river and groundwater levels.

There was no predictable pattern to these diverse low rainfall events. Hence, when estimating the minimum yield of a source it is necessary to bear in mind the types of drought that past experience shows are possible.

An estimate of drought rainfall for catchment modelling can be made either by using a knowledge of recorded minimum rainfall for a region, expressed as a percentage of the average, or by carrying out a statistical analysis (Tabony, 1977) of available rainfall measurements.

3.12 MINIMUM RATES OF RUN-OFF

In temperate climates with variable rainfall, when minimum run-offs are expressed as rates per unit catchment area, it can often be seen that catchment geology and topography are the major influences, except where human activity has interfered. Clearly the dry weather flow of many small catchments is zero; and *bournes*, which are streams flowing strongly when the water table is high, dry out gradually from their headwaters as the water table level falls away from the stream bed. In temperate zones where rainfall occurs throughout the year, large rivers do not dry up.

Several studies have attempted to predict minimum flows of specified severity after regional analysis of flow records. Notable examples include those for: Malaysia (Enex, 1976), Europe (Gustard, 1989), New York State (Darmer, 1970) and that produced in 1992 by the IoH, now CEH for the UK (Gustard, 1992). The IoH provided formulae for estimating drought flows based on a detailed classification of 29 types of soil systems. A summary national equation for the UK, which accounted for 62% of the variations encountered, is:

$$\text{1-day mean flow exceeded 95\% of the time, } Q_{95(1)} = 44B^{1.43}S^{0.033}A^{0.034}$$

where $Q_{95(1)}$ is the 1-day flow, expressed as a percentage of the long-term average daily flow, B is the *Base Flow Index* (BFI) (Table 3.4); S is the Standard Average Annual Rainfall 1941–70 in mm; A is the catchment area in km^2.

$$\text{Mean annual 7-day minimum flow, } MAM(7) = 6.40\ Q_{95(1)}{}^{0.953}S^{-0.0342}$$

where $MAM(7)$ is expressed as a percentage of the long-term average daily flow.

The BFI represents the proportion of river flow which it is estimated to be derived from underground storage. IoH developed a procedure for evaluating BFI from an analysis of a long sequence of daily flows by locating the minima of consecutive non-overlapping 5-day flow totals. The analysis then searches for the ‘turning points’ in the sequence of minima, connecting them together to form the estimated base flow hydrograph. Figure 3.2 shows a graphical representation of the procedure. The BFI is the volume of flow below the base flow hydrograph, divided by the total flow for the same period.

Table 3.4 Typical base flow indices for various rock types

Dominant characteristics			
Permeability	**Storage**	**Example of rock type**	**Typical BFI range**
Fissured	High storage	Chalk	0.90–0.98
		Oolitic limestones	0.85–0.95
	Low storage	Carboniferous limestone	0.20–0.75
		Millstone Grit	0.35–0.45
Intergranular	High storage	Permo-Triassic sandstone	0.70–0.80
	Low storage	Coal measures	0.40–0.55
		Hastings Beds	0.35–0.50
Impermeable	Low storage at shallow depth	Lias	0.40–0.70
		Old Red Sandstone	0.46–0.54
		Metamorphic-Igneous	0.30–0.50
	No storage	Oxford and London Clay	0.14–0.45

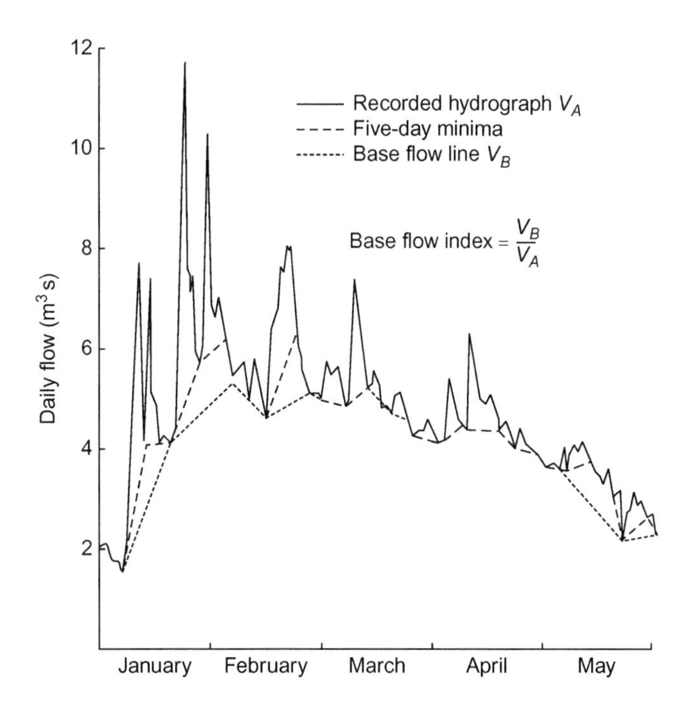

FIGURE 3.2

Derivation of base flow index.

The BFI values for over 1300 gauged catchments in the UK are given in the latest issue of the CEH/British Geological Survey (BGS) publication UK Hydrometric Register (Marsh, 2008). For any ungauged catchment an estimate of the BFI can be obtained either by interpolation between the values for upstream and downstream gauging stations, quoted in the Hydrometric Register for the same river, or by transposing the value for a gauged catchment with a similar average annual rainfall, surface geology and soil type. The approach has been adopted by a number of countries for analysis of their river low flows, the advantage being that annual values of BFI tend to be more stable than other low flow variables. Details of the method are given in the IoH's Low Flows Study (IoH, 1980), the predecessor to IoH Report 108 (Gustard, 1992). The range of BFI values typically applying is shown in Table 3.4.

Many of the formulae presented in IoH Report 108 have now been incorporated into the software package 'Low Flow 2000' (Young, 2003), a GIS-based decision support software system that has been adopted by EA and SEPA (the Scottish Environment Protection Agency) as a best practice tool for estimating low flows in ungauged catchments throughout the UK. The system includes extensive data on artificial influences to enable these to be taken into account but, because of the size and complexity of the software, it is not generally available to the average practitioner.

3.13 MAXIMUM RAINFALLS

Figure 3.3 is a plot of the maximum measured rainfalls recorded at individual points in the world. It must be stressed that the highest falls are precipitated only in the most unusual hill areas in certain climatic zones, once the duration exceeds 1 day. In addition, the lines do not represent any continuous event, except possibly for a storm lasting for up to, say, 4 hours. The maximum figures for Britain are seen to be about 20–30% of world maxima, with the lowland easterly part of the country suffering less severely in long duration storms. The greatest is the Martinstown, Dorset, storm of 1955 in which 280 mm (11 in) of rain were officially noted in 18.5 hours, with an unofficial estimate at the heart of the storm claiming 350 mm (14 in).

Maximum rainfalls vary with the season of the year because thunderstorm intensities are associated with high sea and air temperatures. However, high rainfalls over a day or so may occur at any time of year, whenever the weather system can bring in moist air steadily and where conditions (often orographic) exist to cause precipitation.

It is quite possible for many years to elapse without any outstanding maximum rainfall event and then for several unusual maximums to be clustered close together, probably due to high concurrent sea temperatures and dominant weather system movement routes. Many countries have compilations of extreme meteorological events and these should not be overlooked. Many national meteorological agencies make available long period rainfall measurements and duration–intensity–frequency estimates. These are often adequate for estimating storm magnitudes that can be expected more frequently than once every 50 years; but possible rarer storms need to be investigated by thorough regional studies. In the UK the most up-date information is contained in Volume 2 of the *Flood Estimation Handbook* (FEH, 1999); this shows how rainfall frequency calculations can be applied to UK catchments, leading to estimation of either the rainfall depth for

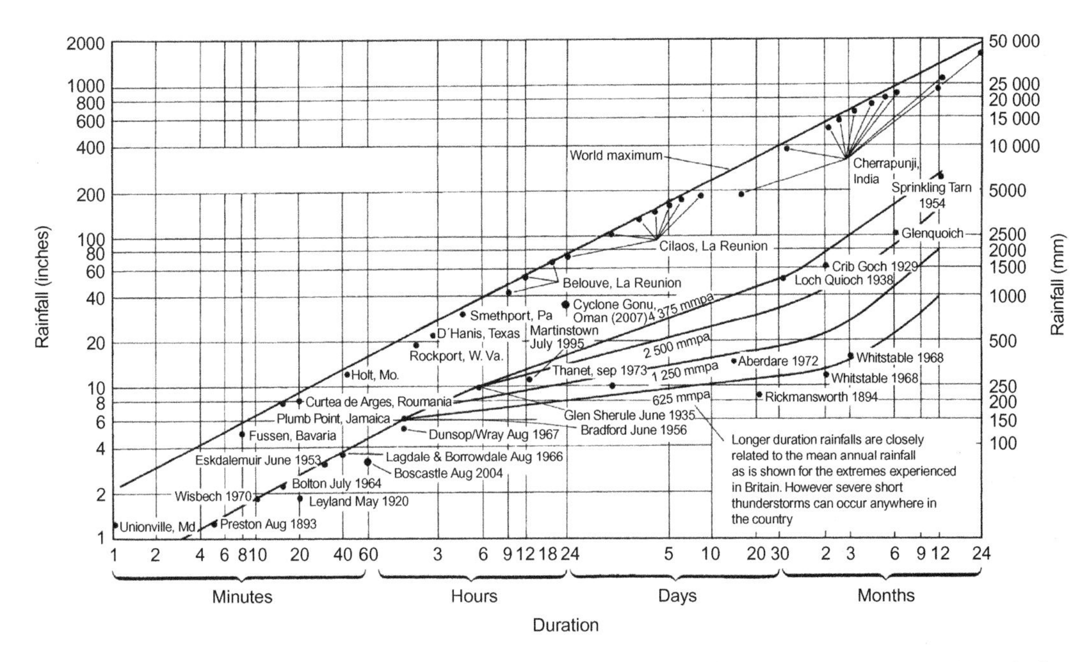

Sources of information: *Paulhus, J. L. H. (1965). Indian Ocean and Taiwan Rainfalls set new record.* Monthly Water Review, ***93****(5), pp. 331–335; Dhar, O. N. and Farooque, S. M. T. (1973) A study of rainfalls recorded at Cherrapunji.* International Association of Hydrological Sciences, ***XVIII****(4), pp. 441–450 and Black & Veatch.*

FIGURE 3.3

Maximum recorded world and British rainfall.

a given return period and duration, or of the return period corresponding to a given depth and duration of rainfall. The parameters of the rainfall frequency model are provided digitally on FEH CD-ROM to a 1 km grid (NERC, 2009).

Improved rainfall depth–duration–frequency values for use in UK flood studies have recently been published by Defra, while a review of rainfall frequency and probable maximum precipitation estimation methods for both the UK and several other countries was published in an earlier report by Svensson et al. (2006).

3.14 MAXIMUM RUN-OFFS

Peak flood discharges are notoriously difficult to measure directly because of their transient nature, high velocity, large debris content and the difficulty in gaining access to the river during times of flood. The problem is exacerbated when flows come out of bank and significant discharge occurs across the flood plain. Peak discharges can, however, often be estimated indirectly by means of hydraulic calculations using sediment or debris marks left behind by the flood to indicate peak water level, cross-section area and water surface slope.

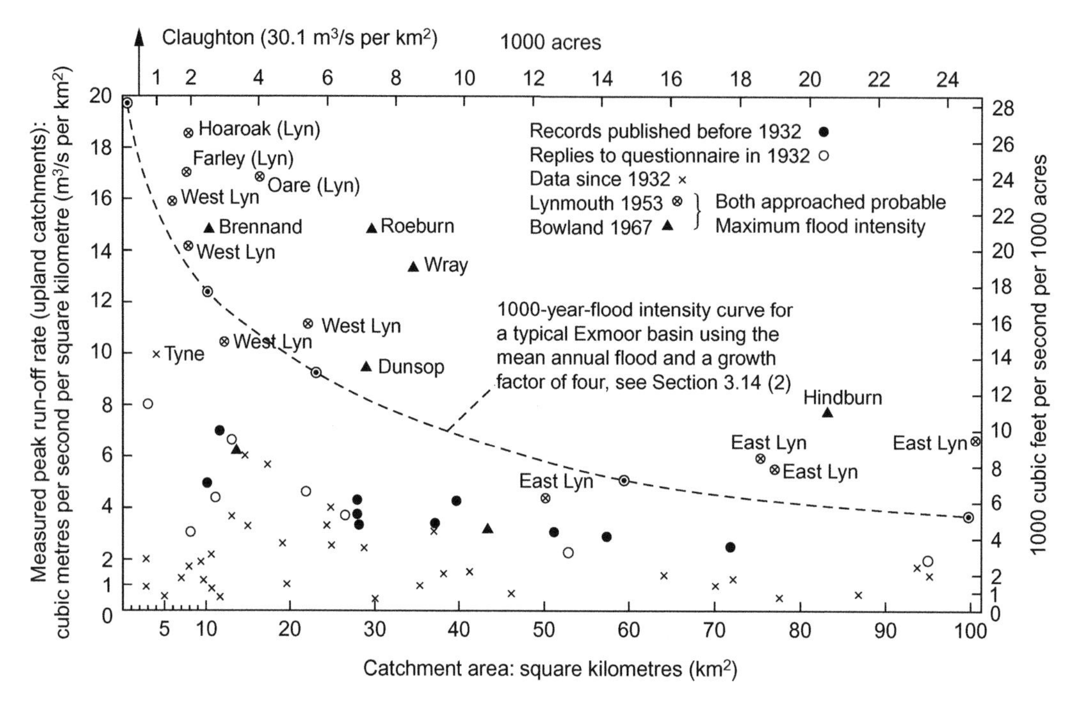

Sources of information: *ICE (1960) and others.*

FIGURE 3.4

Flood data in Britain in and since the ICE 1933 report.

Figure 3.4 gives some recorded maximum run-off data experienced in the UK. In lowland areas, peak run-off rates are normally much lower than Figure 3.4 suggests because of the temporary storage available in side channels and drains or in adjacent low lying land and the greater likelihood of drier soil conditions before a storm. For example in the fenland area of Eastern England entire catchments may be drained quite adequately by pumping stations capable of pumping no more than 13 mm of run-off from their catchment per day. This is very much less than the maximum precipitation rates shown in Figure 3.3 where rainfalls up to about 250 mm in 24 hours can occur in Eastern England, as shown by the plot for Thanet. These figures are not however, large compared with experience elsewhere in the world where 10 times higher rates can be experienced.

There are four methods of estimating maximum run-offs; the fourth applies specifically to the UK.

1. ***Probability analysis of existing flood records.*** This method plots recorded peak floods against a probability distribution, as described for the analysis of droughts in Sections 3.18 and 3.19, in order to estimate the probability of occurrence of a flood of given magnitude. The type of probability distribution used is that which gives a best straight line plot for the recorded floods. Usually the peak flood for each year of record (the *AMAX series*) is plotted; but sometimes all peak flows above a given level are plotted (a peaks-over-a-threshold

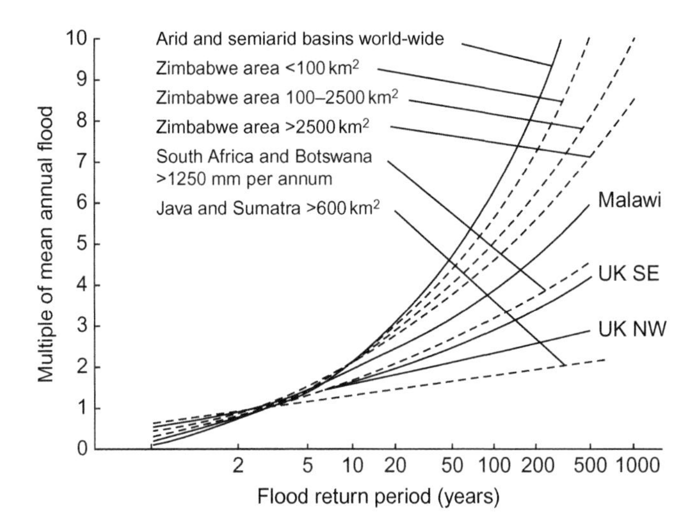

Sources of information: *UK – Law, F. Inst. of Hydrology. Elsewhere – Meigh, J. R., Farquharson, F. A. K. and Sutcliffe, J. V. (April 1997)* Hydrological Sciences Journal, ***42****(2), pp. 225–244.*

FIGURE 3.5

Examples of regional dimensionless flood frequency curves.

analysis). In the latter case care has to be taken to ensure that the events plotted are truly independent and that they are counted per 'water year' (i.e. summer plus winter) and not per calendar year.

There are limitations to the value of this method because a probability plot of past annual maximum floods can only be extrapolated reliably to estimate return periods up to about twice the length of the period of record. Thus, a 30-year record of annual maximum floods cannot be safely used to estimate the magnitude of a 100-year return period flood. In addition, the accuracy of past flood records cannot be checked and catchment conditions may have altered since records were taken or may alter in the future.

2. ***Use of regional flood probability curves.*** A different approach, applicable worldwide, is to use regional flood probability curves which are published for many regions (Fig. 3.5). The mean or median annual flood for a given catchment, termed the '*index flood*', is obtained from the period of historic record. A '*growth factor*' (i.e. multiplier) taken from the appropriate regional flood probability curve is then applied to the index flood to give the probable magnitude of a flood of given return period T. Published flood probability curves are derived from analyses of flood magnitude frequencies for catchments in the same region possessing similar characteristics. In the absence of an adequate historic record, the mean annual flood may be estimated using formulae based on regression analyses with various catchment characteristics such as catchment area, average annual rainfall, soil type or average catchment slope.

   In the UK, the current approach is to adopt the median annual flood as the index flood, rather than the mean. The *Flood Estimation Handbook* (FEH) also introduces the concept of a '*pooling group*' of catchments with similar characteristics to replace geographical regions. Full details of the method are given in the Volume 3 of the handbook (FEH, 1999).

3. ***Flood estimation derived from maximum precipitation.*** In this method the maximum precipitation that has occurred, or may occur, on a catchment is converted to consequent flood flow using either a unit hydrograph derived from the catchment, or a 'synthetic' hydrograph whose derivation assumes the most unfavourable catchment conditions that are likely to pertain during a probable maximum flood event. The method is of wide application because use of the unit hydrograph permits the calculation of flood flows from a wide range of rainstorms. Both approaches are similar but as the second is principally used for designing impounding reservoir overflows, it is described in Sections 5.17–5.19.
4. ***Flood estimation in the UK.*** The principal guidance for estimating flood magnitudes in the UK and Ireland is now Volumes 3 and 4 of the FEH (1999). Alternative methods are presented for flood frequency estimation on any catchment, gauged or ungauged. Volume 3 provides statistical procedures intended principally for use for return periods between 2 and 200 years. However, these statistical procedures should not be used for assessing the required design capacity of impounding reservoir flood overflow works, except where no loss of life can be foreseen as a result of a dam breach and where very limited additional flood damage would be caused (Section 5.17). Volume 4 uses a unit hydrograph rainfall–run-off method and is applicable to a wider range of return periods than the statistical approach given in Volume 3. However, this method model has now been superseded by an improved rainfall–run-off model – the Revitalised Flood Hydrograph (ReFH) model – for modelling flood events (Defra, 2005; Kjeldsen, 2007). The original rainfall–run-off method was designed for use with FEH Volume 2 procedures for estimating depth and duration of rainfall for a given return period, or vice versa. This permits the estimation of floods for return periods between 2 and 10 000 years and derivation of the probable maximum flood from estimates of the probable maximum precipitation. However, these rainfall depth–duration–frequency relationships have now been updated (Section 3.13).

The procedures set out in Volumes 3 and 4 of the FEH (1999) are relatively complex by reason of the number of influencing and limiting factors that need to be taken into account. Although the FEH provides much useful guidance on the choice and use of method, the application of the procedures to any particular case in the UK or Ireland is best undertaken by an experienced hydrologist. Issues identified with the FEH are mentioned in Section 5.17.

PART II YIELD OF SURFACE SOURCES

3.15 INTRODUCTION, DEFINITIONS AND CONCEPTS

The term 'yield' does not have a precise meaning. Its definition varies according to the context, which is important to understand when interpreting yield values quoted by others. Therefore, it is essential to state clearly the basis of yield calculations when quoting a yield figure to others.

The following terms are used to define yield more precisely or to indicate the way in which yield has been calculated:

- hydrological yield
- operational yield
- source yield
- system yield
- average yield
- historic yield
- probability yield
- failure yield

Terms such as 'deployable output' (DO) or 'water available for use' (WAFU) are sometimes understood to be general yield terms whereas in fact they have precise meanings in legal or regulatory contexts. The meaning of all of these terms is discussed in the following paragraphs while examples of their use are given later in the chapter.

Understanding the difference between terms in the first two rows of the above list is fundamental to appreciating the subtleties in more precise definitions. For surface water sources, *hydrological yield* is limited by the overall availability of rainfall and river flow, perhaps boosted by storage to iron out short-term shortages. However, the water available to satisfy demand may be less than that available in purely hydrological terms. It may be limited by hydraulic constraints such as pump, intake or pipe capacity or by the terms of an abstraction licence or agreement. Hydraulic yield so constrained is referred to as *operational yield*. In 1996, the UK Government published its *Agenda for Action* which introduced a new framework for the assessment of water resources in the newly privatized UK water industry. The paper assigned fairly precise definitions within the regulatory framework to terms like DO and WAFU. Their relationship to other terms like 'hydrological yield' is illustrated in Plate 1(a).

DO is the constant rate of supply that can be maintained from a water resources system except during periods of restriction. DO may be constrained by the specified 'Level of Service'; the historic period for which data are available; the physical capacity of the supply system (pipes, pumps, intakes, etc.); abstraction licences or permits; and water quality and environmental constraints (UKWIR/EA, 2000).

From the above, it can be seen that quoting a DO figure without defining the constraints under which it was calculated is usually rather meaningless. Levels of service (LoS) are discussed below under 'source reliability' while other constraints are discussed in Sections 3.18–3.24 in relation to specific types of sources.

The supply figure that must be relied on is *WAFU*. This can be compromised by '*outage*', which is the temporary loss of DO due to planned events, such as maintenance of source works, or unplanned events such as power failure, system failure and unacceptably high levels of pollution; turbidity, nitrates and algae. Outage must therefore be added to WAFU to get DO.

More often than not, the yield quoted for a source is its *drought yield*; the amount of water which the source can produce in times of drought. However, for much of the time the source can provide more than this. When water is plentiful, higher rates of output can support higher rates of demand or may enable other sources to be rested. As rainfall and river flows decrease in dry periods, output is reduced to ensure that supplies can be maintained throughout a period of drought.

The long-term output of the source through wet and dry periods is called the *average yield.* This is a good indicator of source potential and its overall contribution to satisfying demand. However, drought yields are the figures most often quoted since they can be relied on most of the time.

Droughts vary enormously in their severity and duration so when quoting a drought yield, the drought parameters must be defined. Drought severity can be considered in terms of the quantity of rainfall or run-off that occurs in a given period and the frequency with which this happens. The definition of yield that incorporates this combination of quantity and frequency is called the *probability yield*; the source output that can be maintained throughout a drought of given severity or frequency.

In sparsely populated or remote areas with scattered communities, a single water supply source is often adequate to supply the needs of the local population. The yield of this source is referred to as *source yield.* As the population grows and communities merge into towns and conurbations, these single sources are often connected by distribution systems so that any one community may be served by more than one source. The yield of these linked systems is termed *system yield.*

Linked systems can also realize benefits from the *conjunctive use* (Section 3.25) of different types of sources. Most surface water sources face a reduction in water availability in the summer or dry season. The early summer is a time when groundwater storage is normally at its maximum following winter recharge. The linking of these different types of source may increase security of supply by taking advantage of their differing yield characteristics. This introduces the concept of *critical drought period.*

In essence, the yield of any source is defined by:

$$\text{Yield} = \frac{\text{inflow over the critical period} + \text{any storage that is available}}{\text{Length of the critical period}} \tag{3.1}$$

At a run-of-river intake, there may be no storage available to buffer brief periods of minimum flow. The critical period in this case would be zero and yield is effectively the minimum river flow of specified frequency. Impounding reservoirs provide storage, which can maintain supplies above the rate of minimum flow for a certain period. How long this period is depends on how much storage is available and how long flows remain below the target rate of draw-off; i.e. drought duration (Plate 1(b) and Section 3.19).

The period over which available storage is drawn down is critical to determining the size of the yield, hence *critical period.* The concept of *critical drought period* is fundamental to both yield assessment and water supply system design. The critical period of a single source system is relatively easy to determine. However, once a number of sources of different types are linked together and subject to complex operating rules, the critical period is more difficult, if not impossible, to define as it varies with system configuration and with different operating rules. This is one of the main reasons why yield assessments in the UK are generally required to use system simulation modelling (Section 3.24).

All water utility managers have to be aware that almost every source has a quantifiable risk of partial failure. However, *source reliability* can be viewed from two perspectives:

- climatic or hydrological variability;
- the capacity or operational constraints of the water supply system.

The method used to quantify yield depends on which of these perspectives is the focus of concern.

Hydrological parameters such as rainfall and river flow have a natural variability and ultimately their availability will limit the yield attainable. Individual sources may be developed to abstract all of the water available in times of drought. Their output is therefore related closely to hydrological variability and attempts are made to describe this variability by comparing it to a statistical or probability distribution (Section 3.18).

In regions where water is relatively plentiful and living standards are high, supply systems are developed in stages in response to the growing demand of the population and the needs of the environment. As demand rises, the capacity of source works is increased or new sources are developed to satisfy that demand for a given level of reliability. The required reliability may be specified by a central planning authority as one of the criteria in system design. In a privatized water industry, a reliability standard may be set or imposed by regulation, the standard being determined by a combination of cost, customer preference and environmental implications. Typically, such a standard would be one of a number of LoS to which the utility manages its supply. Relevant typical LoS that have been used in the UK are:

- a hosepipe ban not more than once in 10 years on average;
- a need for a major public campaign requesting voluntary savings of water not more than once in 20 years on average;
- Drought Orders imposing restrictions on non-essential use not more frequently than 1 year in 50 on average;
- Drought Orders authorizing standpipes or rota cuts not more frequently than 1 year in 100 on average.

Such LoS are not statutory requirements but are in essence a contract between customer and supplier setting out the standards of service that customers can expect to receive. Reliability involves a balance between the needs of customers and the needs of the environment and as such, it may change with time. This affects the DO of the system. If DO is set too high, cut-backs in supply will be too frequent and the LoS will not be met; the DO then has to be reduced until the LoS are achieved.

Hydrological yield is usually calculated on the assumption of draw-off to supply at a constant target rate. However, it has always been recognized that water supply sources are not operated in an "on/off" way. If the target rate cannot be maintained, a lesser rate almost certainly can. In reality, water resources managers do not wait until supplies are about to run out before initiating cut-backs. System operation rules normally specify that target draw-offs are reduced when reservoir storage or river flows fall below pre-defined thresholds. Such thresholds may be stipulated in the terms of an abstraction licence or agreement. System operating rules are often linked to LoS.

The types of analyses that are involved in the 'hydrological' and 'operational' approaches to yield estimation are discussed in Sections 3.18–3.24.

3.16 HISTORY OF YIELD ESTIMATION IN THE UK

In England and Wales, the original approach to yield estimation was to adopt historic yields based on experience in major historic droughts. These yields normally referred to single sources or small groups of sources.

In the late 1960s and 1970s there was a change towards probability yields based on statistical analyses of historic drought flow sequences. These analyses allowed synthetic flow sequences to be constructed to represent droughts of varying duration, severity or return period.

As individual sources increasingly became merged into linked systems for which complex operating rules were sometimes devised, the yield of individual sources became less relevant and the difficulty of calculating yield for such systems became more apparent. In 1995–96 there was a notable drought across much of the country and a review was undertaken of the lessons to be learned for water resources and supply arrangements, both for the ongoing drought and for the longer term. The Government consulted widely and the results of the review were published in a paper entitled *Water Resources and Supply: Agenda for Action* (DoE, 1996). Annex E of the paper outlines the framework in which the yield of surface water sources for public water supply are to be assessed in the UK on a consistent basis. The framework embodies two main principals:

1. yield is to be assessed by simulation of the realistic operation of the water resources system in question; and
2. yield is defined as the supply that can be met with a given LoS.

Effectively, this means that a computer simulation model of the water resources system is required (Section 3.24). However, not all water supply sources occur in integrated water supply systems, particularly in sparsely populated or developing countries. Different source types have different characteristics and hence methods of estimating their yield vary. A *Handbook of Source Yield Methodologies* for the UK was published by UKWIR in 2014 (UKWIR, 2014).

3.17 METHODS OF YIELD ESTIMATION – GENERAL

Equation (3.1) describes the basic yield calculation for all single sources. Any yield assessment relies on historical data sufficient for the purpose. However, it must be accepted that available data are unlikely to reveal a drought that would be critical to yield determinations. In arid or semi-arid regions, rivers frequently dry up altogether for at least part of the year. Without storage, the *historic yield* of such streams is zero. However, they may not dry up every year and so some water is available in certain years. The more water that is required, the lower the frequency at which it is available. This combination of quantity and frequency is called the *probability yield* (Section 3.15).

3.18 RIVER INTAKE YIELDS

By far the most reliable estimate of river intake yield is obtained by analysing a long, reliable record of river flows at or near the point of abstraction. If there are no restrictions on the quantity of water that can be abstracted, yield for a given level of reliability (i.e. a probability yield) can be obtained from a statistical analysis of the available flow record. It is important that the flows are natural (i.e. free of any artificial influences such as upstream abstractions, discharges or river regulation effects) so that the observed variability in the flow record can be assumed to conform to an appropriate statistical distribution. This is likely to be true if the flow variability is natural but it is very unlikely to be true if the record contains artificial influences. If such influences are present in the flow record, the record must be naturalized (Section 3.9). Statistical analyses can then be performed on the naturalized flows to derive estimates of flow for a given frequency and level of reliability. Alternatively, a rough assessment may be made by ranking the low flows and plotting them on probability paper (Fig. 3.6). If appropriate, estimates of the artificial flow component are then added back in to give a total flow figure. This is important because in some rivers effluent returns from industry or treatment works are a major component of river flow in the dry season.

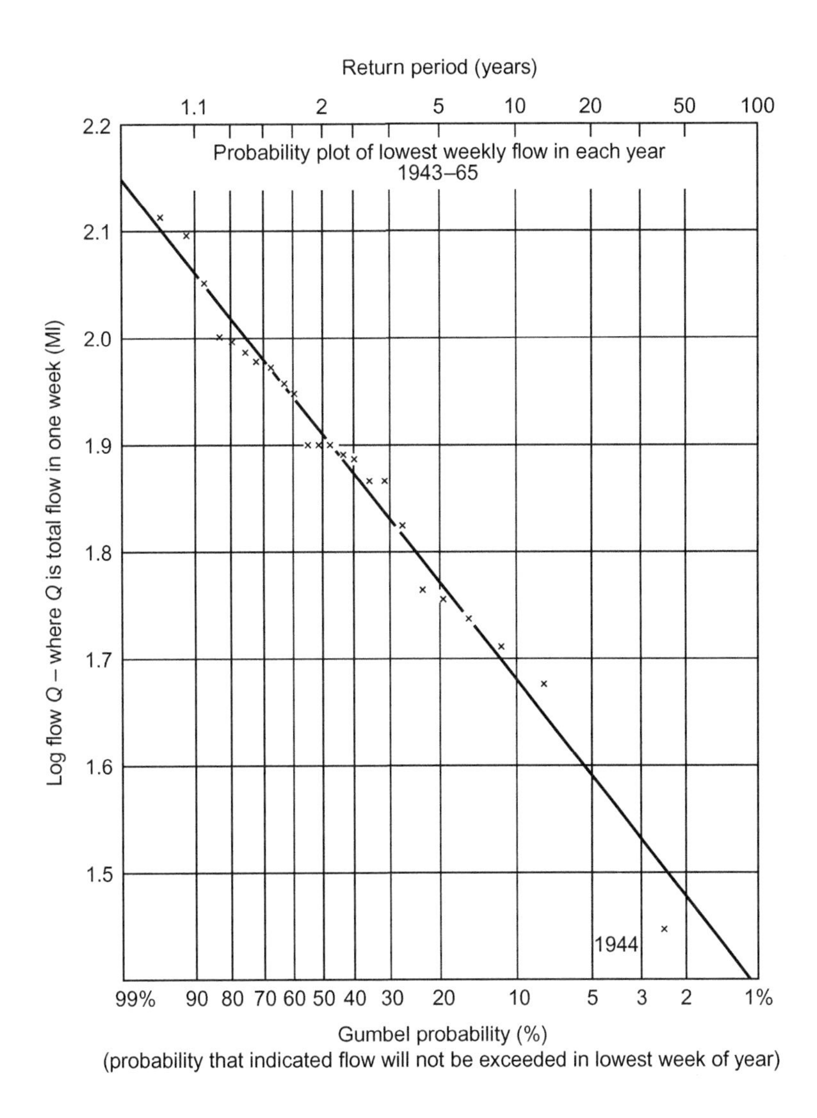

FIGURE 3.6

Naturalized dry weather flow probability plot.

If of sufficient quality, these effluent returns may be a vital resource and so cannot be ignored in any yield assessment.

As with all yield assessments, the question of critical period must be considered. For a water supply system fed from a river intake that has no storage, the critical period is zero. In practice, most water supply systems have at least some storage (e.g. bankside storage at the intake or service reservoirs within the distribution system) that can provide a buffer to maintain supplies through brief shortages. Bankside storage reservoirs may typically store the equivalent of 7-day supply to safeguard the system against pollution incidents in the river when the intake may have to be closed. The critical period for the source would then be 7 days and source yield would be determined by the average flow (of appropriate frequency) over a 7-day period. This is likely to be a much more stable value than daily flow, which can be subject to erratic changes due to gate operation or other artificial influences upstream. The shorter the critical period, the more important it is to base the analysis on daily flows. If only monthly data are available for the location in question, a scaling factor should be applied to the monthly values to reduce them to the relevant shorter period. An appropriate factor can be obtained from an analysis of daily low flow sequences from a flow record on a similar or nearby river in the region.

The above discussion relates to the determination of the *hydrological yield* of a river intake where water availability is limited only by climatic variability and hydrological factors. In reality, the yield of river intakes may be constrained by a number of additional factors including the following:

- Licence or permit constraints: abstraction limits linked to river flows; compensation flows or environmental sweetening flows; and the rights of other users;
- Hydraulic capacity of the intake;
- Water quality constraints: tidal effects and saline intrusion; pollution, etc.

In modern UK parlance, this constrained figure tends to be referred to as the 'deployable output' of a source.

If the *hydrological yield* of the river is greater than the physical capacity of the intake and any pumps involved, these would constrain source yield. Such factors could be inherent in the design or have resulted from post-construction morphological changes in the river channel (erosion, deposition or changes in river course – see Section 6.21).

Many river intakes are located near the tidal limit of rivers in order to be able to exploit the maximum potential hydrological yield of the catchment. Even where water levels at the intake exhibit tidal influence (e.g. a twice-daily cycle) the water itself may remain fresh in all but the most extreme circumstances. These usually occur when extreme high tides coincide with low river flows allowing saline water to intrude further up the estuary than normal and perhaps reach the intake. The saline water (intrusion) travels up the estuary as a wedge (due to its greater density) with freshwater on top. The depth from which water is drawn into the intake may therefore determine the increase in salinity. The frequency of occurrence of salinity higher than that acceptable for supply determines whether it affects yield (WAFU). Rare events are likely to be treated as *source outages*, in which case they would have no effect on quoted yields or DO. Frequent events may mean that for high levels of reliability (low probability of failure) the source has no reliable yield at all. This situation might arise due to continuing increases in upstream abstractions.

Water quality problems at intakes are not restricted to saline intrusion. A serious pollution incident in the river upstream could cause the intake to be closed for a time. As mentioned earlier, some intakes benefit from bankside storage to provide supplies during the period of closure. As with saline intrusion, intake yields may become affected if the water quality problems become more frequent or protracted. Background levels of pollution by chemicals such as nitrates (from fertilizers) are one example. High nitrate levels are often associated with the first period of significant run-off after a long dry spell when nitrate build-up in the soil is washed into rivers. It may be possible to model this effect using long sequences of river flow record if the relationship between nitrates, flows and dry periods can be defined adequately. Water treatment plants are increasingly provided with means of dealing with commonly occurring pollutants such as nitrates (Sections 10.24–10.26). In such cases yield (WAFU) will not be affected until the level of the pollutant exceeds the capacity of the treatment process to deal with it. The risk of a pollution incident affects yield if the river intake is a single source; it is less likely to affect the yield if the source is used conjunctively with other sources.

3.19 YIELD OF DIRECT SUPPLY IMPOUNDING RESERVOIRS

The yield of a direct supply impounding reservoir is defined by the following equation:

$$\text{Yield} = \frac{\text{Storage} + \text{inflow over the critical period}}{\text{Length of the critical drawdown period}} \tag{3.2}$$

The critical period (Section 3.15) is the time from when the reservoir first starts to be drawn down to the time when it is at its maximum drawdown. It is the period over which storage is used to supplement river flows to maintain a reliable yield, the larger the available storage, the longer the critical period. The other main factors determining yield are the length of the dry season, over which flows have to be boosted, and the volume of reservoir inflows during this period. If winter rains are regular and the reservoir fills every year, the critical period is short, perhaps 6–10 months. If the storage is small, it may be even shorter. If winter rains occasionally fail, droughts may extend over 2 years or even longer. In this case, storage has to be eked out over a longer period making the critical period longer. Plate 1(b) illustrates the principles involved. The volume of water that supports yield is A (the volume of inflow over the period) plus B (the volume of storage used to maintain the required supply over the same period). So if the volume is X million litres and the critical period is Y days the reservoir yield is X/Y = V Ml/d.

The historic yield is calculated by identifying the longest period in the historic period that has the smallest volume of inflow (per unit of time). To get a robust yield figure the longest period of record should be used in order to ensure it contains a severe drought that is unlikely to be exceeded

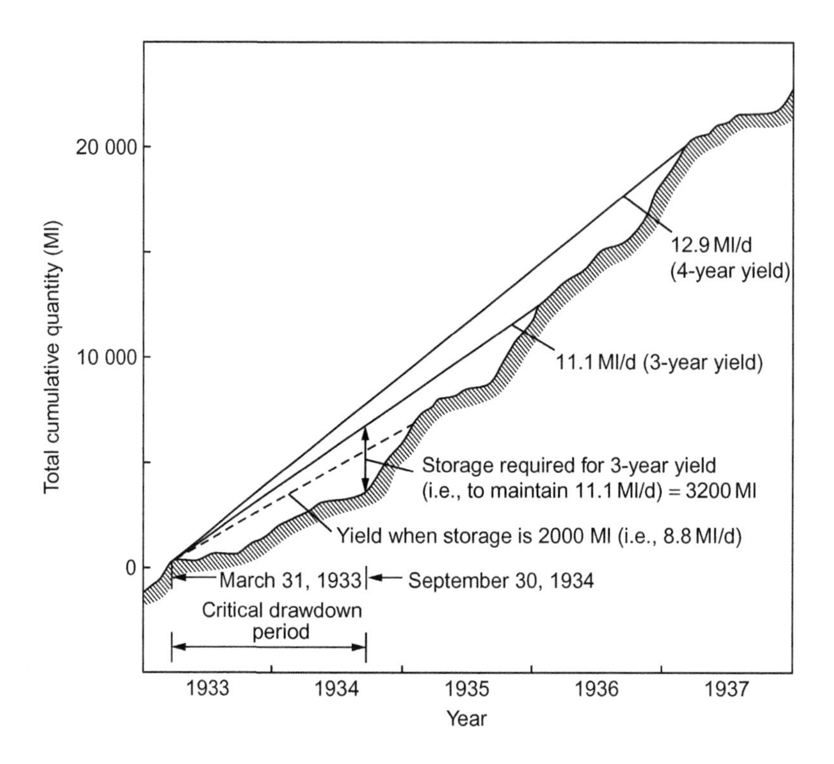

FIGURE 3.7

Five-year mass flow diagram.

often. This may require the reservoir inflow record to be extended back in time (by flow correlation or rainfall/run-off modelling) if the direct record is not long enough. Figure 3.7 shows a method for assessing historic yield graphically; cumulative catchment run-off is plotted over the period of record.

If a *probability yield* is required, the drought of required duration (the critical period) with that probability of recurrence (return period) must be identified in the historic record. The chances of finding one are slight so a *synthetic drought* is sometimes constructed from available run-off records. For the design of a new reservoir, many of the variables such as storage volume, compensation flow, etc. may be undecided and a number of options may need to be assessed. Each option may have a different critical period. It is very unlikely that data will be available for a historic drought for which all the critical periods of interest have the same return period. One way of getting round this is to construct a nested synthetic drought in which run-off of all durations has the same return period so that, in say a 36-month sequence of river flows, the driest 6, 7, 8, 15, 18, 24, etc. periods all have the same probability of recurrence; say 2% (1 in 50 years). An illustration of this is shown in Figure 3.8. Another graphical method, illustrated in Figure 3.9, is to express the minimum yield as a proportion of the mean catchment

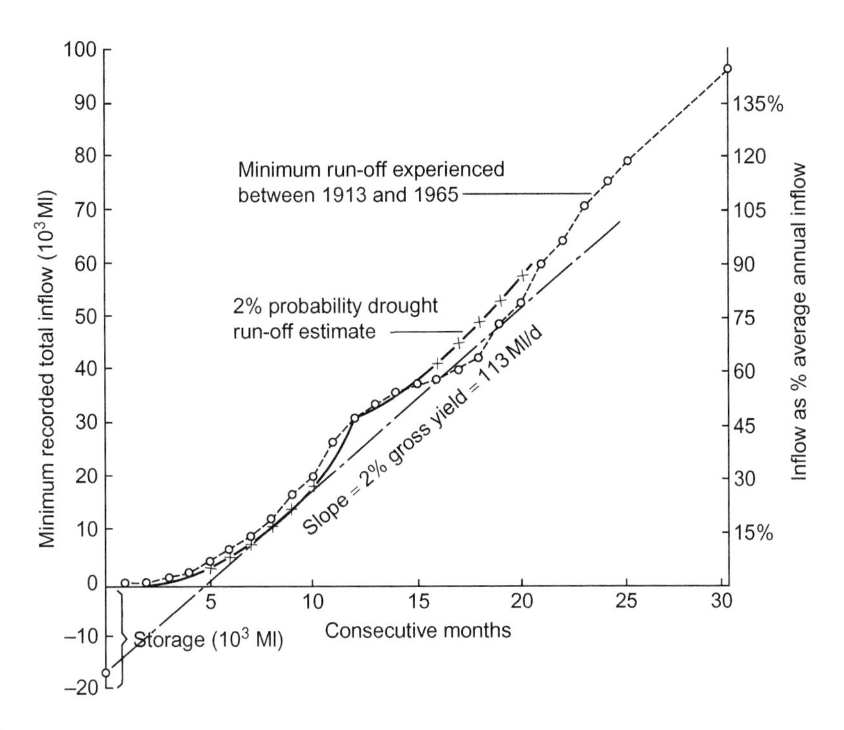

FIGURE 3.8

Minimum run-off diagram.

flow over the period of record and the reservoir storage as a proportion of the mean annual flow volume.

Most of the commercial software available to simulate complex water resources systems summarized in Section 3.24 can also be used to simulate individual reservoirs. Alternatively, a spreadsheet may be used with parameters such as inflow, storage, outflow, etc. arranged in columns across the page and values for each time step arranged in rows down the page. The elevation/storage and, or, elevation/surface area relationships (Section 5.18) can be represented as lookup tables. As the time step for such calculations is rarely less than a day and often as long as a month, reservoir spills, if any, are simply the balance of inflows minus outflows when the storage is full. There are no hydraulic considerations involved.

Evaporation from the surface of a water body, such as a reservoir, is greater than from the equivalent area of land surface. Potential evaporation is higher because the albedo (reflectivity) of the reservoir surface is lower and more of the sun's energy is available to drive evaporation. However, the ratio of actual to potential evaporation is also higher because of the availability of water to evaporate. In temperate regions, the increased losses from the surface of the reservoir may not be significant, especially in the case of a relatively small reservoir in a large catchment. However, in arid or semi-arid zones, such losses can be very large and should be allowed for in the yield calculation. In the simplest calculations, daily or monthly evaporation figures can be applied to the average surface area of the reservoir. However, if a spreadsheet model is used, it is straightforward to apply evaporation figures to the area of the reservoir at each time step, computed using a storage/elevation/area relationship.

Other factors that should be taken into account in a reservoir simulation include draw-off, licence stipulations, compensation flows and emergency storage allowance.

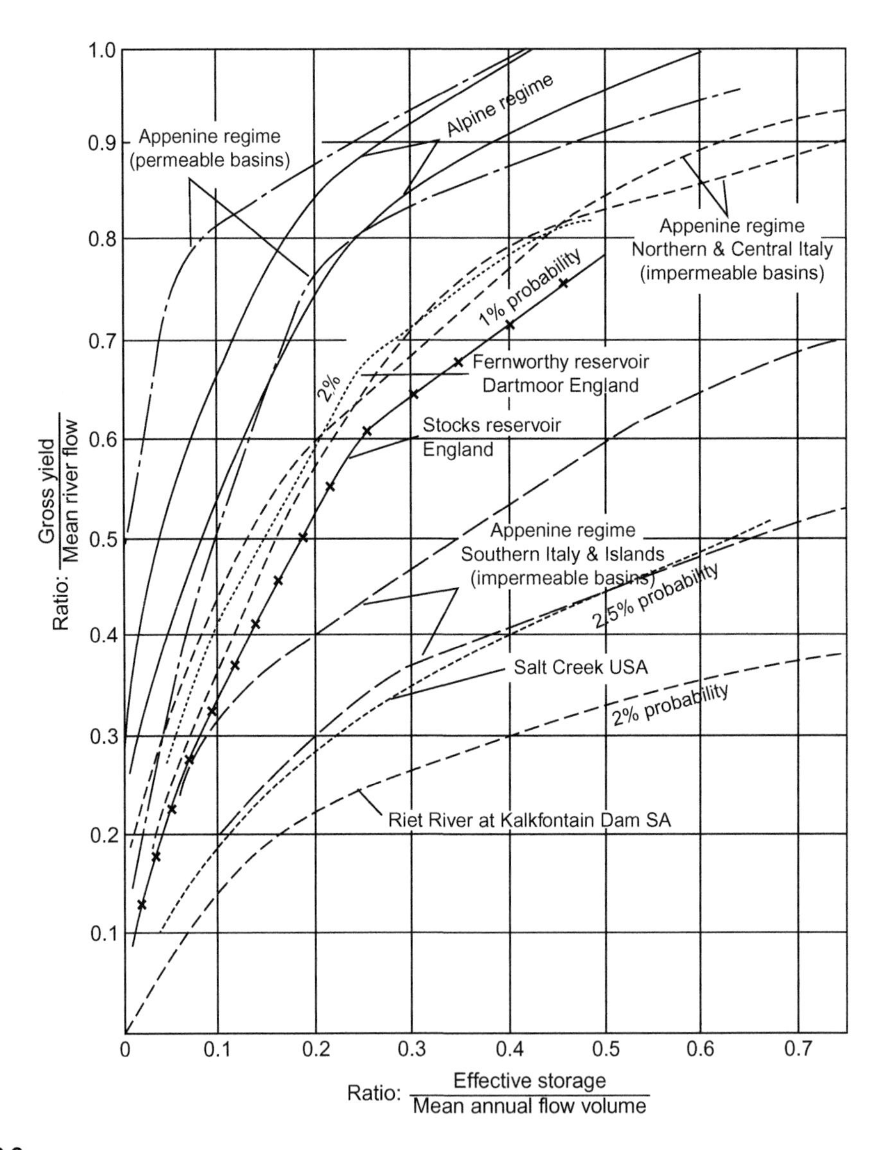

FIGURE 3.9

Typical yield/storage relationships for direct supply reservoirs.

3.20 YIELD OF A PUMPED STORAGE RESERVOIR

To estimate the yield of a storage reservoir fed by pumps at an intake it is more accurate to work on the basis of daily flows. The use of mean monthly flows is inexact and necessitates reducing the computed yield by an arbitrary amount to allow for part of the flows in excess of the mean being uncollectable. If some daily flow records are available, these may assist in estimating what allowance should be made; otherwise a frequent practice is to assume that only 90% of the potential abstraction is possible. The better option is to develop a series of daily flows that represent flows during a period of minimum run-off. The method involves use of minimum monthly flows of a suitable probability (say 2%) and applying to them a factor derived from a relevant, albeit short, record of daily flow for one of the months.

With daily flows available, the basic calculations are as shown in Table 3.5. However, the sub-calculation required to calculate potential abstraction depends on: (1) the rule(s) laying down the abstraction conditions; (2) the assumed maximum pumping capacity; and (3) the time lag between a change of flow and the consequent change of pumping rate.

Factors (2) and (3) need careful consideration before starting the calculation. With appropriate control equipment and variable speed pumps, the abstraction may closely follow flow variations; but other arrangements are often adopted. A range of fixed speed pumps may be started and stopped automatically to give outputs such as *0.5Q*, *1.0Q*, *1.5Q* or *2.0Q* according to their combination. In addition, if pump output is manually controlled, the time lag between change of flow and change of abstraction will depend on the manning pattern and on whether variable or fixed speed pumps are used.

A simplistic calculation can be based on assuming an unlimited volume of reservoir storage is available so the maximum drawdowns occurring each year can be calculated and plotted on probability paper to find the drawdown probable, e.g. once in 50 years. However, this does not solve the problem of what yield is available with a reservoir of a lesser capacity than the maximum drawdown. Calculations then have to proceed for a range of draw-offs to produce a yield–storage relationship with an associated risk in order to be able to determine the yield for a given storage for any degree of risk. This process is difficult to carry through successfully unless the flow record is long enough to make the probability plot one for interpretation only. Points from an unnatural distribution do not permit confident extrapolation.

Table 3.5 Pumped storage reservoir calculation (abstraction condition – two thirds flow above 30×10^3 m^3/d)

					Reservoir contents	
Day	Flow to intake 1000 m^3/d	Quantity available at intake 1000 m^3/d	90% pumped to storage 1000 m^3/d	Amount supplied ex storage 1000 m^3/d	Net change of storage 1000 m^3/d	In store end of day 1000 m^3/d
15.8	42.0	8.0	7.2	5.5	+1.7	151.7
16.8	39.0	6.0	5.4	5.5	−0.1	151.6
17.8	38.0	5.0	4.5	5.6	−1.1	150.5

In view of these difficulties, it is more satisfactory to work directly on an estimate of the daily flows likely during, say, a drought of 2% risk. It is then possible to compute the yield of 2% risk according to the size of storage reservoir adopted and the three factors mentioned above. A further limiting consideration is the need to ensure refilling of the reservoir after some maximum drawdown. The most secure provision is to ensure refilling of the reservoir in any single wet season following a dry season, but this is not always possible. Some pumped storage schemes accept that refilling will only occur once every few years, but they should be tested to make sure they can achieve an initial filling sufficient to meet the anticipated initial demand.

All such problems can be dealt with by applying appropriate computer calculations to the record of daily flows. Among the most useful results is that of finding the most economic size of pumps to install. Increasing pump capacity beyond a certain level may increase the yield by only a small amount when flows fall rapidly and critical drawdown periods are short. On the other hand, the need to ensure refilling of the reservoir during a wet period may be an over-riding factor determining maximum pumping capacity.

3.21 YIELD OF REGULATING RESERVOIRS

A regulating reservoir (Fig. 3.10) impounds water from a catchment A, and releases water to support an abstraction at some location B downstream, when flows at B are not sufficient to meet the required abstraction. This means that the yield obtainable is greater than that provided by catchment A alone. Usually some compensation water has to be released from A to maintain a flow in the stream below it; and at B various abstraction conditions may apply in order to preserve the natural low flow regime of the river. This may require either:

- maintenance of a given flow continuously below the intake B; or
- abstraction at B to be fully supported by equivalent releases from A until the natural flow at B reaches a certain figure; thereafter, as flows at B continue to rise, releases from A are cut back until the natural flow at B is sufficient to support the whole abstraction.

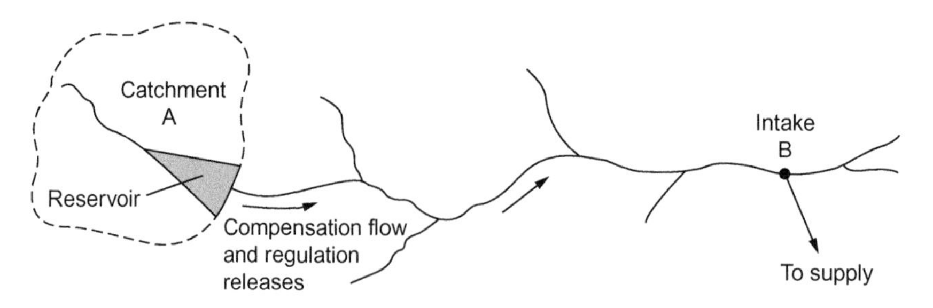

FIGURE 3.10

Regulating reservoir.

As with pumped storage schemes it is best to calculate minimum yield by first producing daily flows for, say, a 2% drought period for the natural flows at the intake point B. The proportion of this flow diverted into the regulating reservoir A has to be assessed and probably bears a varying relationship to the natural flow at B. The calculation can then proceed according to the rules laid down for compensation releases at A and the abstraction conditions at B. An example is given in Table 3.6.

Two points need to be borne in mind. Releases from the reservoir may need to include an allowance for evapotranspiration and other losses en route to the abstraction site (hence the notional 10% addition to Col. (4) in Col. (5) of Table 3.6). Secondly, the time taken for released water to travel to the abstraction point must be taken into account. Depending on the distance involved, the time lag may range from several hours to 1 or 2 days or more. This can mean that an increase of release in expectation that flow at the intake will decline may be wasted if rain should come and increase the flow. Hence, a further allowance of 10% or 15% may have to be added to the releases to cover discrepancies between actual and theoretical release requirements. A dry weather recession curve for the natural flow at the intake, converted into a guiding rule, can be used to aid release decisions. Use of such a decision rule in the UK has shown that actual releases tend to be up to 20% more than the theoretical requirement in wet years and about 3% more in dry years.

Table 3.6 Regulating reservoir calculations. Conditions: abstraction required 28 Ml/d. Reservoir compensation release 2.0 Ml/d. Residual flow below intake 10 Ml/d

	(1)	(2)	(3)	(4)	(5)	(6)	
Day	**Natural flow at intake 1000 m^3**	**Reservoir inflow 1000 m^3**	**Net flow at intake 1000 m^3**	**Intake flow deficiency 1000 m^3**	**Reservoir release 1000 m^3**	**Reservoir change/day 1000 m^3**	**Cum 1000 m^3**
1	45.0	3.3	41.7	Nil	2.0	+1.3	+1.3
2	41.0	3.0	38.0	Nil	2.0	+1.0	+2.3
3	38.0	2.8	35.2	0.8	2.9	−0.1	+2.2
4	35.5	2.6	32.9	3.1	5.4	−2.8	−0.6
5	33.0	2.4	30.6	5.4	7.9	−5.5	−6.1
6	etc.						

Col (3) = Col (1) − Col (2)
Col (4) = Abstraction + Residual flow − Col (3) − Compensation release
Col (5) = Col (4) × 110% + Compensation release
Col (6) = Col (2) − Col (5)

3.22 YIELD OF CATCHWATERS

A catchwater is usually a channel which leads water from some remote catchment into an impounding reservoir. The remote catchment would not otherwise contribute any flow to the reservoir so the flow of the catchwater increases the yield of the reservoir. The yield of the catchwater is the amount of water the catchwater can provide with specified frequency or reliability over a given period. This period would normally be the critical drawdown period of the reservoir so catchwater yield is effectively the amount by which the catchwater raises the yield of the reservoir.

It is very unusual for flows in catchwaters to be measured directly. The contribution made by the catchwater therefore has to be estimated. Unless it is extremely large, the catchwater will not be able to divert all of the run-off reaching it. Flows in excess of catchwater capacity will be lost (unless the overflow occurs within the catchment of the reservoir). Catchwaters often originate at an intake on another stream, or capture flows from streams they cross along their route. Intake capacity and compensation flow requirements may therefore be other limiting factors. In view of the above, catchwater yield requires the evaluation of two main parameters:

1. the volume of water reaching the catchwater in the critical period; and
2. the proportion of total flow that the catchwater can capture.

Parameter (1) depends in turn on the capacity of the catchwater and the variability of potential inflows. One way of determining catchwater capacity is to install a temporary gauge and measure it directly during a period of heavy rainfall. If this is not feasible, the capacity may have to be estimated by hydraulic analysis (according to channel dimensions and gradient).

If the catchwater is to be incorporated into a simulation model of the overall system, parameter (1) must be provided as a long, historic flow series while parameter (2) is calculated by the model for each time step. The time step must be short enough (15 minutes to an hour) to represent the variability of flows in the inflow sequence. If the time step is too long, average flow over the time step may be less than the capacity of the catchwater, suggesting that all of the run-off can be captured.

If the catchwater is not to be modelled using a long sequence of historic flows, the volume of water reaching the catchwater and the proportion of run-off captured have to be calculated separately. The former can be obtained from a flow duration curve (Fig. 3.11) for an appropriate period; e.g. an average year or a 2% drought year. The proportion of run-off captured can then be obtained from a catchwater transfer curve such as that shown in Figure 3.12. This shows a relationship between catchwater size and percentage of run-off collectable, derived from hydrographs observed on small British catchments with average rainfalls of 1500 mm or more. However, the relationship was found to hold true against data from a tropical catchment, such as one in Singapore. Although Figure 3.12 strictly applies to average annual flows, it can be used without too much error to assess yield during a given season of the year, provided the appropriate average daily run-off for that season is used in place of the average annual flow.

The catchwater contributions to the reservoir are added to its direct catchment inflow, so that the minimum yields for various consecutive months can be plotted on a diagram such as Figure 3.13. This shows the extra yield a catchwater provides for a given reservoir storage and the change of critical drawdown period.

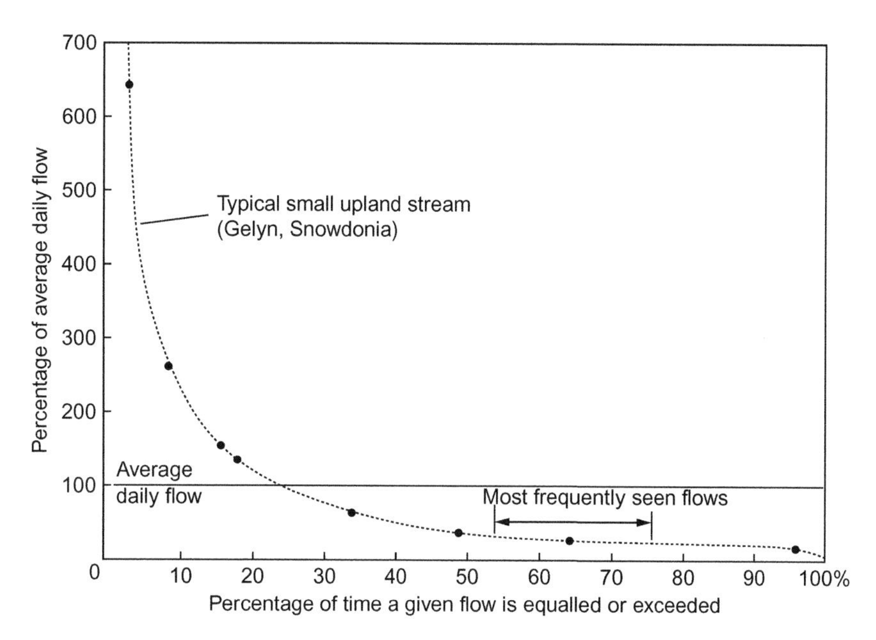

Note: *The curve is derived by measuring the number of hours the stream flow exceeds given level of flow.*

FIGURE 3.11

Flow duration curve.

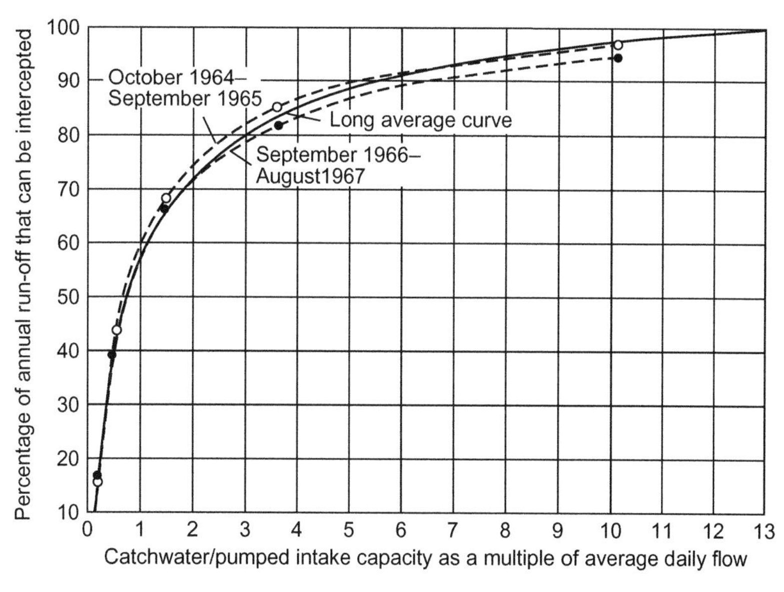

Note: *The design curve is based on hourly flow duration data from small mountain catchments in England and Wales.*

FIGURE 3.12

Catchment transfer curve.

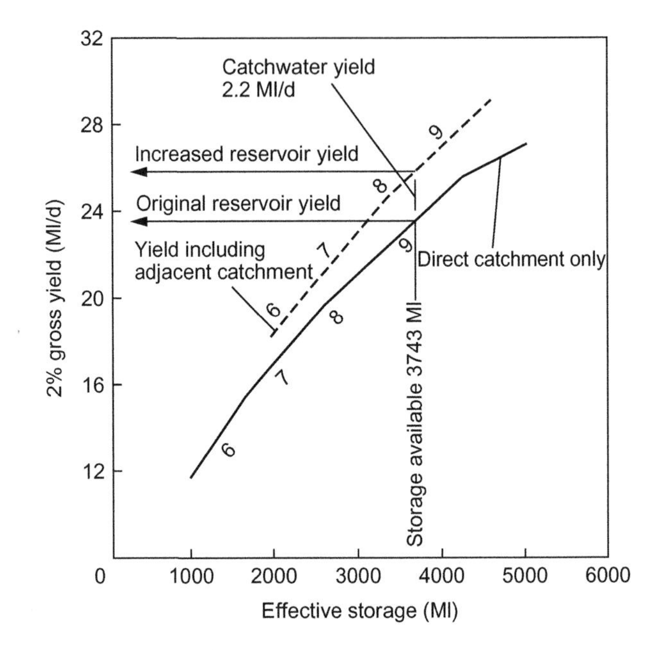

Note: *The figure by each straight line denotes the length of the critical period in months.*

FIGURE 3.13

Catchwater yield/storage diagram.

Since many catchwaters are simply open unlined channels cut to a gentle gradient in a hillside, they often contribute little or no inflow during the dry season, and do not contribute flow to the reservoir until initial precipitation is sufficient to wet their bed and banks to saturation level. Therefore, their main contribution to the reservoir is during the wet season and their impact on yield is more on average yield rather than drought yield.

3.23 COMPENSATION WATER

Compensation water is the flow that must be discharged below an impounding reservoir to maintain the water rights of riparian owners and other abstractors downstream. Each country tends to have its own water law to preserve water rights and setting quantities of compensation water can involve extended legal dispute. In Britain, the compensation water from most impounding reservoirs is set by some Parliamentary Act. In the early 1900s compensation water was often set at one third of the gross yield of the reservoir, but this proportion tended to reduce to one quarter in later years (Gustard, 1987). Nowadays compensation water is often required to be varied seasonally and extra discharges as 'spates' may be stipulated at certain times of the year to meet fishing interests.

The discharge of a fixed amount every day has been criticized on environmental grounds as being 'unnatural' and not conducive to the maintenance of fauna and flora, which need periods of varying flow (see also Section 6.8). Considerable progress has been made in quantifying the water requirements at different stages in the life cycle of fish, invertebrates and macrophytic vegetation (Bullock, 1991). American studies of physical habitats have been followed in France, Norway, Australia and the UK as a means of defining environmentally acceptable flows. The software calculations with the PHABSIM program, available in the public domain, depend on field measurement of river velocity, depth, substrate and tree cover. They determine ecological preferences and hence seasonal variation of compensation water but, inevitably, not all the requirements can be met if a reasonably economic yield is to result. Hence, some compromise solution has to be found. Nevertheless, the technique gives a satisfactory means of engineering water resource developments to achieve minimum environmental damage.

3.24 YIELD OF WATER RESOURCES SYSTEMS

In relatively sparsely populated regions, water supplies are typically drawn from single sources, particularly where settlements are separated by large distances and it is impractical to link them. However, in more densely populated regions, water supply systems are generally supplied from multiple sources of varying types. There are numerous advantages in having an integrated water supply system: individual sources can be rested or taken out of commission for maintenance without undue interruptions to supply; and water can be transferred from a part of the system with a surplus to an area experiencing a drought. The lack of a fully integrated system contributed to the shortages that were felt in the Yorkshire region of the UK in 1995. However, estimating the yield of an integrated system is more complicated than for a single source, largely as a result of the differing critical drought periods for each source type; the only practical way of calculating the yield of an integrated system is by computer simulation.

In UK Environment Agency Guidelines (EA, 2012) for the estimation of the yield of surface water and conjunctive use schemes, the recommended approach includes behavioural analysis; a model is used to simulate the realistic operation of a water resource system as currently configured using historic data corrected for current catchment conditions. The method further requires:

- the derivation of long, naturalized historic sequences for all inflows to the water resource system being modelled;
- system control rules linked to LoS which allow for the introduction of demand management practices to cut supplies during periods of drought;
- yield estimates to allow for the provision of emergency storage to accommodate the operational uncertainty regarding the duration of a particular drought.

The hydraulics of water movement around the system is not normally simulated by the model, but realistic physical limitations such as pump and mains capacity are taken into account, as are the limits and conditions imposed by licence agreements.

Inflows to the system need to be for as long a period as possible in order to represent the full range of hydrological variability in the modelled sequence and to allow the company's target LoS to be tested. If necessary, naturalized historical flow records need to be extended back in time, either by correlation with other long-term records or by the use of rainfall–run-off modelling.

Demand is defined at model nodes, which coincide with major demand centres. Base demand is the average demand over the year, to which a seasonal demand profile is applied.

The DO of the system is calculated as the output to supply which can be met over the whole period of the simulation with the required LoS. It is defined by system performance throughout the worst drought in the record and takes account of the varying critical drought periods for different sources and parts of the system.

Simple water resource simulation models can be developed using spreadsheets. They are quick and easy to develop and have the advantage that numerical and graphical output can be tailored to the user's requirements. Commercial programs are available including:

- WRMM Alberta Environment, Calgary, Alberta
- HEC-ResSim www.hec.usace.army.mil/
- Aquator http://www.oxscisoft.com/aquator/index
- MIKE-Basin www.dhi.di/mikebasin
- Miser www.tymemarch.co.uk

Some of these, such as WRMM and HEC-ResSim, are freely available while others are designed for use by water companies with large, permanent water supply systems and may include automatic source optimization routines.

3.25 CONJUNCTIVE USE AND OPERATION RULES

When a water utility has several sources, conjunctive or integrated use of them may be a means of improving the total yield or of reducing costs, or both. Thus extra water from an underground source when the water table is high, or from a river in flood, may permit a cut-back in the supply from an impounding reservoir with a 2-year critical drawdown period, enabling it to store more water. Similarly, it may be possible to keep storages with short critical periods in continuous full use to avoid overspills and so maximize their supply, at the same time reducing draw-off from a larger reservoir with a longer critical period, thereby gaining larger reserves to meet critical drawdown conditions. Similarly, it may be possible to reduce costs if the source producing the cheapest water can be overrun for part of the year, whilst dearer sources are cut back (Lambert, 1992; Parr, 1992).

However, physical conditions may limit possibilities for conjunctive use, for example:

- isolation of sources and their supply areas;
- supply areas at different elevations;
- incompatibility of one source water with another;
- the need of certain manufacturers to use only one type of water.

It is not always advisable to change frequently from one type of water to another, particularly if one is a 'hard' water from underground and the other a 'soft' river or impounding reservoir supply.

To ascertain potentials for conjunctive use the whole system of sources needs drawing out in diagrammatic form, showing:

- source outputs (average day critical yield; maximum day plant output);
- impounding or pumped storage capacity; length of critical drawdown period;

- area served, line of trunk feeders, key service reservoirs fed;
- elevation of supply at sources; high ground areas in the supply area;
- any legal or other restrictions on source outputs.

It is helpful to allocate a different colour for each source and its associated works. The possibilities can then be examined for conjunctive use. Key factors will probably be the need for major interconnecting pipelines and extension of treatment works capacity. The cost of these must be roughly assessed to see whether they are likely to be worthwhile for the possible gain in yield. Once a possible scheme for conjunctive use has been clarified, this can be tested by computer calculations on a month-by-month basis to check the combined yield during a chosen critical dry period.

Operating rules can be developed to assist in judging when storage reservoirs can supply more than their minimum yield for a given risk. Monthly reservoir drawdowns over a long period of simulated inflows can be used to develop a control curve as shown in Figure 3.14. This shows the minimum storage required at the beginning of every month to ensure maintenance of a given supply rate. To produce a control curve of this type involves calculating backwards in time, from an assumed zero storage at the end of each month of the year. By applying this process to droughts of every duration, it is possible to locate the maximum storage required at any time of the year to ensure the reservoir never quite empties at the assumed abstraction rate. Different abstraction rates will require different control curves. A family of such curves is therefore produced, each for a different level of supply. Hence, reference to the curves and a storage/water level chart, can show whether the water level in the reservoir permits an increased abstraction or not.

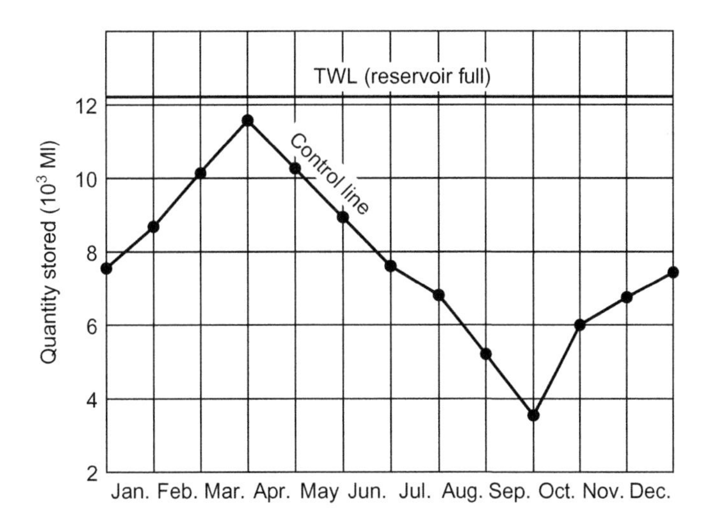

Note: *The control line represents the levels to which the reservoir could have been drawn down during any month in the period 1910–64 (55 years) and still maintain a total outflow of 46 Ml/d to supply and 13 Ml/d to compensation. When the contents are above the control line the draw-off may exceed the 46 Ml/d to supply (up to the limit of the treatment works capacity). If the contents fall to or below the control line the draw-off must be limited to 46 Ml/d to supply.*

FIGURE 3.14

Reservoir control rule.

Reservoir control curves have to be based on the most severe droughts historically recorded, or on a 'design' drought of specified probability. Neither can forecast the magnitude of some drought. Hence a control curve tends to be of more practical use in permitting extra water to be supplied when storage is high, than when storage is low. If a reservoir is three parts empty with the dry season not yet ended, most engineers would attempt to restrain demand, in preference to relying solely on a control curve, which poses a significant risk that it might not apply to what the future may bring.

3.26 RAINWATER COLLECTION SYSTEMS

Rainwater tanks collecting run-off from roofs or impervious surfaces form a useful source of drinking water where daily rainfall is frequent, as in equatorial climates, or for monsoon periods in monsoon climates. The supply is particularly useful if local sources are polluted, since only simple precautions are necessary to keep the rainwater free of pollution.

For individual house rainwater collecting tanks the principal constraint is usually the size of tank which can be afforded or which it is practicable to install. Tanks up to 600 or 800 litres are generally formed of one piece of material and are transportable. Tanks of 1000-litre capacity or more are usually more economical if constructed in situ, e.g. in reinforced concrete. Such tanks have the advantage they are repairable if they leak. Tanks made of plastic plates bolted together tend to fracture under the repeated bending caused by changing water levels and the fracture is usually unrepairable. Steel tanks made of plates bolted together tend to rust at the joints.

Roof areas are not usually a principal constraint provided houses are one storied. There is usually an area of roof, which can discharge to guttering along one side, perhaps with a return along another side. Traditional roofing for low income communities is galvanized iron, but corrugated asbestos cement roofing, clay tiling and asphalt may be found. With house occupancies varying from six to 12 people, about 12–22 m^2 of developable roof area per person is usually available.

Daily rainfall records need to be tabulated. The daily run-off to the collecting tank is usually calculated according to the roof area available per person. The run-off is taken as *95% × (daily rainfall − 1 mm)*, the 1 mm deduction being for initial evaporative loss on the roof, and the 95% allowing for guttering overspill. Some field tests need to be undertaken to assess the size of guttering required; where it should be located; and what allowance should be made for overspill during intense rainfall. Local practice and experience, where available, can act as a guide. Although UK rainfall conditions are not likely to apply overseas where roof rainfall collection systems are mainly used, some UK publications can provide guidance (BS EN 12056:2000; Sturgeon, 1983). For large roofs, e.g. of commercial premises, etc. the use of siphonic outlets to guttering can achieve economy in downpipe sizing. The siphons become primed by air-entrainment of initial flows, thus making the whole head between gutter outlet and the bottom of the downpipe available for the discharge, reducing the size of downpipe required (May, 1982, 1997).

Calculations proceed by using trial abstractions over typical recorded dry periods, commencing when the collecting tank can be assumed full. It is convenient to work on 'units of roof area' and 'storage available per person', sizing this up later according to the average number of people per household for which the design should cater. Operating rules for householders should be simple, e.g. either a fixed amount X per person per day, reduced to $0.5 \times X$ when the tank is half empty. An appropriate minimum draw-off rate would be 8 litres per person per day, the basic minimum requirement for direct consumption (drinking and cooking purposes) (Section 1.4). It is necessary to add an allowance for the *bottom water* and *top water* to the calculated theoretical tank capacity. The top water allowance, about 150 mm, means the householder does not have to restrict his or her take until the water level is that much below overspill level. The level can be marked clearly inside the tank. The advised abstraction X has to be expressed in terms of commonly available vessels. A standby supply must also be available to cover longer dry periods than assumed for the calculation and also for rescuing householders who run out of water for any reason.

Although house tanks will not give a sustainable amount in dry seasons of the year, they will be well used in the wet season because they relieve the householder of having to carry water from a distance. The rainwater in the tanks does not need to be chlorinated, but tanks do need cleaning out annually. Mosquito breeding in tanks can prove a nuisance in some climates; gaps in covers should be sealed and vents covered with insect mesh (Section 20.15). Water quality and treatment aspects of rainwater collection are discussed in Section 8.23.

3.27 THE LIKELY EFFECTS OF CLIMATE CHANGE

The assessment of water availability for supply inevitably has to assume *climatic stationarity* when considering hydrological variability; that is wet years, dry years, floods and droughts continue to vary about a constant long-term mean much as they have in the past. This assumption is necessary in order to describe the variability in terms of a statistical or probability distribution which in turn facilitates the estimation/prediction of extreme events (e.g. the 1 in 100-year drought) that perhaps have not yet been observed in rainfall and run-off records. *Climate change* is a process which involves non-stationarity and therefore has to be considered separately. The term climate change is frequently used imprecisely and it is essential to understand the difference between climate change and climate variability.

Climate variability is the natural variation of climatic parameters such as temperature or rainfall about the long-term average. Large variations in these parameters occur, both seasonally and from year to year. Without climate change, these variations occur more or less symmetrically about the long-term mean, which is effectively constant. The words ‘long term’ are important here because clusters of hot or cold, wet or dry years might produce short-term trends or cycles. These do not provide evidence of climate change.

The climate can be said to be changing when there is a clear trend in the long-term average of one or more climatic parameters over a long period. Trends might be seen in maximum and minimum values as well as the mean, or may be observed in event frequency such as the period between floods or droughts of a given magnitude.

Few hydrological records extend back more than 50 years and so most are too short to identify long-term trends with much confidence. Where trends in river flows can be identified, it is difficult to separate the effects of climate change from changes in land use, farming practice or urbanization. Nevertheless, conventional wisdom states that our present climate is changing and the possible impact of this change on future source yields needs to be considered.

The Intergovernmental Panel on Climate Change (IPCC) was established in 1988 by the WMO and the United Nations Environment Program (UNEP) to prepare assessments of the current state of scientific knowledge concerning climate change. Their Fifth Assessment Report (AR5) provides a summary of the situation up to 2014 and strongly links future climate to current and future emissions of greenhouse gases, particularly CO_2 (IPCC, 2013, 2014). Many of their conclusions are fairly broad and qualified by '*probability of occurrence*' or '*risk*' indicators. This is partly because of uncertainties in the science but also because efforts are being made by governments and international organizations to reduce greenhouse gas emissions in the future. Our future climate will therefore depend at least in part on the success of these efforts. This would make precise predictions of future climate impractical even if the science was perfectly understood.

Highly complex models of the global climate system (oceans as well as atmosphere) have been run using a range of scenarios, each representing a different level of greenhouse gas emissions in the future. The models are used to translate possible emissions scenarios into resulting climatic conditions. No likelihood of occurrence is attached to these scenarios which should not therefore be regarded as future predictions, more a constrained form of sensitivity analysis.

AR5 paints a picture of steadily rising air and sea temperatures and rising sea levels. Changes in the pattern of precipitation will not be uniform but it is thought that extreme precipitation events will become more intense and frequent in many regions. In most dry subtropical regions the report concludes that "...climate change is projected to reduce renewable surface water and groundwater resources...intensifying competition for water among sectors" (IPCC, 2014). Clearly such changes need to be taken into account in the design of water supply schemes for the future.

The IPCC sets the scientific and mitigation framework for the world as a whole. However, as indicated above there are significant regional differences in the impacts of climate change and the vulnerability of communities to it. For this reason, government or regional policy on how to incorporate climate change impacts into future planning is likely to vary widely. Guidance may also change as the results of new research become available. It is therefore advisable to check the publications and websites of national organizations for the latest planning advice or policy. The situation in the UK is described below by way of example; some other countries may not have quantified projected changes in as much detail.

A comprehensive set of climate projections for the UK was published in 2009 in the form of maps covering the whole of the UK and its territorial waters (Met Office, 2009; Jenkins, 2009). The scientific background to these maps was published in a series of supporting reports. The projections are known as UKCIP09 and supersede earlier projections made in 2002 (UKCIP02). They are based on climate modelling carried out mainly at the Met Office Hadley Centre but include contributions from over 30 organizations and sponsored principally by Defra (Department for the Environment, Food and Rural Affairs) and DECC (Department for Energy and Climate Change), the two government departments responsible for policy and regulations in the fields of energy, greenhouse gas emissions and the environment.

The maps show projected changes from the 1961–90 baseline for 16 climatic parameters related to temperature, precipitation, relative humidity and cloud cover. They present climatic conditions in three future periods (30-year periods centred on 2020, 2050 and 2080) for three different scenarios of greenhouse gas emissions and five levels of probability. This enables the user to make risk based decisions for future planning in different parts of the country. The maps are available at (http://ukclimateprojections.metoffice.gov.uk/), the UKCP09 website managed by the

UK Environment Agency with the Met Office. In summary, the projections suggest that by the 2080s under a medium emissions scenario:

- average temperatures will rise across the UK;
- summers warm more than winters, particularly in Southern England;
- mean daily maximum and minimum temperatures increase across the UK in both summer and winter;
- average annual precipitation changes little across the UK but winter precipitation increases in western regions while summer precipitation decreases in many, but not all parts of the UK.

The range in estimates is large. For example, the range across the country for likely changes in mean summer temperature is from +1°C to 8.1°C for different emissions scenarios. The corresponding range for the most likely change is from 2.3 °C to 3.0 °C for a low emissions scenario and from 3.7 °C to 5.0 °C for the high emissions scenario. The picture for projected changes in precipitation is less clear. In the Channel Islands, e.g. the likely variation in mean summer precipitation is from a 16% increase (low emissions scenario) to a 65% decrease (high emissions) although the most likely outcome is for a decrease in all scenarios.

The demand for water is linked directly to temperature, particularly maxima, and to lack of precipitation. This would suggest a steady increase in future demand. However, reality is rather more complex than this as recent experience in the UK demonstrates significant increases in water use efficiency, and therefore suppression of demand, in response to drought and water shortages (UKWIR, 2013).

The yield of surface and groundwater sources is directly linked to river flow and recharge and only indirectly linked to temperature and precipitation. A further stage of analysis is therefore required to evaluate the impact of climate change on water supply sources. EA guidance to UK water companies (EA, 2012) suggests that the method of analysis used should be proportionate to the potential risk of a loss of source yield and this in turn should be evaluated by a vulnerability assessment.

In high-risk zones it is anticipated that rainfall/run-off modelling will be required to derive flow sequences from rainfall data perturbed to allow for the effects of climate change. The generated flow sequences are then used in catchment or water resource system models to calculate the yield of river intakes and reservoirs under future conditions. Similarly, soil moisture accounting models can be used to estimate groundwater recharge under future climatic conditions using perturbed precipitation and evapotranspiration sequences. Wetter winters may increase recharge while drier summers are likely to shorten the recharge period and may give rise to lower groundwater levels at the start of the recharge season. The overall impact on recharge may therefore be hard to predict and requires detailed modelling to derive robust estimates of future groundwater yield.

In the UK, detailed time series have been produced by the 'Future Flows Project' for 282 rivers and 24 boreholes throughout England, Wales and Scotland using the latest UKCIP projections and base data generated by the Hadley Centre Regional Climate Model (HadRM3). The data span the 150-year period 1950–2099 and are available under licence from CEH and BGS (Prudhomme, 2012): http://www.ceh.ac.uk/sci_programmes/water/future%20flows/ffgwlproductsanddatasets.html.

Projected changes to river flow in the 2050s are summarized on the 'National Changes in River Flow' web page:

http://www.ceh.ac.uk/sci_programmes/water/future%20flows/ffriverflowchanges-2050s.html.

In low-risk zones, yield might be calculated using historic flow or rainfall time series perturbed by monthly factors designed to represent the likely impact of climate change. Such factors are presented in the UKWIR (2009) assessment report of the significance of UK climate projections to water resource management plans.

In developed countries like the UK, water supply systems usually involve the linking of a number of different sources and the creation of Water Resource Zones (WRZs). The conjunctive use of different types of source (river intakes, surface water reservoirs and boreholes) can help to increase the cumulative yield of individual sources (Section 3.25). Climate change impacts on WRZs are likely to be less than on individual sources, partly because of conjunctive use benefits but also because the yield of WRZs is not always resource-limited but may depend on limits imposed by licences and abstraction permits.

In the UK, water companies are required by government to produce Water Resource Management Plans (WRMPs) every 5 years outlining how they propose to maintain a reliable source of supply to satisfy projected demand over the succeeding 25 years. These Plans are submitted for approval by the Environment Agency who is responsible to Defra for managing water abstractions and protecting the environment. The Environment Agency (EA, 2011, 2012) has produced detailed guidelines on how to produce a WRMP which include allowances for any changes in demand and supply due to climate change. Many water companies have commissioned specific studies to review the impact of climate change on the yield of their sources and on the balance between supply and demand in their WRZs. Table 3.7 shows the climate change allowances made by a selection of UK water companies in the current reporting period which extends to 2040.

Table 3.7 Examples of climate change allowances (dry year annual average)

	Company	Anglian Water	Yorkshire Water	Thames Water
Resource zone		Ruthamford South	Grid surface water	London
Chosen level of service		Reference	Note[a]	Preferred
Change in deployable output due to climate change – (8.1BL; % of DO)	2019–20	0%	–3%	–1%
	2029–30	0%	–8%	–3%
	2039–40	0%	–10%	–4%
Target headroom (climate change component) – (14BL; % of Total TH)	2019–20	26%	15%	35%
	2029–30	39%	11%	43%
	2039–40	31%	71%	34%
% of consumption driven by climate change	2019–20	0.34	0.08	0.18
	2029–30	0.85	0.17	0.41
	2039–40	1.37	0.26	0.65

Sources: Anglian Water: Water Resources Management Plan 2014, December 2013; Yorkshire Water: Water Resources Management Plan, Yorkshire Water Services, August 2014; Thames Water: Final Water Resources Management Plan, 2015–2040, September 2015.

[a]*Notes*: 1 in 25-year TUBs, 1 in 80-year Drought Order.

BL=baseline case references; DO=deployable output (Section 3.15 et seq.); TH=target headroom (the minimum buffer between supply and demand).

If rainfall–run-off or recharge models are not available, the guidelines provide factors to perturb historic flow and recharge sequences directly. These perturbed sequences are then used as input to groundwater or surface water system simulation models to determine yield at specified future time horizons. The impact on yield in intervening years is determined by interpolation between these time horizons and the present-day baseline.

REFERENCES

Ackers, P., White, W. R., Perkins, J. A. and Harrison, A. J. M. (1978). *Weirs and Flumes for Flow Measurement.* John Wiley & Sons.

Allen, R. G., Periera, L. S., Raes, D. and Smith, M. (1998). Irrigation and drainage Paper 56. *Crop Evapotranspiration – Guidelines for Computing Crop Water Requirements.* FAO, Rome.

Bosch, J. M. and Hewlett, J. D. (1982). A review of catchment experiments to determine the effects of vegetation changes on water yield and evapotranspiration. *Journal of Hydrology* **55**(1/4), pp. 3–23.

BS 1100:1998 (R2007). *Measurement of Liquid Flow in Open Channels. Determination of Stage-Discharge Relation.* BSI. Also available Draft BS ISO 1100-2 *Hydrometry Measurement of Liquid Flow in Open Channels Part 2: Determination of the Stage-Discharge Relation.* BSI.

BS EN 748:2007. *Hydrometry. Measurements of Liquid Flow in Open Channels Using Current-Meters or Floats.* BSI.

BS EN 6416:2005. *Hydrometry. Measurement of Discharge by the Ultrasonic (Acoustic) Method.* BSI.

BS EN 12056:2000. *Part 3: Gravity Drainage Systems Inside Buildings: Roof Drainage, Layout and Calculation.* BSI.

BS ISO 1438:2008. *Hydrometry – Open Channel Flow Measurement Using Thin-Plate Weirs.* BSI.

Bullock, A., Gustard, A. and Grainger, E. S. (1991). *Instream Flow Requirements of Aquatic Ecology in Two British Rivers, Report 115.* IoH.

Calder, I. R. (1992). *Evaporation in the Uplands.* John Wiley & Sons.

Creutin, J. D. and Obled, C. (1982). Objective analysis and mapping techniques for rainfall fields: an objective comparison. *Water Resources Research* **18**(2), pp. 413–431.

Darmer, K. I. (1970). *A Proposed Streamflow Data Program for New York.* USGS Water Resources Division, Open File Reports, Albany, NY.

Dean, T. J., Bell, J. P. and Baty, A. J. B. (1987). Soil moisture measurement by an improved capacitance technique. Part 1. Sensor design and performance. *Journal of Hydrology* **93**, pp. 67–78.

Defra (2005). *Revitalisation of the FSR/FEH Rainfall-Runoff Method.* R&D Technical Report FD1913/TR, Defra, July 2005.

DoE (1996). *Water Resources and Supply: Agenda for Action.* HMSO. (including Annexes D and E contributed by EA)

Doorenbos, J. and Pruitt, W. O. (1976). *Crop Water Requirements. Irrigation and Drainage Paper 24.* FAO.

EA (2001). *Good Practice in Flow Naturalisation by Decomposition.* EA, National Hydrology Group, Version 2.0.

EA (2011). *Climate Change Approaches in Water Resources Planning – Overview of New Methods.* EA report SC090017/SR3.

EA (2012). *Water Resources Planning Guideline; The Technical Methods and Instructions.* Guideline developed jointly by EA, Ofwat, Defra and the Welsh Government.

Enex (1976). *Magnitude and Frequency of Low Flows in Peninsular Malaysia, Hydrological Procedure No. 12.* Enex/Drainage and Irrigation Dept, Malaysia.

FEH (1999). *Flood Estimation Handbook, 1999. Vol. 1 – Overview; Vol. 2 – Rainfall Frequency Estimation; Vol. 3 – Flood Frequency Estimation: Statistical Procedures; Vol. 4 – Flood Estimation: FSR Rainfall-Runoff Method; Vol. 5 – Catchment Description*. IoH.

Gustard, A., Cole, G., Marshall, D. and Bayliss, A. (1987). *A study of compensation flows in the UK*. Institute of Hydrology, Wallingford, 86pp. (IH Report No.99).

Gustard, A. and Gross, R. (1989). *Low flow regimes of Northern and Western Europe*. FRIENDS in Hydrology Edited by Roald, L., Nordseth, K and Hassel, K. A. Proceedings of International conference convened by the Norwegian National Committee for Hydrology FRIENDS: Flow Regimes from International Experimental and Network Data Sets. International Hydrological Programme of UNESCO, and the International Association of Hydrological Sciences (IAHS).

Gustard, A., Bullock, A. and Dixon, J. M. (1992). *Low flow estimation in the United Kingdom*. Institute of Hydrology, Wallingford, 88pp. (IH Report No.108).

Hall, J. and Nott, M. (1994). The naturalisation of flow records by decomposition. *Presentation at the National Hydrology Meeting of the British Hydrological Society, London, 17th March*.

Herschy, R. W., White, W. R., and Whitehead, E. (1977). *The Design of Crump Weirs, Technical Memo No.8*. DoE Water Data Unit.

Holford, I. (1977). *The Guinness Book of Weather Facts and Feats*. Guinness Superlatives Ltd.

Hough, M. N. and Jones, R. J. A. (1997). The United Kingdom Meteorological Office rainfall and evaporation calculation system: MORECS version 2.0 – an overview. *Journal of Hydrology and Earth System Sciences* **1**, pp. 227–239.

Hudleston, F. (1933). A summary of seven years experiments with rain gauge shields in exposed positions 1926–1932 at Hutton John, Penrith. *British Rainfall* **73**, pp. 274–293.

IoH (1980). *Low Flow Studies Report*. IoH.

IPCC (2013). *Climate Change 2013: The Physical Science Basis. Contribution of Working Group I to the Fifth Assessment Report of the Intergovernmental Panel on Climate Change*. Cambridge University Press.

IPCC (2014). *Climate Change 2014 – Synthesis Report, Summary of the Fifth Assessment Report of the Intergovernmental Panel on Climate Change for Policymakers*. Cambridge University Press.

IWE (1937). Report of Joint Committee to consider methods of determining general rainfall over any area. *Transactions of the IWE XLII*. pp. 231–299.

Jenkins, G. J., Murphy, J. M., Sexton, D. M. H., Lowe, J. A., Jones, P. and Kilsby, C. G. (2009). *UK Climate Projections: Briefing Report*. Met Office Hadley Centre, Exeter, UK.

Kirby, C., Newsom, M. D. and Gilman, K. (1991). *Plynlimon Research: The First Two Decades. Report No. 109*. IoH.

Kjeldsen, T. R. (2007). The revitalised FSR/FEH rainfall-runoff method. *Flood Estimation Handbook Supplementary Report No. 1*. CEH.

Lambert, O. A. (1992). An introduction to operational control rules using the ten component method, Occasional Paper No. 1, British Hydrological Society, 1990. *Water Resources and Reservoir Engineering*. (Ed. Parr N. M., et al.). BDS. Thos. Telford, pp. 11–40.

Lapworth, C. F. (1965). Evaporation from a reservoir near London. *Journal IWE* **19**(2), pp. 163–181.

Lau, J., Onof, C., and Kapeta, L. (2006). Comparisons between calibration of urban drainage models using point rain-gauge measurements and spatially varying radar-rainfall data. *7th International Workshop on Precipitation in Urban Areas, 7 December, Switzerland*.

Marsh, T. J. and Hannaford, J. (Eds) (2008). *UK Hydrometric Register. Hydrological Data UK Series*. CEH.

Manley, R. E. (1978). Simulation of flows in ungauged basins. *Hydrological Sciences Bulletin* **3**, 85–101. (See also Hysim user manual, Manley R.E. and Water Resources Association Ltd, 2006).

May, R. W. P. (1982). *Report IT 205, Design of Gutters and Gutter Outlets*. Hydraulics Research Station.

May, R. W. P. (1997). The design of conventional and siphonic roof drainage systems. *Journal CIWEM*, pp. 56–60.

McCulloch, J. S. G. (1965). Tables for the rapid computation of the Penman estimate of evaporation. *East African Agricultural and Forestry Journal* **30**(3), pp. 286–295.

Met Office (2009). *UK Climate Projections Science Report: Climate Change Projections*. Met Office Hadley Centre, Exeter (Version 3, Updated December 2010).

NERC (2009). *FEH CD-ROM, Version 3.0*. Centre for Ecology & Hydrology, UK.

Parr, N. M., et al., (1992). *Water Resources and Reservoir Engineering*. British Dams Society, Thomas Telford, pp. 11–40.

Penman, H. L. (1963). *Vegetation and Hydrology. Technical Communication No. 53*. Commonwealth Bureau of Soils.

Prudhomme, C., Dadson, S., Morris, D. and Williamson, J., et al., (2012). *Future Flows Climate Data* NERC-Environmental Information Data Centre.

Rodda, J., Downing, R. A. and Law, F. M. (1976). *Systematic Hydrology*. Newnes-Butterworth, pp. 88–89.

Sargent, R. J. and Ledger, D. C. (1992). Derivation of a 130 year run-off record from sluice records for the Loch Leven catchment, south-east Scotland. *Proceedings of the ICE Water, Maritime Energy* **96**(2), pp. 71–80.

Simpson, R. P. and Cordery, I. (1987). A review of methodology for estimating the value of streamflow data. *Institution of Engineers, Australia, Civil Engineering Transactions* **24**, pp. 79–84.

Smith, L. P. and Trafford, B. D. (1975). *Technical Bulletin 34, Climate and Drainage*. MAFF. HMSO.

Smith, M. (1993). *Irrigation and Drainage Paper 49. CLIMAT for CROPWAT: A Climate Database for Irrigation Planning and Management*. FAO, Rome.

Sturgeon, C. G. (Ed.) (1983). *Plumbing Engineering Services Design Guide*. Inst. of Plumbing.

Svensson, C., Jones, D. A., Dent, E. D., Collier, C. G. et al. (2006). *Review of Rainfall Frequency and Probable Maximum Precipitation Methods. Report No.2 of Defra project Number WS194/2/39 Reservoir Safety – Long Return Period Rainfall*. Defra, UK.

Tabony, R. C. (1977). *The Variability of Long Duration Rainfall Over Great Britain, Scientific Paper 37*. Meteorological Office.

Thompson, N., Barrie, I. and Ayles, M. (1981). *The Meteorological Office Rainfall and Evaporation Calculation System: MORECS*. The Meteorological Office.

Tor, S. M. (1998). Strategies towards sustainable development in Nigeria's semi-arid regions. *Journal CIWEM* **12**, pp. 212–215.

UKWIR/EA (2000). *Report 00/WR/18/2, A Unified Methodology for the Determination of Deployable Output From Water Sources, Vol. 2*. UKWIR, UK (EA R&D Technical Report Ref. No. W258).

UKWIR (2009). *Report 09/CL/04/11, Assessment of the Significance to Water Resource Management Plans of the UK Climate Projections 2009*. UKWIR, UK.

UKWIR (2013). *Report 13/CL/04/12, Impact of Climate Change on Water Demand – Main Report*. UKWIR, UK.

UKWIR (2014). *Report 14/WR/27/7, Handbook of Source Yield Methodologies*. UKWIR, UK.

USACE (2000). *Hydrologic Engineering Center Hydrologic Modeling System, HEC-HMS, Technical Reference Manual, Report CPD-74B*. USACE, March 2000.

USACE (2010). *Hydrologic Engineering Center Hydrologic Modeling System, HEC-HMS Users Manual, Report CPD-74A*. USACE.

USGS (1968). *Techniques of Water-Resources Investigations of the USGS. Book 3. Chapters A6 (1968) and A7 (1968) on Stream Gauging Procedure*. United States Geological Survey. US Government Printing Office.

USGS (1969). *Techniques of Water-Resources Investigations of the USGS. Book 3. Chapter A8 (1969) on Stream Gauging Procedure*. United States Geological Survey. US Government Printing Office.

USGS (1985). *Techniques of Water-Resources Investigations of the USGS. Book 3, Chapter A16, Measurement of Discharge Using Tracers*. United States Geological Survey. US Government Printing Office.

Wilson, E. M. (1990). *Engineering Hydrology*. 4th Edn. Macmillan.

WMO (1966). *Measurement and Estimation of Evaporation and Evapotranspiration. Note No. 83*. WMO.

WMO (2008). *Guide to Meteorological Instruments and Observing Practices No. 8*. 7th Edn. WMO.

Young, A. R., Grew, R. and Holmes, M. G. R. (2003). Low Flows 2000: a national water resources assessment and decision support tool. *Water Science & Technology* **48**(10), pp. 119–126.

CHAPTER

Groundwater Supplies

4

4.1 GROUNDWATER, AQUIFERS AND THEIR MANAGEMENT

In many countries more than half of the groundwater withdrawn is for domestic water supplies and globally groundwater provides over 45% of the world's drinking water (UNESCO-WWAP, 2009). Half the world's megacities and hundreds of major cities on all continents rely upon or make significant use of groundwater (Vrba, 2004). Small towns and rural communities rely on it for domestic supplies. Even where groundwater provides lower percentages of total water used, it may serve local areas where no alternative accessible supplies exist and usually does so with relatively low cost and good water quality. Groundwater is also used to bridge water supply gaps during long dry seasons and drought periods.

The most prolific sources of underground water are the sedimentary rocks, sandstones and limestones, the latter including the chalk. They have good water storage and transmissivity, cover large areas with extensive outcrops for receiving recharge by rainfall and have considerable thickness but are accessible by boreholes and wells of no great depth.

The porosity of a rock is not an indicator of its ability to give a good water yield. Clays and silts have a porosity of 30% or more but their low permeability, due to their fine grained nature, makes them unable to yield much water. Solid chalk has a similar high porosity and low permeability but is a prolific yielder of water because it has an extensive network of fissures and open bedding planes, which store large quantities of water and readily release it to a well, borehole or adit. Limestone, on the other hand, is so free draining that, though it may yield large quantities of water in wet weather, the aquifer rapidly drains away in dry weather and so has low yield.

In England the main aquifers used for public supply (Rodda, 1976) comprise:

- The Chalk and to a lesser extent the Upper and Lower Greensands below it, both of the Cretaceous Period.
- The Sherwood Sandstones of the Triassic–Permian Period, previously named the Bunter and Keuper Sandstones.
- The Magnesian and Oolitic Limestones of the Jurassic Period and, to a lesser extent, the Carboniferous Limestones and Millstone Grits of the Carboniferous Period.

Twort's Water Supply. DOI: http://dx.doi.org/10.1016/B978-0-08-100025-0.00004-1

The chalk and greensands, and the limestones of the Jurassic Period, are widely spread over southern and eastern areas of England; the Triassic sandstones and Carboniferous limestones occur in the Midlands. Similar formations occur in northern France and across the lowland northern plains of Europe where they give good yields.

Some of the largest aquifers worldwide are listed in Table 4.1. The formations most extensively used for water supply are the following:

1. The shallow alluvial strata, sands and gravels of Tertiary or Recent age, which are so widespread and of easy access that, despite their varying yields, they form the principal source of supply in many parts of the world.
2. The many areas of Mesozoic to Carboniferous age sedimentary rocks – chalk, limestones and sandstones that mostly give good yields, subject to adequate rainfall.
3. The hard rock areas of Paleozoic to Cambrian or pre-Cambrian age. This includes igneous and metamorphic rocks which are relatively poor yielders of water unless well fractured and fissured, but which have to be used in many parts of the world because better sources are scarce.

Over vast areas of Africa and India, reliance has to be placed on the relatively small yields that can be drawn from the ancient hard rocks under (3) above. The Deccan Traps of India, for example cover almost 0.5 million km^2 (Singhail, 1997) and are used by a large rural population. Almost one half of the African continent is underlain by hard basement rocks. Although these provide relatively poor yields, they are the main source of underground water for rural populations (Wright, 1992; MacDonald, 2000). These formations, typically yielding 10–100 m^3/d per borehole, are sufficient for basic domestic use by small populations but are normally inadequate for any large urban, agricultural or manufacturing development.

The advantages of groundwater are substantial. The wide area typically occupied by aquifers makes it possible to procure water close to where it is required. Many aquifers provide water that requires no treatment other than precautionary disinfection though in many developing countries even disinfection is not adopted. The cost of borehole installation and associated pumping is relatively modest if the water depth is not great; and the supply can be increased if necessary by drilling additional boreholes subject to the availability of groundwater resources.

However, this ease of exploitation often leads to failure to conserve and protect underground supplies. For example, the large cities of Bangkok, Jakarta, Calcutta and Manila were initially able to gain supplies by drilling and pumping from boreholes close to, or even within, their urban areas but yields have reduced due to spread of paved and built-on areas. Subsidence can result from over-pumping; intensive pumping of groundwater from the thick series of alluvial aquifers beneath Bangkok has resulted in their partial dewatering and consolidation and consequential ground subsidence at the surface.

There is often a failure to monitor a groundwater resource and the effects of overdraw may remain unseen until the resource is seriously depleted. The European Environment Agency (EEA) estimates that about 60% of European cities with more than 100 000 inhabitants are located in or near areas with groundwater over-exploitation, as shown by severe supply problems in parts of Spain and Greece (Stanners, 1995). There is also a common failure to understand the importance of aquifer protection, which is particularly important in urban areas overlying an aquifer where poor sanitation methods or badly maintained sewerage systems result in shallow aquifers becoming polluted.

Table 4.1 Major aquifer systems worldwide

			Estimated volumes		
	Strata	**Basin area km^2 × 10^6**	**Reserves MCM[a]**	**Recharge MCM/year**	**Use MCM/year**
Nubian Aquifer System[1,2] Egypt, Libya, Chad, Sudan	Cambrian and Tertiary	2.0	150×10^6	Small	460
Great Artesian Basin Australia[3,4]	Triassic- Cretaceous sandstones	1.7	20×10^6	1100	600 (1975)
Hebei Plain, China[3]	Quaternary alluvium	0.13	-0.75×10^6	−35 000	10 000
Algeria, Tunisia North Sahara[3,5]	Lower – Upper Cretaceous alluvium	0.95	Very large	Very small	>Recharge
Libya[6]	Cambrian – Cretaceous sandstones	1.8	24×10^6	Small	>Recharge
Ogallalah Aquifer, USA[7,8]	Jurassic sediments	0.075	350 000	60	3400
Dakota Sandstone Aquifer USA[8,9]	Upper Mesozoic sandstones	>0.4	$>4.0 \times 10^6$	>315	725
Umm er Radhuma Aquifer, Saudi Arabia[10]	Tertiary – Palaeocene sediments	>0.25	25 000	1048	n.a.

Notes: [a]MCM, million cubic metres.

Sources of information: [1]Idris, H. and Nour, S. (1990). Present groundwater status in Egypt and the environmental impacts. *Environmental Geology Water Science* **16**(3), pp. 171–177.
[2]Lamoreaux, P. E. et al. (1985). Groundwater development, Kharga Oasis Western Desert of Egypt: a long term environmental concern. *Environmental Geology Water Science* **7**(3), pp. 129–149.
[3]Margat, J. and Saad, K. F. (1984). Deep-lying aquifers; water mines under the desert. *Nature & Resources* **20**(2), pp. 7–13.
[4]Habermehl, M. A. (1980). The Great Artesian Basin, Australia. *BMR Journal of Australian Geology and Geophysics* **51**(9).
[5]De Marsily, G., et al. (1978). Modelling of large multi-layered aquifer systems: theory & applications. *Journal of Hydrology* **36**, pp. 1–34.
[6]Mayne, D. (1991). *The Libyan Pipeline Experience.* Brown & Root Ltd.
[7]ASCE (1972). *Groundwater Management.* ASCE.
[8]Johnson, R. H. (1997). Sources of water supply pumpage from regional aquifer systems. *Hydrogeology Journal* **5**(2), 54–63.
[9]Helgeson, J. O., et al. (1982). Regional study of the Dakota aquifer. *Ground Water* **20**(4), 410–414.
[10]Bakiewicz, V., et al. (1982). Hydrogeology of the Umm er Rhaduma aquifer Saudi Arabia, with reference to fossil gradients. *Quarterly Journal of Geology* **15**, 105–126.

Sometimes the physical or chemical quality of a groundwater can act as a constraint on its use. Proximity to the sea poses a salinity hazard affecting boreholes in coastal areas and on oceanic islands. Development of freshwater resources then requires great care because, once seawater is drawn in, it may prove difficult or impossible to reverse the process (Section 4.15). In areas where water circulates to great depth, the groundwater that eventually emerges at the surface may be warm or even hot. Whilst air cooling methods can be applied, the warm or hot groundwater may have taken up an undesirably high concentration of minerals or gases that may be difficult to deal with.

Cessation of large scale abstractions of groundwater can cause groundwater levels to rise. In Paris, London, Birmingham and Liverpool, rising groundwater levels have occurred because major water-consuming industries, that historically abstracted water from on-site boreholes, are being replaced by other lower water demand activities and connections to the public water supply. These rising groundwater levels threaten tunnels and deep basements with flooding, can cause chemical attack on the structural foundations of buildings and decrease the ability of drainage systems to dewater surface areas. Similarly, rising groundwater levels can occur where excessive irrigation is applied or where large scale river impounding schemes for hydropower development or irrigation raise water levels upstream, causing soil salinization and sometimes land slope instability.

If groundwater is to provide sustainable resources for the future, aquifers have to be managed. The techniques described in this chapter allow aquifer behaviour to be assessed and thereby managed. However, conservation and pollution control require a suitable legal and regulatory framework (Chapter 2). In England and Wales the Environment Agency (EA) is the statutory body responsible for the protection and management of groundwater resources and provides guidance on the requirements through various documents including *Managing Water Abstraction* (EA, 2013a) and *Groundwater Protection: Principles and Practice* (GP3) (EA, 2013b).

4.2 YIELD UNCERTAINTIES AND TYPES OF ABSTRACTION WORKS

The yield of a well, borehole or adit is dependent on the following:

- aquifer properties of the strata from which the water is drawn, the thickness and extent of the aquifer and area of its outcrop;
- extent to which the water storing and transmitting network of fissures, cracks and open bedding planes in the aquifer are intercepted;
- depth, diameter and construction details of a borehole or well, and the type of screen or gravel packing;
- recharge of the aquifer (Section 4.3) and the other abstractions already taking place;
- limitations imposed by existing use of the same aquifer by subterranean abstraction, springs or, indirectly, via discharges to watercourses.

Uncertainty arises because a borehole intercepts only a small volume of the strata. In fissured rock, it may not intersect fissures or planes large enough to give a good supply; a boring in one location may give a poor yield while a second boring only a few feet away yields four or five times as much. Clearly the more fissures or cracks there are, the more likely it is that a reasonable yield will be obtained. Thus, boreholes in the UK are favoured in the well-fissured layers of the Upper Chalk or in the looser Pebble Beds of the Bunter Sandstone. In alluvial gravels and sands there is less uncertainty with yield but lenses of low permeability can reduce transmissivity.

Estimation of the probable yield of a proposed underground development is difficult. Knowledge of the hydrogeology of the area and records of the yield of other boreholes or wells in a similar formation can be of help. However, where little is known about the hydrogeology of a proposed borehole site, it is advisable to sink a 'pilot hole', usually about 150 mm diameter, to gather information from the samples withdrawn. If the results seem promising and it is decided to adopt the site, the pilot hole can be reamed out to form a borehole large enough to accommodate a pump. However, it must be borne in mind that, if the small pilot bore appears to give a good flow for its size, this may not be a reliable indicator that a larger yield could be obtainable from a larger hole in the same or at a nearby location. A method of approach (Sander, 1997) for locating a borehole for optimum yield uses mapping data from lineations, bedrock geology, vegetation and drainage to produce probabilities of yield. The method is used to assist rural groundwater development in poor aquifers. Practical guidance for rural water supply in sub-Saharan Africa is available in MacDonald (2000), based on research carried out by BGS with funding from the UK DFID. The manual presents a number of methods for quick assessment of groundwater resources.

Need for Hydrogeological Survey

Where an aquifer is already supporting abstractions and feeds rivers and streams flowing through environmentally sensitive and recreationally important areas, it is essential to conduct a full hydrogeological investigation of the aquifer before new abstraction is proposed. An environmental assessment of the possible effect of a proposed new abstraction for public supply is often a requirement of an abstraction licensing authority or government department. Section 4.19 lists the key steps required in the UK. A hydrogeological investigation may also be needed to ensure that sources of pollution, such as leachates from solid waste tips, pumping from mines or abandoned contaminated industrial sites, will not render water from a proposed abstraction unsuitable for public supply. Such an investigation is complex and may be extensive; it is best carried out by an experienced hydrogeologist who is familiar with the range of geophysical and other techniques available (Section 4.7).

Types of Abstraction Works

The range of abstraction works which are used for different ground conditions include:

1. Shallow or deep *boreholes*, sunk by different methods, with different methods of screening to exclude fine material and to keep the bore stable and drawing water from a horizon selected for best quality or quantity.
2. Large diameter *wells*, 15–25 m deep, sometimes with a borehole sunk from their base to reach water-bearing strata, the well lining being sealed to prevent entry of surface waters.
3. Large diameter *wells with adits* to intercept water-bearing fissures or bedding planes. Adits are unlined tunnels, approximately 1 m wide by 2 m high and of length up to several kilometres. The manual excavation required renders this type of works undesirable where other, safer, options of abstraction exist (Section 4.17).
4. A '*well field*' comprising several moderately sized boreholes, spaced apart in some pattern; their yields being collected together.
5. *Collector wells* with porous or unjointed pipes laid in the river bed or in shallow bankside deposits of sand and gravels.
6. *Galleries* driven into a hillside to tap the water table (Section 4.16).

These methods of development are described in detail in the following sections.

4.3 POTENTIAL YIELD OF AN AQUIFER

It is often necessary to quantify the limiting yield of an aquifer for water supply purposes. Formerly this used to be taken as equal to the 'long-term' average recharge from rainfall percolation, provided the storage capacity of the aquifer is large enough to even-out year to year variations of the recharge. However, studies of climate variability have shown there can be long runs of years of below-average recharge so that aquifer yields rarely exceed 90% of the 'long-term' average recharge. However, when allowances are made to conserve aquifer-fed springflows and to limit saline intrusion a lower proportion will be available; for an initial estimate 75% of the average 'long-term' recharge may be assumed but the likely maximum sustainable development should be subject to detailed study.

To assess the potential yield of an aquifer, the groundwater catchment must first be defined from the contours of the water table. It is often assumed that the water table reflects the ground surface to a reduced scale, but variations from this can occur in asymmetric scenery containing features such as escarpments, or where one valley has cut down deeper than its neighbours; the underground water table catchment 'divide' may not coincide with the topographical surface catchment divide above. It is also often assumed, in the absence of any evidence to the contrary, that underground flow is in the direction of the major slope of the water table; however, this may not always be true, for instance where there are karst limestone fissures.

Assessing the Amount of Recharge. Sometimes the aquifer catchment outcrop may be remote from the point at which the wells tap the strata concerned. The average recharge is estimated as average rainfall minus evapotranspiration and other losses over the outcrop area (Sections 3.6–3.8). Recharge can also occur as leakage where the catchment includes clay-covered areas, surface run-off subsequently leaking directly into the aquifer through river beds. Even with mathematical modelling (Section 4.5) of the aquifer hydrology, leakage is difficult to quantify until there is a long record of successful pumping. Recharge from river bed leakage is usually only significant in arid areas. In temperate wetter areas the hydraulic gradient between the water level in the river and the adjacent groundwater table level is usually too small to cause significant river bed leakage.

Percolation formulae (Summers, 1987) can be of use to assess aquifer recharge, but they need confirming before being used outside the area for which they were derived. Formulae for semi-arid areas like Jordan or Western Australia are based on a simple percentage of 'long-term' average rainfall, say 3% or 5%. Research by tracking the tritium contents of chalk–limestone pore water and fissure water has demonstrated that, whereas water passing through fissures may travel down to the water table at more than 0.3 m/d, the pore water recharge front may move down at only 1–2 m/year. Distinguishing between the volumes travelling by these alternative routes is fraught with difficulty because it depends on the size of the recharge event and the possibility that pore water will drain out through fissures during any major drawdown of the water table. A more helpful analytical approach lies in recession analysis of dry-weather flows from groundwater catchments.

River Flow Recession Curves can be used to determine that part of river flow, termed '*base flow*', which is fed from underground aquifer storage (Section 3.12). In prolonged dry weather the natural flow of a river comprises only aquifer drainage through springs. At any instant of time, spring flow Q is related to the volume of stored water S in the aquifer by the relationship:

$$Q = kS$$

If Q_o is the spring flow at time $t = 0$, and Q_t is the flow at later time t:

$$Q_t = Q_o e^{-kt}$$

where k has the unit day^{-1} if t is in days. Typical values for k lie between 0.01 per day in a good aquifer, to 0.10 per day in a relatively impermeable aquifer.

Consequently if log Q_t is plotted against t (days) for a prolonged dry period, this should give a straight line of slope k as shown in Figure 4.1. If a single long period free of rain is not available for analysis, it is possible to link together shorter dry period flow recessions which, plotted as shown in Figure 4.2, will be asymptotic to the natural recession curve. Section 3.12 shows how this recession curve can be used to estimate the base flow for a given period of record.

Base flow is percolation routed through storage. Hence, by summing the baseflows over a period and adding a correction for the change of storage between the beginning and end of the period, the percolation for the period can be estimated. The correction for change of storage is obtained by taking the change of water table level between the beginning and end of the period and multiplying this by the aquifer storage coefficient (Section 4.4). Ranked and plotted on probability paper (Fig. 4.3), these seasonal percolation values can be used to estimate recharge probabilities.

Percolation Gauge or Lysimeter Drainage Readings can also be used, but it is difficult to be sure that they represent actual average catchment conditions. Their rims prevent surface run-off and,

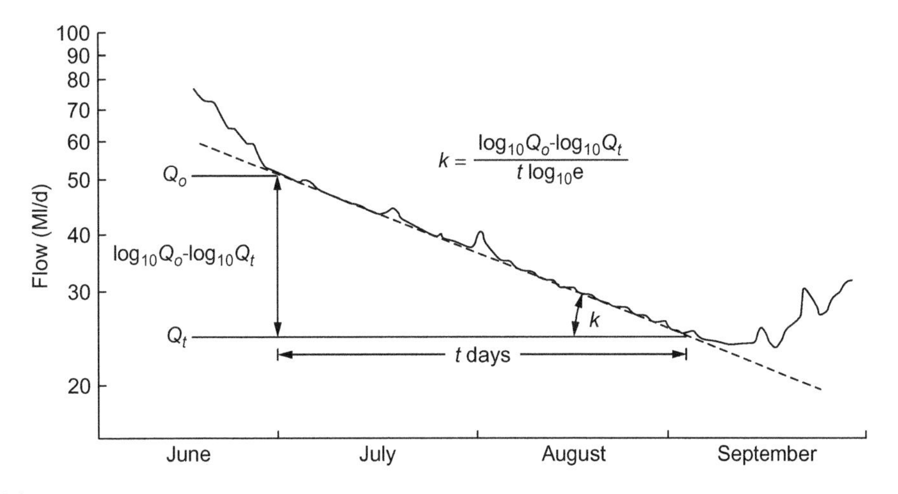

FIGURE 4.1

Groundwater recession graph.

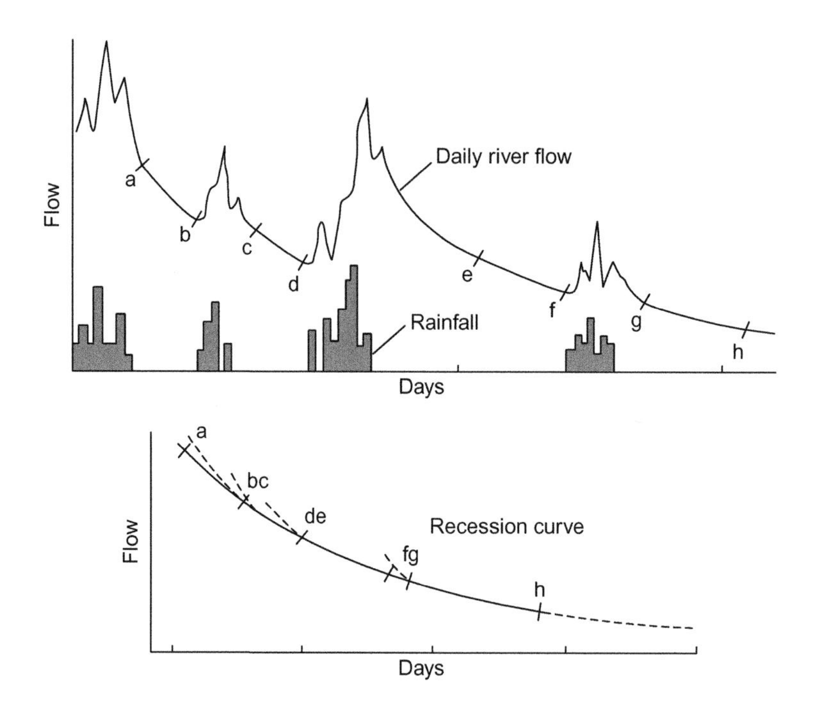

FIGURE 4.2

Recession curve drawn from parts of daily flow record.

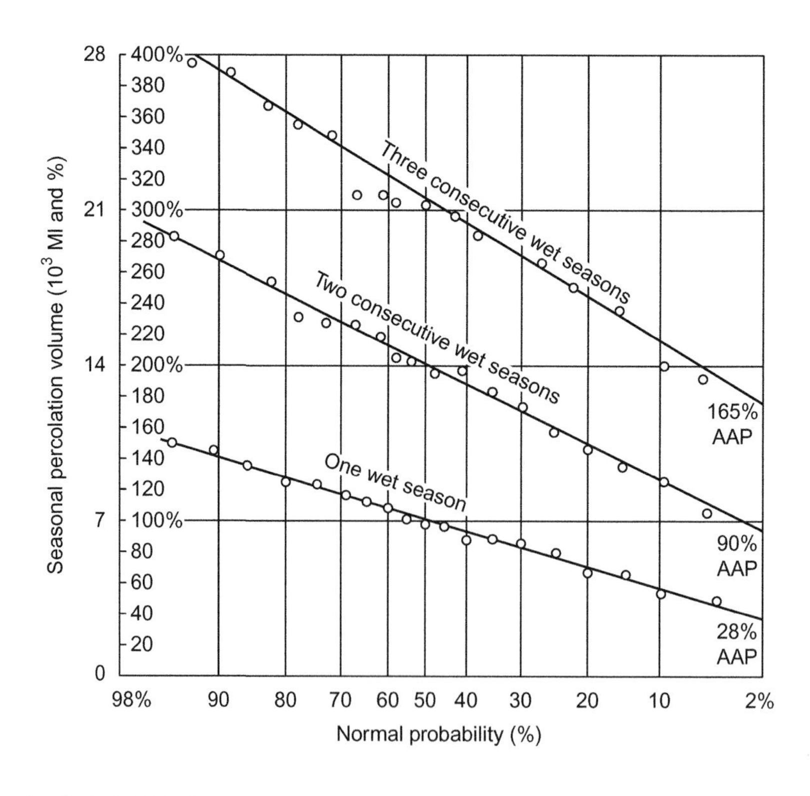

FIGURE 4.3

Seasonal percolation probability plot.

being on level ground, local run-off does not occur; therefore, the measured percolate may be an optimistic estimate of the amount that reaches the water table under natural conditions. Practical difficulties occur when such gauges are kept in use for a long period. They can become moss covered or the soil may shrink away from the edge of the container. Alternatively, percolation can be estimated by a soil moisture storage balance method of the type demonstrated by Headworth (1970). Once the readily available moisture in the root zone is used up by evapotranspiration in a dry period, the subsequent build-up of soil moisture deficit must be made up before excess rainfall can percolate to the aquifer once more. The average percolation produced by this method depends strongly on the amount of moisture stored within the root zone. Storage or '*root constant*' values range from 25 mm for short-rooted grassland to over 200 mm for woodland and need to be calculated for each specific climatic regime, preferably with daily rainfall data.

Yield Constraints. Having found the potential yield of a groundwater catchment, it is possible to compare it with the total authorized (or licensed) existing groundwater abstractions. This gives the upper limit to what extra yield may be obtained or, if too much is already being taken, the extent of groundwater mining that must be taking place. It should be noted that mining can occur in one part of an aquifer whilst elsewhere springflows may indicate no mining is apparent. This indicates low transmissivity within the catchment or parts of it, suggesting a better siting of abstraction wells or a different approach to estimating the maximum yield.

Where a group of wells exists within a catchment, the average maximum drawdown they can sustain without interfering with their output can be assessed. Using the average storage coefficient S' for the aquifer supplying the wells, the amount of storage that can be draw upon to produce this drawdown can be estimated. From this it is possible to calculate the minimum yield for various time periods for a given severity of drought, i.e.:

Abstraction possible = storage available for given pumping level + recharge during the given time period − loss from springflows

This calculation would be carried out for say a 2% drought (50-year return period) for periods of 6–7 months (first summer); 18 months (2 summers + 1 winter); and so on. This can reveal what is the length of the dry period that causes the lowest yield and whether pumps are sited low enough to average out percolation fluctuations.

Seawater Intrusion into wells close to the sea may limit abstraction. The pumping has to be kept low enough to maintain a positive gradient of the water table to the mean sea level. However, since seawater is 1.025 times denser than freshwater, under equilibrium conditions the freshwater–saltwater interface is $40h$ below sea level, where h is the difference in height between the freshwater surface level and the sea level. Thus if the rest water level in a well is 0.15 m above mean sea level, the saline water interface will be about 6.0 m lower, or 5.85 m below mean sea level. However, in practice, the interface will be a zone of transition from fresh to salt water rather than a strict boundary. This means that a pump suction can be sited slightly below mean sea level without necessarily drawing in seawater; but this is heavily dependent on the aquifer's local characteristics, because a stable interface may only be formed in certain types of ground formations. A more reliable policy is to use coastal wells conjunctively with inland wells. In the wet season the coastal wells are used when the hydraulic gradient of the water table towards the sea is steepest, reducing the risk of drawing in seawater. In the dry season the inland wells are used. By this method a higher proportion of the water that would otherwise flow to the sea is utilized than if only the coastal wells were used. Small low-lying oceanic islands face special seawater intrusion problems described in Section 4.15.

4.4 ASSESSMENT OF AQUIFER CHARACTERISTICS

The two principal characteristics of an aquifer are its horizontal transmissivity T, which is the product of its permeability times its wetted depth; and the storage coefficient S'. Transmissivity T is the flow through unit width of the aquifer under unit hydraulic gradient. Its units are therefore m^3/m per day, often abbreviated to m^2/d.

The storage coefficient S' is defined as the amount of water released from an aquifer when unit fall in the water table occurs. Where free water table conditions occur it is the volume of water that will drain from a unit volume of an aquifer (expressed as a percentage of the latter) by gravity with a unit fall of the water table level; sometimes known as *specific yield* S_y. However, when the aquifer is confined under pressure because of some impervious layer above, it is the percentage of unit aquifer volume that must be drained off to reduce the piezometric head by unit depth. The difference between these two meanings, although subtle, is vital. Whereas the former may be in the range 0.1–10.0%, the latter may be 1000 times smaller (demonstrating the incompressibility of water). In the following formulae S can be S' or S_y depending on the specific aquifer conditions.

By considering a well as a mathematical 'sink' which creates a cone of depression in the water table, it has been shown by Theis (1935) that drawdown in a homogeneous aquifer due to a constant discharge Q initiated at time $t = 0$ is:

$$h_o - h = \frac{Q}{4\pi T}\left(-0.5772 - \log_e u + u - \frac{u^2}{2.2!} + \frac{u^3}{3.3!} + \frac{u^4}{4.4!}\cdots\right)$$

$$h_o - h = \frac{Q}{4\pi T} W(u) \tag{4.1}$$

where $u = r^2 S/(4Tt)$, h_o is the initial level and h is the level after time t in a well at distance r from the pumped well. Any consistent set of units can be used. For example, if Q is in m^3/d and T is in m^2/d, then h_o and h must be given in metres. S is a fraction. *W(u)*, the '*well function*' of the Theis equation, can be obtained from tables (Walton, 1970). Although both this equation and the Jacob (1950) simplification of it are derived for a homogeneous aquifer, they are found to work in well-fissured strata, such as Upper and Middle Chalk. The essential need is to be able to assume reasonably uniform horizontal flow on the spatial scale being considered, together with the absence of any impervious layer that interferes with drawdown.

In the Theis method of solution a type curve of *W(u)* against u is overlaid on a plot of the pump test drawdowns versus values of log (r^2/t). Where a portion of the type curve matches the observed curve, coordinates of a point on this curve are recorded. With these match point values the equations can be solved for S and T.

However, Jacob's less exact method is easier to apply and meets most situations that confront an engineer. He pointed out that if time t is large, as in most major pumping tests, then u is small, (say less than 0.01); therefore, the series in the Theis equation (4.1) can be shortened to:

$$h_o - h = \frac{2.30Q}{4\pi T}\log_{10}\left(\frac{2.25Tt}{r^2 S}\right) \tag{4.2}$$

Plotting drawdown against time at an observation borehole within the cone of depression thus produces a straight line as illustrated in Figure 4.4.

If the drawdown for one logarithmic cycle of time is read off, the value of $\log_{10}(2.25Tt/r^2S) = 1.0$ in equation (4.2). Hence, where $h_o - h$ is in metres and Q is in m^3/d:

$$h_o - h = \frac{(2.30Q)}{(4\pi T)}$$

$$\text{i.e. } T = (2.30Q)/(4\pi(h_o - h))$$

Reading off the time intercept, t_o days, for zero drawdown:

$$S = 2.25Tt/r^2, \text{ where } T \text{ is in m}^2\text{/d and } r \text{ is in metres.}$$

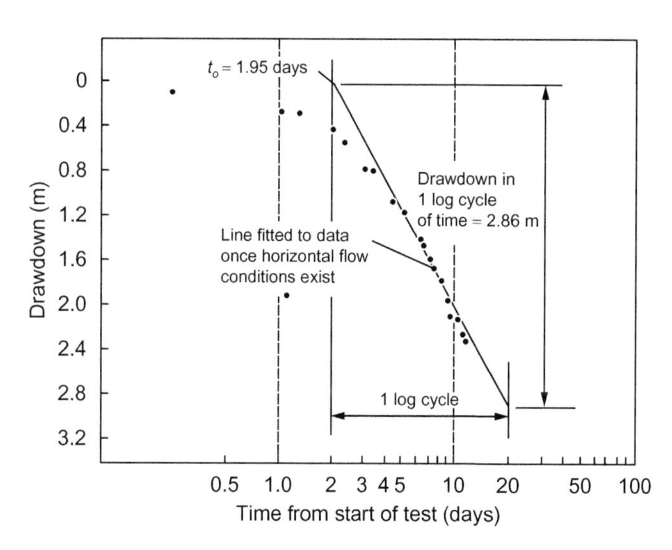

Note: *Data plotted from observation well 200 m from test well.*

FIGURE 4.4

Jacob's pump test analysis for test pumping of a well at 4000 m^3/d.

The following qualifications apply:

1. If the regional water table rises or falls as a whole during the test then the drawdown should be adjusted by the equivalent amount.
2. Early time data should not be used because there will be substantial initial vertical flow as the storage is evacuated. Boulton (1963) suggests that the necessary horizontal flow conditions exist when $r > 0.2d$ and $t > 5dS/K_t$ days; where d is the wetted aquifer depth and K_t is the vertical hydraulic conductivity which can be taken as T/d if K_t is not otherwise known. (In horizontally layered strata, K_t may be one third or less of the horizontal conductivity.) Some reiteration with values of S and T is required to use this guide.
3. Where the pumped well only partially penetrates the aquifer (Kruesman, 1990), it is necessary to adjust the drawdown values for the resulting non-standard flow lines unless the observation well is at least 1.5 times the aquifer depth away from the source. Rather than make complex adjustments it is preferable to use the Jacob rather than the Theis method on a long-term pumping test.
4. Where drawdown is large compared with the aquifer thickness and the aquifer is unconfined, the measured drawdowns should be corrected (Jacob, 1950) by subtracting from them a correction for large drawdown:

$$\text{Correction} = (\text{Drawdown})^2/2 \times \text{wetted aquifer depth}$$

Data observed during the recovery part of a test can also be analysed to derive S and T. If only levels in the pumped well are obtainable, the information that can be gained is usually limited to an estimate of T on recovery (Driscoll, 1986).

Once S and T are established, it is possible to predict with the above equations what drawdown below current rest water level will result at different pumping rates, different times and other distances. Where more than one well can create a drawdown at a point of interest, the total effect can be calculated by the principle of superposition, i.e. the drawdowns due to individual well effects can simply be added. Analytical solutions (Walton, 1970; Kruesman, 1990) exist for many aquifer conditions, including boundary effects from impermeable faults and recharge streams. Care is needed to adopt the solution appropriate to the lithology and recharge boundary.

It will be appreciated that the engineer has no control over the values of S and T found at a well site. The water drawn from the hole may come from local aquifer storage after a long residence time. However, re-siting, deepening and duplicating a bore are all options that may be called upon once S and T are known.

4.5 GROUNDWATER MODELLING

The use of numerical models in hydrogeology has increased dramatically in the last 30 years with the easy availability of computing power and groundwater modelling software. The development of a useful groundwater model is a multistage process. Anderson and Woessner provide an excellent review of groundwater modelling techniques and, in their introduction, they give a concise guide to modelling protocol (Anderson, 1991).

Early groundwater movement models used an electrical analogue of Darcy's law of groundwater flow, which states:

$$\text{Flow } Q = Tiw$$

where T is transmissivity and i is hydraulic gradient through an aquifer cross section of width w. Transmissivity was defined in Section 4.4 as flow per m width i.e. m^3/m per day under unit hydraulic gradient. With the introduction of computers, mathematical modelling of groundwater flows became possible. The most common models use the finite difference or finite element approach. Under the former, and assuming flow is near enough horizontal, a grid is superimposed on a plan of the aquifer to divide it into 'nodes'. Between nodes the flow is related to the hydraulic gradient and the transmissivity of the aquifer in directions 'x' and 'y' (the transmissivity sometimes being taken the same in each direction). The finite difference method adjusts the calculations so that 'boundary conditions' between each flow stream and its neighbours match. Finite difference models in general require fewer input data to construct the finite difference grid. The finite element method is better able to approximate irregularly shaped boundaries and it is easier to adjust the size of individual elements as well as the location of boundaries, making it easier to test the effect of nodal spacing on the solution. Finite elements are also better able to handle internal boundaries and can simulate point sources and sinks, seepage faces and moving water tables better than finite differences.

Equations (with many terms) can therefore be set up connecting node-to-node flows and head changes, and to which overall limiting boundary conditions apply, such as the assumed (or known) upstream initial head applying, the lateral boundaries of the aquifer and the downstream conditions applying, such as the outcrop of the aquifer. The equations can then be solved by the computer to match the boundary constraints by reiterative methods or matrix formulation, to a specified degree of accuracy, thereby providing outputs of water table contours and field flows. Three-dimensional flows and both transient and steady-state condition of flows can be dealt with, together with matters such as consequent aquifer-fed springflows and the effect of pumped abstractions.

There are many computer models available for simulating groundwater flows; a summary of their characteristics and capacities is given by Maidment (1992). MODFLOW developed by US Geological Survey in 1988 is quoted as 'popular and versatile'. It can deal with two- and three-dimensional flows and incorporates numerous ancillary facilities. Another useful finite difference model is PLASM, originally written for the Illinois State Water Survey in 1971 and described as 'recommended as a first code for inexperienced modellers because it is interactive and easiest to operate'. Groundwater modelling requires a sound knowledge of hydrogeology and practical experience of modelling techniques. A hydrogeological investigation of the aquifer is also essential to derive the data for the model (Sections 4.2 and 4.7).

Data Requirements

Setting up and running a model requires a large amount of data. The advantages that numerical models have over analytical models are that they allow for spatial and temporal variations in many parameters. The disadvantage is that values for these parameters must be specified by the modeller.

Most of the parameters listed below must be defined for every element of a numerical model. Generally the area modelled is divided into smaller regions, which are allocated the same values, but this can lead to loss of flexibility.

- *Boundaries* must be defined for all models, the area modelled being completely surrounded, so that the flow equations can be solved; for resource modelling it is sufficient to ensure that the boundaries are far enough removed from the area of interest in the model that they do not significantly affect the solution; however, this is not true for protection zone modelling;
- *Recharge rates* must be specified for every node and, since they are critical to model credibility, they should allow for spatial variation due to differences in cover, soil type, vegetation, rain shadows caused by hills and leakage from distribution systems;
- *Hydraulic conductivity* or, in some cases transmissivity, must be defined for every node;
- *Storage parameters* are required for each formation in transient models and in models used for particle tracking;
- *Transport parameters* are difficult to obtain but, if required for a particular model for contaminant transport (as opposed to particle tracking), should include dispersion coefficients, adsorption isotherms and decay coefficients;
- *Interaction parameters* for defining aquifer relationships with rivers are important for regional hydrogeological models; for rivers two parameters are, in most cases, needed: river elevation (easily ascertained) for each node affected and river bed permeability (usually not well known);
- *Leakance parameters* are needed where the aquifer level is known, or expected, not to be the same (in hydraulic continuity) as the surface level in a connected water body and where the connection is categorized as 'leaky'; leakance can be determined by careful head measurements but is usually obtained during model calibration and can, therefore, be a source of error; it varies significantly along a river.

Model construction is often an iterative procedure. It is unusual for all of the parameters required for the model to be well quantified at the start. The development of a useful and useable model involves fine tuning of the parameters in order to achieve calibration.

Model Calibration

Once the data sets are acquired and the model run, it is necessary to assess if the parameter set and the conceptualization reproduce the observed performance of the aquifer. The output from the model is compared with measured data, generally of water levels from boreholes and flow in rivers. The model parameters are varied until a good correlation is achieved between the model and measured data. To verify interaction, measurements of base flow are often compared with model estimates of river flows to calibrate the model; therefore it is important that the rivers are correctly represented. However, there is no unique solution to the flow equations. Similar head patterns can be achieved with quite different combinations of input parameters. This is especially true for regional models where there are large numbers of parameters that can be varied.

4.6 TEST PUMPING OF BOREHOLES AND WELLS

In most countries concerns about the need to preserve the natural aquatic environment and the rights of existing abstractors mean that authorization to abstract more underground water will only be obtained if the engineer is able to show, from records taken during test pumping, that no unacceptably harmful effect will be caused (Boak, 2007). Therefore, test pumping a borehole and monitoring its impact on the aquifer and the catchment area are essential. The aim of test pumping is to find how much water a well or borehole will presently yield and to determine:

1. the effect, if any, of the pumping on adjacent well levels, spring and surface flows;
2. the sustainable amount that can be abstracted during dry periods of different severities;
3. the drawdown/output relationship whilst pumping in order to decide the characteristics required for the permanent pumps installed;
4. the quality of the water abstracted;
5. data sufficient to derive estimates of the key characteristics of the aquifer penetrated, i.e. its transmissivity and storage coefficient.

An initial problem is to decide what capacity of pump to use for test pumping: if much less than well yield the test will not prove the yield and the effect of such abstraction on adjacent sources; but if larger than well yield it means a waste of money. A major cost of test pumping is the temporary discharge pipeline required; for a large output it may need to be laid a considerable distance to minimize the risk of the discharged water returning to the aquifer and affecting water levels in the test and observation wells.

The test pump size should be estimated from experience in drilling the well and on what other holes in similar formations have given. Water level recovery rates after bailing out (Section 4.9) and on stopping drilling are important indicators of possible yield; the drill cores can indicate where well-fissured formations have been encountered, and an experienced driller should be able to notice at what drilling level there have been signs of a good ingress of water to the hole. Electrical submersible pumps are most commonly used for test pumping, output being adjusted by valve throttling. In rare instances a suction pump (i.e. one above ground having its suction in the well) may be used if the water level in the well is near ground level; for small holes an airlift pump is a cheap but rarely used alternative. The discharge must be accurately measured and continuously recorded. A venturi meter, orifice meter, or vane meter can be used for measuring discharge or, for best accuracy, the discharge may be turned into a stilling tank equipped with a V-notch (Section 14.14) with a float recorder to log the water level over the notch. The advantage of a stilling tank and weir is that it provides a nearly constant outlet head for all outputs and is a positive method of measurement. If meters in the discharge pipeline are used, the pump outlet head will vary with the flow making it more difficult to maintain a constant test pump output; the accuracy of the meter must be checked before test pumping and at least once during the test.

Test Pumping Regimes

Every effort should be made to carry out the test pumping when groundwater levels are near their lowest seasonal decline towards the end of the dry season. Whilst dry weather during the test is an advantage, it cannot be guaranteed in a variable climate, and some rainfall during the test period does not invalidate the observations taken, provided it is not extreme. Test pumping outside the dry season when groundwater levels are high should be avoided if at all possible, because it gives uncertain estimates of the yield during the dry season.

Test pumping should run for at least 3 weeks, with three stages of increasing output, such as one third, two thirds and full output without stoppage, each for at least 24 hours. The pump is then stopped and the recovery rate of the well water level is carefully measured. When sufficient of the recovery curve against time has been obtained to show its trend towards starting rest level, the pump should be restarted at maximum output and kept at that rate for 14 days, the recovery again being measured when pumping is stopped.

If pump output is greater than well yield, drawdown will increase continuously; the output must be throttled back to achieve a steady drawdown. In other cases there may be a quick initial drop of water level followed by a slower decline, or a diminishing rate of decline asymptotic to some maximum drawdown for the given output. It is not always possible to reach stable conditions in a 14 day pumping test.

Problems can occur. If the pump fails in the middle of the test, it is important to record the well water level recovery; after re-starting, the flow rate should be adjusted to the previous value. It is not easy to keep the pump output constant because it reduces as drawdown increases and is affected by electrical supply voltage; adjustment of pump output by valve operation has therefore to be done carefully since over-adjustment and re-correction produce readings that are difficult to interpret. Options for measuring the water level accurately in a pumped borehole are: dipping with an electrical contact device to detect the water level (manual); measuring pressure below the water level with an air pressure instrument (not very accurate); and recording with a submerged electronic pressure transducer (instrument must be calibrated and readings corrected for barometric variation).

The pumping water levels and pump outputs should be plotted as shown in Figure 4.5 to derive a '*type curve*' of output against drawdown. It is also useful to re-plot the stepped drawdown results as a Bruin and Hudson curve (Bruin, 1955) as shown in Figure 4.5. This can reveal the proportion of the drawdown due to the characteristics of the aquifer, and the proportion due to hydraulic characteristics of the well or borehole, the latter being distinguished by the turbulent flow losses, the former by the laminar flow losses. Examination of the turbulent loss coefficient c can sometimes suggest high entry losses to the well due to poor design of the well screens or gravel packing. Good (i.e. low) figures for c in cQ^2, where Q is in m^3/min, are below 0.5. The laminar loss coefficient b describes the relative permeability of the aquifer(s) feeding the well.

Use of Observation Wells and Monitoring

To determine aquifer characteristics (Section 4.4) it is desirable to have two observation boreholes sited within the likely cone of depression of the water table around the borehole when test pumping takes place. The distance of these holes from the test well should be such that the drawdown they experience is large enough to give a reasonably accurate measure of the drawdown but small relative to the depth of saturated aquifer which contributes flow to the pumped well, so that the flow through the aquifer at the observation bore site is essentially horizontal. Experience has shown that these requirements frequently result in observation holes being sited between 50 and 200 m of the pumped well. The larger this distance is the greater is the 'slice' of aquifer brought under observation.

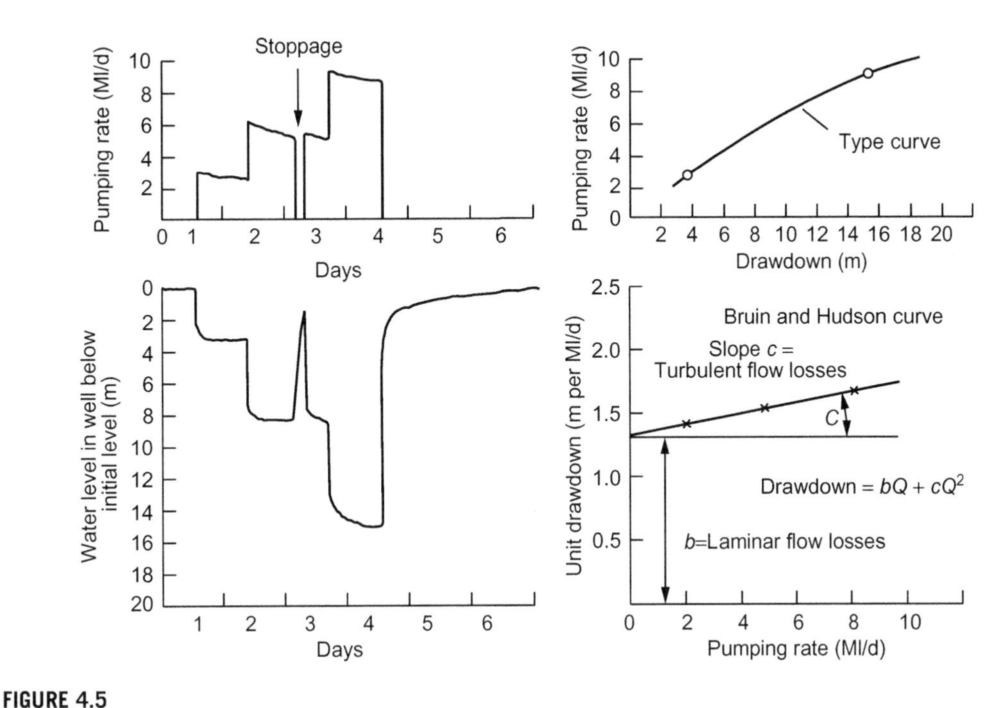

FIGURE 4.5

Step-drawdown test results (Bruin and Hudson, 1955).

Similarly, more information will be gained if the two observation holes are on radials 90° apart. The radial distance of each bore from the well should be measured.

Water levels in the observation boreholes should be measured regularly before, during and after test pumping, the frequency being increased after starting, stopping or altering the pump output sufficient to log rapid changes of water level. Preferably, a water level recorder should be used. An existing well, such as a household well, can be used for observation purposes, provided the quantity of water being drawn is too small to affect its standing water level. Usually, however, no conveniently placed existing well will be available and, if observation bores are required, they will have to be drilled. Observation boreholes are often omitted because of their installation cost, with reliance being placed solely on results from the pumped borehole and the effect, if any, on existing boreholes and wells nearest to the pumped site.

It is also essential to monitor water levels in all wells and ponds, and the flows from springs and in streams, that could be affected by the test. The area covering features to be monitored is a matter of judgment, but would often be about 2 km radius from the test well. However, all important abstractions that could be sensitive to flow diminution, such as watercress beds and fisheries, should be monitored even if they appear to lie outside any possible range of influence of the proposed abstraction. The paramount need is to have proof that the features were not affected by the test pumping. The measurements of key indicator wells and springs may be started 6–9 months before the planned timing of the test pumping. Temporary weirs may be needed to measure springs and stream flows. On streams or rivers, measurements of the flow both upstream and downstream of the test pump site may be necessary. The aim is to get a sufficiently long record of water table levels and stream flows that their trend before and after the test pumping can show whether there is any interruption or change caused by the test pumping.

There is also a need to check that the underground water table catchment area (*yu*), which contributes to the test pumping, is the same as the topographical catchment. Increased abstraction from one underground catchment area can cause the water table divide to migrate outwards, reducing the area of an adjacent catchment feeding other sources, which would, therefore, at least notionally suffer a reduction of potential yield.

Catchment rainfall measurements must be taken daily; if no rain gauge stations exist in the catchment, one or more temporary gauges must be installed.

4.7 GEOPHYSICAL AND OTHER INVESTIGATION METHODS

Samples of strata will be retrieved during the sinking of a borehole. However, they may not be properly representative of aquifer water quality because inevitably the samples will be disturbed and the groundwater quality will have been disrupted also, being only a mixture of all the flows entering the bore. More precise information can be derived from a wide variety of subsurface 'down-hole' geophysical methods. Direct physical observation or measurement can be obtained by use of:

- a television camera to view the walls of a boring;
- a caliper device to measure and record bore diameter with depth;
- recording instruments to obtain profiles of water temperature, pH and conductivity with depth;
- small sensitive current meters to detect differences of vertical flow rates in the borehole, thus indicating changes in inflow or outflow rate from strata.

Electrical and nuclear instruments can provide information concerning the strata penetrated. Resistivity measurements taken down the hole, using an applied electrical potential through probes, can locate the boundaries of formations having different resistivities. This can aid identification of the type of strata penetrated and distinguish between fresh and saline waters in the formation. Measurement of the '*self*' or '*spontaneous*' electrical potential existing between strata at different depths of the formation can assist in detecting permeable parts of the aquifer. Nuclear downhole tools used in aquifers are mainly (in water) of three types:

- natural gamma detectors which pick up natural radioactivity from potassium in clays (so correlating with clay content);
- gamma–gamma tools which use a gamma radiation source to bombard the formation and measure back scattered radiation produced, which is inversely proportional to the formation density and can be used to indicate degree of cementation or clay content also;
- neutron tools which are used to give an indication of water content and porosity.

All these measures, co-ordinated and supplementing each other, can give additional information about an aquifer. In particular they can indicate the best locations for inserting well screens or gravel packs (Section 4.8). More information is available in IWES *Water Practice Manual No. 5* (1986) and ISO/TR 14685:2001 *Hydrometric Determinations – Geophysical Logging of Boreholes for Hydrogeological Purposes – Considerations and Guidelines for Making Measurements* (ISO, 2001).

Many aquifer systems comprise different aquifer formations at different levels. These can be investigated by using packers down the borehole to isolate individual aquifers and, for each, measure inflow and piezometric head and take water samples for analysis. Various multi-tube assemblies can be used to withdraw samples of water from the isolated aquifers and, if inserted in an observation borehole, can remain in place for subsequent monitoring purposes. The production bore can, of course, be similarly investigated before a pump is installed. The method can reveal whether it is advisable to seal off inflows of undesirable quality (including signs of contamination), or which come from formations whose catchment outcrop is known to include potential sources of pollution best avoided.

4.8 BOREHOLE LININGS, SCREENS AND GRAVEL PACKS

The components of a borehole are illustrated in Figure 4.6.

The upper part of a borehole is usually lined with solid casing which is concreted into the surrounding ground. This is to prevent surface water entering and contaminating the hole and to seal off water in the upper part of the formation which may be of lower quality than that within the main aquifer.

Where a borehole encounters weakly cemented or uncemented sands and gravels, a perforated or slotted lining may be needed to support such formations, while permitting fine particles adjacent to the screen to be washed out and improve the yield of the borehole. There are many different types of screen; the simplest comprise slots cut in borehole casing, whilst the most sophisticated are stainless steel, wedge wire-wound screens with accurately set apertures. In non-corrosive waters screens are usually made of steel; in corrosive waters, plastic coated steel, phosphor bronze, glass reinforced plastic or rigid PVC slotted or perforated screens may be used. Figure 4.7 shows the aperture and percentage open area for some typical 300 mm diameter screens. The screen aperture required has to be decided according to the size of particles in the formation. Because of unavoidable partial blocking of the screen by sand particles, the effective area of a screen is usually estimated at less than half its initial open area. The water entry velocity through the apertures needs to be limited to about 30 mm/s to avoid high turbulence losses. Working from the required yield, the open screen area and the amount of blockage that might occur over time, the length of screen necessary can be calculated.

If the apertures or slots of a well screen are suitably sized, withdrawal of fines from the formation should result in the coarser material forming a 'naturally graded pack' (filter) against the screen; this prevents further withdrawal of fines when the borehole is pumped. A formation is considered capable of forming a natural pack if the D_{10} size (aperture size through which 10% of the material by weight passes) is greater than 0.25 mm, the D_{90} size (size through which 90% passes) is less than 2.5 mm, and the Uniformity Coefficient (UC) is between 3 and 10.

$$UC = D_{60}/D_{10}$$

where D_{60} is the aperture size through which 60% of material passes (by weight).

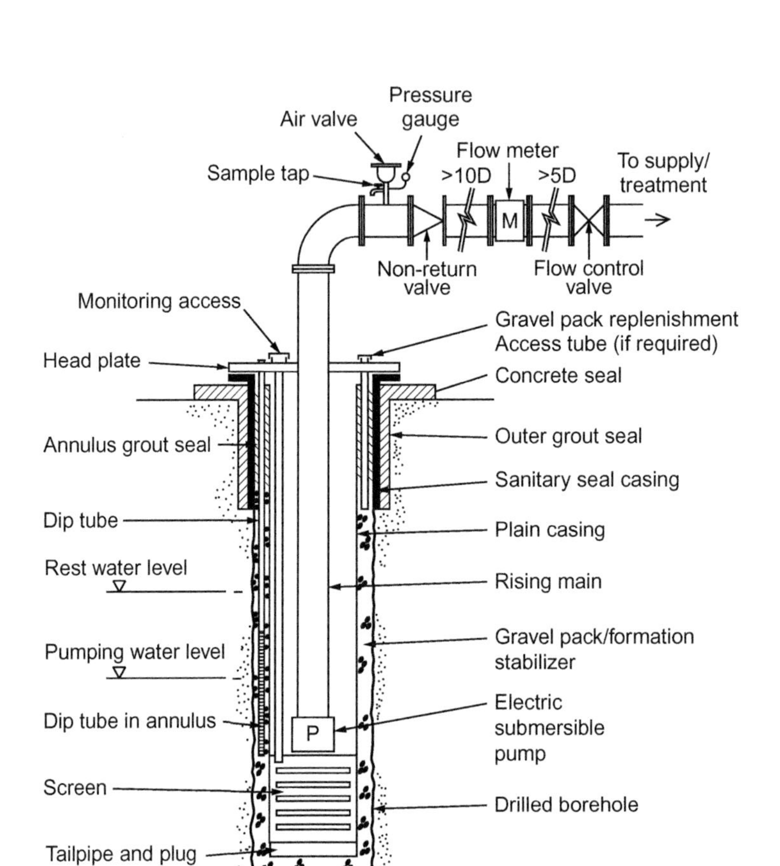

FIGURE 4.6

Components of a borehole.

During test pumping and development of the borehole, abstraction rates should be higher than will occur during the permanent production pumping so that no fines are withdrawn in production.

Where the grading of the formation material is unlikely to form a 'natural pack', a gravel pack should be placed outside the screen. The screen is suspended centrally in the borehole by means of lugs fitted at intervals on the outside and gravel is tremied into the annular space. The gravel pack can be of uniform sized gravel or graded gravel. Uniform sized packs are suitable when the UC of the formation material is less than 2.5, in which case the 50% size of the gravel pack should be four to five times the 50% size of the formation material (Terzaghi, 1943). Otherwise, a graded gravel pack should be used, its grading being selected to parallel the grading curve of the formation material so that the UCs are similar; the D_{15} size of the pack should be four times the D_{15} size of the coarsest sample from the formation, with the D_{85} size of the pack less than four times the D_{85} size of the finest formation material sampled.

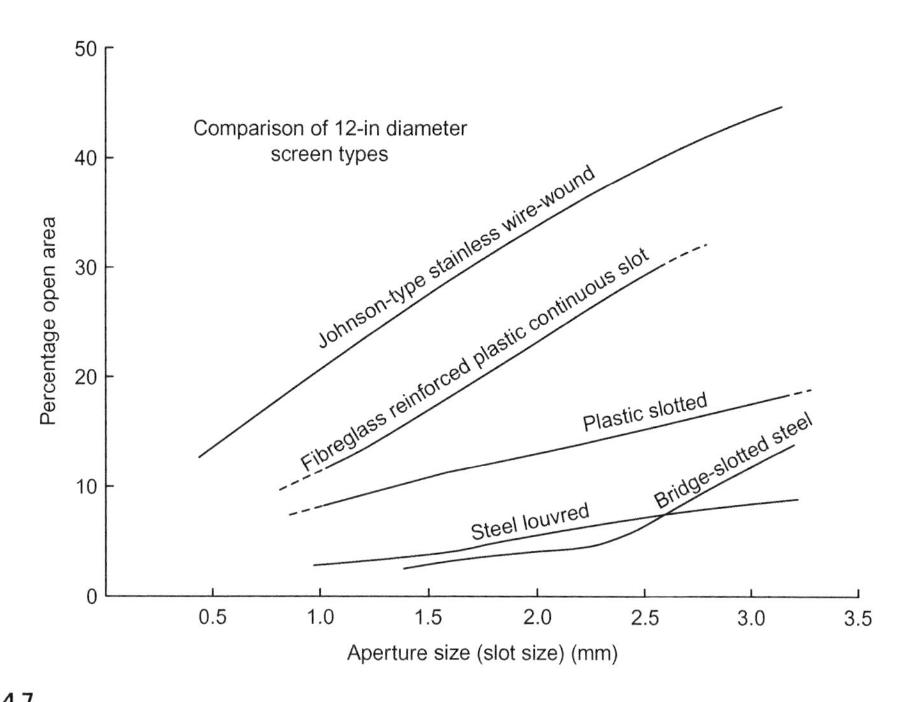

FIGURE 4.7

Comparison of screen open areas.

The minimum thickness of the gravel pack should be 75 mm. Installing a thinner pack becomes uncertain because of imperfect verticality of the boring, variations in its diameter, and the problem of placing the gravel evenly in the annulus at depth. Even with the use of a tremie pipe there may be zones where the pack is virtually absent and others where the thickness is greater than necessary. Part of the development of a newly sunk borehole may involve the need to remove 'mud cake' from the formation face, and a thicker pack tends to reduce the ability to remove the cake. There are many methods for selecting an appropriate gravel pack (Monkhouse, 1974; Campbell, 1973), each usually for a particular type of aquifer formation.

In loose sandy formations such as Greensand, the need for a gravel pack has to be taken into account when deciding on the starting diameter for a boring. The expected water pumping level is also an important factor because it determines where the pump is to be sited, and consequently the diameter required down to pump level to ensure the pump can be accommodated with reasonable flow conditions to its suction. Although it is possible to ream out a borehole to a larger diameter in some cases, it may not be practicable in a loose formation where solid or slotted lining has to be driven down closely following the boring tools to prevent collapse of the hole.

4.9 CONSTRUCTION OF BOREHOLES AND WELLS

Boreholes for public water supply are usually of large diameter; methods for sinking them differ from the smaller diameter holes sunk for oil drilling or for the water supply to a few houses. The *percussive method* is widely used. A heavy chisel is cable suspended and given a reciprocating motion on the bottom of the hole by means of a 'spudding beam', which alternately shortens the cable and releases it at a rate that varies with the size and weight of chisel used. Considerable skill is required by the operator to adjust the rate of reciprocation to synchronize with the motion of the heavy string of tools down the hole and avoid any violent snatch on the cable. This cable has to be paid out gradually and kept exactly at the right length for the chisel to give a sharp clean blow to the base of the hole. The wire cable is left hand lay and, under the weight of the string of tools tends to unwind clockwise, thus rotating the chisel fractionally with each blow. After a while the unwinding builds up a torque so that, when the weight is off the cable just after a blow on the base of the hole, the torque breaks a friction grip device in the rope socket attachment and the cable returns to its natural lay. Thus slow rotation of the chisel continues. Every so often the string of tools must be withdrawn and a bailer – a tube with a flap valve at the bottom – is lowered to clear the hole of slurried rock chippings. Progress is therefore slow and, depending on the hardness of the formation, the drill chisel may require re-sharpening from time to time.

Another percussive method is use of a '*down-the-hole*' hammer operated by compressed air. This is up to 10 times faster in hard rock than a cable-operated chisel, but it is unsuitable for soft formations and cannot be used in substantial depths of water. The chippings have to be ejected to the surface by the exhaust air from the hammer passing through the annular space between the drill rods and the borehole. Unless the boring is of small diameter there may be difficulty in raising the chippings, in which case an open-topped collector tube may be positioned immediately above the tool to collect the chippings. The down-the-hole hammer is predominantly used on small boreholes up to about 300 mm diameter.

For *rotary drilling* of large holes in hard material a roller rock bit is used, equipped with toothed cutters of hard steel that rotate and break up the formation. Water is fed to the cutters down the drill rods and, rising upwards through the annular space, brings the rock chippings with it. In the 'reverse circulation' method the water feed passes down through the annular space and up through hollow drill rods. This achieves a higher upward water velocity making it easier to bring chippings to the surface. Air is sometimes used instead of water, but the efficiency falls off as depth below water increases. Heavy drill collars may have to be used to give sufficient weight on the rock cutters. In addition, a large quantity of water may be required for the drilling of a large diameter hole. Rotary rock bit drilling is widely used by the oil industry, usually for holes of substantially smaller diameter than those needed for public water supply. Clay or bentonite drilling fluids can be used to support unconsolidated formations and assist in raising chippings. However, the use of such suspensions is not always advisable in water well drilling in case they seal off water-bearing formations. Instead polymer based, low solids, biodegradable muds are preferred since these are not so likely to seal off water-bearing formations.

Diamond core drilling is seldom used for sinking water supply holes except for small diameter holes in hard formations or for trial holes where rock cores are needed.

An alternative method of rotary core drilling for large diameter holes is to use chilled shot fed down to the bottom of a heavy core-cutting barrel. The barrel has a thickened bottom edge under which some of the chilled shot becomes trapped and exerts a strong point load on the formation causing it to break up. A small amount of water is added for lubrication and further chilled shot is added so that eventually, with the annular cut deepening, a core of rock enters the barrel. To break off the core and bring it to the surface, some sharp pea gravel or rock chippings are sent down, causing the core to be gripped by the barrel, rotation of which causes the core to be broken off. Progress in hard formations is slow and, if fissures are encountered, problems of loss of shot or water can occur.

When soft or loose material is encountered, a solid casing or slotted lining may have to be pushed down to the level required as the boring proceeds, or after withdrawal of the chisel. Once a lining is inserted, the hole may have to be extended in a smaller diameter. Several such 'step-ins' of diameter may be required if the formation has several separated layers of loose material. Biodegradable muds may be used with rotary drilling to keep the hole open, provided the mud does not seal off water-bearing formations or is not lost through fissures; in which case it may not be necessary to line the hole until it is completed to full depth. However, rotary drill rigs may have difficulty in penetrating ground containing hard boulders and rotary drilling is considerably more expensive than percussion drilling.

A typical specification for verticality is that the hole should be no more than 100 mm off vertical in 30 m of depth but it is not necessary to insist on this at depths below the possible siting for the pump. It is especially difficult to keep a boring vertical when steeply inclined hard strata are encountered. Verticality of a borehole is not as important as are 'kinks' in the boring down to the proposed level for the permanent pump. Kinks above pump level may throw the pump to one side of the boring, making poor pump suction entry conditions and causing difficulties in lowering or removing the pump. If correction is essential it may be necessary to ream out a hole to a larger diameter to get the remainder of the boring in line.

The duties of the borehole driller are:

1. not to lose drilling tools down the hole;
2. to note and log down every change of drilling conditions such as increased or decreased speed of drilling, change of tools necessary, quick or sluggish recovery of water level after bailing out, etc.; and
3. to note and log down the lengths of different casings used, the nature and depth of all cores and chippings produced from the boring and to keep all cores and chipping samples.

Tools lost down a hole, such as the drill bit which is attached to the drilling rods, will cause days of delay in attempts to recover them and, if 'fishing' (with a wide variety of special tools) is unsuccessful, it may be necessary to ream out the hole to larger diameter down to the tool to recover it. Sometimes tool recovery may defeat all efforts and the hole has to be abandoned if it has not been driven to the required depth.

When a water borehole is being sunk it is particularly important to watch for signs of large fissures, indicated, for example by a sudden drop of the chisel or drill, increased flow of water into the hole or fast recovery of water level when bailing out. Borehole progress logs must be meticulously kept; core samples should be stored in wooden core boxes, their depths being clearly marked and samples of chippings and soft material should be stored in properly labelled plastic bags. The samples need to be examined by an experienced geotechnical engineer or hydrogeologist following the identification and classification guidelines set out in BS EN 14688 and BS EN 14689.

Wells up to about 1.2 m diameter can be sunk in hard material by rotary drilling using a heavy core barrel and chilled shot as previously described. In less hard materials, such as chalk, a large percussion chisel may be used to chop up the formation, the excavated material being removed by suction pump or by grab. In very soft or loose material, wells may be sunk *kentledge* fashion; a concrete caisson of the required well diameter, with a lower cutting edge, is sunk into the ground by hand excavation of the core material below the cutting edge. As excavation deepens, further rings are added to the top of the caisson. The excavated material is removed by skips or by a crane grab. In extremely soft material, excavation by grab alone may be sufficient for the caisson to descend by its own weight until harder material is encountered. Early wells were hand dug with linings of brick; later wells had linings of cast iron segments bolted together. Either interlocking precast concrete rings or concrete segments bolted together are now used.

The upper part of a well must be capped and, where there is any risk of direct ingress of contaminated or polluted surface water (Section 4.11), must be sealed by cement grouting on the outside to prevent ingress. Where a well is sunk in river bed deposits, the casing should be extended to above river flood level.

4.10 DEVELOPMENT AND REFURBISHING BOREHOLES AND WELLS

On completion of a borehole or well it should be '*developed*' to maximize its yield. The objective is to remove the clay or finer sand particles from the natural formation surrounding the slotted linings or well screens to improve the flow of water into the boring and to ensure that any gravel packing is properly compacted. 'Surging' is the most common development technique. It consists of pumping from the well at maximum rate, and then suddenly stopping the pump thereby causing flow to wash in and out of fissures in the formation thus dislodging silt and fine sand therein. The piston effect of using a bailer can also be used, but must not be too energetic to avoid screen damage or collapse of uncased parts of the hole. 'Swabbing' and 'surging' with a piston can be used, a swab having a valve in it which allows water upflow. Both tools promote energetic flow through a screen during their up and down movement, the surge plunger in particular causes strong agitation in a gravel pack, encouraging rearrangement of particles in the pack and removal of fines. However, these tools cannot be used with wire-wound screens, which are supported by internal vertical bars; instead they have to be operated in the casing above the screen where they may still be effective but considerably less so.

Airlifting, although an inefficient form of water pumping, can be useful for the development of a hole carrying sand-charged water, which would be highly abrasive to centrifugal pumps. There are several variations of airlift for this purpose; these depend on the vertical movement of the air pipe up or down inside the eductor pipe. It is also possible to seal the top of a boring and inject compressed air at the top, driving the water down into the formation, prior to release of the pressure and starting the airlift.

Jetting can be used. Depending on whether the gravel pack has to be agitated, or the screen or formation cleaned, either low or high pressure can be used. A jetting head with horizontal water jet nozzles is suspended on the end of a drill pipe and slowly rotated whilst jetting. Care has to be taken with non-metallic screens or where the screen may have been weakened by corrosion.

Other methods sometimes adopted comprise chemical treatment or acidization but it is strongly advisable that these should only be undertaken by an engineer experienced in such techniques. The most common chemical treatment uses a dispersant, such as sodium hexametaphosphate, for example, Calgon, to assist in the removal of fine material. Acidization is used in calcareous strata to enhance the size of fissures in chalk or limestone in the immediate vicinity of the borehole. Concentrated hydrochloric acid is usually used, together with an inhibitor to minimize corrosion of any mild steel casing. The acid is applied by pipe below the water level. If the top of the borehole is closed the pressure of the evolved carbon dioxide can enhance penetration of the acid into fissures. However, the handling of acids is dangerous and should be subject to strict safety procedures. In addition, the carbon dioxide produced can be hazardous unless special attention is paid to ventilation.

The performance and condition of water supply boreholes will change with time and performance monitoring is a core activity in good groundwater supply management strategy. Comprehensive data gathering and interpretation of the results enable the optimum maintenance and rehabilitation strategy and implementation plan to be adopted. Best practice guidance on monitoring, maintenance and rehabilitation can be found in a CIRIA report on the subject (Howsam et al., 1995).

4.11 POLLUTION PROTECTIVE MEASURES: MONITORING AND SAMPLING

The dangers of groundwater contamination are widespread and varied. Industry, agriculture and urbanization generate large and varied suites of groundwater pollutants, many of which are of public supply significance. The characteristics of the various pollutants vary in time and location as Table 4.2 indicates. All represent the background addition of some substances to groundwater, which may be controllable somewhere in the water supply cycle, but prevention cannot be managed necessarily within the catchment. The implications for the public supply are very high. Treatment becomes necessary initially and the resources may then become difficult to manage to avoid inducing or accelerating further contamination. Ingress of *Cryptosporidium* oocysts or *Giardia* cysts or other harmful microbiological organisms (Sections 7.60–7.63) to a well from agricultural land via poor well cap seals is often a cause of notices to boil water and can be troublesome and expensive to deal with.

Groundwater pollution can sometimes be 'cleaned up'; many methods have been used in the last 20 years, mainly in the USA where funds have been provided for the purpose but the technology for restoration of contaminated groundwater and aquifers remains costly and imprecise. Remedial action may be impracticable where predicted clean-up time is decades or longer. The aquifer may become completely unusable so that substitute resources, such as imported supplies or treated surface water, are needed and a valuable and inexpensive resource is lost.

Table 4.2 Major sources of potential groundwater pollution

Occurrence	Local/linear mode	Distributed mode
Seasonal/periodic	Road salting Rail and road verge herbicides Silage Combined sewer overflows	Agricultural fertilizers, herbicides and pesticides Sewage sludge spreading
Continuous	Road drainage Cesspool overflows Septic tank effluent disposal to land Solid waste tip and landfill leachates Contaminated abandoned industrial land Influent polluted rivers Run-off from grazing land (cryptosporidium)	Industrial atmospheric discharges
Random	Road/rail tanker spills Pipeline and sewer breakages Fires Defective storage of industrial or agricultural chemicals	Nuclear and industrial accident fallout

The concept of evaluating the vulnerability of groundwater sources to pollution is based on consideration of the lithology and thickness of the strata above the aquifer and the surface soil leaching properties. From this, the size of ***Source Catchment Protection Zone*** (SCPZ) required around a borehole or well may be derived, based on the estimated *travel times* of potential pollutants within the saturated zone to the abstraction point (Adams, 1992). In the UK, work initiated by the National Rivers Authority and later taken over by the EA, suggested three zones of protection should be investigated for protection of underground abstraction works:

1. ***Inner zone (Zone 1)*** defined by a 50-day travel time from any point below the water table to the source, based principally on biological decay criteria, and has a minimum radius of 50 m;
2. ***Outer zone (Zone 2)*** defined by a 400-day travel time from any point below the water table, based on the minimum time required to provide dilution and attenuation of slowly degrading pollutants and has a minimum radius of 250 or 500 m around the source, depending on the size of the abstraction;
3. ***Total catchment (Zone 3)*** defined as the area around a source within which all groundwater recharge is presumed to be discharged at the source. In confined aquifers, the source catchment may be displaced some distance from the source. For heavily exploited aquifers, the final SCPZ can be defined as the whole aquifer recharge area where the ratio of groundwater abstraction to aquifer recharge (average recharge multiplied by outcrop area) is >0.75. There is still the need to define individual source protection areas to assist operators in catchment management.

To delineate such zones about any particular source, a series of field investigations are necessary; initially to create a conceptual model which later, as more data is obtained, is sufficient to produce a calibrated model of the catchment so that the zones can be more accurately defined. In the UK, zones are not statutory but are used as a guide for the control of catchment activities by water utilities. However, SCPZ1 has been noted in statutory guidance as the minimum area under the former Groundwater Directive that is identified for the protection of drinking water. SCPZs are also recognized in the Environmental Permitting Regulations (EPR) as zones where certain activities cannot take place, for example in certain standard rule permits (EA, 2013b). Elsewhere, as in Germany, protection zones can be statutory, being set up under local or Federal legislation.

Continual vigilance over the surface catchment to an underground source should be exercised. All potential sources of pollution (cesspools, septic tanks, farm wastes, industrial waste and solid waste tips, farm or industrial storages of chemicals, etc.) should be identified and recorded on a map of the catchment area. All these risks should be regularly monitored to ensure they are properly controlled. Poorly functioning household sewage disposal works may have to be replaced by more efficient plant and farmers may have to be advised of improvements to their waste disposal practices.

Monitoring water quality will identify signs of contamination and whether it is a recent and local incident or more likely to be from a distant part of the catchment. The chemical constituents of the water can show evidence of the types of pollution listed in Table 4.2. If a pumped borehole penetrates several water-bearing strata, depth specific sampling devices can be used below the pump with isolating inflatable packers. This can be useful for ascertaining at what level a contaminant is entering the borehole. Gas operated ejection devices or small diameter piston pumps can be used to withdraw samples from sections of the borehole using the isolating packers. Otherwise samples from the pumped water will be a composite sample of the water entering the hole from the full thickness of the formation in contact with the well.

Groundwater sampling needs a slightly different approach to surface water sampling. The chemistry of groundwater tends to be relatively stable, but dissolved oxygen and redox (Section 7.54) potential may be more important than for surface waters. Quality changes during sampling have to be considered, mitigated as much as possible by taking measurements of parameters such as dissolved oxygen, pH and conductivity in the field. The groundwater temperature may also be a useful indicator, because surface percolation to the upper levels of an aquifer will show a substantially greater seasonal temperature variation than water drawn from depth. Long-term groundwater abstraction may cause an increase in dissolved solids, which may imply that abstraction is drawing water from more distant parts of the aquifer, possibly indicating that the catchment area monitored may need to be extended.

4.12 RIVER FLOW AUGMENTATION BY GROUNDWATER PUMPING

Regulating river flow by pumping from groundwater is practised as a valid means of developing water resources to their full potential while still protecting the environment. Water is pumped from an aquifer and discharged to a river to augment its flow during a low flow period. Usually the abstraction points lie within the river catchment basin. The Shropshire

Groundwater Scheme is an example in the UK where water is abstracted from the sandstone aquifer underlying North Shropshire and discharged into the River Severn to meet peak dry weather demands. To be successful river regulation by groundwater pumping has to meet certain basic criteria:

- the scheme should provide a satisfactory increase in yield above that which could be obtained by direct abstraction from the aquifer;
- discharge of the pumped groundwater to the river should result in economies in pipeline costs;
- the abstraction pumping capacity must be not less than the desired gain in river flow during the design drought period, plus any reduction of springflows feeding the river caused by the pumping;
- if recirculation losses through the river bed occur, it must be practicable to adopt equivalent extra abstraction capacity.

Prolonged groundwater abstraction will almost certainly reduce some springflows contributing to the river flow. To avoid iterative calculations it may be better, in a simple approach, to assume that all springflows within a given distance from the groundwater abstraction points will be reduced to zero by the pumping during a prolonged drought. Pilot well tests and a detailed knowledge of the catchment hydrogeology are necessary to assess which springflows will be affected. A better approach is to adopt groundwater mathematical modelling. This permits the cumulative effect of pumping to be traced for droughts of different severity and length. The method can reveal the effect on local wells used for private supplies, farming and other interests, whether adjacent catchments are affected; and to check if percolation is sufficient to restore the aquifer storage during later wet periods.

In general river regulation by groundwater pumping is most successful where the pumping has a delayed, minimal, short-lived effect on natural river flows. To achieve this, several carefully sited abstraction points, pumped according to a specific programme, may be necessary. Where a regulated flow substantially in excess of the natural low river flow is required, prolonged pumping tests at the proposed abstraction points during a dry period are essential. This can be an expensive operation. In addition, caution is required in estimating the possible yield during some extreme drought conditions not previously experienced, because aquifer drawdown conditions during some future critical event cannot be known with any accuracy.

Augmenting river baseflows as a consequence of groundwater recharge has the following advantages:

- it allows better management of water resources by recharging with surplus water during high flow periods; and
- it raises low flows in the summer, provided that the timing and movement of recharge water is appropriate, with consequent environmental benefits.

However, the disadvantages include:

- it is only applicable where recharging an unconfined aquifer in hydraulic continuity with the surface water system (springs or river channel inflow);
- additional studies and trials are required to determine feasibility and possible effects.

4.13 ARTIFICIAL RECHARGE AND AQUIFER STORAGE RECHARGE

Artificial recharge (AR) supplements the natural infiltration of water into the ground to augment groundwater yields. AR can be used strategically where an aquifer is overexploited such that no further abstraction would be allowed or where lack of natural recharge prevents its utilization. The water abstracted is not necessarily the same water that was recharged and abstraction boreholes may be some distance apart. It is often practised for water treatment reasons, as in the Netherlands where the dune sands are used to store potable water (Jones, 1999).

The two basic methods adopted are the use of spreading areas or pits and injection through boreholes. The former technique has dominated because of its simplicity and the ease with which clogging problems can be overcome. The relatively unknown performance of wells and the need to pass only pre-treated water down them limits their usefulness except where land for recharge pits is at a premium. Untoward events can happen during recharge operations: at an Israeli site (Sternau, 1967) the injection of water down a well caused unconsolidated sands around it to settle and, in less than 2 days after pumping ceased, the borehole tubes and surface pump house sank below ground level. In a British experiment (Marshall, 1968) even the use of a city drinking water supply for recharge did not prevent the injection well lining slots from becoming constricted with growths of iron bacteria.

At times when existing surface water treatment works are working below capacity, for example when there is a seasonal fall in demand, it may be possible to make water available for well recharge. This is being practised in the Lee Valley (O'Shea, 1995) north of London. Complete efficiency cannot be expected because of the relatively uncontrollable and 'leaky' nature of aquifer storage. Because the recharged source may not be used for some months, there is a loss of resources as the 'recharge mound' decays outwards and down the hydraulic gradient. Even where abstraction facilities have been located specifically to minimize this, the losses are considerable. Special AR operations in coastal areas have been used to prevent saline intrusion from the sea, as for example in coastal Israel (Harpaz, 1971; Aberbach, 1967) and southern California (Bruington, 1965).

Recharge as a way of improving wastewater quality has been used in the Netherlands, Germany and Scandinavia for many years (IAHS, 1970). Alluvial sand aquifers of the coastal Netherlands are recharged with heavily contaminated River Maas water after primary filtration. The abstracted water still has to be treated using activated carbon to remove heavy metals but can be used for public supply. The strata that provide the natural treatment during infiltration have to be 'rested' and allowed time to recover and re-oxygenate. At Atlantis in South-West Africa a successful AR scheme is part of the local water resources management system. Treated domestic sewage effluent and urban storm run-off from the 67 000 population is recharged through lagoon systems and re-abstracted from specially sited production wellfields (Wright, 1996).

Recharge boreholes are vulnerable to various types of clogging due to accumulation of suspended solids, gas bubbles that come out of solution, and microbial growth filling interstices or screen apertures. Borehole recharge systems using polluted water are normally not acceptable. Hence, only recharge pits are considered below.

Although dimensions vary, a recharge pit bears resemblance to a slow sand filtration bed (Sections 9.12 and 9.13) because replaceable filter media normally covers the base. The rating of a pit depends on the rate at which the raw water will pass through the filter media and the rate at which the underlying aquifer will accept it. The former depends upon raw water quality, any pre-treatment of the water (Section 9.15) and the depth of water kept in the pit. Little, if anything, can be done to improve the rate at which the aquifer accepts the filtrate, but a decline in the infiltration rate must be guarded against by tests to ensure the filter media is working satisfactorily. Published results for infiltration rates vary widely.

Pilot tests are always required at a new site, as are initial investigation bores to ensure the pit floor will be above the water table by a margin as big as possible to give the best opportunity for water quality improvement and to allow for increased groundwater storage and its 'mounding' below the recharge pit. Iron-pan layers and other impedances to vertical flow should be avoided.

Aquifer storage and recovery (ASR) is a subset of AR and comprises one or more of the following:

- storage of water in a suitable aquifer (through wells) during times when water is available and recovery of water from the same well when needed;
- normally achieved by using dual purpose boreholes;
- injection of water into an aquifer containing non-potable water;
- abstraction of the same water that was injected;
- utilization of a confined aquifer to minimize potential environmental impacts.

The water is stored in the aquifer locally to the borehole; it improves resource management and has operational advantages. Water is recharged into the aquifer during periods of low demand and then recovered during periods of high demand.

ASR could be considered rather than AR in parts of confined aquifers that have not been used for productive water supply because of poor quality (Jones, 1999). The use of ASR in areas of outcrop of potable aquifers is less likely as the aquifer may be fully licensed, natural recharge will occur and the groundwater to surface water interaction will be more immediate and hence have environmental impacts (Williams, 2001).

The key constraints on the development of ASR identified by research (Jones, 1999) include:

- recovery efficiency, a measure of how much is recovered against how much is put in, is less significant where both recharged and native waters are of similar quality;
- clogging issues including: air entrainment, suspended sediment, bacterial growth, chemical reactions, gas production and compaction of clogging layer;
- water quality changes: chemical components of the recharged water reacting with groundwater and aquifer, particularly the organochlorine compounds (Section 7.24) in the treated recharge water;
- hydraulic properties of aquifers;
- operational issues including: variability in volume and quality of water available for recharge, less significant when treated effluent is used; variation in daily and weekly demand in relation to average monthly or annual demand; and site selection;
- regulatory issues.

Of particular environmental concern is the possible impact on water levels in adjacent aquifers or in the target aquifer. ASR schemes are designed to result in no net change in abstraction from the aquifer. In the long term there may be some net change but this should be small in relation to the overall resource. A scheme that injects and then abstracts the same volume of water, as a long-term steady-state average, will have no impact on local or regional groundwater levels. However, the seasonality of the scheme means that local water levels will first be increased and then reduced. The timing and absolute value of these changes are important in assessing the eventual impact and will need to be determined by site specific model studies or field trials.

4.14 GROUNDWATER MINING

In many countries of the Middle East and North Africa groundwater is being used at a rate greatly in excess of the rate at which the resources are being replenished. This tends to occur either because there is no institutional framework for controlling development or because demands for water cannot be ignored while adequate resources appear to exist. Climatic patterns have changed during the last few tens of thousands of years; areas, which are now arid, were once comparatively humid. In that earlier period, high rates of recharge applied where none now takes place, as in the Libyan Desert and much of Saudi Arabia. In both regions there are groundwater developments on a vast scale that 'mine' this water. Many thousands of cubic kilometres of groundwater can be developed, though the cost of doing this and taking the water to where it is needed are high. The Great Man-Made River Project in Libya (Table 4.1) is a modern example, where the capacity now exists to pump many millions of cubic metres of groundwater through 4 m diameter concrete pipelines for over 600 km from inland desert wellfields to the Mediterranean coast. In Saudi Arabia 'fossil' groundwater is being mined for public supply and here, as in some other countries, groundwater exhaustion problems have only been managed through large scale adoption of seawater desalination.

The use of water resources without knowledge of their sustainability, or where it is known that mining is taking place, may be looked upon as an irresponsible course of action in most cases. However, often the use of technology to minimize demand or maximize efficiency in the use of water is too costly to be politically acceptable, especially in respect of agriculture, which remains the greatest and least efficient user of water resources. It is estimated that a 10% improvement in the efficiency of agricultural use would double the resources available for public water supply.

4.15 ISLAND WATER SUPPLIES

Large numbers of islands abound in all the major oceans, particularly off South East Asia and in the Pacific. In the Philippines alone there are over 7000 islands and Indonesia comprises over 13 000 mainly small islands. The Republic of the Maldives in the Indian Ocean is a nation comprised entirely of about 1300 coral atoll islands, of which about 200 are populated. Geologically the islands range from the Maldive type of atoll islands which are less than 2 m above high tide level, to larger islands with a rocky core surrounded by a rim of sediment and reef. Many of these islands are isolated, lack natural resources and are short of water. Such freshwater as exists often occurs as an extremely fragile lens, floating on saline water below. Population densities are high. Malé island, capital of the Maldives, had a population density in 2014 of 26 000 per km^2. The fresh groundwater therefore not only became polluted with sewage but eventually was virtually consumed by over-abstraction so that dependence on desalinated water and external resources is now near total. Similar problems can occur in other places where tourism increases water demand beyond the ability of local resources to supply.

On the larger rocky islands, surface water resources can sometimes be developed from perennial streams or springs, but stream flow is ephemeral on all but the larger islands. Usually there is heavy dependence on groundwater, which exists as a reserve of freshwater, usually described as forming 'a lens' of freshwater within the strata of the island, below which lies denser saline water infiltrated from the sea. The lens is not usually regular, but strongly distorted by complications of geology that influence fresh groundwater flow to the sea. A lens thickness of 20–30 m is common, with a transition zone between the fresh and saline water, the thickness of which is a function of the aquifer properties, fluctuations in the tidal range and due to variation in the rate of rainfall recharge. The amount of water represented by a given lens depends on the specific yield of the strata; a coral sand may have a high specific yield (20% or more) whereas a limestone or volcanic rock may have a much lower capacity to store water (less than 1%).

The simplified elements of an island water balance are illustrated in Figure 4.8 including the transition zone between fresh and saline waters. The dynamic balance between percolation from rainfall and outflow of freshwater to the sea determines the dimensions of the lens of freshwater, mainly its depth because the width is limited. This is the 'classical view' of the lens configuration, in which the height of the lens surface above mean sea level is 1/40th of the depth of the lens below mean sea level (Section 4.3). However, observations of many islands show that a more realistic model incorporates vertical and horizontal flow due to the effect of a typical two-layered sand overlying a reef limestone system (Wheatcroft, 1981; Ayers, 1986; Falkland, 1993). The transmission of the tidal fluctuations to the lens is not necessarily proportional to the distance from the shore, but rather to the depth of the sand aquifer and the transmissivity of the underlying reef limestone. Dispersion is also incorporated in the model to allow for the thickness of the transition zone between the fresh and saline water, which is typically of the order of 3–5 m.

However, other than rainfall, assigning a quantity to all the components in Figure 4.8 is very difficult. For instance the groundwater outflow is usually a significant proportion of the water balance, but it cannot be directly measured; it can only be roughly estimated from the groundwater gradient and the hydraulic conductivity of the aquifer. Similarly, evapotranspiration use by vegetation is difficult to estimate because of the unusual situation that such vegetation can find freshwater

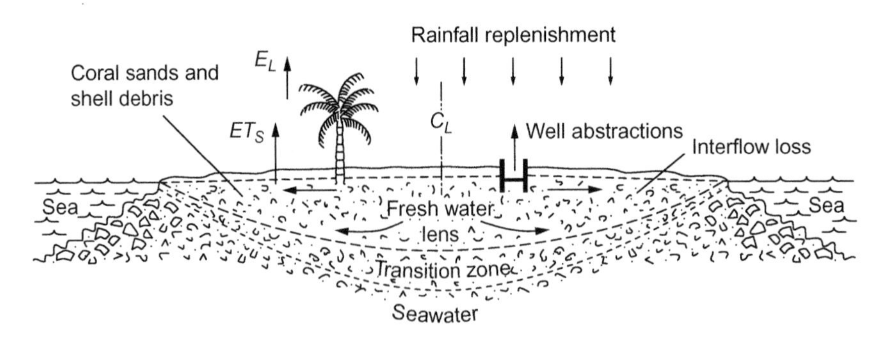

FIGURE 4.8

Diagram of lens of freshwater floating on seawater on a coral island.

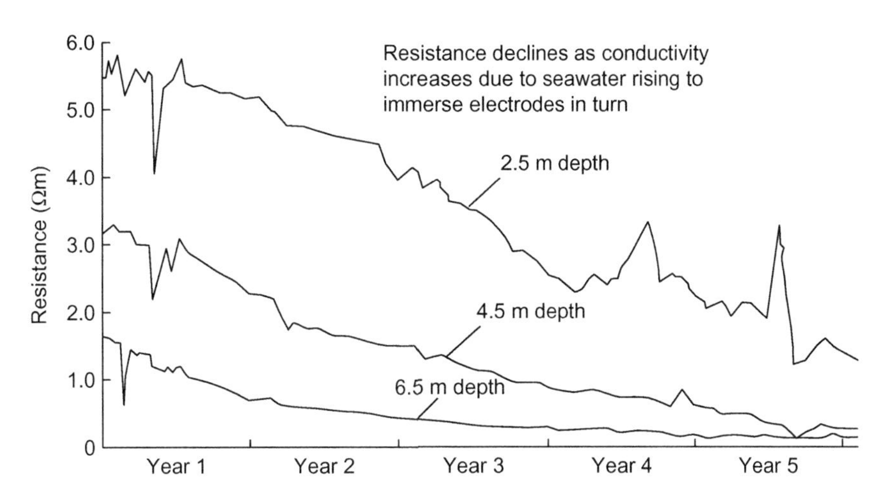

FIGURE 4.9

Graph indicating rise of freshwater/seawater interface as lens of freshwater thins due to over-abstraction – Malé 1983–88.

continuously available at shallow depth. Hence a water balance estimate can only reveal whether the likely replenishment of the lens is obviously less (or more) than the estimated losses from outflow and abstraction. Consequently the only safe way to evaluate the situation is to measure the thickness (and extent) of the lens and to keep it monitored.

Measuring the electrical conductivity at the base of the freshwater lens and across the transition zone is the most sensitive method of monitoring freshwater reserves. Over this zone, small changes in the elevation of the lens are reflected in large changes of electrical conductivity. Figure 4.9 shows measurements taken on Malé Island between 1983 and 1988 at a number of depths. The figure illustrates the virtual demise of the freshwater reserves during the period.

Controlled development of island freshwater is best achieved by shallow abstraction, using groups of shallow wells or collector systems, or even infiltration trenches. The use of boreholes is usually not the best approach because drawdown of the freshwater surface during pumping tends to cause upward movement of the saline water below. The rate of 'up-coning' of saline water depends on the geology, rate of pumping and depth of the borehole; but once contaminated with saline water the aquifer may take many weeks or longer to recover a usable quality of freshwater. The proportion of rainfall that recharges groundwater is subject to many factors. A relationship between rainfall and recharge derived by Falkland (1993) for atolls and larger topographically low islands is shown in Figure 4.10 and is a useful guide for preliminary use.

However, the difficulties of preventing overdraw from an island aquifer has to be recognized. The local population will have been able to use house wells for their supply. However, because the level of the water in wells remains virtually constant (even though over-abstraction is causing the freshwater lens to thin), householders will see no evidence to suggest they should reduce their take. The technical solution required to preserve the lens is to close all the private wells and install a public surface abstraction system, which takes only an amount equal to the average replenishment

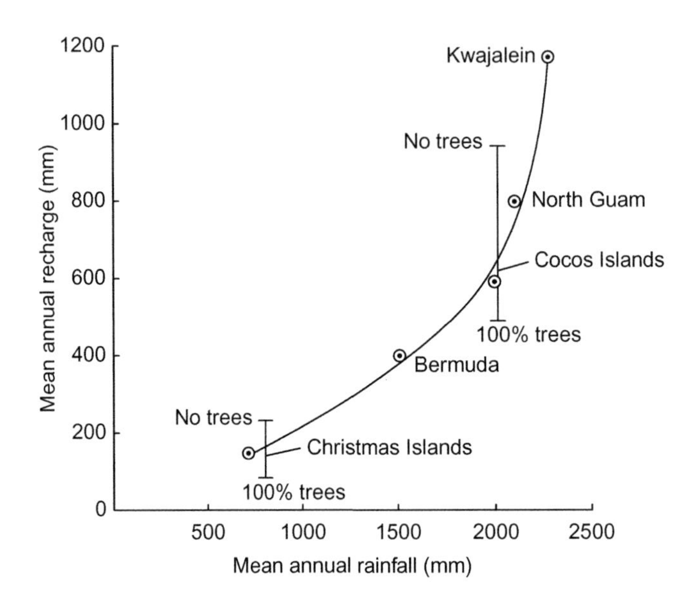

FIGURE 4.10

Relationship between mean annual rainfall and estimated mean annual groundwater recharge.

of the aquifer, supplying householders with a rationed supply via standpipes or metered connections. Such a policy would be difficult to implement because householders may not be persuaded of the technical need.

4.16 COLLECTOR WELLS AND OTHER UNDERGROUND WATER DEVELOPMENTS

Collector wells sunk in the river bed are widely used in alluvial deposits in valleys in arid areas where the dry weather flow of the river is underground. They take the form of large diameter concrete caissons sunk in the river bed, the caisson having to be of sturdy construction to withstand the force of the flood flow in wet weather. Access by bridge is necessary, the bridge and the top of the caisson being sited above maximum flood level. From the base of the caisson, collector pipes are laid out horizontally usually upstream, though sometimes across the stream bed or fanwise upstream. The collectors usually comprise porous, perforated or unjointed concrete pipes, 200–300 mm diameter, laid in gravel filled trenches cut in the river bed sediments, connected to the wet well. It is important to lay the collectors at sufficient depth to avoid their destruction during floods when some of the stream bed can mobilize into suspension. Where the bed is narrow with deep deposits a simple large diameter well may suffice to intercept flow down the valley. The yield of such wells is limited by replenishment flow down the valley. If ordinary centrifugal pumps (which are the cheapest) are used, they have to be sited above maximum flood level with their suctions dipping into the water, which limits the range of water levels from which water can be pumped. Use of submersible pumps can overcome this difficulty.

The ***patented Ranney well*** is an early form of collector well, usually made of cast iron and sunk in river bankside gravels or alluvial deposits fed from the river, with perforated collector tubes jacked or jetted out horizontally from the base in the most appropriate configuration, often parallel to the river. This type of well uses the bankside deposits to act as a filter. Consequently, the well may have a short productive life if suspended sediment is drawn into the bankside deposits; but back flushing of the collectors can prolong the useful life, sometimes very effectively. The advantage of a Ranney well is its relative cheapness. See also Section 6.27 for radial wells.

Galleries are similar to the pipes of collector wells and are common in some areas of the world where seasonal watercourses cross thick and extensive alluvial strata that offer large storage potential. Galleries are comparatively large diameter perforated collector pipes buried 3–5 m depth or below the minimum dry season groundwater level, and surrounded by a filter medium. It is essential to locate galleries to avoid those parts of a watercourse where active erosion of the bed could occur when the river is in spate.

Qanats are believed to be of ancient Persian origin and are widespread on the flanks of mountain ranges in Iran. They are also widespread in the Maghreb where they are known as '*foggaras*'; in the Arabian Peninsula where they are known as '*aflaj*' (singular '*falaj*'); and are also found in Afghanistan and China. The source or 'mother well' is dug to some depth below the water table relatively high on the flanks of a mountainous area (Fig. 4.11). A series of additional wells is dug in a downhill line towards the area where the water is needed and the wells are connected by tunnels below the water table so that the groundwater is drained down-gradient. The tunnel gradient is slightly less than the groundwater gradient and much less than the topographical gradient;

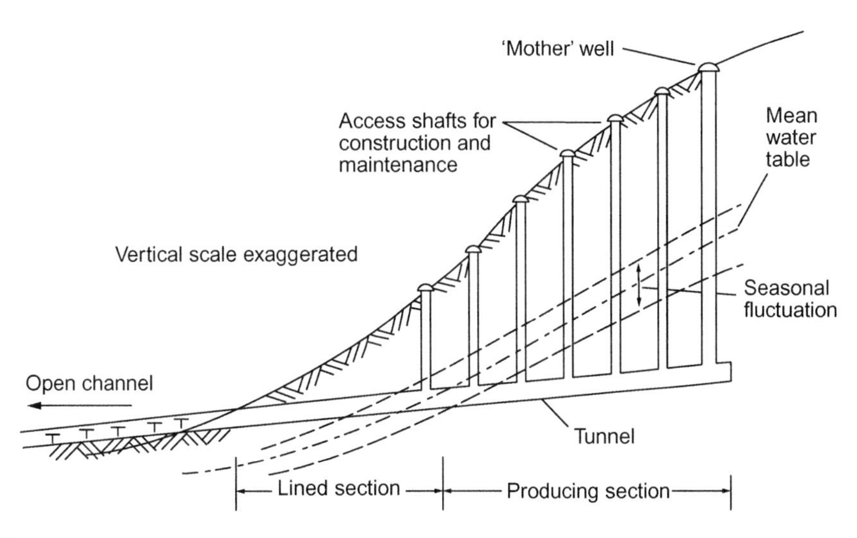

FIGURE 4.11

Section through typical Qanat system.

therefore the tunnel subsequently emerges above the water table and eventually at the ground surface where it acts as a canal to deliver the water where required.

The systems were hand dug and hence tunnels are comparatively large for ease of construction and to allow regular maintenance. The tunnel section from its point of emergence from the water table has to be clay-lined to minimize losses. In an increasing number of cases, competing large scale pumped groundwater abstraction is causing serious interference with these ancient gravity systems. As a result they are slowly going out of use through neglect as their discharge declines and the necessary maintenance is no longer done.

4.17 BOREHOLE AND WELL LAYOUTS

For most public supply purposes, a borehole capable of giving upwards of 5 Ml/d is required and, for this, a boring of at least 300 mm diameter is necessary to accommodate the pump. In practice a more usual size would be 450 or 600 mm diameter to provide good flow characteristics in the boring and the rising main, to allow for the diameter of the boring to be reduced with depth or for installing screens, and to allow for any lack of verticality in the hole. If a boring is to be sunk at a site where there are no boreholes in the vicinity to provide information on the strata to be penetrated it is advisable to sink a pilot borehole first. Choice of the size of this boring can present a problem. If it is small, starting from 150 mm and reducing to 100 mm, it will be cheap and give the strata information required, but should it strike a good yield of water it may turn out too small for insertion of a pump to develop its yield. It can, perhaps, be reamed out to a larger diameter or a larger borehole may be sunk nearby to accommodate the pumps. However, there is no certainty in fissured formations such as chalk and sandstone that a nearby hole will yield as much, or more. In one case where two 300 mm diameter borings were sunk in the chalk of southern England, the first hole gave just over 3 Ml/d, which was considered not enough and a second hole was sunk only 5 m away and gave more than 18 Ml/d.

A single boring of 450–600 mm diameter will permit only one pump to be installed. If the borehole is to be the only source of supply, a single pump would not safeguard the water supply adequately; the time required to replace a defective pump, even a submersible, could interrupt the supply longer than can be tolerated. If the pumping level is not more than about 20 m below ground level it may be economic to construct a well sufficiently large to accommodate two pumps. The well can be sunk over a trial boring, if the latter gives a good yield; or a second boring can be sunk a few feet away and a connection between the two can be 'blown through' by using a directional charge (Fig. 4.12). If, however, the trial borehole fails to give a satisfactory yield, but it is nevertheless thought the site should be capable of giving more, a well may be sunk elsewhere on the site, a second boring being sunk from its base. By use of a pump in this second boring, the well may be dewatered so that an adit can be driven towards the trial boring (also dewatered) connecting the two bores together. This makes it possible to utilize the combined output of the two borings.

Adits are seldom constructed now, but in the past many miles of adit were driven from wells sunk in chalk or sandstone formations to intercept more water-bearing fissures. The adits were generally unlined, about 2 m high by 1.2 m wide, driven to a slight upward grade from

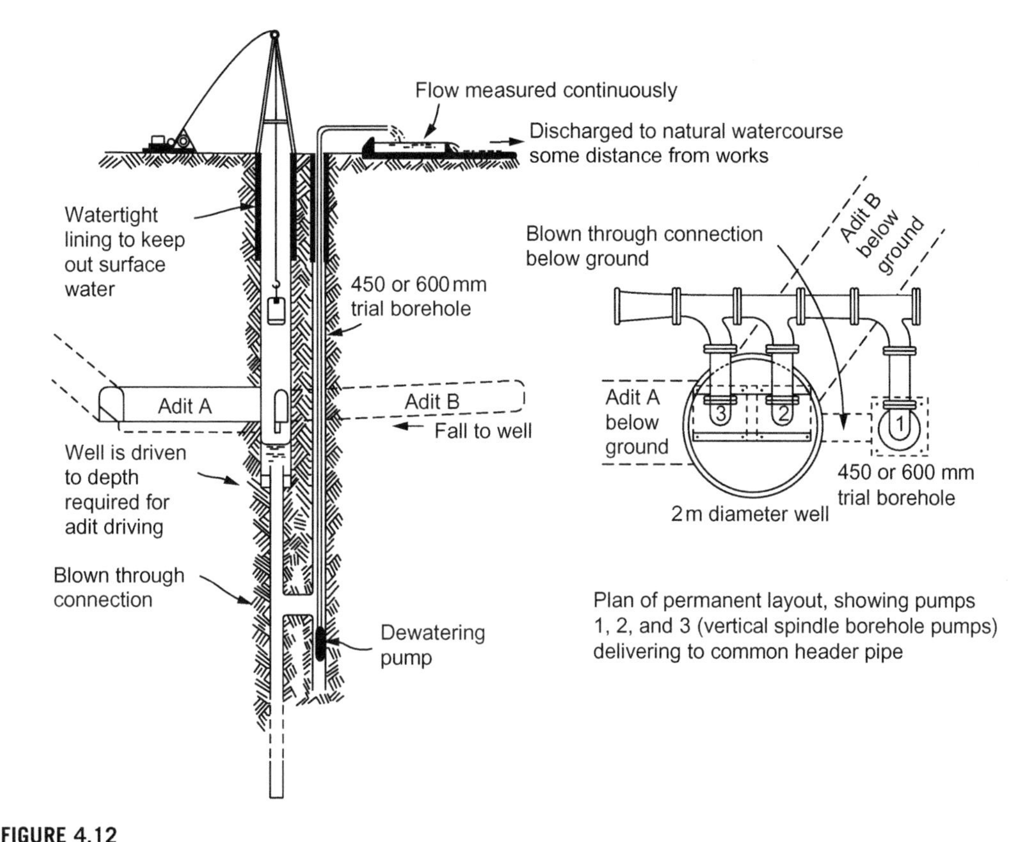

FIGURE 4.12

Layout using trial borehole, deep well and adits.

the well and with a channel on one side to drain incoming water to the well shaft when driving. The risk of meeting a large water-bearing fissure meant that quick, reliable escape means had to be assured for the miners; dual power supplies for the dewatering pumps and for the crane operating the access bucket were therefore essential. Excavation was slow due to lack of space for muck haulage. The high cost and residual risk of adit driving compared with other means (well development) of improving yields from boreholes effectively ended adit construction.

A hand pump is the simplest device to lift groundwater to the surface. Hand-operated village pumps are considered the most suitable low cost option for safe water supply in resource-poor settings common in rural areas in developing countries. The hand pump gives access to deeper groundwater, up to 100 m; a deeper well offers more protection from pollution due to use of contaminated buckets.

During the International Drinking Water Supply and Sanitation decade (1981–90) boreholes, hand dug wells and tube wells were constructed and water pumps provided to developing countries. Unfortunately this top-down approach led to the installation of pumps that were difficult to maintain. Studies showed that about 90% of most hand pumps break down within 3 years due to worn out or broken components. In 1987 the World Bank completed a technical review of a large number of hand pumps (WB, 1987a) and provided guidelines for the selection of water supply technology and systems that best meet the needs of a community (WB, 1987b). The guidance introduced the *Village Level Operation and Maintenance* (VLOM) approach, also referred to as *Management of Maintenance*, which has led to the development of hand pumps which require minimal maintenance. These 'VLOM pumps' allow villagers in remote locations to maintain the pumps themselves and are part of a larger strategy to reduce dependency of villages on government and donor agencies. The main types of traditional lever action hand pumps are the India Mark III for shallow to medium depth wells and the Afridev for deep wells. Direct action pumps include the Nira AF-85 for shallow wells. Typical performances of the more common VLOM and other hand pumps are detailed in the WaterAid *Technical Brief Handpumps* (WaterAid, 2013).

4.18 CHOICE OF PUMPING PLANT FOR WELLS AND BOREHOLES

The yield–drawdown characteristic curve for the source will be obtained from the test pumping of the completed well (Curve A in Fig. 4.13). The curve shows the pumping water level expected when the rest level is 'average'. However, allowance has to be made for the fluctuation of the water table level during the wet and dry seasons of the year and the lowest rest level that might occur in an extreme dry weather period. There may also be deterioration of the yield with time due to such unpredictable matters as silting of fissures, change of aquifer recharge due to changed conditions on the outcrop, or increased abstraction from the aquifer elsewhere.

The lift to be specified for the borehole pump(s) should take account of the normal seasonal fluctuation of the water table level; this should be available from data collected before the pumping test takes place. Historical records for existing wells in the same aquifer may indicate the lowest water table level likely in an exceptionally prolonged dry season. The specified lift should also allow for deterioration in well flow characteristics with time. Again, experience of boreholes in the same area or in similar formations elsewhere, can act as a guide but, in its absence an arbitrary allowance for increase in drawdown should be made (say 33%) or that future yield will decline by 33% for a given drawdown.

If an arbitrary reduction of yield is assumed, as shown by Curve B on Figure 4.13, curves b–b′ and c–c′ show its variation in the wet season and dry season, respectively. The characteristic curve of a possible pump has also to be added onto the diagram, shown as curve x–x′ for a given fixed speed pump. The curve x–x′ applies to the head necessary to lift the water to the top of the well and to any additional lift to some storage tank. The pump characteristic has to be such that it is capable of giving the maximum output required (10 Ml/d in Fig. 4.13) if the well characteristics should decline to curve c–c′. However, if the well yield–drawdown characteristics remain as Curve A, it will be necessary to throttle the output by part closure of the delivery valve, or to use of a motor with a variable speed drive (Section 19.24). Curve y–y′ represents the pump output at a lower speed of rotation, when only about 6 Ml/d, say, is the required output and the pumping water level in the well is higher. Another alternative would be to use a pump with a dummy stage, into which an impeller would be fitted later when drawdown increases (Section 19.3).

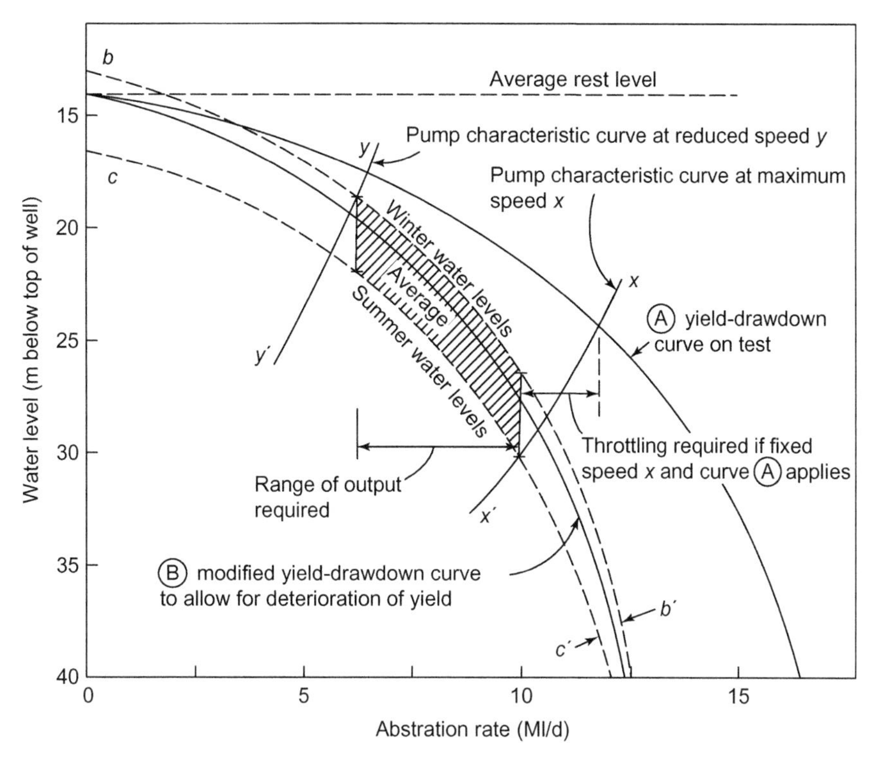

FIGURE 4.13

Yield–drawdown curves for a well with pump characteristics curves superimposed.

The choice of pumping arrangement depends on cost. Throttling a fixed speed pump wastes energy. A pump with a variable speed drive will be more expensive than a fixed speed pump and its average efficiency over the range of operation may not be as high as that of a fixed speed pump, which can spend most of its time running at its design duty. It is advisable to define the range of expected operational conditions and discuss the most suitable arrangement with the pump manufacturer. The data provided should include:

- the well yield–drawdown characteristics;
- the additional lift required from the well head (and its characteristics);
- the expected time periods per annum for which the drawdown will be in the upper, middle and lower range of values;
- what increase of drawdown should be assumed for the first 10 or 15 years;
- the average, maximum and minimum rates of pumping that will be required.

An important point to note is that the licensed abstraction is often stated as a quantity that must not be exceeded in any 24 hour period, whereas generally the actual pumping hours will be less, for example, 22 hours to allow for shutdowns to attend to routine maintenance matters. Similarly demands less than the licensed abstraction would normally be met by pumping for reduced hours, which implies that storage must be available to maintain the supply to consumers when the pump is not operating.

A mixed flow, fixed speed submersible pump (Plate 32 (c)) with a fairly flat efficiency curve about the design duty and a reasonable slope to the head-output curve may be the most suitable for well pumping if the drawdown range is not excessive (Section 19.3). Vertical spindle centrifugal pumps would only be installed for large outputs because of their much greater cost.

4.19 ENVIRONMENTAL IMPACT ASSESSMENTS

An assessment of the hydrogeological impact of groundwater abstractions is required for licence applications in the UK under the Water Act 2003. A methodology for Hydrological Impact Assessment (HIA) is outlined by the Environmental Agency in line with their abstraction licence process (Boak, 2007). The methodology identifies 14 steps necessary in carrying out an assessment. They are:

1. Establish the regional water resource status.
2. Develop conceptual model for the abstraction and the surrounding area.
3. Identify all potential water features that are susceptible to flow impacts.
4. Apportion the likely flow impacts to the water features.
5. Allow for mitigating effects of any discharges, to establish net flow impacts.
6. Assess the significance of the net flow impacts.
7. Define the search area for drawdown impacts.
8. Identify all features in the search area that could be impacted by drawdown.
9. For all these features, predict the likely drawdown impacts.
10. Allow for the effects of measures taken to mitigate the drawdown impacts.
11. Assess the significance of the net drawdown impacts.
12. Assess the water quality impacts.
13. If necessary, redesign the mitigation measures to minimize the impacts.
14. Develop a monitoring strategy.

These steps are not intended to be prescriptive; the detail required for each step should match the situation. The methodology depends on the development of a good conceptual model of the aquifer and the abstraction itself. Advice is also given on how to undertake an HIA in karstic aquifers and fractured crystalline rocks.

Other UK legislation and regulation relevant to the assessment of the impacts of groundwater abstraction on water resources and the water-related environment include the Habitats Directive, the Water Framework Directive (Section 2.4), and Catchment Abstraction Management Strategies (CAMS).

The EA is responsible for the protection and management of groundwater resources in England and Wales. Their guidance document *Groundwater Protection: Principles and Practice (GP3)* describes their approach to the management and protection of groundwater in England and Wales (EA, 2013b). It provides a framework within which to manage and protect groundwater and takes account of the government's sustainable development strategy and the water strategies of both the Department for Environment, Food and Rural Affairs (Defra) and the Welsh Government.

REFERENCES

Aberbach, S. H. and Sellinger, A. (1967). A review of artificial groundwater recharge in the coastal plain of Israel. *Bulletin of International Association of Scientific Hydrology* **12**(1), pp. 75–77.

Adams, B. and Foster, S. S. D. (1992). Land-surface zoning for groundwater protection. *Journal of IWEM,* June pp. 312–320.

Anderson, M. P. and Woessner, W. W. (1991). *Applied Groundwater Modelling: Simulation of Flow and Advective Transport.* Academic Press Inc.

Ayers, J. F. and Vacher, H. L. (1986). Hydrogeology of an atoll island; a conceptual model from a detailed study of a Micronesian example. *Ground Water* **24**(2), pp. 185–198.

Boak, R. and Johnson, D. (2007). *Science Report – SC040020/SR2, Hydrogeological Impact Appraisal for Groundwater Abstractions.* EA.

Boulton, N. S. (1963). Analysis of data from non-equilibrium pumping tests allowing for delayed yield from storage. *Proceedings of the ICE* **26**, pp. 469–482.

Bruin, J. and Hudson, H. E. (1955). *Selected Methods for Pumping Test Analysis, Report 25.* Illinois State Water Supply.

Bruington, A. E. and Seares, F. D. (1965). Operating a seawater barrier project. *Journal of the Irrigation and Drainage Division, ASCE* **91**(IR-1 Pt. 1), pp. 117–140.

BS EN ISO 14688. *Geotechnical Investigation and Testing. Identification and Classification of Soil. Part 1:2002 + A1:2013 Identification and Description, Part 2:2004 + A1:2013 Principles of Classification.* BSI.

BS EN ISO 14689. *Geotechnical Investigation and Testing. Identification and Classification of Rock. Part 1:2003 Identification and Description.* BSI.

Campbell, M. D. and Lehr, J. H. (1973). *Water Well Technology.* McGraw Hill.

Driscoll, F. G. (1986). *Groundwater and Wells.* 2nd Edn. Edward E. Johnson Inc.

EA (2013a). *Managing Water Abstraction.* EA.

EA (2013b). *Groundwater Protection: Principles and Practice (GP3).* EA.

Falkland, A. (1993). Hydrology and water management on small tropical islands. *Yokohama Symposium, Hydrology of Warm Humid Regions.* IAHS Publication No. 216, pp. 263–303.

Harpaz, Y. (1971). Artificial ground-water recharge by means of wells in Israel. *Hydraulics Division, Proc. ACE.* HY12. 1947–1964.

Headworth, H. G. (1970). The selection of root constants for the calculation of actual evaporation and infiltration for chalk catchments. *Journal of IWE* **24**, p. 431.

Howsam, P., Misstear, B. and Jones, C. (1995). *Monitoring, Maintenance and Rehabilitation of Water Supply Boreholes, Report 137.* CIRIA.

IAHS (1970). *International Survey on Existing Water Recharge Facilities.* Publication No. 87. IAHS.

ISO (2001). ISO/TR 14685. *Hydrometric Determinations – Geophysical Logging of Boreholes for Hydrogeological Purposes – Considerations and Guidelines for Making Measurements* (2001). ISO.

IWES Water Practice Manual No.5 (1986). *Chapter 9, Downhole Geophysics, Groundwater Occurrence, Development and Protection.* IWES.

Jacob, C. E. (1950). *Flow of Groundwater.* John Wiley & Sons.

Jones, H. K., Gaus, I., Williams, A. T., Shand, P. and Gale, I. N. (1999). *A Review of the Status of Research and Investigations, Report WD/99/54.* ASR-UK. BGS.

Kruesman, G. P. and de Ridder, N. A. (1990). *Analysis and Evaluation of Pumping Test Data*. 2nd Edn. Int. Inst. for Land Reclamation and Improvement, Publication 47. Wageningen.

MacDonald, A. M. and Davies, J. (2000). *A Brief Review of Groundwater for Rural Water Supply in Sub-Saharan Africa, BGS Technical Report WC/00/33*. British Geological Survey. NERC.

Maidment, D. R. (Ed.). (1992). *Handbook of Hydrology, Ch. 22, Computer models for sub-surface water*. McGraw-Hill Inc.

Marshall, J. K., Saravanapavan, A. and Spiegel, Z. (1968). Operation of a recharge borehole. *Proceedings of the ICE* **41**, pp. 447–473.

Monkhouse, R. A. (1974). The use of sand screens and filter packs for abstraction wells. *Water Services,* 78.

O'Shea, M. J., Baxter, K. M. and Charalambous, A. N. (1995). The hydrology of the Enfield-Haringey artificial recharge scheme, north London. *QJEG* **28**, 115–129, The Geological Society.

Rodda, J. C., Downing, R. A. and Law, F. M. (1976). *Systematic Hydrology*. Newnes-Butterworth, Table 3.10.

Sander, P. (1997). Water well siting in hard rock areas; identifying probable targets using a probabilistic approach. *Hydrogeology Journal* **5**(3), pp. 32–43.

Singhail, B. B. S. (1997). The hydrogeological characteristics of the Deccan Trap formation in India. *Proc. Rabat Symposium, Hard Rock Hydrosystems*. IASH Publications No. 241, pp. 75–80.

Stanners, D. and Bourdeau, P. (Eds) (1995). *Europe's Environment, Chapter 5, Inland waters*. European Environment Agency, pp. 57–108.

Sternau, R. (1967). Artificial recharge of water through wells: experience and techniques. *P Haifa Symposium, Artificial Recharge and Management of Aquifers*. IAHS.

Summers, I. (1987). *Estimation of Natural Groundwater Recharge*. Reidel Publishing Co., p. 449.

Terzaghi, K. (1943). *Theoretical Soil Mechanics*. John Wiley & Sons.

Theis, C. V. (1935). The relation between lowering the piezometric surface and the rate and duration of discharge of a well using groundwater storage. *Transactions of the American Geophysical Union* **16**, pp. 519–614.

UNESCO-WWAP (2009). *Water in a Changing World. The United Nations World Water Development Report 3*. UNESCO, Paris.

Vrba, J. and van der Gun, J. (2004). *The World's Groundwater Resources, Contribution to Chapter 4 of WWDR-2 Draft*. International Groundwater Resources Assessment Centre Report IP 2004-1.

Walton, W. C. (1970). *Groundwater Resource Evaluation*. McGraw-Hill.

WaterAid (2013). *WaterAid Technical Brief Handpumps*. WaterAid, <www.wateraid.org/technologies>.

WB (1987a). *Rural Water Supply Handpumps Project. Handpumps Testing and Development: Progress Report on Field and Laboratory Testing*. The World Bank.

WB (1987b). *Community Water Supply. The Handpump Option*. The World Bank.

Wheatcroft, S. W. and Buddemeier, R. W. (1981). Atoll hydrology. *Ground Water* **19**(3), pp. 311–320.

Williams, A. T., Barker, J. A. and Griffiths, K. J. (2001). *Assessment of the Environmental Impacts of ASR Schemes. Report CR/01/153*. BGS. NERC.

Wright, A. and du Toit, I. (1996). Artificial recharge of urban wastewater; the primary component in the development of an industrial town on the arid west coast of South Africa. *Hydrogeology Journal* **4**(1), pp. 118–129.

Wright, E. P. and Burgess, W. G. (1992). *The Hydrogeology of Crystalline Basement Aquifers in Africa*. Special Publication 66. Geological Society.

CHAPTER

Dams and Reservoirs

5

5.1 INTRODUCTION

Reservoirs are provided to regulate raw water resources and may be required to store rainfall in wet periods for use in dry periods. Such storage typically requires an impounding dam across a natural river valley. Service reservoirs provide a similar function but related to treated water (Chapter 20). A dam across a natural river valley is sometimes called an impounding dam, as it impounds a natural watercourse.

The earliest dam Smith was able to report in his book *A History of Dams* (Smith, 1971) was the 11 m high Sadd el-Kafara dam, built between 2950 and 2750 BC, the remains of which lie 20 miles south of Cairo. It had upstream and downstream walls of rubble masonry each 24 m thick at the base, with a 36 m wide, gravel-filled space between; it appeared to have had a short life because it suffered from the two principal defects that continued to plague many dams for the next 4500 years – it leaked and was probably overtopped. However, the nearby Ma'la dam, located in Giza, operated successfully for over 3000 years. The remains of the dam can still be seen today. The dam was constructed using a greater volume of material than that contained in the Great Pyramid at Giza; the irrigation water storage that it provided was one of the keys to Egypt's prosperity.

There are many materials of which a dam can be made – earth, concrete, masonry or rockfill. The choice depends on the geology of the dam site and what construction materials are nearest to hand. Concrete and masonry dams require hard rock foundations; rockfill dams are normally built on rock, but have been built on alluvial deposits; earth dams can be built on rock and also on softer, weaker formations such as firm clays or shales.

Masonry dams are still built in developing countries where labour costs are low. Where labour costs are high, masonry has generally been replaced by mass concrete compacted by using immersion vibrators or, in recent years, by roller compaction. It was not until the middle of the 19th century that concrete and masonry dams began to be designed according to mathematical analysis of the internal and external forces coming upon them; nor until the first quarter of the 20th century was the behaviour of earth dams sufficiently understood for mathematical design procedures to be applied to them also. Recent advances in dam design are such that only the salient principles involved can be given here.

Twort's Water Supply. DOI: http://dx.doi.org/10.1016/B978-0-08-100025-0.00005-3

5.2 ESSENTIAL RESERVOIR CONDITIONS

For a successful reservoir the following conditions need to be fulfilled:

1. The valley sides of the proposed reservoir must be adequately watertight to the intended top water level of the reservoir and must be stable at that level.
2. Both the dam and its foundations must be sufficiently watertight to prevent dangerous or uneconomic leakage passing through or under the dam.
3. The dam and its foundations must withstand all forces coming on them.
4. The dam and all its appurtenant works must be constructed of durable materials.
5. Provision must be made to pass all floodwaters safely past the dam.
6. Provision must be made to draw water from the reservoir for supply and compensation purposes and for lowering the reservoir water level in emergencies.

To fulfil these conditions, any proposed reservoir site must be subjected to detailed geological investigation to ensure that there are no fault zones, hidden valleys or permeable strata through which unacceptable leakage would take place and that the dam foundations are adequate (ICOLD, 2005). In addition, the nature and stability of the hillsides surrounding the reservoir must be investigated (ICOLD, 2002). At high reservoir levels areas of hillside will become saturated. Any rapid lowering of the reservoir may leave such areas with high internal pore pressures, higher (saturated) unit weights and reduced cohesion. Should this reduce the stability sufficient to cause a large volume to slide into the reservoir surge waves would result, threatening the dam and also any people on the lake shore.

5.3 WATERTIGHTNESS

All dams leak to some extent; but the risk of that leakage carrying with it material from the dam or its foundations must be avoided. Hence, as much as possible of all such leakage should be collected by an underdrainage system and delivered to a collecting basin where it can be continuously measured to reveal any signs of increase and inspected to see it is not carrying material with it.

To reduce seepage beneath the foundations of a dam it is usually necessary to construct some form of 'cut-off' (ICOLD, 2012b). Types of work to reduce seepage under (and around) dams include the following, which may be combined in any way:

- the construction of a cut-off across the valley below the dam;
- grouting the foundations beneath the dam to reduce their permeability;
- provision of an upstream blanket of low permeability.

A ***concrete cut-off***, typical of those used prior to about 1960, is illustrated in Figure 5.1. It is taken down sufficiently far to connect into sound rock or clay at the base and is usually filled with concrete, the trench being about 2 m wide. Many earlier dams had the cut-off backfilled with puddle-clay.

Where a deep cut-off is necessary, it is now more common to use a plastic concrete diaphragm wall. The trench is excavated under bentonite slurry using a grab in soils or a 'hydrophraise' machine in rock. After excavation, the slurry is replaced by a plastic concrete mix, the slurry being

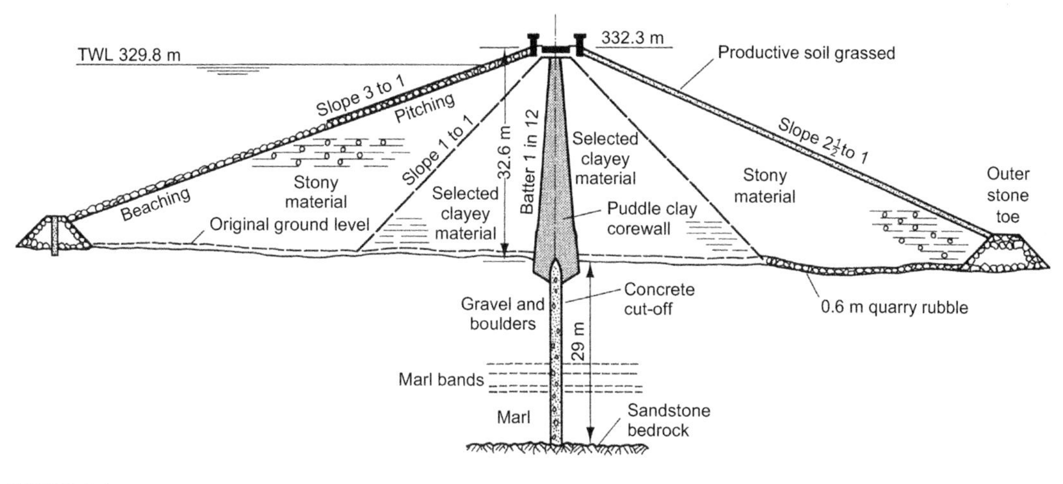

FIGURE 5.1

Deep concrete cut-off at Taf Fechan dam, Wales (Engineer: Binnie & Partners).

de-sanded and recirculated for further use. Alternatively, particularly for temporary works, such as cofferdams, a self-setting slurry can be used for both trench support and provision of the cut-off.

Whatever form of deep cut-off is used, the junction of the top of the cut-off with the core of the dam requires the most careful design, to ensure that the differential stiffness does not allow settlement to open up a leakage path around the top of the cut-off, or act as a propagator for cracks.

A wide, shallow cut-off is shown in Figure 5.2; this is used where a sound foundation material exists not far below ground surface. The corewall material – in this case 'rolled clay', which is a mixture of clay and coarser materials – is taken right down to the bottom of the cut-off.

Some open excavation cut-offs have been extraordinarily deep and therefore difficult and expensive to construct. The classic case is the cut-off for the Silent Valley dam for Belfast (McIldowie, 1934–35), which reached a maximum depth of 84 m before sound rock was met. Diaphragm wall cut-offs have been taken down to depths of more than 120 m, e.g. Mud Mountain dam in the USA.

As a 'rule of thumb', cut-offs are generally taken down to between 50% and 100% of the depth of retained water above. Two thirds the impounded water depth is an often used criterion, but may need to be adjusted locally for particular geology. It is also important to extend the cut-off wall into the abutments of the dam, to reduce seepage around the ends of the dam.

Other forms of cut-off have been used in poor ground conditions; for example, Linggiu dam in Malaysia has two rows of small-diameter secant piles through the permeable alluvial deposits, with jet grouting of the soils between them.

Diaphragm walls have been used in several cases for the repair of leaking dams. Diaphragm walls were constructed within the puddle-clay corewalls of Lluest Wen dam in South Wales and Balderhead dam in Teesdale, UK.

Grouting is normally adopted to improve the watertightness of a rock which is already basically sound, reasonably impermeable, and not liable to decompose, even with some leakage through it. In practice, it is difficult and generally uneconomic to grout rock where the packer permeability

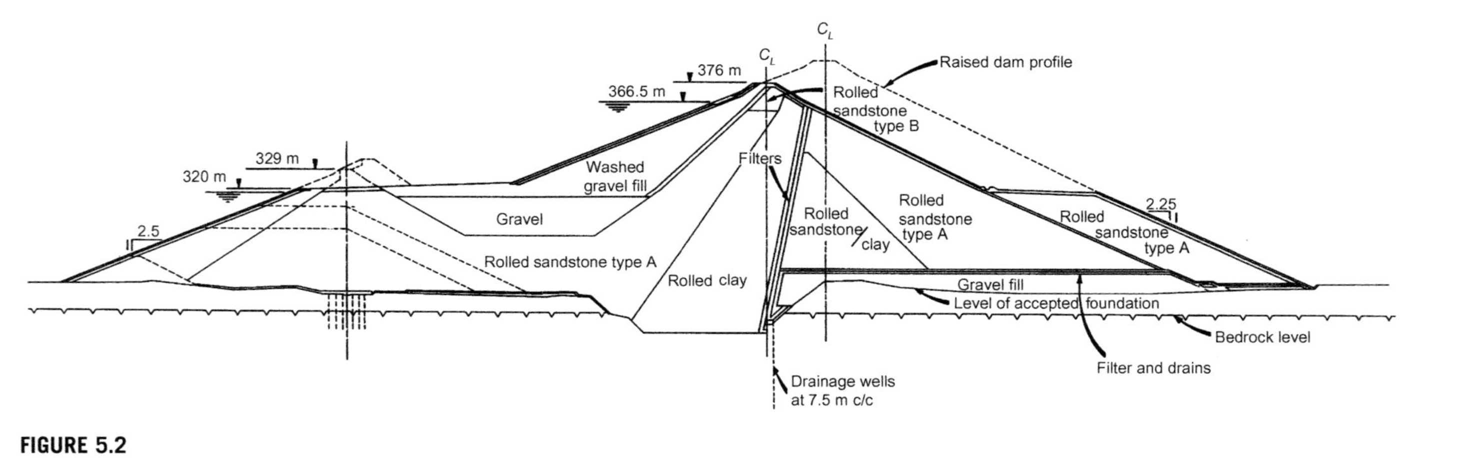

FIGURE 5.2

Shallow, wide clay cut-off at Mangla dam, Pakistan (Engineer: Binnie & Partners).

test results show a value of less than 5 Lugeon units (about 5×10^{-7} m/s). In exceptional cases it may be economically justified to attempt permeation grouting of materials other than fissured rock, such as sands, gravel, silts, clay and mixtures of these materials. These are often difficult to grout, despite use of a variety of methods; complete success is not ensured and additional precautions may be needed in the design of the dam to allow for unavoidable leakage (Ischy, 1962; Geddes, 1972). Guidance on grouting is given in the CIRIA guide (Rawlings, 2000).

After completing foundation grouting it is usual to ensure that the contact zone between the dam material and the excavated ground surface is sealed by carrying out 'contact' or 'blanket' grouting, through a line of grout pipes with packers set above and below foundation level. After sufficient height of the dam has been constructed to provide weight, grout is injected into the pipes between the packers.

Grouting can be highly successful, but tales are legion concerning the damage that can be done if the procedure is not carefully controlled. Risks of causing heave and un-wanted 'hydrofracture' are great.

A fully impervious cut-off is often not possible, but water pressures beneath the dam need to be controlled for stability and to prevent erosion. This can often be achieved using a 'partial' cut-off, together with pressure-relief wells downstream. Partial cut-offs do not significantly reduce flow volumes; many dams have been built with both a full cut-off and pressure-relief wells.

Upstream blankets have been used to lengthen the seepage path, under earth dams in particular, if the foundation is suitable. This is constructed of low permeability material and is usually used in conjunction with a cut-off beneath the dam. An example is Tarbela Dam, Pakistan. Considerable care is needed with blanket thickness and use of graded material, in layers if necessary, since the hydraulic gradients are high.

5.4 STRENGTH AND DURABILITY OF A DAM

Every dam must be secure against failure by sliding along its base; this applies whether the dam is of concrete, masonry, rock or earth. Although arch dams rely less on base friction for their stability, it is still a contributing factor. Resistance to failure against shearing along any surface is another criterion to be applied. Sliding is resisted by the weight of the dam acting in combination with friction and cohesion along its base contact with the foundations. Usually the resistance of rockfill and earthfill embankments dams against sliding or direct shear is ample, on account of their large widths. Stability checks on such dams tend to focus instead on internal failure mechanisms, such as the development of slip circle instabilities or erosion and piping failures (Section 5.9).

Driving forces are static (hydraulic pressure and gravity) and dynamic (seismic inertia and hydrodynamic) – see Section 5.22. Sliding is a more crucial failure mechanism to examine when designing a masonry or concrete dam; these, being constructed to much steeper slopes than earth dams, should also be checked against overturning. Where concrete dams are arched, the abutments must be strong enough to take the end thrusts of the arch, and the foundations must also be strong, as they are more highly stressed than would be the case for an equivalent height concrete gravity dam.

Durability is of the utmost importance since dams are one of the few structures built to last almost indefinitely. Their design life is sometimes described as 'monumental'. The durability of an earth dam is primarily dependent upon the continuance of its ability to cope with drainage requirements. It has to contend with seepage, rainfall percolation and changing reservoir levels. The movement of these waters must not dislodge or take away the materials of the dam; otherwise its condition and stability will deteriorate at a progressively increasing rate. It must perform this function despite long-term settlement of itself and its foundations and may be required to withstand seismic movements.

5.5 TYPES OF DAM

The most economic forms of dam construction are those that make maximum use of natural materials available nearby. In the lower reaches of a river there may be an abundance of soils such as clay, sand and gravels. Use of such materials would usually be economically and environmentally preferred and the resulting dam would be compatible with the foundations, which are likely to be of similar materials. The upper reaches of a river may be in much steeper and more rocky terrain. There may even be a dearth of soils. In such a terrain a rockfill or concrete dam is likely to be more appropriate.

Dams constructed of earthfill or rockfill are termed embankment dams. The simplest example is a homogeneous embankment made entirely of the same material; this would need to be sufficiently impermeable to retain the reservoir without excessive leakage, while being sufficiently strong so as to be self-supporting. More advanced designs might feature a central core of clay or silt as the impervious element, with supporting shells, or shoulders, of stronger soils or even rockfill. In between the clay and shell materials, graded sand and gravel filters are provided as a transition between the different materials. Clay-core earth embankments and clay-core rockfill embankments are amongst the most common forms of dam construction.

Where clay is not economically available, the central waterproofing element may be of another material, such as asphalt. A number of rockfill dams with a central asphalt core have been constructed in Norway. In the case of rockfill embankments, it is also possible to make the dam watertight by providing a suitable upstream facing, instead of an impermeable central core. Such upstream facings are usually either of asphalt or reinforced concrete. The latter have grown in popularity in recent years.

Where the foundations are good and significant flood passage structures are required, it may be economical to form the whole dam of concrete. The various forms of concrete dam are generally named after their geometry or function. A simple concrete dam, approximately triangular in cross section and relying on its own dead weight for stability, is termed a concrete gravity dam. Similar stability can result by providing an upstream wall, generally sloped in order to attract a vertical component of the water load, supported on the downstream side by a series of propping walls or buttresses. This arrangement is termed a buttress dam. Concrete arch dams take the form of a shell spanning across a narrow valley, such that the load is conveyed to the valley sides as well as to the valley base.

There are many variants to those described above. The adaptation of local materials to provide a specific function remains an area where dam engineers can use their experience and talents to be imaginative and creative. In some cases, it might be advantageous to combine two different types of dam in one location. The main types of dam and some considerations in their design and construction are discussed in the following sections.

EMBANKMENT DAMS

5.6 TYPES OF DESIGN

To avoid a misunderstanding, it is necessary to mention that the term 'earth' does not mean that tillable soil can be used in an earth dam. All such 'soil' must be excluded, because it contains vegetable matter which is weak and, in its decomposition, could leave passages for water percolation. The first operation when undertaking earth dam construction is therefore to strip off all surface soil containing vegetable matter. The dam can be constructed, according to the decision of the designers, of any material or combination of materials such as clay, silt, sand, gravel, cobbles and rock.

Early designs of earth dams usually had a central core of impermeable 'puddle clay', supported on either side by one or two zones of less watertight but stronger material (Fig. 5.1). The central puddle-clay core is impermeable but its structural strength is low. This clay core must join to a cut-off in the ground below. The earliest dams had clay-filled cut-off trenches but use was soon made of concrete to fill the cut-off trench, although the corewall continued to be made of clay. The inner zones of the shoulders of the dam contained a mixture of clay and stones, so was impermeable to some extent, but the stony material added to its strength. The outer zones would contain less clay again and more stones, perhaps boulders and gravel. The main purpose of the shoulders was to hold the inner core of clay and add strength to the dam. Boulder clay (a glacier-deposited mixture of clay, silt and stones with boulders) was a favourite material for this outer zone in the UK, because it had the right characteristics and there were extensive deposits of it in those areas where impounding schemes developed.

Many such dams have been and continue to be successful. Their outer slopes were decided by accumulated experience and were often 3:1 (horizontal:vertical) upstream and 2:1 or 2.5:1 downstream, with flatter slopes for the higher dams. There were few measurement tests that could be applied. Although many new kinds of tests are now available, earth dam construction still involves judgment from experience and constant supervision as the work proceeds to ensure sound construction. Only a minute proportion of the variable earth materials in a dam can be tested, so construction must always come under continuous detailed supervision by an experienced engineering team. To assist with fill placement and compaction quality control, trials are usually carried out to develop procedures giving the desired results with the compacting plant proposed. Provided the established procedures are strictly followed, satisfactory construction can be achieved with less field and laboratory testing.

In modern dams, a puddle-clay core is no longer used. Instead the rolled-clay core is made of a mixture of compacted clay and coarser materials. The core usually has a base width of 25–50% of the height of the dam, to limit the hydraulic gradient across it and reduce the risk of leakage at foundation level. A large dam of this sort is shown in Figure 5.2.

Sometimes, a dam may consist wholly of rolled clay with only an outer protective zone of coarser material on the water face for protection against wave action. However, dams with the same core and shoulder materials normally include internal drainage to improve stability and reduce the risk of erosion.

Although clays of moderate to high plasticity are preferred for core material, dams have been constructed successfully using clays of very low plasticity and silt for the core. In the latter case, the filters protecting the core against internal erosion are of great importance. An alternative

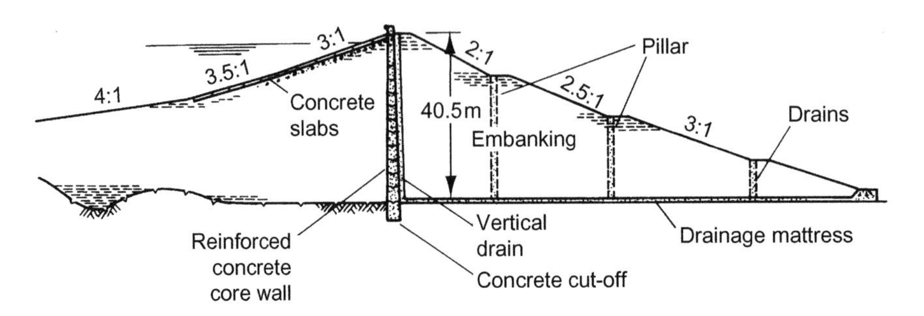

FIGURE 5.3

Reinforced concrete corewall in Gyobyu dam, Rangoon, Myanmar (Engineer: Binnie & Partners).

used is the concrete corewall; Figure 5.3 shows one of the earliest examples. Generally a concrete corewall is limited to use with stiffer shoulder materials such as sand, gravel or rockfill and to good foundations.

The principal aim is to use as much locally available material as possible, because this helps minimize cost and, often, environmental impact. The challenge for the designer is how to incorporate the available materials in the dam in a manner which uses their different characteristics efficiently and safely.

5.7 PORE PRESSURE AND INSTRUMENTATION IN EARTH DAMS

An earth dam is built in layers of material, 150–450 mm thick, which are compacted with vibrating or heavy machinery. The material has to have a moisture content close to its 'optimum' value to achieve maximum in-situ density. Soils that need to be flexible, such as the core material are usually compacted slightly wet of the optimum. If the soil is too wet, the compaction is ineffective and the plant cannot operate. Soil used in the shoulders is often stiffer and, to achieve this, the placement water content is at or just below the optimum value. At the other end of the range, if the soil is too dry, effective compaction cannot be achieved. Control testing of the fill materials, both at the borrow pit and after placement is necessary to ensure compliance.

The water within the soil ('porewater') is affected by the construction process. When the soil is compressed by the compaction process and the added load, water pressure in the pores increases. The air in the soil is either pushed out or goes into solution, increasing the degree of saturation. As the effective strength of the soil is reduced by high porewater pressure, the dam can become unstable. Clays of low to medium plasticity such as Glacial Till (Boulder Clay) tend to compress when loaded, and high construction pore pressures can often occur. Heavily over-consolidated clay, such as the Lias Clay or Oxford Clay dilate when excavated and at low stresses (less than 10–15 m of fill) the compaction porewater pressures are often low or even negative (suction).

The maximum allowable construction porewater pressures to ensure an acceptable factor of safety against failure can be calculated. During construction, these porewater pressures can be monitored using piezometers. Some examples are shown in Figure 5.4. The Casagrande type is

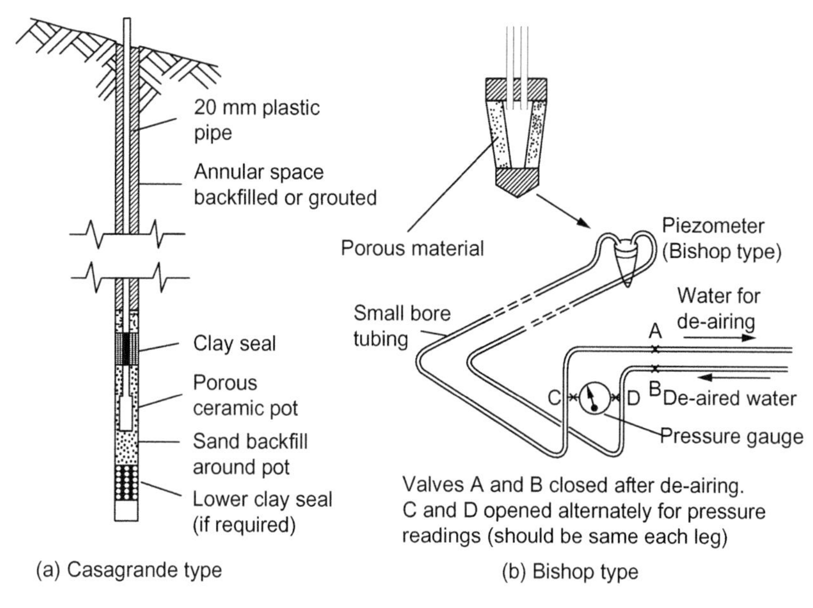

FIGURE 5.4

Porewater pressure sensing devices.

suitable for dam foundations and the Bishop type suitable for installation in the embankment fill. Other types of piezometers using different methods of measuring the water pressure at the tip are available. These include vibrating wire and pneumatic systems. Details of such instruments are given in Dunnicliff (1993). In unsaturated fill materials only piezometers that can be de-aired, such as the Bishop type, are likely to give reliable long-term readings. The piezometer readings can be used to control the rate of construction so allowing excess pore pressure to dissipate.

During first impounding, piezometers placed in the upstream shoulder of an earth dam can register how the water level in the shoulder responds to the reservoir water level; piezometers in the foundation and downstream shoulder can check the performance of the core and cut-off. During operation piezometers can show how fast the water drains out of the upstream shoulder with a fall in reservoir water level. This is important in order to avoid a slip failure due to ‘rapid drawdown’. Piezometers at low level just downstream of the corewall can show pressures which indicate leakage through the corewall, especially so if their pressure rises and falls with reservoir water level. It is thus possible to monitor the behaviour of an earth dam through the use of piezometers, both during and immediately post-construction and long-term operation. In many soils, the end-of-construction porewater pressures are a worst case, but in some – particularly the high plasticity clays – the dam stability deteriorates as the pore pressures come to equilibrium. This process may take tens of years.

Settlement gauges, extensometers and inclinometers can measure the deformation of a dam both during and after construction. Surface markers permit movement and settlement to be detected by precise surveying, which must be based on benchmarks established outside the limits of any land settlement caused by the presence of the dam and by the weight of water in the reservoir.

5.8 STABILITY ANALYSIS IN DAM DESIGN

The stability of the earth dam against slope failure is an essential element of the design. There are many limit equilibrium slope stability analysis methods and a significant number of commercial slope stability computer packages are available. For the majority of embankment dams, a circular slip analysis based on Bishop's rigorous method (Bishop, 1955) is satisfactory. However, the geology of the site and the shape of the dam may indicate that a non-circular slip surface may be more critical. Simple non-circular methods include two-part and three-part wedge analyses; more sophisticated methods include Morgenstern and Price's method (Morgenstern, 1965).

Computer packages may offer a number of different analyses methods but some methods are not reliable for embankment dam design, particularly where there are a variety of materials in the dam and its foundation. The Ordinary or Fellenius method, which ignores inter-slice forces, should not be used and methods which do not fully define the inter-slice forces such as Janbu's method or Bishop's simplified method should only be used with caution.

The dam needs to be checked for a number of different modes of slope failure. The most important of these situations modes are:

- the end of construction (prior to impounding for the upstream slope and with the reservoir impounded for the downstream slope);
- long term, with steady-state seepage conditions established within the dam;
- after the rapid drawdown of the reservoir, which can destabilize the upstream slope; and
- in most countries including Britain, seismic conditions.

In practice, the required factor against failure (factor of safety) used in design is related to the risks and consequences of a failure, so differs for the various modes. Natural materials such as soils are inherently variable and, even with good quality ground investigation and strength testing, there will always be some uncertainty over the choice of suitable design parameters. Although a moderately conservative set of design parameters should be chosen from the test results, the factor of safety still needs to cover these uncertainties.

Figure 5.5 shows the output from a rapid drawdown analysis for Mangla Dam. This figure shows the reservoir not fully drawn down. There is normally a critical pool level for which the minimum factor of safety is lower than that for complete drawdown.

Finite element analyses of embankment dams can be very helpful, particularly for those embankments that have wide intermediate berms. The strains required to mobilize the resistance in some parts of a dam may be sufficient to cause failure elsewhere. Elasto-plastic or strain softening soil models are necessary to obtain realistic results. Dynamic finite element analyses are also used to assess changes in porewater pressures and deformations under seismic conditions.

Failures can occur on pre-existing sheared surfaces. These are usually in the foundation soils and are a result of tectonic shearing or periglacial solifluction. An example is the failure of Carsington Dam during construction, where some sheared foundation soil was not removed (Fig. 5.6). There are a few cases of failure within the embankment where shear surfaces were formed by previous movement and not removed. Figure 5.7 shows a section of the failure at Tittesworth Dam.

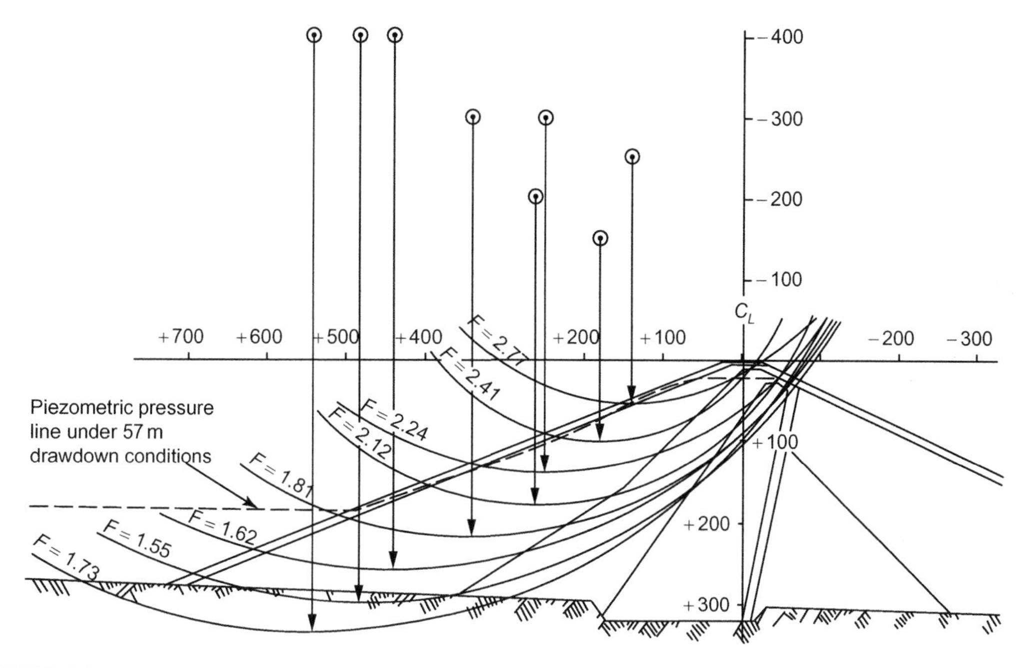

FIGURE 5.5

Slip circle stability analysis results for upstream slope for Mangla dam for 57 m drawdown and static conditions. Lowest factor of safety is 1.55.

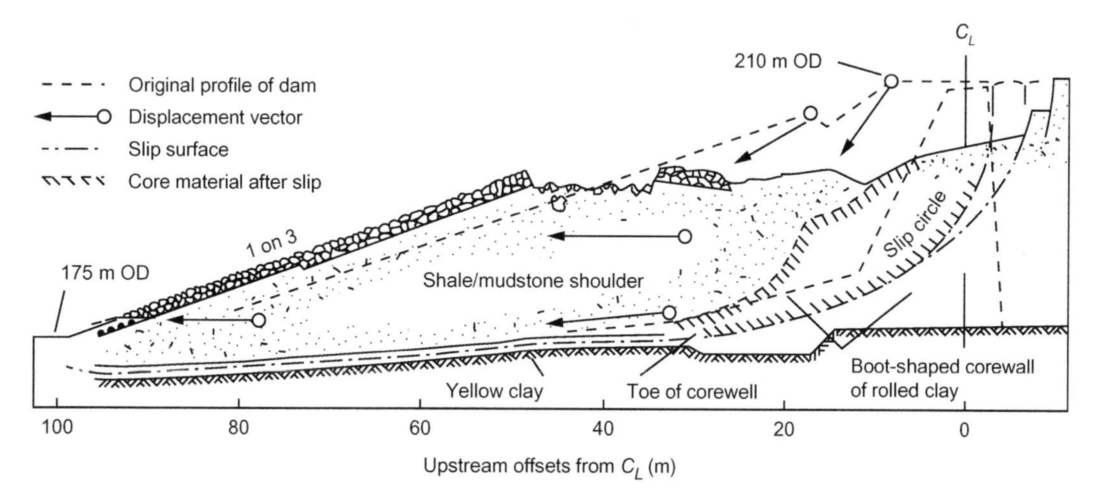

FIGURE 5.6

Failure of Carsington dam 1984, UK. Overstressing of material near toe of core, as construction was nearing completion, caused progressive slip circle failure.

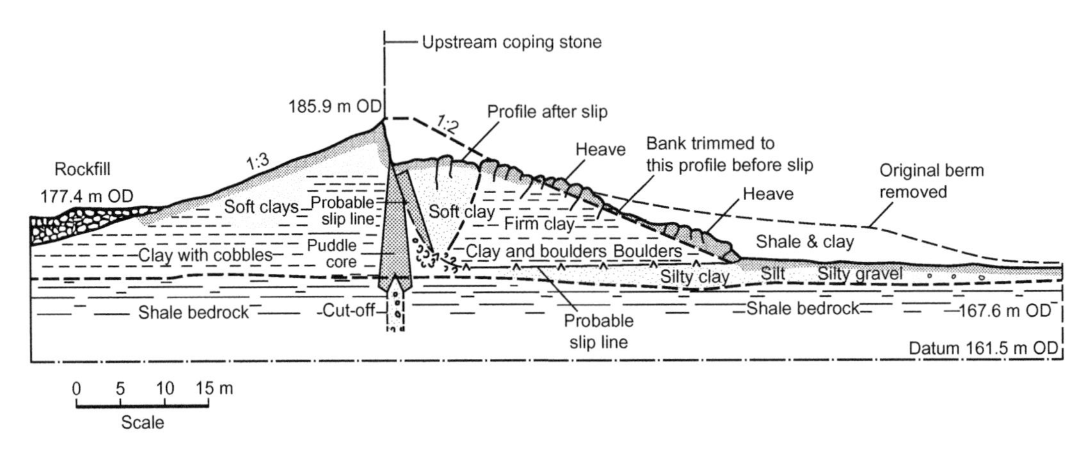

FIGURE 5.7

Failure of Tittesworth dam 1960, UK. Removal of part of small berm to downstream shoulder caused wedge-type failure through puddle-clay core. Later investigations showed that the embankment had failed in the same location but had not been properly rebuilt.

5.9 DRAINAGE REQUIREMENTS FOR AN EARTH DAM

Much emphasis has been put on the slope stability aspects of dam design but dam failure due to inadequate assessment of slope instability is extremely rare. Failures are more likely to be due to poorly designed drainage or to poor detailing of the contact between the dam and its foundation.

The preceding sections illustrate that an earth dam is subject to continuous water movements within it. Water levels in the upstream shoulder rise and fall with reservoir level; the downstream shoulder has a moisture content and often a positive porewater pressure which rise and fall with rainfall, and it may have to accept some inflow from seepage through the core and from springs discharging to its base and abutments. The principal requirement is that all these water movements must not dislodge or carry away any material of the fill, especially of the core.

Drains are therefore necessary within a dam to accommodate the water movements that occur. However, internal erosion must be guarded against since it is a factor in a significant number of embankment failures (ICOLD, 2014). A drainage layer of coarse material, such as gravel, cannot be placed directly against fine material such as clay, or water movement would carry clay particles into and through the drainage layer. The removal of the clay would progressively enlarge leakage paths until there is a serious danger of disruption of the fill and breakthrough of water from the reservoir. Every drainage layer of coarse material has therefore to be protected by a 'filter'. The filter material must have a particle size small enough to prevent the fine material moving into it, and coarse enough to prevent its own movement into the larger material of the drain. Sometimes two filters have to be placed in succession when there is a large difference in size between the fine material to be drained and the coarse material of the drainage layer. All such drainage layers must convey their water individually to inspection traps where the flows should be regularly

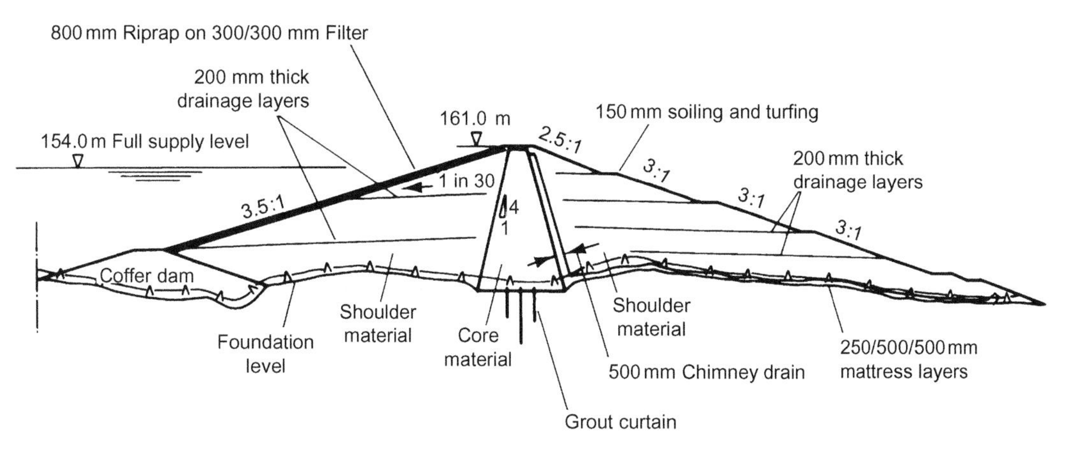

FIGURE 5.8

Chimney and blanket drains in 58 m high Upper Muar dam, Malaysia, constructed of residual clay soils on weathered rock foundations (Engineer: Binnie & Partners, in association with SMHB).

measured and inspected to make sure they are not carrying suspended material from the dam. Any flow of cloudy drainage water from a dam should be tested for the amount and size of suspended solids it carries.

The literature includes a number of sets of filter rules to design the grading of filter materials to protect soils from internal erosion. The work by Sherard and Dunnigan (Sherard, 1989) gives simple and effective rules for most soil types. Special care needs to be exercised for gap-graded and widely graded soils that are not internally stable.

An example of a drainage layout is shown in Figure 5.8. The main drainage system comprises a 'chimney' drain, against the downstream slope of the core, to transfer seepage flows to a 'drainage mattress' laid at formation level, discharging to an outlet chamber at the downstream toe of the dam. Drainage 'blankets' are laid at intervals in the upstream shoulder to relieve pore pressures that build up during construction and for drainage of the upstream shoulder during rapid drawdown. In some embankment dams drainage galleries have been used, but they are principally included for interception of pressure-relief wells in the dam foundations.

5.10 SURFACE PROTECTION OF EARTH DAMS

The upstream face of an earth dam must be protected against erosion by water and wave action. Common forms of protection are:

- stone pitching;
- riprap;
- concrete blocks (various shapes);
- concrete slabs; and
- asphaltic concrete.

Stone pitching, either grouted or open-jointed, comprises stones of fairly uniform thickness and size, generally about cuboid, which are laid by hand with relatively close jointing. Pitching is generally found on older dams. Riprap is formed by the random packing of pieces of angular rock with a range of sizes and shapes, which are usually placed by machine; although it is generally advantageous to hand-place small pieces of rock between the large pieces to assist in locking them together. Beneath the rock must lie layers of material which act as filters between the large rock and the finer material of the dam that must be prevented from being drawn up through the rock surface by wave action.

Where large waves are anticipated and large rock is unavailable, concrete blocks can be used. These are shaped to dissipate the wave energy and a number of proprietary designs are available, such as tetrapods and dolosse. When concrete slabs are used for protection they need to be placed without open joints to avoid dislodgement by wave action, but drainage of the fill through them should be possible. They should be laid on one- or two-graded gravel filter layers.

For many years slope protection of dams has been designed empirically, with satisfactory results in most cases (Johnston, 1999). A conventional rule used by many engineers required the depth of riprap or concrete blockwork on dams in the UK to be not less than one sixth of the 'significant' wave height in the design event. Experience at a number of reservoirs suggests that this is not always adequate, although a factor in some cases has apparently been the adoption of an insufficiently rare windstorm event for the design of the wave protection works.

The key current reference on the stability of riprap under wave attack (and on the appropriate detailing and under layers) is CIRIA *Rock Manual* (CIRIA, 2007). For blockwork and slabbing, a design procedure was recommended by Yarde et al. (1996), based on physical model studies undertaken at HR Wallingford. Adoption of this procedure, however, may be hampered by limitations in the ranges of experimental parameters used to derive the equations. An experience-based approach using data provided by Herbert et al. (1995) has been proposed (Besley, 1999). Further information on the subject of wave protection at UK embankment dams is given by Johnston et al. (1999).

Erosion of the downstream face of a dam by rainfall–run-off must be prevented. It is usual to achieve this by turfing the surface of the dam or by soiling and seeding it with a short-bladed strongly rooting grass. If the dam is more than about 15 m high then, according to the type of rainfall climate, berms should be constructed on the downstream slope, each berm having a collector drain along it so that surface run-off is collected from the area above and is not discharged in large amounts to cause erosion of the slope below. Cobbles or gravel are sometimes used to surface the downstream slope, particularly in arid zones where rainfall is insufficient to maintain a good grass growth throughout the year. Surface drains on the downstream slope should discharge separately from the internal drains to the shoulder.

It is unwise to allow any tree growth on, or at the foot of, the downstream slope of an earth dam. Some tree roots can penetrate deeply in their search for water, and drainage layers could be penetrated by such roots. The presence of trees may obscure proper observation of the condition of the slope. A damp patch, or development of a depression on the slope, may indicate internal drainage systems not working properly or some leakage through the core. Any signs of this sort need to be investigated. It is best to treat the downstream slope uniformly, with the same type of covering (gravel or grass) throughout, e.g. Jari Dam, Mangla (Plate 3(a)), since this helps to disclose any discrepancies that might need investigation.

5.11 ROCKFILL AND COMPOSITE DAMS

Rockfill dams are appropriate for construction at locations where suitable rock can be quarried at or near the dam site, and where the foundations will not be subject to material settlement due to loading or to erosion from any seepage through or under the dam. The design must, of necessity, incorporate a watertight membrane, which is generally sited either centrally in the dam or in the form of an upstream facing.

Early concrete-faced dams of dumped rockfill featured embankment slopes of about 1 (vertical) on 1.5 (horizontal) and were satisfactory in service with heights up to about 75 m. Above that height face cracks and excessive leakage tended to occur due to the compressibility of the dumped rockfill. This problem was alleviated somewhat when it was discovered that sluicing rockfill with water improved compaction, but the modern rockfill dam really dates from the 1960s, when vibrating rollers became standard equipment for compaction (Plate 3(b)).

A variety of rocks can be used with modern methods of compaction; even relatively weak rocks such as sandstones, siltstones, schists and argillites have been used. It is usual to examine all possible sources of locally available material, to carry out laboratory tests on samples and pilot construction fills and to base construction procedures and zoning of materials on the results of such tests. The specification for construction should set out the required layer thickness, normally 1–2 m for sound rock and 0.6–1.2 m for weak rock, and the number of passes required by a 10 tonne vibratory roller to achieve adequate rockfill breakdown and strength (Cooke, 1990).

Centrally located waterproofing can typically comprise a core of clay, silt, asphalt or concrete. Central cores of clay or silt are thicker than asphalt or concrete cores. An earth core would be protected by zones of transitional material between the clay and the rockfill (Section 5.9). Two or more layers of transitional material are necessary, grading through from clayey sand against the corewall, to a crushed rock with fines against the rockfill.

Where upstream facings are provided, these are generally of either reinforced concrete or asphalt. In both cases the final facing is screeded up the slope over one or more layers of fine bedding material, compacted both horizontally and up the slope. Concrete facings are discussed in International Commission for Large Dams (ICOLD) Bulletin 141 (ICOLD, 2011b). However, a more recent development has been to place the rockfill against upstream kerbs of roller-compacted concrete (RCC), as construction proceeds. This produces a relatively impermeable face, which is a useful safeguard against construction floods. It also provides a good base on which to cast the subsequent waterproofing membrane. Watertightness of a concrete face is achieved by sealing vertical joints with waterbars. Horizontal joints are not required, construction joints being formed with reinforcing steel passing through them.

In the case of an upstream facing, the associated cut-off trench (or plinth) must necessarily follow the curve of the upstream toe of the dam in order to connect with it. Such upstream membranes need to be sufficiently flexible to be able to accommodate long-term settlement of the rockfill without unacceptable leakage. Concrete-faced rockfill dams are now among some of the highest dams in the world. Their essential simplicity and construction economy has made them very attractive where good rockfill is plentiful. However, problems with a few such dams occurred in 2006. These were high dams, greater than 140 m, in steeply sided valleys, where vertical expansion joints in the facing had been omitted and the rockfill was basalt. In these cases, settlement deformations caused central shearing, or crushing, failures of the concrete facings over almost their entire heights.

The leakages produced were easily accommodated by the internal rockfill and so the dams were not in structural danger; however, the matter illustrates the need for care when design concepts are taken beyond the range of previous experience.

The rockfill body of an upstream-faced rockfill dam is inherently stable. The water load is applied more vertically downwards than horizontally, the rockfill is totally drained and sliding resistance is over the whole of the base. As a result no concrete-faced rockfill dams have experienced structural failure and stability analyses on assumed failure planes are not generally done. Instead the upstream and downstream slopes of the dam are decided based on the quality of the rockfill and on precedent. Slip circle analyses are, however, necessary when an earth core is incorporated, to determine the extent to which the core strength and pore pressures affect the dam's stability. In the case of a central-core dam the water load is applied in a mainly horizontal direction and only the downstream shoulder resists sliding forces.

An asphalt or bituminous concrete facing can also be used. It has an inherent flexibility and may be able to absorb larger deformations than a reinforced concrete facing without cracking. However, it can still be vulnerable to differential settlement at its connection with the plinth. Cold weather affects the flexibility of bituminous materials, hot weather causes creep, frequent variations in temperature can cause fatigue; and ultraviolet radiation can damage a coating that is not submerged. Hence the standard practice is to provide a protection layer and a membrane sealer. Modern practice is to use a single 100–200 mm thick impervious bituminous concrete layer spread by a vibratory screed and then rolled by a winched vibratory roller.

A centrally installed bituminous diaphragm wall avoids problems with extreme temperatures and ultraviolet radiation, is less vulnerable to damage, and permits the cut-off to be constructed along the centreline of the dam. An example of an internal asphaltic core to a rockfill dam is shown in Figure 5.9. At 107 m high, this is one of two dams at High Island, Hong Kong (Vail, 1975), which are, unusually, founded below the sea bed across a sea inlet. The core consists of 1.2 m thick

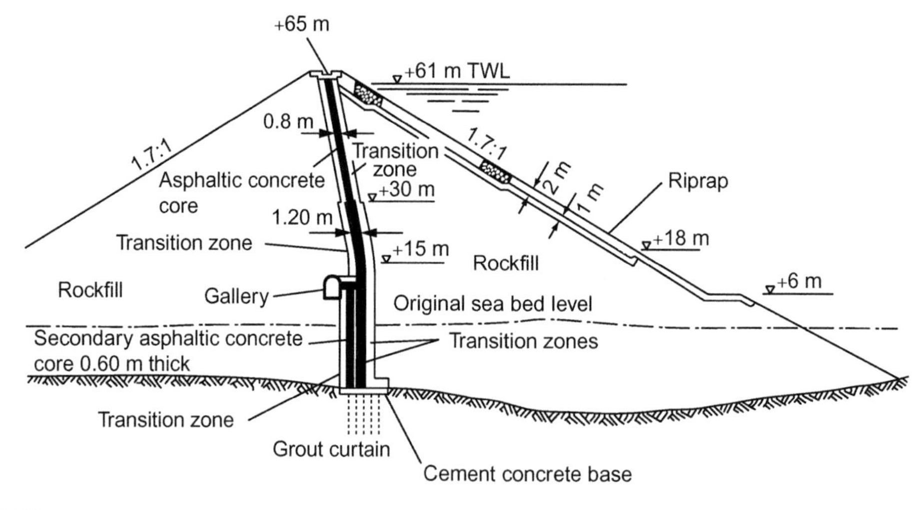

FIGURE 5.9

Rockfill shoulders with asphaltic concrete core in High Island West dam, Hong Kong, China (Engineer: Binnie & Partners).

asphaltic concrete, made of 19 mm aggregate mixed with cement and water, preheated and then mixed with about 6% of hot bitumen. Below a certain level, the core is duplicated and above a certain level it is reduced to 0.8 m thick.

CONCRETE AND MASONRY DAMS

5.12 GRAVITY DAM DESIGN

The weight of a gravity dam (and the width of its base) prevents it from being overturned when subjected to the thrust of impounded water (Kennard, 1995). To prevent sliding the contact with the foundation and the foundation itself must have appreciable shear strength. In carrying out stability calculations for any proposed section of a dam, the 'uplift' must be taken into account. Uplift is the vertical force exerted by seepage water which passes below a dam or which penetrates cracks in the body of the dam. Maximum or '100% uplift' on any section through a dam would be a triangular pressure diagram as shown in Figure 5.10; this assumes the upstream value equals the

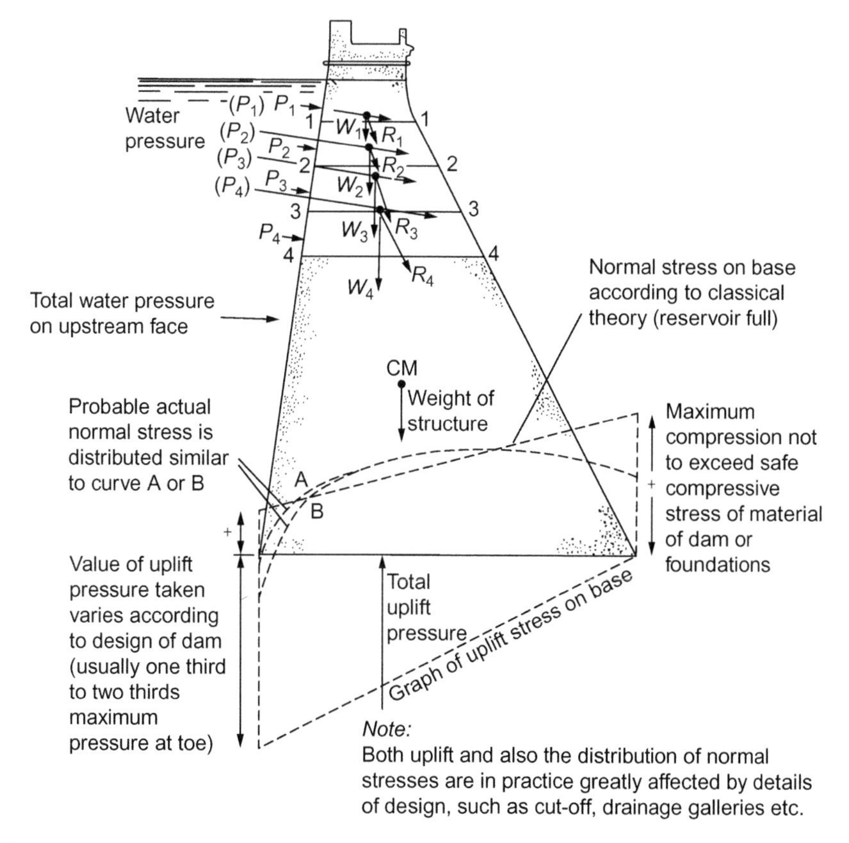

FIGURE 5.10

Simple stability analysis of gravity dam showing effect of uplift on base of dam and at other sections.

pressure exerted by the maximum water level during flood conditions above the section. Below the base of a dam with no cut-off, the pressure diagram is trapezoidal if, at the downstream toe, there is a potential uplift from any tailwater level.

Taking these three forces into account – the uplift, the water thrust and the weight of the dam – the accepted rule is that the resultant of these forces should pass within the middle third of the section being analysed. This ensures that no tension is developed at the upstream face of the dam. At first sight the foregoing appears illogical since the assumption of 100% uplift assumes a crack exists, which the design should prevent. However, perfect construction everywhere is not likely to be achieved and cracks might progressively develop if there is any tension at the face. Even if no crack exists, there can be uplift pressure in the pores of concrete of the dam due to the passage of seepage. Figure 5.11 shows pore pressures measured inside the concrete of the Altnaheglish dam in Northern Ireland, where concrete deterioration had occurred, before remedial work was undertaken. This pore pressure acts, of course, only on the proportion of the concrete

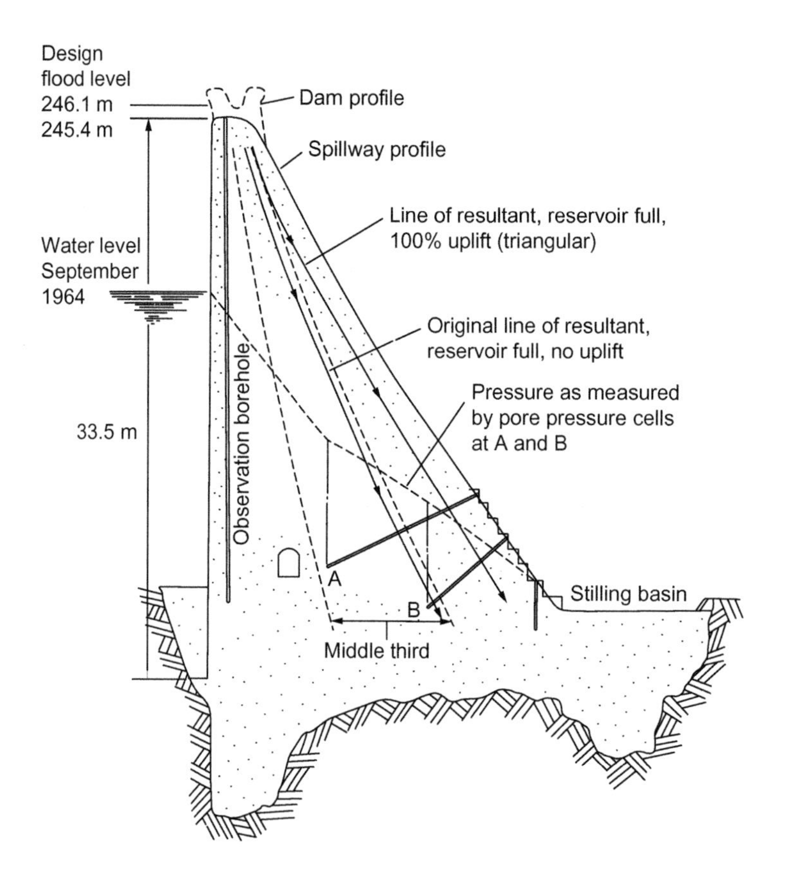

FIGURE 5.11

Effect of uplift on 'resultant' at Altnaheglish dam, Northern Ireland.

which consists of voids, so that the total uplift is only some proportion of the concrete area multiplied by the pore pressure.

In order to reduce uplift from seepage below a dam, a concrete-filled cut-off trench may be sunk at the upstream toe, as shown in Figure 5.11. Even if this cut-off does not make the foundation rock wholly watertight, it lengthens any seepage path and thus reduces uplift below the base. Relief wells or drainage layers can be inserted at intervals below the downstream half of the dam base; these also reduce the total uplift. With these provisions the amount of uplift assumed in the design can be less than '100% uplift'.

Ice thrust may have to be taken into account in the design of gravity dams in cold climates. Estimates (Davis, 1969) of the thrust force vary according to anticipated ice thickness, and range from 2.4×10^2 to 14.4×10^5 N/m^2 of contact with the vertical face of the dam. Seismic forces from earthquakes also have to be considered. It is now accepted that minor seismic damage can occur in areas, such as the UK, previously thought to be free of earthquake risk (Section 5.22).

The design of gravity dams appears deceptively simple, if only the overall principles described above are considered. In fact, controversy raged for more than half a century over the subject of how stresses are distributed within a gravity dam, until computer programs permitted computation of stresses using two- or three-dimensional finite element analysis. The method determines displacements at each node, and stresses within each element of the structure, the latter being considered for analysis as an assemblage of discrete elements connected at their corners.

5.13 GRAVITY DAM CONSTRUCTION

Concrete. The construction of concrete gravity dams is relatively simple. Plates 4(b) and 5(c) show the Wadi Dayqah curved concrete gravity dam and the Tai Lam Chung concrete gravity dam respectively. Most of the remarks concerning the preparation of foundations and cut-off trenches for earth dams apply also to gravity dams. The key problem with mass concrete dams is to reduce the amount of shrinkage that occurs when large masses of concrete cool off. Heat is generated within the concrete as the cement hydrates and as this heat dissipates, which may take many months, the concrete cools and shrinks. In order to reduce the effects of this shrinkage the concrete is placed in isolated blocks and left to cool as long as possible before adjacent blocks are concreted. 'Low-heat' cement is frequently used to reduce shrinkage, or sometimes a blend of cement and pozzolan. Pozzolans, such as pulverized fuel ash, also have cementitious properties, but a much slower and lower rate of heat development. Shrinkage is also reduced if the content of cementitious materials is kept as low as possible and if the aggregate size and proportions are as high as possible. Ice flakes may be added to water used in the concrete mix, while in the larger installations, water cooling pipes may be laid within the concrete to draw-off excess heat as it is produced. Concrete dams are best constructed in areas where rock suitable for the making of concrete aggregates is abundant.

Some concrete dams have been raised or strengthened by installing post-tensioned cables through the concrete mass near the upstream face, and grouting them into the foundation rock to provide additional vertical forces to counterbalance the overturning moments introduced by increased water loads.

Masonry. The construction of masonry dams is expensive because of the large amount of labour required to cut and trim the masonry blocks. In the UK, masonry dams are no longer built, but elsewhere dams have had masonry facing and concrete hearting, primarily to improve the appearance of the dam and also to avoid the need to provide shuttering for the concrete. Overseas, particularly in India, construction of masonry gravity dams continued, but seldom for large structures, due to the unpredictability of masonry behaviour under seismic forces.

Great care must be taken with masonry dams to fill all the joints and beds completely with a watertight mortar mix and to pay special attention to the quality and watertightness of the work on the upstream face. It was Dr G. F. Deacon's insistence on this when constructing the 44 m high masonry Vyrnwy dam for Liverpool in 1881 that has left the dam still in excellent condition over 130 years later. Up to then, he remarked 'there was probably no high masonry dam in Europe so far watertight that an English engineer would take credit for it' (Deacon, 1895–96).

5.14 ROLLER-COMPACTED CONCRETE DAMS

RCC dams were developed to combine the best attributes of a fill dam (offering a plant-intensive method of construction) and a concrete dam (offering a small volume of erosion-resistant material). Although there are reported to have been a number of RCC dams constructed in the 1960s and 1970s, it was not until the early 1980s that take up of the methodology really started. A sound foundation similar to that required for a traditional concrete gravity dam is needed, and then RCC can be placed in successive layers to form a monolithic dam structure. One of the main advantages of the RCC type of methodology is the very rapid rate of placement. RCC dam stability requirements are similar to those for concrete gravity dams. The main differences between the two types are the concrete mix design and method of placing and the design of spillways, outlet works, drainage and inspection galleries to facilitate RCC construction.

Initially, different approaches were made to RCC dams (Dunstan, 1994). One approach was to use a relatively lean and porous mix for the body of the dam, relying on upstream waterproofing for watertightness. This was usually achieved by using a thickness of richer, impermeable concrete on the water face. The other approach was to use a 'high-paste' concrete containing more cementitious material (i.e. cement and pozzolan) that is sufficiently impermeable not to need upstream waterproofing. Its greater tensile strength permits economy in the profile of the dam by the use of steeper side slopes. The lean-mix dam can suffer seepage if the upstream waterproofing is not effective, whereas the high-paste RCC dam has a greater propensity to suffer thermal cracking, so the thermal conditions in the dam have to be studied carefully. In recent years, some low-paste RCC dams have been constructed using a geomembrane to ensure a waterproof upstream face.

The Japanese have developed their own form of RCC dams with thick upstream and downstream faces of traditional concrete protecting the RCC in the interior of the dam. The latter has a cementitious content of 120 or 130 kg/m^3 depending on the height of the dam. This form of RCC dam has not been found to be economic outside Japan. Another interesting development is the faced symmetrical hardfill dam (FSHD) (Londe, 1992). This has a low-strength body made of cement-stabilized fill but with a waterproof reinforced concrete upstream face. Both upstream and downstream faces are sloped, with the upstream slope therefore attracting vertical water load.

The FSHD is generally economic where there is a lower quality foundation, a high dynamic loading and/or a large design flood.

Recent trends have been towards use of high-paste RCC which has been found to be the economic solution under most conditions. A further factor is the increasing size of RCC dams; during the 1980s the average height was only 40 m, whereas in the first decade of the 21st century the average height of RCC dams is approaching 100 m. Plate 4(a) shows the RCC being placed at Olivenhain dam in the USA. The main part of this 97 m high dam was placed in only 6 months.

ICOLD Bulletin 126 includes case histories and provides guidance for the design and construction of RCC dams (ICOLD, 2003). Several RCC dams over 200 m in height have been constructed in the last decade.

5.15 ARCH DAM DESIGN

The principle of design of an arch dam is greatly different from that of a gravity dam. The majority of the strength required to resist the water thrust is obtained by arching the dam upstream and transferring the load to the abutments. The abutments must therefore be completely sound. The theory of design is complex, with the dam resisting the water thrust partly by cantilever action from the base and partly by arching action from abutment to abutment. An arch dam acts in compression and can be much thinner than a gravity dam – e.g. Mudhiq (Plate 5(b)). Early designs were based on the 'trial load' procedure. The dam was assumed to consist of unit width cantilevers one way and unit width arches the other. The water load at each point was then divided between the 'cantilevers' and 'arches' so that their deflections at every point matched. The modern method is to construct a three-dimensional finite element model of the dam to evaluate stresses under various loadings. Physical models have also been used to measure strains and to give a first approximation to the likely distribution of stresses, and to then act as a check on the mathematical calculations.

There are many variations from the simple uniform arch shape; the most economic section is curved both vertically and horizontally and results in the horizontal arches varying in radii with level. A dam of this kind is called a double curvature dam and is especially economical in the use of concrete.

River and flood diversions are usually taken in tunnels through the abutments; flood overspill may be passed over a central spillway. Arch concrete dams are among the highest in the world and are inherently stable when the foundations and abutments are solid and watertight. However, the stresses in the concrete of the dam and in the foundations and abutments can be very high, so that the utmost care needs to be taken in the site investigations and in the design and the construction. Whatever may be the results of the theoretical analyses of forces and stresses on the dam, sufficient reserve strength must be included in the design to meet unknown weaknesses.

5.16 BUTTRESS OR MULTIPLE ARCH DAMS

Where a valley in rock is too wide for a single arch dam, a multiple arch dam may be used. This comprises a series of arches between buttresses, as shown in Plate 4(c) for Nant-y-Moch dam. Each section of the dam, consisting of a single arch and its buttresses, may be considered as achieving stability in the same manner as a gravity dam. A similar design uses a 'diamond head' at the upstream end of each buttress. In this case there is no arch action and seals are needed between each pair of 'diamond heads'. The buttress dam provides considerable saving of concrete compared with a mass concrete dam; on the other hand this saving may be offset by the extra cost of the more complicated shuttering required and the more extensive surface areas requiring a good finish for appearance. A buttress dam has the advantage that uplift is negligible, because the space between the abutments allows uplift pressure to dissipate. Occasionally a buttress dam is advantageous where the depth to sound rock would make a mass concrete gravity dam more expensive.

Buttress dams (or barrages) may also be used in connection with river control and irrigation works, where many gates have to be incorporated to control the flow. In this case there are no arches: the buttresses are to gravity design and have to take the extra load from the gates when closed.

FLOOD AND DISCHARGE PROVISION

5.17 DESIGN FLOOD ESTIMATION

The assessment of flood risk is a vital element in the safe design, maintenance and operation of impounding reservoirs. Earth dams are inherently erodible and uncontrolled overtopping can lead to catastrophic failure. Overtopping of a rockfill, masonry or concrete dam needs to be avoided except where the design specifically provides for this. It is therefore necessary to specify a design flood, in combination with wave action, which the dam must be capable of withstanding. Greater security is required against dam failure where there is a major risk of loss of life and extensive damage and a lower security where the threat is less severe.

A wide range of methods are used for computing reservoir design floods. These methods include:

- empirical and regional formulae;
- envelope curves;
- flood frequency analysis;
- various types of rainfall–run-off and losses models, including the unit hydrograph, the US Soil Conservation Service, the transfer function and the Gradex methods.

The Design Flood (*Guidelines*) produced by the International Commission on Large Dams (Committee on Design Flood) (ICOLD, 1990) contains useful summaries of the more commonly used methods and their limitations. The guidelines also contain some important cautions about reservoir design flood estimation, for example:

1. The determination of design flood is a complex problem requiring the contributions of specialist engineers, hydrologists and meteorologists whose involvement must be sought through the whole process.
2. The choice of the design flood involves too many and varied phenomena for a single method to be able to interpret all of them.

3. All methods available are based on meteorological and hydrological records both for the river basin under investigation and often for other comparable basins. The results derived from the application of a particular method will essentially depend on the reliability and applicability of the data adopted.
4. The exceptional circumstances causing extreme floods are often such that the records of historical, usually moderate, floods provide only a poor indication of conditions during a Probable Maximum Flood (PMF) or a 10 000-year event.

The reader may like to refer to ICOLD Bulletin 82 – *Selection of Design Flood – Current Methods* (ICOLD, 1992).

For dams in the UK the key advisory document is the fourth edition of *Floods and Reservoir Safety* (ICE, 2015) which recommends the levels of protection for different categories of dam. The recommended floods standards in the UK are summarized in Table 5.1. Chapter 3 of *Floods and Reservoir Safety* guides the engineer on the application of methods in the 1975 *Flood Studies Report* (FSR) (NERC, 1975) and the *Flood Estimation Handbook* (FEH, 1999), involving the use of a unit hydrograph rainfall–run-off and losses model to derive reservoir design flood inflows. Chapter 4 covers reservoir flood routing and Chapters 5 and 6 of *Floods and Reservoirs Safety* contain methods for estimating wave overtopping and determining appropriate wave freeboard allowances. An update on the derivation of the FEH catchment parameters and depth–duration–frequency calculations via a web-based service is given by Stewart and Young (2015).

Table 5.1 Design flood standards by dam category adapted from *Floods and Reservoir Safety*, ICE, 2015

Dam category	Potential effect of a dam breach	Safety check flood[a]	Design flood[b]
A	Where a breach could endanger lives in a community	Probable maximum flood (PMF)	10 000-year flood
B	Where a breach 1. could endanger lives not in a community or 2. could result in extensive damage	10 000-year flood	1000-year flood
C	Where a breach would pose negligible risk to life and cause limited damage	1000-year flood	150-year flood
D	Special cases where no loss of life can be foreseen as a result of a breach and very limited additional flood damage would be caused	150-year flood	150-year flood

Notes: [a]The wave freeboard allowance (if any) required in conjunction with the safety check flood is based on the acceptability of the amount of overtopping resulting from the mean annual maximum hourly wind speed.
[b]The wave freeboard allowance in conjunction with the design flood should prevent virtually all wave overtopping resulting from the mean annual maximum hourly wind speed, with a minimum value of 0.60 m applying for Categories A and B dams, 0.40 m for Category C and 0.30 m for Category D.

The statistical analysis of flood events has a very limited role in reservoir design flood estimation in the UK. The reason for this is that extrapolation of statistical flood estimates to the high return periods relevant to freeboard and spillway design can lead to gross under- or over-design, given the relatively short period for which flood data are typically available. Flood estimates become increasingly unreliable when the return period of the design event is greater than about twice the length of the record. It is not possible here to provide a guide to each of the methods of flood estimation currently employed. The following paragraphs, however, provide some basic guidance on the unit hydrograph rainfall–run-off method which has been used for design flood estimation in many countries.

Unit Hydrograph Approach

There are numerous different forms of unit hydrograph and loss models, but all are similar in that they convert a rainfall input to a flow output using a deterministic model of catchment response. These deterministic models have three common elements:

- the unit hydrograph itself which often has a simple triangular shape defined by time to peak;
- a losses model which defines the amount of storm rainfall which directly contributes to the flow in the river; and
- baseflow (i.e. the flow in the river prior to the event).

Where possible the model parameters should be derived from observed rainfall–run-off events. If no records exist the model parameters may be estimated from catchment characteristics.

A unit hydrograph for a given catchment shows the flow resulting from unit effective rainfall in unit time on the catchment, for example the flow following, say, 10 mm effective rainfall falling in 1 hour. It assumes the rainfall is uniform over the catchment and that run-off increases linearly with effective rainfall. Thus the run-off from 20 mm of effective rainfall in 1 hour is taken as double that due to 10 mm and so on, and the ordinates of the hydrograph are doubled. Similarly if rainfall continues beyond the first hour, the principle of superposition can be used, the resulting hydrograph being the sum of the ordinates of the hydrographs for the first, second, third, and so on, hours.

Due to the complexities involved, the derivation of a catchment unit hydrograph from storm rainfall and run-off records is best carried out by an experienced hydrologist. Losses by evaporation, interception and infiltration have to be deducted from rainfall, and the baseflow has to be deducted from the measured streamflow. A complication is that rainfall events are seldom of unit duration, so further analysis is required to produce a unit hydrograph for unit time and unit rainfall. It is usual to derive several unit hydrographs from separate rainfall events for comparative purposes and, where a catchment is too large for uniform rainfall intensity over the whole of it to be assumed, it is necessary to treat individual tributaries separately and assess their combined effect.

One relatively straightforward method of obtaining an adequately severe unit hydrograph is suggested by the US Bureau of Reclamation (USBR, 1977) as follows:

(1) Time T_c of concentration is given by:

in US units	in metric units
$T_c = (11.9\ L^3/H)^{0.385}$ hours	$T_c = (0.87\ L^3/H)^{0.385}$ hours

where: L = length of longest tributary (miles – km); H = fall of this tributary (feet – metres); T_c = time elapsing between onset of storm and time when all parts of the catchment begin contributing to flow to measuring point.

(2) Time to peak T_p for the hydrograph for one unit of rainfall over the catchment is given by:

$$T_p = 0.5\ (\text{rainfall duration}) + 0.6T_c$$

where T_p, T_c and rainfall duration are in hours.

(3) Peak run-off rate R_p:

in US units	in metric units
$R_p = 484\ AQ/T_p$ ft³/s (cusec)	$R_p = 0.2083\ AQ/T_p$ m³/s (cumec)

where A is catchment area (sq. miles – km²) and Q is the unit rainfall (inch – mm). If A is in acres and Q is in inches, the peak run-off in US units is given by $R_p = 0.756\ AQ/T_p$ ft³/s.

Use of this formula tends to over-estimate flood peaks in temperate, flat and permeable areas; for these, a simple triangular unit hydrograph, with a base of 2.5 times the time to its peak, should suffice. Its peak flow is $2.2/T_p$ m³/s for each km² of catchment for every 10 mm of effective rainfall. The value of T_p often lies in the range $1.3(\text{area})^{0.25}$ to $2.2(\text{area})^{0.25}$ where the area is in km².

A variety of alternative loss models can be incorporated into the unit hydrograph approach. The two most common are to:

- derive a single percentage run-off value applicable throughout the whole storm; or
- adopt an initial loss x mm at the start of the design storm followed by continuing losses of y mm/h throughout the event.

Once each of the model elements has been defined for a catchment, the unit hydrograph method may be used to estimate the total run-off from any rainfall event. The design rainfall is in the form of a hyetograph defined by a depth/duration characteristic of the area and arranged into a selected storm profile. A 'bell-shaped' profile is most commonly used for storm events up to about 24 hour duration, whereas the recorded profiles in severe historic events are used to define more realistic design storm profiles for longer duration events.

Table 5.2 Some probable maximum precipitation (PMP) estimates

Location	Country	PMP in mm				
		20 min	1 h	6 h	15 h	24 h
South Ontario	Canada	–	–	410	–	445
Guma	Sierra Leone	81	183	–	580	630
Selangor	Malaysia	–	160	300	–	460
Shek Pik	Hong Kong	101	220	–	915	1200
Garinono	Sabah	81	162	420	620	675
Brenig	Wales					
	May–Sep	74	109	183	–	254
	Oct–Apr	38	72	165	–	272
Tigris (50 000 km^2)	Iraq	–	–	60	–	167
Jhelum, Mangla	Pakistan					
(2500 km^2 sub-area)	Dec–May	–	–	185	–	295
	Jan–Nov	–	–	365	–	575

The design storm rainfall may be a statistically derived design event to produce a flood of specific return period (the *T*-year event), or may be a probable maximum precipitation (PMP) to produce a PMF. Rainfall depth–duration–frequency values, sometimes including estimates of the PMP, are available for many regions. Table 5.2 gives some examples of the PMP estimates adopted for the areas draining to a number of major dam sites; Figure 3.3 also gives information on maximum precipitations that have been recorded.

5.18 SPILLWAY FLOOD ROUTING

If a reservoir is full to top water level, a flood inflow causes the reservoir water level to rise and this causes increasing rates of discharge over the spillway. In a conventional ungated spillway, with a fixed weir crest, the temporary ponding of water in the reservoir results in a maximum spillway discharge that is less than the maximum rate of flood inflow to the reservoir, with the peak outflow occurring after the peak inflow, as illustrated in Figure 5.12. (It may be noted that, when the outflow is at its maximum, it is equal to the inflow.)

This phenomenon is referred to as 'flood attenuation' and the calculation to process the inflow hydrograph and produce an outflow hydrograph is termed 'flood routing'. In the past, graphical approaches were sometimes adopted to assist in flood routing but are no longer necessary, as computer programs or spreadsheets can be used much more conveniently. In most practical circumstances the reservoir water surface can be treated as horizontal at each time step and the flood routing approach is therefore described as 'level pond'.

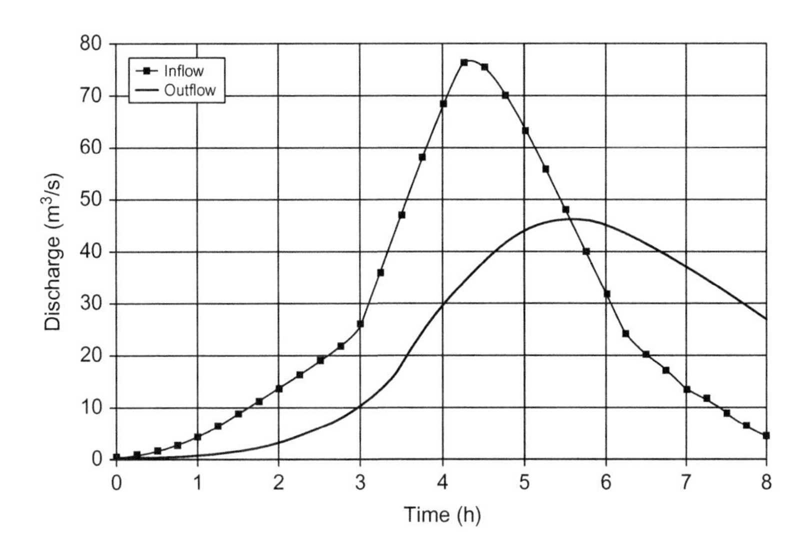

FIGURE 5.12

Example flood routing through a reservoir.

Suitable software for undertaking reservoir flood routing is included in industry-standard open-channel flood modelling software. The principal data required to carry out a reservoir flood routing are:

- the inflow hydrograph, at a suitable time step (preferably no longer than about a 10th of the time to the peak of the incoming flood);
- the spillway rating (that is the relationship between discharge and reservoir level); and
- the relationship between the water level and surface area in the reservoir.

In practice, a storage volume relationship for the reservoir may be available, rather than a surface area relationship, and some software may accept the volume relationship, rather than require an area relationship to be provided. However, it is always best practice to convert a volume relationship to the corresponding area relationship, to check that it is consistent with the known surface area of the reservoir and increases appropriately as the water level rises.

The user needs to define the initial water level in the reservoir, e.g. whether it is assumed that the reservoir is initially just full, or is discharging the mean discharge or some other amount. There may be further complexities, such as the presence of a bywash, which intercepts part of the incoming flow and carries it past the reservoir. Such cases should be evaluated by an experienced hydraulics engineer.

Level-pond flood routing is undertaken for each time step equating the difference between the inflow and outflow to the change in the volume of water contained. The basic flow balance equation (normally in metre-second units) is:

$$\frac{I_1 + I_2}{2} - \frac{O_1 + O_2}{2} = (H_2 - H_1)\frac{(A_1 + A_2)}{2\Delta t}$$

where I is the incoming discharge, O is the outflow, A is the reservoir surface area and H is the water level in the reservoir, and the subscripts 1 and 2 refer to the beginning and end of the time step, Δt. Theoretically it is necessary to iterate to a solution, and this is what the best software does. However, if a short enough time step is adopted, it can generally be assumed that, within a time step, $O_2 = O_1$ and $A_2 = A_1$, in which case the basic equation simplifies to:

$$H_2 = H_1 + \frac{(I_1 - O_1)\Delta t}{A_1}$$

In this case, the values of O_2 and A_2 are calculated only after H_2 has been determined, ready to be used as the values of O_1 and A_1 in the next time step. If this simplified approach is adopted, the user should beware of using too small a time step Δt. If the calculations proceed with values of ΔH that are too small to be evaluated accurately within the available precision of the calculator or computer, the accuracy of the simulation could be severely prejudiced.

5.19 DIVERSION DURING CONSTRUCTION

When a dam is constructed in a natural river valley, some means of temporarily diverting the river flow is needed in order for the dam foundations to be properly prepared and for the construction of the works to proceed. Where flows are seasonal, advantage can be taken of this, ensuring that especially critical periods of construction are carried out in low-flow periods. Plate 5(a) shows closure of the River Chira at Poechos Dam, Peru.

In some cases, it may be possible to divert the river via tunnels in the abutments and to work within an effectively dry river bed. In wide valleys it may be more economic to divert the river onto one side of the valley while starting construction on the other side, then to divert the river though temporary openings in the partially completed structure and start construction on the remaining side. In exceptional cases, where rivers are subject to regular high floods, it may be necessary to allow the partially completed dam to periodically over-top and to design the works accordingly to accommodate this.

Diverting through a temporary opening left in the main dam body or allowing the partially completed structure to be overtopped by floodwater are techniques especially suitable for concrete or masonry dams. These are essentially non-erodible and therefore can be easily recovered after inundation.

It is not possible to follow this procedure with an earth dam which, if overtopped at any stage during construction, would probably be destroyed. In such cases diversion is generally via a tunnel or culvert. Culverts and tunnels used initially for river diversion can later be used for housing the permanent draw-off pipes from the reservoir. Care needs to be taken with culverts to ensure that they do not cause differential stress concentrations in the earthfill lying above. Unless an earth dam is founded on sound rock (which is often not the case), it is undesirable to have a central culvert left through the body of an earth dam. If a culvert is unavoidable it is better to recess the culvert as much as possible into a trench, to minimize the tendency for it to generate stress differentials in the main dam body. Differential settlement can also cause the culvert to fracture or open up a path for impounded water to leak through the junction where the culvert passes through the corewall of the dam (Section 5.28).

The flood capacity of temporary diversion works may well be laid down in the construction specification but is also of key importance to the contractor building the works, since he would normally have to shoulder any risks associated with an event of greater severity occurring during the construction of the scheme. The flood discharge capacity of the diversion works required during construction generally reflects the consequences of damage should a larger flood occur. In the early stages of construction the consequences of under-capacity may be slight. Later, insufficient capacity may entail loss of much of the dam. It is not uncommon therefore to see the design return period for diversion increase as construction develops. Diversion works may initially be required to accommodate the 1 in 5 year flood, but eventually be required to take the 1 in 50 or 1 in 100 year event immediately prior to dam closure.

Each case has to be considered individually, taking all the relevant circumstances into account. These include the nature of the formation, the type of dam, the proposed method of construction and the volume stored behind the embankment, where this could become large enough before diversion ceases to pose a risk to life and property downstream. In planning a diversion scheme, rainfall, run-off and flood routing studies need to be carried out to find the critical circumstances applying at each stage of the construction.

Some risk is unavoidable and should be the subject of insurance provision. In the case of the Oros dam in Brazil (W&WE, 1960), a diversion tunnel of 450 m^3/s capacity had been built, but 600 m of rainfall fell on the catchment in a week. Despite all efforts to raise the dam ahead of the rising waters, it was overtopped on 25 March 1960 and required major re-construction.

5.20 FLOOD SPILLWAYS

Spillway works can generally be classified into three types:

- integral overflows that convey floodwater over the crest of the dam;
- open ‘chute’ spillways, which convey floodwater via a channel constructed on natural ground at a dam abutment or at a low point on the reservoir perimeter; and
- structures that discharge floodwater via an opening or culvert through or under a dam, or via a tunnel driven through an abutment.

The crest of a dam has generally to be high enough above the highest water level in the reservoir under flood conditions to prevent waves and spray passing onto the crest and the downstream face of the dam. For an earth dam, the usual practice in the UK results in the top of the embankment being up to about 2 m higher than the crest of the overflow weir, with a substantial wavewall capable of protecting the bank against the run-up from wave action also being provided. For dams in other parts of the world, where reservoirs and floods can be of much greater magnitude than in the UK substantially higher freeboards may be necessary as for Mangla dam (Fig. 5.2), where almost 10 m of freeboard was provided for flood surcharge and wave run-up (Binnie, 1968).

Integral Spillway

Direct discharges are permissible over concrete (including RCC) and masonry dams and some rockfill dams, if provided with a suitably designed spillway. The practice can be economical, because very little alteration to the profile of such a dam is necessary to accommodate the overflow section. Typical examples are shown in Plates 4(b), 5(b) and 5(c). The water usually cascades down a smooth or stepped face of the dam. At the foot of the dam the water must be turned into a stilling basin to dissipate some of its energy, because the main danger with this type of overflow is scour at the toe of the dam during an extreme flood. Sometimes the water is ejected off the face of the dam by a ski jump or other means (Plate 5(b)), which throws the water some distance away from the toe of the dam, thereby lessening the danger to the toe. It is essential that all the construction is massive and soundly founded on solid rock. The design is normally not appropriate to a valley where good hard rock does not appear in the river bed at the dam and for some distance downstream.

Sometimes an additional feature of a concrete or masonry dam is the inclusion of automatic crest gates at the overflow weir. The gates permit water to be stored above the level of the overflow weir crest but, after a certain level is reached, the gates are opened automatically and permit progressively increasing discharges to pass over the weir. This permits increased storage without raising the flood overflow level. Except in the case of some concrete dams on hydro electric schemes in Scotland, gates have seldom been used on dams in the UK, in case storm conditions might interfere with the power supplies for operating the gates at the crucial time. Elsewhere their use is widespread, even where there is risk of failure to operate due to meteorological, security or even political circumstances.

Another method is the use of a siphon hood over the weir crest, in order to achieve a greater discharge per unit length of crest than would occur for the same water head over a fixed weir. Unit discharges of between 6 and 20 m^2/s have been achieved for design heads, relative to the spill level, of between 0.12 and 0.9 m (Ackers and Thomas, 1975). Modern spillway siphons are usually of the air-regulated type, being designed to provide a progressive increase in discharge at increasing heads, in comparison with older 'make-and-break' designs that tend to cause sudden fluctuations in discharge and hence unstable conditions both upstream and downstream. Ultimately, if the upstream water level continues to rise and the discharge increases, an air-regulated siphon reaches 'blackwater' flow, in which there is little or no air entrained in the flow. Beyond this point, the siphon behaves like a short culvert, with the discharge being approximately proportional to the square root of the operating head, so further increases in discharge capacity require substantial rises in the upstream pool level. For this reason, spillway siphons should normally be designed with a generous factor of safety on discharge, to allow for the uncertainties inherent in the determination of design floods, or provided with supplementary emergency spillway capacity.

Siphons require careful design, either adopting a previously proven design or verification by model testing. A key design consideration is the avoidance of cavitation, and this limits the operating head to about 8 m. Siphons are not normally suitable for use in cold climates, where ice formation in winter might block or decrease the capacity of the siphon. Booms may be needed to protect siphon inlets from trash.

Chute Spillway

At most major reservoirs in the UK floodwaters are discharged via a spillway channel around the end of the dam (Plate 6(a)). Such spillways should be constructed on natural ground and not on the embankment dam itself where settlement will occur. The construction has usually to be massive, of concrete or masonry, because of the need to prevent dislodgement of any part under flood conditions, when significant pressure fluctuations can occur in the flow. Plates 7(a) and 7(b) show larger examples where the spillway gated structure is sited on original ground between massive monoliths against which the dam embankments abut.

At the upstream end of the channel there is an overflow weir, often with a rounded crest to increase the discharge coefficient and minimize damage by floating debris. A common configuration is a 'side-channel' (Plate 6(a)), in which the overflow weir forms one side of a channel that runs along the reservoir shoreline for a short distance upstream of the dam abutment. The channel downstream of the weir must be designed to take the water away without any backing up of water on the weir. Flow here is usually subcritical and the channel gradient is gentle, until it reaches a point near the abutment where it starts to fall more steeply and critical flow occurs (Sections 14.7 and 14.8). This acts as a control point for the flow of water in the spillway and determines whether the flow conditions upstream allow the overflow to perform up to its full discharge capacity, or is 'drowned out' and behaves as a submerged weir. In many cases, flows in this area are further constricted by the presence of a bridge, affording access to the dam crest; this can create a 'control' based on orifice-type flow conditions.

A common design problem is that the floor of the channel upstream of the corewall of the dam is subject to uplift when the reservoir is full, so must be designed accordingly. The channel must also be effectively sealed against the corewall of the dam, to prevent leakage passing below the floor or behind the walls of the spillway.

Downstream of the hydraulic control point near the abutment of the dam, the overflowing water becomes supercritical and passes at increasing speed down the steeper part of the channel. Attempts have sometimes been made to destroy some of its energy by constructing piers or steps in the channel, but these are normally not particularly effective and can cause problems with air bulking of the flow and with spray, which can erode soft material adjacent to the walls of the channel. At the end of the spillway channel a 'stilling basin' is normally constructed to dissipate some of the energy of the fast-moving water. The most usual method is to induce the water to form a hydraulic jump within the stilling basin (Section 14.11).

Historically, spillway channels have often been curved, especially in the vicinity of the abutment, where the channel steepens and when the channel reaches the base of the valley, approaching the stilling basin. It was also common to accompany the steepening gradient with a reduction in channel width. These curves and transitions generally result in complex patterns of cross waves in supercritical flow and often result in overtopping of the channel walls at locations that were not anticipated by the designer. Wherever possible, new spillways should be designed so that they are straight and with no change in width or channel shape throughout sections where supercritical flow occurs. Unless designed in a manner that ensures virtual certainty of the flow conditions, the designs of new channel spillways should normally be subjected either to physical model testing or (if appropriate) analysis by CFD (Section 14.18).

Bellmouth Spillway

A bellmouth (or morning-glory) spillway (Plate 6(b)) may be adopted for small to moderate discharges where the expense of cutting a spillway channel is great, or where a tunnel has to be driven through one of the abutments for the diversion of the river during construction of the dam. The bellmouth must be constructed on firm ground, preferably either clear or almost clear of the toe of the dam. If built within the body of the dam, differential settlement of the bellmouth and embankment fill may cause disruption of the fill, or the different levels of fill against it might cause it to tilt. The bellmouth shaft may be vertical or sloping and it may join the tunnel with a smooth or sharp bend.

Three primary flow modes are usually possible with a bellmouth spillway:

- crest control, where the rim of the bellmouth acts as a weir;
- orifice control, in which the throat of the shaft (normally at the base of the bellmouth, where the diameter of the shaft becomes uniform) acts as a control and drowns out the crest;
- tunnel control, in which the shaft and tunnel flow full and the discharge is a function of hydraulic friction and form losses.

These three flow modes would usually apply in the above sequence, at increasing upstream pool levels but in some cases there may be a direct transition from crest control to tunnel control, without any region of orifice control. A potential disadvantage of the bellmouth spillway is that there is little increase in discharge capacity once the weir is drowned and orifice or tunnel flow control applies. Unless a supplementary emergency spillway is to be provided, it is, therefore, normally best practice to design a bellmouth spillway so that crest control applies over the entire discharge range, with a generous margin provided before either orifice control or tunnel control would occur.

The profile adopted for the rim and interior of the bellmouth is normally similar to that for an ogee type of spillway. *Design of Small Dams*, Section 212 (USBR, 1977) gives profiles and discharge relationships for bellmouth spillways, based on research by the US Bureau of Reclamation in the 1950s. See also *Morning Glory Spillways* (USACE, 1987) for a simpler comprehensive presentation.

Radial vanes are generally incorporated in the top of a bellmouth shaft to minimize the occurrence of vortex action, which can otherwise reduce the discharge capacity and cause problematic flow conditions in the shaft and tunnel. If a bellmouth spillway is located close to the reservoir shoreline or the upstream face of the dam, the limited depth and width of the approach on that side of the rim can lead to 'starvation' of that part of the weir, reducing the overall discharge capacity. Normally, the bellmouth should be positioned so that there is a minimal starvation effect.

A key requirement in the design of bellmouth spillways is to avoid (or severely limit) the occurrence of siphonic conditions at the base of the shaft and the upstream end of the tunnel when tunnel control applies. This is primarily to avoid cavitation and consequent damage to concrete (Section 15.12). The damage done by cavitation can be catastrophic. At the Tarbela dam in Pakistan in 1974 (Kenn, 1981) a thickness of over 3 m of concrete tunnel lining was eroded within about 24 h. This damage is believed to have been caused by erosion due to the collapse of vapour pockets against the concrete surface after being generated at the interface between the very high velocity jet emerging from a part open gate and adjacent slower moving water. Great care should be taken in the interpretation of model test results in cases where sub-atmospheric pressures occur.

The junction of the base of the shaft with the tunnel requires careful design, as the falling water brings with it appreciable quantities of air. The design should be such that this air can be safely evacuated, avoiding a phenomenon known as 'blow-back', in which pockets of air force their way up the shaft, temporarily reducing discharge. Although siphonic conditions at the base of the shaft should be avoided, siphons may be included or added at the bellmouth lip as for weir spillways.

An advantage of the bellmouth overflow and tunnel, apart from use of the tunnel for river diversion during construction, is that it may also accommodate the draw-off functions. The supply pipe and compensation water pipe may be led through the tunnel, either fully encased in concrete or passing through a separate compartment, whilst the scour valve may discharge directly to the base of the bellmouth shaft.

Bellmouth spillways are not suitable in climates where substantial ice may form on the reservoir during winter. Also, even for small reservoirs, the size of the bellmouth and its throat must be large enough to allow passage of any debris that might be brought down by the river. The bellmouth presents a danger to sailing and fishing boats when the reservoir water level is high, so guards must be fixed around its perimeter; these must also be designed not to hold back floating debris that could restrict the bellmouth discharge capacity.

Emergency Spillways

It is sometimes not feasible to provide sufficient capacity at a single spillway to accommodate the whole design flood flow. This may be for reasons such as space limitation or cost. In any case, if the main spillway is gated, it may be held that there should some back-up spillway provision in case the main spillway gates cannot be opened.

Operation of an emergency spillway should be automatic and reliable and so should not be reliant on any human intervention or any systems. It will be required to operate in the most extreme circumstances when normal operation procedures are unlikely to be possible. Often limited space means that a simple weir cannot be used, so designs usually involve triggering the opening of ample water passages. Several arrangements are commonly used.

Fuse plug embankments. These consist of erodible embankments constructed on a non-erodible (concrete or rock) base, often in bays with different crest levels to allow increments in discharge. The embankments consist of a clay or silt core sloping downstream and supported by shoulders of non-cohesive sandy gravel. Sand filters are used each side of the core for the usual reasons (Section 5.9) and as their presence downstream of the core facilitates erosion. The embankment must be designed and constructed as a dam but the grading of the downstream shoulder material is critical. Plate 8(a) shows a four bay fuse plug spillway of capacity 14 000 m^3/s for Poechos Dam, Peru.

Arrangements in which a whole fuse plug section is designed to fail at once may not be suitable where there is concern about the effects of the consequent significant flood wave passing downstream. In such cases a pilot channel may be used. This is a section of width about half the height of the fuse plug and with crest level about 1 m below the main plug crest (for plug height of 3–9 m). Model tests carried out by USBR (1985) showed that the pilot channel encourages the initial breach and allows the plug to be eroded away laterally at a rate of between 1.4 and 2.8 m/h for heights of 3 and 9 m, respectively. A 100 m wide fuse plug 6 m high with a central pilot channel would thus take about 25 min to erode away fully, thus reducing the steepness of the flood wave considerably.

Tipping gates. These may be of two types, both allowing progressive flow increase as gates tip:

- reusable 'folding' steel or concrete walls 'hinged' at the base and designed to topple into a recess in the spillway channel floor; Plate 8(b) shows concrete tipping gates at a reservoir in the UK;
- disposable steel or concrete fabrications designed to topple and be carried away by the resulting flow; these may be made with straight or labyrinth weir sections (Plate 8(c)) such as those designed by Hydroplus – used on over 30 dams worldwide since 1990.

Another arrangement sometimes used, where appropriate resources (perhaps military) are assured locally, is to arm sections of the dam embankment with explosives or facilities for their quick installation.

Operation of emergency spillways is intended to be rare, if at all, but some arrangements produce a sudden large discharge – even if the full flow is delivered in steps. The consequences of the sudden increases in flow downstream should be carefully considered and controls may have to be put in place to avoid the worst damage or loss of life that sudden flood flows could cause.

5.21 DRAW-OFF ARRANGEMENTS

Draw-off works at water supply reservoirs can be for water supply, reservoir drawdown and for providing downstream river compensation flows. A summary of requirements and UK practice is provided in the paper by Scott (2000).

Except when a reservoir is shallow, water supply draw-off pipes are usually designed for withdrawing water at several different levels from within the reservoir. A common provision may be simply upper, middle and lower, although more may be used in deep reservoirs. The choice of levels depends on the depth/volume relationship of the reservoir and the expected variation of water quality (and perhaps temperature) with depth, which may change seasonally (Section 8.1). The upper draw-off must be a sufficient distance below top water level to avoid constant changing from it to mid-level draw-off with normal reservoir level fluctuations. The bottom draw-off level may need to allow for sediment.

Such arrangements are normally sited in draw-off towers (Figure 6.6). Wet-well towers feature draw-offs discharging into a flooded tower and a single extraction point from the tower at low level. Such towers are less common nowadays owing to the difficulty of maintaining the underwater valves and pipework. A more common option is the dry-well tower, in which all extracted water is contained by valved pipework accessible by means of internal ladders and platforms. At each draw-off there are usually two valves, or else an outer sluice-gate and an inner valve. The inner valve is used for normal operation, the outer valve or sluice-gate is used to permit the inner valve to be maintained.

The draw-off tower at an embankment dam would typically be sited close to the upstream toe of the embankment with access by footbridge from the dam crest or shoreline (Plate 6(b)). In the case of concrete dams they are more economically tied to the main dam or sited within it. In some cases draw-off arrangements have been located in sloping galleries on the reservoir abutments. This can be a cheaper and less prominent option as well as attracting less seismic loading. The draw-offs transfer water to one or more outlet pipes. It is not good practice to site such pipes directly in an embankment, but rather they should be sited in culverts or tunnels and with some means of upstream isolation. They can, of course, be directly embedded in the body of a concrete dam.

A scour pipe is required for a reservoir to ensure that the approach channel to the lowest draw-off is kept free of silt, and to make it possible to lower the water level in the reservoir at a reasonably fast rate in case of emergency. Recommended drawdown rates for reservoirs are not standardized and are to a large extent site-specific. Typical criteria can be a drop of a quarter of the reservoir height in 14–21 days in conjunction with an average wet month inflow or, alternatively, half of the reservoir volume in 30 days in conjunction with mean inflow. In both cases the rate of reservoir drop should not be so as to threaten the stability of the dam or reservoir margins and the rate of release should not be so great as to cause excessive distress downstream.

Those pipes that release flow to the downstream watercourse (including the scour pipe, any scour branch off the main draw-off pipes and the outlet for compensation water) normally discharge either into a stilling basin, designed to dissipate the energy, or via a disperser valve (Section 18.17) that turns the flow into a spray.

5.22 SEISMIC CONSIDERATIONS

Earthquakes are severe and common in some countries, especially those bordering the Pacific Ocean. In other countries, such as the UK, they are less common and less severe. Nevertheless, dams are major items of infrastructure with the potential to cause considerable damage should they fail. In view of this, just as dam and reservoir works are designed to accommodate rare flood events, they have also to be designed to be able to accommodate very rare seismic events.

Seismic loadings need to be considered in the design of both new dams and the appraisal of old dams. In the UK, the key references are *An Engineering Guide to Seismic Risk to Dams in the United Kingdom* (Charles, 1991) and an 'application note' covering aspects of the use of the guide (ICE, 1998). These documents provide a structured framework for the consideration of seismic risk similar to that adopted for floods. The reader may also refer to ICOLD Bulletin 148 which provides guidance on the selection of seismic parameters for dam design (ICOLD, 2010).

DAM REGULATION, SUPERVISION AND INSPECTION

5.23 STATUTORY CONTROL OVER DAM SAFETY IN THE UK

Regulations are adopted in many countries to ensure that dams are regularly inspected and are constructed or altered only under the charge of properly qualified engineers. Reservoir safety in the UK is governed primarily by the Reservoirs Act 1975, as amended by the Water Act 2003 and the Flood and Water Management Act 2010 (ICE, 2000, 2014). The Reservoirs (Scotland) Act 2011 is in the early stages of implementation and, in the meantime, the unamended Reservoirs Act 1975 continues to apply in Scotland. At the time of writing, a Reservoirs Bill is progressing through the Northern Ireland Assembly. Reservoir safety regulations in a number of other countries follow the model of the UK Reservoirs Act 1975.

The 1975 Act applies to 'large raised reservoirs' which are currently defined as having a capacity of 25 000 m^3 or more above the level of any part of the adjacent land. Provision has been made to reduce the qualifying capacity to 10 000 m^3, including in the Scottish and Northern Ireland legislation, but implementation of this has been deferred in England, Wales and Scotland. Specifically excluded from the 1975 Act are lagoons covered by the Mines and Quarries (Tips) Act 1969 (together with the corresponding 1971 Regulations and the Quarries Regulations 1999) and navigation canals.

The 1975 Act includes roles for the Secretary of State, for the 'undertaker' (the owner and/or operator of the reservoir), for the 'enforcement authority' and for 'qualified civil engineers'. A series of 'panels' of such engineers have been set up under the advice of the Institution of Civil Engineers to perform the particular functions under the Act, namely:

Construction Engineer	Responsible for supervising the design and construction or the alteration of a reservoir to increase or reduce its capacity
Inspecting Engineer	Carries out periodic inspections, normally at intervals of 10 years
Supervising Engineer	Has a continuous appointment to 'supervise' the reservoir, watching out for problems and keeping the undertaker informed

Three panels have been set up to cover the duties of Construction Engineer and Inspecting Engineer, the panels being defined according to the type of reservoir on which the engineer is qualified to exercise their duties, as follows:

All Reservoirs Panel	All reservoirs covered by the Act
Non-impounding Reservoirs Panel	All reservoirs, except impounding reservoirs
Service Reservoirs Panel	Service reservoirs only

The Scottish and Northern Ireland legislation have some differences in terminology, e.g. 'reservoir manager' instead of 'undertaker', but make similar provisions for panels, and it is expected that the panels will be run in parallel, with coordinated appointment processes.

All engineers on the above panels are also entitled to act as Supervising Engineer for all reservoirs covered by the Act. In addition, the Supervising Engineers Panel covers those qualified to act only as Supervising Engineers. Panel appointments are made for a period of 5 years and are renewable.

The 1975 Act and corresponding Scottish and Northern Ireland legislation contain provisions for the registration of reservoirs and for the enforcement of the legislation, both of which functions, for England and Wales, were transferred to the Environment Agency in October 2004. These duties in Wales now reside with Natural Resources Wales. In Scotland, the duties are being transferred to SEPA under the Reservoirs (Scotland) Act 2011. In Northern Ireland, they are due to reside with the Rivers Agency. Enforcement is facilitated by requirements for the issue of certificates by the qualified civil engineers under specific circumstances, in particular in association with the construction or alteration of a reservoir, the issue of inspection reports and the implementation of 'measures taken in the interests of safety'. Appointments of qualified civil engineers by the undertakers also have to be notified to the enforcement authority.

A Guide to the Reservoirs Act 1975, 2nd edition (ICE, 2015) contains the full text of the 1975 Act as amended and the relevant regulations, a commentary on the application of the Act and flowcharts to illustrate the duties of the various parties. The 2nd edition applies specifically to England and the original guide (ICE, 2000) remains applicable in Scotland until the Reservoirs (Scotland) Act is implemented. Anyone engaged on work involving reservoirs in the UK should refer to these guides and to the current guidance on the websites of the British Dam Society and respective governments:

http://www.britishdams.org/reservoir_safety/default.htm
https://www.gov.uk/reservoirs-owner-and-operator-requirements
https://www.gov.uk/government/publications/reservoir-safety-a-guide-for-reservoir-owners
https://naturalresources.wales/water/registering-a-reservoir/?lang=en
http://www.gov.scot/Topics/Environment/Water/16922
http://www.sepa.org.uk/regulations/water/reservoirs/
http://www.dardni.gov.uk/index/rivers/reservoirs-bill-ni-2.htm

Particular points worth noting in relation to the Reservoirs Act 1975 are:

- responsibility for the construction or enlargement of a reservoir remains with the Construction Engineer for between 3 and 5 years from first filling;
- a Supervising Engineer is not required until the end of the Construction Engineer's involvement, upon issue of their final certificate;
- the first inspection under Section 10 of the Reservoirs Act 1975 is required within 2 years of the issue of the final certificate; and
- the Inspecting Engineer must be 'independent' of the Construction Engineer and the undertaker.

There are some differences with regard to certain of the above points in the legislation for Scotland and Northern Ireland.

The provisions of this Act and the earlier Reservoirs (Safety Provisions) Act of 1930 have worked well, in that they have assisted in preventing any dam failures involving a loss of life in the UK for over 80 years. A particular feature of the 1975 Act is that it imposes a personal responsibility on the engineer who issues a certificate. Such responsibility can only be effectively exercised by an engineer having adequate experience of dam design and construction and who has access to the specialist services frequently required in making a proper inspection. Many dams in the UK are 100–150 years old and require careful attention to ensure their continued safety.

5.24 DAM DETERIORATION SIGNS

Routine observations to ensure a dam remains in good condition are too numerous to list here. The International Commission for Large Dams (ICOLD) publishes a number of bulletins which include guidance on dam surveillance (ICOLD, 2000, 2009, 2013a, b). Reference may also be made to guides on the safety of embankment dams (Johnston, 1999) and concrete and masonry dams (Kennard, 1995), both of which provide detailed checklists, but each dam requires its own specific programme of monitoring. All instrumentation of a dam, such as settlement or tilt gauges and instrument for measuring porewater pressures and underdrain flows need regular monitoring and checking for accuracy. Only some matters of importance, principally related to signs of leakage, can be mentioned here.

Good access to a dam is important and should be suitable for heavy construction plant that might be needed for repairs. The situation arising if emergency work should be required has to be envisaged: night work may be necessary under heavy rainfall and in high winds and with a flood overflow. All gates, valves and other mechanical controls should be in easily operational condition, and access to shafts, galleries and inspection pits should be safe, properly ventilated and lit to safeguard against accidents to personnel adding to the troubles of an emergency.

The catchment needs monitoring for important changes of use and for any evidence of hillside movement that could indicate instability.

For earth dams leakage may be evidenced by damp patches on the downstream face, by areas with unusually luxuriant plant growth or by increased underdrain flows or porewater pressures. However, leakage may also cause settlement. When properly built, an earth dam should have smoothly regular upstream and downstream slopes; the crest should either be straight or to a pre-formed curve. Normally the dam crest should initially show an even rise towards the centre or highest part of the dam because a settlement allowance should have been incorporated during construction, usually of about 1% of the dam height. Irregularities of line and level of the crest that are not explained by the anticipated post-construction settlement may be evidence of erosion due to leakage, as could a localized depression in the embankment.

The upstream facing of the dam, whether of pitching, riprap, concrete blockwork or slabbing, should be checked for slope movement, sometimes visible from distortions in the water line. Concrete slabbing of the upstream face should be checked for signs of settlement or damage caused by wave action pulling out the supporting material.

The ground at the toe of an earth dam is often wet, sometimes supporting a growth of rushes. This is a natural collecting point for surface run-off but undue wetness may be caused by seepage due to a hidden defect such as cracking, erosion or settlement of the corewall. All underdrain inspection pits and drainage water should be examined for evidence of silt or clay being carried out of the dam. The puddle-clay corewalls of old dams are particularly vulnerable to erosion through leakage, and thought must be given to the possibility that the dam has suffered leakage or settlement, in the past, which has not been effectively dealt with.

Concrete dams can show signs of leakage by damp patches on the downstream face or by the growth of moss and lichen at joints. Signs of stress or movement in a concrete dam are spalling of concrete at joints, opening up of joints and cracks, or displacement irregularities, both on the surface of the dam and in any internal gallery or shaft. Since displacement of shuttering when the dam was constructed, or initial settlement which has ceased, can cause irregularities, it is important to log them and the areas where no irregularities occur, so that any new signs of movement can be detected. In masonry or masonry-faced dams, leakage will be through joints; cracking or fallout of pointing needs investigation to see whether it is caused by weathering, mortar softening, increased stress, or possibly acid water seepage attack.

The flood discharge works to any dam need regular inspection for signs of settlement and for any other irregularities which could induce scour that might damage the works during a flood overflow.

A 'trained eye' is necessary for the inspection of dams, especially for old earth dams, where a small surface defect may eventually prove to be related to a dangerous condition of the dam developing internally, hence the need to investigate the cause of anything that seems to be 'not quite right' or 'not as it should be'.

5.25 RESERVOIR SEDIMENTATION

There is a vast range in the degree to which reservoirs worldwide are affected by sedimentation. Figure 5.13 shows the range of suspended sediment loads experienced in rivers of different size, as reported by Fleming (1965). However, the design and operational practices at the reservoir

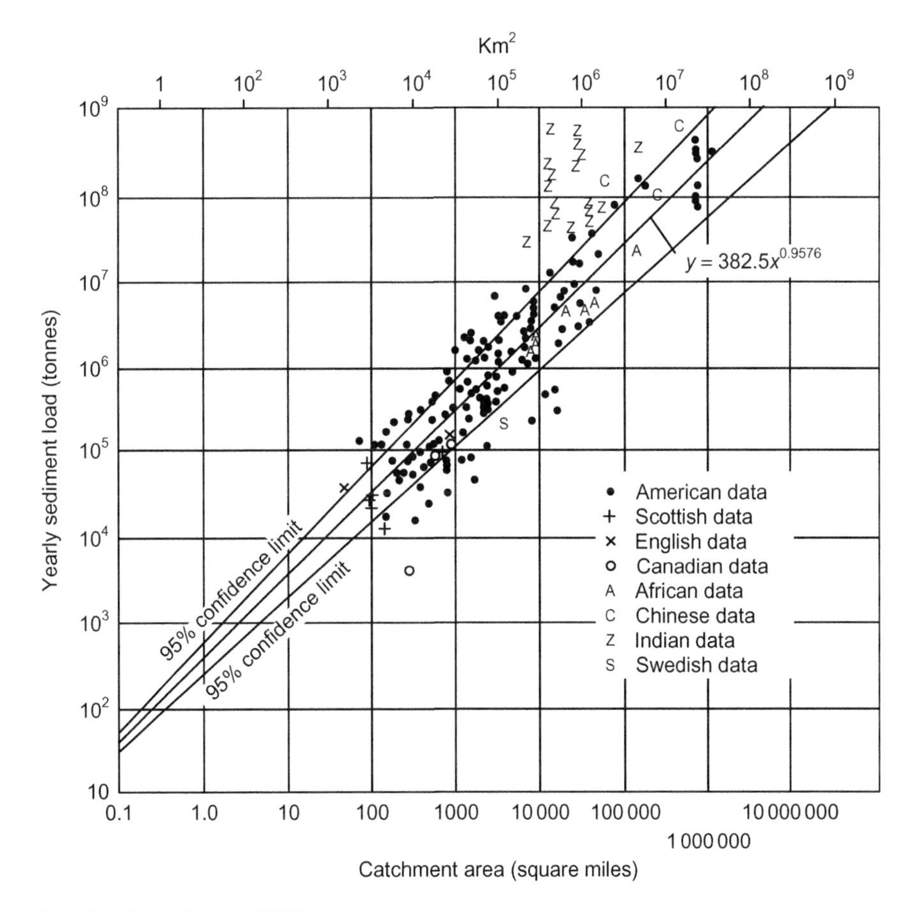

Based on information from: *Fleming (1965).*

FIGURE 5.13

Average suspended sediment concentrations.

can significantly affect the degree to which the incoming sediment load is trapped. Morris and Fan (Morris, 1998) cite a reference to an average worldwide rate of loss of storage through sedimentation of up to 1%, and White (2001) quotes the case of a reservoir in Austria that was completely filled with sediment within 1 year of commissioning. On the other hand, a survey of 95 reservoirs in the UK found an 'almost negligible loss of storage', of about 0.1% per annum.

The most comprehensive current source of information specifically applicable to water supply reservoirs in the UK is by Halcrow Water (Halcrow, 2001). This points out that, although sedimentation rates in the UK are generally not high by global standards, they can be significant in some locations. Historic approaches to sediment management for British reservoirs include:

- provision of a silt trap ('residuum lodge') where the main watercourse enters at the upstream end of the reservoir; and
- provision of a bypass channel ('bywash') to divert the most sediment-laden flows.

These measures have generally been successful, although there can be problems in the disposal of trapped sediment, including the operation of bottom outlets to release sediment accumulated in the main reservoir basin into the downstream watercourse.

In many parts of the world, the sustainability of reservoirs is threatened by high sediment loads in the incoming rivers. These loads may be the result of many factors, such as:

- natural erodibility of the catchment, for example during rainfall–run-off;
- landslides, occurring naturally, or triggered by man's activities;
- earthquakes;
- deforestation, agricultural and urban development; and
- fallout of volcanic dust and ash.

ICOLD Bulletins 140 and 147 provide guidance on sedimentation and its management (ICOLD, 2011a, 2012a). In recent years, attention has been focussed on how to design and operate reservoirs in those parts of the world that are subject to high sediment loads. White (2001) concentrates on the feasibility of regular flushing of reservoirs to pass the sediment downriver and minimize the loss of useful storage capacity. Information is given on the factors that make sediment flushing practicable, which include:

- the shape of the reservoir basin – flushing is more effective in comparatively narrow steep-sided reservoir basins;
- the ability to lower the water level in the reservoir, which depends on the discharge capacity of the low-level outlets provided;
- the availability of sufficient water for flushing, which depends mainly on the reliability of seasonal patterns of inflow to the reservoir and the ease of refilling after flushing;
- the capacity of the reservoir in relation to the annual inflow; and
- the mobility of the deposited sediments.

Morris and Fan (1998) provide many case studies and examples of techniques that have been deployed to manage sediments successfully, including the removal of sediments that have been deposited over many years.

5.26 ENVIRONMENTAL CONSIDERATIONS AND FISH PASSES

Raw water storage reservoirs represent major investments, so are constructed only where there is a strong, demonstrable need. However, the construction of such major works and the blockage of natural rivers, inevitably have environmental impacts and some of these can be seen as negative. Typical impacts include the need to relocate people displaced by the reservoir, loss of heritage features, destruction or change of local flora and fauna and changes to local water quality and groundwater regimes. The experience with resettlement has been mixed, in spite of great efforts to plan and build suitable replacement accommodation and facilities. ICOLD Bulletin 146 sets out lessons learnt (ICOLD, 2011c).

No such reservoir project can now proceed unless environmental issues, such as the ones above, are addressed at the planning stage. All major funding agencies and most countries now have rules and/or laws to deal with associated environmental impacts. At the planning stage, the benefits of the project are assessed as well as any negative impacts. A focus at this stage is on trying to eliminate negative aspects or where elimination is not possible, to mitigate the effects. Projects can only proceed if positive outcomes exceed any negative impacts; on occasion a negative impact may be so severe (e.g. the loss of a unique species) that it may be sufficient to stop a project proceeding.

An overview of the environmental issues involved with respect to dams and reservoirs is given by Carpenter (2001). The issues are too numerous to be dealt with here, but a special mention is made of fish passage, as fish pass structures are often incorporated into dams, so that fish migration can continue along rivers after the dam and reservoir have been constructed.

The primary emphasis for the provision of fish passes in the past was to allow the upstream passage of migratory salmonid species (salmon and sea trout). These species are diadromous, meaning that their life cycle involves both the sea and freshwater. Historically little or no attention has been paid to other species or indeed to downstream fish migration, except with regard to their exclusion from water intakes. The safe passage of freshwater fish and diadromous species such as eel and shad is now also considered.

There are a number of approaches to the design of fish passes, which are mostly variations on the themes of steps, slopes and lifts. The 'step' approach splits the height difference into a series of smaller drops, with various forms of traverse separating pools in which the fish can rest. In the 'slope' approach, the water passes down a relatively steep slope that contains various forms of baffle to dissipate the energy and limit the water velocity, the best-known versions of these devices being due to Denil (1909) and Lariner and Miralles (Lariner, 1981). In a 'lift', the fish are attracted into a confined space and then lifted either mechanically or hydraulically before being released upstream.

The design of fish passage arrangements for adult migratory salmonids has been developed over many years but knowledge is still being gained on the needs and appropriate passage arrangements for other species. The design of fish passage arrangements is a specialist matter, requiring appropriate expert advice. In the UK, the key reference on the subject is the EA's *Fish Pass Manual* (Armstrong, 2004). See also Section 6.9.

5.27 STATUTORY CONSENTS AND REQUIREMENTS

Reservoirs and associated structures require the consent of the national and local planning authorities. Regulations vary from country to country but, in the UK, reference must be made under the Town and Country Planning Act 1990 for consent. This action normally attracts comment from various statutory bodies, local organizations, environmental groups and other stakeholders depending on the sensitivity of the location. Early consultation with these bodies is necessary to avoid delay or refusal of the planning application.

In the UK, where nationally or internationally designated nature conservation sites could be affected, consultation with Natural England, the Countryside Commission for Wales, the Environment and Heritage Service (Northern Ireland), or Scottish Natural Heritage is necessary. This may lead to the modification of the reservoir's design, method of construction, or operation to ensure that its impacts are moderated and concerns allayed. The requirements are set out in the Wildlife and Countryside Act 1981, Countryside and Rights of Way Act 2000, and the Conservation (Natural Habitats, &c) Regulations 1994 (as amended).

If other designated features, such as archaeological sites, recreational assets or important landscapes might be affected, there would be similar requirements under other legislation to consult with the relevant statutory authority and potentially modify the proposals to accommodate their concerns.

Larger reservoirs – or those likely to have a significant effect on the natural or human environment – require the preparation of a formal environmental impact assessment (EIA) under the Town and Country Planning (Environmental Impact Assessment) (England and Wales) Regulations 1999 (or the equivalent legislation in Scotland and Northern Ireland). Typically, the EIA draws together, in a systematic way, an assessment of the project's likely significant environmental effects. This helps to ensure that the importance of the predicted effects, and the scope for reducing any adverse effects, are properly understood by the public and the relevant regulators. Dams and other installations designed to impound or permanently store water always require an EIA if the new or additional amount of water impounded or stored exceeds 10 million cubic metres. Smaller structures likely to have an adverse effect on sensitive sites also require an EIA.

Similar requirements for an EIA apply to major reservoir projects outside the UK, either under the relevant national law or to meet the requirements of international funding agencies.

5.28 DAM INCIDENTS

Dams are monumental structures, designed to rigorous standards and for a prolonged, some might say indefinite, life. Many thousands of dams have been built over the years, the majority in the last 100 or so years, and the vast majority have served society well. However, very occasionally failures have occurred and it is useful for practicing engineers to learn from such events as are described below.

Inundation of hillside materials increases their unit weight and decreases their cohesion, so that, if the water level in the reservoir should be rapidly lowered, the loosened wet material may slide into the reservoir. This happened in the Vaiont dam disaster of 9 October 1963 in Italy (Jaeger, 1965) when a landslide of gigantic proportions fell into the reservoir, causing a 100 m high flood wave to pass over the crest of the 206 m high arch dam causing the deaths of 3000 people in the valley below. The dam was not destroyed but the reservoir was afterwards abandoned.

The Dolgarrog disaster (ENR, 1926) in Wales on 25 November 1925 (the last to cause any loss of life in the UK) was caused by continuing leakage below a low concrete wall only 3 m high that had been used to heighten the level of water in Lake Eigiau. At one point the wall had been taken only 0.5 m deep into clay foundations and leakage at this point so widened its passage that there was ultimately a sudden breakthrough of the lake waters. The wave of water destroyed another small dam below and engulfed the village of Dolgarrog, causing 16 deaths.

An early failure in 1884 was that of the Bouzey Dam in France (ICE, 1896), a 19 m high masonry dam which moved 0.35 m downstream under the static force of the water. It was cemented back on its foundations, but 11 years later in 1895, it split horizontally about the middle. The 60 m high concrete St Francis Dam feeding Los Angeles failed on 13 March 1928 (ENR, 1928). It was placed on weak foundation material, so that a section of the dam broke out and a flood wave 40 m high travelled down the valley at 65 km/h, causing 426 lives to be lost.

The failure of the Malpasset Dam in France (Jaeger, 1963) on 2 December 1959 illustrates the catastrophic nature of an abutment failure. The arched concrete dam was 60 m high but only 6.5 m thick at the base. It was judged afterwards that, owing to rock joint pressurization, the left abutment moved out ultimately as much as 2 m, causing rupture of the arch in a few seconds, instantaneously releasing the whole contents of the reservoir which engulfed the Riviera town of Fréjus 4 km downstream. The dam was almost completely swept away.

In Italy, the multiple-arch 43 m high Gleno Dam suddenly failed on 1 December 1923 (ENR, 1924) when one of the buttresses cracked and burst in a matter of a few minutes. The cause was attributed to poor quality workmanship: the concrete in the arches was poor and inadequately reinforced with scrap netting that had been used as hand grenade protection in the Great War, and there was evidence of lack of bond with the foundations.

The near failure of Lluest Wen dam in South Wales in 1970 (Twort, 1977) illustrates the dangers resulting from placing a culvert through the centre of an earth dam, even when the foundation is rock. The dam was built in 1896 in a coal mining area. A 'pillar' of un-mined seams of coal was left below the dam, but land settlement occurred outside this pillar and this, combined with the weight of the upstream draw-off tower, caused the plug of concrete in the culvert to fracture within the zone of the puddle-clay corewall. A small 150 mm drain pipe through the plug was fractured and seepage occurred into this, bringing with it clay from the corewall. After an unknown length of time, probably several years, a 2 m deep hole appeared in the crest of the dam 20 m above and, shortly afterwards, clay slurry began emerging from the drain pipe. The matter was rectified by lowering the reservoir and inserting a concrete cut-off wall through the clay corewall into bedrock below.

REFERENCES

Ackers, P. and Thomas, A. R. (1975). Design and operation of air-regulated siphons for reservoir and head-water control. *Symposium on Design and Operation of Siphons and Siphon Spillways*, May. London. BHRA.

Armstrong, G., Aprahamian, M. and Fewings, A. (2004). *Fish Passes – Guidance Notes on the Legislation, Selection and Approval of Fish Passes in England and Wales*. Version 1.1. EA.

Besley, P., Allsop, N. W. H. and Ackers, J. C. (1999). Waves on reservoirs and their effects on dam protection. *Dams & Reservoirs* **9**(3), December.

Binnie, G. M., Gerrard, R. T. and Eldridge, J. G. (1968). Mangla. Engineering of Mangla. *Proc. ICE*, 1967, 38 (November) pp. 337–576 and 1968, 41 (September) pp. 119–203. ICE.

Bishop, A. W. (1955). The use of the slip circle in the stability analysis of slopes. *Géotechnique, ICE* **5**(1), 7, March.

Carpenter, T. G. (Ed.) (2001). *Environmental, Construction and Sustainable Development*, Vol. 2. John Wiley & Sons Ltd, Sustainable Civil Engineering.

Charles, J. A., Abbiss, C. P., Gosschalk, E. M. and Hinks, J. L. (1991). *An Engineering Guide to Seismic Risk to Dams in UK*. ICE.

Cooke, J. B. (1990). *Rockfill and Rockfill Dams, Stan Wilson Memorial Lecture*. Univ. of Washington, USA.

CIRIA, CUR, CETMEF (2007). *The Rock Manual – The Use of Rock in Hydraulic Engineering*. 2nd Edn. CIRIA, London.

Davis, C. V. (1969). *Handbook of Applied Hydraulics*. McGraw-Hill.

Deacon, G. F. (1895–96). The Vyrnwy works for the water supply to Liverpool. *ICE Trans*, CXXVI Pt. IV, pp. 24–125.

Denil, G. (1909). Les echelles a poissons et leur application aux barrages de Meuse et d'Ourthe (Fish ladders and their use at the Meuse and Ourthe dams). *Bull. Acad. Belg.*, pp. 1221–1224.

Dunnicliff, J. (1993). *Geotechnical Instrumentation for Monitoring Field Performance*. John Wiley & Sons Inc., New York.

Dunstan, M. R. H. (1994).The state-of-the-art of RCC dams. *International Journal of Hydropower and Dams*, March.

ENR (1924). *Engineering News Record*. 7 August, pp. 213–215.

ENR (1926). *Engineering News Record*. 7 January 1926 and 25 November.

ENR (1928). *Engineering News Record*. 29 March p. 517, and 12 April p. 596.

FEH (1999). *Flood Estimation Handbook*. 5 Vols. IoH.

Fleming, G. (1965). Design curves for suspended load estimation. *Proc. ICE*, 43, pp. 1–9.

Geddes, W. E., Rocke, S. and Scrimgeour, J. (1972). The Backwater Dam. *Proc. ICE*, 51, March p. 433.

Halcrow Water (2001). *Sedimentation in Storage Reservoirs*. Final Report. DETR. London.

Herbert, D. M., Lovenbury, H. T., Allsop, N. W. H. and Reader, R. A. (1995). *Performance of Blockwork and Slabbing Protection for Dam Faces*. HR Wallingford Report SR 345.

ICE (1896). The Failure of the Bouzey dam (Abstract of Commission Report). *Proc. ICE* CXXV, p. 461.

ICE (1998). *Application Note to 'An Engineering Guide to Seismic Risk to Dams in UK'*. ICE.

ICE (2000). *A Guide to the Reservoirs Act 1975*. ICE. Thomas Telford.

ICE (2014). *A Guide to the Reservoirs Act 1975*. 2nd Edn. ICE Publishing.

ICE (2015). *Floods and Reservoir Safety: An Engineering Guide*. 4th Edn. ICE Publishing.

ICOLD (1990). *The Design Flood – Guidelines*. Committee on Design Flood. ICOLD.

ICOLD (1992). *Bulletin 82*. Selection of Design Flood – Current Methods. ICOLD.

ICOLD (2000). *Bulletin 118*. Automated Dam Monitoring Systems – Guidelines and Case Histories. ICOLD.

ICOLD (2002). *Bulletin 124*. Reservoir Landslides: Investigation and Management. Guidelines and Case Histories. ICOLD.

ICOLD (2003). *Bulletin 126*. Roller-Compacted Concrete Dams – State-of-the-Art and Case Histories. ICOLD.

ICOLD (2005). *Bulletin 129*. Dam Foundations: Geological Considerations, Investigations, Treatment, Monitoring. ICOLD.

ICOLD (2009). *Bulletin 138*. General Approach to Dam Surveillance. ICOLD.

ICOLD (2010). *Bulletin 148*. Selecting Seismic Parameters for Large Dams – Guidelines. ICOLD, 2010 Revision.

ICOLD (2011a). *Bulletin 140*. Sediment Transport and Deposition in Reservoirs. ICOLD.

ICOLD (2011b). *Bulletin 141*. Concrete Faced Rockfill Dams – Concepts for Design and Construction. ICOLD.

ICOLD (2011c). *Bulletin 146*. Dams and Resettlement – Lessons Learnt and Recommendations. ICOLD.

ICOLD (2012a). *Bulletin 147*. Sedimentation and Sustainable Use of Reservoirs and River Systems. ICOLD.

ICOLD (2012b). *Bulletin 150*. Cutoffs for Dams. ICOLD.

ICOLD (2013a). *Bulletin 157*. Small Dams: Design, Surveillance and Rehabilitation. ICOLD.

ICOLD (2013b). *Bulletin 158*. Dam Surveillance Guide. ICOLD.

ICOLD (2014). *Bulletin 164*. Internal Erosion of Existing Dams, Levees and Dykes and their Foundations. ICOLD.

Ischy, E. and Glossop, R. (1962). An introduction to alluvial grouting. *Proc. ICE* 21 March, p. 449.

Jaeger, C. (1963). The Malpasset Report. *Water Power* 55–61, 15 February.

Jaeger, C. (1965). The Vaiont Rock Slide. *Water Power*, 17 March pp. 110–111 and 18 April pp. 142–144.

Johnston, T. A., Millmore, J. P., Charles, J. A. and Tedd, P. (1999). *An Engineering Guide to the Safety of Embankment Dams in the United Kingdom*. Report 171, BRE, 1990 new Edn.

Kenn, M. J. and Garrod, A. D. (1981). Cavitation damage and the Tarbela Tunnel collapse of 1974. *Proc. ICE* February, p. 65.

Kennard, M. F., Owens, C. C. and Reader, R. A. (1995). *An Engineering Guide to the Safety of Concrete and Masonry Dams in the UK*. Report 148, CIRIA.

Lariner, M. and Miralles, A. (1981). *Etude hydraulique des passes a ralentisseurs* (Hydraulic Study of Denil Fishways). Unpublished report, CEMAGREF.

Londe, P. and Lino, M. (1992). The faced symmetrical hardfill dam: a new concept for RCC. *Water Power and Dam Construction* **44**(2), 19–24, February.

McIldowie, G. (1934–35). The construction of the Silent Valley Reservoir. *Trans. ICE* 239, p. 465.

Morgenstern, N. R. and Price, V. E. (1965). The analysis of the stability of general slip surfaces. *Géotechnique* **15**, pp. 79–93.

Morris, G. L. and Fan, J. H. (1998). *Reservoir Sedimentation Handbook: Design and Management of Dams, Reservoirs, and Watersheds for Sustainable Use*. McGraw-Hill, New York.

NERC (1975). *Flood Studies Report*. NERC, London.

Rawlings, C. G., Hellawell, E. E. and Killkenny, W. M. (2000). *Grouting for Ground Engineering*. CIRIA C514, London.

Scott, C. W. (2000). Reservoir outlet works for water supply. *Proc. ICE, Water & Maritime Engineering*, Vol. 142, December.

Sherard, J. L. and Dunnigan, L. P. (1989). Critical filters for impervious soils. *Journal for Geotechnical Engineering, ASCE* **115**(7), pp. 927–947.

Smith, N. (1971). *A History of Dams*. Peter Davies. London.

Stewart, E. J. and Young, A. (2015). *Flood Estimation Handbook – Update 2015*. Circulation No 127, British Hydrological Society.

Twort, A. C. (1977). The repair of Lluest Wen dam. *Journal of IWES*, July, p. 269.

USACE (1987). *Hydraulic Design Criteria. Morning Glory Spillways*. HDC 140-1 to 140-1/8. Publication of the Headquarters. USACE.

USBR (1977). *Design of Small Dams*. US Bureau of Reclamation, US Govt Printing Office, Washington.

USBR (1985). *Hydraulic Model Studies of Fuse Plug Embankments*. REC-ERC-85-7. Engineering and Research Center, Denver. USBR.

Vail, A. J. (1975). The high island water scheme for Hong Kong. *Water Power and Dam Construction*, January, p. 15.

White, W. R. (2001). *Evacuation of Sediment From Reservoirs*. Thomas Telford.

W&WE (1960). The breaching of the Oros Dam, North East Brazil. *Water & Water Engineering*, August, p. 351.

Yarde, A. J., Banyard, L. S. and Allsop, N. W. H. (1996). *Reservoir Dams: Wave Conditions, Wave Overtopping and Slab Protection*. HR Wallingford Report SR 459.

CHAPTER

Intakes

6

6.1 INTAKE FUNCTION, TYPES AND MAIN CONSIDERATIONS

The primary function of an intake structure is to safely withdraw a required quantity of water from a water source over a pre-determined range of water levels at a selected location and to transfer it to the conveyance conduit for treatment or use. The water source where the intake structure is required may be natural or it may be artificially created or altered; potential sources include streams and rivers, canals, lakes, reservoirs, estuaries or the sea.

Various types of intakes have been used historically for different applications. Table 6.1 shows the main types and key features.

Table 6.1 Types of intake and their application

Type	Location	Capacity
Cylinder with wedge wire or perforated plate fine screen (multiple for larger flows)	Submerged, above bed of river, lake or sheltered sea	Low to medium
Radial well (from collector chamber)	In permeable strata in the banks of rivers or lakes or in ocean shores	Low to medium
Stream bottom grid	In the bottom of small upland streams	Low to medium
Bankside intake	In the bank or a river	Medium to very large
Direct pumping	Pumps on pontoon or jetty	Low to medium
Tower intake	Away from the shore in water depth up to about 100 m. Access via a bridge if practicable	Low to medium
Velocity cap intake	Submerged offshore	Medium to large
Shoreline	On the sea shore, estuary bank or reservoir	Medium to very large

Twort's Water Supply. DOI: http://dx.doi.org/10.1016/B978-0-08-100025-0.00006-5

The intake design flow should suit the needs of the system it supplies but will be limited by the hydrology of the water source and by environmental constraints. Abstraction for potable water use competes in many locations with the demands for other uses such as agriculture, industry, hydropower, thermal and nuclear power cooling and navigation (canals). Licences for abstraction, water rights and priority uses are discussed in Chapter 2 and Section 6.5.

Other considerations affecting the location, type and design of intakes include: access, sediment management, river morphology, geology, water level variations and flooding, water quality and pollution, fish exclusion and protection, biogrowth, trash and ice. These aspects are discussed in the following sections.

GENERAL CONSIDERATIONS

6.2 INTAKE CAPACITY

An intake must have sufficient capacity to satisfy the demand of components of the transfer system and other facilities downstream in all normal circumstances. This may be described as the normal maximum instantaneous demand. This is determined as the combination of all flows, including those for screen backwashing, feeds to filter backwash tanks and other process offtakes at treatment works downstream. The intake must be able to abstract and deliver the resulting aggregate demand under all design conditions, considered together, such as low-source water level, partial screen obstruction and accumulated biogrowth in conduits.

At many locations, construction of an intake involves work in water or excavation below the water table next to a water body. This type of work requires the mobilization of special construction plant which, for offshore marine intakes, may be very costly, or construction of a cofferdam at some cost and at the penalty of temporary environmental impacts. Where there is sufficient water for an extension to the abstraction licence to be granted and to avoid having to mobilize expensive equipment or build expensive temporary works a second time, it may make sense to build the intake structure large enough to facilitate increased abstraction in the future. Therefore, the design of the intake should take into account increases in demand which could materialize in the future.

6.3 HYDROLOGY, BATHYMETRY AND HYDRAULICS

Before the intake site is confirmed and the type of intake is selected data on hydrology (Chapter 3), bathymetry and hydraulics of the source need to be collected and assessed. Historical data on flows, water quality, floods, maximum and minimum water levels, currents, suspended load and bed load and likelihood of freezing or ice accumulation must be assembled. In addition, for estuaries and where an intake is to be located at sea it will be necessary to obtain information about tides and waves that would apply at the location.

Bathymetry of the area normally plays an important role in the selection of the most appropriate type of intake structure and the details need to be obtained at an early stage of the selection process. At locations where the bathymetry may change as a result of flows, floods, tidal currents and other naturally occurring events, it may be necessary to obtain information on the variability of the bathymetry at different times of the year and at different stream flows (Section 6.21).

6.4 WATER QUALITY AND PROTECTION FROM POLLUTION

The selection of site for an intake structure must ensure that the source water in the area of the intake is not polluted and that measures are in place to ensure that future pollution is not likely to occur or that systems are in place to allow shut down of the intake if pollution of the catchment upstream is detected early enough. However, the aim should be to ensure that enforced temporary shutdown of the intake facility as a result of pollution is not necessary. Unless there is spare capacity at other sources serving the same distribution system, intake shut down can be tolerated only if there is sufficient raw water storage to cover the shutdown period. Otherwise, as far as practicable, the selected location should be such that the risk of pollution from substances that are difficult or expensive to deal with is minimized.

If a sufficient history of water quality data is not available a sampling and monitoring programme should be implemented, of duration sufficient to cover periods of low and high flows, but at least over 1 year. In drought, discharges are much less diluted while in floods there may be appreciable discharges of untreated wastewater if sewerage systems get overloaded, as well as agricultural run-off.

Pollution of a watercourse from agricultural discharges and releases of phosphates and nitrates is generally of concern and requires particular attention in selecting a location of an intake for potable water supply. Spillage of fuel is also a risk in some catchments. However, some treatment clarification processes such as flotation can deal with hydrocarbons better than others (Section 8.18).

Catchment management plays an important part in the control of pollution upstream of an intake and can help reduce turbidity and suspended sediment in the abstracted water. In the EU, countries have to implement River Basin Management Plans (Section 2.4); the principal aim being to achieve 'good status' of the water in the basin. In all catchments conservation of vegetation and soil, containment of contaminants, interception of pollutants and treatment of discharges play an important role in preserving good water quality for both abstraction purposes and for the benefit of river ecology.

6.5 ABSTRACTION RESTRICTIONS

In most countries construction of intake works and abstracting water from a source are subject to regulation affecting aspects, which are often interlinked, such as environmental impact, planning and permitting. Some form of environmental impact assessment is likely to be required for water supply intakes. Some of the environmental issues which need to be considered are discussed in Sections 6.8 and 6.11. The regulatory framework that applies for the intake location must be understood early in the process of intake planning.

The Water Resources (Abstraction and Impounding) Regulations 2006 (UKSO, 2006) is the statutory instrument which covers the abstraction of water for England and Wales. Separate requirements apply in Scotland where the role of the regulatory body comes under The Scottish Office. In Northern Ireland the legislation comes under The Fisheries Act (Northern Ireland) 1966 and the overall responsibility for regulation comes under the Department of Agriculture for Northern Ireland (DANI).

In England and Wales abstraction of 20 m^3/d or more from most water sources (river or stream, reservoir, lake or pond, canal, spring, underground source, dock, channel, bay, estuary or arm of sea) requires an abstraction licence from the Environment Agency. Currently over 48 000 licensed abstractions of varying sizes are registered.

The United States Environmental Protection Agency (US EPA) regulates water abstractions in the USA under the provisions in the Clean Water Act (Section 316b). The Act requires that the location, design, construction and capacity of a water intake reflect the best available technologies for minimizing adverse environmental impact. A considerable amount of background research and consultation with interested parties was carried out before releasing these regulations and form a very useful guide for applications elsewhere outside the USA where the environmental guidelines may not be fully defined.

The licensed abstraction may vary at different times of year and may also depend on the available flow at the location at any particular time so as to maintain a minimum flow downstream. The limitations may require controls which affect the design of the works so need to be considered early. The consents may include requirements during construction of the intake and its subsequent operation and maintenance.

In Europe, member states have incorporated the EIA Directive and the Strategic Environmental Assessment (SEA) Directive into their laws. Projects are classed as Annex 1 projects which will always require an Environmental Impact Assessment (EIA) and Annex 2 projects for which member states determine environmental assessment requirements. In the UK environmental assessment follows stages of: *screening* to determine if the project requires assessment; *scoping* to define the issues to be covered in the assessment; and preparation of the Environmental Statement.

In the US, requirements for study of environmental impact come under the National Environmental Policy Act (NEPA). An Environmental Assessment (EA) is produced to determine the need for an Environmental Impact Statement (EIS) which is a very onerous undertaking. Even the EA is now considered quite onerous.

6.6 TRASH AND COARSE SCREENS

Floating or submerged trash, such as timber, tree branches, reed, plastic or other material, is likely to be present at an intake, particularly after storms or in floods. Trash which could damage downstream equipment or obstruct openings is prevented from entering the intake by trash racks located in front of the intake.

Trash racks are sized to deal with the likely amount of trash that may reach the location. They require regular cleaning to remove any trash that has been retained. This can be carried out manually where the amount of trash is small. Where a large amount of trash is likely to be present, mechanized trash rack cleaning equipment will be needed and can be operated periodically as required (Plate 9(a)). If a very large amount of trash is expected it may be necessary to install continuous chain-operated raking devices; initiation of operation can be controlled automatically on pressure drop across the racks if suitable instruments are installed.

The bar spacing at a trash rack should be selected to ensure that any material that passes through the screens will not damage mechanical equipment in the intake system such as pumps. A bar spacing of 40–150 mm clear gap is often used. Screen bars may be of rectangular flat section or round bars; they are fixed in a frame with intermediate supports if required. Biological growth on trash racks is a common problem and requires periodic inspection and cleaning. Excessive growth will restrict flow, increase headloss and may cause malfunction of downstream equipment such as band screens or pumps. A common material for trash screens is coated or galvanized carbon steel. For estuarial waters or the sea it may be necessary to use stainless steel; however, cupro-nickel is an alternative material which minimizes biofouling although it is not compatible with cathodic protection systems if they are needed for other reasons.

The average flow velocity immediately in front of screens needs to be low enough to ensure that impingement (Section 6.10) of fish does not occur. A mean velocity of 100–150 mm/s is required to minimize entrainment of small fish (Schuler, 1975). Where large flows are involved this can result in a very large screen area. The hydraulic headloss (Section 6.15) across the screen when partially blinded by trash must be taken into account in the hydraulic calculations for the intake system.

Metallic intake trash racks located in fresh water lakes and rivers in the northern regions are subject, in cold weather, to the formation of frazil ice (AWWA, 2012) which may result in complete blockage and shut down of the intake. Such blockages tend to occur at night and mostly under water and may not be noticed. Nevertheless, increased load due to the ice can result in collapse of the trash rack if the rack is unable to support the additional load. A headloss alarm would allow the intake system to be shut down manually or, preferably, automatically before damage occurs. The problem is reduced by use of low thermal conductivity materials, such as wood or fibre reinforced plastic for rack components.

6.7 ICE

Design of intake structures located in areas where the water body can freeze during winter periods need to consider the forces induced by ice sheets and accumulations of floating ice against the structure. Ice loading can be severe and must be evaluated at sites in rivers, lakes and the sea in North America, Northern Europe and Northern Asia. Moving ice sheets at the surface of an open water body can induce large foundation loads and cause failure.

The magnitude of the forces induced by the ice should be estimated based on the ice conditions, the levels at which it can form in relation to the structure and the movement of the ice. The following ice forces should be considered:

- Vertical forces that arise as a result of water level change after the ice has formed around (and gripped) the structure;
- Horizontal load from moving ice floes;
- Horizontal load (ice expansion) due to temperature fluctuations;
- Pressure from ice that has accumulated within passages in the structure.

Several design approaches, codes, recommendations and guidelines (some draft) are in use in various situations. The reader is referred to guidance in API RP 2N, ISO 18806, Lloyd (2005), IEC 61400 and Frannsson (2009).

If ice is likely to form against a structure, information on its formation, particularly thickness, should be collected. If a structure is located near the shoreline ice loading may be less but the presence of the ice may render the intake temporarily unusable. At inland locations in lakes and rivers where formation of ice of significant thickness is less likely, alternative measures, such a bubble curtains, can be taken to prevent its formation close to the structure.

6.8 ENVIRONMENTAL ISSUES

Water intakes directly affect the flow, water level and sediment behaviour and may indirectly affect infiltration to or from aquifers, water temperature, nutrients and weed growth, all of which are likely to impact the biology and ecology of the source. Other environmental aspects include visual amenity, flora and fauna, noise and recreation. Work will be required to assess these aspects early in the intake planning process with possible further detailed work done, if required, at the design stage. Stakeholders with an interest should be consulted as early as possible; options for avoiding the need for an intake or for its location or configuration will need to be considered.

Intakes and their screens (Sections 6.12 and 6.13) have the potential to kill or damage animals by entrapment or impingement and affect fish and other marine organisms (including micro-invertebrates, fish eggs and larvae) in the area of the intake. Entrapment occurs when the organism is drawn into the intake and is not able to swim away or otherwise escape. Impingement occurs when an organism is held in contact with the intake structure or screen by the flow. The severity of entrapment or impingement will depend on the abundance of the aquatic life at a given time, its size and behaviour in the vicinity of the intake, and on the abstracted flow, screen-mesh size and the physical arrangement of the intake.

Changes to flow may affect spawning grounds or nursery areas where juvenile fish shelter in gravel, reeds or weed which invertebrates inhabit. Flow change may also affect the reproduction of insects. Other biological issues include water quality, sedimentation and temperature since these can affect the egg, larval and juvenile fish stages as well as plants.

Clearly the smaller the ratio of abstracted flow to minimum downstream flow the less the impact on the ecology of the water source. Therefore, abstracted flows should be minimized as far as possible but often the abstracted flow will represent a significant proportion of minimum flow downstream. Compensation flows will need to be allowed for (Section 3.23) and, in addition, mitigation should be considered so that the effects of the flow reduction are reduced or compensated by ancillary works.

If the basic information about the aquatic biota present at a location is not available a site-specific assessment study should be carried out to check the species present, obtain baseline information and for intake design. This will help ensure that a proper assessment of the impact of constructing an intake at the site can be made and assist in selecting the appropriate measures to minimize any adverse impacts on the aquatic biota (Fedorenko, 1991) when considering the construction of an intake.

6.9 FISH BEHAVIOUR

Early concerns about the impact of intakes on fish were limited to commercial species, including those that migrate upstream such as salmon and sea trout. Water ecology and the extent of environmental concerns and legislation now require consideration of all species of plant and animal that may be affected by the construction and operation of works such as intakes.

The migration of anadromous species such as salmon and sea trout is very well known and has been allowed for in the construction of dams, barrages and intakes for at least 50 years. More recently, attention has focussed on less common catadromous species such as the eel which spawns in remote oceanic waters. Juvenile eels (elvers) migrate into coastal and inland waters to grow to adulthood and then return to the sea. Eel stocks in rivers have depleted over the years and the European Commission has implemented an eel recovery plan (Council Regulation No 1100/2007) which aims to return eel stocks to sustainable levels. Guidelines on measures to be taken at intakes are set out in *The Eel Manual* (Sheridan, 2014).

Fish are equipped with a number of senses which serve principally for feeding and locating prey, spawning and escape from predators. Senses include sight, touch, smell and taste, hearing and pressure and voltage (along the lateral line). Fish may be shoaling (swimming in large groups for protection) or solitary. They may congregate in the shelter of plant growth (reeds, mangrove or weed) or structures or in shadow in order to avoid predators.

Some fish remain in the same fresh water system (*potamodromous*) and some migrate seasonally between fresh and salt water in search of food (*amphidromous*) or as part of their life cycle (*diadromous*). The latter are either *anadromous* (salmon and sea trout) which return to fresh water for spawning or the less common *catadromous* species (such as eel) which migrate to the sea for breeding. Some species move up and downstream on a daily basis.

Fish are mostly cold-blooded (exceptions are some tuna and shark) and are adapted to a certain water temperature range. Their activity generally reduces with temperature. This can lead to the existence of a temperature barrier for upstream passage of migrating fish in water of temperature below about $5 \pm 3°C$, the variability being due to species and differences in local stream velocity. Fish breathe oxygen dissolved in the water but there is a limit (oxygen saturation) which decreases with increasing temperature. Clough (2001) states that muscle in most fish is of two types: red, contracting in the presence of oxygen and white, using stored glycogen (which is replaced on availability of oxygen over a period of hours). Red muscle is used for sustained swimming and white for burst swimming.

Some fish need to move to be able to breathe (shark) but most are able to take rest periods between bursts of activity. This pattern is useful for evasion from predators and other dangers and for upstream progress in rivers where shelter from higher stream velocities may be found in the lee of rocks or in pools. Much work has been done to investigate fish swimming capacity since this is important for the design of fish passes (Section 5.26) and intake systems. However, results are often affected by differences between the experimental and the natural environment and by behavioural and motivation factors. Swimming capabilities of fish depend on their size but also the species; e.g. 50 mm long sockeye salmon can swim twice as fast as lake sturgeon of the same length (100 and 50 cm/s, respectively) (Schuler, 1975).

Data gathered in fish swim speed experiments has been utilized in the formulation of algorithms which are used in swim speed software such as that developed by Clough (2001) for some fresh water fish in UK rivers and developed for transit of fish through culverts in North America. Fish swim speed is a function of species, stage (juvenile or adult), length, water temperature and duration. Burst swim speeds may be as much as 10 times sustained swim speeds.

Fish tend to avoid areas where their senses suggest that predators or other dangers may be present. This aspect of behaviour can be utilized to keep them away from intakes as discussed in Section 6.10.

The environmental studies that are required for intakes will include assessment of the species of vertebrate and invertebrate animals that may be present as well as other biological life, and then assessment of the impact of different options for the intake. These studies require expert knowledge and consultation is best commenced early while all options are still on the table.

6.10 FISH EXCLUSION

Legislation and guidance provided by regulating authorities in Europe, North America and other countries reflect the increasing requirements for measures that must be incorporated at water abstraction facilities to minimize the 'entrapment' and 'impingement' (Section 6.8) of aquatic species. The best available technology and good environmental practices increasingly require use of guidance systems that help guide the fish clear of an intake and thus avoid entrapment or impingement. The provision of such facilities is particularly important where large volumes of water are to be abstracted. Use of an intake with horizontal flow with approach velocity less than the swim speed of the fish provides an environment away from which the fish can escape. Whether approach velocity should be compared with the sustained or burst swim speed depends on the approach geometry; for a long channel approach sustained swim speed should be used.

Knowledge of the fish species that are likely to be present at a particular location is necessary for selecting particular fish exclusion systems. Specialist advice needs to be sought from suppliers of fish guidance equipment before selecting any particular combination of systems for a location. Options include behavioural barriers such as bubble curtains, acoustic fish fences, intense flashing lights, electrical charge and flow altering arrangements. Physical barriers such as net barriers are also used. Behavioural barrier systems are more appropriate at locations where physical screening is not practical or not desirable on other grounds including the species or life stages of fish present (USBR, 2006).

Bubble curtains: A bubble curtain is a simple type of behavioural barrier; the bubble curtain is generated bypassing compressed air into a perforated pipe running along the base of the barrier. Depending on the depth of water, more than one compressed air pipe may be required to create an effective air bubble curtain. The bubble curtain is aligned to guide the fish to a bypass or away from the intake.

The design of the air pipes and supports needs care to ensure stability in water particularly where the current is high or in tidal water where there is wave action. Submerged pipes carrying compressed air normally require weighting for stability and support. A continuous compressed air supply will consume energy and will require a power source which can be problematic in remote locations. Bubble curtains are more effective when used with other methods such as acoustic fences or lights.

Acoustic fish fences: Suitably positioned sound projectors in the area of a water abstraction point can help to guide fish away from an intake. The acoustic stimulus is produced by electromechanical sound projectors which generate sound of frequencies between 5 and 600 Hz at source levels of around 160 dB re 1 μPa @ 1 m. The sound projection system can either be mounted directly on the river/lake bed or can be located on a horizontal framework supported by columns.

When used with a bubble curtain the sound generated by the projectors is contained by the air bubbles, allowing a precise linear wall of sound to be developed. The trapping of the sound signal within the air bubble curtain prevents contamination of the surrounding area by sound which typically falls to ambient levels at a range of few metres (2 or 3 m) from the bubble curtain axis.

The acoustic fence is a widely used behavioural system for guiding anadromous species, resident coarse fish or estuarine marine fish which are sensitive to sound stimulus. The system depends on availability of a reliable power supply and requires regular maintenance to ensure proper functioning.

In the UK an acoustic deflection system when used with intensive light screen is considered as the best available technology for large intakes for coastal power plants, when used in conjunction with travelling fine screen (either drum screen or band screen) fitted with proven fish recovery and return systems.

Intense flashing lights: Linear arrays of intense flashing lights are used to generate a visual stimulus. The light is generated by LED powered devices that create white light in a vertically orientated beam. When used with a bubble curtain, the air bubbles reflect the light beam and improve the visibility from the direction of approaching fish and deflect them away from the abstraction point. Irregular flashes are more effective than regular ones; modulated intense light is more effective than continuous steady light and is particularly effective in repelling eels. However, light screens can be visible above water at night and may be objectionable from a light pollution point of view.

Electric charge barrier: Electrical charge screens have been used at many locations in the past as a barrier to fish entry but have gone out of use partly because of safety concerns with human and other animals present near an intake. These screens use alternating current. Some newer systems are available (e.g. Geiger fish repelling system), where widely spaced electrically charged electrodes are used to generate an electrical field. The system uses direct current and is arranged to suit site-specific conditions of water conductivity, temperature and depth of flow. Water quality (salinity) and environmental conditions at the water abstraction point need to be carefully examined before considering use of electrical barriers for fish exclusion.

Flow altering behavioural barriers comprise devices such as louvres and changes in depth, width and direction of flow to produce velocity changes or turbulence. Louvre systems do not physically exclude fish from the intakes but create hydraulic flow conditions that help to guide fish along the line of louvres and away from intake.

Net barriers: Net barriers can be utilized if the location of the intake is appropriate and leaves ample space for fish to go past the intake facility. They can be particularly suitable for excluding jellyfish from a marine intake provided the net barrier covers the whole depth of water. Selection of appropriate mesh size is important to ensure that gilling and resulting mortality of fish does not occur. Net barriers will not be effective for juvenile fish or larval/egg stages of fish. They can be relatively inexpensive to install but have high maintenance requirements.

6.11 FISH PROTECTION SYSTEMS

Fish protection systems are available to help minimize fish entrainment. Impingement is avoided by using a sufficiently low flow velocity in the vicinity of an intake screen to enable the fish to swim away. Apart from behavioural barriers and nets (Section 6.10) there are two forms of fish protection systems available: passive screens and active screens with fish return systems.

The aim is to avoid fish of all sizes and stages of development from being drawn into the intake as far as possible. The screen opening size is selected to be small enough so that fish (or larvae or eggs) of the target size cannot pass. Further information on screening at intakes may be found in ASCE (1982), EA (2005) and Turnpenny (2014).

Passive screening requires no provision for collection or return of fish.

Active screening provides for screening equipment to safely collect any fish that arrives at the fine screening equipment and to return them to the water source downstream of the intake with minimum mortality and damage.

Fine screening can be regarded as the first stage in the treatment of raw water. It is best done near to the intake and upstream of mechanical equipment such as pumps and in a location where fish may be recovered and returned to the water source. However, the screening reduces the load on treatment processes downstream.

6.12 SCREENING – PASSIVE

Passive screens are static devices using wedge wire (Johnson type cylindrical screening units similar to their well screens) or perforated cling free plates mounted on cylindrical cans or polyhedron units (Tabrogge/Tapis type) and provide robust means of excluding aquatic life and floating debris from intake water. Closely spaced wedge wires or small perforations can provide a friendly choice for excluding juvenile fish, eggs and larval/post-larval stages of aquatic life from an intake by reducing screen opening size as appropriate. Field trials indicate that the overall effectiveness of cylindrical wedge wire screens depends on the biological characteristics of the species present and physical parameters of the screens (slot width). In general entrainment decreases with longer larval length (very little entrainment for larval lengths over 7 mm for many species); a 0.5 mm slot width is required to reduce entrainment of eggs (Dixon, 2005). The opening size of passive screens is usually small enough to avoid the need for secondary screening. Hydraulic losses are discussed in Section 6.15.

Passive screening systems have no moving parts below water and provide a relatively simple method of excluding debris and aquatic life. In static water or where flow velocity is small, trash may accumulate around the screens. At such locations, provision of an air backwash clearing facility should be considered to remove any debris that collects on the external face of the screens. Debris is thrown back into the source water so no disposal arrangements are necessary. Such cleaning will not totally remove weed or fibrous material which tends to get wrapped around the screen wedge wires nor will it remove growth on the inside of the screen (Gille, 2003). Therefore, provision should be made for periodic removal of screen units for cleaning.

Air backwash cleaning of passive screens can be more effective for smaller units than large ones. The process may be objectionable in some locations due to noise, safety or local recreational use of the water, particularly for large screens. Therefore, use of multiple units should be considered, together with means of isolating each unit to allow continued use of the intake while one unit is undergoing cleaning or is removed for repair. Air bubbles dispersed in water reduce the buoyancy of bodies immersed in it or floating on it; therefore, it would be necessary to design an exclusion zone around the affected area.

Some species and stages are more active at night and so more likely to be impinged on the screen. This tendency can be reduced by reducing abstraction at night, when the water demand may be lower or by employing a behavioural barrier.

Passive screening offers simplicity and many advantages (economic and environmental) but presents some practical issues, not necessarily apparent from the screen supplier's technical material, which need to be considered at the design stage:

- Stainless steel of a grade appropriate for the source water reduces metallic corrosion of screen units but does not control biofouling and marine growth. Cupro-nickel should be considered as an alternative for screen materials, particularly for intakes in the sea or estuaries, since it reduces biofouling but at an increased cost.
- The air backwash rate should be controlled so that flow is gradually increased at the start of each wash sequence. This allows fish to move away and avoid harm and reduces the stress on the air pipework which is initially full of water.
- Any chlorination of the water to control biogrowth in downstream conduits should be carried out at a location downstream of the screen to ensure that no chlorine enters the water source. Chlorine dosing lines should be laid inside the intake conduit wherever possible and other measures taken, such laying in a duct to avoid damage during conduit cleaning.
- Where chlorination within the screen units is used marine organisms in the area will be stunned or killed and some may enter the water conveyance system further downstream. Facilities for periodic cleaning will therefore be required beyond the screen units.
- Screening units installed in estuaries or open sea are at risk of damage by large fish or mammals unless they are very robust or are protected by a netting. The latter is beneficial where smarms of jellyfish can be expected.

6.13 SCREENING – ACTIVE

Active screening uses wire mesh or perforated plate screen panels which move through the incoming water using a form of mechanical drive. The three basic systems are:

Drum screen (Plate 9(b))
Travelling band screen (Plate 9(c))
Multi-disc screen

Water passes through the submerged parts of the screen panels and the retained material is washed off with water jets. The screen panels move through the water at a speed which is pre-set or controlled automatically according to headloss across the screen. The approach velocity of the flowing water should be selected to suit the requirements of the material to be retained by the screening system. To minimize damage to juvenile fish and other living organism, an approach velocity of about 50 cm/s is recommended for times when fish are likely to be present.

All three systems can be provided with troughs to retain fish, fish eggs or larval fish depending on the mesh size selected. When the screen panels emerge above the water the collected fish are washed off the panels using low-pressure water sprays and retained in water troughs at the bottom of each screen panel. The contents of the fish water troughs are emptied into a collection channel and returned back to the water source downstream of the intake with minimum harm or damage to the fish. Other rubbish retained by the screen panels is washed off the screen panels, after the fish have been collected and removed, by high pressure water sprays and disposed to collection baskets.

A supply of clean water under pressure is needed for screen washing; this has to be pumped from the screened flow, through debris filters if necessary. The total flow required for washing may be about 1% of the maximum throughput; more may be needed where there is provision for fish recovery.

The three types of screen have been in use for several decades and are well proven. Band screens have a smaller footprint and require a smaller structure than drum screens. However, they have more moving parts and require more maintenance. As is increasingly required for environmental reasons, fish collection systems can be retrofitted to any of the screen types if they were not provided with this facility originally.

The travelling band screens and the drum screens are available in through flow or dual flow configurations and in a variety of different materials including epoxy-coated or galvanized carbon steel, stainless or non-metallic mesh and UV resistant polymer modules according to the water source. For seawater applications stainless steel should have a Pitting Resistance Equivalent Number (PREN) of at least 40 to avoid crevice corrosion, erosion corrosion and pitting. PREN depends on the proportion of noble metals (chromium (Cr), molybdenum (M), tungsten (W)) and nitrogen (N) present in the alloy:

$$\text{PREN} = 1 \times \%\text{Cr} + 3.3 \times (\%\text{M} + 0.5 \times \%\text{W}) + 16 \times \%\text{N}$$

6.14 BIOLOGICAL FOULING OF SCREENS AND INTAKE CONDUITS

Biofouling affects the hydraulic performance of intake systems and needs to be allowed for to ensure that downstream facilities such as pumping stations perform as intended. Organisms that colonize underwater surfaces, such as mussels, barnacles, larvae and spores of algae, use a diverse array of biological glues to attach themselves to wet surfaces; the result is biofouling. Two forms occur: microfouling and macrofouling.

Microfouling occurs within a short time after a clean surface is immersed in water. Bacteria colonize the surface within hours and form a biofilm which is an assemblage of attached cells, commonly referred to as slime, and may grow to 500 μm in thickness.

Macrofouling comprises soft or hard fouling which develops and overgrows the microfouling slime. Soft macrofouling comprises algae and invertebrates such as soft corals, sponges, anemones, tunicates and hydroids; hard macrofouling comprises invertebrates such as barnacles, mussels (including widely found quegga and zebra mussels in fresh water) and tubeworms. The specific organisms that develop in the fouling community depend on the water chemistry, substratum, geographical location, season and factors such as predation (Callow, 2002).

Macrofouling is a highly dynamic process and can achieve appreciable thickness on engineered surfaces of concrete or metal. Over extended periods the fouling builds up in multiple layers and results in a significant reduction in available flow area; it can have a detrimental effect on the hydraulic performance of the affected facility. In general the effective roughness of the surface on which macrofouling has occurred will be much higher.

Various technologies for preventing or reducing biofouling are available and can be used effectively depending on the use of the abstracted water. The technologies available are anti-fouling toxic coatings, non-toxic coatings, anti-fouling cupro-nickel alloys and chlorination.

Toxic coatings, although effective, are currently restricted in their use because of environmental concerns. Non-toxic coatings are currently available but their application to civil engineering structures, particularly where the coatings cannot be renewed periodically, requires further evaluation.

Continuous or intermittent low-level chlorination can be effective if it is distributed uniformly on the intake surfaces. Use of sodium hypochlorite as a means of controlling biofouling is widely adopted particularly where seawater is used on a large scale for desalination. Cathodic protection is ineffective as far as biofouling is concerned.

Use of copper or copper-nickel alloy (90%Cu-10%Ni) for metallic elements in an intake structure and screens offers both high resistance to seawater corrosion and biofouling but at an added cost in comparison with other traditional materials.

An intake facility where marine growth can occur will need to be cleaned periodically to ensure that it does not interfere with the hydraulic performance of the facility. The amount (and frequency) of cleaning will be reduced by use of anti-fouling techniques but cannot be eliminated.

6.15 HEADLOSS THROUGH SCREENS

Hydraulic headloss (ΔH) (m) through a clean coarse screen can be determined using the Kirschmer formula:

$$\Delta H = \beta (W/b)^{4/3} (V_a^2/2g) Sin\vartheta$$

where β is a dimensionless shape factor, W is the width of bar (m), b is the clear space between bars (m), ϑ is the angle of bars to the horizontal and V_a is the approach velocity (m/s). Values of β are 1.67, 1.83 and 2.42 for rectangular bars with semi-circular shape at both edges, upstream edge only and flat edged bars, respectively. For round bars β is 1.79. Values of β less than 1 can be obtained with specially streamlined bars but their benefit is questionable if hydraulic design assumes partially blocked conditions (IS 11388:2012). The loss through fouled screens may be estimated by modifying bar width and spacing according to the thickness of fouling.

The headloss through clean fine screens is given by:

$$\Delta H = \frac{1}{2g}(Q/CA)$$

where Q is flow (m^3/s), A is effective mesh opening (m^2) and C is the coefficient of discharge (typically 0.6) which should be confirmed with the screen manufacturer.

6.16 COPING WITH WATER LEVEL VARIATIONS

The water level in all water bodies (river, lake or reservoir, estuary or sea) varies to some extent and an intake facility will need to take into account the expected variations. At times of drought, river levels will be low; so low that abstraction might not be permitted. Even if abstraction is permitted the water level may not be high enough to drive the required flow into the intake; a weir or barrage may be required to sustain an adequate water level (Section 6.25).

Intake structures on lakes and rivers also need to cope with the maximum flood levels which may be much higher than the levels prevalent most of the time. All the electrical equipment should be mounted above the maximum flood level. Mechanical equipment may be located below maximum flood level if it is adequately protected from damage but access for maintenance should also be considered; long periods of inaccessibility are not desirable.

Lake and particularly reservoir water levels can be expected to vary over a wide range; if an intake is to be effective at most times it will need a draw-off near the bottom. In reservoirs it is an advantage to be able to select the draw-off level for water quality purposes (Section 8.1). Draw-off towers are described in Sections 5.21 and 6.30.

Intakes located in estuaries and on the sea shore will be subject to daily tidal water level variations and variations due to atmospheric pressure and wind set up. These factors may have a significant influence on the arrangement of the intake, particularly where the tidal range is large. It is usual to design tidal intakes so that the normal maximum instantaneous required flow can be abstracted at the lowest astronomical tide, taking into account the hydraulic losses which may be expected in normal circumstances. However, even lower water levels can be experienced due to high atmospheric pressure. The coincidence of occasional demand higher than the normal instantaneous maximum flow, abnormally high losses (before the intake system is cleaned) and unfavourable meteorological effects may need to be considered together in a risk analysis. It is then possible to make an informed choice of intake level which allows the supply system to meet its objectives without having to design for the worst combination of all variables.

Intakes located on tidal waters should be designed for the highest astronomical tide plus the effect of storm surge (low atmospheric pressure and wind set up – both of which may be very significant) and an allowance for future sea level rise. All the operating equipment will need to be located above the highest water level predicted, including an allowance for free-board and waves. In seismic areas a tsunami risk assessment should be made and the intake arranged so that no damage occurs in such an event, or that it is limited to equipment that can be easily and quickly repaired or replaced.

6.17 INTERFERENCE WITH OTHER USERS

A new intake or modifying an existing intake, particularly in relation to abstracting more water, is very likely to affect other users of the source. The licensing authority will want to be satisfied that any increased abstraction will be acceptable so it is essential to investigate all other uses of the water body, including recreational use. Adverse impacts will need to be investigated and stakeholders will need to be consulted and a way found to render the proposal acceptable.

Where an intake is located in a water body that is used for recreation it will require particular care to ensure the safety of the public and at the same time ensure that the intake and the water quality will not be adversely impacted by recreational activities which may include sporting activities in direct contact with water (swimming, diving, water skiing) or activities such as boating, sail boarding or fishing. At locations where the intake facilities are completely submerged and not visible care will be required to make sure that inadvertent damage to the facilities does not occur, for example by anchors being dropped by boats, and it may be necessary to provide markers or set up exclusion zones.

6.18 STABILITY AND CONSTRUCTION

By their very nature intake structures are frequently located where lateral forces are not in balance. This applies particularly to earth pressures on structures in river banks or on the shores of lakes and the sea; it also applies to groundwater pressures which may not be in balance with water levels in the water body, especially as water levels recede. Other lateral forces include ice (Section 6.7), waves, currents, loads from floating trash and water level differentials due to hydraulic losses. In seismic areas, several different (horizontal and vertical) forces need to be considered: inertia of structural elements, inertia of contained water, sloshing in chambers where there is a free water surface and seismic hydrodynamic forces on the outside of the structure, applying to totally immersed elements as well as those protruding above the water surface. Lateral stability must be considered at various stages of construction to guard against movement.

The intake structure should have sufficient resistance to uplift to ensure that it does not float with the highest flood level that may occur at the selected location. Additional measures may be necessary to ensure stability during construction. The design and arrangement of the intake must also ensure that it will not float when the structure is partially or wholly dewatered for maintenance and cleaning.

As described in Section 6.21, sediment movement poses a risk to intakes. Sediment adjacent to the intake may be eroded due to currents at bends in a river, local effects at obstructions or due to waves. The intake must be arranged, with sheet pile or other cut-off elements, to ensure that the structure cannot be undermined by erosion in extreme floods or storms. Equally, measures may be needed to ensure the approach to the intake does not silt up.

6.19 ACCESS FOR OPERATION AND MAINTENANCE

Access to intake structures is needed for operation and maintenance. Components that may need access include trash racks (screens), chlorine dosing injectors, stop logs if included and lifting gear. Access not requiring the use of divers is preferred for safety reasons; it should also not require the use of boats if possible since conditions may prevent access just when it is needed. For marine intakes which may be located well offshore in order to reach water depths desirable for water quality and fish exclusion (20–40 m), access by boat or diver cannot be avoided. If underwater work is necessary it should be restricted to few tasks and made as simple as possible to minimize diver hours underwater. Screens, gates and stop logs should be configured so that they can be installed and removed with minimum intervention.

For larger intakes, or anywhere total shutdown cannot be accepted, the intake layout should permit sections to be taken out of service for cleaning and maintenance while other sections remain in service. It may sometimes be necessary to provide one isolatable intake bay, with its own gates and screens, for each pump, but one section for each pair, or small group of pumps is normally sufficient.

Even with the use of anti-fouling materials and chlorine dosing, biofouling of intake conduits can be expected (Section 6.14). Therefore, spaces behind screens and conduits should be accessible safely. Provided part of the intake structure is above the water surface at all times, gates and stop logs can be provided to seal the intake. Due to the hazards involved means of double isolation should preferably be provided if personnel entry is necessary. Equipment for operating the gates and stop logs has to be mounted on the top of the intake and will itself need maintenance. An alternative to provision for personnel entry is to provide facilities for the entry and retrieval of robotic inspection and cleaning machines such as remote-operated vehicles and material handling equipment. It is helpful if there is a shaft or chamber with land access for servicing inspection and cleaning operations.

6.20 INTAKE LOCATION AND SITING

Typical locations for intakes are listed in Table 6.1 along with some of the types of intake appropriate to those locations. The factors described in Sections 6.2–6.8 and 6.16–6.19 must be taken into account in the selection of the location and arrangement for all intakes. Some factors relevant to intakes of various types on streams and rivers and for open water are covered in Sections 6.21–6.31.

Locations where disturbance to the natural environment should be avoided include nursery areas in estuaries, reefs or shores which serve anadromous migrating or other fish; sub-tidal rock reefs with growth of aquarium and other macrophites (microscopic algae) as these areas serve as significant nurseries for juvenile stages of prawns and rock fish; protected rocky shoreline habitats sheltering intertidal fish larvae which lack the capacity of planktonic dispersal; and areas where herring spawning is known to occur (intakes need to be at least 2 km away from such areas).

Foundations conditions must be assessed by a geological desk study and confirmed by ground investigation and associated laboratory testing. The desk study will assist in choice of a suitable site for the structure while the ground investigation will support calculations of the stability of the intake at the site finally selected (Section 6.18).

Other physical factors to be considered in the siting of an intake include depth of water and variation in level; direction and strength of currents; local scouring near the structure; and navigation.

INTAKES ON STREAMS AND RIVERS

6.21 SEDIMENT TRANSPORT

Rivers carry and erode or deposit sediment; the processes involved are complex and the reader is referred to Graf (1971) for an understanding. Finer material is carried in suspension and coarser material is carried as bed load but there is usually a mix of sizes carried. At lower discharges, the coarser material may act as armour on the river bed until a threshold flow is reached. Raudkivi (1993) suggests division of sections of a river into four types: mountain streams; steep rivers with slope (S) steeper than 1 in 1000; rivers on plains ($10^{-3} > S > 10^{-4}$) and large rivers ($S < 10^{-4}$); which correspond to the largest size of sediment carried: boulders, gravel, sand and silt, respectively. In upper reaches rivers are typically in the process of cutting down by bed erosion. In lower reaches accretion is typically taking place. In the transition, which may be abrupt or extended, a river is said to be 'in regime', neither eroding not accreting on average.

In a given reach sediment will be present in the flow if it is brought in from upstream (or a tributary) or has been picked up from the bed or banks of the channel. As flow increases the concentration of sediment carried in increases and the amount picked up from the bed also increases. The latter effect can produce lowering of the bed in floods. Conversely, when flow decreases, as happens downstream of abstractions, sediment deposition will increase.

Within the same reach sediment may be being deposited or eroded due to variations in velocity, turbulence and shear stress at non-linear features such as bends and obstructions. The erosion which occurs on the outside of bends often leads to movement of the channel in time. The past behaviour of the river channel and tendency for future movements should be assessed in a study of the river morphology.

The amount of sediment transported by a river may be measured against its catchment area. Annual average sediment yields range typically between 100 and 10 000 tons/km^2, affected by rainfall, terrain steepness, rock and soil type, vegetation cover and disturbance by man. Sediment concentration at a given location is very roughly proportional to flow and typically ranges between 0 and 10 000 mg/l. Very high concentrations (over 100 000 mg/l) may be experienced in rivers flowing through highly erodible material, such as loess in China, at times of flood.

It is important that reliable information on bed load and suspended sediment is available for rivers or streams where intakes are being considered. This will assist in the selection of intake type and of any measures to reduce the amount of sediment entrained in the abstracted flow. Data on sediment concentration and bed material can be collected as part of the hydrological data collection (Section 3.3) and should allow a sediment rating curve to be developed alongside the flow rating curve for the reach under study. It should be noted that sediment transport is greater on the rising part of a flood hydrograph than for the same flow on flood decay. This hysteresis effect is largely due to limitations in supply of sediment from upstream (Raudkivi, 1993).

6.22 SEDIMENT EXCLUSION AND REMOVAL

Intakes located on rivers will abstract sediment suspended in the flow and may also take in bed load depending on the level of the intake sill with respect to the river bed. Raudkivi (1993) and Razvan (1989) suggest ways of eliminating withdrawal of bed load and reducing abstraction of suspended sediment.

It is often beneficial to locate an intake on the outside of a bend for several reasons: the surface with less sediment in suspension is moving towards the outside of the bend; the bed flow is moving away from the outside of the bend and the higher scouring effect at the outside of the bend tends to prevent accretion of bed load in the approach to the intake. If a bend is not available similar effects can be induced artificially, one method being a weir (Section 6.25) with a bed load flushing passage. By such means it should be possible to prevent entry of bed load and to reduce, but not eliminate, the amount of suspended load abstracted.

A settlement basin in an approach channel just upstream of the intake can be used to settle out the coarser suspended sediment. Alternatively, the abstracted flow may be passed through a desanding basin or even settlement ponds in which finer material may be settled. Settlement is achieved by reducing the flow velocity and providing space for the sediment to accumulate for later removal (SEPA, 2008). However, removal of very fine and light particles from suspension by such means is not usually practicable without flocculation (Sections 8.5 and 8.10). The settlement basin will require periodic maintenance cleaning of the sediment trapped. This can be facilitated by inclusion of sediment traps with scour valves for gravity discharge or by use of 'jet-type' dredge pumps (FWR, 2002).

Some finer particles will tend to settle out where not wanted and will need to be removed. Facilities should be provided within the intake system to collect and remove the sediment. This will generally require sections of the intake structure to be emptied behind closed gates and the accumulated sediment removed using manual or mechanical means.

6.23 FLOOD PASSAGE

Intake facilities located on watercourses and reservoirs will need to be designed to deal with the passage of floods without being damaged or undermined. Intakes on rivers should also be capable of accommodating possible changes in river bed profile during passage of floods; this requires understanding of the river morphology and the temporary river bed changes that may occur during floods (Section 6.21). An intake location on the outside of a bend, selected to minimize sediment abstraction (Section 6.22) is more prone to local bed lowering in floods and may require a deeper sill foundation.

Any electrical equipment must be located above the highest flood level to be expected during the life of the project, or should be suitable for immersion.

6.24 UPLAND INTAKES

Upland streams and rivers are fast flowing high energy environments in confined and steep valleys with average river bed slope of 1% or more, sometimes with stepped pools along the watercourse. Sediment size is much larger than that in the lowland rivers and bedrock is often exposed and strewn with large cobbles and boulders. Water depths are low and velocities are high.

For less steep slopes a full (Section 6.25) or partial weir may be employed to hold water levels up sufficient for a lateral intake; a wide channel is needed to encourage the large bed load to pass or it must be accepted that bed maintenance will be required. The essential features of a typical partial weir or groyne intake are shown in Figure 6.1.

For very steep (and narrow) streams the most suitable arrangement is an overpass type of intake. A transverse channel is built across and below the bed and a grating of concrete beams or metal slats placed over the top but sloping steeply downstream (Fig. 6.2). The grating acts as a coarse screen with the openings sized to prevent larger material and trash from dropping through; it is instead rolled off downstream by the force of the flow. Finer material does enter but can be deposited in a debris trap before the water is led off to an aqueduct or pipe.

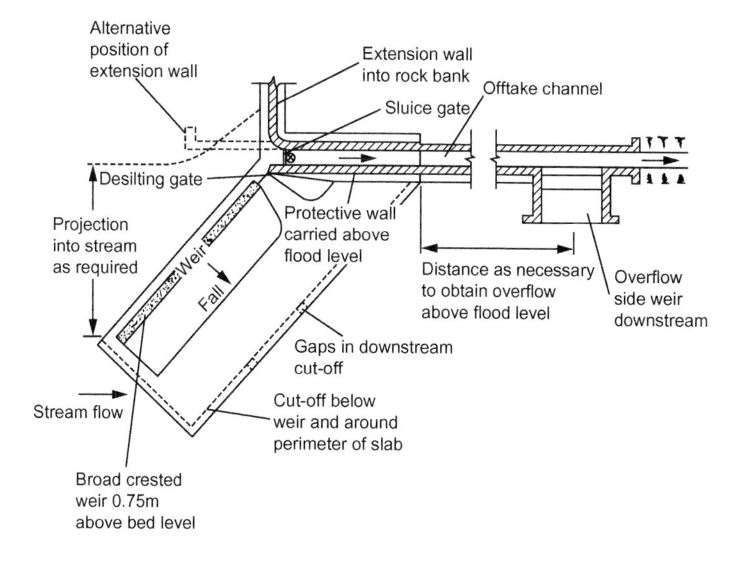

FIGURE 6.1

Groyne intake for small abstraction on a 'flashy' river, Cyprus.

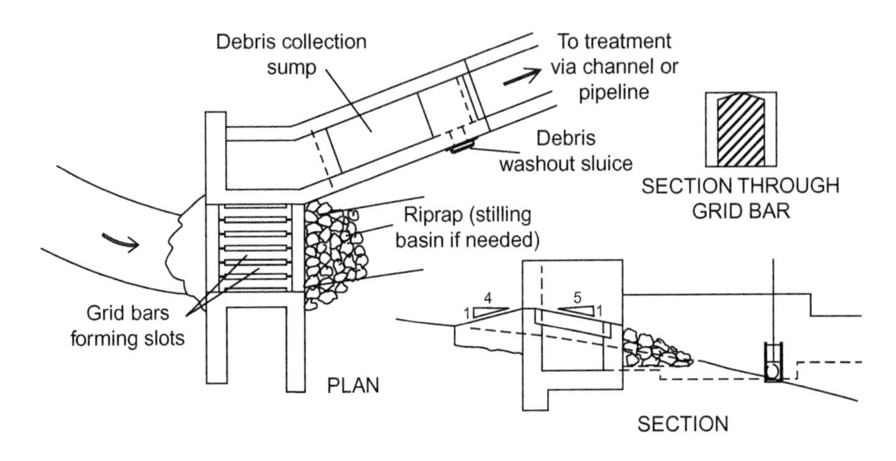

FIGURE 6.2

Upland bottom intake for steep streams.

6.25 BARRAGES AND WEIRS

Barrages and weirs can be built if necessary to raise the water level at the intake sufficient for the required flow to be abstracted. So that upstream water levels are increased as little as possible at high flows and floods, barrages and weirs are often equipped with control gates; these may be automated to maintain the required water level. Such discharge arrangements also need to meet compensation flow requirements (Section 3.23).

Excess hydraulic energy downstream of the weir or barrage must be dissipated or controlled to ensure that undue erosion does not occur. If adequate measures are not taken the stability of the structure can be at risk. If the structures are located on permeable strata a cut-off may be needed and suitable measures may be required to ensure stability due to uplift or overturning (Chapter 5).

Barrages and weirs form an obstruction to the movement of sediment transported by the river flow and some of the sediment may be deposited upstream of the barrage or weir. This needs to be allowed for in the design and suitable provision should be made for periodic flushing of the sediment, particularly on the side adjacent to an intake (Fig. 6.3). Similarly the reduced supply of sediment from upstream will tend to cause the bed downstream to lower in time.

Passage of floating vegetation, branches and trees needs to be considered in the design of weirs to make sure that trash will not obstruct the weir. Floating trash should be kept clear of barrage gates by guiding it towards a bank where it can be safely removed or by making suitable provision for passing through a gate opening.

Barrages and weirs form obstructions to the upstream migration of fish; depending on their location it may be necessary to provide fish passes with continuous controlled flow (Clay, 1961; SEPA, 2008).

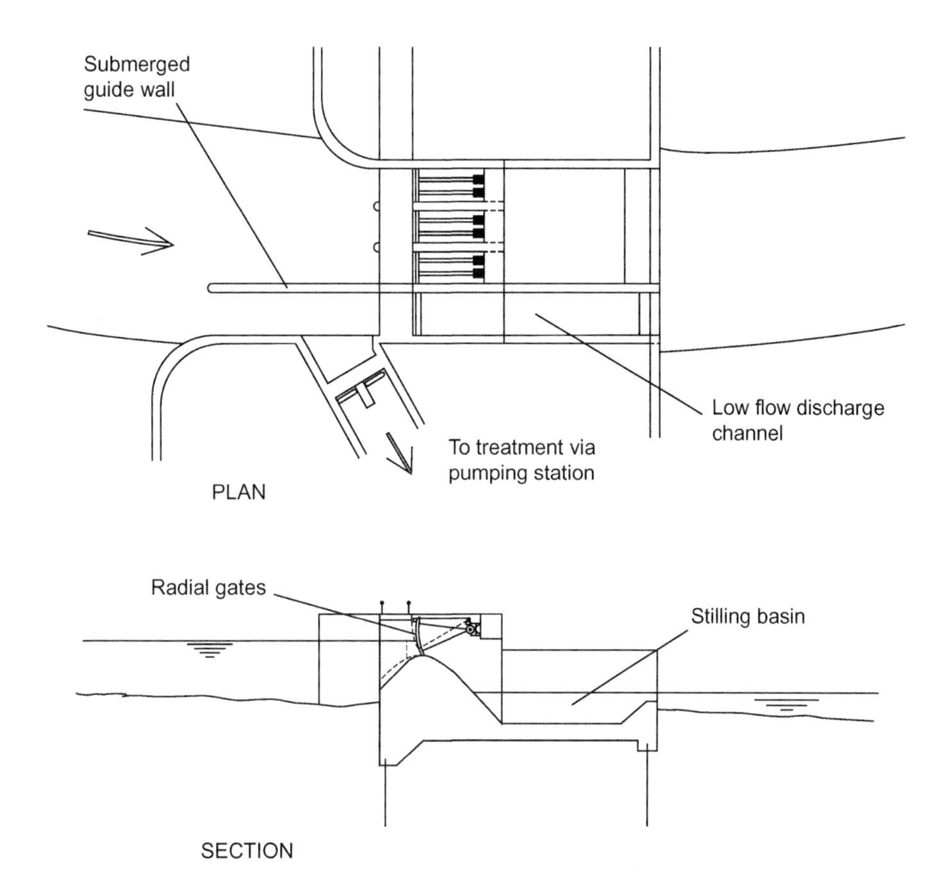

FIGURE 6.3

Side intake with small barrage.

6.26 BANKSIDE INTAKES

Bankside intakes are constructed within a river or canal bank, preferably where the bank is stable and where study of the morphology indicates that the approach will not silt up. If bank erosion is likely, local river training and bank defences should be considered. River training can also help reduce siltation in front of an intake. However, in a river which has been subject to major abstraction, for irrigation for example, reduced average flows can make siltation unavoidable; dredging may be inevitable.

A typical bankside intake for larger flows (Fig. 6.4) will include: trash racks (Section 6.6) and associated cleaning equipment; fine screens (Section 6.13) to remove smaller debris and fish (and fish return channels); collection and disposal facilities for trash and rubbish from

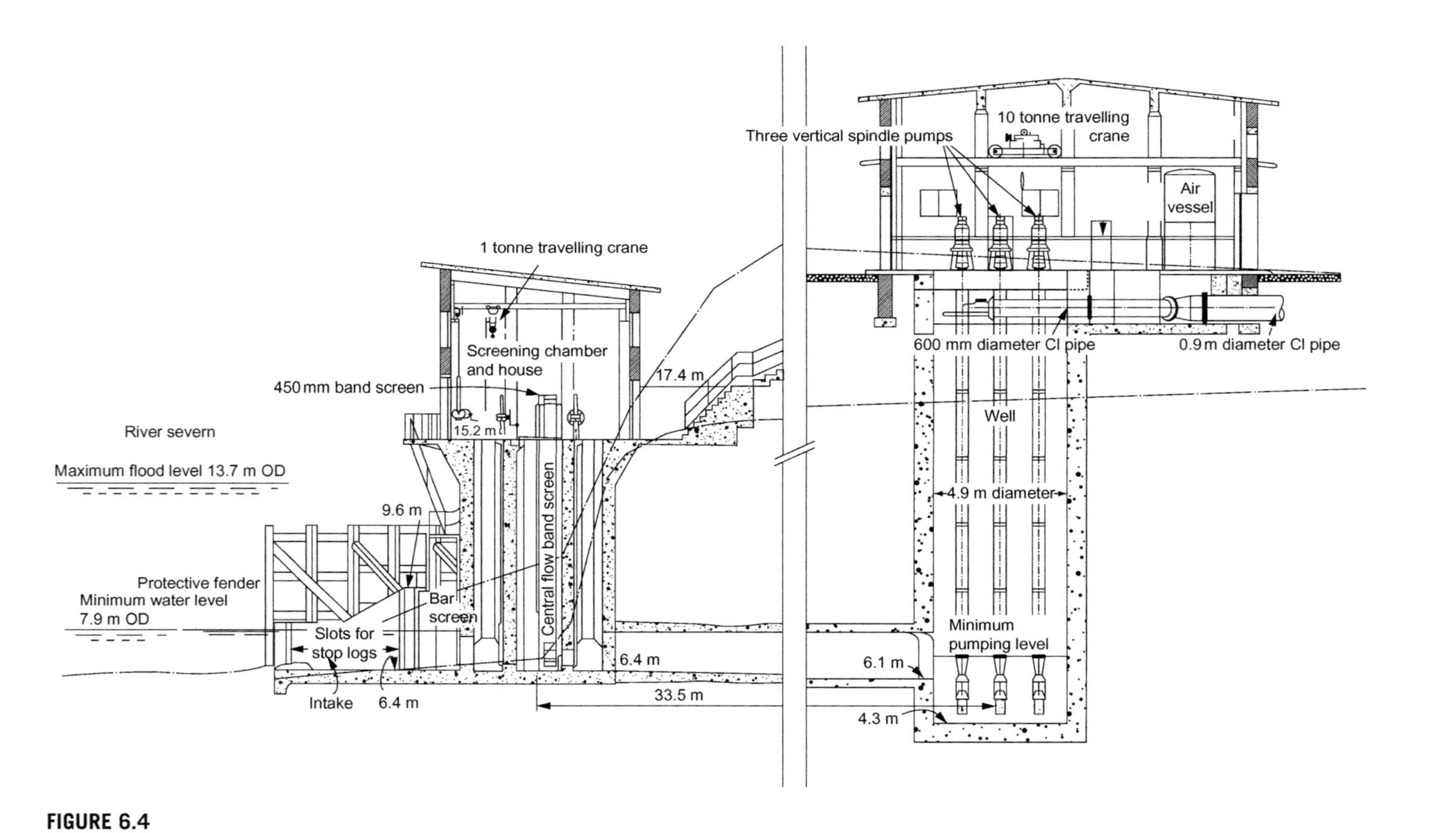

FIGURE 6.4

Bankside intake on River Severn (Engineer: Binnie & Partners).

screens; isolating facilities for dewatering whole or parts of the structure for maintenance; and pumps and associated equipment for transferring water for treatment or use. Chlorination equipment should be installed if biofouling, particularly from mussel growth, is expected to be a problem.

A forebay may be needed to avoid the front of the structure at the trash rack line from protruding into the river channel. The forebay acts as a transition and approach to the intake but it should be kept as short as possible to reduce the area from which sediment has to be cleaned periodically. Where the channel is used for navigation, intakes should be protected from damage by impact. It may be necessary to construct dolphins on piles just off the intake; these would be designed to take impact loads determined as for bridge piers.

If the screens are of the passive type (Section 6.12) facilities will be required for air blowback equipment for backwash, lifting out and replacing screen units for cleaning and markers or other measures to indicate the position of underwater screen units. Active fine screens (Section 6.13) will require collection and return facilities for fish and disposal arrangements for washings.

For smaller capacity intakes a siphon intake may be used. A suction pipe is laid on piles or a jetty above water level but with its inlet submerged. Vacuum equipment is needed for priming. The siphon pipe can discharge to a pumping wet well or directly to the pump inlet. The siphon intake can only be used where the water level variation is small.

6.27 RADIAL WELL INTAKES

Subsurface intakes using horizontal or inclined directionally drilled radial wells provide a method of abstracting water from permeable strata beneath or beside a river bank or the sea. The strata need to be hydraulically connected to the water in the river and not blinded by silt or other fine material. Visible above ground works are minimal. The collected water is filtered through the ground and requires less treatment than water drawn directly from the same river. Entrapment and impingement of fish in the river are avoided completely and there is no issue with trash. Radial well intakes are, therefore, an environmentally friendly option in rivers with sand beds and slope between about 1 in 1000 and 1 in 10 000. However, this type of intake is normally limited to relatively small abstracted flows.

The radial wells used to collect the water need to be located in an area where they will not suffer damage during floods when the bed material can go into suspension. The number of drilled wells will depend on the rate of percolation and the quantity of water to be collected; site investigation and testing are likely to be necessary as for a borehole (Sections 4.2 and 4.6). The top of the collecting well shaft should be located above flood level for ease of access and to house any electrical equipment for the pumps. A typical arrangement is illustrated in Figure 6.5. Other forms of subsurface intake are described in Section 4.16 and in *Overview of Desalination Plant Intake Alternatives* (WaterReuse, 2011).

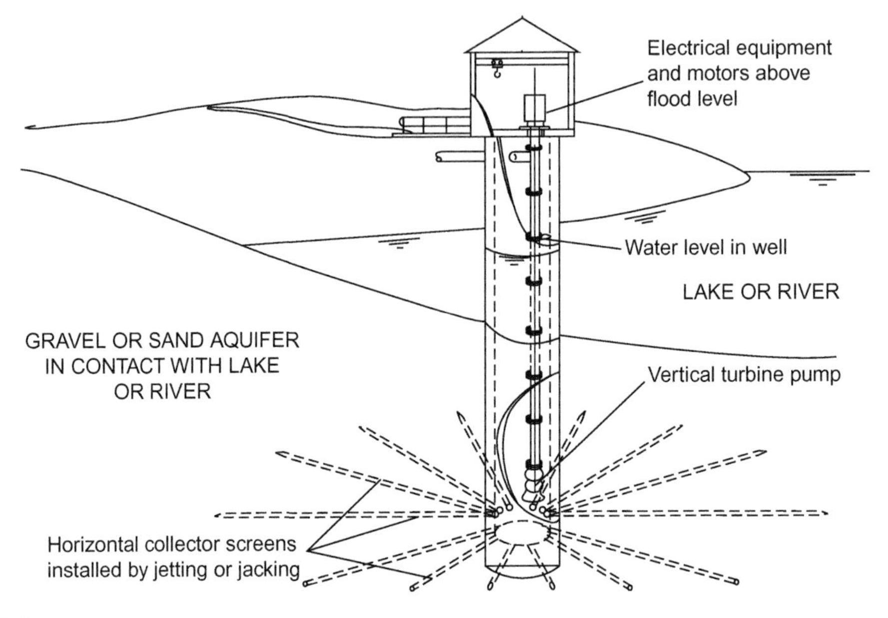

FIGURE 6.5

Radial well intake for lakes and large rivers.

INTAKES FOR OPEN WATER

6.28 SHORELINE INTAKES

A shoreline or surface intake is one located near the water on the shore of a lake, estuary or the sea. It may incorporate dykes or bunds to form an inlet approach channel to provide some protection against waves and strong currents in places where fast flowing water may be present at times of high flows or floods. This type of intake is feasible only where sufficient water depth is available near the shoreline.

The main features of a shoreline intake will be similar to those described in Section 6.26 for bankside intakes. However, shoreline intakes should be located outside of the littoral zone; therefore, a long forebay may be necessary.

Although marine intakes may be constructed on the shore, submerged intakes are preferred due to lower environmental impact. These are described in Section 6.31.

6.29 INTAKES ON JETTIES OR PONTOONS

Where the shore shelves gently a shoreline intake may be impracticable. Other options include using a jetty or a pontoon as a base for submersible intake pumps located below water level or for centrifugal pumps on the deck with suction pipes below water. Pumps supported from a jetty will draw water from a fixed elevation above the bed while those attached to a pontoon will draw water from a fixed depth below the surface. Screens for pump inlets can be of the passive type (Section 6.12).

Where there is a large range of water level, for example in estuaries with a large tidal range, intake pumps can be mounted on a pontoon riding on piles. The pontoon can be moored away from the shore or berthed to a jetty which can carry the delivery pipework; in the latter case the pipework has to be configured to accommodate tidal movement. For small intakes this can be achieved using flexible hoses but for larger intakes the options are a floating pipeline of HDPE, in an arrangement similar to that used for dredger discharge, or a pivoting coated steel pipework with a special ball joint at each end of the pivot. A floating pipeline would be needed for an intake pontoon moored away from the shore but would pose a large obstruction to navigation and would not be acceptable in some situations.

For smaller water level ranges the pumps can be sited on land with the suction pipework laid on a jetty and with its inlet below lowest water level. This arrangement requires a means of priming the suction pipework, for example using a vacuum pump.

The position of the pump inlet with respect to the bed and to water surface requires a compromise. There should be sufficient submergence at all times for pump suction conditions and to exclude floating trash but the inlet should be as high as possible above the bed to avoid intake of damaging solid matter stirred up in rough weather.

6.30 TOWER INTAKES

Tower intakes are often used in reservoirs, usually near a dam, where they provide a means of withdrawing water for supply using multiple intake draw-off pipes. This arrangement enables water to be drawn from the most appropriate level to give optimum quality in relation to sediment content, biological conditions and temperature. It is usual to include an access bridge from the top of the dam or abutment.

The water from each level is conveyed to a vertical stand pipe located in the tower and controlled using valves provided on each draw-off pipe. Screens are normally provided at the entrance to each draw-off pipe to exclude floating objects. The vertical stand pipe continues in a tunnel to carry the water to the downstream side of the dam. Controls for all the valves are usually located near the top of the tower where access is provided. Draw-off arrangements for dams are discussed in Section 5.21. A typical arrangement showing the basic facilities is shown in Figure 6.6.

6.31 OFFSHORE INTAKES

Offshore or submerged intakes are used where the water depth at or near the shore is not sufficient to abstract the flow required or where construction at or near the shoreline is unacceptable on environmental or other grounds. At some locations it may be necessary to locate the intake structure at an offshore location to ensure that the water quality meets the requirements and that fish

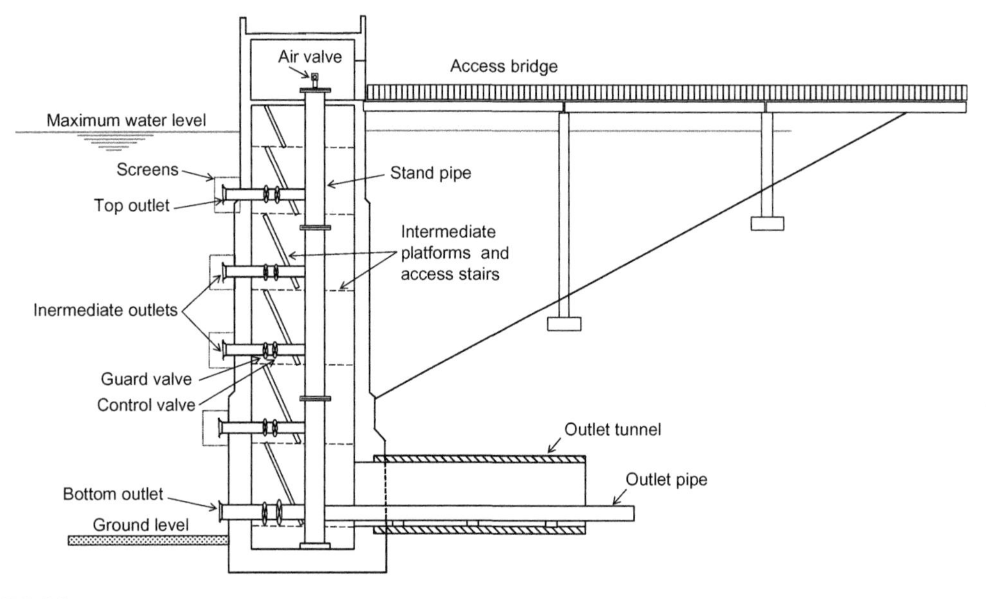

FIGURE 6.6

Tower intake for reservoirs and lakes.

entrapment and impingement are minimized. The most usual type of structure for an offshore intake is the velocity cap type intake.

The offshore intake structure will need to be connected to the onshore facilities by a tunnel, pipeline or culvert to transfer the water to the remaining intake facilities located onshore. The structures required will include:

- offshore velocity cap intake including coarse screens;
- pipeline (or tunnel or culvert) to connect to onshore facilities;
- bed protection around the intake structure;
- small dosing pipeline for carrying chlorine solutions for biofouling control;
- onshore reception chamber for incoming water;
- coarse and fine screening;
- isolating facilities for dewatering the onshore structure for maintenance;
- pumping station to house pumps and associated equipment for transferring water for treatment or use.

Where the design intake flow is large it may be necessary to use more than one intake structure to keep the physical size of the structure and the connecting conduits to manageable proportions. Two intakes for a single facility may also be required to enable flow through one to be stopped for periodic inspection and maintenance without complete system shutdown.

With an offshore velocity cap type intake structure, fish entrainment is minimized purely by selecting the proportions and arrangement of the inlet areas and by achieving low uniform flow velocity through the coarse screens. The threshold of the waterway opening needs to be kept above the bed level by a sufficient distance to ensure that stirred up bed sediment is not drawn in with the flow during floods or storms. Where necessary the bed should be suitably protected with gravel or rock.

Incoming flow at the opening needs to have velocity less than 0.15 m/s and to pass over a threshold platform with width at least 1.5 times the height of the opening. Tests carried out using live fish confirmed that this arrangement achieved a lower fish entrapment than other geometries that were in use previously and gave fish near the entrance sufficient signs to avoid being drawn into the intake (Schuler, 1975). A typical arrangement of velocity cap intake is illustrated in Figure 6.7. With this arrangement it is not possible to exclude the smaller life forms of fish and the location of the intake needs to be selected to ensure that small and juvenile fish are not present in the offshore area as far as practicable.

Velocity cap intakes have to be pre-fabricated on land, transported and lowered into position in the water. The weight of precast concrete units can be an advantage for stability under wave lateral loading but makes it necessary to use large lifting equipment mounted on a barge or ship. However, under seismic loading structural lightness is helpful, particularly for the intake upper works. Fabrications in glass or fibre reinforced plastic can be handled and installed with less expensive equipment and can be used, at least for the upper part of a velocity cap intake.

Foundations for offshore intakes can be of the buried raft type with rock armour for erosion protection and ballast. Rock or ground anchors are often considered; but protection from corrosion and reliable anchor grouting underwater are issues which need careful evaluation. If needed for other

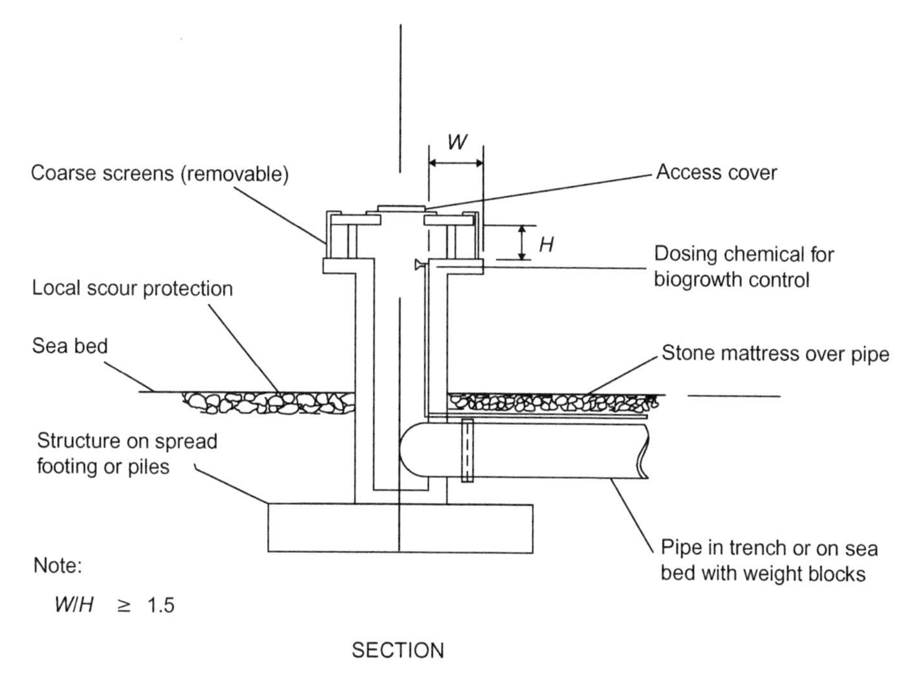

FIGURE 6.7

'Velocity cap' type intake for offshore use.

installations on the project, large diameter drilling or piling equipment mounted on a jack-up barge may be used to achieve anchorage to a rock foundation. However, mobilization of such equipment for only one or two intakes can be a major cost factor.

The whole of the offshore intake structure is normally located completely under water and some form of marking or exclusion zone above the water surface is necessary to ensure it is not affected by navigation or other activities in the area. The coarse screens at the intake will require regular inspection and maintenance to ensure that they are free from marine growth or damage and the arrangement selected should be such that the screens can be removed with ease for maintenance when required.

Flow from the offshore intake to the remaining facilities on shore is by gravity. The connecting conduit therefore needs to be placed at a low enough level to ensure that the system is always full of water with no air trapped and that the top of the conduit is below the hydraulic gradient. The conduits need to be arranged such that air does not accumulate within the transfer conduits. Ideally the velocity in the conduit should be sufficient to prevent settlement of any solids carried in. However, where the range of intake flows is appreciable this is not feasible and provision for cleaning must be made.

Offshore velocity cap intakes are used for large flows. It is preferable not to locate passive fine screens at such intake because of the inefficiency and impracticality of the air cleaning arrangements that would be required. For intakes nearer the shore and in relatively shallow water an alternative is to install an array of smaller 'can' type intakes, each with passive fine screens and with a supply of air for back flushing and of chlorine for control of biofouling. The array is connected to the intake conduit via manifolds. Such an arrangement is complicated and requires underwater maintenance but avoids the need for large expensive equipment for installation.

REFERENCES

API RP 2N. *Recommended Practice for Planning, Designing and Constructing Structures and Pipelines in Arctic Conditions*. API Recommended Practice 2N. (Reaffirmed 2007). API.

ASCE (1982). *Design of Water Intake Structures for Fish Protection*. ASCE.

AWWA (2012). Chapter 5. Intake Systems, *Water Treatment Plant Design*. 5th Edn. AWWA with ASCE.

Callow, M. E. and Callow, J. A. (2002). Marine bio fouling: a sticky problem. *The Biologist* **49**(1), 10, University of Birmingham, UK.

Clay, C. H. (1961). *Design of Fishways and Other Fish Facilities*. Dept. of Fisheries, Canada.

Clough, S. C. and Turnpenny, A. W. H. (2001). *Swimming Speeds in Fish: Part 1*. R&D Technical Report W2-026/TR1. Environment Agency.

Dixon, D. (2005). *Field Evaluation of Wedge Wire Screens for Protecting Early Life Stages of Fish at Cooling Water Intakes*. Electric Power Research Institute (EPRI), California, Final Report No 1010112.

EA (2005). *Screening for Intake and Outfalls: A Best Practice Guide*. Science Report SC030231. EA.

Fedorenko, A. Y. (1991). *Guidelines for Minimising Entrainment and Impingement of Aquatic Organisms at Marine Intakes in British Columbia*. Habitat Management Division of Fisheries and Oceans, Vancouver, British Columbia.

Frannsson, L. and Bergdahl, L. (2009). *Recommendations for Design of Offshore Foundations Exposed to Ice Loads*. Elforsk Rappor 09.55.

FWR (2002). *An Investigation into the Removal of Sediment From Water Intakes on Rivers by Means of Jet-Type Dredge Pumps*. Report No. 1187/1/02. FWR.

Gille, D. (2003). Seawater intakes for desalination plants. *Proc. Eur. Conf. on Desalination and the Environment*, IWA.

Graf, W. H. (1971). *Hydraulics of Sediment Transport*. McGraw Hill.

IEC 61400-3. *Recommendations for Design of Wind Turbine Structures With Respect to Ice Loads*. IEC.

IS 11388. *Recommendations for Design of Trash Racks for Intakes*. Bureau of Indian Standards. 6.2012.

ISO 18806. *Arctic Offshore Structures*. ISO.

Lloyd (2005). *Guideline for the Construction of Fixed Offshore Installations in Ice Infested Waters*. Part 6. Germanischer Lloyd Oil and Gas GmbH.

Raudkivi, A. J. (1993). *Sedimentation. Exclusion and Removal of Sediment From Diverted Water*. Hydraulic Structures. Design Manual – 6. IAHR.

Razvan, E. (1989). *River Intakes and Diversion Dams*. Elsevier.

Schuler, V. J. and Larson, L. E. (1975). *Improved Fish Protection at Intakes systems*. ASCE Environmental Engineering Division Paper No 11756.

SEPA (2008). *Engineering in the Water Environment. Good Practice Guide – Intakes and Outfalls*. SEPA.

Sheridan, A. (2014). *Screening at Intakes and Outfalls: Measures to Protect Eel – The Eel Manual*. GEH00411BTQD-E-E. EA.

Turnpenny, A. W. H. and Horsfield, R. A. (2014). *International Fish Screening Techniques*. WIT Press.

UKSO (2006). *Water Resources, England and Wales*. The Water Resources (Abstraction and Impounding) Regulations 2006, The UK Stationary Office Limited.

USBR (2006). *Fish Protection at Water Diversions*. USBR.

WaterReuse (2011). *Overview of Desalination Plant Intake Alternatives*. Desalination Council. WaterReuse Association, USA.

CHAPTER

Chemistry, Microbiology and Biology of Water

7

7.1 INTRODUCTION

This chapter is divided into six parts:

- Part I lists alphabetically 50 parameters or groups of parameters that make up some of the more usual physical and chemical characteristics of water and describes their significance. Some of these parameters constitute a risk to human health, others affect the aesthetic quality of the water supplied and others relate to treatment issues.
- Part II considers the standards relating to the above parameters and their derivation. It also looks at levels of monitoring and the analytical requirements.
- Part III looks at the microbiology of water and the most common waterborne diseases. It also looks at the associated standards, levels of monitoring and the testing of water for pathogenic organisms.
- Part IV considers water biology in terms of the significance of macroorganisms on water quality.
- Part V reviews emerging areas of concern in respect of drinking water quality.
- Part VI discusses Drinking Water Safety Plans, which are based on risk assessment and management of water supplies from source to tap.

Guideline Values and Standards associated with the various physical, chemical and microbiological parameters discussed in Parts I and III can be found in Tables 7.1(A)–(E) and 7.2.

PART I SIGNIFICANT CHEMICAL AND PHYSICO-CHEMICAL PARAMETERS IN WATER

7.2 ACIDITY

An acidic water is one which has a pH value of less than 7.0 (Section 7.38). The acidity of many raw waters is due to natural constituents, such as dissolved carbon dioxide or organic acids derived from peat or soil humus. These are unlikely to lead to pH values much below 5.5. Some apparently unpolluted moorland waters may have pH values below 4.5 due to 'acid rain' which is formed

Twort's Water Supply. DOI: http://dx.doi.org/10.1016/B978-0-08-100025-0.00007-7

when atmospheric sulphur dioxide, derived from the burning of fossil fuels, combines with water vapour to form dilute sulphuric acid.

Surface waters sometimes become contaminated with acidic industrial effluents. Acidic wastes from disused mines can also provide a significant source of acid contamination in some parts of the world. One of the worst incidents in the UK occurred in 1992 when 50 million litres of acidic, metal laden water was accidentally released, on failure of a tailings dam at Wheal Jane tin mine, into the River Carnon in Cornwall.

Acidity is an important factor in water treatment, especially when optimizing the coagulation process and in ensuring that the treated water entering supply is non-corrosive or aggressive to materials in the distribution network.

7.3 ACRYLAMIDE

Polyacrylamides are used extensively in water treatment as coagulant aids, filter conditioners and in sludge treatment. There is a potential health risk from free residual acrylamide monomer in the treated water but this is controlled by dose and product specification rather than direct measurement.

7.4 ALGAL TOXINS

Algae are discussed in detail in Part IV.

Certain algae produce toxins when they die off and decay. Cyanobacteria, otherwise referred to as blue-green algae, are photosynthetic bacteria that have similar properties to algae. The species *Microcystis*, *Oscillatoria* and *Anabaena* produce hepatotoxins. Microcystin-LR is the most studied of these and has been shown to cause liver damage. Other species, including *Oscillatoria* and *Anabaena*, produce anatoxin, a neurotoxin that attacks the nervous system (Section 8.13). Following an intense bloom of these algae, the water may develop a greenish hue and scums may form in the shallow margins of a lake or reservoir. This scum may prove fatal to fish or to animals drinking at the water's edge (Carmichael, 1993) and humans may be affected either through contact with contaminated water, or consumption of contaminated fish or other species taken from such waters (Hunter 1991). In July 2014 there was a major bloom of cyanobacteria in Lake Erie, USA. This resulted in unacceptable levels of microcystin in the drinking water supplied to Toledo, Ohio, and some 400 000 people were advised not to drink their tap water for several days. WHO (2015) has produced a technical brief providing information on the management of cyanobacteria in drinking water supplies.

Activated carbon has been used successfully to remove the toxins from raw waters and ozone has also been shown to be effective at breaking down the toxins into less toxic by-products (Falconer, 1989; Lahti, 1989).

7.5 ALKALINITY

In a general sense 'alkalinity' is taken to mean the opposite of 'acidity' i.e. as the pH value increases (Section 7.38) alkalinity increases. More accurately, the alkalinity of a water mainly comprises the sum of the bicarbonates, carbonates and hydroxides of calcium, magnesium, sodium and potassium. Calcium and magnesium bicarbonates predominate in waters that are associated with chalk or limestone and comprise the temporary hardness of a water (Section 7.29). Where the alkalinity is less than the total hardness, the excess hardness is termed permanent hardness. Conversely, where the alkalinity is greater than the total hardness, the excess alkalinity is usually due to the presence of sodium bicarbonate, which does not affect the hardness of the water. Because bicarbonate ions can exist at pH values below pH 7.0, a measurable alkalinity is still obtained with 'acidic' waters down to pH values of 4.5.

Alkalinity provides a buffering effect on pH, which is an important factor in many water treatment processes. It is also a key factor in determining the corrosive or aggressive nature of a water.

7.6 ALUMINIUM

Aluminium can occur in detectable amounts in many natural waters as a consequence of leaching from the substrata. It is also found in the run-off from newly afforested areas. However the most usual source of aluminium in public water supplies comes from incomplete removal of aluminium-based coagulants during treatment (Section 8.20). Current thinking is that the residual aluminium in the water leaving large water treatment works using aluminium-based coagulants should be less than 0.1 mg/l, and for small treatment works it should be less than 0.2 mg/l (WHO, 2011a). However, by taking a holistic approach to optimizing all stages of the treatment process, it should be possible to keep the residual aluminium in water entering supply after clarification and filtration to less than 0.05 mg/l (or 50 μg/l) as Al. A percentage of this aluminium is likely to settle out in the distribution system and accumulate as flocculant material, especially in areas where the flow is low. Any disturbance of these sediments, either through flow reversals or sudden changes in flow due to burst water mains, for example, may result in consumer complaints of discoloured or 'dirty' water.

Standards relating to aluminium are based on the aesthetic quality of the water supplied. However, public concerns have been expressed about the possible neurotoxic effects of aluminium in drinking water. It has been established that the aluminium content of water used for renal dialysis should be no greater than 0.01 mg/l to avoid neurological problems in dialysis patients. Concerns have also been expressed that aluminium in drinking water might be a risk factor in the development and acceleration of more general neurological disorders (Kawahara, 2011). A number of epidemiological studies have been carried out that demonstrate a positive relationship between aluminium in drinking water and Alzheimer's disease. However, the 1997 WHO Environmental Health Criteria document for aluminium concludes that while the positive relationship identified in these studies cannot be totally dismissed, strong reservations about inferring a causal relationship are warranted (WHO, 1997, 2010, 2011a).

In 1988 there was a major incident at an unmanned treatment works in Cornwall, UK, when 20 tonnes of aluminium sulphate were inadvertently pumped into the treated water tank. This resulted in contaminated water being supplied to a large number of consumers in the surrounding area, causing considerable concern. Subsequent studies carried out into the possible risks to human health from short-term exposure to the water supplied concluded that it was unlikely that there would have been any persistent or delayed harm to health from any of the contaminants involved in the incident (COT, 2013). However, further studies were needed as a precautionary measure into any long-term effects on infants at the time of the incident and also those *in utero*. The incident also resulted in a significant tightening of procedures for accepting chemical deliveries at all water treatment works, including the need for dedicated filling points.

7.7 AMMONIACAL COMPOUNDS

Ammonia is one of the forms of nitrogen found in water (see also Section 7.35). It exists in water as ammonium hydroxide (NH_4OH) or as the ammonium ion (NH_4^+), depending on the pH value, and is usually expressed in terms of mg/l 'free' ammonia (or 'free and saline ammonia'). 'Albuminoid ammonia' relates to the additional fraction of ammonia liberated from any organic material present in the water by strong chemical oxidation. 'Kjeldahl nitrogen' is a measure of the total concentration of inorganic and organic nitrogen present in water.

Ammoniacal compounds are found in most natural waters. They originate from various sources, the most important being decomposing plant and animal matter. Increased levels of free ammonia in surface waters may be an indicator of recent pollution by sewage, agriculture or industrial effluent. However, some deep borehole waters, which are of excellent organic quality, may contain high levels of ammonia as a result of the biological reduction of nitrates. The source of any substantial amount of ammonia in a raw water should always be investigated, especially if it is associated with excessive bacterial pollution.

The level of free ammonia in a raw water is important in determining the chlorine dose required for disinfection. Chlorine reacts with any ammonia present in the water to form monochloramine, dichloramine and finally nitrogen trichloride. This reaction has to be completed before a free chlorine residual can be achieved (Section 11.9). Ammonia removal may, therefore, have to be considered (Sections 10.27 and 10.28), if the ammonia concentrations in the raw water are high enough to create a chlorine demand.

There is no health risk associated with the levels of ammonia typically found in drinking water, although the water may become unacceptable on taste and odour grounds if high concentrations of dichloramine or nitrogen trichloride are allowed to develop during chlorination.

7.8 ARSENIC

Arsenic is toxic to humans and, if detected in water, its origins should always be investigated. It is a natural water contaminant in many parts of the world, particularly in areas of geothermal activity, and is naturally present at high concentrations in groundwater in several countries. There have been significant health problems in parts of Bangladesh caused by high levels of arsenic in drinking water supplies (Smith, 2000) and recent epidemiological studies in Bangladesh, India and Chile (Ahsan, 2006; von Ehrenstein, 2006; Hopenhayn, 2006) all indicate correlations between arsenic in drinking water and skin cancer, reproductive effects and anaemia.

Arsenic can be found in surface run-off from mining waste tips or in areas where there are certain types of metalliferous ore. Its presence in water may also be the result of pollution from weedkillers and pesticides containing arsenic, or from chemicals used in wood preservation.

There are a number of treatment processes to remove arsenic (Section 10.14) and in the UK several water companies have had to install additional treatment at groundwater sites in order to meet the required standard for arsenic.

7.9 ASBESTOS

The widespread use of asbestos cement pipes in distribution systems has raised concerns that the fibres found in water may be a danger to health. A study carried out in the UK found that some drinking waters can contain up to 1 million asbestos fibres per litre, with more than 95% of these being less than 2 μm in length (WRc, 1984). The UK Department of Health CASW Committee reported in 1986 that there was substantial evidence to show that asbestos, as found in drinking water, did not represent a hazard to health (DoE, 1990a). WHO likewise considers that there is no consistent evidence to show that ingested asbestos is hazardous to health. Thus, there is no health-based guideline value for asbestos in drinking water. This is supported by a report commissioned by Defra (2002), based on evidence from epidemiological studies and from results of laboratory animal feeding.

Few countries continue to install asbestos cement pipes, mainly because of the inhalation risks associated with working with the dry material. The use of new asbestos cement pipes has been banned in the UK since 2000, because of this health and safety risk.

7.10 BORON

Boron exists in the aquatic environment primarily as boric acid and borate. The main sources of boron in surface waters are from industrial discharges or detergents in treated sewage effluents. However, naturally occurring boron may be found in groundwater sources in areas with boron rich deposits. Furthermore, seawater contains up to 6 mg/l of boron, so its presence in a source water near the coast may indicate saline intrusion.

Boron is not significantly removed by conventional treatment processes but can be removed by reverse osmosis (RO) if the process is operated at pH values greater than 10 (Section 10.47). The concentrations of boron in most drinking waters tend to be very low.

7.11 BROMIDE AND IODIDE

Seawater contains 50–60 mg/l bromide, so the presence of bromide in well or borehole sources near the coast could be evidence of seawater intrusion. Ozone treatment of water containing bromide may result in the formation of bromate as a disinfection by-product (DBP; Sections 7.24 and 11.19). In March 2004, the producer of a bottled table water in the UK had to withdraw supplies because of elevated levels of bromate. The bottled water was produced from tap water, which was first desalinated. Calcium chloride was then added to meet a minimum standard for calcium and the product water disinfected with ozone. Traces of bromide present in the calcium chloride were oxidized to bromate, giving concentrations of up to 22 μg/l.

Many natural waters contain trace amounts of iodide, usually at levels less than 10 μg/l, although higher concentrations are found in brines and brackish water. The levels found in drinking water are too low to contribute significantly to dietary requirements.

7.12 CADMIUM

The main sources of cadmium in the aquatic environment are from industrial discharges or diffuse pollution from air pollution or certain types of fertilizer. It is occasionally found in drinking water as an impurity in zinc used to galvanize pipes, or in solders, or other metal fittings.

7.13 CALCIUM

Calcium is found in most waters, the level depending on the type of rock through which the water has permeated. It is usually present as calcium carbonate or bicarbonate, especially in waters that are associated with chalk or limestone, and as calcium sulphate associated with gypsum. Calcium chloride and nitrate may also be found in waters of higher salinity. Calcium bicarbonate forms temporary hardness, the other salts being linked to the permanent hardness (Section 7.29).

Calcium is an essential part of human diet, but the nutritional value from water is likely to be minimal when compared to the intake from food. The main concern of calcium in drinking water is the potential for scale formation.

7.14 CARBON DIOXIDE

Free carbon dioxide in a water (as distinct from that existing as carbonate and bicarbonate) depends on the alkalinity and pH value of the water. It is an important factor in determining the corrosive properties of a water (Section 10.41).

Surface waters usually contain less than 10 mg/l free CO_2 but some groundwaters from deep boreholes may contain more than 100 mg/l. A simple means of reducing free CO_2 in a water is by aeration (Section 10.18).

7.15 CHLORIDE

Chloride is found in nearly all waters and is derived from a number of sources, including natural mineral deposits; seawater intrusion or airborne sea spray; agricultural or irrigation discharges; urban run-off due to the use of de-icing salts; or from sewage and industrial effluents. It is usually combined with sodium and to a lesser extent with potassium, calcium and magnesium, which makes chloride one of the most stable components in water.

Most rivers and lakes have chloride concentrations of less than 50 mg/l Cl and any marked increase may be indicative of sewage pollution or, if the increase is seasonal, urban run-off linked to the application of rock salt ('grit') to roads. The chloride content of a sewage effluent under dry weather flow could increase the chloride content of the receiving water by as much as 70 mg/l. Brackish or estuarine waters can contain several hundred mg/l of chloride, with seawater typically containing around 20 000 mg/l.

Excessive chlorides can give rise to corrosion and also to taste problems. High concentrations of chloride tend to enhance the corrosion rates of iron, steel and plumbing metals, especially when coupled with low alkalinity (Section 10.41). A sensitive palate can detect chloride in drinking water at as low a level as 150 mg/l and concentrations above 250 mg/l may impart a distinctly salty taste (Section 7.49). However, in arid or semi-arid areas, people may have to drink water containing much higher levels of chloride if no alternative supply is available.

Conventional water treatment processes do not remove chlorides. If the chloride content of a water has to be reduced then some form of desalination has to be applied (Section 10.43).

7.16 CHLORINATED HYDROCARBONS

The term 'chlorinated hydrocarbons' covers a wide range of volatile organic chemicals used as solvents, metal cleaners, paint thinners, dry cleaning fluids and as a raw material in the synthesis of other organic chemicals. They tend to be found as microcontaminants in groundwaters that have been subjected to industrial pollution and can continue to contaminate the soils and substrata long after the original source of pollution has been removed. Some may also be found in surface waters as a result of effluent discharges.

The three most common groups to be detected in drinking water are the chlorinated alkanes, which include tetrachloromethane (or carbon tetrachloride); the chlorinated ethenes, which include trichloroethene (TCE) and tetrachloroethene (PCE); and the chlorinated benzenes. Many chlorinated solvents are considered potential carcinogens and some, such as TCE and PCE, may degrade in anaerobic groundwaters to produce more toxic substances such as vinyl chloride. There is little commonality between the recommended standards and, in the case of the chlorinated benzenes, the taste and/or odour thresholds are usually well below any health risk levels.

7.17 CHLORINE RESIDUAL

Chlorine remains the principal biocide and disinfectant used in water treatment in most countries (Chapter 11). Chlorine gas tends to be used at major water treatment works, although many works in urban areas now use sodium hypochlorite, either as a liquid or generated on-site by electrolysis of salt, on safety grounds. Chlorine-containing powders, crystals and solutions tend to be used at smaller works or at sites where the procurement of chlorine gas presents difficulties. The effectiveness of chlorine as a disinfectant arises from its high chemical reactivity. However, it is recognized that the use of chlorine for disinfection can result in the formation of a wide range of undesirable chlorinated compounds, known collectively as DBPs (Section 7.24). Although usually present at very low concentrations, some of these compounds are potentially hazardous to health and others produce objectionable tastes and odours. Notwithstanding these problems, which can be minimized by appropriate treatment control (Section 11.7), the use of a suitable disinfectant such as chlorine is essential for ensuring that a water is bacteriologically safe to drink, unless other reliable means of disinfection (e.g. boiling) can be used.

There is no evidence that the levels of chlorine residual normally found in drinking water are harmful to health. Most consumers become familiar with the levels that are normal for their local water supply. However, sudden increases are likely to be noticed and generate taste complaints, particularly from consumers with sensitive palates. Taste and odour complaints can also arise from the reaction of chlorine with other trace substances present in the water (Section 11.11) or with certain plumbing materials (Section 7.49).

7.18 CHROMIUM

Chromium is found in the aquatic environment in predominantly two forms, trivalent chromium (Chromium III or Cr^{3+}), which occurs naturally, and hexavalent chromium (Chromium VI or Cr^{6+}), which also occurs naturally due to erosion of chromium deposits but is found more commonly as a result of industrial pollution. Chromium can also be found in drinking water as a corrosion product of chrome-plated taps.

Chromium VI salts tend to be more soluble than those of chromium III and are also more harmful to human health. Because of the uncertainties of the toxicology, WHO currently has a provisional guideline value for total chromium. Most other regulatory standards are also based on total chromium. However, the presence of chromium VI in drinking water is becoming something of an issue in the USA and, in July 2014, the State of California introduced a specific standard of 0.01 mg/l for chromium VI in drinking water.

7.19 COLOUR

The colour of a water is usually expressed in Hazen units, which are the same as TCU (true colour units) or mg/l on the platinum cobalt (Pt–Co) scale. Water can often appear coloured because of colloidal or other material present in suspension, which means that true colour must be determined only after filtration, usually through a 0.45 μm filter. The colour in unpolluted surface waters is caused by the presence of humic and fulvic acids, which are derived from peat and soil humus. These chemicals, known as chromophores, absorb light of particular wavelengths, giving a characteristic yellow/brown colour. In some waters the colour is enhanced by the presence of iron and manganese, which is often organically bound. Waters subject to industrial pollution may also contain a wide variety of coloured materials.

The level at which colour becomes unacceptable depends largely upon consumer perception, with most consumers noticing a colour of 15 TCU in a glass of water.

7.20 COPPER

Copper is rarely found in unpolluted waters, although trace amounts can sometimes be found in very soft, acid moorland waters. The most usual source of copper in drinking water is from corrosion of copper, and copper-containing alloys, used in domestic plumbing systems.

Newly installed copper pipework may give rise to 'blue' water, especially if there is prolonged stagnation in the domestic distribution system. Water containing as little as 1 mg/l of copper can cause blue/green stains on sanitary fittings. Much lower concentrations can accelerate corrosion of other metals in the same system (Campbell, 1970) making it inadvisable to use galvanized steel piping or storage tanks downstream of copper piping.

Copper concentrations above 2.5 mg/l can impart an unpleasant and astringent taste to a water and some individuals may suffer acute gastric irritation at concentrations above 3 mg/l.

7.21 CORROSIVE QUALITY

Ideally, a treated water entering supply should not be corrosive or aggressive to plumbing materials or to concrete (Section 10.41). Many factors determine whether a water will be corrosive but the three principal characteristics that exacerbate corrosion are:

- a low pH value, i.e. acidity;
- a high free carbon dioxide (CO_2) content; and
- an absence, or low amount, of alkalinity.

Free chlorine residual is also a factor in the dezincification of brasses containing more than 15% zinc.

Waters that tend to be corrosive include soft moorland waters; shallow well waters of low pH with low temporary hardness but high permanent hardness; waters from greensands and other iron bearing formations; chalk and limestone waters with a high CO_2 content; waters from coal measures; and waters with a high chloride content. Desalinated water is also very corrosive unless suitably treated.

One measure of assessing the corrosive nature of a water is the so-called saturation pH (pHs). This can be determined by placing the water in contact with powdered chalk or marble for a defined period. The final pH value is referred to as the stability or saturation pH and if this is greater than the initial pH value, the water is likely to be corrosive to iron, steel and cement. If the pHs value is less than the initial pH value, then the water is likely to deposit calcium carbonate as a protective layer on the interior of metal pipework. If the two pH values are the same, the water is said to be in equilibrium. The stability pH value can also be calculated using the Langelier formula (Section 10.41) (Langelier, 1936). There are a number of other indices with wider capabilities that can be used to calculate the corrosion potential of a water (Rossum, 1983).

Another aspect of corrosion potential is whether a water will dissolve lead or copper (i.e. be plumbosolvent or cuprosolvent) (Section 7.32). Such waters may require the addition of orthophosphate or silicate as well as pH and alkalinity adjustment, either individually, or in various combinations, to reduce the corrosion potential.

7.22 CYANIDE

Cyanide and cyanide complexes are only found in waters polluted by effluents from industrial or mining processes involving use of cyanide. Most cyanides are biodegradable and should be removed by chemical treatment before an effluent is discharged into a receiving water.

Chlorination to a free chlorine residual under neutral or alkaline conditions effectively decomposes any remaining cyanide that may be present in a raw water.

7.23 DETERGENTS

There are several substances that can cause foaming in water, the largest group being synthetic detergents or surfactants. Many surface waters downstream of urban areas contain detergent as a result of sewage effluent discharges. In recent years, however, the increased use of biodegradable detergents, which can be removed by normal sewage purification processes, has led to a reduction in the residual detergent in sewage effluent. The main reason for removing detergents from water is to prevent foaming, although some components of anionic surfactants are toxic to aquatic life. The presence of the more common anionic surfactants can be detected using the methylene blue colorimetric method. An anionic surfactant detected by this reaction is called a methylene blue active substance (MBAS).

7.24 DISINFECTION BY-PRODUCTS (DBPs)

Chemical oxidants, such as chlorine and ozone, are traditionally used as disinfectants to inactivate pathogenic organisms present in raw waters (Chapter 11). They are also used to assist in the oxidation of iron and manganese and to break down taste and odour forming compounds. Ozone is used extensively to break down pesticides. Side reactions with organic and inorganic constituents present in a raw water can give rise to low concentrations of a number of compounds, collectively known as DBPs. In the case of organic constituents, the rate of DBP formation can be significantly reduced by optimizing the treatment process to remove or substantially reduce precursors prior to the final disinfection stage. Furthermore, the disinfection process should be designed, maintained and operated in such a way as to minimize the potential for DBP formation but without compromising the effectiveness of the disinfection. In 1991 a major cholera epidemic occurred in Peru after water officials were put under pressure to reduce chlorination because of a perceived hypothetical cancer risk from chlorination by-products.

Over the last 30 years much research has been carried out into the significance and control of DBPs, especially those associated with chlorination. Routine monitoring tends to focus on the regulated DBPs, namely trihalomethanes (THMs) and haloacetic acids (HAA), and it is considered likely that adequate control of these compounds will also result in adequate control of most other chlorinated DBPs. Some of the more commonly found DBPs are discussed in more detail below.

Bromate. The presence of bromate in drinking water tends to be associated with disinfection. However, early in 2000, a UK water company carrying out checks in advance of the new Drinking Water Directive requirements for bromate found unacceptable levels at one of its groundwater sources. The site of a former chemical works was identified as the source of contamination. The affected source had to be taken out of supply and there were increasing concerns for other sources in the area as the plume of contamination moved across the aquifer.

In terms of DBP formation, ozone reacts with any bromide present in a raw water to form bromate (Section 11.19) (DoE, 1993a). The rate of bromate formation in drinking water depends on the amount of natural organic matter present in the water, the alkalinity and the ozone dose applied (Siddiqui, 1995). Bromate may also be present in commercially available sodium hypochlorite or may be formed during on-site electrolytic generation of sodium hypochlorite (Section 11.14) due to bromide present in the salt feedstock. The concentrations found in hypochlorite solution can range from 2.8 mg/l to more than 20 mg/l (DoE, 1993b).

Chloramines. The process of chloramination involves dosing a controlled amount of ammonia to chlorinated water (Section 11.8). Although less powerful than free chlorine, the resultant monochloramine retains disinfectant properties and provides a more stable residual within the distribution system. Dichloramine and nitrogen trichloride are associated DBPs which can also be formed if the process is poorly controlled, with resultant taste and odour problems. Similar problems arise if chloraminated water is mixed with water containing a free chlorine residual within a distribution system.

Chlorate and chlorite. Although not strictly DBPs, chlorate and chlorite can be found in water that has been treated with chlorine dioxide (Section 11.16). Chlorate, like bromate, may also be present in commercially available sodium hypochlorite and in sodium hypochlorite produced by electrolysis.

The UK drinking water regulations specify a maximum chlorate concentration of 0.7 mg/l as ClO_3 in the treated water as a condition of use for plants producing sodium hypochlorite by electrolysis. The UK drinking water regulations also apply British Standards BS EN criteria for the use of chlorine dioxide and sodium chlorite, such that the combined concentration of chlorine dioxide, chlorite and chlorate must not exceed 0.5 mg/l as chlorine dioxide in the water entering supply.

Chlorophenols may be formed from the chlorination of trace levels of phenolic compounds present in the raw water or as degradation products from the breakdown of phenoxy acid herbicides. They have very low organoleptic thresholds and, if present, are immediately noticed by consumers as an antiseptic taste. The taste thresholds for the most commonly found chlorophenols in drinking water are well below any health-related value.

Haloacetic acids (chloro- and bromoacetic acids) and haloacetates may be formed when surface waters are chlorinated, depending on the organic precursors present. WHO sets guideline values for two individual HAAs and a provisional value for a third, whilst the US EPA has a cumulative standard for five HAAs. Trichloroacetic acid can be used as a herbicide and has been found in detectable concentrations in some raw waters.

Halogenated acetonitriles. Chloro- and bromo-halogenated acetonitriles may be formed when surface waters containing algae are chlorinated or chloraminated. Dichloroacetonitrile is the most commonly found.

Sodium dichloroisocyanurate, which is widely used as a disinfectant for swimming pool water, can also be used as an emergency disinfectant of drinking water. It may also be used as source of chlorine for point-of-use water treatment. Although not strictly a DBP, WHO has guideline values for both residual sodium dichloroisocyanurate and associated cyanuric acid.

Trihalomethanes (THMs) may be formed when surface waters containing naturally occurring organic compounds, such as humic and fulvic acids, are chlorinated. They are also formed by the reaction of chlorine with some algal derivatives (Section 11.7). The group comprises four compounds: trichloromethane or chloroform, bromodichloromethane, dibromochloromethane and tribromomethane or bromoform. Some epidemiological studies have suggested weak correlation between incidences of certain cancers with consumption of water containing THMs, but the level of exposure has not always been adequately characterized and, in some cases, no account has been taken of other cancer risk factors or other routes of exposure. Likewise, studies in the USA claim a correlation between THM intake and early pregnancy terminations and low birth weight. A study into the relationship of THM concentrations in public water supplies to stillbirth and birth weight in three water regions in England suggested a significant association of stillbirths with maternal residence in areas with high total THM exposure but it also concluded that further research was needed (Toledano, 2005). A further study in 2007 commissioned by Defra (2007a) found little evidence for a relationship between THM concentrations in drinking water and risk of congenital anomalies.

WHO sets guideline values for each of the four individual THMs and uses a fractionation approach (the sum of the ratios of the concentration of each to its respective guideline value should be equal to or less than 1) to derive a standard for total THMs.

Nitrites (see also Section 7.35) can be formed as a DBP where ultraviolet radiation is used to disinfect waters containing moderate to high concentrations of nitrate (Table 11.7). Undesirable levels of nitrite can also be formed at the end of long and complex distribution systems where disinfection by chloramination is practised (Section 11.8).

7.25 ELECTRICAL CONDUCTIVITY AND DISSOLVED SOLIDS

Conductivity measures the ability of a solution to carry electrical current. This depends on the presence of ions in solution and therefore provides a useful indication of the total dissolved solids, or salts, present in a water. For most waters a factor in the range 0.55–0.70 multiplied by the conductivity in μS/cm gives a close approximation to the dissolved solids in mg/l. The factor may be lower than 0.55 for waters containing free acid and greater than 0.70 for highly saline waters. Conductivity is temperature dependent and a reference temperature (usually 20°C or 25°C) is used when expressing the result. One of the advantages of the conductivity determination is that it can be easily measured in the field or used in continuous monitoring.

High levels of dissolved salts can result in taste complaints, as well as causing excessive scaling to domestic and industrial water systems. Water with a dissolved solids content of more than 1000 mg/l could be unpalatable, although many natural mineral waters exceed this level. Water with low dissolved salts is desirable for many industrial processes but may be unacceptable to consumers, again on taste grounds, and may also be corrosive to domestic plumbing.

7.26 ENDOCRINE DISRUPTING SUBSTANCES

It has been long recognized that exposure to certain chemicals such as polychlorinated biphenyls, tributyl tin, certain pesticide residues and some pharmaceutical residues can have an adverse effect on the endocrine (or reproductive) system of animals in the aquatic environment. Some of these substances are already subject to environmental quality standards. The full impacts of others, including steroids such as oestrogen, are still being researched. Most of these substances enter the aquatic environment through industrial discharges and sewage effluents, or via diffuse pollution.

The UK Environment Agency (EA) has investigated the impact of synthetic and natural oestrogen residues on the aquatic environment as part of its environmental strategy, especially in terms of the feminization of male fish (EA, 1998). In 2012 a literature review into endocrine disrupting substances in drinking water (Defra, 2012) identified 325 substances for which concerns had been expressed with regard to potential endocrine disrupting activity. Of these, only six were identified as warranting more detailed consideration (*p*-benzylphenol, dibutylphthalate, 4-nitrophenol, digoxin, fluticasone and salbutamol). The report concluded that the predicted worst-case concentration of each of these chemicals would not constitute a significant risk to human health when considered in terms of the equivalence to the natural hormone oestradiol.

The impact of such residues is of more significance to the aquatic environment than to drinking water. There is usually significant dilution of sewage effluent discharges before downstream water is abstracted as a source of drinking water. Furthermore, advanced treatment processes installed for pesticide removal are likely to be equally effective at removing any remaining endocrine disrupting residues. Traces of oestrogens can also be removed by chlorination at concentrations typically used for final disinfection.

7.27 EPICHLOROHYDRIN

Epichlorohydrin is a chemical intermediate used in the manufacture of polyamine coagulants and also in some ion exchange resins used in water treatment and softening. Traces may be found in drinking water but concentrations are controlled by product specification and use.

7.28 FLUORIDE

Naturally occurring fluoride is found in varying concentrations in most drinking waters. Some deep groundwaters in the UK contain between 2 and 5 mg/l of fluoride and much higher concentrations are found in other parts of the world, especially in areas associated with fluoride-containing minerals.

The levels of fluoride in drinking water have to be closely controlled as excessive amounts can lead to fluorosis, with resultant mottling of the teeth, and in extreme cases even skeletal damage. The maximum concentration also has to be related to climatic conditions and the amount of water likely to be consumed. Blending is the preferred option if there is sufficient low fluoride water available; otherwise specialized treatment is required to remove excess fluoride (Section 10.16).

Low levels of fluoride are dosed to public supplies in the UK (MoH, 1969), the USA, Australia and elsewhere as an effective means of reducing dental caries. The greatest reduction of dental decay occurs if fluoridated water is drunk in childhood during the period of tooth formation. Although consumers accept the presence of naturally occurring fluoride in the drinking water, many object to the deliberate addition of fluoride to public water supplies in spite of the stringent controls that are attached to the process. In the UK the decision to fluoridate water supplies has to be taken by the relevant health authority and must be evidence based. It is also subject to public consultation.

7.29 HARDNESS

The term 'hardness' originally referred to the tendency of a water to form scum or curd when it reacts with ordinary soap. Hardness also relates to the scale that precipitates in kettles and utensils when water is boiled. The form of hardness produced on heating is the temporary, or carbonate, hardness consisting of calcium and magnesium bicarbonates. Permanent, or non-carbonate, hardness (which is not precipitated by heating) is due to other minerals or salts of calcium and magnesium, such as sulphates, that may be present in the water. Hardness, like alkalinity (Section 7.5), is usually expressed in mg/l as $CaCO_3$ although other notations may be used, such as French or German degrees, in association with domestic appliances such as dishwashers.

The descriptive terms commonly applied are as follows:

Hardness description	Hardness as $CaCO_3$ mg/l
Soft	0–50
Moderately soft	50–100
Slightly hard	100–150
Moderately hard	150–200
Hard	200–300
Very hard	>300

Problems caused by excessive hardness relate mainly to the formation of scale in boilers and hot water systems. Consumers in hard water areas also complain of scale deposition on kitchen utensils and increased soap usage, with associated scum formation. Conversely, waters containing less than 30–50 mg/l total hardness tend to be corrosive and may need additional treatment to reduce the risk of plumbo- and cuprosolvency. Desalinated water has virtually zero hardness and is highly corrosive, requiring treatment to render it non-aggressive to metallic plumbing materials (Section 10.49).

A number of studies have been carried out into relationships between the hardness of water and the incidence of cardiovascular disease. In 2005 a literature review was carried out in the UK (Defra, 2005a), which concluded that there was evidence that hard water gave a protective effect. A further study, however, carried out in 2008 (Defra, 2008a) found no evidence of an association between step changes in drinking water hardness, calcium or magnesium and cardiovascular mortality. Many other characteristics vary with the hardness of a water and it is not clear whether the relationship observed might be due to a protective factor in hard water or a harmful factor in soft water. Very few UK water companies now soften public water supplies following advice issued by the then Department of Health and Social Security (DHSS, 1971).

7.30 HYDROCARBONS

Hydrocarbons include petroleum, mineral oils, coal and coal tar products, and many of their derivatives such as benzene and styrene, which are produced by the petrochemical industry for industrial processes. In oil producing parts of the world, background concentrations of hydrocarbons may be present in surface and groundwaters from natural sources. However their presence is more usually the result of pollution.

On rare occasions, traces of benzene and other solvents have been reported in domestic water supplies. This has usually occurred where there has been localized permeation of plastic water pipes (Section 17.25) from contaminated ground conditions or other external sources of hydrocarbons, such as petroleum spillages. Although the practice of lining water mains with coal tar has not been practised in the UK since the 1970s, some mains lined with coal tar still exist and these can result in chemicals such as benzopyrene leaching into the water (Section 7.41).

A number of hydrocarbons have health-related standards, but most are well above the thresholds for taste or odour.

7.31 IRON

Iron is found in most natural waters and can be present in true solution, or in suspension as a colloid, or as a complex with other mineral or organic substances. Iron in surface waters is usually in the ferric (Fe^{3+}) form, but the more soluble ferrous (Fe^{2+}) form can be found in deoxygenated conditions that may occur in some deep boreholes, especially in greensand areas, or in the bottom waters of lakes and reservoirs. On exposure to air, such waters rapidly become discoloured as the iron oxidizes to the ferric form and precipitates out. Severe corrosion can occur where unlined iron is used in rising mains and pumps at boreholes that contain aggressive water. This gives rise to very iron rich water, especially when there are long periods of stagnation between pumpings.

Iron salts are extensively used as coagulants in water treatment (Section 8.21). With good process control, it should be possible to keep the residual concentration of iron in the water entering supply to less than 0.05 mg/l. However, incomplete removal during treatment can lead over a period of time to significant deposits of iron in the distribution system. Corrosion products from unlined cast iron mains also contribute to the build up of ferruginous material in a distribution system. Regular flushing, providing it is adequately controlled, can help to control this build up of deposits which otherwise gives rise to discoloured water problems when disturbed (Section 16.3). The deposits can also contribute to the growth of iron bacteria, which in turn can cause further water quality deterioration by producing slimes or objectionable odours.

There is no health-related standard for iron in drinking water but consumers are likely to reject discoloured water on the grounds of appearance. Excess iron can cause brown stains on laundry and plumbing fixtures and can also impart a bitter taste to the water at levels above 1 mg/l.

7.32 LEAD

Lead is a cumulative poison and the hazards of exposure to lead in the environment have been well documented over the years. With the decrease in levels of lead in atmospheric emissions and with reduced use of paint-containing lead, lead in drinking water is now considered to be the main source of controllable exposure. Recently there has been growing concern, and an increasing body of evidence, that quite low levels of exposure to lead can affect learning ability and behavioural problems in children and correlations have been found between lead levels in drinking water and lead levels in blood (Fertmann, 2003; Brown, 2012). Within the European Union, the maximum allowable lead concentration in drinking water has been reduced twice in the last 10 years, firstly in 2003 from 50 to 25 μg/l and then again in 2013 to a final standard of 10 μg/l. Schools and childcare centres are of particular interest, as are public buildings where drinking water is made available to the public.

Lead is rarely found in detectable concentrations in the aquatic environment, except in areas where soft acidic waters come into contact with galena or other lead ores or where rivers are subject to pollution from disused lead mines. The main source of lead in drinking water is the dissolution of lead service pipes and internal domestic plumbing, along with any original lead solder, which may still be present in older properties in the UK and elsewhere. Although plumbosolvency tends to be associated with soft, acidic waters from upland and moorland catchments, it can also occur in hard water areas, especially where the hardness is mainly non-carbonate. Thus, the concentrations of lead in drinking water should always be monitored closely in areas where lead pipes are known to be present. Treatment for plumbosolvency (Section 10.14) by stabilizing the final water pH value and/or dosing orthophosphate has been shown to be very effective in many parts of the UK and the USA.

In England and Wales, the benefits of treatment are shown by the significant decrease in lead concentrations in regulatory drinking water samples since 1990. By the end of 2013, 99.91% of regulatory samples contained less than 25 μg/l of lead and 99.31% contained less than 10 μg/l of lead (DWI, 2013a). Although these results are very promising, it is recognized that in some areas the ongoing failures are due to particulate lead, which sloughs off the interior of the pipe. In such situations the only viable solutions are pipe replacement or, if the lead pipe is still structurally sound, possibly lining with a thin plastic liner. Lead pipe replacement is the only permanent solution but this may be difficult where the water company does not own the pipework. In 2013, DWI issued guidance to the industry on ways of addressing the situation (DWI, 2013b).

Unacceptable levels of lead have also been found with some new unplasticized PVC pipes, where lead compounds used as stabilizers in the pipe manufacture are leached out. Although the use of solder with a high lead content for copper pipe joints is now banned in the UK, occasional high levels of lead are still found where the wrong type of solder has been used.

7.33 MANGANESE

Manganese can be found in detectable concentrations in both surface and groundwaters. The concentration of soluble manganese rarely exceeds 1.0 mg/l in a well aerated surface water but much higher concentrations can occur in groundwaters subject to anaerobic conditions. Manganese can also re-dissolve from the bottom sediments in impounding reservoirs if the bottom water becomes deoxygenated. This leads to an increase in the overall manganese content of the water when the reservoir water 'turns over' (Section 8.1).

Manganese is an essential element in the human diet but, in excess, it can act as a neurotoxin. A study carried out in Canada (Bouchard, 2011) suggested a link between intellectual impairment in children and high levels of manganese (up to 2.7 mg/l) in local groundwater supplies. However, current guidelines and standards are based on aesthetic quality, as levels of manganese as low as 0.1 mg/l can cause staining of laundry and sanitary ware. It is also undesirable even in small quantities in water supplies as it can precipitate out in the presence of oxygen or after chlorination. The precipitate forms a black slime on internal surfaces in the distribution system and, if disturbed, gives rise to justifiable consumer complaints. The tolerable concentrations of manganese in a distribution system are generally lower than those for iron for, although deposition of manganese is slow, it is continuous. Thus the onset of serious trouble may not become apparent for some 10–15 years after putting a manganese rich supply into service.

7.34 NICKEL

Nickel occurs widely in the environment and is occasionally found at low levels in drinking water. Increased concentrations may occur when the plating on nickel- or chrome-plated taps becomes damaged or starts to break down. Increased concentrations may also be found in the water boiled in electric kettles with exposed elements, although these are becoming increasingly uncommon (WRc, 2007). Nickel is known to cause skin sensitization and the ingestion of high levels via food or drinking water may cause dermatitis in some, but not all, nickel sensitive individuals.

7.35 NITRATE AND NITRITE

These parameters are considered together because both are oxidation states of ammonia and conversion from one form to the other can occur naturally within the aquatic environment.

Nitrate

Nitrate is the final stage of oxidation of ammonia and the mineralization of nitrogen from organic matter. Most of this oxidation in soil and water is achieved by nitrifying bacteria and can only occur in a well oxygenated environment. The same bacteria are active in percolating filters at sewage treatment works, resulting in large amounts of nitrate being discharged in sewage effluents from such works. The use of nitrogenous fertilizers on the land can also give rise to increased nitrate concentrations in both surface and underground waters. Nitrate levels in surface waters often show marked seasonal fluctuations, with higher concentrations occurring in winter when run-off increases due to winter rains at a time of reduced biological activity. During summer, the nitrate levels are likely to be reduced by biochemical mechanisms and by algal assimilation in reservoirs. Bacterial denitrification and anaerobic reduction to nitrogen at the mud interface can, in addition, substantially reduce the nitrate levels in reservoirs. The EC Nitrates Directive of 1991 (CEU, 1991) required member states to take measures to control farming practices that are likely to cause increased nitrates in surface and groundwaters. In the UK a designated action programme required Nitrate Vulnerable Zones to be set up and the promotion of best practice to farmers in the use and storage of fertilizers and manure. By 2002 55% of England was designated a Nitrate Vulnerable Zone, with similar designations in Scotland, Wales and Northern Ireland. The focus now is on diffuse pollution, with farmers being encouraged to adopt catchment sensitive farming (Section 7.55). Implementation of the 1998 EC Directive on Urban Waste Water Treatment (CEU, 1998b) has also resulted in the setting up of Sensitive Areas across many parts of England. These are intended to reduce the level of eutrophication in inland or coastal waters (Section 7.75) and to protect surface waters used for the abstraction of drinking water by reducing the levels of phosphorus (Section 7.40) and nitrates in discharges from sewage treatment works.

Waters containing high nitrate concentrations are thought to be potentially harmful to infants. At the neutral pH of the infant stomach, nitrate can undergo bacterial reduction to nitrite, which is then absorbed into the bloodstream and converts the oxygen carrying haemoglobin into methaemoglobin. Whilst the methaemoglobin itself is not toxic, the effects of reduced oxygen carrying capacity in the blood can be serious, especially for infants having a high fluid intake relative to body weight. However, it has become apparent that methaemoglobinaemia is a problem associated with rural shallow wells subject to microbial contamination, rather than public water supplies. There are many examples of public water supplies with relatively high nitrate levels in Europe and North America where no problems with methaemoglobinaemia have been reported. In 1997, an epidemiological study carried out by Leeds University indicated that there was a small but statistically significant correlation between the incidence of childhood diabetes and nitrate in drinking water. The study was restricted to a very small geographical area and the correlation could have arisen by chance. A follow-up study, commissioned by the then DETR examined the incidence of childhood diabetes over large areas of England and Wales and the results failed to demonstrate any correlation with nitrate concentrations in drinking water (DETR, 1999a).

Nitrite

Nitrite is the intermediate oxidation state between ammonia and nitrate and can be formed by the reduction of nitrates under conditions where there is a deficit of oxygen. Surface waters, unless badly polluted with sewage effluent, seldom contain more than 0.1 mg/l nitrite as N. Thus, the presence of nitrites in surface waters in conjunction with high ammonia levels indicates pollution from sewage or sewage effluent. The presence of nitrites in groundwater may also be a sign of sewage pollution. Conversely, it may have no hygienic significance as nitrates in good quality groundwaters can be reduced to nitrite under anaerobic conditions, especially in areas of ferruginous sands. New brickwork in wells is known to have a similar effect. Nitrite concentrations may increase in distribution systems receiving chloraminated water.

Ingested nitrites, some of which may be formed by bacterial reduction of nitrates, can react with secondary and tertiary amines found in certain foods to give nitrosamines. This has given rise to concern as nitrosamines are potentially carcinogenic. However, there is no epidemiological evidence of an association between nitrite levels in drinking water and cancer incidence.

There are no simple methods for treating a water to reduce its nitrate–nitrite concentration, other than by blending it with another supply with low or negligible concentrations of the same. Thus improved catchment management is the most sustainable approach to dealing with point source or diffuse nitrate pollution, especially for groundwaters. Treatment processes currently used for nitrate removal include ion exchange and biological removal under controlled conditions using denitrifying bacteria (Sections 10.24–10.26).

7.36 ORGANIC MATTER, BIOCHEMICAL OXYGEN DEMAND (BOD) AND CHEMICAL OXYGEN DEMAND (COD)

Surface waters, and groundwaters affected by surface waters, contain organic matter from a variety of sources such as plant and animal material, including partially treated domestic waste, and industrial effluents. The amount of natural organic matter (NOM) present in a water can be estimated from the oxygen absorbed permanganate value (PV), from the total organic carbon content (TOC), from the biochemical oxygen demand (BOD) or from the chemical oxygen demand (COD). All are gross measures of the total concentration of organic substances and are usually dominated by the concentration of NOM, i.e. humic and fulvic acids. (Humic and fulvic acids are complex molecular structures resulting from the decay of humus, itself the product of decay of plant, animal and microbial matter.)

The proportion of TOC that passes a 0.45 μm filter is referred to as dissolved organic carbon (DOC) and the residue is particulate organic carbon (POC) (Section 10.34). BDOC (biodegradable dissolved organic carbon) and AOC (assimilable organic carbon) are the biodegradable fractions of DOC (Section 10.35) and are indicators of the potential for bacterial growth. TOC/DOC is an important parameter in water treatment as it relates to the DBP formation potential of a water. TOC can be measured relatively easily in the laboratory. The analysis allows simple differentiation between particulate and dissolved fractions and is more useful for process control than BDOC/AOC measurements. BDOC/AOC is used to assess the effectiveness of biological processes and bacterial growth in distribution systems.

UV absorbance at 254 nm is used as a surrogate measure of selected organic constituents (usually the DBP precursors, for example, humic acids, tannin, lignin, etc.). It provides an indication of the aggregate concentration of UV absorbing organic constituents in a water, rather than individual constituents. It is easily measured using on-line instruments and is sometimes used in automatic control algorithms to identify the coagulant dose for coagulation (Section 12.4). UVa (m^{-1}) is used together with DOC (mg/l) to calculate SUVA (specific UV absorbance in l/mg.m = $UV_{254} \div DOC$). SUVA is a good indicator of the humic fraction of DOC which is effectively removed in coagulation and can therefore be used to assess the nature and reactivity of DBP precursors in water (Section 8.12).

The oxygen absorbed (PV) test provides an indirect, but by no means complete, measurement of TOC by determining its oxidizability. The 5-day BOD test gives an indication of the oxygen required to degrade any organic matter present biochemically, as well as the oxygen needed to oxidize inorganic materials such as sulphides. The test provides an empirical comparison of the relative oxygen requirements of surface waters, wastewaters and effluents. For example, if a sewage effluent with a high BOD is discharged into a stream, the oxygen required by organisms to break down the organic matter in the effluent is taken from the overall oxygen content of the receiving water. This depletion could potentially destroy fish and plant life.

The COD test is less specific, as it measures all chemically oxidizable material, rather than just biologically active organic matter. BOD and COD are usually found in low concentration, are more difficult to measure and are not sensitive as a design tool in water treatment.

Organic Micropollutants

The application of gas chromatography and mass spectrometry to water analysis since the 1970s has revealed the presence of many hundreds of different organic compounds. These are derived from naturally occurring substances in the environment, as well as from materials produced, used or discarded by industry and agriculture and, when present, are usually at very low concentrations of less than 1 μg/l. Some are known to be toxic or carcinogenic to animals at concentrations far higher than detected in drinking water; others are known to be mutagenic (i.e. capable of making heritable changes to living cells) under laboratory testing. A number have yet to be identified fully, although ongoing improvements in analytical capability mean that more substances are being detected at ever decreasing concentrations.

Assessing the risk that these compounds might present to human health is difficult, given the very low levels found in drinking water. Different methods of extrapolating data obtained from observing the effect of high concentrations on animals, in order to deduce the risk to human health of ingesting far smaller quantities, have given results in some cases varying by several orders of magnitude (Hunt, 1987). Epidemiological studies to discover some statistical relationship between cancer mortality and water type (e.g. comparing the effect of using surface waters likely to contain organic matter, with the effect of using groundwater likely to contain less) show 'some association' in some studies, whilst other studies have given inconsistent results. Thus, any risk to health due to the presence of organic micropollutants in water is likely to be very small, difficult to measure, and to be manifest only after a long period of exposure. However, the potential risk has led to the view that, where possible, the concentration of certain pollutants should be kept as low as possible in drinking water.

7.37 PESTICIDES

Pesticides cover a wide range of compounds used as insecticides, herbicides, fungicides and algicides. The term can also refer to chemicals with other uses such as wood preservation, public hygiene, industrial pest control, soil sterilization, plant growth regulation, masonry biocides, bird and animal repellents and anti-fouling paints.

Pesticides find their way into natural waters from direct application for aquatic plant and insect control, from percolation and run-off from agricultural land, from aerial drift in land application, and from industrial discharges. Organic pesticides are often toxic to aquatic life even in trace amounts and some, particularly the organo-chlorine compounds, are very resistant to chemical and biochemical degradation. Where pesticides and algicides are used for aquatic control they can cause deoxygenation as a result of the decomposition of the treated vegetation. This in turn can cause other problems such as the dissolution of iron and manganese and the production of tastes and odours.

Even with controlled application, traces of pesticides are still found in sewage effluents and in a number of groundwater and river sources used for public water supplies. Pesticides used in non-agricultural situations, particularly on hard surfaces, pose a higher risk of contaminating water sources. Accidental discharges of pesticides in bulk to watercourses occasionally occur and can have serious implications, causing fish death and making it necessary for a temporary shutdown of any water intakes. In the UK the use of pesticides is regulated by the Health & Safety Executive (HSE); similar arrangements exist in the USA. Even so, some pesticides that have been banned or restricted for many years remain persistent in the environment and have very occasionally been found in groundwater sources long after the original contamination occurred.

Many pesticides are insoluble in water or have limited solubility; others degrade readily, depending on the soil type. Many are rapidly adsorbed onto sediment or suspended material and this property can be utilized in treatment processes involving coagulation followed by sedimentation and filtration. Oxidation using ozone, chlorine, chlorine dioxide or potassium permanganate effectively breaks down some organic pesticides, although in some cases the use of chlorine or ozone may result in degradation products that are more toxic than the original pesticide, or give rise to odorous compounds (Section 7.49). Other options for pesticide removal are presented in Sections 10.35 and 10.36. In general it is preferable to restrict the levels of pesticides in the raw water, through catchment management (Part VI), rather than having to remove them in the treatment process.

In certain situations, selected pesticides may actually be applied to sources of drinking water to control disease-carrying insects. Such pesticides have to be approved for use by the national authorities and applied under very strictly controlled conditions.

Overall, the amounts of pesticides which are likely to be consumed from drinking water tend to be a very small fraction of those likely to be consumed from foodstuffs. WHO and US EPA set health-related standards for a number of individual pesticides. The EC Directive does not take into account the wide variation in toxicity of the various pesticides that might be detected in drinking water but applies a single standard of 0.1 μg/l to most pesticides and relevant metabolites and degradation products.

7.38 pH VALUE OR HYDROGEN ION

The pH value, or hydrogen ion concentration, determines the acidity of a water. It is one of the most important determinations in water chemistry as many of the processes involved in water treatment are pH dependent. Pure water is very slightly ionized into positive hydrogen (H^+) ions and negative hydroxyl (OH^-) ions. In very general terms a solution is said to be neutral when the numbers of hydrogen ions and hydroxyl ions are equal, each corresponding to an approximate concentration of 10^{-7} moles/l. This neutral point is temperature dependent and occurs at pH 7.0 at 25°C. When the concentration of hydrogen ions exceeds that of the hydroxyl ions (i.e. at pH values less than 7.0) the water has acidic characteristics. Conversely, when there is an excess of hydroxyl ions (i.e. the pH value is greater than 7.0) the water has basic characteristics and is described as being on the alkaline side of neutrality.

The pH value of unpolluted water is mainly determined by the inter-relationship between free carbon dioxide and the amounts of carbonate and bicarbonate present (Section 10.41). The pH values of most natural waters are in the range 4–9, with soft acidic waters from moorland areas generally having lower pH values and hard waters which have percolated through chalk or limestone generally having higher pH values.

Most water treatment processes, but particularly clarification and disinfection, require careful pH control to optimize the efficacy of the process fully. The pH of the water entering distribution must also be controlled to minimize the corrosion potential of the water (Section 7.21).

7.39 PHENOLS

Phenolic compounds found in surface waters are usually a result of pollution from trade wastes such as petrochemicals, washings from tarmac roads, gas liquors and creosoted surfaces. Decaying algae or higher vegetation can also release natural phenols into the aquatic environment, whilst traces of phenols and other phenol-like compounds may be found in good quality groundwaters, especially in areas with coal- or oil-bearing strata. Most phenols, even in minute concentrations, produce chlorophenols on chlorination (Section 7.24). Even trace amounts can render the water unacceptable to consumers because of objectionable taste and/or odour (Section 10.33).

Phenols can be effectively removed by superchlorination at pH 7−10, with chlorine to phenol ratios of between 6:1 and 10:1, by oxidation with ozone or chlorine dioxide, or by adsorption onto activated carbon. However, it is preferable to seek to eliminate the source wherever possible.

7.40 PHOSPHATES

Phosphates in surface waters mainly originate from sewage effluents containing phosphate-based synthetic detergents, from industrial effluents, or from agricultural run-off following the use of inorganic fertilizers. Groundwaters usually contain insignificant concentrations of phosphates, unless they become polluted. Phosphorus is one of the essential nutrients for algal growth and can contribute significantly to eutrophication of lakes and reservoirs (Section 7.75).

Orthophosphates may be added during water treatment for plumbosolvency control (Section 7.32). The applied dose is initially around 1 mg/l as P, gradually decreasing to around 0.7 mg/l as P as the treatment takes effect and the system becomes optimized. Although phosphate treatment is effective for lead, phosphate dosing of waters with high alkalinity and low pH may slightly increase cuprosolvency. Orthophosphates, along with polyphosphates, can also be used as corrosion inhibitors to reduce the risk of supplying 'red water' from corroding iron pipes.

Phosphate dosing has played an important role in meeting the tighter lead standards. However, concerns have been expressed about the impact of such dosing on the environment. Typically up to 40% of the phosphate in sewage effluent comes from detergents and about 50% from human excretion, with the phosphate added to water supplies to control plumbosolvency contributing no more than 10%. In areas of low intensity agriculture, sewage effluents represent the main source of phosphate in the aquatic environment but the impact from plumbosolvency treatment is still very low. High intensity agriculture contributes significantly more amounts of phosphates to the environment, making the contribution from plumbosolvency treatment even lower. Thus the overall impact of phosphate dosing for plumbosolvency control is likely to be small, even in situations where it is deemed necessary to reduce phosphate levels in order to further protect the aquatic environment.

7.41 POLYNUCLEAR AROMATIC HYDROCARBONS (PAHs)

Polyaromatic hydrocarbons (PAHs) are a group of organic compounds that occur widely in the environment as the result of incomplete combustion of organic material. Trace amounts of PAHs have been found in industrial and domestic effluents. Their solubility in water is very low but can be enhanced by detergents and by other organic solvents which may be present. Although PAHs are not very biodegradable, they tend to be taken out of solution by adsorption onto particulate matter. If present in raw waters, they are usually removed during coagulation, sedimentation and filtration. However, they can be re-introduced in the distribution system from mains that have been lined with coal tar pitch. In the UK, coal tar, which can contain up to 50% of PAHs, was used until the 1970s to line iron water mains to prevent rusting. In some situations this lining may eventually break down, releasing PAHs in solution and as particulates into the water. There tends to be a seasonal variation in the concentrations found, with solubility linked to water temperature. There is also evidence that chlorine dioxide, when used as a disinfectant or for taste and odour control, can result in elevated concentrations of PAHs at consumers' taps (DoE, 1990a).

Several PAHs are known to be carcinogenic at concentrations considerably higher than those found in drinking water, with the main routes of exposure being from food and cigarette smoke. Drinking water is normally monitored for five indicator parameters, namely benzo(b)fluoranthene, benzo(k)fluoranthene, benzo(a)pyrene, benzo(ghi)perylene and indeno(1,2,3-cd)pyrene, of which benzo(a)pyrene is considered to be the most harmful and has a separate standard.

7.42 RADIOACTIVE SUBSTANCES

Many water sources contain very low levels of radioactive substances, which are mainly naturally occurring radionuclides in the uranium and thorium decay series. These tend to be alpha particle emitters, although some (e.g. radium-228 and potassium-40) are beta particle emitters. Other radionuclides may be found in water sources as a result of pollution from human activities, for example, the nuclear fuel and power industries, or where radioactive isotopes are used in medicine or industry, or other similar activities. Most of these man-made radionuclides are beta emitters, including tritium which is usually man-made but may also occur naturally. The impact of man-made radionuclides on the environment is subject to tight regulatory control. If there is any likelihood of a water source or a drinking water supply becoming contaminated then these regulatory controls would be used to ensure that remedial action was taken.

The contribution of drinking water to total radiological exposure is generally extremely small. The International Commission on Radiological Protection (ICRP) provides detailed advice and recommendations on the control of exposure to radiation. WHO uses these recommendations, along with advice and recommendations from the International Atomic Energy Agency (IAEA), in formulating its guidelines on radionuclides in drinking water. These guidelines are based on a recommended reference dose level (RDL) of the committed effective dose, equal to 0.1 mSv from 1 year's consumption of drinking water and the dose coefficients for adults; mSv (millisievert) is the 'effective dose equivalent'. It is a measure of the effect produced on a person by different types of radiation, taking into account the nature of the radiation and the organs exposed. This RDL represents less than 5% of the average effective dose each year from natural background radiation. These recommendations apply only to existing operational water supplies and to new supplies. Other advice would apply in emergency situations where radionuclides have been accidentally or deliberately released into the environment.

It is neither feasible nor practical to monitor water supplies routinely for all individual radionuclides. WHO recommends a screening approach, with no further action being required if the gross alpha activity is $\leq$0.5 Bq/l and the gross beta activity is $\leq$1.0 Bq/l. 1 becquerel (B) corresponds to one nuclear transformation per second; 1 curie (Ci) = 37×10^9 Bq. Specific radionuclides need only be identified if either of these screening levels is exceeded. In the event of this happening, a dose estimate has to be made for each radionuclide likely to be present, using activity-to-dose conversion factors (based on 1 year's consumption of 2 litres of water per day). The sum of the individual dose estimates determines whether the RDL (or total indicative dose) is likely to be exceeded. If the screening relates to a single sample, the RDL would only be exceeded if consumers were exposed to the measured concentrations for a full year. A single exceedance does not, therefore, necessarily mean that the water supply is unsuitable for consumption, but signals the need for further investigations. In the event of the RDL being exceeded, appropriate medical advice should be sought and remedial measures taken to reduce the level of exposure.

The more commonly found nuclides in drinking water include:

- Potassium-40, which occurs naturally in a fixed ratio to stable potassium;
- Radium-226 and Radium-228;
- Radon, a radioactive gas released as a decay product of radium;
- Tritium, which occurs naturally at low levels but, at higher concentrations, is indicative of man-made pollution, thereby providing an indication of other potentially more harmful radionuclides; and
- Uranium-238.

In November 2013, the European Council published a Directive under the Euratom treaty (CEU, 2013b). This introduced a standard for radon in drinking water and set out associated monitoring frequencies.

7.43 SELENIUM

Selenium is naturally present in soil in many parts of the world. It is also naturally present in water, particularly in areas of geothermal activity. However in some areas the concentrations of selenium in surface waters can be significantly increased due to irrigation returns.

Selenium is an essential element at low levels and is included in some mineral supplements – particularly in Europe, where selenium levels in soil are low. However, in excess selenium is toxic and cumulative in the body.

7.44 SILICA

Silica can be found in water in several forms, as a result of degradation of silica-containing rocks such as quartz and sandstone. Natural waters can contain between 1 mg/l of silica, in the case of soft moorland waters, and up to about 40 mg/l in some hard waters. Much higher levels are found in waters from volcanic or geothermal areas.

Silica levels in water supplies do not constitute a risk to health. However they are important in a number of industrial processes as silica can form a very hard scale, which is difficult to remove.

7.45 SILVER

Trace amounts of silver are occasionally found in natural waters but it is rarely found in detectable concentrations in drinking water. However, silver can be used as a disinfectant in domestic water treatment units or point-of-use devices (e.g. ceramic filter candles or granular activated carbon impregnated with silver) and this may result in elevated levels in the treated water (Butkus, 2003).

7.46 SODIUM

Sodium compounds are abundant in the environment and are also very soluble in water. The element is present in most natural waters at levels ranging from less than 1 mg/l to several thousand mg/l in seawater. The taste threshold for sodium in drinking water depends on several factors, such as the predominant anion present and the water temperature. If sodium is present as sodium chloride, the taste threshold is around 150 mg/l as sodium but is higher, at around 220 mg/l as sodium, if the sodium is present as sodium sulphate.

The use of base exchange or lime-soda processes to soften hard waters can result in a significant increase in the sodium concentration of the softened water. This situation also applies to domestic softeners and, ideally, there should always be a separate tap available for drinking water that is supplied directly off the mains supply when a domestic softener is installed. It is advisable that consumers on low sodium diets do not drink point-of-use softened water and the water should not be used for making up baby food.

7.47 SULPHATES

The concentration of sulphate in natural waters can vary over a wide range from a few mg/l to several thousand mg/l in brackish water and seawater. Sulphates come from several sources such as the dissolution of gypsum and other mineral deposits containing sulphates; seawater intrusion; the oxidation of sulphides, sulphites and thiosulphates in well aerated surface waters; and from industrial effluents where sulphates or sulphuric acid have been used in processes such as tanning and paper pulp manufacturing. Sulphurous flue gases discharged to atmosphere in industrial areas often result in acid rain water containing appreciable levels of sulphate.

High levels of sulphate in water can impart taste and, when combined with magnesium or sodium, can have a laxative effect (e.g. Epsom salts). Consumers tend to become acclimatized to high sulphate waters and, in some parts of the world, waters with very high sulphate contents have to be used if no alternative supplies are available. Although there is no health-based guideline for sulphate, WHO recommends that health authorities be notified if concentrations exceed 500 mg/l because of the gastrointestinal effects of high sulphate levels.

Bacterial reduction of sulphates under anaerobic conditions can produce hydrogen sulphide, which is an objectionable gas smelling of bad eggs. This can occur in deep well waters but the odour rapidly disappears with effective aeration. It can also occur if there is seawater intrusion to a shallow aquifer which is polluted by sewage.

7.48 SUSPENDED SOLIDS

The suspended solids content or filter residue of a water quantifies the amount of particulate material present and includes both organic and inorganic matter such as plankton, clay and silt. The suspended solids content of a surface water can vary widely depending on flow and season, with some rivers under flood conditions having several thousand mg/l of material in suspension (Section 6.21).

The measurement of suspended solids is usually on a dry weight-volume basis and gives no indication as to the type of material in suspension, the particle size distribution or the settling characteristics. Thus there is no direct correlation between suspended solids and turbidity (Section 7.50). However, in terms of drinking water quality, it is important that suspended solids are adequately removed from the raw water prior to final disinfection so that the efficacy of the disinfection process is not impaired.

7.49 TASTE AND ODOUR

There are four basic taste sensations, namely sweet, sour, salt and bitter. What is regarded as taste is in fact a combination of these sensations with the sensation of smell. In examining water samples, the odour rather than the taste of a sample is often evaluated as it avoids putting a possibly suspect sample into the mouth. However, it is still desirable to check the taste of a final treated water. A subjective or qualitative assessment of the taste or odour is often carried out at the time of sampling. This can then be supplemented by a quantitative measurement, reported as the Dilution Number (or the Threshold Number), which is carried out under controlled laboratory conditions. The quantitative test is based on the number of dilutions of the sample with taste- and odour-free water necessary to eliminate the taste or odour.

Many tastes and odours in drinking water are caused by natural contaminants such as extracellular and decomposition products of plants, algae and microfungi. Certain of the blue-green algae (Section 7.4) and actinomycetes (Section 7.73), when present in a raw water, can give rise to very distinctive earthy and musty tastes and odours. Raw waters contaminated by agricultural and industrial discharges may also give rise to serious taste and odour problems, which are often exacerbated by the use of chlorine for disinfection. Chlorine itself can give rise to extensive taste and odour complaints, especially if the applied dose has been increased suddenly for operational reasons. However, some tastes and odours may be caused by the condition of the domestic plumbing system for example, chlorophenolic tastes/odours associated with old washing machine hoses; astringent tastes from the dissolution of plumbing materials such as zinc or copper; and poorly sited pipes resulting in the build-up of biofilms.

Taste and odour of drinking water tends to be very subjective, with consumers becoming familiar with the taste or odour associated with their local water supply. Thus any complaint of an unusual taste or odour should always be investigated immediately, in case it relates to more serious changes.

7.50 TURBIDITY

The measurement of turbidity, although not quantitatively precise, is a simple and useful indicator of the condition of a water. Turbidity is defined as the optical property that causes light to be scattered and absorbed rather than transmitted in straight lines through a sample. Although turbidity is caused by material in suspension, it is difficult to correlate it with the quantitative measurement of suspended solids in a sample, as the shape, size and refractive indices of the particles in suspension all affect their light-scattering properties. For the same reason turbidity measurements can vary according to the type of instrument used. Nephelometers measure the intensity of light scattered in one particular direction and are highly sensitive for measuring low turbidity. Other instruments measure the amount of light absorbed by particles when light is passed through a water sample. Early measurements were made using the Jackson Candle Turbidimeter, with results reported in Jackson units (JTU). However, this was a comparatively crude form of measurement and not suitable for measuring low turbidities. Current methods use a primary standard based on formazin which, if used with nephelometric instruments, gives an equivalent turbidity measurement in FTU (i.e. FTU = NTU).

Raw water turbidity can vary over a very wide range, from virtually zero to several thousand NTU, but effective treatment should consistently be able to produce a final water with turbidity levels of less than 1 NTU. Turbidity meters are therefore an essential tool in optimizing and controlling water treatment processes from the raw water to the final water leaving the water treatment works. An important aspect of turbidity monitoring is the significant impact of particulate material on the efficacy of the disinfection process. Current thinking is that turbidities should be no more than 1 NTU, and ideally much lower, in order to ensure effective disinfection under normal operational conditions.

Turbidity affects the aesthetic quality of the water supplied. Levels have to be acceptable to consumers, who will notice significant changes, for example when sediments are disturbed following a burst main. Excessive dissolved air in tap water can cause consumer concerns over apparent turbidity even though there is no risk to health. In such situations, the water appears milky when in a glass but gradually clears from the bottom upwards.

7.51 ZINC

Zinc tends to be found only in trace amounts in unpolluted surface waters and groundwaters. However, it is often found in the water at consumers' taps as a result of corrosion of galvanized iron piping or tanks, or dezincification of brass fittings.

Zinc is beneficial in moderation in diets but is harmful in excess, producing copper deficiency and other adverse effects. There are no health-related standards but zinc can cause an astringent taste and also cause opalescence in some waters at concentrations of more than 3 mg/l.

PART II WATER QUALITY GUIDELINES AND STANDARDS FOR CHEMICAL AND PHYSICAL PARAMETERS

7.52 DRINKING WATER GUIDELINES OR STANDARDS (CHEMICAL AND PHYSICAL)

Guideline values or enforceable standards apply to many of the parameters discussed in Part I and these are listed in Tables 7.1(A)–(E). Care should be taken when comparing the numerical values in these tables, as there are inconsistencies between the authorities over the units of measurement and notations for some parameters. Reference is made in the tables to the following documents and the units and notations quoted for each parameter are as they appear in the respective documents.

Table 7.1(A) Inorganic chemical parameters of health significance

	WHO Guidelines 4th Edn. 2011	UK Water Supply (Water Quality) Regulations Prescribed conc. or value	EC Directive 98/83/EC November 1998 Parametric value	US EPA Primary Drinking Water Regulations Maximum contaminant level (MCL)
Antimony (Sb)	0.02 mg/l	5 μg/l	5 μg/l	0.006 mg/l
Arsenic (As)	0.01 mg/l^P	10 μg/l	10 μg/l	0.010 mg/l
Asbestos (fibres >10 μm in length)	No consistent evidence			7×10^6 fibres/l
Barium (Ba)	0.7 mg/l			2 mg/l
Beryllium (Be)				0.004 mg/l
Boron (B)	2.4 mg/l	1.0 mg/l	1.0 mg/l	
Bromate (BrO_3)*	0.01 mg/l^P	10 μg/l	10 μg/l	0.010 mg/l
Cadmium (Cd)	0.003 mg/l	5 μg/l	5 μg/l	0.005 mg/l
Chloramines (Cl_2)*	Mono- 3 mg/l			4 mg/l[1]
Chlorate (ClO_3)*	0.7 mg/l^P	700 μg/l[2]		
Chlorine (Cl_2)	5 mg/l^C			4 mg/l[1]
Chlorine dioxide/ Chlorite (ClO_2)*	0.7 mg/l^P	0.5 mg/l[3]		0.8 mg/l[1,4]
Chromium (Cr)	0.05 mg/l^{P,T}	50 μg/l	50 μg/l	0.1 mg/l^T
Copper (Cu)	2 mg/l^S	2.0 mg/l	2.0 mg/l	1.3 mg/l[5]
Cyanide (CN)		50 μg/l	50 μg/l	0.2 mg/l free cyanide
Fluoride (F)	1.5 mg/l	1.5 mg/l	1.5 mg/l	4 mg/l
Lead (Pb)	0.01 mg/l	10 μg/l	10 μg/l[6]	0.015 mg/l[5]
Manganese (Mn)		50 μg/l		
Mercury (Hg)	0.006 mg/l	1.0 μg/l	1.0 μg/l	0.002 mg/l
Nickel (Ni)	0.07 mg/l	20 μg/l	20 μg/l	
Nitrate	50 mg/l as NO_3^{E}	50 mg/l as NO_3	50 mg/l as NO_3	10 mg/l as N
Nitrite	3 mg/l as NO_2^{E}	0.50 mg/l as NO_2 0.10 mg/l ex works	0.50 mg/l as NO_2 0.10 mg/l ex works	1 mg/l as N
Nitrate + Nitrite	$NO_3/50 + NO_2/3 \leq 1$	$NO_3/50 + NO_2/3 \leq 1$	$NO_3/50 + NO_2/3 \leq 1$	
Selenium (Se)	0.04 mg/l^P	10 μg/l	10 μg/l	0.05 mg/l
Sodium dichloroisocyanurate	50 mg/l as sodium dichloroiso-cyanurate, 40 mg/l as cyanuric acid			
Thallium (Tl)				0.002 mg/l

(Continued)

Table 7.1(A) (Continued)

	WHO Guidelines 4th Edn. 2011	UK Water Supply (Water Quality) Regulations Prescribed conc. or value	EC Directive 98/83/EC November 1998 Parametric value	US EPA Primary Drinking Water Regulations Maximum contaminant level (MCL)
Turbidity – see also Table 7.1(E)	<1 NTU[7]			1 NTU to 5 NTU[8]

Notes:
1 μg/l = 0.001 mg/l.
[P]Provisional guideline value.
*Usually present in drinking water as a disinfection by-product.
[C]Concentrations of the substance at or below the health-based guideline value may affect the appearance, taste or odour of the water, leading to consumer complaints.
[T]For total chromium.
[S]Staining of laundry and sanitary ware may occur below guideline value.
[E]Short-term exposure.
[1]Comprehensive Disinfectants and Disinfection Byproducts Rules (Stage 1 and Stage 2) Maximum Residual Disinfectant Level (MRDL) for Regulated Disinfectants – The highest level of disinfectant allowed in drinking water.
[2]Set as a condition of use under Section 31(4) of the Regulations for on-site electrolytic generation of chlorine.
[3]Set as a condition of use under Section 31(4) of the Regulations for on-site generation of chlorine dioxide. The combined concentration of chlorine dioxide, chlorite and chlorate and chlorate must not exceed 0.5 mg/l as chlorine dioxide in the water entering supply.
[4]1.0 mg/l as chlorite when present as a disinfection by-product for works using chlorine dioxide as a disinfectant.
[5]The Lead and Copper Rule – Action level – additional steps must be taken to control the corrosiveness of the water if more than 10% of tap water samples exceed this level.
[6]Based on a weekly average value ingested by consumers.
[7]To ensure effectiveness of disinfection.
[8]Performance standard for treatment works using conventional and direct filtration – the maximum turbidity should be <1 NTU, with turbidities of ≤ 0.3 NTU in at least 95% of samples taken in any month. Works using other filtration to conventional or direct filtration must follow state limits, which must include turbidity at no time exceeding 5 NTU.

The WHO *Guidelines for Drinking-water Quality* (WHO, 2011a)

The 4th edition of the WHO *Guidelines for Drinking-water Quality* integrates the 3rd edition, published in 2004, with both the first addendum from 2006 and the second addendum from 2008. It focuses on the public health aspects of drinking water and provides health-related guideline values that can be applied worldwide. Many countries therefore use the WHO guideline values as a basis for setting national guideline values or legally enforceable standards. The guideline values for chemical parameters are based on the potential for such parameters, or substances, to cause adverse health effects after long periods of exposure. Some parameters have provisional guideline values on the grounds that available data on health effects are either limited or at present poorly defined. The Guidelines also provide guidance on the aesthetic quality of drinking water, whereby a water may be unacceptable to consumers on the grounds of appearance, or taste and odour, but there is no associated health risk.

Table 7.1(B) Organic chemical parameters of health significance

	WHO Guidelines 4th Edn. 2011	UK Water Supply (Water Quality) Regulations Prescribed conc. or value	EC Directive 98/83/EC November 1998 Parametric value	US EPA Primary Drinking Water Regulations Maximum contaminant level (MCL)
Aromatic hydrocarbons				
Benzene	0.01 mg/l	1.0 μg/l	1.0 μg/l	0.005 mg/l
Ethylbenzene	0.3 mg/l[1]			0.7 mg/l
Styrene	0.02 mg/l[1]			0.1 mg/l
Toluene	0.7 mg/l[1]			1 mg/l
Xylenes	0.5 mg/l[1]			10 mg/l total
Chlorinated alkanes				
Tetrachloromethane (Carbon tetrachloride)	0.004 mg/l	3.0 μg/l		0.005 mg/l
Dichloromethane	0.02 mg/l			0.005 mg/l
1,2-Dichloroethane	0.03 mg/l	3.0 μg/l	3.0 μg/l	0.005 mg/l
1,1,1-Trichloroethane				0.2 mg/l
1,1,2-Trichloroethane				0.005 mg/l
Chlorinated benzenes				
Monochlorobenzene				0.1 mg/l
1,2-Dichlorobenzene	1 mg/l[1]			0.6 mg/l
1,4-Dichlorobenzene	0.30 mg/l[1]			0.075 mg/l
Trichlorobenzene				0.07 mg/l
Chlorinated ethenes				
1,1-Dichloroethene				0.007 mg/l
1,2-Dichloroethene	0.05 mg/l			*cis* 0.07 mg/l *trans* 0.1 mg/l
Tetrachloroethene	0.04 mg/l	)	)	0.005 mg/l
(PCE) and		) Sum 10 μg/l	) Sum 10 μg/l	
Trichloroethene (TCE)	0.02 mg/l[P]	)	)	0.005 mg/l
Pesticides				
Total pesticides		0.50 μg/l[2]	0.50 μg/l[2]	
Individual pesticides *except for*:	See Table 7.1(C)	0.10 μg/l[2]	0.10 μg/l[2]	See Table 7.1(C)
-Aldrin	Sum of aldrin +	0.030 μg/l	0.030 μg/l	
-Dieldrin	dieldrin: 0.00003 mg/l	0.030 μg/l	0.030 μg/l	0.030 μg/l
-Heptachlor		0.030 μg/l	0.030 μg/l	0.0004 mg/l
-Heptachlor epoxide		0.030 μg/l	0.030 μg/l	0.0002 mg/l

(Continued)

Table 7.1(B) (Continued)

	WHO Guidelines 4th Edn. 2011	UK Water Supply (Water Quality) Regulations Prescribed conc. or value	EC Directive 98/83/EC November 1998 Parametric value	US EPA Primary Drinking Water Regulations Maximum contaminant level (MCL)
Polycyclic aromatic hydrocarbons (PAHs)		0.10 μg/l[3]	0.10 μg/l[3]	
Benzo(a)pyrene	0.0007 mg/l	0.01 μg/l	0.01 μg/l	0.0002 mg/l
Other organic compounds				
Acrylamide	0.0005 mg/l	0.10 μg/l[4]	0.10 μg/l[4]	*(TT)*
Di(2-ethylhexyl) adipate (DEHA)				0.4 mg/l
Di(2-ethylhexyl) phthalate (DEHP)	0.008 mg/l			0.006 mg/l
1,4 Dioxane	0.05 mg/l			
Dioxin				3×10^{-8} mg/l
Edetic acid (EDTA)	0.6 mg/l (free acid)			
Epichlorohydrin	0.0004 mg/l[P]	0.10 μg/l[4]	0.10 μg/l[4]	*(TT)*
Microcystin-LR	0.001 mg/l[5]			
Nitrilotriacetic acid (NTA)	0.2 mg/l			
Polychlorinated biphenyls (PCBs)				0.0005 mg/l
Vinyl chloride	0.0003 mg/l	0.50 μg/l	0.50 μg/l	0.002 mg/l
Disinfection by-products – *see also* ***Table 7.1(A)***				
Haloacetic acids:				0.060 mg/l as a total of 5 acids
-Monochloroacetate	0.02 mg/l			
-Dichloroacetate	0.05 mg/l[P]			
-Trichloroacetate	0.2 mg/l			
Chlorophenols:				
-2,4,6-Trichlorophenol	0.2 mg/l[1]			
Halogenated acetonitriles:				
-Dichloroacetonitrile	0.02 mg/l[P]			
-Dibromoacetonitrile	0.07 mg/l			

(*Continued*)

Table 7.1(B) (Continued)

	WHO Guidelines 4th Edn. 2011	UK Water Supply (Water Quality) Regulations Prescribed conc. or value	EC Directive 98/83/EC November 1998 Parametric value	US EPA Primary Drinking Water Regulations Maximum contaminant level (MCL)
Trihalomethanes:				
-Bromoform	0.1 mg/l[6]	)	)	)
-Chloroform	0.3 mg/l[6]	) 100 μg/l sum of	) 100 μg/l sum of	) 0.080 mg/l total[8]
-Dibromochloromethane	0.1 mg/l[6]	) concentrations[7]	) concentrations[7]	)
-Bromodichloromethane	0.06 mg/l[6]	)	)	)
N-Nitrosodimethylamine (NDMA)	0.0001 mg/l	>10 ng/l[9]		

Notes:
1 μg/l = 0.001 mg/l.
[P]Provisional guideline.
[1]Concentrations of the substance at or below the health-based guideline value may affect the appearance, taste or odour of the water, leading to consumer complaints.
[2]'Pesticides' means organic insecticides, herbicides, fungicides, nematocides, acaricides, algicides, rodenticides, slimicides and related products (inter alia, growth regulators) and their related metabolites, degradation and reaction products. 'Total pesticides' means the sum of all individual pesticides detected and quantified in the monitoring procedure.
[3]Total PAHs – Sum of concentrations of benzo(b)fluoranthene, benzo(k)fluoranthene, benzo(ghi)perylene, indeno(1,2,3-cd)pyrene.
[4]Residual monomer concentration calculated according to the specification of the maximum release from the polymer in contact with the water. This is controlled by product specification. (Acrylamide derives from use of polyacrylamide flocculants in water treatment; epichlorohydrin from its use in the manufacture of water treatment resins.)
Also (TT). Each water system must certify, in writing, to the state that when acrylamide and epichlorohydrin are used to treat water, the combination (or product) of dose and monomer levels must not exceed the levels specified, as follows:
Acrylamide – 0.05% dosed at 1 mg/l (or equivalent).
Epichlorohydrin – 0.01% dosed at 20 mg/l (or equivalent).
[5]Total microcystin-LR (free plus cell-bound).
[6]The sum of the ratios of concentration for each to their respective GV not to exceed 1.
(Note: bromoform is tribromomethane; chloroform is trichloromethane.)
[7]Where possible a lower value should be aimed for but without compromising the effectiveness of the disinfection.
[8]Action level – additional steps must be taken if more than 10% of tap water samples exceed this level.
[9]A multi-tiered approach to be taken with Tier 1 triggered as a potential hazard in the risk assessment; Tier 2 triggered at a concentration >1 ng/l with further monitoring; Tier 3 triggered at a concentration >10 ng/l on the basis of wholesomeness; and Tier 4 triggered at a concentration >200 ng/l as a notifiable event.

The 4th edition further develops concepts, approaches and information introduced in the 3rd edition (WHO, 2004), including the comprehensive preventative risk management (drinking water safety plan) approach for ensuring drinking water quality, as introduced in the 3rd edition.

Since 1995, the WHO Guidelines have been kept up to date through a process of rolling revision, which leads to the regular publication of addenda that may add to, or supersede, information in the previous volumes.

The European Commission Directive on the Quality of Water Intended for Human Consumption (CEU, 1998a)

The 1998 Drinking Water Directive updates the original Directive of 1980 (CEU, 1980). It is a legal document that applies only to EU Member States and is intended to enable consistent water qualities and obligations to be achieved throughout the European Union. Member States were required to adopt the Directive into national law within 2 years of it coming into force and to achieve compliance with most of the standards specified by 25 December 2003. These standards are, for the most part, based on guideline values in the 3rd edition of the WHO Guidelines. The Directive contains 43 numerical standards for chemical and radiological parameters, of which 30 are mandatory on the grounds of health risks. The remaining 13 standards, along with five parameters that do not have numerical values, are termed indicator parameters. Most of the indicator parameters relate to the aesthetic and organoleptic characteristics of a water and have no associated health risk. Their purpose is rather to indicate problems or potential problems with the treatment or distribution of the water. Member States are allowed to set values for additional parameters not included in the Directive. They can also adopt more stringent standards than those specified, although the Commission has to be notified. Any failure of a standard has to be investigated but the water supplier may then be granted a special dispensation, which permits non-compliant water to be supplied for up to 3 years. Such dispensations or 'derogations' are time limited and can only be authorized if there is no associated risk to human health in continuing to supply the water. They are also conditional on appropriate remedial action being taken. There is a requirement in the Directive to review the standards at least every 5 years, 'in the light of scientific and technical progress'.

The UK *Water Supply (Water Quality) Regulations*

Public water supply arrangements are different in England and Wales compared to Scotland and Northern Ireland, as is the role of the regulatory authorities. The requirements of the 1998 EC Directive have been transposed into the respective national laws by way of the Water Supply (Water Quality) Regulations. This was done in England in 2000 (UK, 2000), with subsequent amending Regulations in 2001 (UK, 2001a); in Wales in 2001 (UK, 2001b); in Scotland in 2001, with subsequent amending Regulations (UK, 2001c, d); and in Northern Ireland in 2002 (UK, 2002), with subsequent amending Regulations in 2003 (UK, 2003).

More recently in 2007, new Regulations were introduced in Northern Ireland (UK, 2007a) to reflect changes in the delivery of water services and the English and Welsh Regulations were subject to further Amendment Regulations (UK, 2007b, c). The Regulations in Scotland were amended in 2010 and subsequently re-enacted, with some modifications, in 2014 (UK, 2010c; UK, 2014).

A consolidated version of the Water Supply (Water Quality) Regulations incorporating all amendments made to the 2000 Regulations (SI No. 3184) was made available in 2010 (UK, 2010a). This currently exists as an unofficial consolidated version in England, but as an official consolidated version in Wales (UK, 2010b).

Table 7.1(C) Pesticides – parameters of health significance

	WHO Guidelines 4th Edn. 2011	US EPA Primary Drinking Water Regulations Maximum contaminant level	UK DoE Guidance on Safeguarding the Quality of Public Water Supplies 1989 Advisory value
PESTICIDES – see also Table 7.1(B)			
Alachlor	0.02 mg/l	0.002 mg/l	
Aldicarb	0.01 mg/l		
Atrazine	0.1 mg/l[1]	0.003 mg/l	2 μg/l
Carborfuran	0.007 mg/l	0.04 mg/l	
Chlordane	0.0002 mg/l	0.002 mg/l	0.1 μg/l – total isomers
Chlortoluron	0.03 mg/l		80 μg/l
Chlorpyrifos	0.03 mg/l		
Cyanazine	0.0006 mg/l		
DDT + metabolites	0.001 mg/l		7 μg/l – total isomers
2,4-Dichlorophenoxyacetic acid (2,4-D)	0.03 mg/l	0.07 mg/l	1000 μg/l
2,4-DB	0.09 mg/l		
2,4,5-T	0.009 mg/l		
1,2-Dibromoethane (ethylene dibromide)	0.0004 mg/l[P]	0.00005 mg/l	
1,2-Dibromo-3-chloropropane (DBCP)	0.001 mg/l[2]	0.0002 mg/l	
1,2-Dichloropropane (1,2-DCP)	0.04 mg/l[P]	0.005 mg/l	
1,3-Dichloropropene	0.02 mg/l		
Dalapon		0.2 mg/l	
Dichlorprop	0.1 mg/l		40 μg/l
Dimethoate	0.006 mg/l		3 μg/l
Dinoseb		0.007 mg/l	
Diquat		0.02 mg/l	
Endothall		0.1 mg/l	
Endrin	0.0006 mg/l	0.002 mg/l	
Fenoprop (2,4,5-TP)	0.009 mg/l	0.05 mg/l	
Glyphosate		0.7 mg/l	1000 μg/l
Hexachlorobenzene (HCB)		0.001 mg/l	0.2 μg/l

(Continued)

Table 7.1(C) (Continued)

	WHO Guidelines 4th Edn. 2011	US EPA Primary Drinking Water Regulations Maximum contaminant level	UK DoE Guidance on Safeguarding the Quality of Public Water Supplies 1989 Advisory value
Hexachlorobutadiene (HCBD)	0.0006 mg/l		
Isoproturon	0.009 mg/l		4 μg/l
Lindane	0.002 mg/l	0.0002 mg/l	
MCPA	0.002 mg/l		0.5 μg/l
MCPB			0.5 μg/l
Mecoprop (MCPP)	0.1 mg/l		10 μg/l
Methoxychlor	0.02 mg/l	0.04 mg/l	30 μg/l
Metolachlor	0.01 mg/l		
Molinate	0.006 mg/l		
Oxamyl (Vydate)		0.2 mg/l	
Pendimethalin	0.02 mg/l		
Pentachlorophenol	0.009 mg/l[P]	0.001 mg/l	
Permethrin[a]			
Picloram	0.3 mg/l	0.5 mg/l	
Pyriproxyfen[b]			
Simazine	0.002 mg/l	0.004 mg/l	10 μg/l
Terbuthylazine (TBA)	0.007 mg/l		
Toxaphene		0.003 mg/l	
Trifluralin	0.02 mg/l		

Notes: 1 μg/l = 0.001 mg/l.
[P]Provisional guideline value.
[1]Including its chloro-*s*-triazine metabolites; hydroxyatrazine 0.2 mg/l.
[2]Concentrations of the substance above the health-based guideline value may affect the taste or odour of the water, leading to consumer complaints.
[a]Not recommended for direct addition to drinking water as part of WHO's policy to exclude the use of any pyrethroids for larviciding of mosquito vectors of human disease.
[b]Not considered appropriate to set guideline values for pesticides used for vector control in drinking water.

The Regulations contain a number of national standards, most of which are indicator parameters under the Directive. They are also statutory instruments, with a legal obligation for water suppliers to supply wholesome water as defined under the Regulations. The most recent amendment of the Regulations in England and Wales includes a specific regulatory requirement for adequate treatment and disinfection of water supplies. Failure to comply is deemed a criminal offence and liable to prosecution. The UK legislation fully embraces the risk-based approach set out in the WHO guidelines, requiring drinking water safety plans to be in place for all public water supplies. In the 10 years since their instigation, these have evolved to become central to the process of managing water quality and determining investment solutions needed to mitigate residual risks.

Private water supplies (i.e. any supply of water intended for human consumption that is not provided by a public water supplier) in the UK also have to meet the requirements of the European Directive. They are subject to a similar set of Regulations as public water supplies and are enforced by local authorities in England, Scotland and Wales and, in Northern Ireland, by the Drinking Water Inspectorate, assisted by the environmental health departments of local councils.

Table 7.1(D) Radioactive substances – parameters of health significance

	WHO Guidelines 4th Edn. 2011	**UK Water Supply (Water Quality) Regulations. 2010 Prescribed conc. or value**	**EC Directive 98/83/EC November 1998 Parametric value**	**US EPA Primary Drinking Water Regulations Maximum contaminant level (MCL)**
Reference dose level/total indicative dose	0.10 mSv/year	0.10 mSv/year[#,1]	0.10 mSv/year[#,1]	
Screening levels –				
Gross alpha activity	≤0.5 Bq/l[2]	0.1 Bq/l		15 pCi/l[3]
Gross beta activity	≤1 Bq/l[2]	1 Bq/l		4 mrem/year[4]
Radium-226 + -228				5 pCi/l
Radon	100 Bq/m^3 [5]		100 Bq/l[6]	
Tritium[7]		100 Bq/l[#]	100 Bq/l[#]	
Uranium-238	10 Bq/l			30 μg/l

Notes:
[#]Indicator parameter, with standard set for monitoring purposes only.
[1]Total indicative dose excludes tritium, potassium-40, radon and radon decay products.
[2]Further action required above these levels.
[3]1 pCi/l = 0.037 Bq/l.
[4]Beta particles and photon emitters. 1 mrem/year = 0.01 mSv/year (rem = roentgen equivalent in man).
[5]The WHO reference level for radon concentration in indoor air. Screening levels in water should be set on the basis of the national reference level for radon in air and the distribution of radon in the national housing stock.
[6]To be implemented by EU Member States by 28 November 2015. Member States may set a level higher than 100 Bq/l but lower than 1000 Bq/l. Remedial action must be taken where radon concentrations exceed 1000 Bq/l.
[7]Effectively a screening parameter for the presence of artificial radionuclides. Monitoring only required if there is a source of tritium in the catchment and it cannot be shown by other means that the level of tritium is well below 100 Bq/l.

Table 7.1(E) Physical characteristics and substances undesirable in excess

	WHO Guidelines 4th Edn. 2011 Values based on aesthetic quality and acceptability to consumers	UK Water Supply (Water Quality) Regulations. 2010 Indicator parameters specification concentration or value	EC Directive 98/83/EC November 1998 Indicator parameters – parametric value	US EPA Secondary Drinking Water Regulations Secondary standards
Aluminium (Al)	0.2 mg/l	200 μg/l*	200 μg/l	0.05–2.0 mg/l
Ammonia (NH_4)		0.50 mg/l	0.50 mg/l	
Chloride (Cl)	250 mg/l	250 mg/l[1]	250 mg/l	250 mg/l
Colour	15 TCU	20 mg/l Pt/Co scale*	Acceptable to consumers and no abnormal change	15 colour units
Conductivity		2500 μS/cm at 20°C	2500 μS/cm at 20°C	
Copper (Cu)				1 mg/l
Corrosivity				Non-corrosive
Dissolved solids	1000 mg/l			500 mg/l
Fluoride (F)				2 mg/l
Foaming agents				0.5 mg/l
Hydrogen sulphide	0.05 mg/l as H_2S			
Iron (Fe)	0.3 mg/l	200 μg/l*	200 μg/l	0.3 mg/l
Manganese (Mn)	0.1 mg/l	50 μg/l*	50 μg/l	0.05 mg/l
Oxidizability (O_2)			5 mg/l[2]	
pH (Hydrogen ion)		6.5–9.5*	$\geq$6.5 and $\leq$ 9.5[1]	6.5–8.5
Silver (Ag)				0.10 mg/l
Sodium (Na)	200 mg/l	200 mg/l*	200 mg/l	
Sulphate (SO_4)	250 mg/l	250 mg/l[1]	250 mg/l	250 mg/l
Surfactants (foaming agents)				0.5 mg/l
Taste/odour		Acceptable to consumers and no abnormal change	Acceptable to consumers and no abnormal change	3 as Threshold Odour Number
Total organic carbon		No abnormal change	No abnormal change	
Turbidity – see also Table 7.1(A)	0.3 NTU[3]	1 NTU ex works 4 NTU at the consumer's tap*	Acceptable to consumers and no abnormal change <1 NTU ex works[4]	
Zinc	3.0 mg/l			5 mg/l

Notes:
1 μg/l = 0.001 mg/l.
*National requirements.
[1]The water should not be aggressive.
[2]Does not have to be measured if TOC is being measured.
[3]For works treating surface water, and groundwater under the influence of surface water, this value, prior to disinfection, demonstrates that there are significant barriers against pathogens that adsorb to particulate matter.
[4]For works treating surface waters.

The US EPA National Primary Drinking Water Regulations

Current regulations are as specified in the 1996 amendments to the Safe Drinking Water Act (US EPA, 1996). In the USA, drinking water quality regulations were mandated by law under the Safe Drinking Water Act of 1974 and are issued by US EPA. The regulations apply to public water supply systems and the primary standards are enforceable by the US EPA. They are mandatory on all public water systems that supply water to at least 15 connections, or 25 consumers, for at least 60 days in a year. Some 15% of Americans receive their drinking water from private wells, which are not covered by the federal regulations but by local rules set by states and local governments.

There are two categories of standards: the National Primary Drinking Water Regulations (NPDWRs or primary standards), which are health-related; and the National Secondary Drinking Water Regulations (NSDWRs or secondary standards), which are non-enforceable guidelines relating mainly to aesthetic parameters. The starting point for the primary standards are Maximum Contaminant Level Goals (MCLGs), which are based on the level of a contaminant below which there is no known or expected risk to human health. These goals contain a margin of safety and are not enforceable but they form the basis of the Maximum Contaminant Levels (MCLs) which are enforceable. US EPA is required to carry out a cost–benefit analysis on each standard and may, if necessary, adjust the standard for a particular supply to a level that 'maximizes health risk reduction benefits at a cost that is justified by the benefits'. So MCLs are set as close as is feasible to MCLGs, using the best available treatment technology and taking cost into consideration. Several of the primary standards are linked to treatment or monitoring requirements, which have been promulgated as a series of Rules.

The Safe Drinking Water Act requires US EPA to review and revise, as appropriate, each NPDWR at regular intervals; the last update was published in May 2009 (US EPA, 2009). US EPA is also required to identify and prioritize contaminants that are currently not subject to any proposed or promulgated primary drinking water regulation but may require regulation in the future. This is done by way of the Drinking Water Contaminant Candidate List, which was last published in October 2009. Some, but not all of the contaminants listed are also included in the WHO forward work programme.

Full details of the above Regulations and the associated treatment rules can be found on the US EPA website (http://www.epa.gov/safewater).

7.53 COMMENT ON THE APPLICATION OF HEALTH-RELATED STANDARDS

Drinking water may be one of several sources of exposure for the health-related standards listed in Tables 7.1(A)–(D). In some cases, it may be a very minor source, with little impact on the overall exposure. In other cases, such as lead and DBPs, drinking water may play a significant role in the overall level of exposure.

Occasionally it may be possible to determine the impact of exposure and the associated health effects from studies involving humans. More usually studies are carried out on laboratory animals (e.g. rats and mice), especially if carcinogenic or potentially carcinogenic substances are involved. The dose at which there is no observed adverse effect (NOAEL) is determined and 'uncertainty factors' varying from 10^{-2} to 10^{-4} are applied to take account of species differences, nature and severity of adverse effects, and the quality of the data. Further factors are then applied to give a tolerable daily intake, which is the amount that can be ingested without any appreciable risk to health over an average human life span (usually taken to be 70 years), adjusted for the average human body weight (usually taken to be 60–70 kg). Consideration has to be given to the contribution of drinking water to the overall exposure, compared to the intake from food and other sources, and the volume of water consumed (usually taken to be an average of 2 litres per day). Account also has to be taken of the impact on infants and children who, because of their higher intake and lower body weight, may be more susceptible than adults to exposure from certain substances. In the case of carcinogens or potential carcinogens, mathematical models may be used to extrapolate the effect to dose levels low enough to reduce the risk of cancer development in human cells to an 'acceptable level', such as one case per lifetime per 100 000 individuals. Problems may arise over interpretation whichever procedure is used, for example limited term tests at high dosage rates may not be representative of life-duration ingestion of much smaller quantities; the extrapolation relationship may be assumed wrongly; and different effects may arise with combinations of substances or different forms of a substance.

The above is a very simplified overview of the way in which health-related drinking water quality standards are developed. The derivation of chemical guideline values is detailed fully in Chapter 8 of the 4th edition of WHO *Guidelines for Drinking-water Quality* and there are numerous papers that give a more scientific account, for example Fawell (1991). Suffice to say that all the health-related chemical standards listed in Tables 7.1(A)–(D) have large safety margins built in to ensure that the results achieved are no worse than those estimated on the basis of an 'acceptable degree of risk'. Failure to meet a health-related standard means that the water is unwholesome but it does not necessarily mean that the water is unsuitable or unfit for human consumption.

7.54 SAMPLING FOR PHYSICAL AND CHEMICAL PARAMETERS

Sampling Frequencies to WHO, EC, UK and US EPA Requirements

WHO provides guidance both on operational monitoring and on the level of sampling needed to verify the quality of the water supplied. Rather than specifying sampling frequencies, emphasis is placed on the need for clearly defined objectives that take account of the local conditions, including facilities available for sampling and analysis, and the variability and likely levels of contamination that may be present in the water.

The EC Drinking Water Directive sets out a minimum frequency for sampling and analysis based on the volume of water distributed each day within a water supply zone. Where water is supplied via a distribution system, the point of compliance is deemed to be the point in a building where drinking water is normally made available to consumers. There are two levels of monitoring under the Directive, audit monitoring and check monitoring. Audit monitoring is carried out for the health-related parameters that have mandatory standards, whereas check monitoring is intended to provide more regular information on the general microbiological quality of the water and the overall effectiveness of the treatment process. Check monitoring, which relates mainly to the indicator parameters, is carried out at higher sampling frequencies than those required for audit monitoring. For a typical zone supply receiving 10 000 m^3/d and serving a population of around 45 000 consumers, 34 samples a year would be required for parameters on check monitoring and four samples a year would be required for parameters on audit monitoring. In all cases the samples have to be representative of the quality of the water supplied throughout the year in each water supply zone.

Drinking water regulations in the UK follow the same pattern of check monitoring and audit monitoring, taking into account the need for both check and audit monitoring for those parameters that are also national standards. Sampling frequencies at treatment works are based on the volume supplied, whereas sampling frequencies at consumers' taps are based on zonal population, with a maximum allowable population of 100 000 per water supply zone. Zonal samples are usually taken at randomly selected properties, although some parameters may be sampled from a fixed supply point within the supply system, such as the outlet of a treatment works or service reservoir, provided the water is representative of the zone as a whole. Supply points can only be used for certain parameters, such as pesticides, that are not subject to change and remain at the same concentration or value throughout the distribution system. Samples for parameters such as lead, copper and trihalomethanes have to be taken from consumers' taps. There are also requirements on the regularity of sampling, in that the number of samples specified per annum has to be spread evenly throughout the year.

US EPA sampling requirements are complex. The number of samples to be taken for routine monitoring of the primary standards varies according to the parameter being monitored, the type of source water, whether the source is defined as 'vulnerable' or 'non-vulnerable' to specific contaminants, and the size of population supplied. Reduced monitoring is also permitted in some situations provided approval has been granted. Samples for the secondary standards should be taken at similar intervals to those for the primary standards. There is also an unregulated contaminant monitoring programme for parameters on the Contaminant Candidate List, for which final standards have yet to be set. During the current monitoring period (2012–15), 30 contaminants will be monitored, 28 chemical parameters and two viruses.

Minimum Sampling Requirements Where No Regulations Apply

In countries where there are no guidelines or legal requirements for sampling, the water company/supplier should establish a risk-based sampling programme that also takes into account local conditions and competencies. The programme should include taking samples of source waters, the treated water leaving water treatment works and the water at key points within the distribution system for basic chemical analysis. Furthermore, samples of the treated water leaving a water treatment works should be taken for routine microbiological analysis preferably daily or, if this is not possible, at a frequency that demonstrates that the treatment processes and, in particular, disinfection have not been compromised.

Sampling Techniques for Physical and Chemical Parameters

It is of paramount importance that correct procedures are always followed whenever water samples are taken, to ensure that the samples are representative of the water being supplied. The order of sampling is also important, with physical–chemical samples being taken first and before any samples for microbiological analysis. Samples for copper, zinc and lead taken at consumers' taps should be taken as the first litre of water drawn from the tap, before flushing. Likewise, samples for PAHs and other organic compounds should always be taken before the tap is disinfected or flamed.

It is also very important that samples are not subject to contamination or change during the sampling process or during transportation to the laboratory. Therefore, whenever possible, samples should be taken by trained and experienced personnel, using dedicated sampling bottles and equipment. There should also be a chain of custody between the sample being taken and its receipt by the laboratory.

General information on sampling procedures and methods of sampling for chemical and microbiological parameters are fully documented in a series of UK publications under the title of Methods for the Examination of Waters and Associated Materials (EA, 1976 et seq.). Individual water companies in the UK have developed their own sampling manuals based on these publications. Another useful publication is *Standard Methods for the Examination of Water and Wastewaters* (AWWA, 2012) jointly produced by the American Water Works Association, the American Public Health Association and the American Water Pollution Control Federation. The International Organisation for Standardisation (ISO) also provides guidance on the design of sampling programmes and sampling techniques (ISO, 2006).

On-Site Testing and Field Analysis

Ideally some parameters, such as temperature, pH, conductivity and residual disinfectant, should be measured at the time of sampling as significant changes in concentration or value may occur even over a short period of time. Other parameters that should be carried out as on-site measurements include redox potential, dissolved oxygen and carbon dioxide. These are particularly important for groundwater samples where there can be a rapid change in concentration (or value) due to pressure changes as the sample is taken. The loss of carbon dioxide from a deep borehole sample may result in an increase of pH and, if present, soluble iron and manganese can precipitate out as the water comes into contact with atmospheric oxygen. Special sampling techniques are required in such situations.

Under some circumstances it may be necessary to carry out a more comprehensive analysis in the field, especially if the site is remote and without ready access to a laboratory. There are numerous test kits available for field analysis, covering a wide range of parameters, and varying greatly in complexity and accuracy. The simplest are 'test strips' which are dipped into the water sample. The intensity of the subsequent colour development is then compared against a strip of standard colours for specific concentrations of the parameter under test. These provide a fairly crude but objective result. There is a wide range of pre-calibrated test discs also available for monitoring most of the common water quality parameters. Reagents, usually in tablet form, are added to a standard volume of sample contained in a glass sample cell. The resultant colour development is compared visually against a blank sample, using a coloured glass test disc in the appropriate range of concentrations. Most of these test kits are user friendly and can be used with minimal training but, as they involve visual comparisons, the results should always be treated with a degree of caution. It is also advisable to introduce an element of analytical quality control, as far as is practicable, to such systems, especially if the results are intended for regulatory reporting.

A broad range of parameters can be monitored in the field by means of electronic meters, for example pH, redox, dissolved oxygen, turbidity, conductivity and temperature. Multi-parameter meters are also available which can measure a number of parameters with a single instrument. Such instruments are available with an integral data logging system and are extremely useful for continuous monitoring surveys. Comprehensive 'field laboratories' are also commercially available. These typically provide reagents and apparatus for measuring single parameters, such as residual chlorine, or multiple parameters within a single unit. A pre-calibrated spectrophotometer can be used for colorimetric tests, with electronic meters for pH and conductivity. A digital titration system may also be included for parameters such as hardness and chloride. Gas and liquid chromatographs can also be used to test for organic parameters such as pesticides. A higher level of skill is required to operate such instruments and a higher level of analytical quality control should be applied to ensure the validity of the results.

7.55 RAW WATER QUALITY

Classification

WHO no longer provides specific guidance on raw water classification. Instead, it recommends effective catchment management and source protection as part of the drinking water safety plan approach (see Part VI). However the relationship between treatment and the levels of faecal contamination developed by WHO (1993), and summarized in Table 7.2, remains relevant today, as does the conclusion that adequate disinfection should produce at least 99.99% reduction of any enteric viruses present in the raw water.

Although such raw water classifications provide useful guidelines in assessing whether a source of water is suitable for public supply, in many parts of the world it may be necessary to take account of other circumstances. Consideration should always be given to the unlisted or unquantifiable risks that apply in a catchment, such as the desirability of choosing a source that is least likely to be affected by domestic and industrial wastes, or subject to dangerous accidental pollution. Account also needs to be taken of physical and financial constraints that may make it impossible for the water supplier to provide water that complies fully with the quality requirements of the WHO Guidelines or other standards. For example, problems may arise in maintaining consistent treatment and disinfection because of lack of resources, inadequate supplies of materials, or lack of skills in the local labour force. However, every effort should be made to provide consumers with a palatable and aesthetically pleasing water supply that is free of bacteria, harmful chemicals and objectionable tastes or odours, especially if the alternative is an untreated source of doubtful quality.

The 1975 EC Directive on the quality required of surface water intended for the abstraction of drinking water (CEU, 1975) was repealed at the end of 2007 as part of the implementation of the Water Framework Directive (CEU, 2000). However, the requirement to prevent deterioration of raw water quality still remains relevant. The Water Framework Directive is intended to deliver real benefits to the aquatic environment and so improve the quality of raw water abstracted for drinking water purposes. Its aim has been to bring all surface water bodies within the European Union up to good (ecological and chemical) status by 2015, although this deadline may be extended in respect of individual water bodies to 2021 or 2027 in specific circumstances. Groundwaters should achieve good quantitative and chemical status by the same deadlines.

The Water Framework Directive established a regime under which environmental quality standards are set for specific substances. These are designated priority or priority hazardous substances, the latter being a subset of particular concern, which pose a significant risk to the aquatic environment, including risks to waters used for the abstraction of drinking water. The aim under the Directive is to reduce or eliminate pollution of surface water by these substances. The initial list identified 33 substances or groups of substances shown to be of major concern for European Waters, with a further 14 substances being subject to later review. Subsequent amendments have been made by way of a daughter Directive (CEU, 2013a).

Table 7.2 Classification of water sources according to bacterial quality and the recommended level of treatment

Source	Level of contamination	Treatment
Groundwater [a]		
Deep protected wells[b]	Free of faecal contamination *E. coli* 0/100 ml	Disinfection[c] for distribution purposes only
	Evidence of faecal contamination *E. coli* ≤ 20/100 ml	Disinfection[c]
Unprotected groundwater e.g. shallow wells	Faecal contamination *E. coli* ≤ 2000/100 ml	Filtration[d] and disinfection[c]
	Evidence of faecal contamination *E. coli* > 2000/100 ml	Not recommended as a source of drinking water[e]
Surface waters [a]		
Protected impounded waters	Essentially free of faecal contamination *E. coli* ≤ 20/100 ml	Disinfection[c]
Unprotected impounded upland water or upland river	Faecal contamination *E. coli* 20–2000/100 ml	Filtration[d] and disinfection[c]
Unprotected lowland river	Faecal contamination *E. coli* 200–20 000/100 ml	Long-term storage or pre-disinfection, filtration[d], additional treatment[f] and disinfection[c]

Notes:
Table is based on information in WHO *Guidelines for Drinking-water Quality. Volume 1, Recommendations.* 2nd Edition 1993.
[a]If the sources are contaminated with *Giardia* cysts or *Cryptosporidium* oocysts they must be treated by processes additional to disinfection (Sections 9.20 and 9.21).
[b]Water must comply with the WHO guideline criteria for pH, turbidity, bacteriological and parasitological quality.
[c]WHO conditions for final disinfection must be satisfied (Section 11.5).
[d]Filtration must be either rapid gravity (or pressure) preceded by coagulation-flocculation and where necessary clarification or slow sand filtration. The degree of virus reduction must be >90%.
[e]Water from these sources should be used only if no higher quality source is available. Drinking water from such sources carries a risk of inadequate virological quality.
[f]Additional treatment may consist of slow sand filtration, ozonation with granular activated carbon adsorbtion or other processes demonstrated to achieve >99% virus reduction.

The Directive also requires EU member states to identify all water bodies (groundwater and surface water) that are used for the abstraction of water for human consumption and provide more than an average of 10 m^3/d in total, or serve more than 50 consumers, along with water bodies that might be so used in the future. The main objectives are to ensure that, with the correct treatment regime in place, the treated water entering supply will meet the requirements of the Drinking Water Directive and that the raw water quality is unlikely to deteriorate to the point where additional treatment is required. The catchments associated with these water bodies are designated Drinking Water Protected Areas.

Monitoring Raw Water Quality

The purpose and objectives of monitoring surface waters and groundwaters have to be taken into account when developing a monitoring programme, along with the various factors that might influence the outcomes, such as spatial and temporal variations. A useful document is *Planning of Water-Quality Monitoring Systems*, produced by the World Meteorological Organisation (WMO, 2013).

In the UK, the various environmental bodies (Environment Agency in England; Natural Resources Wales in Wales; Scottish Environment Protection Agency in Scotland; and Environment and Heritage Service in Northern Ireland) have developed monitoring strategies for surface waters and groundwaters as part of the requirements of the Water Framework Directive (DWI/EA, 2012). Furthermore, the UK drinking water regulations require water companies to identify every abstraction point from which water is drawn for drinking water and to carry out a risk assessment for each one and set up a monitoring programme accordingly. The Regulators then have the power to specify the number of raw water samples to be taken and the nature of the analysis to be carried out. Minimum frequencies have been set for surface waters providing on average more than 100 cubic metres of water per day, which are linked to the size of population served. There is no minimum sampling frequency for groundwater sources but historical water quality has to be taken into consideration. The results then link into the monitoring for the designated Drinking Water Protected Areas.

Catchment Assessment (See Also Part VI) and Catchment Management

As part of the risk assessment approach advocated by WHO, catchments to new sources should be surveyed and mapped fully, with any potential sources of pollution noted, before the source is brought into use. Subsequent catchment surveys should be carried out at regular intervals and any changes noted. Where chemical tests show that toxic substances are present in a raw water, their source should be traced and, if possible, eliminated. If this is not possible, additional monitoring must be set up for such substances in both the raw and treated waters. The use of waters having consistently high coliform and *Escherichia coli* counts, or dangerously sited with respect to any waste discharge, should be avoided if at all possible. Attempts should also be made to divert any sources of pollution that are close to an intake used for drinking water supplies. If this is not possible then they should be kept under frequent observation and monitored accordingly.

US EPA likewise advocates identifying and controlling potential contaminants in the catchment and the need for source protection. It provides guidance to individual states on conducting source water assessments for public water supply systems. However, each programme of assessments is tailored to the individual state's water resources and drinking water priorities.

Catchment management is playing an increasingly important role in reducing the levels of potential contaminants in raw waters. An efficiently managed scheme will help to reduce pollution from agriculture and also help to control urban and chemical pollution from sites within a catchment. However, in order to operate successfully, the scheme must involve close liaison and co-operation between all interested parties, including water companies/suppliers, communities, landowners, farmers, highway authorities, industrialists, the various regulators and any other interested parties. In 2012, the UK Water Industry Research Limited published a benefit assessment framework for identifying and quantifying the benefits of catchment management (UWIR, 2012).

7.56 BASIC PRIORITIES IN WATER QUALITY CONTROL

In some parts of the world, the availability of well-equipped laboratories and resources for water quality testing may be limited. The level of testing in such circumstances must concentrate on the most essential parameters. The following is a suggested list of priorities for testing which, for the sake of completeness, includes microbiological testing as defined in Sections 7.68–7.70.

New Sources

Ideally any new source should be sampled and subjected to a full chemical analysis, including toxic substances, and a full microbiological analysis before it is put into supply. Similar checks should be carried out whenever new sources of pollution are suspected in the catchment or treatment processes are being significantly altered.

Simple Checks at Source Works

Simple but important checks at works treating surface waters include twice daily visual checks at the intake to the works and spot checks for any unusual odours in the raw water. Daily samples of the raw water should also be tested for the more easily measurable parameters such as turbidity, colour, conductivity and pH. Although groundwaters are inherently more stable and the water quality is less variable than surface waters, daily checks on the same parameters, plus odour, should be carried out on the raw water. Significant changes in these parameters may indicate a breach of integrity of the boreholes or contamination of the source. Samples of raw water should also be analysed for microbiological parameters at least monthly.

Where coagulation, clarification and filtration are applied, the coagulant dosage should be checked daily. The pH and turbidity of the clarified and filtered water should be checked at least daily and also after any change has been made to the coagulant dose. If a chlorine-based disinfectant is being used, at least twice daily checks should be carried out on the dosage rate and on the residual disinfectant in the water entering supply, along with daily checks for turbidity, colour, taste, odour, conductivity and pH.

Where possible, analysis for the recognized indicator organisms, coliform bacteria and *Escherichia coli* should be carried out preferably daily, but at least weekly, on samples of treated water leaving the water treatment works. If no suitable laboratory is available for microbiological analyses, consideration should be given to on-site testing by a trained operative visiting the works on a routine basis.

All results should be recorded on a day sheet and reviewed at regular intervals by a suitably qualified person. Any trends should be identified and acted upon.

Water Quality in the Distribution System

Samples of water should be taken for full chemical analysis from strategic locations within the distribution system at quarterly, six-monthly, or yearly intervals, depending on the type of water and the size of the population supplied. If there is any cause for concern then more frequent checks should be carried out for parameters of health significance, e.g. trihalomethanes, pesticides, PAHs and the heavy metals. However, this would depend on the level of risk, especially where local laboratory capabilities are limited.

Samples for microbiological analysis should be taken at least monthly from strategic locations within the distribution system, including any service reservoirs and from consumers' taps.

7.57 MONITORING AT WATER TREATMENT WORKS

The basic priorities listed in Section 7.56 should, as far as possible, be applied to all drinking water supplies. However, a higher degree of monitoring is desirable at water treatment works that have variable quality source waters and/or involve multiple treatment processes and/or supply a large number of consumers.

In general, a given raw water tends to have characteristics that fall within a certain range, often according to rainfall or other seasonal conditions. As soon as sufficient microbiological and chemical data have been obtained to establish this range and after appropriate treatment has been adopted, routine testing can concentrate on monitoring parameters which are most likely to indicate any abnormal change or deterioration in quality. Additional monitoring should always be undertaken if the source is of variable quality or is subject to sudden and dramatic changes that could affect the optimization of the treatment processes. Such monitoring should include routine chemical and microbiological parameters, along with any other potential contaminants that might be present in the raw water (e.g. substances that could be precursors of the formation of DBPs). However, it must always be remembered that the results obtained only apply to the very short window of time when the samples were taken.

On-Line Monitoring

On-line monitoring systems (see also Sections 8.1 and 12.31) can be useful alternatives to manual checks, especially as they give a fast response to sudden changes in raw water quality, provided the associated power supply can be assured. However, devising a reliable and comprehensive system can prove difficult as it is impossible to monitor for every potentially harmful substance that may be present. Surrogate monitoring systems such as fish monitors have been used with some success. The simplest form is a tank, containing a number of suitably sensitive fish such as salmonids, which is supplied continuously with the water being monitored. Any fish deaths may indicate the presence of a pollutant that needs to be investigated further. A more sophisticated version involves continuously monitoring the gill movement or electrical responses of a number of fish. Organic monitors that detect changes in the absorbance of the raw water at a specific ultraviolet frequency provide another alternative. In an ideal situation the raw water quality should be monitored upstream of the intake, with alarm systems connected either to the treatment works control room or to a central control room that is manned permanently, and with sufficient time built in to shut the intake in the event of a major pollution incident. Bankside storage helps to even out fluctuations in raw water quality and also helps to maintain supplies if the intake has to be closed to allow a plume of pollution to pass by (Section 8.1). Ideally, such storage should be divided into at least two separate compartments, so that the raw water in each can be tested alternately before being used.

On-line monitoring systems are also widely used to check the efficacy of the treatment processes. Turbidity, pH and conductivity monitors provide basic but important water quality information. More sophisticated on-line monitors that can measure residual coagulant (iron and/or aluminium) are available, as are UV absorption monitors which provide a surrogate for DOC. These provide additional information on the effectiveness of the clarification and the filtration stages of treatment and are also useful for checking the quality of the treated water entering supply. Similar systems are used for monitoring the residual disinfectant level at various stages during the disinfection process and in the treated water entering supply (Section 12.35).

Where possible, statistical process control should be applied to enable identification of 'out of control' conditions or development of trends. Appropriate trigger values can then be applied (based on statistical interrogation of historical data sets) that would prompt appropriate action and response that could include emergency shutdown of the supply. On-line monitoring systems are ideal tools for process control but they have to be properly maintained and calibrated by trained operators. This may make them too sophisticated for small rural supplies where a more hands-on approach to monitoring might be more appropriate.

7.58 METHODS OF CHEMICAL ANALYSIS

It is important that methods of analysis be standardized as far as possible in order to achieve comparability of results. In 1972 in the UK, the then DoE established the Standing Committee of Analysts to set up working groups to produce suitable methods for water analysis. The committee, which now comes under aegis of the EA, represents a wide range of interests in the water industry and has produced detailed guidance in a series of publications under the title of *Methods for the Examination of Waters and Associated Materials* (EA, 1976 et seq.). Each publication looks at a single analytical method or linked group of methods. Another useful publication is *Standard Methods for the Physical and Chemical Examination of Water and Wastewaters* (AWWA, 2012). This comprehensive book, which is updated at regular intervals, provides a valuable reference and the methods described in it have had a wide influence on standards adopted in other countries. The International Organisation for Standardisation also provides a series of methods of analysis (ISO, 1983 et seq.). *Field Testing of Water in Developing Countries* (Hutton, 1983) provides a further useful reference.

All analytical methods for chemical parameters should be fully evaluated and validated by the laboratory carrying out the tests before a method is adopted for routine use. The increasing use of sophisticated analytical equipment means that many methods have to be adapted for use with a particular instrument. The initial performance tests should demonstrate that the analytical system is capable of establishing, within acceptable limits of deviation and detection, whether any sample contains the parameter under analysis at concentrations likely to contravene the required standard. Such performance testing should cover the entire analytical procedure, including sample preparation and any concentration steps. The precision and trueness or accuracy of the test in terms of maximum tolerable values for total error and systematic error and limit of detection should be ascertained, along with checks for recovery and resilience against possible interferences. The method should also be validated for each type of sample matrix on which it is to be used (e.g. hard, soft or surface water). The required performance characteristics for chemical parameters are specified both in the EC Drinking Water Directive and in the US National Primary Drinking Water Regulations. The UK DWI has issued guidance on analytical systems to water companies in England and Wales as part of more general guidance on the implementation of the Regulations (DWI, 2010a), which is updated at regular intervals. However, it is recognized that the required level of performance cannot always be met with the current methods of analysis available, especially for some of the more obscure organic parameters.

Each laboratory should also have established and documented procedures for routine analytical quality control as applied to each validated method. External quality control schemes or inter-laboratory proficiency testing schemes should be used where available, as these provide further useful information on a laboratory's capabilities for carrying out an acceptable level of analysis. Guidance on setting up such schemes is available from ISO (ISO, 2007).

Under the US National Primary Drinking Water Regulations, all analyses, other than turbidity, chlorine residual, temperature and pH, have to be carried out by a certified laboratory. The UK drinking water regulations require that all analysis on samples for regulatory purposes be conducted by or under the supervision of a person who is competent to perform the task. The DWI has issued guidance on the level of training considered necessary to meet these requirements. Furthermore, water companies in England and Wales are required to use laboratories that have UKAS accreditation under ISO/IEC 17025 to Drinking Water Testing Specification (DWI, 2013c). This covers sampling, transportation and analysis of all regulatory drinking water samples.

7.59 QUALITY ASSURANCE OF WATER TREATMENT CHEMICALS AND MATERIALS IN CONTACT WITH DRINKING WATER

Some water treatment chemicals and materials that come into contact with drinking water may have an adverse impact on the quality of the water supplied. WHO refers to the need for verification protocols as part of its Water Safety Plan approach to ensuring the safety of drinking water supplies (Section 7.82). The EC Drinking Water Directive requires member states to take appropriate measures to ensure that such chemicals and materials or associated impurities do not pose a risk to human health. The majority of chemicals and filter materials used by member states in water treatment are covered by appropriate European Standards; the European Commission's (DG Enterprise) Expert Group – Construction Products Drinking Water is providing advice on the development of a revised European Acceptance Scheme for drinking water construction products.

Under the UK drinking water regulations all chemicals used in the treatment of drinking water and materials which come in contact with drinking water (such as pipe linings, etc.) have to bear an appropriate CE marking, or conform to an appropriate harmonized standard or European technical approval or an appropriate British Standard. All other products and materials have to be approved by the appropriate authorities. Approval is based upon consideration as to whether the use of a substance or product will adversely affect the quality of the water supplied, or cause a risk to the health of consumers. The DWI publishes a list of approved products for use in public water supplies throughout the UK (DWI, 2015). The list is updated at least once a year and covers items such as: water treatment chemicals; filtration, including membrane filtration and electrodialysis systems; ion exchange media; systems used to generate disinfectants in situ; construction products used in water treatment processes, including pipework and storage installations; in situ applied repair material; construction products and coatings used in raw water and treated water installations; and water retaining vessels and pipework used for the provision of drinking water in emergencies. Maximum permissible dosages are stipulated for some proprietary products, together with other conditions of use as appropriate (Sections 7.3, 7.24 and 7.27). If a water supplier fails to meet these requirements, then it is guilty of an offence.

There are similar requirements in the US (NSF/ANSI, 2013).

PART III WATER MICROBIOLOGY

7.60 DISEASES IN MAN THAT MAY BE CAUSED BY WATERBORNE BACTERIA AND OTHER ORGANISMS

Humans are subject to a number of diseases which can be waterborne and may be caused by the presence of pathogenic bacteria and other organisms, such as protozoa and viruses, in drinking water supplies, in water used for bathing or immersion sports, or via other routes. It has long been recognized that the ingestion of water contaminated with excrement can result in the spread of diseases such as cholera and typhoid and that adequate treatment and other measures are required to prevent outbreaks of such diseases. The WHO Guidelines state that 'The potential health consequences of microbial contamination are such that its control must always be of paramount importance and must never be compromised'.

This section considers some of the intestinal diseases which have commonly, although not invariably, been waterborne. In all cases, the associated organisms are present in large numbers in the excreta of an infected host and are relatively resistant to environmental decay. Many can cause illness even when ingested in small numbers. Schistosomiasis and other parasitic diseases of the tropics are reviewed as part of water biology in Part IV.

7.61 BACTERIAL DISEASES

Cholera is caused by the bacterium *Vibrio cholerae* and is characterized by acute diarrhoea and dehydration. Many strains have been identified and those in serogroups O1 and O139, which produce cholera toxin, are associated with outbreaks of epidemic cholera. Infection is usually contracted by ingestion of water contaminated by infected human faecal material, but contaminated food and person to person contact may also be sources. The link between the disease and drinking water was made by John Snow in 1854, following an outbreak in London.

The disease is endemic in many countries. However, the bacteria are short lived in water and have low resistance to chlorine so it is unlikely to spread in communities with controlled water supplies and effective sewage disposal. Vulnerable communities with inadequate sanitation systems and limited access to clean drinking water are particularly at risk, especially following a natural disaster. In 2010 Haiti was hit by a severe earthquake and the subsequent cholera outbreak killed more than 6000 people.

Typhoid fever is caused by the bacterium *Salmonella* Typhi. Infection is usually contracted by ingestion of material contaminated by human faeces or urine, including water and food (e.g. milk, shellfish). The bacteria are moderately persistent in water but have low resistance to chlorine. However, in 1937 there was a large waterborne outbreak of typhoid fever in Croydon, UK, killing 43 people. Investigations revealed that it was caused by a combination of circumstances including a person who was a carrier of *Salm.* Typhi working down a well, which was pumping into supply, coincidental with the filtration and chlorination plants being bypassed (Suckling, 1943).

There have been a number of more recent outbreaks, which are also believed to have been due to water contamination coinciding with inadequate disinfection. *Paratyphoid fevers* are also caused by *Salmonella*, in this case Paratyphi *A, B* or *C* and infection may exceptionally be via contaminated water.

Bacillary dysentery is caused by bacteria of the genus *Shigella*, with *Sh. dysenteriae* 1, *Sh. flexneri*, *Sh. boydii* and *Sh. sonnei* being the four recognized species. Infection can occasionally be contracted via water contaminated by human faeces but more commonly it is due to ingestion of food contaminated by flies or by unhygienic food handlers who are carriers. *Traveller's diarrhoea* is a term applied generally and may have several potential causes, but it may be that some forms of pathogenic *Escherichia coli* or, more rarely, *Shigella* are responsible. It is probably transmitted in the same way as bacillary dysentery and water may sometimes be the vehicle. However, the bacteria are short lived in water and have a low resistance to chlorine. The problems associated with enterovirulent *E. coli* are reviewed in Section 7.81.

Leptospirosis is caused by species of the bacterium *Leptospira*. There are now 16 recognized species of which *Leptospira interrogans* is the most important. Leptospires are motile, spiral-shaped organisms and pathogenic strains can cause symptoms ranging from mild fever to severe jaundice in the case of Weil's disease. The organisms are shed in the urine of infected rats, dogs, pigs and other vertebrates, and are often present in ponds and slow-flowing streams visited by such animals. People who bathe in, fish in or sail on these waters are at risk, becoming infected via the mouth, nasal passages, conjunctiva or skin abrasions. The bacteria are very sensitive to disinfectants and are eliminated by normal water treatment processes but sewer workers remain at risk.

Legionnaires' disease is an acute form of pneumonia caused by the bacterium *Legionella pneumophila*. Fifty-two species are now described in the genus *Legionella*, of which quite a few have been implicated in a range of infections in humans generally termed Legionellosis (EA, 2005). The bacterium *L. pneumophila* is regarded as the most dangerous, having been identified in all outbreaks of Legionnaires' disease. (The name comes from an outbreak of the disease causing high mortality in war veterans attending a Legion Convention in Philadelphia, USA in 1976). Outbreaks of the disease can be sudden and result in significant mortality rates. Other forms of Legionellosis, such as Pontiac fever, are less severe and exhibit no particular age distribution.

Legionella bacteria are widely present in low numbers in surface waters and possibly in some groundwaters. They are persistent in water and will multiply, but have a low resistance to chlorine. However, if present within protozoa they can survive conventional water treatment, including disinfection with chlorine, and retain an ability to colonize water systems such as cooling towers, evaporative condensers and even domestic hot water tanks, showers and other plumbing system components. They are thermotolerant and ideally suited to grow in the warm water systems of buildings at 30–55°C but do not survive sustained temperatures above 60°C. Infection occurs from the transport of the bacterium by aerosols or airborne water droplets, which are inhaled. Several of the outbreaks of Legionnaires' disease have been caused by aerosols blown from exposed cooling towers of air-conditioning plants at hotels, hospitals and similar large buildings. Whirlpools and jacuzzis, where recirculated warm water is sprayed, have also been implicated. Official advice is available (HSE, 2013, 2014) on preventive measures, including designing and maintaining water systems in buildings to minimize the risk of colonization; minimizing the accumulation of sediments and slimes; and maintaining hot water systems above 60°C and cold water systems below 20°C. Biocides are effective at controlling *Legionella* in air-conditioning systems using wet evaporative cooling towers but are generally less effective in water distribution systems within buildings. There is no evidence of transmission by ingestion, so infection is not attributable directly to drinking water supplies.

Campylobacteriosis is caused by bacteria of the genus *Campylobacter*, which includes 14 species. They are endemic worldwide and are the most common cause of food-related bacterial gastroenteritis in England and Wales, with *Campylobacter jejuni* and *C. coli* the most common human pathogens detected. These bacteria are also associated with the faeces of a wide range of wild and domesticated animals and birds, including poultry and gulls. *Campylobacters* are widespread in the environment and frequently found in sewage. They have been detected in surface waters, where they can survive for several weeks at cold temperatures. They are sensitive to disinfectants used in water treatment and conventional water treatment should eliminate their presence in drinking water supplies. However, there have been a number of reported outbreaks associated with unchlorinated or inadequately chlorinated surface water supplies or with contaminated storage facilities. Problems can also occur with private water supplies where there is no treatment.

7.62 OTHER BACTERIA

The ***Pseudomonas*** group is commonly found throughout the environment. Some species may be present in human and animal excrement and many can multiply in water containing suitable nutrients. Their presence in drinking water may result in an overall deterioration in microbiological quality and lead to consumer complaints of taste and odour. *Pseudomonas aeruginosa* is an opportunist pathogen which can cause infection particularly in people whose natural defence mechanisms may be impaired, for example the very old, the very young or the immunosuppressed. It is capable of colonizing taps and plumbing fittings in buildings, particularly where cold and hot water supplies are adjacent to each other. Under static conditions, the cold supply can warm up, creating suitable conditions for growth. This has resulted in the pathogen being responsible for a number of hospital-acquired infections. This reinforces the importance of well-designed and maintained distribution systems in buildings. Other members of the pseudomonas group may be associated with the growth of biofilms within a distribution system, being detected as part of the colony count or heterotrophic plate count (Section 7.69).

Members of the ***Aeromonas*** group are naturally present in the aquatic environment, with methods available for their specific detection. Their presence in drinking water does not necessarily indicate faecal pollution but highlights possible inadequacies in the treatment process or ingress within a distribution system, where there is also the potential for regrowth and biofilm development. Most treatment processes effectively remove Aeromonads and free chlorine residuals of 0.2–0.5 mg/l are considered sufficient to control the organisms in distribution systems. When present their numbers depend on the residence time of the water, its organic content and the residual disinfectant level.

7.63 PROTOZOAL DISEASES

Amoebic dysentery is caused by the microscopic parasite *Entamoeba histolytica*. The parasite is distributed throughout the world and exists in two stages, only one of which, the cyst, is infective. The parasite infects mainly primates and, following infection, resides in the large intestine in humans where it reproduces and generates further cysts, which are passed in the faeces. The cysts can survive for several days in water at temperatures of up to 30°C and are relatively resistant to disinfection. As infection occurs when the cysts are ingested, outbreaks of the disease are likely if water supplies are contaminated with domestic sewage containing viable cysts. Coagulation, clarification and filtration are the most commonly employed methods for physical removal of parasites, while ozone or ultraviolet light are used for inactivation. Combinations of treatment technologies can result in parasite removal/inactivation greater than 6-log, resulting in reliable public health protection. More commonly the disease is transmitted by person-to-person contact or via food contaminated by carriers.

Cryptosporidiosis is an acute self-limiting diarrhoeal disease caused by the protozoan parasite *Cryptosporidium*. Improvements in water treatment technology and practice within the developed world, especially in relation to chlorine disinfection, have seen a significant decline in the number of outbreaks of disease attributable to consumption of public water supplies over the last several decades. However, of the 31 outbreaks reported in the UK during the last 25 years or so, 29 have involved *Cryptosporidium*. Therefore, this organism remains a realistic microbial threat to the safety of drinking water.

There are currently around two dozen recognized species of *Cryptosporidium* and more than 60 reported genotypes. Although the great majority (>98%) of reported cases of human illness involve just three species (*C. hominis*, *C. parvum* and *C. meleagridis*), there are published reports of human infection with a further 13 species and at least five genotypes. The faecal–oral route is the usual means of exposure, either by direct contact with infected animals or humans or via contaminated water or food. Not all infected persons necessarily develop the symptoms of the disease but the illness is likely to be serious or even life threatening for patients who are immunologically compromised (Smith, 1992).

The parasite has a complex life cycle, which takes place within the body of the host and can include repeated cycles of autoinfection. Infective oocysts, which are 4–5 μm in diameter, are then shed in vast numbers in the faeces of infected animals and persons. Oocysts have been found in water sources that are subjected to faecal contamination originating from humans and/or livestock (Sturdee, 2007). Their presence in surface waters is often associated with intensive animal grazing. Oocysts are also occasionally found in some groundwater sources, either via fissures in the overlying strata or through compromised borehole structures (Carmena, 2006). Treated sewage effluent can, on occasion, also contain large numbers. The oocysts can survive and remain infective in water and moist environments for several months and are resistant to high concentrations of chlorine (Section 11.6). This, coupled with the small size of the oocysts, can result in low numbers of oocysts occasionally penetrating conventional water treatment processes.

In recent years there have been an increasing number of outbreaks of cryptosporidiosis associated with public water supplies in both the UK and North America. Following an outbreak in Swindon and Oxfordshire during the winter of 1988–89 involving over 500 cases of cryptosporidiosis, an expert committee was set up by the UK Government under Sir John Badenoch, to examine the problem. The Badenoch Report (DoE, 1990b) made many recommendations concerning water treatment practices, monitoring, and the role of various authorities in the event of a suspected waterborne outbreak. There were a number of other outbreaks during the 1990s, resulting in the expert committee being reconvened and issuing a second report (DoE, 1995) providing further guidance. A third report was produced after the Group was reconvened under Professor Ian Bouchier (DETR, 1998), following an outbreak of cryptosporidiosis in North London in which groundwater supplies were implicated. Since 2000 there has been a significant reduction in the number of outbreaks of cryptosporidiosis associated with public water supplies in the UK, although there have been some notable exceptions in North West England, Scotland, North West Wales and the East Midlands. In the Scottish outbreak and one of the North West outbreaks, the raw water catchments were shown to be at risk from grazing sheep and the raw water aqueducts were shown to be vulnerable to ingress. One of the largest historical outbreaks of cryptosporidiosis associated with public water supplies occurred in the City of Milwaukee in 1993, when more than 400 000 consumers became ill and a number died.

Changes in the seasonal incidence of cryptosporidiosis in the UK point to the effectiveness of focusing on treatment measures. Recent trends show a sustained reduction in the number of cases of cryptosporidiosis reported during spring, coincidental with an increased occurrence in young farm animals, especially lambs. However, an autumn peak has been reported for most years between 2000 and 2012 (PHS, 2013). Although other factors such as overseas holidays, swimming pools and other recreational water-based activities may play an important part in this peak, there is no room for complacency (Sopwith, 2005; Goh, 2005).

Knowledge has moved on significantly in recent years, with ongoing research into the methodology for the detection of oocysts in water, methods for assessing oocyst viability and infectivity and genetic fingerprinting. Both ozone and ultraviolet irradiation treatment have been shown to inactivate the oocysts (Sections 9.20, 11.21 and 11.23), but multiple barrier systems, including suitable membrane systems, are still needed to remove them. The water safety plan approach (Section 7.82) helps to mitigate the risk, especially in terms of monitoring of activities in catchments, introducing control measures aimed at reducing the oocyst challenge in the raw water, and ensuring that the treatment processes are not compromised.

Giardiasis is also a diarrhoeal disease and is caused by the protozoan parasite, *Giardia duodenalis*. Like cryptosporidiosis, the disease is usually self-limiting and is caused by the ingestion of cysts by a susceptible host. The faecal–oral route is the usual means of exposure, either by direct contact with infected animals or humans, or via contaminated water or food. Infected animals can contaminate surface waters and, in North America, beavers are frequently blamed for associated outbreaks. The cysts are larger than *Cryptosporidium* oocysts, being 7–10 μm wide and 8–12 μm long, and can survive for many days in a cool aqueous environment. Being larger, the cysts are effectively removed from drinking water by physical methods of treatment, such as filtration. They are also much more susceptible to disinfection than *Cryptosporidium*.

Giardiasis is a worldwide disease and there have been a number of reported waterborne outbreaks in the USA. Evidence of waterborne infection in the UK has been confined to situations where there has been direct faecal contamination of water used for drinking, usually with untreated private supplies.

7.64 VIRAL DISEASES

Viruses differ from bacteria in that they are very much smaller and can multiply only within suitable host cells, in which they produce changes that give rise to a range of diseases. They are usually quite host-specific. Human enteric viruses are a diverse group of more than 140 different viruses that are known to be shed in human faeces. The level of enteric viruses in domestic sewage is generally orders of magnitude lower than that of faecal bacteria. However, viruses are more environmentally resistant and can migrate considerable distances. Consequently, they are widespread in surface waters and have been detected in groundwaters that are remote from direct sources of contamination.

Contaminated water has been associated with outbreaks of *Hepatitis* in the Far East and elsewhere resulting from various deficiencies, including treatment failure, compromises in the distribution system, badly constructed wells allowing contamination from adjacent cesspits, and heavy rainfall. There are also risks of infection when using sewage-polluted waters for recreational purposes, or in the recycling of wastewater for domestic use without adequate treatment and disinfection.

Little direct information is available on the removal of viruses by water treatment processes, mainly because of the limitations in analysis, but information gained using cultured viruses indicates that effective treatment, particularly disinfection, will, if applied properly, produce effectively virus-free drinking water. Membrane systems based on nanofiltration and RO are also very effective at removing viruses.

Enteric viruses that are emerging as important risks in the transmission of waterborne viral diseases are reviewed in Part V.

7.65 MICROBIOLOGICAL STANDARDS FOR DRINKING WATER

Pathogenic bacteria and other organisms are often difficult to detect in water that has been treated effectively. If present, their numbers are likely to be very small. Their presence even in sewage effluent or polluted river waters may be only infrequent or may only occur at irregular intervals, depending on the level and source of contamination. Analysing directly for pathogenic bacteria is not therefore a practical safeguard for a water supply and, indeed, routine monitoring for process control purposes would be both impracticable and unnecessary. Instead, evidence of any pollution with excreta from man or animals should be sought using simpler and more accessible tests. If the evidence is positive, it should be assumed the water may also contain pathogenic bacteria and must therefore be regarded as unsuitable for supply purposes.

Table 7.3 sets out the WHO Guidelines, the EC Drinking Water Directive standards, the current UK standards and US EPA standards for the microbiological quality of drinking water. Coliform bacteria should not be detected in the water leaving a water treatment works provided the treatment processes, particularly disinfection, are adequate. The principal requirement common to all the standards for drinking water at the point of supply, including the consumer's tap, is that there should be no total coliforms or *Escherichia coli* (faecal coliforms) detected in any 100 ml of sample.

Microbiological standards should be rigorously adhered to and any failure should be investigated immediately and suitable corrective action taken as necessary (Section 7.69).

In addition to specific microbiological standards, the EC Drinking Water Directive requires that water intended for human consumption should not contain any microorganisms and parasites in numbers that constitute a potential danger to human health. The Directive also sets microbiological standards for water offered for sale in bottles and containers. These are more stringent than the standards that apply at consumers' taps and include additional numerical standards for colony counts and *Pseudomonas aeruginosa*.

There are MCLGs of zero for *Cryptosporidium*, *Giardia* and enteric viruses in public water supplies in the USA. The Surface Water Rules, which apply to systems treating surface water sources or groundwater sources under the direct influence of surface water, enables US EPA to issue a system of 'log removal credits' for *Cryptosporidium*, *Giardia* and viruses. The Long Term 2 Enhanced Surface Water Rule, which applies to public water supplies serving 100 000 or more people, targets additional treatment requirements for *Cryptosporidium* at higher risk sites. It also contains provisions to reduce risks from any uncovered final water storage facilities. There are four classifications for treatment works with filtration and these depend on the average *Cryptosporidium* concentration in the source water. The log removal required is then based on the type of filtration used. Works without filtration are required to provide additional treatment such as chlorine dioxide, ozone or UV to achieve 2- or 3-log inactivation depending on the number of oocysts/litre in the source water. Works with ≤ 3.0 oocysts/litre in the source water and operating with direct filtration are required to provide up to 3-log removal (Section 9.19). Water suppliers are also expected to achieve 3-log or 99.9% removal or inactivation of *Giardia lamblia*, with 4-log or 99.99% removal or inactivation required for viruses, either by disinfection alone or by a combination of filtration and disinfection. Under the Groundwater Rule, one of the options for corrective action at groundwater sources identified as being at risk is to install treatment to achieve at least 4-log inactivation or removal of viruses.

Table 7.3 Microbiological standards

	Colonies/ml at 22°C	Colonies/ml at 37°C	Total coliforms	*E. coli* (faecal coliform)	Enterococci (faecal streptococci)	*Clostridium perfringens*	Cryptosporidium
WHO Guidelines 4th Edition 2011							
Treated water entering the distribution system				ND in any 100 ml sample[a]			
Treated water in the distribution system				ND in any 100 ml sample[a]			
All water directly intended for drinking				ND in any 100 ml sample[a]			
EC Directive 1998							
The point where water emerges from taps that are normally used for human consumption for water supplied from a distribution network	No abnormal change[b]		0/100 ml[b]	0/100 ml	0/100 ml	0/100 ml[b]	
UK Water Supply (Water Quality) Regulations							
Water leaving a water treatment works	No abnormal change[b]	No abnormal change[b]	0/100 ml	0/100 ml			
Water in service reservoirs	No abnormal change[b]	No abnormal change[b]	0/100 ml[c]	0/100 ml			
Water at consumers' taps	No abnormal change[b]	No abnormal change[b]	0/100 ml[b]	0/100 ml	0/100 ml	0/100 ml[b]	
US EPA National Primary Drinking Water Regulations (MCLs)							
At sites representative of the water throughout the distribution system	Treatment requirement for surface water <500 colonies/ml	Treatment requirement for surface water <500 colonies/ml	Treatment requirement for surface water MCL <5.0% positive samples in a month[d] (MCLG – zero)				TT – 99% removal (MCLG – zero)

The National Primary Drinking Water Regulations also require treatment processes to be in place to achieve:

- 99.9% removal/inactivation of *Giardia lamblia*
- 99.99% removal/inactivation of enteric viruses

Notes:
ND = Not Detectable; MCL = Maximum Contaminant Level (US, mandatory standard); MCLG = Maximum Contaminant Level Goal (US, non-mandatory); TT = Treatment requirement – unfiltered supplies are required to include *Cryptosporidium* in catchment control provisions.
[a] *E. coli* or thermotolerant coliform bacteria.
[b] Indicator parameter.
[c] 95% of the last 50 samples taken must meet the standard.
[d] Including faecal coliform and *E. coli*. No more than one sample should be total coliform positive for water supply systems that take less than 40 samples per month.

Standards of microbiological testing can vary in different parts of the world. Terminology is often inconsistent and confusing, with the same term being used to cover one test procedure in one country and a different test procedure in another. In hot climates, many waters give total bacterial counts substantially higher than in temperate climates, and certain types of organisms may be more abundant. It can be difficult for the engineer to interpret bacterial test results and the advice of an experienced water microbiologist should always be sought in obtaining a definitive interpretation.

7.66 USE OF COLIFORMS AS AN INDICATOR OF MICROBIOLOGICAL POLLUTION

Coliform bacteria are widespread throughout the environment and have long been used as indicator organisms by water microbiologists. This is because tests for them are relatively simple and they can be detected in low numbers. The coliform group contains many species that can multiply in water and are not of faecal origin. Some coliforms are able to grow at higher temperatures, giving rise to the terms thermotolerant and 'faecal' coliforms. Emphasis is usually placed on hygienic importance, for which the occurrence in faeces is the most significant.

Current standards tend to be more specifically based, focusing on *Escherichia coli.* This is a thermotolerant coliform and is consistently present in very large numbers in the faeces of warm-blooded animals, including man, where it is a natural inhabitant of the intestine. Many coliform bacteria, including *E. coli*, can survive for a considerable time in water, making them a good indicator for the presence of other pathogenic bacteria. Thus, the detection of *E. coli* in drinking water supplies provides clear evidence of faecal contamination. If coliform bacteria are detected, but no *E. coli*, it is likely that the contamination may be from soil or vegetation, or it may provide a warning that more serious contamination could follow, especially after heavy rain. However, the presence of any coliform bacteria in treated water indicates either deficiencies in the treatment process or some form of post-treatment contamination and the circumstances should always be investigated immediately.

Although the absence of coliform bacteria, and more particularly *E. coli*, implies that the water is unlikely to be contaminated, it cannot be guaranteed that other intestinal pathogens are absent. This is because other pathogens, such as viruses and protozoa, although less likely to be present, may be more resistant to disinfection.

7.67 FREQUENCY OF SAMPLING FOR MICROBIOLOGICAL PARAMETERS

The WHO Guidelines do not specify sampling frequencies. The EC Drinking Water Directive sets minimum frequencies for samples taken from consumers' taps, based on the volume of water supplied within a water supply zone. Colony counts at 22°C, coliform bacteria, *E. coli* and *Clostridium perfringens* are all subject to check monitoring, with samples for *Clostridium perfringens* only being required if the water originates from or is influenced by surface water. For a zone supply of up to 1000 m^3/d the minimum number of samples required per year is 4, the number increasing by three samples for every 1000 m^3/d or part thereof for larger volumes. Enterococci are monitored at the audit frequency, which is one sample per year for volumes of up to 1000 m^3/d, increasing by one sample for each 3300 m^3/d or part thereof up to 10 000 m^3/d, then three samples plus one sample for each 10 000 m^3/d or part thereof up to 100 000 m^3/d, and 10 samples plus one sample for each 25 000 m^3/d or part thereof up to volumes greater than 100 000 m^3/d.

Similar minimum frequencies are specified in the UK drinking water regulations, with additional samples being required to meet the national standards for coliform bacteria and *E. coli* in the water leaving water treatment works and in service reservoirs.

US EPA sets out routine monitoring requirements for coliform bacteria and *E. coli* in the Total Coliform Rule. The minimum number of samples is based on the population supplied and range from one sample per month for a population of up to 1000, to 480 samples per month for populations greater than 3 960 000. Samples must be taken at regular intervals throughout the month, apart from sites sampled once a month. Where untreated water is supplied without disinfection, for example with non-piped or community supplies, a minimum sampling frequency should be established based on local conditions.

Raw Water Sampling

In the UK, there are no specific set frequencies for taking microbiological samples from raw water sources. However, water companies are required to identify each abstraction point from which water is drawn for drinking water and to carry out a risk assessment to inform the degree of monitoring required (Section 7.55).

When initially published, the US EPA Long Term 2 Enhanced Water Treatment Rule required up to 2 years of monitoring for *Cryptosporidium* in the source water to water treatment works treating surface water or groundwater affected by surface water. This applied to works both with and without filtration, to determine whether any additional treatment was required to meet the standard. However, works with filtration that provided at least 5.5-log treatment for *Cryptosporidium* and works without filtration that provided at least 3-log treatment for *Cryptosporidium* were exempt, as were works that intended to install that level of treatment. Treatment works with filtration were also required to monitor the source water for *E. coli* and turbidity. US EPA issued a guidance manual on these requirements in February 2006 (US EPA, 2006). The Groundwater Rule established a risk-based approach to identify groundwater sources susceptible to faecal contamination triggered by monitoring for *E. coli*, Enterococci or coliphages.

7.68 SAMPLING FOR ROUTINE MICROBIOLOGICAL PARAMETERS

The sample should be representative of the water in supply and care should always be taken to avoid accidental contamination either during or after sampling. Personnel taking microbiological samples should be adequately trained and aware of the responsibilities of their role. They should be aware of the need to avoid crosscontamination between raw and treated water samples. Microbiological samples should be transferred immediately to dark storage conditions and kept at temperatures between 2°C and 8°C during transit to the analysing laboratory. They should be analysed as soon as practicable on the day of collection. In exceptional circumstances, commencement of the analysis may be delayed by up to 24 hours after a sample is taken but only if the sample has been stored under the above conditions.

Raw water samples should be taken at the inlet to the treatment works and preferably from a dedicated sampling point. Dip samples should be avoided, unless there is no alternative means of taking the sample, and due consideration should be given to health and safety.

Water leaving a water treatment works should be sampled from a point that is representative of the water entering supply, which may mean having more than one sample point. Samples should be taken from a dedicated sample tap that should not be used for any other purposes. The tap should be of an approved design and made of metal. It should be kept clean and also be prominently labelled with a dedicated reference number. Delivery pipework to the tap should be as short as possible, with the pipe made of a suitable material and opaque. Where possible the pipe between the sample point and the sample tap should be a single length of pipe, with no joints and minimal bends. It is generally not advisable to take microbiological samples from constantly running taps as the action of turning the tap off to disinfect it and then turning it on again to run to waste can dislodge particulates or biofilm from the sample line.

Sample taps and pipework at service reservoirs should be as described above. The system should be designed to ensure that water sampled is representative, as far as possible, of the reservoir as a whole and take account of the inlet/outlet arrangements and any hydraulic arrangements within the service reservoir, such as separate compartments (Section 20.6).

Sampling from consumers' taps should ideally be carried out at a tap supplied directly from the service pipe, so as to be representative of the water in supply. The tap should be in good repair and free of attachments. Many mixer taps currently available in developed countries have components or are made of materials that make them very difficult to disinfect. Extra care is therefore always needed to ensure that the tap is adequately prepared and disinfected before being sampled.

Method of Sampling

As soon as any chemical samples have been taken, the sample tap should then be disinfected using a blowtorch in the case of a metal tap, or, in the case of most consumers' taps, by swabbing the outside and as much of the inside of the tap as is possible with a chemical disinfectant, such as 10% w/v sodium hypochlorite solution. Disinfectant wipes can also be used to wipe the outside of taps that would not withstand flaming with a blowtorch. A few minutes contact time should be allowed for the disinfectant to work before the tap is run to waste until the water is cool or all the chemical disinfectant has been removed. During running to waste and sampling the flow rate of the water should remain steady to reduce the risk of any biofilm being dislodged into the sample.

Only sterilized sample bottles should be used for microbiological sampling, which ideally have been subject to a batch sterility quality control check before releasing for use. If the water being sampled contains residual chlorine, the bottle should contain sufficient 1.8% w/v sodium thiosulphate to dechlorinate the sample. The sample bottle should not be rinsed but filled in one action, with the cap or stopper being removed for the minimum time possible. The lip of the bottle should not be allowed to come into contact with the tap and the bottle should be filled without splashing, leaving a small air space below the cap or stopper. Care should always be taken not to contaminate the cap or stopper and if accidental contamination is suspected the sample should be discarded and taken again in a new bottle.

In some cases it may be necessary to take samples from hydrants or standpipes, particularly for new and repaired mains. The hydrant box should always be cleared of any accumulated debris and water and the outlet should be dosed with sodium hypochlorite before the standpipe, which should be kept in a clean condition, is attached. The hydrant should then be cracked open to fill its outlet and the standpipe and allowed to stand for at least 5 min before flushing. Flushing should continue until the residual chlorine level is that of the mains supply and the microbiological sample should be carefully taken without turning the water off.

Occasionally dip samples may have to be taken for investigational purposes. Special sterilizable sampling cans can be obtained for this purpose; otherwise, wide mouthed sterile sample bottles securely attached to sterile wire can be used for taking the sample. It requires special care to obtain a sample representative of the body of water being sampled whilst avoiding contamination of the sample.

7.69 ROUTINE TESTS FOR BACTERIAL CONTAMINATION OF WATER

A typical suite for routine microbiological monitoring usually includes the following tests:

Colony counts (or heterotrophic plate counts) at 20–22°C and 37°C. Large numbers of micro-organisms occur naturally in both ground and surface waters, many of which are associated with soil and vegetation and can survive for long periods in the environment. Counts of such organisms, grown as colonies on or in nutrient agar, yeast extract agar or other suitable non-selective agar, provide a useful means of assessing the general bacterial content of a water. The colony count, or heterotrophic plate count, following incubation at 20–22°C gives an indication of the diversity of bacteria present at normal environmental temperatures.

Although the result does not have any direct health significance, it provides a useful means of assessing the efficacy of the various water treatment processes in terms of overall bacterial removal. It also gives an indication of the general microbiological state of a given distribution system. Incubation at the higher 37°C temperature encourages the growth of bacteria that can thrive at body temperature and which, therefore, may be of animal origin. The main value of both tests is to provide a background level or reference for a particular source water, treatment works or distribution system. A sudden marked increase, particularly in the 37°C count, could be indicative of treatment deficiencies or of a more serious problem developing, such as ingress within the distribution system. Marked changes over and above the normal seasonal trends for colony counts at both temperatures could indicate longer term changes in the microbiological quality of the water. The counts are also of value where the water is used in the manufacture of food and drink as they could be an indication of a potential spoilage problem.

Coliform bacteria count. As discussed in Section 7.66, the tests for coliform bacteria make it easy to detect and enumerate these bacteria when present in water. The term 'coliforms' traditionally referred to bacteria capable of growing at 37°C in the presence of bile salts and of fermenting lactose at this temperature, producing acid and gas after 24–48 hours incubation. They are also gram and oxidase-negative and non-spore forming. Improved understanding of the coliform group has enabled more rapid and direct test methods to be developed. These are based on the expression of the β-galactosidase gene, without the requirement for gas production. Methods using membrane filtration provide at least a 'presumptive' result under normal laboratory conditions after 18 hours incubation. Further tests are then carried out to confirm the result.

Escherichia coli (Faecal coliform) count. Traditionally coliforms of faecal origin, characterized by *E. coli*, have been considered capable of growth and of expression of their fermentation properties at the higher temperature of 44°C. This has resulted in them sometimes being referred to as thermotolerant. Standards now tend to require the application of a stricter taxonomic definition that is based on the detection of *E. coli*, and recognizing that strains that are not thermotolerant have the same sanitary significance. As with coliform bacteria, a 'presumptive' result is available after 18 h of incubation. Methods incorporating expression of the β-glucuronidase gene, which is highly specific for *E. coli*, are now available and these can provide a confirmed result after 18 hours. Alternatively, specific confirmation based on production of indole from tryptophan is required. A number of confirmatory test kits are commercially available.

The detection of *E. coli* provides a reliable indicator of recent faecal contamination. The presence of other thermotolerant coliform bacteria, particularly when detected in warmer tropical or sub-tropical waters, provides a less reliable indication of such contamination. However, for many routine monitoring purposes an acceptable correlation between *E. coli* and faecal coliforms, as defined by thermotolerance, is often assumed.

Enterococci and Clostridium perfringens. Tests for the presence of Enterococci and *Clostridium perfringens* are required under the EC Directive to verify the effectiveness of treatment of surface derived sources and also the quality of the water at the consumer's tap. These bacteria are commonly used as secondary indicators of faecal pollution and can be used to extend the scope of the testing, especially if there is a need to confirm whether there is a problem.

The test for Enterococci can be used to assess the significance of coliform organisms in the absence of confirmed *E. coli*, as they are more persistent indicators of faecal contamination than *E. coli*. The ratio between the numbers of *E. coli* and Enterococci present in a sample may in some circumstances provide some indication as to whether the source of the contamination is human or animal. However, careful interpretation of any such ratios is necessary and even then the outcome may be unreliable.

The presence of spore-forming, sulphite-reducing anaerobes, such as *Clostridium perfringens*, is also associated with faecal contamination. The presence of such bacteria, especially in well or borehole supplies, can indicate remote or intermittent contamination. During treatment, the presence of *Clostridium perfringens* in the filtered water and/or the final water may indicate deficiencies in the filtration process (e.g. filter breakthrough) or in the disinfection process. At some works this may correlate with a potential for the breakthrough of protozoan cysts such as *Cryptosporidium*. However, this has been the subject of debate and the correlation is not universally true.

Investigation of a Microbiological Failure

Any microbiological failure, either in the water leaving a water treatment works or within the distribution system, must always be investigated immediately. Action should always be taken on the presumptive result, even if this does not subsequently confirm, with repeat samples being taken from the tap giving the original failure as well as additional samples, as appropriate to provide information on the extent of potential contamination.

If the failing sample is on water leaving a water treatment works then the integrity of the treatment processes should be checked and further samples taken both at the works and within the distribution system. If the failing sample is on water within the distribution system or from a consumer's tap, samples should be taken from at least two other taps in adjacent properties and on the same supply to rule out potential wider contamination, as well as from any water assets upstream (service reservoirs, water treatment works). All samples should be tested additionally for Enterococci and *Clostridium perfringens*.

It is also advisable to take additional samples following any interruptions to supplies or any repair work that might have compromised the integrity of the distribution system. In the event of a suspected waterborne outbreak of illness, sampling for a variety of other microbiological parameters may be appropriate.

7.70 METHODOLOGY FOR MICROBIOLOGICAL EXAMINATION

The methods for the routine microbiological examination of water are well documented in a series of booklets published in the UK by the EA, under the auspices of the Standing Committee of Analysts (EA, 2002 et seq.). Equivalent test procedures are given in the *American Standard Methods* (AWWA, 2012) and also in international standards published by ISO (ISO, 1985 et seq.). Established methods should always be used for routine analyses. Good laboratory practice should be adopted at all times, with special precautions being taken to avoid accidental contamination of samples once they are in the laboratory. Appropriate quality control procedures should be used at all stages of the analysis and should include the use of positive and negative controls. Laboratories should also participate in external quality control schemes, such as inter-laboratory tests, where these are available.

The choice of method for coliform analysis depends to a certain extent on the number of organisms likely to be present. Methods based on membrane filtration and colony counting and Most Probable Number, such as the defined substrate 'Colilert' method, are widely used for the analysis of drinking water. Membrane filtration is a good approach for treated waters but it may not be suitable for highly turbid waters, or for waters containing only a small number of indicator organisms in the presence of large numbers of other bacteria capable of growing on the media used, thereby masking the result. Multiple tube or most probable number methods can also provide good results for treated waters and, in some instances, may be more suitable for highly contaminated samples or samples containing appreciable quantities of sediment or particulate matter.

Presence–absence tests for coliform bacteria. This is a modification of the traditional multiple tube procedure, based on a single 100 ml volume of appropriate medium, such as minerals-modified glutamate, instead of a series of tubes of different volumes. It can be used to determine whether coliform bacteria are present in a sample but the drawback is that a positive sample gives no indication of the relative number of organisms present in the sample. There are now a variety of proprietary formulations, such as 'Colilert', which can be used in this format. These proprietary formulations have additional advantages in being easier to set up than the traditional method, with confirmed results, including the presence of *E. coli*, available after incubation at 37°C for 18 hours, rather than 18–24 hours.

Field test kits. There are a number of field test kits available. Most are based on membrane filtration or presence–absence testing and consist of a small portable incubator, which can if necessary be plugged into a car battery, and a means of aseptically preparing and processing the sample.

The most simplistic form consists of a 'dip slide' with a small area of sterile agar medium, which is selective for coliforms, at the end of a stick attached to a screw cap for a dedicated tube or container. The 'slide' is dipped in the water sample and incubated in its tube. The result is essentially a presence–absence test, although a very crude semi-quantitative estimate of numbers may be possible. However, the test does not relate to a specific volume of sample, only to the magnitude of contamination. Whilst it would not be applicable to treated drinking water, a 'dip slide' test can assist in initial assessments of raw water. The more sophisticated test kits provide the equivalent of a portable laboratory. Some test kits offer rapid initial results, especially where significant numbers of *E. coli* may be present, after only 8–12 hours incubation. However, the full 18 hours incubation is essential to provide reliable results.

Considerable care has to be taken with all field test kits to ensure that samples do not become contaminated during analysis and that adequate sample dilutions have been prepared to cover the expected concentration ranges. Colony identification and counting must be carried out by an experienced person, familiar with techniques of identifying coliform bacteria, since non-coliform organisms may also be visible on the membrane. This is particularly so in tropical or sub-tropical climates. A basic level of analytical quality control should also be adopted with these test kits, including positive and negative control samples and a record of the incubator temperature.

7.71 PROTOZOAL EXAMINATION

Sampling methods for *Cryptosporidium* and *Giardia* include continuous capture by filtration, large volume grab samples or composite samples. For general surveillance purposes, a large volume of water sampled through a filter over a long period of time, usually a minimum of 1000 litres over 24 hours, is likely to give the best result. Although this works well for treated waters, sampling by filtration may not be possible for raw waters that are very turbid or contain high numbers of algae, as both are likely to block the filter. Dedicated sampling equipment is required for continuous filtration, whilst grab samples, usually 10–20 litres, can provide a more accessible means of obtaining operational samples. Grab samples will always give a less representative result and liaison with the analysing laboratory is required to avoid overloading analytical capacity.

Standard methods of analysis are available for both protozoa. Although greatly improved in recent years, the biggest problem associated with methods for *Cryptosporidium* is still the maintenance of analytical consistency through a multi-stage procedure to ensure acceptable percentage recovery efficiencies. Furthermore, the methods detect both live and dead oocysts, resulting in problems in the application of treatment technologies that do not physically remove the oocysts, for example ozone or ultraviolet irradiation (Sections 11.21 and 11.23).

Although recent research has developed reliable methods to determine viability and genotyping of oocysts, these are still regarded as either too expensive or require too high a level of technical competence to be useful for routine examination. Laboratories should also partake in external quality control schemes, such as inter-laboratory tests, where these are available to provide an indication of recovery rates for the methods used.

7.72 VIROLOGICAL EXAMINATION

Routine testing of water for viruses is not generally practised due to the complexity of the methods and the requirement for specialist laboratory facilities with highly trained analysts. Bacteriophages, which are viruses that can infect bacterial cells, have been proposed as possible indicators for enteric viruses in drinking water since it has been found that some bacteriophages are inactivated at similar rates to enteroviruses during treatment. Furthermore, the isolation of bacteriophages is relatively straightforward and rapid.

However, as molecular techniques for the detection of enteric viruses develop, there is increasing evidence of a divergence in inactivation or removal rates during water treatment between different virus types, and associated evidence that bacteriophages are likely to be unsuitable indicators for enteric viruses under these conditions (Defra, 2013). Nevertheless, it is generally considered that a well operated conventional water treatment process, with a disinfection stage, will be effective in addressing any risk posed by viruses in drinking water.

7.73 OTHER PROBLEM ORGANISMS

Iron bacteria. There are several groups of iron bacteria, all of which are capable of abstracting and oxidizing any ferrous and manganous ions present in a water. The process is continuous with a large accumulation of rust coloured or black deposits developing over time. These deposits tend to accumulate in storage tanks and on the walls of pipes in low flow areas of the distribution system. Pipes may become blocked or the flow seriously impaired and any disturbance of the deposits results in badly discoloured water.

The growth of iron bacteria also results in an increase in the organic content of the water that could in turn encourage the growth of other nuisance organisms. In combination with sulphur bacteria, they also contribute to the corrosion of iron and steel pipelines (Section 10.39).

The development of the organisms can only occur where there is sufficient iron or manganese present in a water at the ideal oxidation–reduction potential. Thus, iron bacteria are widely found in the bottom sediments of raw water reservoirs where the depletion of oxygen provides adequate ferrous ions. They can also be associated with ferruginous groundwaters containing high levels of free carbon dioxide and low levels of oxygen.

Sulphur bacteria. There are two groups of sulphur bacteria with implications for the water industry. Sulphate-reducing bacteria grow in anaerobic conditions and reduce any sulphate present in the water to hydrogen sulphide. They contribute to galvanic corrosion of water mains and can cause taste and odour problems (Section 10.39). Sulphur oxidizing bacteria grow in aerobic conditions and produce sulphuric acid from any sulphides present, for example in sewers.

Nematodes. Nematodes are small worm-like organisms and a number of non-pathogenic free-living varieties are occasionally found in drinking water. Their presence does not necessarily indicate a health risk; rather they are regarded as an aesthetic problem in terms of consumer acceptability. They are more likely to be found in supplies derived from nutrient rich surface waters and in the 'dead zones' of distribution systems. If large numbers are detected then a strategic and comprehensive flushing programme, utilizing flushing nets to examine the effectiveness of the remedial action, should be undertaken in the first instance.

Actinomycetes are a diverse group of bacteria which, together with some genera of microfungi, can give rise to earthy or musty tastes and odours, particularly in water derived from nutrient rich lowland sources. Actinomycetes occur particularly at the margins of surface water bodies wherever decomposing organic material is present. Two compounds commonly formed as metabolites by actinomycetes are geosmin and 2-methylisoborneol (MIB), both of which have very low threshold taste/odour values. Both compounds can give rise to major consumer acceptability issues when present in drinking water.

PART IV WATER BIOLOGY

7.74 INTRODUCTION

Rivers and lakes support a wide range of freshwater organisms (plants and animals), collectively referred to as biota. A similar range of organisms, the marine biota, is adapted to saline waters in estuaries and seawater. It is important to recognize that while some organisms only affect temperate environments, others, both freshwater and marine, occur worldwide. Climate changes are increasingly bringing tropical organisms into areas once specific to temperate biota. Until recently marine biota have seldom affected the water supply industry but the need to extend water resources to brackish and seawater has meant that the effect of a range of marine organisms on engineering structures and treatment technologies may now need to be taken into account.

A large number of freshwater organisms have been reported as causing problems in water supply systems. Consumers regard the presence of living organisms in potable water as aesthetically unacceptable, as they indicate an impure product. Apart from their mere presence, such biota can cause deteriorating water quality by introducing turbidity and tastes and odours as a result of their metabolism, as well as greatly increasing the chlorine demand.

Only one animal, whose adult life is free living, has been positively identified as a health hazard in drinking water. This is a water flea, which is the host for a human parasite, the guinea worm, which can infest man if ingested. There are other animals which are parasitic in man and which produce life stages that are free living and could be transmitted (e.g. Schistosomiasis) if not prevented, killed or removed by treatment. Finally there are animals which if not removed would proliferate in the distribution systems in biofilms, where their activities could facilitate the survival of disease organisms such as bacteria or viruses.

7.75 SOURCE WATERS AND STORAGE RESERVOIRS

There are two main groups of plants in rivers and standing waters, algae that are largely microscopic and the larger, readily visible, macrophytes – commonly named water weeds. Algae are simple green plants, most of which are free floating (planktonic) forms ranging in size from single-celled species of 2–5 μm diameter to larger colonial forms up to several millimetres. There are, however, some larger species of algae which grow as attached fronds in rivers or as upright forms on lake beds. The number of algae and the species found depend on environmental conditions such as temperature, the concentrations of dissolved salts and particularly the available nutrients such as nitrogen and phosphorus which the plants using energy from sunlight, build into their biomass.

Given the right conditions, plants will grow until the available nutrients are exhausted, and if not controlled may cause a variety of nuisances. For example, water weeds or larger attached algae can cause major engineering problems, such as mechanical breakages to control structures and even river channel blockages. Surface 'blooms' (sudden outbursts of rapid growth) of floating algae can affect slow moving and standing waters, causing fish kills through de-oxygenating the water and the sliming of fish gills as the algae die and decay. In recent years, excessive growth has become more frequent (Rodgers, 2008). The term Hazardous Algal Bloom (HAB) has been adopted for these events. In 1998 a major fish kill of over 150 tonnes of trout and coarse fish occurred in a fish farm fed from the Kennet and Avon canal in the UK due to this effect.

In recent years many rivers have shown an increase in nutrients, a feature which also occurs as rivers flow downstream. Starting as clean upland streams they often become polluted lowland waters from releases of wastes and agricultural run-off. This tendency to nutrient enrichment has become known as eutrophication and is outlined in Chapter 8. Whilst additional nutrients may improve plant and animal growth up to a certain point, beyond that point the excess nutrients can have a deleterious biological effect. This is because rapid growing biota are favoured and these species tend to be tolerant of most forms of pollution. Their growth results in a loss of biodiversity. Many of the clean water biota which support animals such as salmon and trout, dragonflies and mayflies are lost in this way. By comparison, the more tolerant species cause nuisances such as midge swarms or noxious algal events like HABs, sometimes with foul odours.

The macrophytes may be present as floating plants or as emergent or submerged water weeds. These plants maintain healthy ecosystems by providing food and shelter for numerous animals. They also create water quality improvements by allowing settlement of particulates in quiescent areas, by incorporation of nutrients from the water and by oxygenation. However, if present in excess, macrophytes can prove troublesome by reducing the carrying capacity of watercourses, blocking intakes and water control structures, whilst sudden die off can lead to foul water.

Although lakes and reservoirs vary considerably in size and other conditions, the range of species of algae is small and there is a characteristic makeup of species of algae that is found in similar water bodies anywhere in the world. As the waters leave the uplands and enter lowland plains there is an increase in dissolved solids, in temperature, and, in many developed countries, in nutrients from sewage discharges. There is commonly a transition in the algae from diatoms that tend to dominate in cold upland waters to green and to blue-green algae that thrive in warm, often shallow and nutrient rich lowland waters.

Nutrient poor upland waters are known as oligotrophic (poorly fed) whilst the richer nutrient waters are eutrophic (well fed) waters. Oligotrophic waters tend to have a few species of flagellate algae such as the Chrysophyte genera *Synura*, *Uroglena* and in the spring there may be blooms of diatoms, such as *Cyclotella*, *Tabellaria* or *Asterionella*. Green algae are common in many ponds and shallow water bodies, common species being *Chlamydomonas, Chlorella, Euglena, Chlorococcus, Coelastrum, Cosmarium, Oocystis, Pediastrum, Scenedesmus and Staurastrum*. The term blue-green algae is retained here for convenience, although now becoming known as Cyanobacteria, being more closely related to bacteria than algae. Many blue-green algae are colonial and able to grow very rapidly in warm waters. A number of the commonest blue-green have flotation 'devices' such as gas vacuoles in the cells or mucus which binds the colonies into large rafts; examples are *Anabaena, Microcystis, Aphanizomenon* and *Oscillatoria*. As noted in Section 7.4, some of the blue-green algae also produce toxins present in the mucopolysaccharides that make up the mucus released by the cells. The toxins are not always present and their presence cannot currently be predicted except that high concentrations of the named blue-green species give rise to a higher risk. They also form scum as the cells die off and cause taste and odour. There have been numerous accounts of the ecology and biology of algae but the works of Palmer (1977) in the USA and Bellinger (1992) in the UK provide a good general introduction. Illustrations of algae genera of significance in water supply taken from Bellinger (1992) are shown in Figure 7.1.

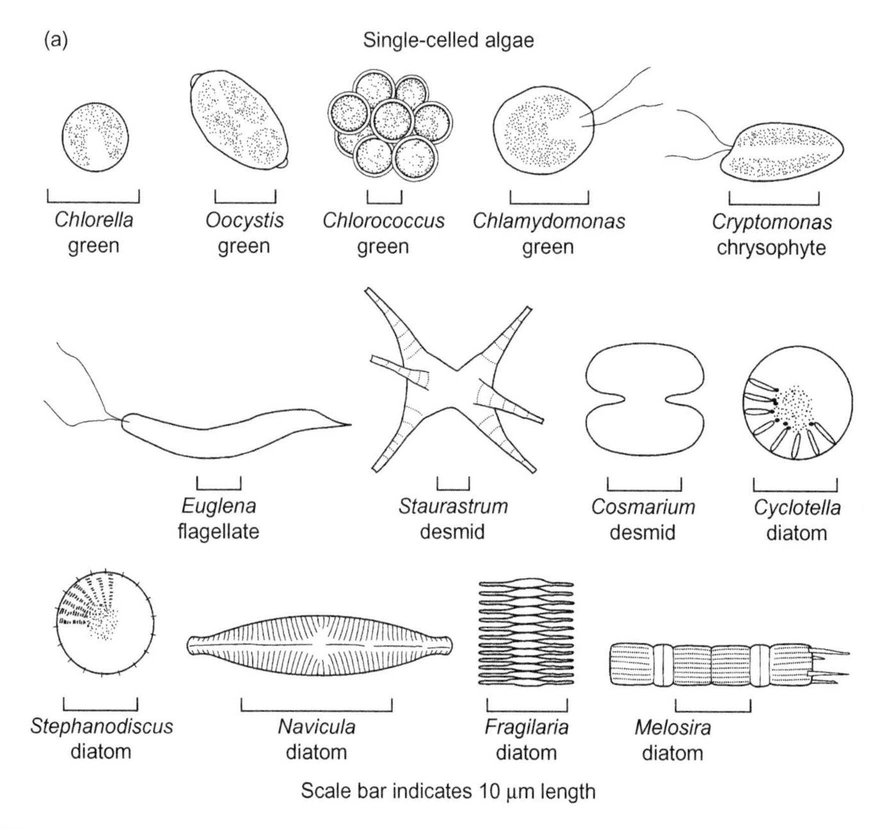

FIGURE 7.1

(a) Algae genera of significance in water supply – single-celled algae. (b) Algae genera of significance in water supply – colonial algae.

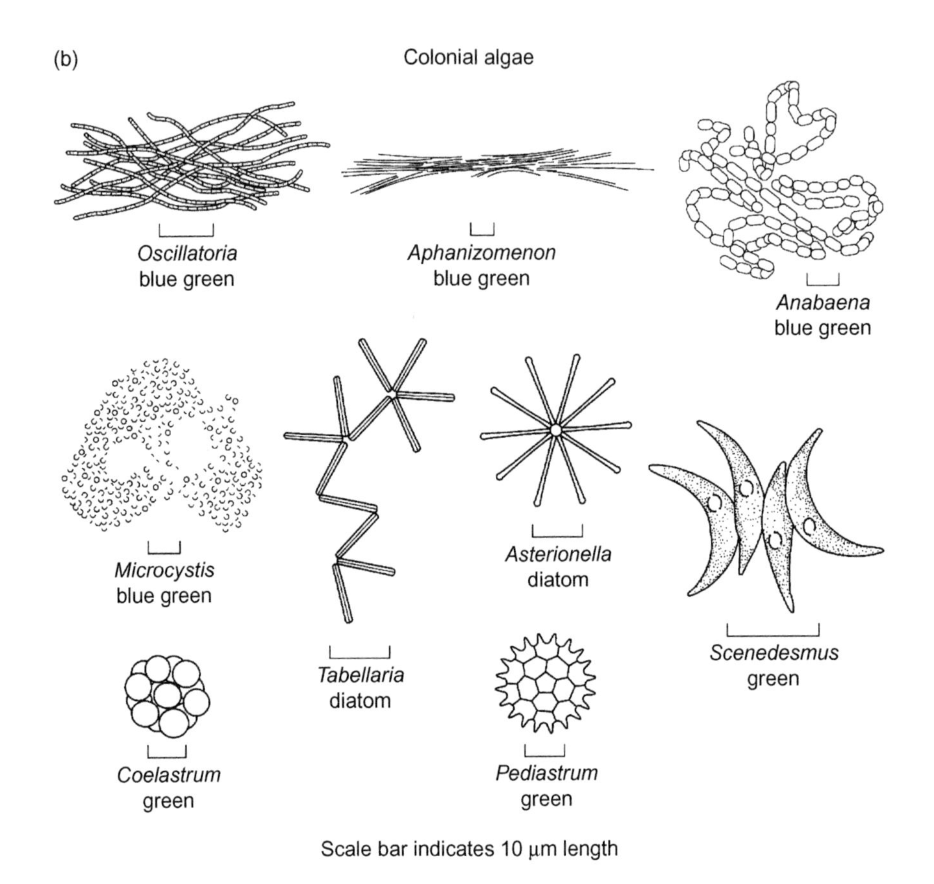

FIGURE 7.1

(Continued)

There has been a great deal of research and development of methods to reduce the adverse effects of eutrophication, varying from control of nutrient contribution from agriculture and from sewage works using a range of phosphorus and nitrogen removal technologies (Anonymous, 1997) to in-lake methods of lake management. Recent eutrophication management in Germany, where blue-green algae (Cyanobacteria) and toxins production were serious issues risking the safety of supplies, has shown that reducing total phosphorus concentrations to below 50 μg/l as phosphorus as an annual mean was successful in ensuring the safety of the supply (Chorus, 2011). It should be emphasized that this step needs to be examined carefully, using full evaluation with the development of a Water Safety Plan, before such investment is undertaken. The traditional algal control method, of direct intervention by dosing copper sulphate (IWEM, 1969) to the water at times of blooms to kill the algae, is not now acceptable to most authorities due to the toxicity of copper to other biota including humans. The reduction by precipitation of soluble phosphates by dosing aluminium or iron salts in a contained volume at the entry to the water body has been used successfully (Croll, 1992). Typically, 1–5 moles of Fe or Al to 1 mole of phosphate is required, with a pH range for precipitation from 5.5 to 7.0 depending on the metal salt used. Hayes (1984) reported that a ferric sulphate dose of 3.0–5.4 mg/l as Fe on the inflow to Foxcote reservoir reduced soluble phosphorus concentration to $<10\ \mu g/l$. The phosphate precipitated should be regularly removed by dredging. Introducing species or biological communities as a form of biomanipulation or adding barley straw, which decays in the water releasing an algal toxin, have been employed and, whilst attractive in large water supply reservoirs, could also prove useful and cost effective in small impoundments (Purcell, 2012).

The use of ultrasound for disruption of algal cells is being developed (Lee, 2002) and has been used in the UK where units of modest power have achieved worthwhile operational gains. Units available in 2015 operate at frequencies of 15 kHz to 10 MHz and have an effective operating radius of 200 m. They can be arranged to traverse the reservoir or fixed by mooring and can be operated from solar panels thereby giving greater flexibility in their deployment. It is critical to keep the transducer face clean for effective performance. The use of ultrasound in this context is reported as being harmless to fish and other aquatic life. Schneider (2015) reported successful use in a New Jersey reservoir which helped to control geosmin and methyl isoborneol and achieve a reduction in coagulant dose and improved performance of dissolved air flotation and filters. This book cannot cover all the literature on the control of algae and other biota in eutrophic waters but the reader may find reviews by Purcell (2012), Sutcliffe (1992) and Moss (1996) useful.

Animals in natural waters causing problems include Chironomid midges, causing nuisance with the risk of its larvae penetrating media filters, and sponges, barnacles, mussels and pipe mosses causing pipeline and conduit blockage (biofouling), reducing hydraulic transfer capacity. Animals such as water fleas, *Rotifers*, *Cyclops*, *Asellus*, *Gammarus*, *Daphnia* and Chironomid larvae lead to contamination and tastes and odours in the distribution system and in service reservoirs. Most organisms can readily be removed during treatment but some such as the chironomid midge larvae may be difficult to prevent from passing through to the distribution system and to the consumer.

There are marine biota that can cause problems in RO desalination plants, swimming pools and oceanaria. The most common problem has been HABs, which in this instance are known as 'Red Tides'. These are dense blooms of dinoflagellate algae which commonly tint seawater a reddish brown colour and are transported over long distances by currents and wind. These particular organisms are associated with nutrients from agricultural run-off and human waste disposal and are capable of fast growth (Dela-Cruz, 2003). In the sea, the strong movement of waters along coasts then tends to aggregate the rapidly growing cells, forming red tides in shallow bays of the sort commonly chosen by man for economic use and positioning of intakes. These blooms release toxins causing fish and other fixed marine organisms in the area to die, resulting in foul odours and unsightly appearance.

7.76 TRANSFER STAGES

Plants can seriously reduce the crosssection of open channels and also increase the frictional resistance to flow. This is usually a minor problem in the UK but can be an important issue in tropical countries where plant growth can be substantial. It is also more often a problem associated with bulk water transfers as in irrigation canals, in which velocities are usually low.

Plant growth in watercourses, whilst essential for a good aquatic habitat, can at the same time promote problems for the water manager. In the Ely Ouse scheme in the UK, the good habitat provided by plants has encouraged coarse fish. These shoal and block the intake screens for this inter-river transfer scheme. In many tropical countries snails, carrying the intermediate vector for schistosomiasis (bilharzia), thrive in clean water canals and can create a major health hazard. Molluscicides are needed to control the snail, combined with adequate education of the local population with regard to sensible personal hygiene practice.

Plants do not in themselves cause problems in pipelines, as without light they cannot grow. However, other organisms may affect raw water transfer pipelines. Biofilms of bacteria provide some food for larger organisms which foul the water by removing oxygen and releasing their metabolic products to it. These biofouling organisms are usually filter-feeding animals such as sponges, moss animalcules (these are colonial forms resembling the common hydroid *Hydra*), mussels and, where sediments have built up in major transfer systems, even cockles. These forms thrive upon bacteria, algae and other fine plant debris carried into the pipelines and in part therefore can benefit water quality but uncontrolled growths can rapidly result in increased headloss in the system, with tastes and odours imparted to the water, and blockages when dead animals fall from the pipe walls. Adequate control of the Zebra Mussel, *Dreissena polymorpha* by seasonal low level chlorination to kill aquatic larvae has been successfully used but formation of chlorination by-products now precludes this form of control. A new control method based on feeding mussels with pellets containing material high in organically bound potassium, which is toxic to them but not to other wildlife or humans, has been proposed (Aldridge, 2006).

The effects of marine biofouling of intakes by mussels and other seawater organisms have increasingly created problems for power and desalination plants. Intake screens are usually subject to regular cleaning (Section 6.6), usually by physical means, or made of material toxic to biofouling (e.g. cupro-nickel), but culverts, pipelines and tunnels have also been colonized and in many cases chemical control with chlorine has had to be adopted, usually based on continuous dosing for power plants and shock dosing for RO plants.

In addition to the filter-feeding animals, which are sedentary or fixed to the walls of pipelines and tunnels, there are mobile animals such as water louse, shrimps and fish. These animals can swim in the water and thrive in transfer systems with low water velocities such as tunnels for bulk water transfer systems. Unfortunately, these occasionally cause water quality issues due to seasonal dying off and may on occasion block screens or pump sumps.

7.77 TREATMENT STAGES

The seasonality and speciation of algae have become important issues in water supply as many of the algae have highly specific effects on the treatment process (Section 8.13). In investigating problems with algae in the treatment process, it is important to be able to measure algae and the specific removal rates accurately. This has been difficult as the techniques of counting or measuring the plants have to vary to suit the organisms being counted. The standard measurement in the past has been direct counting under a light microscope with numbers being reported as cells or organisms or colonies per unit volume, with some agencies preferring to report the plan area or even the volume of the cells per unit volume. The cost and subjectivity of counting has led to the use of chemical determinands, such as the measurement of POC or the green pigment chlorophyll a, which is measured fluorimetrically. Particle counting using modified optical Coulter counter techniques has been used more recently in special studies.

There are no limits set for animals or plants in drinking waters though there is a presumption that they should be absent. In practice, it is difficult to prevent some carryover of algae and small animal components from treatment processes though these are killed by the final chlorination before leaving the treatment works. Values for an algal content of between 100 to 1000 cells/ml, which is a range close to the lowest discernible limit for direct optical counting, have been used as a guide over the years. The lower number (100 cells/ml) has been advocated as a practical limit for the treated water prior to final chlorination (Mouchet, 1984).

The toxins that can be associated with algae both in freshwaters and in seawater and which have led to the concept of HABs are removable by modern comprehensive treatment process trains (Section 10.43).

7.78 SERVICE RESERVOIRS AND DISTRIBUTION SYSTEMS

Water in service reservoirs and distribution systems will be chlorinated and, in the majority of cases, kept in the dark as most service reservoirs are covered to avoid UV-oxidation of the chlorine residual or aerial contamination of the water entering supply zones. It is therefore often surprising how many living organisms are found in such areas. Plants are seldom found in such systems as light levels are too low, but where brick or concrete conduits have been laid too near the surface the roots of trees can penetrate and cause blockage and a pathway for other organisms to enter.

Service reservoirs should have midge and mosquito proof mesh (Section 20.15) fitted to all the essential air vents provided near the roof of the reservoir as such gaps have been shown to allow ready colonization by the flying adults which lay their eggs on cool damp surfaces.

Apart from contamination at the surface contact points, the main source of animals found in the distribution system is the treatment process. In a number of cases in the past, it has been found that the source may not simply be the bypassing of the clarification and filter stages by animals in the raw water, but rather that animal communities build up in the sedimentation tank and filters and provide a regular slippage of animals to the final water. These animals include a wide variety of freshwater forms including nematode worms, water fleas, the water louse and Chironomid midge larvae. Granular activated carbon adsorbers operating in the biological mode provide ideal conditions for animals to colonize (Section 10.35). Membrane filters are effective in stopping passage of animals. Pre-treatment oxidation by chlorine or ozone of raw water from a eutrophic water body and post-ozonation/chlorination have been shown to be effective in controlling such nuisance animals.

PART V NEW AND EMERGING ISSUES

7.79 INTRODUCTION

Under the forward work programme for the ongoing revision of the WHO *Guidelines on Drinking-water Quality*, addenda are issued when guideline values are set following the evaluation of new contaminants, or as new scientific information becomes available for existing guidelines. Similar processes of review and revision are built into the EC Directive and the US EPA National Primary Drinking Water Regulations. WHO has collaborated with US EPA and other bodies on a number of emerging issues, especially those relating to water and infectious disease (WHO, 2003). However, a number of topics have emerged recently that have given rise to consumer concern. Some of these issues are discussed below.

7.80 CHEMICAL ISSUES

Disinfection by-products. A number of DBPs are discussed in Section 7.24. Epidemiological studies continue to suggest associations between adverse health effects and exposure for several DBPs that do not currently have health-related standards. Concerns have been raised over haloacetic acids and also nitrogen-containing disinfection by-products (N-DBPs) including N-Nitrosodimethylamine. N-DBPs are believed to be more toxic than the DBPs that are currently regulated. However, there is limited data available on their toxicity and the precursors and treatment conditions that give rise to their formation.

N-Nitrosodimethylamine (NDMA). This potent carcinogenic compound is formed as a DBP when dimethylamine (DMA) reacts with monochloramine or hypochlorous acid (HOCl) in the presence of ammonia or nitrite (WHO, 2002). It may also be present in surface waters subjected to industrial or sewage effluent discharges and as a microcontaminant of certain pesticides. NDMA can also be in water that has been treated with polyDADMAC (polydiallyldimethylammonium chloride) or Epi-DMA products as coagulants or coagulant aids and when ion exchange resins containing the quaternary ammonium group come into contact with chlorinated water (Wilczak, 2003).

NDMA is included on the US EPA Contaminant Candidate List 3 and has been detected in low concentrations in some drinking waters in North America. A study carried out for Defra (2008b) looked at 41 selected treatment works in England and Wales and NDMA was detected at concentrations of up to 0.058 μg/l in the final waters at three of the works. Although well below the WHO guideline value of 0.1 μg/l, DWI issued guidance on trigger levels relating to wholesomeness (DWI, 2008). NDMA can be reduced to acceptable levels by ultraviolet irradiation.

Endocrine disrupting chemicals. Endocrine disrupting chemicals are discussed in Section 7.26. They may disrupt human or animal endocrine systems and can either be oestrogenic or androgenic. It has been claimed that they may be responsible for increases in breast and testicular cancer and for decreases in sperm counts. This is neither a new nor an emerging issue but media interest in the impact of hormone residues associated with the use of contraceptive pills can create significant consumer concern. Furthermore a number of oestrogenic hormones used in human and veterinary pharmaceuticals are included in the EPA Contaminant Chemical List 3.

A study commissioned by DETR in the UK (DETR, 1999b) investigated a number of approved water supply products as potential sources of endocrine disrupters. Transient and low levels of leaching were observed in some construction materials. Higher and more persistent levels were noted in some in-situ glues and additives but exposure is likely to be very low. These products were all used within buildings rather than the public water supply system.

Research is also continuing on other substances, such as bisphenol a, nonyl phenol and nonyl ethoxylates, which have been shown in bioassays to possess endocrine disrupting properties. Analytical methods have been developed to determine the presence of oestrone, 17α-ethynyl oestradiol and 17 β-oestradiol at the concentrations of interest and further research is ongoing.

Pharmaceuticals and veterinary medicines. Sewage effluent tends to be the main source of pharmaceuticals and related substances, including recreational drugs, in the aquatic environment. Surveys on the levels of pharmaceutical residues in treated sewage effluent and the associated receiving waters have been carried out by the EA. A study of the lower River Tyne (Roberts, 2006) identified traces of a number of pharmaceutical compounds. A desk-based review carried out for Defra (2007b) found reports of several compounds being detected in drinking water in different parts of the world, with levels of up to 0.1 μg/l, being found in some treated waters in Europe and the USA. In 2011, further research was carried out in the UK to determine the levels of pharmaceuticals in source and treated waters in England (Defra, 2011a). The study looked at 17 compounds, including cocaine, detected at very low levels (sub μg/l) in source waters at four sites. Six of these compounds were also detected at lower levels in the treated drinking water and a subsequent study evaluating the toxicological risk associated with these compounds (Defra, 2014) concluded that the levels measured were unlikely to pose an appreciable risk to public health.

The structure and nature of the individual compounds are key to the efficacy of removal during water treatment. Advanced treatment processes, such as ozone and activated carbon, give removal rates of greater than 90% in many cases, with RO being very effective at removing a wide range of pharmaceutical residues.

Similar concerns have been raised over the possible presence of veterinary medicines in source waters. Studies in the UK and New Zealand have indicated that hormones present in animal wastes from livestock farms may also be having an impact on the aquatic environment (Matthiessen, 2006; Sarmah, 2006). A desk study carried out in 2011 (Defra, 2011b) on the current knowledge of veterinary medicines in drinking water considered that such medicines were unlikely to be a potential risk to health but more research was needed.

Household and personal care products. Recently there has been media interest in the fate of domestic cleaning products, cosmetics and toiletries in the aquatic environment. Improvements in analytical capabilities mean that the chemicals associated with these products are now being detected at trace levels in some surface waters. Currently very little is known about the consequences of their presence in the environment and more research is needed to ascertain any potential risks to human health.

Metaldehyde is a molluscicide that is widely used for slug control by the domestic market and by agriculture. In the UK between 6000 and 10 000 tonnes of slug pellets are estimated to be applied annually and increasing levels of metaldehyde are now being found in surface water sources, especially after prolonged rainfall.

A number of UK water companies have detected metaldehyde in concentrations above the standard for individual pesticides in the treated water leaving their works. Advanced oxidation processes using UV in conjunction with hydrogen peroxide have been shown to be effective at reducing levels, but incur high operating costs (Holden, 2013). Pilot scale investigations using high doses of powdered activated carbon have also shown promising results (Veolia, 2013). However, a key strategy for reducing metaldehyde levels in drinking water is controlling its use in the catchment rather than having to resort to costly end of pipe treatment solutions.

Methyl tertiary butyl ether (MTBE) is the main fuel oxygenate for unleaded petrol. It is highly soluble and highly mobile in the aquatic environment; not significantly absorbed, and not significantly attenuated or biodegraded. There is a possibility of it contaminating groundwaters via leaking storage tanks and pipes.

MTBE is considered to be more of a problem in the USA than in Europe and is included on the US EPA Drinking Water Contaminant Candidate List. In the UK, water companies carry out some limited operational monitoring at sites considered to be at risk. If present in drinking water, it would be detected by taste/odour at levels much lower than any perceived health risk.

Perchlorates can be used in the manufacture of explosives but are more generally used as an additive to petroleum products to increase the octane number. They have been detected in groundwater sources in California and other parts of the US. Although a number of epidemiological studies have been carried out, there remain significant gaps in knowledge of health effects, levels in drinking water and effective removal during treatment. Perchlorate is on the US EPA Contaminant Candidate List.

Perfluorooctanesulphonate (PFOS) was a key ingredient in the manufacture of 'Scotchgard' and is also a component of firefighting foams. It is a toxic substance and, although produced and used in very limited quantities these days, it remains persistent in the environment. It excited interest in the UK following the Buncefield oil depot explosion and fire in December 2005, when there were concerns that the local aquifer had become contaminated. A survey was carried out for Defra (2008c) into the prevalence of PFOS and associated perfluorooctanoic acid (PFOA) in drinking water, which indicated that PFOS and PFOA did not appear to be a widespread contaminant in raw and drinking water in England. Concentrations of PFOS, where detected, were very low and the source involved was considered at higher risk either from local contamination or a specific incident. DWI subsequently issued guidance on trigger levels for PFOS and PFOA in relation to wholesomeness (DWI, 2009). Trace levels of PFOS and other perfluoroalkyl surfactants have been detected in a number of surface waters in the USA.

Uranium is widespread throughout the environment. In some countries it is naturally present in drinking water and this can constitute the major source of intake. Although the radiological aspects of uranium are widely recognized, little information is available on chronic health effects caused by uranium as a chemical. The British Geological Society (BGS) has carried out a survey of uranium in British groundwaters (Defra, 2006) and found concentrations of <0.02 to 48.0 μg/l. Much higher concentrations have been found in groundwaters in other parts of the world and it is of increasing concern in some Eastern European countries.

7.81 MICROBIOLOGICAL ISSUES

Aeromonas. In addition to being widely distributed in the aquatic environment there are now two major distinct groups and 19 recognized species of *Aeromonas*, some of which have been associated with health effects. The organisms are widely found in the aquatic environment and have been associated with regrowth and biofilm problems in water distribution systems. The potential for a waterborne route of infection has been proposed but is yet to be conclusively demonstrated. Aeromonads are causative agents of some wound infections and septicaemia and have been associated with cases of diarrhoeal illness. Most treatment processes effectively remove Aeromonads and free chlorine residuals of 0.2–0.5 mg/l are considered sufficient to control the organisms in distribution systems.

Arcobacter. This is a group of bacteria, related to *Campylobacter*, for which the pathogenicity for humans is uncertain. There are currently nine recognized species of *Arcobacter*, of which *Arcobacter butzleri* has been associated with diarrhoeal illness. There have been isolated incidents when water supplies may have been implicated as a potential source of infection but the epidemiology is not well understood and a food-borne route is most likely as there is a recognized association with raw meat. No data are currently available on removal efficiency during treatment but, as *Arcobacter* species are similar to *Campylobacter*, they are likely to be readily inactivated by disinfectants such as chlorine and ozone.

Enterovirulent Escherichia coli. Certain strains of *E. coli* are an important cause of diarrhoeal illness through the possession of distinct virulence factors. These are the enteropathogenic, enterotoxigenic, enteroinvasive and Vero cytotoxigenic *E. coli*. Most data currently available relates to the enterohaemorrhagic strain *E. coli* 0157, which does not ferment sorbitol. The organisms can cause a wide range of symptoms including vomiting, fever, bloody and mucoid stools and, in severe cases, acute renal failure. Most outbreaks have been associated with the consumption of contaminated food. However, these organisms may be found in water sources where specific techniques are required to detect them.

Conventional water treatment processes provide effective removal as they do for the non-pathogenic *E. coli* strains. Minor outbreaks have been linked to small private water supplies with no treatment and usually associated with farms, or to recreational water use. However, in May 2000 there was a major outbreak in Walkerton, Ontario, which involved a municipal water supply. This resulted in 2300 cases of illness and seven deaths. The cause of the outbreak was contamination of a well with cattle excreta, following heavy rain, coupled with inadequate treatment and disinfection.

Mycobacterium avium subsp. Paratuberculosis (MAP) is one of a group of related species of bacteria collectively referred to as the *M. avium* complex (MAC) and generally referred to as environmental mycobacteria. Water is an important potential source of human exposure to environmental mycobacteria, some of which have been implicated as causing disease in humans. These include chronic inflammatory disease of the intestines such as Johne's disease and Crohn's disease.

These organisms are excreted in large numbers by infected animals and are common in the environment. Thus it is likely that human exposure is common, and transmission is thought to be through the faecal–oral route. It has been demonstrated that some members of the mycobacterium group can colonize distribution and hot and cold plumbing systems and water features such as whirlpool baths and showers where they may proliferate. Exposure in these settings may include respiratory and skin contact. There are other potential sources, including contaminated food, and other risk factors, such as an individual's state of immunocompetency, are involved in the establishment of disease. Environmental mycobacteria such as MAP are difficult to recover from water and environmental samples.

Members of the MAC, other than MAP, have been isolated from drinking water samples (Defra, 2005b). These organisms should be removed by conventional water treatment processes, although they are comparatively resistant to chlorine. While there is clearly a potential role for waterborne transmission, there remains considerable uncertainty about its contribution.

Microsporidia are single cell obligate intracellular fungal parasites. They proliferate inside the cell and are released in the faeces as spores. They are common in nature and rarely cause infection in humans, although they do represent a significant risk for immunodeficient individuals. They are readily inactivated by UV irradiation but are more resistant to chlorination than many bacteria and viruses.

Cyclospora are protozoan parasites similar but larger, at 8–10 μm, to *Cryptosporidium.* They cause diarrhoeal illness and infection is understood to be through consumption of contaminated food or water. Water-associated outbreaks have been reported in some parts of the world. Treatment processes effective at removing or inactivating *Cryptosporidium* will also deal with *Cyclospora.*

Adenoviruses. There are 57 serotypes of Adenovirus, grouped into seven sub-groups. Mammalian, including human adenoviruses, belong to the *Mastadenovirus* genus and can cause upper respiratory infections, conjunctivitis, gastroenteritis and febrile illness. Their broad host range ensures wide occurrence in natural waters throughout the year. Traditionally identified by electron microscopy there are now cell culture, molecular and ELISA techniques available for detecting some sub-groups from water samples. Adenoviruses can be transmitted by water but, while they are reported to be more resistant than some viruses, they should be inactivated during conventional water treatment processes, including disinfection.

Norovirus is the most common cause of viral gastroenteritis in England and Wales. Molecular characterization has demonstrated the Norovirus to be a distinct genus of the *Calicivirus* family. There are currently five genogroups, with genogroups I and II responsible for the majority of illness in humans. Multiple strains are recognized and different strains can be circulating in the community at any one time. These viruses can be detected by electron microscopy, using polymerase chain reaction (PCR) and ELISA techniques in clinical samples. Detection in environmental samples is more difficult but is possible by quantitative PCR after appropriate concentration. They are highly infectious, spreading readily in confined communities, such as care homes, hospitals and on cruise liners, causing nausea and diarrhoea. Noroviruses are susceptible to chlorination and should be inactivated by the disinfection strategies used at water treatment works. The most common means of spread is human-to-human but it can also be food- or waterborne.

Other Enteroviruses. The term enterovirus encompasses a range of different virus groups broadly related by their ability to infect the intestinal tract, causing a wide range of symptoms often including fever, meningitis and abdominal pain. They are usually present in large numbers in the faeces of infected individuals and, therefore, in sewage. Research in recent years has improved understanding of these viruses, methods of detection and differentiation. Prominent among the enteroviruses are Coxsackieviruses, Caliciviruses and Echoviruses. Conventional water treatment processes, particularly disinfection, should be designed with the removal or inactivation of such viruses in mind.

Biofilms and associated problems. There is an increasing number of bacteria that are associated with biofilms and which are able to survive in otherwise adverse environments (e.g. the presence of chlorine) by virtue of being internalized by protozoans such as *Acanthamoeba* spp. Such bacteria provide a potential challenge in distribution and plumbing systems. Among the bacteria potentially implicated in this respect are *Helicobacter pylori*, a cause of gastritis and duodenal ulcers, and *Legionella pneumophila* (Section 7.61).

Quantitative Microbiological Risk Assessment (QMRA) is increasingly being used for the more detailed evaluation of pathogen loading in abstracted waters and the efficacy of treatment systems, with particular emphasis on disinfection, for removing or inactivating them.

Antibiotic resistance. There is recognition that global overuse of antibiotics is linked to the emergence of multiple resistant bacteria, including pathogens, and so poses a potentially serious threat to public health. Recent research has identified the presence of antibiotic resistant bacteria in urban wastewater river sediments downstream of a large wastewater treatment plant (Novo, 2013; Amos, 2014). This has resulted in an awareness that the presence of resistant bacteria in the aquatic environment may provide opportunities for transfer and dissemination of resistance genes, even though the consumption of drinking water from properly operated supply systems does not represent a significant risk.

Rainwater harvesting/grey water recycling systems within buildings. There is a growing interest in water conservation and associated sustainable solutions, including rainwater harvesting and/or grey water recycling for toilet flushing and garden watering. Properly installed, maintained and operated systems present very little risk to consumers. However, there is always the possibility of backflow of the reclaimed water into the drinking water supply and/or illegal or accidental crossconnections between the reuse system and the drinking water system. Concerns have been raised in the UK after drinking water supplies to a number of 'eco-houses' on a housing development were found to be contaminated with *E. coli*.

Climate change and extreme weather events. Climate change is already having an impact in terms of extreme weather events associated with extensive flooding or prolonged droughts associated with water scarcity (Section 3.27). Both are likely to present problems for drinking water supplies, sanitation and hygiene, especially in areas of increasing population. Related issues of making more efficient use of water through rainwater harvesting and grey water systems, and treatment for effluent reuse, continue to present design challenges to manage inherent microbiological risks.

PART VI WATER SAFETY PLANS

7.82 INTRODUCTION

The concept of drinking water safety plans goes beyond the need for source water protection and good water supply management. It builds on good practice, but also takes a comprehensive risk assessment and risk management approach to the whole water supply chain from catchment to consumer. The key objectives are to:

- minimize the contamination of source waters;
- reduce or remove any contamination by using appropriate treatment; and
- prevent contamination as the water passes through the distribution system to the point of supply.

Although the water supplier has the primary responsibility for the development and implementation of such plans, other parties such as drinking water quality regulators, authorities responsible for monitoring public health and the environment, and all users of water also have an important role to play.

The key components of a water safety plan, and the steps involved in developing one, are set out in Chapter 4 of the 4th edition of the WHO *Guidelines for Drinking-water Quality* (WHO, 2011a) and the WHO *Water Safety Plan Manual* (WHO, 2009). WHO has provided additional specific guidance and supporting documentation on water safety planning for:

- distribution systems (WHO, 2014a);
- small community water supplies (WHO, 2012, 2014b);
- chemical safety of drinking water (WHO, 2007).

7.83 STRUCTURE OF A WATER SAFETY PLAN

A water safety plan is essentially made up of three components, namely:

- *system assessment* of the whole water supply chain to determine whether water of a quality which is required to meet the health-related targets and which is aesthetically acceptable, can be delivered up to the point of consumption; this includes the assessment of design criteria for new plant, systems and facilities;
- operational monitoring to ensure that the *control measures* are effective in meeting the required health-based or acceptability criteria and that any deviations from the required system performance are detected in a timely manner; and
- *management and communication* procedures documenting the system assessment, the level of monitoring and the actions required under normal operational and incident conditions.

The system assessment is intended to identify potential hazards in every part of the supply system from source to tap. A level of risk is then assessed for each hazard, along with the measures needed to control that risk to ensure that the water as supplied is safe to drink and that the required standards are being met. The likelihood of occurrence of each hazard and severity of the consequences determines the level of risk, which can then be ranked in priority for action. The control measures in turn collectively determine the nature and frequency of the operational monitoring. Management procedures should include appropriate actions for investigating a failure to meet the required health-based or acceptability criteria or an operational control; proposed remedial action following such a failure; and appropriate levels of communication and reporting both internally and to external stakeholders. Validation processes also need to be developed to ensure that the plan is effective, based on sound science and technical information, and benchmarked against other similar systems. Plans need to be supported by training programmes, and the development of standard operational and maintenance procedures and processes for ensuring timely corrective actions if necessary.

Ideally, a water supplier should develop a water safety plan for each individual water system in its area of supply. This may not be practicable in the case of very small supply systems, and a generic model may have to be developed for supplies of a similar nature and with similar levels of risk.

Although the water supplier is ultimately responsible for preparing and implementing its water safety plans, a multi-disciplinary team of experts on drinking water supply should ideally be convened, with representatives from the authorities responsible for the protection of public health in the area of supply. The team would then draw up the plan, with each step being documented, for that supply chain. Once agreed, the plan should be implemented as soon as possible. However, it should be reviewed on a regular basis and kept up to date to ensure that all components are still relevant. It should also be reviewed if any one component changes significantly or if there has been a problem with the quality of the water supplied.

There are four elements to a complete water safety plan. The water supplier can deal with only three of these, namely the catchment (resource and source protection), treatment and piped distribution systems; these are discussed in more detail below. The fourth key component is the consumer, over whom the water supplier has little or no influence once the water passes into the consumer's pipework. The supplier is generally limited to advising the consumer of the water quality aspects and also the implications of storing and piping water within their premises.

Catchment

The factors that could impact the quality of a source water include: type of catchment, in terms of its geology and hydrology; the impact of local weather patterns; the nature of the land within the catchment and how it is used; any existing catchment controls or source protection; other water uses (e.g. irrigation) and how they could affect the raw water; and any other known or planned activities within the catchment. The source of water could be a deep borehole, a shallow well, an upland catchment or a lowland surface water catchment. Each will have different hazards that need to be identified.

Control measures could be applied to the catchment, or at the point of abstraction. Where a water supplier has little or no influence over the catchment, they may look to treatment as the best means of controlling the risk, although stakeholder engagement with landowners, land managers and land users should also be considered. If the source is a trans-boundary river that flows through more than one country, there is a need for good communication and co-operation between the various governments involved.

Treatment

The team assessing the treatment processes should have a complete understanding of the processes involved and their capability to remove the contaminants or potential contaminants that might be present in the raw water. The efficacy and reliability of each stage of process needs to be assessed, with particular focus on disinfection. The appropriateness of the treatment chemicals being used, or proposed, should be included in the assessment, along with the levels of process monitoring and control.

Consideration also needs to be given to any identified hazards that cannot be controlled in the catchment and that may not be removed or reduced to an acceptable level by the existing treatment. Suitable control measures should then be identified. These may include further optimization of the treatment processes, with new or upgraded processes being installed as necessary, or stopping or restricting abstraction during periods when the raw water quality is seriously compromised.

Piped Distribution Systems

Once the water leaves the treatment works, its quality is likely to deteriorate. The rate of deterioration will depend on the condition of the distribution system and the residence time in each component. Distribution systems are likely to include storage facilities (such as service reservoirs), pumping stations and pipe networks, with either constant or intermittent supplies. The hazards associated with the condition of the network, its performance and the relationship with the associated operation and maintenance strategies and practices need to be understood in order to identify suitable control measures.

Non-Piped, Community and Household Systems

Intermittent supplies and non-piped supplies significantly increase the complexity of the problem and risks to public health, over which the water supplier may have little control. Similarly, the supplier is unlikely to have any control over the point of supply in terms of hygiene, or the condition of the plumbing materials used. In such situations, the relevant authorities need to work with the supplier to develop surveillance and education programmes, and to provide advice, so that consumers are aware of the risks involved.

Plumbing within buildings can represent a significant potential source of deterioration in both the microbiological and the chemical quality of the water supplied. Risks arise through poor design and the type of materials used, or from later alterations to the plumbing system. There is also a risk that backflow contamination could affect public water supplies. The control of hazards at consumer premises requires the involvement of other stakeholders and the development of appropriate supporting programmes, including building regulations and water fittings regulations. However, control measures can only be as effective as the policing by the enforcement agencies.

7.84 DEVELOPMENT OF A WATER SAFETY PLAN

To date, water safety plans are not universally available for public water supplies. However, it is important to recognize that each water supplier will be producing plans from different starting points. National drinking water standards apply in many countries and water suppliers in the North America, Western Europe and some Australasian countries already have comprehensive management practices that incorporate safety plan principles or can be used as a basis for developing such plans.

In the UK the water safety plan approach has been incorporated into the current drinking water regulations for England and Wales (UK, 2010a, b). This requires water companies and suppliers to carry out an assessment of hazard and risk mitigation for each individual water system in its area of supply and submit the outcome of the assessment to the Regulator.

A similar approach has to be taken for private supplies in England and Wales. Through the Private Supply Regulations 2009, local authorities have the duty to produce risk assessments for all *'private supplies of water intended for human consumption'* by 2015. The legislation covers all sources of water and qualities of water intended for human consumption including water used for food production and, if requested by the owner or occupier, for single dwellings. The Drinking Water Inspectorate has issued technical guidance on water safety plans (DWI, 2005) for private water supplies (DWI, 2010b) and guidance on reporting the outcomes of the assessments (DWI, 2014). The methodology may not be practical for very small supply systems and a generic model may have to be developed for supplies of a similar nature and with similar levels of risk.

Future drinking water legislation in Europe is likely to incorporate the principles of water safety plans. However, in countries where there are no national standards or trade associations through which consistency can be achieved, individual water suppliers will need to draft their standards, procedures and documentation based on their individual circumstances and resources, using the guidance provided by WHO. For developing countries, plans should be appropriate to the situation and be attainable but at the same time should provide for continuous improvement. For example, it would not be cost effective to attempt to produce very high quality water with sophisticated treatment processes only to distribute it to consumers on an intermittent basis and via a distribution system in poor condition. Risks to public health would be better managed by first improving the reliability of supply and then by addressing the treatment processes.

WHO guidance on water safety plans extends beyond drinking water to include:

- *Safe Drinking-water from desalination* (WHO, 2011b);
- *Guidance for the safe use of wastewater, excreta and grey water* (WHO, 2006), covering policy and regulation, wastewater use in agriculture, wastewater and excreta use in aquaculture and excreta and grey water use in agriculture.

REFERENCES

Ahsan, H., Chen, Y. and Parvez, F., et al., (2006). Arsenic exposure from drinking water and risk of premalignant skin lesions in Bangladesh: baseline results from the health effects of arsenic longitudinal study. *American Journal of Epidemiology* **163**(12), pp. 1138–1148.

Aldridge, D. C., Elliott, P. and Moggridge, G. D. (2006). Microencapsulated BioBullets for the control of biofouling zebra mussels. *Environmental Science & Technology* **40**(3), pp. 975–979.

Amos, G. C. A., Hawkey, P. M., Gaze, W. H. and Wellington, E. M. (2014). Waste water effluent contributes to the dissemination of CTX-M-15 in the natural environment. *Journal of Antimicrobial Chemotherapy* **69**(7), pp. 1785–1791.

Anonymous (1997). Biogrowth. *Water Quality International*, March/April.

AWWA (2012). *Standard Methods for the Examination of Water and Wastewaters*. 22nd Edn. AWWA.

Bellinger, E. G. (1992). *A Key to Common Algae: Freshwater, Estuarine and Some Coastal Species*. 4th Edn. IWEM.

Bouchard, M. F., Sauve, S. and Barbeau, B., et al., (2011). Intellectual impairment of school-age children exposed to manganese from drinking water. *Environmental Health Perspectives* **119**, pp. 138–143.

Brown, M. J. and Margolis, L. (2012). Lead in drinking water and human blood levels in the United States. *Morbidity and Mortality Weekly Report* **61**(4), pp. 1–9.

Butkus, M. A., Edling, L. and Labare, M. P., et al., (2003). The efficacy of silver as a bactericidal agent: advantages, limitations and considerations for future use. *Journal of WSRT-Aqua* **52**(6), pp. 407–416.

Campbell, H. S. (1970). Corrosion, water composition and water treatment. *Journal of SWTE* **19**, pp. 11–15.

Carmena, D., Aguinagalde, X. and Zigorraga, C., et al., (2006). Presence of *Giardia* cysts and *Cryptosporidium* oocysts in drinking water supplies in Northern Spain. *Journal of Applied Microbiology* **102**, pp. 619–629.

Carmichael, W. W. (1993). Harmful algal blooms: a global phenomenon. *Proceedings of the 1st International Symposium on Detection Methods for Cyanobacterial (Blue-Green Algal) Toxins*. Royal Society of Chemistry, Bath.

CEU (1975). Directive on the quality of surface water intended for the abstraction of drinking water. 75/440/EC, *EC Official Journal L194*.

CEU (1980). Directive on the quality of surface water intended for human consumption. 80/778, *EC Official Journal L229/11*.

CEU (1991). Directive concerning protection of waters against pollution caused by nitrates from agricultural sources. 91/676/EC. *EC Official Journal L375*.

CEU (1998a). Directive on the quality of surface water intended for human consumption. Directive 98/83. *EC Official Journal L330/41*.

CEU (1998b). Directive concerning urban waste water treatment. 91/271/EC, *EC Official Journal L135*: as amended by Directive 98/15/EC, *CELEX-EUR, EC Official Journal L67*.

CEU (2000). Directive establishing a framework for community action in the field of water policy. 2000/60/EC. *EC Official Journal L327/1*.

CEU (2013a). Directive amending directives 2000/60/EC and 2008/105/EC as regards priority substances in the field of water policy 2013/39/EU. *EC Official Journal L 226/1*.

CEU (2013b). Directive laying down requirements for the protection of the health of the general public with regard to radioactive substances in water intended for human consumption 2013/51/EURATOM. *EC Official Journal L 296/12*.

Chorus, I. and Schauser, I. (2011). *Oligotrophication of Lake Tegel and Schlachtensee, Berlin Analysis of System Components, Causalities and Response Thresholds Compared to Responses of Other Waterbodies*. Report 45/2011. Federal Environment Agency (Umweltbundesamt), Germany.

COT (2013). Subgroup Report on the Lowermoor Water Pollution Incident. *Committee on Toxicity of Chemicals in Food, Consumer Products and the Environment*.

Croll, B. T. (1992). *Phosphate Reduction in Reservoirs*. IWEM Scientific Section, Huntingdon.

Defra (2002). *Asbestos Cement Drinking Water Pipes and Possible Health Risks*. Defra.

Defra (2005a). *Review of Evidence for Relationship Between Incidence of Cardiovascular Disease and Water Hardness*.

Defra (2005b). *Further Analysis of the Incidence of Mycobacterium avium Complex (MAC) in Drinking Water Supplies (Including the Detection of Helicobacter pylori in Water and Biofilm Samples)*.

Defra (2006). *Uranium Occurrence and Behaviour in British Groundwater*.

Defra (2007a). *Chlorination Disinfection By-Products and Risk of Congenital Anomalies in England and Wales*.

Defra (2007b). *Desk Based Review of Current Knowledge on Pharmaceuticals in Drinking Water and Estimation of Potential Levels*.

Defra (2008a). *A Study Into the Possible Association Between Step Changes in Water Hardness and Incidence of Cardiovascular Disease in the Community*.

Defra (2008b). *NDMA – Concentrations in Drinking Water and Factors Affecting Its Formation*.

Defra (2008c). *Survey of the Prevalence of Perflourooctane Sulphonate (PFOS), Perfluoroactanoic Acid (PFOA) and Related Compounds in Drinking Water and Their Sources*.

Defra (2011a). *Targeted Monitoring for Human Pharmaceuticals in Vulnerable Source and Final Waters*.

Defra (2011b). *Desk-Based Study of Current Knowledge on Veterinary Medicines in Drinking Water and Estimation of Potential Levels*.

Defra (2012). *A Review of Latest Endocrine Disrupting Chemicals Research Implications for Drinking Water.*

Defra (2013). *Viruses in Raw and Partially Treated Water: Targeted Monitoring Using the Latest Methods.*

Defra (2014). *Toxicological Evaluation for Pharmaceuticals in Drinking Water.*

Dela-Cruz, J., Middleton, J. H. and Suthers, I. M. (2003). Population growth and transport of the red tide dinoflagellate, *Noctiluca scintillans*, in the coastal waters off Sydney Australia, using cell diameter as a tracer. *Journal of Limnology and Oceanography* **48**(2), pp. 656–674.

DETR (1998). *Cryptosporidium in Water Supplies*. Third Report by Group of Experts.

DETR (1999a). *Nitrate in Drinking Water and Childhood-Onset Insulin-Dependent Diabetes Mellitus in Scotland and Central England.*

DETR (1999b). *Exposure to Endocrine Disrupters via Materials in Contact With Drinking Water.*

DHSS (1971). *Circ.71/150*. 17 August.

DoE (1990a). *Guidance on Safeguarding the Quality of Public Water Supplies*. HMSO.

DoE (1990b). *Cryptosporidium in Water Supplies*. Report by the Group of Experts. HMSO.

DoE (1993a). *Formation of Bromate During Drinking Water Disinfection*. Report No DWI0137. FWR, DWI contract 0137.

DoE (1993b). *Formation of Bromate During Electrolytic Generation of Chlorine*. Report No. DWI0136. FWR, DWI contract 0136.

DoE (1995). *Cryptosporidium in Water Supplies*. Second Report by the Group of Experts. HMSO.

DWI (2005). *A Brief Guide to Drinking Water Safety Plans.*

DWI (2008). *Guidance on the Water Supply (Water Quality) Regulations 2000 Specific to N-nitrosodimethylamine (NDMA) Concentrations in Drinking Water.*

DWI (2009). *Guidance on the Water Supply (Water Quality) Regulations 2000 Specific to PFOS (Perfluorooctane Sulphonate) and PFOA (Perfluorooctanoic Acid) Concentrations in Drinking Water Information Letter 10/2009.*

DWI (2010a). *Revised DWI Guidance on the Water Supply (Water Quality) Regulations Information Letter IL 9/2010.*

DWI (2010b). *Legislative Background to the Private Water Supplies Regulations 2009, Section 9 (E&W): Private Water Supplies: Technical Manual*. October 2010.

DWI (2013a). *Drinking Water 2013 – A Report by the Chief Inspector*. July 2014.

DWI (2013b). *DWI PR14 Guidance – Lead in Drinking Water*. July 2013.

DWI (2013c). *Information Letter 5/2013 on Drinking Water Analysis and Regulatory Requirements*. November 2013.

DWI (2014). *Private Water Supplies, Protection of Drinking Water Sources: Roles Responsibilities and Pollution Prevention Advice*. October 2014.

DWI (2015). *List of Approved Products for Use in Public Water Supply*. January 2015.

DWI/EA (2012). *The Contribution of the Water Supply (Water Quality) Regulations to the Implementation of the Water Framework Directive in England & Wales*. June 2012.

EA (1976 et seq.). *Methods for the Examination of Waters and Associated Materials*. A series of booklets 1976 onwards. Standing Committee of Analysts.

EA (1998). *Endocrine-Disrupting Substances in the Environment: What Should be Done?* Consultative Report.

EA (2002 et seq). *The Microbiology of Drinking Water: Parts1–14*. The Standing Committee of Analysts – Methods for the Examination of Waters and Associated Materials.

EA (2005). *The Determination of Legionella Bacteria in Waters and Other Environmental Samples: Part 1 Rationale of Surveying and Sampling*. The Standing Committee of Analysts – Methods for the Examination of Waters and Associated Materials.

Falconer, I. R., Runnegar, M. T. C. and Buckley, T., et al., (1989). Using activated carbon to remove toxicity from drinking water containing cyanobacterial blooms. *Journal AWWA* **18**, pp. 102–105.

Fawell, J. K. (1991). Developments in health-related standards for chemicals in drinking water. *Journal IWEM* **5**, pp. 562–565.

Fertmann, R., Hentschel, S., Dengler, D., Jansen, U. and Lomnel, A. (2003). Lead exposure by drinking water: an epidemiological study in Hamburg, Germany. *International Journal of Hygiene and Environmental Health* **207**, pp. 235–244.

Goh, S., Reacher, M. and Casemore, D., et al., (2005). Sporadic cryptosporidiosis decline after membrane filtration of public water supplies, England, 1996–2002. *Emerging Infectious Disease* **11**(2), pp. 251–259.

Hayes, C. R., Clark, R. G. and Stent, R. F., et al., (1984). The control of algae by chemical treatment in a eutrophic water supply reservoir. *Journal of Institute Water Engineering Science* **38**, pp. 149–162.

Holden, B. (2013). What links slugs, thistles and UV? (ultraviolet advanced oxidation processes (AOPs): international experience and perspectives). *London Conference*, December 2013.

Hopenhayn, C., Bush, H. and Bingkang, A., et al., (2006). Association between arsenic exposure from drinking water and anaemia during pregnancy. *Journal of Occupational & Environmental Medicine* **48**(6), pp. 635–643.

HSE (2013). *Legionnaire's Disease: The Control of Legionella Bacteria in Water Systems*. Approved Code of Practice and Guidance L8, 4th edition. HSE.

HSE (2014). *Legionnaire's Disease: Technical Guidance Series*. HSE Code HSG 274.

Hunt, S. M. and Fawell, J. K. (1987). The toxicology of micropollutants in drinking water; estimating the risk. *Journal IWES* **41**, pp. 276–284.

Hunter, P. R. (1991). An introduction to the biology, ecology and potential public health significance of the blue green algae. *PHLS Microbiology Digest* **8**, pp. 13–15.

Hutton, L. G. (1983). *Field Testing of Water in Developing Countries*. WRc.

ISO (1983 et seq.). *Physical, Chemical and Biochemical Methods*. ISO/TC 147/SC2.

ISO (1985 et seq). *Microbiology of Water*. ISO/TC 147/SC4.

ISO (2006). *Water Quality – Sampling – Part 1: Guidance on the Design of Sampling Programmes and Sampling Techniques*. ISO 5667-1:2006 (Part of a series of guidance on sampling (general methods) ISO/TC 147/SC6).

ISO (2007). *Water Quality – Interlaboratory Comparisons for Proficiency Testing of Analytical Chemical Laboratories*. ISO/TC 20612:2007.

IWEM (1969). *Manual of British Water Engineering Practice*. 4th Edn. IWEM.

Kawahara, M. and Kato-Negishi, M. (2011). Link between aluminium and the pathogenesis of Alzheimer's disease: the integration of the aluminium and amyloid cascade hypotheses. *International Journal of Alzheimer's Disease* **2011**, Article 276393.

Lahti, K. and Hissvirta, L. (1989). Removal of cynaobacterial toxins in water treatment processes: review of studies conducted in Finland. *Water Supply* **7**, pp. 149–154.

Langelier, W. F. (1936). Analytical control of anti-corrosion water treatment. *Journal AWWA* **28**(10), pp. 1500–1521.

Lee, T.-J., Nakano, K. and Matsumura, M. A. (2002). Novel strategy for cyanobacterial bloom control by ultrasonic irradiation. *Water Science and Technology* **46**(6–7), pp. 207–215.

Matthiessen, P., Arnold, D. and Johnson, A. C., et al., (2006). Contamination of headwater streams in the United Kingdom by oestrogenic hormones from livestock farms. *Science of the Total Environment* **367**, pp. 616–630.

MoH (1969). *Fluoridation Studies in the UK and Results Achieved After Eleven Years*. Report of Research Committee of the MoH, No.22. HMSO.

Moss, B., Madgwick, F. J. and Phillips, G. L. (1996). *A Guide to the Restoration of Nutrient-Enriched Shallow Lakes*. EA and Broads Authority.

Mouchet, P. (1984). Potable water treatment in tropical countries: recent experiences and some technical trends. *Aqua* **3**, pp. 143–164.

Novo, A., et al., (2013). Antibiotic resistance, antimicrobial residues and bacterial community composition in urban wastewater. *Water Research* **47**(5), pp. 1875–1887.

NSF/ANSI (2013). NSF/ANSI Standard 60: *Drinking Water Treatment Chemicals – Health Effects*.

Palmer, C. M. (1977). *Algae and Water Pollution*. EPA-600/9-77-036.

PHS (2013). *Cryptosporidium: Statistics 2000–2012*. Public Health England.

Purcell, D., Parsons, S. A. and Jefferson, B., et al., (2012). Experiences of algal bloom control using green solutions barley straw and ultrasound, an industry perspective. *Water and Environment Journal* **27**(2), pp. 148–156.

Roberts, P. H. and Thomas, K. V. (2006). The occurrence of selected pharmaceuticals in wastewater effluent and surface waters of the lower Tyne catchment. *Science of the Total Environment* **356**(1–3), pp. 143–153.

Rodgers, J. H. (2008). *Algal Toxins in Aquaculture*. Southern Regional Aquaculture Centre, Publication No. 4605.

Rossum, J. R. and Merrill, D. T. (1983). An evaluation of the calcium carbonate saturation indexes. *Journal AWWA* **75**(2), pp. 95–100.

Sarmah, A. K., Northcott, G. L. and Leusch, F. D. L., et al., (2006). A survey of endocrine disrupting chemicals (EDCs) in municipal sewage and animal waste effluents in the Waikato region of New Zealand. *Science of the Total Environment* **355**(1–3), pp. 135–144.

Schneider, O. D., Weinrich, L. A. and Brezinski, C. (2015). Ultrasonic treatment of algae in a New Jersey reservoir. *Journal AWWA* **107**, p. 10.

Siddiqui, M. S., Amy, G. L. and Rice, R. G. (1995). Bromate formation: a critical review. *Journal AWWA* **87**(10), pp. 58–70.

Smith, H. V. (1992). Cryptosporidium and water: a review. *Journal IWEM* **6**(5), pp. 443–451.

Smith, A., Lingas, E. and Rahman, M. (2000). Contamination of drinking-water by arsenic in Bangladesh: a public health emergency. *Bulletin of the World Health Organisation* **78**(9), pp. 1093–1103.

Sopwith, W., Osborn, K. and Chalmers, R., et al., (2005). The changing epidemiology of cryptosporidiosis in North West England. *Epidemiology and Infection* **133**(5), pp. 785–793.

Sturdee, A. P., Foster, I. D. L. and Bodley-Tickell, A. T., et al., (2007). Water quality and *cryptosporidium* distribution in an upland water supply catchment, Cumbria, UK. *Hydrological Processes* **21**(7), pp. 873–885.

Suckling, E. V. (1943). *The Examination of Water and Water Supplies*. 5th Edn. Churchill.

Sutcliffe, D. W. and Jones, J. G. (1992). *Eutrophication: Research and Application to Water Supply*. Freshwater Biological Association.

Toledano, M. B., Nieuwenhuijsen, M. J. and Best, N., et al., (2005). Relation of individual trihalomethane concentrations in public water supplies to stillbirths and birth weight in three water regions in England. *Environmental Health Perspective* **113**(2), pp. 225–232.

UK (2000). *The Water Supply (Water Quality) Regulations 2000*. Statutory Instrument 2000 No.3184.

UK (2001a). *The Water Supply (Water Quality) (Amendment) Regulations 2001*. Statutory Instrument 2001 No.2885.

UK (2001b). *The Water Supply (Water Quality) Regulations 2001*. Welsh Statutory Instrument 2001 No.3911.

UK (2001c and d). *The Water Supply (Water Quality) (Scotland) Regulations 2001*. Scottish Statutory Instrument 2001 No.207 and Amendment Regulations 2001, Scottish Statutory Instrument 2001 No.238.

UK (2002). *The Water Supply (Water Quality) Regulations (Northern Ireland) 2002*. Statutory Rule 2002 No.331.

UK (2003). *The Water Supply (Water Quality) (Amendment) Regulations (Northern Ireland) 2003*. Statutory Rule 2003 No.369.

UK (2007a). *The Water Supply (Water Quality) (Northern Ireland) Regulations 2007*. Statutory Rule 2007 No.147.

UK (2007b). *The Water Supply (Water Quality) Regulations 2000 (Amendment) Regulations 2007*. Statutory Instrument 2007 No.2734.

UK (2007c). *The Water Supply (Water Quality) Regulations 2001 (Amendment) Regulations 2007*. Welsh Statutory Instrument 2007 No.3374.

UK (2010a). *The Water Supply (Water Quality) Regulations 2000 SI No.3184 (Unofficial Consolidated Version 2010)*.

UK (2010b). *The Water Supply (Water Quality) Regulations 2010 No.994 (W.99)*.

UK (2010c). *The Water Quality (Scotland) Regulations 2010*. Scottish Statutory Instrument 2010 No. 95.

UK (2014). *The Public Water Supplies (Scotland) Regulations 2014*. Scottish Statutory Instrument 2014 No. 364.

UKWIR (2012). *Quantifying the Benefits of Water Quality Catchment Management Initiatives. Volume 1 - A Benefit Assessment Framework*. UK Water Industry Research Limited Report Ref: No. 12/WR/26/10.

US EPA (1996). *Public Law 93-523*, Safe Drinking Water Act (December 1974) and amendments under the Safe Drinking Water Act 1996. USA.

US EPA (2006). *Source Water Monitoring Guidance Manual for Public Water Systems for the Long Term 2 Enhanced Surface Water Treatment Rule*. Office of Water (4601M) EPA 815-R06-005.

US EPA (2009). *National Primary Water Regulations*. EPA816-F-09-004.

Veolia Water (2013). *ActifloCarb Pilot Plant Report – Metaldehyde Removal*. Internal Report, March 2013.

von Ehrenstein, O. S., Guha Mazumber, D. N. and Hira-Smith, M., et al., (2006). Pregnancy outcomes, infant mortality and arsenic in drinking water in West Bengal, India. *American Journal of Epidemiology* **163**(7), pp. 662–669.

WHO (1993). *Guidelines for Drinking-water Quality. Volume 1, Recommendations*. 2nd Edn. WHO.

WHO (1997). *Environmental Health Criteria 194 – Aluminium (International Programme on Chemical Safety)*. WHO.

WHO (2002). *N*-Nitrosodimetheylamine. Concise International Chemical Assessment Document 38. WHO.

WHO/US EPA (2003). *Emerging Issues in Water and Infectious Disease*. WHO.

WHO (2004). *Guidelines for Drinking-water Quality*. 3rd Edn. WHO.

WHO (2006). *Guidelines for the Safe Use of Wastewater, Excreta and Greywater*, **Vol. 1–4**. WHO.

WHO (2007). *Chemical Safety of Drinking-Water: Assessing Priorities for Risk Management*. WHO.

WHO (2009). *Water Safety Plan Manual (WSP Manual), Step-by-Step Risk Management for Drinking-Water Suppliers*. WHO.

WHO (2010). *Aluminium in Drinking-Water Background Document for Development of WHO Guidelines for Drinking-water Quality*. WHO.

WHO (2011a). *Guidelines for Drinking-water Quality*. 4th Edn. WHO.

WHO (2011b). *Safe Drinking-Water From Desalination*. WHO/HSE/WSH/11.03. WHO.

WHO (2012). *Water Safety Planning for Small Community Water Supplies: Step-by-Step Risk Management Guidance for Drinking-Water Supplies in Small Communities*. WHO.

WHO (2014a). *Water Safety in Distribution Systems*. WHO.

WHO (2014b). *Water Safety Plan: A Field Guide to Improving Drinking Water Safety in Small Communities*. WHO.

WHO (2015). *Management of Cyanobacteria in Drinking-Water Supplies: Information for Regulators and Water Suppliers*. WHO/FWC/WSH/15.03.

Wilczak, A. J., Assadi-Rad, A. and Lai, H. H., et al., (2003). Formation of NDMA in chloraminated water coagulated with DADMAC cationic polymer. *Journal AWWA* **95**(9), pp. 94–106.

WMO (2013). *Planning of Water-Quality Monitoring Systems*. Technical Report Series No.3 WMO No. 113.

WRc (1984). *Asbestos in Drinking Water*. Technical Report TR 202. WRc.

WRc (2007). *Further Study on the Potential for the Release of Nickel From Kettle Elements*. Report Ref: UC 7341/1, prepared for Scottish Executive.

CHAPTER

Storage, Clarification and Chemical Treatment

8

8.1 RAW WATER STORAGE

This may be regarded as a first stage in treatment as it may involve a complex combination of physical, chemical and biological changes. Raw water storage has been regarded as a 'first line of defence' against the transmission of waterborne diseases; this aspect is still of major importance if the unstored water is liable to excessive bacterial pollution from sewage, even though such pollution may only occur occasionally, for example if storm-water sewage overflows discharge into a river. A few days' storage improves the physical and microbiological characteristics of a surface water through the effect of a combination of actions including sedimentation, natural coagulation and chemical interactions, the bactericidal action of ultraviolet radiation near the water surface and numerous biotic pathways which help to reduce enteric microorganisms (Sykes, 1971). Storage in a reservoir for a period of from 1 to several months produces a substantial decrease in the numbers of bacteria of intestinal origin; the specific organisms of typhoid and cholera also disappear. The die-off rate for enteric coliforms, designated here as the time to achieve a 90% loss of bacteria or T_{90}, in lakes and other open waters varies from 2–3 hours in strong sunlight in clear waters, to 10 hours in more turbid waters (Kay, 1993). An additional benefit of short-term storage is that it allows a river intake to be shut down to avoid or investigate any pollution which might, for example, be indicated by the death of fish or by other information (Young, 1972) such as changes to physical chemical characteristics of the water.

The UK DoE recommended in circular No. 22/72 (DoE, 1971) that water supplies at risk from accidental spillages of industrial chemicals on roads and manufacturing sites should be protected by at least 7 days' storage. This was to allow closure of the intake until the pollution risk was over, to dilute any polluted water entering the intake with clean stored water and to allow further self-purification to take place. By 1979 rather more than two thirds of river derived supplies in England and Wales had received some storage in a reservoir prior to treatment (DoE, 1979). Such buffer storage is still desirable but alternative strategies are now considered, particularly when storage is neither economic nor practical. Such strategies (Chapter 7, Part VI) might include catchment management, regulations to avoid or reduce the risk of industrial chemicals being spilt in locations where they might enter an aquifer or surface source, or protection of abstraction sites by continuous water quality monitoring systems coupled with the provision of an alternative source of supply (DoE, 1993).

Twort's Water Supply. DOI: http://dx.doi.org/10.1016/B978-0-08-100025-0.00008-9

The Water Framework Directive (Section 2.4) will have an impact on the need for management of raw water storage.

It may be beneficial to have an emergency bypass so that water may be taken directly from a river instead of from a reservoir. This could be used in the case of exceptionally high algal growth in the water or of pollution having occurred, or being suspected in water in the reservoir. However, this action can have operational and hence economic consequences in that waters exhibiting eutrophication are very likely to be at risk of carrying oocysts of *Cryptosporidium* and appropriate processes need to be included in the treatment plant.

Instruments for on-line monitoring of source water quality parameters such as temperature, pH, conductivity, ammonia or dissolved oxygen are used by many water undertakings on their surface water supplies (Section 7.57). Incidents caused by accidental spillage of compounds such as phenols, which cause taste and odour problems at very low concentrations and were able to pass through conventional process works and enter distribution before being detected, have led to the development of increasingly sophisticated raw water monitoring. Often there is a need for monitoring but no one compound could be generally said to be the main risk, as for example at an intake downstream of a road bridge at which accidental spillage might occur. In such instances fish monitors based on behavioural and metabolic activity monitoring of sensitive fish can be used (Moldaenke, 2009). There are several instruments capable of measuring very low concentrations of soluble organic substances such as UV absorption (at 254 nm) for dissolved organic carbon and fluorescence for hydrocarbons by fluorometry (at 360 nm) (Westaby, 2010). Chlorophyll a can also be monitored by fluorometry (at 685 nm). Most instruments are provided with facilities for alarms and trending; these give notice of sudden peaks and allow identification of the compounds within a few minutes. This is usually sufficient time to close down an intake before a water supply is threatened by irretrievable contamination.

Potential Problems in Raw Water Storage

There are potential disadvantages in the prolonged storage of raw waters, which should be taken into account when considering adoption of storage and its management. The most obvious of these is the likelihood of growth of various forms of plants, either rooted aquatic types (macrophytes) which may choke shallow waters, or free floating water weeds such as duckweed and, in the tropics, water hyacinth or *Salvinia* as well as planktonic types such as algae (phytoplankton), which may increase the difficulties of treatment. The main issues for water supply raised by the presence of these plants and also animals are discussed in Part IV of Chapter 7. Storage reservoirs that are less than about 10 m deep can allow light to reach the bottom; this may encourage the growth of rooted plants unless the stored waters are sufficiently turbid to reduce light penetration. Shallow reservoirs are therefore generally avoided if there is any likelihood that plant growth could be high.

Waters which contain sufficient nutrient materials to support prolific growths of aquatic plants are usually described as eutrophic. Lakes and water bodies exhibit a range of concentrations, from low nutrient conditions in upland lakes on igneous rocks (oligotrophic), through moderate or mesotrophic lakes where a balanced ecology with some shoreline macrophytes and a wide range of planktonic algae occur, to lowland water bodies which tend to be eutrophic and even hypertrophic waters, where prolific growths of plants commonly occur due to enrichment by sewage and where in temperate climates there are usually seasonal peaks. In most reservoirs that are more than about

10 m deep, and many are designed in this way to avoid excessive plant growth, an additional complication is that thermal layering or stratification may occur on a seasonal basis in temperate climates. As temperature rises in the spring warmer water tends to remain at the surface due to its lower density. In the absence of any strong wind induced circulation the colder and now denser water below remains and ceases to mix with the surface water. The upper and lower layers are known respectively as the epilimnion and hypolimnion; in between there is a zone known as the thermocline in which there is a relatively steep change in temperature with depth. In reservoirs averaging less than 10 m deep thermal stratification in temperate lands is often only temporary, lasting for a few days at most. In colder climates there may be winter stratification due to ice formation which ensures the water body and its biological life is protected from lethal frosts (Moss, 2010).

Thermal stratification can clearly affect retention time of incoming water and is often of major importance with reference to water quality. In many large reservoirs, there are facilities for withdrawing the water for treatment at several different levels which can be chosen as circumstances dictate. Multiple draw-off facilities are discussed in Section 5.21. In the case of eutrophic reservoirs, the ability to avoid drawing from surface water with high concentrations of algae is particularly useful. The draw-off tower can be sited some way from the shore in order to avoid the build up of surface aggregations of algae that onshore winds can cause.

In some large upland reservoirs, where nutrients and temperatures constrain plant growth, the water in the hypolimnion can have a high standard of purity, remaining cold and well oxygenated. This is typical in Scandinavian countries and is reported in Lake Constance (Bodensee) (Lang, 2010). In eutrophic reservoirs, however, organic impurities, either in the incoming water or released by leaching from bottom muds and inundated soils, may accumulate in the bottom water. As a result of plant and animal respiration, and of bacterial activity, the concentration of dissolved oxygen falls and may approach zero. Under such conditions major chemical changes take place in the transition from anoxic conditions, when only combined oxygen compounds such as sulphate (SO_4^{2-}) and nitrate (NO_3^-) are present, to full anaerobic conditions when even these compounds have been reduced to sulphide and nitrogen. This transition can be measured by redox potential which falls from +200 mV when free oxygen is present to −200 mV in anaerobiosis.

Redox potential or oxidation–reduction potential (ORP) is the potential developed in a cell between a metal electrode (e.g. platinum) and a reference electrode during an oxidation–reduction reaction; it is reported with respect to the potential of the hydrogen electrode which is zero. The typical range measured is −1000 mV to +1000 mV. Oxidizing agents (e.g. Cl_2, O_2) increase the ORP, whereas reducing agents (e.g. sulphites) lower the ORP. Thus, redox potential is used to determine the oxidizing or reducing characteristics of a solution.

At high ORP ions such as sulphate, nitrate or ferric predominate whilst the presence of sulphites, ammonia or ferrous ions in significant concentration are indicative of low ORP. In aquatic redox reactions, ORP is related to bacterial activity (e.g. oxidation of iron in biological filters and reduction of ferric hydroxides or oxides in lakes). The important consequence for water quality is that under negative redox conditions numerous other chemical compounds can re-dissolve from the sediments into the overlying water. Iron, which is present as insoluble oxide or hydroxide, is usually the most prominent of these substances but if manganese is present (usually as insoluble oxides), it may also become soluble. Both iron and manganese often exist in combination with organic colouring matter. The concentrations of plant nutrients, phosphates and ammonia commonly also increase in bottom water close to anaerobic sediment. Actinomycete fungi often proliferate in these conditions and lead to tastes and odours in the raw water (Asquith, 2013). Under such circumstances, abstraction from the hypolimnion for water treatment should be avoided. Some authorities prefer to abstract this water for compensation flows to 'bleed' the higher concentrations of dissolved contaminants from the stored water.

When the surface water cools down in the autumn and wind action becomes effective, the reservoir water mixes and, under normal circumstances, water from mid depth or deeper is of good quality. Rapid mixing caused by strong autumnal winds can cause sudden 'turnover' resulting in rapid deterioration in the quality of the water at the surface in respect of colour, iron and manganese. A more gradual effect of the resumed mixing is that the plant nutrients carried up from the bottom may increase the growth of algae in the water and cause an autumnal bloom of algae. This is not uncommon in eutrophic lakes and reservoirs and may render the surface water difficult to treat.

Reservoir management to control stratification has been established practice since the 1960s. Mechanical pumps have been used with varying degrees of success both to oxygenate the bottom layers of a reservoir and to control stratification. Pioneering studies in the UK (Pastorok, 1980) showed that algal populations as well as water quality could be managed by water movement control. This approach led to arrangements for carefully controlled raising of cool, low oxygen water from below the thermocline using jetted inlets in those reservoirs which had pumped inflows. Jets entrain the bottom waters (also low in algae) in the flows induced to the surface and increase the amount of satisfactory water above the thermocline. The system is suitable for large flat-bottomed reservoirs where the bottom water volume is large relative to the surface epilimnion, but is not effective in narrow natural valley impoundments where the hypolimnion is small and where more vertically oriented mixing is needed. Airlift pumping systems such as the 'Bubble Gun' (Henderson-Sellers, 1984) and 'Helixor' (Ridley, 1966), which made use of rising air bubbles in tubes to entrain bottom water so that it rose to the surface whilst being aerated by the bubbles were shown to be more effective than mechanical pumps. Later in the 1970s simpler and more economical methods were developed such as the use of compressed air plumes from perforated airlines laid along the bottom of the reservoir (Davis, 1980; Mitrakas, 2013). Other aeration systems include the 'Speece' cone, where air is introduced into the top of a cone, air bubbles rise counter-current to the downward water flow and aerated water is discharged at the bottom of the cone into a pipe diffuser system (Dominick, 2009), SolarBee® where a solar-powered floating mechanical axial flow mixer and an intake hose draw water up from below and distribute it across the surface and 'ResMix' which uses a slow-rotating axial flow impeller moving water from the surface of the reservoir to the bottom (Elliot, 2013).

Growth of plants can deplete nutrients from the water particularly if the growth is not limited by light or grazing. Management of stored water to make use of this characteristic has been carried out by Thames Water Utilities (TWU) for many years (Steel, 1975). Water from the River Thames in autumn and winter frequently has nitrate nitrogen concentrations which exceed the EC Directive value (Section 7.35) of 50 mg/l as NO_3, due to pollution from agricultural run-off. Toms (1981) showed that an empirical relationship could be used to predict the reduction of nitrate by storage and that long-term stored water with low nitrate could be used to reduce nitrate concentrations entering supply from other sources by blending. TWU use storage in one of their larger reservoirs for prolonged periods to reduce nitrates by the combination of algal growth and bacterial denitrification.

8.2 GENERAL CONSIDERATIONS FOR WATER TREATMENT PLANTS

The degree and type of treatment required for surface water derived from rivers, lakes and reservoirs, groundwater and brackish and sea water depends on the raw water quality and the treatment objectives. Suitable processes include chemical treatment (e.g. coagulation, pH adjustment, disinfection), flocculation, chemical or biological oxidation, solid–liquid separation, ion exchange, adsorption and desalination as necessary. Many of these processes generate a waste stream that itself requires treatment before it can be discharged to the environment. The following sections and Chapters 9–12 provide details of these processes and associated chemical facilities.

Chapter 14 considers the subject of hydraulics, which is key to the mechanical performance of water treatment and transmission works. Once water enters a treatment works, flow should be by gravity wherever possible: it is inadvisable to re-pump water between clarifiers and media filters as this would break up floc. Hence a site having a gentle gradient of 1 in 10 to 1 in 15 is favoured. The typical headloss across a treatment plant comprising clarifiers and filters (inlet chamber to the treated water reservoir), assuming mixing is by hydraulic means, is generally in the range 5.5–6.5 m, of which filters (underdrain, clean media and an allowance for clogging) account for 2–2.5 m. Inter-stage pumping may be unavoidable if there are two filtration stages in series or membrane filtration is included in the process; additional head of about 2.5–3 m and 7–30 m, respectively should then be allowed.

When siting works adjacent to a river it is important to avoid locating any structure below the highest flood level because of the difficulty and cost of countering uplift problems. A works should be provided with a safe means for disposing of overflows caused by fault or mal-operation. Overflow arrangements are typically provided at the works inlet, filters and disinfection contact tank and/or treated water reservoir. All water retaining structures should be provided with means for dewatering.

8.3 MICROSTRAINERS

These are revolving drums mounted in open tanks with a straining medium which is usually a stainless steel wire fabric of a very fine mesh, fitted to the periphery of the drum. The drum is submerged for about 75% of its diameter (66% of area) and rotates at about 0.5–5 rpm (peripheral drum speeds of 3–50 m/min). Water to be treated enters the drum axially under gravity and flows out radially through the fabric, depositing particulate matter. Cleaning is accomplished by a row of water jets along the full length of the drum operating at about 2.5 bar pressure. Particulate matter intercepted by the fabric rotates to the top of the drum where it is backwashed into a hopper running the full length of the drum and conveyed by a pipe which also acts as the axle for the drum assembly, to a point outside. Water jets use about 1–1.5% of the total quantity of water strained but this washwater should be filtered and chlorinated.

Total headloss through a microstrainer unit including inlets and outlets varies from about 150 to 200 mm. Single units have capacities of 10 m^3/h to a maximum of 4000 m^3/h for a 3.2 m diameter × 5 m wide drum.

Microstrainers improve the physical quality of a water by removing particles down to about 20 μm, but there is no change in the chemical characteristics of the water. The ideal water for microstraining is a lake or large reservoir supply which does not contain a large amount of suspended matter but which contains moderate quantities of zooplankton, algae and other microscopic-sized particles. Mouchet (1998) has reported total algae removal in the range 40–70% with removal up to 100% for some species, confirmed by Sethunge (2009). For the removal of zooplankton, microstrainers are located either at the beginning or end of the treatment process. Fabrics commonly used with stored waters are made of woven stainless steel wires of 0.05 mm diameter with apertures of 23 and 35 μm. However, a coarser mesh at 200 μm aperture is sometimes used after granular activated carbon (GAC) filters to remove eroding particles of the carbon and any bacterial flora or zooplanktons that sometimes develop in GAC filters. Attempts to use plastic mesh materials have not been successful in water supply.

Microstrainer operation is fully automatic. A microstrainer screen can easily be damaged if too great a loading is placed upon it; hence, head across the screen should be monitored and an alarm initiated if it approaches the maximum desirable value. The rotational speed of the drum assembly can be adjusted so that the optimum differential headloss across the fabric can be maintained to achieve maximum removal efficiency irrespective of the raw water flow or quality. Most installations have an automatic fail-safe bypass weir which diverts unstrained water when the screens become overloaded. This causes deterioration of the treated water at times of peak loading and therefore is not generally acceptable as an alternative to media filtration for potable waters. Their most extensive and successful role in water supply has been to lighten the loading on rapid or slow sand filters so that the length of run of these filters between cleaning is extended, thereby increasing their output by as much as 50%. In comparison with roughing filters (Section 9.10), microstrainers (which have surface loading rates up to 80 $m^3/h.m^2$) require less space and produce lower headloss; their capital and running costs are lower. Chlorine should not be added before water enters the microstrainer.

SEDIMENTATION AND SETTLING TANKS

8.4 GENERAL DESIGN CONSIDERATIONS

Sedimentation tanks are designed to reduce the velocity of water to permit suspended solids to settle out by gravity. There are many different designs of tanks and most are empirical. The effluent quality for suspended solids or turbidity depends on the design and can be in the range 1–5 mg/l and 1–5 NTU, respectively. The quantity, size, shape, density and nature of suspended solids in a water, the water temperature and the effluent quality required all influence the performance of a tank design. Laboratory jar tests are performed on samples of raw water and a suitable tank design has to be selected. Designs which have been successful before under similar conditions are a useful start but alternatives that could provide a more economical and efficient solution should be considered.

8.5 PLAIN SETTLING

In plain settling (or sedimentation), suspended solids in a water are permitted to settle out by gravity alone; no chemicals are used. For this purpose the water can be left to stand in a tank, although with continuous supply at least two such tanks have to be used alternately. Such fill-and-draw tanks are seldom used in modern plants, except for filter washwater recovery. Instead plain sedimentation tanks are designed for continuous throughput, the velocity of flow through the tank being sufficiently low to permit gravitational settlement of a portion of the suspended solids to occur. In practice the application of plain sedimentation in waterworks is very restricted because impurities such as algae, aquatic plant debris and finely divided mineral matter do not settle at a rate sufficient for a tank of reasonable size to be utilized. Plain settling is most frequently used as a preliminary treatment for fast flowing river waters carrying much suspended solids such as occur in the tropics or at intake works on water transfer schemes in order to minimize the amount of suspended material passing into the system. Under most circumstances chemically assisted sedimentation, which is a more complex process, is adopted (Section 8.8).

The velocity with which a particle in water falls under gravity depends on the horizontal water velocity, the size, relative density and shape of the particle and the temperature of the water. The theoretical velocity V (mm/s) of falling spherical particles in slowly moving water (Reynolds numbers of less than 0.5) is given by:

$$V = \frac{g}{1.8 \times 10^4}(r-1)\frac{d^2}{\gamma}$$

where $g = 9.81\ \text{m/s}^2$, r is the relative density (γ_s/γ_w) of the particles, d is the diameter of the particles in mm and γ is the kinematic viscosity of water in m^2/s, which varies with the temperature of the water as given in Table 8.1 (Camp, 1946). The coefficient of kinematic viscosity (μ in m^2/s) = coefficient of absolute viscosity (Ns/m^2) divided by density (kg/m^3) where N (Newton) is kg.m/s^2.

A number of different (mainly empirical) formulae have been given for the settlement of sand and soil particles in still water; some of the values derived are given in Table 8.2.

Aluminium and iron flocs have a specific gravity of about 1.002, particle size as large as 1 mm and a settling velocity (at 10°C) of about 0.8 mm/s (Fair, 1968). Clay particles generally have a grain diameter of 0.01 mm to less than 0.001 mm (1 μm) so that it is impracticable to remove them from a water by simple sedimentation, or even by filtration, without prior chemical coagulation treatment (Sections 8.10–8.17).

Maximum velocity to prevent bed uplift or scour. Apart from the settling rate in still water it is, of course, essential that once a particle has reached the base of the tank it shall not be

Table 8.1 Kinematic viscosity of water

Temperature (°C)	0	5	10	15	20	25
Value γ (m^2/s) $\times 10^{-6}$	1.79	1.52	1.31	1.15	1.01	0.90

Table 8.2 Settling speeds of particles of relative density r

Diameter of particle (mm)	Settling speed[1] (mm/s) – sand[a] (r = 2.65@ 10°C)	Settling speed[2] (mm/s) – sand[a] (r = 2.65@ 10°C)	Settling speed[2] (mm/s) – coal (r = 1.5@ 10°C)	Settling speed[2] (mm/s) – sewage solids (r = 1.2@ 10°C)
1.0	100	140	40	30
0.6	63	–		
0.5	–	70	20	17
0.4	42	–		
0.2	21	22	7	5
0.1	8	6.7		
0.06	3.8	–		
0.05	–	1.7	0.4	0.3
0.04	2.1	–		
0.02	0.62	–		
0.01	0.15	0.08	0.02	0.008

Sources of information: [1]AWWA (1969); [2]Imhoff (1971).
Notes: [a]Particle sizes 0.1 mm and below are classed as silt.

re-suspended by the velocity of flow of water over the bed. Camp (1946) gives the channel velocity V_c (m/s) required to start motion of particles of diameter d (mm) as:

$$V_c = \left(\frac{8\beta g}{10^3 f}(r-1)d\right)^{1/2} \tag{8.1}$$

where r is the relative density, β is about 0.04 for sand in a smooth bed, 0.10–0.25 for sand in a rippled bed, or in the range 0.04–0.06 for sticky flocculent materials, $g = 9.81\ \text{m/s}^2$ and f is the friction factor in the Darcy–Weisbach formula, in which the hydraulic gradient is given by:

$$i = \frac{f}{4R}\frac{V^2}{2g}$$

where V is velocity in m/s and R is the hydraulic radius (cross-sectional area divided by wetted perimeter) which can be taken as the flow depth in a relatively wide and shallow tank. The value of f is typically in the range 0.02–0.03.

Maximum horizontal velocity of flow. A third flow measure which must be taken into account is that the horizontal velocity of flow must not be so great as to prevent, by turbulence, the settling of particles under gravity. There is general agreement that this velocity should not be more than 0.3 m/s to allow sand grains to settle. This is, of course, too high a velocity for the settling of particles of light relative density (1.20 and less), but this is the figure normally used for sewage grit chambers where the heavier material is to be deposited and the lighter material left to carry over. At 0.2 m/s faecal matter, i.e. organic matter, begins to settle. These velocities are in contrast to 1.0–1.5 m/s to minimize suspended solids settlement in pipelines.

8.6 THEORY OF DESIGN OF TANKS

The flow Q through a rectangular sedimentation tank of length L, water depth y and width b is shown in Figure 8.1. The time of fall (y/V) of a particle of silt of vertical falling speed V from entry to the tank to reach the bottom before the water leaves the tank must equal the time of horizontal flow (Lby/Q). Therefore $V = Q/A$, where A ($=Lb$) is the surface area of the tank. Q/A is known as the surface loading rate and is expressed as $m^3/h.m^2$, m/h or mm/s.

In a tank with uniform flow distribution all particles with a falling speed greater than Q/A reach the bottom before the outlet end of the tank. Particles with a speed less than Q/A are removed in the same proportion as their speed bears to Q/A, for example if the speed V is only half Q/A then only half the particles falling at this speed reach the bottom. Q/A is thus a measure of the effective removal of the particles in any tank. For example, in a tank of 300 m^2 surface area with inflow of 1.2 m^3/s, $Q/A = 14.4$ $m^3/h.m^2$ (4 mm/s) so that, theoretically, all particles with V of 4 mm/s or more are removed, 50% of those having V of 2 mm/s, 25% of those with V of 1 mm/s and so on. Therefore, the performance of a tank is independent of depth and retention time. This concept is the basis for the design of multi-tray horizontal flow tanks and inclined plate and lamella settlers (Sections 8.14 and 8.17).

The foregoing theory, however, assumes that the falling particles do not hinder each other; but McLaughlin (1959) showed in laboratory experiments with clay and aluminium sulphate that the faster particles, settling through the slower ones, gather some of the latter up and drag them out of suspension. This is the case with flocculated suspensions where, during settlement, agglomeration of the particles takes place and so particles settle much faster. The performance of a tank in which

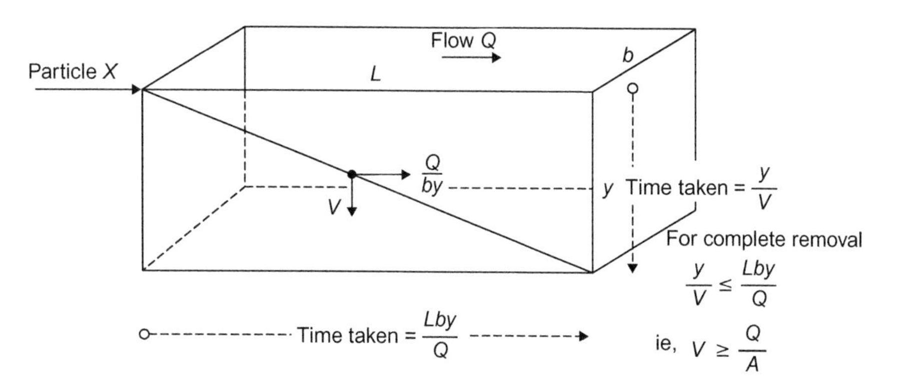

FIGURE 8.1

Theoretical flow in a horizontal flow rectangular tank.

this phenomenon occurs is related not so much to its surface loading rate as to the time of residence. These findings relate to still water. Thus, dependent upon the nature and size of the settling particles, the range of sizes, the degree of concentration of the suspension and the amount of turbulence, the performance of a sedimentation tank may relate to its surface loading rate or to its residence time, or partly to both.

8.7 GRIT TANKS

In waterworks, grit primarily constitutes sand, gravel and other abrasive material and is present mostly in water abstracted from rivers. If allowed to enter the works it could damage intake pumps and settle in raw water pipelines and inlet process units. Intakes should therefore be sited and designed to minimize the uptake of grit (Section 6.22) and tanks should be provided upstream of pumps to trap the grit. Grit tanks are also used as traps for sand, anthracite, GAC and other filter media carried over in used filter washwater to protect any pumping equipment downstream.

Grit tanks operate as plain settling tanks (unaided by coagulants) and are sized to capture particles with diameter larger than 0.1 mm. From Table 8.2 a silt particle of 0.1 mm falls in still water at 8 mm/s. The residence time in the tank should be at least 1.5 times the time taken for 0.1 mm particles to settle to the floor of the tank. Grit tanks for sand removal usually have surface loading rates of 10–25 $m^3/h.m^2$, a water depth of 3–4 m and length to width ratio of at least 4:1 (Kawamura, 2000). To prevent scour of settled material the horizontal velocity must be smaller than the scour velocity calculated from equation (8.1) in Section 8.5. This is usually about 0.3 m/s for sand.

CHEMICALLY ASSISTED SEDIMENTATION OR CLARIFICATION

8.8 CHEMICALLY ASSISTED SEDIMENTATION

This comprises several separate processes of treatment which together make up the complete system known as 'clarification'. This system is designed to remove from a water suspended materials, colour and other soluble material such as that of organic origin and soluble metals such as iron, manganese and aluminium ahead of filtration processes. It is a delicate and chemically complex phenomenon having three stages: (1) the addition of measured quantities of chemicals to water and their thorough mixing; (2) coagulation and flocculation, or the formation of a precipitate which coalesces and forms a floc; and (3) solid–liquid separation.

8.9 CHEMICAL MIXING

Most chemical reactions in water treatment applications are completed within 5 seconds and therefore the principal objective in chemical mixing is to obtain rapid and uniform dispersion of the chemical in the water to ensure that chemical is mixed and reactions are completed in the shortest possible time. Inadequate mixing of a coagulant such as aluminium sulphate can impair the formation of a good floc and would result in poor plant performance which can only be corrected by using excess of the chemical. The addition and mixing of chemicals to the water is a continuous process and is frequently described as either rapid or flash mixing. The design of mixers is often based on the concept of velocity gradient, which was first developed by Camp and Stein (Camp, 1943) for flocculation. It is an inadequate parameter for design of mixers but, in the absence of a better design approach, it is still being used for mixer design and its value is used to express the degree of mixing at any point in the liquid system. The velocity gradient G (s^{-1}) is defined in terms of power input by the following relationship developed by Camp and Stein for flocculation:

$$G = \left(\frac{P}{\mu V}\right)^{1/2} \tag{8.2}$$

where P is the useful power input (W), V is the volume (m^3) and μ is the absolute viscosity (Ns/m^2 or kg/m.s) at the water temperature. V can be taken as the flow rate multiplied by the residence time in the mixer. Mixing efficiency is directly related to the local flow turbulence created and should give a high degree of chemical-in-water homogeneity within a short time, with low absorption of power. The methods used for mixing can be hydraulic or mechanical. In hydraulic mixers all elements of liquid within the mixer are subjected to the same retention time, akin to plug flow and they are suitable for many mixing applications. Mechanical mixers are mostly of the backmix reactor type and when applied to a continuous flow system the elements of liquid within them have a distribution of residence time. This is not ideal for coagulant mixing. They require long residence time to make allowance for short-circuiting and the headloss across the mixing chamber is as much as that required for hydraulic mixing.

Hydraulic mixing makes use of the turbulence created due to the loss of head across an obstruction to flow, such as an orifice plate, pipe expansion or valve or by the sudden drop in water level when water flows over a weir or hydraulic jump. The latter is usually formed at a flume (Chow, 1959) (Section 14.8) in a channel with a local width constriction and change in floor level designed to produce supercritical flow under all operating flows. The ratio of the depth just upstream of the jump y_1 (m) to the downstream depth y_2 (m) is given by the following equation (Section 14.11):

$$\frac{y_1}{y_2} = \frac{1}{2}(1+8F^2)^{1/2} - 1$$

where F (Froude number) $= V_1/(gy_1)^{1/2}$, V_1 is the velocity upstream of the jump (m/s) and g is the acceleration due to gravity (m/s^2). For the hydraulic jump to form, the ratio (y_2/y_1) should be >2.4 and therefore $F > 2$. For F between 4 and 9 about 40–70% of energy is available for mixing (Schulz, 1992). Hydraulic mixer efficiency is flow dependent. Mixers should be designed for the operating range of plant flow rates, if necessary with facilities for taking sections of the mixer out of service at low flow rates.

When pipelines are used for mixing, a pipe of equivalent hydraulic length of at least 20 pipe diameters should be allowed. Static mixers are pipeline hydraulic mixers employing a repeated pattern of stationary, shaped diverters (elements) which induce rotational and radial flow and subsequent shear, some making use of vortex shedding. The fixed elements are installed in a housing, usually of the same diameter as the pipe. The most efficient designs combine the shortest mixing length with lowest headloss and should be used where mixing times required are short, for example coagulation. Plate 10(b) shows a static mixer for a pipeline; performance of such a mixer is demonstrated using computational fluid dynamics (CFD) modelling (Section 14.18) as shown in Plate 10(a). In raw water applications, debris such as weeds and twigs could obstruct this type of unit, which may not be suitable for unscreened river abstraction. If good dispersion followed by rapid mixing is not achieved when mixing alkalis (e.g. lime, caustic soda) in waters with alkalinity greater than 10 mg/l as $CaCO_3$ calcium carbonate scale may form on the elements due to localized softening; also ferric salts could form hydroxide precipitate on the elements. Gases should ideally be injected as solutions using a static mixer in a sidestream of flow. For carbon dioxide and ozone, the sidestream flow should be about 1–2% and 7.5–10% of the plant flow, respectively. Static mixers can also be installed in channels (Plate 10(c)), for which the headloss required is in the range 10–150 mm. Lower water depth with reduced flow has little or no effect on the performance of channel mixers. Channel mixers are also more suitable for raw water applications than units installed in pipework since they can be accessed for cleaning.

Chemicals should be injected at about twice the velocity of pipe flow at a point within the mixer housing. Mixing should be completed within the mixer and samples of the mixed solution should be taken at the outlet of the mixer. The location of a static mixer in a pipe should be discussed with the supplier as clear straight lengths are required both upstream and downstream of some types of mixer. The ratio of pipe flow to chemical flow should be in the range 10^2–10^4 depending on the design of the injection system. Mixing is rapid when viscosities of the chemical and the receiving water are similar and is, therefore, improved by applying the chemical in a diluted form. Mixer units should be designed for easy removal for cleaning although a bypass is considered unnecessary.

The useful power input of hydraulic mixers is related to headloss by the equation:

$$P = Q\rho gh$$

where P is the useful power input (W), Q is the flow (m^3/s), h is the headloss (m), ρ is the density of water (kg/m^3) and g is the acceleration due to gravity (m/s^2). It is generally found that adequate mixing is obtainable in a free fall weir with a headloss of between 300 and 400 mm. The power input for a headloss of 350 mm is about 3.5 kW per m^3/s of water flow and, with a mixing time of 5 seconds, gives a G value of about 700 s^{-1}. For mixing polyelectrolytes (Section 8.22) as a coagulant aid following coagulant mixing the headloss should be kept to about 150 mm, with associated G values in the range 300–500 s^{-1} to minimize the damage to the floc already formed. Static mixers require a headloss in the range 100–1000 mm although headloss as much as 1500 mm is not uncommon where a high turndown in water flow is required. The residence time is typically 2–3 seconds and the associated G value could be as high as 5000 s^{-1}. The mixing efficiency of all types of mixer is defined by the coefficient of variation (*CoV*) which, for good mixing, should be less than 0.05. $CoV = \sigma/x$ where σ is the standard deviation of the tracer concentration in each sample and x is the mean concentration of the tracer. For example, for a *CoV* of 0.05 (i.e. '95% mixed') there is a 99.9% probability that all samples taken downstream will be within $\pm 3\sigma$ of the mean mixed value. This is rarely achieved except in static mixers.

Hydraulic mixers are usually simple and particularly suitable where some headloss can be tolerated. They have no moving parts or direct power consumption so maintenance is negligible. A disadvantage is that the efficiency of mixing suffers if works throughput is lowered outside the operating range of the mixing device. Mechanical mixing is achieved in purpose built chambers equipped with mechanical rotary impellers such as radial flow turbines or axial flow propellers. Ideally mixer manufacturers should be consulted for guidance on sizing. Typical residence times are of the order of 15–30 seconds and the velocity gradient value G varies between about 300 and 600 s^{-1}. This gives a power input range equivalent of about 4–10 kW per m^3/s of water flow at 20°C and the value selected depends upon the raw water quality, chemical to be mixed and degree of short-circuiting in the chamber. In recirculation pump jet mixing, about 2.5–5% of the plant flow is drawn upstream of the mixing chamber and returned in a pipe through an orifice plate on to a plate in the mixing chamber (Skeat, 1969). The residence time and the power input are the same as for the impeller type mixers. This concept has been extended to pipelines where the water is returned into the pipe against the direction of plant flow through an injection nozzle selected to give a full cone spray (Kawamura, 2000). Due to a tendency for nozzle blockage in raw water mixing applications, clarified, filtered or plant service water should be used as the motive water for this type of mixer. Mixing time and velocity gradient are similar to those for hydraulic mixers. In both types of pumped jet mixers the chemical may be injected into the return pipe or at the point of turbulence. Mechanical mixers have the advantage that they are not affected by flow variations. To maintain a uniform velocity gradient at varying works throughputs the power input can be varied by fitting the mixer with a variable speed motor.

Pumps, in particular the centrifugal type, are good mixers. The lower the efficiency of the pump the better the mixing; for example in a pump of 75% efficiency, a significant part of the 25% energy lost is due to turbulence (Section 19.2) which can be used for mixing. However, consideration should be given to possible volatilization of the chemical due to low pressure on the suction side and hence corrosion of pump internals by the concentrated chemical. For this reason gases such as chlorine in solution should not be mixed in pumps and even for other chemicals, materials of wetted parts should be carefully selected. If the alkalinity of the water is greater than 10 mg/l as $CaCO_3$, alkalis should not be dosed into pumps because of the potential to form calcium carbonate scale in the pump. When chemicals are injected into pumps, the manufacturer must be consulted to verify the suitability of the pump materials.

The method of injecting chemicals also contributes to the performance of the mixer. For dosing chemicals into pipelines of diameters (D) up to 600, 1000 and 2500 mm, injection tubes one of length $0.33D$, two of $0.15D$ and four of $0.15D$, respectively, should be used. Injection tubes used for slurries or with chemicals likely to form precipitates in contact with water (e.g. alkalis) should be withdrawable. The injection tube could be a pipe fitted with a nozzle or a perforated diffuser. The nozzle velocity should be about 0.75 m/s or 50% of the velocity of flow in the large pipe, whichever is the greater. In all cases, the injection tubes are mounted in a plane perpendicular to the direction of flow and in the case of horizontal pipes the tubes are inserted with their axes at 45° to the horizontal. For static mixers the injection criteria vary according to the mixer manufacturer.

Injection of chemicals at weirs or flumes is best achieved via perforated pipes (for clear solutions) or channels (for slurries) running the full length of the weir about 50 mm above the water surface and just upstream of the point of turbulence. Submerged perforated pipes similarly located are used for gases such as chlorine in solution.

8.10 CHEMICAL COAGULATION AND FLOCCULATION

Coagulation and flocculation are essential processes in the treatment of most surface waters; one exception being the application of slow sand filtration (Section 9.13). The two processes, operating in conjunction with solid–liquid separation processes, remove turbidity, colour, cysts and oocysts, bacteria, biological matter, viruses and many other organic substances of natural and industrial origin. Some are removed directly and some indirectly through attachment or adsorption onto particulate matter.

The terms coagulation and flocculation are two separate processes, contrary to common usage. In coagulation the coagulant containing the aluminium or iron salt is mixed thoroughly with the water and various species of positively charged aluminium (3+) or iron (3+) hydroxide complexes are formed. These positively charged particles adsorb onto negatively charged colloids such as colour, clay, turbidity and other particles through a process of charge neutralization. Flocculation is the process in which the destabilized particles are bound together by hydrogen bonding or Van der Waal's forces to form larger particle flocs during which further particulate removal takes place by entrapment. Flocculation is usually achieved by a continuous but much slower process of gentle mixing of the floc with the water in one of numerous types of plant. In the theory of flocculation the rate at which it takes place is directly proportional to the velocity gradient (Camp, 1943) and equation (8.2) used in Section 8.9 for mixing is also used for determining the velocity gradient G for flocculation. If the residence time in a flocculation chamber is t seconds then the extent of flocculation which takes place, or the number of particle collisions which occur, is a function of the dimensionless expression Gt which is given by the equation:

$$Gt = \frac{1}{Q}\left(\frac{PV}{\mu}\right)^{1/2}$$

where the symbols have the same meaning as in equation (8.2).

For the common coagulants of aluminium and ferric salts the value of G for flocculation is usually in the range 20–100 s^{-1} with the residence times in flocculation chambers varying from 10 to 40 minutes. However, there are cases where flocculation times approaching 60 minutes have been necessary for waters of extremely high colour and low temperature (Adkins, 1997). The value of Gt would be in the range from 20 000 to 200 000. The values of G and t depend on the raw water quality (e.g. colour, turbidity, algae), water temperature and the required floc size (inversely proportional to G). Typical target floc sizes are 2–5 mm for clarifiers in general; 2.5 mm–150 μm for dual media filters in direct filtration (Hudson, 1981) and 25–50 μm for dissolved air flotation (DAF; Edzwald, 2010). Therefore, each application should be individually evaluated by pilot trials unless adequate information is available for almost identical conditions. In direct filtration where the intention is to form a microfloc, G values of the order of 100 s^{-1} and a residence time of about 10 minutes are used. For DAF the values used are typically: G about 50–70 s^{-1} and residence time about 15 minutes for algal laden water and 20 minutes for waters with colour. Amato (2001) has suggested that 10 minutes is adequate for most UK waters. Flocculation for sedimentation including lamella settlers would require G value in the range 30–70 s^{-1} and residence time between 20 and 40 minutes. For optimum flocculation, the coagulated water should be subjected to a decreasing level of energy with time; the so-called 'tapered energy' flocculation provided in two or three equal size compartments. The G values quoted above are the mean values for two or three stage flocculators. The G value applied in the last stage is about 10–30% of the first stage G value. For example, in DAF the first and last stage G values could be 100 and 25 s^{-1}. In the case of the high rate 'Actiflo' process (Section 8.17), in which microsand is used to ballast the floc, typical G values and residence times are 150–300 s^{-1} and 4–8 minutes, respectively, usually applied in a single stage.

8.11 TYPES OF FLOCCULATORS

The agitation required for flocculation is usually provided by either hydraulic or mechanical means. The most common hydraulic flocculator is the baffled basin in which a sinuous channel is equipped with either around-the-end or over-and-under baffles. The flocculation energy is derived primarily from the 180° change in direction of flow at each baffle. For the around-the-end type, which is preferred for their ease of cleaning the minimum water depth is 1 m and the headloss across the flocculator is in the range 500 mm to 1 m. The residence time is 20–25 minutes. The distance between baffles should be at least 500 mm and that between the end of each baffle and the wall should be about 1½ times the distance between the baffles. The baffle spacing should be increased gradually with channel length to achieve tapered flocculation. The floor of the channel should slope towards the outlet. The baffled basins have no mechanical or moving equipment and produces near plug flow with low short-circuiting. The disadvantages are: most headloss occurs at the 180° bends where the G value may be too high, but inadequate in the straight channels; G value varies with flow (but this could be partly overcome by providing removable baffles); and suspended solids tend to settle in the channel. When designing baffled sinuous channels some allowance must, therefore, be made for water quality; suggested velocities to minimize settlement are 0.40, 0.30 and 0.25 m/s for high, moderate and low turbidity waters, respectively.

Over-and-under baffles can be as deep as 3 m and therefore have a smaller footprint. A disadvantage is the removal of settled solids on the floor of the tank; weep-holes and valved drains are used to mitigate the problem. They are often used in the COCO DAFF process. Typically, either a single tank, or two tanks in parallel to accommodate for redundancy, each with four compartments, is provided. The two under baffles are provided with rectangular bottom openings, each to give a headloss of about 50 mm. The over baffles and the inlet and outlet weirs are submerged to minimize the risk of floc damage.

Other hydraulic flocculators are helicoidal-flow, staircase-flow, gravel-bed and Alabama types (Schulz, 1992). Hydraulic flocculation is also used in sludge blanket clarifiers (Section 8.16) which are typically of the flat-bottomed type. Flocculation takes place within the clarifier and is hydraulically induced. Coagulant dosed water enters via orifices in flow distribution pipe laterals or via orifices discharging from low level inlet pipes injecting flow downwards onto the floor of the clarifier. Flocculation is completed within the sludge blanket through a process known as contact flocculation. With cold waters, however, this may need to be supplemented with external flocculation.

There are several types of mechanical device for flocculation, common designs being the paddle type stirrers, mounted either horizontally or vertically in the flocculating chamber, or axial flow turbine type stirrers; the latter being used for high energy flocculation. The power term in the velocity gradient expression for a paddle type is given by the equation:

$$P = F_D V = \frac{1}{2} C_D \rho A V^3$$

where P is the power input (W), F_D is the drag force (m kg/s^2), C_D is the drag coefficient, A is the submerged area of the paddles (m^2), V is the relative velocity of the paddles (m/s) with respect to water, ρ is the density of water (kg/m^3); C_D is about 1.8 for a paddle stirrer. V may be approximated to 0.75 times the peripheral velocity of the paddle or equal to $1.5\pi rn$, where r is the effective radius of the paddle (m) and n is the number of revolutions (s^{-1}). The speed of rotation of the paddles varies from 2 to 15 rpm and the peripheral velocity of the paddles ranges from 0.3 to1.2 m/s. For one- two- or three-stage flocculators the peripheral velocities would be 0.6, 1.2/0.6 and 1.2/0.6/0.3 m/s, respectively. For high energy flocculation the paddle tip speed could be as much as 3 m/s.

Tapered flocculation chambers are usually made up of two or three equal size compartments in series to minimize short-circuiting and the stirrers should be counter-rotating. In order to optimize the velocity gradient the stirrer in each compartment should if possible be fitted with variable speed motors with that in the last compartment having infinitely variable speed facilities. The compartments should be separated by either around-the-end or over-and-under baffles arranged to give diagonal flow in the compartment; the headloss across the baffle walls produces a G value of up to 20 s^{-1}. Flow velocity should be limited to 0.25 m/s, depending on the floc characteristics to minimize floc shear. The tank dimensions (L – length in the direction of flow and b – width) vary according to the type. For the horizontal shaft type, the tank should be long and narrow (L:b of at least 4:1 with a square cross section perpendicular to the direction of flow) and with depth of about 3 m. Tanks incorporating a single vertical stirrer should ideally be square or if rectangular have a ratio L:b not exceeding 1:1.5. Wider tanks can be accommodated by incorporating an additional stirrer, increasing the overall allowable L:b ratio to 3:1. The depth is not critical provided the stirrer can be supported from the gearbox without having to use bottom bearings. This limits the paddle type to a depth of about 4–5 m whereas the turbine type could be as deep as 7 m; the clearance to the floor should not be less than 500 mm.

Turbine flocculators normally use axial flow impellers. Radial flow impellers cause significant floc damage. The blades provide less surface area than a comparable paddle type, but they run at a higher speed; typically tip speed is about three times that of the paddle type, with a maximum of 3 m/s, thus providing high collision rates while using less energy. The operation at a higher speed reduces gearbox size and therefore is more efficient. Griffiths (1996) states that bulk fluid velocity is an essential design parameter and recommends a value of 0.03 m/s. This paper also provides a method of comparison between different impeller types and speeds and gives the following formula for shear stress (τ) in N/m^2:

$$\tau_{\max} \propto N_p^{8/9} V_{tip}^2$$

where N_p is the power number and V_{tip} (m/s) is the tip speed. High energy flocculation is suitable for DAF and direct filtration which require a small floc size. The higher speeds can handle floc especially those strengthened by polyelectrolyte without detriment. Flow patterns developed also reduce floc settlement in the basins. They help to reduce the footprint with width to length ratio of 2:1 and are used in deep tanks (e.g. Mekorot WTW, Israel; tank depth 8.65 m and water depth 6.745 m). They are smaller, lighter and do not require end bearings (which is the case with horizontal paddles and can be the case with vertical paddles) and do not require seals (which is the case for horizontal paddles). They are used widely in all flocculation applications. The velocity gradient as defined by Camp (1943) does not apply to this type of system.

Vertical shaft paddle type stirrers should have a diameter and paddle height greater than two thirds of a plan dimension and should have 500 mm clearance to the walls. For good performance, horizontal shaft type stirrers should be mounted in sinuous channels to reduce short-circuiting and should have at least three paddles on each diametrical arm. When flocculation tanks are dedicated to individual clarifiers, measures should be taken (such as a free fall outlet) to prevent flocculator flow patterns being transmitted to the downstream clarifier and to limit transfer velocity to less than 0.1 m/s.

8.12 FACTORS AFFECTING COAGULATION AND FLOCCULATION

Impurities which require coagulation and flocculation before they can be removed by solid–liquid separation processes can broadly be classified as either inorganic or organic material. The more usual type of inorganic material encountered in water treatment settles easily, especially when it is of particulate size; when chemically unaided settling is used 1 or 2 hours retention usually removes at least 40–50% of particulate matter. A primary coagulant, sometimes assisted by a coagulant aid, usually removes 98% or more of the suspended particulate matter for lightly loaded water with suspended solids not exceeding about 300 mg/l. Such a removal rate may not be adequate for those river waters typical of regions where suspended solids are regularly measured in excess of 1000 mg/l; concentrations over 10 000 mg/l are not uncommon. In such places it may be necessary to introduce a settlement stage unaided by chemicals before chemical sedimentation or, alternatively, to have two separate stages of chemical treatment and settlement in series because the quantity of solids in suspension and the sludge formed by the addition of chemicals are too much to remove in one stage of treatment.

Not all inorganic matter settles out quickly in plain settlement and in some waters, such as are found in Central Africa, it may be weeks or even months before any significant settlement of colloidal material takes place. With such waters there is no option but to use two clarification stages each with chemical coagulation. Organic colour and finely divided mineral matter, including various forms of clay, are included in this category. Clay in water is a hydrophobic colloidal suspension, or 'sol', in which the surfaces of the particles are considered to have a negative charge. This charge contributes to the stability of the sol by helping to prevent the particles coalescing into larger particles which would then have a relatively rapid rate of settling.

The excellent coagulating effect of aluminium or ferric salts may be due to the triple positive charge on the trivalent aluminium or ferric hydroxide complex ion neutralizing the negative charges on the clay particles. Literature on the coagulation of clay and similar materials is reviewed by Packham (1962, 1963); he describes his own investigations which have developed the theories in a somewhat different direction. His experiments confirmed that it is not primarily the aluminium or ferric ions in solution which react with the clay particles, but it is the mass of rapidly precipitating hydroxides of aluminium or ferric which enmesh them. In subsequent work by other researchers this is identified as one of two mechanisms in coagulation and is called 'sweep coagulation'. It requires excess coagulant and occurs in about 1–7 seconds. The other mechanism is one of neutralization of negatively charged colloids by positively charged hydrolysis products which are soluble hydrated hydroxide complexes of aluminium or ferric salts.

The reaction is completed within a second (Amirtharajah, 1978). The presence of some anions such as sulphate (SO_4^{2-}) helps the coagulation process (Fair, 1968). Sweep coagulation uses more chemicals and produces more sludge than for charge neutralization, but removal of trace contaminants is better.

pH is an important factor in coagulation. The optimum pH range corresponds to that over which minimum solubility of the hydrolysed coagulant products occurs and maximum turbidity and colour removal is achieved. Typically for ferric salts the coagulation pH is 5 or higher and for aluminium salts it is between 6.5 and 7.2. For coagulation of natural organic matter such as colour, coagulation pH values for ferric and aluminium salts are about 4.5 and 5, respectively, whilst optimal turbidity removal typically occurs at pH 6–7.5.

US EPA (1999) includes in the Disinfection By-Products Rule (DBPR) a requirement for 'enhanced coagulation' for greater removal of natural organic matter to minimize disinfection by-product (DBP) formation. It is defined as the addition of excess coagulant, a change in coagulant type, or a change in coagulation pH for improved total organic carbon (TOC) removal. US EPA proposes that enhanced coagulation is applied to surface waters treated by conventional treatment unless (1) the concentration of TOC in the raw or treated water is <2 mg/l; or (2) the concentration of TOC in the raw water is <4 mg/l, the alkalinity is >60 mg/l as $CaCO_3$ and the concentrations of total trihalomethane and haloacetic acid (five regulated species) in the distribution system are below 50% of the maximum contaminant levels (MCLs); or (3) for treatment employing only chlorine for disinfection, the concentrations of total trihalomethane and haloacetic acid (five regulated species) in the distribution system are below 50% of the MCLs. The DBPR also lists the TOC removal percentages required at TOC >2 mg/l for various ranges of alkalinity values. All values are based on quarterly running averages.

SUVA (Section 7.36) is used to assess the ability to remove TOC by coagulation. SUVA values less than 2 are indicative of hydrophilic organic material of non-humic origin, with low chlorine demand and low trihalomethane formation potential (THMFP). SUVA values greater than 4 are indicative of hydrophobic organic material mostly of aquatic humic origin, rich in highly aromatic compounds, with high chlorine demand and high THMFP. SUVA values in the range 2–4 are indicative of a water containing a mixture of hydrophobic and hydrophilic material, representing both humic and non-humic substances. TOC in water with low SUVA (<2) is difficult to remove with coagulation while TOC with high SUVA (>2) is easier to treat by coagulation; the higher the SUVA the greater the TOC removal. According to the US EPA Disinfectants and Disinfection By-Products Final Rule mandates, it may not be necessary to comply with the indicated TOC removal requirement if raw or treated water SUVA is ≤2. Filtered water SUVA values in the range 2–4 in an existing works are likely to indicate potential for elevated THM formation and should be investigated to identify options for treatment process optimization. SUVA values >4 suggest reappraisal of the process may be necessary.

Polyelectrolytes (Section 8.22) can be used as coagulants for waters containing high turbidities (and for sludge conditioning). For example, in the 1365 Ml/d Al Karkh water treatment works for Baghdad a cationic polyacrylamide was used as a coagulant to reduce the raw water suspended solids from 30 000 mg/l to about 500 mg/l in pre-settlement tanks. In the subsequent clarification stage the same polyelectrolyte was used as the coagulant aid to aluminium sulphate. For waters of low turbidity, polyelectrolytes are ineffective as coagulants because, at the very low dosages applied, the residence time provided for flocculation is insufficient to produce large floc. They are, however, effective in producing fast settling, larger, denser floc with coagulated floc particles formed when an aluminium or ferric salt has been used as the primary coagulant, resulting in increased settlement rates. In this application, polyelectrolytes are called coagulant aids and are added between 60 seconds (warm water) and 5 minutes (cold water) after the mixing of the coagulant, preferably close to the clarifier to minimize the damage to the floc. This time delay is shown to be more important for ferric than aluminium coagulants (Brejchova, 1992) and found to be applicable in particular to polyacrylamide type polyelectrolytes. Coagulation by polyelectrolytes follows charge neutralization, bridging, or both of these. Therefore, they would be effective even when the polyelectrolyte carries a charge of the same sign as that of the particles to be coagulated. Hence, the best type of polyelectrolyte for an application should be determined by the laboratory jar tests.

8.13 EFFECT OF ORGANIC CONTENT AND ALGAE

The variety of impurities present in raw waters and their varying concentrations explain the dictum that all waters are different and that water treatment is an 'art' as well as a science. This particularly applies to waters whose primary natural impurity is organic in origin, as these are very often the most difficult to treat. Miscellaneous fragments of animal and vegetable matter contribute towards the organic content, as does the organic colouring matter derived from peat and similar sources consisting largely of humic and fulvic acids and of more complex compounds, partly in true solution and partly in colloidal form. The optimum chemical treatment for such waters is sometimes very difficult to achieve, particularly when it is necessary to remove dissolved iron or manganese. Laboratory jar tests are an essential tool in the formulation of treatment for all waters. Where possible these should be followed by testing at pilot scale.

Planktonic animals and plants, particularly the plants (phytoplankton) which are algae, can cause a variety of problems in treatment. Their removal in microstrainers has been mentioned in Section 8.3. Table 8.3 lists some common organisms and their implications for treatment in temperate climates. Algae may occur in large quantities in lowland and in eutrophic waters. Their removal by flocculation, sedimentation and filtration is basically similar to, but more variable than, that of other forms of suspended material because of their widely diverse types, densities and shapes. Some algae, for example diatoms which have silica in the cell walls, are notably denser than water. These algae and many other algae with large or simple shaped cells are removed effectively by chemical treatment. On the whole, however, plankton do not settle very readily when growing rapidly, as they are buoyant either because of oxygen gas in their cells or as they actively swim, and the coagulant dose usually has to be increased to be effective. Many small-celled types, and those which swim using their motile flagellae, are often not well retained in floc, and are liable to pass through subsequent rapid filters. Conversely, problems often arise because of the tendency of some algae to form dense blooms in lake waters; blue-green algae (*Cyanobacteria*) are typical of this group. Such blooms can cause a rapidly increasing loss of head in the filters. The removal of planktons by coagulation and settlement can be improved by the use of a polyelectrolyte as a coagulant aid, and either killing them or simply inactivating them using pre-treatment with chlorine, ozone (Bauer, 1998) or an algicide (Parr, 1992; House, 2002). However, these methods have problems due to toxicity of the residues and, in the case of chlorine, the formation of DBPs (Hoehn, 1980) or

Table 8.3 Organisms reported as having caused difficulties in British waterworks

Group	Genus	Difficulty experienced (✓)					
		Blocking of filters	**Penetration of filters**	**Taste or odour**	**Growth in pipes, channels**	**Unsightly scums**	**Potentially toxic substances**
Bacillario-phyceae (Diatoms)	*Asterionella*	✓					
	Cyclotella	✓					
	Fragilaria	✓					
	Synedra	✓		✓			
	Nitzschia	✓					
	Stephanodiscus	✓	✓				
	Melosira	✓					
	Diatoma	✓					
Chlorophyta	Green unicells		✓				
	Chlamydomonas		✓				
	Cladophora				✓	✓	
	Spirogyra	✓					
Cyanobacteria 'Blue-green algae'	*Anabaena*			✓		✓	✓
	Aphanizomenon	✓				✓	
	Lyngbya	✓	✓				
	Microcystis			✓		✓	✓
	Oscillatoria	✓	✓	✓	✓	✓	✓
Xanthophyceae	*Tribonema*	✓					
Chrysophy-ceae	*Synura*		✓	✓			
	Dinobryon	✓					
	Mallomonas	✓	✓				
Euglenophyta	*Euglena*		✓	✓			
Dinophyceae	*Peridinium*	✓					
Rotifers	*Diglena*		✓				
Crustacea	*Cyclops*	✓	✓				
	Diaptomus		✓				
	Daphnia	✓	✓				
	Asellus				✓		
	Bosmina	✓					

(*Continued*)

Table 8.3 (Continued)

Group	Genus	Difficulty experienced (✓)					
		Blocking of filters	Penetration of filters	Taste or odour	Growth in pipes, channels	Unsightly scums	Potentially toxic substances
Nematodes					✓		
Oligochaetes	*Nais*				✓		
Insects	*Chironomus*		✓		✓		
	Tanypus		✓		✓		
Filamentous bacteria	*Sphaerotilus* (*Cladothrix*)				✓	✓	
Iron bacteria	*Crenothrix*				✓		
Polyzoa	*Plumatella*				✓		
	Cristatella				✓		
Porifera (sponges)	*Ephydatia*				✓		
	Spongilla				✓		
Molluscs	*Dreissena*				✓		

other side effects such as taste and odour formation (Section 10.32). Without killing algae their removal by sedimentation is about 50–75% (Markham, 1997). Some algae such as blue-green algae are especially adapted to live in the top layers of water; the cells of these forms contain minute air vacuoles and are particularly buoyant and difficult to settle. The DAF process is therefore particularly suited to blue-green algae removal because of the natural tendency for the algae to rise to the surface (Bare, 1975; Edzwald, 1990). The process can also be effective with other less buoyant algae, such as some diatoms and is described in more detail in Section 8.18. Based on UK experience, ozone pre-oxidation, coagulation, flotation and filtration remove about 96% of influent cells while rapid gravity filtration alone removes 63–75% (Henderson, 2008).

Many freshwater blue-green algae produce toxins harmful to mammals, birds and fish. These include hepatotoxins (e.g. *Microcystis*) and neurotoxins (e.g. *Anabaena*). Of the toxins, WHO (2011) has set a Provisional Guideline Value of 0.001 mg/l (1 μg/l) for total microcystin-LR (free plus cell bound). Harmful toxin concentrations occur when the plant populations increase due to high summer growths, lack of natural predation and high nutrient concentrations in the water. Control measures include good watershed management and nutrient control. Treatment methods include oxidation (chlorine, ozone), advanced oxidation (UV/peroxide), powdered activated carbon treatment, GAC adsorption, biological treatment (ozone/GAC) and nanofiltration or reverse osmosis (Alvarez, 2010). For a detailed review of toxic cyanobacteria in water the reader is referred to the publication by WHO (1999).

Microorganisms such as cysts and oocysts are readily removed by coagulation and flocculation followed by filtration with or without clarification (Section 9.20); a removal of 99.9% should be feasible for bacteria (WHO, 1996), although it is reported that such removal efficiencies are achieved mostly in the summer months, reducing to 70% in the winter months (O'Conner, 1997). Removal mechanisms include direct removal by charge neutralization and entrapment in floc and through removal of particles to which bacteria are attached.

Of the other impurities which are encountered in raw waters, it is far less easy to be precise as to removal efficiency by chemical coagulation. In many lowland waters there are to be found small but appreciable concentrations of substances in the run-off from farmland, drainage from agricultural land and sewage and trade effluent discharges. Many of these are organic substances and are usually classified as volatile or non-volatile with many more sub-divisions, such as biodegradability. Whilst it is known that chemical coagulation, flocculation, clarification and rapid filtration are at least partially effective in their removal, in particular those proportions adsorbed on to particles, different processes can offer a higher degree of removal of organic substances. These are described in Chapter 10.

CLARIFIERS

8.14 HORIZONTAL FLOW CLARIFIERS

Some simple types of settlement tank are in use for the clarification of flocculated waters. For large volumes of water containing a relatively heavy load of suspended solids a relatively dense floc is formed which settles easily and in warm climates, where the viscosity of the water is lower thereby permitting more rapid settlement of floc, the large horizontal flow sedimentation tank can be an economical solution for clarification. Although a large tank is necessary in order to keep velocities sufficiently low to permit settlement of floc to the base of the tank, its construction is simple because it need not be very deep and few internal walls are required. In the simplest design of horizontal flow tank, floc is allowed to accumulate on the floor of the tank until such time as the increasing velocity of water above the accumulated sludge begins to stir it up, thereby affecting the clarity of the effluent. When this occurs the tank should be cleaned out. Alternatively, moving scrapers operated continuously or intermittently add to the overall efficiency by pushing the settled floc to outlets in the base of the tank where it may be drawn off as sludge.

Accepting their initial low rating and correspondingly high civil construction cost, horizontal tanks are nevertheless versatile clarifiers; modifications to an original simple design to meet changing conditions of raw water can be accommodated if provision has been made at the design stage. For example, rotating flocculators and sludge scrapers can be added and some tanks have had their rate of flow increased by the addition of inclined tube or plate modules (Section 8.17). Since the depth of a tank does not influence its performance (Section 8.5) some horizontal flow tank designs use the principle of shallow depth sedimentation, with a sloping tray so that the direction of flow of the water reverses up and over the tray and exits near the inlet. Yet another variation is where the depth of the rectangular tank is divided by as many as four inclined trays and flow takes place in parallel streams between the trays in a concept similar to inclined plate settlers. In this way the capacity of a tank can be increased three- to fourfold on the same footprint. Circular tanks have a centre-feed and, after chemical flocculation which usually takes place in a central compartment, water flows radially and upwards to peripheral collecting launders or to a combination of radial and peripheral launders. Circular sedimentation tanks are fitted with rotating sludge scrapers, either with the drive mechanism mounted on a central platform or bridge or, with the drive unit mounted on the outside wall and the scraper bridge pivoted on a central support.

In the design of rectangular or circular tanks for treating heavily silted water, the feed water should be conveyed in channels rather than pipes to permit easy access for cleaning of settled solids. The advantages of these tanks are greater tolerance to hydraulic and quality changes, ideal for stop/start operation; infinite turndown; simplicity of operation; suitability for water containing high silt loads; and performance which is largely unaffected by diurnal temperature change. The primary drawback is their low surface loading rate and hence the large footprint and associated capital costs. Compared to rectangular tanks, circular tanks do not lend themselves to a compact layout. One such circular tank design known as the 'Centrifloc' is used for the treatment of R. Tigris water at the 1365 Ml/d Al Karkh water treatment works in Baghdad and is shown in Figure 8.2.

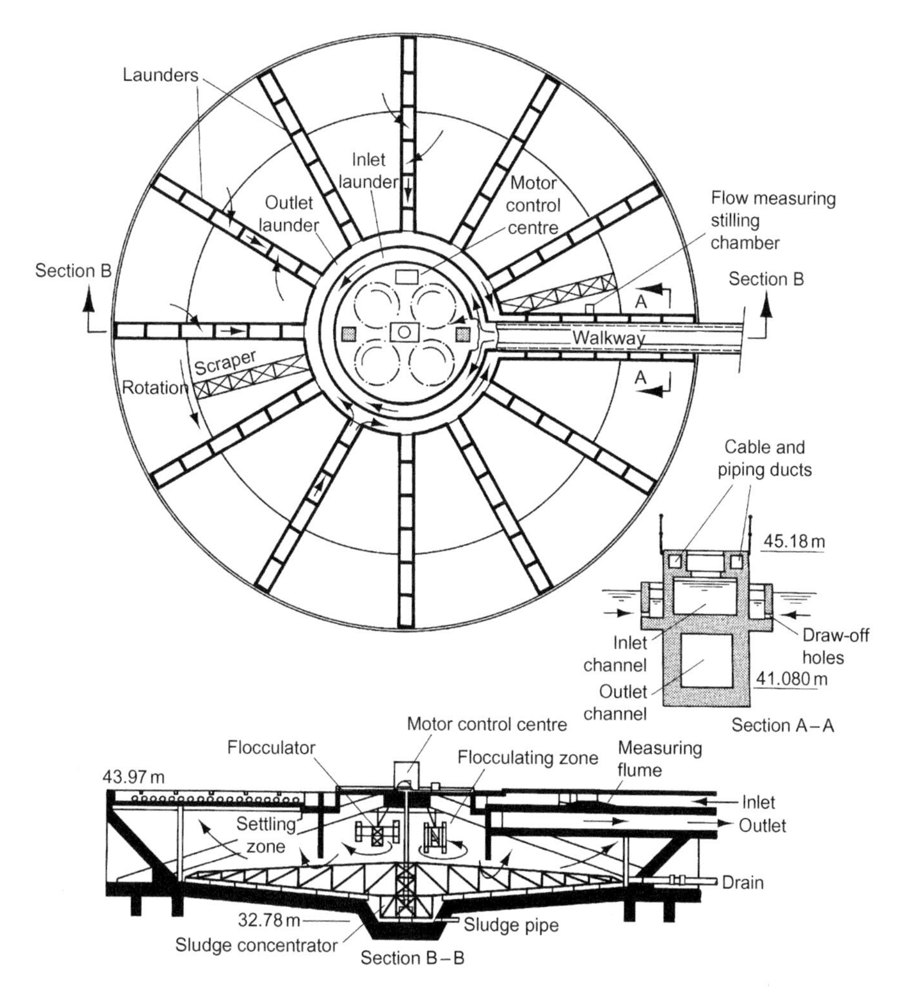

FIGURE 8.2

Circular Centrifloc® clarifier, 51 m diameter, unit flow 76 Ml/d @ 1.7 $m^3/h.m^2$, at Al Karkh for treatment of R. Tigris water for Baghdad (Engineer: Binnie & Partners; Contractor: Paterson Candy Ltd).

8.15 DESIGN CRITERIA

Without the aid of a coagulant, a rectangular or circular tank only works at a surface loading up to about one third the rate shown by jar tests (33% efficiency). With coagulant assisted sedimentation, the actual efficiency is generally much closer to the results obtained by laboratory jar tests. However, theoretical settling rates cannot be directly translated to tank design because of short-circuiting in the tank. The safety factor can be limited to between 1.3 and 1.5 by careful design of the inlet and outlet arrangements. These design features for rectangular tanks include multiple inlets at about 1.5 m centres sized to give an inlet velocity of about 0.5 m/s and perforated baffles, with orifice diameters of 100–200 mm to give a headloss of less than 10 mm to minimize floc shear, at the inlet and outlet ends across the whole cross section of the tank to create a more uniform flow pattern through the tank. Whilst these measures help, they may have little effect when the temperature of the incoming water differs from that in the tank and so creates turbulence which must subside before the particles can resume settling under quiescent conditions.

In horizontal flow tanks the horizontal flow of water should be designed to be laminar (Reynolds number < 2000) and short-circuiting and instability of flow should be minimized (Froude number $> 10^{-5}$). For tanks of dimensions L, b and y (m), surface loading rate S (m^3/h.m^2) and flow Q (m^3/s), Huisman (1970) defined Reynolds and Froude numbers as follows:

$$\mathrm{Re} = \frac{Q}{b+2y}\frac{1}{\gamma} \quad \text{and} \quad F = \frac{S^2}{g}\frac{L^2(b+2y)}{1.3\times 10^7 by^3}$$

where γ is kinematic viscosity (m^2/s) and $g = 9.81$ m/s^2.

Chemically assisted sedimentation tanks for the removal of organic matter and light flocculated particles, have length to width ratios greater than or equal to 4:1 and surface loading rates between 0.75 and 1.75 m^3/h.m^2, which, with a coagulant aid, may increase to 2.5 m^3/h.m^2. The mean horizontal flow velocity should be in the range 0.25–1.0 m/min. The flow over the outlet weir should be maintained below 50 m^3/h per m of weir length and should preferably be about a quarter of this value. Alternatively, 90° V-notches with a water depth of about 75 mm and spacing less than 0.6 m may be used. Submerged orifice outlets are useful in minimizing the passage of floating material to filters or where freezing is likely. Orifices should be sized to give a flow velocity of about 0.6–0.7 m/s with an orifice diameter greater than 30 mm and headloss 35–40 mm. They should be about 75 mm below the water surface or more (0.1–0.2 m) below the water surface where freezing is a problem, depending on the thickness of the ice layer. To achieve the outlet flow, double sided launders are usually used. These are placed about 0.5 m from the end wall (rectangular tanks) or peripheral wall of circular tanks. Alternative arrangements include 'finger' launders of length $<20\%$ of the tank length (rectangular tanks) or radial launders (circular tanks) spaced at less than 8 m on the circumference feeding a collector launder. The depth of the rectangular tanks should be adequate for sludge deposits and storage; a minimum of 3 m is recommended for scraped tanks and 5 m recommended for manually cleaned tanks (Section 8.19). Circular tanks typically have a side water depth of 4.5 m. Detention times for particles of low relative density, i.e. the theoretical time of travel of water in the tank, varies from a minimum of 1½ hours to an average of 4 hours, but is more usually about 2½–3 hours.

Surface loading rates in clarifiers are often obtained by dividing the daily or hourly flow by the gross surface area of the tank. This can be misleading as it may not take into account the area occupied by the mixing and flocculation compartments (where these are provided), and the effluent launders, which usually account for at least 12% of the gross area. A better guide to the true loading rate is to take the horizontal clarification area available at a depth of about 1.25 m below the surface.

8.16 SLUDGE BLANKET OR SOLIDS CONTACT CLARIFIERS

Probably the earliest type of sludge blanket tank is the so-called 'hopper-bottomed tank'. The sludge blanket effect is obtained simply by allowing the chemically treated water to flow upwards in an inverted pyramid type of tank with the angle of slope usually 60° to the horizontal and terminating in a vertical section of 1.5 m water depth. They are usually square in plan (occasionally circular); although rectangular multiple hopper tank variations have been constructed for large capacity works. As the water rises in the tank of increasing area, its velocity progressively decreases until, at a given level, the reduced upward drag on the particles counterbalances the weight of the particles, leaving them suspended in the water. Their presence forms a kind of blanket in the water in which chemical and physiochemical reactions and flocculation can be completed and in which a straining action to remove some of the finer particles may also take place. While the top of the blanket is usually well defined there is no distinguishable bottom to the blanket because of the varying density of the particles forming the blanket.

The concentration of the sludge blanket is controlled by allowing it to bleed off via suitably placed hoppers or lightweight suspended cones. Sludge is also removed from the bottom of the hopper. The hopper-bottomed tank is not cheap to construct as the total water depth, governed by geometry, is usually about the same as a dimension of the plan. During construction care must be taken to ensure that the inverted pyramid is reasonably accurate and the inlet pipe discharges at the geometric centre; if it does not, streaming of the water to one side can occur and upset the stability of the sludge blanket. The inlet velocity (m/s) should be maintained at around a quarter of the rise rate in the vertical section of the clarifier (m/h), to avoid 'boiling' at the blanket surface. The surface loading of hopper-bottomed tanks is within the range 2–3.5 $m^3/h.m^2$, the lower velocity may sometimes be needed for floc formed in the removal of colour from a soft reservoir water.

In order to reduce construction costs there has been a continual striving towards simplification of the tank shape. This has led to the development of the flat-bottomed sludge blanket clarifier, occasionally circular but usually rectangular in plan, as illustrated in Figure 8.3. In this design chemically treated water is directed downwards onto the base of the tank through orifices discharging from low level inlet pipes before passing upwards through the blanket to the surface collecting launders. The sludge is collected in hoppers with their lips placed at the top level of the blanket and is removed under hydrostatic head.

A proprietary design of upward flow flat-bottomed sludge blanket tank is the Pulsator® (Fig. 8.4) where a proportion of the incoming chemically dosed water is lifted into a chamber built onto the main inlet channel, by applying a vacuum of about 0.65 m water gauge using a centrifugal fan, and released into the tank. This creates a pulsing effect in the blanket. The frequency and

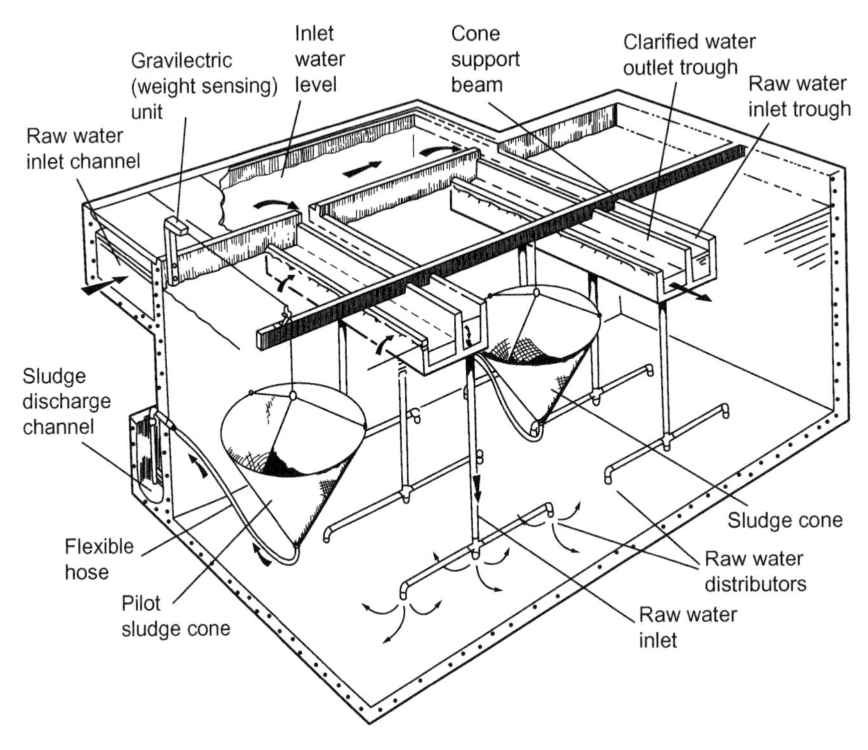

FIGURE 8.3

Flat-bottomed sludge blanket clarifier with Gravilectric® sludge cone (Paterson Candy Ltd).

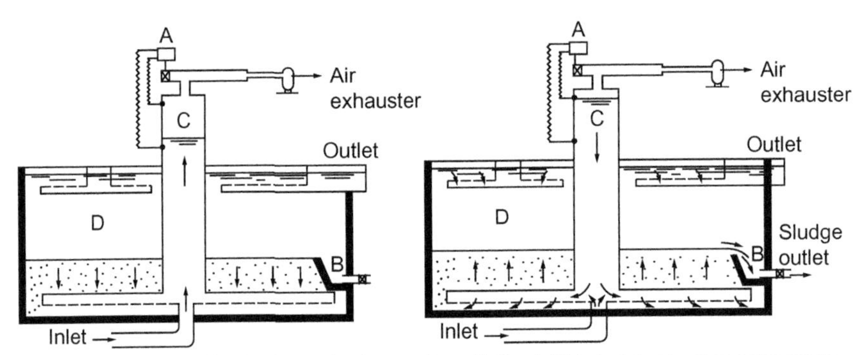

FIGURE 8.4

Pulsator® clarifier (Degremont, France).

duration of pulsation is varied according to the flow and water quality and typically would be of the order of 30–50 seconds and 8–10 seconds, respectively. These would normally be set to achieve a loading during the pulse in the tank of about 7–8 $m^3/h.m^2$ for waters containing low settleable solids and 10–12 $m^3/h.m^2$ for waters containing high settleable solids.

Sludge blanket clarifiers are suitable for many types of waters, including turbid ones, provided the particulate matter is of low density. A safe upper limit for turbidity is about 500 NTU but, depending on the nature of the particulate matter, much higher (up to 1000 NTU) peaks can be accommodated. Heavy particulate matter tends to settle on the bottom of the clarifier. When used in such applications these clarifiers should be provided with scrapers or other methods for intermittent removal of bottom sludge. Alternatively they should be drained down and cleaned after the rainy season, or about once or twice a year. In this respect, hopper-bottomed tanks which have bottom drains are more appropriate for highly turbid waters. For heavily silt laden waters sludge blanket clarifiers should be preceded by simple rectangular or circular sedimentation basins as described in Section 8.14. The surface loading rates of flat-bottomed sludge blanket clarifiers typically vary from 2 to 5 $m^3/h.m^2$. They have a side water depth of about 4.5–5 m, which is typically made up of a bottom distribution zone of 0.6–1.0 m, sludge blanket of 2.15–2.25 m and clarified water depth of 1.75 m. In some designs the clarified water depth could be as low as 1.0 m. In sludge blanket tanks, the blanket concentration is typically 20–25% v/v (after 10 minutes settlement in a 250 ml cylinder) and 0.1–0.2% w/v, although with some waters containing high settleable solids due to high colour and turbidity it could be as much as 30–35% v/v and 0.25–0.5% w/v.

The operation of sludge blanket tanks is somewhat sensitive to sudden changes in flow and raw water quality, but very tolerant to gradual changes which would be detrimental to the operation of many other types of clarifier. They require greater operator skill than simple settling tanks. They are not suitable for stop/start operation. Restart following a lengthy shut down may take more than 24 hours. However this could be reduced if sludge from a similar clarifier is available for seeding the blanket. Stoppages of 3–6 hours, can however be accommodated. There is also a constraint on flow turndown, which is normally limited to about two thirds of the maximum flow. The Pulsator clarifier, because of its high intermittent flow, can be operated down to about one third of the maximum flow.

The performance of sludge blanket clarifiers (and horizontal flow settling tanks) is known to be influenced by temperature (Hudson, 1981). The temperature effect is normally diurnal and is caused by the creation of thermal gradients within the clarifier due to the walls of the tank being heated by the sun or by warmer water entering from an open raw water storage tank or an exposed raw water main, giving rise to density currents within the clarifier. The result is disturbance of the blanket and carry over of floc towards the evening. Sometimes similar effects have been attributed to the release of gases due to bacterial activity in sludge. Measures taken to minimize carry over of floc include use of polyelectrolyte as a coagulant aid; inclusion of tube modules in the clarified water zone; or perhaps both techniques. Intermittent chlorination would help to overcome bacterial activity in sludge. So long as their sensitivity is appreciated and they are operated intelligently sludge blanket clarifiers will produce a good quality effluent (turbidity of about 1 NTU).

There are many other designs which endeavour to achieve the same high level of performance as sludge blanket tanks by providing extra mixing energy for flocculation or by the recirculation of sludge, or by a combination of both methods. Nearly all such variations have circular configurations and are usually equipped with bottom sludge scrapers, but operate as solids contact clarifiers. The essential feature is that the settled floc is used to seed the incoming dosed water, thereby accelerating the flocculation process. This is achieved internally (i.e. 'Accentrifloc' clarifier) by feeding the chemically dosed raw water into the flocculation zone where it is mixed with sludge recirculated by the aid of a rotor impeller or externally (i.e. 'Pre-treator' clarifier or Densadeg® clarifier) where a pump is used to recycle sludge to mix with dosed water. Clarifiers of these designs can be operated at surface loading rates about two to three times the loading of those without recirculation. Densadeg®, which incorporates tube modules, can operate at much higher ratings. Optimization of performance requires some skill as there are several operating variables such as: rotor impeller speed (peripheral speed 0.5–1.5 m/s); recirculation flow (up to 20% of tank throughput for external recirculation or up to five times or, for softening, up to 15 times the tank throughput for internal circulation); blanket depth; and sludge withdrawal rate. Such clarifiers have a high mechanical plant content and are comparatively high in capital and operating costs.

8.17 HIGH RATE CLARIFIERS

Commercial competition between treatment plant manufacturers and limited space for construction of treatment facilities have provided the main incentive for the development of new techniques for obtaining higher flow ratings in clarifiers. The maximum rate for a sludge blanket tank using coagulant alone is about 2 $m^3/h.m^2$; the upper limit with polyelectrolytes is about 5 $m^3/h.m^2$. In proposals for clarification it is important to check the probable performance at the lowest water temperatures likely to apply, when the viscosity of the water is highest. This can be the limiting criterion for performance.

Tube or Plate Settlers

Some success with increasing flow rates through existing clarifiers has been achieved by the use of tube modules or plates, which make use of the principle established long ago by Hazen (1904), that a settling tank should be as shallow as possible in order to shorten the falling distance for particles. The use of all-plastic tube modules or plates is a logical development of shallow depth settlement by multiple trays (Section 8.14), which helps to lessen the sludge removal problem of wide shallow trays. The 'tubes' are circular, hexagonal or square in cross section, of hydraulic diameters in the range 50–80 mm and made of plastic, usually polystyrene loaded with carbon black to protect against ultraviolet radiation. The best shape for a tube of a given area is a shape with the largest perimeter. The order of preference based on projected area is hexagonal (regular or chevron), then square followed by circular (Degremont, 2007). The tubes have a large wetted perimeter relative to the cross-section area and thereby provide laminar flow conditions which theoretically offer optimum conditions for sedimentation. Laminar flow is not achieved immediately when water enters the tubes, but after a transition length L_t (m) given by the Schiller formula (Coulson, 1999), $L_t = 0.0288\ Re.D$, where Re is the Reynolds number which should be less than 280 and D is the hydraulic diameter (m) which is equal to $4s/p$, where s is the cross-section area (m^2) of the tube and p its wetted perimeter (m). In the design calculations this transition length must be deducted.

Sedimentation takes place in the length following the transition length and it will retain all particles with a settling velocity less than V_s (m^3/h.m^2) (Yao, 1973) which is given by:

$$V_s = \frac{Vk}{\text{Sin}\,\theta + \frac{L_s}{y}\text{Cos}\,\theta}$$

where, V is average velocity of flow in the tubes (m^3/h.m^2) which is equal to U sin θ, where U is the average upward velocity (m^3/h.m^2) which is the surface loading rate of the settling tank (i.e. rate of flow÷plan area); L_s is settling length (m) which is equal to $(L - L_t)$ where L is the total tube length (m); k is a coefficient (1.0 for parallel plates, 1.33 for circular and hexagonal tubes and 1.38 for square tubes); θ is angle of inclination of tubes to the horizontal; and y is perpendicular distance between two adjacent plates (m).

Flow in the tubes or plates should be laminar with $Re < 200$, but <50 is preferred. Froude number should be $>10^{-5}$. V should be <10 m/h, detention time in the tubes should be <10 minutes and in the plates should be <20 minutes and surface loading rate for the tank should be <7.5 m^3/h per m^2 plan area of the tank depending on the raw water quality. The clear water depth above the plates or tube modules should not be less than 300 mm. The vertical depth occupied by the tubes or plates is 500–1000 mm. The velocity under tubes or plates should be less than 10 mm/s and the clarified water removal rate should not be greater than 15 m^3/h per m equivalent launder length. In tube (or plate) settler design at least one third of the tank length should remain tube (or plate) free. When the angle of inclination of the tubes to the horizontal is between 55° and 60° the solids settled on the inclined wall of the tube slide downwards under gravitational forces along its lower side, the clarified water flowing in the counter-current direction. Most tube settlers have been used in circumstances where it was desired to increase the flow through or improve the performance of an existing clarifier, particularly horizontal and radial flow clarifiers and some sludge blanket clarifiers, so that the increased clarification rates achieved were still in the range 2–5 m^3/h.m^2 (Galvin, 1992).

Sludge blanket clarifiers are uprated by incorporating tube or plate modules in the clear water zone. The clear distance between the tube pack and the top water level and the sludge layer below is usually maintained at about 500 mm in each direction. The vertical depth occupied by the tube pack is about 650–750 mm. An example of a sludge blanket clarifier with tube modules in the clarified water zone is illustrated in Figure 8.5. The use of tube packs allows up to twofold increase in the surface loading rate; in essence the tube packs, which provide a much larger settling area than the clarifier plan area, trap the floc carried over from the blanket when subjected to such high rates. Major disadvantages with tube settlers particularly those in sludge blanket clarifiers are that the floc carried over to the filters tends to be fine and may not be retained well in the filters and that the clear water depth of only about 500 mm encourages algal growth on the tube modules which, along with slime growth if allowed to form, would partially clog them and affect performance. It is desirable to use anthracite-sand filters downstream of these clarifiers. However, the use of the tubes helps to minimize the effect of thermal 'boiling' and wind on the clarifier performance. There are few examples of proprietary clarifiers where tube packs are incorporated into the design of new tanks.

An example of a sludge blanket clarifier which has made practical use of the plate system within the sludge blanket is the Super Pulsator®. The plates are about 300 mm apart, at an angle of

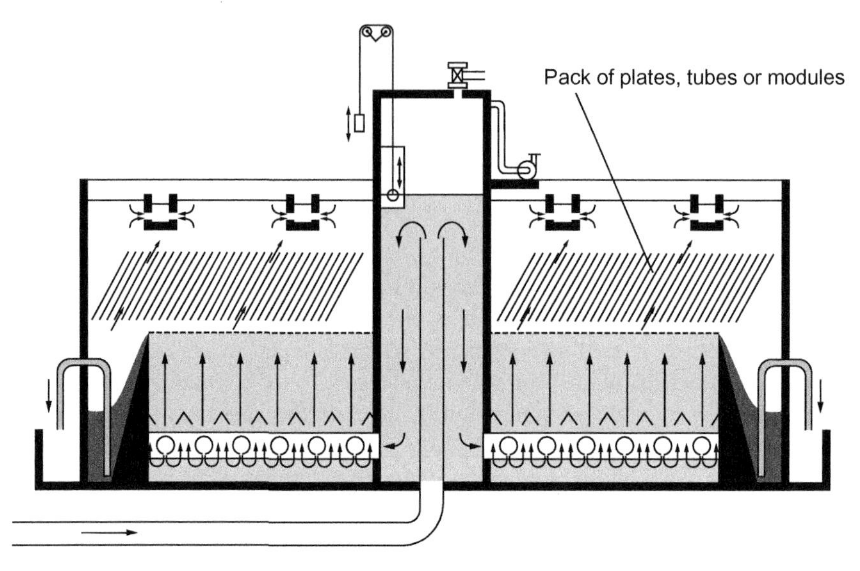

FIGURE 8.5

Pulsatube Pulsator® clarifier (Degremont, France).

60° to the horizontal and perpendicular to the sludge collection hoppers; coagulated water passes upwards and sludge travels between the plates into the hoppers. It is reported to achieve clarification rates in the range 5–10 m^3/h.m^2 and sludge concentration about twice that in a Pulsator operating at the same upward flow rate.

Lamella Clarifiers

The principle of shallow depth sedimentation has been extended to the design of a parallel plate system, sometimes referred to as lamella clarifiers. Clarifiers using the plate system are usually purpose built to take advantage of the high settlement rates which can be obtained and the greater density of sludge provided. They require an efficient flocculation stage which is critical for successful operation. The flocculated water enters at the base of lamella plates and travels upward between the plates counter-current to settled sludge moving down (Plate 11(a)). In some designs the uneven distribution to the inlet of the lamellas is corrected by introducing the flocculated water flow individually into each lamella space via slotted openings in the side walls of channels running on both sides of the plate pack along the length of the tank. Each space between the lamella plates therefore acts as an independent settling module. The lamella plates extend the full depth of the tank and rise about 125 mm above the top water level. Clarified water is collected in decanting launders running along each side of the plate pack by submerged orifices or V-notches one between each pair of plates (Plate 11(b)). About 1.5 m is allowed in the bottom of the tank for the collection of sludge which is removed by a circular scraper or a scraper of the chain and flight or reciprocating type to a central hopper or a series of small hoppers at one end of the tank. For small tanks sludge could be collected in hoppers placed underneath the plate pack. The plates are inclined at 55–60° to the horizontal. The turbidity of the clarified water is about 1–2 NTU.

With this arrangement, the settling area available is equal to the sum of the projections of plates in a horizontal plane. Thus the settling area is very large on account of the overlapping of plates but occupies a relatively small plan area. The total settling area is equal to $(n-1)\ LW \cos\theta$ where n is the number of plates, L the plate length in water (m) after deducting the transition length, W the plate width (m) and θ the angle of inclination of the plates to the horizontal. The value of n should be determined taking plate thickness and spacing between plates into consideration. The plates should be flat and not corrugated and they are usually made of stainless steel but sometimes of plastic. Plate width is about 1.25–1.5 m and plate length is about 2.5–3.25 m including the length of 125 mm above the normal water surface; plate thickness is usually 0.7 mm for stainless steel. The horizontal spacing between plates is varied according to the application and is normally in the range 50–80 mm. Lamella clarifiers are designed on the basis of settling rates related to the projected area of the plates. Depending on the settling velocity of the flocculated particles, the settling rate (also known as the Hazen velocity) could be in the range of 0.6–1.5 m/h, with the higher rates (>1.0 m/h) being used for waters containing high levels of turbidity. Surface loading rates equate to about 7.5–25 $m^3/h.m^2$ (based on plan area of the clarifier) and can therefore give a much reduced plan area (up to 95%) compared with more conventional horizontal flow clarifiers. This also means that the retention time within the clarifier is low, sometimes 20 minutes or less, so that control of chemical treatment becomes more exacting. The tendency for algae to grow on the plates is a problem with lamella clarifiers. Such clarifiers should therefore be enclosed in a building. The use of chlorine for algae control could corrode the length of the stainless steel plates above water (due to air above the water containing moist chlorine) and the plates should therefore be protected by a plastic laminate down to about 300 mm below the water surface.

Other High Rate Clarifiers

In ballasted flocculation clarifiers (e.g. the Actiflo® process) a suspension of fine quartz sand (d_{10} of 100–150 μm and uniformity coefficient of 1.6) is used as a ballasting agent to form a weighted floc of density in excess of 2.5 kg/l, resulting in very high settling velocities. In the process coagulant is first mixed in the water using turbine stirrers (mixing time 1–2 minutes), followed by sand in a similar mixer (1–2 minutes) and high energy flocculation (4–8 minutes). Polyelectrolyte is added to any one of the three tanks depending on the water quality. Settlement takes place in flat-bottomed or hopper-bottomed tanks fitted with inclined parallel plates or tubes (at 60° and vertical height about 1.25 m) operating at rates in the range of 25–60 $m^3/h.m^2$ (Cailleaux, 1992). Even higher rates are possible in the so-called Actiflo®-Turbo process, which incorporates a shroud around the sand turbine to enhance mixing of the coagulated water with sand and polyelectrolyte. The sludge containing the sand falls to the bottom and in the case of flat-bottomed tanks, is removed by scrapers to a series of hoppers located at one end. The sludge is then recycled at 3–6% of works throughput via hydrocyclones to separate the sand from the sludge. Hydrocyclones are operated at constant rate and multiple units are required for clarifier output turndown, which may be over 5:1. The sludge discharge from the hydrocyclones (overflow) is about 80% of the recycle rate. The recovered sand (underflow 20%) is then made up with fresh sand to account for losses in the sludge stream and recycled to 'seed' the incoming water again. The process requires about 3–6 g/l of sand which is equivalent to about 0.08% of the volume of the settling tank. The make up sand is of the order of 1–2 mg/l and is injected either continuously or in larger quantities intermittently (about once a week). The process removes *Cryptosporidium* oocysts and *Giardia* cysts like any other clarification process, which follows coagulation and flocculation. There is however the potential risk of returning some oocysts and cysts back to the process along with the recycled sand. This aspect should be investigated by pilot trials. The process is highly dependent on polyelectrolyte and the control of the dose is paramount as over dosing could harm the downstream filters. The proportion of water lost as sludge can be about 2.5% of works throughput consisting of about 0.1% w/v solids. The energy consumption of the process can be in the range 0.01–0.02 kWh/m^3 of water treated.

The principal advantage of high rate clarifiers is that they occupy a small footprint, but they need close attention and optimization of chemical treatment. Most of them depend on polyelectrolyte dosing. The retention time in some of them is very short and therefore tolerance to changing water quality is significantly reduced and the sensitivity to optimum operating parameters is significantly increased.

8.18 DISSOLVED AIR FLOTATION

DAF operates on the principal of the transfer of floc to the surface of water through attachment of air bubbles to the floc. The floc accumulated on the surface, known as the 'float', is skimmed off as sludge (Section 8.19). The clarified water is removed from the bottom and is sometimes called the subnatant or 'floated' water. The primary components of the process are illustrated in Figure 8.6. Since rain, snow, wind or freezing could cause problems with the float, flotation tanks must be fully enclosed in a building; some users enclose the flocculation tanks as well. The process

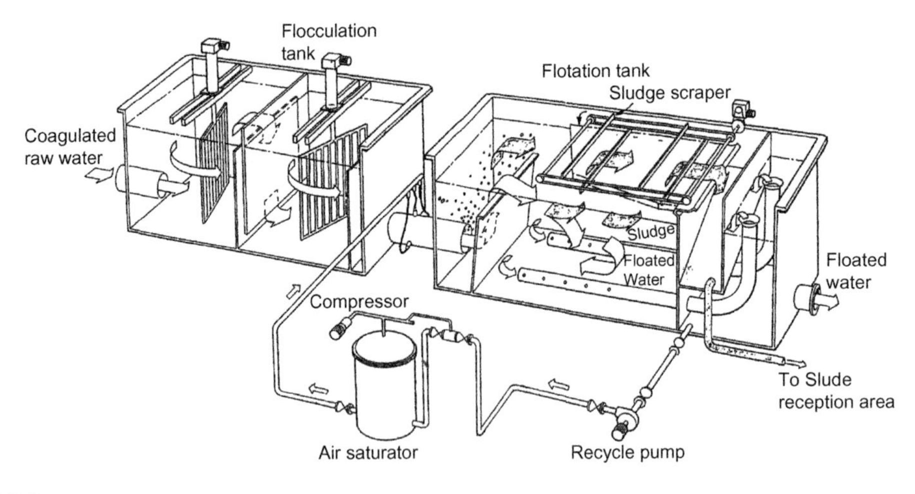

FIGURE 8.6

Diagram of dissolved air flotation plant (Purac Sweden, AB).

is particularly suited to treatment of eutrophic, stored lowland or otherwise algae laden waters and soft, low alkalinity upland coloured waters (Longhurst, 1987; Rees, 1979). Like all clarification processes flotation performance depends on the effectiveness of coagulation and flocculation. Polyelectrolyte dosing is often included to compensate for reduced performance at low water temperature or if the floc is fragile. Although the process has been successfully used for some directly abstracted waters other clarification methods tend to be more suitable for treatment of such waters especially when the turbidity consistently exceeds about 100 NTU (Gregory, 1999; Valade, 2009). Table 8.4 shows some typical results when treating algal laden waters.

There is, however, some experience with eutrophic waters with very high counts of algae where DAF has not been successful, so that caution is necessary when choosing the process. It should be noted that sedimentation can achieve degrees of removal comparable to flotation, if algae are first inactivated by chlorination. This would however result in the formation of DBPs by the action of chlorine on algal metabolic products.

Flotation is preceded by a flocculation stage of the hydraulic or mechanical type usually dedicated to each flotation cell. The flocculation tank should normally have at least two compartments in series (Section 8.11). Flotation is normally carried out in rectangular tanks designed with surface loading rates (based on total basin surface area) between 8 and 12 $m^3/h.m^2$ but rates as low as 5 $m^3/h.m^2$ or as high as 15–20 $m^3/h.m^2$ have been used on some plants (Pfeifer, 1997; Shawcross, 1997; Edzwald, 2012). At rates >20 $m^3/h.m^2$ (depending on water temperature) there is a risk of air entrainment in the clarified water which could cause negative head due to air binding in downstream filtration processes (Section 9.3). This can be overcome by minimizing the high velocity that can cause bubble entrainment at the DAF outlet or as in some proprietary designs by withdrawing clarified water uniformly over the entire floor of the tank or by using deeper tanks. When air entrainment is a problem, a de-aeration tank could be provided at the outlet.

In flotation the solids loading can vary in the range 0.25–1.0 kg dry solids/$h.m^2$. Typical tank depth is 2–3 m and the preferred length:width ratio is 1.33–2.5:1 with lengths up to 15 m using end-feed of air or 20 m with centre-feed of air. Width is limited to about 6 m for scraped tanks. The retention time in the flotation tank is between 10 and 20 minutes. The velocity in the subnatant opening should not exceed 0.05 m/s. The flow over the clarified water discharge weir should be less than 100 m^3/h per m of weir length.

For effective flotation the quantity of air required is about 5–10 g/m^3 or 4–8 l/m^3 of water treated and requires a recycle flow rate of about 6–15% (typically 8–10%) depending on temperature

Table 8.4 Comparison of algal cells in the raw water and % removal after coagulation by a ferric salt and sludge blanket clarifiers at 1 $m^3/h.m^2$ or flotation at 12 $m^3/h.m^2$ (Parr, 1991)

Alga	Raw water	Sludge blanket clarifiers	Flotation
Aphanizomenon	179 000	87%	98%
Microcystis	102 000	76.5%	98%
Stephanodiscus	53 000	58.7%	82.8%
Chlorella	23 000	84.3%	90.4%

and dissolved oxygen concentration of the incoming water (Edzwald, 1992). The recycle flow should be included in the flow used for the computation of the rates for the flotation unit. Recycle water should preferably be filtered water. Clarified water, if used, should be strained to prevent recycle nozzle blockage. Oil-free compressors are preferred but not essential for the air supply. Air is dissolved in recycle water under pressure either in pressure vessels equipped with an eductor on the inlet side for adding air or in a packed column; the operating pressures of the two respective saturator systems are 6–7 and 3.5–6 bar. In packed columns a packing depth of 0.8–1.2 m of 25–37.5 mm Pall or Raschig rings of polypropylene (unsuitable for chlorinated water) or polyvinylidene difluoride are used. The hydraulic loading rate of the air dissolving units lies in the range 50–90 $m^3/h.m^2$. Saturator efficiency for packed column type is about 90–95% whilst that for unpacked type is about 65–75% (Amato, 1997, 2001). Saturator efficiency is 100 times the amount of air measured in the recycle water divided by the amount of air that could be dissolved theoretically. Air saturated water is returned to the flotation tank through a series of nozzles or needle valves to give a sudden reduction in pressure and release of air bubbles in a white water curtain. Typically bubble size ranges from 10 to 100 μm with a mean diameter of 40 μm (Zabel, 1984). The outlets are usually spaced at 0.3–0.6 m for needle valves and 0.1–0.3 m for nozzles (Dhalquist, 1997). A typical nozzle density is about 10 per m^2 provided in two or three manifolds which could be isolated independently to facilitate greater turndown of recycle flow without loss of pressure. The contact time in the riser section should be about 100–120 seconds.

In plants where there is a need for raw water ozonation and flotation, the two processes could be combined with air in the flotation process being replaced by an ozone–air or ozone–oxygen mixture (Boisdon, 1994).

High rate flotation processes are finding application as they require smaller footprint. These include proprietary designs AquaDAF® (Plate 12(a)), Clari-DAF® (Plate 12(b)) and Enflo-DAF™, for which surface loading rates up to 50 $m^3/h.m^2$ are claimed. AquaDAF® comprises a pre-fabricated perforated false floor with distribution of holes of different sizes across the floor, designed for uniform withdrawal of flow over the whole area of the tank which is said also to maintain a deeper bubble blanket throughout the float area. The holes are also thought to act as bubble collectors and allow for bubble coalescence minimizing carry over to the filters. The combination of these effects is believed to provide performance comparable to conventional flotation (where clarified water is collected at one end and the bubble blanket is concentrated at the inlet end and grows shallower along the length of the tank) but at much higher rates. The surface loading rates are between 25 and 50 $m^3/h.m^2$. The width of the tank is greater than the length in the ratio 1.5–2:1 and the depth is about 4 m. The other design parameters (such as flocculation requirements, bubble size, recycle ratio and air dose) are similar to conventional flotation. Clari-DAF tank geometry is similar to conventional flotation design, but deeper (about 4.5 m) and clarified water is removed through a pipe lateral system located on the floor of the tank. Enflo-DAF™ also uses a deep tank (up to 5 m) (Amato, 2012).

Since the clarified water is taken from the bottom of the tank in the flotation process, it could be combined with rapid gravity filtration in the same tank with the filtration section placed underneath (DAFF) for example Enflo-Filt™ and Flofilter®. In this case, the surface loading rates of the two processes need to be the same and should include the recycle flow. COCO DAFF® is an innovative combined flotation–filtration design in which air and water flow counter-current as against co-current in the conventional DAF process (Fig. 8.7). Air is introduced with recycle water across

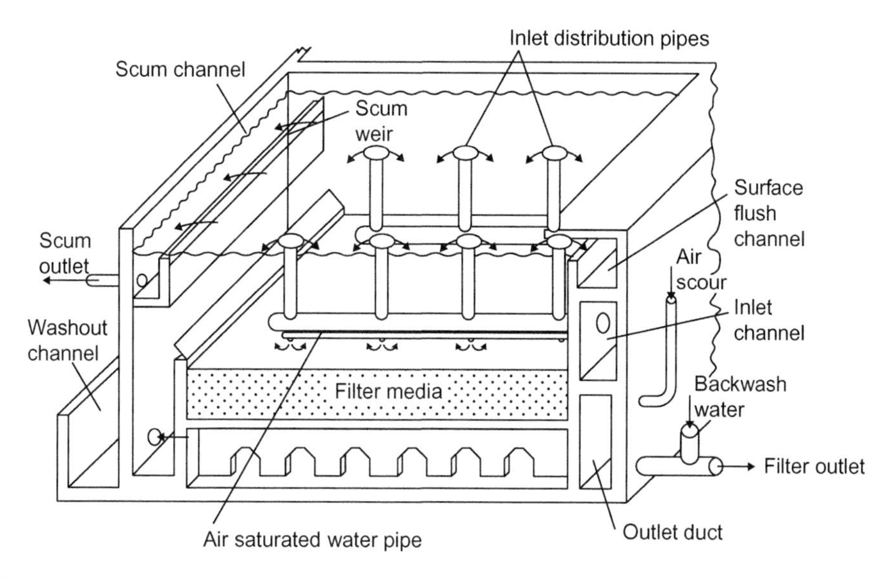

FIGURE 8.7

Typical arrangement of a COCO DAFF tank (Black & Veatch).

the total tank sectional area below the flotation zone and therefore only the filter surface loading rate should include the recycle flow. COCO DAFF gives more efficient particle–bubble interaction, and therefore increase in turbidity during desludging is minimized (Officer, 2001). The process combines flotation and gravity filtration in one tank and uses a group of flocculation tanks common to all of the flotation cells. Flocculation is usually hydraulic (Section 8.11) and continues within the bubble blanket. Since the recycle flow is dissipated into the clarified water and not to the flocculated water as in conventional DAF, floc damage is minimized. The process requires far fewer recycle nozzles.

The flotation process is suitable for stop/start operation and has a flow turndown of about 2:1 or greater depending on the design of aeration manifolds. The former is one of its advantages when dealing with a water subject to high algal loadings; a plant can be 'switched in' as and when needed and will give a steady quality treated water within 45 minutes (Rees, 1979). Apart from the drawbacks common to all high rate clarifiers, the flotation process has high energy requirements (about 0.05–0.075 kWh/m^3 of water treated).

8.19 SLUDGE REMOVAL FROM CLARIFIERS

Effective removal of sludge is very important for the efficient operation of clarifiers. With a raw water having suspended solids less than about 250 mg/l (most waters used in many countries fall in this category) the sludge volume to be removed from the tank should not exceed about 2.5% of inflow. For raw waters having high suspended solids ($>$500 mg/l) the sludge may be as high as

5–10% by volume of inflow; for suspended solids >1000 mg/l the sludge may have to be removed continuously in order to keep the tanks in operation, even at a reduced output, and to maintain an acceptable water quality. Under such conditions output may have to be reduced and the sludge volume can be as much as 20–30% of inflow. Special measures for sludge removal must obviously be provided for heavily silted waters and, for those having above 1000 mg/l solids or possibly less, depending on the nature of the solids, it is necessary to provide scraping equipment for all designs if throughput and quality are to be maintained. Scrapers move sludge to a series of hoppers located at the inlet end of rectangular tanks or usually in the centre or periphery of circular tanks. The tank floor should slope towards the hoppers. Hoppers are of an inverted pyramid shape with an included angle of about 60°. Sludge is removed from hoppers individually under hydrostatic head using the full water depth in the tank.

Rectangular tank scraper design depends on the tank geometry; they are either of the travelling bridge type with or without suction headers (for tanks up to 25 m × 75 m) or of the chain and flight or cable hauled type (for tanks up to 6 m × 50 m). Bridge scrapers with suction headers (speeds varying from 1.0 to 2.0 m/min) and chain and flight type (speeds less than 0.5 m/min) are suitable for tanks treating heavily silted waters. The speed of bridge scrapers without suction headers is 0.5–1.0 m/min for scraping and about 2.5 m/min for the return.

Circular tanks have radial or diametrical scrapers with the bridge supported from the centre and driven with a central or peripheral drive unit. In larger tanks, support is also provided by a travelling wheel on the outside wall. The peripheral speed of the scraper is about 1.0–2.0 m/min and should not exceed 3 m/min. In some square tanks, corners are curved on the bottom and the scraper arm is equipped with a pivot type corner blade extension, which reaches out to the corners and then folds back on itself when traversing the four sides (pantograph mechanism).

Some circular tank designs include a suction header similar to the rectangular tanks. Tanks employing suction headers draw sludge at about 2 l/s.m of tank width or radius from points just in front of the scraper blades or squeegees using pumps mounted on the bridge; others are aided by submersible pumps or down pipes with eductors. The floors of scraped circular tanks have slopes of about 1:10–1:20 and those of scraped rectangular tanks have slopes of about 1:300–1:500. Rectangular tanks equipped with suction headers do not require a slope except for drainage. Unscraped rectangular tanks usually have a cross fall of about 1:10 to a central channel running the length of the tank and a longitudinal fall of 1:200. By including high pressure water jets (at 3.5–4 bar) for cleaning the slope can be reduced to about 1:250. Valves for sludge removal are always better placed outside the walls of the tank. Both valves and pipework should be adequately sized to pass the maximum sludge withdrawal rate, which can be 400% or 500% greater than the average rate. The valves should be of the full bore type like plug or pinch valves (Sections 18.12 and 18.14).

Sludge removal from blanket tanks is generally easier than with other designs, although the same rules for valve and pipe sizing must be used. The positioning of sludge hoppers is less critical than with other designs as a sludge blanket is in continual movement and migrates towards the space left by evacuated sludge. Hoppers usually occupy about 10–15% of the total settling area of the tank. The hydrostatic head available for sludge removal depends on the discharge level of the sludge pipe. Sludge removal is usually operated by an automatic system having adjustable timers for varying the duration of opening of sludge discharge valves at pre-selected but adjustable time intervals. When suspended solids are high continuous removal of sludge may be necessary. Sludge should be withdrawn from hoppers individually; manifolding hopper outlets to allow for simultaneous withdrawal is not recommended.

A method for initiating sludge removal, which has met with considerable success is a design that relies on sensing the differential weight of concentrated sludge in water. The equipment is shown in Figure 8.3 and consists of several flexible sludge cones suspended in water and one of which (called the master or pilot cone) is connected by a cable to a load cell. 'Slave cones' are normally positioned at a level 150 mm below the master cone so that they are always full of sludge when the master cone reaches the pre-set weight. The load cell is sufficiently sensitive that when the weight of the sludge reaches a pre-set value (usually when the cone is about two thirds full of sludge) the load cell initiates the opening of the desludging valves on all the cone outlets.

In flotation tanks sludge or 'float' collects on the water surface and is removed by mechanical or hydraulic means or a combination of the two. For highly coloured waters, the float should not be allowed to accumulate in the tank for more than 30 minutes, whereas much longer periods can be tolerated for algal laden or turbid waters. Mechanical units are usually scrapers of the travelling type for example chain and flight, reciprocating or bridge and the choice is made primarily on the tank dimensions. The flight speed should be about 0.5 m/min or greater (up to 2 m/min) if assisted by hydraulic desludging. Another design is a scraper with blades fixed to a cylinder which rotates over the beach. It produces a thin sludge. All types are known to cause 'knockdown' of float solids to some degree, a process where clarified water is contaminated with the sludge as a result of de-aeration of the float due to the disturbances caused by the scraper. In hydraulic desludging, the clarified water draw-off is restricted intermittently to raise the water level in the cell until the sludge layer overflows into a collection trough. The overflow rate should be greater than 6 l/s per m weir length. Water spray bars should be used to reduce friction at the walls and the float area should be free of any fixtures that would restrict the movement of the float. In COCO DAFF, where there is no natural cross-flow velocity, filtered or flocculated water is pumped to a side channel (central channel in duplex units) and overflows a ski-jump shaped weir, pushing the float towards the sludge collection trough. This method produces sludge of low concentration compared to mechanical methods but with reduced 'knockdown'. The full-depth air blanket across the full process area in the COCO DAFF process improves the chance of any sludge knockdown being re-floated.

COAGULANTS AND COAGULANT AIDS

8.20 ALUMINIUM COAGULANTS

Aluminium sulphate is the most widely used aluminium coagulant. It is available in a number of solid forms such as block, kibbled or ground and is also available as a solution. In waterworks practice aluminium sulphate is frequently but incorrectly referred to as 'alum'. The solid form has the composition $Al_2(SO_4)_3xH_2O$ where x may range from 14 to 21 containing 14–18% w/w Al_2O_3 (alumina) or 7.5–9% w/w Al (aluminium), depending on the number of molecules of water (x). The liquid form contains 8% w/w Al_2O_3 or 4.2% w/w Al. The amount of Al_2O_3 or Al in any solid form of aluminium sulphate containing x moles of water is given by $y\%$ w/w $Al_2O_3 = [5.67 \div (19 + x)] \times 100$ and $z\%$ w/w $Al = [3 \div (19 + x)] \times 100$, respectively.

The aluminium sulphate dose is therefore normally expressed in mg/l as *y*% w/w Al_2O_3 depending on the form of aluminium sulphate, or more usefully in mg/l as Al; 1 mg/l as *y*% w/w Al_2O_3 is equal to $5.29\ y \times 10^{-3}$ mg/l as Al.

When dosed into water, the formation of an aluminium hydroxide floc is the result of the reaction between the acidic coagulant and the natural alkalinity of the water, which usually consists of calcium bicarbonate. A dose of 1 mg/l of aluminium sulphate as Al reacts with 5.55 mg/l of alkalinity expressed as $CaCO_3$ and increases the CO_2 content by 4.9 mg/l. Thus if no alkali is added the alkalinity would be reduced by this amount with a consequent reduction in pH. If a water has insufficient alkalinity or 'buffering' capacity, additional alkali such as hydrated lime, sodium hydroxide, or sodium carbonate must therefore be added; the alkalinity expressed as $CaCO_3$ produced by 1 mg/l of each chemical (100% purity) is 1.35, 1.25 and 0.94 mg/l, respectively. The aluminium hydroxide floc is insoluble over relatively narrow bands of pH, which may vary with the source of the raw water. Therefore pH control is important in coagulation, not only in the removal of turbidity and colour but also to maintain satisfactory minimum levels of dissolved residual aluminium in the clarified water. The optimum pH for coagulation of lowland surface waters is usually in the range of 6.5–7.2, whereas for more highly coloured upland waters a lower pH range, typically 5–6, is necessary. Lowland waters usually contain higher concentrations of dissolved salts, including alkalinity and may therefore require the addition of an acid in excess of that provided by the coagulant. Under these circumstances, it is usually more economic to add sulphuric acid rather than excess aluminium sulphate to obtain the optimum coagulation pH value. The use of excess coagulant to depress the pH would result in more sludge production.

There are some coagulants, such as polyaluminium chloride (PACl), aluminium chlorohydrate (ACH), polyaluminium chlorosulphate (PACS) and polyaluminium silicate sulphate (PASS), which are formulated to contain high basicity, a measure of hydroxyl ions present in the coagulant. A definition of basicity is given by Letterman (1999). Aluminium sulphate has zero basicity whereas that of the polymerized aluminium salts varies from about 50% for $Al_2(OH)_3Cl_3$ (i.e. 10% w/w Al_2O_3) to about 85% for ACH (24% w/w Al_2O_3). The H^+ ions produced by aluminium sulphate, PACl (10% w/w Al_2O_3) and ACH are 6, 3 and 1, respectively. High basicity coagulants therefore depress the pH of the treated water less than aluminium sulphate, thereby minimizing the need for coagulation pH adjustment and reducing the alkali dose required for subsequent final pH correction. These salts also have a number of additional benefits: they limit aluminium residuals whilst maintaining optimum coagulation properties; they produce stronger and more readily settleable floc than aluminium sulphate, thus reducing the need for polyelectrolytes as coagulant aids; coagulation is less affected by low temperature and they produces less sludge than aluminium sulphate. In some waters such salts can be used in lower doses than aluminium sulphate and over a broader optimum pH range (6–9). There are several grades of PACl containing 10%, 18% or 24% w/w Al_2O_3; the 10% and 18% w/w grades being the most commonly available. The other polymeric aluminium salts PACS and PASS although less common, behave in a similar manner to PACl. The properties of the most commonly used aluminium coagulants are summarized in Table 12.3.

The presence of aluminium in drinking water can be harmful to users of renal dialysis (Section 7.6). It has also been suggested that aluminium in drinking water may be associated with neurological disorders and Alzheimer's disease (Section 7.6). Although no definitive link has been established, as a precautionary measure, some water undertakings prefer ferric to aluminium coagulants (Carroll, 1991). When making a change in coagulant type, care should be taken to clean process units free of all accumulated floc, which would otherwise dissolve and increase the aluminium concentration in the water if the ferric coagulant is used outside the optimum pH range for the aluminium coagulant.

8.21 IRON COAGULANTS

Iron coagulants in the ferric form behave similarly to aluminium sulphate and form ferric hydroxide floc in the presence of bicarbonate alkalinity. A dose of 1 mg/l of ferric sulphate or chloride as Fe neutralizes 2.7 mg/l alkalinity expressed as $CaCO_3$ and increases the CO_2 content by 2.36 mg/l. Ferric hydroxide floc is insoluble over a much broader pH range (4–10) than aluminium hydroxide floc. The lower end of the pH range (4–5.5) is useful for treating highly coloured moorland waters.

Iron coagulants are available as ferric sulphate, ferric chloride and ferrous sulphate. Ferric salts are very corrosive acidic liquids. Ferric sulphate is usually preferred to ferric chloride since the introduction of chloride ions may increase the corrosivity of a water (Section 10.41). Ferrous sulphate, traditionally referred to in its hydrated form ($FeSO_4.7H_2O$) as 'copperas', is used as a coagulant usually in conjunction with chlorine; in practice excess chlorine is used. Chlorinated ferrous sulphate is not used due to the potential risk of DBP formation by the action of excess chlorine with DBP precursors in the raw water. Ferrous sulphate on its own is used as a coagulant in processes utilizing high pH values such as lime softening (pH 10–11) and manganese removal (pH 9). Iron coagulants have the advantage of producing a denser floc than that produced by aluminium sulphate thereby producing improved settlement characteristics but at the expense of about a 40% increase in the weight of hydroxide sludge when compared to aluminium coagulants.

Polymeric ferric sulphates are now available; they contain about 12.5% w/w Fe and are claimed to perform better and at lower doses than ferric sulphate (Jiang, 1996). There are ferric-aluminium sulphate coagulants; one such product contains approximately 10% w/w of the metal oxides made up of 7% w/w Al_2O_3 and 3% w/w Fe_2O_3. The properties of commonly used of iron coagulants are summarized in Table 12.3.

Many iron coagulants contain approximately 2–6 g of manganese per kg of iron as an impurity depending on the product specification. This contributes to the manganese concentration in the water.

8.22 COAGULANT AIDS AND POLYELECTROLYTES

Coagulant aids are used to improve the settling characteristics of floc produced by aluminium or iron coagulants. The coagulant aid most used for a number of years was activated silica; other aids included sodium alginates and some soluble starch products which are still in use. These substances had the advantage of being well-known materials already used in connection with the food industry and were thus recognized as harmless in the treatment of water. Polyelectrolytes came later into use and were more effective. They now comprise numerous synthetic products: long chain organic chemicals, which may be cationic, anionic or non-ionic. The theory of their action has been reviewed by Packham (1967).

Polyacrylamides are the most effective of the synthetic group of polyelectrolytes, but for their safe use the toxic acrylamide monomer residue (the raw materials used in their manufacture) which is not adsorbed by the floc, should be virtually absent from the product. The UK Water Supply (Water Quality) Regulations (UK, 2010) state that no batch may contain more than 0.02% w/w of free acrylamide monomer based on the active polymer content; the dose used must average no more than 0.25 mg/l and never exceed 0.5 mg/l. US EPA allows a maximum dose of 1 mg/l on a regulated acrylamide limit of 0.05% w/w in the polyelectrolyte.

Polyelectrolyte doses used are very small in relation to the dose of the primary coagulant. Natural polyelectrolyte (starch based) doses vary between 0.5 and 2.5 mg/l whereas polyacrylamide doses vary between 0.05 and 0.25 mg/l. Polyelectrolytes are added as a coagulant for turbid waters or after the primary coagulant as a coagulant aid (Section 8.12). Sometimes they are added just prior to filtration in very small doses (about 0.01 mg/l) to flocculate microfloc particles carried over from the clarifiers and filter-passing algae; care in control of the dose is necessary because excess polyelectrolyte could result in 'mud ball' formation and other problems in the filters. The cationic polyelectrolyte PolyDADMAC (polydiallyldimethylammonium chloride) is known to reduce the dose of inorganic coagulants when the two are used together. This has been observed in waters of high colour with both low and high alkalinity and conductivity. The reduction in the aluminium sulphate dose could be as much as 50% (Gebbie, 2005). The effect this reduction in the coagulant dose has on the TOC removal should be tested in the laboratory. PolyDADMAC could form N-nitrosodimethylamine (NDMA) (Section 7.80).

Most polyelectrolytes are powders and a solution must be prepared for dosing. For successful preparation the powder must be wetted properly by using a high energy water spray before dissolving; the solution should be allowed to age for about an hour in cold water or 30 minutes in warm water conditions before use. For polyacrylamide the solution should be prepared at about 2.5 g/l, whereas for natural polyelectrolytes the solution concentration could be as high as 25 g/l. Following metering the solution should be diluted 10-fold to assist transfer in the pipe and dispersion at the point of application. Once a batch of stock solution is prepared it should be used preferably within about 24 hours.

The practical effect of introducing polyelectrolytes in many existing waterworks has been to increase the settling rate and hence allow substantially greater output through the clarifiers; an additional use is to assist in the recovery of used filter washwater and thickening and dewatering of sludges (Chapter 13). Laboratory tests should be carried out to identify the most appropriate polyelectrolyte for the application. Historically, when using organic polyelectrolytes, cationic types have been used as a coagulant aid and anionic types for sludge treatment, with non-ionics being tried as an alternative. Cationic polymers are more toxic to aquatic life than anionic or non-ionic polyelectrolytes, however, and their use is discouraged if there is a risk of discharge to a local watercourse from sludge treatment plant, especially for low alkalinity and low pH receiving waters. Besides being less toxic to aquatic life, anionic polyelectrolytes are also less prone to foul downstream membranes when residuals are present in water from pre-treatment processes or supernatant returned from sludge treatment processes. A review of organic polyelectrolytes is given by Bolto (2007).

8.23 RAINWATER HARVESTING

Rainwater is typically devoid of all dissolved solids but contains dissolved gases (oxides of carbon, nitrogen and sulphur) which result in pH values of about 5.5 or lower. Rainwater is therefore devoid of alkalinity; it is acidic (low pH), low in mineral content and aggressive towards calcium bearing materials such as concrete and some of the metals typically used in domestic plumbing. In coastal areas rainwater may also contain up to 15 mg/l of sodium chloride from sea spray. The presence of chlorides exacerbates corrosivity. Rainwater is also unpalatable in taste, again due to the low solids content (Meera, 2006). A study carried out in Bangladesh (Karim, 2010) found that harvested rainwater had very low colour and turbidity but microbiological quality did not comply with local and WHO Guidelines. When stored in brick-, concrete- or cement-lined tanks the acidity of the water resulted in pH in excess of 8.5. The low mineral content of the water was considered to be an adverse health concern.

Rainwater is harvested from roofs (Section 3.26) via guttering into storage tanks and may be contaminated, particularly in the 'first flush' after a dry period, by detritus that has collected on the roof, for instance leaves, insects or more importantly bird faeces. In coastal locations, salt deposits from sea spray may also be present.

Water falling on roofs or other surfaces designated for the purpose of collection is directed, via screens to remove leaves, bird feathers and other coarse debris, to a collection tank. The first flush, typically about 0.5–2 l per m^2 of collection area (Krishna, 2005), is discharged to waste by means of 'stand pipes' or other proprietary designs and may be used for non-potable purposes such as gardening or toilet flushing.

The roof surfaces over which the rain is collected should be made of materials from which no undesirable compounds may be dissolved or leached if rainwater is intended for human consumption; examples are lead flashings and bituminous roof material.

A code of practice for rainwater harvesting is given in BS 8515 and an information guide has been issued by the EA (2008).

REFERENCES

Adkins, M. F. (1997). Dissolved air flotation and the Canadian experience. *Proc International Conference on Dissolved Air Flotation*. CIWEM, London.

Alvarez, M., Rose, J. and Bellamy, B. (2010). *Treating Algal Toxins Using Oxidation, Adsorption, and Membrane Technologies*. Water Research Foundation, Denver, Colorado.

Amato, T. (1997). DAF: its place in an integrated plant design. *Proc International Conference on Dissolved Air Flotation*. CIWEM, London.

Amato, T., Edzwald, J. K. and Tobiason, J. E. (2001). An integrated approach to dissolved air flotation. *Water Science and Technology* **43**(8), pp. 19–26.

Amato, T., Nattress, J. and Mackay, D. (2012). Seawater v. surface water treatment applying Enflo-DAF™ technology. *Proceedings AWWA Water Quality Technology Conference, Toronto*.

Amirtharajah, A. (1978). Design of rapid mix units. *Water Treatment Plant Design* (Ed. Sanks, R. L.). Ann Arbor Science.

Asquith, E. A., Evans, C. A., Geary, P. M. et al. (2013). The role of *Actinobacteria* in taste and odour episodes involving geosmin and 2-methylisoborneol in aquatic environment. *Journal WSRT* **62**(7), pp. 452–467.

AWWA (1969). *Water Treatment Plant Design*. AWWA.

Bare, W. F. R., Jones, N. B. and Middlebrooka, E. (1975). Algal removal using dissolved air flotation. *Journal WPCF* **47**(1), pp. 153–169.

Bauer, M. J., Bayley, R., Chipps, M. J. E. A. et al. (1998). Enhanced rapid gravity filtration and dissolved air pretreatment of River Thames Reservoir Water. *Water Science and Technology* **37**(2), pp. 35–42.

Boisdon, V., Bourbigot, M. M., Nogueria, F. et al. (1994). Combining of ozone and flotation to remove algae. *Water Supply* **12**(3/4), pp. 209–220.

Bolto, B. and Gregory, J. (2007). Organic polyelectrolytes in water treatment. *Water Research* **41**, pp. 2301–2324.

Brejchova, D. and Wiesner, M. R. (1992). Effect of delaying the addition of polymeric coagulant aid on settled water turbidity. *Water Science and Technology* **26**(9–11), pp. 2281–2284.

BS 8515 (2009). *Rainwater Harvesting Systems – Code of Practice.*

Cailleaux, C., Pujol, E., De Dianous, F. et al. (1992). Study of weighted flocculation in view of a new type of clarifier. *Journal WSRT* **41**(1), pp. 28–32.

Camp, T. R. (1946). Sedimentation and the design of settling tanks, Paper 2285. *Trans ASCE*, III (3), pp. 895–958.

Camp, T. R. and Stein, P. C. (1943). Velocity gradients and internal work in fluid motion. *Journal of Boston Society of Civil Engineers* **30**, pp. 219–237.

Carroll, B. A. and Hawkes, J. M. (1991). Operational experience of converting from aluminium to iron coagulants at a water supply treatment works. *Water Supply* **9**, pp. 553–558.

Chow, V. T. (1959). *Open Channel Hydraulics*. McGraw Hill, New York.

Coulson, J. M. and Richardson, J. F. (1999). *Chemical Engineering*, **Vol. 1**, 6th Edn. Pergamon Press.

Dahlquist, J. (1997). State of DAF development and application to water treatment in Scandinavia. *Proc International Conference on Dissolved Air Flotation*. CIWEM, London.

Davis, J. M. (1980). Destratification of reservoirs, a design approach. *Water Services* **84**, pp. 497–504.

Degremont (2007). *Degremont Water Treatment Handbook*, **Vol. 1**, 7th Edn. Lavoisier.

DoE (1971). *First Annual Report of the Steering Committee on Water Quality*. Circular No. 22/72.

DoE (1979). *Second Biennial Report: February 1977–February 1979.*

DoE (1993). *Bankside Storage and Infiltration Systems*. UK Water Research Centre Report.

Dominick, S. and DiNatale, K. (2009). Mile-high decisions: Denver. *Lakeline*, Fall.

EA (2008). *Harvesting Rainwater for Domestic Uses: An Information Guide*. Environment Agency.

Edzwald, J. K. (2010). Dissolved air flotation and me. *Water Research* **44**, pp. 2077–2106.

Edzwald, J. K. and Haarhoff, J. (2012). *Dissolved Air Flotation for Water Clarification*. AWWA & McGraw Hill.

Edzwald, J. K. and Wingler, B. J. (1990). Chemical and physical aspects of dissolved air flotation. *Journal WSRT* **39** (1), pp. 24–35.

Edzwald, J. K., Walsh, J. P., Kaminski, O. O. et al. (1992). Flocculation and air requirement for dissolved air flotation. *Journal AWWA* **84**(3), pp. 92–100.

Elliot, S. and Swan, D. (2013). Source water management – deep reservoir circulation. *7th Annual WIOA NSW Water Industry Operations Conference.*

Fair, G. M., Geyer, J. C. and Okun, D. A. (1968). *Water and Wastewater Engineering*, **Vol. 1**. John Wiley.

Galvin, R. M. (1992). Lamella clarification in floc blanket decanters: case study. *Journal WSRT-Aqua* **41**(1), pp. 28–32.

Gebbie, P. (2005). Dummy's guide to coagulants. *68th Annual Water Industry Engineers and Operators' Conference*, Bendigo.

Gregory, R., Zabel, F. F. and Edzwald, J. K. (1999). Sedimentation and flotation, *Water Quality and Treatment*, 5th Edn. McGraw-Hill.

Griffiths, S. (1996). The effect of agitator impeller design on maximum shear stress and the resulting impact on the flocculation process. *Journal CIWEM* **10**(5), pp. 324–331.

Hazen, A. (1904). On sedimentation, Paper No. 980. *Trans ASCE* **53**(2), 45–71.

Henderson, R., Chips, M. and Cornwell, N. (2008). Experience of algae in UK waters: a treatment perspective. *Water and Environment Journal* **22**, pp. 184–192.

Henderson-Sellers, B. (1984). *Engineers Limnology*. Pitman Advanced Publishing Program, Pitman.

Hoehn, R. C., Barnes, D. B., Thompson, B. C. et al. (1980). Algae as a source of trihalomethane precursors. *Journal AWWA* **72**(6), pp. 344–350.

House, J. and Burch, M. (2002). Using algicides for the control of algae in Australia. *CRC for Water Quality and Treatment, Australia.*

Hudson Jr, H. E. (1981). *Water Clarification Processes*. Van Nostrand Reinhold.

Huisman, L. (1970). *Lecture Notes: European Course in Sanitary Engineering*. Delft University of Technology.

Imhoff, K. (1971). *Disposal of Sewage*. Butterworth.

Jiang, J. Q., Graham, N. J. D. and Harward, C. (1996). Coagulation of upland coloured water with polyferric sulphate compared to conventional coagulants. *Journal WSRT* **45**(3), pp. 143–154.

Karim, M. R. (2010). Quality and suitability of harvested rainwater for drinking in Bangladesh. *Water Science and Technology: Water Supply* **10**(3), pp. 359–366.

Kawamura, S. (2000). *Integrated Design of Water Treatment Facilities*. John Wiley.

Kay, D. and Hanbury, R. (1993). *Recreational Water Quality Management*, **Vol. 2**. Ellis Horwood.

Krishna, J. H. (2005). *The Texas Manual on Rainwater Harvesting*, 3rd Edn. Texas Water Development Board.

Lang, U., Schic, R. and Schroder, G. (2010). The decision support system Bodensee online for hydrodynamic and water quality in lake constance (Chapter 6). *Decision Support System Advances* (Ed. Devlin, G.). Intech.

Letterman, R. D., Amirtharajah, A. and O'Melia, C. R. (1999). *Coagulation and Flocculation, Water Quality and Treatment*, 5th Edn. McGraw Hill.

Longhurst, S. J. and Graham, N. J. D. (1987). Dissolved air flotation for potable water supply. *Public Health Engineer* **14**(6), pp. 71–76.

Markham, L., Porter, M. and Schofield, T. (1997). Algal and zooplankton removal by dissolved air flotation at Severn Trent Ltd. *Proc Int Conf on Dissolved Air Flotation*, London.

McLaughlin, R. T. (1959). The settling properties of suspensions. *Journal ASCE* **85**(HY12), pp. 9–41.

Meera, V. and Ahammed, M. M. (2006). Water quality of rooftop rainwater harvesting systems; a review. *Journal WSRT* **55**(4), pp. 257–268.

Mitrakas, M., Samaras, P., Stylianou, S. et al. (2013). Artificial destratification of Dipotamos reservoir in Northern Greece by low air injection. *Journal WST* **13**(4), pp. 1046–1055.

Moldaenke, C., Baganz, D. and Staaks, G. (2009). *Monitoring of Toxins in Drinking Water Quality by the ToxProtect64 Fish Monitor*. Techneau 3.6.3.4, 2009.

Moss, B. (2010). *Ecology of Freshwaters: A View for the Twenty-First Century*. Wiley-Blackwell.

Mouchet, P. and Bonnelye, V. (1998). Solving algae problems: French expertise and world-wide applications. *Journal WSRT* **47**(3), pp. 125–141.

O'Conner, J. T. and Brazos, B. J. (1997). Evaluation of rapid sand filters for control of microorganisms in drinking water. *Public Works* **128**(3), pp. 52–56.

Officer, J., Ostrowlski, J. A. and Woollard, P. J. (2001). The design and operation of conventional and novel flotation system on a number of impounded water types. *Water Science and Technology: Water Supply* **1**(1), pp. 63–69.

Packham, R. F. (1962). The theory of coagulation process. (1) The stability of colloids. *Journal SWTE* **11**(1), pp. 53–63.

Packham, R. F. (1963). The theory of coagulation process. (2) Coagulation as a water treatment process. *Journal SWTE* **12**(1), pp. 106–120.

Packham, R. F. (1967). Polyelectrolytes in water clarification. *Journal SWTE* **16**(2), pp. 88–102.

Parr, W. and Clarke, S. J. (1992). *A Review of Potential Methods for Controlling Phytoplankton, With Particular Reference to Cyanobacteria, and Sampling Guidelines for the Water Industry*. Water Research Foundation Report FR0248.

Parr, W., Clarke, S. J. and Gourlay, G. L. (1991). *Method of Controlling Phytoplankton*. UM 1223. WRc.

Pastorok, R. A. (1980). Review of aeration/circulation for lake management in restoration of lakes and inland waters. *Proc International Conference*, Portland, Maine.

Pfeifer, B. J., Harris, D. I. and Adkins, M. F. (1997). DAF and catalytic filtration for treatment of a challenging water supply in a small northern community. *Proc AWWA Annual Conference*, Atlanta, Georgia.

Rees, A. J., Rodman, D. J. and Zabel, T. (1979). Water clarification by flotation. *TR 114*, WRc.

Ridley, J. E., Cooley, P. and Steel, J. A. P. (1966). Control of thermal stratification in Thames Valley Reservoirs. *Journal SWTE* **15**, pp. 225–245.

Schulz, C. R. and Okun, D. A. (1992). *Surface Water Treatment for Communities in Developing Countries*. John Wiley.

Sethunge, S. S., Manage, P. M. and Jayasinghe, S. (2009). Microstrainer for removal of algae in drinking water treatment. *Proceedings of the 14th International Forestry and Environment Symposium*, Sri Lanka.

Shawcross, J., Tiffany, T., Nickols, D. et al. Pushing the envelope: dissolved air flotation at ultra-high rate. *Proc International Conference on Dissolved Air Flotation*. CIWEM, London.

Skeat, W. O. and Dangerfield, B. J. (Eds) (1969). *Manual of British Water Engineering Practice*, **Vol. III**, 4th Edn. W. Heffer & Sons.

Steel, J. A. P. (1975). Management of Thames Valley reservoirs. *Effects of Storage on Water Quality*. WRc.

Sykes, G. and Skinner, F. A. (Eds) (1971). *Microbial Aspects of Pollution*. Academic Press.

Toms, I. P. (1981). Reservoir management. *Proc Water Industry '81*, CEP Consultants.

UK (2010). *The Water Supply (Water Quality) Regulations*. Statutory Rule 2010 No. 128 (Northern Ireland).

US EPA (1999). *Microbial and Disinfection Byproduct Rules Simultaneous Compliance Guidance Manual*. EPA-815-R-99-015.

Valade, M. T., Becker, W. B. and Edzwald, J. K. (2009). Treatment selection guidelines for particle and NOM removal. *Journal of Water Supply: Research and Technology – Aqua* **58**(6), pp. 424–432.

Westaby, C. (2010). Hydrocarbon in water monitors using fluorescence. *Power Plant Chemistry* **12**(9), pp. 534–539.

WHO (1996). *Guidelines for Drinking Water Quality*, Vol. 2, 2nd Edn.

WHO (1999). *Toxic Cyanobacteria in Water: A Guide to Their Public Health Consequences, Monitoring and Management* (Ed. Chorus, I. and Bartram, J.). E & F N Spoon, London.

WHO (2011). *Guidelines for Drinking-water Quality*. 4th Edn.

Yao, K. M. (1973). Design of high rate settlers. *Journal ASCE Environmental Engineering Division* **99**(EE5), pp. 621–637.

Young, E. F., Wallingford, F. E. and Smith, A. J. E. (1972). Raw water storage. *Journal SWTE* **21**(2), pp. 127–152.

Zabel, T. (1984). Flotation in water treatment. *Scientific Basis of Flotation* (Ed. Ives, K. J.). NATO ASI Series, Boston.

CHAPTER

Water Filtration

9

GRANULAR MEDIA FILTRATION

9.1 RAPID FILTRATION – INTRODUCTION

Clarification is usually followed in a waterworks by solid–liquid separation processes which usually include rapid filtration; the basic principles determining the removal of particles by filtration are discussed below. Usually rapid filtration is preceded by chemical treatment of the water; rapid filtration without chemical treatment is used in 'primary' filtration before slow sand filtration.

Rapid gravity filters are commonly the last solid–liquid separation stage in the treatment of fresh water for human consumption. The objective with all designs of filters is to reduce the solids content measured as turbidity to less than 0.3 NTU with an upper limit of 1.0 NTU, these being the usual target values. More recently however, the need to ensure the removal of *Giardia* cysts and *Cryptosporidium* oocysts has led to turbidity targets of less than 0.1 NTU (95 percentile value) with a maximum of 0.3 NTU being applied to filtered waters.

9.2 RAPID FILTRATION – MECHANISMS

In rapid filtration, the removal of particles is largely by physical action, although physicochemical processes may also occur. The size of grain of the filter media, usually sand, is normally within the range of 0.4–1.5 mm whereas particles which may be removed by simple filtration, for example mineral particles or diatoms, may be at least 20 times smaller. In fact, a proportion of particles even several hundred times smaller than the size of the sand grains may be removed. To achieve effective removal of the smaller particles, the addition of a coagulant to form a floc, containing aluminium or iron hydroxides, is usually necessary; but even the floc particles may be very small compared with the size of a grain of the filter media. Therefore, filtration is more than a simple straining action such as that provided by microstrainers (Section 8.3). There may be some straining action due to a coating on the surface of a filter but in general filtration is a process in which some depth of the filter media is utilized. During filtration the flow within the filter bed is laminar; the loss of head through the media is proportional to the velocity of flow of water.

Twort's Water Supply. DOI: http://dx.doi.org/10.1016/B978-0-08-100025-0.00009-0

Extensive research has been carried out to determine the mechanisms of removing relatively small particles in filtration. Some of this work has been carried out and summarized by Ives (1967, 1969, 1971). The general conclusion is that the principal mechanisms of filtration are physical and they may be considered under the headings of gravity (or sedimentation), interception, hydrodynamic diffusion, attraction and repulsion. The contribution of each mechanism depends on the nature of the water and its chemical treatment.

Research on filtration has also considered mechanisms of attraction (or repulsion) between particles and filter grains; in the absence of any attraction, some particles would tend to become detached. Such considerations are, however, complex and any clear conclusions are of limited application. Van der Waals forces are well known as attractive forces between molecules and theoretically they apply to nearly all materials in water, but their range is usually limited to minute distances of less than 0.05 μm.

Finally, it should be mentioned that in some water treatment processes physico-chemical or chemical reactions occur in contact with filter grains. One example is the deposition of calcium carbonate from waters having a positive calcium carbonate saturation index. Another is the chemical oxidation and deposition of compounds of iron and manganese. After such reactions have commenced the coated grains may provide active surfaces to catalyse further oxidation reactions (Section 10.11). In some instances the biological oxidation of ammonia, iron or manganese or other biochemical action takes place when passing a water through a rapid filter, due to the development of the necessary bacterial flora in organic impurities on the filter grains (Sections 10.12 and 10.28).

9.3 DESIGN AND CONSTRUCTION OF RAPID GRAVITY FILTERS

The part of a rapid gravity filter that removes the solids from the incoming water is the filtering medium which is usually sand.

There are a number of variations to the design of rapid gravity filters: for instance the sand size can either be (nominally) constant, i.e. monograde sand, or it can vary from fine to coarse i.e. graded sand; the depth of the sand can be shallow or deep; the direction of flow of water is usually downflow but could be upflow. Alternatively, the filter media could use three or more layers of sand and pebbles or, other materials altogether, having differing grain size and specific gravity. When a fine sand is used, the collection of solids during filtration, and hence the build-up of headloss, tends to be in the top layers. In coarser sands the solids penetrate to a greater depth. So long as there is an adequate factor of safety in bed depth against complete dirt penetration it makes good sense to utilize at least some of the bed depth for solids capture provided the backwashing system can be relied on to remove accumulated solids and achieve thorough cleansing of the sand before its next working cycle.

Filter Media

Sand filters use either graded sand (fine to coarse or heterogeneous) or coarse monograde sand (uniform size or homogeneous). No single media specification (size and depth) can be applied universally for all waters; the choice depends on the water quality and upstream processes, filtered water quality objectives, cleaning method, filtration rate and length of filter runs. In graded sand filters the bed depth typically comprises 0.7 m of 0.6–1.18 mm fine sand (effective size 0.63–0.85 mm), 0.1 m of 1.18–2.8 mm coarse sand, 0.1 m of 2.36–4.75 mm fine gravel and 0.15 m of 6.7–13.2 mm coarse gravel. The effective size, d_{10}, is defined as the size of aperture through which 10% by weight of sand passes. For applications requiring a finer sand, the two upper layers are changed to 0.7 m of 0.5–1.0 mm sand ($d_{10} = 0.54–0.71$ mm) and 0.1 m of 1.0–2.0 mm coarse sand, the gravel layers remaining the same. Depending on the slot size of the underlying filter nozzles, the bottom gravel layer can be omitted and replaced by more of the next layer. The homogeneous sand filter has a 0.9–1 m deep bed of typically 0.85–1.7 mm sand ($d_{10} = 0.9$ mm) placed on a 50 mm layer of 4–8 mm or 75 mm of 6.7–13.2 mm gravel. Homogeneous sand of effective size up to 1.3 mm has also been used. The stated size ranges for sand and gravel are generally 5 and 95 percentiles. For estimating the sand depth some employ the rule that the depth of sand should be ≥ 1000 times its effective size (Kawamura, 2000). Pilot studies may be done to confirm sand depth, for large plants in particular.

Some filter plant designers use the term 'hydraulic size' in place of effective size (Stevenson, 1994). This is defined as the size particles would have to be, if all were the same size, in order to match the surface area of a sample covering a range of sizes. For media with size range 1:2, the hydraulic size is approximately 1.36 × the lower size in the range; e.g. for 0.85–1.7 mm sand the hydraulic size is 1.16 mm.

Other filter media such as anthracite (Section 9.7), granular activated carbon (GAC; Section 9.9), garnet, pumice (Farizoglu, 2003), expanded clay particles and glass are also used in filtration applications. Garnet is a dense (s.g. 3.8–4.2) medium which is used as the bottom layer of multimedia filters containing anthracite and sand. It occupies about 15% of the bed depth and the effective size could be as low as 0.35 mm; being dense, it requires about three times the wash rate as anthracite to give the same bed expansion. Pumice and expanded clay are porous media and could be used in biological filtration (Section 10.28). Glass is a suitable filter medium of similar specific gravity to sand.

The sand should be of the quartz grade with a specific gravity in the range 2.6–2.7. Bulk density is about 1.56 g/cc. The uniformity coefficient (UC) is expressed as:

$$UC = d_{60}/d_{10}$$

where d_{60} is the size of aperture through which 60% of sand passes. UC values should be less than 1.6 and usually lie between 1.3 and 1.5. Lower UC values would make the medium costly as a high proportion of fine and coarse medium is discarded and higher values would reduce the voidage. Typically sand has a voidage of 37–40%, defined as: 100 × (particle density − bulk density)/ particle density. Loss in weight on ignition at 450°C should be <2% and the loss in weight on acid washing (20% v/v hydrochloric acid for 24 hours at 20°C) should be <2%. The sand should be tested for friability (BW, 1996) to ensure that washing operations do not produce fines.

Underdrain Systems

The filter media is placed on a system to collect water from the underside of the bed in an even manner and to spread air and water uniformly through the bed during cleaning. There is a choice of collector system design. One system comprises nozzles set in PVC pipe laterals, the spacing between laterals being infilled with concrete (Fig. 9.1). The design is commonly adopted to apply air and water separately during cleaning and therefore finds application in graded media, dual or triple media or GAC filters. In another design, nozzles are set in a reinforced concrete false floor with a plenum (space below) (Fig. 9.2). The floor is either constructed in situ on plastic formwork or made up of pre-cast concrete slabs supported on concrete sills. The design allows water and air to be applied simultaneously or separately and the system provides for better distribution of air and water than the pipe lateral systems. In separate air–water designs, the supporting layers beneath the media are used to prevent sand from penetrating down to the floor as nozzles often have slot sizes larger than the sand size. In combined air–water washing the gravel layer limits erosion of the nozzle dome caused otherwise by the localized fluidized sand; the nozzles, which usually have slot sizes in the range 0.25–0.5 mm, also minimize the risk of sand penetration. Nozzle density depends on the type of nozzles but is typically 30–35 nozzles/m^2. Several other underdrain systems, mostly of proprietary design, have been successfully used in many parts of the world. An example is the design by Xylem (formerly Leopold), comprising high-density polyethylene underdrain blocks, which clip together to form longer laterals (Fig. 9.3). The blocks incorporate a dual lateral design with a water recovery channel that ensures uniform distribution of air and washwater over laterals up to 12 m in length. The design incorporates reverse graded (coarse-to-fine-to-coarse) gravel to prevent disruption caused by the air during air–water backwash. The floor can be fitted

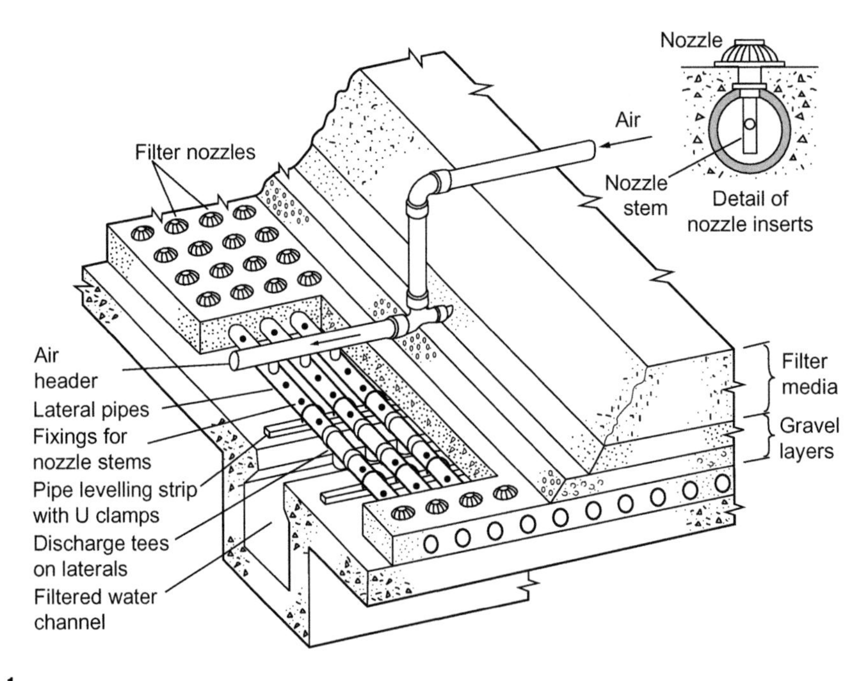

FIGURE 9.1

Pipe lateral filter floor arrangement (Paterson Candy Ltd).

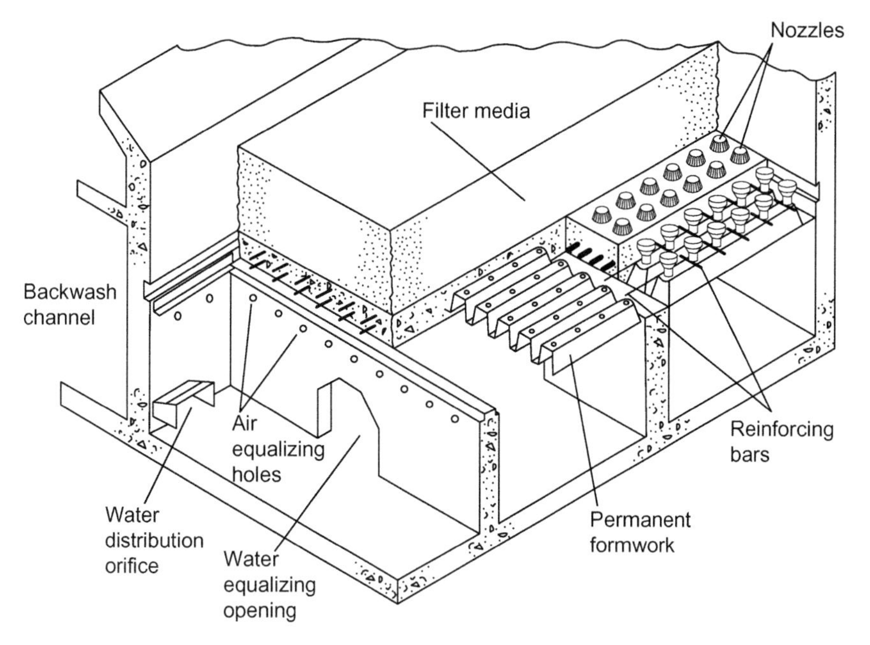

FIGURE 9.2

Plenum floor (Paterson Candy Ltd).

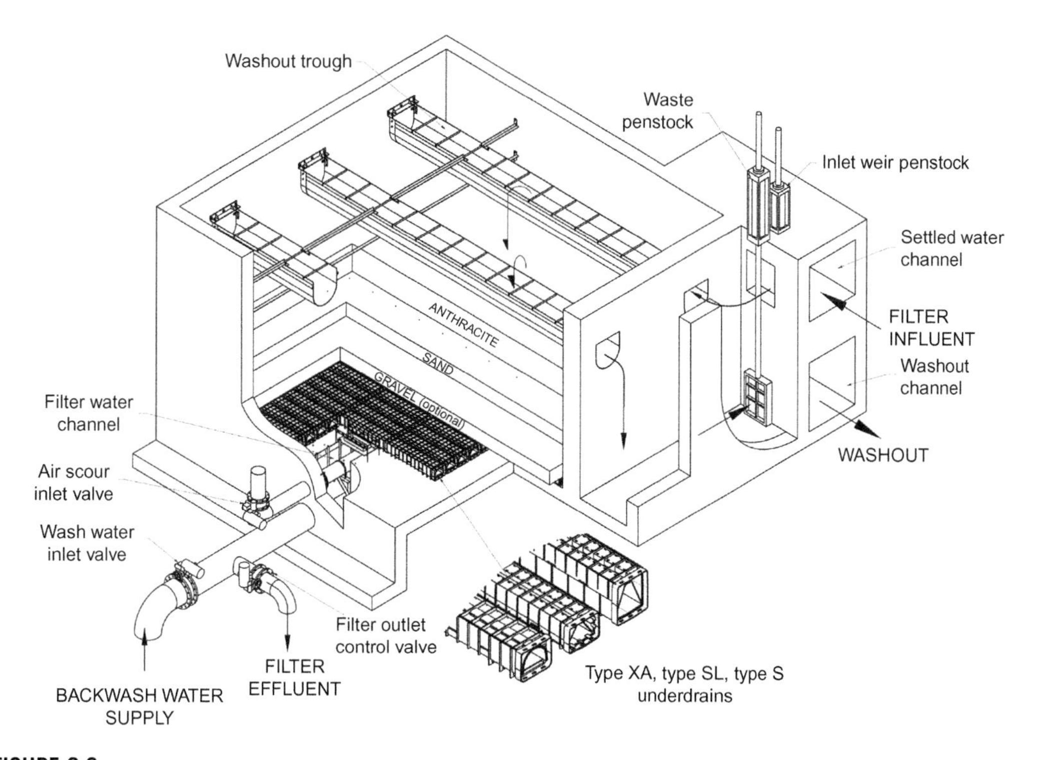

FIGURE 9.3

Block underdrain filter arrangement.
Courtesy Xylem Inc., USA.

with a porous plate that helps to eliminate the need for support gravel and reduce the overall bed depth. This is ideal for GAC filters as it simplifies media removal for regeneration. Care is required to avoid clogging of the porous plate as a result of manganese oxide deposits, biological growth in biological activated carbon contactors, or in softening plants which produce waters with a high calcium carbonate precipitation potential. Existing graded media filters with pipe lateral underdrains designed for separate air and water washing can be converted to monograde media filters with combined air–water washing by adopting the block underdrain system (e.g. Xylem) or by placing separate air laterals above the gravel layer with orifices covered in gauze and pointing downwards.

Filter Configuration

The overall number and size of filters vary. The number of filters is selected to minimize the sudden increase in the filtration rate when removing a filter from service for washing as this would dislodge retained deposits and increase the filtered water turbidity in the other filters. Ideally it should be possible to take two filters out of service simultaneously (one draining down and one washing) although it would be adequate to design for one filter out for washing. In general, the more filters the better but a minimum of six filters is usually desirable; four filters may be used with at least three operating in filtration mode provided plant throughput is reduced during maintenance of another filter. Up to five filters could be arranged in one bank; six filters or more would be arranged as even numbers in two banks. The limiting factors for size are the uniform collection of filtered water, even distribution of washwater and air and the travel length of washwater to the collection channel during washing. Typically, filter sizes vary from 25 to 100 m^2 with lengths in the range 8–20 m and widths 3–5 m. The washwater collection channel is usually located on one side along the length of the filter. Filter beds up to twice these sizes can be constructed by providing two identical beds separated by the washwater collection channel in the middle, thus limiting the distance travelled by dirty washwater to a maximum of 5 m. For some applications, such as combined air–water backwashing of anthracite–sand filters, washwater collecting troughs are placed above the filter bed (Section 9.7).

Filtration Rates

Filtration rates are selected to suit the application. Filters with deep bed homogeneous sand for iron removal are rated at 6–7.5 $m^3/h.m^2$ and for manganese removal at about 15–18 $m^3/h.m^2$. When used downstream of clarifiers homogeneous sand filters are rated at about 6–12 $m^3/h.m^2$ with the higher rate being used when water upstream of the clarifiers is treated by a combination of a coagulant and a polyelectrolyte. At filtration rates above 15 $m^3/h.m^2$ the quality of filtrate tends to deteriorate and at rates in excess of about 20 $m^3/h.m^2$ the rate of headloss development becomes too rapid. Shallow bed, graded sand filters are usually rated at about 75% of the rates for deep bed, homogeneous sand filters. The rates achieved with multimedia filters (Section 9.7) are similar to those achieved with deep bed homogeneous sand filters. Where there is concern over *Cryptosporidium* oocysts and *Giardia* cysts in raw water supplies filtration rates are limited to about 6–7 $m^3/h.m^2$ to minimize the risk of particulate breakthrough.

Headloss, Air Binding and Negative Head

In the downflow filter design with upflow washing it is usual for the filter to operate with about 1.5–2 m or more of water depth over the bed. However, there are some proprietary designs which operate with a much smaller depth of water (down to 0.5 m) and even with negative head conditions, but the latter is not regarded as good practice as difficulties can occur with cracking and mudballing in the filter bed and air binding (also called air blinding), especially with high filtration rates. The pressure distribution in a filter bed is illustrated in Figure 9.4. Negative head can occur when 'clogging head' (total headloss less the clean media loss) at any depth exceeds the static head (water depth) at that point. The point of negative head development varies with the filter media; it is nearer the media surface for graded sand, about one third way down the media in coarse homogeneous sand and just below the anthracite layer in dual media filters. Under negative head conditions dissolved gases in water are released into the space between sand grains restricting the water flow, increasing the headloss and prematurely terminating filter runs (Scardina, 2004). It can also result in 'mudball' formation and poor filtrate quality when air binding is restricted to part of the filter and due to channels formed by escaping gases. Negative head is particularly a problem in filters treating cold waters because of greater solubility of gases in water at low temperature, surface waters or well-aerated groundwaters. It could also be a problem when filters are preceded by dissolved air flotation (DAF) clarifiers operating at high rates typically greater than about

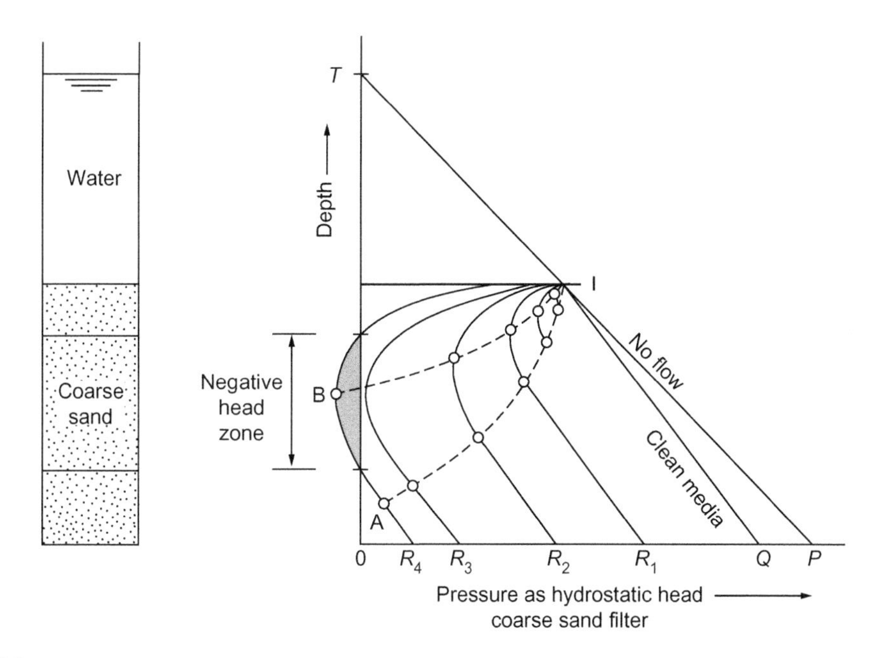

FIGURE 9.4

Pressure distribution in a rapid gravity filter.

Note: Scales of axes are same so OP = OT. IQ is pressure with clean media, IR_{1-4} are pressures with increasing dirt loss.

12–15 $m^3/h.m^2$ unless the water is deaerated or the design allows for efficient air separation (Section 8.18). Negative head can be overcome by providing sufficient water depth above the top of the media or by washing the filters at a headloss less than the static head down to the point of negative head development.

The headloss in a clean filter is directly proportional to the viscosity of the water hence the temperature and is made up of clean media loss and underdrain loss (nozzles contribute about 0.05 mm). Clean media loss at a filtration rate of 6 $m^3/h.m^2$ for sand of 0.85–1.70 mm, 1.0 m deep and voidage of 40% is 0.178 m at 10°C and 0.115 m at 25°C. Typically, when a filter is returned to service after washing, the loss of head through the bed, underdrain system and the filter outlet should be less than about 0.35 m. The rate of headloss development is a function of solids retention capacity of the filter and is lower for coarse homogeneous sand filters than for graded sand filters; in the latter it is improved by the use of an anthracite layer (Section 9.7). The headloss allowed for retention of suspended solids (clogging head) with all the filters in service should not be less than 1.5 m on the basis of 24-hour filter runs.

Solids Retention

The maximum solids retention capacity of a filter is a function of the voids which is approximately equal to 40–45%. In practice, only about a quarter of this space is available for solids removal. It is considered that the solids retention capacity of a gravity filter is limited to 10 and 35 g dry solids/l of voids for light hydroxide floc and suspended solids in river water, respectively. For example, a filter of 0.9 m sand bed, with a voidage of 45% and filtration rate of 7.5 $m^3/h.m^2$, washing every 24 hours, cannot accept more than 5.6 mg/l of suspended turbidity or 20 mg/l of light hydroxide floc in the influent over the run length. The respective equivalent filter loadings are 1012 and 3544 g/m^2 filtration area. Cleasby (1999) quotes a value less than 1580 g/m^2 per m of media. Some designers tend to use a more conservative loading of 350 g/m^2 for sand filters, whilst others use a value of 1000 g/m^2. Solids loading on to anthracite–sand filters would be higher and is of the order of 700 g/m^2. When anthracite–sand filters are used for direct filtration a solids loading of about 1500 g/m^2 is used. The objective is to achieve a minimum filter run length of 24 hours. These values should be used for guidance only. Actual values for design purposes should be determined by pilot plant trials.

Flow Control

As the bed becomes clogged the headloss through it increases resulting in a reduction in the filtration rate if available head remains constant. For best filter performance it is desirable to have a constant filtration rate and any changes in filtration rate, such as when removing a filter for washing, should be as smooth as possible. Therefore, control systems which divide flow equally between filters and allow filtration without fluctuations in rate are essential for good filtration results. Equal flow may be achieved by using weirs to proportion the flow equally to all filters or it may be achieved by sizing the outlet pipework and valves of each filter to limit the maximum flow hydraulically. In such cases, after a filter is backwashed, the level of water in the filter rises to such a level above the outlet head on the filter that it is sufficient to overcome the headloss through a clean filter and its underdrain system. As the sand becomes progressively clogged during a filter run, increasing the headloss through the filter, so the water level in the filter box rises to the maximum possible. The simplicity of the rising level filter design is attractive and has many advantages (Cleasby, 1969), but the free fall of water into the filter (typically about 2 m) can break the fragile floc into small fragments which may not be effectively removed in the filters; to minimize floc damage the maximum fall should be kept below 300 mm. Furthermore, the absence of any flow control on the outlet valve prevents any form of slow start being incorporated when a filter is brought back on-line after being backwashed.

A commonly used control regime is the constant flow/constant level method, in which a common inlet channel feeds all the filters via submerged penstocks. The inlet channel is therefore hydraulically linked with all of the filters, ensuring a common water level. Each filter outlet is provided with a flow meter and a modulating valve, the latter being controlled to divide the total inlet flow equally between the filters in service. The set-point for the flow control loop for each filter is adjusted as necessary to maintain the level measured in the common inlet channel within a pre-set range.

One of the drawbacks of constant flow filters is that turbidity breakthrough occurs towards the end of the filter run. This is less likely in declining rate filters where the filtration rate decreases as the headloss develops. Declining rate filters are based on a simple design (Hudson, 1981; Cleasby, 1993a, b). The system is best suited to a group of six or more filters so that the additional flow to be shared when one filter is taken out of service for washing is not excessive. It is reasonable to design on the basis of a maximum flow range through each filter of ±35% of the average filtration rate, the average being taken as the total output divided by the total sand bed area provided. The filter inlet is submerged and, to restrict the filtration rate to the maximum, it is necessary to install some form of restricting orifice or valve on the outlet. Filters are washed in a fixed sequence and individual filter instrumentation for loss of head or quality of filtrate, i.e. turbidity, is used only to detect a filter whose behaviour is out of line with the rest for some reason. In terms of hardware, the system is very simple and this is its chief merit. A detailed analysis of rapid gravity filters and their hydraulics is given by Stevenson (1998).

9.4 BACKWASHING

Rapid gravity filters employing graded sand are washed by separate application of air and water through the bed in the reverse direction; the used washwater is removed by a washwater collection channel. After taking the filter out of service, the filter is allowed to drain down until the water lies a few centimetres above the top of the bed. Air is then introduced through the collector system at a rate of about 6.5–7.5 mm/s. The air breaks up the surface scum and dirt is loosened from the surface of the sand grains. This is followed by an upward flow of water at a carefully selected velocity to expand and fluidize the bed. Under this condition, the voids between grains of sands are increased and the resulting rotation of grains and consequent attrition between grains produces a scouring action to remove attached deposits. The wash rate should be just sufficient to achieve fluidization velocity (incipient fluidization) with little bed expansion. Increasing the backwash rate beyond this state would be counterproductive because as the distance between sand grains increases, the scouring action is reduced. High backwash rates may result in loss of sand and wastage of water and energy. The washwater collection channel sill is usually placed about 100–150 mm above the sand.

In the UK, typical practice is to use wash rates to produce 1–5% bed expansion. The rates are viscosity dependent and therefore are affected by water temperature, with higher rates used at warm water temperatures. Typical wash rates to give 2% bed expansion are given in Table 9.1.

An empirical relationship has been developed to express bed expansion of graded media in terms of temperature (Tebbutt, 1984) which is the ratio of expansions at T°C and 20°C; this is $1.57\ e^{-0.02255T}:1$.

Applying this equation to typical UK conditions shows that if summer wash rates are used throughout the year, a 40% increase in the degree of expansion occurs in the winter. This would lead to wastage of water and could result in loss of sand. Therefore, in temperate climates, means for seasonal adjustment of wash rates is advisable.

Filters comprising deep bed coarse homogeneous sand rely upon the simultaneous application of air and water in the wash phase, followed by a water rinse. This regime is also referred to as 'collapsed pulse' and is more effective than separate air–water washing. In both phases the water rate is well below the fluidization velocity and does not cause the bed to expand. This prevents hydraulic grading and maintains the homogeneity of the filter bed. The air rate is typically 16 mm/s of free air. Various combinations of water rates are used. Examples are: 2 mm/s in the combined air/water wash phase and 4 mm/s in the rinse phase; 2 mm/s in the combined air/water wash phase and 15 mm/s in the rinse phase; or a water rate of 4–5 mm/s in both the combined air/water wash and rinse phases. In the combined air–water wash method, the influence of temperature on wash rate is less pronounced; for every 10°C rise in water temperature wash rate increases by about 0.4–0.6 mm/s. In such filters, the washwater collection channel sill is about 500 mm above the sand and the sill has a large forward chamfer to allow locally suspended sand to drop out as flow approaches the weir crest. Simultaneous application of air and water is not used with graded sand media as this would intermix the media and could also result in loss of media into the washwater collection channel. Simultaneous air and water washing can be used with fine media if this is placed on block-type underdrains without the supporting gravel layers used in graded media. Media loss in this case can be avoided by providing elevated washout troughs, terminating the air flow before the rising washwater reaches the troughs and continuing with water only.

Table 9.1 Wash rates (mm/s) for sand filters to give 2% bed expansion at varying water temperatures[+]

		Water temperature (°C)					
Sand size range (mm)	**Effective size (mm)**	**5**	**10**	**15**	**20**	**25**	**30**
0.5–1.0	0.54–0.71	3.1*	3.5*	4.0*	4.5*	5.0	5.5
0.6–1.18	0.6–0.85	4.4	5.0	5.6	6.3	6.9	7.5

Notes: [+]These rates should be increased by up to 20% where possible, to take account of possible variations in media size and voidage.

*In practice, a minimum rate of 5 mm/s is recommended for 0.5–1.0 mm sand.

The duration of the wash phases depends on the method of wash and filter influent quality. For designs with air and water applied separately, air scour lasts about 3–4 minutes and the water wash lasts about 4–6 minutes; typically for designs with the air and water applied simultaneously, air is first introduced about 1.5–2 minutes before the water to allow the air flow to become established, after which the combined air–water wash proceeds for about 6–8 minutes. The air flow is then stopped and the water flow continues to rinse the bed for another 8–10 minutes. The total period a filter remains off-line for washing is about 30–45 minutes, which includes about 15–30 minutes for draining the filter down depending on the headloss in the filter. The total water consumption per wash amounts to about 2.5 bed volumes.

To remove as much of the dirty backwash water as possible from the top of the filter before it is refilled and put to use, it is common to allow the filter influent (clarified water) into the filter and to flow across the top of the bed from the side remote from the washwater collection channel as the last stage of backwashing. This is called 'surface-flush' or 'cross wash' and increases the total washwater consumption to about three bed volumes. In one filter design the filter is washed as usual, but the used washwater is contained above the media by allowing the level to rise. At the end of the wash the used washwater is rapidly discharged to waste via flap gates in one side wall. An advantage of this design is that high wash rates may be applied without fear of media loss. Typical free air and water rates used in this design are 14–22 mm/s and 10–18 mm/s, respectively and can be applied either concurrently or separately. The water usage is about 1–2.5% of works throughput.

When the filter is returned to service after backwashing there is a short period, lasting about 15–60 minutes, when the filtrate turbidity is high. This is due to the displacement of residual washwater containing solids loosened from the bed during backwashing and also the lower solids removal efficiency of the freshly washed media. The options available for reducing this effect and allowing time for the filter to 'ripen' before returning it to normal service are to return the first 15–60 minutes of filtrate to waste (or usually to the works inlet), to allow the filter to stand for up to about 30 minutes (so-called 'delayed start'), to start the filter at a slow rate (so-called 'slow start') or a combination of these. All of these features can be controlled automatically as part of the wash sequence provided that the filters are appropriately designed.

Water used for backwashing should be filtered and preferably chlorinated, particularly if there is no pre-chlorination at the works, to reduce the risk of biological growth in the underdrains. In works using aluminium coagulants washwater is best taken upstream of any final pH correction (if >7.8) to minimize the risk of dissolving aluminium hydroxide floc retained in the filter. The total amount of washwater used has an important bearing on the economy of a treatment works, especially in relation to the net yield of a source. The total washwater used should normally not exceed 2–2.5% of the treated water output on the basis of 24-hour filter runs.

9.5 OPERATION OF FILTERS

When the maximum permitted headloss has been reached the filter run is terminated. Typically, the headloss allowance for dirt is 1.5 m with all filters in service; this should be increased to take account of filters out of service for backwashing or maintenance. It is usual to design for a minimum filter run length of 24 hours. The length of the run is typically 24–60 hours. Long run lengths permit savings on backwash water but encourage bacterial growth in the filter bed, in particular if the raw water contains organic matter or is of poor bacteriological quality and is not treated with a disinfectant upstream. It is therefore desirable to restrict maximum run lengths to 48 hours in warm water temperatures and to 60 hours in cold water temperatures. Any appreciable increase in backwash frequency above the design value would reduce output below the design level. The sequence of backwashing including the drain down usually takes about 30–45 minutes so that, for a bank of filters requiring to be washed more frequently than every 24 hours, it could prove necessary for one filter to be draining whilst another is being backwashed. In difficult circumstances, when filter runs are as short as 6–12 hours, it is often the practice to discharge the total contents of the filter to waste instead of draining to supply in order to speed up the wash cycle: this is sometimes referred to as 'dumping' the filter.

The primary parameters used to initiate a filter backwash are duration of filter run, elevated headloss, or deterioration of filtrate quality measured by turbidity. In most installations all three parameters are monitored and can be used to initiate the washing cycle automatically. More often than not, only the first two parameters are used for automating the start of a filter washing cycle. It is useful to monitor filtered water turbidity both in the combined filter flow as well as in the effluent from each filter, with an alarm initiated on a high value. The individual filter turbidity helps to identify filter breakthrough which may be used to give early warning of *Giardia* cysts or *Cryptosporidium* oocysts (for sources with risks) in the treated water, although particle counters on individual filter outlets would be more appropriate. In automatic filter plants, manual washing should always be possible as an alternative so that special cleansing measures can be taken when necessary.

For air scouring, compressors working in conjunction with air storage vessels or Roots type positive displacement blowers are used. Air is usually applied at a pressure of about 0.35 bar g at the air inlet valve. Once the air scour is established, the pressure falls to around 0.2 bar g. For backwashing it is usually most economic to use gravity flow from a large elevated storage tank, since the rate of flow required is large and electrical demand charges are minimized by keeping such a tank topped up by a relatively small pump drawing from the filtered water supply or from the treated water supply pumps. The tank should have two compartments with each sized for at least one filter wash. To ensure that the backwash rate does not change too much with the head in the tank, either flow rate controllers are used or the tank is arranged to have a large surface area. Pumps used to introduce water directly into filters are usually of the centrifugal type. The head required at the washwater inlet valve of the filter is about 5 m.

A flow meter should be provided on the backwash main, used in control loops to set the optimum wash rate and to change the rate between wash and rinse phases and between winter and summer periods. Butterfly valves (double flanged) should be used for all filter valves. Plug valves are sometimes used for flow control purposes. Penstocks tend to leak after a few years in service and should only be considered for the filter inlet. On large filters or filters designed for automatic washing, power assisted actuators are necessary for operating penstocks and valves. Electric actuators are often used but pneumatically operated actuators are also successfully employed. Primary filters are not normally covered except in very cold climates, to prevent freezing. Once filtered, water should not be exposed to contamination and therefore all chambers and channels downstream should be covered and any access hatches should be sealed water tight. The water from a rapid gravity filter is not completely free of bacteria and, before the water passes into supply, it must be disinfected (Chapter 11). A comprehensive analysis of operation and maintenance (O&M) of filters is given by Logsdon (2002).

9.6 CONSTRUCTION AND OPERATION OF PRESSURE FILTERS

Pressure filters are similar in concept to rapid gravity filters, except that they are contained in a steel pressure vessel. Perforated pipes or a steel plate with nozzles are used for collecting the filtered water and for distribution of the washwater and air scour. The steel pressure vessel is cylindrical, arranged either horizontally (Fig. 9.5) or vertically. With a pipe lateral underdrain system the bottom of the vessel is usually filled with concrete so as to obtain a flat base. In the plenum floor design a steel plate is used, into which nozzles are screwed. In a horizontal vessel, vertical plates are sometimes welded inside to give a rectangular shaped sand bed within the cylinder so that the bed may be washed evenly and there are no 'dead' areas beneath which air scour and water pipes cannot be placed. More commonly, media is placed in the entire vessel such that the depth is equally distributed above and below the centreline of the filter. The whole of the cylinder is kept filled with water under pressure and an air release valve is provided at the highest point for the release of trapped air. To avoid the requirement for special transportation procedures for large loads the maximum diameter of filters is limited to 3–4 m and the length/height is limited to about 12 m.

The backwashing of such filters is very similar to that of an open rapid gravity filter. A bellmouth and pipe can be used for the removal of dirty washwater in a vertical filter; for most horizontal filters a single vertical plate located near to one of the dished ends facilitates washwater removal, but for the larger filters, a central washout channel formed by two vertical plates is necessary.

The advantage of pressure filters is that excess raw water pressure is not lost when the filtration process takes place, as is the case with an open rapid gravity filter system. About 3 m head may be

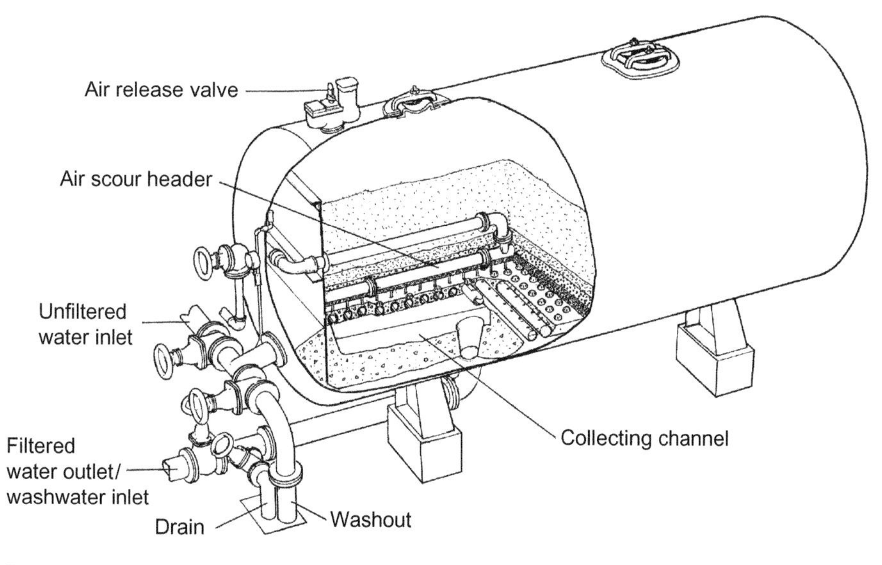

FIGURE 9.5

Sectional view of a pressure filter (Black & Veatch).

lost in friction through the sand bed and the inlet and outlet fittings. This includes an allowance for dirt loss of about 1.5–1.8 m. This combined headloss is between the common inlet and outlet bus mains serving a battery of filters. Pressure filters may be installed in a pumped or gravity pipeline without a large loss of pressure on the supply. Air binding, hence negative head, should not occur in pressure filters if the pressure in the media is always greater than that at the points upstream where air could have gone into solution.

Pressure filters suffer from the disadvantage that the state of the bed under backwashing conditions and when the plant is working cannot be directly observed. It is of vital importance, therefore, that every pressure filter is fitted with an open box or dish in the front of it, into which the washwater is turned so that at least any washing out of the sand may be observed and the backwash rate immediately reduced.

When coagulation or other chemical treatment is required chemicals are injected and mixed under pressure and flocculation must be hydraulically carried out in pressure vessels fitted with baffles. The same applies when a contact tank is needed on the downstream side for disinfection. Pressure filter installations are often not provided with any form of individual flow control. The result is that each operates as a declining rate filter for at least part of its filtration cycle and flow is concentrated through the clean, newly washed filters. An orifice plate may be installed in the outlet to restrict flow. However, a better option is to modulate the outlet valve using signals from a flow meter on each filter so that the flow equals the total flow divided by the number of filters in service. As the headloss through a filter increases the outlet valve opens to maintain this average flow. When employing flow controllers headloss can be monitored on individual filters; the filter run can be terminated on any of headloss, length of filter run or turbidity breakthrough. Care is needed when providing this form of control on pumped systems to minimize the pumping head required.

The equipment required for air scouring pressure filters and for valve control is similar to that used for rapid gravity filters. The filtered water usually has enough pressure for backwashing and this is often used so avoiding the need for backwash tanks or pumps. For a filter washed by separate air and water the required rate of application of water is about four times as great as the rate of filtration which is about 5–6 $m^3/h.m^2$. Thus in a large filter battery of about 15 filters, groups of five filters can be taken out of service at a time and the combined filtrate from four of the filters can be used to wash the fifth filter, and so on, until all filters in the group are washed. In large pressure filter plants, arrangements are usually made to wash filters in groups at a specific time each day. Monitoring of individual filters for loss of head is seldom done except when outlet flow control is used. In fact, because a large battery of filters is usually supplied by a common inlet bus main and the outlets from filters also connect to a common outlet bus main, it is only meaningful to measure the headloss across the battery of filters. Individual filtrates could usefully be monitored for turbidity.

Specific regulations apply to the design, fabrication and testing of pressure vessels. In the UK, BS PD 5500 (2012) is followed. Condensation on the outside of the tanks is a continual nuisance as it causes corrosion of the steel shell and staining of the floor below. The pressure applied to steel pressure filters is not usually in excess of 80 m head of water, which should be adequate for most distribution systems. Above this pressure, the thickness of plates used for the steel shell is likely to cause costs to rise rapidly. The use of pressure filters in place of rapid gravity filters is limited by the common requirement for upstream treatment by clarification processes operating at atmospheric pressure and practical considerations limiting their suitability for high throughputs. For these reasons, pressure filtration is usually restricted to small plants where it is undesirable to break the hydraulic gradient, iron and manganese removal (primarily in groundwaters), treatment of stored water by direct filtration or organics removal in GAC adsorbers.

MULTI-LAYER AND OTHER METHODS OF FILTRATION

9.7 USE OF ANTHRACITE MEDIA

The most efficient form of media grading for maximum capture of solids in a rapid gravity filter would be to have the sand decreasing in size in the direction of flow. This is not possible with a sand media because hydraulic regrading takes place during backwash, so that the finer sand collects at the surface of the bed. This can be countered by using separate layers of filter materials having different density and grain size, the denser materials being at the bottom of the bed and the less dense at the top. One type of multi-layer filter bed in wide use is the two-layer filter using anthracite over sand. It has been found that the filtrate quality from anthracite–sand can be as good as that from conventional sand only filtration. Filter runs can be 1.5–3 times longer than with sand media owing to the ability to retain solids throughout the entire bed depth and the associated reduced rate of headloss development. A layer of anthracite is sometimes incorporated in sand filters to extend filter runs and to improve filtrate turbidity. They are usually operated at filtration rates similar to those used on coarse homogeneous sand filters although high filtration rates, even up to 15 $m^3/h.m^2$, can be achieved, albeit at the cost of short filter runs. The specific gravity of anthracite (1.4–1.45) is lower than that of sand (2.6–2.7), bulk density is about 0.73 g/cc and voidage is about 50%. The apparent bulk density after compaction during filtration is approximately 0.79 g/cc. The lighter anthracite is placed on denser sand. The ratio of media sizes, anthracite:sand, is usually in the range 2:1 to 4:1. The anthracite used is usually of size 1.18–2.5 mm (d_{10} = 1.36 mm) or 1.4–2.5 mm (d_{10} = 1.57 mm) and UC < 1.5. The sand is of size 0.5–1.0 mm (d_{10} = 0.55 mm) but sometimes of 0.6–1.18 mm (d_{10} = 0.75 mm). Where the filters are washed by separate application of air and water, the depth of the anthracite bed is of the order of 0.15–0.3 m with sufficient sand to give a combined depth of about 0.75 m. Backwashing with concurrent application of air and water allows deeper anthracite layers (up to 0.6 m) to be used, providing increased dirt-carrying capacity but with a corresponding increase in the depth of the sand layer (up to 0.8 m). The sand should be supported on gravel (Section 9.3) depending on the type of underdrain system used. For the pipe lateral system, graded sand and gravel should be used and for plenum floor design a shallow gravel layer (50 mm of 4–8 mm or 75 mm of 6.7–13.2 mm) is normally used. Sources of anthracite (carbon content at least 90%) are limited to a few countries in the world. Therefore, there is a tendency to use high grade bituminous coal (carbon content at least 80%) in place of anthracite. This is generally acceptable provided it is non-friable and can meet the standards for friability and acid solubility (Section 9.3) and hardness (BW, 1996; AWWA, 2009; ASTM, 2005).

Table 9.2 Wash rates (mm/s) for anthracite–sand filters at varying water temperatures

	Water temperature (°C)					
Bed expansion (%)	**5**	**10**	**15**	**20**	**25**	**30**
10	4.5	5.3	6.2	7.2	8.0	8.9
15	6.5	7.5	8.3	9.2	10.0	10.7
30	11.4	12.1	12.8	13.5	14.3	15.3

Beds of anthracite can be expanded or fluidized at about the same backwash rate as a sand bed when the size of the anthracite grains is about 1.5–3 times that of the sand grains. Backwash rates are used to give 10–15% expansion of the combined bed, with an occasional wash at higher rate to give up to 30% bed expansion for regrading the media. The rates are viscosity dependent and therefore, affected by the water temperature. Wash rates required for a bed of anthracite ($d_{10} = 1.3$ mm) and sand ($d_{10} = 0.55$ or 0.75 mm) are given in Table 9.2.

Filters are often cleaned by air scour followed by water backwash. The air scour rate when washing in this way is generally in the range 8–12 mm/s. The water rate is as for graded sand filters with temperature compensation. Concurrent application of air and water requires high level suspended washwater collection troughs in place of the conventional low level collection channel as otherwise lighter anthracite would be carried to waste. The sequence in this regime consists of simultaneous application of air at 16 mm/s and water at about 2 mm/s as for coarse homogeneous sand filters; air is turned off before the water level reaches the sill of the trough weir and the wash rate is then increased to 12–16 mm/s for rinsing and regrading the media. The troughs are usually placed with invert level about 600 mm above the unexpanded media surface; this permits about 5 minutes of concurrent wash.

After rinsing at fluidization velocities, the two-layer bed settles down again with regrading of the two layers by density and grain size. In practice the media are mixed at the interface (100–150 mm) and evaluation by Cleasby (1975) on intermixing suggests that the mixed interface is beneficial as it reduces the voidage of the anthracite layer and produces better filtrate quality but at the expense of higher headloss due to greater retention of suspended solids. To prevent intermixing a coarse sand would be required but this would defeat the objective of the dual media filter where a fine size sand must be used to achieve the required filtrate quality.

In some designs, anthracite is used as the sole filter media. In one plant in the Middle East the bed depth used is about 2 m, size range 1.4–2.5 mm ($d_{10} = 1.5$ mm). Filters are washed by air at 20 mm/s followed by water (6.5–16.7 mm/s) to give a bed expansion of 5–10%.

Expanded clay material (e.g. Filtralite) has found increasing use in place of anthracite in dual media filters in recent years, often providing improved filter run times and thus reduced operating costs without compromising filtrate quality. Typically media size would be 1.5–2.5 mm ($d_{10} = 1.7$ mm), with a bulk density of 0.55 g/cc and a voidage in the range 58–65%. Specific gravity is approximately 1.05 g/cc. The media depth is about 500–900 mm supported on 400–800 mm of 0.8–1.2 mm sand. Filters are normally backwashed by simultaneous application of air and water, similar to anthracite–sand filters. Typical rates are water wash at 4.5 mm/s with air at 12.5 mm/s followed by a water rinse at 25 mm/s to give a bed expansion of about 25% at 20°C. The media could also be used in single media filters of one size (600–1200 mm deep, 0.8–1.6 mm, $d_{10} = 0.95$ mm) or two sizes with 300–1000 mm depth of 1.5–2.5 mm media on top of 400–1000 mm depth of 0.8–1.6 mm media. Wash rates are similar to those given above (e.g. Bedrichov direct filtration plant, Czech Republic).

9.8 USE OF ANTHRACITE TO UPRATE FILTERS

In recent years, the output of many existing filter plants has been increased by changing their media from sand to anthracite–sand, increasing the previously conventional filtration rates of 4–6 $m^3/h.m^2$ for graded sand filters to rates of 6–12 $m^3/h.m^2$ where there is no risk from *Cryptosporidium* oocysts (Section 9.20). However, these increases have frequently been associated with the improvement of floc characteristics by the addition of a polyelectrolyte as a coagulant aid. Use of polyelectrolyte may also be necessary to prevent penetration (Crowley, 1979) through the anthracite and sand by small green algae, dominated by minute species such as *Nannochloris* and *Ankistrodesmus* with cell diameters of 4–8 μm.

It is important to pay attention to floc size when adopting anthracite filters: too fine a floc may pass through the anthracite layer and cause too large a load to reach the sand below; too large a floc in relation to the anthracite size may place too large a load on to the anthracite, defeating the object of gaining filtration in depth by using anthracite and sand. Since anthracite is an expensive material compared to sand it is important to ensure a filter is adequately designed hydraulically before anthracite is used for uprating. This means checking that the inlet, outlet and flow control pipes and valves can accept the higher flow rates; that the distribution of water at the inlet to the filter does not cause excessive scouring of the anthracite–sand bed because of the higher flow rate; that the higher backwash rate required does not result in loss of anthracite over the washwater weir; that the washwater discharge channel is of sufficient size and gradient to accept the increased backwash without backing up; and that the filter underdrain system can accept higher upthrust. Expanded clay media could also be used to uprate filters although the hydraulic considerations discussed above for anthracite would remain applicable.

9.9 USE OF GRANULAR ACTIVATED CARBON

Granular activated carbon (GAC) is normally employed for adsorption of dissolved organic matter (Section 10.35). It is also a good filter medium and therefore could be used on its own for filtration of turbidity but, for such filtration, GAC would be subjected to aggressive and frequent washing and, being more friable than sand, would be susceptible to breakdown. Consequently, losses as fines could be substantial. Filters are either of the gravity or pressure type. Filtration rates typically vary in the range 6–7.5 $m^3/h.m^2$ and media depth is about 1–1.2 m. Media depth is a function of the required empty bed contact time which could vary between 5 and 30 minutes; for filtration duties the media depth is about 1–1.2 m while depths up to 2.5 m (for gravity filters) and 3 m (for pressure filters) are used for adsorption. GAC size range is either 0.525–1.70 mm (d_{10} = 0.6–0.7 mm) or 0.85–2.00 mm (d_{10} = 0.9–1.1 mm) depending on the application. GAC should be tested for friability (BW, 1996). Water soluble ash content should be less than 1% w/w.

The adsorption capacity of the GAC gradually becomes exhausted and the media requires periodic reactivation (Section 10.35).

GAC is placed in the filters on 50 mm of gravel or directly on appropriate filter nozzles. Sometimes a 300 mm layer of sand of size 0.5–1.0 mm (d_{10} = 0.55 mm) is used at the bottom to capture any GAC fines. Since most carbons abrade easily, the filters should be washed only by the sequential application of air and water, with the duration of the air scour phase limited to minimize attrition of the GAC media. Most types of underdrain system are therefore suitable. A water rate in the range 5–12 mm/s is used, depending on the effective size, the base material of the GAC (i.e. coal, wood, peat or coconut) and the water temperature, to give 20–30% expansion of the carbon bed. Wash rates for different grades of a coal-based GAC at varying water temperatures to give 20% and 30% bed expansion are given in Table 9.3. Air scour is applied at a rate of around 6.5 mm/s although rates as high as 14 mm/s are used by some.

9.10 UPWARD FLOW FILTRATION

Upward flow filtration with upflow washing has been used for a few potable water treatment plants in the UK, but its use is more appropriate to industrial water applications, as roughing filters ahead of slow sand filters or to tertiary sewage filtration where a high standard of filtrate quality is not so important. The principle used in upflow filters is to have progressively finer sand in the direction of flow, which allows the filter to carry a greater load of impurity before backwashing because the larger particles tend to be held in the lower, coarser part of the filter, leaving the upper layers to deal with the smaller particles. However, unless the finer grades of sand are restrained they would be washed away at higher rates of filtration (as well as during backwashing); to prevent this a filtrate collector pipe system is buried in the top layer of fine sand, with strainers located on the side of the pipes so that filtrate water flow has to change from a vertical to a horizontal direction, thus preventing expansion of the sand. During backwashing the filtrate collector is not used and dirty washwater escapes from an elevated trough. Another method is the use of a grid, square in section, located about 0.1 m below the surface of the sand. During filtration the sand arches between individual members of the grid which prevent expansion of the sand, whilst during backwashing the arches are intentionally broken by successive applications of air and backwash water. Piped lateral floors with large orifice nozzles are used to distribute the incoming water. Screening of the raw water to remove leaves and other debris is essential to prevent blockages.

9.11 DIRECT FILTRATION

Some surface waters can be treated by coagulation, flocculation and rapid filtration (gravity or pressure), eliminating clarification. Such waters need to be carefully selected and pilot tested. In general, a water source is considered to be suitable for direct filtration when average turbidity and

Table 9.3 Wash rates (mm/s) for GAC to give 20% and 30% bed expansions at varying water temperature

Grade	GAC size range (mm)	Effective size range (mm)	Water temperature (°C)					
			5	10	15	20	25	30
F200	0.425–1.70	0.6–0.7	4.5 (5.6)	5.0 (6.4)	5.8 (7.2)	6.3 (7.8)	6.7 (8.3)	6.9 (8.6)
F300	0.600–2.36	0.8–1.0	6.4 (8.1)	6.9 (8.9)	7.8 (10.0)	8.5 (10.3)	9.0 (11.4)	9.3 (11.5)
F400	0.425–1.70	0.6–0.7	3.9 (5.0)	4.7 (6.0)	5.3 (6.7)	5.8 (7.2)	6.4 (8.1)	6.5 (8.2)
TL830	0.850–2.00	0.9–1.1	6.9 (8.9)	7.8 (9.7)	8.6 (10.3)	8.9 (10.8)	9.5 (11.7)	9.7 (11.8)

Source of information: Chemviron Carbon Ltd, UK.
Note: Values for 30% bed expansion shown in brackets.

colour values are less than 10 NTU and 25 Hazen, respectively, with peaks of 40 NTU and 40 Hazen for periods less than 24 hours. Maximum values are likely to be significantly lower for filter run times of 24 hours. The total organic carbon value should be less than about 2 mg/l as it influences the coagulant requirement. The coagulant dose should be no more than 1 mg/l as Al or 1.5 mg/l as Fe, although higher doses for short periods are acceptable. A polyelectrolyte may be used as a coagulant aid, but good dose control is essential to eliminate the risk of mudball formation due to carryover of polyelectrolyte. Algae, both the filter clogging and passing types, can cause problems such as shortened filter runs if numbers are high (Chan Kin Man, 1991); an upper limit of 2000 asu/ml for diatoms is reported (Hutchinson, 1974) (asu/ml = areal standard units/ml; 1 asu is $20 \times 20\ \mu m$ and for a medium-sized algae 1 asu/ml can be approximated to 0.1 μg/l of chlorophyll a). The total flocculated solids load on to the filters should be limited to about 20–25 mg/l, with short-term peaks up to about 60 mg/l. Direct filtration operates well on microflocs and requires untapered high energy flocculation (Section 8.11). The filters used are either deep bed (0.9–1.0 m) monograde sand ($d_{10} = 0.9$ mm) or anthracite–sand containing 0.3–0.4 m anthracite ($d_{10} = 1.3$ mm) and 0.6 m graded sand ($d_{10} = 0.55$ mm) (Sections 9.3 and 9.7). Filtration rates should be maintained below about 7.5 $m^3/h.m^2$.

9.12 FILTER PROBLEMS

Filter problems can occur due to incorrect design (filtration and wash rates, media grading, flow control method, etc.) for the water to be treated, poor hydraulic design, poor performance of upstream processes resulting in overloading or precipitation, use of unsuitable material, poor installation in particular of the underdrain system or mal-operation. These may result in short filter runs, inefficient filter cleaning (dirty filters), very high starting headloss (up to 1 m), media loss, loss in capacity and shortfalls in filtered water quality, and sometimes even ruptured filter underdrain systems. Installation problems include damaged, blocked or poorly fitted nozzles, incorrect levelling of the floor or pipe laterals and nozzles (outside their tolerance limits), construction debris left in the pipe laterals or the plenum, poor sealing of the floor slabs or pipe/duct joints, incorrect grout being used and differential settlement. These flaws can be checked before placing the media by a hydraulic pressure test with all the nozzles plugged and by testing for uniform air distribution with about 150 mm of water in the filter sufficient to cover the nozzles. Observation of the air scour pattern during backwash is an important way of identifying underdrain problems in operating filters.

Blocked nozzles, usually the result of construction debris left in the underdrains, manganese or calcium carbonate deposits, biological growth, or lime or sand particles in the clean washwater, can result in high pressures in the underdrain system leading to its rupture. The risk of damage to the underdrain system can be minimized by incorporating a standpipe with free discharge in the washwater main; a pressure relief valve is not recommended as these may become seized after long periods without operation. Damaged or poorly installed nozzles are a common problem with operating filters and allow sand into the underdrain system reducing its capacity; during washing they can cause channelling, sand boils, high localized velocities with sand ingress into support layers and gravel brought to the surface by jetting and loss of media. Sand boils and consequent upset of gravel layers can also be the result of sudden introduction of backwash water. Some channelling and sand leakage into the underdrain system could be attributed to incorrect sizing of sand and gravel media or use of nozzles with a slot size incompatible with the media size. Slots in nozzles should generally be selected at about 60% of the smallest sand grain size.

Deficient washing results in the build up of floc, oxide deposits and organic matter (algal and detrital) ultimately leading to the formation of mudballs and jetting (build up of columns of support gravel through the media), cracks in the filter bed and bed shrinkage with media pulling away from the walls. Sometimes polyelectrolyte, when dosed in filters or carried over in the clarified water due to overuse, can encourage mudball formation. Early turbidity breakthrough and premature excessive headloss development during filtration should always be investigated. A good analysis of filter problems is given by Hudson (1981), Lombard (1995), Baylis (1971) and Lopato (2012). The problem of air binding is discussed in Section 9.3. It is important that filter washes are frequently witnessed to identify problems early to allow corrective measures to be taken. Observations to be made during backwashing include uniformity of air distribution, undisturbed areas, sand boils, mudballs and media carryover. The wash efficiency of a filter can be checked by taking core samples from several places in the bed after backwashing and analysing 250 ml of media from different depths in each sample for suspended solids by washing it thoroughly with water (up to 250 ml) and measuring the solids by volume using an Imhoff cone. For a good wash, suspended solids should be less than 2% v/v. An alternative criteria defined by Bauer (1997) based on Thames Water Utilities (TWU) experience is that the sand after backwashing should contain particulate organic carbon less than 0.4 g and suspended solids less than 2.4 g per litre of filter medium.

SLOW SAND FILTRATION

9.13 INTRODUCTION AND HISTORY

Slow sand filters were the first effective method devised for the purification in bulk of surface waters contaminated by pathogenic bacteria. They remain equally effective today and they address a growing number of issues for which their use in new works should be encouraged.

'Slow' sand filters are so called because the rate of filtration through them may be only one twentieth or less of the rate of filtration through rapid gravity or pressure filters. Most of London's surface derived supplies are treated by slow sand filters although a number of less efficient works have been closed down. Filters in the remaining works have been uprated, with improvements to pre-treatment and filtration rate; the current total filter area is about 38 ha. Slow sand filters have also been adopted for supplies to cities including Amsterdam, Antwerp, Belfast, Paris, Stockholm and Budapest. A summary of slow sand filter plants operated by TWU is given in Table 9.4.

Slow sand filters are an efficient method of producing water of good bacteriological, physical and organic quality which requires only marginal chlorination before distribution. They achieve 2- to 4-log removal of coliforms, *Escherichia coli*, pathogenic organisms, cercariae of *Schistosoma* (Pike, 1987), ova, cysts such as *Giardia* and oocysts such as *Cryptosporidium*; they remove viruses (Dullemont, 2006), oxidize ammonia and biodegradable natural organic matter ($<80\%$) and reduce turbidity to less than 1.0 NTU.

Table 9.4 Slow sand filter plants operated by TWU, UK

Works	Capacity (Ml/d)	Pre-treatment	Filtration area (m^2)
Ashford Common	690	Pre-ozone, anthracite–sand gravity filters, main ozone	32 × 3121
Coppermills	680	Sand gravity filters, main ozone	34 × 3400
Hampton	790	Anthracite-sand gravity filters, main ozone	112 900 (25 filters)
Kempton	200	Sand gravity filters, main ozone	12 × 3640
Fobney	73	Pre-ozone, coagulation, (PACl), Actiflo clarification, rapid gravity filtration, main ozone	12 × 900

Source of information: Thames Water Utilities (TWU), UK.
Notes: *1.* All slow sand filters have a GAC sandwich of 100–150 mm in a total bed depth of 800–900 mm. In addition there is a gravel layer of 100 mm.
2. Filter underdrain systems are porous concrete floor type.
3. Filtration rate varies in the range 0.3 $m^3/h.m^2$ (average) to 0.5 $m^3/h.m^2$ (maximum).
4. Fobney pre-treatment: PACl - Polyaluminium chloride.
5. Post-treatment is chloramination.
6. All works except Fobney serve London.

9.14 MODE OF ACTION OF SLOW SAND FILTERS

The slow sand filter acts by a combination of straining and microbiological action, the latter being the more important. Purification of the water takes place not only at the surface of the bed but also for some depth below. Van de Vloed (1955) has given a clear account of the purification process. He distinguishes three zones of purification in the bed: the surface coating (the *schmutzdecke*); the 'autotrophic' zone, existing a few millimetres below the *schmutzdecke*; and the 'heterotrophic' zone, which extends some 300 mm into the bed.

When a new filter is put into commission and raw water is passed through it, during the first 2 weeks the upper layers of sand grains become coated with a sticky reddish brown deposit of partly decomposed organic matter together with iron, manganese, aluminium and silica. This coating tends to absorb organic matter existing in a colloidal state. After 2 or 3 weeks a film of algae, bacteria and protozoa develops in the uppermost layer of the sand, to which is added finely divided suspended material, plankton and other organic matter deposited by the raw water. This skin is called the *schmutzdecke* and it acts as an extremely fine-meshed straining mat.

In the autotrophic zone, the growing plant life breaks down organic matter, decomposes plankton and uses up available nitrogen, phosphates and carbon dioxide, providing oxygen in their place. The filtrate thus becomes oxidized at this stage.

In the heterotrophic zone, which extends some 300 mm into the bed, bacteria multiply to very large numbers so that the breakdown of organic matter is completed, resulting in the presence of only simple inorganic substances and unobjectionable salts. The bacteria act not only to breakdown organic matter but also to destroy each other and so tend to maintain a balance of life native to the filter so that the resulting filtrate is uniform.

The biological processes require oxygen and if this is absent anaerobic conditions would set in, resulting in the formation of hydrogen sulphide, ammonia, soluble iron and manganese from their oxides, and taste and odour producing substances. Therefore, to ensure satisfactory operation the feed water must contain sufficient oxygen, indicated by the filtrate oxygen which should not be allowed to fall below 3 mg/l (Huisman, 1974). The efficiency of the process is also temperature dependent. At low water temperatures the rate of biological reactions and activity of bacteria consuming microorganisms reduces rapidly and the rate of reduction in *E. coli* falls sharply, thus requiring the downstream chlorine dose to be increased. The reduction in permanganate value, a measurement of the organic content (Section 7.36), decreases by $(T+11)/9$ where T is the water temperature in °C (Visscher, 1987). Below 6°C ammonia oxidation ceases. If water temperatures less than 2°C persist for prolonged periods, consideration should be given to covering the filters.

9.15 CONSTRUCTION AND CLEANING OF SLOW SAND FILTERS

The bed of sand in a slow sand filter is 0.6–1.25 m thick and is laid over a supporting bed of fine gravel with a filtrate collector pipe or channel system below. Water is introduced via a pipe above the sand and the outlet pipework is arranged to ensure that the sand is always fully submerged. The water passes downwards through the whole arrangement installed in a shallow watertight tank of large area. In some plants a porous concrete floor is used in place of pipes. Filtration rates used are typically 0.1–0.3 $m^3/h.m^2$. Figure 9.6 shows the construction of a typical slow sand filter using the original Coppermills Works (TWU) as an example; the 0.675 m sand bed lies on 75 mm of fine gravel, which, in turn, rests on a bed of porous concrete. Each filter is about 34 m × 90 m and the filtration rate was in excess of 0.3 $m^3/h.m^2$. These filters have since been upgraded by the addition of a GAC sandwich layer for pesticide reduction (Table 9.4). Below the concrete, collector drains direct the filtered water to the main effluent pipes. The sand is ungraded: because the filters are not backwashed, hydraulic grading of the media does not occur and size distribution in the bed is thus random. The sand has a size range 0.21–2.36 mm (d_{10} = 0.30 mm) and UC 1.5–3.5 (less than 2 is preferable). The stated size range generally refers to the 5th and 90th percentiles. The sand should be hard and should not be liable to breakdown when subjected to skimming and washing processes. The suspended solids and particulate organic content of the sand should not exceed 0.5 g and 0.1 g per litre of media, respectively.

The raw water is admitted gently to the filter bed and percolates downwards. Directly after a bed has been cleaned a head of only 50–75 mm of water is required to maintain the design rate of flow through the bed. However, as suspended matter in the raw water is deposited on to the surface of the bed, the *schmutzdecke* builds up on the surface and increases the headloss through the bed. To maintain the flow at a uniform rate as far as possible the outlet valve is gradually opened until the headloss across the bed reaches some predetermined value between 0.6 and 0.9 m, when the bed must be taken out of service and cleaned. The maximum permissible headloss should be kept to about 1.0 m. If fine grain sand is used then the depth should be reduced to minimize the resistance. To prevent negative head (Section 9.3) the maximum permissible headloss must be kept less than the

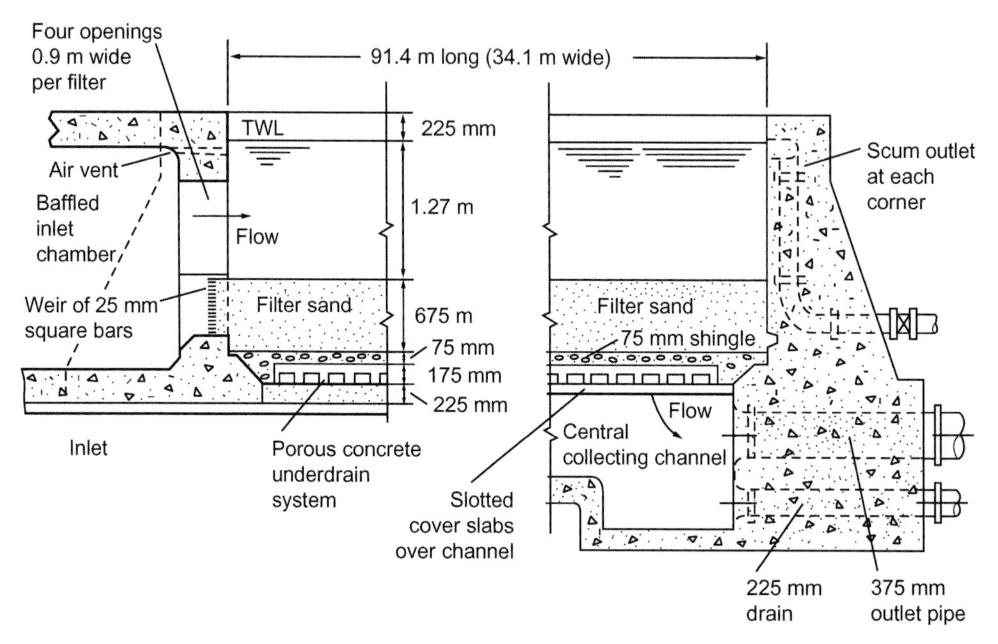

FIGURE 9.6

Slow sand filter.

Source of information: Thames Water Utilities Ltd.

depth of water on top of the sand. Normally the water depth above the sand bed is maintained at about 1.2–1.5 m with a maximum of 2 m. The free board should be greater than 0.3 m. As a further preventative measure the outlet weir (if there is one) should be set above the media surface level.

Each time a sand bed requires cleaning it is drained of water and the top 12–25 mm of the sand surface is carefully scraped off. The filter is then returned to service by gradually increasing the flow over 24 hours, sometimes longer. This process is repeated until the bed is thinned to the minimum practical thickness for efficient filtering which should not be less than 0.5 m (Visscher, 1987). When this stage is reached the bed is then topped up with clean new sand to its original level or the old sand may be returned if it has been adequately washed and cleaned in a sand washing machine.

The interval between successive scrapings may vary from several months during the winter when pre-filtration is installed, to 10 days where no pre-filtration occurs and algal growth is at a maximum. Resanding would only be necessary every 2–5 years depending on the scraping frequency. Originally, slow sand filters were invariably scraped by manual labour (typically for scraping: 5 person-hours/100 m^2 and resanding: 50 person-hours/100 m^2) (ASCE, 1991) but the increased cost and decreased availability of this type of labour led to the introduction of mechanical methods, in particular for large plants thereby reducing labour typically to about 2–4 person-hours/100 m^2 for scraping and 5–8 person-hours/100 m^2 for resanding. Resanding by the wet slurry method is reported to reduce labour requirement further (Kors, 1996). Some of the mechanical methods are fairly simple, for example, using small tracked skimming machines discharging to dump trucks. One of the difficulties of mechanizing the cleaning of the older slow sand filters is that their sizes and geometry were often not uniform since these factors were unimportant in the days of manual cleaning. However, with new installations, provided that the filters are constructed with the correct dimensions, it is possible to span them with a standard sand lifting bridge which runs on tracks along the sides of the filters, thus greatly reducing the labour necessary for scraping (Lewin, 1961). Other cleaning methods used are wet harrowing and flushing (Collins, 1988; Joslin, 1997) which do not disturb the attached bacteria and the use of a suction dredger using laser depth control (Glendinning, 1996) which avoids the need to drain down the filter. Sand scrapings must be washed to remove dirt before being stored for later return to the filter. Washing must be carried out soon after removal. In small plants manual washing by high pressure water jet is used. In larger plants mechanized systems consisting of screens and hydrocyclones are used. The standards aimed for cleaned sand are: silt concentration in the range 0.5–1.0 g/l of sand and particulate organic concentration 0.1 g/l of sand or 0.1% by weight of sand (Toms, 1988; Harrison, 1997). Slow sand filter plants serving the Northern Production Unit in Antwerp are unique in that the upper layer (15–30 cm) of sand is cleaned in situ by water backwash at a rate of about 15–30 cm/h (Huisman, 1974).

Each filter requires five valves: inlet, outlet, backfilling, waste and drain. The valved inlet discharges into the filter over a weir arranged to give a velocity of about 0.1 m/s for good distribution of water and to avoid damaging the *schmutzdecke*. The weir box or the inlet pipe is provided with a valved drain to remove the top water when the filter needs cleaning. The backfilling valve is used to refill the filter (after scraping) by reverse flow to a water depth of about 250 mm to remove air entrained in the voids. The waste valve is used to discharge the filtrate for a period until the filter is 'ripened' after it is brought on line following skimming or resanding; filtered water is discharged to waste or returned to other filters in use until chemical and bacteriological tests prove it to be satisfactory. The filtration rate during such ripening should be similar to that in production since the range of fauna vary with the rate. In operation, it is important to minimize fluctuations in inflow and raw water quality. The plant should be designed to operate at constant rate with a constant level of water above the sand surface, or rising water level in the filter. With the latter, the shallow water depth allows penetration of sunlight, encouraging algae growth and there is always the danger of disturbing the *schmutzdecke* by the incoming water. With the low filtration rate, manual control of filter inlet and outlet is more than adequate and cheaper but automatic control is useful for large plants. Water level in the filter can be kept constant by a level controlled inlet valve and a constant filtration rate is maintained by a flow controlled outlet valve. Each filter should be fitted with a loss of head gauge.

9.16 USE OF PRE-TREATMENT WITH SLOW SAND FILTERS

Slow sand filters may operate as the sole form of filtration for raw water turbidities of up to about 10 NTU. Pre-treatment such as storage or pre-filtration may be used to reduce the load of suspended matter on the slow sand filters and so permit longer intervals between cleanings or a faster rate of filtration, or both. van Dijk (1978) recommended that, if the average raw water turbidity exceeds 50 NTU for more than a few weeks or is in excess of 100 NTU for a day or so, then pre-treatment is indispensable. Pre-treatment can take the form of river bed or bank filtration, Ranney wells, storage, plain settlement, media filtration (horizontal, pebble matrix, rapid gravity or

Table 9.5 Performance of slow sand filters for London water supply (1998–99)

Treatment works and type of pre-treatment currently adopted	Rate of filtration (m^3/h.m^2)	Quantity of water filtered per hectare of sand bed cleaned (Ml/ha.d)[a]
Hanworth Road[b] – no pre-treatment	(0.004–0.059)[c]	(3.15–7.05)[c]
Kempton Park; rapid gravity pre-filtration and ozone	0.3 (0.132–0.152)[c]	33–49.5 (12.6–16)[c]
Ashford Common; pre-ozone, rapid gravity pre-filtration and main ozone	0.3 (0.132–0.137)[c]	41.4–69 (10–15.7)[c]

Source of information: TWU, UK.
Notes: [a]1 Ml = 10^3 m^3. 1 ha = 10^4 m^2. So 1 Ml/ha = 0.1 m^3/m^2.
[b]Hanworth Road installation is now closed.
[c]Values in brackets show the information for the summer seasons in 1959–63; since then ozone treatment and GAC sandwich have been introduced.

upflow) (Galvis, 1993), microfiltration (Section 9.19) or a combination of these depending on the turbidity of the raw water. If the filtrate has persistently low oxygen content (<3 mg/l) then aeration of the raw water would be necessary. Pre-treatment can be more extensive and sometimes includes chemical coagulation and flocculation without clarification or with clarification (Welte, 1996) and filtration (Abrahamsson, 2006). Residual coagulant metal appears not to be detrimental to the biological activity (Dorea, 2006). If raw water *E. coli* counts are in excess of about 10 000/100 ml, the filtered water should be subjected to full disinfection as marginal residual chlorination, without the use of a chlorine contact tank (Section 11.5), would be inadequate. A summary of the effects of pre-treatment is given by Ridley (1967), who compared three London supply works: Hanworth Road with slow sand filtration only; Kempton Park with rapid gravity filters preceding slow sand filters; and Ashford Common with microstraining preceding slow sand filters. Figures for the period 1 April–30 September (when algal growths are usually most prominent) for the 5 years 1959–63 inclusive and the corresponding data for 1998–99 following upgrading of the pre-treatment on operating criteria are shown in Table 9.5.

During the period 1956–63, Kempton Park rapid gravity filters used 1.15–1.62% washwater (as a percentage of the water filtered) while Ashford Common microstrainers used 1.64–2.03%. The extra washwater used and the capital and running costs of pre-filtration and ozonation are usually taken as economically justified in view of the very considerable increase in water filtered through the slow sand filters between cleanings.

9.17 LIMITATIONS AND ADVANTAGES OF SLOW SAND FILTERS

Slow sand filters have tended to be ignored when new plants are being considered because of the amount of land and labour they use and their relatively large capital cost. However, there is still a place in water treatment for slow sand filters, providing their advantages and limitations are carefully weighed in each particular case. They may also play a greater role in future water treatment technologies owing to pressures for less chemical usage in treatment.

Slow sand filters do not significantly reduce the 'true colour' (Section 7.19) of a water. They are thus only suitable for treating waters of relatively low colour. Another factor is that slow sand filters are not effective in removing iron and manganese in solution and, being a biological process, there is a marked reduction in efficiency in the removal of other contaminants at low temperatures (Section 9.14).

A high concentration of algae in the raw water can cause treatment difficulties due to clogging of the filters or anaerobiosis and impartation of taste into the water if the cells die off in the filter, due to excessive respiratory demands. The chlorophyll a concentration of the feed should ideally be limited to 5 μg/l (ASCE, 1991; Bauer, 1995) with a peak of 15 μg/l. Thus, slow sand filters are not the treatment of choice for highly eutrophic waters in which sudden peaks of algae may appear in spring, summer or autumn. Small algae such as diatoms increase the headloss through the *schmutzdecke*. Methods to prevent or control algae include pre-treatment, covering of slow sand filters and chemical treatment. Low to moderate algal concentrations, in particular filamentous species, on the other hand are beneficial to the process.

Slow sand filters are also not very suitable for the removal of any substantial quantity of finely divided inorganic suspended matter. With upstream rapid gravity or upflow filtration, however, they may function successfully for years treating waters which are intermittently laden with fine silt.

Removal of organics by slow sand filters is not complete (typically achieving a reduction in biodegradable dissolved organic carbon of 46–75%) and removal of pesticides is type dependent (Lambert, 1995). Therefore on some waters they are supplemented by additional treatment usually ozone and GAC adsorbers used either upstream or downstream of slow sand filters. At TWU's Ashford Common Works a 0.135 m GAC sandwich layer is incorporated within the slow sand filters; ozonation is carried out upstream of roughing and slow sand filters (Glendinning, 1996). At Ivry and Orly plants near Paris, ozone and GAC are used downstream (Welte, 1996).

Apart from the above, the other stated limitations of slow sand filters often appear of doubtful validity on closer inspection and four examples are given to illustrate this. Firstly, it is stated that slow sand filters occupy a greater area of land than do coagulation and rapid gravity plants. This is true so long as one ignores the question of sludge disposal and chemical dosing, but if adequate sludge disposal and chemical facilities are included the land required for slow sand filters may well be no greater than that for conventional treatment. Secondly, it is maintained that slow sand filters are expensive in capital costs. However, when the cost of chemicals is taken into account, slow sand filters may be cheaper in whole life costs than conventional plants. Thirdly, whereas it was usual to design slow sand filters for low rates shown in Table 9.5 for 1959–63, it has been found that with improvements to pre-treatment (e.g. pre-ozonation, GAC, redesigned primary filters, intermediate ozone) without covering (Glendinning, 1996) or with covering (Wilson, 1996), filtration rates have been almost doubled to about 0.3 $m^3/h.m^2$. Abrahamsson (2006) reported that covering increased filter run time from 6 months to 4 years at Stockholm Works. Other techniques such as the use of replaceable non-woven synthetic fabric layer (geotextiles) on top of sand has helped to reduce the cleaning frequency (Hendricks, 1991) and a sixfold reduction in cleaning frequency is reported by Klein (1994). Fourthly, slow sand filters are said to be labour intensive, but where designed specifically for the latest methods of mechanical cleaning, they may require no more operator input than conventional treatment plants, taking into account the greatly extended run times and the much reduced labour required for sludge treatment and disposal.

In some European countries slow sand filters are used as a final polishing stage of treatment, being preceded by as many as five or six stages of treatment including chemical coagulation and sedimentation (Schalekamp, 1979), ozone, GAC, softening, etc. at River-Lake plant, Loenderveen and Wesperkarspel, Netherlands (Kors, 1996).

MEMBRANE AND MISCELLANEOUS FILTERS

9.18 MF AND UF MEMBRANE FILTRATION

Membrane filtration employs a semi-permeable membrane to separate materials according to their physical and chemical properties when a pressure differential or electrical potential difference (electrodialysis – see Section 10.45) is applied. These processes can be broadly classified according to the membrane pore size, filtration mechanism(s), and size of particles removed (Fig. 9.7). Associated classifications are microfiltration (MF), ultrafiltration (UF), nanofiltration (NF) and

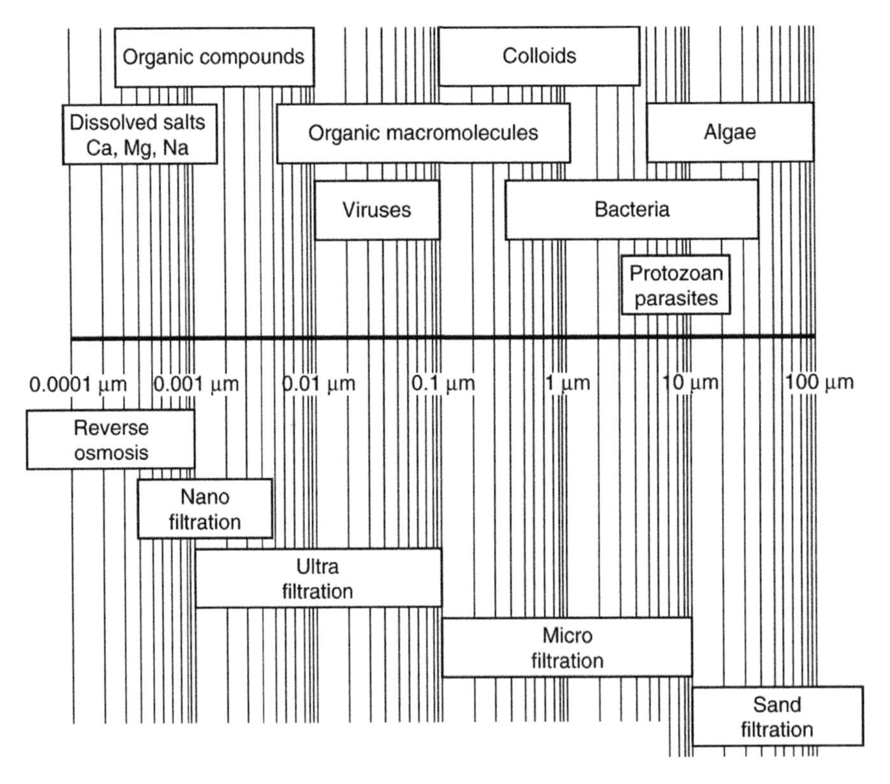

FIGURE 9.7

Comparison of treatment methods with particle size.

reverse osmosis (RO). RO and NF are used to remove many types of dissolved ions and molecules from water (Section 10.46), whereas MF and UF remove particles from water. MF and UF are sometimes called low pressure membrane filtration, because typical MF/UF transmembrane pressures (TMPs) are 1–3 bar compared to NF operating pressures of 6–14, 7–20 bar for low-salinity RO, and 55–80 bar for seawater desalination with RO (not taking credit for any energy recovery).

Filtration with MF/UF removes particles based on size exclusion applying a sieving mechanism. Unlike more traditional granular media filtration, MF/UF does not require physiochemical conditioning, such as addition of a coagulant prior to filtration to enable removal of small particles. MF membranes used for water treatment generally have pore sizes of 0.1–0.5 μm and UF of 0.005–0.1 μm. To define their ability to remove large organic molecules dissolved in water, some UF membranes are also rated based on molecular weight cut-off (MWCO) and nominal values of 100 000–200 000 daltons are typical. (Dalton is a unit of measurement defined to be 1/12 of the mass of one atom of carbon.) MF provides essentially no removal of dissolved materials and UF removes very little of dissolved materials of concern for potable water. For example, neither type removes dissolved manganese, true colour, or most other dissolved organic material or disinfection by-products (DBPs) such as trihalomethanes. While UF removes some large organic molecules, the precursors of DBPs, DBPs themselves, and taste and odour causing compounds generally have much lower molecular weights than the MWCO, so these solutes generally pass through both MF and UF. To augment removal of dissolved contaminants, MF/UF can be integrated with pretreatment processes that convert the dissolved parameter into a particulate form, such as coagulant or powdered activated carbon addition to adsorb dissolved organics, sources of taste and odour and DBPs, or addition of oxidants to convert manganese and iron to particulate or colloidal forms.

MF/UF consistently removes small particles, thereby providing effective and predictable control of turbidity with typical filtrate values of less than 0.1 NTU. Speth and Reiss (2005) summarized a literature review of 122 studies that showed that MF and UF membranes produce extremely high quality water regardless of influent turbidity and that there is no apparent difference in turbidity removal between membrane type, manufacturer, or whether a coagulant was used. Mean filtrate values had a median of 0.06 NTU and the maximum values had a median of 0.08 NTU. Studies by Jacangelo et al. (1991, 1995), Coffey et al. (1993) and others have shown that MF and UF also consistently remove *Giardia* cysts and *Cryptosporidium* oocysts exhibiting removals greater than 4-log and ranging up to 6- or 7-log. Log removal expresses the filtration removal effectiveness for a target organism, particle or surrogate and is calculated as: log [feed concentration] − log [filtrate concentration]; thus, a log removal value (LRV) of 4 = 99.99% removal and a LRV of 2.5 = 99.68%. Cysts are not detected in the filtrate of intact MF/UF membrane systems and the reported values are therefore limited by the sensitivity of the measurements. While there is no observed difference in removal of protozoan cysts (which are 2–15 μm in size) between MF and UF, UF provides greater removal of viruses (which are 1–2 orders of magnitude smaller). One regulatory agency that bases MF/UF ratings on carefully controlled microbial challenge tests, California Department of Public Health, routinely grants 4-log removal credit for *Giardia* and *Cryptosporidium* to accepted MF/UF membranes. For MF membranes, the virus removal credit is typically 0- to 0.5-log, while for UF it ranges from 0- to 4-log depending on performance and viewpoint of the designer or regulator. Theoretically, MF with 0.2 μm pores should not reject virus-sized particles but limited virus removal is observed. It has been speculated that the mechanism for this may be rejection of larger particles on which viruses have become attached, or adsorption of viruses on the cake layer. Even given the high microbial LRVs of MF/UF, chemical disinfection (e.g. chlorination) should follow membrane filtration to protect the public health with a multiple barrier approach.

The use of low pressure membrane filtration has increased rapidly over the past two decades – partly due to concern over microbial contaminants and the well proven ability of MF/UF membranes to provide consistently low turbidity filtrate and high LRVs independent of chemical pre-conditioning.

There are significant differences between the various types of MF/UF systems. The most basic hydraulic difference is that some employ membranes mounted inside pressure vessels (called pressurized or encased systems) (Plate 13(b), a typical arrangement is shown in Plate 13(c)), while others apply membranes mounted in a tank open to atmospheric pressure (called submerged, immersed or vacuum systems) (Plate 15(b)). With encased systems, a positive pressure is applied to the upstream side of the membrane to transport the filtered water through the membrane while for submerged systems a negative pressure is applied to the downstream side to draw filtrate through the membrane. Either type can utilize a site's hydraulic gradient, so selection and equipment arrangement may be influenced by topography. For example, siphon-driven and mixed siphon pump submerged UF membrane systems were selected to fit the topography and hydraulic profile and to reduce costs at the 273 000 m^3/d extension of Chestnut Avenue Waterworks (Freeman, 2004) and the retrofit of membranes into existing granular media filters at the 184 000 m^3/d Choa Chu Kang Waterworks (Ratnayaka, 2008).

The membrane geometry most commonly used in MF and UF is hollow fine fibre with lumen (e.g. internal bore) diameters typically ranging from 0.4 to 1 mm (Plate 13(a)). Most systems use polymeric membranes, but ceramic materials allow a wider range of temperature and pH, and have greater oxidant resistance which can be beneficial during cleaning (Fujiura, 2006). Ceramic membranes for potable water production were expensive and rare outside Japan. However, the design has since been developed to house multiple membrane elements in a single pressure vessel, as opposed to the single membrane element per vessel of the original design. Costs have come down and ceramic membranes are being adopted for an increasing proportion of projects. Their advantages over polymeric membranes include longer membrane life (20 years vs. 10), absence of fibres and related integrity issues, reduced backwashing frequency, improved cleanability and better resistance to aggressive cleaning chemicals. These factors allow increased run times between washes, more consistent permeability and more reliable long-term output. Further developments have allowed costs to be reduced further, including a modified design that houses up to 200 membrane elements in a single pressure vessel (Galjaard, 2012) (Plates 14(a)–(c)). Polymeric membranes are most frequently manufactured from polyvinylidene difluoride, polyethersulfone or polysulfone; although some older systems use polypropylene or cellulosic materials. Ceramic membranes are currently made of aluminium oxide but other materials are being considered. They allow ozone to be used to reduce the potential for fouling by natural organic matter, reduce the TMP development and increase permeability.

While most membranes are based on an out-to-in flow pattern, with the raw water applied on the outside of the fibres and the filtrate flowing inside, some encased systems use an in-to-out pattern. Most MF and UF processes are operated in a dead-end filtration mode, while operating in a cross-flow mode has become rare. In the latter, only part of the inflow is filtered through the membrane; the remainder flows tangentially to the membrane surface, carrying with it the particulates removed on the surface whilst a portion is recirculated back to the feed water supply. If a cross-flow option is compared to dead-end, it is important to include energy usage in the cost comparison, because the recirculation can be a major energy user. In general, an MF/UF system can be characterized by the following key parameters: hydraulic design (encased vs. submerged), chemistry and materials of construction, chemical tolerance (to oxidants, disinfectant and pH) (Table 9.6), physical tolerance (to temperature and pressure), flow pattern (outside-in vs. inside-out), flow type (dead-end vs. cross flow), pore size, packing density, backwash type, pumping and air requirements, energy usage and physical arrangements (horizontal vs. vertical and single or multiple modules per vessel).

Certain features are similar for all types of MF/UF as currently practised. These systems have five basic operating cycles: filtration; backwash, which is sometimes augmented with an air scour; chemically enhanced backwash (CEBW); chemical clean-in-place (CIP); and integrity verification. Typically, a unit is in filtration mode for approximately 85% of the time, producing filtrate at a set flow or flux (filtrate flow/membrane area, generally expressed in $l/m^2.h$ and termed lmh). Periodically, usually two or three times an hour, the particle build-up on the surface is removed by backwashing, generally with filtered water. Often the water backwash is assisted with an air scour and/or air-driven water backwash. Periodically, the membrane is also chemically cleaned with a CEBW or a CIP operation. CEBW is a special backwash operation conducted at typical frequencies ranging from daily to once or twice per week, depending on feed water quality, in response to fouling, with chemicals such as sodium hypochlorite, caustic soda or acid, added to the backwash water supply. CIP uses more concentrated chemical solutions with soak cycles as well as flushing cycles, to clean the membrane. These are typically conducted monthly, or in some cases quarterly or less, as allowed by local conditions, flux, pre-treatment, and related fouling. Terminal TMP, the condition at which cleaning must be conducted, varies in the range from 0.7 to 3 bar. Feed water recovery varies between 90% and 98%, depending on the operating mode and on raw water quality. This includes an allowance for backwash water which can be about 2–10% of product flow.

Table 9.6 Characteristics of selected membrane materials (AWWA, 2005)

Material	Type	Oxidant tolerance	pH range
Polyvinylidene difluoride (PVDF)	MF/UF	Very high	2–11
Polyacrylonitrile (PAN), rarely used	MF/UF	High	2–11
Polyethersulfone (PES)	UF	High	2–13
Polysulfone (PS)	UF	Moderate	2–13
Polypropylene (PP), rarely used	MF	Low to nil	2–13
Cellulosic, rarely used	UF	Moderate	5–8
Ceramic, aluminium oxide	MF	Greatest/widest tolerance	1–12 +

Notes: *1.* PVDF generally has tolerance to up to 5000 mg/l chlorine residual (short-term) and 20 mg/l (continuous), chlorine dioxide to 5 mg/l, $KMnO_4$ to 5 mg/l, and even limited tolerance to ozone if special construction methods are used, while PES can generally tolerate up to 100 mg/l of chlorine. Ceramic material has the greatest tolerance to oxidants.
2. Ceramic membrane information is based on personal communications with ceramic membrane manufacturer, METAWATER and is not from the AWWA (2005) reference.

Critical design parameters for membrane systems are capacity with respect to seasonal demand and related water temperature, sustainable flux, and level of redundancy. Flow through MF/UF is more temperature sensitive than through granular media; the limiting condition typically occurs in winter when water viscosity is at its maximum. Designers should evaluate the relationship between seasonal temperature, demand and MF/UF ability to provide sufficient year-round capacity. The sensitivity of membrane capacity or flux to viscosity can be roughly accounted for by the factor $1.03^{(T-20)}$ with 20°C being the reference temperature. Additional equations are presented in the US EPA's *Membrane Filtration Guidance Manual* (US EPA, 2005).

Flux should be sustainable in the long term; if the flux is too high the facility will experience high operating costs, more fouling which reduces capacity, and stress which leads to excessive fibre breakage in polymeric membranes. Ideally, long term pilot trials should be used to determine the optimum stable flux. Often, however, it is not feasible for a pilot programme to capture the complete range of raw water quality and temperature variations and the cumulative effect of multiple filtration and CIP cycles. In such instances, the performance of full-scale plant in similar applications can be considered, if available. While it is difficult to generalize, in many cases in a temperate climate, a maximum instantaneous flux of 40–50 lmh for polymeric membranes (100–200 lmh with ceramic) is sustainable. For especially challenging applications, such as filtering secondary effluent for water reclamation or treating a coagulated water without an intermediate clarification process, 38–42 lmh is generally sustainable, while for treating low fouling waters a flux of 60–80 lmh might be suitable. If a pilot trial is conducted, it should include seasonal water quality and temperature variations that may be anticipated for the full-scale plant and should use standard full-scale membrane modules to avoid misleading results. Trials should also include full representation of all pre-treatment and recycled streams. Generally, use of polyelectrolyte in pre-treatment or recycled conditioning is avoided or limited to chemicals with a proven record of not causing fouling. Since such fouling can be severe and irreversible, any polyelectrolyte being considered should be included in the pilot programme.

The level of redundancy should also be carefully considered in the planning and design of membrane facilities. The evaluation should consider the frequency and duration of peak-day events as well as provisions for emergencies, availability of alternative water sources, and storage in the distribution system. Operators of MF and UF systems report a significant difference between membrane filtration and granular media. With membranes, low turbidity filtered water is easy to achieve, independent of pre-conditioning; however, it can be impossible to force more filtrate through a membrane, even to address a short-term need. Sometimes one or two spare trains are included to cater for such requirements. To avoid queuing for routine operating cycles such as backwashing, larger facilities may need to consider redundancy within sections of the plant.

One of the benefits of MF/UF membrane technology is the integrity verification test. Generally practised once each day, this simple automated test verifies there are no flaws greater than a set resolution or size, typically 3 μm, to coincide with the low end of the size range of *Giardia* cysts and *Cryptosporidium* oocysts. While there is some variation, the general description of the test is as follows. About 1 bar of air is applied to one side of the membrane, generally the filtrate side to prevent drying fouling material on the working surface; the membrane itself is wet and the pores are full of water. After pressurizing the system with air and isolating it with valves, the air pressure or flow rate is monitored for 5–15 minutes. If the membrane is intact, then all of the pores should be smaller than 3 μm, in which case air leakage would be low. With an intact membrane, the surface tension of the water holds the water in the pores and the low rate of leakage will be due to diffusion of air into the water. However, if the membrane is not intact, the leakage rate is higher. If the automated test indicates a leak, operators can find and repair the leak with a series of similar manual tests. It should be noted that turbidity measurement is not sensitive enough to indicate flaws in full-scale MF/UF systems (Jacangelo, 1991). Some researchers are trying to demonstrate 4-log removal with particle counting, but this has not been widely accepted, partly due to the sensitivity and cost of the instruments, but partly due to the reliability of the air-based membrane integrity test. Guo (2010) provides a useful summary of other options available for integrity testing of membranes.

Both MF and UF systems can be applied directly on raw water with only a pre-filtration stage to reduce particle size to a range 150–500 μm depending on specific requirements. Pre-filtration generally comprises automatic backwashing screens employing weave-wire or wedge-wire media. In addition, membrane processes can be preceded by pre-treatment to augment removal of dissolved materials either to meet water treatment goals (such as for colour, manganese, hardness or DBPs) or to protect the membrane from fouling. Pre-treatment options cover the full range of water treatment technologies, including coagulation, flocculation, sedimentation, DAF, lime softening or activated carbon. Post-treatment such as advanced oxidation can also be applied to address DBP precursors, micropollutants and taste and odour causing compounds.

A typical plant consists of raw water pumps feeding pre-screens, followed by a bank or several banks of membrane modules, with recirculation pumps if needed for cross-flow mode operation. To simplify the descriptions in this paragraph, the text is based on encased membranes. For a submerged membrane system, the order of unit operations is slightly different with pre-screens followed by the membrane tank and then pumps drawing water through the membrane cells and discharging into a common header downstream. A module generally consists of one or sometimes several membrane elements in a pressure shell, complete with feed inlet ports, distributors, outlets and filtrate and permeate removal points. In a bank, several modules may be connected in parallel. Pumps used on a membrane plant are usually of the centrifugal type. Both MF and UF systems can be provided with multiple stages, with used backwash water and bleed from the recirculation water, if any, treated in a second stage. A flow schematic diagram of a typical membrane plant is shown in Plate 15(a).

O&M cost for a membrane plant is primarily made up of energy, membrane replacement and labour. Adham (2005) surveyed 87 full-scale MF/UF plants with capacities greater than 4000 m^3/d; these reported median O&M costs of US$39.8 per 1000 m^3. These operating costs comprise the following categories: labour (32% of the total), energy (30%), parts (10%), chemicals (9%), chemical disposal (2%) and other (8%). Comparing these results to Witcher et al. (2014) shows that O&M costs vary; they reported US$195 per 1000 m^3, not including raw water charges, for an 83 Ml/d membrane-based surface WTP in California with the following breakdown: labour (31%), electric power (39%), chemicals (16%), maintenance (5%), laboratory (1.3%), solids handling (1.1%) and the remainder miscellaneous.

9.19 MISCELLANEOUS FILTERS

Numerous variations of the conventional sand filter have been developed over the years. One type of filter comprises a vertical, hopper bottomed cylindrical vessel filled with sand of grain size 0.9–1.2 mm ($d_{10} = 0.9$ mm). Raw water is introduced into the cylindrical section just above the hopper and flows upward through the sand bed counter-current to the sand moving continually downwards and leaves the filter at the top over an outlet weir (e.g. 'Dynasand'). The sand containing dirt is conveyed from the hopper section by an airlift pump; the turbulent action of the pump cleans the sand with additional hydraulic or mechanical cleaning in a sand washer at the top. The cleaned sand is returned to the top of the filter. Filtration rates can vary in the range 5–10 $m^3/h.m^2$. Such filters are used as primary filters ahead of slow sand filters or in direct filtration mode with coagulation. In the latter case filtrate turbidity is in the range 0.1–0.5 NTU. Water loss for continuous sand cleaning is in the range 5–10% of the inflow.

Filtration of water can be achieved with *diatomaceous earth* filters or *pre-coat* filters, where the filtering medium is formed on septums or 'candles' of metal or other materials inside a pressure vessel. Excellent removal of suspended matter may be achieved. A good account of the process is given by Cleasby (1999) and Marsh (2004).

Cartridge filters with elements of the woven fibre type rated at 5–10 μm are used in reverse osmosis plants preceding high pressure feed pumps. These filters remove particulate matter passing the pre-treatment stage and are used as safety filters to protect high pressure feed pumps and the membrane (Section 10.47). Once used, they are normally discarded, although backwashable elements are now available. Cartridge filters are used in some groundwater treatment plants where waste discharge from the works is not permitted, or on emergency supplies. They can provide 4-log removal of *Cryptosporidium*.

Some microfibre filters are designed to allow backwashing (e.g. AMF by Amiad). These are automatic self-cleaning pressurized units capable of removing up to 98% of suspended solids above 2 μm. The filter medium is a multi-filament textile fibre wound onto a cassette. A large number of these filter cassettes are interconnected within a housing vessel to provide the required flow capacity. Backwashing is completed using water admitted under pressure via nozzles that direct the water onto the outer surface of the fibres.

A range of automatic, self-cleaning mechanical filters variously using screens, multi-filament textile fibre cassettes or discs are available, capable of achieving up to 98% of suspended solids removal in the range 10–800 μm. These find application as one of the pre-treatment stages in many water treatment plants, including membrane desalination plants.

There are so called '*depth filters*' where the filtration efficiency increases with passage as a 'dynamic filtering' medium is deposited on a filtration surface, either from particles in the water being treated or by addition of a suitable fine material such as diatomaceous earth, Fuller's earth (a fine non-cohesive clay) or powdered activated carbon. The filtration surface can be fibres wound on to a cassette or a candle with grooves, fibres arranged in a bundle around a central core or a porous cloth woven to produce a series of filtration tubes (e.g. Kalsep). The fibrous filters are capable of filtering particles down to about 5 μm, and achieving *Cryptosporidium* oocysts removal up to 1-log consistently (O'Neill, 1995). The porous cloth type is known to be capable of removing particles greater than 1 μm and therefore provide 2- to 3-log removal of oocysts.

Other filter designs include the *Green Leaf* and travelling bridge type. The former is a valveless rapid gravity filter based on siphon technology. Multiple filter cells are arranged around a central core and each filter is washed by using the filtrate from the remaining filters. In the latter design the filter is made up of several shallow bed cells, with one cell washed at a time by a 'vacuum cleaner' action using a hood placed on each cell. Washwater is available from the remaining cells. The filter bed could be 450–600 mm deep with sand only or sand-anthracite.

CRYPTOSPORIDIUM OOCYSTS AND *GIARDIA* CYSTS REMOVAL

9.20 *CRYPTOSPORIDIUM*

Removal of *Cryptosporidium* oocysts can be achieved by any process that removes particles down to a size of 3 μm or smaller. However, conventional water treatment processes in current use for drinking water supply cannot guarantee complete removal. Since disinfectants are largely ineffective against oocysts (Sections 7.63 and 11.6) the removal of oocysts in the upstream solid–liquid separation processes must therefore be maximized. Well operated and maintained conventional treatment processes, which include coagulation, flocculation and clarification followed by granular media filtration, can be expected to achieve 2- to 3-log removal consistently (Standen, 1997). Criteria for good clarification performance would be turbidity less than 1 NTU and total coagulant metal ion concentration less than 1 mg/l as Fe or 0.5 mg/l as Al depending on the coagulant used (UKWIR, 1998). Filters should be operated at about 6 $m^3/h.m^2$ with all filters in service. Filtrate turbidity should be less than 0.1 NTU (Logsdon, 1998). Filtered water turbidities <0.1 NTU and particle counts <50/ml are indicators of good treatment for controlling *Cryptosporidium* (Edzwald, 1998). Second stage filtration such as GAC adsorbers or manganese removal filters may provide a further barrier with removal <0.5-log but cannot be relied on for effective removal because oocysts are not enmeshed in residual floc in the same way as upstream of primary filters. The US EPA (2006) adopts an approach that determines the degree of removal required based on the historic challenge from *Cryptosporidium* oocysts and the number of customers supplied. Log removal credits are ascribed to treatment processes as follows: 3-log credits for conventional treatment (coagulation, clarification and filtration) or slow sand filtration (with no pre-chlorination) and 2.5-log credits for direct filtration or slow sand filtration (as secondary filtration with no pre-chlorination). Additional treatment is not required for oocysts concentration <0.075/l. For systems supplying more than 10 000 consumers from raw water containing 0.075–1.0 oocyst per litre, 1.0–3.0 oocyst per litre or ≥3.0 oocysts per litre (based on the maximum value for a 12-month running annual average or 2-year arithmetic mean if twice-monthly monitoring is conducted), plants with conventional treatment (coagulation, clarification and filtration) or slow sand filtration (with no pre-chlorination) must be provided with additional treatment sufficient to provide an additional 1-, 2- and 2.5-log removal or inactivation respectively to comply with a total *Cryptosporidium* removal or inactivation of at least 5.5-log. The additional log reduction or inactivation required for these same raw water quality envelopes downstream of direct filtration or slow sand filtration (as secondary filtration) are 0.5-log greater than required downstream of conventional coagulation, clarification and filtration. The additional treatment could be bankside filtration (respective credits of 0.5-log if 25 feet from source or 1-log if 50 feet from source), cartridge filtration (respective credits 2- or 2.5-log for single or in series), membrane filtration or disinfection by chlorine dioxide, ozone or UV. Challenge testing is required for cartridge and membrane filters to demonstrate the removal efficiency. Systems serving less than 10 000 consumers are not required to be monitored and do not require additional treatment.

Microfiltration and ultrafiltration can achieve consistent removal rates better than 2-log and 4-log respectively (Section 9.18); in some studies higher removals (>6-log) (Jacangelo, 1995) have been reported. *Cryptosporidium* oocysts are resistant to even high concentrations of chlorine (Section 11.6) but ozone, when used in the doses typically applied at water treatment works, is reported to produce 1- to 2-log inactivation of oocysts (Badenoch, 1990). US EPA provides

Ct values for *Cryptosporidium*, at different water temperatures, which are listed in Table 11.4. Chlorine dioxide is less effective against *Cryptosporidium* and *Ct* requirements are high; for example *Ct* values for 1-, 2- and 3-log credits are 179, 357 and 536 mg.min/l, respectively at 15°C (US EPA, 2006). UV disinfection is widely accepted for inactivating *Cryptosporidium* (Section 11.25).

It is important to note that, irrespective of the various degrees of removal/inactivation achieved in different processes, the critical factor is the number of oocysts remaining in the water, which is a function of the number present in the raw water and how this relates to the infectious dose.

In the UK, three reports were produced by a committee of experts (Badenoch, 1990, 1994; Boucher, 1998) and, based on the recommendations made in these reports, the following design and operational guidelines are proposed to minimize the risk of *Cryptosporidium* passing into water supplies:

1. The design and operation of rapid gravity filters should ensure sudden surges of flow are avoided; flow changes should be limited to 1.5–5% of prevailing flow per minute, depending on raw water type and associated floc strength.
2. Rapid gravity filters should not be restarted after shutdown exceeding 24 hours without backwashing.
3. After cleaning, slow sand filters should not be brought back into use without an adequate 'ripening' period.
4. Bypassing part of the water treatment process should be avoided.
5. Turbidity of individual filters should be monitored to detect early turbidity breakthrough. Sudden increases in turbidity should be investigated as that could infer dislodging of oocysts.
6. Recycling supernatant water from clarifier sludge and filter backwash water treatment facilities to the treatment works inlet should not be practised unless no more than 5% return of oocysts to the works inlet can be assured. This may require membrane filtration or ozone dosing. The recycle flow should be between 5% and 10% of raw water flow.
7. 'Slow start' alone is inadequate. Filtrate at the start of the filter run (first filtrate or filter to waste; typically one bed volume) should be diverted to waste, recycled to the works inlet or stored for backwash water. This could be supplemented with a 'delayed start' (standing the filter for about 15–60 minutes). Figure 9.8 illustrates the effects of these three operations.
8. Particle counters are more representative of oocyst breakthrough than turbidity monitors.

Logsdon (1998) suggested for plants treating water from low or medium risk sources, the target quality for the supernatant return from the settling process should be a turbidity of <5 NTU, suspended solids of <10 mg/l and total coagulant metal ion concentration of <5 mg/l as Al or Fe.

9.21 *GIARDIA* CYSTS

Giardia cysts are larger organisms (8–12 μm) than *Cryptosporidium* oocysts (3–6 μm) and are consequently easier to remove by solids–liquid separation processes than *Cryptosporidium*. In most cases *Giardia* removal is about 1- to 2-log better than the corresponding value for *Cryptosporidium* removal. Because of their larger size they can be better correlated to turbidity than *Cryptosporidium* oocysts.

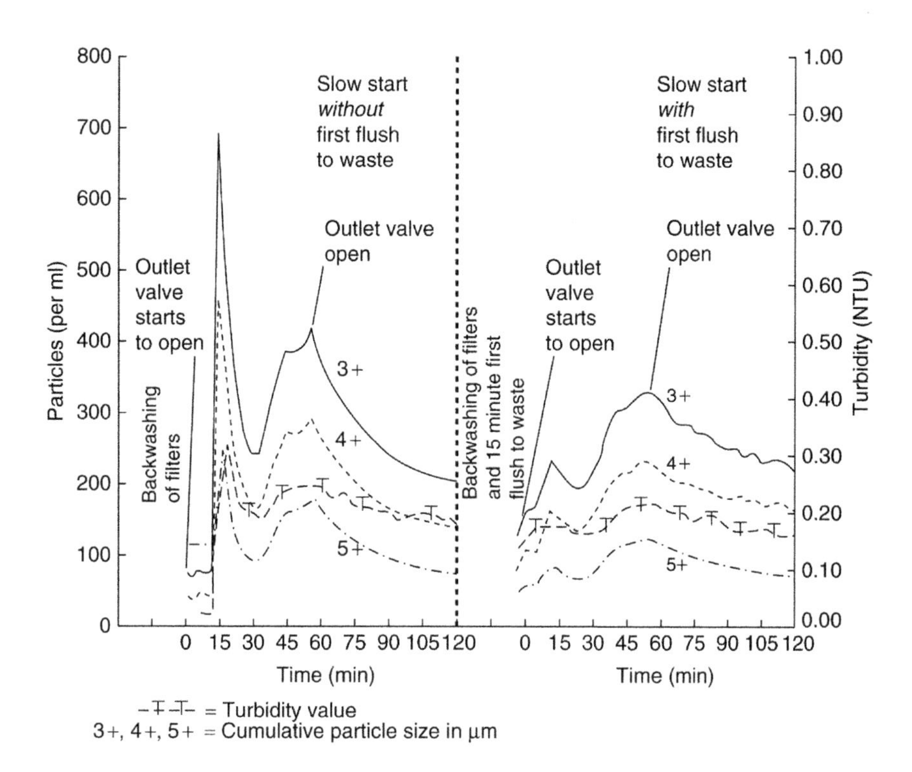

Source of information: *Portsmouth Water Company.*

FIGURE 9.8

Particle count and turbidity on filter start up — effect of first filtrate to waste.

Giardia cysts are less resistant to disinfection than *Cryptosporidium* oocysts, though chlorine is still not very effective (Section 11.6). *Ct* values for 1-log inactivation are given in Table 11.2. For 2-log inactivation with free chlorine residual of 2 mg/l, a *Ct* value of 110 mg.min/l is required at pH 7 and 5°C, improving to about 27 mg.min/l at 25°C. Monochloramine is almost ineffective. Chlorine dioxide performs better with *Ct* values of 17 (at 5°C) and 7.3 (at 25°C) mg.min/l at pH 6–9. Ozone is the most effective; *Ct* values for 2-log inactivation at pH 7–9 are 1.30 mg/l.min at 5°C and 0.32 mg.min/l at 25°C (US EPA, 1991). UV disinfection is widely accepted for inactivating *Giardia* (Section 11.25).

REFERENCES

Abrahamsson, J. and Dromberg, P. (2006). Covering of slow sand filtration; qualitative and operational aspects. *Recent Progress in Slow Sand and Alternative Biofiltration Processes* (Ed. Gimbel, R., Graham, N. J. D. and Collins, M. R.). IWA Publishing.

Adham, S., Chiu, K., Gramith, K. et al. *Development of a Microfiltration and Ultrafiltration Knowledge Base.* AwwaRF.

ASCE (1991). *Slow Sand Filtration* (Ed. Logsdon, G. S.). New York.

ASTM (2005). *Standard Test Method for Ball Pan Hardness*. D3802.

AWWA (2005). *Manual of Practice M53: Microfiltration and Ultrafiltration Membranes for Drinking Water*. 1st Edn. AWWA.

AWWA (2009). *Standard for Granular Filtering Material*. AWWA, B100–09.

Badenoch, J. (1990). *Cryptosporidium* in Water Suppliers. *Report of the Group of Experts*. London.

Badenoch, J. (1994). *Cryptosporidium* in Water Supplies. *Second Report of the Group of Experts*. London.

Bauer, M., Buchanan, B., Colbourne, J. et al. (1995). The GAC/slow sand filter sandwich: from concept to commissioning. *Water supply* **13**(3/4), pp. 137–142.

Bauer, M. J., Bayley, R., Chipps, M. J. et al. (1997). Enhanced rapid gravity filtration and dissolved air pretreatment of River Thames Reservoir Water. *Proc IAWQ-IWSA Joint Specialist Conference on Reservoir Management and Water Supply*, Prague.

Baylis, J. R., Gullans, O. and Hudson Jr., H. E. (1971). Chapter 7a: filtration, *Water Quality and Treatment*. 3rd Edn. McGraw Hill.

Boucher, I. (1998). *Cryptosporidium* in Water Supplies. *Third Report of the Group of Experts*. London.

BS PD 5500. (2012). *Specification for Unfired Fusion Welded Pressure Vessels*. BSI.

BW (1996). *The Specification, Approval and Testing of Granular Filtering Material and Code of Practice for the Installation of Filtering Material*. BW:P.18.96R.

Chan Kin Man, D. and Sinclair, J. (1991). Commissioning and operation of Yau Kom Tau Water treatment works (Hong Kong) using direct filtration. *Journal IWEM* **5**(2), pp. 105–115.

Cleasby, J. L. (1969). Filter rate control without controllers. *Journal AWWA* **61**(4), pp. 181–185.

Cleasby, J. L. (1975). Effect of media intermixing on dual media filtration. *Journal of Environmental Engineering ASCE* **101**(EE4), pp. 503–516.

Cleasby, J. L. (1993a). Status of declining rate filtration design. *Water Science and Technology* **27**(10), pp. 151–164.

Cleasby, J. L. (1993b). Declining rate filtration. *Water Science and Technology* **27**(7–8), pp. 11–18.

Cleasby, J. L. and Logsdon, G. S. (1999). Granular bed and precoat filtration, Chapter 8. *Water Quality and Treatment* (Ed. Letterman, R. D.). McGraw-Hill.

Coffey, B. M., Stewart, M. H. and Wattier, K. L. (1993). Evaluation of microfiltration for metropolitan's small domestic water systems. *Proc. AWWA Membrane Technology Conf.*

Collins, M. R. and Eighmy, T. T. (1988). Modifications to the slow rate filtration process for improved trihalomethane precursor removal. *Slow Sand Filtration* (Ed. Graham, N. J. D.). Ellis Horwood.

Crowley, F. W. and Twort, A.C. (1979). Current strategies in water treatment developments. *Proc. ICE Conf Water Resources: A Changing Strategy*.

Dorea, C. C. and Clarke, B. A. (2006). Impacts of chemical pretreatment in slow sand filtration. *Recent Progress in Slow Sand and Alternative Biofiltration Processes* (Ed. Gimbel, R., Graham, N. J. D. and Collins, M. R.). IWA Publishing.

Dullemont, Y. J., Schijven, J. F., Hinjen, W. A. M. et al. (2006). Removal of microorganisms by slow sand filtration. *Recent Progress in Slow Sand and Alternative Biofiltration Processes* (Ed. Gimbel, R., Graham, N. J. D. and Collins, M. R.). IWA Publishing.

Edzwald, J. K. and Kelly, M. B. (1998). Control of cryptosporidium: from reservoirs to clarifiers to filters. *Water Science and Technology* **37**(2), pp. 1–8.

Farizoglu, B., Nuhoglu, A., Yildiz, E. et al. (2003). The performance of pumice as a filter bed material under rapid filtration conditions. *Filtration & Separation* **40**(3), pp. 41–47.

Freeman, S., Verapaneni, S., Presdee, J., et al. (2004). World's largest membrane filtration plants. *Proc. AWWA Annual Conf.*, Orlando.

Fujiura, S., Tomita, Y., Kanaya, S., et al. (2006). Ceramic membrane microfiltration with coagulation. *Proc. AWWA Annual Conference & Exposition*. San Antonio.

Galjaard, G., Clement, J., Ang, W. S. et al. (2012). Ceramac®-19 demonstration plant ceramic microfiltration at Choa Chu Kang Waterworks. *Water Practice and Technology* **7**(4), pp. 1–7.

Galvis, G., Fernandez, J. and Visscher, J. T. (1993). Comparative study of different pre-treatment alternatives. *Journal WSRT-Aqua* **42**(6), pp. 337–346.

Glendinning, D. J. and Mitchell, J. (1996). Uprating water-treatment works supplying the Thames water ring main. *Journal CIWEM* **10**(1), pp. 17–25.

Guo, H., Wyart, Y., Perot, J., Nauleau, F. and Moulin, P. (2010). Low-pressure membrane integrity tests for drinking water treatment: a review. *Water Research* **44**, pp. 41–57.

Harrison, N. (1997). Dunnore point water treatment works–Phase 1, Extension. *Water & Sewage Journal* Spring, pp. 41–42.

Hendricks, D. (1991). *Manual of Design for Slow Sand Filtration*. AwwaRF.

Hudson Jr, H. E. (1981). *Water Clarification Processes*. Van Nostrand Reinhold.

Huisman, L. and Wood, W. E. (1974). *Slow Sand Filtration*. WHO.

Hutchinson, W. and Foley, P. D. (1974). Operational and experimental results of direct filtration. *Journal AWWA* **66** (2), pp. 79–87.

Ives, K. J. (1969). Theory of filtration. *Proc 4th World Filtration Congress*. IWSA, Ostend.

Ives, K. J. (1971). Filtration: the significance of theory. *Journal IWE* **25**(1), pp. 13–20.

Ives, K. J. and Gregory, J. (1967). Basic concepts of filtration. *Journal SWTE* **16**(2), pp. 147–169.

Jacangelo, J. G., Laine, J. M., Carns, K. E. et al. (1991). Low pressure membrane filtration for removing giardia and microbial indicators. *Journal AWWA* **83**(9), pp. 97–106.

Jacangelo, J. G., Laine, J. M., Cummings, E. W. et al. (1995). UF with pretreatment for removing DBP precursors. *Journal AWWA* **87**(9), pp. 100–112.

Joslin, W. R. (1997). Slow sand filtration: a case study in the adoption and diffusion of a new technology. *Journal New England WWA* **111**(3), pp. 294–303.

Kawamura, S. (2000). *Integrated Design and Operation of Water Treatment Facilities*. 2nd Edn. John Wiley.

Klein, H. P. and Berger, C. (1994). Slow sand filters covered by geotextiles. *Water Supply* **12**(3/4), pp. 221–230.

Kors, L. J., Wind, A. and van der Hoek, J. P. (1996). Hydraulic and bacteriological performance affected by resanding, filtration rate and pretreatment. *Advances in Slow Sand and Alternative Biological Filtration* (Ed. Graham, N. and Collins, R.). John Wiley.

Lambert, S. D. and Graham, N. J. D. (1995). A comparative evaluation of the effectiveness of potable water filtration. *Journal WSRT-Aqua* **44**(1), pp. 38–51.

Lewin, J. (1961). Mechanisation of slow sand and secondary filter bed cleaning. *Journal IWE* **15**(1), pp. 15–46.

Logsdon, G. (1998). Removal of micro-organisms by clarification and filtration processes, special contribution. *Water Supply* **16**(1/2), pp. 208–220.

Logsdon, G. S., Hess, A. F., Chipps, M. J., et al. (2002). *Filter Maintenance and Operations Manual*. AwwaRF.

Lombard, H. K. and Hanroff, J. (1995). Filter nozzle and underdrain systems used in rapid gravity filtration. *Water SA* **21**(4), pp. 281–298.

Lopato, L., Binning, P. J. and Arvin, E. (2012). Review of diagnostic tools to investigate the physical state of rapid granular filters. *Journal WSRT-Aqua* **61**(3), pp. 123–141.

Marsh, J. H. (2004). Water filtration using diatomaceous earth. *Water World* **20**(6), p. 16.

O'Neill, J. G. (1995). *An Evaluation of Fibrous Depth Filters for Removal of* Cryptosporidium *Oocysts From Water in Protozoan Parasites and Water.* RSC Special Publication 168. Royal Society of Chemistry.

Pike, P. G. (1987). *Engineering Against Schistosomiasis Bilharzia; Guidelines Towards Control of the Disease.* MacMillan.

Ratnayaka, D. D., Lee, M. F., Tiew, K. N., et al. (2008). Application of membrane technology to retrofit large-scale conventional water treatment plant in Singapore. *Singapore Water Week Convention Proceedings.*

Ridley, J. E. (1967). Experiences in the use of slow sand filtration, double sand filtration and microstraining. *Proc. SWTE*, **16**, pp. 170–184.

Scardina, P. and Edwards, M. (2004). Air binding of granular media filters. *Journal of Environmental Engineering* **130**(10), pp. 1126–1138.

Schalekamp, M. (1979). The development of the surface water treatment for drinking water in Switzerland. *Proc IWES Symposium: The Water Treatment Scene – The Next Decade.*

Speth, T. F. and Reiss, C. R. (2005). Chapter 2, water quality, Manual of practice M53, *Microfiltration and Ultrafiltration Membranes for Drinking Water.* 1st Edn. AWWA.

Standen, G., Insole, P. J., Shek, K. J. et al. (1997). The use of particle monitoring in the performance optimisation of conventional clarification and filtration processes. *Water Science and Technology* **36**(45), pp. 191–198.

Stevenson, D. G. (1994). The specification of filtering materials for rapid gravity filtration. *Journal IWEM* **8**(5), pp. 527–533.

Stevenson, D. G. (1998). *Water Treatment Unit Processes.* Imperial College Press.

Tebbutt, T. H. Y. and Shackelton, R. C. (1984). Temperature effects in filter backwashing. *Public Health Engineering* **12**(3), pp. 174–178.

Toms, I. P. and Bayley, R. G. (1988). Slow sand filtration, approach to practical issues. *Slow Sand Filtration* (Ed. Graham, N. J. D.). Ellis Horwood.

UKWIR (1998). *Guidance Manual Supporting Water Treatment Recommendations From the Badenoch Group of Experts on Cryptosporidium* (98/DW/06/5).

US EPA (1991). *Guidance Manual for Compliance With the Filtration and Disinfection Requirements for Public Water Systems Using Surface Water Sources.* EPA 570391001.

US EPA (2005). *Membrane Filtration Guidance Manual.* EPA 815-R-06-009.

US EPA (2006). National Primary Drinking Water Regulations: Long Term 2 Enhanced Surface Water Treatment Rule. *Federal Register* **71**(3).

van de Vloed, A. (1955). Comparison between slow sand and rapid filters. *Proc. 3rd Congress of the IWSA*, London.

van Dijk, J. C. and Oomen, J. H. C. M. (1978). *Slow Sand Filtration for Community Water Supply in Developing Countries. Tech. Paper 10.* WHO.

Visscher, J.T., Paramasivam, R., Raman, A., et al. (1987). *Slow Sand Filtration for Community Water Supply, Technical Paper No. 24.* IRC.

Welte, B. and Montiel, A. (1996). Removal of BDOC by slow sand filtration: comparison with granular activated carbon and effect of temperature. *Advances in Slow Sand and Alternative Biological Filtration* (Ed. Graham, N. and Collins, R.). John Wiley.

Wilson, D. (1996). Uprating of barmby water treatment works. *Advances in Slow Sand and Alternative Biological Filtration.* John Wiley.

Witcher, G., Treloar, T., Henderson, R. et al., 2014. Managing membranes to minimize opex: an update 10 years after start-up. *AWWA/AMTA 2014 Membrane Technology Conf, March 10–14, Las Vegas, Nevada.*

CHAPTER

Specialized and Advanced Water Treatment Processes 10

SOFTENING OF WATER

10.1 HARDNESS COMPOUNDS

A description of hardness is given in Section 7.29. A large proportion of waters from underground sources are hard, particularly waters from chalk and limestone aquifers which often have a carbonate hardness of 200–300 mg/l as $CaCO_3$. The hardness compounds are taken into solution because the water becomes acidic as it acquires carbon dioxide from the soil formed by the oxidation of organic matter. A major source of non-carbonate hardness in surface waters is the calcium sulphate present in clays and other deposits. In contrast, many surface waters from the older geological formations are soft or very soft, for example 15–50 mg/l as $CaCO_3$, because the rocks are largely impermeable and insoluble.

10.2 PRINCIPAL METHODS OF SOFTENING

There are three principal methods of softening a hard water. In the first, hardness compounds are precipitated by adding lime (calcium hydroxide, $Ca(OH)_2$) and soda ash (sodium carbonate, Na_2CO_3) and then removed by clarification and filtration. In the second method the nature of the hardness compounds is changed bypassing the water through a bed of 'ion exchange' resin. In the third method, membrane processes such as reverse osmosis (RO) remove all dissolved salts from water at an efficiency of about 95–99%; nanofiltration (NF) removes bi- or tri-valent ions (e.g. Ca^{2+}, Mg^{2+}, Al^{3+}, CO_3^{2-}, SO_4^{2-}) at an efficiency of about 80–85% and monovalent ions (e.g. Na^+, K^+, Cl^-) at an efficiency of up to 40%. The differences in the three methods are important because the chemical and membrane processes reduce the total dissolved solids (TDS) in a water, a feature that is often desirable for industrial applications.

Softening in drinking water treatment is usually applied to a proportion of the flow (split-treatment) to soften the water to a hardness value below the required value and this stream is then blended with the remaining unsoftened water.

Twort's Water Supply. DOI: http://dx.doi.org/10.1016/B978-0-08-100025-0.00010-7

10.3 THE LIME-SODA PROCESS OF SOFTENING

The aim of the lime-soda process is to convert calcium and magnesium compounds to the virtually insoluble forms, calcium carbonate ($CaCO_3$) and magnesium hydroxide ($Mg(OH)_2$). Magnesium carbonate ($MgCO_3$), unlike calcium carbonate, does not precipitate in cold water.

The stages of treatment involved are set out in Table 10.1. Lime is added to remove the temporary hardness and soda ash is added to remove the permanent hardness. Several applications of lime and soda ash are needed to remove magnesium hardness. In practice, complete removal of hardness is undesirable because this renders a water highly aggressive. In most hard waters the calcium temporary hardness forms the major component; therefore, it often suffices to remove only this by the addition of lime. Caustic soda (sodium hydroxide, NaOH) can be used in place of lime for carbonate and non-carbonate hardness removal; alkalinity reduction is only 50% that of lime softening. Soda ash is formed in the reactions and sometimes it may be supplemented by soda ash addition to remove non-carbonate hardness. The advantages of using caustic soda are that it is easier to handle than lime, only one chemical may be required and the quantity of calcium carbonate sludge produced is less (Section 13.8). The drawbacks are its higher cost, the hazardous nature of the chemical and the associated addition of sodium to the treated water. As a result of the high operating pH (9.5–10.5), chemical softening also removes many of the heavy metals (Section 10.14), including arsenic, iron and manganese, by precipitation.

Table 10.1 Lime-soda softening processes

Reaction
To remove carbon dioxide in water add LIME (not a softening reaction):
$H_2CO_3 + \mathbf{Ca(OH)_2} = \mathit{CaCO_3} + 2H_2O$
To remove calcium temporary hardness add LIME:
$Ca(HCO_3)_2 + \mathbf{Ca(OH)_2} = 2\mathit{CaCO_3} + H_2O$
To remove calcium permanent hardness add SODA ASH:
$CaSO_4 + \mathbf{Na_2CO_3} = \mathit{CaCO_3} + Na_2SO_4$; $CaCl_2 + \mathbf{Na_2CO_3} = \mathit{CaCO_3} + 2NaCl$
To remove magnesium temporary hardness add LIME + more LIME, Stage 1:
$Mg(HCO_3)_2 + \mathbf{Ca(OH)_2} = MgCO_3 + \mathit{CaCO_3} + 2H_2O$
The calcium carbonate precipitates but the magnesium carbonate does not, so further LIME is added, Stage 2:
$MgCO_3 + \mathbf{Ca(OH)_2} = \mathit{Mg(OH)_2} + \mathit{CaCO_3}$
The magnesium hydroxide and calcium carbonate precipitate
To remove magnesium permanent hardness add LIME and SODA ASH:
$MgCl_2 + \mathbf{Ca(OH)_2} = \mathit{Mg(OH)_2} + CaCl_2$; $MgSO_4 + \mathbf{Ca(OH)_2} = \mathit{Mg(OH)_2} + CaSO_4$
The addition of soda ash then converts the calcium chloride and calcium sulphate to calcium carbonate as in Section 10.3

Notes: 1. Compounds in **bold** are those being added and compounds in *italic* are those precipitating.
2. Compounds are: H_2CO_3 – carbonic acid (carbon dioxide in water); $Ca(HCO_3)_2$ – calcium bicarbonate; $Ca(OH)_2$ – calcium hydroxide (hydrated lime); $CaCO_3$ – calcium carbonate; $CaSO_4$ – calcium sulphate; $CaCl_2$ – calcium chloride; H_2O – water; Na_2CO_3 – sodium carbonate (soda ash); $Mg(HCO_3)_2$ – magnesium bicarbonate; $Mg(OH)_2$ – magnesium hydroxide; $MgCO_3$ – magnesium carbonate; $MgCl_2$ – magnesium chloride; $MgSO_4$ – magnesium sulphate.

10.4 SOFTENING PLANT

Lime is used in the hydrated form. This is a dry powder with a low solubility which is dosed as a slurry (Section 12.9). Although also delivered as a powder, soda ash is significantly more soluble in water than hydrated lime and is dosed as a solution. The concentration of a saturated solution varies with temperature (Table 12.1). Usually a solution is prepared with about 60% of the saturated concentration at the lowest anticipated water temperature. Clarification is in hopper-bottomed clarifiers of the sludge-blanket type or solids recirculation clarifiers (Section 8.16).

An excess of up to 10% of both lime and soda ash is added over the dose stoichiometrically required in order to complete the reactions in reasonable time. If magnesium is present, a portion of it may also cause precipitation but the reaction is slower than that of calcium. Lime-soda softening reduces the carbonate hardness value to 35–50 mg/l as $CaCO_3$ and the total hardness to 75–125 mg/l as $CaCO_3$. If the water to be softened contains suspended solids and organic matter such as colour, these can be removed concurrently although coagulants and coagulant aids have to be added. The coagulants used are usually of the iron type, because of their high coagulation pH values, but aluminium sulphates can also be used; at high pH, insoluble magnesium aluminate and not aluminium hydroxide is formed. Normally softeners are operated at surface loading rates in the range 3–4 $m^3/h.m^2$; when magnesium is to be removed, the rates are about 2–2.5 $m^3/h.m^2$ because of the gelatinous nature of the precipitate.

The dosages of lime required for softening are high, being of the order of 100–200 mg/l. The process produces a large amount of liquid sludge due to the precipitation of hardness and coagulation of suspended solids and colour (Sections 13.10–13.12 for treatment and disposal).

10.5 WATER SOFTENING BY CRYSTALLIZATION

Softening reactions can be accelerated by using sand grains for seeding the crystallization of calcium carbonate. Softening takes place in a cylindrical reactor partially filled with filter sand (0.2–0.6 mm). The water is injected with the softening chemical and passed upwards at a surface loading rate of 50–120 $m^3/h.m^2$ to fluidize the sand bed. Calcium carbonate deposits on the sand grains which grow to form pellets of about 1–2 mm in diameter and accumulate at the base of the reactor, from where they can be periodically removed. Make up sand is added either at the top or at the base. The reaction tanks, called pellet reactors, are typically about 6 m deep and up to 4 m in diameter. Lime is used when the ratio of carbonate hardness to total hardness is high; in the intermediate hardness range caustic soda is used; and when carbonate hardness is very low, soda ash is used. van Dijk (1991) reported that crystal growth is adversely affected when the phosphate content of a water exceeds 0.5 mg/l as PO_4 (0.15 mg/l as P). Fluffy pellets are formed when iron is present in the water above about 1 mg/l as Fe (van der Veen, 1988).

The advantages of the pellet reactor over the conventional softening process are its high surface loading rate, the improved handling characteristics of pellets compared with sludge and the requirement for only a small excess of softening chemical. The disadvantages of the process are that it does not remove magnesium, the hardness after softening is in the range 50–100 mg/l as $CaCO_3$ (van Honwelingen, 1994) and is therefore unsuitable for most industrial uses, suspended solids concentrations in the product are high (up to 30 mg/l), and the removal of turbidity and colour in raw water by coagulants cannot be performed in the same reactor. However, it is ideal for softening groundwaters.

Lime is used as a slurry containing 10–100 g $Ca(OH)_2$/l. The use of lime water (saturated lime) is not usually practical because of its low lime content (e.g. 1.76 g/l at 10°C); the volume to be added would be over 10% of the volume of water to be softened. Nevertheless, there are plants operating with lime water. Use of lime slurry results in carry over of calcium carbonate, undissolved lime and inert impurities in the lime, amounting in total to about 20–30 mg/l as suspended solids. These are removed by dosing the softened water with acid and iron coagulants followed by filtration. By improving the quality of lime and its solubility, the carry over of suspended solids can be reduced by up to 80% (van Eekeren, 1994).

Various alternatives to chemical crystallization are available, including physical processes using magnetic, ultrasonic, electrolytic, electrostatic or electronic devices fitted to pipelines carrying hard water. These processes do not change the chemical properties of the water but modify the crystal structure so that crystals are less likely to agglomerate. This results in an increase in particles in suspension and a decrease in the formation of scale. The devices are used in waterworks on sample and lime dosing lines and on domestic hot water systems with some success but appear to be most suited to applications where water is continuously recycled rather than once-through systems.

10.6 STABILIZATION AFTER SOFTENING

The softening reactions of precipitation are not usually wholly completed in the clarification tanks and therefore, the water leaving is usually supersaturated with calcium carbonate and tends to form further deposits, mainly of calcium carbonate, in the later stages of the treatment plant for example media filters. The water, therefore, needs to be stabilized by injecting carbon dioxide (CO_2) into the water to achieve an alkalinity of about 30–50 mg/l as $CaCO_3$ (1 mg/l as $CaCO_3$ = 1.22 mg/l as HCO_3 = 0.4 mg/l as Ca) and a slightly positive Langelier Index (Section 10.41). Sulphuric acid is used in some cases but, as it converts carbonate to sulphate, care must be taken to ensure that water is not rendered corrosive due to the low calcium carbonate concentration. The objective of adding CO_2 or sulphuric acid is to prevent after-precipitation while producing a non-aggressive water. Another method of avoiding after-precipitation is to add about 0.5–2 mg/l as P of a sequestrant such as a polyphosphate (e.g. sodium hexametaphosphate). This method is usually preferred for industrial applications. Softening is usually followed by anthracite–sand filters.

10.7 BASE EXCHANGE SOFTENING

In the ion exchange (IX) process of softening, when water containing hardness salts is passed through a bed of strong acid cation exchange resin in the sodium form, calcium and magnesium are substituted by sodium. The hardness of the water is reduced to almost zero but the TDS concentration undergoes little change; alkalinity and pH values are unaffected. When the resin's capacity to exchange calcium and magnesium for sodium is exhausted, the bed is regenerated bypassing a concentrated sodium chloride solution through it. The reverse action then takes place, with the calcium and magnesium ions held in the resin being released in the effluent and the sodium from sodium chloride being substituted. The regeneration wastewater is very hard with a high concentration of dissolved salts and its disposal may present problems (Sections 10.25 and 13.9).

The IX resins used are cross-linked polystyrene spherical particles or 'beads'. There are many different types of resin distinguished by the chemical functional group attached to the polystyrene matrix. Those used for base exchange softening have strong acid functionality. The total number of functional groups in the resin determines its exchange capacity. The exchange capacity of a resin for softening is most frequently stated in terms of the hardness removed by a specific volume of resin, for example *x* g eq per litre of resin, and is specific to each resin; typical values quoted vary in the range 1.5–2 g eq/l (g eq = gram equivalent, that is equivalent weight in grams; 1 g eq of hardness = 50 g as $CaCO_3$). The actual operating capacity is less than the reported value due to leakage and incomplete regeneration. A second measure of performance in base exchange softening is the amount of salt that must be used in regeneration per unit of hardness removed. The theoretical figure for regeneration is 117 g of salt per 100 g of $CaCO_3$ (or its equivalent) removed but, in practice depending on the resin used, up to 400 g of salt per 100 g of $CaCO_3$ removed may be required. This can be reduced by as much as 50% by using more efficient, counter-current regeneration techniques.

10.8 PLANT FOR ION EXCHANGE (IX) SOFTENING

The plant required for IX softening is similar to that for pressure filtration. At least two vessels arranged in lead/lag configuration should be provided. The vessels are typically rubber-lined steel vertical type of diameter up to 4 m. The media bed would usually be about 1.2 m deep. Typical operating modes comprise in service (i.e. softening); backwashing with softened water; regeneration; and rinsing to remove excess regenerant. Flow during both softening and regeneration is normally downwards and operation is automated. The surface loading rate for softening varies in the range 12–20 $m^3/h.m^2$ and that for regeneration is about 2.5 $m^3/h.m^2$ to give at least 30-minute contact time. A saturated solution of sodium chloride containing about 26% w/w NaCl (33% w/v) is initially prepared and diluted to about 5–10% w/v before use. Backwash water is applied at a rate to produce a bed expansion between 50% and 100%; rates vary with the water temperature and resin type. The plant is normally operated under pressure so that repumping after softening can be avoided; the loss of head is usually 4–5 m.

In general, IX plants are rarely used for softening public supplies; they are not suitable for softening turbid water (e.g. suspended solids should be $<$1 mg/l) or water that contains significant concentrations of iron and manganese. Chlorine and chloramine in the water cause degradation of the resin: recommended upper limits are 0.3 mg/l for cation and 0.1 mg/l for anions resins. Specially formulated resins could be used for higher concentrations. The process also adds an equivalent concentration of sodium to the water for the hardness removed and is therefore not recommended for drinking or cooking. It is used as dilution/carrier water for alkalis not containing calcium and for domestic water softeners to prevent scaling of washing machines and dish washers. In these applications IX vessels are typically fabricated in GRP (glass or fibre-reinforced plastic). In industry, the tendency has been for IX softening to be superseded by demineralization plant, which can produce a water more exactly tailored to the type of process water required (Section 10.10).

10.9 HARDNESS AND ALKALINITY REMOVAL BY ION EXCHANGE

When used for the removal of hardness and alkalinity, the process uses a weak acid hydrogen form cation exchange resin and therefore hydrogen, instead of sodium, is exchanged for calcium and magnesium equivalent to the alkalinity content of the water; sodium and potassium are not removed unless the alkalinity exceeds the hardness. Alkalinity reacts with the exchanged hydrogen ions producing carbon dioxide which is removed in a degasser, leaving a residual carbon dioxide concentration of about 5–10 mg/l in the water depending on the efficiency of the degasser. Sulphuric or hydrochloric acid is used as the regenerant instead of sodium chloride. Alternatively, alkalinity alone can be removed using chloride from anion exchange resins (dealkalization). Chloride is exchanged for bicarbonates and sulphates in the water and sodium chloride solution is used as the regenerant.

Water dealkalized to give <14 mg/l of $CaCO_3$ with lime is sometimes used in the preparation of lime slurry and as carrier water for lime slurry since it minimizes the risk of scaling in tanks and pipework (Section 12.10).

10.10 DEMINERALIZATION OF WATER BY ION EXCHANGE

The IX process of softening is only a particular example of ion exchange treatment and is more specifically an example of strong cation exchange with the resin in sodium form. In demineralization a strong acid hydrogen form cation exchange resin replaces calcium, magnesium, sodium and potassium by hydrogen ions. This is followed by a second stage of treatment using a strong base anion exchange process in which chloride, sulphate and nitrate are removed. The cation exchange resin is regenerated with sulphuric or hydrochloric acid. Use of hydrochloric acid has advantages in that it can be applied at higher concentration than sulphuric acid (which is often limited to avoid calcium sulphate precipitation in the bed) resulting in higher regeneration efficiency, lower resin inventory and reduced waste volumes. Anion exchange resin is regenerated using sodium carbonate or caustic soda solution. Carbon dioxide formed in the first stage is often removed by 'degassing' or by aeration as an intermediate stage between the cation and anion exchange vessels. The product water has a pH of 7–9 and TDS concentration is very low, with a conductivity less than 20 μS/cm. Such a treatment is therefore called demineralization and is now most frequently adopted in industry for the production of special quality process waters.

In 'mixed bed' IX, both cation and anion exchange resins as described above are mixed in one vessel. During backwashing the resins are hydraulically separated by virtue of their density difference. This allows separate regeneration of the two components (Fig. 10.1) after which the resins are remixed with an upflow of low pressure air. A mixed bed gives a water of neutral pH and conductivity of less than 0.2 μS/cm. Demineralization can, in theory, be applied to brackish waters but, because of the regenerant chemical consumption, the process only finds application for waters having less than about 500 mg/l dissolved solids. For waters with higher dissolved solids, the IX process would not be economic. For public supply purposes other processes should be considered (Section 10.43).

In potable water treatment, the IX process also finds application in the removal of arsenic and chromium (Section 10.14), radionuclides (Section 10.15), nitrate (Section 10.24), ammonia

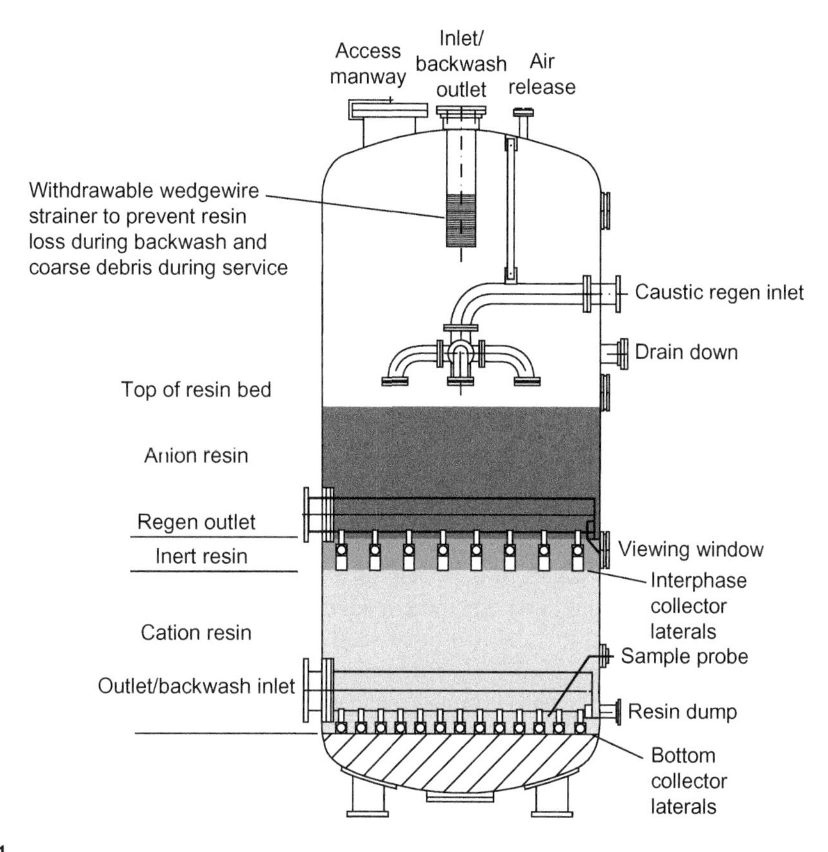

FIGURE 10.1

Internal arrangement of a mixed bed demineralization unit.

(Section 10.27) and organics (Section 10.34). One of the major drawbacks of IX is the problem of the disposal of wastewater which is highly saline and non-biodegradable (Section 13.9).

In the production of drinking water from seawater by reverse osmosis (RO) desalination (Section 10.47), boron removal may not be complete, leaving an unacceptable concentration of boron in the permeate. IX processing can be used for further treatment of the RO permeate either from the first or second membrane pass for boron elimination (Jacob, 2007). Weak base IX resins with methyl glucamine functionality can selectively absorb boron (present as borate in solution) from practically any salt background. The total exchange capacity for boron is 0.7 g eq/l. Removal efficiency is very high with boron leakage usually <0.1 mg B/l. The treatment capacity is in the range 20–30 bed-volumes/h; surface loading should not exceed 40 $m^3/h.m^2$.

Regeneration is co-current flow and is carried out in two steps: first, the absorbed borate is displaced with mineral acid, then the resin is converted into the free base form with sodium hydroxide solution. The regeneration levels should be 50 and 25 g of sulphuric acid and caustic soda per litre of resin, respectively.

REMOVAL OF IRON, MANGANESE AND OTHER METALS

10.11 IRON AND MANGANESE – GENERAL

Traces of iron and manganese are found in many water sources, both surface and underground. Concentrations may occasionally range up to 20 mg/l of iron and 5 mg/l of manganese but at such high values most of the metals, in particular iron, are in particulate form so that they may be relatively easily removed by solid–liquid separation methods. It is the dissolved fractions of iron and manganese which can be troublesome and the disadvantages arising from their presence above certain concentrations are described in Sections 7.31 and 7.33. Treatment for the removal of these metals is therefore often necessary.

10.12 REMOVAL OF IRON AND MANGANESE FROM UNDERGROUND WATERS

When iron occurs in underground waters, it is usually in solution in the ferrous form in a water that is devoid of oxygen. Such waters are fairly common in aquifers that underlie an impermeable stratum for example the greensand beds in the UK and other sand formations underlying clays. Manganese occurs in appreciable amount in only a minority of those raw waters which contain iron. Many waters from deep boreholes in sandstone contain iron and manganese in their lowest state of oxidation, Fe (II) and Mn(II), and are present in solution. When oxidized, the ferrous Fe(II) and manganous Mn(II) are converted to the next common oxidation state, that is ferric Fe(III) and Mn(IV). Under certain circumstances Mn(IV) can be oxidized further to permanganate Mn(VII). Values in brackets denote the corresponding valencies. When a sample of groundwater is first drawn it may appear perfectly clear, but after exposure to air for a short time it acquires a turbid appearance and after a further period a red-brown precipitate of ferric hydroxide is formed. Oxidation of Mn(II) is very slow under most conditions although it may co-precipitate with iron, if present.

Removal of iron and manganese is effected by oxidation, followed by the separation of Fe(III) and Mn(IV) as ferric hydroxide ($Fe(OH)_3$) and manganese dioxide (MnO_2) precipitates by filtration. For high concentrations of iron (>5 mg/l) filtration may be preceded by settling; a coagulant or a coagulant aid may be necessary to assist the process. Solids contact type clarifiers are suitable since oxidation is catalysed by the oxides already present in the sludge blanket or in the recirculated sludge.

Oxidation can be by oxygen in air or by the use of a strong oxidant such as chlorine, potassium permanganate, chlorine dioxide or ozone. In most cases, oxidation is influenced by pH value. In oxidation reactions hydrogen ions are produced which, in turn, react with alkalinity. However, since the concentrations of iron and manganese present are generally low there is usually sufficient alkalinity in the water to buffer the effect of hydrogen ions and prevent a consequent reduction in pH which would otherwise reduce the reaction rates. Table 10.2 gives the stoichiometric quantities of oxidant (in mg) required to oxidize 1 mg of iron or manganese, the corresponding reduction in alkalinity and the optimum pH range for the oxidation reaction. Lower water temperature reduces the rate of oxidation.

Table 10.2 Oxidation of iron and manganese

Metal	Oxidant	Stoichiometric quantity of oxidant (mg/mg Fe or Mn)	Reduction in alkalinity (mg $CaCO_3$/mg Fe or Mn)	Optimum pH
Fe(II)	Oxygen	0.14	1.80	>7.5
Mn(II)	Oxygen	0.29	1.80	>10.0[a]
Fe(II)	Chlorine	0.63	2.70	>7.0
Mn(II)	Chlorine	1.29	3.64	>9.0[a]
Fe(II)	Potassium permanganate	0.94	1.49	>7.0[b]
Mn(II)	Potassium permanganate	1.92	1.21	>7.0[b]
Fe(II)	Chlorine dioxide	0.24[1]	1.96	>7.0
Mn(II)	Chlorine dioxide	2.45[1]	3.64	≤7.0
		0.49[2]	2.18	≥7.5
Fe(II)	Ozone	0.43	1.80	c
Mn(II)	Ozone	0.87	1.80	c

Sources of information: [1]Knocke (1991a); [2]Faust (1998).
Notes: [a]The use of a catalytic filter medium may reduce the pH to 7.5–8.5.
[b]Reaction is known to proceed at pH >5.5.
[c]pH value at which the reaction occurs is less dependent than for other oxidants. Low pH values are preferred as ozone performs better under acidic conditions.

Oxygen is added by aeration. In practice, larger volumes of air are required to compensate for inefficiencies in the aeration system used. Although the oxygen requirements are small, the rate of reaction is slow and pH dependent: 90% oxidation of iron would require about 40 minutes reaction time at pH 6.9, but only 10 minutes at pH 7.2. In some waters, aeration alone is adequate for complete iron oxidation because the removal of free carbon dioxide in the aeration process raises the pH above 7.5. The oxidation of manganese is much slower and also requires elevated pH for successful oxidation: at pH 9.5 about 1 hour for 90% oxidation is required. Therefore, a strong oxidant such as chlorine, potassium permanganate, chlorine dioxide or ozone is usually necessary. These are also effective in oxidizing iron. The application of strong oxidants to water containing iron (II) and manganese (II) results in rapid oxidation, although the rates are pH dependent. In practice, excess oxidant is used to satisfy the demands due to organic matter, hydrogen sulphide and, in the case of chlorine, ammonia when present in water. The use of chlorine may be inadvisable when treating waters containing organic substances due to the possibility of disinfection by-products (DBPs) being formed (Section 7.24).

Potassium permanganate is a stronger oxidizing agent than chlorine and does not form DBPs in the presence of organic substances. It can, however, result in the formation of colloidal manganese which may be more difficult to remove. Manganese dioxide formed in the reaction adsorbs Mn(II) and catalyses its oxidation, which brings about an improvement in Mn(II) removal and a reduction in the amount of potassium permanganate required. The potassium permanganate dose applied must be carefully controlled to minimize any excess passing into supply which could give a pink colour to the water. Products of potassium permanganate oxidation can 'muddy' the water and form 'mudballs' in filters.

Chlorine dioxide is particularly useful as an oxidant in the presence of high ammonia concentrations (Section 11.16) that would otherwise react with chlorine. A disadvantage of chlorine dioxide is the limitation on the dose that can be applied and the formation of disproportionation products.

Ozone reacts readily in the absence of organic matter to oxidize soluble Fe(II) and Mn(II) to the insoluble Fe(III) and Mn(IV) forms. When both metals are present, iron is oxidized first, followed by manganese. Excess ozone can oxidize Mn(II) to its highest oxidation state Mn(VII) which gives a pink colouration to the water. When present, organic matter is oxidized by ozone before iron and manganese and the dose would be much higher than that required for oxidation in the absence of organics. Iron and manganese are more receptive to oxidation by ozone once organics have been removed by coagulation.

In the case of iron, oxidation is followed by settling and filtration or filtration alone, depending on the concentration of iron in the water. In the presence of turbidity (and colour) and when the Fe (II) concentration is greater than about 5 mg/l, settling or flotation could be assisted by a coagulant with or without a coagulant aid or by a coagulant aid alone. Direct filtration is used when the iron concentration is less than about 5 mg/l. Sand ($d_{10} = 0.6$ mm) or anthracite–sand filters with filtration rates of 5 (Fe $<$ 5 mg/l), 7.5 (Fe $<$ 3 mg/l) and 10 (Fe $<$ 2 mg/l) $m^3/h.m^2$ are suitable for the application.

Manganese is usually found in low concentrations compared to iron and, following oxidation, the water is subjected to direct filtration. In the absence of turbidity and upstream coagulation, filtration rates of 10 $m^3/h.m^2$ or higher are used. Filtration assists in the oxidation of manganese in the presence of an oxidant (e.g. chlorine) through adsorption of Mn^{2+} ions from the water. This can either be on to $MnO_x(s)$ previously deposited on the filter media, or onto a media containing manganese oxide, for example pyrolusite or a proprietary material such as 'Polarite'. Once adsorbed onto the surface, the oxidation reaction between the oxidant and adsorbed Mn^{2+} is catalysed by the $MnO_x(s)$. The number of sites available for such removal is a direct function of pH; as pH is increased there are fewer H^+ ions in the water to compete with Mn^{2+} ions for these sites. Oxidation is effective at pH $>$6.5, with the highest sorptive capacity at about 8.0 (Knocke, 2006). The proprietary media usually contain at least 65% by weight MnO_x. Its specific gravity is in the range 3.5–4.0 and its size range is usually 0.35–0.85 mm, with d_{10} of 0.425 mm. It is used in the ratio sand, or anthracite–sand, to oxide media of 4:1. The retention capacity in filters is about 0.1–0.7 kg of Mn/m^2 of surface area. It is reported that the effective size and density of the media are not altered by the presence of the oxide coating (Knocke, 1991b). Catalysed oxidation can reduce manganese to values $<$0.01 mg/l.

An alternative filter media is manganese greensand, formed by treating greensand (glauconite), which is a sodium zeolite, with manganous sulphate followed by potassium permanganate. Mn-greensand removes soluble iron and manganese by a process of IX, frequently with the release of hydrogen ions. The process is, therefore, pH dependent, being virtually ineffective below pH 6.0 and very rapid at pH values above 7.5. When the Mn-greensand is saturated it is regenerated by soaking the filter bed with potassium permanganate (intermittent regeneration–IR). This procedure oxidizes manganese on the surface of Mn-greensand to MnO_2 thereby reactivating the exchange sites. It is reported that the exchange capacity is 1.45 g of Fe or Mn/l of Mn-greensand and that 2.9 g of potassium permanganate (as a 1% w/v solution) per litre of Mn-greensand is required for regeneration (Benefield, 1982). Alternatively, potassium permanganate is continuously applied to the bed by dosing it at the filter inlet, which maintains Mn-greensand active (continuous regeneration–CR) and catalyses the oxidation reaction. Oxidation is rapid at pH values in the range 6.0–7.0. CR also oxidizes some iron and manganese before the water reaches the filter. Mn-greensand then acts as a filter medium in addition to catalytic oxidation of any residual soluble manganese. Mn-greensand has an effective size in the range 0.30–0.35 mm with a uniformity coefficient of 1.4–1.6 and is usually capped with a layer of anthracite to achieve longer filter runs. In the CR process chlorine can be used in place of potassium permanganate. The benefits are longer filter runs, no risk of pink water and lower cost.

It is reported that manganese dioxide filter media in filters behaves in a similar manner to Mn-greensand with operation being carried out in either the IR or CR mode (Merkle, 1997), with chlorine being the most suitable oxidant. In the IR mode, filter washwater is dosed with chlorine; the minimum quantity (mg) can be calculated from $Q \times c \times 1.3$ mg Cl_2/regeneration where Q is the total quantity of water treated between regenerations and c is the average concentration of Mn in the water. An allowance of about 30–40% should be made for chlorine demand within the filter media.

Stronger oxidants, such as ozone, chlorine dioxide and potassium permanganate, tend to form colloidal precipitates, as mentioned earlier, which may not be well retained by the filters unless a coagulant is added prior to filtration. The use of catalytic filtration media is usually limited to manganese which is otherwise difficult to remove. In most plants iron is oxidized ahead of filtration.

Organic substances, such as humic, fulvic and tannic acids, when found in groundwaters, can form soluble complexes with iron and manganese which are not easily oxidized by oxygen (aeration). These and any soluble inorganic complexes such as silicates, sulphates and phosphates are removed by oxidation using strong oxidants or sometimes by coagulation. However, when iron is strongly complexed in the presence of significant concentrations of humic and fulvic acid, strong oxidants are sometimes ineffective. The oxidation of manganese is not similarly influenced by the presence of dissolved organics because it is not strongly complexed by organic matter (Knocke, 1990, 1991b).

Fe(II) and Mn(II) can be removed biologically by utilizing the ability of certain bacteria to produce enzymes and/or polymers that, by catalytic action, promote oxidation in the presence of oxygen in the water (Sharma, 2005). Those which promote iron oxidation are generally considered to be autotrophs but the physiology of those promoting manganese oxidation is poorly defined (Rittmann, 1989; Nealson, 2006). Bacteria are usually present in groundwaters which contain these metals: e.g. *Gallionella ferruginea* (specific to iron), *Leptothrix* sp., *Crenothrix polyspora* and *Sphaeratilus natans*. If absent, they can be introduced from a suitable source, rapid gravity or pressure sand filters being used as biological reactors. When both metals are present in water two filtration stages are usually necessary, manganese being removed in the second stage because manganese removal bacteria require a completely aerobic environment (Mouchet, 1992). For iron removal aeration should be controlled at $<10\%$ saturation (Morris, 2001), particularly at pH values greater than 7, to prevent the chemical process competing with the biological process and therefore to minimize the risk of a chemically formed precipitate breaking through the coarse media used in high rate biological filters.

Based on a number of plants operating in France it is reported that Fe(II) oxidation takes place with pH in the range 6.0–7.5 and dissolved oxygen concentration of 0.25–1.5 mg/l. Mn(II) removal needs a pH greater than 7.5 and a dissolved oxygen concentration in excess of 5 mg/l (Mouchet, 1992). The optimum water temperature is in the range 10–15°C depending on the predominant bacteria. Toxic elements such as heavy metals (e.g. zinc), and/or compounds such as hydrogen sulphide, hydrocarbons and chlorine must be absent. Ammonia interferes with the process and, when present in excess of about 0.20 mg/l as N, the manganese removal stage must be designed for a simultaneous, slower nitrification process. When ammonia is present in excess of about 1.0 mg/l as N, a separate biological nitrification stage using biological aerated filters (BAFs) should be included between the iron and manganese removal stages (Section 10.28). Heard (2002) however, observed in pilot trials a significant manganese removal in the presence of ammonia 0.01–1.10 mg/l as N. Due to the oxidation kinetics, filters can be operated at rates ranging from 10 to 40 $m^3/h.m^2$ using coarse sand media (0.95–1.35 mm) with a solids retention capacity of 1–4 kg Fe or Mn/m^2 filter surface area (Mouchet, 1992). Unlike the chemical process where a clarification stage is necessary when iron concentration exceeds 5 mg/l, the biological iron removal process allows for direct filtration even when iron concentration is as high as 25 mg/l. Backwashing is by combined air–water wash, using raw or treated but unchlorinated water. Filters can be of the open gravity or enclosed pressure type; the filtered water requires aeration (in iron removal only) and disinfection. Biological filters require a seeding period before they operate at their optimum removal efficiency; this is reported to vary from 1 week for iron to 3 months for manganese (Bourgine, 1994). For related reasons flow variations should be minimized. Sludge produced in biological treatment is well suited to thickening (thickened sludge concentrations of 3–8% w/v) and dewatering (Mouchet, 1992).

Other merits of biological treatment, compared to conventional physical–chemical processes are: better treated water quality (Fe and Mn residuals generally not detectable and no interference with dissolved silica which otherwise forms iron–silica complexes); longer filter runs; easier operation; lower capital cost (plants are much more compact) and operation cost (no chemicals for Fe and Mn oxidation or coagulation–flocculation); less washwater losses; and reduced manpower.

10.13 REMOVAL OF IRON AND MANGANESE FROM RIVER AND RESERVOIR WATERS

Since surface waters frequently receive treatment which includes rapid sand filtration, often preceded by coagulation and sedimentation, the removal of iron and manganese is normally included, when necessary, in the same plant. Most river waters used as sources for water supplies are well oxygenated, if not saturated with oxygen. Usually significant proportions of iron and manganese in such river waters are therefore present in insoluble forms which are removed by sedimentation and filtration treatment. When water is drawn from the bottom of a reservoir, however, it may be deoxygenated and the iron and manganese dissolved (Section 8.1), although in such cases the iron Fe(II) is usually fairly easily oxidized. Occasionally the iron, and more often the manganese Mn(II), may be present as soluble organic complexes in a very stable form. A proportion of these soluble organics are removed by coagulation but the remainder require strong oxidants to release them from complexed organics and to oxidize them to Fe(III) and Mn(IV) (Section 10.12). The lime-soda softening process, which operates at a pH value of 10.6, removes both Fe(II) and Mn(II).

With a soft reservoir water, which requires only direct filtration, manganese can be removed by raising the pH above 9.0 before filtration; the coagulant should be an iron salt. The use of an oxidant and/or a catalytic filter medium would permit operation at a lower pH (7.5–8.0). More often however, when both iron and manganese are found complexed with organics in a reservoir water, a satisfactory treatment is oxidation of Fe(II) using a strong oxidant followed by coagulation using an iron salt and clarification for removal of colour, precipitated Fe(III) and turbidity, then manganese removal at pH 9.0 or at 7.5–8.0 in downstream filters, depending on the presence of a catalytic medium. If an aluminium coagulant must be used, manganese should be removed in a second filtration stage after all aluminium floc has been removed in the first stage filter, otherwise aluminium floc carried over from clarifiers would dissolve at pH values above about 7.6–7.8. The use of chlorine at the works inlet for Fe(II) oxidation may not be desirable due to the possibility of DBP formation. In that case Fe(II) could be removed by oxidation in the first stage filter after DBP precursors have been removed by coagulation and clarification, although any turbidity carry over from the clarifiers could still contribute to DBP formation if clarified water turbidity is poor. Biological oxidation could also be used on surface waters in second stage filters.

In some cases where a trace of manganese (less than the WHO Guideline value of 0.05 mg/l as Mn) has passed through a filtration plant, final chlorination can cause precipitation of the metal to form objectionable deposits in the mains. Typically, therefore, works operators seek to achieve manganese concentrations of less than 10 μg/l in the treated water. Precipitation of iron and manganese in distribution systems can be controlled through sequestration (also called chelation). Sequestering agents such as polyphosphates and sodium silicates increase the solubility of the metal ion by forming a bond with it, thereby preventing precipitation.

10.14 REMOVAL OF OTHER METALS

There are several other metals that can be present in a raw water and some might be added in the distributed water due to the corrosion of water mains and plumbing systems. Metals present in dissolved form in raw water can usually be removed by precipitation as the metal hydroxide. This involves adjustment of the raw water pH, usually by adding an alkali, to a value at which the metal precipitates and can be removed by coagulation and filtration. In most cases the pH value required for precipitation can be achieved during coagulation by using aluminium or iron salts; the latter are favoured because they have a wider coagulation pH range. Metals which show good removal during coagulation are arsenic (pH 6–8), cadmium (pH >8), chromium (pH 6–9), lead (pH 6–9) and mercury (pH 7–8) (AWWA, 1988). Other metals which require high pH values such as those experienced in lime-soda softening include barium (pH 10–11), copper (pH 10) and zinc (pH 10).

Arsenic

Arsenic (As) occurs in the soluble form as As(III) (arsenite) under anaerobic conditions predominantly in groundwaters and as As(V) (arsenate) under aerobic conditions, more commonly in surface waters. Both the forms can be effectively removed by coagulation followed by solid–liquid separation processes. Arsenic As(V) is removed with equal efficiency by aluminium and iron coagulants at pH <7.5 (Cheng, 2002) but iron coagulants are more effective than aluminium coagulants in removing As(III) and As(V) at pH >7.5 (Edwards, 1994). As(III) can be oxidized to As(V) by chlorine (may react with organic matter to form trihalomethanes), ozone, chlorine dioxide or potassium permanganate; the stoichiometric requirements per mg of As(III) being 0.95, 0.64, 1.8 and 1.40, respectively (Hoffman, 2006). The oxidation is rapid: 95% conversion is achieved in five seconds with 1 mg/l of free chlorine in the pH range 6.5–9.5 (Gottlieb, 2005). Significant removal of As(V) can also be achieved during the oxidation of Fe(II) by co-precipitation when the two coexist (Cheng, 2002). Lime softening at pH 11 to remove all magnesium also removes arsenic (McNeill, 1997).

Arsenic removal methods applied to groundwater include adsorption on to activated alumina (AA) or granulated ferric hydroxide, IX, RO, nanofiltration and biological oxidation. The AA process is more effective in removing As(V) than As(III) (Wang, 2002), but adsorption is reduced by silicate and phosphate at pH >7. The effect of silicates (at pH >5) and phosphate (at pH >6) is much more adverse on As(III) removal (Simms, 1998). Other ions which reduce the removal efficiency are sulphates and bicarbonates. The capacity for arsenic removal is pH dependent and is optimum in the range 5.5–6.0 over which As(V) is in ionic form and AA is protonated. At lower pH AA tends to dissolve (Chwirka, 2000). Arsenic removal capacity drops rapidly as pH approaches 8.2 (Clifford, 1999) and pH adjustment may be necessary for alkaline waters. Because of the effect of sulphate on arsenic removal; hydrochloric acid is the preferred chemical. Ideally a plant should consist of an equal number of duty/standby vessels (Sections 10.8 and 10.25 give configurations) operating in the downflow mode. Rubel (2003) suggested a media size of 0.3–0.5 mm and an empty bed contact time (EBCT) defined as the bulk volume of adsorbent bed (m^3) divided by the water flow rate (m^3/min), in the range 6–10 minutes. Bed depth should be 0.9–1.8 m. Following exhaustion the bed is backwashed to give about 50% bed expansion (equivalent to 17 $m^3/h.m^2$) using raw water and then subjected to two upflow regeneration steps at 3 $m^3/h.m^2$ using 5% w/w caustic soda with a raw water rinse phase in between at 6 $m^3/h.m^2$. After regeneration, media (at pH >13) should be returned to pH 5.5–6.0 progressively in downflow neutralization steps using raw water dosed with sulphuric acid starting at pH >2.5. During operation, AA could leach out aluminium and this should be regularly measured in the treated water.

The IX process removes As(V) to less than 1 μg/l, but not As(III) unless it is oxidized to As(V). Many anions, in particular sulphate, interfere with As removal and IX should only be applied to waters containing dissolved solids <500 mg/l and sulphate <25 mg/l as SO_4. The process is not economic when sulphate is >150 mg/l as SO_4 (Clifford, 1999). When sulphate is present in the water, a bed which has become saturated with arsenate will begin to release it on take up of sulphate which results in a higher concentration of arsenate in the treated water than in the source water. The runs must therefore be terminated early. The process uses a strong base anion exchange resin of the chloride form and is independent of pH; regeneration is by sodium chloride. EBCT is 1.5–3 minutes. The quoted exchange capacity of the resins is low (1.4 g eq/l) but does not exhaust rapidly because arsenic concentrations in most groundwaters are very low. The presence of Fe(III) can complex arsenic and affect removal (Clifford, 1998).

Arsenic can be removed by adsorption in filters on to granulated ferric hydroxide (Thirunavukkarasu, 2003; Driehaus, 1998) which is available as proprietary media: e.g. Bayoxide

33® and GFH®. Bayoxide 33® media has a size range 0.3–2 mm, adsorption capacity of 4–12 g As/kg, bulk density of 320 kg/m^3 and a specific gravity of 3.6. Typical media depth used is 0.7–1.1 m with an EBCT of 2.8–5 minutes. The typical flow rate is 14–22 BV/h. The feed pH should be in the range 5.5–8.5. The media is said to reduce As concentrations from up to 200 to 2–4 μg/l. The media is very effective for removing As(V) but As(III) is also removed if iron concentration in the water is low; otherwise it requires pre-oxidation. The bed is backwashed at 25–30 m/h. The bed is not regenerated; once exhausted it is discarded. (Source of information: Severn Trent Water Purification Inc., Tampa, Florida, USA.) As(V) is removed by co-precipitation in biological iron removal filters (Katsoyiannis, 2004; Lehimas, 2001).

Nanoparticle agglomerated media such as the proprietary products Adsorbsia® and MetsorbG® (titanium dioxide), ArseneXnp® (iron oxide) and Isolux® (zirconium oxide) are used in arsenic removal (Westerhoff, 2006). Electrodialysis removes about 80% of arsenic.

Lead

Lead is rarely a contaminant of any significance in natural water. In polluted waters, the total concentration of lead could be as high as 10 mg/l with the dissolved fraction usually less than 0.01 mg/l (Galvin, 1996). Particulate lead is effectively removed by settlement with the aid of coagulants and soluble lead is removed at pH >9 such as would be used in lime softening. Lead in drinking water is mainly introduced through corrosion of plumbing systems containing lead pipes and fittings, brass fixtures and lead compounds used in pipe jointing materials. Except in very low alkalinity waters, lead enters the water not so much from direct contact with the lead pipe wall as from contact with lead-rich corrosion products formed as a scale on the pipe wall. The overall solubility of lead in drinking water, referred to as plumbosolvency, is controlled by the solubility of these corrosion products and this in turn is strongly influenced by alkalinity and pH. The waters which are most corrosive towards lead have a low alkalinity (<50 mg/l as $CaCO_3$) and pH <7. Except for very low alkalinity waters (<5 mg/l as $CaCO_3$), increasing pH above 7 dramatically reduces the lead solubility, with a theoretical minimum at a pH of approximately 9.8. At alkalinity levels above around 100 mg/l, lead solubility becomes increasingly less sensitive to changes in pH value. The least plumbosolvent water has a pH greater than 8.5 and alkalinity of between 10 and 80 mg/l as $CaCO_3$ (Schock, 1996; Sheiham, 1981).

Corrective measures are applied at the treatment works to minimize dissolution of lead in distribution. This is accomplished by appropriate control of pH and alkalinity of the water to assist in the formation of relatively insoluble lead compounds consisting principally of carbonates and oxides as a film on lead surfaces or by the dosing of orthophosphates to form a coating of sparingly soluble lead orthophosphate. For waters of low alkalinity (<50 mg/l as $CaCO_3$), lead solubility can be greatly reduced by increasing the pH into the range 9–10 (Schock, 1996) but this would contravene most national and international drinking water standards. In practice, therefore, elevation of pH value is restricted to about 8.5 and not less than 8.0 at the consumer's tap. There is generally no need to increase alkalinity (Sheiham, 1981) although this may be desirable if the buffering capacity of the water is inadequate to maintain the pH unaltered in the distribution system. In a well-buffered water pH variation would be limited to about 0.5 units whereas if buffering is inadequate the variation could be as much as 2.5 units. Soft and aggressive waters with alkalinity less than 50 mg/l as $CaCO_3$ could also be treated by lime and carbon dioxide. While control of pH is generally sufficient to reduce lead levels to around 50 μg/l for moderate to low alkalinity waters, any further reduction normally requires the addition of orthophosphate to form a less soluble lead phosphate film on the pipe wall. The use of pH elevation is in any case generally not suited to higher alkalinity waters where the formation of calcium carbonate scale would be a problem; orthophosphate is required in such waters.

Orthophosphate dosing provides a simple and effective treatment for achieving significantly lower lead concentrations than is possible by pH adjustment. Lead solubility in the presence of orthophosphate is comparatively insensitive to pH more than about 7, especially for low alkalinity waters. For higher alkalinity waters the most effective pH range is in the region 7.2–7.8. There are suggestions that raised dissolved organic carbon levels in the water can result in increased plumbosolvency, possibly as a result of a sequestering mechanism; in such cases phosphate dosing may be less effective than elsewhere.

The formation of protective lead phosphate films is considerably quicker for new lead pipes than for older material, particularly where older pipe has extensive carbonate scale protecting the elemental lead from the phosphate. The rate of initial film formation is thus often accelerated by using a high initial dose of 1.5–2 mg/l as P (4.5–6 mg/l as PO_4), particularly in higher alkalinity waters. Thereafter a maintenance dose in the range 0.8–1.2 mg/l would normally be sufficient to achieve optimum reduction, although higher doses may continue to be used for higher alkalinity waters. Further reductions in lead concentrations often continue to be seen for 2 years or more after phosphate dosing commences as the system reaches a new equilibrium, particularly in areas of high water alkalinity. Lead solubility increases with increasing water temperature and some water companies operate with increased phosphate doses in summer and reduced doses in winter.

The film formation process is reversible and interruptions to treatment should be kept to a minimum (<1 week); damage or repair to lead plumbing systems would also affect the protection (Colling, 1992). Older pipe may also be more susceptible to lead-rich deposits flaking off from the pipe wall, giving rise to particulate lead in the water; this may on occasion exceed 200 μg/l. Such episodes may be initiated by mechanical vibration or water hammer in the supply pipework. Phosphate dosing is of little value in eliminating such flaking.

Orthophosphate is added as one of its sodium salts or as orthophosphoric acid (Table 12.2). Polyphosphates have no effect in reducing plumbosolvency and are known to increase lead concentrations, on occasions, by sequestering lead into soluble compounds. Where polyphosphate appears to be achieving reductions in plumbosolvency, this is likely to be the result of orthophosphate formed by reversion of a portion of the polyphosphate. Monitoring is critical to the success of plumbosolvency treatment. Parameters to be monitored should include pH, alkalinity, orthophosphate and water temperature, and should be carried out both at the treatment works and in the distribution system.

The application of plumbosolvency treatment could have adverse or beneficial effects in the following areas depending on the characteristics of the distribution system: discolouration of water due to iron release; corrosion of other metals and cement lining of pipes; biofouling; scaling; and DBP formation. In addition, other practices may interfere with the effectiveness of plumbosolvency control strategies, including chloramination and mixing of waters in the distribution system. In particular, changes in disinfection regime from free chlorine to chloramination have been observed to cause significant increases in soluble lead concentrations (Renner, 2004). Research suggests this is caused by changes in the oxidation potential, changes which are exacerbated by the presence of natural organic matter (NOM) in the treated water (Valentine, 2009). Recent studies (UKWIR, 2007) suggest that phosphate dosing in the presence of fluoride, either naturally present or dosed artificially for dental caries control, can result in the formation of fluorapatite, a hard crystalline substance that is highly insoluble. The mechanism for its formation is not fully understood but appears to be initiated by local high temperatures, such as may be found in combination boilers or commercial calorifiers. This can result in solids deposition on heat exchange surfaces, ultimately leading to equipment failure.

Aluminium

Aluminium (both soluble and particulate) in surface waters is readily removed by coagulation using aluminium or iron coagulants in the pH range 6.5–7.2 (Section 8.20). Since aluminium hydroxide is soluble at pH greater than about 7.3, it cannot be removed together with manganese. When manganese is present, therefore, aluminium is first removed by coagulation and filtration (with or without clarification) followed by elevation of pH (>8), oxidation and secondary filtration for manganese removal (Section 10.12). A similar treatment regime is applied when treating raw water containing manganese with aluminium coagulants. There are exceptions to this rule as in some waters manganese can be removed at pH <7.5. Aluminium is rarely found in groundwaters in excess of 10 μg/l.

Chromium

When present in surface waters, chromium exists as Cr(III) and Cr(IV). The former is an essential dietary element. Cr(VI) is toxic, but because the two species are inter-convertible in the environment, the provisional WHO Guideline Value of 0.05 mg/l is for total chromium (WHO, 2011). Cr(III) occurs as various species of hydroxide complexes while Cr (VI) occurs as $HCrO_4^-$ (pH <6.5) and CrO_4^{2-} (pH >6.5). Cr(VI) is readily reduced to Cr(III), which is insoluble in the pH range 7–10 (McNeill, 2012). Treatment is therefore by coagulation using ferrous salts where Cr(VI) is reduced to Cr(III), which is adsorbed on to the iron floc and removed in clarifiers. Excess iron is removed by oxidation using aeration or chlorination in filters (Blute, 2014). Blute also demonstrated that weak base anion exchange resins with amine functionality operating at pH 6.0 were effective in removing Cr(VI). Treatment should be conducted with at least two vessels in parallel (Sections 10.8 and 10.25 give suitable configurations). A single charge can treat up to 320 000 BVs and therefore it is not economical to regenerate. Once breakthrough occurs, the media is discarded. Both methods reduced Cr(VI) to <0.05 mg/l. Use of strong base anion exchange resins with a quaternary amine functional group to remove Cr(VI) was demonstrated in test work carried out for the Water Research Foundation (2014). Resins tested were able to achieve a Cr(VI) reduction from 18 to 8 μg/l with 15 000–30 000 BVs treatment capacity over a wide range of pH. The spent resin is regenerated with sodium chloride (Section 13.9).

10.15 REMOVAL OF RADIONUCLIDES

An introduction to radionuclides in water is given in Section 7.42. Radon, one of the most common nuclides in drinking water, is a water soluble gas primarily found in groundwaters in soil containing granite. Its solubility is about 230 cm^3/l at 20°C (USDHHS, 2012). It has a high Henry's constant (2.26×10^3 bar at 20°C, 1 atm) and is therefore easily stripped-off by aeration. Packed tower aerators (Section 10.20) show removal efficiencies of about 98%, closely followed by the diffused air system (Section 10.23). Removal in a spray aerator (Section 10.21) was found to be less than 75% (Dixon, 1991). For a packed tower the packing height is about 3 m with a surface loading rate of about 75 $m^3/h.m^2$. The packed tower performance is not influenced by the air-to-water flow ratio. Problems with packed towers are carbonate scaling and iron deposits if Fe(II) is present in the water (Section 10.20). Radon discharged to air from packed towers is diluted to insignificant ground level concentrations. Granular activated carbon (GAC) adsorbers are also known to remove radon but not as effectively as aeration. EBCT required is more than an hour and GAC adsorbers are therefore suitable only for small capacities (Haberer, 1999).

Uranium, as well as all particulate radionuclides, can be removed by coagulation using iron or aluminium salts followed by solid–liquid separation. Methods used for dissolved radionuclides are: lime softening at pH >10.5 for uranium and radium-226; strong base anion exchange of the chloride form for uranium, particularly in low sulphate water (Zhang, 1994); strong acid cation exchange of the sodium or calcium type for radium (Snoeyink, 1987); and mixed bed IX for β-particles (US EPA, 2002), radium and uranium. Both strong base and strong acid resins are regenerated with sodium chloride. RO could be used for uranium, radium, α- and β-particles.

DEFLUORIDATION AND FLUORIDATION

10.16 DEFLUORIDATION

Some groundwaters contain high levels of fluoride with concentrations well in excess of 1.0 mg/l as F (Section 7.28). Levels in excess of 1.5 mg/l as F may cause dental fluorosis leading to mottling of the teeth; reduction of fluoride may therefore be necessary. Defluoridation can be achieved by chemical precipitation, adsorption or by membrane desalination processes. Lime softening at pH >10 followed by soda ash dosing removes fluoride in the presence of magnesium by adsorption on to magnesium hydroxide. The water may be dosed with magnesium sulphate or dolomitic lime; 1 mg/l of fluoride requires 50 mg/l of magnesium (Degremont, 2007). Precipitation by aluminium sulphate requires doses up to 750 mg/l and is not economical. High levels of fluoride can be reduced down to the solubility of calcium fluoride using lime, leaving about 8 mg/l fluoride in water.

Adsorption onto AA is successfully used for fluoride removal. It is highly selective to fluoride in the presence of sulphate and chloride when compared to synthetic IX resins. In the presence of bicarbonates, although the fluoride concentration is reduced, the adsorption capacity shows a major decline. Silica is also known to interfere with the adsorption of fluoride. The adsorption process is best carried out under slightly acidic conditions (pH 5–6); the lower the pH the more effective the removal. There are proprietary AA media used for defluoridation. One such is SORB 09™, which has a size range 0.6–1.5 mm, with an adsorption capacity of 1.2–2.5 g F/100 g and bulk density of 320 kg/m^3. At pH below 6.0, treatment ratios are stated to be 1000–2000 BV per unit volume of media with service cycles of 5–25 days between regenerations, depending on feed water fluoride. At least two vessels operated out of phase should be provided. Media depth used is 1.2–1.5 m with an EBCT of 4.5–5 minutes. The typical flow rate is 12.2–17.0 $m^3/h.m^2$. The media is said to reduce fluoride concentrations from up to 20 mg/l to less than 1.5 mg/l. The bed is backwashed followed by regeneration using 1–1.5% w/v caustic soda. AA is then reactivated with acid conditioning (about 0.25% w/v H_2SO_4). Beds should be rinsed after each stage to displace residual NaOH and H_2SO_4. The total waste generated is less than 1% of the plant flow. (Source of information: Severn Trent Water Purification Inc., Tampa, Florida, USA). Breakthrough occurs slowly and the operating cycle should be terminated before the alumina is fully exhausted. During operation, aluminium could leach out and this should be regularly measured in treated water. The plant and its operation are similar to that described for arsenic removal (Section 10.14).

Rejection of fluoride by RO (Section 10.46) is pH dependent. Rejections at alkaline pH values can be greater than 99% due to fluoride being in the salt form. Rejection is less than 50% at acidic pH values due to fluoride being present as dissociated ions. Electrodialysis (Section 10.45) also removes fluoride.

10.17 FLUORIDATION

Many waters contain only trace quantities of natural fluoride and some regional or national health authorities consider that it should be added to reduce the incidence of dental caries. Fluoridation of water is carried out by using either hexafluorosilicic acid, disodium hexafluorosilicate (also called sodium hexafluorosilicate) or, for small water supplies, sodium fluoride which is a powder (Section 12.25 and Table 12.2).

The fluoride dose required varies with ambient temperature (T°C); higher temperatures require lower dosages, since more water is consumed in a hot climate. The approximate dose required can be calculated by: $F = 0.34/E$ (Gallagan, 1957), where $E = -0.38 + 0.0062(T \times 1.8 + 32)$.

AERATION

10.18 PURPOSE

Aeration has a large number of uses in water treatment; the more common are to:

- increase the dissolved oxygen content of the water;
- reduce tastes and odours caused by dissolved gases in the water, such as hydrogen sulphide, which are then released;
- decrease the carbon dioxide content of a water and thereby reduce its corrosiveness and raise its pH value;
- oxidize iron and manganese from their soluble to insoluble states and thereby cause them to precipitate so that they may be removed by clarification and filtration processes; and to
- remove certain volatile organic compounds.

According to Henry's law, the equilibrium solubility of a gas in water is given by $C_s = kp$ where C_s is the saturation concentration of the gas in water (mg/l), p is the partial pressure in bar and k is the coefficient of absorption which is equal to $(55\ 600 \times M) \div H$ where M is molecular weight of the gas and H is Henry's constant, expressed in units bar. The higher the value of H, the easier the desorption.

The purpose of aeration is to speed up these processes and there are two main types of aerators in general use: those in which water is allowed to fall through air, for example free-fall aerators, packed tower aerators and spray aerators, and those in which air is injected into water, for example injection aerators, diffused air-bubble aerators and surface aerators. For them to be effective it is essential they provide not only a large water–air interface (high area:volume ratio) but, at the same time, a high degree of mixing and rapid renewal of the gas–liquid interface to facilitate transfer of oxygen.

10.19 CASCADE AERATORS

Cascade aerators are of linear, rectangular or circular shape and consist of a series of steps over which water flows (Plate 16(b)). Figure 10.2 shows the design of a cascade aerator. Such aerators are widely used as water features: they take large quantities of water in a comparatively small area at low head, they are simple to keep clean and they can be made of robust and durable materials

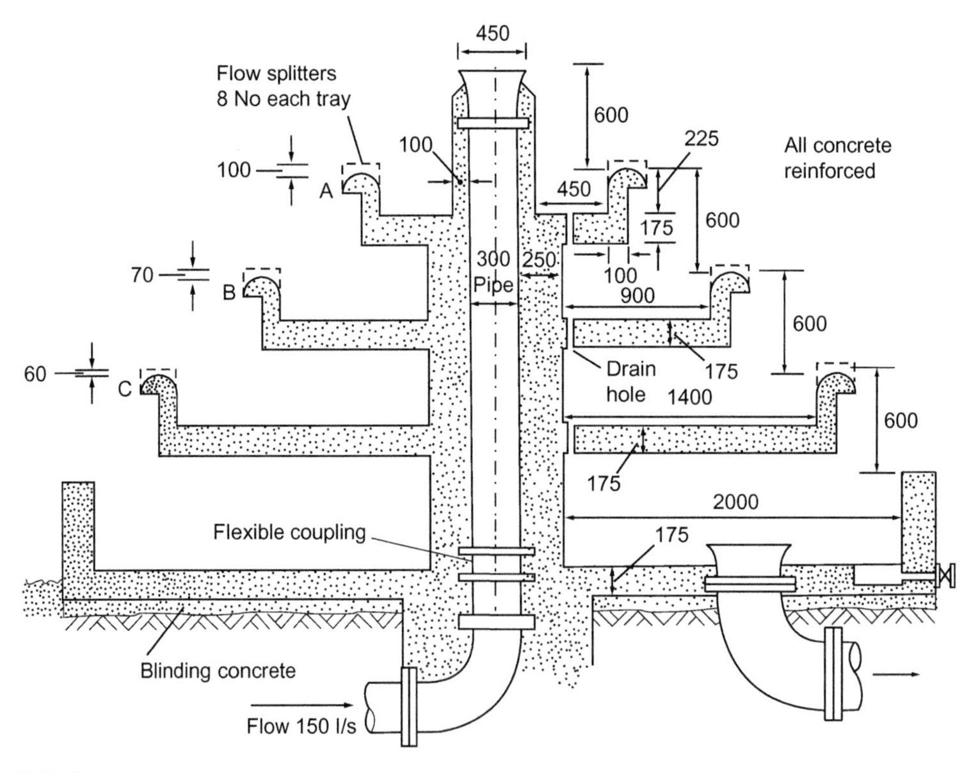

FIGURE 10.2

Cascade aerator.

giving a long life. The steps are usually made of reinforced concrete. The aerator should preferably be in the open air or, for protection against air borne pollution, freezing and algal growth, in a small house which has plenty of louvred air inlets. They are efficient for raising the dissolved O_2 content but not for CO_2 removal; reduction of CO_2 content is usually in the range of 60–70%. From work carried out by the former UK Water Pollution Control Laboratories (DoE, 1973) an empirical relationship has been developed for the ratio r of the oxygen deficit just above a weir to that just below which is:

$$r = \frac{C_s - C_u}{C_s - C_d} = 1 + 0.38abh(1 - 0.11\,h)(1 + 0.046T)$$

where a is 1.25 in slightly polluted water and 1.00 in moderately polluted water; b is 1.00 for a free-fall weir and 1.30 for a stepped weir; h is the height of the fall (difference in water levels) in metres, and T is the water temperature in °C, C_u and C_d are oxygen concentrations upstream and downstream and C_s is the saturation concentration at T°C. This equation is valid only for falls up to about 0.6 m. For falls up to 3 m more representative results are obtained by replacing the term $0.38abh(1-0.11h)$ by: $ab(0.65 + 0.25h)$.

In the design of such aerators certain hydraulic criteria should also be met. These include the depth of the receiving pool y (m) defined in terms of the Drop number (D) as:

$$y/Y = 1.66D^{0.27} \quad \text{and} \quad D = q^2/gY^3$$

where Y (m) is the height of the fall from crest to pool floor and q is the unit discharge defined as the flow Q (m^3/s) per unit length L (m) of weir. The horizontal component x (m) of the trajectory velocity of the water fall over the weir assuming no air resistance, is defined for a weir coefficient of 1.705 and where H is the head over the weir (m) and y is the vertical component of trajectory (m) by:

$$x = [1.33H(Y - y)]^{0.5}$$

The number of steps varies from three to 10 and the fall in each step from 0.15 to 0.6 m. The rate of flow should be limited to about 125 m^3/h.m weir length. To allow entrained air to mix in the water, each step should have a pool of water of depth at least 0.3 m. Weirs with serrated edges perform better (van der Kroon, 1969) as they help to break water flow into separate jets. If the water is allowed to cling to the steps especially at low discharge rates, the efficiency is reduced. This could be avoided by using a sharp edged thin plate weir or by providing a protruding lip using a steel plate. The space requirement is typically of the order of 0.5 m^3 per 1 m^3/h water treated. The oxygen transfer efficiency for 50% saturation at 10°C and atmospheric pressure, starting at zero is calculated to be 1.56 kg O_2/kWh.

10.20 PACKED TOWER AERATORS

Packed tower aerators consist of a vertical steel, plastic or concrete cylinder filled with plastic or ceramic packing, such as pall rings, berls or saddles (Fig. 10.3). Water is applied at the top through a distributor and a counter-current flow of air is usually blown (forced-draught) but sometimes drawn (induced-draught) using an oil-free blower or fan; air must be filtered to prevent contamination of water. The air-to-water volumetric flow ratio is in the range 25:1 to 75:1. The water becomes distributed in a thin film over the packing so providing the large water–air interface required for good mass transfer. Surface loading rates are of the order of 50–100 $m^3/h.m^2$. Packed towers constitute the most efficient form of aeration and are used primarily for groundwaters; suspended solids in surface waters can rapidly clog the packing. They give over 90% oxygen transfer and 85% CO_2 removal efficiencies. They are also used for the removal of hydrogen sulphide, ammonia and volatile hydrocarbons and can deal with varying throughputs with no decrease in efficiency. Typically, the tower needs up to about 10 m head of water.

Drawbacks of the system are that CO_2 removal increases the tendency for scale deposition, whilst in the presence of Fe(II), ferric hydroxide precipitate tends to foul the packing. A sequestering agent such as sodium hexametaphosphate can be used to eliminate scale deposition. The packing can also foster growth of iron bacteria which can be problematic, although normal

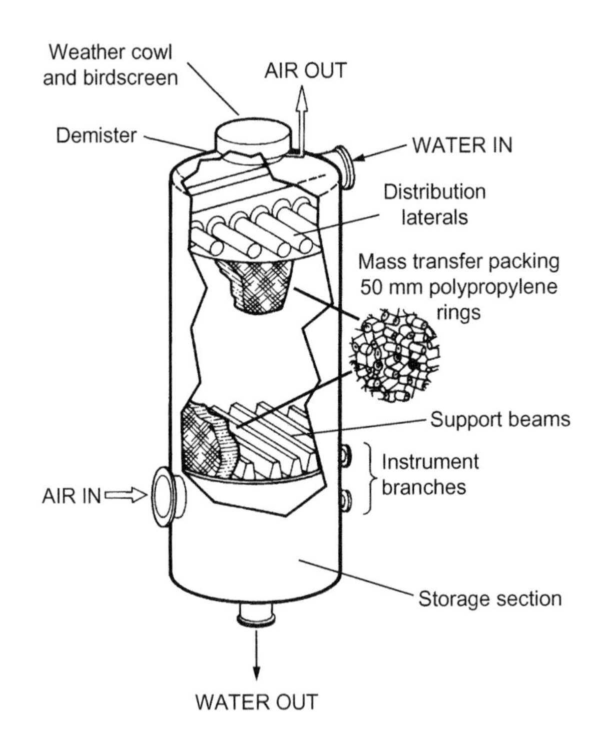

FIGURE 10.3

Packed tower aerator.

bioslimes can be accommodated on some types of packing. After a number of years, the packing needs replacing, depending on the deposits that have accumulated on the surfaces of the material. Capital and operating costs are high but the packed tower aerator is effective for dealing with high CO_2 content, ammonia (Section 10.27) or volatile hydrocarbons (Section 10.30). The oxygen transfer efficiency for 50% saturation at 10°C and atmospheric pressure, starting at zero oxygen is calculated to be 5.0 kg O_2/kWh.

10.21 SPRAY AERATORS

Spray aerators comprise nozzles installed on a pipe grid at 1–3 m spacing. Water sprayed from the nozzles forms fine streams of small droplets which produce a large contact area with the air. They can be used on hard waters and those containing iron and manganese without risk of scale deposition or fouling problems. About 70% CO_2 removal and 80% oxygen transfer efficiencies can be obtained with the best type of spray nozzles. Nozzle diameter ranges from 25 to 35 mm with flow rates up to 30 m^3/h and is a compromise to prevent clogging but avoid the need for excessive pressure. Up to 20 m head of water may be required. A large collecting area (typically 0.5–1 m^2 per m^3/h water treated) is necessary to accommodate the large number of nozzles required and to minimize overlapping of jets. The receiving pool depth should be about 150 mm. Extra efficiency is obtained in some types of plant in which the spray is broken up by impinging on a plate. The sprays require protection from wind and freezing and should be enclosed in a louvred building. They are normally used on groundwaters; suspended solids in surface waters can block the nozzles. The oxygen transfer efficiency for 50% saturation at 10°C and atmospheric pressure, starting at zero oxygen is calculated to be 1.0 kg O_2/kWh.

10.22 INJECTION AERATORS

Injection aerators avoid the need to break the water pressure if this is particularly inconvenient or wasteful of energy. The water may be sprayed into a compressed air space at the top of a closed pressure vessel under pressure. The air has to be circulated by a compressor as for dissolved air flotation (Section 8.18).

Alternatively, compressed air may be injected directly (for pipe diameters up to about 300 mm) or by a side stream static mixer (for pipe diameters 300–500 mm) into the flowing water in a pipe upstream of an inline static mixer; or air at atmospheric pressure may be drawn into the pipe where a constriction, such as the throat of a venturi tube, reduces the water pressure below atmospheric. In the latter case, the venturi tube has a much narrower throat and a much longer divergence cone downstream than a venturi designed for flow measurement. Aeration under pressure does not remove CO_2; its effect is to increase the oxygen content of the water and to saturate it with nitrogen at the operating pressure. The latter can be a disadvantage because air can appear as bubbles when the pressure is released and can lead to air binding and consequently spurious head loss in a downstream filtration process. In the case of pressure filters, an air release valve should be provided. For static mixer designs with direct air injection, the expected oxygen transfer is about 40–50% with a maximum oxygen concentration of 4 mg/l, whereas for side stream transfer it is about 65% with a maximum oxygen concentration of 6 mg/l. Oxygen transfer efficiencies of the two systems will depend more on the energy consumption of the side stream pump and air supply compressor compared to that contributed by the headloss across the static mixer.

10.23 OTHER TYPES OF AERATORS

Mechanical surface aerators operate on the basic principle of entraining air by agitation of the water surface. There are several forms of surface aerators which are successfully used in sewage treatment works. Those where the axis of rotation of the impeller is vertical are used for aerating water in large open tanks. The oxygen transfer efficiency for a mechanical surface aerator is about 0.5 kg O_2/kWh.

Diffused air bubble aeration is usually carried out bypassing air through some form of a diffuser placed at the bottom of a tank through which the water flows horizontally. The diffuser may be a perforated pipe or a porous ceramic plate. Aeration efficiency is a function of bubble size and tank depth; with smaller bubbles, efficiency increases with decreasing depth. Tank depth varies from 3 to 4.5 m; tank width should not exceed twice the depth. Residence time is about 10–20 minutes. Air (at atmospheric pressure) to water volumetric ratio is usually between 1:1.25 and 1:2.5. Air must be filtered to minimize contamination of water and clogging of the diffusers and the compressors used must be of the oil-free type. The oxygen transfer efficiency for 50% saturation at 10°C and atmospheric pressure, starting at zero oxygen is calculated to be 2.5 kg O_2/kWh. This method is also widely used in sewage treatment. In water treatment, the perforated pipe arrangement finds application in the destratification of reservoirs (Section 8.1).

NITRATE REMOVAL

10.24 GENERAL

Nitrate is found in undesirable concentrations in some water sources (Section 7.35). Blending of sources is the simplest method for achieving low nitrate supplies. For surface waters long-term storage can be used to absorb the seasonal peaks and to bring about natural biological denitrification (Section 8.1). Where such methods are not feasible, treatment of the water for nitrate reduction is necessary. The two nitrate reduction processes predominantly used are IX and biological denitrification. Electrodialysis reversal (EDR) and RO processes can be used for nitrate removal and are discussed in Sections 10.45 and 10.46. Both are desalination processes and are substantially non-selective. This can be an advantage as it results in the removal of multiple contaminants in a single process step. They bring about a nitrate removal in excess of 80%. Therefore, treatment is applied only to a small part of the flow so that the desired concentration is achieved after blending with the remainder.

10.25 ION EXCHANGE (IX) PROCESS FOR NITRATE REMOVAL

The IX process is similar to that used for water softening (Section 10.8). In nitrate removal a strong base anion exchange resin in the chloride form is used. As the water passes through the bed of resin contained in a pressure vessel, nitrate and other anions in water are exchanged with chloride in the resin, thus releasing chloride into the water. When the resin is saturated with respect to nitrate (indicated by nitrate breakthrough) the resin is regenerated with sodium chloride solution (brine) and the exchange process is reversed; anions absorbed on the resin are replaced by chloride ions. To ensure that flow is maintained during regeneration ideally 100% standby units should be provided but a total capacity of at least 3 × 50% flow for small works and 4 × 33% for large works may be considered. Either 'classical' anion exchange (Section 10.10) or nitrate selective exchange resins can be used in the process. The former has been used with success except for waters with significant sulphates. Nitrate is less well absorbed than sulphate and nitrate uptake falls off rapidly as raw water sulphate increases. If any sulphate is present in the water, a bed which has become saturated with nitrate will begin to release it on take up of sulphate; this results in a higher concentration of nitrate in the treated water than in the source water. This drawback can be overcome by terminating runs early. High levels of sulphate reduce the resin capacity for nitrate and increase regenerant (sodium chloride) consumption by reason of removal of anions other than nitrate and, as a result of sulphate removal, more chloride is released from the resin to the water. Therefore, for waters containing mass ratio of sulphate (as SO_4) to nitrate (as N) in excess of 3.43:1 or nitrate (as NO_3) in excess of 0.77:1 (1 g of nitrate as N = 4.43 g of nitrate as NO_3), it may be more economical to use nitrate selective exchange resins. These preferentially absorb nitrate and therefore do not show many of the drawbacks inherent to classical resins.

The surface loading rate for nitrate removal is about 30 $m^3/h.m^2$. Allowing about 20% of the time for regeneration, this corresponds to a mean throughput of 576 $m^3/d.m^2$. Volumetric flow rate should be less than 40 bed volumes (BV)/h; 20 BV/h is regarded as a suitable design value with all columns in service. The resin bed would be about 1.5 m deep and vessel diameters would usually be in the range 1–4 m. The length of a run (time between two consecutive regenerations) is a function of operating conditions and the resin and is usually 8–12 hours. The regeneration sequence on completion of a run comprises: backwashing of the bed to remove suspended solids; counter-current regeneration using 6–10% w/w sodium chloride solution at 3–4 BV/h for 1.5 BVs; upflow slow rinse at 3–4 BV/h for 2 BVs; a fast rinse in the direction of service flow at 6 BV/h for 4 BV; return the bed to service. Frequency of backwashing is about every 10–25 cycles depending on the raw water quality. It is normal to use raw water for backwash and treated water for rinse. The quantity of salt required is about 160 g NaCl/l of resin for classical resins and 125 g NaCl/l for nitrate selective resins. The nitrate removal capacity for the two types of resin is about 0.25 g eq/l of resin. When treating waters containing high alkalinity, provision for acid washing of the bed using hydrochloric acid is recommended to remove calcium carbonate deposits.

Drawbacks with the use of IX for nitrate removal are that the process increases the chloride concentration and reduces alkalinity of the product water; and the need for disposal of the spent regenerant. Although the increase in chloride level of the water has no health implication, it can increase the corrosivity of the water (Section 10.40). The problem is less severe with nitrate selective resins. At the beginning of the run chloride ions increase, accompanied by a reduction in the alkalinity. The product water would therefore have a high chloride to alkalinity ratio initially; classical resins produce higher ratios than nitrate selective resins. This problem can be overcome by either or both the following methods: mixing the output of a run in a treated water reservoir before forwarding it to supply; dividing the flow between two or more parallel units and operating them out of phase. A further but more costly option is to use a bicarbonate solution in place of the chloride solution in the last 10–15% of the regeneration phase, to replace the chloride ion in the resin with bicarbonate ion. Thus at the start of the run bicarbonate ions and not the chloride ions are released into the water.

10.26 BIOLOGICAL PROCESS FOR NITRATE REMOVAL

Biological denitrification of drinking water is based on the heterotrophic process that occurs in the anoxic zone in sewage treatment. In this process an organic carbon source is used to sustain bacterial growth, using oxygen bound in nitrate for respiration, reducing it to nitrogen. In water treatment, bacteria first consume any dissolved oxygen in water before the oxygen in nitrate and create an additional demand for carbonaceous matter. Organic carbon should be added to the water as most water supplies contain relatively low concentrations; the sources of organic carbon (substrate) are methyl alcohol, ethyl alcohol or acetic acid. Stoichiometric quantities of methyl alcohol, ethyl alcohol or acetic acid required for each 1 mg of dissolved oxygen and nitrate (as N) are 2.57, 1.85 and 3.62 mg, respectively; the actual demand could be up to 1.5 times greater. Optimum pH is in the region of 7.5. The process has little effect on the alkalinity of the water. A trace concentration of phosphate of less than 0.5 mg/l as P is also needed to assist bacterial growth, and should be added if this is not present in the water. The process is sensitive to temperature and reaction rates decrease markedly below about 8°C. The autotrophic denitrification process which uses hydrogen (Gross, 1988) or sulphur compounds (Soares, 2002) to sustain bacterial growth, is also used in some full-scale plants. The reaction with sulphur produces acidity and should be neutralized with calcium carbonate.

The biological process is usually carried out in fluidized bed or up or down flow fixed bed reactors where the biological growth is physically supported on a medium. Fine sand is commonly used in fluidized bed reactors, whilst a porous medium such as expanded clay is used in fixed bed reactors. In a fluidized bed, water mixed with the substrate flows upward at 20–30 $m^3/h.m^2$ to provide 40–50% bed expansion and giving a detention time typically of 5–10 minutes. Before start up, the bed requires seeding with bacteria. This could take up to 1 month. As the biomass builds up, a proportion of the sand is periodically removed from the bed, the bacterial film is stripped in a sand cleaning plant and the sand returned to the bed. Fixed bed reactors are usually based on conventional sand filter principles, the media used being a porous medium, such as expanded clay, of coarse grain size. An EBCT of 10–15 minutes is usually used. The backwashing regime is similar to that used in conventional sand filters. The treated water from the biological reaction is devoid of oxygen and contains DOC and bacterial floc carried over from the reactor. Hence, the water needs to be reoxygenated and filtered through GAC and sand filters before passing to supply.

REMOVAL OF AMMONIA

10.27 CHEMICAL AND PHYSICAL METHODS

Ammonia is present in water as saline or free form (Section 7.7). The most common method used for ammonia removal is 'breakpoint' chlorination (Section 11.9) where ammonia nitrogen is completely oxidized to nitrogen leaving a residual of free chlorine. To minimize DBP formation breakpoint chlorination should be practised after DBP precursor removal. Organic nitrogen is not destroyed by chlorine. Ozone does not normally oxidize ammonia. Neither chlorine dioxide nor potassium permanganate affects ammonia.

When present as free ammonia or the ammonium ion, ammonia can be removed in packed tower aerators (Section 10.20). In the latter case the pH of the water needs to be raised to 10.5–11.5 to convert all ammonium ions to free ammonia. A design for a packed tower is given by Short (1973). Due to the large air:water ratio (3000:1), however, the height of the tower and the chemicals required for pH adjustment the process is usually uneconomical.

Other options for ammonia removal include strong acid cation exchange resins, as used in base exchange softening (Section 10.7) and RO, which rejects about 99.5% of the ammonium ion (NH_4^+) at pH 8.5 (Section 10.46).

10.28 BIOLOGICAL METHODS

Biological oxidation of saline ammonia to nitrate, known as nitrification, takes place in two steps (Richard, 1978): initially conversion to nitrite by *Nitrosomonas* bacteria, followed by oxidation to nitrate by *Nitrobacter* bacteria. Both steps require oxygen. Carbon dioxide is the carbon source; 1 mg/l ammonia (as N) consumes about 7.2 mg/l alkalinity (as $CaCO_3$). When treating some soft waters, therefore, alkalinity may have to be added. The process also requires about 0.3 mg/l phosphates (as PO_4) (Lutle, 2013) to allow nitrifying bacteria to develop. The seeding period for this is about 1–2 months. The optimum pH range for the reaction is between 7.2 and 8.2. In practice, nitrification is almost 100% complete (Lytle, 2007). The optimum temperature for bacterial growth is in the range 25–30°C. The water temperature should be greater than 10°C; there is no biological activity below 4°C. The process requires oxygen at a rate of about 4.57 mg per mg of ammonia (as N) (Richard, 1978). At high ammonia concentrations, simple saturation of the water with oxygen by aeration may therefore prove inadequate and oxygen must be continually added.

In biological removal of ammonia in drinking water treatment, nitrification is usually carried out in rapid gravity filters. A maximum ammonia concentration of 1.5 mg/l as N can be removed in conventional rapid gravity filters, depending on the temperature and the dissolved oxygen concentration of the influent water. Filtration rates could be in the range 5–10 $m^3/h.m^2$ and, to remove 1 mg/l ammonia (as N), the necessary EBCT is about 20, 10 and 5 minutes at 5, 10 and 30°C, respectively. After nitrification, the water could be devoid of oxygen depending on the ammonia concentration and may need aeration. For higher ammonia concentrations BAFs have been used.

These are similar in principal to the trickling and/or aerated filters used in sewage treatment where there is a continuous flow of air through the media of the filter bed. The ammonia loading of a BAF filter is typically 0.25–0.6 kg of ammonia (as N)/day.m^3 (of media), depending on the type and effective size (d_{10}) of media.

In drinking water treatment the BAF consists of a bed of coarse media (d_{10} between 1.5 and 3.0 mm); filtration is either downflow counter-current or upflow co-current with air flow injected continuously into the bottom of the filter bed using either an independent pipe lateral system or special nozzles in a plenum floor design. The volumetric ratio of air:water is in the range 0.3–1.0 (Degremont, 2007). Upflow filters are generally 15–25% more efficient than downflow filters; to remove 1 mg/l of ammonia as NH_4^+ (0.78 mg/l as N) at pH 7.2 and water temperature of 10°C, the EBCT required for upflow and downflow filters using d_{10} 2 mm 'Bioloite™' (an expanded clay medium) is about 3 and 4 minutes, respectively. Pilot plant work has demonstrated that, based on first-order reaction kinetics, the reaction rate constant for upflow filter was 60% higher than that for the downflow filter (Heard, 2002). The depth of the filter medium would be a function of EBCT and filtration rate and is dependent on the raw water ammonia concentration: a water containing 2.5 mg/l ammonia (as N) would require a depth of about 2.5 m for an upflow filter compared to 3 m for a downflow filter. Upflow BAFs have a surface loading rate of 10–12 $m^3/h.m^2$ and a coarse medium (d_{10} 1.5–2.0 mm) while downflow BAFs operate at surface loading rates of 8–10 $m^3/h.m^2$ (depending on the suspended solids loading) and use a coarser media. Such surface loading rates are feasible with expanded mineral filter media mostly of proprietary makes. For example, Filtralite of size 2.5–5.0 mm (d_{10} 2.7 mm) removes about 90% of raw water ammonia using an EBCT of 12 minutes. Filters using naturally occurring media such as pozzolana or carbon (2–5 mm or even larger to give high specific surface per unit volume) operate at about 5 $m^3/h.m^2$ with EBCT of about 20–30 minutes (Lacamp, 1990).

BAFs are washed by concurrent application of air and water at about 16 and 4 mm/s, respectively followed by a high rate rinse at 12.5 mm/s. The minimum process air rate is about 0.8 mm/s. Washwater should be free of chlorine. When treating surface water BAFs are best used after the clarification stage and downflow filters are generally preferred. Upflow filters are generally used for groundwaters particularly those containing high ammonia concentrations that require high air: water flow rates and low turbidity. BAFs are best followed by conventional rapid gravity sand or anthracite–sand filtration in order to produce a water free of suspended matter. The biological process is adversely affected by chlorine, hydrogen sulphide, heavy metals and precipitates from iron and manganese oxidation and other suspended solids in the water. When ammonia is present together with iron and manganese, the order of biological removal is iron, ammonia followed by manganese (Mouchet, 1992). Iron is best removed separately by chemical oxidation. If the ammonia concentration is high manganese is not removed in the same filter unless adequate EBCT is provided. BAF beds also remove organic carbon effectively. Manganese dioxide coated sand filters have been successfully used to oxidize low concentrations of ammonia biologically and manganese by catalytic oxidation (Janda, 1994).

The biological reaction principle can also be applied to sedimentation tanks of the sludge blanket type. The sludge acts as the medium for the growth of nitrification bacteria. Oxygen for the nitrification reaction is limited to that which can be contained in the feed water and the ammonia removal is limited to about 0.5 mg/l as N.

REMOVAL OF VOLATILE ORGANIC COMPOUNDS FROM GROUNDWATER

10.29 GENERAL

Volatile organic compounds (VOC) (Section 7.16) are not removed by conventional water treatment processes, being stable towards most oxidants including ozone, and they are not biodegradable. The most cost effective treatment is considered to be packed tower aeration; but adsorption onto GAC may be more economical in hard waters (Booker, 1998) (Section 10.35). Spray aerators in an enclosed storage tank could be used to reduce trihalomethanes formed in the disinfection process (Brooke, 2011), although less volatile DBPs would remain.

10.30 PACKED TOWER AERATORS

In packed towers (Section 10.20), also known as air stripping towers, the contaminated water flows downwards through a packing, counter-current to an air flow which strips the VOCs into the gas phase and discharges them through the top of the tower. The treated water is collected at the bottom of the tower. Since VOCs have high Henry's constants (e.g. carbon tetrachloride – 2.04×10^{-2} atm/mol.m^3 at 20°C) the process removes over 99.99% of VOCs (Hess, 1983).

The design parameters are air-to-water volumetric flow ratio, surface loading rate, type and size of packing and depth of packing. These are influenced by temperature, chemistry of the water and the mass transfer characteristics of the packing. Towers are usually constructed in polyethylene, glass-reinforced plastic or rubber-lined mild steel and should be provided with a good water distribution system and mist eliminators on the air discharge. Packing types are generally pall rings, Rashig rings or Berl saddles manufactured from plastic or ceramic materials; packing height should be limited to about 6 metres. For greater packing heights two or more towers in series should be considered. The surface loading rate of the tower needs to be selected to prevent flooding of the packing and is usually about 60–75 m^3/h.m^2. The tower diameter should be limited to 3–4 metres. High water flow rates should be divided between two or more towers in parallel. The air-to-water volumetric ratio is usually about 25:1 to 30:1 to limit the air pressure drop across the packing to 10–40 mm of water per metre of packing.

In packed towers, dissolved oxygen is increased and carbon dioxide is removed while the VOCs are stripped from the water, resulting in iron precipitation and calcium carbonate scaling (Section 10.20). Although the quantities are small, air stripping results in release of VOCs to atmosphere, which may be of some concern for plants located in urban areas.

10.31 ADSORPTION AND CHEMICAL OXIDATION

Adsorption of organic compounds by GAC is well known (Section 10.35) and the technique can be used to remove VOCs. The adsorbers are of the fixed bed type and can be conventional rapid gravity or more commonly, pressure filters. The principal design parameters are: EBCT; type of GAC used; bed depth and hydraulic loading. These parameters should be evaluated by pilot scale tests.

For pressure filters, typical values are EBCT in the range 10–30 minutes, surface loading rates between 10 and 20 $m^3/h.m^2$ and bed depths of 2.5–3 m. The adsorption capacity of GAC for VOCs is generally small and the media needs frequent regeneration if VOCs are present in concentrations in excess of 100 μg/l (Foster, 1991). Experience in the USA shows 12–18 month regeneration frequency (Dyksen, 1999).

VOCs can be oxidized by hydroxyl free radicals formed when ozone is used in combination with hydrogen peroxide (H_2O_2) or with UV radiation (Glaze, 1987) (Section 10.36). Ozone (O_3) is first injected into a reaction chamber consisting of two or three compartments to satisfy the ozone demand; hydrogen peroxide is injected into the second or the third compartment together with further addition of ozone. Laboratory or pilot trials are required to optimize the ratio H_2O_2:O_3 and the contact time which depends on the concentration and the nature of the organic matter. The ratio can vary between 0.3 and 1. The oxidation stage should be followed by GAC adsorbers to remove the by-products of the oxidation and excess hydrogen peroxide; the latter dechlorinates water.

TASTE AND ODOUR REMOVAL

10.32 CAUSES OF TASTES AND ODOURS

The source of a taste or odour in a water is often difficult to identify, but the following list includes the most likely causes.

1. Most of the taste and odour caused by biological action on rotting debris are due to the production of geosmin and 2-methylisoborneol (2-MIB) (Westerhoff, 2005; Asquith, 2013). These are responsible for earthy, musty or mouldy tastes and odours in drinking water detectable at concentrations as low as 5–10 ng/l. They are frequently associated with actinomycetes fungi and some species of blue-green (*Cyanophyceae*). These organisms have the most favourable conditions for their growth in stagnant waters and commonly occur in deeper, poorly mixed reservoirs and also in long lengths of pipeline left standing in warm surroundings, such as the plumbing system of a large building, and the first water drawn in the morning may have an unpleasant taste or odour of the kind mentioned.
2. Decaying vegetation or algae may give rise to grassy, fishy, woody, 'pharmaceutical', fruity and cucumber type odours. Algae mostly cause offensive odours as they die and cell contents leach into the water.
3. Sulphate-reducing bacteria (*Desulphovibrio* and *Desulphotomaculum* in particular) give rise to the hydrogen sulphide (rotten egg) smell. Hydrogen sulphide is naturally present in some groundwaters in concentrations up to 10 mg/l. The taste and odour threshold for hydrogen sulphide depends on the olfactory response of the individual. Respective values in water are 0.06 mg/l and 0.01–0.001 μg/l; in air, the odour threshold is 0.003–0.03 ml/m^3 (Pomeroy, 1969). Iron bacteria (e.g. *Gallionella* and *Leptothrix*) often produce unpleasant tastes and odours with their death and decomposition and they also create an environment for sulphate-reducing bacteria to grow.
4. Iron above a concentration of about 0.3 mg/l imparts a bitter taste to a water.

5. Excessive chloride and sulphate impart a brackish taste to a water; taste thresholds for sodium and calcium chloride are 200 and 300 mg/l as Cl, respectively; those for sodium, calcium and magnesium sulphate are from 250, 250 and 400 mg/l as SO_4, respectively.
6. Industrial wastes are a prolific source of taste and odour of all kinds, of which those produced by phenols are the most frequently experienced. In the presence of free residual chlorine the phenols form a 'chlorophenol' medicinal taste which is quite pronounced; a quantity as small as 0.001 mg/l phenol may react with chlorine to form an objectionable taste. Methyl tertiary-butyl ether (MTBE), a petrol additive, has an odour threshold of 15 μg/l.
7. Chlorine does not, by itself, produce a pronounced taste except at high residuals, but many taste troubles occur due to reactions between chlorine and a number of organic substances. These tastes are usually described as 'chlorinous'. Reaction of chlorine with ammonia can produce tastes and odours; the taste threshold for monochloramine, dichloramine and nitrogen trichloride being 5, 0.8 and 0.02 mg/l, respectively (Black & Veatch, 2010). This is a common problem where waters containing free chlorine and combined chlorine are mixed in a distribution system (Section 11.8).

10.33 METHODS OF REMOVAL OF TASTES AND ODOURS

Tastes and odours of biological origin are geosmin and MIB. Those of industrial origin such as phenol are also invariably of organic nature. Oxidation and adsorption onto activated carbon are considered to be the most effective methods available for reduction of tastes and odours associated with organic compounds. Of the oxidants, ozone is very effective in destroying some of the taste and odour producing compounds. In theory, saturated hydrocarbon compounds, such as geosmin and MIB, which are responsible for musty and earthy tastes, and some of the chlorinated hydrocarbons are not oxidized by ozone. In practice however ozone is known to be effective in removing geosmin and MIB (Glaze, 1990; Chen, 1998) with up to 80% MIB removal using 4 mg/l. Biological removal in slow sand filters is reported to be >95% (Lundgren, 1988) and Nerenberg (2000) reported up to 65% removal in BAC following ozonation. Ozone-hydrogen peroxide (Section 10.36) can be more effective in reducing tastes and odour (Ferguson, 1990), including those due to geosmin and MIB (Duquet, 1989). UV-hydrogen peroxide (Section 10.36) is also effective against geosmin and MIB; 90% MIB removal is achieved with a peroxide dose of 8 mg/l but at a UV fluence of 1000 mJ/cm^2 (Rosenfeldt, 2005).

Other oxidants such as potassium permanganate and chlorine dioxide have been successfully used for taste and odour removal. Chlorine dioxide is useful when phenols are present as it does not form chlorophenols (Section 11.16).

Activated carbon is the most effective method of removing taste and odour compounds of organic nature. It can be used in either the powdered activated carbon (PAC) (10–40 μm) or GAC form. GAC is normally used when taste and odour removal is required continuously for long periods; for seasonal occurrence lasting several days at a time or dealing with pollution incidents it may be economical to use PAC. PAC dosages can vary between 25 and 50 mg/l with a maximum of 100 mg/l (Snoeyink, 1999). MIB removal with PAC varies from 40% at 25 mg/l to 90% at

100 mg/l. The dose is usually divided between raw water or inlet to clarifiers and downstream filters, with dosages to the filters maintained below about 5 mg/l. When applied to raw water the PAC dose may need to be increased to accommodate natural organic matter (NOM) which normally would be removed by coagulation. A Water Research Foundation funded study (2013) has shown that a moderate dose of superfine coconut shell PAC achieved meaningful reductions in concentrations of DBP precursors and NOM. Addition of PAC with the coagulant can reduce the adsorption efficiency because PAC is incorporated into the floc particles and organics must therefore diffuse through the floc (AwwaRF, 2000). When PAC is dosed after coagulation, organic removal is improved (Huang, 1995) because PAC adheres to the outer surface of the floc. Good mixing and sufficient contact time are the important design parameters. When dosed at the works inlet, a contact time of 15 minutes is sufficient. Sludge blanket and solids recirculation tanks provide adequate contact time; however, organics contained in sludge are known to utilize some of the adsorption capacity of PAC. For filter inlet dosing the water depth above the media provides the necessary contact time. PAC and oxidant addition (e.g. for manganese removal) should be separated by either the contact time required for adsorption or sufficient time to ensure that there is no oxidant residual at the PAC dosing point. PAC is stored as a powder in bags or in silos or as a slurry in tanks and dosed as a slurry (Section 12.22).

GAC is used in adsorbers of the rapid gravity or pressure filter type (Sections 9.3 and 9.6). It is sometimes used in place of sand as a medium in rapid gravity filters for turbidity removal (Section 9.9), but better overall performance is achieved when GAC adsorbers are used after rapid gravity filtration. For taste and odour removal only, an EBCT of 10 minutes is appropriate. pH values in the acidic range enhance the adsorption process. The life of a GAC bed between regenerations may be 2–3 years or about 10 years if ozone or ozone-hydrogen peroxide oxidation is used to enhance biological activity in GAC by oxidizing organic matter to biodegradable forms (Section 10.35).

Aeration can sometimes improve the palatability of water which is made poor due to stagnation, such as the bottom waters of reservoirs. It is also effective for taste and odour caused by compounds with high Henry's constants such as chlorinated solvents and some hydrocarbons. Taste complaints resulting from iron, etc. in water that has been left to stagnate in the ends of mains can be overcome by flushing.

Hydrogen Sulphide Removal

In natural waters hydrogen sulphide (H_2S) is in equilibrium with hydrosulphide (HS^-) and sulphide (S^{2-}); H_2S is predominant up to pH 7, between pH 7 and 13 more than 50% is HS^- and above pH 13, S^{2-} predominates. Aeration is effective in removing H_2S at pH <6 when over 90% is free H_2S; a packed tower aerator (Section 10.20) removes $>95\%$. Aeration removes CO_2 in preference to H_2S thus raising the pH. Acid dosing may therefore be necessary to maintain the pH <6.

H_2S can also be removed by chemical oxidation using chlorine, hydrogen peroxide or ozone. In the oxidation reactions, elemental sulphur is formed initially and then further oxidized to sulphuric acid. The latter reaction takes place at low pH. At pH values between 6 and 9, elemental sulphur reacts with residual sulphides, even in the presence of an oxidant, to form obnoxious polysulphides and a milky blue suspension of colloidal sulphur. Treatment consists of converting colloidal sulphur and polysulphides formed in the oxidation reaction to thiosulphate by the addition of sulphur dioxide (or sodium bisulphite) and then converting the thiosulphate formed to sulphate by chlorination (Black and Veatch, 2010; Monscwitz, 1974). Polysulphides are not formed at pH >9. Potassium permanganate oxidizes H_2S only to elemental sulphur; it is also known to introduce soluble manganese into the water. The stoichiometric quantities of oxidant required for the oxidation of 1 mg of H_2S to elemental sulphur are 2.1 mg of chlorine, 1 mg of hydrogen peroxide, 1.41 mg of ozone or 6.2 mg of potassium permanganate. Edwards (2011) reported that at pH <7.5, converting H_2S into elemental sulphur rather than sulphate using potassium permanganate in excess of the stoichiometric quantity resulted in a faster reaction, lower sludge production and complete removal of H_2S. The resulting particles of manganese dioxide (MnO_2) and elemental sulphur can be removed with a greensand filter (Section 10.12), which serves as a catalyst for oxidation and removal of hydrogen sulphide as well as any soluble manganese. Due to high cost, chemical oxidation is commonly used to remove the residual H_2S remaining in the water after the majority has been removed by aeration.

NATURAL ORGANIC MATTER AND MICROPOLLUTANTS REMOVAL

10.34 GENERAL

Natural organic matter (NOM) consists mostly of anionic polymeric compounds and can be either hydrophobic or hydrophilic. It is measured as total organic carbon (TOC), made up of particulate (POC) and dissolved (DOC) fractions. NOM includes humic and fulvic acids derived from vegetation and soils in the catchment, which are responsible for the colour typical of upland waters (Section 10.37). A small fraction of these compounds, known as DBP precursors, react with chlorine used for disinfection to form trihalomethanes (THMs) and haloacetic acids (HAAs) although, rarely, these can also occur naturally in the raw water. Other precursors include algal metabolic products and cell debris (Section 11.7).

Micropollutants are described in Chapter 7 and include pesticides, algal toxins (Sections 7.4 and 8.13), volatile organic compounds (VOC) such as chlorinated organic solvents, taste producing geosmin and MIB, polynuclear aromatic hydrocarbons (PAHs), polychlorinated biphenyls, endocrine disruptors (ECDs), 1,4-dioxane, N-nitrosodimethylamine (NDMA) and methyl tertiary-butyl ether (MTBE). Micropollutants occur in concentrations of the order of micro- or sub-microgram per litre and in most instances their removal is necessary to comply with standards set for drinking water quality.

Conventional chemical coagulation (followed by solid–liquid separation processes) removes NOM. In most instances a significant reduction in the concentration of DBP precursors is achieved by enhanced coagulation (Section 8.10). The removal of micropollutants by coagulation, however, is generally small. For effective removal of micropollutants, conventional treatment processes alone are usually insufficient and should be supplemented with advanced treatment processes including advanced oxidation processes (AOPs) (Kruithof, 2013) (Sections 10.35 and 10.36).

It is well known that activated carbon is effective in removing organic compounds from water. PAC is dosed in powdered form, normally at the inlet to the treatment works, permitting high doses that are then removed from the process with clarifier sludge. When dosed in this location any

NOM present in the raw water will compete with organic micropollutants, decreasing the adsorption capacity towards the micropollutant. PAC is frequently used to address seasonal occurrence of pesticides lasting several days at a time. The removal efficiency increases with dose and contact time. An allowance should be made in the PAC dose to raw water to satisfy demand from competing organic matter, particularly when elevated levels of NOM are present. Complete removal of several pesticides has been reported in direct filtration plants treating groundwaters using PAC doses of 5–10 mg/l supplemented by a small coagulant dose to minimize PAC breakthrough; PAC accumulated in the filter is then available for adsorption throughout the duration of the filter run (Haist-Gulde, 1996).

10.35 ADVANCED TREATMENT PROCESSES

Granular Activated Carbon (GAC) Adsorbers

PAC is costly and is discarded after one use whereas GAC, although about two to three times the cost of PAC, can be reactivated after exhaustion and reused. Since the introduction of more rigorous standards for drinking water quality the use of GAC has become the predominant process for the removal of organic matter including micropollutants. GAC adsorbers are commonly installed downstream of rapid gravity filters used for turbidity removal (Section 9.9).

Activated carbon can be made from wood, coal, coconut shells or peat. The material is first carbonized by heating and then is 'activated' by heating to a high temperature whilst providing it with oxygen in the form of a stream of air or steam. Sometimes chemical activation by phosphoric acid is used. It is then ground to a granular or powdered form. It is a relatively pure form of carbon with a fine capillary structure which gives it a very high surface area per unit of volume. The adsorption capacity of GAC is described by various parameters including Iodine Number and BET surface area (Table 10.3).

GAC adsorbers are of conventional rapid gravity or pressure filter design (Section 9.9) and the basic design parameters are the EBCT and bed depth or hydraulic loading ($m^3/h.m^2$). Bed depths up to 2.5 m for rapid gravity filters and 3 m for pressure filters are used (Section 9.9).

Table 10.3 Typical adsorptive capacity data for GAC

Grade	GAC size range (mm)	Effective size range (mm)	Iodine No.[a]	BET surface area (m^2/g)[b]
F200	0.425–1.70	0.6–0.7	850	900
F300	0.600–2.36	0.8–1.0	950	1000
F400	0.425–1.70	0.6–0.7	1050	1100
TL830	0.850–2.00	0.9–1.1	1050	1050

Source of information: Chemviron Carbon Ltd, UK.

Notes: [a]Iodine number: It indicates a GAC's ability to adsorb organic compounds and be regenerated. It should be greater than 500 mg/g of carbon (AWWA B604-12).

[b]BET surface area: It indicates the surface area available for adsorbates in water. Measured by N_2-BET method (Brunauer, 1938).

GAC characteristics vary according to the base material used. For example, the adsorptive capacity for the pesticide atrazine varies in the order wood > coconut shell > peat > coal (Paillard, 1990a). However, coal-based GAC finds wide use for most water treatment applications as it has a distribution of both mesopores (2–50 nm diameter) and micropores (up to 2 nm diameter), a structure suitable for medium to large (colour, taste and odour) and small organic molecules (micropollutants), respectively. Pilot plant work or Rapid Small Scale Column tests (RSSCT) should be used to optimize the GAC type and other design parameters such as adsorption capacity (by Freundlich adsorption isotherm) and to determine the life of carbon between reactivation (by RSSCT). EBCT varies for different micropollutants and is usually in the range 5–30 minutes; for pesticides it is 15–30 minutes and for DBPs and VOCs it is about 10 minutes.

Although GAC removes most micropollutants efficiently, the adsorption capacity towards some is low, so that frequent reactivation may be necessary which makes the adsorption process costly. For example, using an EBCT of 10–30 minutes most pesticides or DBPs may show breakthrough in 6–12 months and VOCs in 12–18 months. If only taste and odour removal is required, breakthrough normally occurs in 2–3 years when using an EBCT of about 10 minutes. Breakthrough of TOC generally occurs in about 3 months. In one UK works it has been shown that the TOC removal efficiency reduced from 90% to 10% in 14 weeks, but this did not have any adverse effect on the final water THM concentration which was significantly less than that before the installation of GAC (Smith, 1996). For TOC removal a rule-of-thumb used for estimating GAC life is 50 m^3 water treated per kg of GAC (Langlais, 1991).

Backwash requirements are similar to those used for GAC as a filter medium (Section 9.9). The frequency of backwashing of GAC adsorbers at groundwater sites can vary between once every 2 and once every 8 weeks depending on the raw water quality; wash is normally by water only. Frequent air scour tends to break down GAC and where air scour is necessary it should be limited to every 5–10 washes.

At surface water sites, in order to maintain low bacterial counts in the filtered water, the backwashing should include sequential application of air scour and water wash, with the wash frequency being every 2–3 days. This can also help to control the growth of microanimals (zooplanktons such as nematodes, and chironomid midge larvae) because the wash frequency is shorter than their reproductive cycle. The problem of microanimals can also be overcome by chlorinating the backwash water or taking a filter out of service for a period sufficient to produce anaerobic conditions in the filter, to kill the microanimals (Weeks, 2003). This should be carefully controlled so as not to lose biological activity in the filter and to prevent the formation of ammonia and nitrite in the filter. Alternatively, microanimals in the filtered water could be removed by the use of microstrainers (which would also remove any carbon fines). A combination of ultrasound and sand filtration was found to be successful in a demonstration plant (Matsumoto, 2002).

GAC should be reactivated as and when it is exhausted with respect to organic compounds. Reactivation can be achieved (either on or off-site) by heating to 800°C in steam or CO_2, or chemically. Up to 25% of the carbon may be lost in these processes. The loss in adsorption capacity after 1, 4 and 7 reactivations is about 5%, 10% and 20%, respectively (Marc, 1998). The addition of virgin carbon to make up for carbon lost after each reactivation helps restore the GAC to almost its original capacity. Typically the iodine number should not be allowed to fall below about 60% of its initial value (Table 11.3), as reactivation recovers only about 300 points.

When reactivation of the carbon is required it is usually removed from the adsorber by means of a water operated eductor (5 volumes of water to 1 volume of GAC) or by recessed impeller centrifugal pumps (3 volumes of water to 1 volume of GAC) operating at less than 1000 rpm. The removal is only about 90–95%. In some designs, adsorbers are provided with a sloping floor or recessed drain in the floor discharging to a collector system, which helps to improve GAC removal efficiency. All pipework, in particular bends, should be in stainless steel. Straight lengths could be in ABS or PVC-U. Bend radii should be 5–10 pipe diameters. The pipeline velocities should be maintained between 1.5 and 2.0 m/s. The same equipment and design parameters should be used for carbon placement in adsorbers.

Virgin GAC (coal-based, as used in water treatment) contains contaminants such as aluminium (0.65%), iron (0.35%), copper (0.0025%) as well as traces of manganese and arsenic, and reactivated GAC contains additional chemicals adsorbed in the process and not completely removed in the reactivation. Materials that could leach into the filtrate when GAC is placed in adsorbers include sulphides, sulphites and bisulphites (causing chlorine demand and odours), alkali (resulting in high pH), phosphates (if phosphoric acid is used in the activation process) and metals such as aluminium, iron, manganese and copper (Lambert, 2002). Repeated backwashing with water followed by running to waste of the filtrate should be carried out until tests confirm that water is of acceptable quality for supply. For GAC activated by phosphoric acid the phosphate content after reactivation should not exceed 1% w/w.

The impact of reactivated GAC on water quality can be minimized by pre-acid and post acid wash in the reactivation process. Pre-acid wash is useful for manganese, aluminium hydroxide and calcium carbonate and post acid wash is useful for manganese.

Biological Activated Carbon Reactors

Dissolved organic carbon (DOC) present in water is primarily in the form of refractory (i.e. poorly biodegradable) compounds with some biodegradable dissolved organic carbon (BDOC) or assimilable organic carbon (AOC). BDOC and AOC are complementary parameters used to characterize biodegradable organic matter but give different values, with AOC normally less than BDOC for the same sample. All organics are adsorbed onto GAC irrespective of their biodegradability but the presence of BDOC leads to some biological activity occurring after a few weeks, depending on water temperature. Although a high proportion of the organics is removed by adsorption, GAC life is ultimately determined by the refractory organic breakthrough.

Ozone, ozone-peroxide or UV-peroxide treatment of the water upstream of GAC adsorbers can, however, oxidize a large proportion of high molecular weight refractory organics into smaller more assimilable forms (BDOCs). This promotes biological activity in the adsorber, helping to remove organic matter which is otherwise unaffected or only marginally affected directly by oxidation. VOCs and some pesticides are examples of such compounds. In the adsorbers the GAC acts more as a biomass substrate and less as an adsorber; hence GAC beds used in this fashion are called biological activated carbon (BAC) reactors. In the biological reaction the BDOCs are converted to carbon dioxide, thus more of the adsorption capacity of GAC is available to deal with a smaller proportion of unoxidized refractory organics. The life of the GAC between reactivations is therefore increased manyfold and breakthrough occurs when GAC is exhausted with respect to refractory organics. For example, for pesticide removal an increase in running time to 1.5 years is reported (Graveland, 1996). DBPs once formed cannot be removed with ozone or BAC treatment.

BAC reactors are of similar design to GAC adsorbers and filters (Sections 9.9). Because of the biological matter present zooplankton (microanimals) can grow in BAC reactors and filtered water can exhibit high heterotrophic plate counts (HPCs) resulting from this biological activity. Good reactor management is therefore required, with efficient and frequent washing (Section 9.9). In some cases the filtered water may be devoid of oxygen due to oxygen consumption during biological activity, therefore oxygenation of the filtered water may be necessary (Graveland, 1996). For most applications an EBCT of 10–15 minutes is generally considered adequate. Increased EBCT helps to increase the service life of carbon between reactivations. DOC is more effectively removed in BAC reactors than in GAC adsorbers; following a very high initial removal it stabilizes at about 30–40% thereafter (Armenter, 1996; Bauer, 1996). A high proportion of the remaining DOC could be non-biodegradable refractory organic matter; therefore the use of ozone with GAC produces a biologically stable water thus reducing the risk of aftergrowth in the distribution system and minimizing DBP formation in subsequent chlorination for disinfection. Biologically stable water would have an AOC value in the range 50–100 μg/l for water carrying a chlorine residual or <10 μg/l for water without a chlorine residual (LeChevallier, 1993). Chlorination increases the AOC concentration in water in distribution, probably by oxidation of large organic molecules. Therefore, DOC should be removed by oxidation (e.g. ozone) processes followed by biological filtration (e.g. BAC).

While ozonation significantly reduces numbers of pathogenic microorganisms, the water from BAC reactors contains a higher level of non-pathogenic bacteria (HPC), thus requiring an effective disinfection stage such as chlorination for their removal.

Magnetic Ion Exchange Process

The MIEX® (magnetic ion exchange) process is used to remove DOC from water and is based on a conventional IX mechanism. The resin, which is proprietary and can only be purchased from a single supplier, contains a quaternary amine functional group and incorporates a magnetic component. The particle size is typically 150–180 μm. This is significantly smaller than conventional fixed bed IX resins and provides improved reaction kinetics. The surface of the resin is positively charged which allows adsorption of negatively charged organic material.

Process resin is added at a typical dose of 10–30 ml resin/l. In the original design the resin was contacted with the water in two or three continuously stirred tank reactors in series with a total retention time of about 10–30 minutes, followed by settlers where 85–95% of the settled resin was separated and returned to the contactors. The remaining resin was regenerated in a batch process and returned to the reaction tank along with make-up resin to replace any losses (<0.1%). The magnetic properties of the resin result in heavy agglomerates, which facilitate over 99.9% recovery in the settlement tanks operating at a surface loading rate of about 10 $m^3/h.m^2$. More recently, the functions of the reactor and the settler tanks have been combined into a single vessel, from which a portion of the resin is continuously removed by pumps to a settling hopper from which it is removed for regeneration. The resin is regenerated at pH 10 with sodium chloride at a concentration of 100–120 g/l NaCl and dosed at 360 g NaCl per litre of resin (Slunjski, 2000). Sodium bicarbonate can also be used for regeneration if required to reduce the salinity of the waste stream. Shortcomings of the process are that part of the resin is reused without regeneration, high resin concentrations are required in the contactor vessel and the resin only has a single source of supply. In addition, resin 'blinding' by NOM results in large regenerant volumes being required. Some of these shortcomings have been addressed by a high performance resin, MIEX®-Gold, which allows reduced inventories, improved DOC removal and reduced regenerant volumes.

The SIX® process (Suspended Ion Exchange) is a relatively new development of the MIEX® process, developed by PWN Technologies of Holland. In this process the water is dosed with resin at a concentration of 4–15 ml/l and admitted to two vertical contactors followed by a lamella settler to separate the resin from the treated water. The settled resin is then completely regenerated in brine solution and returned to the dosing tank. The SIX® process is claimed to be more compact, with a lower resin inventory, reduced salt usage, improved control of the adsorption process without 'blinding' the resin or biomass production, and improved effluent quality compared to the MIEX® process (Galjaard, 2011). Its major advantage over MIEX® is that the resin can be obtained from a range of suppliers.

MIEX® removes predominantly hydrophilic organic matter (Singer, 2007) whereas coagulation removes predominantly hydrophobic organic matter. Thus the combined operation of the MIEX® process and coagulation can result in increased overall removal of DOC and lower coagulant doses (Budd, 2005).

Disposal options for the saline waste stream produced in these processes are included in Chapter 13.

10.36 ADVANCED OXIDATION PROCESSES

Advanced oxidation processes (AOPs) involve the production and application of highly reactive free radical molecules; in water and wastewater treatment this is most commonly the hydroxyl free radical ($OH^{\bullet}$). Radicals contain unpaired valence electrons which react to oxidize a wide range of microcontaminants, making them useful in water treatment. One of the earliest methods of producing hydroxyl radicals in industrial waters was the use of Fenton's reagent, using ferrous iron (Fe^{2+}) to decompose hydrogen peroxide (H_2O_2) to form $OH^{\bullet}$ before being reduced back to ferrous iron to produce another radical and water.

$$Fe^{2+} + H_2O_2 \rightarrow Fe^{3+} + OH^{\bullet} + OH^{-}$$

$$Fe^{3+} + H_2O_2 \rightarrow Fe^{2+} + HO_2^{\bullet} + H^{+}$$

Hydroxyl free radicals are strong oxidants with an oxidation potential of 2.80 V compared to ozone (2.07 V) and chlorine (1.36 V), and can oxidize organics including those resistant to ozone such as some pesticides and VOCs. Methods commonly used to generate hydroxyl radical now include combinations of ozone and ultraviolet radiation with hydrogen peroxide. Ozone can react with hydrogen peroxide or, through a series of reactions involving hydroxide ions, form the hydroxyl radical, as indicated below.

Hydrogen peroxide induced radical formation:

$$H_2O_2 \Leftrightarrow HO_2^- + H^+$$

This dissociation of H_2O_2 is a function of the prevailing pH and may be assessed from its pK_a value where:

$$pK_a = 11.7$$

$$HO_2^- + O_3 \rightarrow HO_2^{\bullet} + O_3^-$$

Hydroxide ion induced radical formation:

$$OH^- + O_3 \rightarrow HO_2^{\bullet} + O_2^{\bullet -}$$

$$HO_2^{\bullet} = H^+ + O_2^-$$

$$O_2^- + O_3 \rightarrow O_2 + O_3^-$$

$$O_3^- + H^+ \rightarrow HO_3$$

$$HO_3 \rightarrow O_2 + OH^{\bullet}$$

$$OH^{\bullet} + O_3 \rightarrow HO_2^{\bullet} + O_2$$

Ozone can also absorb UV radiation and react with water molecules to form hydrogen peroxide which can then react to form hydroxyl radicals.

$$O_3 + hv + H_2O \rightarrow H_2O_2$$

In practice, the UV-O_3 process is a complex means of forming H_2O_2 in situ. A simpler way to form hydroxyl radicals is the direct photochemical reaction of hydrogen peroxide absorbing UV radiation. In this process, water is dosed with H_2O_2 followed by UV irradiation (Sections 11.23–11.25). The fluence (dose) required for this process is high (500–750 mJ/cm^2) compared to the UV fluence for *Cryptosporidium* inactivation (20–70 mJ/cm^2) (Ijpelaar, 2007). Key advantages of the UV-H_2O_2 process compared to ozone based oxidation are that it does not produce bromate, requires less space, and typically has a lower capital cost.

$$H_2O_2 + hv \rightarrow 2OH^{\bullet}$$

The hydrogen peroxide dose is typically in the range of 5–15 mg/l. Pilot work by Hofman (2004) showed that organic contaminants can be controlled by a UV dose of 540 mJ/cm^2 and a H_2O_2 dose of 6 mg/l. H_2O_2 is photoreactive over the wavelength range 185–400 nm; the highest hydroxide free radical yields are over the UV range 200–280 nm. Since the output of low pressure UV lamps is set at 254 nm with medium pressure UV lamps also having emission in that range, both are effective for generating hydroxyl radicals.

The use of AOPs involving the hydroxyl radical can increase oxidation of contaminants due to the radical's higher reactivity when compared to ozone or other conventional oxidants.

For example, this process is known to improve atrazine removal, from 15–40% using ozone alone, to 70–85% (Mouchet, 1991). Removal is better at low alkalinities and with pH in the range 7–7.5. The H_2O_2:ozone ratio used depends on the application and varies from less than 0.3:1 for controlling taste and odour compounds (Ferguson, 1990) to 0.5:1 for controlling pesticides and VOCs (Paillard, 1990b). The AOP process is best used downstream of coagulation and solid–liquid separation processes after a significant portion of the organic load has been removed. Like ozone, the 'peroxone' process converts refractory organic substances to readily biodegradable compounds which can be removed in BAC reactors. These reactors also remove the by-products of oxidation and any unreacted hydrogen peroxide which would otherwise dechlorinate the water when final residual chlorination is adopted. Although AOP can be used for the micropollutants listed in Section 10.34, certain micropollutants can be treated by UV photolysis alone. These include ECDs (Rosenfeldt, 2004) and NDMA (Lobo, 2007; Wong, 2000).

There are other processes that can be used to generate hydroxyl radicals. Much research has been conducted into the use of titanium dioxide as a catalyst in conjunction with UV radiation. In part due to the difficulties of feeding hydrogen peroxide, researchers are also looking at using UV and chlorine as a means of generating hydroxyl radicals. However, the hypochlorite ion is a strong radical scavenger. Therefore, the process is best used at a pH less than 6. Nevertheless, elimination of H_2O_2 dosing makes this an area of attractive research.

10.37 COLOUR REMOVAL

Natural colour is primarily due to the presence of humic substances classified as humic and fulvic acids. These are colloidal, hydrophilic, anionic polymeric compounds which are refractive (non-biodegradable). Fulvic acid is more soluble than humic acid. Both can be removed by coagulation; low coagulation pH values are considered to be most effective. Iron salts tend to be most effective at pH values of about 5–5.5 but aluminium salts are successfully used with lowland waters in the pH range 6.5–7.2 and upland waters in the pH range 5.5–6.2. Since colour is organic in nature it is susceptible to oxidation by chlorine, chlorine dioxide and ozone. Although chlorine is very effective at oxidizing colour, it is not recommended due to the formation of THMs (Rook, 1976) predominantly caused by reaction with fulvic acid. The ozone dose required to breakdown colour can be high and it is therefore beneficial to use it after chemical coagulation, sedimentation and filtration, the ozone being used to oxidize any residual colour. Pilot work has shown that up to 5.5 mg/l ozone with 15 minutes contact reduced colour from 70 to less than 20 Hazen and the dose varied proportional to the colour value.

Nanofiltration (NF) membranes of the spirally wound cellulose acetate type are used to remove colour in several plants in Scotland. It is reported that true colour removal from 100 Hazen to less than 1 Hazen has been achieved (Merry, 1995). This was accompanied by a reduction in iron (mostly organically complexed) from 1.35 to less than 0.01 mg/l as Fe. Membranes were operated in the cross-flow mode (Section 9.18) with recycle. Dual media filters followed by 5 μm cartridge filters were used in the pre-treatment. A high degree of fouling of the membranes was observed, caused by a combination of organic colour complexed with iron and biological films. Frequent cleaning by a weak acid and detergent and periodic washing with sodium hypochlorite were necessary, with membrane replacement every 3 years. The Loch Fyne process which utilizes tubular membranes also of cellulose acetate has been found to be effective with minimal pre-treatment. The use of mechanical foam-ball cleaning every 4–6 hours minimizes the use of chemicals and therefore allows the membranes to be operated with a high fouling rate. Performance was slightly inferior with respect to iron and manganese removal. These membrane designs are suitable for small capacity plants (Irvine, 2000).

GAC with a high proportion of mesopores and macropores, that is high Molasses Number (e.g. lignite-based GAC) adsorbs colour effectively.

CORROSION CAUSES AND PREVENTION

10.38 PHYSICAL AND ELECTROCHEMICAL CORROSION

High velocity water flow can cause deep pitting and erosion of surfaces due to a phenomenon known as cavitation. Section 15.12 describes how cavitation occurs and Sections 18.20 and 19.15 discuss its effects on valves and pumps, respectively. Cavitation erosion can be avoided by hydraulic design to avoid high velocity, low pressure flow impinging on surfaces where an increase in pressure causes collapse of any vapour pockets. The use of hard corrosion resistant materials can help resist cavitation damage but care must be taken with some stainless steels whose corrosion resisting chromium oxide layer is easily removed by cavitation (and erosion).

Electrochemical corrosion occurs where two metals of differing electropotential are immersed in a common body of water. All natural waters can act as an electrolyte, but the degree to which they do so depends on the dissolved salts present. The metal which loses ions to the electrolyte and therefore corrodes is termed the 'anode'; the other metal is the 'cathode' which gains ions. Of the metals commonly used in water supply systems, zinc is anodic to iron and steel, which are themselves anodic to copper. Thus, in zinc-galvanized pipes or tanks zinc provides sacrificial cathodic protection to steel, but it may corrode if fed by water which has passed through copper pipes. Corrosion of lead joints can occur in lead-soldered copper pipework, because lead is anodic to copper. Electrochemical corrosion can also occur because the pipe metal may change along its length, creating sections of differing metal which have an electropotential between them. Corrosion processes are complicated by additional chemical reactions that can occur. Thus anodic corrosion of iron piping can produce tubercles of rust, which alter the rate of corrosion and beneath which other forms of corrosion (e.g. anaerobic) can occur. Ions deposited on the cathode may reduce its cathodic effect, so the electrochemical reaction dies away; or the ions may be constantly removed by flowing water or combine with other substances in the water so the anodic corrosion continues or may even increase.

Crevice corrosion is a form of electrochemical corrosion where a corrosion cell is created in a narrow crevice between metal elements of the same material and grade when immersed in an electrolyte such as seawater. Examples include bolting and other close fitting components. The more common and cheaper grades of stainless steel do not provide sufficient resistance to crevice corrosion (or pitting) in such situations. Super-duplex or super-austenitic stainless steels with Pitting Resistance Equivalent Number (PREN) >40 provide adequate resistance to crevice corrosion, as well as to pitting, erosion corrosion and other phenomena, such as stress corrosion cracking, for components in contact with seawater or brine.

$$PREN = a + 3.3 \times (b + 0.5c) + 16 \times d$$

where a, b, c and d are the percentages by weight of chromium (Cr), molybdenum (Mo), tungsten (W) and nitrogen (N), respectively.

Unprotected metal pipes corrode externally when buried in wet soil which acts as an electrolyte. Corrosion can be increased if the pipes become anodic to other buried metallic structures in the soil, particularly if these structures are acting as 'earths' to electrical apparatus which introduces stray currents into the soil. Soil surveys can locate conditions likely to cause external corrosion which may include (Tiller, 1984):

- soils with a moisture content above 20%;
- acid soils with a pH of 4 or below;
- poorly aerated soils containing soluble sulphates (Section 10.39);
- soils with a resistivity less than 2000 ohm.cm.

Sleeving, wrapping and cathodic protection of pipelines are measures widely adopted to prevent external corrosion of steel and iron pipelines as described in Sections 17.15, 17.16 and 17.20.

Material compatibility should be carefully considered during the design stage of new water supply works. Such consideration should include the following specific aspects:

- aluminium in contact with concrete should be coated with epoxy or similar;
- aluminium should be protected from direct contact with incompatible metals;
- limiting concentrations of chloride for stainless steels are 200 mg/l for Grades 304 or 304L and 1000 mg/l for Grades 316 and 316L;
- for seawater super-duplex or super-austenitic stainless steels should be used;
- ventilation should be provided to minimize risk of condensation concentrating chlorine vapours from open tanks of chlorinated water (such as filters) causing corrosion of unpainted pipework (e.g. stainless).

10.39 BACTERIAL CORROSION

Sulphate-reducing bacteria (i.e. *Desulphovibrio desulphuricans*) existing in anaerobic conditions, can cause corrosion of iron. These bacteria are capable of living on a mineral diet and, as a result of their metabolism, they produce hydrogen sulphide which attacks iron and steel to form ferrous sulphide. Thus steel becomes pitted and cast iron becomes 'graphitized' (Cox, 1964), in which state it becomes soft. Sulphate-reducing bacteria may be most numerous in waterlogged clay soils in which oxygen is absent and where sulphur in the form of calcium sulphate is likely to occur. In such clays sulphate-reducing bacteria can produce one of the most virulent forms of attack on iron and steel. Backfilling pipe trenches with chalk, gravel or sand to prevent anaerobic conditions arising, in addition to cathodic protection and sleeving, are used to prevent this form of corrosion.

Iron bacteria (e.g. *Crenothrix*, *Leptothrix* and *Gallionella* types) may be present in a water which is deficient in oxygen (as evidenced by the presence of sulphates and hydrogen sulphide in the water) and may cause internal corrosion. These bacteria have the ability to absorb oxygen and then oxidize the iron in water or from iron and steel pipes and to store it. The bacteria are aerobic and, under favourable conditions, can form large deposits of slime which are objectionable, giving rise to odours and staining. The potential for growth of such bacteria is ever present in a water with a high iron content.

Tuberculation on the inside of an iron main is sometimes initiated by sulphate-reducing bacteria or, more commonly, by organic substances in the water, low pH or high oxygen content of the water. The external surface of a tubercle or nodule consists of a hard crust of ferric hydroxide, often strengthened by calcium carbonate. Below this crust conditions tend to be anaerobic so that sulphate-reducing bacteria can flourish creating further products of corrosion. Hence two remedies for the prevention of internal tuberculation are the aeration of a sulphate or hydrogen sulphide containing water, and lining the interior of the pipeline with an epoxy resin or with cement mortar.

10.40 CORROSION CAUSED BY ADVERSE WATER QUALITY

Desalinated water which is devoid of dissolved substances is highly corrosive to metals and in the presence of dissolved carbon dioxide (CO_2) is also highly aggressive to concrete and mortar (Section 10.49). Some waters have a tendency to corrode metals due to a high content of dissolved solids (e.g. chloride or sulphate). In some cases, free residual chlorine tends to cause corrosion due to its oxidation potential. The degree of corrosion depends largely on the acidity or CO_2 content of the water and the extent to which this is countered by the presence of calcium bicarbonate alkalinity in the water and a sufficiently high pH. Where bicarbonate alkalinity is present, an excess of dissolved CO_2 is necessary to prevent the decomposition of bicarbonate ions (HCO_3^-) back to carbonate ions (CO_3^{2-}) and subsequent precipitation of calcium carbonate by the following equation:

$$CO_3^{2-} + CO_2 + H_2O \leftrightarrow 2HCO_3^-$$

The relationship between alkalinity and carbon dioxide at equilibrium is given in Table 10.4, which is based on a graph by Cox (Miles, 1948).

Table 10.4 Relationship between alkalinity, equilibrium pH and dissolved carbon dioxide

Alkalinity as $CaCO_3$ (mg/l)	25	50	75	100	125	150	175	200	250	300
pH at equilibrium, i.e. pH_s	8.8	8.1	7.7	7.6	7.5	7.4	7.35	7.3	7.2	7.0
Free CO_2 at equilibrium (mg/l)	0.0	1	2.5	4.5	7.5	12	18.5	27	32	60

When the concentration of carbon dioxide exceeds the requirement to maintain the equilibrium between carbonate and bicarbonate ions the water tends to dissolve any coating containing carbonates (such as concrete or mortar) with which it is in contact. The excess of CO_2 over that required for equilibrium is termed *aggressive* carbon dioxide. For example, from Table 10.4, a water having an alkalinity of 125 mg/l as $CaCO_3$ at pH 7.5 is aggressive if the free CO_2 concentration exceeds 7.5 mg/l. Conversely, a water with less free carbon dioxide deposits some of the bicarbonate as calcium carbonate. Where the water is only slightly over-saturated this deposition occurs slowly with the calcium carbonate often forming a protective coating on the surface of metals. If however, the water is significantly over-saturated the precipitation occurs rapidly in the amorphous soft form and a protective coating is unlikely to be formed.

A useful method of determining whether or not water is in equilibrium with calcium carbonate is the 'chalk test'. This requires taking a stoppered glass bottle to which some powdered chalk is added, filling the bottle carefully and completely with the water of interest, and replacing the stopper so as to exclude all air. After some 12 hours or more the final pH should be measured. This pH value, denoted as pH_s, is termed the 'saturation pH value' of the water. Alternatively, the pH_s of a water could be calculated from:

$$pH_s = (pK_2 - pK_s) - (\log_{10}[Ca^{2+}] + \log_{10}[2Alk])$$

where, terms in [] are concentrations expressed in g moles/l and values for the term $pK_2 - pK_s$ as a function of temperature and TDS are given in Table 10.5.

Table 10.5 Relationship between pK_2–pK_s, total dissolved solids and temperature

	pK_2–pK_s at TDS (mg/l)										
Temp (°C)	**20**	**40**	**80**	**120**	**160**	**200**	**240**	**280**	**320**	**360**	**400**
0	2.45	2.58	2.62	2.66	2.68	2.71	2.74	2.76	2.78	2.79	2.81
10	2.23	2.36	2.40	2.44	2.46	2.49	2.52	2.54	2.56	2.57	2.59
20	2.02	2.15	2.19	2.23	2.25	2.28	2.31	2.33	2.35	2.36	2.38
30	1.86	1.99	2.03	2.07	2.09	2.12	2.15	2.17	2.19	2.20	2.22

Source of information: Water Works Engineering, Qasim, S. R. et al., Prentice-Hall, New Jersey, USA, 2000.

The numerical difference between the pH of the water and its pH_s is known as Langelier Saturation Index (LSI):

$$LSI = pH - pH_s$$

When the LSI is negative, the water is liable to be aggressive and if LSI is positive the water tends to deposit calcium carbonate. However, Langelier Index does not quantify the degree of under- or over-saturation of water. This can be measured in the 'chalk test' by determining the total alkalinity prior to the test and again, immediately afterwards. The difference between the two measurements is termed the Calcium Carbonate Precipitation Potential or 'CCPP'.

Values of CCPP between +3 and +10 mg/l suggest that the water is capable of precipitating a protective coating on the surface of metals. Values much in excess of 10 mg/l indicate that protective coatings are unlikely to be formed due to the nature of the precipitate formed. If the CCPP is negative, the water is aggressive and will attack concrete and mortar, particularly when the CCPP is lower than −10 mg/l (as $CaCO_3$). Consequently, treatment for an aggressive water consists of chemical addition, sufficient to produce a thin uniform protective coating of calcium carbonate. The treatment must be continuous and usually consists of dosing an alkali such as hydrated lime to ensure that the pH of the treated water is maintained only slightly above the pH_s value (i.e. 0.2–0.3), the resulting deposition being in the form of a smooth hard scale. Table 10.4 is a useful practical guide for determining the saturation pH (pH_s) of a water.

The concentration of both chloride and sulphate ions in the water can have a significant effect on the corrosive nature of the water. Both the rate of corrosion and the dissolution of iron increase markedly as the concentration of chloride and/or sulphate increases. Work by Larson and Skold (Larson, 1957) has suggested the use of the following ratio between chloride plus sulphate and the alkalinity (Larson ratio, *LR*) where [] is the concentration expressed in g moles/l:

$$LR = ([Cl^-] + 2[SO_4^{2-}]) \div [HCO_3^-]$$

Water with a $LR > 0.2–0.3$ is considered corrosive.

10.41 CORROSIVENESS OF VARIOUS WATERS

There are a wide variety of naturally occurring waters with differing characteristics which determine their aggressiveness to metals and concrete. The characteristics of the principal types are discussed below with observations on protection of materials.

Hard surface waters, with moderately high alkalinity. A large proportion of lowland river waters associated with sedimentary geological formations have a positive Langelier Index. Generally, therefore, such waters are satisfactory for use in contact with concrete and mortar and for the avoidance of general corrosion of metals, and are not usually plumbosolvent (Section 10.14).

Hard groundwaters with high alkalinity. Although the mineral composition of many groundwaters from chalk, limestone and other sedimentary formations may often be similar to that of many surface waters, the carbon dioxide content of the water is usually much higher. The excess carbon dioxide should be removed, usually by aeration, to give a slightly positive Langelier Index. Although water with a positive Index should, theoretically, deposit calcium carbonate, in practice it will probably not do so to any appreciable extent in water mains or metal tanks. Its anti-corrosion effect, however, probably depends in part upon the rapid deposition of calcium carbonate on cathodic areas where electrochemical corrosion is initiated, the effect being to stifle this type of incipient localized corrosion. Groundwater with moderately high alkalinity due to calcium bicarbonate are plumbosolvent (Section 10.14).

Groundwaters with low alkalinity but high free carbon dioxide. Many small sources from some of the older geological formations and some from gravel wells and springs comprise waters of this type. Many of the gravel sources can have a high non-carbonate hardness. Aeration can be adopted to reduce the carbon dioxide content and lime can be added to achieve a further reduction. If the water is already hard it may be preferable to neutralize the free carbon dioxide by adding caustic soda. For the correction of the corrosive characteristics of very soft waters, passage through a filter containing granular limestone ($CaCO_3$) may be useful.

Soft waters from surface sources. Certain lake waters are very pure, with a bacteriological quality approaching that of a groundwater. These waters do not, however, contain any appreciable quantity of carbon dioxide. They are therefore almost neutral in reaction and many may also have a very low alkalinity, for example 10 mg/l as $CaCO_3$. From the point of view of corrosion protection and minimizing plumbosolvency, however, they need careful treatment (Section 10.14). In some instances an economical and satisfactory form of treatment is simply to add a small dose of lime. In cases where a greater amount of alkalinity is required, dosing with carbon dioxide followed by lime is adequate. Alternatively, use of calcium chloride or sulphate with soda ash or sodium bicarbonate may be convenient.

The majority of surface waters derived from upland catchments however, are moderately to highly coloured by peaty matter, which contains organic humic and fulvic acids that render the water acidic in reaction with a pH of 6 or lower. To counter the general corrosiveness of this type of water and its plumbosolvent action, the organic acids must be removed (Turner, 1961). This is usually achieved by chemical coagulation followed by filtration. Aluminium or iron coagulants, being acidic, destroy alkalinity: the carbonate or bicarbonate is converted to free carbon dioxide (Section 8.21). Hence lime should be added after filtration and the carbon dioxide is then largely converted back to calcium bicarbonate.

A soft water (low alkalinity) will be aggressive to any calcium bearing material including concrete. It is therefore critical that concrete should be fully and properly compacted in the construction process. The surface of a smooth concrete surface will, after a few years of attack by low alkalinity water, take on a rough 'sandpaper-like' feel. This is due to the water attacking the cement paste on the surface. If this attack then exposes honeycombing resulting from poor construction, water can then readily penetrate further into the structure where it can continue to leach out cementitious material, causing significant weakening.

For the protection of concrete surfaces, benefits are obtained from use of proprietary shuttering designed to draw excess water to the surface during curing, resulting in a denser outer layer with improved resistance to water penetration. Additives can also be used for such situations. These are generally designed to provide improved compaction of the concrete as they provide lubrication of the cement, sand and aggregate particles. They do not, of themselves, confer any protection against low alkalinity water attack. For the most severe attack, barrier systems are available, for application to the surfaces of concrete. Among the best of these are glass-flake coatings of approximately 1–1.5 mm thick. Such barrier systems require a good concrete substrate to which they can bond. Protection of concrete and its reinforcement from attack can therefore be achieved by a combination of measures affecting the concrete mix and its placement, supplemented, where necessary, by application of additional protective measures such as coatings.

10.42 DEZINCIFICATION

Galvanizing on iron and steel is susceptible to corrosion by water containing excess free carbon dioxide, which dissolves zinc, and by any water in which the pH is excessively high. The term 'dezincification', however, usually refers to the effect on brasses (alloys of copper and zinc), when the zinc is dissolved. A particular form of this type of corrosion is called 'meringue' dezincification because of the bulky white layer of corrosion product which appears, the effect of which is to cause failure of fittings, mainly hot water fittings when they are constructed of hot-pressed brass. The effect, investigated by Turner (1961) was found to occur with waters having a pH >8.2 and a chloride to alkalinity ratio greater than those shown in Table 10.6.

Subsequent investigations (AwwaRF, 1996) showed that this relationship could be applied to dezincification in general, although at pH <7.6 this type of corrosion did not occur. These findings reinforce the need for the addition of alkalinity (lime and CO_2 dosing) to very soft waters to reduce their aggressiveness to metals. If a water contains an appreciable amount of chloride, then simply raising its pH without increasing its alkalinity may increase its tendency to cause dezincification and may also fail to control its corrosiveness towards other metals.

Table 10.6 Limiting values of chloride-to-alkalinity ratio (at pH values greater than 8.2) for dezincification

Chloride (mg/l as Cl)	10	15	20	30	40	60	100
Alkalinity (mg/l as $CaCO_3$)	10	15	35	90	120	150	180
Chloride:alkalinity ratio	1:1	1:1	1:1.75	1:3	1:3	1:1.5	1:1.8

DESALINATION

10.43 INTRODUCTION

Desalination is the term that describes processes used to reduce the concentration of dissolved solids in water, which are usually referred to as total dissolved solids (TDS) and expressed in mg/l. Sometimes electrical conductivity, which is expressed as microSiemens per cm (μS/cm), is used as a surrogate measurement of TDS because it is easier to measure. While the exact ratio varies depending on the concentrations of specific constituents, TDS can be approximated by multiplying the conductivity by 0.66. Natural waters may be classified in broad terms as summarized in Table 10.7.

The need for desalination is increasing due to population growth and limited new sources of fresh, low TDS, water. According to the findings in its 2016 survey, Black & Veatch notes that water utilities are increasingly looking at alternative sources of water such as reuse and desalination (B&V, 2016). Cumulative desalination capacity increased from $5 \times 10^6\ m^3/d$ in 1980 to $15 \times 10^6\ m^3/d$ in1990, $28 \times 10^6\ m^3/d$ in 2000 and $56 \times 10^6\ m^3/d$ in 2010 (estimated) (Veerapaneni, 2011). Several methods have been commercially developed for the desalination of high TDS waters. Selection of the correct process requires evaluation of process efficiency and costs, capital and operating. As a general guide, the most frequent application of desalination processes based on TDS concentration is given in Table 10.8.

Table 10.7 Classification of waters according to TDS

Type of water	TDS value (mg/l)
Sweet water	0–1000
Brackish water	1000–5000
Moderately saline water	5000–10 000
Severely saline water	10 000–25 000
Seawater	Above 25 000

Table 10.8 Classification of desalination processes according to TDS

Process	Approximate feed TDS value (mg/l)
Ion exchange (IX)	Up to 100
Electrodialysis (ED)	Up to 1000
Nanofiltration (NF)	Up to 1000 (but depends on divalent/monovalent ratio)
Low-pressure reverse osmosis (RO)	Up to 10 000
High-pressure RO	10 000–40 000
Distillation	Above 30 000

RO is the preferred method up to at least 10 000 mg/l TDS because it is significantly more energy efficient than any of the distillation methods at these concentrations. For higher salinity applications, such as seawater desalination, the choice between RO and distillation depends on site-specific issues. In the past, larger facilities utilized distillation, but that has changed. The installed capacity and percentage of desalination plants applying RO has increased due to improvements in pre-treatment effectiveness, energy efficiency and economy of scale. Examples of these improvements include improved granular media filtration or MF/UF membrane filtration, more permeable (lower energy) RO membranes, and more efficient and reliable energy recovery devices (ERDs).

10.44 ION EXCHANGE

The use of IX processes for reducing mineral or saline constituents in water is discussed in Section 10.10. IX rarely finds application for waters containing more than 1500 mg/l of dissolved solids. Even at lower concentrations, it is usually more cost effective to apply RO as pre-treatment to IX to produce demineralized industrial process water.

10.45 ELECTRODIALYSIS

In electrodialysis (ED), a DC electrical potential is applied between electrodes and the ionic constituents in the water are thus caused to migrate through semi-permeable membranes which are selective to cations and anions. Desalination occurs in a series of cells separated by alternate pairs of cation and anion membranes (Fig. 10.4). A number of membrane pairs can be installed in parallel between a pair of electrodes to form a stack and several stacks can be connected hydraulically together either in parallel to increase output or in series to increase salt removal. Each electrodialysis pass removes about half of the remaining ionic material; therefore, three and four pass systems are typical for brackish water applications. ED is not used for seawater desalination. The reduction in dissolved solids obtained by ED is directly related to the electrical energy input and therefore to the cost of electricity. ED membranes, like RO membranes, are subject to fouling and require suitable pre-treatment.

In the ED process, the current flow is uni-directional and the desalinated and concentrate compartments remain unchanged. In electrodialysis reversal (EDR), the electrical polarity is reversed periodically which results in a reversal in direction of ion movement and provides 'electrical flushing' of scale forming ions and fouling matter. This self-cleaning can allow operation at higher levels of supersaturation of sparingly soluble salts, thereby achieving higher water recoveries.

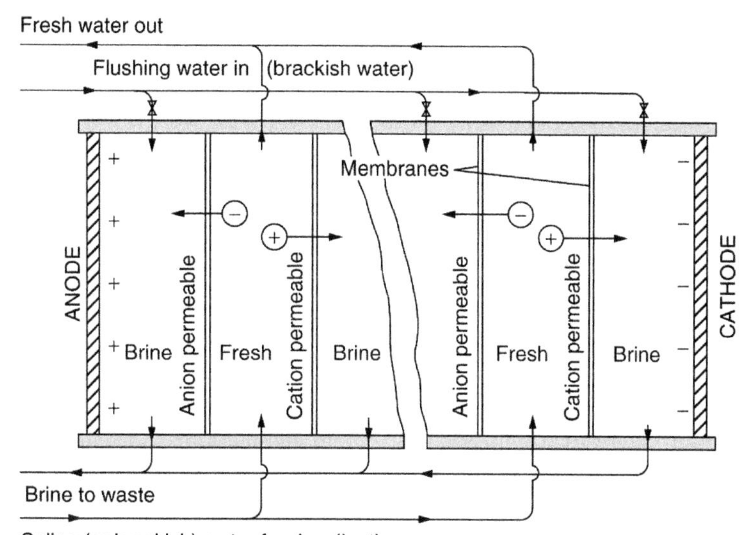

A d.c. voltage is applied across a battery of cells. The anions in the saline (or brackish) water, induced by the e.m.f. applied, pass through the anion-permeable membrane into the brine cells, but cannot pass the cation-permeable membrane. Similarly the cations are also collected in the concentrate or brine compartments which discharge to waste. The saline (or brackish) water in alternate chambers loses ions and reduces in salinity. A proportion of the saline feed is directed to the concentrate (brine) compartments to flush out the brine.

FIGURE 10.4

Principle of electrodialysis plant.

When comparing to RO it is important to note that ED/EDR product water is not filtered. The ions being removed pass through the ED/EDR membrane whilst the water which becomes product flows tangentially along the membrane surface. Only charged material is removed; uncharged material, such as dissolved silica, organic material, colour, viruses, bacteria, protozoan cysts and other microbes are not removed by ED/EDR.

A newer type of ED has been developed called electrodialysis metathesis (EDM), which can be used in concert with RO to concentrate the waste solution further, lowering the waste flow rate (Bond, 2015). EDM achieves higher concentrations without precipitation by placing the removed cations and anions into different channels. Ideally, EDM can be applied as part of a zero liquid discharge approach to allow use of RO in locations that do not have lower cost options for disposal of concentrate (Bond, 2007, 2011).

10.46 REVERSE OSMOSIS AND NANOFILTRATION

In osmosis, a salt solution is separated from pure water by a semi-permeable membrane and there is a net flow of water across the membrane to the higher concentration side until equilibrium is achieved. This equilibrium occurs when the pressure on the pure water side is offset by the osmotic pressure of the salt solution. In reverse osmosis (RO), the osmotic pressure is overcome and reversed. Applying pressure, greater than the transmembrane osmotic pressure, to the higher concentration solution results in a net flow of water through the membrane into the lower concentration side. Osmotic pressure is a physical property that is a function of concentration and type of solute. As an approximation, there is about 0.7 bar of osmotic pressure for every 1000 mg/l of TDS. An operating pressure much higher than the average transmembrane osmotic pressure is required to achieve an economically feasible flow (by a factor of about 2 for seawater and more than 10 for brackish water).

Nanofiltration (NF) is a type of RO. Both are pressure driven, cross-flow filtration processes that utilize semi-permeable membranes to remove dissolved solutes from water. These membranes also remove particulate material but that is a misapplication, because even small (i.e., less than 1 μm) particles foul the membrane yielding resistance to flow and lowering the permeability. Therefore, RO/NF processes require high quality pre-treatment, especially for surface water applications.

On visual examination, NF and RO membranes and their system components look identical. The main difference is that NF exhibits greater selectivity in rejecting dissolved material. NF provides lower rejection of monovalent ions, such as sodium and chloride (typically 10–40% depending on the membrane and operating conditions), while providing higher rejection of multivalent ions, such as calcium and sulphate (typically 80–85%), compared to about 95–99% rejection of most ionic material by RO. Both NF and RO provide greater than 98% rejection of dissolved organics with molecular weight greater than 200 Daltons and both provide some rejection of smaller organic molecules. (Dalton is a unit of measurement defined to be one twelfth of the mass of one atom of carbon.) AOC removal of over 97% has been reported for NF, the efficiency depending on pH and the molecular weight cut-off of the membranes (Hofman, 2004). Neither NF nor RO remove dissolved gasses, such as radon, carbon dioxide (CO_2) or hydrogen sulphide (H_2S). For some ions, such as ammonium ion (NH_4^+), rejection is a function of pH. Ammonium, and hence ammonia, is rejected better at pH higher than 8. With pore size no larger than about 1 nm

(0.001 μm), both NF and RO reject microbial material including bacteria and viruses. NF applications include softening (Section 10.2), colour removal (Section 10.37), control of DOC and removal of DBP precursors. The range of RO applications include these, as well as desalination of brackish and seawaters and removal of specific contaminants (such as nitrate, fluoride and radium). An advantage of both RO and NF is the ability to remove multiple contaminants in a single treatment step. Many communities using RO/NF benefit from softening while also receiving potable water treated for regulated parameters, such as colour, DBP precursors and nitrate.

NF should not be viewed as a less effective version of RO. NF rejection is more selective than RO, which can be an advantage in cases where the goal is partial removal of organics or hardness but where the concentrations of TDS, sodium and chloride in the source water do not require treatment. NF is sometimes chosen instead of RO because of the perception that NF consumes less electricity; however, with the development of modern, ultra low pressure RO membranes, the difference in operating pressures between these membrane types has narrowed. In cases where either type could be used, a detailed capital and operating cost comparison is needed to determine the most economic option. The comparison should include operating costs for electricity, cleaning chemicals and an allowance for future membrane replacement. Due to the higher rejection, RO tends to require less membrane area than NF, which in turn lowers initial capital cost and the membrane replacement budget. This may offset the lower operating pressure of NF that may occur in some cases.

When used to desalinate seawater, RO feed pressure is typically 55–70 bar range. On typical low salinity applications, RO pressure is 7–20 bar while NF is 6–14 bar. Higher operating pressure, and therefore higher energy consumption, is needed if the temperature is decreased or if any of the following are increased: flux, feed concentration, recovery or fouling.

All desalination processes consume energy. The total energy consumption of thermal desalination ranges from 5.2 kWh/m^3 of desalinated water to more than 12 kWh/m^3 while the consumption of an entire seawater RO (SWRO) facility, operating at 50% recovery, is in the range 3.2–4.5 kWh/m^3 (Veerapaneni, 2011). Considering the RO step alone, for a seawater with a feed TDS of 35 000 mg/l and a recovery of 50%, actual facilities can operate at about 2.5 kWh/m^3, which represents a thermodynamic efficiency of about 42% when compared to the theoretical energy requirement of 1.05 kWh/m^3 (Mistry, 2011). Energy use for thermal methods is relatively independent of feed water TDS, while for RO it is a function of TDS, temperature, recovery and type of ERD used.

Almost all modern SWRO facilities apply ERDs and increasingly RO plants treating lower TDS feeds also employ ERDs (Plate 18(c)). These first became popular with SWRO owing to the potential to recover significant energy from the concentrate stream. For example, at 50% recovery, SWRO has a concentrate stream representing half the feed flow rate at a pressure that is only slightly less than the feed pressure, allowing between 40% and almost 50% of the energy to be recovered (Moch, 2002). With a typical feed pressure of 65 bar, the concentrate pressure upstream of the concentrate control valve is greater than 60 bar.

Overall ERDs can be classified as either centrifugal or positive displacement devices (Hauge, 1999). Centrifugal devices are rotating devices in which the pressurized concentrate stream impacts a surface, causing rotation of a shaft, whose energy is linked to the feed pump's shaft. Centrifugal ERDs include Francis turbines, Pelton wheels and hydraulic turbochargers. Francis turbines are the oldest type and are essentially reverse running pumps, but due to disadvantages of high maintenance, vibration and cavitation issues, are not widely used today for such high pressure applications. A Pelton turbine is essentially a wheel with cups on the perimeter that turns under the impact of high pressure water from one or more adjustable nozzles. Pelton wheels are fairly widely used but their limitations include lower efficiency (about 85–88%) than the more recently developed turbochargers and positive displacement devices (90 to >95%) as well as the fact that concentrate leaving a Pelton wheel is at atmospheric pressure and so must exit by gravity or be repumped. A turbine may be direct coupled to a pump so that the energy recovered can be transferred to a portion of the membrane feed flow, or may drive a generator whose energy is used to power pumps indirectly. A turbocharger is an example of a direct coupled ERD system (Plate 18 (b)). Positive displacement ERDs, including work exchangers and pressure exchangers (Plate 18 (a)), directly transfer the hydraulic energy of the concentrate to the feed stream. While they exhibit high energy efficiency, and are being widely applied, positive displacement ERDs allow some leakage between the feed and concentrate streams that flow through the device; this results in an increase in feed TDS.

The most commonly used RO/NF membrane configuration for potable water production is the spiral wound type as is illustrated in Figure 10.5.

The hollow fibre configuration was used in the 1970s and 1980s but is rarely used today, although there is at least one manufacturer still using this design. With spiral wound elements, layers of membrane formed on flat sheets are wrapped around a central product water tube between layers of permeate and feed-concentrate spacer material; this forms a cylindrical element that is

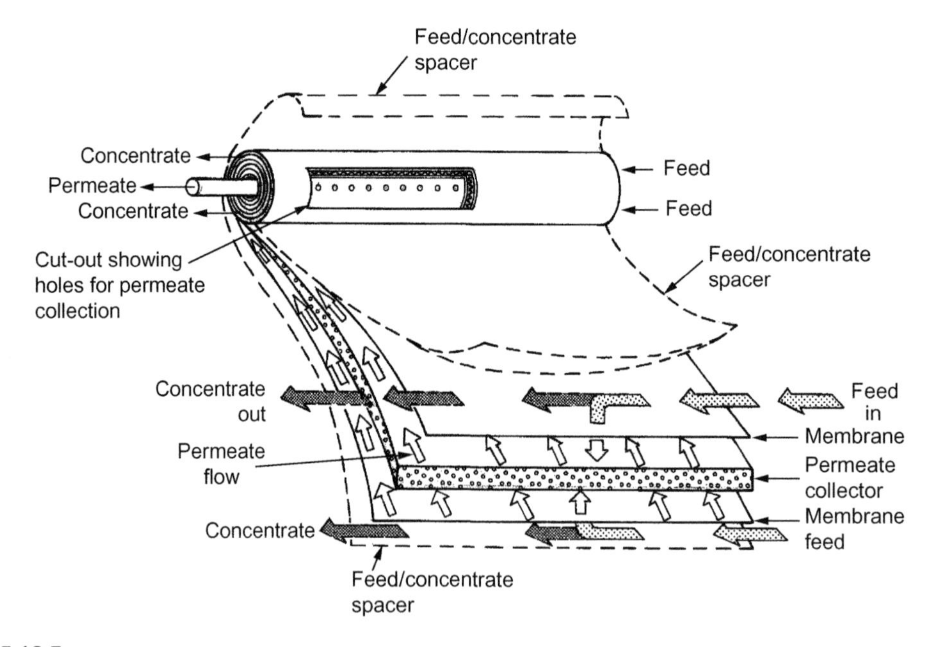

FIGURE 10.5

Spiral wound reverse osmosis membrane.

mounted inside a pressure vessel. As feed water flows from one end of the cylinder to the other, it flows along the membrane surface and a portion passes through the membrane to the permeate side and then to the product water tube. Typically, large systems generally have six or seven 40-inch long (or sometimes four 60-inch long) elements in series inside each pressure vessel. Essentially all RO manufacturers present the length of spiral elements in inches, typically either 40 inches (1.016 m) or 60 inches (1.524 m). Standard diameters are 4 inch (0.102 m) and 8 inch (0.203 m), but 16 inch (0.406 m) and 18 inch (0.457 m) are also available.

Almost all municipal RO applications employ polyamide membrane chemistry, rather than the older style cellulose acetate derivatives. The main reasons are that polyamide membranes are productive at much lower operating pressures, provide higher salt rejection, do not require the same level of pre-acidification and have longer service life. Cellulose acetate membranes can tolerate continuous exposure to residual chlorine up to about 1.0 mg/l but essentially require a chlorine residual to control biological degradation; they are sensitive to pH outside the range 5.5–6 and water temperatures >30°C. Polyamide membranes have no tolerance towards chlorine or other oxidants but are not degraded by bacteria. Polyamide spiral wound elements have a wide pH operating range (3–10 for most types and up to 1–12 for short-term cleaning exposure for some types) and temperature range (up to 45°C); however, operating at the extremes does shorten service life.

The main controlled parameters in an RO/NF system are feed pressure and recovery, which is the ratio of treated water ('permeate' or 'product') flow to feed flow. This is typically accomplished by an automated control system that adjusts the feed pump to yield a permeate flow set point, and a modulating concentrate control valve to yield a recovery set point. Generally, the feed pump is adjusted by a variable speed drive (VSD), also known as an adjustable or variable frequency drive (AFD or VFD). Some older systems utilize a modulating control valve on the pump discharge, rather than a VSD; however, this increases energy consumption. Since the permeate and recovery controls influence each other, programming tolerances and response times should be selected to avoid cycling or 'hunting'. Lower TDS applications also have a blending control loop to bypass a portion of the feed to the finished water. The blend ratio is usually based on a flow ratio or an online conductivity measurement, which can indirectly indicate parameters such as TDS or hardness based on local experience.

The purified stream leaving an RO/NF plant has passed through the membrane; therefore, it contains essentially no particulate material and only a small portion of the original dissolved material. The material rejected by the membrane is concentrated into a stream called reject or concentrate (sometimes also referred to as brine or retentate). RO/NF plants are arranged in stages and passes. While sometimes these terms are used interchangeably, with precise usage the meanings differ. Within a multi-stage train the concentrate from one stage flows via internal piping to feed the next stage. Stages are used to increase recovery. For example, a seawater plant, which would typically operate at 45–50% recovery, would employ a single stage. That stage would consist of multiple vessels in parallel which would each house six or seven elements per vessel (assuming 40-inch long elements). For a low TDS (e.g. NF or brackish water RO) plant operating at 75–85% recovery, two stages with six or seven elements per vessel would typically be used; for 85% and greater, three stages are generally used. There are fewer vessels in each subsequent stage to maintain sufficient velocity across the membrane surface to achieve good mixing and hence prevent localized areas of high concentration, which in turn can result in inorganic scaling on the membrane. In a two-stage system about 67% of the vessels are in the first stage; in a three-stage system about 57% of the vessels are in the first stage and about 28% in the second stage.

More than one pass is employed when additional RO or NF treatment is needed to meet finished water quality requirements. In a two-pass plant, permeate from the first pass is repressurized as feed to a downstream RO/NF system (Plate 17(a)). In this case there are two RO/NF plants in series. Sometimes multiple passes are used to produce ultra-pure water for industrial facilities. Sometimes a partial or full second pass is applied to seawater desalination to meet boron or other special limits, such as chloride or bromide. Long Beach, California has considered two passes of NF for seawater desalination (Harrison, 2005; Leung, 2005).

10.47 RO AND NF PLANT DESIGN

While detailed calculations are frequently conducted using computer modelling programs, an understanding of basic equations can assist designers as well as operators. A preliminary design can be conducted after selecting recovery and flux. Recovery and vessel arrangement are discussed in Section 10.46. RO and NF concentrate dissolved material and therefore achievable water recovery is a function of the presence of sparingly soluble salts, such as calcium carbonate, calcium sulphate, barium sulphate and silica. When the products of their molar concentrations on the feed-concentrate side exceed the related solubility product, precipitation occurs, causing scale to form on the membrane surface which lowers permeability. Solubility product (K_{sp}) is the product of the equilibrium molar concentrations of the ions (moles/l) of a sparingly soluble salt in a saturated solution in water. For $CaCO_3$ it is defined as:

$$K_{sp} = [Ca^{2+}] \times [CO_3^{2-}] = 3.8 \times 10^{-9} \text{ at } 25^{\circ}\text{C}$$

If identified early enough, most scales can be removed with chemical cleaning, although silica scale can be irreversible. However, it is better to avoid scaling and the attendant higher operating costs and risk of irreversible damage. Therefore, precipitation should be prevented by limiting the recovery to the maximum allowable. For brackish RO and NF an antiscalant is applied to optimize the recovery. For seawater RO/NF, recovery is limited by the maximum allowable operating pressure, which generally allows 45–55% recovery with only sulphuric acid addition to control calcium carbonate scaling. For NF and brackish RO applications, commercially available antiscalants, frequently based on polyphosphates and chelating agents, are added to the feed to allow higher recovery by slowing the reactions sufficiently that the concentrated stream leaves the system before precipitation forms. While detailed calculations are needed to determine maximum allowable recovery for a specific case, frequently 75–85% is achievable when a modern antiscalant and sulphuric acid are added to brackish RO/NF feed water. In general, antiscalant is typically dosed at 3–5 mg/l and feed pH is adjusted to yield an LSI of less than 2 in the concentrate, depending on the type of antiscalant and other conditions.

Setting the correct maximum flux is critical for avoiding excessive fouling or decreased plant capacity. While it is impossible to be definitive in a text book, some suggested RO/NF fluxes that are generally achievable assuming good pre-treatment are as follows: seawater 12–14 l/m^2.h (lmh); brackish groundwater 20–25 lmh; surface water 16–20 lmh; water reclamation 18–20 lmh. Higher flux may be successfully applied in some cases but cost–benefit studies tend to favour lower flux.

Even if performance is stable over the long-term, higher flux consumes more electricity since the driving pressure is proportional to flux and higher flux requires more frequent chemical cleaning in almost all cases. A preliminary design based on a single vessel containing six 8-inch diameter × 40-inch long RO/NF elements can be approximated with the following rules of thumb: 4.6 m^3/h for brackish groundwater and 3 m^3/h for seawater.

Three main parameters are monitored as a function of time to determine if a system is operating correctly and to plan maintenance activities: salt passage, differential pressure, and normalized permeate flow. Salt passage can be calculated for TDS or specific solutes. Conductivity passage can be monitored as a surrogate measurement. A gradual increase in salt passage (possibly up to 5% per year) may be expected and should be accounted for in the initial design calculations; however, a more rapid increase indicates damaged o-rings or membrane. With RO/NF, differential pressure refers to the feed-concentrate pressure drop. If this exceeds a manufacturer specified value (~4 bar for a 6-element vessel) the force can irreparably damage the element by telescoping which causes salt leaks at the glue lines. The most complex calculation is normalized permeate flow (or the related normalized net driving force). To account for changes in operating conditions, permeate flow is normalized for feed concentration, temperature and recovery. Simplified versions of the equations are summarized in Table 10.9.

Additional major issues regarding the design of RO/NF include pre-treatment, post treatment, concentrate disposal and proper materials selection to avoid corrosion. Pre-treatment is needed to protect the membrane from fouling and plugging. After the feedwater has been well pre-treated, cartridge filtration with disposable elements, typically rated at 5 μm, is applied just upstream of the RO/NF as a guard filter (Section 9.19). The only exception is that systems with MF/UF membrane filtration pre-treatment sometimes do not include cartridge filters if steps are taken to be certain that no particulate material is present or enters the feedwater. For groundwater sources, pre-treatment can be as simple as cartridge filtration and addition of antiscalant and sulphuric acid to control scaling. Sometimes groundwater also requires removal of materials that can scale the membrane, such as oxidized iron, manganese and hydrogen sulphide (H_2S) gas. H_2S gas does not scale the membrane surface per se but, if oxidized, the resulting elemental sulphur sticks to the membrane, impedes permeate flow and causes damage. For some groundwaters, those with reduced and soluble forms of iron, manganese or hydrogen sulphide, it may be best to prevent oxidation of the feed upstream of RO/NF by excluding air ingress of the feed. Ionic iron and manganese will be rejected just as other ions are. The H_2S will pass through the membrane just as other dissolved gases do; the resulting H_2S in the permeate stream can be subsequently removed by aeration or chemical oxidation (Section 10.33).

One of the major issues with seawater desalination has been the boron concentration in the finished water. The boron concentration in seawater is generally in the range of 5–7 mg/l. Until recently, the WHO Guideline value for drinking water had been 0.5 mg/l; however, in 2011 WHO revised the boron guideline to 2.4 mg/l. Boron ions are weakly charged so boron rejection by RO is low compared to most ions and rejection by NF is even lower. The charge density on boron ions increases with pH. Busch et al. (2003) showed that boron rejection at pH 8 is 45–75% using standard brackish water elements and 85–92% with standard seawater elements, while at pH 11 rejection is increased to 98–99%. Frequently, it is not practical to operate the first pass of a facility at elevated pH due to the risk of scaling by calcium carbonate; therefore, to meet boron limits a partial or full second RO pass or selective IX (Section 10.10) may be required.

Table 10.9 Family of equations that describe the RO process

Description	Equation
Flux (*F*)	$F = Q_p/A$
Recovery (*R*)	$R = (Q_p/Q_f) \times 100$
Salt Passage (*SP*)	$SP = (C_p/C_f) \times 100$
Salt Rejection (*SR*)	$SR = 100 - SP$
Differential Pressure (ΔP_{fc})	$\Delta P_{fc} = P_f - P_c$
Normalized Permeate Flow (*NPF*)	$NPF = Q_p \times TCF_{25} \times \left(\frac{NDP_{initial}}{NDP_{today}}\right)(MC)$
Net Driving Pressure (*NDP*)	$NDP = P_{fcavg} - P_p - P_{osm}$
Temperature Correction Factor (*TCF*)	$TCF_{approx} = 1.03^{(25-T)}$

where:
F = flux, l/m^2.h
Q_p = permeate flow measured at ambient temperature, l/h
A = active membrane area, m^2
R = recovery, percent
Q_f = feed flow measured at ambient temperature, l/h
SP = salt passage, percent
C_p = permeate concentration, mg/l
C_f = feed concentration, mg/l
SR = salt rejection, percent
ΔP_{fc} = feed-to-concentrate pressure differential
P_f = feed pressure, bar
P_c = concentrate pressure, bar
NPF = normalized permeate flow, l/h
TCF = temperature correction factor, dimensionless
$NDP_{initial}$ = net driving pressure during initial operation
(i.e., typically based on readings taken during the first 24–48 hours of operation), bar
NDP_{today} = net driving pressure when Q_p was measured, bar
MC = membrane compaction factor or aging factor; for most modern membranes manufacturers consider $MC = 1$
NDP = average net driving pressure, bar
P_{fcavg} = average feed-concentrate pressure, bar
$= (P_f + P_c)/2$
P_f = feed pressure, bar
P_c = concentrate pressure, bar
P_p = permeate pressure (for simplified calculation, assume ~0), bar
P_{osm} = osmotic pressure, bar
~approximately $= \frac{C_f + C_c}{2000} \times (0.7)$
C_f = feed concentration of total dissolved solids, mg/l
C_c = concentrate concentration of total dissolved solids, mg/l

For surface water, more extensive pre-treatment is required to remove bacteria, turbidity, colour, organic matter, oil and grease. This usually includes a combination of disinfection, coagulation, single- or two-stage dual media filtration or clarification followed by dual media filtration and GAC adsorption depending on the parameters to be removed. Dissolved air flotation can be useful where oil and grease or algae, in particular harmful algal blooms (HABs) (e.g. 'red tides') (Section 7.75), are a concern (Amato, 2012). HABs can cause operational problems and therefore their removal is necessary

(Caron, 2010). Disinfection upstream of the RO/NF may be needed for control of biofouling. If needed, intermittent application of disinfectant or even intermittent pH adjustment may be more effective than continuous pre-disinfection in some cases. If pre-disinfection is used then residual disinfectant should be removed by the addition of sodium bisulphite or other reducing agent upstream of the RO/NF to protect the membranes from unwanted oxidation. The use of MF/UF, instead of granular media is applied in some cases. MF/UF provides low turbidity and low Silt Density Index (SDI) feedwater for RO/NF and may be cost effective given the downward trend of membrane prices. In some seawater applications pre-treatment requirements can be reduced by using beach wells or bank filtration rather than open intakes (Missimer, 1994; Wright, 1997; Voutchkov, 2005) but the cost and practicality will depend on the specific site.

The main pre-treatment goals are to maintain feed water turbidity less than 0.1 NTU and 15-minute SDI (ASTM, 2002) less than 5, but less than 3 is the preferred value. In addition, the concentrations of free chlorine and oil and grease should be nil, and TOC and microbial activity may need to be controlled, but there are no firm quantified values for these. As an approximation, TOC up to 3 or 4 mg/l is probably acceptable, but it depends on the attraction of the organic material to the membrane.

In general, a pilot program should be seriously considered for surface water RO/NF applications, including seawater desalination, and even for many groundwater applications. The pilot program should closely mimic the planned full-scale facility, including pre-treatment conditions and recycle streams within the facility. Full-scale, commercially available RO/NF modules should be piloted since it is difficult to scale-up results accurately. Multiple RO/NF operating cycles are needed to project longer term results. Given a stable groundwater source, a pilot program may consist of 3 or 4 months of testing; for a variable surface water a year or longer may be advised to account for the impact of seasonal changes.

The flow diagram in Plate 17(a) illustrates the issues discussed in this section. This figure shows the process steps for a two-pass SWRO facility but the example can also be generalized to brackish water RO as well as NF. For further study, readers should refer to AWWA's *Manual of Practice* (AWWA, 2007). Plates 17(b) and 17(c) show equipment in a large SWRO plant in India.

10.48 THERMAL PROCESSES

Distillation works on the principle that the vapour produced by evaporating seawater is free from salt and the condensation of the vapour yields pure water. The majority of modern plants use the multi-stage flash thermal process (MSF) or the thermal vapour compression system multi-effect desalination (TVC-MED). Mechanical vapour compression MED (MVC-MED) units are less common; due to current limitations in compressor technology the maximum capacity of MVC-MED units is 3000 m^3/d per unit. Recent practice has tended towards installation of MSF units for distillers in the capacity range 30 000–90 000 m^3/d with TVC covering the 10 000–68 000 m^3/d range. The feed water is usually deaerated and chemically treated to reduce the pH and to control calcium carbonate and sulphate and magnesium hydroxide precipitation/scaling. In some plants, NF is used to reduce calcium sulphate scaling potential. Most large thermal distillation plants are constructed as dual-purpose stations for both desalination and the generation of power. The trend towards higher operating temperatures means that greater attention has to be paid to the reduction of corrosion and the use of cost effective materials and chemicals to combat corrosion.

One of the major design parameters for all distillers is the performance ratio – a measure of the efficiency of energy utilization. The amount of energy required to desalt a given brine concentration varies according to the degree of sophistication of the plant installed, that is annual energy costs reduce as capital costs increase. Other factors to be taken into account include size of units, load factor, growth of demand, interest rate on capital, and technical matters concerning the auxiliary services, repairs and maintenance. For detailed design some 70 design parameters have to be settled. Many of these are concerned with the safe or most economic limits for the temperatures, velocities and concentrations of the coolants, brines, brine vapour, steam, steam condensate and boiler feed water. Other aspects that require careful attention include the prevention of scaling, corrosion, erosion, the purity of the distillate, the efficiency of heat exchangers and the nature and cost of the auxiliary plant involved.

Major problems, which formerly occurred with seawater distillers, were scale formation on heat transfer surfaces due to the presence of carbonates and sulphates of calcium; internal corrosion due to hot sodium chloride and the presence of dissolved gases such as oxygen, ammonia and hydrogen sulphide; plant start-up problems and running at low capacity. These problems have been largely overcome and continuous unit operation in excess of 8000 hours at variable load conditions in the range 60–100% of full load are commonplace.

To augment this brief introduction to thermal desalination, the authors refer readers to the publications by El-Dessouky (2002) and Sommariva (2004).

10.49 POST TREATMENT

Post treatment of the permeate is very important to avoid corrosion of downstream piping and equipment. The RO product water is slightly acidic (pH 5–6) and soft with little alkalinity, while distilled water approaches zero hardness with an alkalinity not likely to exceed 2 mg/l. These waters are aggressive to metal and asbestos cement pipes and take up calcium from mortar-lined pipes (Section 10.40). They are unsuitable for distribution and are unpalatable, being flat and insipid. Generally, NF product is less aggressive since the rejections are lower than with RO, but the risk of corrosion in the distribution piping should be evaluated for both RO and NF. The product water should therefore be remineralized to produce a minimum alkalinity of 40 mg/l as $CaCO_3$, CCPP of 3–10 (Section 10.40), a positive Langelier Index (0–0.5) (Lahav, 2007) and Larson–Skold Index <1 (Section 10.40). Suitable treatment methods include hydrated lime and carbon dioxide dosing; carbon dioxide dosing followed by filtration through limestone and subsequent lime or caustic dosing to raise the pH above the saturation value (Migliorini, 2005); sodium bicarbonate and calcium chloride dosing (note that high chloride to alkalinity ratio is unsuitable for duplex brass fittings); or blending with a suitable brackish water from an artesian source (Applegate, 1986). For MSF plants, carbon dioxide could be recovered from the first stage of the process (recovery decreases from the first to the last) and the recovery rate increased with increasing top brine temperature, bicarbonate content and salinity of seawater and decreased with seawater pH value. Seawater itself cannot be used as a means of remineralization as it does not contain a sufficiently high ratio of alkalinity to TDS to permit this. A concise account of remineralization is given by Withers (2005). Although RO/NF removes bacteria and viruses, disinfection (e.g. chlorination) is generally practised both to provide a multiple barrier to safeguard against leakage due to membrane imperfections or leaking seals and for security in distribution.

10.50 EFFLUENT DISPOSAL

Effluents from desalination processes have a higher density than seawater as a result of the concentration of TDS being approximately double that of seawater. RO/NF processes generate a concentrated waste stream which must be disposed of in a manner that minimizes its impact on the environment (Mauguin, 2005). While this may be more of an issue for inland RO and NF facilities, available options should also be evaluated for coastal plants. Generally the order of preference (increasing cost) is: discharge to sea either direct or when co-located with a power plant, blending with cooling water discharge from the power plant (typically suitable for SWRO); discharge to the outfall of a wastewater treatment plant (WWTP); sewer discharge to a WWTP; deep well injection; disposal to evaporation ponds; or discharge to a special concentration process. In the past, the main high-concentration process has been seeded slurry evaporation with MVC, which has very high capital and operating costs. Newer methods are being developed to augment evaporation and reduce energy costs, such as EDM (Section 10.45) and selective precipitation to reclaim concentrated waste material for beneficial use (Bond, 2007, 2011). For coastal plants, sea discharges should be designed to achieve suitable localized dilution and dispersal, with the design ideally verified by modelling.

Distillation process effluents also contain corrosion products, for example copper, zinc, iron, nickel and aluminium and any additives, for example corrosion inhibitors. Distillation process effluents have elevated temperatures. Disposal of such hot, hyper-saline waters may give rise to adverse ecological effects on the environment. However, the possible environmental effect of the concentrated hot effluent discharge is largely mitigated by mixing with the cooling water needed to convey the low grade rejected heat from the distiller. This cooling water is normally not concentrated and is at ambient temperature. The use of a long sea outfall provided with discharge ports designed to achieve further dilution of the effluent, followed by subsequent tidal dispersion usually makes it possible to meet a predetermined water quality criteria. It is often necessary to carry out tailored studies at each site to enable prediction of dispersal. Discharge to the marine environment should be well designed and managed (UNEP, 2003).

10.51 THE COSTS OF DESALINATION

The subject of desalination costs can be discussed only cursorily because of its diversity and complexity. The purpose of this brief review is therefore to indicate the principal items contributing to the production costs of desalinated water and to point out how these are likely to be influenced by local conditions. The major factors to be considered in any desalting application include: type and characteristics of the saline feed; type of desalination process to be used; local infrastructure; local costs of primary energy source (e.g. oil, gas or electricity); and availability and costs of chemicals needed for feed pre-treatment, product conditioning and plant cleaning. Also important, in the case of distillation plants, is the choice of performance ratio and whether dual-purpose operation is proposed (Section 10.48). Except for the smallest plants, staffing costs are not usually a major item. Table 10.10 summarizes the principal factors affecting the cost of the desalination processes currently popular as applied to single purpose plants (i.e. producers of water only). In the case of brackish waters (Table 10.8) the primary energy and power demands are usually less than half those for seawater desalination, whether electrodialysis or RO is used. Thermal processes are not usually adopted for brackish water desalting because of their unfavourable energy demands. The feed water and chemical consumption for brackish water desalination are also only about one third of those for SWRO.

The costs of desalted water are very variable and site specific. Amortization of the initial capital investment, together with energy costs, usually accounts for up to 80% of the total water cost. The energy components can be calculated from the data in Table 10.10 if local primary energy unit costs are known. The components of operation and maintenance (O&M) costs for a 40 000 m^3/d brackish water RO plant are typically: labour 27%, chemicals 10%, energy 48%, membrane replacements 8% and miscellaneous 6%. For a 40 000 m^3/d SWRO facility the O&M breakdown is typically labour 8%, chemicals 4%, energy 81%, cartridge filters 1%, membrane replacements 4% and miscellaneous 2%.

Table 10.10 Cost contributory factors for major types of seawater desalination processes

Process	Usual maximum size of unit to date (m^{3*}/d)	Total primary fuel energy demand (MJ/m^{3*})[a]	Components of energy demand: Power (kWh/m^3)	Components of energy demand: Heat (MJ/m^3)	Chemical consumption (g/m^{3*})	Feed water consumption (m^3/m^{3*})	Annual cost of spares and replacements as % of initial capital cost of desalter
Multi-stage flash evaporation (MSF)	90 000	120–400	3–4	185–300	4–6	5–10	2%
Thermal vapour compression (TVC-MED)	68 000	120–300	2–3	170–250	5–7	5–10	2–3%
Mechanical vapour compression (MVC-MED)	3000	88–130	8–15	Nil	5–7	2.0–2.5	2–4%
Seawater reverse osmosis (SWRO)	10 000	45–65	4–6	Nil	10–20	2.0–2.5	4–6%
Brackish water reverse osmosis (BWRO)	15 000	22–33	2–3	Nil	6–12	1.2–1.5	4–6%

Notes: *Output.
[a]Lower value for dual-purpose plant.
[1]Data refer to stand-alone plants. Heat and power assumed generated on site.
[2]Thermal efficiencies assumed: 80% for steam boilers for MSF plant; 33% for diesel generators for RO and MVC plant.
[3]Maximum capacity for RO plant refer to one train within a possible multi-train arrangement.
Higher energy value of diesel fuel taken as 37.85 MJ per litre (1 MJ = 0.278 kW h).

Table 10.11 Capital and water costs for large desalination plants (US $)

Type of plant	Plant capital cost ($ per m^3/d output)	Water cost ($/$m^3$)
Multi-stage flash evaporation (MSF)	1000–1900	0.4–1.5
Thermal vapour compression (TVC-MED)	1100–1800	0.4–1.5
Mechanical vapour compression (MVC-MED)	670–990	1.5–2.2
Reverse osmosis (seawater)	700–1200[1]; 1500–4000[2]	0.5–1.2; 2.0–3.0[3]
Reverse osmosis (brackish water)	1200–2000 (one plant at 4000)[4]	0.25–1.8

Notes: [1]Plants in West Asia, North Africa and Europe.
[2]Plants in USA and Australia.
[3]Plants in Australia.
[4]Influenced by pre-treatment applied.

Only the briefest indication of capital costs can be given in the present context for the larger sizes of the five major types of desalination plant listed in Table 10.10. Plant costs, erected and commissioned, including typical costs of civil works, local product storage and all other site-specific costs (such as cost of intake, discharge of effluents, fuels storage and handling, etc.) in 2015 were as given in Table 10.11.

REFERENCES

Amato, T., Park, K., Kim, W. et al. (2012). Application of Enflo-DAF™ Technology to SWRO pre-treatment design and operation. *Proc. of the 6th International Conference on Flotation for Water and Waste Water Systems*, New York.

Applegate, L. E. (1986). Post treatment of reverse osmosis product waters. *Journal AWWA* **78**(5), pp. 59–65.

Armenter, J. L. L. and Canto, J. (1996). Filtration by GAC and on-site regeneration in the treatment of water for the city of Barcelona. Review of 15 years in operation. *Water Supply* **14**(2), pp. 119–127.

Asquith, E. A., Evans, C. A., Geary, P. M. et al. (2013). The role of Actinobacteria in taste and odour episodes involving geosmin and 2-methylisoborneol in aquatic environments. *Journal WSRT-Aqua* **62**(7), pp. 452–467.

ASTM (2002). *Standard Test Method for Silt Density Index (SDI) of Water*. D4189-95.

AWWA (1988). Committee report on: a review of solid solution interactions and implications for controlling trace inorganic materials. *Journal AWWA* **80**(10), pp. 56–64.

AWWA (2007). *Manual of Practice M46, Reverse Osmosis and Nanofiltration*. 2nd Edn. (Ed. Bergman, R.).

AWWA B604-12. *Standard for Granular Activated Carbon*. AWWA.

AwwaRF (1996). *Internal Corrosion of Water Distribution Systems*, 2nd Edn.

AwwaRF (2000). *Optimization of Powdered Activated Carbon Application for Geosmin & MIB Removal*.

B&V (2016). *2016 Strategic Directions: Water Industry Report*, Black & Veatch Insights Group.

Bauer, M., Buchanan, B., Colbourne, J. et al. (1996). The GAC/slow sand filter sandwich – from concept to commissioning. *Water Supply* **14**(2), pp. 159–175.

Benefield, L. D. and Judkins, J. F. (1982). *Process Chemistry for Water and Wastewater Treatment*. Prentice-Hall.

Black & Veatch Corporation (2010). *White's Handbook of Chlorination and Alternative Disinfectants*. 5th Edn. Wiley.

Blute, N., Wu, X., Cron, C. et al. (2014). Hexavalent chromium treatment implementation in Glendale, Calif. *Journal AWWA* **106**(3), pp. E160–E174.

Bond, R. G. and Veerapaneni, V. (2007). *Zero Liquid Discharge for Inland Desalination*. AwwaRF and AWWA, Denver, CO.

Bond, R. G., Klayman, B., Spencer, C. et al. *Zero Liquid Discharge Desalination*. Water Research Foundation, Denver, CO.

Bond, R.G., Davis, T., DeCarolis, J. et al. (2015). *Demonstration of a New Electrodialysis Technology to Reduce the Energy Required for Salinity Management*. California Energy Commission Report CEC-PIR-11-020.

Booker, N.A., Hart, J., Hyde, R.A. et al. (1998). *Removal of Volatile Organics in Groundwater*. WRc.

Bourgine, F. P., Gennery, M. and Chapman, J. L. (1994). Biological processes at saints hill water treatment plant, Kent. *Journal IWEM* **8**(4), pp. 379–391.

Brooke, E. and Collins, M. R. (2011). Post-treatment aeration to reduce THMs. *Journal AWWA* **103**(10), pp. 84–96.

Brunauer, S., Emmet, P. H. and Teller, E. (1938). Adsorption of gases in multimolecular layers. *Journal of the American Chemical Society* **60**, pp. 309–319.

Budd, G. C., Long, B. W., Edwards-Brandt, J. C. et al. *Evaluation of MIEX Process Impacts on Different Source Waters*. AwwaRF.

Busch, M., Mickols, W.S., Jons, S. et al. (2003). Boron removal in seawater desalination. *Proc. International Desalination Assoc*, World Congress, Bahamas.

Caron, D. A., Garneau, M., Seubert, E. et al. (2010). Harmful algae and their potential impacts on desalination operations off southern California. *Water Research* **44**(2), pp. 385–416.

Chen, T., Huddleston, J.I., Atasi, K.Z. et al. (1998). Factor screening for ozonating the taste- and odor-causing compounds in source water at Detroit, USA. *Taste and Odour Preconference Seminar*.

Cheng, A. S. C., Fields, K. A., Sorg, T. J. et al. (2002). Field evaluation of as removal by conventional plants. *Journal AWWA* **94**(9), pp. 64–77.

Chwirka, J. D., Thomson, B. M. and Stomp III, J. M. (2000). Removing arsenic from groundwater. *Journal AWWA* **92**(3), pp. 79–88.

Clifford, D. (1999). Ion exchange and inorganic adsorption, *Water Quality and Treatment: A Handbook of Community Water Supplies*. 5th Edn. McGraw-Hill.

Clifford, D., Ghurye, G. and Tripp, A. (1998). Arsenic removal by IX with and without brine reuse. *Proceedings of AWWA Inorganic Contaminants Workshop*, San Antonio, Texas.

Colling, J. H., Croll, B. T., Whincup, P. A. E. et al. (1992). Plumbosolvency effects and control in hard waters. *Journal IWEM* **6**(3), pp. 259–268.

Cox, C. R. (1964). *Operation and Control of Water Treatment Processes*. WHO.

Degremont (2007). *Degremont Water Treatment Handbook*. 7th Edn. Lavoisier.

Dixon, K. L., Lee, R. G., Smith, J. et al. (1991). Evaluating aeration technology for radon removal. *Journal AWWA* **83**(4), pp. 141–148.

DoE (1973). *Notes on Water Pollution*. DoE (UK), Note No. 61, June.

Driehaus, W., Jekel, M., Hildebrandt, U. et al. (1998). Granular ferric hydroxide – a new adsorbent for the removal of arsenic from natural water. *Journal WSRT-Aqua* **47**(1), pp. 30–35.

Duquet, J. P., Bruchet, A. and Mallevialle, J. (1989). New advances in oxidation processes: the use of ozone/hydrogen peroxide combination for micropollutant removal in drinking water. *Water Supply* **7**(4), pp. 115–124.

Dyksen, J., Raczko, R. and Cline, G. (1999). Operating experience at VOC treatment Facilities, Part 1: GAC. Opflow. *AWWA* **25**(1), pp. 11–12.

Edwards, M. (1994). Chemistry of arsenic: removal during coagulation and Fe-Mn oxidation. *Journal AWWA* **86**(9), pp. 64–78.

Edwards, S., Alharthi, R. and Ghaly, A. E. (2011). Removal of hydrogen sulphide from water. *American Journal of Environmental Science* **7**(4), pp. 295–305.

El-Dessouky, H. T. and Ettouney, H. M. (2002). *Fundamentals of Salt Water Desalination*. Elsevier.

Faust, S. D. and Aly, O. M. (1998). *Chemistry of Water Treatment*. 2nd Edn. Ann Arbor Press.

Ferguson, D. W., McGuire, M. J., Koch, B. et al. (1990). Comparing 'Peroxone' and ozone for controlling taste and odour compounds. Disinfection by-products and micro-organisms. *Journal AWWA* **80**(4), pp. 181–191.

Foster, D. M., Rachwal, A. J. and White, S. L. (1991). New treatment process for pesticides and chlorinated organics control in drinking water. *Journal IWEM* **5**(4), pp. 466–476.

Galjaard, G., Martijn, B., Koreman, E. et al. (2011). Performance evaluation SIX®-Ceramac® in comparison with conventional pre-treatment techniques for surface water treatment. *Water Practice & Technology* **6**(4), pp. 1–2.

Gallagan, D. J. and Vemillion, J. R. (1957). Determining optimum fluoride concentration. *Public Health Reports* **72** (6), pp. 491–493.

Galvin, R. M. (1996). Occurrence of metals in waters: an overview. *Water SA* **22**(1), pp. 7–18.

Glaze, W. H. (1987). Drinking water treatment with ozone. *Environmental Science and Technology* **21**(3), pp. 224–230.

Glaze, W. H., Schep, R., Chauncey, E. C. et al. (1990). Evaluating oxidants for the removal of model taste and odor compounds from a municipal water supply. *Journal AWWA* **82**(5), pp. 79–84.

Gottlieb, M. C. (2005). Ion exchange application in water treatment, Chapter 12. *Water Treatment Plant Design* (Ed. Baruth, E. E.). 4th Edn. AWWA/ASCE.

Graveland, A. (1996). Application of biological activated carbon filtration at Amsterdam water supply. *Water Supply* **14**(2), pp. 233–241.

Gross, H., Schooner, G. and Rutten, P. (1988). Biological denitrification process with hydrogen oxidizing bacteria for drinking water treatment. *Water Supply* **6**, pp. 193–198.

Haberer, C. and Raff, O. (1999). *Removal of Naturally Occurring Radionuclides From Drinking Water – An Overview*. Vom Wasser, 93, September.

Haist-Gulde, B. and Baldauf, G. (1996). Removal of pesticides by powdered activated carbon – practical aspects. *Water Supply* **14**(2), pp. 201–208.

Harrison, C.J., Amy, E.C., Yann, A. et al. (2005). Bench-scale testing of seawater desalination using nanofiltration. *AWWA Membrane Tech Conf*, Phoenix, US.

Hauge, L. J. (1999). Pressure exchanger. *Desalination & Water Reuse* **9**(1), p. 54.

Heard, T. R., Hoyle, B. G. and Hieatt, M. J. (2002). Aerated biological filtration for the removal of ammonia and manganese in a major new water treatment works under construction in Hong Kong. *Water Supply* **2**(1), pp. 47–56.

Hess, A. F., Dyksen, J. E. and Dunn, H. J. (1983). *Control Strategy – Aeration Treatment Technique – Occurrence and Removal of VOCs From Drinking Water*. AwwaRF.

Hoffman, G. L., Lyttle, D. A., Sorg, T. J. et al. (2006). *Design Manual: Removal of Arsenic From Drinking Water Supplies by Iron Removal Process*. EPA/600/R-06/030.

Hofman, J. A., Ijpellaar, G. F., Heijman, S. G. J. et al. (2004). Drinking water treatment in Netherlands: outstanding and still ambitious. *Water Science and Technology; Water Supply* **4**(5–6), pp. 253–262.

Huang, W. J. (1995). Powdered activated carbon for organic removal from polluted raw water in Southern Taiwan. *Journal WSRT-Aqua* **44**(6), pp. 275–283.

Ijpelaar, G. F., Harmsen, D. J. H. and Heringa, M. (2007). *UV Disinfection and UV/H_2O_2 Oxidation: By-Product Formation and Control*. Techneau, D2.4.1.1.

Irvine, E., Grose, A. B. F., Welch, D. et al. (2000). Nanofiltration for colour removal – 7 years operational experience in Scotland. *Membrane Technology in Water and Wastewater Treatment* (Ed. Hills, P.). Royal Society of Chemistry.

Jacob, C. (2007). Seawater desalination: boron removal by ion exchange technology. *Desalination* **205**(1–3), pp. 47–52.

Janda, V. and Rudovsky, J. (1994). Removal of ammonia in drinking water by biological nitrification. *Journal WSRT-Aqua* **43**(3), pp. 120–125.

Katsoyiannis, I. A. and Zouuboulis, A. I. (2004). Application of biological processes for the removal of arsenic from groundwaters. *Water Research* **38**, pp. 17–26.

Knocke, W. R. (1991b). Removal of soluble manganese by oxide-coated filter media: sorption rate and removal mechanism issues. *Journal AWWA* **83**(8), pp. 64–69.

Knocke, W. R. (2006). Personal communication.

Knocke, W. R., Van Benschoten, J. E., Kearney, M. et al. *Alternative Oxidants for the Removal of Soluble Iron and Manganese*. AwwaRF.

Knocke, W. R., Van Benschoten, J. E., Kearney, M. et al. (1991a). Kinetics of manganese and iron oxidation by potassium permanganate and chlorine dioxide. *Journal AWWA* **83**(6), pp. 80–87.

Kruithof, J. C. and Martijn, B. J. (2013). UV/H_2O_2 treatment: an essential process in a multi barrier approach against trace chemical contaminants. *Water Science and Technology: Water Supply* **13**(1), pp. 130–138.

Lacamp, B. and Bourbigot, M. M. (1990). Advanced nitrogen removal processes for drinking and wastewater treatment. *Water Science and Technology* **22**(3), p. 54.

Lahav, O. and Birnhack, L. (2007). Quality criteria for desalinated water following post-treatment. *Desalination.* **207**, pp. 286–303.

Lambert, S. D., Guillermo, S. M. and Graham, G. (2002). Deleterious effects of inorganic compounds during thermal regeneration of GAC: a review. *Journal AWWA* **94**(12), pp. 109–119.

Langlais, B., Reckhow, D. A. and Brink, D. R. (Eds) (1991). *Ozone in Water Treatment – Application and Engineering*. AwwaRF.

Larson, T. E. and Skold, R. V. (1957). Corrosion and tuberculation of cast iron. *Journal AWWA* **49**(10), pp. 1294–1302.

LeChevallier, M. W., Shaw, N. E., Kaplan, L. A. et al. (1993). Development of a rapid assimilable organic carbon method for water. *Applied and Environmental Microbiology* **59**(5), pp. 1526–1531.

Lehimas, G. F. D., Chapman, J. I. and Bourgine, F. P. (2001). Arsenic removal in groundwater in conjunction with biological-iron removal. *Journal CIWEM* **15**, pp. 190–192.

Leung, E., Trejo, R., Rohe, D. L. et al. (2005). Prototype testing facility for two-pass nanofiltration membrane seawater desalination process. *AWWA Memb Tech Conf*, Phoenix.

Lobo, W. S. and Reid, A. (2007). Advanced oxidation processes (AOP) The next 'Silver Bullet'... But which process is best? A comparison of treatment processes combining ozone, UV and Hydrogen peroxide. *World Congress on Ozone and Ultraviolet Technologies*. Los Angeles.

Lundgren, B. V., Grimvall, A. and Sävenhed, R. (1988). Formation and removal of off-flavour. *Water Science and Technology* **20**(8/9), pp. 245–253.

Lutle, D. A., White, C., Williams, D. et al. (2013). Innovative biological water treatment for the removal of elevated ammonia. *Journal AWWA* **105**(9), pp. E524–E539.

Lytle, D. A. (2007). Biological nitrification in a full-scale and pilot-scale iron removal drinking water treatment plant. *Journal WSRT-Aqua* **56**(2), pp. 125–136.

Marc, J.-F. and Pinker, B. (1998). Reactivation of granular activated carbon for drinking water treatment. *Journees Information Eaux 98*, Portiers.

Matsumoto, N., Aizawa, T., Ohgaki, S. et al. (2002). Removal methods of nematodes contained in the effluent of activated carbon. *Water Science and Technology. Water Supply* **12**(3), pp. 183–190.

Mauguin, G. and Corsin, P. (2005). Concentrate and other waste disposals from SWRO plants: characterisation and reduction of their environmental impact. *Proceeding of Conference on Desalination and the Environment*, Santa Margherita Ligure, Italy. *Desalination*, **182**(1–3).

McNeill, L. S. and Edwards, M. (1997). Arsenic removal during precipitative softening. *Journal of Environmental Engineering, ASCE* **123**(5), pp. 453–460.

McNeill, L. S., McLean, J. E., Parks, J. L. et al. (2012). Hexavalent chromium review, Part 2: chemistry, occurrence and treatment. *Journal AWWA* **104**(7), pp. E395–E405.

Merkle, R. B., Knocke, W. R., Daniel, L. et al. (1997). Dynamic model for soluble Mn^{2+} removal by oxide-coated filter media. *Journal of Environmental Engineering, ASCE* **123**(7), pp. 650–658.

Merry, A. et al. (1995). Membrane treatment of coloured water. *Proc of International Symposium of Wastewater Treatment and 7th Workshop on Drinking Water*, Montreal.

Migliorini, G. and Meinardi, R. (2005). 40 MIGD potabilization plant at Ras Laffan: design and operating experience. *Desalination* **182**, pp. 275–282.

Miles, G. D. (1948). The action of natural waters on lead. *Journal of the Society of Chemical Industry* **67**, pp. 10–13.

Missimer, T. M. (1994). *Water Supply Development for Membrane Water Treatment Facilities*. Lewis Publishers/ CRC Press.

Mistry, K. H., McGovern, R. K., Thiel, G. P. et al. (2011). Entropy generation analysis of desalination technologies. *Entropy* **13**(10), pp. 1829–1864.

Moch, I. and Harris, C. (2002). What seawater energy recovery system should I use? A modern comparative study. *International Desalination Association (IDA) World Congress on Desal and Water Reuse*, Bahrain.

Monscwitz, J. T. and Ainsworth, L. D. (1974). Treatment of hydrogen sulphide. *Journal AWWA* **66**(9), pp. 537–539.

Morris, T. and Siviter, C. L. (2001). Application of a biological iron removal process at grove water treatment works. *Journal CIWEM* **15**(2), pp. 117–121.

Mouchet, P. (1992). From conventional to biological removal of iron and manganese in France. *Journal AWWA* **84** (4), pp. 158–167.

Mouchet, P. and Capon, B. (1991). *Recent Evolution in Drinking Water Treatment Technology*. Extract Revue Travaux, Degremont, July/August.

Nealson, K. H. (2006). The manganese-oxidising bacteria. *The Prokaryotes* **5**, pp. 222–231.

Nerenberg, R., Rittmann, B. E. and Soucie, W. J. (2000). Ozone/biofiltration for removing MIB and geosmin. *Journal AWWA* **92**(12), pp. 85–95.

Paillard, H., Partington, J. and Valentis, G. (1990a). Technologies available to upgrade potable waterworks for triazines removal. *IWEM Scientific Section, Pesticide Symposium*, London, 11–12 April.

Paillard, H., Legube, B., Gilbert, M. et al. (1990b). Removal of nitrogonous pesticides by direct and radical type ozonation. *EC Annual Conference on Micropollution*, May, Lisbon.

Pomeroy, R. D. and Cruse, H. (1969). H_2S odour threshold. *Journal AWWA* **61**(12), p. 677.

Renner, R. (2004). Plumbing the depths of DC's drinking water crisis. *Environmental Science and Technology* **38** (12), pp. 224A–227A.

Richard, Y., Brener, L., Martin, G. et al. (1978). Study of the nitrification of surface water. *Progress in Water Technology* **10**(5/6), pp. 17–32.

Rittmann, B. E. and Huck, P. M. (1989). Biological treatment of public water supplies. *CRC Critical Reviews* **19**(2), pp. 119–184.

Rook, J. J. (1976). Haloforms in drinking water. *Journal AWWA* **68**(3), pp. 168–172.

Rosenfeldt, E. J. (2005). UV and UV/H_2O_2 treatment of methylisoborneol (MIB) and geosmin in water. *Journal WSRT-Aqua* **54**(7), pp. 423–434.

Rosenfeldt, E. and Linden, K. (2004). Degradation of endocrine disrupting chemicals bisphenol A, Ethinyl estradiol and estradiol during UV photolysis and advanced oxidation processes. *Environmental Science and Technology* **38** (20), pp. 5476–5483.

Rubel, Jr. F. (2003). *Design Manual: Removal of Arsenic From Drinking Water by Adsorptive Media*. EPA/600/R-03/019.

Schock, M. and Oliphant, R. J. (1996). *The Corrosion and Solubility of Lead in Drinking Water, Internal Corrosion of Water Distribution Systems*. 2nd Edn. AwwaRF.

Sharma, S. K., Petrusevski, B. and Schippers, J. C. (2005). Biological iron removal from groundwater. *Journal WSRT-Aqua* **54**(4), pp. 239–247.

Sheiham, I. and Jackson, P. J. (1981). The scientific basis for control of lead in drinking water by water treatment. *Journal IWES* **35**(6), pp. 491–515.

Short, C.S. (1973). *Removal of Ammonia From River Water*. TP101, WRc.

Simms, J. and Azizian, F. (1998). Pilot-plant trials on the removal of arsenic from potable water using activated alumina. *Proc. AWWA Water Technology Conference*, Denver.

Singer, P. C., Schneider, M., Edwards-Brandt, J. et al. (2007). MIEX for removal of DBP precursors: pilot plant findings. *Journal AWWA* **99**(4), pp. 128–139.

Slunjski, M., O'Leary, B., Tattersall, J. et al. (2000). MIEX water treatment process. *Proc. Aquatech*, Amsterdam.

Smith, D. J., Pettit, P. and Schofield, T. (1996). Activated carbon in water treatment. *Water Supply* **14**(2), pp. 85–98.

Snoeyink, V. L. and Summers, R. S. (1999). Adsorption of organic compounds, Chapter 13. *Water Quality and Treatment: A Handbook of Community Water Supplies*. (Ed. Letterman, R. D.). 5th Edn. McGraw-Hill.

Snoeyink, V. L., Cairns-Chambers, C. and Pfeffer, J. L. (1987). Strong acid IX for removing barium, radium and hardness. *Journal AWWA* **79**(8), pp. 66–72.

Soares, M. I. M. (2002). Denitrification of groundwater with elemental sulfur. *Proc. Water Quality Technology Conference*, San Diego.

Sommariva, C. (2004). *Desalination Management and Economics*. Faversham House Group.

Tiller, A. K. (1984). Corrosion induced by bacteria. *The Public Health Engineer* **12**(3), pp. 144–147.

Thirunavukkarasu, O. S. (2003). Arsenic removal in drinking water using granular ferric hydroxide. *Water SA* **29**(2), pp. 161–170.

Turner, M. E. D. (1961). The influence of water composition on dezincification of duplex brass fittings. *Proc SWTE* **10**, pp. 162–178.

UKWIR (2007). *Investigation into the Emerging Issue of Fluorapatite Formation*. UK Water Industry Research Limited Report. Ref: No. 08/DW/04/11.

UNEP (2003). *Guidelines for the Environmental Sound Management of Seawater Desalination Plants in the Mediterranean*. United Nations Environment Programme, Sngemini, Italy.

USDHHS (2012). *Toxicological Profile for Radon: Chapter 6.* US Department of Health and Human Services, Public Health Service, Agency for Toxic Substances and Disease Registry, Atlanta, Georgia.

US EPA (2002). *National Primary Drinking Water Regulations; Radionuclides.* Office of Ground Water and Drinking Water, Section I-C.10.

Valentine, R. L. and Lin, Y.-P. (2009). *The Role of Free Chlorine, Chloramines, and NOM on the Release of Lead into Drinking Water.* Water Research Foundation, Denver, CO.

van der Kroon, G. M. and Schram, A. H. (1969). Weir aeration – Part 1: single free fall. *H_2O* **22**(2), pp. 528–537.

van der Veen, C. and Graveland, A. (1988). Central softening by crystallisation in a fluidized-bed process. *Journal AWWA* **80**(6), pp. 51–58.

van Dijk, J. C. and Wilms, D. A. (1991). Water treatment without waste material – fundamentals and state of the art of pellet softening. *Journal WSRT-Aqua* **40**(5), pp. 263–280.

van Eekeren, M. W. M., van Paassen, J. A. M. and Merks, C. W. A. M. (1994). Improved milk-of-lime for softening of drinking water – the answer to the carry-over problem. *Journal WSRT-Aqua* **43**(1), pp. 1–10.

van Honwelingen, G. A. and Nooijen, W. F. J. M. (1994). Water softening by crystallization. *European Water Pollution Control* **3**(4), pp. 33–35.

Veerapaneni, S. V., Klayman, B., Wang, S. et al. *Desalination Facility Design and Operation for Maximum Efficiency*. Water Research Foundation, Denver, CO, No. 4038.

Voutchkov, N. (2005). Use of large beach well intakes for large desalination plants. *EDS Newsletter* **22**.

Wang, L., Chen, A. S. C., Sorg, T. J. et al. (2002). Field evaluation of As removal by IX and AA. *Journal AWWA* **94** (4), pp. 161–173.

Water Research Foundation (2013). *DBP Precursor and Micropollutant Removal by Powdered Activated Carbon.* Report 4294.

Water Research Foundation (2014). *Hexavalent Chromium Treatment With Strong Base Anion Exchange.* Report 4488.

Weeks, M. A., Olsen, A., Leadbetter, B. S. C. et al. Invertebrate infestation in granular activated carbon adsorbers in potable water treatment. *Proc. Developments in Activated Carbon processes.* Society of Chemical Industry.

Westerhoff, P. (2006). *Arsenic Removal With Agglomerated Nanoparticle Media.* AwwaRF.

Westerhoff, P., Rodriguez-Hernandaez, M., Baker, L. et al. (2005). Seasonal occurrence and degradation of 2-methylisoborneol in water supply reservoirs. *Water Research* **39**, pp. 4899–4912.

WHO (2011). *Guidelines for Drinking-Water Quality*, 4th Edition. WHO.

Withers, A. (2005). Options for recarbonation, remineralisation and disinfection for desalination plants. *Desalination* **179**(1–3), pp. 11–24.

Wong, J. M. (2000). Treatment technologies for the removal of NDMA from contaminated groundwater. *Groundwater Resources Association of California* **11**(2).

Wright, R. R. and Missmer, T. M. (1997). Alternative intake systems for seawater membrane water treatment plants. *Proc. International Desalination Assoc. World Congress*, Madrid.

Zhang, Z. and Clifford, D. A. (1994). Exhausting and regenerating resins for uranium removal. *Journal AWWA* **86** (4), pp. 228–241.

CHAPTER

Disinfection of Water

11

11.1 DISINFECTANTS AVAILABLE

The term 'disinfection' is used to mean the destruction of infective organisms in water to such low levels that no infection of disease results when the water is used for domestic purposes including drinking. The term 'sterilization' is not strictly applicable because it implies the destruction of all organisms within a water and this may not be either necessary or even achievable. Nevertheless, the word is often used loosely, as in 'domestic water sterilizers'.

On a municipal scale, the following disinfectants are in common use:

- Chlorine (Cl_2)
- Chloramines (NH_2Cl, $NHCl_2$)
- Chlorine dioxide (ClO_2)
- Ozone (O_3)
- Ultraviolet (UV) radiation.

Sodium hypochlorite (NaOCl), which forms the same hydrolysis products as chlorine when dosed to water, is increasingly used in the UK to avoid risks associated with chlorine storage in populated areas. For small plants or under special circumstances, other compounds which release chlorine when dissolved in water may be used, including calcium hypochlorite ($Ca(OCl)_2$).

The organisms in water which may require killing or inactivating by disinfection include bacteria, bacterial spores, viruses, protozoa and protozoan cysts, worms and larvae. The efficacy of disinfection depends on numerous factors: the type of disinfectant used; the amount applied and the time for which it is applied; the type and number of organisms present; and the physical and chemical characteristics of the water.

DISINFECTION USING CHLORINE AND CHLORAMINES

11.2 ACTION OF CHLORINE

The precise action by which chlorine kills bacteria in water is uncertain but it is believed that the chlorine compounds formed when chlorine is added to water rupture bacterial membranes and

Twort's Water Supply. DOI: http://dx.doi.org/10.1016/B978-0-08-100025-0.00011-9

inhibit vital enzymic activities resulting in bacterial death. Chlorine is also a strong oxidizing agent that will break up organic matter in a water; in so doing, because it is a highly reactive chemical, it can form a wide range of chlorinated compounds with the organic matter present. Among these are the trihalomethanes (THMs) and haloacetic acids (HAAs) for which limits have been set for health reasons (Sections 7.24 and 11.7). Chlorine can also restrain algal growth, react with ammonia and convert iron and manganese in the water to their oxidized forms that may then precipitate. Hence, there are a number of factors to be taken into consideration when using chlorine as a disinfectant.

11.3 CHLORINE COMPOUNDS PRODUCED

When chlorine is added to water which is free from organic matter or ammonia, hypochlorous acid HOCl is formed which is further dissociated to H^+ and OCl^-:

$$Cl_2 + H_2O = HOCl + HCl$$

$$HOCl \Leftrightarrow H^+ + OCl^-$$

The dissociation is favoured by high pH and temperature of the water as shown in Table 11.1. The sum of the hypochlorous acid HOCl and hypochlorite ion OCl^- concentrations is together known as 'free chlorine'. These are the most effective forms of chlorine for achieving disinfection. Of the free chlorine, hypochlorous acid is a far more powerful bactericide than the hypochlorite ion. Thus free chlorine acts more rapidly in an acid or neutral water. Therefore, when final pH correction is practised, the alkali should be added after the disinfection process has been completed.

Table 11.1 Variation of HOCl as percentage of free chlorine with pH and temperature values of water

	Percent HOCl at water temperatures (percent OCl^- = 100 − percent HOCl)						
pH	0°C	5°C	10°C	15°C	20°C	25°C	30°C
6.0	98.5	98.3	98.0	97.7	97.4	97.2	96.9
6.25	97.4	97.0	96.5	96.0	95.5	95.1	94.6
6.5	95.5	94.7	94.0	93.2	92.4	91.6	91.0
6.75	92.3	91.0	89.7	88.4	87.1	86.0	84.8
7.0	87.0	85.1	83.1	81.2	79.3	77.5	75.9
7.25	79.1	76.2	73.4	70.8	68.2	66.0	63.9
7.5	68.0	64.3	60.9	57.7	54.8	52.2	49.9
7.75	54.6	50.5	46.8	43.5	40.6	38.2	36.0
8.0	40.2	36.3	33.0	30.1	27.7	25.6	23.9
8.25	27.4	24.3	21.7	19.5	17.6	16.2	15.0
8.5	17.5	15.3	13.5	12.0	10.8	9.8	9.1
8.75	10.7	9.2	8.0	7.1	6.3	5.8	5.3
9.0	6.3	5.4	4.7	4.1	3.7	3.3	3.0

Any ammonia present in the water, either as a contaminant or as a result of chemical dosing, will react with chlorine to form chloramines. These compounds are formed in a stepwise manner, with the successive formation of monochloramine NH_2Cl, dichloramine $NHCl_2$ and trichloramine NCl_3 (nitrogen chloride) (Section 11.8). Of these compounds, the sum of the monochloramine and dichloramine concentrations is known as the 'combined chlorine'; total chlorine is the sum of combined chlorine and free chlorine.

Free chlorine is many times more powerful as a bactericide than combined chlorine. Butterfield (1943–46) estimated that 25 times as much combined chlorine is needed to achieve the same degree of kill of bacteria as free chlorine in the same time. Of the chloramines, dichloramine is more powerful than monochloramine requiring only about 15% of the monochloramine dose for inactivation of *Escherichia coli*. Since ammonia is often naturally present in a water, it is usual to add sufficient chlorine to react with all the ammonia present and produce an excess of free chlorine sufficient to achieve speedy disinfection. As a consequence, the efficacy of chlorine as a disinfectant is influenced by a number of conditions.

11.4 FACTORS RELATING TO THE DISINFECTION EFFICIENCY OF CHLORINE

The following factors have to be taken into account when treating water with chlorine.

The stage at which chlorine is applied. Chlorine is often applied at more than one stage in the treatment of a water. 'Pre-chlorination' refers to the application of chlorine to a water (often raw water) at the works inlet, that is before clarification and filtration. 'Intermediate chlorination' refers to chlorine added between stages of treatment. 'Final chlorination' refers to the final disinfection of a water before it is put into supply. The purposes of pre-chlorination and intermediate chlorination are often partly biological so as to reduce bacterial content, prevent bacterial multiplication and restrain algal growth; and partly chemical, so as to assist in the precipitation of iron and manganese and achieve other oxidation benefits. Final chlorination is always for the purpose of disinfecting the water and to maintain a residual in the distribution system so that it is safe for drinking.

Effect of turbidity. The effect of turbidity in a water is to hinder the penetration of chlorine: bacteria can be shielded in particles of suspended matter and thus be protected from the effect of the chlorine. It is always necessary therefore, that final disinfection by chlorine is applied as a final stage of treatment in water which contains low turbidity. For effective disinfection, WHO (2011) suggests a guide level value for turbidity of less than 1 NTU.

Consumption of chlorine by metallic compounds. A substantial amount of chlorine may be used to convert iron and manganese in solution in the water into products which are insoluble in water (Section 10.12). Reduction of these parameters by upstream processes is therefore essential. Typically, iron and manganese should be less than 0.1 mg/l as Fe and 0.05 mg/l Mn, respectively. If at the point of chlorine application their levels are too low to justify removal, the dose must take their demand into account.

Reaction of chlorine with ammonia compounds and organic matter. Ammonia compounds may exist in organic matter present in the water or separately from organic matter (Section 7.7); in either case they will react to form combined chlorine which is not as effective a bactericide as free chlorine (Section 11.8). Chlorine may be used in the oxidation of some organic matter, but at the risk of forming disinfection by-products (DBPs; Section 11.7). Ammonia in water presented for disinfection should not exceed 0.01 mg/l as N. When this value is exceeded or when organic matter is present, an allowance should be made both in the chlorine dose and contact time to satisfy the chlorine demand prior to disinfection. Therefore, the substances that are causing a chlorine demand must be removed prior to disinfection by upstream treatment or an allowance for them must be made in the chlorine dose, otherwise disinfection could be compromised.

Low temperature causes delay in disinfection. The rate of disinfection is significantly affected by temperature, reducing as the temperature falls. The difference in kill rate of bacteria between the temperatures of 20°C and 2°C is noticeable both with free and combined chlorine. This must be borne in mind when determining the required contact period. The reduction in the rate of disinfection with falling temperature is offset to some extent by a small increase in the equilibrium concentration of the hypochlorous ion at lower temperatures.

Increasing pH reduces effectiveness of chlorine. In free chlorine, hypochlorous acid is formed in greater quantities at low pH values than at high values. Thus disinfection by free chlorine is more effective at low pH values; the guide value suggested by WHO (2011) is less than 8.

The number of coliforms presented for disinfection. This has an influence on the disinfection efficiency. To be confident of achieving 100% compliance with the requirement of zero coliforms after the disinfection stage, the water subjected to disinfection ideally should not contain more than 100 coliforms/100 ml. Most groundwaters satisfy this criterion. In surface waters, coagulation followed by solid–liquid separation processes, achieves up to 99.9% bacteria removal (Section 8.13). Consequently pre-disinfection in addition to conventional treatment is only required for heavily polluted surface waters.

Time of contact is important. The disinfecting effect of chlorine is not instantaneous and sufficient time must be allowed for the chlorine to kill organisms. This important factor is dealt with in the next section.

11.5 CHLORINE RESIDUAL CONCENTRATION AND CONTACT TIME

Of all the factors influencing the disinfection efficiency of chlorine discussed above, the most important are free residual concentration, contact time, pH and water temperature. The term 'free residual' refers to the amount of free chlorine remaining after the disinfection process has taken place. Given adequate chlorine concentration and contact time, all bacterial organisms and most viruses can be inactivated. Thus, a useful design criterion for the disinfection process is the product of contact time (t in minutes) and the free chlorine residual concentration (C in mg/l) at the end of that contact time. This is known as the 'Ct value' or 'exposure value' (WHO, 2011). On this basis the guide level of 0.5 mg/l free residual concentration after 30 minutes contact proposed by WHO would have a Ct value of 15 mg.min/l. This is shown to provide a 12.5-fold factor of safety so that a degree of inefficiency in the contact tank performance can be tolerated (Stevenson, 1998).

The WHO *Ct* criterion of 15 mg.min/l is for faecally polluted water and therefore may be varied according to the bacteriological quality of the source water. For example a groundwater free of *E. coli* and containing no more than 10 coliforms/100 ml could have a *Ct* value of 10 mg.min/l with *t* not less than 15 minutes and for groundwaters where coliforms and *E. coli* are completely absent, chlorination sufficient to maintain a residual in the distribution system with no contact at the treatment works could be acceptable. On the other hand for surface waters a higher *Ct* value would be used, for example 30 mg min/l, with *t* being not less than 30 minutes, and *C* being 0.5–1.0 mg/l depending on the degree of bacterial pollution. If chlorine demand is to be satisfied in the disinfection stage *C* and *t* should be increased.

Giardia lamblia cysts and enteric viruses are more resistant to disinfection compared to *Legionella*, heterotrophic bacteria and coliforms. For this reason they are used in the US Surface Water Treatment Rule (SWTR) (US EPA, 1991) to define the required disinfection effectiveness for a treatment process incorporating solid–liquid separation processes and inactivation by disinfection. The rule requires that for raw water containing an average of 1 *Giardia* cyst/100 l, the processes should achieve 3-log (99.9%) removal for *Giardia lamblia* cysts and 4-log (99.99%) removal of enteric viruses. According to the SWTR, direct filtration treatment is given 2-log credit for *Giardia* removal and 1-log for virus removal, whilst conventional treatment of clarification and filtration is given 2.5-log credit for *Giardia* and 2-log credit for viruses. These requirements increase for higher raw water *Giardia* concentrations. This leaves 0.5- to 1-log inactivation of *Giardia* and 2- to 3-log inactivation of viruses to be achieved by disinfection. Tables 11.2 and 11.3 give the *Ct* values stated in the SWTR to achieve 1-log inactivation of *Giardia* and 2- and 3-log inactivation of viruses.

The WHO *Ct* criterion is applicable to the inactivation of bacteria and most viruses and therefore cannot be directly compared with the *Ct* values for inactivation of cysts given in Table 11.2. Comparison with *Ct* values in Table 11.3 confirms that the WHO criterion has a high factor of safety, which is desirable to ensure complete inactivation.

Table 11.2 *Ct* values for achieving 1-log inactivation of *Giardia lamblia*

		***Ct* value at water temperature**					
Disinfectant	**pH**	**0.5°C**	**5°C**	**10°C**	**15°C**	**20°C**	**25°C**
Free residual chlorine of 2 mg/l	6	49	39	29	19	15	10
	7	70	55	41	28	21	14
	8	101	81	61	41	30	20
	9	146	118	88	59	44	29
Ozone	6–9	0.97	0.63	0.48	0.32	0.24	0.16
Chlorine dioxide	6–9	21	8.7	7.7	6.3	5	3.7
Chloramines	6–9	1270	735	615	500	370	250

Source of information: US EPA (1991).

Table 11.3 *Ct* Values for achieving 2- and 3-log inactivation of enteric viruses at pH values 6–9

Disinfectant	Log inactivation	*Ct* values at water temperature					
		0.5°C	5°C	10°C	15°C	20°C	25°C
Free residual chlorine	2	6	4	3	2	1	1
	3	9	6	4	3	2	1
Ozone	2	0.9	0.6	0.5	0.3	0.25	0.15
	3	1.4	0.9	0.8	0.5	0.4	0.25
Chlorine dioxide	2	8.4	5.6	4.2	2.8	2.1	1.4
	3	25.6	17.1	12.8	8.6	6.4	4.3
Chloramines	2	1243	857	643	428	321	214
	3	2063	1423	1067	712	534	356

Source of information: US EPA (1991).

In the expression *Ct*, the contact time *t* is the time the water remains in the contact tank. This tank is fitted with baffles and, if these are cast in concrete, the volume of the baffles can significantly reduce the volume within the tank. The theoretical residence time (t_T) in the contact tank is the volume of water in the tank divided by the rate of flow. This would be achieved with perfect plug flow conditions through the tank. In practice, eddies and short-circuiting in the tank result in some water passing through the tank in less time than t_T. The main causes of short-circuiting are non-uniform distribution at the inlet (difficult to avoid as the water generally enters the tank as a jet), redirection of flow at the end of baffled lanes, and having to lift all the flow to the surface at the weir outlet. Any contraction, expansion or bend in the flow path due to a non-ideal layout will give further inefficiencies. Chlorine contact tanks are typically designed so that at least 90% of water passing through the tank remains in the tank for more than the required contact time at the design flow rate. This is referred to as the t_{10} time, that is the time it takes for the first 10% of water to pass through the tank, or the minimum period that 90% of the water will remain in the tank. It should be noted that this 10% criteria is not a universal standard and sometimes a different threshold such as 5% is used. The ratio of t_{10}/t_T is a measure of short-circuiting in the contact tank and varies in the range 0–1. A value of 1 is an indication of perfect plug flow but in practice the value typically varies between 0.6 and 0.8 for a rectangular tank divided into a series of baffled lanes.

An optimum design with a t_{10}/t_T of 0.7–0.8 can be achieved either by physical modelling or computational fluid dynamics (CFD) models (Plate 21(b) and Section 14.18). In the absence of these tools, the following basic design parameters are suggested: the inlet jet should be baffled to disperse the flow; the tank should be divided into long straight out and return channels of length: width ratio greater than 10 and depth:width ratio less than 1.5; and a weir outlet should be provided to maintain the required volume of water under all flow conditions. With these provisions, a value of t_{10}/t_T of 0.6–0.7 may be achieved. The actual time of contact provided by newly designed or existing tanks can be checked by timing the passage of a slug (pulse input) or a continuous (step input) dose of a tracer chemical such as lithium chloride or sodium fluoride (Teefy, 1996). CFD models can be used to simulate a tracer test (Plate 21(a)).

Pipelines are ideal for contact as they provide good plug flow characteristics and the ratio of t_{10}/t_T can exceed 0.95. However, in practice it should be reduced to about 0.9 to allow for bends and exit and entry conditions (possibly 0.8 if there are a large number of bends). The pipeline should remain within the treatment plant site boundary so that control and monitoring of the residual chlorine at its downstream end is possible and convenient.

The free residual chlorine concentration of the water leaving the contact tank, after the requisite time of contact, can be reduced if desired by partial dechlorination (Section 11.12) to suit the needs of the distribution system.

11.6 EFFICIENCY OF CHLORINE IN RELATION TO BACTERIA, ENTERIC VIRUSES AND PROTOZOA

Bacterial kill. The work of Butterfield (1943–46) has shown that under nearly all conditions the typhoid bacillus and other enteric pathogenic bacteria are at least as susceptible to chlorination as *E. coli*. Due to the far greater concentrations of *E. coli* present in pollution of human or animal origin (Section 7.66) it is reasonable to assume that, if *E. coli* are absent in a 100 ml sample of disinfected water, then the water is also free of pathogenic bacteria. The spores of bacteria are, however, more resistant to the action of chlorine than are the bacteria; fortunately the bacteria causing most waterborne diseases are not spore formers. The spore-forming *Clostridium perfringens* (*Cl. welchii*) used as an indicator of pollution (Section 7.69) is not considered significant for health.

Virus kill. The pathogenic enteric viruses, described in Section 7.64, occur in far smaller numbers than *E. coli* in a polluted water. However, they can survive for long periods in water and the minimum dose causing human infection is believed to be very low. Test methods available for detecting the presence of viruses (Section 7.72) cannot be used for routine monitoring. The enteric viruses have also been shown to be more resistant to chlorine than *E. coli*. Poynter (1973) reported Russian experiments had indicated that higher levels of residual chlorine and longer periods of contact were required to eliminate viruses than were required to destroy *E. coli*; similar results were obtained by Scarpino (1972).

The consequence of these difficulties is that tests showing the absence of *E. coli* in 100 ml samples of disinfected water do not give the same level of confidence that viruses are absent as they do for the absence of pathogenic bacteria. However, the *E. coli* test remains the only practicable means at present for routine monitoring and the WHO (2011) state that: 'it has been demonstrated that a virus-free water can be obtained from faecally polluted source waters' if the following chlorine disinfection conditions are met:

- the water has a turbidity of 1 NTU or less;
- its pH is below 8.0;
- a contact period of at least 30 minutes is provided; and
- the chlorine dose applied is sufficient to achieve at least 0.5 mg/l free residual chlorine during the whole contact period.

Protozoa resistance to chlorine. The principal protozoa pathogenic to humans – *Entamoeba histolytica, Giardia lamblia* and *Cryptosporidium parvum* – were described in Section 7.63. Like viruses, the cysts can remain viable in the environment for long periods; the ingestion of a very few may be sufficient to cause human infection. However, detection of their presence by routine testing is impracticable.

Protozoa are more resistant to chlorine than viruses. WHO (2011) describes the cysts of *E. histolytica* as 'among the most chlorine-resistant pathogens known'. Because of their size – *E. histolytica* 10–20 μm; *Giardia* spp. 8–12 μm; *Cryptosporidium* 4–6 μm – protozoa can be removed by conventional processes of coagulation, clarification and media filtration (Section 9.1) and membrane filtration (Section 9.18).

No completely reliable disinfection procedure to eliminate these protozoan cysts and oocysts has yet been found. Irrespective of the treatment methods available (Section 9.20), it is paramount that the sources of pollution likely to give rise to the presence of such pathogenic protozoa in a water are minimized. If their presence is detected in a water an intensive search for the source is necessary (Section 7.83). When they are detected or suspected in treated water, it is necessary to advise consumers to boil water used for drinking until evidence of their elimination from the supply is obtained (Section 11.28).

11.7 CHLORINATION AND THE PRODUCTION OF DBPs

When chlorine is applied to water containing precursors, which arise from the presence of natural colour and algal metabolic products, DBPs for example THMs and HAAs are formed (Section 7.24). The reaction rate is favoured by elevated free chlorine residuals and precursor concentrations. Typically, when total organic carbon (TOC) >4 mg/l at the works outlet, the chlorine dose required to achieve a satisfactory free residual at the tap after 2–3 days transit in the distribution system will result in THM values >100 μg/l. The reaction rate is also favoured by alkaline pH: at pH 9 about 10–20% more THM will form than at pH 7; increase in temperature: below 10°C there is no appreciable increase; and extended contact time (UKWIR, 2000). On the other hand, HAA formation generally decreases with increase in pH while dichloroacetic acid formation is independent of pH. Limits on DBP concentrations in drinking water have been set because of their possible health effects. Emphasis is now placed on avoiding or limiting pre-chlorination of a raw water containing organic matter in order to minimize the formation of DBPs. If a pre-disinfection stage is necessary, chlorine dioxide or ozone may be used instead of chlorine (Sections 11.16 and 11.21). Alternatively, since chloramine does not react with organic matter to produce DBPs to the same extent as free chlorine, the chlorine dose may be kept low enough to produce only chloramines by making use of ammonia naturally present in the water, or by applying chloramination of raw water (Section 11.8). If relatively high levels of DBPs are expected at the treatment works outlet, the degree of removal of organic substances prior to chlorination must be maximized. In some instances, it has been sufficient to move the pre-chlorination dosing point further downstream in the treatment process, for example after coagulation and clarification has removed a high proportion of DBP precursors. If that is insufficient, advanced treatment processes, described in Section 10.36, can be used to prevent DBP formation by the removal of precursors before final chlorination is applied.

While DBP formation continues throughout the distribution system, there is some evidence that HAAs undergo some biodegradation in distribution. DBP formation reaction is initially fast, with up to 50% being formed in the first hour or so, but takes several hours or even days to complete and therefore its concentration at the consumer's tap could be much higher than in the treated water leaving the works. The problem can be exacerbated in nutrient-rich waters when 'booster chlorination', that is addition of further chlorine at some key point(s) in the distribution system, has to be adopted to limit biological aftergrowth within the system. The formation of THMs and HAAs in the distribution system can be minimized by controlled dosing of ammonia to convert free chlorine residual to chloramine, which effectively halts the THM reaction. Hong (2007) reported that HAA formation with chloramination is slow but, because chloramine in analytical samples is not quenched by the AWWA recommended quenching agent, false high HAA concentrations have historically been measured in laboratory determinations.

Whatever DBP control measures are adopted, the WHO Guidelines emphasize that the disinfection process must not be compromised and that 'inadequate disinfection in order not to elevate the DBP level is not acceptable'.

11.8 THE AMMONIA–CHLORINE OR CHLORAMINATION PROCESS

Combined chlorine, that is mono- and dichloramines resulting from the reaction of chlorine with ammonia in water, is not commonly used as a primary disinfectant because it is a much weaker and slower acting disinfectant than free chlorine. In some instances, a theoretical contact period of several hours would be required for chloramine to achieve adequate disinfection of certain difficult waters (Smith, 1990). However, ammonia is sometimes deliberately added after final chlorination to produce a chloramine residual in the water passing into the distribution system. The primary reasons for using chloramines rather than chlorine are: greater persistence of chlorine residual in distribution; reduction in THM and HAA formation; superior control of biofilm growth (aftergrowth of bacteria or low forms of animal life) in the distribution system; and the possibility of higher doses (about 2.5 mg/l) with less risk of producing chlorinous tastes. The weight ratio of chlorine to ammonia (as N) is usually in the range 4:1 to 5:1; when the stoichiometric ratio of 5:1 is exceeded monochloramine is destroyed. Ammonia is added after final chlorination when the free chlorine has acted for the requisite contact time. The reaction is fast, taking only a few seconds to form a combined residual. The most effective pH range for chloramination is 7.5–8.5 when monochloramine is predominant. Excess ammonia can be used to ensure monochloramine is predominant. Dichloramine is considered to be a stronger disinfectant than monochloramine but it decays faster. Trichloramine has no disinfection properties and is volatile. Di- and trichloramines have very low taste and odour threshold concentrations (Section 10.32).

Monochloramine can be preformed as a solution of 1.5 g Cl_2/l by reacting ammonium sulphate with sodium hypochlorite in the presence of caustic soda at pH of about 10. About 5% excess ammonia should be maintained to ensure monochloramine is predominant. The product should not be stored for more than about 12–18 hours depending on the ambient temperature.

An advantage of maintaining residual chloramine in the distribution system is that, if routine examinations show chloramine is present at the ends of the distribution system, this should indicate that no serious pollution has entered the pipes en route. The residual cannot be high enough to disinfect pollution entering the system, but it is useful for monitoring the state of the distribution system. The principal disadvantage of chloramination is the nitrification of any excess ammonia or free ammonia released by the decay of chloramines. This process results in the formation of nitrite in the distribution system by nitrifying bacteria. Nitrite are toxic (Section 7.35). Methods of controlling and preventing nitrite formation include: optimizing the chlorine-to-ammonia ratio, typically 5:1 (Lieu, 1993); flushing out the affected sections; decreasing residence time in service reservoirs; removing excess ammonia locally by breakpoint chlorination (Section 11.9); chlorite addition (McGuire, 2009); reducing natural organic matter; periodic changes to free chlorine; pH control; and re-chloramination of the affected sections to eliminate bacterial growth (AwwaRF, 2003). Chloramine also forms N-nitrosodimethylamine (NDMA) (Section 7.80) and cyanogen chloride (CNCl).

Blending of chloraminated water with water containing free residual chlorine in distribution systems can result in breakpoint chlorination or in the formation of dichloramine and nitrogen trichloride, which are well known for causing taste problems. If blending is necessary, it should be carried out upstream of a service reservoir. Barrett (1985) developed a blending model which could be used to predict acceptable blends.

Chloramines are not removed by reverse osmosis or deionization which are the processes incorporated in conventional dialysis units (Ward, 1996). Relevant health professionals should therefore be advised whenever a change is proposed from free chlorine to chloramines. Chloramines can be removed either by granular activated carbon (GAC) installed upstream of the dialysis unit or by chemical reduction with ascorbic acid.

11.9 BREAKPOINT CHLORINATION

Where a water already contains natural ammonia the production of chloramine is unavoidable when chlorine is added. Substantially more chlorine may then have to be added to ensure the production of free chlorine to enhance bacterial kill. Initially chlorine reacts with ammonia present to form monochloramine. Increasing the chlorine-to-ammonia ratio successively leads to the conversion from monochloramine to dichloramine, trichloramine and ultimately nitrogen by chemical oxidation. Only when this reaction is complete does the addition of further chlorine produce free chlorine. Stoichiometrically the breakdown of ammonia to nitrogen commences at a chlorine: ammonia as N ratio of 5:1 and completes at a ratio of 7.6:1. (Chlorine:ammonia as NH_3 or NH_4^+ is 6.26:1 or 5.92:1, respectively.) In practice, the ratio for complete breakdown could be as much as 10:1 and is pH dependent. The point at which the free chlorine begins to form is called the 'breakpoint' for the water, and adding enough chlorine to exceed this is called 'breakpoint chlorination'. This is illustrated in Figure 11.1. In laboratory experiments Palin (1950) observed that for neutral to slightly alkaline pH when the ratio of chlorine to ammonia (as N) is less than 5:1 (by weight) the residual was mainly NH_2Cl; breakpoint occurred at 9.5:1 for pH 6; between 8.2:1 and 8.4:1 for pH 7–8; and 8.5:1 for pH 9. As the ratio increased to 10:1 and above there was a decrease in combined chlorine accompanied by increases in NCl_3 and free Cl_2. Apart from the advantage of producing free chlorine, breakpoint chlorination can sometimes reduce taste and odour

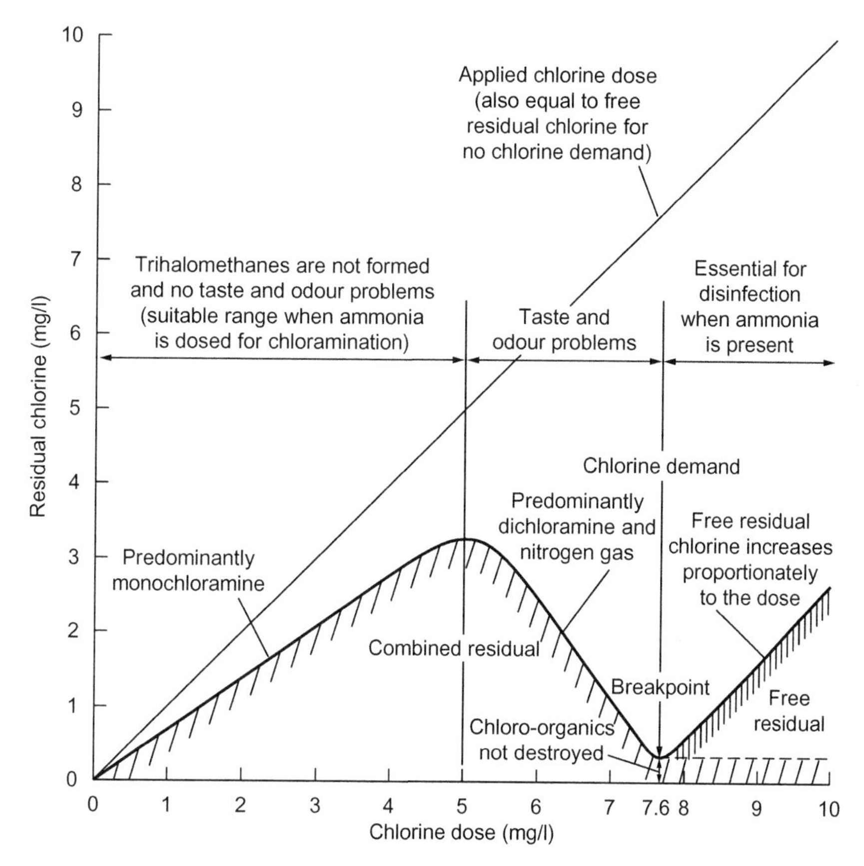

FIGURE 11.1

Theoretical breakpoint curve for a water containing 1.0 mg/l ammonia (as N).

problems resulting from di- and trichloramines. The foregoing reactions are complex, being dependent on numerous factors such as temperature, pH and contact time. The breakpoint reaction could take about 20 minutes to complete and depends on the water quality. In some waters the ammonia content may be so high (0.5 mg/l) that the amount of chlorine required to achieve breakpoint is uneconomic and other means to reduce the ammonia first should be adopted (Section 10.27).

11.10 SUPERCHLORINATION

'Superchlorination' refers to the practice of applying a high dose of chlorine, often much higher than necessary to meet the normal chlorine demand. The method is most often used on borehole or well waters which, though normally free of pollution, may be subject to unpredictable levels of pollution following heavy rainfall or some other circumstances. The normally unpolluted water may only require a small protective dose of chlorine of the order of 0.2 mg/l. To wait for the pollution to occur, detect it, and then increase the dose is impracticable since inevitably some of the pollution would pass into supply. Superchlorination is also practised on heavily polluted river waters, where the pre-treatment process does not include a pre-disinfection stage. In these circumstances, a continuous high dose of chlorine and adequate contact time is given sufficient to counter the worst conditions likely. After the contact period the water is partially dechlorinated by the injection of sulphur dioxide or sodium bisulphite (Section 11.12), leaving the desired residual to enter supply.

11.11 TYPICAL CHLORINE DOSE

Typically chlorine doses to final treated waters are in the range 0.2–2.0 mg/l of free chlorine to give a residual of about 0.02–0.3 mg/l at the consumer's tap. The lower doses (0.2–0.5 mg/l) tend to be used on clear groundwaters not subject to pollution; the higher doses relate to treated surface waters or to well or borehole supplies which are liable to sudden pollution, where superchlorination followed by partial dechlorination after adequate contact time may be advised. Some raw waters, particularly surface waters, can have a high chlorine demand of 6–8 mg/l (Smith, 1990).

Free chlorine has a taste threshold concentration of 0.6–1.0 mg/l. Chlorine dosing therefore often causes taste and odours, principally by the reaction of chlorine with some of the many trace compounds in the water (Section 10.32).

Equipment for storing and dosing chlorine is described in Section 12.13.

11.12 DECHLORINATION

Dechlorination can be achieved with sulphur dioxide (Section 12.18), sodium bisulphite ($NaHSO_3$) (Section 12.18) or sodium metabisulphite ($Na_2S_2O_5$) (Section 12.19) and, if required following superchlorination (Section 11.10), should be carried out after chlorine has had adequate contact time for disinfection. In this case, only partial dechlorination is required. In pre-treatment for reverse osmosis, complete dechlorination is the objective (Section 10.47). Chloramines can be similarly dechlorinated. Stoichiometrically 0.9 mg of sulphur dioxide or 3.6 mg of sodium bisulphite (25% w/w SO_2) or 1.38 mg of sodium metabisulphite (65% w/w SO_2) removes 1.0 mg of chlorine or 0.742 mg of monochloramine; in practice at least 15% more sulphur dioxide or twice the stoichiometric quantity for sulphites would be required; the dechlorination reaction is generally fast, taking less than 1 minute to complete. Sodium thiosulphate ($Na_2S_2O_3$), which is sometimes used to dechlorinate chlorinous waste discharges to the environment, takes longer, typically up to 5 minutes.

Other dechlorination methods include filtration through GAC or use of ammonia (Section 11.9) or hydrogen peroxide. GAC is sometimes used to protect chlorine sensitive reverse osmosis membranes; flow rates of 10–20 $m^3/h.m^2$ are typical with EBCT between 10 and 15 minutes. GAC catalyses the dechlorination reaction and in theory GAC is not consumed by chlorine. However, it should be replaced on a regular basis due to adsorption of other material. GAC used for dechlorination is not suitable for reactivation as it becomes weak in this service and breaks down during reactivation.

11.13 USE OF AMMONIA

Ammonia (anhydrous or aqueous) or ammonium sulphate is used for chloramination and should be added after disinfection and partial dechlorination (if applicable). Ammonia is typically dosed to provide a ratio in the range 3:1–5:1 of chlorine:ammonia (as N). Details of storage and dosing plant are provided in Section 12.20.

11.14 HYPOCHLORITE PRODUCTION ON SITE BY ELECTROLYSIS

Sodium hypochlorite can be produced by the electrolysis of brine (a solution of sodium chloride) in a fully mixed cell. A direct current passed through a solution of sodium chloride (common salt) containing Na^+ and Cl^-, produces chlorine at the anode, and hydrogen at the cathode. With mixing of the catholyte, anolyte and sodium ions in the solution, sodium hypochlorite (Na^+OCl^-) is produced. The principal reactions are as follows:

At the anode	$2Cl^- - 2e \rightarrow Cl_2$
At the cathode	$2H_2O + 2e \rightarrow 2OH^- + H_2$
On mixing	$2Na^+ + 2OH^- + Cl_2 \rightarrow Na^+OCl^- + Na^+Cl + H_2O$

The process has the advantage that the hypochlorite solution can be manufactured on site, thus avoiding the risks of transporting, storing and handling liquid and gas chlorine and the difficulty of meeting all the associated safety measures required. On-site generation (Plate 19(c)) produces a hypochlorite solution that is easy to handle, with favourable running costs as compared with the use of purchased liquefied chlorine. The process can be used on large or small supplies.

A diagram of the process used is shown in Figure 11.2; a plant is illustrated in Plate 19(c). The brine for electrolysis is prepared by withdrawing saturated brine from a salt saturator at a concentration of approximately 360 g/l and diluting this to 20–30 g/l. About 3 kg of sodium chloride is required to form 1 kg of chlorine; approximately 50% of the sodium chloride is converted to hypochlorite. Suitable salt storage arrangements are described in Section 12.15. The feed water temperature should be maintained above 5–10°C to maximize anode coating life; the product solution temperature should be below 30°C to inhibit chlorate formation. The temperature rise in the electrolyser is limited to about 10–15°C. In cold climates it may be necessary to warm the feed water usually using the product in the secondary circuit, and in hot climates to cool the dilution water using chillers.

The chlorine content of the hypochlorite produced is in the range 6–9 g Cl_2/l. The electrical consumption is 4–5 kWh per kg of chlorine produced; a low voltage direct current (e.g. 40 V) is used. Hydrogen gas is produced at the cathode at the rate of 330 litres at 15°C and 1 bar pressure, per kg of chlorine produced. The sodium hypochlorite solution with the hydrogen is fed to an enclosed hydrogen disentrainment tank of retention time >5 minutes, which can also act as a solution storage tank. The hydrogen degasses very rapidly to an air space at the top of the tank, which is force-ventilated with air at a rate 100 times the maximum hydrogen production rate. This reduces

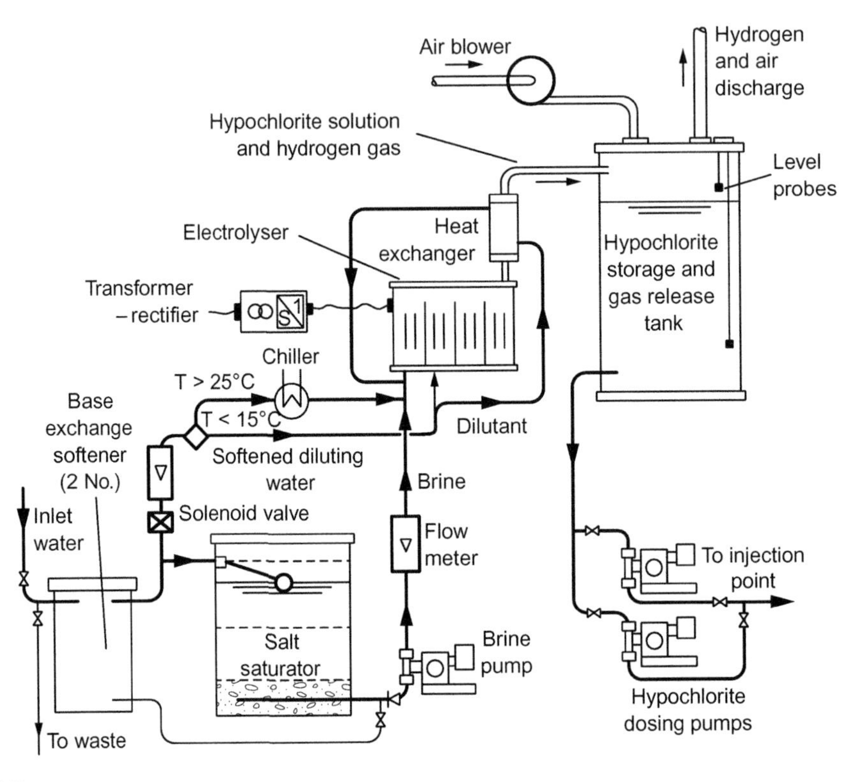

FIGURE 11.2

On-site hypochlorite production by electrolysis of brine.

the hydrogen concentration to one quarter of the 'lower explosion limit' of 4% of hydrogen in air by volume. The diluted gas is then vented to the atmosphere.

At coastal installations, hypochlorite can be produced from seawater. Due to the hardness of seawater with the presence of heavy metals and suspended matter, a special cell design with appropriate anodes and cathodes is used; the chlorine content of the hypochlorite solution is about 1–2 g/l and the respective feed water requirement is 1–0.5 m^3/kg of chlorine. The feed water should be free of pollution and have a relatively constant salinity and must be passed through a strainer to remove coarse suspended particles. Hypochlorite produced from seawater is used primarily for biological growth control in raw water, or cooling water treatment at desalination, or coastal power generation plants. The hypochlorite solution is unstable with a decay rate of 2–3% per hour due to the presence of heavy metals (Black & Veatch, 2010) and is not suitable for storage.

Safety measures. Electrochlorination plants present a potential fire and explosion hazard due to the hydrogen produced. Precautions should therefore be taken in the siting, layout and design of the plant (HSE, 1987). The plant should conform to the Explosive Atmosphere Directive (ATEX) 94/9/EC and electrical apparatus and instrumentation for installation in potentially explosive atmospheres should be selected and installed in accordance with BS EN 60079. It would also be necessary to prepare a hazardous area classification for the parts of the plant producing and storing sodium hypochlorite in accordance with BS EN 60079.

By-products. Several side reactions take place during the electrolysis of brine. One such produces both chlorate (ClO_3^-) (Section 7.24) which has a WHO Guideline value of 0.7 mg/l, and perchlorate. Commercial units are known to produce less than 50 μg of chlorate as ClO_3^- per 1.0 mg of chlorine, although this could almost double after 24 hour storage. The process also produces bromate from bromide present as an impurity in the salt: 1 mg of bromide forms 1.6 mg of bromate (BrO_3^-). All the bromide in the salt and in process water will be converted to bromate (Stanford, 2011).

All the sodium chloride used in the electrolyser feed is added to the water being treated as sodium and chloride. Chloride coupled with low alkalinity in a water can contribute to corrosion of metals (Section 10.41). The quantities of sodium and chloride produced are approximately 1.2 and 1.8 mg, respectively per mg of chlorine produced.

11.15 TESTING FOR CHLORINE

Common methods for on-line measurement of the chlorine content of a water are presented in Section 12.35.

Manual testing for chlorine requires the addition of reagents, in the form of tablets, powders or liquid to a measured volume of water sample. The most commonly used reagent for measuring chlorine content is diethyl-*p*-phenylenediamine (DPD) to produce a pink colouration, as described for on-line monitors in Section 12.35. The subsequent addition of potassium iodide to the same sample produces a colour intensity proportional to the total chlorine concentration (i.e. free plus combined chlorine); the difference between the two readings (i.e. total chlorine less free chlorine) gives the combined chlorine concentration.

The DPD-chlorine colour may be measured on a bench-top spectrophotometer at a wavelength of 515 nm, a handheld photometer or, less commonly nowadays, by comparison against a standard colour disc. Photometers may be either single parameter at a fixed wavelength or multi-parameter with a series of pre-set wavelengths. A typical measuring range for such instruments would be 0–5.0 mg/l Cl_2 with a resolution of 0.01 mg/l. The sample cell path length should be at least 10 mm. Standard discs containing a series of pre-calibrated glass filters are used in a handheld comparator in which the discs available most commonly cover the range 0–1.0 mg/l or 0–4.0 mg/l Cl_2.

11.16 USE OF CHLORINE DIOXIDE

Chlorine dioxide (ClO_2) is produced on site due to its relatively short half-life, commonly by the reaction between a solution of chlorine (in water) and sodium chlorite ($NaClO_2$) in a glass reaction chamber packed with porcelain Rashig rings. Sodium chlorite is supplied as a liquid containing 26% w/w of $NaClO_2$ in small containers or in bulk or as a powder containing 80% w/w $NaClO_2$ from which a solution of up to 31% w/w is made for use (Table 12.4). Spillages of sodium chlorite should be washed quickly as evaporation leads to deposits of highly flammable sodium chlorite powder.

The proportion of chlorine to sodium chlorite (100% w/w $NaClO_2$) will vary from the stoichiometric ratio of 0.39:1 to as much as 1:1, depending on the alkalinity of the water. Excess chlorine, above the stoichiometric requirement, should be limited to that required to neutralize the alkalinity, otherwise any further excess chlorine will promote chlorate (ClO_3^-) production and cause DBPs to form if precursors are present. Some of the alkalinity also reacts with the hydrochloric acid produced by the action of chlorine on water which otherwise would have reduced the chlorine dioxide yield by about 20%. The pH value should be about 4 and therefore for most waters the chlorine concentration needs to be over 500 mg/l. In practice, chlorine dioxide solution concentrations are maintained at less than 1 g/l in open systems, and 10 g/l in fully enclosed pressurized systems.

Other processes for producing chlorine dioxide include (1) reaction between $NaClO_2$ and hydrochloric acid; (2) reaction between $NaClO_2$, sodium hypochlorite and hydrochloric acid; (3) reaction between Purate® (a mixture of sodium chlorate and hydrogen peroxide) and sulphuric acid; and (4) electrolysis of a $NaClO_2$ solution (e.g. Pureline®).

The acid process uses about 1.25 times more sodium chlorite than the chlorine process to produce the same amount of chlorine dioxide. Stoichiometrically 1.67 g of sodium chlorite (100% w/w $NaClO_2$) and 0.54 g of HCl are required to produce 1 g of chlorine dioxide. In practice about 50% more sodium chlorite is required. Furthermore, between 300% and 350% of the stoichiometric quantity of acid is required to lower the pH (to ≤ 0.5) and neutralize alkalinity and maximize the yield. The electrolysis process is claimed to produce 99.5% ClO_2.

Chlorine dioxide is most commonly used as a disinfectant in cases where problems of taste and odour arise with chlorine, particularly those due to the presence of phenols (Walker, 1986). It is a powerful oxidant but at the limited permitted dose levels (limited because of by-products formed), its oxidation potential is not fully utilized. It is known to oxidize iron (II), manganese (II) (Knocke, 1991) (Section 10.12), colour (Aieta, 1986) and certain types of tastes and odours. It does not produce THMs or oxidize DBP precursors, nor does it react with ammonia or phenols in water. In the USA, it is primarily used as a substitute pre-oxidant for chlorine at the inlet to the works for taste and odour control, colour removal, pre-disinfection and iron and manganese oxidation because, unlike chlorine, it does not produce DBPs. Its bactericidal efficiency is comparable with that of free chlorine in the neutral pH range (Bernard, 1965) but, unlike chlorine, its efficiency increases with increasing pH (Aieta, 1986). Chlorine dioxide therefore has particular advantages for disinfecting waters liable to produce chlorophenol tastes, have a high pH or which contain substantial concentrations of ammonia.

The principal drawback with chlorine dioxide use is the formation of chlorate (ClO_3^-) and chlorite (ClO_2^-), both in the generation process and subsequently in the dosed water. Chlorate is produced when generating chlorine dioxide at too low pH and with high excess chlorine. At concentrations less than 10 g/l, chlorine dioxide disproportionates under alkaline and acidic conditions to form chlorate and chlorite, respectively. Disproportionation is the transformation of a compound into two dissimilar compounds by a process involving simultaneous oxidation and reduction. In practice, the disproportionation products are kept to a minimum by maintaining the pH of the solution in the range 3.5–7.5. Chlorate is also known to form by the exposure of water dosed with chlorine dioxide to sunlight, by increased pH (such as in softening) or by the action of chlorite ions with free residual chlorine in the contact tank or distribution system. Chlorite originates from the reactants in the generation process and from the disproportionation of chlorine dioxide. Typically, about 65–75% of the chlorine dioxide dose will disproportionate to chlorite. The chlorite so formed is effective in minimizing nitrite formation in the distribution system where chloramination is practised.

In the UK, a limit has been set for the combined residual levels of chlorine dioxide, chlorite and chlorate of 0.5 mg/l as ClO_2 in the water entering supply (UK, 2010), which effectively restricts the chlorine dioxide dosage to about 0.75 mg/l. US EPA has a maximum residual disinfectant level of 1.0 mg/l as ClO_2 which would allow a dosage of about 1.5 mg/l.

When used as a pre-oxidant it is seldom applied at a dose greater than 1.0 mg/l. At a dose of 1.0 mg/l, chlorate ion in the treated water would be in the range 0.2–0.4 mg/l. Chlorate can be removed by ferrous chloride and it is reported that 6–7 mg/l ferrous chloride as Fe per 2.5 mg/l of chlorine dioxide dose was effective in reducing the combined species concentration to 0.2 mg/l (Hurst, 1997). Sodium bisulphite or sulphur dioxide is effective in removing chlorite ions, achieving some 95% removal efficiency in the pH range 5–6.5. Due to the restrictions on its by-products in the finished water, chlorine dioxide is rarely used as the primary disinfectant.

11.17 SODIUM HYPOCHLORITE SOLUTION

Sodium hypochlorite is available for waterworks purposes as a straw-coloured solution containing 14–15% w/w of available chlorine. It is frequently used in place of chlorine gas for safety reasons. It can be supplied in small containers or in bulk, but loses its chlorine strength when exposed to atmosphere or sunlight. For properties and storage see Section 12.13 and Tables 12.4 and 12.5.

11.18 CALCIUM HYPOCHLORITE

Calcium hypochlorite is available in both powder and granular forms. The powder, commonly known as 'bleaching powder', is widely used in less developed countries of the world for disinfection of water supplies. It contains 30–35% w/w of releasable chlorine and excess lime. Calcium hypochlorite granules contain 65–70% w/w chlorine. Storage and dosing arrangements for both forms are provided in Section 12.16.

DISINFECTION USING OZONE

11.19 ACTION OF OZONE

Ozone (O_3) is a powerful oxidizing agent widely used for primary disinfection and oxidation. The bactericidal effect of ozone is rapid, usually requiring a contact time between 4 and 10 minutes with dosages of 1–3 mg/l. It is also known to be more effective than chlorine in killing viruses, cysts and oocysts. Ozone is often used to provide multiple benefits beyond disinfection:

- Oxidation of inorganic compounds such as iron (II), manganese (II) or hydrogen sulphide;
- Oxidation of organic compounds such as endocrine disrupting compounds, pharmaceutical and personal care products (PPCPs), algal toxins, colour and taste and odour-causing compounds (AwwaRF, 2007);
- Increased organics removal during coagulation;
- Reduced formation of chlorination DBPs; and
- Improved particle removal during sedimentation and filtration processes.

Ozone has a half-life in water between 0.5 and 5 minutes and its decay can be modelled using first order kinetics. The half-life is a function of water quality, temperature and pH. As the concentration of organics, temperature and pH decrease, the half-life increases. The half-life of ozone is too short for it to be effective as a residual disinfectant in the distribution system. Consequently, a free chlorine, monochloramine or chlorine dioxide residual is provided at the final stage of treatment.

Ozone does not react readily with ammonia and thus it is useful for the disinfection of waters containing ammonia which would otherwise require a large dose of chlorine. Ozone, in conjunction with GAC filtration, has been found useful for the removal of some pesticides in water (Foster, 1991). Pre-ozonation results in improved coagulation of particles and turbidity (Tobiason, 1995) and improved particle removal in sand and GAC filtration (Bourgine, 1998). To enhance coagulation, the ozone dose required is about 0.4 mg/mg TOC (Langlais, 1991).

Experiments have shown that ozone does not remove organic matter but breaks it down into smaller, more biodegradable organic matter (BOM) (Huck, 2000). These organic by-products, being more biodegradable than their precursors, can provide nutrients for biological growth and thus promote aftergrowth (regrowth) in the distribution system (Langlais, 1991). Therefore, BOM needs to be removed to prevent aftergrowth (Section 10.36).

The use of ozone has been found to limit the formation of chlorination DBPs, specifically THMs and HAAs. The mechanism for this is through oxidation of organic precursors and reduction in chlorine demand (Langlais, 1991). However, some of these ozonated precursors react more readily with chlorine, leading to an increase, rather than a decrease, in the formation of DBPs such as chloroacetic acid and chloroaldehydes and ketones during final chlorination (Foster, 1991; Hyde, 1984). The potential health significance of the many by-products of ozone is still not well understood. Among the identified by-products are formaldehyde, organic peroxides, unsaturated aldehydes, epoxides, HAAs and the inorganic by-product bromate. The addition of biological activated carbon reactors (Section 10.36) after ozonation can remove a high proportion of the organic by-products, but not the inorganic by-product bromate.

Bromate (BrO_3^-) is formed during ozonation of waters containing bromide. The current drinking water standard for bromate set by the EU, US EPA and WHO is 10 μg/L.

Bromate formation can be minimized through several control strategies. Bromide is present in water and is first oxidized to the weak acid-base pair of hypobromous acid (HOBr) and hypobromite (OBr^-), which has an acid dissociation constant (pKa) of 8.68. Bromate is then formed through the reaction of ozone with hypobromite. Hypobromous acid does not react with ozone; therefore one control strategy is to limit the concentration of hypobromite by reducing the pH.

Depression of pH to less than 7 minimizes bromate formation but may not be economical for high alkalinity waters (Siddiqui, 1995). A second control strategy is to add ammonia before ozonation, which reduces bromate concentration through monobromoamine (NH_2Br) formation (Ozekin, 1998). A third strategy is addition of chlorine to oxidize the bromide to hypobromous acid prior to ozone addition and then add ammonia to form monobromoamine (Buffle, 2004). A fourth strategy is addition of chlorine dioxide (ClO_2) prior to the addition of ozone although, following ozonation, any unreacted chlorine dioxide will form another DBP, chlorate (ClO_3^-) (Buffle, 2006). Hydrogen peroxide (H_2O_2) dosed after ozonation progressively decreases bromate formation with increasing dose. For maximum reduction, the weight ratio of H_2O_2:O_3 would be greater than 2:1 (Kruithof, 1995). Hydrogen peroxide is usually added immediately after the oxidation stage in the ozone contactor and would therefore prevent disinfection because it reacts with the residual ozone. Bromides are usually found in very small concentrations (<0.2 mg/l) in water and their removal (by ion exchange or reverse osmosis) is not cost effective. There are no proven methods in full-scale operation for the removal of bromate after its formation.

11.20 PRODUCTION OF OZONE

Ozone is unstable and must be generated on site. In small capacity, low-concentration generators, ozone is formed using air and a lamp emitting UV light. In commercial scale ozone generators, it is produced by passing air or oxygen through a silent electrical discharge. A high voltage alternating current is applied between two electrodes separated by a dielectric material and a narrow gap through which the gas containing oxygen is passed. The gap width is a function of the oxygen concentration, the dielectric material and the characteristics of the power supply. Typical gap widths are between 0.3 and 3 mm. Electrical power options commonly available are low frequency (50–60 Hz) in air fed ozone generators and medium frequency (350–6000 Hz) in oxygen fed (and some air fed) ozone generators. Medium frequency generators have a smaller discharge gap and allow higher ozone production per tube (Plate 20(a)). Low frequency systems have the simplest power supply.

In practice, the two electrodes are generally concentric tubes but may also be arranged as plates. In concentric tube designs, the outer electrode is a stainless steel tube. The dielectric is a glass or ceramic material, which can be plated onto the outside tube or the inner electrode. Historically a glass tube with an inner metallized coating forms the inner electrode. A typical ozone generator consists of several hundred to several thousand of such tubes assembled in a large vessel. The high voltage is applied to the inner tube; the stainless steel outer tube is connected to ground.

Approximately 90–95% of the energy input is converted to heat and must be removed through a cooling system. Small generators may utilize air whereas large generators rely on water. Manufacturers recommend that cooling water be non-corrosive, non-plating, free of settleable solids and have a low chloride content to minimize corrosion of stainless steel. Depending on the water quality, the cooling water system could be once-through (open loop) type using filtered plant water or closed loop type that uses a water-cooled heat exchanger or refrigerated water chillers. The system capacity should be adequate to limit the increase in water temperature to less than 5°C. The feed (air or oxygen) to the ozonizer must be oil-free (hydrocarbon content less than 40 ppm by volume as methane) and filtered through a 10 μm nominal (40 μm absolute) filter to reduce electrode fouling. Feed gas must be very dry; the normal requirement is that the feed should have a dew point below −62°C to prevent fouling. To achieve this dryness, it is usually necessary to refrigerate the feed gas and pass it through a desiccant. The feed gas should be cool with a temperature below 25°C and any compressors or blowers that deliver gas to the generators must be of the oil-free type. A schematic diagram of an ozone plant is shown in Figure 11.3.

With air as the feed, typical production is up to 4% w/w ozone in air (52.3 g O_3/m^3 of air at 0°C, 1.013 bar). Air feed systems can be either low pressure (<2 bar g) or high pressure (>4 bar g). In a high pressure system, most of the moisture can be removed in the aftercooler, which eliminates the need for a refrigerant dryer. High pressure systems are commonly used in packaged ozone generators.

With oxygen as the feed gas, typical production is about 10–12% w/w ozone in oxygen (148–179 g O_3/m^3 of oxygen at 0°C, 1.013 bar) although ozone generators capable of 16% w/w (240 g O_3/m^3) are entering the marketplace. Methods of supplying oxygen include use of liquid oxygen (LOX), delivered in bulk to site by tanker or pipeline, on-site production of LOX, or

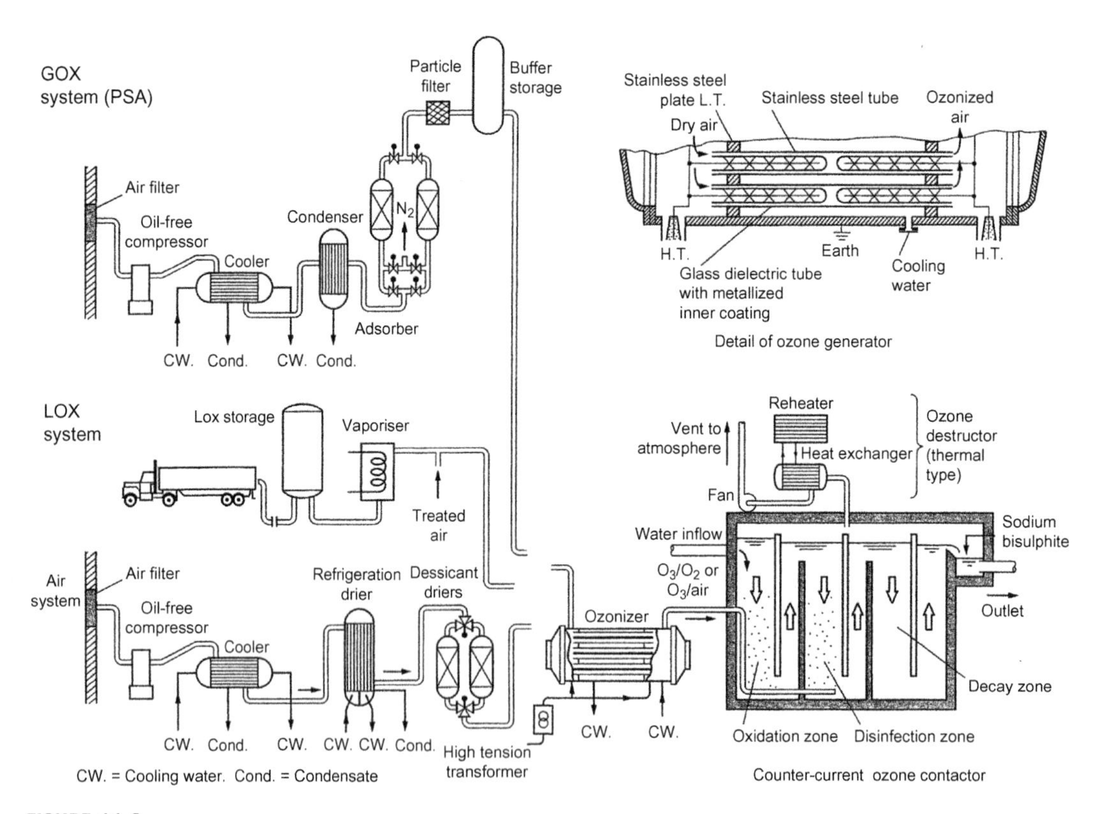

FIGURE 11.3

Ozone plant.

on-site generation of gaseous oxygen (GOX). LOX delivered in bulk is almost 100% pure oxygen with a dew point lower than $-80°C$ and therefore does not normally require further treatment. It is stored on site in a vacuum-insulated storage tank. Vaporizers, which use ambient air, warm water, steam or electrical energy, convert LOX to GOX for use in the ozone generators. Ozone generators have been found to be as much as 10–20% more efficient when a small amount of nitrogen is present (Rackness, 2005). Typical ranges for nitrogen addition are 0.1–4% w/w and vary by ozone system supplier. It is now common to install a supplemental air system that removes the moisture and hydrocarbons to the required limits and blends air (composed of nearly 75.5% w/w nitrogen) with the vaporized LOX.

LOX may be produced on site, in pre-engineered packaged plants with capacities greater than 20 t/d, by cryogenic air separation where liquefaction of air is followed by fractional distillation to separate oxygen, nitrogen and argon. This method is commonly used in commercial gas production plants where oxygen, nitrogen and argon are required as products. Oxygen concentration in the product gas is usually greater than 95% w/w.

More commonly, GOX is produced on site by pressure swing adsorption (PSA) or vacuum swing or vacuum-assisted pressure swing adsorption (VSA/VPSA) processes. These comprise one to three vessels containing a synthetic zeolite material which selectively adsorbs nitrogen from the air at elevated pressures allowing oxygen to pass through. A second adsorbent also removes moisture, carbon dioxide and hydrocarbons. Once the bed is saturated, it is regenerated by subjecting the bed to a lower pressure to release the by-products: nitrogen, moisture, carbon dioxide and hydrocarbons. A continuous stream of oxygen is maintained by switching the beds periodically at 30 second to 2-minute intervals. In PSA, adsorption is at 1–3 bar and regeneration is at atmospheric pressure whilst in VSA/VPSA, adsorption is at 0.2–1 bar and regeneration is under vacuum. The processes produce oxygen of purity 90–94% w/w with low hydrocarbon content (less than 10 ppm as methane) and dew point lower than $-80°C$. If the system has less than three adsorbers, an oxygen buffer storage vessel is usually provided. For VSA/VPSA systems, an oxygen booster compressor is required for feeding the ozone generator.

In general, bulk delivered LOX is the most economic choice for oxygen capacities up to 20 t/d. Adsorptive processes are suitable for small to intermediate capacities (PSA: 0.1–10 t/d and VSA/VPSA: 10–60 t/d) and cryogenic plants are best for capacities of more than 20 t/d. The source of oxygen should be determined through an economic evaluation that compares the cost and availability of delivered liquefied oxygen with the electrical and maintenance requirements of on-site generation alternatives.

The output of ozone generators increases with the increase in oxygen concentration, improved quality of the feed gas (i.e. dryness, hydrocarbon content and dust content), decrease in the gas discharge gap between the electrodes and increase in the frequency of the current applied to the dielectric.

The specific energy consumption for ozone generation is dependent on a number of factors and is in the range 20–27 kWh per kg O_3 produced for low pressure air feed plant, and about 8–12 kWh per kg O_3 for oxygen feed plants. Specific energy consumption for on-site oxygen production processes are about 0.3–0.35, 0.3–0.4 and 0.4–0.45 kWh/kg O_2 for cryogenic, VSA/VPSA and PSA, respectively. Ozone production plants require cooling water between 1.5 and 3 m^3/kg O_3 produced, the quantity being a function of maximum cooling water temperature and heat transfer within the generator shell.

11.21 OZONE DISSOLUTION AND CONTACT

Ozone must be transferred from the gas phase to water efficiently to minimize losses. Therefore, the design of the ozone–water contacting units is critical to the transfer process. In ozone contactors for disinfection, fine bubble diffusers are commonly used to transfer ozone into water. They are either porous rod, dome or disc type and require submergence depths between 5.5 and 7.5 m. They are operated under the ozone generator discharge pressure. Transfer efficiencies achieved are usually greater than 90%. Transfer efficiency is a function of the diffuser pore size and submergence. The number of diffusers may be calculated based on the rated gas flow for each diffuser. Usually an additional 10–30% may be installed to account for fouling, and blow off diffusers are required to release water from the submerged section of the ozone gas conveyance pipeline. Systems that produce low concentrations of ozone require more diffusers than systems capable of producing ozone at 8–12% w/w. The number of diffusers controls the dimensions of the ozone contactor transfer cell. Clogging of the diffuser pores due to the precipitation of iron and manganese is a disadvantage and tendency to foul precludes their use in pre-ozonation applications if raw water suspended solids concentrations are high. The gaskets and diffusers require annual inspection and periodic replacement based on the deterioration of the gaskets and the fouling.

If a pressurized water flow is available, a vacuum can be created through the throat of a venturi to draw ozone gas into solution, resulting in excellent ozone transfer, but at the cost of energy required for pressurizing the side stream. These injection systems commonly achieve 93–98% transfer efficiency but the side stream must be pressurized to 3–6 bar and the gas to liquid ratio limited to less than 0.3–0.5 volume gas/volume liquid. Transfer is maximized by applying the ozonated side stream at locations with backpressure exceeding 5 m, minimizing the side stream pipe length, and providing good dispersion and mixing of the side stream with the bulk flow. The side stream flow is usually less than 10% of the main stream flow. Pipeline static mixers can provide mixing with the bulk flow and are similar to those described in Section 8.9. The main line static mixer may require about 0.5 m headloss. The static mixer should be followed by a contactor for oxidation and disinfection reactions to complete.

Alternative transfer methods include turbine mixers and static radial diffusers. The turbine mixers aspirate ozone feed gas into the water and eject the resulting mixture into the contactor in a manner which encourages mixing with the bulk liquid (Langlais, 1991). In some designs, the mixer motor is submerged. In static radial diffusers, about 10% of the water flow is pumped to a submerged radial diffuser where it meets the ozone gas stream, dividing this up into very fine bubbles when injected into the contactor through the diffuser head (Anonymous, 1994).

Ozone contactors with diffusers are commonly designed to incorporate at least two transfer cells in series. In the first transfer cell, the initial ozone demand is satisfied, and in the second, a sufficient quantity of ozone is added for disinfection. US EPA (2006) provides guidance on *Ct* values for *Giardia*, virus and *Cryptosporidium* inactivation (Tables 11.2–11.4). *C* is the concentration of dissolved ozone in mg/l. In ozone contactors, the residual is measured at several locations because of its short half-life. Therefore, *Ct* is calculated for the contact time upstream of each measuring point and the summated *Ct* values for the contactor is used to determine the disinfection credits (Black & Veatch, 2010). A further stage is usually included to provide retention for ozone to decay. At the outlet of the ozone contactor, excess ozone may be removed from the water by physical or chemical methods. One such method involves cascading the water over weirs. Ozone is a highly volatile gas and is easily stripped from water (Henry's Constant = 100 000 atm.m^3/mol (AwwaRF, 1991)). A second more widely accepted method for removing ozone is by dosing an ozone scavenger such as sodium bisulphite, hydrogen peroxide or calcium thiosulphate.

According to US EPA (1990), counter-current ozone contactors with diffusers may receive flat inactivation credits of 1-log for viruses and 0.5-log for *Giardia*, if ozone residuals of 0.1 and 0.3 mg/l, respectively, are maintained at the outlet of the first cell of the contactor. Therefore, maintaining an ozone residual of at least 0.3 mg/l at the outlet of the first cell may reduce the Ct required in subsequent stages. The procedure for calculation of Ct values in the subsequent cells is described in the US EPA design criteria (US EPA, 1990; Langlais, 1991). Ideally t should be the t_{10} value which for baffled ozone contactors could vary in the range of 0.5–0.7 of the t_T (Section 11.5).

The number of cells in a baffled contactor is determined by the need to achieve uniform flow and to minimize short-circuiting. The greater the number of baffles, the closer the flow regime is to plug flow. The preferred flow configuration in the transfer cell is one in which the ozone gas injected by diffusers in one cell flows upwards counter-current to the water flowing downwards. In the second cell, the flow is reversed using underflow baffles to enable counter-current flow to be re-established in the second transfer cell (Fig. 11.3). This provides good liquid–gas contact and helps to minimize short-circuiting. To minimize capital costs, transfer may occur in adjacent counter-current and co-current cells. Where large flows are treated, the flow is split into several equal parallel streams. Contactors are often designed using CFD modelling (Section 14.18) to minimize short-circuiting (Plate 21(c)).

With the advent of side stream injection, ozone may be injected into a new or existing pipeline, provided it is constructed or coated with ozone compatible materials. Pipeline contactors approach plug flow more closely than rectangular baffled contactors and may serve a necessary purpose of conveying water between processes. Side stream injection systems typically consist of only one location through which ozone can be delivered to the process flow.

Table 11.4 *Ct* values (mg.min/l) for *Cryptosporidium* inactivation by ozone[1] (US EPA, 2006)

	Ct value at water temperature (°C)										
Log credit	**≤0.5**	**1**	**2**	**3**	**5**	**7**	**10**	**15**	**20**	**25**	**30**
0.25	6.0	5.8	5.2	4.8	4.0	3.3	2.5	1.6	1.0	0.6	0.39
0.5	12	12	10	9.5	7.9	6.5	4.9	3.1	2.0	1.2	0.78
1.0	24	23	21	19	16	13	9.9	6.2	3.9	2.5	1.6
1.5	36	35	31	29	24	20	15	9.3	5.9	3.7	2.4
2.0	48	46	42	38	32	26	20	12	7.8	4.9	3.1
2.5	60	58	52	48	40	33	25	16	9.8	6.2	3.9
3.0	72	69	63	57	47	39	30	19	12	7.4	4.7

Note: [1]Systems may use this equation to determine log credit between the indicated values:
Log credit = $(0.0397 \times (1.09757)^{Temp}) \times Ct$.

About 5–10% of the ozone introduced into the water remains in the spent gas at the top of the contactors. A destructor is required to convert unused ozone to oxygen for safe discharge. This conversion can be achieved by heating the gas to 350°C at which temperature decomposition takes place in 5 seconds, or by heating the gas 5–10°C above the inlet temperature (to decrease relative humidity) and passing it through a catalyst to accelerate decomposition. The higher temperature thermal system usually incorporates a heat recovery system. Its main drawbacks are a longer start-up time, increased energy use and larger footprint. The catalytic system on the other hand can be prone to fouling by moisture. Destructors are usually of stainless steel construction and any chlorine from pre-chlorination could cause severe corrosion at high operating temperatures.

Ozone contactors are normally of concrete construction, which is resistant to attack by ozone. The specification should be the same as for other water retaining structures (Mrazek, 1981); a good concrete mix (BS EN 206, 2013) and compaction are essential. Special concrete admixtures and 50 mm of rebar cover are used to protect the rebar from corrosive conditions. Maximum design crack width should be limited to 0.2 mm.

11.22 OZONE SAFETY

Ozone is toxic and a dangerous gas to handle; its odour perception threshold is less than 0.02 ml/m^3. The 15 minute exposure limit is 0.2 ml/m^3 (HSE, 2011). Ozone leak detectors must be installed to shut down ozone generating equipment in the event of a leak. Other precautions must be taken in the design and layout to minimize hazards to health (HSE, 2014). In addition, ozonized air is highly corrosive in the presence of moisture; hence piping and other equipment must be of special materials, mostly stainless steel grade 316L.

DISINFECTION USING ULTRAVIOLET RADIATION

11.23 UV DISINFECTION

UV radiation is a form of electromagnetic radiation of wavelength between 100 and 400 nanometres (nm), between X-rays and visible light in the electromagnetic spectrum. There are four classes of UV radiation: UV-A (315–400 nm), UV-B (280–315 nm), UV-C (200–280 nm) and vacuum UV (100–200 nm) (Meulemans, 1986).

The main mechanism of UV disinfection occurs via absorption of UV light in the germicidal spectrum (200–300 nm) by the cellular nucleic acids within a microorganism. When UV light penetrates bacteria, viruses and protozoan cysts, the energy is absorbed by the deoxyribonucleic acid (DNA) or ribonucleic acid (RNA) within the microbe resulting in photochemical damage. The damage physically inhibits the enzymes used for nucleic acid synthesis and thus blocks copying of the damaged DNA or RNA during replication. This renders the microorganism inactive.

Of the pathogens of concern in drinking water treatment, viruses are the most resistant to UV disinfection, with the highest UV dose requirements associated with Adenovirus Types 40 and 41. Bacteria are less resistant to UV than viruses with *Cryptosporidium* oocysts and *Giardia* cysts being the most susceptible pathogens to UV disinfection. The required UV doses to achieve various inactivation levels for *Cryptosporidium*, *Giardia* and Adenovirus, as defined in the US EPA's Long Term 2 Enhanced Surface Water Treatment Rule (LT2ESWTR) (US EPA, 2006) and UV Disinfection Guidance Manual (UVDGM) (US EPA, 2006a), are presented in Table 11.5.

Some microorganisms have the ability to repair UV-induced damage to their genome through the processes of photorepair and dark repair. In practice, the UV doses applied in the disinfection of drinking water result in damage to the microbial genome that is so significant that its reactivation mechanisms are rendered ineffective. As a result, microbial reactivation in drinking water disinfected with UV is not of significant concern if operated in accordance with recognized guidelines and dose requirements.

Table 11.5 UVDGM dose requirements (in mJ/cm^2)

	Log inactivation							
Target pathogen	**0.5**	**1.0**	**1.5**	**2.0**	**2.5**	**3.0**	**3.5**	**4.0**
Cryptosporidium	1.6	2.5	3.9	5.8	8.5	12	15	22
Giardia	1.5	2.1	3.0	5.2	7.7	11	15	22
Virus	39	58	79	100	121	143	163	186

11.24 UV EQUIPMENT

UV systems used for the disinfection of drinking water typically consist of the reactor shell, UV lamps, lamp sleeves, UV sensors, ballasts, electrical/control enclosures and cleaning systems. UV systems may also include equipment such as on-line UV transmittance (UVT) monitors, temperature sensors and water level probes. Figure 11.4 illustrates a simplified UV reactor; Plate 20(b) shows a typical UV installation.

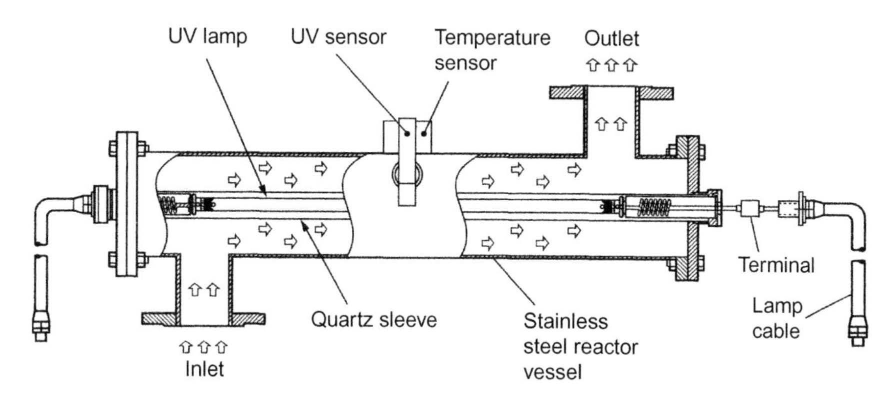

FIGURE 11.4

UV treatment chamber.

Based on information from: Hanovia Ltd.

UV reactors used in drinking water applications are closed inline vessels, with lamps arranged parallel, perpendicular or at an angle to the direction of flow. Current commercially viable UV disinfection systems employ low pressure (LP), low pressure high output (LPHO) or medium pressure (MP) lamps. The low or high 'pressure' refers to the mercury vapour pressure inside the lamp. LP and LPHO lamps emit near-monochromatic output at 254 nm while MP lamps emit a polychromatic spectrum, including the germicidal wavelength range of 200–300 nm. While MP lamps have a much higher UV output per unit length than LP and LPHO lamps, they are much less efficient since much of the radiation that they emit is outside the germicidal range that is effective for disinfection. A comparison of the operating characteristics of LP, LPHO and MP lamps is presented in Table 11.6.

The lamps inside the reactor are encased in sleeves, typically constructed of high purity quartz, which isolate the electrical components from the water, help maintain optimum operating temperatures and serve to protect the lamps from liquid-generated forces and thermal shock. The intensity of UV light is measured at a fixed point within the reactor by UV sensors. UV sensors are installed in quartz sleeves or in ports with quartz monitoring windows, providing a barrier against the wetted interior of the reactor while allowing passage of UV radiation for measurement of the UV intensity.

Lamps are powered by ballasts which regulate the electrical power and control the level of energy to operate the lamps. Ballasts that are commonly used in UV disinfection systems consist of electronic, electromagnetic or magnetic ballasts. Electronic ballasts and electromagnetic ballasts allow continuous adjustment of lamp output (i.e. intensity); transformer-based magnetic ballasts allow only step adjustment or fixed lamp output. During system operation, the ballasts can vary the power to the lamps from the minimum power setting (approximately 30–50%) to 100% power which enables the UV system to operate efficiently at a wide range of flow rates and UVT values. The ballasts are mounted in electrical enclosures and are connected to the lamps by dedicated cables.

Table 11.6 Operating characteristics of mercury vapour lamps

Parameter	LP	LPHO	MP
Germicidal UV radiation	Monochromatic (254 nm)	Monochromatic (254 nm)	Polychromatic, (inc. 200–300 nm germicidal range)
Mercury state	Liquid	Solid amalgam	Liquid
Mercury vapour pressure (Pa)[1]	Approx. 0.93	0.18–0.6	40 000–4 000 000
Operating temperature (°C)[1]	Approx. 40	60–100	600–900
Electrical input (W/cm)[1]	0.5	1.5–10	50–250
Electrical to germicidal UV conversion efficiency (%)	35–40	30–40	10–20
Relative number of lamps[1]/UV equipment capital cost	High	Intermediate	Low
Lifetime (hrs)	8000–10 000	8000–15 000	4000–9000

[1]*Source of information*: US EPA (2006a).

Three general methods are used to clean lamp sleeves and UV sensors to remove fouling: (1) off-line chemical cleaning, which consists of flushing or spraying the interior of a drained UV reactor with a cleaning solution; (2) on-line mechanical cleaning, using wipers (Teflon rings, stainless steel brush) that physically remove fouling as they are driven up and down the lamp and sensor sleeves by electric motors or pneumatic piston drives; and (3) on-line chemical/mechanical cleaning, using a collar filled with cleaning solution to remove fouling by physical contact with the sleeve in combination with chemical cleaning. The chemical cleaning solutions usually consist of phosphoric or citric acid.

The majority of the energy input to UV lamps is not converted to light, but is wasted as heat. The water flowing through a UV reactor constantly cools the lamps, avoiding premature lamp ageing or lamp failure as a result of overheating. Lamp overheating may result from low or no flow conditions in a reactor or from exposure of UV lamps to air in reactors that are not completely flooded. As a result, UV reactors typically include water level probes and temperatures sensors to ensure that they are completely flooded and that the water temperatures remain below levels that can cause overheating. Medium pressure reactors often incorporate a cooling circuit for use on reactor start-up, when forward flow is inhibited until the lamps reach operating temperature.

Some UV systems may be equipped with on-line transmittance monitors, which allow continuous, real-time measurement of the UVT. Depending on the dose-control approach employed by the UV reactor, the on-line UVT measurement may be used for system control or solely as a monitoring device. The most common type of on-line UVT monitor is a flow through spectrophotometer that measures the UV absorbance (UVA) and/or UVT at 254 nm from a side stream that is diverted from the main influent or effluent flow to the UV disinfection facility.

Disinfection of water by UV radiation has enjoyed a spike in popularity as a result of its many advantages over conventional disinfection methods by chemicals such as chlorine. However, UV, like any other disinfection technology, also has its share of disadvantages. A comparison of UV disinfection versus chemical disinfection is presented in the Table 11.7.

Table 11.7 Comparison of UV disinfection versus chemical disinfection

Advantages	Disadvantages
• Short contact time	• Higher power consumption
• Minimal space requirements	• May require uninterruptible power supply and standby power source
• No transportation, storage or handling of hazardous chemicals	• Effectiveness reduced by suspended solids, colour, soluble organic matter and turbidity
• No significant formation of toxic disinfection by-products[1]	• Risk of microbial reactivation by light and dark repair mechanisms
• Simple equipment; easy to operate and maintain	• No disinfectant residual
• Good reliability	• Need for periodic replacement of lamps
• Effective against *Giardia* and *Cryptosporidium*	• High UV doses required for inactivation of Adenovirus Types 40 and 41
• Performance not affected by pH of water	

Notes: [1]Includes formation of THMs, HAAs and AOC at UV doses applied in the disinfection of water (Ijpelaar, 2007). Nitrite formation can occur via exposure of nitrate to wavelengths ranging between 200 and 240 nm, which can be significantly reduced in MP systems by incorporating quartz sleeves that block wavelengths below 240 nm when high levels of nitrate are present.

11.25 UV DOSE AND PROCESS DESIGN CONSIDERATIONS

UV dose is the product of the UV intensity to which a microbial population is exposed and the time of exposure. The theoretical exposure time within a UV reactor can be calculated by dividing the volume of a UV reactor by the flow rate. UV intensity is typically measured in milliwatts per square centimetre (mW/cm^2) or watts per square metre (W/m^2) and time in seconds (s). The resulting units of UV dose are $mW.s/cm^2$ or $W.s/m^2$, which are equivalent to millijoules per square centimetre (mJ/cm^2) and joules per square metre (J/m^2), respectively. However, the UV dose provided by a UV reactor cannot be calculated simply by multiplying the UV intensity by the theoretical exposure time since all real world reactors have a dose-distribution. Various UV doses are experienced by different microbial populations as they flow through the reactor as a consequence of a non-uniform UV intensity distribution, coupled with non-ideal hydraulics resulting in various velocities and flow paths throughout the reactor. As a result, UV reactors must be tested to validate the dose that is provided under various operating conditions, as reviewed in the following section.

The principal parameters used for the design of UV disinfection systems include the disinfection requirements (target pathogen, required level of inactivation and associated UV dose requirement), flow rate, lamp ageing, fouling and associated water quality constituents, UVT and possibly turbidity or suspended solids (for unfiltered supplies). The design dose for the UV system should be based on the target microorganism, the desired level of inactivation and the applicable guidelines, as enforced by the local regulatory authority.

The dose requirements of the UVDGM, as presented in Table 11.5 have been accepted in numerous countries besides the United States, including Canada, Australia, the UK and other European countries. The UV disinfection requirements of the Austrian Standards Institute (ÖNORM) and the German Technical and Scientific Association for Gas and Water (DVGW) have also been widely applied around the world. In the UK, the DWI have issued guidance on the use of UV for disinfection which draws heavily from the UVDGM and requires installations to be validated to one of the UVDGM, DVGW or ÖNORM protocols (DWI, 2010). The guidance also includes details of monitoring and other regulatory requirements. The DVGW and ÖNORM regulations require UV systems to provide a *Bacillus subtilis* reduction equivalent dose (RED; Section 11.26) of 40 mJ/cm^2 to provide a general disinfection barrier, achieving a 99.99% (4-log inactivation) of human pathogens including bacteria, protozoan cysts and a majority of viruses. These requirements are quite different from those provided in the UVDGM which allow for selection of the required dose based on a specific target pathogen and required log inactivation.

The distribution of UV light throughout a UV reactor, and thus the disinfection efficiency of a UV system, is a function of the UV output of the lamp, the absorptive properties of the quartz sleeve and impact of sleeve fouling, as well as the UVT of the water. UV lamps degrade as they age and exhibit reduced output over time. The rate at which lamps degrade is specific to each lamp and ballast technology, with typical lifetimes presented in Table 11.6. Lamp ageing tests are typically conducted in order to assess the reduced intensity achieved at the end of the lamp life.

The disinfection efficiency of a UV system can be greatly affected by deposits that build up on the lamp sleeves, resulting in a reduction in UV light reaching the target pathogens in the water. In addition, fouling can occur on the UV sensor sleeves, resulting in falsely low UV intensity measurement. In either case, fouling results in UV system operation at elevated power levels to compensate for the reduction in the UV light. Fouling can occur from accumulation of deposits as a result of thermal effects, photochemical processes, or natural precipitation of minerals and typically consists of iron, aluminium, manganese and/or calcium and magnesium (hardness). Fouling may also be organic in nature, such as biofilms, which may occur within inactive reactors that are left flooded for long periods.

UVT is a measurement of the ability of UV radiation to penetrate the water and is reported as a percentage for a sample of a defined path length (typically 1 or 10 cm). As UV light travels further away from a UV lamp, the surrounding water absorbs more UV, with less energy available for disinfection. UVT is a function of the physical (suspended solids, turbidity), dissolved organic (humic/fulvic acids, phenols) and dissolved inorganic (iron and manganese) characteristics of the water. Typical UVT values for drinking water presented for disinfection range between 70% and 99%. The attenuation of UV light through water can also be expressed in terms of UVA, which has a direct mathematical relationship with UVT as demonstrated in the following equation:

$$UVT = 100 \times 10^{-UVA}$$

Suspended solids and turbidity also affect the efficiency of UV disinfection due to their ability to scatter the radiation and shade or shield the organisms. Research by Qualls et al. (1983) concluded that particles larger than 8–10 μm have an adverse effect on UV disinfection efficiency. Passantino et al. (2004) and Christensen and Linden (2002) found that the effects of turbidity on UV disinfection are accounted for in UVT measurements up to 10 NTU. In the event that the turbidity is in excess of 10 NTU, the dose requirements outlined in Table 11.5 may not apply, and site-specific dose–response testing is likely to be required.

11.26 UV REACTOR VALIDATION

Due to the dose-distribution occurring within UV reactors, full-scale systems must be validated in order to demonstrate their disinfection performance. UV systems can be validated on site at a water treatment facility or off site at a validation test centre. There are four internationally recognized UV validation test centres, one in Austria, one in Germany and two in the USA, each of which employs validation protocols dictated by the country's regulatory agency.

The validation procedure is a two stage process. In the first stage, the UV sensitivity of a selected surrogate microbe is accurately determined in the laboratory using a collimated beam apparatus. The test is repeated at a series of different UV doses to derive a dose–response curve for the organism. In the second stage, the same surrogate microbe is introduced at a known concentration to the inlet of the UV reactor under test. Typical surrogate microbes include *Bacillus subtilis* spores, MS2, Qβ, T1UV and T7 bacteriophage. The reactor is operated over a wide range of conditions by adjusting parameters such as the water flow rate, water UVT (altered by the addition of additives such as lignin, sulphonic acid, humic acid, instant coffee) and reactor power and the reduction in viable organism numbers for each test condition is analysed. Relating these values back to the dose–response curve derived in the first stage allows an equivalent UV dose to be determined for each condition. Since the

UV dose is measured indirectly and is based on the measured reduction of the surrogate microbe, this is referred to as the Reduction Equivalent Dose (RED) for the associated operating condition.

Two general UV system control philosophies can be developed from validation data, referred to as the intensity set point and calculated dose-monitoring approaches. The intensity set point approach is the only monitoring method approved by ÖNORM and DVGW, while the UVDGM allows for UV system operation with both control philosophies. If operated using the intensity set point approach, delivery of the required dose by a UV system is ensured by maintaining the UV intensity at or above an intensity set point while maintaining the flow through the reactor at or below a maximum flow rate. The benefit of this approach is that on-line UVT monitoring is not required since the manner in which the intensity set point is determined accounts for the impact of UVT on sensor readings. However, the increased simplicity of this monitoring approach results in decreased operating efficiency. As a result, this approach is typically best applied to smaller installations with relatively low power demand that will benefit from a less complex UV system with reduced maintenance requirements. Additional operating flexibility and efficiency can be obtained in these systems by utilizing multiple intensity set points or an algorithm for live calculation of the intensity set point based on the measured flow rate.

UV reactors validated in accordance with the UVDGM's calculated dose approach are operated using dose-monitoring algorithms that calculate the RED of the surrogate microbe as a function of the operating variables of the UV system, typically consisting of UVT, flow rate (Q), measured UV intensity (S), maximum UV intensity (S_o) and the number of banks of lamps ('*Banks*' in the equation below) in operation. A typical format for an *RED* monitoring algorithm is presented in the following equation. Constants A, B, C and D are derived from validation test results.

$$RED = 10^A \times UVT^B \times \left(\frac{S/S_o}{Q}\right)^C \times Banks^D$$

Recent advances in validation techniques have led to the use of multiple surrogate microbes having UV sensitivities that bracket that of the target pathogen. UV sensitivity (D_L) is a measurement of the RED required to achieve a specific log inactivation of a microbe, based on the microbial dose–response with a low pressure UV lamp (i.e. 254 nm). A minimum of two surrogate microbes are required for this validation approach, with one surrogate being more sensitive to UV disinfection (lower D_L value) than the target pathogen and the other surrogate being more resistant (higher D_L value). This validation approach allows for the development of an algorithm that directly calculates the log inactivation or RED of the target pathogen, rather than the RED of the microbial surrogate. Typical formats for the log inactivation (Logi) and RED algorithms developed with this validation approach are presented in the following equations.

$$Logi = 10^A \times UVT^B \times \left(\frac{S/S_o}{Q \times D_L}\right)^C \times Banks^D$$

$$RED = Logi \times D_L = 10^A \times UVT^B \times \left(\frac{S/S_o}{Q \times D_L}\right)^C \times Banks^D \times D_L$$

UV reactor validation is subject to various uncertainties associated with the testing procedures, data analysis, model development and the accuracy of the UV equipment. In addition, biases exist as a result of differences in UV sensitivity between the microbial surrogate and the target pathogen (known as RED bias) as well as potential sources of polychromatic bias that are specific to MP systems. The benefit of advanced Logi or RED algorithms developed from multiple surrogates is that the RED bias can be removed from the algorithm since the UV sensitivity is included as a variable in the calculation.

For UV systems operated using the UVDGM intensity set point or calculated dose-monitoring strategy, a validation factor (VF) is applied to the RED to account for the biases and uncertainties associated with a specific reactor's validation. The *RED* is divided by the *VF* to calculate a validated dose (D_{Val}) as presented in the equation below.

$$D_{Val} = \frac{RED}{VF}$$

In order for a UV facility to be granted disinfection credit for *Cryptosporidium*, *Giardia* or virus inactivation, the D_{Val} must be greater than or equal to the required dose values presented in Table 11.5.

11.27 EMERGING UV ISSUES

Recent research has demonstrated that action spectra bias, a type of polychromatic bias, may be significantly greater in some MP UV systems than that reflected by the procedures in the UVDGM (Townsend, 2013). Action spectra bias occurs when a MP UV system is validated with a surrogate microbe that is more sensitive to germicidal wavelengths than the target pathogen. Inactivation of the challenge microbe during the validation may be partially due to wavelengths that will not have a significant impact on the target pathogen, resulting in an overestimation of the validated dose and log inactivation of the target pathogen at a water treatment facility. Action spectra bias is reactor, validation and site-specific and not all MP reactors will be affected. For those MP systems that are affected an action spectra bias factor (CF_{AS}) can be applied in the calculation of the VF. Guidance on procedures for the determination of CF_{AS} for MP systems is currently in development by the Water Research Foundation.

11.28 BOILING WATER

Boiling water is an extremely useful process for disinfecting water because boiling kills bacteria, viruses, ova and cysts present in polluted water and WHO (2015) gives the log reduction achieved for these organisms. It is reported to be equally effective whether the water is clear or cloudy, relatively pure or highly contaminated (though obviously contaminated or cloudy water should be used only as a 'last resort'). Turbid water should preferably be filtered through a clean cloth before boiling. Alternatively, the water should be boiled for up to 5 minutes. If a supply that is normally safe becomes contaminated with microorganisms, it may be necessary for a water utility to advise consumers to boil all water used for drinking or cooking or brushing teeth. In this situation, it is sufficient to bring water to the boil and the use of electric kettles with automatic switch-off is acceptable (Boucher, 1990). In the UK, immunocompromised individuals are advised to boil all drinking water from any source (Boucher, 1990). For the complete sterilization of a polluted water the WHO recommends bringing the water to a 'rolling boil' (large bubbles continuously coming to the surface) and then maintaining this for at least one minute for a clear water. At high altitude one minute extra time should be given for every 1000 m above sea level because of the lower temperature at which boiling takes place (WHO, 2011). Boiled water can become re-contaminated once it has cooled as it has no residual disinfectant and therefore should be stored in a clean closed container.

DISINFECTION OF WATERWORKS FACILITIES

11.29 DISINFECTION OF WATER MAINS AND TANKS

The UK recommended practices for disinfecting mains have been set out in Technical Guidance Notes 2, 3 and 4 published by Water UK (2011). Before a new or renovated pipeline is put into service, it should first be swabbed clear of dirt and debris with a foam swab and flushed with water. It should then be filled with water containing at least 20 mg/l of free chlorine and allowed to stand for 16 hours, after which it should be flushed and recharged with mains water and allowed to stand for a further 16 hours. Samples should then be taken from a number of points along the main and at its extremities and, if samples are found to be free of coliform organisms and give satisfactory results for residual chlorine, taste, odour and appearance, the main can be brought into service. All service pipes connected to the main should be flushed out. Alternatively, a renovated main may be disinfected for a minimum period of 30 minutes with 50 mg/l of free chlorine followed by flushing, filling with mains water and sampling. For very large diameter pipelines and treated water tunnels filling with water with 20 mg/l of free chlorine may be impracticable and consideration should be given to passing a slug of more highly chlorinated water through the system. Alternatively, manual swabbing or spraying, possibly using a remote-operated vehicle, may be considered.

A similar treatment is advised for an operation which involves cutting the main, and where there is a risk of contamination from water in the trench or other foul water (e.g. from sewer). For repairs to a live main which has to be cut with minor soiling at opening, all surfaces which come in contact with the drinking water must be cleaned down with water containing 1000 mg/l free chlorine. In all cut main repairs, following disinfection and flushing the main must only be returned to service when bacteriological and other qualitative tests have proved satisfactory. US practice is set out in AWWA standard C651-05 (AWWA, 2005).

Disinfection of reservoirs is set out in AWWA standard C652-11 (2011). Alternatively, the following variation could be used. Tanks and service reservoirs are usually hosed down with strong jets of clean water and the walls and floor are then brushed down with a chlorine solution containing not less than 20 mg/l chlorine as against 200 mg/l recommended in the AWWA standard. The reservoir is then half filled with water containing at least 0.5 mg/l chlorine and, after standing for 24 hours, a sample of the water is tested for coliform organisms. If the sample fails, the chlorine dose of the inlet water is increased to 1.0 mg/l and the reservoir is filled, the water being retested after a further 24 hours. If coliforms are absent, the reservoir can be filled with water having 0.5 mg/l chlorine, which is resampled and tested when the reservoir is full. Personnel entering a service reservoir which has been emptied for inspection or maintenance should scrub their footwear in a tray containing 1000 mg/l chlorine immediately before entering the reservoir. Normally chlorine tablets, bleaching powder, calcium hypochlorite granules or sodium hypochlorite are used as chlorine sources, but for large new transmission mains and reservoirs, chlorine water from a gas chlorinator might be injected into the filling water. All test water containing chlorine should be dechlorinated before disposal; chemicals available include sodium bisulphite, sulphur dioxide, sodium metabisulphite and sodium thiosulphate (UK Water, Technical Guidance Note 14 (UK Water, 2010)).

11.30 CONTROL OF AFTERGROWTH (REGROWTHS) IN DISTRIBUTION MAINS

Considerable aftergrowth may occur in distribution mains especially with nutrient-rich waters. Its cause and control is discussed in detail in a WHO publication (Levi, 2004). Organic matter, iron and manganese deposits, algae and corrosion products may foster growth in the mains of bacteria and other forms of life, such as biofilms, which are resistant to control measures (Flemming, 1998). Problems caused by biofilms include taste and odour, brown water, corrosion of mains, bacterial contamination, increased disinfectant demand, reduction of dissolved oxygen, greater headloss, creation of habitats for bacteria, viruses, protozoa and fungi and sloughing of biomass. In systems with biofilms it is often difficult to retain any residual chlorine in the water. No practicable level of residual chlorine prevents problems of aftergrowth if algae, residual iron or aluminium floc, suspended solids or substantial amounts of BOM (Section 10.34) are allowed to remain in a treated water. Phosphates used in plumbosolvency control can also contribute to aftergrowth. Swabbing followed by slug dosing with a heavy dose of chlorine passed slowly along the affected mains is a short-term remedy, but the permanent cure lies in optimizing the treatment process and thus improving the quality of the water distributed (Mouchet, 1992). In extensive distribution systems carrying nutrient-rich waters it may be necessary to adopt 'booster chlorination', that is the addition of further chlorine or chloramination of the water at some key distribution point or points in the system, usually at the inlet or outlet of a service reservoir. Chloramination has also become more common as a means of controlling biofilms and avoiding exceeding chlorination by-product concentrations (Section 11.7). This has become essential in some utilities (Norton, 1997). As noted in Section 7.68, bacteriological testing of water in service reservoirs is required under the UK Water Regulations.

11.31 DISINFESTATION OF DISTRIBUTION MAINS, WELLS AND BOREHOLES

Most distribution systems can become infested with small aquatic animals (Evins, 2004), especially systems with old mains carrying treated lowland surface waters which contain considerable organic matter. The most commonly experienced animals are *Asellus* (water louse), *Gammarus* (freshwater shrimp), nais worms and nematode worms. Occasionally the larvae of midges and flies may be found, having passed through filter beds or gained entry to a service reservoir. Many other small aquatic organisms can occasionally be present. The flushing of mains and chlorination is largely ineffective where animal growth is prevalent. The ability of the animals to leave reproductive spores or to reproduce from fragments means that reinfestation tends to be rapid: hence disinfestation is necessary. This can be achieved using approved chemicals such as Permasect WT which contains pyrethin or permethrin. The application has to be controlled: under Regulation 31 of the UK Water Regulations (DWI, 2014), the average and maximum concentrations of pyrethin and permethrin must not exceed 10 and 20 μg/l, respectively, nor may its use continue for more than 7 days.

In cases where corrosion of the main is the cause of high turbidity or coloured water, flushing and pigging with polyurethane foam swabs have been shown to be effective (Smith, 1986).

Iron bacteria may develop in mains and other types of bacteria may occasionally multiply into substantial colonies. One of the aeromonas strains, *Aeromonas hydrophila*, is believed capable of causing gastroenteritis (Edge, 1987). Bacterial infestation of mains can usually be dealt with by flushing, swabbing and then chlorinating. Typically this has to be done in sections when, as is likely, a whole system is affected. Maintenance of an enhanced chlorine residual can assist in preventing regrowth, but this practice may be inhibited by the need to keep DBPs below the permitted maximum (Section 7.24). In the case of iron bacteria in mains, the proper remedy is to revise the treatment of the water to remove the iron or, if the bacteria grow because of deterioration of old iron mains, to reline the mains (Section 16.11).

Iron bacteria are particularly prone to develop in wells and boreholes drawing water from ferruginous formations. These bacteria are sessile, that is attach themselves to a surface. The majority produce large masses of extracellular covering material in the form of slime which clings to well screens and borehole linings. Often the presence of large iron bacterial growths remains undetected until part of the slime detaches and is discovered in the water, or the yield of the well or borehole falls off because of clogging of screens. The slimes protect the bacteria against any biocide; hence physical removal of the slimes is necessary. Pumps and rising mains in wells and boreholes can be withdrawn and cleaned. However, to clean well screens, surging, jetting, chemical applications or steam injection may have to be adopted (Section 4.10). Prevention of regrowth in well screens and borehole linings after physical cleaning may have to comprise slug dosing with chlorine and then flushing. Chlorine should never be applied direct to a well or borehole as this allows the extent of the problem to remain unknown and may encourage corrosion of the well or borehole lining and of the pump inserted. The continuous application of chlorine may also be inadvisable since it tends to precipitate the iron and manganese in solution.

REFERENCES

Aieta, E. C. and Berg, J. A. (1986). Review of chlorine dioxide in water treatment. *Journal AWWA* **78**(6), pp. 62–72.

Anonymous (1994). *Ozonia Radial Diffuser*. Ozonia Product Information, Zurich.

AWWA C651-05. *Disinfection of Water Mains*. AWWA.

AWWA C652-11. *Disinfection of Water Storage Facilities*. AWWA.

AwwaRF (1991). *Ozone in Water Treatment: Application and Engineering*. Lewis Publishers Inc.

AwwaRF (2003). *Ammonia From Chloramine Decay: Effects on Distribution System Nitrification*. Report 90949.

AwwaRF (2007). Removal of EDCs and Pharmaceuticals in Drinking and Reuse Treatment Processes. Report 91188.

Barrett, S. E., Davis, M. K. and McGuire, M. J. (1985). Blending chloraminated and chlorinated waters. *Journal AWWA* **77**(1), pp. 50–61.

Bernard, M. A., Israel, B. M. and Olivieri (1965). Efficiency of chlorine dioxide as a bactericide. *Applied Microbiology* **13**(5), pp. 776–780.

Black & Veatch Corporation (2010). *White's Handbook of Chlorination and Alternative Disinfectants*. 5th Edn. Wiley.

Boucher, I. (1990). *Cryptosporidium in Water Supplies*. Third Report of the Group of Experts, London.

Bourgine, F. P., Chapman, J. I., Bastment, R. J. et al. (1998). The effect of ozonation on particle removal in drinking water. *Journal CIWEM* **12**(3), pp. 170–174.

BS EN 206:2013. *Concrete – Specification, Performance, Production and Conformity*. BSI.

BS EN 60079. *Explosive Atmospheres*. BSI.

Buffle, M. -O. (2006). *Enhanced Bromate Control During Ozonation: Pre-oxidation With ClO_2* (on line). http://e-collection.ethbib.ethz.ch/ecol-pool/diss/fulltext/eth16266.pdf. (January 2008).

Buffle, M.-O., Gailli, S. and von Gunten, U. (2004). Enhanced bromate control during ozonation: the chlorine-ammonia process. *Environmental Science and Technology* **38**, pp. 5187–5195.

Butterfield, C. H., Wattie, E. and Megregian, S. (1943–46). Influence of pH and temperature on the survival of coliform and enteric pathogens when exposed to free chlorine. *US Public Health Reports*, **58**(51), pp. 1837–1866; **59**, pp. 1661–1671; **61**(6), pp. 157–193; also *Journal AWWA* (1948), **40**, pp. 1305–1512.

Christensen, J. and Linden, K. (2002). New findings regarding the impacts of suspended particles on UV disinfection of drinking water. *Proceedings of the Annual Conference of the American Water Works Association*. New Orleans, LA, June 16–20.

DWI (2010). *Guidance on the Use of Ultraviolet (UV) Irradiation for the Disinfection of Public Water Supplies*. February 2010.

DWI (2014). *List of Approved Products*, (updated annually), London.

Edge, J. C. and Finch, P. E. (1987). Observations on bacterial aftergrowth in water supply distribution systems. *Journal IWEM* **1**, pp. 104–110.

Evins, C. (2004). Chapter 6: Safe piped water: managing microbial water quality in piped distribution systems. *Safe Piped Water* (Ed. Ainsworth, R.). IWA Publishing.

Flemming, H. C. (1998). Biofilms in drinking water system. *GWF Wasser/Abwasser* **139**(13), pp. 65–71.

Foster, D. M., Rachwal, A. J. and White, S. L. (1991). New treatment processes for pesticides and chlorinated organics control in drinking water. *Journal IWEM* **5**(4), pp. 466–476.

Hong, T., Liu, S., Song, H. et al. (2007). HAA formation during chloramination – significance of monochloramine's direct reaction with DOM. *Journal AWWA* **99**(8), pp. 57–69.

HSE (1987). *Fire and Explosion Hazards at Electrochlorination Plant*. Document No. Nov. HSE 490/11.

HSE (2011). *EH40/2005 Workplace Exposure Limits* (updated regularly), 2nd Edn.

HSE (2014). *Ozone Health Hazards and Precautionary Measures*. EH 38. 3rd Edn.

Huck, P. M., Coffey, B., Amirtharajah, A. et al. *Optimising Filtration in Biological Filters*. AwwaRF.

Hurst, G. H. and Knocke, W. R. (1997). Evaluating ferrous ion for chlorite removal. *Journal AWWA* **89**(8), pp. 98–105.

Hyde, R., Zabel, T. and Green, L. (1984). *Proceedings, Seminar on Ozone in UK Water Treatment Practice*. IWES, September.

Ijpelaar, G. F., Harmsen, D. J. H. and Heringa, M. (2007). *UV Disinfection and UV/H_2O_2 Oxidation: By-Product Formation and Control*. Techneau, D2.4.1.1.

Knocke, W. R., Van Benschoten, J. E., Kearney, M. J. et al. (1991). Kinetics of manganese(II) and iron(II) oxidation by potassium permanganate and chlorine dioxide. *Journal AWWA* **83**(6), pp. 80–87.

Kruithof, J. C. and Meijers, R. T. (1995). Bromate formation by ozonation and advanced oxidation and potential options in drinking water treatment. *Water Supply* **13**(2), pp. 93–103.

Langlais, B., Reckhow, D. A. and Brink, D. R. (Eds.) (1991). *Ozone in Water Treatment, Application and Engineering*. AwwaRF. Lewis Publication.

Levi, Y. (2004). Chapter 2: Minimising potential for changes in microbial quality of treated water. *Safe Piped Water* (Ed. Ainsworth, R.). IWA Publishing.

Lieu, N. I., Wolfe, R. L. and Means, E. G. (1993). Optimising chloramine disinfection for the control of nitrification. *Journal AWWA* **85**(2), pp. 84–90.

McGuire, M. J., Wu, X., Blute, N., Askenazier et al. (2009). Prevention of nitrification using chlorite ion: results of a demonstration project in Glendale, Calif. *Journal AWWA* **101**(10), pp. 47–59.

Meulemans, C. C. E. (1986). *The Basic Principles of UV-Sterilisation of Water*. Ozone + Ultraviolet Water Treatment. International Ozone Association, Aquatec Amsterdam, Paris.

Mouchet, P. (1992). Physico-chemical degradative changes in water while in the distribution system. *Techniques, Sciences, Methods* **6**, pp. 299–306.

Mrazek, L. G. (1981). Resistance of concrete structures to ozone penetration. *Concrete International* **3**(4), pp. 69–74.

Norton, C. D. and Le Chevallier, M. W. (1997). Chloramination, its effect on distribution water quality. *Journal AWWA* **89**(7), pp. 66–77.

Ozekin, K., Westerhoff, P., Amy, G. et al. (1998). Molecular ozone and radical pathways of bromate formation during ozonation. *Journal of Environmental Engineering ASCE* **124**(5), pp. 456–462.

Palin, A. T. (1950). Chemical aspects of chlorine. *Journal IWE* **IV**(7), pp. 565–581.

Passantino, L., Malley, J., Knudson, M., Ward, R. and Kim, J. (2004). Effect of low turbidity and algae on UV disinfection performance. *Journal AWWA* **96**, pp. 128–137.

Poynter, S. F. B., Slade, J. S. and Jones, H. H. (1973). The disinfection of water with special reference to viruses. *Journal SWTE* **22**, pp. 194–206.

Qualls, R. G., Flynn, M. P. and Johnson, J. D. (1983). The role of suspended particles in ultraviolet disinfection. *Journal of the Water Pollution Control Federation* **55**, pp. 1280–1285.

Rackness, K. (2005). *Ozone in Drinking Water Treatment: Process Design, Operation, and Optimisation*. AWWA.

Scarpino, P. V., Berg, G., Chang, S. L. et al. (1972). A comparative study of the inactivation of viruses in water by chlorine. *Water Research* **6**, pp. 959–965.

Siddiqui, M. S., Amy, G. L. and Rice, R. G. (1995). Bromate ion formation: a critical review. *Journal AWWA* **87** (10), pp. 58–70.

Smith, D. J. (1990). The evolution of an ozone process at Littleton water treatment works. *Journal IWEM* **4**(4), pp. 361–370.

Smith, G. A. (1986). A case study in pipeline maintenance. *Water and Wastewater International* **1**(3), pp. 4–5.

Stanford, B. D., Pisarenko, A. N., Snyder, S. A. et al. (2011). Perchlorate, bromate, and chlorate in hypochlorite solutions: guidelines for utilities. *Journal AWWA* **103**(6), pp. 71–83.

Stevenson, D. G. (1998). *Water Treatment Unit Processes*. Imperial College Press.

Teefy, S. (1996). *Tracer Studies in Water Treatment Facilities: A Protocol and Case Studies*. AwwaRF.

Tobiason, J. E., Edzwald, J. E. and Edzwald, J. K. (1995). Effects of ozonation on optimal coagulant dosing in drinking water treatment. *Journal WSRT-Aqua* **44**(3), pp. 142–150.

Townsend, B., Hawley, T. and Coggins, J. (2013). This way forward: addressing action spectra bias concerns in medium pressure UV reactors. *Proceedings of the North Carolina American Water Works Association and Water Environment Association 93rd Annual Conference*, Concord, NC, November 10–13.

UK (2010). *Water Supply (Water Quality) Regulations*.

UK Water (2010, 2011). *Technical Guidance Notes 2, 3, 4 and 14*.

UKWIR (2000). *Toolboxes for Maintaining and Improving Drinking Water Quality*. Report 00/DW/03/11, Section 3. Toolbox for Trihalomethane Formation.

US EPA (1990). *Guidance Manual for Compliance With Filtration and Disinfection Requirements for Public Water Systems Using Surface Water Sources*. Appendix O. Guidance to predict performance of ozone disinfection system.

US EPA (1991). *Guidance Manual for Compliance With the Filtration and Disinfection Requirements for Public Water Systems Using Surface Water Sources*.

US EPA (2006). National primary drinking water regulations: long term 2 enhanced surface water treatment rule. *Federal Register* **71**(3), pp. 653–786.

US EPA (2006a). *Ultraviolet Disinfection Guidance Manual for the Final Long Term 2 Enhanced Surface Water Treatment Rule*. EPA 815-R-06-007.

Walker, G. S., Lee, F. P. and Aieta, E. M. (1986). Chlorine dioxide for taste and odour control. *Journal AWWA* **78** (3), pp. 84–93.

Ward, D. M. (1996). Chloramine removal from water used in hemodialysis. *Advances in Renal Replacement Theory* **3**(4), pp. 337–347.

WHO (2011). *Guidelines for Drinking Water Quality*, 4th Edn.

WHO (2015). *Technical Brief: Boil Water*. WHO/FWC/WSH/15.02.

CHAPTER

Chemical Storage, Dosing and Control

12

12.1 CHEMICALS USED IN WATER TREATMENT

Chemicals are used for a range of different purposes in water treatment but the principal uses can be divided into basic functions as below. Their use is described in the chapters indicated; aspects of their storage and dosing are described in this chapter:

- Coagulation and coagulation aids (Chapter 8)
  - Aluminium salts
  - Ferric salts
  - Polyelectrolytes
- pH adjustment (Chapters 8, 10 and 11)
  - Acid (including CO_2)
  - Alkali
- Disinfection and dechlorination (Chapter 11)
  - Chlorine based
  - Dechlorination
  - Ammoniation
- Oxidation (H_2O_2, $KMnO_4$) (Chapter 10)
- Organics removal (PAC) (Chapter 10)
- Plumbosolvency control (Chapter 10)
- Fluoridation (caries control) (Chapter 10)

Chemicals used in water treatment must be approved for use by the appropriate governing authority (Section 7.59). These include EN standards in the European Union and NSF/ANSI Standard 60 in the USA. In the UK, chemicals conforming to the appropriate BS EN standard may be used without seeking further approval from governing authorities (DWI, 2014). Other chemicals are required to be approved individually for use to ensure they do not impart taste, odour, colour or toxicity to the water or otherwise be objectionable on health grounds. ANSI/NSF Standard 60 is a health-based standard which details minimum requirements and evaluation processes for the control of potential adverse human health effects from use of chemicals in drinking water treatment.

Twort's Water Supply. DOI: http://dx.doi.org/10.1016/B978-0-08-100025-0.00012-0

There is no official certification or approval of products and the onus is on the supplier to demonstrate product conformance with the standard. This standard is widely used by other countries. In Australia, guidelines for chemicals used in water treatment are set out in the *Australian Drinking Water Guidelines* (ADWG, 2011).

12.2 CHEMICAL DOSING EQUIPMENT

Chemical storage and dosing plant requirements depend on the form of the chemical. With the exception of certain chemicals which are unstable and are therefore unsuitable for transport or prolonged storage, most chemicals used in water treatment are delivered to site as liquefied gases, solutions or solids. Solids in turn may be delivered in block form, as crystals or as powders. Chemicals that are unsuitable for storage, such as ozone (Section 11.20) or chlorine dioxide, are produced on site from more stable chemicals that are amenable to on-site storage.

Chemical dosing plant typically comprises storage facilities, solution or slurry preparation tanks and chemical metering and conveying systems. Table 12.1 indicates the form of a number of chemicals used in water treatment and summarizes the associated storage and dosing plant arrangements.

Properties of some of the commonly used chemicals in water treatment are given in Table 12.2.

It is usual for storage facilities to be sized for 28 days' demand at average dose and maximum flow rate, or the size of one consignment plus the demand for the period between placing the order and receiving a delivery, allowing for public holidays. Increased storage may be required for locations where access is affected by bad weather or where chemicals have to be imported.

Most chemicals delivered in solid form, as well as some delivered in concentrated solution, are made up in batches into solutions or suspensions of known concentration. At least two batching tanks are required for each chemical in order to maintain continuity of dosing; additional tanks would allow maintenance and cleaning without interruption to dosing. Each tank is normally sized so that one or two batches are prepared in a work shift. Accurate batching and dilution, with proper mixing, is required to maintain uniform concentrations because, in most cases, metering to the point of application is volumetric. For soluble solids, the solution strength should be well below the solubility at the lowest water temperature. For powders, such as lime and powdered activated carbon (PAC), suspensions need to be maintained at a value of less than 10% w/v and must be continuously stirred. The concentration of batches should be checked periodically for accuracy by hydrometer, conductivity meter or chemical analysis.

Liquid chemical storage vessels should be located in separately bunded areas, dedicated to each individual chemical (Cassie, 2003). Bunds should be designed to hold 110% of the contents of the combined storage volume served by the bund. All chemical drainage including that from bunded and hardstanding areas should be collected, neutralized and disposed of separately and should not be allowed to contaminate watercourses. Chemical plant located in basements or other areas which could flood in the event of a burst pipe should be avoided where possible. Easy access, including turning circles, should be provided for chemical delivery vehicles.

All chemical storage and dosing systems should be designed to minimize risks to operators when in use and should take into account all relevant health and safety guidance available from chemical suppliers and associated safety committees. Hazards should be eliminated through design as far as practicable and safety equipment, such as safety signs, safety showers, eye baths, first-aid boxes, protective goggles and clothing and breathing apparatus, as appropriate for the chemicals in use, should be provided.

Table 12.1 Physical form of chemicals used in water treatment and associated dosing plant design

Chemical type	Typical storage and dosing arrangement
Liquefied gases	
Ammonia, carbon dioxide, chlorine, oxygen, sulphur dioxide	Stored as liquefied gas under pressure in bulk or in delivery drums/cylinders. Vaporized prior to metering. Dosed as gas under pressure or, more commonly, as aqueous solution employing a water-driven eductor system
Solutions/suspensions	
Aluminium sulphate, ammonium hydroxide, ammonium sulphate, caustic soda, citric acid, ferric chloride, ferric sulphate, hexafluorosilicic acid, hydrated lime slurry, hydrochloric acid, hydrogen peroxide, monosodium phosphate, orthophosphoric acid, polyaluminium chloride, polyaluminium chlorosulphate, polyaluminium silicate sulphate, polyelectrolytes, polymeric ferric sulphate, sodium bisulphite, sodium hypochlorite, sulphuric acid	Delivered in bulk, intermediate bulk containers (IBCs), drums or carboys. Stored in as-delivered form. Dosed either directly from storage tanks or via day tanks by metering pumps. Chemical may be diluted in day tank, especially when chemical usage is low. Carrier water often added downstream of dosing pump to improve mixing at point of application
Solids (blocks, crystals, granules, powders)	
Aluminium sulphate, ammonium sulphate, calcium hypochlorite, hydrated lime, limestone, monosodium phosphate, potassium permanganate, polyelectrolytes, powdered activated carbon, quicklime, sodium carbonate, sodium chloride, sodium chlorite, sodium hexafluorosilicate, sodium metabisulphite, sodium thiosulphate	Delivered in bulk, flexible IBCs (FIBCs), drums or bags and transferred to storage silos, or else kept in delivered packaging until required for use, when transferred to hopper. As delivered chemical transferred from bulk silo or hopper pneumatically or by screw conveyor to duty/standby batching tanks for solution/slurry preparation. Batched chemical delivered to point of application from batching tank by metering pump, with or without carrier water
Prepared on site	
Chlorine dioxide, monochloramine, ozone, sodium hypochlorite	Sodium hypochlorite is stored in tanks prior to dosing by metering pumps

Table 12.2 Properties of some chemicals commonly used in water treatment

Chemical	Function	Form	Density (for solids: bulk density; for gases: at 1 atm, 20°C)	Materials	Freezing point/ solubility	Storage	Dosing concentration
Ammonia (BS EN 12126: 2012)	Chloramination	Flammable liquefied gas	Lighter than air; 0.71 g/l	Carbon steel PTFE Polypropylene	380 g/l at 15°C 340 g/l at 20°C 280 g/l at 30°C Henry's constant 1.6×10^{-5} atm. m^3/mol at 25°C	29 kg, 59 kg cylinders; 530 kg drums	
Ammonium hydroxide (25% w/w[b] NH_3) (BS EN 12122: 2005)	Chloramination	Hazardous liquid	0.916 g/l at 10°C 0.910 g/l at 20°C	Carbon steel, stainless steel, Al	−55°C	Vertical or horizontal tanks fitted with safety relief valve and vacuum breaker and earthed	Neat or diluted to suit[a]
Ammonium sulphate (26% w/w NH_3) (BS EN 12123: 2013)	Chloramination	Crystalline powder	1770 kg/m^3	Thermoplastics, stainless steel (304, 316) (**iron, Cu, Zn, Tin and their alloys*)	727 g/l at 10°C 754 g/l at 20°C 781 g/l at 30°C	Bags (20 kg, 50 kg)	10% w/v[f]
Ammonium sulphate 40% w/w $(NH_4)_2SO_4$ (10% w/w NH_3) Also available as 38% w/w (BS EN 12123: 2012)	Chloramination	Solution pH 5.2–6.0	1.231 g/l at 10°C 1.228 g/l at 20°C 1.224 g/l at 30°C	Thermoplastics, stainless steel (304, 316) (**iron, Cu, Zn, Tin and their alloys*)	−14°C	Bulk or IBCs	Neat
Carbon dioxide (BS EN 936: 2013)	pH correction, rehardening	Liquefied gas	Heavier than air; 1.90 g/l	Carbon steel, stainless steel	1.7 g/l at 20°C	34 kg, bulk in refrigerated or cryogenic vessels	Neat (as gas) or solution
Caustic soda 47% w/w NaOH	pH correction, softening	Corrosive liquid	1.504 g/ml at 10°C 1.497 g/ml at 20°C	Steels, thermoplastics, rubber, Ni and Ni	8°C (−25°C when diluted to 20% w/w NaOH)	Carboys (45 l) steel horizontal pressure vessel	Neat or 20% w/w NaOH

Also available as 50% w/w (BS EN 896: 2013)			1.490 g/ml at 30°C	alloys (T <150°C) (**Al, Tin, Zn, galvanized steel, brass*)		or steel or thermoplastic, PVC/GRP vertical tanks (heated and lagged as applicable)	
Citric acid	Membrane cleaning	Crystalline powder	769 kg/m^3	Stainless steel (316L), GRP. (Epoxy coated or PVC-lined) stainless steel 304 or steel	1170 g/l at 10°C 1450 g/l at 20°C 1800 g/l at 30°C pH (1% w/w solution) 2.2	FIBCs (500 kg) or bags (25 kg, 50 kg)	10% w/v (100 g/l)
Hexafluorosilicic acid 20% w/w H_2SiF_6 (15.8% w/w F) (BS EN 12175: 2013)	Fluoridation	Corrosive liquid. Highly toxic pH 1.2	1.18 g/ml at 25°C	Thermoplastics (PE, PP, PVC-U), Neoprene rubber/steel (**Glass, stainless steel (304, 316), Al, brass, bronze, carbon steel*)	−11.6°C	Horizontal or vertical rubber/mild steel or thermoplastic vertical tanks	Neat or diluted to suit
Hydrated lime 96% w/w $Ca(OH)_2$ (BS EN 12518: 2014)	pH correction, softening	White fine powder	[c]480–540 kg/m^3 [d]400 kg/m^3 [e]1.81 m^3/t	Steels, thermoplastics (**Al, tin, Zn, brass, galvanized steel*)	1.76 g/l at 10°C 1.65 g/l at 20°C 1.53 g/l at 30°C pH 12.4	Bags (25 kg, 50 kg) on pallets, steel or concrete silos	<10% w/v
Hydrated lime 18% w/w $Ca(OH)_2$	pH correction	Milky white liquid pH 12.4	1.11 g/ml at 15°C	(As for hydrated lime)	0°C	Vertical steel or thermoplastic tanks with mixers or recirculation pumps	2.5% w/v to neat
Hydrochloric acid (BS EN 939:2016)	Membrane cleaning, IX regeneration	Corrosive liquid	28% w/w – 1.14 g/ml at 15.5°C, 36% w/w – 1.18 g/ml at 15.5°C	Glass, PVC, PP, PE, Teflon, Hastelloy, rubber/steel (**Al, steels, Ni, Ni-alloys*)	– 57°C (28% w/w) – 33°C (36% w/w)	PVC-U reinforced with GRP, HDPE, PP, rubber/steel	10% w/w to neat

(*Continued*)

Table 12.2 (Continued)

Chemical	Function	Form	Density (for solids: bulk density; for gases: at 1 atm, 20°C)	Materials	Freezing point/ solubility	Storage	Dosing concentration
Hydrogen peroxide 35% w/w H_2O_2 (BS EN 902: 2016)	Oxidation	Hazardous liquid pH 2–4	1.132 g/ml at 20°C	Aluminium (99.5%), Al-Mg alloys, stainless steel (304, 316), HDPE, PVC (**Fe, Cu, Ni, Cr, brass*)	−33°C	PE carboys (50 kg), stainless steel or Al horizontal or vertical tanks or HDPE or PP/GRP, PVC/GRP	Neat
Orthophosphates (i) mono sodium $2H_2O$ (20% w/w P) (BS EN 1198: 2005); (ii) di sodium $2H_2O$ (17% w/w P) (BS EN 1199: 2005); (iii) tri sodium $12H_2O$ (8% w/w P) (BS EN 1200: 2005)	Plumbosolvency control	Crystalline powders	(i) 1200 kg/m^3 (ii) 1200 kg/m^3 (iii) 900 kg/m^3	Thermoplastics, stainless steel (304, 316), rubber/steel (**carbon steel, Al*)	(i) 650 g/l at 10°C, 850 g/l at 20°C, 1010 g/l at 30°C; (ii) 38 g/l at 10°C, 78 g/l at 20°C; (iii) 125 g/l at 10°C, 215 g/l at 20°C, 330 g/l at 30°C	Bags (50 kg)	(i) 40% w/v[a] (ii) 15% w/v[a] (iii) 20% w/v[a]
Orthophosphoric acid 75% w/w H_3PO_4 (24% w/w P) (BS EN 974: 2003)	Plumbosolvency control	Corrosive liquid	1.579 g/ml at 20°C 1.572 g/ml at 30°C	Stainless steel (316), thermoplastics (**carbon steel, cast iron, Al, Al-alloys, brasses, tinned or galvanized*)	−18°C	Lined steel drums (45 l, 200 l), horizontal or vertical stainless steel or rubber/steel or vertical HDPE or PVC/GRP tanks	Neat or diluted to suit[a]
Oxygen	Ozone generation	Liquefied gas	1.355 g/ml at 15°C 1.105 g/ml at 25°C	Stainless steel (316L)		Bulk in cryogenic vessels	100% oxygen
Potassium permanganate (BS EN 12672: 2008)	Oxidation	Granular powder or crystals	1600 kg/m^3	Steels, thermoplastics (**Zn, Cu, Al, galvanized steel, rubber*)	44 g/l at 10°C 65 g/l at 20°C 90 g/l at 30°C	Kegs (50 kg) Drums (150 kg)	1.5–3% w/v

Powdered activated carbon (BS EN 12903: 2009)	Organics removal dechlorination	Powder	[c]410–600 kg/m^3 (depending on the grade) [d]375–500 kg/m^3	Stainless steel (304, 306) rubber/mild steel (for slurry), thermoplastics	–	Bags (25 kg, 50 kg) on pallets, 450 kg bags, 1000 kg bins or steel silos (epoxy coated & earthed)	<10% w/v
Quicklime 95% w/w CaO (BS EN 12518: 2014)	pH correction	Hygroscopic powder	[c]1230 kg/m^3	(As for hydrated lime)	Highly reactive with water[g]	Bags (25 kg, 50 kg) on pallets, steel silos	Slaked to form hydrated lime 2.5–10% w/v
Sodium bisulphite 33% w/w $NaHSO_3$ (20% w/w SO_2) (BS EN 12120: 2013)	Dechlorination/ deoxygenation	Hazardous liquid	1.28 g/ml at 15.5°C pH 3.9–4.4	Thermoplastics (PP, PVC/GRP, PP/GRP), stainless steel (304, 316) (**carbon steel*)	10°C	Drums (45 l, 210 l), stainless steel (316) or PP/GRP, PVC/ GRP, HDPE or rubber/mild steel vertical tanks (heated/lagged as applicable)	Neat or diluted to suit
Sodium carbonate (light grade) 95% w/w Na_2CO_3 (BS EN 897: 2013)	pH correction	Anhydrous crystalline powder	[c]550 kg/m^3	(As for hydrated lime)	110 g/l at 10°C 180 g/l at 20°C 280 g/l at 30°C	Bags (25 kg, 50 kg) on pallets or steel silos	5% w/v (temperate) 15% w/v (tropics)
Sodium chloride (pure dried vacuum grade) 100% w/w NaCl (BS EN 14805: 2008)	Regeneration of ion exchange resin. On-site generation of sodium hypochlorite	Crystalline powder	1200–1360 kg/m^3 s.g. 1.2	Thermoplastics, rubber/carbon steel, stainless steel (316), aluminium alloy NS4 (**stainless steel, carbon steel for moist or salt solutions*)	358 g/l at 10°C 360 g/l at 20°C 363 g/l at 30°C	PE bags (25 kg), 1 t containers, saturators of reinforced concrete of rich mix (1:1.5:3) with 40 mm cover or GRP or FRP	Saturated solution or diluted to suit
Sodium chlorite (i) 26% w/w $NaClO_2$ (ii) 80% w/w $NaClO_2$ (BS EN 938: 2016)	Chlorine dioxide generation	(i) Hazardous liquid; (ii) Hazardous powder	(i) 1.27 g/ml at 20°C; (ii) 1105 kg/m^3	Thermoplastics (PE, HDPE, PVC) PVC/GRP (**Zn and combustibles*)	(i) −7°C (ii) 400 g/l at 20°C	PE kegs (50 kg, 70 kg), Steel drums with PE lining, HDPE vertical tanks	(i) 12.5–20% w/v (ii) 25–30% w/v[a]

(*Continued*)

Table 12.2 (Continued)

Chemical	Function	Form	Density (for solids: bulk density; for gases: at 1 atm, 20°C)	Materials	Freezing point/ solubility	Storage	Dosing concentration
Sodium fluoride 98% w/w NaF (44% w/w F) (BS EN 12173: 2012)	Fluoridation	Crystalline powder, highly toxic	1440 kg/m^3 s.g. 2.558	Same as sodium hexafluorosilicate (**aluminium and glass*)	40.5 g/l at 20°C 41.0 g/l at 25°C pH 7.4	Bags, drums	Saturated solution or diluted to suit[a]
Sodium hexafluorosilicate 98% w/w Na_2SiF_6 (59.4% w/w F) (BS EN 12174: 2013)	Fluoridation	Crystalline powder. Highly toxic	1440 kg/m^3 s.g. 2.679	Thermoplastics, rubber-lined carbon steel	5.5 g/l at 10°C 6.8 g/l at 20°C 8.5 g/l at 30°C pH 3.6	Bags (20 kg, 50 kg) on pallets, steel silos	0.2% w/v[a]
Sodium metabisulphite 96% w/w $Na_2S_2O_5$ (65% w/w SO_2)	Dechlorination/ deoxygenation	Granulated powder	1.48 g/ml at 20°C	Thermoplastics (PP, PVC) GRP, stainless steel (304, 316) (**carbon steel*)	385 g/l at 10°C 390 g/l at 20°C 405 g/l at 30°C pH 4.0–4.8	25 kg PP bags with PVC liner	10% w/v
Sodium thiosulphate $Na_2S_2O_3$ (40% w/w SO_2) $Na_2S_2O_3.5H_2O$ (26% w/w SO2) (BS EN 12125: 2012)	Dechlorination	Coarse crystalline powder	1.667 g/ml (anhydrous) 1.685 g/ml (pentahydrate)	Thermoplastics (PP, PVC) PP/ GRP, stainless steel (304, 316) (**carbon steel*)	330 g/l at 0°C (anhydrous) 520 g/l at 0°C (pentahydrate)	25 kg paper sacks	5–10% w/v
Sodium silicate (wt. ratio SiO_2/Na_2O: 3.22) (BS EN 1209:2003)	Mains corrosion protection	Corrosive liquid pH 11.3	1.38 g/ml at 20°C	Steel, thermoplastics, rubber, Ni and Ni alloys ($T < 150$°C) (**Al, Tin, Zn, galvanized steel, brass*)	0°C	25 l carboys, 210 l drums. Steel horizontal pressure vessel, or steel or thermoplastic, PVC/ GRP or FRP vertical tanks	Neat (viscosity 100 mPa.s at 20°C) or diluted to 4–25 mg/l SiO_2

Sulphuric acid (i) 98% H_2SO_4 w/w[h] (ii) 96% w/w H_2SO_4[h] (BS EN 899: 2009)	pH correction	Corrosive liquid	1.846 g/ml at 10°C 1.836 g/ml at 20°C 1.826 g/ml at 30°C pH < 1.0	Steels, PTFE (* *most other metals*)	(i) 3°C (ii) − 14°C	Carboys (45 l), steel horizontal pressure vessels or steel vertical tanks (lagged for 98% w/w as applicable)	Neat or 10 w/w H_2SO_4
Sulphuric acid 50% w/w H_2SO_4 (BS EN 899: 2009)	pH correction	Corrosive liquid	1.403 g/ml at 10°C 1.395 g/ml at 20°C 1.387 g/ml at 30°C pH <1.0	Thermoplastics, rubber/steel	−37°C	Carboys (45 l), PVC/GRP, PP/GRP, HDPE or rubber/ mild steel vertical tanks	10% w/w to neat
Sulphur dioxide (BS EN 1019: 2005)	Dechlorination	Hazardous liquefied gas	Heavier than air; 2.72 g/l	Carbon steel	170 g/l at 10°C 120 g/l at 20°C 75 g/l at 30°C	30 kg, 65 kg cylinders; 865 kg, 1016 kg drums	Up to 5 g/l

Source of information: Physical data from Perry (2007) and chemical manufacturer's data sheets. Grades given are those available in the UK.
Notes: GRP – Glass-reinforced plastic is a fibre reinforced plastic (FRP); HDPE – High density polyethylene; PE – Polyethylene (Polythene); PP –Polypropylene; PP/GRP– PP lined GRP; PTFE – Polytetrafluoroethylene; PVC – Polyvinyl chloride; PVC/GRP – PVC lined GRP; Rubber/steel – rubber lined carbon steel.
All stainless steel grades are to BS 970 or 1449.
BS EN refers to British/European standards, current edition.
* *Unsuitable materials.*
[a]Softened water should be used for solution preparation and dilution to prevent scaling.
[b]*x*% w/w is *x* percent weight per weight = *x* grammes of the chemical in 100 g of the product.
[c]For calculating silo capacity.
[d]When aerated during bulk delivery.
[e]When stacked in bags.
[f]*y*% w/v is *y* percent weight per volume = *y* grammes of the substance in 100 ml of water (10 y g/l).
[g]Quicklime gives off considerable amount of heat (1140 kJ/kg) during slaking.
[h]Sulphuric acid gives off considerable amount of heat during dilution. Therefore when diluting, acid should be added to a large quantity of water.

12.3 CHEMICAL DOSING

Chemical dosing must be accurate and related to the flow of water to be treated to maintain the required dose. Positive displacement pumps of the reciprocating type with mechanical or hydraulic diaphragm heads are most frequently used for injection but, for lime and PAC suspensions or viscous solutions, such as polyelectrolytes, progressive cavity type positive displacement pumps are sometimes used. For suspensions, peristaltic pumps also find application. Pumps should be provided with a calibration vessel on the suction side, a pressure relief valve (venting to waste), pulsation dampener, and a back pressure valve on the delivery side. In all dosing pumps, parts in contact with the chemical must be of appropriate materials. The maximum stroking speed of reciprocating pumps should be about 120 strokes per minute (spm), in particular for viscous or abrasive chemicals. The motor speed of progressive cavity and peristaltic pumps should be kept to less than 500 and 50 rpm, respectively. In plants where low levels of technology are adopted, a centrifugal pump may be used to pump the chemical to a constant head tank with gravity discharge to the dosing point (WHO, 1997). A control valve and a variable area flow meter are then used for manual dose control.

In general, all chemicals should be diluted in-line after metering, to provide sufficient flow in associated dosing lines to maintain a velocity of 1.0–1.5 m/s and improve mixing at the point of injection. Dilution is usually carried out in the dosing line, downstream of the metering pump. Particular precautions are required for certain chemicals, as indicated under the specific chemicals below.

In some plants it is necessary to apply the same dose to two or more equal streams, for example dosing to individual clarifiers; it is then vital to ensure equal division of the metered chemical flow and for systems operating at atmospheric pressure this is economically achieved by use of a splitter box with equally set V-notch weirs. Alternatively, a recirculation system supplied by centrifugal type pumps can be used, with control valves and flow meters fitted on off-takes to each dosing point. Such systems are particularly suitable for slurries, allowing appropriate velocities to be maintained in dosing lines, minimizing settlement.

Dosing pumps and associated equipment should be surrounded by a low bund wall, ideally arranged to overflow back into the main storage tank bund.

Generally, chemical dosing lines should be provided in duplicate (one duty, one standby). Chemical lines should not be laid in positions where any leakage could damage other lines or cause injury to personnel (e.g. over access ways and above cables). Consideration should be given to using double-contained pipework for chemical lines carrying corrosive or dangerous chemicals, especially where these pass close to access routes in buildings. Toxic gas under pressure or solution lines should be laid outside buildings in separate ducting. Chemical pipes should be laid in trenches in the ground, provided with removable covers for better access in preference to buried ducts. Ducts and trenches for chemical delivery lines should have drainage outlets and should be separate from those carrying electrical or instrumentation cables. Where possible, toxic gases should be conveyed under vacuum and mixed with water just before injection.

Delivery lines for slurries, such as lime, are difficult to keep clean and should be of the flexible hose type. They should be laid flat. Water-flushing of dosing lines should be provided. Instrumentation and electrical cables and sampling and dosing lines form a complex network inside buildings and around the site: allowance should be made for chemical pipes, cable trays, ventilation ducts and other services such as site water supply in the early stages of planning the building and site layouts.

Appropriate safety equipment for operational staff should be provided for all chemical storage and dosing facilities. This should include the provision of safety signs, safety showers, eye baths, first-aid boxes, protective goggles and clothing and breathing apparatus. Safety screens should be provided around pumps, particularly those used for hazardous chemicals such as sulphuric and phosphoric acids and coagulants.

Additional details, including typical batching and dose-control strategies specific to individual chemicals and treatment functions, are provided in the following paragraphs.

COAGULANTS AND COAGULANT AIDS

12.4 COAGULANTS

The properties of the most commonly used coagulants are summarized in Table 12.3.

Aluminium sulphate is the most widely used aluminium coagulant (Section 8.20). It is available in a number of solid grades such as block, kibbled or ground and is also available as a solution typically containing 8–8.3% w/w Al_2O_3. In waterworks practice aluminium sulphate is frequently but incorrectly referred to as 'alum'.

When delivered in solid form, aqueous solutions are usually prepared in suitably lined concrete tanks, equipped with a collector system of perforated pipe laterals in a bed of gravel. The tanks are usually built in the ground so that the material delivered in bulk can be tipped directly into the tanks. Aluminium sulphate is aggressive to concrete so such surfaces are protected with a liner. The inside surfaces of block-receiving tanks are usually protected from damage to the lining by a grid of timber. Penetrations of the lining for fixings to the concrete are inadvisable so the grids are often held in place by wedges. Flotation can be avoided by use of greenheart timber which is heavier than water. Mixing is by pumped recirculation. Solubility, expressed as g $Al_2(SO_4)_3$ per 100 g of water, increases with water temperature and is approximately 33.5 (10°C), 36.4 (20°C) and 40.4 (30°C). A saturated solution is prepared at these concentrations and is subsequently diluted about four- to sixfold in stock tanks before dosing. When block and kibbled forms are delivered in bags, a 200–300 g/l solution is prepared in tanks containing two compartments separated by a timber grid to prevent solid in one compartment damaging the top-entry turbine mixers in the other. Solution (200–300 g/l) from powder grades is prepared in dissolving tanks equipped with similar type mixers. Solution strength is controlled by a hydrometer.

Aluminium coagulants are also available in various high basicity and polymeric forms, formulated to depress the pH of the treated water less than aluminium sulphate (Section 8.20). High basicity coagulants include polyaluminium chloride (PACl), aluminium chlorohydrate (ACH), polyaluminium chlorosulphate (PACS) and polyaluminium silicate sulphate (PASS). There are several grades of PACl containing 10%, 18% or 24% w/w Al_2O_3, the 10% and 18% w/w grades being the most commonly available. These are delivered in liquid form and are stored and dosed in the same way as liquid aluminium sulphate. Polymeric coagulants should not be stored after dilution, to avoid premature hydrolysis.

Table 12.3 Properties of coagulants

Chemical	Physical form	Typical commercial grade	Fe/Al content (% w/w) of the commercial product	pH	Materials	Storage	Specific gravity	Freezing point	Viscosity at 20°C	Coagulation pH range
Aluminium sulphate $Al_2(SO_4)_3.xH_2O$ $x = 14–21$ (BS EN 878: 2016)	Solid	Blocks/slabs (14% w/w Al_2O_3), kibbled, granulated or powdered (17% w/w Al_2O_3)	300/(19 + x)	1.5 for a saturated solution (Section 13.4 for solubility)		50 kg bags, bulk – loose Acid resistant epoxy coated or brick-lined concrete or thermoplastic tanks for solutions or epoxy coated concrete or timber bunkers for solid forms	Bulk density (Loose) m^3/t: slabs/blocks – 1.4, kibbled – 1.2 granulated/ powder – 1.1	–	–	5.5–7.5
Aluminium sulphate $Al_2(SO_4)_3$ (BS EN 878: 2016)	Liquid	8–8.3% w/w Al_2O_3	4.2	1.8–2.2	Thermoplastics, stainless steel (*most common metals and their alloys)	1000 l FIBCs, bulk liquid Vertical or horizontal tanks in stainless steel, rubber/steel. Vertical thermoplastic, PP/GRP, PVC/GRP tanks	1.32 at 15°C	– 15°C	20 m.Pa.s	5.5–7.5
Polyaluminium chloride $Al_x (OH)_y Cl_z$ (BS EN 883: 2004)	Liquid	10%, w/w Al_2O_3	5.3	2.3–2.9	Thermoplastics (*stainless steel, most common metals and their alloys)	1000 l FIBCs, bulk liquid Vertical or horizontal tanks in stainless steel, rubber/steel. Vertical thermoplastic, PP/GRP, PVC/GRP tanks	1.20 at 20°C	– 12°C	3.5–4.5 m. Pa.s	6–9
Polyaluminium chloride $Al_x (OH)_y Cl_z$ (BS EN 883: 2004)	Liquid	18% w/w Al_2O_3	9	1	Thermoplastics (*stainless steel, most common metals and their alloys)	1000 l FIBCs, bulk liquid Vertical or horizontal tanks in rubber/steel. Vertical thermoplastic PP/GRP, PVC/GRP tanks	1.37	– 20°C	25–35 m. Pa.s	6–9
Polyaluminium chlorosulphate $Al_2(SO_4)_x.Cl_y.(OH)_z$ (BS EN 883: 2004)	Liquid	8.3% w/w Al_2O_3	4.4	2.8–3.0	Thermoplastics (*stainless steel, most common metals and their alloys)	1000 l FIBCs, bulk liquid Vertical or horizontal tanks in rubber/steel. Vertica thermoplastic, PP/GRP, PVC/GRP tanks	1.16 at 20°C	– 12°C	4.5 m.Pa.s	6.5–7.8

Polyaluminium silicate sulphate $Al_w(OH)_x(SO_4)_y(SiO_2)_z$ (BS EN 885: 2004)	Liquid	8% w/w Al_2O_3	4.4	3.6–3.8	Thermoplastics (*most common metals and their alloys)	1000 l FIBCs, bulk liquid Vertical or horizontal tanks in stainless steel. Vertical thermoplastic, PP/GRP, PVC/GRP tanks	1.28 at 15°C	0	11 m.Pa.s	6.5–.8
Ferric sulphate $Fe_2(SO_4)_3$ (BS EN 889: 2004)	Liquid	40–42% w/w $Fe_2(SO_4)_3$	12	<1.0	Thermoplastics, stainless steel (*most common metals and their alloys)	1000 l FIBCs, bulk liquid Vertical or horizontal tanks in stainless steel, rubber/steel. Vertical thermoplastic, PP/GRP, PVC/GRP tanks	1.52 at 15°C	– 15°C	30 m.Pa.s	4.0–9.0
Ferric chloride $FeCl_3$ (BS EN 888: 2004)	Liquid	40–42% w/w $FeCl_3$	14–14.5	<1.0	Thermoplastics, tantalum, titanium, Hastelloy C (*common stainless steels, most common metals and their alloys, ABS)	1000 l FIBCs, bulk liquid Vertical or horizontal tanks in rubber/steel. Vertical thermoplastic, PP/GRP, PVC/GRP tanks	1.45 at 15°C	– 2°C	7.5 m.Pa.s	4.0–9.0
Polymeric ferric sulphate $Fe_2(SO_4)_3$	Liquid	48–50% w/w $Fe_2(SO_4)_3$	13.5–14	1	Thermoplastics, stainless steel (*most common metals and their alloys)	1000 l FIBCs, bulk liquid Vertical or horizontal tanks in rubber/steel. Vertical thermoplastic, PP/GRP, PVC/GRP tanks	1.58–1.63 at 15°C	– 20°C	55 m.Pa.s	>4.5

Notes: 1. All liquid coagulants are dosed neat, and solid aluminium sulphate is dosed as a saturated solution. They can be diluted to suit (typically 10–20% w/v for metering), but any further in-line dilution following metering (typically 1-5% w/v) should be limited to a level that does not neutralize more than 2.5% of the coagulant. Diluted solutions of some polymerized coagulants gradually hydrolyse with time with subsequent loss of effectiveness.
2. For coagulants, materials of construction are thermoplastic material such as polyvinylchloride (PVC), polyethylene, HDPE, polypropylene (PP), PVC-lined GRP (PVC/GRP), PP-lined GRP (PP/GRP), rubber-lined mild steel (rubber/steel), stainless steel (316) except for those containing chloride and concrete with suitable linings, for example acid resistant bricks, fibre glass or resin coated. Unsuitable materials are most common metals such as Fe, Al, Zn, Cu and their alloys and concrete. For other notes, refer to Table 12.2.
* *Unsuitable materials.*

Iron coagulants are available as ferric sulphate, ferric chloride and ferrous sulphate as well as various polymeric grades (Section 8.21). Ferric salts are very corrosive acidic liquids. Ferric sulphate is usually preferred to ferric chloride since the introduction of chloride ions may increase the corrosivity of a water.

Liquid coagulants are typically stored in vertical thermoplastic or lined glass-reinforced plastic (GRP) cylindrical tanks (Table 12.3) and metered by pump in the delivered form.

Coagulant dosing is critical to treatment works operation and storage facilities should ideally include a minimum of two storage tanks. This allows continued operation in the event that a tank fails catastrophically or that a contaminated delivery is received. Provision of multiple tanks also allows a tank to be taken off-line for periodic cleaning to remove sludge or precipitated material. Attention is drawn to the increased density of ferric coagulants compared to alum-based chemicals when considering changing from an aluminium-based coagulant to a ferric-based coagulant.

Where natural organic matter (NOM) is present in the raw water (Section 7.36), this tends to be the key parameter determining the required coagulant dose. Sufficient coagulant must be added to satisfy the charge demand of the NOM for effective treatment to occur (Pernitsky, 2006). For raw water where the NOM concentration is stable or only changes slowly, the appropriate coagulant dose can be determined by regular jar tests and set manually. Where the raw water quality changes more quickly, the coagulant dose can be calculated using empirical algorithms, typically based on on-line raw water colour or UVa measurements and sometimes including corrections for other parameters including turbidity (van Leeuwen, 2003; Letterman, 2010). These algorithms are specific to individual treatment works and should be derived from jar tests conducted over the full range of raw water qualities. The resulting algorithm provides a feed-forward control strategy (Section 12.30). Feed-back control (Section 12.28) can be achieved using streaming current detectors (SCD) (Sibiya, 2013) or SCD in combination with computer models (Yavich, 2013). These sample water from downstream of the coagulant dosing point and measure the net surface charge or 'zeta potential' of particles in the dosed water. It is this charge which determines the stability of a colloidal suspension: positively charged ions formed on hydrolysis of the coagulant neutralize this charge (Section 8.10). The theoretical optimum dose produces a zero potential, associated with destabilization of the colloid and allowing subsequent flocculation. In certain applications where the organic content is relatively stable, satisfactory coagulation can be achieved by varying the coagulant dose rate to maintain a pH set-point.

Coagulants should generally be diluted using carrier water, to improve mixing at the point of application. The diluted concentration is usually in the range 1–5% w/v, depending on the alkalinity of the carrier water, and should not neutralize more than about 2–2.5% of the coagulant, to avoid premature hydrolysis of the chemical. In the case of polymeric coagulants, advice should be sought from the supplier before carrier water is adopted, as these chemicals are more prone to premature hydrolysis than the monomeric products. In the absence of specific supplier's information, carrier water should not result in greater than about fivefold dilution.

It is good practice to include a flow meter on the coagulant dosing line, upstream of carrier water addition, to confirm satisfactory pump operation and to compute and verify the dose applied.

12.5 COAGULANT AIDS AND POLYELECTROLYTES

Coagulant aids are used to improve the settling characteristics of floc produced by aluminium or iron coagulants. The most commonly used coagulant aids are synthetic polyelectrolytes, although some soluble starch products are still in use. Polyelectrolytes are long-chain organic chemicals, which may be cationic or anionic with varying charge densities, or non-ionic.

Polyacrylamides are the most effective of the synthetic group of polyelectrolytes (Section 8.22), but restrictions are placed on the maximum quantity of monomer allowed in the treated water. These are met by imposing stringent controls both on the quantity of monomer present in the chemical product after manufacture and on the maximum allowable doses.

Most polyelectrolytes are powders and a solution must be prepared for dosing. For successful preparation the powder must be wetted properly by using a high energy water spray before dissolving; the solution should be allowed to age for about an hour in cold water or 30 minutes in warm water conditions before use. Polyelectrolyte batching plants are typically proprietary skid-mounted systems. These are fully automated and incorporate a vacuum powder transfer system to charge a small storage hopper from a storage bin, a metering screw conveyor to deliver powder to a proprietary wetting head where the powder is contacted with water, a mixing and ageing tank and a stock tank. For polyacrylamide, the solution should be prepared in the range 0.5–2.5 g/l, whereas for natural polyelectrolytes the solution concentration could be as high as 25 g/l. Polyelectrolyte solution tends to be viscous and following metering, the solution should be diluted to less than 0.25 g/l, preferably including a static mixer at the point of dilution, to assist transfer in the pipe and dispersion at the point of application. Once a batch of stock solution is prepared it should be used within about 24 hours.

Some polyelectrolytes are supplied in liquid form, delivered in carboys, drums or flexible intermediate bulk containers (FIBCs). These chemicals should be diluted to around 0.25–0.5% w/v in batching tanks before metering. Further dilution downstream of the dosing pumps to $<0.05\%$ w/v (0.5 g/l) using carrier water, is usually provided to improve mixing at the point of application.

pH ADJUSTMENT AND WATER CONDITIONING CHEMICALS

12.6 pH ADJUSTMENT

Control of pH is an essential aspect of coagulation (Section 8.12) and also plays a key role in the efficiency of disinfection with free chlorine (Section 11.4) as well as the oxidation of metals such as manganese (Sections 10.11–10.13) to allow subsequent removal by filtration. Water conditioning refers specifically to adjustments to pH and alkalinity to reduce the corrosive and aggressive tendencies of the water (Sections 10.40 and 10.49).

Sulphuric acid is the most commonly used chemical for reducing pH in water treatment, although carbon dioxide is also sometimes used under certain circumstances, particularly in conjunction with lime to increase alkalinity. Hydrochloric and citric acid find use for membrane cleaning, with the former also being used for regeneration of certain ion exchange resins. Orthophosphoric acid and hexafluorosilicic acid are used as sources of phosphate and fluoride, respectively rather than for their acidic properties and are discussed elsewhere (Sections 12.24 and 12.25).

Commonly used alkalis include hydrated lime (calcium hydroxide), sodium hydroxide (caustic soda) and sodium carbonate (soda ash). Quick lime (calcium oxide) is also occasionally used.

12.7 SULPHURIC ACID

Sulphuric acid (H_2SO_4) is a strongly corrosive liquid. It is available in various concentrations but normally stored at either 96–98% w/w or 50% w/w. The latter, known as 'battery acid' is less hazardous.

The 96–98% acid is usually stored in horizontal pressure vessels of carbon steel construction. Vertical steel tanks could be used provided they are properly designed and maintained. The iron sulphate formed by the action of acid on steel coats the steel and forms a 'passivation' film, which then protects the carbon steel from further corrosion. 96–98% w/w sulphuric acid is a desiccant and absorbs moisture. Thus, the top acid layer in storage tanks is always dilute, making it very corrosive to steel. Under normal acid usage, this layer moves continually and does no harm to the tank. For installations with low usage, the dilute layer remains static for prolonged periods and therefore could accelerate the corrosion rate of the tank at the acid–air interface. There is usually some gradual corrosion on carbon steel, for which an allowance should be made in the thickness of the steel plate. This corrosion results in the evolution of hydrogen gas and the formation of an iron sulphate precipitate, most of which settles to the bottom of the tank. It is therefore good practice to provide two tanks to allow for periodic cleaning. This also permits internal inspection for periodic integrity assessment.

The 50% w/w acid is highly corrosive towards steel and requires the use of rubber lined steel or vertical tanks from suitable plastics such as polypropylene (PP) or polyvinyl chloride (PVC) lined glass-reinforced plastic/fibre reinforced plastic (GRP/FRP), high density polyethylene (HDPE) or polyethylene (PE).

Concentrated sulphuric acid (96–98% w/w) is diluted downstream of the dosing pumps to $<10\%$ w/v. The reaction is highly exothermic with a resulting temperature $>50°C$ and care is required when using carrier water; static mixers fabricated from solid polytetrafluoroethylene (PTFE) block are suitable at the point of dilution. Precautions are required to ensure an adequate carrier water flow is maintained to ensure heat dissipation and to prevent backflow of water into the acid storage tank. Heat produced on dilution of 50% w/w acid is minimal.

12.8 HYDROCHLORIC ACID

Hydrochloric acid (HCl) is supplied commercially at concentrations mainly within the range 28–36% w/w. At these concentrations it evolves hydrogen chloride vapour with a sharp irritant odour. Both the acid and the vapour are highly corrosive to most common metals. Suitable materials for storage tanks are indicated in Table 12.2. To avoid corrosion to adjacent plant and equipment from any escaping vapour, storage and dosing plant should preferably be located in dedicated rooms. Vent lines from storage tanks should feed into a scrubber unit, designed to cope with the fumes given off and the pressures generated during the filling of the tank. Water, sodium hydroxide solution or dilute acid solution can be used as the scrubbing medium.

12.9 CARBON DIOXIDE

Carbon dioxide (CO_2) forms carbonic acid when dissolved in water. It is usually used for stabilization of water after chemical softening (Section 10.6) and, in conjunction with lime dosing or limestone filters, for increasing the alkalinity of water. Increasing the alkalinity may be desirable either to enhance the water's resistance to subsequent change in pH or to assist in reducing its aggressive and corrosive tendencies (Section 10.41). Carbon dioxide can be delivered in cylinders but is usually delivered in bulk and stored on site as a liquid, either in insulated pressure vessels of the single walled type with refrigeration or as a cryogenic liquid in vacuum-sealed double walled type vessels. In the particular case of thermal desalination plants, CO_2 is often recovered from the process (Section 10.49). It can also be produced on site in packaged units by combustion of fuel oil or natural gas and is recovered by using monoethanolamine (MEA) and purified using potassium permanganate. It is then dried to a dew-point of $-60°C$ and filtered through activated carbon columns. When delivered in bulk, it is drawn off as required through ambient or electrically heated vaporizers and reduced in pressure before being passed through a control valve that regulates the rate of gas flow. The gas can be dosed to the point of application under direct pressure or combined with carrier water and dosed as a solution. For the latter approach, gas may be drawn through a vacuum regulating valve using a water-driven eductor system (Section 12.13) or mixed with water in a side stream static mixer. The dose rate is typically either fixed in proportion to flow (Section 12.27) (when dosed in conjunction with lime, which is controlled to achieve a pH set-point (Section 12.28)) or else controlled to obtain a pH set-point, as measured by a downstream pH monitor (Section 12.28).

Carbon dioxide gas is 1.5 times heavier than air: it will collect in basements, ducts, drains and low-lying areas, where it displaces air and can cause asphyxiation. It is therefore hazardous, with an 8-hour exposure limit of 5000 ppm (by volume) and 15-minute exposure limit of 15 000 ppm by volume (HSE, 2011). Suitable gas leak detection systems should be provided in areas where the gas could collect.

Liquid carbon dioxide plant including storage tanks and vaporizers should be located outdoors, away from underground rooms and ventilation plant intakes. Dosing plant should be located in fully mechanically ventilated, segregated buildings or rooms, and separated from inhabited buildings by a distance of at least 25 m.

12.10 HYDRATED LIME

Hydrated lime $Ca(OH)_2$, is a very finely divided powder resulting from the hydration of quick lime (CaO). It is available commercially at various purity levels ranging from 96% w/w to less than 80% w/w $Ca(OH)_2$. It is usually delivered to site in bulk and stored in vertical steel or concrete silos but is also available in 500 or 1000 kg bags (FIBCs) and 25 kg sacks. It is normally transferred from storage to batching tanks using screw conveyors, although pneumatic conveyors are also used, particularly for more convoluted plant layouts. Water-driven eductor systems have also been used with some success, drawing a metered quantity of powder into the eductor directly beneath the storage silo and conveying the resulting slurry to downstream batching/holding tanks. Lime is only sparingly soluble and it is normal to prepare a slurry at concentrations between 5% w/v and 10% w/v, although lower concentrations may be used on smaller treatment works. Lime transfer during the batching process is controlled by means of volumetric or gravimetric conveyors or by use of load cells. Batching tank mixers are operated continuously to keep lime in suspension. Lime slurry may be dosed using reciprocating diaphragm, progressive cavity or peristaltic pumps. Slurry tends to settle out and this can result in frequent blockage of dosing lines. Settlement can be reduced by maintaining minimum velocities in dosing lines, ideally greater than 1 m/s but not less than 0.3 m/s. The pulsating action of peristaltic pumps can help reduce such blockage. In some countries lime is also available from suppliers as a pre-prepared slurry, typically containing around 18% w/w $Ca(OH)_2$. These proprietary slurries are produced using finely divided lime powder with a comparatively narrow size range and are designed to reduce storage and handling problems on site as well as to provide increased rates of reaction after dosing.

As well as promoting mixing at the point of application, carrier water can be helpful in achieving desired velocities in the associated dosing lines but care must be taken to avoid precipitation of calcium carbonate. Carrier water for lime should, therefore, have an alkalinity less than 14 mg/l as $CaCO_3$. This will usually require de-alkalization, either by ion exchange (Section 10.9) followed by degassing, or by acidification with sulphuric or hydrochloric acid (HCl) to destroy the alkalinity followed by degassing to remove the carbon dioxide produced in the reaction; 1 mg/l of alkalinity as $CaCO_3$ requires 1.0 mg/l 100% sulphuric acid or 0.73 mg/l 100% hydrochloric acid and produces 0.88 mg/l carbon dioxide. Alternatively, lime slurry can be diluted in tanks with at least 15 minutes residence time to allow any softening reactions to reach completion.

Lime dosing downstream of filters can result in an increase in the turbidity of the filtered water depending on the lime dose, the proportion of impurities in the lime and the formation of calcium carbonate precipitate due to localized softening caused by poor mixing. This can be overcome either by using caustic soda or a saturated solution of lime (lime water). Lime water is usually prepared in continuous flow saturators which are upward flow conical hopper bottomed tanks of steel construction comprising a bed of lime charged with a lime slurry (5–10% w/v). Tanks similar to hopper-bottomed sludge blanket clarifiers (Section 8.16) could also be used. Water is fed from the bottom through the bed of lime with lime water drawn from the top as a saturated solution. Surface loading rates are in the range 1–1.2 $m^3/h.m^2$. In some designs, surface loading rates up to 2.0 $m^3/h.m^2$ are achieved by using a turbine mixer to improve contact between lime and water. Adding a polyelectrolyte to improve settling rate or placing lamella plates in the clear solution zone (to increase settling area) would also help to increase the loading rate. Saturators dissolve only about 80% of the lime in the feed and an allowance for this should therefore be made in sizing the saturators. Saturators are also useful for producing a clear solution of lime when the purity of the lime powder is low. Irrespective of the lime purity, regular desludging is required from the base of the tank to remove undissolved lime, calcium carbonate formed by softening grit and other insoluble impurities.

To minimize nuisance from escaping lime dust, lime storage may be located outdoors if suitably weatherproofed or else in a segregated room. Similarly, slurry/solution preparation and dosing plant should be housed in fully segregated rooms.

12.11 SODIUM HYDROXIDE

Sodium hydroxide (NaOH), also known as caustic soda, is available in a range of solution strengths from 10% w/w to 47% w/w and is delivered in bulk, IBCs, drums or carboys. The 47% w/w grade is stored in unlined horizontal pressure steel tanks. All grades, including 47% w/w, are suitable for storage in vertical tanks of unlined steel, PVC-U lined GRP/FRP or HDPE construction. Aluminium, aluminium alloys, tin, zinc (galvanizing), lead, bronze and brass should not be used on plant handling caustic soda. Although the use of 47% w/w product minimizes haulage costs and is thus the cheapest form, it has a relatively high freezing point (8°C). Therefore, tanks and dosing lines should be trace heated to prevent freezing. For this reason, lower concentrations are often used, as these have a much lower freezing point (e.g. 20% w/w freezes at −25°C). The chemical is often dosed in the as-delivered form. The use of carrier water to dilute chemical to around 0.5% w/w can help reduce precipitation of calcium carbonate at the point of injection, by improving mixing and reducing local high concentrations. This is particularly important when dosing into pipes or static mixers which could otherwise quickly become blocked. Carrier water should be softened to eliminate scaling of the dosing lines (Section 10.7).

12.12 SODIUM CARBONATE

Sodium carbonate (Na_2CO_3), also known also as soda ash, is a white powder available as 'light' or 'granular' grade. While the light grade is more readily dissolved, the granular grade is easier to handle in mechanical conveying plant and tends to be favoured. It is supplied in bulk, 750 kg FIBCs or 25 kg bags. It is stored on site in silos or charged as required from bags into smaller hoppers. During transport and storage, soda ash undergoes a slow change, due to the absorption of water and carbon dioxide from the atmosphere, to form sodium bicarbonate and hydrates of soda ash, which form as a cake on exposed surfaces. This reduces the available soda ash, makes mechanical transfer more difficult and hinders subsequent dissolving. It is non-corrosive and is batched using similar equipment to lime powder, described above. Soda ash, however, is relatively soluble and is batched as a 3−5% w/v solution rather than as a slurry. In addition to its use in the lime-soda process (Section 10.3), it is particularly useful for pH correction of low alkalinity waters where its properties, of being a weak alkali that provides both hydroxide and bicarbonate alkalinity, can provide improved pH control.

CHLORINE-BASED CHEMICALS

12.13 CHLORINE

Chlorine (Cl_2) (for properties and storage see Table 12.4) is contained as a liquid under pressure in drums or cylinders. The liquid occupies about 95% of the volume; the remaining 5% is occupied by gas. Most cylinders are designed to draw gas only, but the drum design allows either gas or

Table 12.4 Properties of disinfectants

Chemical	Physical form	Materials	Storage	Specific gravity	Freezing point/ solubility	Dosing concentration
Chlorine (Cl_2) (BS EN 937: 2009)	Liquefied gas	Liquid, dry gas under pressure: mild steel; Dry gas under vacuum: PVC-U Solution: PVC-U, natural rubber	33 kg, 71 kg cylinders; 864 kg, 1000 kg drums	Heavier than air; 2.99 g/l	7 g/l	1–3.5 g/l
Chlorine dioxide (ClO_2) (BS EN 12671:2009)	Aqueous solution Typically 1% w/v ClO_2 (10 g/l)	Thermoplastics (PE, HDPE, PVC) GRP (**Zn and combustibles*) FPM, PTFE, PVC (**Iron, copper and their alloys, chloride and other reducing agents*)	Generated on site (Section 12.16)	1.0 g/ml	0°C	As generated
Sodium hypochlorite (BS EN 901: 2013) Also available in low bromate grades and grades with additives for scale control	Liquid 14/15% w/w Cl_2 pH 11–13	Thermoplastics (PE, PVC, HDPE), GRP or FRP (PVC-U lined), rubber/steel EPDM, PTFE, PVC (**carbon steel, Al, Zn, Cu and their alloys, PP*)	Carboys (25 l, 45 l), 1000 l IBCs, bulk	1.27 g/ml	−17°C	Neat or diluted to 0.5% w/w with softened carrier water
Sodium hypochlorite (generated on site)	Aqueous solution 0.6–0.9% w/w Cl_2	GRP or FRP (PVC-U lined), HDPE	Generated on site (Section 11.14)	Approx. 1 g/ml	Determined by sodium chloride concentration in product	As generated
Calcium hypochlorite $Ca(OCl)_2$ (BS EN 900: 2014)	Granules, tablets 65–70% w/w Cl_2	EPDM, polypropylene, PVDF	25 kg, 40 kg, 50 kg lined steel or plastic drums	Bulk density (granules) 0.8 g/ml	215 mg/ml at 0°C$^+$ 234 mg/ml at 40°C pH 10.4	≤1% w/w Cl_2, as required
Calcium hypochlorite (Bleaching powder, chloride of lime) $CaCl(ClO).4H_2O$	Powder, tablets 30–35% w/w Cl_2	EPDM, polypropylene, PVDF	25 kg, 40 kg, 50 kg lined steel or plastic drums	Bulk density (powder) 0.4 g/ml	215 mg/ml at 0°C$^+$ 234 mg/ml at 40°C	≤1% w/w Cl_2, as required

Notes: $^+$Maximum insoluble matter 15–25% by weight. For other notes refer to Table 12.2.

liquid to be withdrawn. Typically, cylinders are available in 33, 50, 65 and 71 kg capacities whilst drums are available in 864, 966 and 1000 kg capacities. The maximum continuous gas withdrawal rate at 15°C and 2 bar back pressure is 1.0 kg/h for 33 and 50 kg and 1.5 kg/h (or 2.3 kg/h at 25°C) for 71 kg cylinders; that for a drum is 10 kg/h at 20°C and 2 bar back pressure. Higher rates can be achieved by connecting more than one container in parallel. A practical limit to the number of containers connected to a header is about six for cylinders and four for drums. The larger the number of containers connected the greater the number of connecting joints and hence the risk of a leak and also the greater the risk of liquefaction in the pipe. When higher rates are required, the contents should be withdrawn as liquid and vaporized in evaporators before use. The liquid should not be drawn from more than one container and the withdrawal rate should not exceed 180 kg/h. In a few large works where chlorine usage is high, chlorine is stored as a liquid between 4 and 8 bar pressure at 5–30°C in bulk storage vessels, which are filled from road tankers. Chlorine storage facilities must be designed to very high standards of safety (HSE, 1990). Plate 19(a) shows a drum store with evaporator.

A water-driven ejector is used to develop a vacuum and draw chlorine gas through a vacuum regulator to the ejector. The resulting solution contains up to 3500 mg/l Cl_2 for dosing to the water supply. The vacuum regulator reduces the gas supply pressure from up to 10 bar down to a vacuum of about 170 mb. Interruption or failure of the operating water supply will immediately shut down the flow of gas. In the unlikely event that positive pressure should reach beyond the vacuum regulator a pressure relief valve vents a small quantity of gas to outside and operates an internal check valve until such time as normal vacuum conditions are restored.

The flow of gas is metered by maintaining a constant differential pressure drop across a manually or automatically adjusted variable area orifice through which the gas passes, the rate of flow being indicated by a glass tube flow meter. Should an excess vacuum condition occur within the system, as when the gas supply is exhausted, a relief valve operates downstream of the metering section and a no-flow condition is indicated on the flow meter. Depending upon the type and capacity of the equipment the ejector may form an integral part of the unit, or be separately mounted either adjacent to the chlorinator or near the point of application. The chlorine dose rate may be automatically controlled to accommodate variations in water flow or quality, or both, by means of an electric positioner fitted to the variable area orifice (Fig. 12.1).

Safety Precautions

Chlorine containing greater than 0.002% w/w water causes rapid corrosion of steel. Pipework carrying liquid or gaseous chlorine is of carbon steel and must be designed to exclude water or moist air; PVC-U pipework can be used for chlorine gas under vacuum and for chlorine solution applications. Chlorine gas reacts explosively with many organic compounds.

Chlorine is also very toxic with a 15-minute exposure limit of 0.5 ml/m^3 (HSE, 2011). Liquid chlorine leaks are far more dangerous than chlorine gas leaks since, on evaporation, 1 kg of liquid chlorine yields about 335 litres of gas at 15°C. Chlorine gas is heavier than air. A chlorine solution leak releases chlorine fumes laden with moisture and is particularly dangerous as it seems more tolerable to the respiratory tract than inhalation of dry chlorine (Black & Veatch, 2010). Safety precautions must be taken in the design and layout of chlorine installations to safeguard the operators

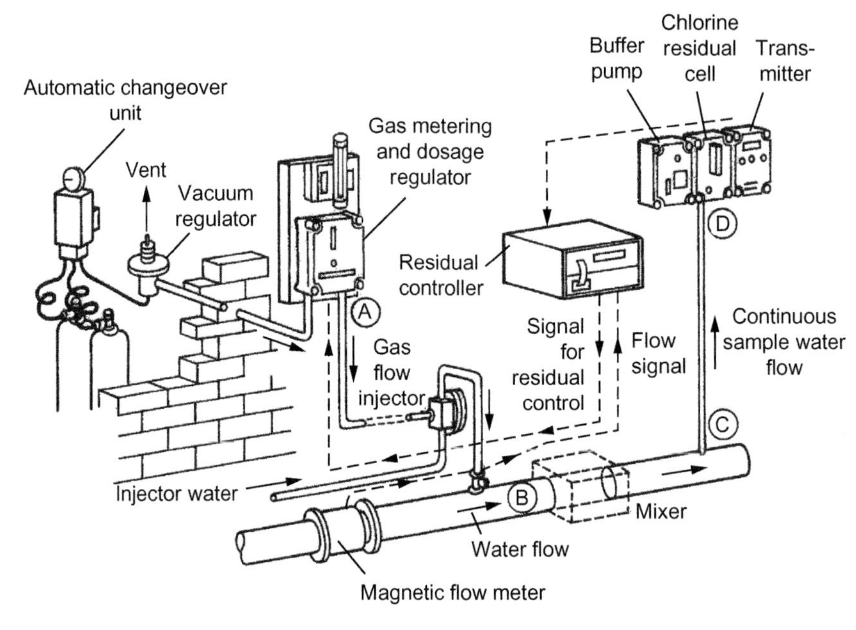

FIGURE 12.1

Vacuum-operated gas metering unit for automatic dosing with flow and pre-set chlorine residual control. Suitable for chlorination, partial dechlorination or chloramination.

and the public (HSE, 1990, 1999). For sites in the European Union countries, it is also necessary to comply with the Control of Major Accident Hazards Regulations (COMAH, 2015).

Chlorine facilities should be designed to minimize leaks and to contain them if they should occur. Storage and dosing equipment should be located within a dedicated building which should be separated from any ventilation intakes of other buildings by at least 25 m and from the site boundary by at least 20 m (for cylinder installations), 40 m (for drum installation using gas) and 60 m (for drum installations using liquid). Chlorine containers should always be installed in a separate store constructed in substantially fire-resistant material. The store should be provided with direct access to outdoors. The store should be designed to be gas-tight as far as is reasonably practical. All doors should open outwards and emergency exit doors should have a push bar operated panic bolt. External windows should be avoided, artificial illumination being used throughout.

A useful safety measure at chlorine installations is to incorporate pneumatically operated actuators on containers connected to the gas header pipework (e.g. ChlorGuard; Plate 19(b)). These allow remote isolation of the connected containers using a motive air system, for effective isolation of the chlorine supply at the container in the event of a downstream leak. Vacuum regulators should be located in the container store to restrict any leaks in the pressurized pipework to the store. Similarly, ejectors should be mounted remote from the chlorinators, local to the point of application so as to restrict the risk of a chlorine solution leak to a short length of pipe at the point of application. Heaters (non-radiant type) may need to be installed in the container store to maintain a temperature greater than 10°C. The ventilation system in the container store should have high level fresh air inlets and low level (500 mm above the floor) extraction of air discharging to outside at high level. Ventilation systems should be designed to give not less than 10 air changes per hour and should be arranged to start at a low leak level and to shut down at a high leak level to contain major leaks. Leak detectors with a 0.5 ml/m^3 detection limit should be installed in enclosed areas where chlorine is handled.

At chlorine sites in urban areas or near housing developments or with inventories of 10 or more tonnes, the use of a chlorine absorption plant for neutralizing a leak should be considered. The absorption plant should be designed to treat the contents of one container (e.g. 1000 kg for a drum installation). The leakage rate could vary from 1.5 kg/min to about 35 kg/min. A typical system consists of a packed tower where the contaminated air flows counter-current to a flow of dilute caustic soda (10–20% w/w NaOH) which is continuously recycled back to the top of the tower. Stoichiometrically, 1000 kg of chlorine (about 340 m^3 as gas) requires 1127 kg of caustic soda (100% w/w NaOH); in practice at least 10% more caustic soda is recommended by the World Chlorine Council (WCC, 2011). An alternative absorbent is ferrous chloride, which could be regenerated by adding iron filings to the tank.

12.14 SODIUM HYPOCHLORITE

Sodium hypochlorite (NaOCl) is available for waterworks purposes as a straw coloured solution containing 14–15% w/w of available chlorine. It is frequently used in place of chlorine gas for safety reasons. It can be supplied in small containers or in bulk, but loses its chlorine strength when exposed to atmosphere or sunlight (for properties and storage see Table 12.4). The rate of decomposition increases with increased temperature as shown in Table 12.5. Therefore, for stability under high ambient temperature and long storage the solution should be chilled to about 10°C using an external cooling circuit.

Table 12.5 Decomposition of sodium hypochlorite after standing in the dark due to age and temperature (based on tests carried out in the dark)

	Initial concentration as available chlorine								
	185 g Cl_2/l			160 g Cl_2/l			120 Cl_2/l		
Weeks	1	2	4	1	2	4	1	2	4
10°C	183	180	176	158	157	154	120	119	118
15°C	180	176	167	157	155	150	119	118	116
20°C	175	166	151	155	150	141	118	116	113
25°C	166	151	127	149	139	123	116	113	106
30°C	151	128	97	139	123	100	112	104	92

Source of information: Product Booklet B303/85, Hayes Chemicals Ltd, UK. 1985

Dosing of hypochlorite can be by positive displacement reciprocating diaphragm pumps. Sodium hypochlorite dissociates to form oxygen which can lead to gas locking of dosing lines. Measures to minimize problems with trapped oxygen include: keeping the pump suction pipework short and simple, with a minimum of valves; providing vents or drains where entrapment of liquid could occur, including vents on pump suctions to above tank liquid level; drilling the internal ball plugs in ball valves in the upstream direction; and arranging pipework with flow in an upward direction. Attention has to be paid to limiting the introduction of chlorate and bromate to the treated water. Commercial hypochlorite of 15% w/w chlorine, produced using the membrane process, contains chlorate (as ClO_3^-) and bromate (as BrO_3^-) of about 0.25% w/w and 0.035% w/w, respectively. This limits the dose that may be applied to about 4.3 mg/l as Cl_2, provided hypochlorite is the sole contributor to bromate in the water.

Commercial hypochlorite contains excess caustic soda which can cause precipitation at the point of application or in the dosing lines downstream of carrier water addition. This can be avoided by using softened carrier water.

12.15 SODIUM HYPOCHLORITE PRODUCTION BY ELECTROLYSIS

Sodium hypochlorite can be produced by the electrolysis of brine (a solution of sodium chloride) (Section 11.14). On-site generation (Plate 19(c)) produces a weak hypochlorite (<10 g/l) solution that is easy to handle and avoids many of the safety hazards associated with the use of purchased liquefied chlorine and bulk sodium hypochlorite.

Brine is prepared from sodium chloride crystals delivered to site in bulk tankers or, for smaller sites, in 25 kg bags. When delivered in bulk, the chemical is usually blown across from the delivery tanker by air compressors into saturator vessels where it is stored under water. Saturators may be concrete tanks constructed in the ground or more usually GRP tanks located above ground. The salt should be of high purity grade, free from calcium, magnesium and heavy metals, in particular manganese (<10 μg/l). Pure dried vacuum salt containing at least 99.7% w/w NaCl is most suitable (Table 12.2). Lower quality salt should be treated to remove hardness, turbidity, iron and manganese. Water for saturation and dilution purposes must have a hardness less than 15 mg/l as $CaCO_3$ and may therefore need to be softened by ion exchange (Section 10.7). The process water requirement is about 125–150 l/kg of chlorine. Additional detail is provided in Section 11.14.

The chlorine content of the hypochlorite produced is in the range 6–9 g Cl_2/l. Hypochlorite solution tanks can be constructed in high density polyethylene or GRP/FRP with a PVC lining. Storage tanks should be designed for hydrogen venting. The hypochlorite solution is injected by a positive displacement metering pump into the water to be treated.

12.16 CALCIUM HYPOCHLORITE

Calcium hypochlorite ($Ca(OCl)_2$) is available in both granular and powder form. Calcium hypochlorite as granules contain 65–70% w/w chlorine and can be supplied in 45/50 kg drums with plastic liners. The granules are readily soluble and solutions of concentration up to 10% w/v (100 g/l) can be made up for dosing. A 1% w/v Cl_2 (10 g Cl_2/l) solution is prepared by mixing 15 kg of granules containing 65% w/w Cl_2 in 1000 litres of water. The chlorine content reduces substantially in a few days if left exposed to the air.

Calcium hypochlorite powder, commonly known as 'bleaching powder', contains 30–35% w/w of releasable chlorine and excess lime. The powder has the advantage that sealed drums of it can be held in store for long periods without serious loss of chlorine. It is best to make up a solution in batches prior to dosing. Assuming the powder to contain 33% w/w of chlorine, a 1% w/v Cl_2 (10 g Cl_2/l) solution would be made up by mixing 30 kg of powder in 1000 litres of water. A 100 litre 'batch' of this solution would be sufficient to give 1.0 mg/l in 1000 m^3 of water. It should be allowed to stand before use to settle out excess lime. The supernatant containing chlorine is drawn off and diluted in a storage tank to the dosing concentration, which is injected into the water to be treated by means of a positive displacement reciprocating pump of the diaphragm type or by constant head feeders. In large plants, saturators may be used (Section 12.10). The resulting chlorine solution rapidly loses its chlorine content if exposed to air or sunlight; hence it needs to be made up daily or every second day. Likewise a drum of bleaching powder begins to lose its chlorine content once opened.

12.17 CHLORINE DIOXIDE

Chlorine dioxide (ClO_2) is produced on site due to its relatively short half-life, commonly by the reaction between a solution of chlorine (in water) and sodium chlorite ($NaClO_2$). Sodium chlorite is supplied as a liquid containing 26% w/w of $NaClO_2$ in small containers or in bulk or as a powder containing 80% w/w $NaClO_2$ from which a solution of up to 31% w/w is made for use (Table 12.2). This and other processes are described in Section 11.16.

The product, typically with a concentration in the range 1–10 g/l, is stored in a day tank and dosed using conventional dosing pumps. Typically, level controls in the day tank initiate starting and stopping of the associated generation system.

Spillages of sodium chlorite should be washed away quickly as evaporation leads to deposits of highly flammable sodium chlorite powder.

DECHLORINATION CHEMICALS

12.18 SULPHUR DIOXIDE

Sulphur dioxide (SO_2) similar to chlorine, is contained in cylinders (30 and 65 kg) or drums (865 and 1016 kg) and is drawn for use as a gas or liquid. The gas withdrawal rates at 15°C and 2 bar back pressure are 0.3 kg/h and 0.45 kg/h from 30 kg and 65 kg cylinders, respectively, and 2.3 kg/h from a drum. Higher withdrawal rates can be achieved by connecting several containers in parallel or withdrawing sulphur dioxide as a liquid and vaporizing it in evaporators. The liquid should not be drawn from more than one container and the withdrawal rate should not exceed 140 kg/h. The equipment used for sulphur dioxide is very similar to that used for chlorine. The dose rate may be automatically controlled proportional to water flow with feedback from a downstream chlorine residual monitor to maintain a pre-set level of chlorine residual.

Safety Precautions

Like chlorine, sulphur dioxide is corrosive; materials of construction used for chlorine can also be used for sulphur dioxide. Sulphur dioxide is a toxic gas with an 8-hour exposure limit of 0.5 ml/m^3 and a 15-minute exposure limit of 1 ml/m^3 (EC, 2009). Liquid sulphur dioxide leaks are far more dangerous than a corresponding gas leak since, on evaporation, 1 kg of liquid sulphur dioxide yields about 370 litres of gas at 15°C and 380 litres at 25°C. It is heavier than air. It is compatible with chlorine and therefore can be stored in the same room as chlorine, with the same advice relating to building segregation and separation being applicable. Sulphur dioxide leaks are treated in absorbers of a similar design to those described for chlorine, using caustic soda as the absorbent.

12.19 OTHER DECHLORINATING CHEMICALS

Sodium bisulphite (Na_2SO_3) is available as a solution containing up to 33% w/w SO_2 (Table 12.2) and is typically stored in vertical tanks fabricated from thermoplastics (PP, PVC/FRP, PP/FRP). Stainless steel (316) is also suitable. Higher concentrations tend to crystallize at relatively high temperatures ($\leq$6°C), causing blockages in dosing pump suction and delivery lines and lower concentrations of around 20% w/w are sometimes preferred for this reason. Any spillages should be neutralized with soda ash, to prevent sulphur dioxide emission, and then be oxidized to neutral sulphate with sodium hypochlorite. Sodium metabisulphite should be used if dechlorinating chemical is required in powder form. Sodium thiosulphate ($Na_2S_2O_3$), is also sometimes used to dechlorinate waste discharges to the environment.

CHLORAMINATION CHEMICALS

12.20 AMMONIA

Ammonia (NH_3) is available as a liquid in cylinders (49 kg and 65 kg) or drums (530 kg). It is withdrawn as a gas at 0.5 kg/h from a cylinder or 2 kg/h from a drum at 15°C, or as a liquid to evaporators. Apparatus used is very similar to that used for chlorine. Ammonia is very soluble in water and is corrosive. Steel piping is suitable for conveyance of ammonia liquid and dry gas. Iron, copper, zinc and aluminium are attacked by ammonia solution, but PVC-U is suitable. Motive water for ammonia dosing units should be softened to a hardness value of less than 25 mg/l as $CaCO_3$ to prevent calcium carbonate scaling of fittings.

Ammonia is toxic with an 8-hour exposure limit of 25 ml/m^3 and a 15-minute exposure limit of 35 ml/m^3 (HSE, 2011) and is lighter than air. It is flammable in air in mixtures between 15.5% and 27% by volume. Electrical apparatus in areas where ammonia is handled should therefore be suitably protected. Ammonia gas forms explosive mixtures with chlorine and sulphur dioxide gases and should therefore be stored away from them. Storage room design requirements are similar to those for chlorine, except for the ventilation system which requires low level fresh air inlets and air extraction and disposal at high level.

Absorbers for treating ammonia leaks are of a similar design to those used for chlorine. Absorbent used for ammonia is usually sulphuric acid (10% w/w H_2SO_4), although water can also be used, owing to its high affinity for ammonia.

Ammonia is also available as aqueous ammonia of concentration up to 40% w/w NH_3, referred to as aqua ammonia. Ammonia vapour may evolve from the solution, which, therefore, requires the same safety considerations as anhydrous ammonia.

Ammonium sulphate is also sometimes used as a source of ammonia for chloramination. It is a crystalline powder (Table 12.2) although it is also available as a solution at various strengths (24–40% w/w). It does not evolve ammonia vapour, so is safer than anhydrous or aqueous ammonia. When supplied as a powder this is usually batched to a 5–10% w/v solution prior to dosing. It does not require the use of softened water for batching or carrier water purposes. Dose control is similar to gaseous ammonia.

OXIDATION CHEMICALS

12.21 POTASSIUM PERMANGANATE

Potassium permanganate, $KMnO_4$, is delivered in crystal form, typically in 25 or 50 kg kegs or 150 kg drums. The crystals do not deteriorate and may be stored indefinitely but should be kept away from organic chemicals such as polymers and activated carbon.

The chemical is prepared as a solution of around 4% w/v, usually in batching tanks with mixers. The product is free flowing, allowing metering conveyors to be used to transfer crystals from a storage hopper to the batching tanks.

Prepared solution is metered to the dosing point using reciprocating diaphragm pumps. Dose rate is normally controlled in direct proportion to flow past the dosing point. Potassium permanganate should not be dosed in the same location as PAC, as the PAC may consume the permanganate (US EPA, 1999), although it can be dosed with coagulants or polyelectrolyte solutions.

12.22 HYDROGEN PEROXIDE

Hydrogen peroxide is rarely used in drinking water treatment for its oxidative properties but is usually dosed in conjunction with either ozone or UV light as part of an advanced oxidation process (Section 10.36).

Hydrogen peroxide is usually delivered as a 35% w/w or 50% w/w solution in carboys, FIBCs or bulk tankers. It readily decomposes to liberate oxygen, which then increases the flammability of neighbouring materials. The rate of decomposition is increased by exposure to sunlight and heat, both of which should be avoided in storage. Decomposition is accelerated by the presence of even very low concentrations of impurities, such as metal salts, particularly multivalent ions, in a process referred to as homogeneous decomposition. Heterogeneous decomposition is catalysed by the presence of solid surfaces in contact with the solution. This can result in explosive reactions with many common materials including iron (including rust), copper, chromium and nickel. It is important therefore to select all materials used for storage and associated instrumentation carefully and to locate the storage and dosing facility away from other flammable materials. Heat is also produced during the decomposition process. In dilute solutions, the heat evolved is readily absorbed by the water present. In more concentrated solutions, the heat raises the temperature of the solution and accelerates the decomposition rate. Stabilizers are added in the manufacturing process to inhibit the catalytic decomposition effect of metals and other impurities that may contaminate the chemical during storage and handling but these are not effective when excessive contamination is present. Hydrogen peroxide can cause spontaneous combustion of many organic materials, such as paper, cloth and wood, especially in hot climates. Under certain conditions, solutions above about 44% w/w can give rise to explosions caused by release of oxygen and associated pressure increases, particularly in the presence of organic initiators.

Storage tanks are usually of polyethylene or stainless steel. Metering pumps are typically of the reciprocating diaphragm type of stainless steel or PTFE, provided with double diaphragms or with pressure relief on the non-peroxide side of the diaphragm, to avoid pressure bursts should the diaphragm fail. Dosing pipework is typically of stainless steel.

ORGANICS REMOVAL

12.23 POWDERED ACTIVATED CARBON (PAC)

PAC is normally delivered to site in bulk or in 500–1000 kg bags (FIBCs), to minimize manual handling on site, although it is also available in 25 kg bags. When delivered in bulk the chemical is blown across to vertical storage silos. PAC delivered in FIBCs is normally stored on site in the delivery packaging until required for use. PAC is prepared and dosed as a slurry in much the same way as lime powder (Section 12.10), using conveyors and batching tanks. If not wetted properly, PAC will tend to float in the batching tank. Wetting can be achieved by feeding PAC into the vortex of a mixer in the tank or via a wetting head consisting of a funnel with a continuous supply of water placed directly under the discharge of the screw feeder. Additional mixing is carried out in the batching tank. Batching tanks may be of rubber-lined or epoxy-coated steel, to resist abrasion. Plant for batching and dosing from FIBCs is commonly supplied as a skid-mounted system incorporating a frame from which the FIBC is suspended. PAC is drawn from the bottom of the FIBC and fed by a volumetric or gravimetric screw feeder into a water-driven eductor system. The resulting PAC slurry is either conveyed directly to the dosing point or else held in batching tanks from where progressive cavity or peristaltic pumps meter it to the dosing point.

PAC is a finely divided powder which can be difficult to contain. To minimize nuisance from escaping dust, storage and handling plant should be housed in fully segregated rooms. Any dust that escapes quickly forms a black covering on all adjacent plant and equipment. PAC is an electrical conductor and should not be allowed to accumulate as dust on open electrical circuits, to avoid potential for short-circuits. Some suppliers indicate that this dust should be considered as potentially explosive and appropriate precautions taken, although others state that this is not necessary. The lower explosive limit for activated carbon dust is variously reported as 50 (Cabot-Norit, 2014) and 140 g/m^3 (Avantor, 2011). This suggests there is a small risk of dust explosions inside silos, conveyors, dust filters and other items of plant handling dry PAC, as well as in the area immediately adjacent to such equipment. In practice, most PAC is manufactured using steam activation. The resulting PAC contains only a very low level of volatile matter as a result of the high temperatures involved and is not considered to present an explosion risk. The so-called 'Deflagration Index' (K_{St}) for steam activated PAC places it in the lowest class of combustible dusts, Class ST1.

To minimize any risk of explosion, dust filters should be provided and sources of ignition should be avoided. Electrical equipment in the immediate vicinity of PAC plant should be protected to IP65 as a minimum and where possible should be located in a separate room. All equipment and piping used for dry handling of PAC should be earthed for static electricity.

Typically, carrier water is added downstream of the metering pump to obtain dilution down to 1% w/v (10 g/l), to maintain velocities in the dosing line and improve mixing at the point of application.

PLUMBOSOLVENCY CONTROL

12.24 PHOSPHATE DOSING

Orthophosphate for plumbosolvency control (Section 10.14) is dosed either as orthophosphoric acid (H_3PO_4) or as one of its sodium salts, typically monosodium phosphate (NaH_2PO_4).

Both chemicals are normally delivered in liquid form, either in bulk or in IBCs, monosodium phosphate (MSP) as 32% w/w and orthophosphoric acid as 75% w/w chemical. The chemical is metered in 'as-delivered' form. MSP is also sometimes delivered in granular form in 25 kg bags and prepared as a solution in batching tanks using metering conveyors. Although more expensive, MSP tends to be used for low alkalinity waters to minimize impact on treated water pH or for applications where chemical demand is low. Historically, chemical was transferred to a day tank before being dosed, to minimize the risk of inadvertent overdosing, but increasingly the day tank is omitted.

A phosphate monitor can be used in closed loop control for phosphate dosing, although, as phosphate is a 'conservative' parameter that does not decay, it is usual to add a fixed dose in proportion to flow with downstream on-line monitoring. Phosphate injection points should be separated from any lime dosing points, to avoid possible formation of insoluble calcium phosphate. Recently some water undertakers have adopted mass flow meters to monitor and control the quantity of orthophosphoric acid applied. These take account of varying concentrations in commercially supplied orthophosphoric acid and ensure a consistent dose is added, rendering on-line analysers unnecessary.

CARIES CONTROL

12.25 FLUORIDATION

Fluoridation of water (Section 10.17) is usually carried out using either hexafluorosilicic acid which is a solution usually containing 20% w/w H_2SiF_6 (15.8% w/w F), or disodium hexafluorosilicate (sodium hexafluorosilicate) which is supplied as a powder usually containing at least 98% Na_2SiF_6 (59.4% w/w F) (Table 12.2). Sodium fluoride (NaF), a powder containing 98% w/w NaF (44% w/w F), is sometimes used for small water supplies. Both the acid and powders are highly toxic and the acid is very corrosive. Acid vapour and dust from the powders should be contained to prevent inhalation and ingestion by operators. Splashes of acid on skin should be washed with copious amounts of cold water.

Disodium hexafluorosilicate is batched into a solution using metering conveyors prior to being dosed. The water used for dissolving the powder or diluting the acid should be softened to <75 mg/l as $CaCO_3$ using a base exchange softener (Section 10.7) to prevent formation of calcium and magnesium fluoride deposits. Alternatively a sequestering agent such as a polyphosphate could be used to minimize scaling. Dilution of fluorosilicic acid between 10:1 and 20:1 often forms an insoluble silica precipitate. Softening the water will not prevent this precipitation. Precipitation can be eliminated by using an acid fortified with 0.1% w/v hydrofluoric acid.

Excessive doses of fluoride can cause mottling of the teeth and skeletal fluorosis. The dosing system should therefore be designed to ensure that the quantity of fluoride injected to supply over 24 hours does not exceed the maximum dosage allowable. Acid should be transferred from a bulk storage tank by pump to an intermediate storage tank (a day tank) which holds not more than 24 hours maximum usage. For the batching system, a day tank or a system to limit the number of batches per day is essential. For powder systems, a hopper with one day's capacity should be used.

Dosing pumps should be sized so that they operate close to their maximum capacity most of the time. Where the works flow varies frequently by more than 5%, the pump motor speed should be arranged to vary automatically with flow. Any fluoride naturally present in the raw water should be taken into account when determining the required dose. Where the raw water fluoride concentration varies then closed-loop control of the output based on a signal from a fluoride monitor could be used, provided a reliable on-line fluoride analyser is available.

Fluoride chemicals are normally added to the final filtered water. Hexafluorosilicic acid can be dosed near to the point of final chlorination as the two chemicals are compatible. Hexafluorosilicic acid should not be added with lime as calcium fluoride can precipitate, although it may be added after the lime and water have been mixed thoroughly. Technical aspects on fluoridation are published in a *Code of Practice* by UK DWI (2005).

CONTROL TECHNIQUES

12.26 CHEMICAL DOSE CONTROL

Historically, chemicals have generally been dosed, under a level of automatic control, using reciprocating pumps, in which dosage adjustment is normally achieved by altering the pump stroke length. Where plant throughput is variable (greater than $\pm 5\%$) the pump motor speed is automatically controlled in proportion to the bulk flow rate measured near the chemical injection point. This type of control is called 'open-loop': it has no feedback or corrective action and the applied dose rate is strictly proportional to the flow (Section 12.27). In a 'closed-loop' system, the pump output is corrected to maintain a given water quality value (such as pH) within a narrow pre-set band, measured downstream of the injection point after the chemical has been well mixed with the water ('feedback' control, Section 12.28). A process controller, working in conjunction with an appropriate water quality measuring instrument, sends a 4–20 mA signal (Section 19.38) back to the pump to adjust its stroke length.

Some pump manufacturers now offer so-called digital dosing pumps, controlled by a digital signal rather than an analogue 4–20 mA signal. These are essentially reciprocating pumps which incorporate a brushless DC stepper motor in place of conventional AC motors (Section 19.20). These allow the pump to operate at a fixed full stroke length while the volume of chemical delivered is regulated by close control of the rotation of the motor, which proceeds in a series of precise steps. This mechanism provides much greater turn-down than is available from a conventional diaphragm pump. For these pumps, as well as for progressive cavity and peristaltic types which can only accommodate a single control input, the water quality signal is combined with the rate of flow signal to provide a single control signal. The same approach is used with chlorinators and similar equipment to control the orifice positioner and so maintain a pre-set residual chlorine concentration in the water (Section 12.13).

12.27 FLOW PROPORTIONAL CONTROL

The simplest form of automatic chemical dose rate control is flow proportional control, in which the chemical dose rate is varied, by changing the speed of the dosing pump or adjusting the position of a regulating valve, in direct proportion to changes in flow past the dosing point. A flow meter provides the required flow signal. This method of control maintains a constant dose (mg/l) and does not make any allowance for variations in the incoming water quality. Flow proportional control is thus suitable for controlling chemicals applied at a fixed dose, such as phosphate or, in many applications, fluoride, where the dose does not change with changes in water quality. For chemicals where the required dose does vary with changes in water quality, the control should incorporate a signal from a water quality monitor.

12.28 FEEDBACK CONTROL

In feedback control, the desired value for the associated water quality parameter is entered as a set-point in the controller by the operator. The signal from a water quality monitor downstream of the chemical dosing point is compared with this set-point and the controller adjusts the dosing rate to maintain the set-point.

Feedback process control may be continuous using conventional PID control, described below, or it may be implemented at regular pre-set intervals. This latter method makes corrections to the process at specific intervals, to take account of the time taken for adjustments made to the chemical dose rate to be observed on the downstream water quality monitor. This period is referred to as the control loop time and includes the transit time from the chemical dosing point to the sampling point, the sample transit time from the sampling point to the water quality monitor, the response time of the water quality monitor and the time taken for the changed chemical dose rate to reach the dosing point (Fig. 12.1). The interval between adjustments to the chemical dosing rate is varied with the flow past the dosing point, to take account of the change in time taken for dosed water to reach the sample point. Flow proportional control is maintained during this interval, to maintain the prevailing dose. It is important to minimize the overall control loop time, particularly where PID control is used (Section 12.31). Control loop times should normally be no more than 4 minutes and should not exceed 10 minutes under any circumstances, unless as part of a cascade system (Section 12.29).

12.29 CASCADE CONTROL

Cascade control is a method of control combining two feedback loops, with the output of one controller (the primary controller) adjusting the set-point of a second controller (the secondary controller). It is particularly useful on processes with a long process time, such as chlorine dose control across a chlorine contact tank (Section 11.5), where loop times in excess of 30 minutes can be expected. The primary controller receives a signal from the water quality monitor at the outlet from the contact tank and compares this with the set-point entered by the operator. The output from this controller varies the set-point of the secondary controller. The secondary controller then adjusts the chemical dose rate to maintain this internal set-point, as measured by a water quality monitor at the inlet to the chlorine contact tank. The secondary controller thus acts on the dosing pump or regulating valve to bring both the residual at the inlet to the chlorine contact tank and the residual at the outlet from the chlorine contact tank back to their desired values. In this case, it is important that the set-point for the primary controller is changed at an interval no less than the overall process loop time.

12.30 FEED FORWARD CONTROL

In a few processes, the required chemical dose can be derived directly from the incoming water quality. Examples are coagulant dosing, where the coagulant dose can sometimes be estimated as a direct function of the raw water colour or UV absorbed value, and chloramination, where the required ammonia dose is often derived directly from the free chlorine residual at the ammonia dosing point. In such cases, feed forward control may be applied, where the signal from a water quality monitor upstream of the chemical dosing point is manipulated in an algorithm in the controller to derive the required chemical dose. This is combined with the prevailing flow rate to derive a control signal to the dosing pump or regulating valve.

12.31 PID CONTROL

A commonly used feedback control regime is PID control. In its simplest form, this combines three separate control actions, proportional, integral and derivative, to provide a single control signal to the dosing pump or control valve. Each of these actions is referred to as a 'term' and PID control is sometimes referred to as three term control.

The proportional term generates a control signal in direct proportion to the difference between the measured water quality parameter and the desired set-point. If used on its own, this form of control results in considerable deviation around the set-point and ultimately results in an offset between the measured value and the set-point.

The integral term modifies the proportional term in proportion to the length of time that the difference between the measured value and the set-point has persisted. Inclusion of the integral term eliminates the offset associated with proportional-only control but also reduces the speed of response of the control system.

The derivative term alters the control output to reflect the rate of change of the measured variable and is used to speed up the response of the control system. Both the integral and derivative actions are dynamic functions which are optimized to match the characteristics of the process. In practice, changes in water quality in water treatment are usually relatively slow and the derivative term is not normally required.

SAMPLING AND ON-LINE ANALYSIS

12.32 SAMPLING

Water is sampled to provide spot checks of water quality (e.g. raw and final/finished) (Sections 7.55–7.57 and 8.1), to assess the performance of individual processes, to permit automatic chemical dose control and, if necessary, to enable plant to shut down. Sample water should be representative of the bulk fluid flow from which it is taken in terms of physical, chemical and microbiological parameters.

It is critical that samples used for automatic control should be representative of the process conditions. This can only be achieved if sufficient care is taken to design the sampling system correctly.

Samples should be taken from a point as close as possible to the chemical dosing point but located after complete mixing has taken place. The sampling point should take into account time required for any chemical reaction to complete (e.g. pH adjustment, chloramination and ozonation). This is particularly true for pH adjustment, where sufficient time should be allowed for complete mixing, dissolving (if lime slurry is being used) and equilibration. In the case of chlorine disinfection, the control sample should be taken after disinfection has been completed.

The sample should be taken from the body of water with a sampling probe and not from the pipe or channel wall. Sample withdrawal tubes should project into the bulk fluid flow in a manner which avoids vibration and the collection of entrained gas bubbles or sediment and should extend into the pipe by at least one third of the pipe diameter, ideally at 45 degrees from the vertical in a horizontal flowing pipe.

If a pump is required to supply the monitor, the pump should be as close as possible to the sampling point. Pumps are normally of the centrifugal or progressive cavity type, with the progressive cavity type preferred as these cause less agitation of the sample, with less risk of chlorine gassing off, for example or turbidity measurements being affected.

Transit time in sample pipework should be kept to a minimum, particularly where it forms part of an automatic control loop. This can be achieved by locating the monitor as close as practically possible to the sample point and by using small bore pipework, or by using a 'fast loop' bypass system. Fast loop systems deliver excess sample to the monitor down a larger diameter pipe, only a small portion of which is delivered to the monitor, with the remainder either being returned to the main works flow or discharged to waste.

Sampling pipework is usually of PVC-U or, where it passes through buried ducts, polyethylene, although stainless steel is also used. Pipe for final water sampling used for regulatory purposes is often of copper with chrome-plated brass taps, to minimize the risk of bacteriological growth and to allow flaming for sterilization when taking samples for bacteriological analysis. The flow rate to sampling taps is typically restricted to around 2 l/min.

12.33 WATER QUALITY MONITORS

Water quality monitors are available for on-line analysis of a wide range of parameters. Those commonly used in water treatment plants are listed in Table 12.6, which indicates the principle of measurement and the primary function of these monitors.

Some measurements can be made using probes inserted directly into the process flow, either through a pipe wall or dipped into an open tank. Care must be taken in these instances to ensure the location is representative of the bulk flow and that easy access is provided for calibration and maintenance. However, most measurements are made in cells mounted remotely from the process, supplied with sample from a remote sample point.

The most common methods of measurement are based on colorimetric or electrochemical principles. Colorimetric methods are based on the addition of chemicals which react with the parameter being analysed to form a coloured complex. The intensity of the colour that develops is determined by measuring its optical density photometrically. Typically, colorimetric monitors operate sequentially on discrete samples of known volume to which measured quantities of the required reagents are added. After allowing an appropriate delay for the associated reactions to complete, the intensity of the colour is measured and the resulting value is displayed. The sample chamber is then emptied and rinsed before the process is repeated with the next sample. Most colorimetric monitors include an automatic calibration sequence which uses standard solutions of known colour intensity to allow automatic adjustment of the instrument. This sequence is normally repeated every 24 hours or other interval as recommended by the supplier.

Electrochemical methods generally incorporate a cell consisting of two dissimilar metal electrodes immersed in the sample solution. The cell can either be galvanic, generating its own potential (amperometric), or be subjected to an externally applied voltage (voltametric). In both cases the

Table 12.6 Water quality monitors in treatment plants

Parameter	Insertion or bypass	Method	Typical location in treatment process	Function
Aluminium	Bypass	Colorimetric	Post coagulation, post filtration, final water	Monitoring
Ammonia	Bypass	ISE or colorimetric	Raw water, post chloramination	Monitoring/ chloramination control
Chlorine	Bypass	Amperometric, colorimetric or polarographic	Manganese oxidation, disinfection, final water	Monitoring/chlorine dose control
Colour	Bypass	Colorimetric	Raw water, final water	Monitoring/coagulant dose control
Conductivity	Bypass or insertion	Conductance (twin electrode)	Raw water, final water	Monitoring
Dissolved organic carbon	See UVa: DOC is inferred from UVa measurements			
Dissolved oxygen	Bypass	Galvanic	Raw water, biological processes	Monitoring
Fluoride	Bypass	ISE	Post fluoridation	Monitoring/fluoride dose control
Iron	Bypass	Colorimetric	Post coagulation, post filtration, final water	Monitoring
Manganese	Bypass	Colorimetric	Raw water, post manganese removal, final water	Monitoring
Nitrate	Bypass	ISE	Raw water, post nitrate removal, final water	Monitoring/blending control
pH	Bypass or insertion	Potentiometric	Raw water, coagulant dosing, manganese removal, disinfection, final water	Monitoring/acid and alkali dose control
Phosphate	Bypass	ISE or colorimetric	Post phosphate addition, final water	Monitoring/phosphate dose control
Turbidity	Bypass or insertion	Nephelometric or surface scatter	Raw water, clarified water, filtered water, pre-disinfection, final water, recycle water	Monitoring/coagulant dose control
UVa/UVT	Bypass	Photometric	Raw water, post clarification, post filtration, pre-UV disinfection	Monitoring/coagulant dose control/UV disinfection control

resulting redox reactions cause a current to flow through the solution; the current flow has a direct relationship with the concentration of the parameter being analysed.

Some analyses are based on the use of ion-selective electrodes (ISEs). These are based on potentiometric measurement using ISEs in a similar way to pH measurement (Section 12.34). In use, a potential difference occurs between the measuring electrode and the reference electrode, as a result of charged ions of the parameter being analysed 'migrating' to an ion-selective membrane. This difference in potential is measured and is proportional to the ion concentration.

Where water quality monitors are used as part of an automatic control loop it is common to provide multiple instruments to validate the signal used for control purposes. Ideally, three instruments are used, with the output from each instrument combined in an algorithm to provide a signal to the controller. If the output from all three instruments lies within a pre-set band then the algorithm transmits the average value to the controller. If one instrument deviates by more than a pre-set amount from the other two, then an alarm is raised and control continues based on the average of the remaining measurements. This is termed triple validation. More recently, to reduce both capital costs and ongoing maintenance requirements, triple validation has been succeeded by so-called dual validation, in which the output from two instruments is compared. Normally one instrument from the pair is designated for use for control while the other instrument provides a policing function. If the output from the two instruments differs by more than a set value then an alarm is raised. Control may continue based on the duty unit or else the process may be shut down, depending on the degree of variation and the criticality of the process.

Sample volumes for continuous on-line monitors are generally around 30–60 l/h. The sample delivery rate to the monitoring cell and past the measuring sensors should be kept constant to avoid creating disturbances and upsetting monitoring performance. Samples which have had no reagents added during the monitoring process can be discharged back to the main process flow, whereas samples that contain added chemicals, either as reagents or for calibration purposes, should normally be discharged to waste. On sites where waste streams are discharged to the local watercourse it is important to take account of the possible impact of these reagents on any regulatory discharge consent, particularly for pH and chlorine residual.

The most commonly monitored parameters on a water treatment plant are pH, turbidity and chlorine. Details of monitors for these three parameters are presented in more detail below.

12.34 pH MONITORS

In simple terms, pH is a measurement of the relative quantities of hydrogen ions and hydroxyl ions in an aqueous solution. pH measurement is based on an electrochemical cell formed between a pH sensitive electrode (usually glass), a reference electrode, and the process sample being measured. A temperature sensor is also included to provide temperature compensation. Often the three probes are combined into a single polypropylene housing.

The pH electrode is typically formed from a glass or plastic tube with a small glass bulb at its base. The tube and bulb are filled with a buffered solution of chloride in which silver wire covered with silver chloride is immersed. The glass bulb at the base of the tube is formed with a thin wall of about 0.1 mm, from a specially formulated, pH sensitive glass. In use, the surface of the special glass adsorbs hydrogen ions on the inside from the buffered chloride solution and on the outside from the process solution being monitored, until equilibrium is achieved. Both sides of the glass are charged by the adsorbed hydrogen ions, creating a potential difference (voltage) proportional to the pH of the sample. So-called 'low-resistance' glass electrodes offer quicker response times than standard glass for low temperatures ($<10°C$) and low conductivity ($<100\ \mu S/cm$) applications.

The reference electrode is designed to maintain a constant potential at any given temperature, and serves to complete the pH measuring circuit within the process solution. It provides a known reference potential for the pH electrode. The difference between the potentials of the pH and reference electrodes provides a millivolt signal proportional to pH. To complete the electrical circuit between the pH electrode and the reference electrode, a small continuous flow of electrolyte is required from the reference electrode into the process solution. This is achieved via a porous membrane or ceramic wick at the base of the electrode, known as a junction, through which the electrolyte seeps very slowly. Sometimes the internal solution is replaced with a gel, which reduces the leakage rate, and so extends the period of operation between refilling. Eventually the gel becomes contaminated by back-diffusion of process solution and the electrode must be replaced. Typically, gel electrodes are less responsive than those that use a liquid electrolyte.

pH instruments require regular maintenance, to remove any fouling of the glass bulb on the pH electrode, to take account of clogging of the junction and, in the case of gel electrodes, contamination of the electrolyte. These factors will contribute to slow response times and instrument drift. For this reason, it is necessary to clean electrodes regularly, particularly when immersed in dirty process solutions, and calibration using buffer solutions of known pH is generally required at least monthly. Gel reference electrodes will usually require annual replacement.

12.35 CHLORINE ANALYSERS

There are three types of on-line chlorine monitors in common use on water treatment plants, employing either electrochemical or colorimetric technologies:

- Amperometric
- Polarographic membrane
- Colorimetric

The basic amperometric method uses two bare electrodes from dissimilar metals, one typically formed from a noble metal such as gold and the other formed from copper. When connected in an external circuit, a current flows between the two electrodes, approximately proportional to the chlorine concentration. The cell is more sensitive to the hypochlorous ion than the hypochlorite ion and is therefore dependent on pH, as the form the chlorine adopts in water depends on the pH. To address this dependence, an acid buffer, usually based on either phosphate or acetate, is added to reduce and stabilize the pH, with the incidental benefit of cleansing the cell. Alternatively, the pH dependence is accounted for by measuring the pH of the sample separately and compensating the chlorine measurement. Cells often include a mechanical cleaning aid, typically using fine grit or small plastic spheres, maintained in suspension by the sample flow or mechanical agitators, which help reduce fouling of the electrodes. Sometimes the electrodes in an amperometric cell are covered with a membrane, which can provide better selectivity of the analysis by only permitting a single chlorine species to pass to the electrodes.

When combined chlorine is present (Section 11.3) total chlorine can be measured by adding potassium iodide to the sample. Iodine is displaced in proportion to the total chlorine present in the sample, which then allows current to flow through the cell in a similar way to chlorine.

Polarographic monitors normally incorporate a membrane and consist of a pair of noble metal electrodes immersed in a common electrolyte. The electrodes are isolated from the sample by a proprietary chlorine-permeable membrane, which can be selected to allow either free or combined chlorine to diffuse through. A voltage is applied across the electrodes which generates a flow of electrons between the two electrodes. The current generated is proportional to the concentration of chlorine present.

The colorimetric method is based on the oxidation of diethyl-*p*-phenylenediamine (DPD) by chlorine to generate a pink colour whose intensity is proportional to the concentration of free or total chlorine present in the sample. The analysis process proceeds sequentially. First, the sample is pumped to the measuring cell and its background light absorption is measured. The reagents are then added, comprising buffer, potassium iodide and an indicator containing DPD. Iodine is released, which reacts with the indicator and turns the sample to magenta. The intensity of the colour is proportional to the iodine concentration which, in turn, is proportional to the total chlorine concentration. The light absorption of the reacted sample is measured and the difference between the two absorption measurements is used to determine the total chlorine concentration. Each measurement cycle takes between 2 and 3 minutes to complete.

12.36 TURBIDITY MONITORS

Most turbidity monitors are based on the nephelometric method, which measures the amount of light scattered at right angles to an incident light beam by particles present in a sample. Measured values are indicated in nephelometric turbidity units, NTU. The basic instrument incorporates a single light source and a photodetector to sense the scattered light. Internal lenses and apertures focus the light onto the sample, while the photodetector is set at 90 degrees to the direction of the incident light to monitor scattered light.

Various light sources can be used in the instrument. In the European Community, regulatory samples analysed in the laboratory are required to be monitored using infrared light at 860 nm (BS EN ISO 7027). In the USA, US EPA regulations require the use of lamps with a colour temperature between 2200 and 3000 K and a detector with a spectral response peak between 400 and 600 nm, the primary wavelengths of natural 'white light'. Typically, tungsten lamps are used to meet the US EPA requirements. The voltage applied to the tungsten lamp determines the spectral output characteristics produced, making a stable power supply a necessity. In addition, being incandescent, the output from the lamp decays with time, so that frequent calibration of the instrument is necessary and annual lamp replacement is required. Instruments with an 860 nm output overcome some of the incandescent lamp limitations, and provide longer service between replacements. Suitable monochromatic light sources include light emitting diodes, lasers and mercury lamps.

As the concentration of particles increases, more particles reflect the incident light, which increases the intensity of the scattered light. Eventually the particles themselves begin to block the transmission of the scattered light, resulting in a decrease in its intensity. The surface scatter design was developed to monitor waters with high turbidities to avoid this problem. This design focuses a light beam on the sample surface at an acute angle. Incident light strikes particles in the sample and is scattered towards a photodetector also located above the sample surface. As turbidity increases, the light beam penetrates less of the sample, thus shortening the light path and compensating for interference from multiple scattering. These instruments are best suited for measuring high turbidities such as are present in raw water and recycle streams (Hach Corporation, 1995).

Air bubbles are the primary source of interference for water turbidities of less than 5 NTU, contributing to increased scattering of the incident light. These can be formed by air coming out of solution due to agitation, changes in pressure or changes in temperature. Most turbidity monitors include bubble traps to ensure they do not interfere with the measurement. For turbidity levels greater than 5 NTU, interference from colour, particle absorption, and particle density become more significant (Sadar, 2002). Instrument fouling can be a significant problem, especially for units monitoring wastewater handling or supernatant return flows. Some instrument designs include automatic wipers to help address this problem.

REFERENCES

ADWG (2011). *Australian Drinking Water Guidelines 6*. Version 3.0 (Updated December 2014).

Avantor (2011). *MSDS – Powdered Activated Carbon*. Performance Materials Inc. (J. T. Baker).

Black & Veatch Corporation (2010). *White's Handbook of Chlorination and Alternative Disinfectants*. 5th Edn. Wiley.

BS EN ISO 7027 (2000). *Water Quality – Determination of Turbidity*.

Cabot-Norit (2014). *MSDS – Powdered Activated Carbon*.

Cassie, S. and Seale, L. (2003). *Chemical Storage Tank System – Good Practice*. CIRIA.

COMAH (2015). *Control of Major Accident Hazards Regulations 2015*. HSE.

DWI (2005). *Code of Practice on Technical Aspects of Fluoridation of Water Supplies*. DWI, UK.

DWI (2014). *List of Approved Products* (updated annually). London. http://dwi.defra.gov.uk/drinking-water-products/approved-products/soslistcurrent.pdf. DWI. UK.

EC (2009). *Health and Safety at Work – Scientific Committee on Occupational Exposure Limits for Sulphur Dioxide*.

Hach Corporation (1995). *Excellence in Turbidity Measurement*.

HSE (1990). *Safety Advice for Bulk Chlorine Installations*. HS/G28.

HSE (1999). *Safe Handling of Chlorine From Drums and Cylinders*. HS/G40.

HSE (2011). *EH40/2005 Workplace Exposure Limits* (updated regularly). 2nd Edn.

Letterman, R. D. and Yiacoumi, S. (2010). Coagulation & flocculation, *Water Quality & Treatment*. 6th Edn. McGraw-Hill.

Pernitsky, D. J. and Edzwald, J. K. (2006). Selection of alum and polyaluminum coagulants: principles and applications. *Journal WSRT-Aqua* **55**(2), pp. 121–141.

Perry, R. H. and Green, D. W. (2007). *Perry's Chemical Engineers' Handbook*. 8th Edn. McGraw Hill.

Sadar, M. J. (2002). Turbidity instrumentation – an overview of today's available technology. *Turbidity and Other Sediment Surrogates Workshop*, April 30–May 2, 2002, Reno, NV.

Sibiya, S. M. (2013). Evaluation of the streaming current detector (SCD) for coagulation control. *Proceedings of the 12th International Conference on Computing and Control for the Water Industry, Perugia, Italy*.

US EPA (1999). *Alternative Disinfectants & Oxidants – Guidance Manual*. EPA 815-R99-014.

Van Leeuwen, J., Holmes, M., Heidenreich, C. et al. (2003). Modelling the application of inorganic coagulants and pH control reagents for removal of organic matter from drinking waters. *Proc Modsim*, pp. 1835–1840.

WCC (2011). *Chlorine Safety Scrubbing Systems*. World Chlorine Council.

WHO (1997). *Guidelines for drinking-water quality*, **Vol. 3**. Surveillance and Control of Community Supplies. WHO.

Yavich, A. A. and Van De Wege, J. (2013). Chemical feed control using coagulation computer models and a streaming current detector. *Water Science and Technology* **67**(12), pp. 2814–2821.

CHAPTER

Energy Use, Sustainability and Waste Treatment

13

PART I ENERGY AND SUSTAINABILITY

13.1 ENERGY USE

After manpower, energy is the highest operating cost item for most water and wastewater companies. Over the last decade, energy consumption by the sector has increased with the implementation of new technologies to meet new potable water and effluent treatment quality standards. During the same period the price of energy has increased substantially. Proposed changes to regulations and standards and objectives to remove emerging contaminants will require additional energy intensive processes and technologies to achieve more exacting requirements.

High energy consumption which is inextricably linked to the issue of carbon emissions and the Climate Change debate is affecting the water industry worldwide. It is therefore important to minimize the use of energy by optimizing efficiency across the water cycle.

Globally the water industry consumes between 1% and 2% of the energy generated nationally of which about 45% is required for the production and delivery of safe drinking water and the rest is used to treat and dispose of wastewater. In the USA '*Nationwide, about 3–4% of USA power generation is used for water supply and treatment ... Electricity represents approximately 50–70% of the cost of municipal water processing and distribution*' (US DOE, 2006).

These statistics need to be put into context; between 10% and 15% of energy demand is used to heat domestic water and 18–30% or more is used for domestic space heating and air conditioning. Reducing domestic hot water temperature by 2°C would have more impact on carbon emissions than can be delivered by water efficiency measures. Furthermore, because there is a linear relationship between energy usage for water production/delivery and water usage, 'real' water conservation will deliver 'real' energy and water savings in addition to a reduction in use of other natural resources (Brandt, 2010).

Energy usage for drinking water is split between raw water abstraction and transfer (25%), treatment (10%) and distribution (65%). The percentage split will vary with the geographical characteristics and performance of the catchment and water supply area, for example, some areas with boreholes may supply direct to distribution without treatment. The dominant energy use in these activities is pumping to transfer water between sources, facilities and centres of demand and in treatment, air compressors for treatment processes (Brandt, 2012).

Twort's Water Supply. DOI: http://dx.doi.org/10.1016/B978-0-08-100025-0.00013-2

13.2 ENERGY EFFICIENCY

Adopting the definition of energy efficiency as being '*using what we need to use efficiently*', potential energy optimization benefits from interventions include:

- Between 5% and 10% improvement to existing pump performance.
- Between 3% and 7% improvement from new and innovative pump technology.
- Gains of between 5% and 30% from modifying pump operating regime to return it closer to the original design condition.
- More complex and large scale pumping energy savings are feasible but frequently show marginal payback using current financial analyses.
- Up to 20% improvement in clean water treatment processes, but the energy use in this category is low.
- Water conservation and demand management can deliver at least 5–10% water and energy savings and greater savings where a utility is resource constrained. These interventions are included here because they both represent key components of any energy use reduction strategy.

The above savings are indicative only, suggesting order of magnitude gains for companies that either have not started or have implemented only limited energy efficiency measures to date. Some companies have made significant progress installing new and refurbishing existing equipment with energy use reduction in mind. However, further progress towards the upper end of the savings ranges indicated will be more difficult to achieve and may not be deliverable in some physical and performance situations. Black & Veatch's 2016 Strategic Directions: Water Industry Report (B&V, 2016) includes the results of a survey of water service providers on their energy reduction goals. Half of respondents had goals for reductions of 10% or less, with a quarter having goals for reductions of between 10% and 25%.

When evaluating energy efficiency interventions, as for all investments, the assessment should include a Whole-Life Cost–Benefit Analysis (WLCBA) (Section 2.10) using electricity prices projected to about half the design life of the plant; that is, to at least 10 years into the future, in order to include future as well as current cost avoidance. The analysis should include a carbon balance that can be drawn up to compare the carbon footprint of installing new, more energy efficient, equipment with that of the energy saved. It should be clear from the analysis that the benefits from equipment replacement will be greater the nearer it is to the end of its life.

Pumping

Since pumping represents up to 90% of energy consumption for clean water production and transfer, improving pumping performance can deliver significant reductions in energy demand.

Modern pumps are better than 80% efficient and their drives, usually by electric motor, are better than 95% efficient. Recent regulations demanding energy efficient motors have resulted in 97% efficiency becoming common but pump selection is critical to realizing such improvements in practice. For example, a high profile aspect of a pump's selection may be its peak flow so it is selected on this basis. However, it may only be required to perform as such for limited periods and most of its working life may be spent at lower flows. The Best Efficiency Point (Section 19.7) should therefore be selected for the most common duty, not a short duration peak flow.

It is still common for pumps to be selected on the basis of lowest capital cost even though numerous studies have shown that this outlay represents less than 10% of pump whole-life cost. Most of the cost of owning a pump, usually over 85%, comes from meeting its energy demand. Maintenance is also significant, more frequently when a cheap pump is selected. Models are available, some of them water company generated, for selecting pumps on the basis the lowest Whole-Life Cost (WLC) analysis; the models accounting for energy, financial depreciation, maintenance and capital costs. The change from 'Capex' and 'Opex' to 'Totex' in investment planning is discussed in Chapter 2.

A pump motor is commonly direct coupled to the pump which is therefore driven at the speed of the motor. The most common type of motor used for driving pumps is the induction motor which, if used without a variable speed drive (VSD), has a speed (synchronous) which is determined by the number of poles of the motor (Section 19.21). Flexibility of pump selection is afforded by indirect drives, usually belts. These allow pumps to be run at non-synchronous speeds by conventional motors. Historically 'Vee belts' were used but they slip and require adjustment so toothed belts are now used; being more reliable, requiring less maintenance and being more energy efficient since there is no slippage.

Electronic VSDs are very popular but are often used in the design process to avoid committing to an accurate pump selection. However, full load electrical losses in a VSD are about 3%, with the best available being about 2% (Section 19.24). This represents a significant operational cost, as well as capital cost. Where pumps are required for variable duties; for example, if there is significant diurnal variation for a single pump's duty, or when precise pressure control is required over different flow ranges, the additional expense of VSDs may be warranted. Otherwise, it is more energy efficient to install a number of fixed speed units with trimming provided by a limited number of variable speed units. Caution should be exercised running pumps at low speeds as their characteristics can change drastically. Stall can occur which results in churning and heating water rather than moving it and pump efficiency can fall dramatically at low speeds. VSDs have other assets including control functions, interrogation facilities and power factor correction. The latter may not actually save much energy but could offer significant financial savings against power factor targets from electricity suppliers.

Motors for borehole pumps typically have lower voltage withstand levels than the short-term voltage spikes with VSD drives (Section 19.24). This, with the need to allow in the cable design for the eddy currents, has meant that use of VSDs for borehole installations has been problematic. The situation is improving with availability of VSD types with smoother outputs and borehole pump motors with higher voltage withstand levels.

One attribute often quoted as a reason for selecting VSDs is soft starting and stopping of pumps, typically in pressure sensitive zones to avoid surges. However, it should be remembered that power failure is usually the critical scenario for surge transients and a VSD will not help in this situation. Flywheels or air/water pressure accumulators are effective solutions.

Pipework design is important to energy efficient pumping and can also minimize surge potential. Pipeline velocities should be selected on the basis of minimum whole-life cost of the whole pumping and transmission system (Section 15.2). Station pipework should, in general, be sized so that velocities are no higher than pipeline velocities and should be arranged to minimize losses employing, where possible, radius bends rather than straight butted or mitred bends and swept rather than mitred tees. Exceptions may need to be made at isolating valves in order to reduce cost.

It may be cost effective to make the structure slightly larger to allow space for a more hydraulically efficient pipe layout and reduce risk of resonance due to pressure wave reflection.

Energy wastage through control valves, where present, should be reduced or, if possible, eliminated. In gravity systems excess head should be used to generate electricity through hydroturbines. In pumped systems a more appropriate, perhaps multi-pump fixed speed installation, or a combination of fixed speed drives with one or more VSDs, may achieve the required flexibility. The capital cost of such replacement technology will often be saved from reduced operating and maintenance costs on the control valves. Sump design and location relative to the pump is also critical, particularly noting that provision of a positive suction head is always beneficial, avoiding priming problems, reducing air entrainment and maximizing efficiency.

It is difficult to quantify potential savings from adopting energy efficient measures at pumping installations because the individual circumstances always vary but, if an old installation requires updating, 5–10% savings may be expected. Pump replacement is usually expensive in operational terms even if capital costs are acceptable. Under these circumstances pumps can benefit from refurbishment including impeller and bowl coatings, seal and bearing replacement. On high-speed pumps improvements of 5% are feasible and on a duty/standby system may be achievable without interrupting supply. In general, pump technology is mature but there is scope for improving the energy efficiency of some pumps, particularly small submersibles and multi-stage boosters and of submersible motors. Improvements of 3–15% may be realized if there is sufficient demand from the water industry.

Potable Water Treatment

This part of the water cycle has relatively low energy demand but savings are still feasible. Much of the energy demand in conventional treatment works is used for pumping, although there can be significant demands related to lighting, heating, air conditioning and ventilation. Transfer, recycle and dosing pumps can benefit as above. Flocculator and stirrer motors should be properly sized to avoid them running at low loads, and where ever possible works should be arranged for gravity flow.

On-site generation of sodium hypochlorite (Section 11.14) or ozone (Section 11.20) for oxidation and disinfection and use of UV are energy intensive processes. Clearly any reduction in the demand for such products by improved clarification and filtration or by better pH control will directly reduce the energy consumed at such generation plant. It is likely that the efficiency of such on-site generation plant may improve somewhat in the future. However, in the case of ozone generation, a large proportion of the energy input is lost as heat. Addition of heat exchangers on the cooling circuit could allow some of the waste heat to be recovered.

Advanced treatment options such as ultrafiltration, nanofiltration and reverse osmosis (RO) require the influent to be boosted to high pressures so pumps for these duties should be selected carefully. If there is any waste from the boosted flow, for example, the reject from RO membranes, this should be let down to atmospheric pressure through pressure recovery devices such turbines or pressure exchangers (Section 13.4). The energy demand of packaged plant could benefit from analysis of some aspects (e.g. heating of chemicals) of cleaning-in-place (CIP) technology (Section 9.18) and its application. Such parameters may even affect the selection of a different process on the basis of WLC analyses. Energy recovery is discussed further in Sections 10.46 and 13.4.

The treatment and disposal of clean water treatment sludges are discussed in detail in Sections 13.7 et. seq. There are limited opportunities to reduce energy demand for sludge treatment, dewatering and disposal beyond reducing the demand for water through conservation and demand management and the choice of a process which requires minimum use of energy.

Building Services

Heating, cooling, ventilation and odour control have been seen as targets for energy saving projects but care is required. Frost protection is important for plant maintenance and avoiding failures; lighting levels are critical when performing certain tasks. Cooling control rooms or areas where live maintenance may be required are also important for reliable equipment and operator comfort, particularly in hotter climates. Task lighting for regular or critical maintenance may also be important so operators and maintenance staff must be consulted on any proposed changes. Solar gain can be compounded by electromechanical plant such as VSDs which generate significant heat. Thermal insulation as part of imposed energy intensive changes such as fans or air conditioning can achieve additional benefits such as noise reduction. Lighting also offers scope for savings at plants where energy efficient sources are not already used or where sensor-controlled lighting is not already in place.

Increasingly new domestic and non-domestic premises are being built with energy conservation in mind. The types of technology employed in their design are also applicable to waterworks buildings. In addition to the use of renewable energy generation technologies, wider adoption of heat exchangers and heat pumps should reduce external demands for energy and could, in some cases, be returned to treatment processes that could benefit.

If changes to existing building services are envisaged it could be beneficial to gain operator support as well as consult with Safety and Risk experts; their suggestions should be practical and their cooperation critical to ensure the new systems work as designed. As above, a modern design should already incorporate best practice but old buildings or those intended for new uses could benefit by up to 15%.

A check list in the *Association of Electrical Equipment and Medical Imaging Manufacturers Journal* (NEMA, 2013) suggests the following:

- LEDs (light emitting diodes) should be first choice for lighting and their light quality is improving;
- VSDs can save 30–40% of power to motors;
- drivers are mainly cost reduction, including future cost avoidance;
- management tools are available;
- standards are changing to reflect new technology;
- smart plugs on some appliances can give significant benefits;
- green walls can help building energy efficiency;
- data centres (servers) have major energy demands and can benefit from energy efficiency analyses.

The indicated savings from use of VSDs apply where there is significant wastage of head due to throttling.

13.3 WATER CONSERVATION AND DEMAND MANAGEMENT

Water conservation and demand management reduce the demand for water and energy resources. Reducing per capita usage and wastage and leakage from the distribution system all contribute to reducing the pumping and treatment volumes. Any reduction in the volumes of fresh and waste water transported and treated within the water cycle will have a direct and proportional reduction in energy consumption.

Water utilities find the consumer responsive to publicity to conserve water and reduce demand during droughts or other events. However, the savings are often temporary, with demand reverting to pre-publicity quantities once the emergency is over. Some long-term savings are possible; for example, by consumers installing water butts and other rainwater collection systems to irrigate gardens and public areas (see also rainwater harvesting in Section 8.23). Long-term water conservation and demand management are largely social and political issues which are not responsive to technical solutions but still need to be resolved for water supplies in their existing state to be sustainable, for example, the extended drought in Australian starting in 2002/03 has led to consumer self-imposed and regulated changes in consumer water usage. Exceptions requiring technical input from the utility include leakage control and influence over the domestic and industrial fittings market.

Distribution system water losses can be reduced and managed by leakage monitoring, detection and repair, system pressure management and pipe replacement and maintenance programmes (Section 16.9). For example reducing the pressure in supply zones with high measured leakage or which are susceptible to high pipe burst rates, can reduce the incidence of leaks and burst especially at night and during periods of lower demand and higher pressure.

Water conservation, demand management and water loss reduction measures should deliver a minimum of 5–10% energy and water resource savings. Where a utility is constrained by the availability of water resources, cost benefit analyses of loss reduction intervention options generally demonstrate that significantly greater reductions are economically achievable.

13.4 RECOVERY OF ENERGY AND CHEMICALS

The major opportunities for energy recovery in water supply systems include in-line turbines installed in raw and potable water pipes where there is a significant hydraulic head difference between the pressure in the pipe and the required pressure immediately downstream. Energy can also be recovered from treatment processes involving high pressure such as membrane plants (see below), but currently the majority of energy recovered by the water industry is through CHP technology dealing with wastewater sludges.

Both raw water and distribution systems can have surplus energy. Surplus pressure is often dissipated through pressure or flow control valves. Such valves are expensive, high maintenance items because the energy they absorb results in hydraulic noise and vibration. The valves can be replaced by energy recovery turbines that will control flows or pressures, will minimize hydraulic issues and generate income from electricity. If the installation is within water company pipework and land ownership there should be no environmental impact and minimal regulatory restraints. However, obtaining 'tie-in' agreements with the local electricity distributor can involve complex and extended negotiations. Recently energy recovery turbines were installed on the inlet to Glencorse water treatment works in Scotland and within distribution pipes by Welsh Water (DCWW).

RO plants have, for some time, been built with energy recovery on the waste (reject) stream. However, energy recovery devices (ERDs) have developed considerably in the last decade and efficiencies are much improved. Therefore, older RO and nanofiltration plants present an opportunity to save energy by improving or introducing energy recovery by retrofitting up-to-date ERDs. The use of ERDs in the waste stream of such plants is discussed in Section 10.46.

If renewable energy is seen as part of an energy recovery strategy then solar photovoltaic (PV) is often feasible. Panels can be mounted on buildings, preferably with south facing (in the northern hemisphere) pitched roofs, on flat roofs or even on spare land. An existing power supply is ideal to enable a convenient connection with metering to enable direct accounting for energy generated. Solar thermal panels may be more efficient at collecting energy but heat is rarely required in clean water processes and is usually more conveniently and reliably collected from CHP systems for wastewater processes. Wind energy is usually intermittent and low power so not commonly applicable to water engineering needs. However, for remote sensors or instruments where a conventional power supply may be difficult, there are packaged units combining solar PV and small wind turbines with batteries to allow continuous output.

There are limited opportunities for energy generation from drinking water treatment processes since waste and sludges usually have no biological energy and can be chemically contaminated (Section 13.7). However, with several previously common materials causing concern due to limited resources, the economics and incentives may change in future and it may become beneficial to extract some chemicals from waste, either for reuse or simply to prevent their discharge to the environment.

13.5 CARBON ACCOUNTING AND EMBEDDED CARBON

Carbon accounting is a means of measuring the direct and indirect emissions to the Earth's biosphere of carbon dioxide and its equivalent gases from industrial activities. Hence if a water turbine is installed in a pipe to recover energy from excessive pressure the generated energy represents a carbon reduction benefit through the reduced carbon output from a conventional generation source. There will be a capital carbon cost, mainly due to the embedded carbon in the materials and workmanship used to manufacture and install the turbine and associated equipment. A Life Cycle Analysis (LCA) would show how long it would take for the capital carbon cost of the water turbine to be repaid by the outputs. LCAs are sometimes complicated by environmental benefits which may accrue beyond the site boundaries or beyond the conventional project life, such as into the demolition phase, so it is important to draw boundaries consistently in time, financial, commercial and geographic terms. For consistency, values used for direct and embedded carbon for different materials are usually taken from approved databases such as the *Inventory of Carbon and Energy (ICE), Version 2.0* (Hammond, 2011).

For water treatment processes energy is a major direct carbon cost but the normal fossil fuel element can be effectively reduced by buying electricity from renewable sources, such as wind power companies.

It is sometimes difficult to find carbon information about chemicals since the production methods vary, however, a material such as granular activated carbon, as used for adsorption, will obviously have a high carbon cost, both from its manufacture and its periodic generation; liquid oxygen is another example.

Chemicals introduced in one part of the treatment process incur direct carbon and financial cost, but can also have a detrimental effect on another part of the process, often requiring further treatment, more chemicals and energy to address as well as the consequences of increased waste generated. The overall multiple effects of adding chemicals should, therefore, be taken into account.

Carbon values for construction materials are also available so comparisons can be made for different building techniques. Some surprises have resulted: glass fibre reinforced plastic kiosks and control cabins were frequently used but their end-of-life disposal is difficult as they cannot readily be recycled due to the hazards of raw glass fibres. In contrast, timber framed and insulated buildings, or even bricks, can be easily demolished and the materials reused or recycled.

When assessing process carbon costs it is not usual to include fugitive emissions unless specifically requested. These include odorous air from aeration or sludge treatment; however, any boiler stack emissions should be accounted for. For clean water treatment and pumping, the main fugitive emissions would relate to diesel powered generators primarily used for standby generation (resilience measure). However, like all standby systems, they need to be run regularly under load and maintained in order to minimize emissions and reduce the risk of failure.

In financial terms a project is sometimes considered viable if the capital cost is paid back within about 5 years; however, the acceptable carbon payback period is much longer. Regulation is tending towards longer term planning, typically 25 years, with renewed focus on balancing capital and operating costs (WLC and Totex; Section 2.10); new principles need to be established on carbon capital cost payback since this can be significantly different. There are also wider potential benefits to carbon accounting since changes in metrics can lead to completely new thinking on project concepts and execution. It has been found that such approaches can have significant financial benefits over conventionally managed projects. Broad scope and early adoption are critical. A predefined project scope presented with a carbon reduction objective at detail design stage is not likely to yield many benefits. If project objectives are considered within a broad brief at concept stage, then the risk is that the initial concept may change but the potential benefits could be much greater. For example a pipeline project will realize some benefit from arranging deliveries of pipes to small stockpiles along the route since this will reduce on-site transport and double handling. With a wider brief and perhaps a change of materials the pipes could be manufactured on site with huge savings in long distance transport costs, cheaper installation and major carbon credits. Different attitudes also encourage staff and contractors at all levels to make carbon and cost savings in their daily tasks so the gains can become cumulative. One benefit across the industry has been the increased use of recycled crushed concrete, since cement is a major carbon emitter due to quarrying and firing cement kilns. The financial benefits are apparently marginal but the carbon saving is significant. Careful reuse of trench excavation material has also reduced landfill volumes and costs significantly. Separation of waste material on site for recycling has also become an accepted norm; rather than all waste being dumped in one skip for discharge to landfill.

At a project level carbon accounting has led to evaluating and managing projects against a set of parameters designed to result in carbon credits. Options may be selected on the basis of least carbon expenditure and/or best carbon payback over whatever term is chosen. In a regulated industry this allows a utility's investment plan to be assessed and scored against other companies with, in some cases, financial penalties for unsatisfactory performance. However, energy is the main source of carbon emissions and the water industry only uses about 1–3% of national energy demand so the industry's contribution to national carbon reduction targets is relatively small. It is nevertheless an important example for other industries and one with a close public connection.

To help establish carbon accounting as an alternative to financial accounting for evaluating and prioritizing projects a nominal price of carbon was established by international agreement under the Kyoto Protocol in 1997. The hope was that this would raise the profile of climate change and the carbon emissions so that, through regulation, an artificial shortage would cause an effective carbon price rise. This would make projects with carbon benefits, such as renewable energy devices and their large scale deployment, more attractive. Some countries, such as the UK, committed to reducing carbon emissions by other means but full international agreement was never achieved. The price has not risen enough to influence commercial decisions; its value on the open market has not reflected the long-term importance of the issue and commercial decision making remains dominated by the habitual short-term focus of the financial markets.

13.6 SUSTAINABILITY AND THE FUTURE

The Brundtland Report (1987) definition of sustainable development is still sufficiently far from normal business practice to exercise the mind. The concept of meeting '*... the needs of the present without compromising the ability of future generations to meet their own needs...*' is comprehensive, inter-generational and beyond normal planning horizons. Infrastructure should normally be planned on a long-term basis anyway but the need to respond to financial and environmental regulation can introduce political influences which are invariably short term.

For the water industry sustainability should mean that all of the resources and facilities that are required to supply potable water should be available for the foreseeable future. The inter-generational timescale requires links with regulatory requirements for the rolling (25 years) investment plans required. If an item of mechanical plant lasts 30 years and civil structures last in excess of 100 years this should allow time to adapt the thought processes to changing circumstances.

The challenges facing a sustainable water industry include:

- sufficient suitable (sustainable) raw water resources; the quality of existing sources has been declining in many parts of the World but, in Europe, should be improved by application of Water Safety Plans (Sections 7.82 and 7.83);
- political acceptance of inter-catchment transfers;
- consumer acceptance of potable water from alternative sources;
- consumer acceptance of the equitable use of water – reduce demand for water in all phases of the water cycle;
- the availability of energy to treat and move water to where it is needed;
- appropriate standards for '*safe drinking water*'; not over regulation leading to increased demands on natural resources;
- appropriate processes (including technology development) to optimize use of energy and chemicals for the treatment and disinfection processes;
- local facilities versus large central works and complex distribution networks; environment reuse or disposal (sinks) for any waste by-products;
- people with sufficient skills at all levels;
- finances and materials to enable whole-life planning, construction and decommissioning of fixed assets;
- affordability of energy beyond the short-term period of an investment analysis.

The availability of raw water resources is by no means ensured. In some areas of the UK where proposed economic development has not flourished as envisaged water infrastructure may seem under-utilized. However, the ecological condition of the rivers is under scrutiny and this may put pressure on existing abstraction regimes. In other areas, notably cities such as London, there is concern that over-abstraction is putting river conditions at risk and depleting aquifers. To some extent local problems can be alleviated by linking catchment areas with pipelines and transfer schemes to improve resilience but if city populations continue to rise then so will the risks of supply shortages in dry years. Desalination of sea water can help bridge the gap but is carbon expensive; it consumes about 10 times the energy of a conventional water treatment plant. Water transfer systems on the scale required are also energy hungry and human behaviour will need to adapt to using less water and adopting recycling practices. Australia is an example of adaptation during their recent 7-year drought, but archaeology is rich with examples of abandoned cities that ran out of water, even in wet climates.

For sustainable energy efficiency the long-term plans require that personnel resources should be assured as well as reliable energy supplies. As can be seen from the above sections a mix of competent engineering skills and experience is essential to design, build, operate and maintain plant and processes in an efficient state. Such assets take time to acquire and build into a balanced team and people are too easily lost. Sustainability requires that long-term plans be drawn up and followed; they should be reviewed on at least a rolling 5-year basis, as well as after supply crises and at the time of management change when short-term issues tend to override softer concerns.

PART II WATERWORKS WASTE AND SLUDGE DISPOSAL

13.7 TYPES OF WASTE

The primary wastes from conventional treatment plants are the sludge from clarifiers, used washwater from rapid gravity or pressure filters and waste from slow sand filter sand washing plants. Sources of waste from membrane filtration plants are membrane washwater and discharges containing chemicals used in chemically enhanced backwash (CEBW) and clean-in-place (CIP) regimes. Other sources include waste regenerant streams from ion exchange (IX), adsorption and radionuclide removal processes.

Secondary sources include chemical wastes (from chemical delivery, storage and dosing facilities); overflows (from the main and secondary treatment processes); water quality monitoring sample flows; waste generated during commissioning (e.g. filter washwater, water used in the disinfection of water retaining structures and pipes, water used in hydraulic tests and membrane preserving chemicals); and water retaining structure draindown prior to maintenance.

There may be discharges from treatment works which need to be properly managed (e.g. gas leaks, chemical dust and waste arising from cleaning processes) as well as wastes arising from chemical deliveries such as packaging, bags, drums, carboys and pallets and also workshop waste (oil and grease), laboratory wastes (hazardous and non-hazardous) and domestic wastewater. Typical sources of waste and the associated disposal requirements are summarized in Table 13.1. A detailed review of secondary waste generation and disposal is reported by Li (2002).

Table 13.1 Waste produced at treatment works and its treatment

Waste source	Waste	Treatment/disposal
Commissioning		
Hydraulic testing water	Free of contaminants	Discharge to a watercourse
Treated water	Chlorinated and may not be compliant with treated water quality standards for supply during start up and process adjustment	Discharge to watercourse after dechlorination if necessary. Reuse if feasible
Water used for disinfecting water retaining structures	High chlorine residual	Discharge to watercourse after dechlorination. Reuse if feasible
Filter backwash water	Media fines	Settlement and discharge to a watercourse. Fines to landfill
Normal operation		
Clarification sludge	Regular batch discharges; 0.1–3% w/v solids	Discharge to sewer where available. Otherwise dewatered sludge to landfill (occasionally to agricultural land if suitable)
Media filter or membrane filter washwater	Regular batch discharges; typically 300 mg/l solids	Recycle with/without settlement. Settled solids to sludge treatment plant
Slow sand filter sand washing plant washwater	Biological matter	To sewer or settlement and disposal
Membrane plant clean-in-place or chemically enhanced backwash water	Acids, chlorine and alkali	Neutralize and dispose to sludge plant or sewer
Nanofiltration (colour removal) reject stream	Highly coloured effluent with very high organic content	Disposal to sewer or sea outfall
Spent ion exchange regenerant and associated backwash and rinse water and adsorption media	High TDS. Very high organic content from DOC removal processes. Toxic waste from arsenic removal process	Neutralize pH and treatment where applicable. Disposal to sewer or sea outfall where available. Discharge to licensed disposal site
Process tank drainage	Contains settled sludge	Remove clear water to site drain. Sludge to sludge treatment plant
Tank under-drainage	May contain dosed chemicals	To surface drain
Main process overflows	Could contain low residual chlorine and/or typically pH in the range 5–7 depending on the location	Discharge to watercourse (subject to discharge consents) If necessary dechlorinate and/or neutralize
Secondary process overflows (sludge treatment plant)	Suspended solids; concentration depends on the location	Watercourse (subject to discharge consents). Sewer or discharge to downstream process units

(*Continued*)

Table 13.1 (Continued)

Waste source	Waste	Treatment/disposal
Chemical wastes (dross, flushings, drainage washings, spillages)	Care necessary to prevent mixing of incompatible materials (NB inadvertent chemical reactions)	Neutralize and discharge to sewer or hold for tanker removal
Chemical tank leakage and spillage on delivery	Concentrated chemicals	Contain in bund and remove for off-site disposal. Use 3-way diversion valves in delivery area to allow rainwater discharge
Water quality sample flows	Some do not contain chemical reagents (e.g. turbidity monitors), others contain reagents	Those with reagents to sewer or sludge plant
Domestic water	Sewage and grey water	Sewer or treatment on site
Laboratory drainage	Hazardous and non-hazardous chemicals	Collect hazardous waste for off-site disposal. Non-hazardous to site drain
Laboratory solid waste	Contaminated waste including hazardous chemical and biological waste	Off-site disposal
Ventilation/dust handling	Dust associated with chemical handling and storage	Hazardous: Off-site disposal
Membrane preserving chemical	Glycerine, sodium bisulphite, etc.	To sewer or off-site disposal
Road drainage	May be contaminated with fuel oil and other spillages	Site drain with oil traps
Process gas from generation plants or leakage	Chlorine, ammonia, sulphur dioxide, ozone, hydrogen, etc.	Scrubbers (Cl_2, SO_2, NH_3), destructors (O_3), dilution and venting (H_2)
Chemical packaging	Bags, drums, pallets, carboys, etc.	Reuse or off-site disposal by skips
Workshop wastes	Oil and grease	Collect for off-site disposal

13.8 TYPES AND QUANTITIES OF SLUDGE

Sludge may be classified according to the type of water treatment process adopted.

Non-chemical sludge arises from microstrainers, pre-settlement unaided by chemicals, backwash water from membrane filtration unaided by coagulant, sand washings (biological) from slow sand filters and associated roughing filter backwash water. These sludges are for the most part relatively innocuous and in some countries it may still be permissible to discharge them to a watercourse or water body without treatment; exceptions are when the raw water has high populations of algae or biological sludge when treatment is required.

Coagulant sludge derives from treatment plants using coagulation. It is the most difficult to treat for disposal because of the relatively large volumes involved and the difficulties of dewatering due to their gelatinous nature. It is dealt with in detail below.

Softening sludge is generated by lime or lime-soda softening. This sludge, being of granular consistency, is comparatively easy to dewater but the quantities are much larger than from the other two sources.

Quantities

The quantities of dry solids in sludge produced in a treatment works employing coagulation, flocculation, clarification and filtration are a function of raw water quality and the chemical treatment applied. They are made up of coagulant hydroxide, suspended solids, precipitated colour, algae, iron, manganese and other chemicals contributing to solids, such as powdered activated carbon (PAC), impurities in lime and polyelectrolyte. Dry solids produced ahead of clarifiers or filters (in direct filtration) can be calculated as follows:

$$\textit{Sludge dry solids}\ (\text{mg/l}) = X + S + H + C + Fe + Mn + P + L + Y$$

where X is coagulant hydroxide (mg/l) $= f \times$ coagulant dose (mg/l as Al or Fe), $f = 2.9$ for Al and 1.9 for Fe; (note that US practice is to include 3 molecules of hydration water in the hydroxide, increasing these factors to 4.9 for Al and 2.9 for Fe (Cornwell, 2006)); S is suspended solids (mg/l); (when suspended solids data is not available a value can be approximated to $2 \times$ turbidity (NTU)); H is $0.2 \times$ colour in Hazen (Warden, 1983); C is $0.2 \times$ chlorophyll a in μg/l; Fe is $1.9 \times$ iron in water in mg/l as Fe; Mn is $1.6 \times$ manganese in water in mg/l as Mn; P is PAC dose (mg/l); and L is lime dose in mg/l as 100% $Ca(OH)_2 \times (1/w - 1)$ where w is the purity of lime expressed as a fraction; and Y is polyelectrolyte dose (mg/l).

In a well operated treatment works using clarification and filtration all but about 1–5 mg/l of the calculated dry solids will be removed in the clarifiers. When softening follows clarification, the sludge dry solids produced in the softener can be calculated as follows (see Sections 10.2–10.4 for softening).

For lime softening to remove carbonate hardness:

$$\textit{Sludge dry solids}\ (\text{mg/l}) = 2CaCH + 2.54MgCH + CO_2 + L = LSSDS$$

where $CaCH$ is calcium carbonate hardness removed as $CaCO_3$ (mg/l); $MgCH$ is magnesium carbonate hardness removed as $CaCO_3$ (mg/l); CO_2 is carbon dioxide removed as $CaCO_3$ (mg/l), L is as above and $LSSDS$ is lime softening sludge dry solids.

For lime-soda softening to remove both carbonate and non-carbonate hardness:

$$\textit{Sludge dry solids}\ (\text{mg/l}) = LSSDS + 2CaNCH + 2.54MgNCH$$

where $CaNCH$ is calcium non-carbonate hardness removed as $CaCO_3$ (mg/l) and $MgNCH$ is magnesium non-carbonate hardness removed as $CaCO_3$ (mg/l).

For caustic soda softening to remove carbonate hardness:

$$Sludge\ dry\ solids\ (\text{mg/l}) = CaCH + 0.54MgCH$$

For caustic soda softening to remove both carbonate and non-carbonate hardness:

$$Sludge\ dry\ solids\ (\text{mg/l}) = CaCH + 0.54MgCH + CaNCH + 0.54MgNCH$$

If softening and clarification processes are carried out together the clarifier and softener dry solids should be summated.

When coagulants are used, the water is usually filtered by rapid gravity or pressure filters. The coagulated water may pass directly to the filters (direct filtration), in which case there is only the sludge content of the filter backwash water to deal with, which should be estimated as for the clarifiers assuming total removal in the filters. In direct filtration with coagulation, the backwash water volume can range from 2–5% of plant input, with an average of 3%. With clarification, irrespective of the wash regime, a filter wash will use a washwater volume equivalent to 2.5 bed volumes for sand or three bed volumes for dual media; this is about 1.5–2.5% of works inflow depending on the frequency of filter washing. Filter backwash water usually contains 0.01–0.05% w/v dry solids with an average of 0.03% w/v and up to 0.2% w/v for direct filtration with coagulation. Note: x% w/v and x% w/w = x g of dry solids in 100 ml or 100 g of liquid containing the solids, respectively. Dry solids are defined as the residue that remains after evaporation of a sample to a constant weight at 103–105°C.

Sludge withdrawn from clarifiers may amount to about 1.5–2.5% of works inflow and contains 0.1–1.0% w/v solids with an average of 0.3% w/v. Dissolved air flotation (DAF) tanks employing full length scrapers for float removal produce sludge flows of the order of 0.5–1% of works inflow, with dry solids concentrations ranging from 2–4% w/v, although removal at high dry solids concentrations may require sparge water to assist conveying sludge to the sludge treatment plant. Beach scrapers produce a weak sludge with flows up to 3% of works inflow and a solids concentration of about 0.1% w/v. Hydraulic float removal methods increase the sludge quantity to about 1–2% of works inflow with solids concentrations of less than 0.5% w/v and usually about 0.1% w/v (Schofield, 1997). For COCO DAFF or DAFF combined losses are about 4% of inflow at a concentration of about 0.5–1% w/v (DAF) and 0.03% w/v (F – filtration). For softeners the quantity discharged is usually in the range 0.5–2% of works throughput, although in some plants it can be as high as 5%, whilst the sludge is withdrawn at concentrations between 5% w/v and 10% w/v; about 85–95% w/w of the solids is calcium carbonate depending on whether the softener is used for combined clarification and softening or for softening only.

For pre-settlement tanks and clarifiers which have to treat waters with heavy silt loads of over 1000 mg/l, the sludge volume removed can range from 5–10% of works inflow and may even reach as high as 30%. The concentration of such a sludge can range from 2–5% w/v solids with a maximum of 10% w/v. These are exceptional cases in which special removal methods are necessary, but it is as well to be aware that they can occur with some raw waters, particularly in overseas countries. At the 1365 Ml/d Al Karkh water treatment works treating river Tigris water for Baghdad, the raw water suspended solids loading could reach 30 000 mg/l, at such times pre-settlement tank desludging accounted for 28% of works inflow and the concentration of the sludge was 10% w/v; the downstream clarifiers accounted for a further 7.5% of inflow at 2.5% w/v concentration. Losses due to filter backwashing remained below 1.5% of inflow. Total water losses were therefore some 37% of works inflow.

Membrane filtration plants have both sludge and chemical waste streams. The former is made up of the blow-down or used backwash water and has a composition similar to the feed, albeit at slightly higher concentration. When coagulation and other chemicals such as PAC are used, quantities should be estimated as for clarifiers and filters. Hypochlorite and an acid or an alkali will be present in CEBW and CIP waste (Section 9.18). The quantity varies between 2% and 10% of inflow, depending on the feed water quality, chemical pre-treatment and whether a clarification stage precedes the process. The sludge concentration is in the order of 0.025% w/v, similar to that in used washwater from rapid gravity filters. The quantity of chemical cleaning waste is a function of the cleaning frequency and varies from several times a week for CEBW to once every few weeks for CIP. It is primarily made up of the chemicals used in cleaning and the rinse water, and is neutralized with sodium bisulphite, acid or alkali as appropriate before discharge.

Waste from IX (Sections 10.7–10.10 and 10.25) and other adsorption processes (Sections 10.14 and 10.16) has high total dissolved solids (TDS) concentration made up of excess chemicals used in regeneration (e.g. Na, Cl, SO_4) and the ions removed in the exchange process (e.g. Ca, Mg, As(v), NO_3, SO_4) and dissolved organic carbon (DOC) (in the magnetic ion exchange (MIEX®) process, Section 10.35). They also have waste streams containing suspended solids from backwashing the beds prior to regeneration and high TDS from rinsing the bed before returning to service.

In slow sand filter installations, used washwater quantities and solids concentrations from sand washing are similar to those of direct filtration plants although the solids do not contain coagulants. The sand washing plant produces between 10 and 20 m^3 of water/m^3 sand. The used washwater contains solids, primarily consisting of organic debris and typically has a volume of 100–200 litres per 100 m^3 of water treated, assuming 50 mm is scraped off every 2 months.

13.9 FILTER, ION EXCHANGE AND RADIONUCLIDE WASTE DISPOSAL

Used filter washwater is very dilute compared to clarifier sludge and it is preferable to dispose of it separately, or keep the two waste streams segregated until the used washwater has undergone settlement to concentrate its sludge solids. There are instances, however, where used washwater and clarifier sludge are combined and treated together, especially in smaller treatment works.

Discharging used washwater to the river downstream of the point of abstraction of the raw water is the simplest and most economical disposal method subject to local discharge standards. In the absence of such standards, it could be practised if dilution is available to reduce the coagulant metal concentrations to less than their respective drinking water quality standards. In such circumstances, an environmental impact assessment would be required to confirm there are no adverse environmental consequences.

Used washwater could be recycled with the solids it contains to the treatment process at the works inlet, where it is mixed with raw water. The main benefit is that solids are subsequently removed in the coagulation and clarification process and can be withdrawn as clarifier sludge at a steady rate and consistent solids concentration. Recycling also helps to maximize washwater recovery. Used washwater is however generated intermittently in relatively large amounts over short periods, with the solids concentrations declining continually until it is almost clear. It contains up to about 10% of the solids removed in the treatment plant. Uncontrolled return of used washwater as shock loads to the main treatment process may adversely affect the coagulation and downstream clarification processes. It should therefore be added, uniformly at a steady flow of up to 10% of the works inflow (Section 9.20), from a flow balancing tank, upstream of the coagulant dosing point and should be well mixed with the raw water flow. A graphical method for sizing a flow balancing tank is given by Warden (1983).

Recycling can return undesirable material (which had already been removed) to the plant. The contaminants of primary concern are microanimals (e.g. *rotifers, cyclopoids, copepods, chironomid midge larvae and nematode worms*) and protozoan organisms such as *Cryptosporidium* oocysts and *Giardia* cysts. If the numbers of these organisms in the raw water are high, then recycling of washwater (which would contain a large number removed in the filter) would exacerbate associated problems. Holding tanks used for washwater can also provide breeding grounds for some of these organisms, in which case the used washwater returns could potentially carry them in increased numbers to the treatment process. This seeding can result in infestation of clarifiers and filters, increasing the risk of passing them to the treated water.

Treatment of used filter washwater by settlement reduces the risk of recycling microorganisms. With 80% settlement efficiency, the increased loading of *Giardia* cysts and *Cryptosporidium* oocysts to the plant through recycling the supernatant is only about 1.2 times the source loading; settlement also helps to reduce manganese, aluminium and iron present as particulates as well as DBP precursor concentrations (Cornwell, 1994). High *Giardia* cyst and *Cryptosporidium* oocyst risk sites should have facilities in the form of a lagoon or similar to divert recycled water in the event of an incident.

Settlement is usually preceded by a trap to remove any filter media carried over in used washwater to minimize damage to downstream pumps. The settlement process is aided by a small dose of polyelectrolyte (0.02–0.2 mg/l of polyacrylamide) and can be by batch or continuous flow sedimentation. Flocculation has shown to be beneficial in the settlement process (Arendze, 2014). Batch tanks take the shape of hopper-bottomed clarifiers (Section 8.16) or shallow rectangular tanks (typically $L{:}w = 5{:}1$ see Section 8.15), with floor sloping towards the outlet end, and usually consist of three or more tanks each sized for at least one filter wash and arranged to operate in rotation with one filling, one settling and one being emptied. Supernatant is decanted using a floating arm draw-off arrangement, with sludge drawn off from the bottom. Continuous flow sedimentation uses lamella plate settlers (Section 8.17) operating at about 0.75 m^3/m^2. h based on the projected area with plates at 60° to the horizontal and horizontal plate separation of 50–55 mm. Mechanical flocculation should be provided upstream. The settlement process concentrates the solids 10-fold to 0.3–1% w/v. The supernatant recovered will have a turbidity usually less than 10 NTU and can be either discharged to a watercourse or returned to the works inlet at less than 10% of the works raw water flow. DAF has been evaluated for treating used washwater (Eades, 2001) and Clari-DAF® (Section 8.18) has been selected for Betasso WTP following an assessment of six alternative technologies (US EPA, 2002). Chemical treatment was limited to low doses of polyelectrolyte and the flocculation time was about 10 minutes. Turbidity of <1 NTU was achieved at rates up to 17 $m^3/m^2 \cdot h$. The recycle rate over the range 5–20% had little impact on performance. The performance of the full scale plant however, did not meet that of the pilot plant at rates $>5\ m^3/m^2 \cdot h$ (Cornwell, 2010). There are very few DAF plants in use in this role to date.

The risk of returning *Cryptosporidium* or *Giardia* to the works can be further reduced by treating the supernatant by membrane filtration (Section 9.18), cartridge or 'depth filters' (Section 9.19), ozonation (Section 11.21) or UV treatment (Section 11.25). In works where pre-ozonation is practised, the high ozone dose required to inactivate *Cryptosporidium* oocysts and *Giardia* cysts can be achieved by injecting a proportion of ozone required for main works pre-ozonation into the supernatant return. An allowance should be made in the applied dose for the ozone demand of the supernatant. The minimum *CT* value required is typically15 mg.min/l with about 2–3 minutes contact time.

The sludge stream from membrane filters can be treated in a manner similar to washwater from media filters. The chemical cleaning wastes from membrane plants can either be treated on site or removed for off-site disposal, depending on the quantity. On-site treatment usually consists of flow balancing followed by neutralization for acids and alkalis, and sodium bisulphite dosing to remove chlorine. Membrane preserving chemicals, such as glycerine (high in biochemical oxygen demand (BOD)) and sodium bisulphite (oxygen scavenger) (about 1 litre per pressure vessel), both used to store the membranes prior to installation, can be discharged to the sewer or removed for off-site disposal.

The volume of regeneration waste in IX processes varies depending on the process. For softening, it is about 0.25% of plant throughput with backwash and rinse water waste amounting to an additional 1.5%. In denitrification plants, depending on the degree of removal required, the volume can be about 1.5–2.0% of the throughput. Waste regenerant volumes from the MIEX® process are <0.05% of plant throughput, with slightly higher volumes from the SIX® (suspended ion exchange) process. Waste may be safely discharged to a sewer or, for sites in coastal areas, to the sea. In other cases, removal by tanker may be necessary as its discharge to a watercourse would not be acceptable. For large plants, tanker removal may need to be preceded by a volume reduction process such as electrodialysis reversal or RO; in some countries solar evaporation may be used. The Eliminate® system used on nitrate removal plants regenerates the resin with potassium chloride and nitrate in the waste stream is converted to nitrogen gas by electrolysis.

The ISEP® nitrate removal process utilizes nitrate selective resin in a carousel with a multiport distribution valve arranged to feed about 30 columns of which about 20 are operating in parallel in different stages of exhaustion with the remainder in stages of rinsing (with the effluent being used to dilute the regenerant brine), counter-current regeneration and displacement. Softened water is used to minimize precipitation of calcium sulphate and calcium carbonate during rinsing and displacement respectively. Recovery is about 99.7% compared to 95% for the conventional processes. The advanced Amberpack® design utilizes a fractal distribution system to ensure uniform flow across the resin bed, thereby increasing the efficiency of the process and improving recovery to >99%.

The saline waste stream from the regeneration system in MIEX® processes used for natural organic matter (NOM) removal has a high organic content. This waste stream should be disposed of either to sewer or, for coastal installations, to the sea. Where these options are not available, alternatives include nanofiltration, to recover about 80% of the sodium chloride, and biological denitrification.

The spent regenerants from adsorption and IX processes used for arsenic removal can be treated by aluminium or iron salts to form insoluble arsenates or by lowering the pH to 5–6.5 to form insoluble hydroxides. Sludge and spent media disposal is a problem as arsenic has the potential to leach back into ground and surface waters. Treatment processes available are reviewed in a US EPA report (MacPhee, 2001).

Chromium waste (Section 10.14) present in dewatered clarifier sludge and exhausted resin should be disposed of in hazardous waste landfills. Spent filter backwash water from the process could be recycled. Sodium chloride in spent brine from regeneration of strong base anion exchange resins is recovered by reduction of Cr(VI) to Cr(III) followed by coagulation and solid–liquid separation with ferrous sulphate.

The spent regenerant produced when reactivating activated alumina used for defluoridation (Section 10.16) can be treated with lime to precipitate calcium fluoride, discharged to sewer or dried in evaporation ponds.

In plants treating radionuclides, these become concentrated in the waste streams and need careful disposal. US EPA (2005) has suggested disposal alternatives based on the concentration of radionuclides in the waste stream. These include disposal of liquid wastes to watercourses, sewer or deep injection wells and treatment by evaporation or precipitation with disposal of solid wastes (dewatered sludge) to various types of landfill facilities.

13.10 SLUDGE THICKENING AND DISPOSAL

Clarifier sludge is not recycled or discharged to a watercourse untreated. It is mixed with settled sludge from used filter washwater settlement tanks and concentrated in continuous flow thickeners where the residence time of the supernatant and the sludge can be varied independently of each other. The thickeners must be preceded by flow balancing tanks to contain and mix intermittent sludge discharges and feed the thickeners at a consistent concentration and uniform rate (Fig. 13.1). In applications where used washwater (without settlement) is mixed with clarifier sludge, the sizing of the flow balancing tanks becomes critical because of the large surges of dilute used washwater. The sludge concentration achieved in the thickener is independent of the feed concentration. However, the greater the feed volume, the larger the thickener required. There are several thickener designs in use; most are of the settlement type developed for industrial applications, which uses heavy duty scrapers with a picket fence attachment. A design developed in the UK by the WRc (Warden, 1983), for waterworks sludge thickening applications, consists of a cylindrical tank of water depth 2–3.5 m with a shallow sloping floor (1 in 20). Sludge is introduced at a central feed well and the supernatant overflows a peripheral weir. The sludge is thickened by the action of a specially designed rake which also moves the thickened sludge to a central hopper, of included angle 60°, for intermittent discharge under hydrostatic head (Albertson, 1992). Lamella settlers (Section 8.17) are sometimes combined with a thickener in a single tank about 5.5 m deep, with a settlement rate of about 0.5–0.7 m/h on the projected area. The horizontal plate separation is typically 100 mm.

With the aid of polyelectrolytes, coagulant sludge can be thickened to concentrations in the range 2–6% w/v solids. The polyelectrolyte dose can be in the range 0.1–1 g/kg of dry solids and

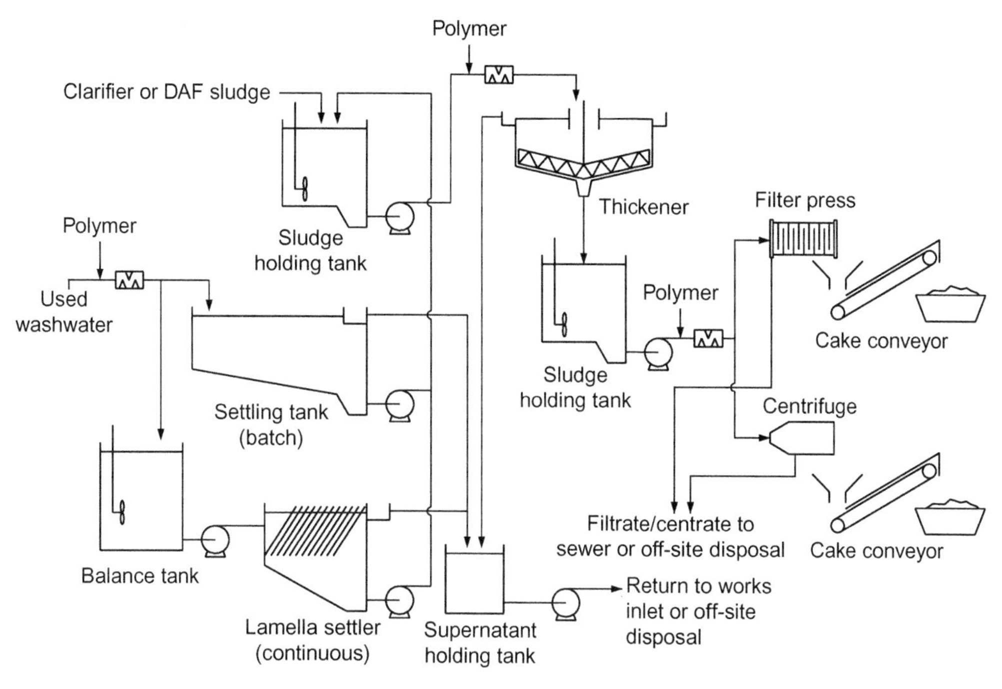

FIGURE 13.1

Schematic diagram of sludge treatment plant.

the dose should be applied proportional to the feed solids concentration and flow. It should be well mixed using an in-line static mixer or similar. The supernatant overflow will have a turbidity less than about 10 NTU. The hydraulic loading should be less than 1.5 $m^3/h.m^2$ and the dry solids loading is less than 4 $kg/h.m^2$. Lower loading rates of up to 1 $m^3/h.m^2$ are often required for thickening wastewater derived from upland water sources, and a limit of 0.8 $m^3/h.m^2$ is usual for algal laden waters. Associated dry solids loading may reduce to 2–2.5 $kg/h.m^2$. Softener sludge is rarely thickened because it concentrates well in the clarifiers. If necessary, it can be further thickened to 20–30% w/w or more at a dry solids loading rate of about 8 $kg/h.m^2$. To prevent the softener sludge becoming too thick, underflow is sometimes recycled to the feed. DAF (Section 8.18) is also sometimes used in waterworks sludge thickening (Haubry, 1983).

Each thickener should have a dedicated feed pump of the progressive cavity type. Desludging can be initiated by sludge blanket level and terminated after a pre-set time, or when the solids concentration in the discharge measured in the outlet pipe falls to a pre-set value. In a well-operated thickener, the supernatant is normally of an acceptable chemical and physical quality for recycling to the treatment process, but the potential risk of microorganism and microanimal return in clarifier sludge is far greater than in the case of recycling used filter washwater. Treatment for microorganism removal similar to that for the supernatant from used washwater settling tanks is therefore required. Chemical quality problems can also arise in thickeners when sludge containing oxides of iron and manganese and NOM are allowed to age in the thickener. Anaerobic conditions can then develop and release iron, manganese, colour and organics that would impart taste and odour into the supernatant, which if recycled, could have an adverse effect on treatment plant performance. Recycling can also lead to accumulation of toxic monomer, derived from polyacrylamide-type polyelectrolytes used in the waste treatment process. This should be calculated by mass balance based on the assumption that all monomer present in the polyelectrolyte will be present in recovered water.

In some cases it is possible to discharge unthickened waterworks sludge to the public sewer, subject to the appropriate consent and provided the proportion of sludge solids is less than 10% of the sewage sludge solids and adequate velocity ($>$0.75 m/s) is maintained in the sewers at all times to prevent silting. The 5-day biochemical oxygen demand (BOD5) of coagulant sludge is in the range 30–100 mg/l (Albrecht, 1972). Thickened sludge may be piped or tankered to a sewage works. Coagulant sludge may aid primary sedimentation of the sewage, thus reducing the solids load to the subsequent biological stages and helping in the removal of phosphate. Thickened waterworks sludge may also be mixed with digested sewage sludge for disposal. Ferric coagulant sludge can prevent hydrogen sulphide gas formation and therefore corrosion of concrete sewers (McTique, 1989).

13.11 SLUDGE DEWATERING

Dewatering of coagulant sludge is difficult because of their gelatinous nature. Materials such as lime, fly ash or diatomaceous earth give body to the sludge and ease dewatering but the quantity of dry solids is increased by 50–100% w/w. At one plant a lime dose of 65 – 90 mg/l as $Ca(OH)_2$ has increased dewatered sludge concentration from 30–40% w/w solids. Lime forms calcium aluminate which is insoluble. The dewaterability of softener sludge depends on the Ca:Mg ratio of the sludge; those with a Ca:Mg ratio less than 2:1 are difficult to dewater while those with a Ca:Mg ratio greater than 5:1 dewater readily.

Lagoons and Drying Beds

Disposal of thickened sludge by lagooning or discharging to drying beds is still widely used in some parts of the world where land is cheap and abundant. Lagoons are shallow structures (0.5–1.5 m sludge depth) excavated from, or formed by, impoundment with earth embankments of slope no steeper than 1 in 2.5, on porous ground above the water table. A complete installation should be provided with a large number of small lagoons, some being filled, some rested for drying and some being emptied of dewatered sludge. Dewatering is by percolation (although with coagulation sludge the base soon becomes impervious), settlement and decanting, and finally evaporation. About 60–80% of the water is removed by settlement and decantation. The average rate of evaporation of wet sludge is about 80% of that for free water. The structure is provided with weir boards or penstocks at several points to allow decanting of the supernatant and rainfall. In cold countries, some dewatering through freezing and thawing can be accomplished, but it is unlikely that more than one application can be dewatered in this manner. In lagooning, sludge is usually placed in layers (0.25–0.5 m deep) in several lagoons in sequence and allowed to dewater before the next layer is placed. This is continued until the total depth is utilized. The coagulant sludge generally consolidates to 10–15% w/w dry solids and softener sludge to greater than 50% w/w solids. The dried sludge is removed by using draglines or front-end loaders. The use of augers to aerate the sludge and expose it to solar evaporation is reported to show a 4-fold increase in the drying rate to give dry solids concentration as high as 70% w/w.

Drying beds, unlike lagoons, have a permeable base of 150–250 mm of sand (effective size of 0.3–0.75 mm, uniformity coefficient less than 4), supported by about 300 mm of graded gravel laid over an under-drainage system of plastic pipes laid with open joints and covered with coarser gravel. As an alternative, porous concrete floors are sometimes used. Dewatering in drying beds is by drainage and evaporation, with the former accounting for about 40–50% of the dewatering. Rainfall delays the drying process as only about 60% of the rainfall is lost by drainage whereas rainfall during the later stages of drying, when the bed is cracked, has little effect on the drying time. However, adjustable decanting facilities help discharge the supernatant and accumulated rainfall. Sludge is usually placed to a depth of about 200–300 mm and coagulant sludge is removed at about 15–25% w/w dry solids.

Drying beds are not extensively used for softener sludge because the sludge penetrates the sand bed during drainage. This could be overcome by polyelectrolyte conditioning and sludge may be dried to over 50% w/w solids. In drying beds, layering is not normally practised as it retards the dewatering by drainage. Otherwise, the operating and emptying of drying beds are similar to those of lagoons. For lagooning and drying beds to succeed as dewatering processes the net evaporation rate must exceed rainfall for a considerable part of the year. Meteorological data should be used for sizing lagoons and drying beds. In sizing, an allowance should be made for storage of sludge during winter and wet months when drying is minimal. The ability of waterworks sludge to dewater by gravity varies a great deal and is very dependent on the characteristics of the individual sludge. Aluminium coagulant sludge usually drain more slowly than do iron coagulant sludge. The rate of draining or settlement for coagulant sludge can be increased by 50% or more by using a conditioning agent such as polyelectrolyte (Novak, 1977). Typical performance for drying beds is about 25 kg/m^2.annum; this could be doubled by the use of polyelectrolytes.

Plate Presses

Dewatering of sludge by plate pressing is becoming increasingly popular for polymer thickened clarification sludge, the recessed plate press being commonly used in waterworks sludge dewatering. A press contains a horizontal stack of vertical rectangular or square plates covered with a filter cloth, to provide a series of chambers, clamped between two fixed end plates. The plates are suspended either from an overhead I-beam or on two side bars (Plate 16(a)). Plates are fabricated in steel, ductile iron or polypropylene, and cloths are of nylon, polypropylene or polyester fabric. Initially the press is closed and sludge is admitted to the press by positive displacement pumps to fill the chambers. The solids are retained on the cloth and the filtrate passes through the cloth and emerges from the drainage ports. The filtrate flow rate is at its maximum at the start, remaining reasonably constant until the pressure begins to build up with the gradual formation of the cake. When the pumping pressure reaches the operating value there is a period of constant pressure filtration (which can last several hours) with a continual decline in filtration rate which eventually ceases when the chambers are full of dry cake. The pump is then stopped, the pressure is released, the press ends are unclamped and the plates are separated from one end, one at a time, to release the cake. The presses are normally operated at pressures from about 7 to 8 bar or sometimes from 14 to 16 bar. For coagulant sludge a 25 mm cake thickness is used; thicker cakes (up to 35 mm) are feasible, but at the expense of increased filtration time.

Recessed plate presses can give a cake of 20–30% w/w dry solids. The volumetric capacity of a press is determined by the chamber depth (25–35 mm), plate dimensions, (0.5 m × 0.5 m up to 2 m × 2 m) and the number of chambers (up to about 160). For example, the capacity of a 30 mm deep, 150 chamber, 2 m × 2 m plate press is 14.25 m^3, which is equal to the volume of the cake. If the sludge is to be dewatered to 30% w/w dry solids, then the weight of dry solids per pressing is about 5.2 t (1 t of wet sludge occupies $(1 - 0.6f)$ m^3 where f is the fraction of dry solids by weight in the sludge, on the assumption that dry solids in waterworks sludge have a specific gravity of about 2.5 (Warden, 1983)).

A drawback with the recessed plate press is that the filtration time can be up to 10 hours or even longer and the cycle time which includes downtime for filling, cake drop and cloth washing, can be 12 hours or longer. This drawback is overcome in the membrane plate press where one of the recessed plates in each chamber is replaced by an inflatable rubber or polypropylene ribbed membrane moulded round a steel insert plate, over which the filter cloth is laid.

Filtration is carried out as in the recessed plate design at about 7–8 bar. The feed pump is then stopped, the membrane is inflated using compressed air or water and the remaining water in the sludge is squeezed out at about 15 bar. The cake thickness is reduced by about 40%. These presses produce thinner and dryer cakes in a much shorter time. Cake dry solids for coagulant sludge range from 25–45% w/w and dry solids for softener sludge is usually greater than 60% w/w. The filtration and compression time can be about 2 hours (made up of 90 and 30 minutes, respectively) and the total cycle time is about 4 hours, thus giving six pressings a day, compared to two for the recessed plate press. In the example for the recessed plate press, the volume of cake after compression to 15 mm is 7.13 m^3. If the sludge is dewatered to 30% w/w dry solids, then weight of solids per pressing is 2.6 t. The sludge processed per day by similar recessed and membrane plate presses will therefore be 10.4 and 15.6 t dry solids, respectively. The press feed pumps are usually piston ram, hydraulic diaphragm or progressive cavity type. Polyelectrolyte is not always added to the press feed; sometimes polyelectrolyte conditioning is limited to thickening only. When added to the press feed, the dose varies in the range 2 and 4 kg/t of solids.

Filter pressing is a batch process. All operations in a cycle can be fully automated. The filtrate produced from a press dewatering coagulant-based sludge initially contains about 50 mg/l suspended solids, reducing to less than 10 mg/l as filtration proceeds. The overall solids capture is better than 98% (99–99.5%). The filter cake is usually discharged into a hopper located underneath the press and removed by screw conveyors for disposal. The energy consumption is about 0.03–0.05 kWh/kg dry solids. The primary advantage of filter presses is the high dry solids achieved. Disadvantages include mechanical complexity, limited degree of automation, high labour requirements, requirements for special support structure (the weight of a press could be up to 100 t), large floor area and building height for the equipment (presses are usually located on the first floor to allow skips to be placed directly beneath), limitations on filter cloth life and high capital cost. Vertical travelling filter cloth presses are used in Japan. It is reported that these are more complex than plate presses, with higher capital and operating costs, but dewatering times are much shorter.

Thermal Processes

Heating dewatered sludge to 40°C, to reduce the viscosity and improve the filtration rate by 1.3–1.5 times that at normal temperatures, has been successfully tried in Japan (Yamane, 2005). The concept of heating sludge is extended in the INOS™ process where sludge is heated up to about 85°C using hot water in the diaphragm of each plate press chamber, followed by application of a vacuum to the sludge to remove moisture as steam. The lowered pressure reduces the boiling point of water in the cake and eases removal of water as steam through the filtrate ports. Ferric coagulant sludge has been dewatered from 3% w/w to 90% w/w dry solids. The dewatering time to achieve 90% solids could be as long as 6 hours. The process can be retrofitted to existing filter presses.

Thermal drying as used in sewage treatment is used to dry waterworks sludge in several plants in Japan. It involves heating the sludge to a very high temperature (up to 800°C) by convection. The dryer works by directly heating the air in a combustion chamber and allowing sludge to cascade through the stream of hot air. The discharge gases are cleaned to remove any particulates carried over. A high proportion of the exhaust air is recycled. Dryers can achieve a solids content of >70% w/w.

Freezing and thawing have also provided an effective method of dewatering coagulation sludge. Natural freezing is more appropriate to countries with more severe winter conditions. There are a few plants where a mechanical freeze-thaw method is used for dewatering (Henke, 1989).

Centrifuges

Centrifuging of sludge has always had a place in the dewatering of softening sludge and, in recent years, has also been used successfully for coagulant sludge. Centrifuges of several designs are available. The solid bowl decanter type, also known as the scroll centrifuge is widely used on waterworks sludge. The bowl is a cylinder on a horizontal axis with a conveying conical section at one end, called the beach, and an inward facing flange or adjustable weir at the opposite end. A helical screw conveyor (scroll) is mounted coaxially inside the bowl with a very small radial clearance. The parameters that affect centrifuge efficiency are bowl speed, scroll differential speed, pond depth, sludge feed rate and polyelectrolyte dose. To increase cake dryness, bowl speed should be increased and feed rate, scroll differential speed, pond depth and polyelectrolyte dose should be decreased. To increase solids recovery, bowl speed, pond depth, scroll differential and polyelectrolyte dose should be increased and feed rate should be decreased. Centrifuges are typically operated at about 1800–3500 rpm to give a centrifugal force at the wall of the bowl of about 1500–2000*g*. The speed differential between the bowl and the scroll is in the range 2–40 rpm but, usually about 10 rpm.

Unlike filter pressing, centrifuging is a continuous process. Its performance on coagulant sludge depends to a considerable degree on polyelectrolyte conditioning of the sludge; polyelectrolyte usage is high and usually in the range 2–6 kg/t of dry solids (Piggott, 1992). The feed to the centrifuges should be maintained at <3% w/v to ensure good mixing of polyelectrolyte with the solids. Coagulant sludge can be dewatered to about 15–25% w/w depending on the nature of the raw water, softening sludge to about 40–50% w/w. Overall solids recovery is normally better than 95% (98–99%). Centrate (i.e. the effluent) water quality from a centrifuge is usually poor with a suspended solids concentration in the range 300–1000 mg/l. It may be spread on land or discharged to a sewage treatment works. The advantages of centrifuging are its enclosed operation and therefore its clean appearance, fast start-up and shutdown; quick adjustment of operating variables; continuous operation if necessary, ready automation and therefore suitability for unmanned operation; low capital cost-to-capacity ratio; and high installed capacity to building area ratio. Disadvantages are low cake dry solids, high demand for energy (about 0.07 kWh/kg dry solids), high polyelectrolyte consumption and poor centrate quality.

Belt, Drum and Screw Presses

Belt presses, where pressure is applied between moving endless belts, are being used successfully for coagulant sludge dewatering in limited numbers in the USA (Migneault, 1987) and France. They are very dependent on polyelectrolyte dosing and the solids content obtained varies from 15–25% w/w.

Vacuum drum filters are successfully used for dewatering softener sludge. A cloth covered drum with horizontal axis above a trough is rotated to pick up the sludge in the trough. Water is removed internally under vacuum and the dewatered sludge is removed by a scraper located just before the drum surface re-enters the trough. With filter loadings in the range 300–450 kg dry solids/h.m^2, cake solids up to 65% w/w have been achieved (Anonymous, 1981).

The screw press is a continuously operating low capacity sludge dewatering unit where sludge is pumped into a cylindrical screened drum within which a screw rotates. The diameter of the screw shaft increases and pitch of the flights decreases towards the outlet end, thus reducing the volume and increasing the pressure of the drum. As the sludge moves towards the outlet it is pressed against the screen and the end plate and the water is squeezed out of the screen and dry sludge cake is discharged. A water spray is periodically applied to clean the screen from outside. The rotational speed is low (<2.5 rpm) and the energy consumption is therefore very low. Solids recovery is >97%. Cake solids can be expected to be lower and polyelectrolyte dose higher than from a centrifuge.

Other Processes

Alternative dewatering processes include the Dehydris Twist process and the Electro Osmosis Dehydrator (ELODE) process. The Dehydris Twist is a mechanical dewatering process that seeks to combine filter press efficiency with the benefits of automation typically associated with centrifuges. Press cloth is supported on a number of flexible drainage filaments. These are potted at each end and encased in a steel chamber, loosely resembling a membrane module. In operation, sludge is pumped into the chamber while one end of the chamber is driven in and out in a horizontal reciprocating motion. The filaments are continuously twisted and untwisted as the piston reciprocates. Water is expressed through the cloth and carried out of the chamber via the drainage filaments. At the end of the dewatering cycle the wall of the chamber is retracted and dewatered sludge drops into collection skips. Dry solids contents of around 40% w/v are claimed.

The ELODE process, widely used in South Korea where it was developed, combines the principles of electrophoresis and electro-osmosis to extract both 'free water' and 'absorbed water' from sludge. In the process, a sludge dewatered by a conventional process to 15–20% w/w is fed between an anode (drum) and a cathode (caterpillar belt) in a belt press. A DC current is passed between the two and the potential difference separates solids from water with sludge particles migrating to the anode and water to the cathode. The combination of the mechanical pressure and electrophoresis/electro-osmosis bring about dehydration of the sludge. The total dewatering time is stated to be about 60 seconds. The dewatered cake is reported to contain up to 60% w/w dry solids. ELODE has been tested on waterworks sludge at two plants in the UK. Both aluminium and iron sludge were tested and the process produced cake between 35% w/w and 45% w/w dry solids compared to 12–15% w/w solids produced by centrifuges. Research is likely to continue into sludge dewatering because of the need in many countries to comply with increasingly stringent conditions for disposal.

Reed beds are an environmentally friendly method developed in Denmark for sewage sludge treatment, and are sometimes used in the UK to dewater waterworks sludge. Long-term dewatering takes place by drainage, evapotranspiration and mineralization. A typical bed consists of several layers of gravel, filter sand and a growth layer (sludge) on top. A geotextile membrane separates the sand and gravel layers. A system of pipes placed in the gravel layer collects water drained through the bed and is pumped to disposal. The total filter bed height is about 600 mm. The feed sludge solids concentration should typically be less than 3% w/v (30 g/l). The reeds' root system provides porosity to the bed and increases drainage. The type of reeds used depends on the sludge to be dewatered and the climate: *Phragmites australis* is commonly used. The loading rate is usually in the range 30–60 kg DS/year . m^2. A minimum of eight beds is considered necessary to achieve the required ratio between loading and resting. In a 10-year operation, a typical cycle is made up of 2 years for planting and two growing seasons, 6 years for operation, followed by resting and emptying over 2 years. Loading is stopped during resting. Dewatered sludge after 10 years' operation could reach an approximate height of 1.2–5 m, with a dry solids content of 40–50% w/w or higher, depending on the climatic conditions. Following emptying, the beds are re-established. Dewatered sludge is either spread on land or disposed in landfills. The total life time of the facility is at least 30 years. Reed beds have lower capital and operating costs compared to mechanical dewatering systems, low carbon footprint and avoid the use of chemicals and associated environmental issues. Trials carried out with a ferric sludge achieved a maximum dewatering level of about 0.025 l/s.m^2, with dry solids (0.16–0.20% w/v) concentrated over 200-fold and 99% reduction in volume. Effluent turbidity was <5 NTU except during loading (Nielsen, 2011).

13.12 BENEFICIAL USES OF SLUDGE

Economic constraints, regulatory requirements and environmental issues are driving water treatment plant operators to examine beneficial uses, alternative disposal methods and resource recovery from waterworks sludge. The UK figures for disposal of waterworks sludge are over 50% to landfill, nearly 30% to sewage treatment works and under 10% spread on land. The respective values for the US are 20%, 24% and 25%, with a further 24% disposed of to watercourses and monofill (Cornwell, 2006). In the UK, of the remaining 10% only 0.5% is used in construction material (brick making and aggregate production), the remainder (9.5%) being lagooned or discharged to a watercourse (Simpson, 2002).

The beneficial uses of coagulant sludge include: land application; the manufacture of cast iron (Henke, 1989); brick making especially for those sludges with high silt content; use as an cement additive as it contains most of the key elements (Si, Ca, Al, Fe) used in cement making (Cornwell, 1990, 2006); and application for the removal of phosphate in sewage treatment (Section 13.10). In agricultural use the potential benefits are nutrient control (increased plant available N and total organic C), pH adjustment, water retention, soil aeration and increased drainage capacity (Elliott, 1991). While it is important to consider the presence of heavy metals in waterworks sludge, these generally only occur in trace concentrations. Therefore, it is not a major concern in land application, but the local regulations for heavy metal loadings on to land must be followed and these may limit the size and the number of applications. Coagulant sludge when used as soil conditioners are known to affect growth of some plants due to the ability of aluminium or iron to fix phosphorus in the soil reducing its availability. Conversely, coagulant sludge can be used in eutrophied lakes to fix phosphorus to minimize algal blooms.

Sludge could be used to supply trace metals such as aluminium, iron and silica required in the manufacture of 'Portland' cement. It is usually added during crushing and homogenization of limestone (raw material). It also finds application in brick making, as coagulant sludge has characteristics similar to clays and shale used in brick manufacture. Sludge with high clay or silt content would be ideal for brick making. The presence of lime, however, would make the sludge unsuitable. Iron in ferric sludge helps to give the red colouration to the bricks. In the USA, coagulant sludge is used in turf grass production farms where it is normally applied during the preparation of the fields for growing the grass. Other applications include use as a bulking agent for composting, in citrus cultivation (iron sludge), as a cover for landfills, land reclamation, manufacture of top soil and fixing hydrogen sulphide (ferric sludge).

Sludge from softening processes contains high purity calcium carbonate with little or no magnesium and is free from coagulant and suspended solids. Consequently, it can be used in flue gas desulphurization and may also be sold for agricultural use or in industry for products such as cosmetics. Such a pure calcium carbonate sludge is only likely to result from the softening of well water derived from a chalk aquifer.

The recovery of aluminium sulphate from sludge has received considerable attention in the past because it offers apparent partial reuse of the coagulant and also reduces the volume of sludge to be handled. In theory, every 1 g of aluminium hydroxide in the sludge could be recovered as 2.2 g of aluminium sulphate by treating the sludge with 1.9 g sulphuric acid at about pH 2. In practice the amount of acid can be much greater if other acid-soluble material is present in the sludge. Typically, recoveries greater than 75% are feasible (Anonymous, 1994). There are also benefits from sludge weight reduction (about 35–40%) and improved dewatering characteristics of acidified sludge. The recovered aluminium sulphate is usually mixed with the commercial product before reuse in the water treatment process. Due to concerns over the possible accumulation of metals and other impurities, such as organic material, the practice has lost favour and has been discontinued in many installations. It is likely to be economical only when the cost of the sulphuric acid is lower than that of purchasing aluminium sulphate.

Lime recovery from softening sludge by recalcination is economically viable since the softening process produces calcium carbonate (Section 10.3). Generally, more lime is produced than is added for treatment. Recalcination is carried out in a furnace at about 900–1000°C. The available lime in the recalcined product may be only about 60–75%, depending on the presence of inerts, such as magnesium, iron, silica and other compounds. Carbon dioxide, which is a by-product, could be used for recarbonation and pH correction of water on the plant.

REFERENCES

Albertson, O. E. (1992). Evaluating scraper designs. *Water Environment and Technology* **4**(1), pp. 52–58.

Albrecht, A. E. (1972). Disposal of alum sludges. *Journal AWWA* **64**, pp. 46–52.

Anonymous (1981). Lime softening sludge treatment and disposal. *Journal AWWA* **78**(11), pp. 600–608.

Anonymous (1994). Coagulant recovery system wins big award. *Water Engineering & Management* **14**(7), pp. 12–13.

Arendze, S. and Sibiya, M. (2014). Filter backwash water treatment options. *Journal of Water Reuse and Desalination* **4**(2), pp. 85–91.

B&V, 2016. *2016 Strategic Directions: Water Industry Report*, Black & Veatch Insights Group.

Brandt, M. J. (2010). *The Challenges for the Urban Water Sector: Future Energy Demand.* IWA World Water Congress.

Brandt, M. J., Middleton, R. A. and Wang, S. (2012). *Energy Efficiency in the Water Industry: A Compendium of Best Practices and Case studies.* UKWIR/GWRC Joint Project CL 11. IWA Publishing.

Brundtland (1987). *Report of the World Commission on Environment and Development: Our Common Future.* Report: A/42/427. UN Documents.

Cornwell, D. A. (2006). *Water Treatment Residuals Engineering.* AwwaRF.

Cornwell, D. A. and Koppers, H. M. M. (1990). *Slib, Schlamm, Sludge.* Report. AwwaRF.

Cornwell, D. A. and Lee, R. G. (1994). Waste stream recycling: its effect on water quality. *Journal AWWA* **86**(11), pp. 50–63.

Cornwell, D. A., Tobiason, J. and Brown, R. (2010). *Innovative Applications of Treatment Processes for Spent Filter Backwash.* Water Research Foundation.

Eades, A., Bates, B. J. and MacPhee, M. J. (2001). Treatment of spent filter backwash water using dissolved air flotation. *Water Science and Technology* **43**(8), pp. 59–66.

Elliott, H. A. and Dempsey, B. A. (1991). Agronomic effects of land application of water treatment sludges. *Journal AWWA* **83**(4), pp. 126–131.

Hammond, G. and Jones, C. (2011). *Inventory of Carbon and Energy (ICE)*, Version 2.0. University of Bath.

Haubry, A. and Fayoux, C. La (1983). Flottation des Boues: Un Avenir Assure, L'Eau, l'Industrie. *Les Nuissances* **79**(1), pp. 20–24.

Henke, H. (1989). Application of freeze-thaw for handling of sludge from the treatment of Great Dhunn Reservoir Water. *KIWA/AwwaRF Experts Meeting*, Nieuwegein.

Li, A., Ratnayaka, D. D. and Hieatt, M. J. (2002). Tai Po water treatment works waste management, management of waste from drinking water treatment. *Proc. CIWEM International Conf.*, London.

McTique, N. E. and Cornwell, D. (1989). Impact of water plant waste discharge on wastewater plants. Proc. *Residuals Management Conference,* AWWA/WPCF. San Diego.

MacPhee, M. J., Charles, G. E. and Cornwell, D. A. (2001). *Treatment of Arsenic Residuals From Drinking Water Treatment Processes.* EPA 600/R-01/033.

Migneault, W. (1987). Santa Clara Valley district sludge concentration and recycling program. California Nevada Section. *Proc. AWWA Fall Conf.*

NEMA (2013). *Association of Electrical Equipment and Medical Imaging Manufacturers (USA).* Vol. 18(8).

Nielsen, S. and Cooper, S. J. (2011). Dewatering sludge originating in water treatment works in reed bed systems. *Water Science and Technology* **64**(2), pp. 361–366.

Novak, J. T. and Langford, M. (1977). The use of polymers for improving chemical sludge dewatering on sand beds. *Journal AWWA* **69**(2), pp. 106–110.

Piggott, G. A., Tse, Y. S. and Ratnayaka, D. D. (1992). Waterworks sludge treatment and disposal options. *Proc. Water Malaysia '92'*, Kuala Lumpur.

Schofield, T. (1997). Sludge removal and dewatering processes for dissolved air flotation system. *Proc International Conference on Dissolved Air Flotation*. CIWEM, London.

Simpson, A., Burgess, P. and Coleman, S. J. (2002). The management of potable water treatment sludge: present situation in the UK. *Water and Environment Journal* **16**(4), pp. 260–263.

US DOE (2006). *Energy Demands on Water Resources – Report to Congress on the Interdependencies of Energy and Water*. US Department of Energy.

US EPA (2002). *Filter Backwash Recycling Rule Technical Guidance Manual*. EPA 816-R-02-014.

US EPA (2005). *A Regulator's Guide to the Management of Radioactive Residuals From Drinking Water Treatment Technologies*. EPA 816-R-05-004.

Warden, J. H. (1983). *Sludge Treatment Plant for Waterworks*. Technical Report 189, WRc.

Yamane, Y. (2005). Water supply sludge treatment in Japan, *Water* 21, p. 26.

CHAPTER

Hydraulics

14

14.1 THE ENERGY EQUATION OF FLUID FLOW

A fluid moves, in accordance with Newton's laws, under the action of external forces. If there is a net force acting on an element of the fluid, then that element will either accelerate or decelerate depending on the direction of that force; or, if the forces are in balance, then the element will remain at rest or at the same velocity. There is a resistance to motion, however, in the form of drag on that element of fluid and, in moving, energy is expended in overcoming that drag. This expenditure of energy appears in the form of turbulence in the water created by the drag of the surfaces of the conduit and by any obstructions or changes to the shape and direction of the conduit. As the eddies of turbulence decay, their kinetic energy is transmitted to the motion of individual water molecules, so the temperature of the water increases slightly; except for the turbulence found in some pumps (Section 19.12) the temperature change is scarcely detectable and too small to be of any practical use. The balance of energy remaining is important because it determines the subsequent level, pressure and kinetic energy of flow.

The energy of a unit mass ρ of water can be expressed as:

$$E = (\rho u^2/2) \quad + \quad (p) \quad + \quad (\rho g z)$$

(i.e. Kinetic energy + Pressure energy + Potential energy)

where u is the velocity of the water, p its pressure and z its height above some given datum. In the civil engineering context it is convenient to divide the expression by ρg so that the energy and pressure are expressed in terms of a height, or 'head' of water to give the expression in the form:

$$\frac{E}{\rho g} = H = \frac{u^2}{2g} + h + z \qquad (14.1)$$

For flow between two points, A and B, as shown in Figure 14.1:

$$H_A = H_B + H_L$$

Twort's Water Supply. DOI: http://dx.doi.org/10.1016/B978-0-08-100025-0.00014-4

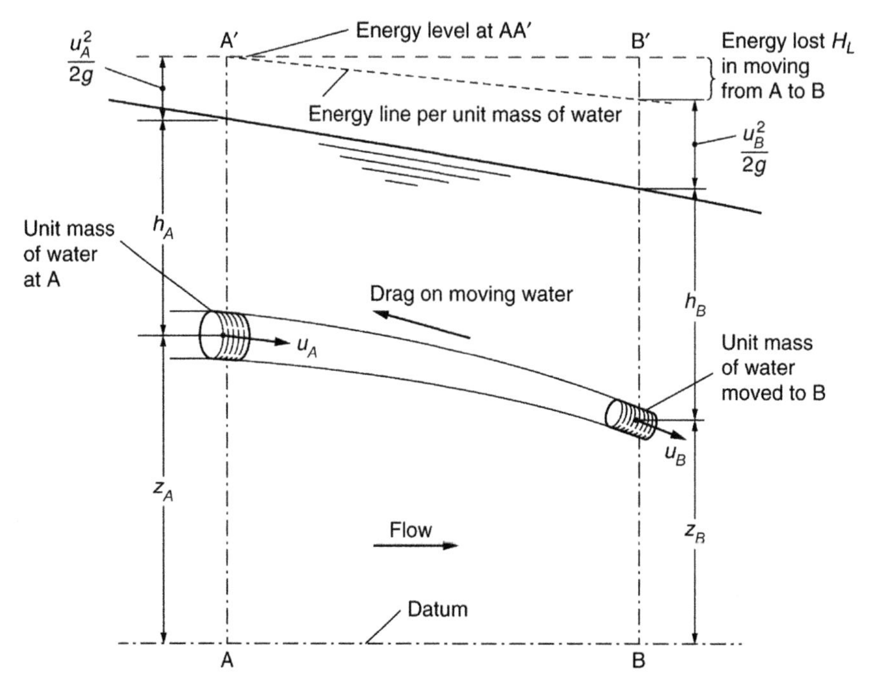

FIGURE 14.1

Variation in energy as unit mass of water moves from position A to B.

where H_L is the energy head lost by the water flowing from A to B. Hence:

$$\frac{u_A^2}{2g} + h_A + z_A = \frac{u_B^2}{2g} + h_B + z_B + H_L \tag{14.2}$$

This is the general form of the energy equation which is fundamental to almost all hydraulic calculations and is referred to several times in this chapter. The Bernoulli equation, one of the most widely quoted equations in fluid dynamics, is the special case of no energy loss (i.e. $H_L = 0$) between points A and B. Bernoulli's equation can be expressed as:

$$\frac{u^2}{2g} + h + z = \text{constant}$$

The limitations of Bernoulli's equation must be carefully noted as it is an idealized case. It applies only to steady flow within a streamline and to flow where no energy is lost through turbulence. Therefore, it can only be applied as an approximation over short distances. In all civil engineering applications energy is lost as flow moves from point A to point B and the challenge is the determination of this loss.

14.2 BOUNDARY LAYERS

The concept of streamlines is most relevant to the idealized condition of flow moving uniformly in a large body of water. Away from the influence of any solid boundary, a particle of water follows its streamline, a condition known as potential flow. However, when a boundary is introduced into the flow the water in immediate contact with the solid surface must be stationary. Away from the surface the velocity increases up to a point where the flow is unaffected by the surface boundary. In this region adjacent to the surface there is a varying velocity gradient, with adjacent streamlines at different velocities. This region is known as the boundary layer. The thickness of the boundary layer depends on a number of factors but the boundary layer can be envisaged as growing from the point of contact of the flow with the solid boundary. In a conduit of finite size the boundary layer increases in the direction of flow until it fills the full depth of flow in an open channel or the full cross section of a pipe.

The velocity profile within a boundary layer is primarily a function of the size of the conduit, the velocity of flow and the density and viscosity of the fluid. These parameters are combined in a single dimensionless grouping known as the Reynolds number Re where: $Re = VD/v$, V being the average velocity of flow, D is some representative dimension of the flow (e.g. the depth of flow in an open channel or the diameter of a pipe) and ν is the kinematic viscosity of the fluid, defined as μ/ρ where μ is the dynamic viscosity and ρ the density.

At low Reynolds numbers the flow is described as *laminar*. In laminar flow viscosity is dominant and the momentum and inertia of the flow have little effect. The boundary layer is small and the velocity gradient within it may be high. A good example is treacle flowing over an object clinging to the surface without separation or turbulence. Laminar flow occurs rarely in water supply systems as Reynolds number would need to be below about 2000; since the value of the kinematic viscosity of water is about 1.1×10^{-6} m^2/s at 15°C, this requires that $V \times D$ be less than about 0.002 m^2/s. Thus, for example, it would apply to a case where the flow velocity was less than 0.02 m/s in a pipe of 100 mm diameter or 0.2 m/s in a pipe of 10 mm diameter.

At a Reynolds number between about 2000 and 4000, depending on the precise conditions and surface roughness, turbulence starts to be generated and there is a sudden transition from laminar to turbulent flow with a marked increase in flow resistance. Viscosity initially remains a major influence but as the Reynolds number increases so the turbulence becomes more dominant until, at some higher Reynolds number depending on the roughness of the boundary, the effect of viscosity becomes negligible and the flow is said to be '*rough turbulent*'. The region where both viscosity and turbulence influence the flow is described as the '*intermediate zone*'. Figure 14.2, which illustrates the variation of flow resistance with Reynolds number, also shows a curve of minimum friction factor or hydraulic resistance as the lower boundary of the intermediate zone. This curve, applicable to very smooth surfaces, is referred to as the '*smooth-turbulent*' boundary.

This relationship between resistance to flow and Reynolds number is best illustrated by the diagram developed by Moody (1944) (Fig. 14.2) for pipe flow. The flow resistance is characterized by the friction factor f, which is explained below. In the laminar flow region f is a function solely of the Reynolds number, but for values of the Reynolds number above the transition from laminar to turbulent flow, f is a function of both Reynolds number and the roughness of the surface. The roughness is usually expressed as the *relative roughness* which for pipes is defined as k_S/d where k_S is a linear measurement of the surface roughness and d is the pipe diameter.

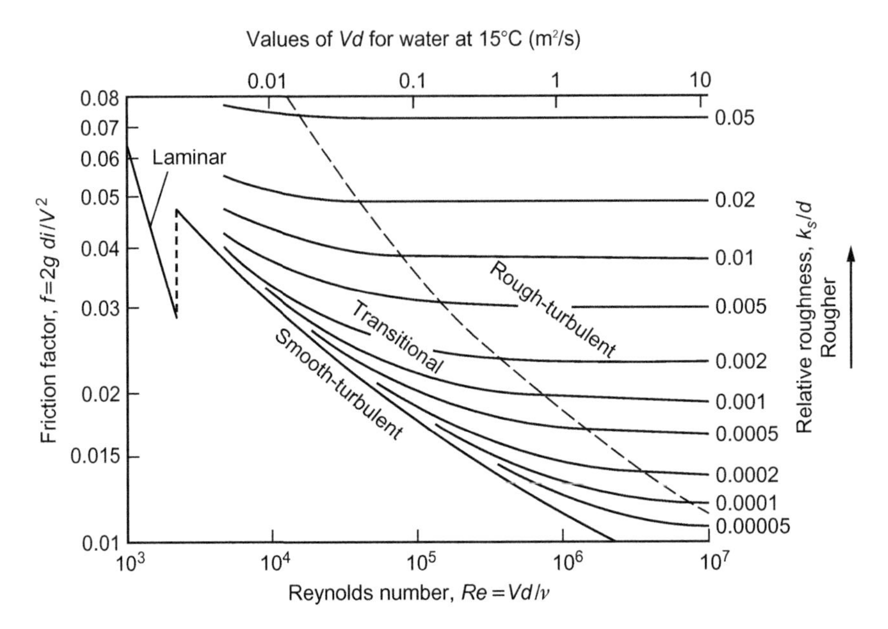

FIGURE 14.2

Moody diagram illustrating the variation of hydraulic resistance with Reynolds number.
Note: i is the gradient of the energy line, the 'hydraulic gradient'.

Turbulence is generated by water moving at different velocities in adjacent streamlines. This generates a shear force between the streamlines tending to retard the faster flow and accelerate the slower flow. Thus, in a boundary layer, the faster flow is dragged down towards the boundary and the slower flow pulled away from the boundary, creating a lateral velocity component and the formation of turbulent eddies. There is a transfer of momentum across the streamlines and the velocity profile is no longer linear but approximately logarithmic with zero value at the boundary.

14.3 PIPE FLOW

Two types of water flow are covered in this chapter: open-surface channel flow and closed conduit flow. In the former the depth of flow can vary; in the latter the area of flow is fixed and for a known flow in a given size of conduit the velocity can be calculated directly.

At the entry into a pipe from a large tank or reservoir, the flow, as it accelerates into the inlet, approximates to the idealized condition of potential flow. However, a boundary layer is generated from the lip of the inlet and, within a relatively short distance downstream of the entry, the boundary layer expands to fill the whole pipe. At Reynolds numbers just above the laminar transition the velocity profile is logarithmic from the boundary wall to the centreline of the pipe. At higher Reynolds numbers the turbulence is such that significant momentum is transferred across the streamlines. Across much of the section the time-averaged velocity is effectively constant but there is a steep velocity gradient near the pipe wall.

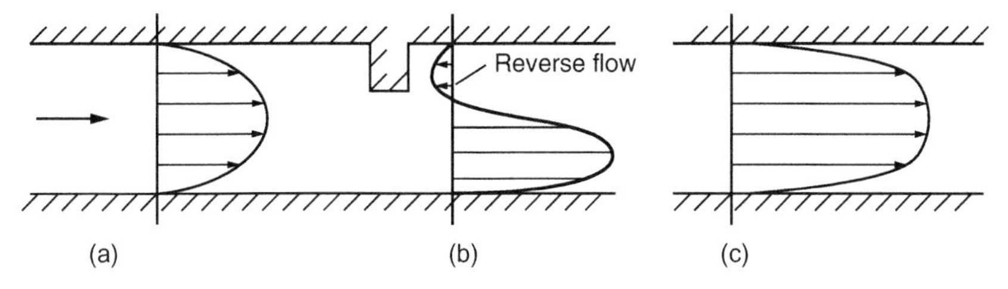

FIGURE 14.3

Velocity profiles in a pipe flowing full.

Figure 14.3(a) illustrates the velocity profile across a pipe operating in the smooth-turbulent zone. At the centre of the pipe the velocity is greatest and is about 1.2 times the average value, although the ratio of the maximum speed of the flow to the average varies as a function of the Reynolds number and the roughness of the walls. Of course, it may vary greatly if an obstacle in the pipe affects the flow as illustrated in Figure 14.3(b). Figure 14.3(c) shows the velocity profile across a pipe operating in the rough-turbulent zone, with a much more uniform velocity across much of the cross section, but with steep velocity gradients close to the walls.

The average velocity in a pipe is simply the rate of discharge divided by the cross-sectional area of the pipe, that is, Q/a, where Q is the discharge and a is the cross-sectional area. This average velocity is denoted as V to distinguish it from the velocities in individual streamlines, u_1, u_2, u_3, etc. The energy equation applies to any streamline flow but to apply the same equation to the whole flow within a pipe the energy equations for the separate streamlines must be summed. Thus the total kinetic energy of the flow is:

$$\sum\left(\frac{u_1^2}{2g}+\frac{u_2^2}{2g}+\frac{u_3^2}{2g}+\ldots\ldots+\frac{u_n^2}{2g}\right)$$

However, this sum does not equal $V^2/2g$. The ratio of $(\Sigma u^2/2g)/(V^2/2g)$ is therefore given the symbol of α and the total kinetic energy of the flow is $\alpha V^2/2g$. In a similar fashion the summation of the pressure energies is βP, where P is the average pressure across the cross section of the pipe. As it is a linear function, the average potential energy in a circular pipe is given by the level of the centre of the pipe, provided the pipe is flowing full. For flow in a straight pipe, away from the influence of bends and obstructions, the pressure distribution is hydrostatic and $\beta = 1.0$. Thus for the whole pipe flow the energy equation is correctly:

$$H = (\alpha V^2/2g) + \beta P + z$$

where V and P are the average velocity and pressure head across the section, respectively, and z is the level of the centre of the pipe above some datum. In practice, the values of α and β are close to unity and in most practical applications the errors involved in ignoring the differences are so small as to be insignificant. Hence, this assumption permits the flow at any section of a pipeline to be related to that at any other point of the same line.

Units Used

As indicated in Section 14.1, for water supply systems the common practice is to express the terms in the energy equation in '*head of water*'. In metric units this is normally expressed in metres. Thus, for example, the difference in level between two reservoirs connected by a pipeline can be considered as a direct measure of the available potential energy between the two locations. In metric units, if the energy, H is expressed in 'metres head of water' then z must also be expressed in metres and the kinetic energy in the same units. This requires that the velocity is in m/s and the gravitational acceleration, g, is in m/s^2 (9.81 m/s^2 at sea level).

The term 'head' is often loosely applied; for example, 'pump head' is used to denote the pump lift, 'head' is also sometimes used in reference to a pressure level or elevation, relative to a datum. In this text 'head' is used simply to mean an equivalent depth of water.

14.4 HEADLOSSES IN PIPES (1) – THE COLEBROOK–WHITE FORMULA

The energy equation (Equation 14.2) for the total flow in a pipe can now be expressed as:

$$(V_1^2/2g) + h_1 + z_1 = (V_2^2/2g) + h_2 + z_2 + H_L$$

where H_L is the energy loss in the pipeline between the two sections 1 and 2. Knowing the total energy level at location 1 (e.g. the level of a reservoir) the energy level at any other point can be calculated if the energy loss H_L can be estimated.

The energy lost through turbulence is caused by two mechanisms: (1) the drag of the pipe walls on the flow and (2) turbulence generated whenever there is a change to the direction or area of the flow. A flowing fluid has momentum and will continue moving in a straight line unless there is an external force acting to divert it; any change to the angle of the boundary walls, particularly if the boundary turns away from the direction of flow, may lead to the flow 'breaking away' from the surface leaving an area of turbulence, the 'wake'. The more abrupt the boundary change the greater is the potential for energy loss. The former mechanism for energy loss is known as the hydraulic resistance or 'friction' loss; the latter is the 'form' loss due to the geometry of the change of cross section or obstruction in the flow. The *pipe friction losses* are continuous over the length of a pipeline; the form losses are localized in the immediate vicinity of the element causing the energy loss and are referred to as *local* or *fitting losses*. (Note that strictly 'friction' relates to the movement of one solid surface against another. In fluid flow the fluid particles against a solid surface are stationary as described earlier and the resistance to motion is caused by shear in the boundary layer. The description of the energy loss as a 'friction' loss is thus not strictly correct but is a convenient way of considering it.)

The loss of energy due to the hydraulic resistance of a pipe is a function of the velocity of the flow – V, the internal diameter of the pipe – d, the length of the pipe – L, the roughness of the surface of the pipe and the characteristics of the flowing fluid, expressed in terms of the kinematic viscosity ν.

Darcy–Weisbach Formula

A dimensionally correct formula for the friction loss is the Darcy–Weisbach equation, which gives the headloss in a length of pipe as:

$$H_L = fLV^2/(2gd) \tag{14.3}$$

where f is a non-dimensional coefficient, known as the *friction factor*, which includes the effects of pipe wall roughness and the fluid viscosity, L is the length of pipe being considered and d is the diameter of the pipe. (Note that the friction factor f is also commonly designated by λ.) Unfortunately f is not constant but varies with the size of pipe and the degree of turbulence of the flow.

Colebrook–White Formula

Colebrook and White showed that f in the Darcy–Weisbach formula is a function of the relative roughness of the pipe surface, the viscosity of the flow and the Reynolds number, Re. From a combination of theoretical analysis and empirical data they showed that:

$$\sqrt{\frac{1}{f}} = -2\log_{10}\left\{\frac{k_s}{3.71d} + \frac{2.51}{Re\sqrt{f}}\right\}$$

which can also be written as:

$$\sqrt{\frac{1}{f}} = -2\log_{10}\left\{\frac{k_s}{3.71d} + \frac{2.51}{(d\sqrt{2gdi})}\right\}$$

where i is the *hydraulic gradient*, H/L, the rate of energy loss along the pipe; k_s is the roughness of the internal surface of the pipe; ν is the kinematic viscosity of the water.

The experimental data showing variation of f with Reynolds number and relative roughness was plotted by Moody and forms the basis of the diagram illustrated in Figure 14.2. Thus the Colebrook–White equation is, in effect, a mathematical representation of the Moody diagram. The formula now widely known as the Colebrook–White equation is a formulation of the Darcy–Weisbach equation with f replaced. It is usually written as:

$$V = -2\sqrt{(2gdi)}\cdot\log_{10}\left[\frac{k_s}{(3.71d)} + \frac{2.51v}{(d\sqrt{2gdi})}\right] \tag{14.4}$$

Equation (14.4) can be solved directly only for V (and hence Q), knowing d and i. More commonly it is required to find i, knowing Q and d. The equation is then not explicit but must be solved using an iterative technique of successive approximations. The Colebrook–White equation, or one of the approximations such as given below allowing a direct solution, is now widely applied and is recommended as the formula to be used to estimate pipeline headlosses. One of the advantages of this equation over the empirical formulae discussed below is that the roughness coefficient, k_s is a function only of the surface roughness of the pipe and does not change with the size of the pipe or velocity of the flow. The factor k_s is sometimes referred to as the *equivalent sand roughness* because the original experiments carried out by Nikuradse utilized sand grains stuck to the inside of the pipes. The value of k_s is meant to represent the equivalent diameter of the sand particles giving that degree of roughness. Although this is only a notional concept, it does provide a physical meaning to the roughness measurement, which does not apply to the coefficients in any of the empirical formulae given in Section 14.5.

A second advantage of the Colebrook–White formula is that it applies over the full range of turbulent flow from the smooth-turbulent condition at Reynolds numbers as low as 3×10^3 to the rough-turbulent flow condition at Reynolds numbers in excess of 1×10^7. The first term in the logarithmic function, $k_s/3.71d$, represents the effect of the pipe roughness and dominates at high Reynolds numbers, whilst the second term includes the dynamic viscosity and dominates at low Reynolds numbers.

There are approximations to the Colebrook–White formula that allow direct calculation of d or i knowing the flow Q and the other of the two parameters. These provide agreement with the Colebrook–White formula within 0.5%, well within the accuracy with which the roughness is known. Two such equations which can be recommended are those of Barr to solve for i, and Pham to solve for D. Thus the three explicit equations can be written in the following similar format.

The Colebrook–White equation, to solve for Q:

$$\frac{V}{\sqrt{(2gdi)}} = \frac{0.9003Q}{d^2\sqrt{(gdi)}} = -2\log_{10}\left[\frac{k_s}{(3.71d)} + \frac{2.51\nu}{d\sqrt{(2gdi)}}\right] \tag{14.5}$$

The Barr approximation (HR Wallingford, 2006), to solve for i:

$$\frac{0.9003Q}{d^2\sqrt{(gdi)}} = -1.9\log_{10}\left[\left(\frac{k_s}{(3.71d)}\right)^{1.053} + \left(\frac{4.932\nu d}{Q}\right)^{0.937}\right] \tag{14.6}$$

The Pham approximation (HR Wallingford, 2006), to solve for d:

$$\frac{0.9003Q}{d^2\sqrt{(gdi)}} = -1.88\log_{10}\left[\frac{0.365(gi)^{0.2}k_s}{Q^{0.4}} + \frac{3.55\nu}{Q^{0.6}(gi)^{0.2}}\right] \tag{14.7}$$

Units must be consistent, for example, d (m); Q (m^3/s); g (m/s^2 – 9.81 m/s^2 at sea level) and k_s must also be in metres (although usually quoted in mm as below). The kinematic viscosity ν of clean water is 1.310×10^{-6} m/s^2 at 10°C and 1.011×10^{-6} m/s^2 at 20°C.

For water mains there is considerable guidance available on the choice of k_s values to adopt for design (HR Wallingford, 2006). Typical values for new clean pipes and indicative values for design purposes that allow for deterioration of interior condition are shown in Table 14.1.

Table 14.1 Values of k_s for design purposes

New clean pipes	**k_s mm**
Steel or ductile iron pipes:	
with spun bitumen or enamel finish	0.025–0.05
with cement mortar lining	0.03–0.1
Concrete pipes	0.03–0.3
Plastic pipes	0.003–0.06
For design with allowance for deterioration:	
Raw water mains	1.5–3.0
Treated water trunk mains	0.3–1.0
Distribution systems	0.5–1.5
For new clean service pipes:	
Galvanized steel	0.06–0.30
Copper	0.002–0.005
MDPE	0.003–0.006
PVC-U	0.003–0.06

Notes: 1. k_s values below 0.01 mm show no significant change in V or i from that at 0.01 mm.
2. Loss at joints, elbows, tees, etc. on the line to the consumer's tap may add 50–70% to pipe losses (see also Table 16.5).
3. It is not possible to quote typical k_s values for old service pipes due to the wide range of interior conditions that can apply.

In raw water mains there may be a tendency for organic slimes to develop on the walls. This slime tends to flatten as the velocity increases and there is evidence that the roughness reduces at higher velocities (HR Wallingford, 2006). The designer of a new pipeline must make a judgement as to whether this effect should be taken into account in the calculation of design headlosses. It may also be necessary in a raw water pipeline to allow for some increase in roughness due to presence of sediment in the invert and possibly fresh water organic growth such as mussels. The latter may greatly increase the roughness. Also if the water quality is such that tuberculation is likely over the life of an iron pipeline, then much higher values may be appropriate.

The availability of sophisticated mathematical models now allows for the hydraulic conditions in single pipes and complex distribution networks to be analysed and for models to be calibrated against measured flow and pressure data by the adjustment of roughness coefficients in each pipe. Once a model using the Colebrook–White formula is calibrated the performance of a system under other flow regimes can be predicted and design improvements can be undertaken with some confidence, without further adjustments to roughness values. However, it is important to remember that measurements are rarely more accurate than $\pm 5\%$ in practice, with errors often greater. Hydraulic modelling software is discussed further in Section 15.15.

Some network programs include the option to use the Hazen–Williams equation (Section 14.5). Given the uncertainties involved in network modelling this does not necessarily lead to greater inaccuracies but it does mean that a model calibrated for one set of flow conditions must be used with greater caution for other flow conditions. This is discussed further in the following sections.

14.5 HEADLOSSES IN PIPES (2) – EMPIRICAL FORMULAE

There are several other formulae for calculating pipe friction headloss; these have been and are still used by water supply engineers. They have to be used with caution as each applies over a limited range of Reynolds numbers and in different areas of the Moody diagram (Fig. 14.2). Thus, for example, the Blasius formula applies at low Reynolds numbers in the smooth-turbulent zone where viscosity dominates; the Hazen–Williams formula applies in the intermediate zone and Manning's formula applies in the rough-turbulent zone where the pipe roughness dominates. The Blasius equation has relatively limited applicability in the civil engineering context but the latter two are discussed in more detail below.

The ***Hazen–Williams formula*** has been used for many years in water supply and is still used widely in the USA. It is well documented and, until the advent of programmable calculators and computers, was considerably easier to use than the Colebrook–White equation. The equation can be expressed in metric units as:

$$H(m) = \left(\frac{6.78L}{d^{1.165}}\right)(V/C)^{1.85}$$

or

$$V(m/s) = 0.355Cd^{0.63}i^{0.54} \tag{14.8}$$

where C is a coefficient, i is the hydraulic gradient (H/L) and d, V, H and L are as defined in Section 14.4.

The coefficient C is not dimensionless; it has units and is therefore a function of the other parameters. As noted above, the Hazen–Williams equation is most accurate for the pipe sizes and velocities typically found in water supply practice. The flow in a pipe of 0.6 m diameter with a velocity of 1.0 m/s has a Reynolds number of about 5×10^5. This is in the intermediate zone in the Moody diagram and the Hazen–Williams formula (Equation 14.8) can be applied with reasonable accuracy provided that velocity and pipe size do not vary greatly from these values. The formula becomes increasingly inaccurate as the Reynolds number varies further away from this mean value. Thus different values of C apply for different pipe sizes and even for the same pipe at different flows.

The value of C can be adjusted to provide a more accurate answer if the parameters do vary significantly. Figure 14.4 shows how the coefficient varies with pipe diameter for a range of pipe roughnesses, and indicates the approximate adjustment needed for velocities varying from 1.0 m/s. Attempts have been made to define the variation of C in terms of the other parameters, d and V, to make the Hazen–Williams formula directly applicable over a greater range. Since the Colebrook–White formula already provides an accurate estimate over the full range of turbulent flow conditions, this seems a pointless exercise and the use of the Hazen–Williams formula should be abandoned for detailed design.

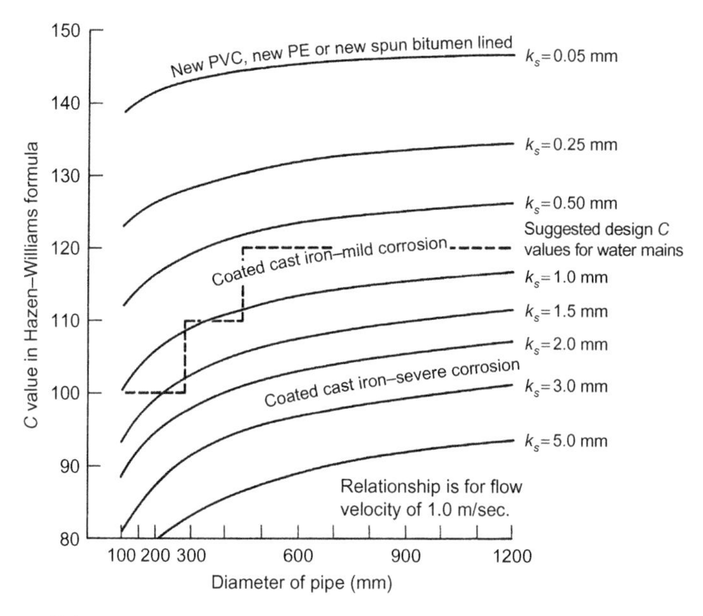

Notes:

1. For 0.5 m/s *add* 5 to *C* value. For 2.0 m/s deduct 5 from *C* value above.
2. For asbestos cement pipes use k_s = 0.50 mm
3. For concrete line pipes the following are Scobey classes:
   Class 4(k_s=0.25 mm) first class finish
   3(k_s=0.50 mm): good finish, all joints filled
   2(k_s=1.25 mm): imperfect finish
   1(k_s=5.0 mm): old concrete pipes
4. For tuberculated cast iron pipes carrying raw water k_s may be 10 → 30 mm and Mannings formula should be used in preference to Hazen Williams

FIGURE 14.4

C values in the Hazen–Williams formula as a function of pipe size and pipe roughness (k_s) for a velocity of 1.0 m/s.

Manning's equation is appropriate for use when the flow is in the fully turbulent range, at either high Reynolds numbers or when the conduit is particularly rough. It is widely used in open-channel flow, for which there are extensive data, and is referred to below in that context. It is not generally recommended for pipeline systems, except possibly in large, rough conduits such as unlined rock tunnels. Manning's equation, in metric units, is normally written in the form:

$$V = \frac{(R^{2/3} i^{1/2})}{n} \tag{14.9}$$

where R is the hydraulic mean depth (also known as the *hydraulic radius*) that is, the area of flow divided by the wetted perimeter, i is the hydraulic gradient (H/L) and n is a roughness coefficient known as the Manning coefficient. For a circular pipe, $R = \pi d^2/4\pi d = d/4$; therefore, for a pipe the equation can be written:

$$V(m/s) = \frac{(0.397d^{2/3}i^{1/2})}{n} \tag{14.10}$$

Again, it should be noted that n has dimensions (in US (fps) units a factor of 1.49 must be introduced into Equation 14.9) and is a function of the size of the conduit. However, it is relatively insensitive to the diameter of a pipe and for many calculations can be assumed to be constant. Some typical values for n are given in Table 14.3.

The use of the Colebrook–White equation (Equation 14.5), or the direct solution approximations of it (Equations 14.6 and 14.7), is therefore strongly recommended for all pipeline calculations.

14.6 LOCAL HEADLOSSES AT FITTINGS

As discussed earlier, headlosses occur at every location where there is a geometric change to the conduit, such as at a bend in the pipeline, a change of section, an obstruction to flow such as a valve, or simply the entry to a pipe or exit from a pipe into a tank. In such cases the headloss is usually expressed as a proportion of the velocity head, or kinetic energy, of the flow, that is, ΔH (or H_L) $= KV^2/2g$, where K is a coefficient depending primarily on the type of fitting or fixture in the pipeline. V is normally taken as the velocity in the upstream pipeline: not, in the case of valves or other obstructions, the velocity through the opening in the fitting itself.

Values of K are almost entirely empirical but there have been extensive experimental measurements on standard fittings on which estimates can be based. Table 14.2 gives values of the headloss coefficients for some standard fittings and suggested values for design. Some standard data is available for valves but, in most cases, it is advisable to obtain values from manufacturers. The suggested design values in Table 14.2 are generally conservative, suitable for estimating losses in short conduits containing several fittings or changes of direction, etc. Whereas in a long pipeline the fittings losses may be a small proportion of the total losses, for example, 5% or less, in a short line with an inlet, several bends and an outlet loss, fittings losses may dominate the total losses. Short conduits conveying water from one tank to another often occur in water treatment works, and it is important not to under-estimate losses when designing weir and overflow levels.

For some calculations such as surge and transient flow analyses, flow distribution or when considering the possible range of pump duties the most severe conditions are often when velocities in the system are high. The greatest velocities will occur when resistance to flow is at a minimum. It may be necessary, therefore, to consider the case of minimum roughness, for example, when the system is new, and to take minimum values for the fittings losses. Table 14.2 gives test measurements which represent such minimum losses. Conversely, the highest pressures may occur when flow resistance is greatest and allowance made for possible higher losses in fittings such as those given as *field values* in Table 14.2 and for increases in roughness that have occurred as the system ages.

Table 14.2 Loss coefficients through pipeline fittings

	K values in $KV^2/2g$	
	Laboratory test	Suggested field[a]
Entrances: (V = velocity through pipe or gate)		
Standard bellmouth pipe	0.05	0.12
Pipe flush with entrance	0.50	1.00
Pipe protruding	0.80	1.50
Sluice-gated or square entrance	–	1.50
Bends – 90°: (45° half values given)		
Medium radius ($R/D = 2$ or 3)	0.40	0.50
Medium radius – mitred	0.50	0.80
Elbow or sharp angled	1.25	1.50
Tees – 90°: (Assumes equal diameters)		
In-line flow	0.35	0.40
Branch to line, or reverse	1.20	1.50
Exits:		
Sudden enlargement: ratio 1:2	0.60	1.00
Sudden enlargement into tank (note: bellmouth does not reduce exit loss)	1.0	1.00
Gradual (long, well tapered) exit	0.20	0.50
Sudden contractions: (Loss on contraction and subsequent expansion; V = velocity through contraction)		
Contraction area ratio:		
1:2	1.00	1.50
2:3	0.65	1.00
3:4	0.40	1.00
Expansion only	–	1.00
Gate valve fully open	0.12	0.25
Butterfly valve fully open	0.25	0.5

Notes: [a]The *K* values in the second column are recommended for assessing loss through short conduits containing several pipe fittings in close proximity. For extensive data on pipe fittings losses refer to Miller (1990) but note that his data is based on accurate laboratory measurements and has no 'design' factors included.

If accurate analysis is required, for example, on an existing installation or where fittings are in close proximity, then reference to more accurate figures may be required (Miller, 1990 is widely regarded as definitive in this regard). For design, consideration should be given to increasing Miller's values for fittings losses as they include no design factors.

It should be noted that the net headloss through a series of adjacent fittings is often less than if they were more widely spaced due to the way the high velocity jet created by one fitting is transmitted through the next. It should also be noted that the losses at some fittings, particularly bends, are a function of the pipeline roughness as well as the geometric shape. For complex pipework arrangements it may be advisable to use a hydraulic model, either a physical model or one developed for Computational Fluid Dynamics (CFD) analysis (Section 14.18) to assess the potential net headlosses.

14.7 OPEN-CHANNEL FLOW

The flow in an open channel follows the same principles that have been developed for pipe flow. Thus the energy equation (Equation 14.1), is still valid and energy is lost in the same way through the resistance of the channel surfaces and locally where the geometry of the channel changes. Whereas, in a given pipe, the area of flow is fixed and hence the velocity is a function only of the flow, in a channel of known dimensions, the velocity is not only a function of the flow but also of the channel geometry and depth of flow. The depth of flow can vary. In the energy equation (Equation 14.1):

$$H = \frac{V^2}{2g} + h + z$$

for open channels the pressure head h becomes the depth of water y; and z is the level of the channel invert above some datum. For a case where the energy is constant any increase in velocity, and hence the kinetic energy must be accompanied by a fall in the potential energy and hence a drop in water surface. Similarly a retardation of the flow and reduction of the kinetic energy must be accompanied by a corresponding rise in the water surface if there is no energy loss. In practice, whilst there may be very little energy loss in a case where the velocity is increased and potential energy is converted to kinetic energy, it is always the case that there is some loss of energy in the reverse process. The latter process, known as *recovery of head*, is never 100% efficient.

The *specific energy head* H_S is defined as the energy head above the channel invert and Equation (14.1) can be written as:

$$H_S = \frac{V^2}{2g} + y$$

where y is the depth of flow.

$V = Q/A$ where Q is the flow and A the area of the flow, thus:

$$H_S = \frac{Q^2}{(A^2 2g)} + y \tag{14.11}$$

Since A is a function of the depth y, Equation (14.11) relates the specific energy of flow (H_S) to a cubic function of y. This is made clearer if a rectangular channel is considered, for which $A = yb$ where b is the width of the channel. Substituting for A gives:

$$H_S = \frac{Q^2}{(y^2 b^2 2g)} + y$$

Substituting q for Q/b, where q is known as the *unit discharge*, that is, the flow per unit width in a rectangular channel, gives:

$$H_S = \frac{q^2}{(y^2 2g)} + y \qquad (14.12)$$

For any value of H_S (above a certain minimum value, to which reference is made later) there are two possible depths y which satisfy Equation (14.12) for the same flow q, that is, H_S can be made up of two different combinations of velocity energy and depth energy. There are a number of common examples that demonstrate this, two of which are shown in Figure 14.5.

The first is the flow over a dam spillway or weir. Upstream, the flow is deep and the velocity very low. Downstream the same flow has a much higher velocity and the depth is much reduced. The second case illustrates the flow under a freely discharging sluice gate. Again the upstream flow is deep and slow; the downstream flow is shallow and fast. In both cases it is assumed that there is negligible energy loss across the structure. For reasons which are explained below, the slow deep flow is known as *sub-critical flow*, whilst the fast shallow flow is known as *super-critical flow*.

Equation (14.12) can be plotted for y as a function of the specific energy head H_S with a constant unit discharge, as illustrated in Figure 14.6. This shows how the equation provides two answers for a particular value of the specific energy. It also shows that there is a minimum value of

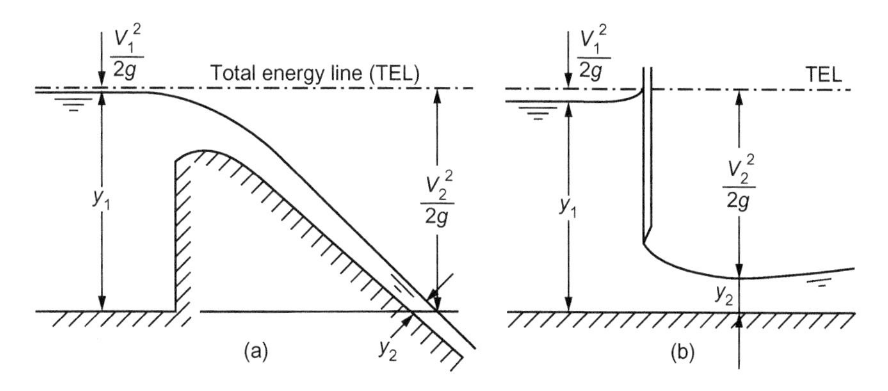

FIGURE 14.5

Examples of alternative flow depth.

the specific energy at which there is only a single value of depth which solves the equation; this is known as the *critical depth*. The critical depth is an important concept in open-channel flow; it marks the boundary between sub- and super-critical flow. If the depth is less than the critical depth then the flow is super-critical; if the depth is greater than the critical depth the flow is sub-critical.

14.8 CRITICAL DEPTH OF FLOW

The graph in Figure 14.6 shows that, for unit discharge, the minimum energy use and critical depth of flow occur when *dH/dy* equals zero. Differentiating Equation (14.12) and putting the differential equal to zero for the minimum value gives:

$$\frac{dH}{dy} = \frac{-2q^2}{(y^3 2g)} + 1 = 0$$

that is, $y^3 = q^2/g$, from which the critical depth:

$$y_c = \sqrt[3]{(q^2/g)} \tag{14.13}$$

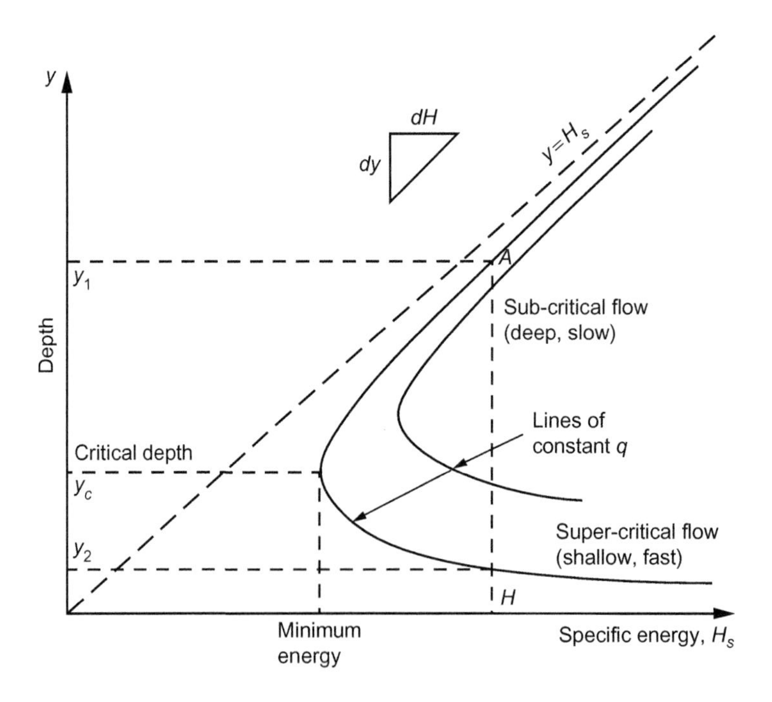

FIGURE 14.6

Relationship between specific energy and depth for a given unit discharge.

Equation (14.13) is important because it demonstrates that the critical depth in a rectangular channel is a function only of the discharge per unit width. Furthermore, substituting y_c^3 for q^2/g in Equation (14.12), and simplifying, the following simple relationship between critical depth and the minimum specific energy is obtained:

$$H_{S(\min)} = 1.5y_c = 1.5(q^2/g)^{1/3} \tag{14.14}$$

Thus the minimum energy is also a function of the unit discharge alone. Hence, given the unit discharge, the critical depth and minimum energy head at any location can be calculated.

Consider the case of flow in a rectangular channel with unit discharge q and assume the flow is sub-critical. As shown in Figure 14.7, the specific energy level (the energy head above the invert) is denoted by the chain dotted line and comprises the depth plus the kinetic energy of the flow – the velocity head. If part of the bed is raised, as indicated in the diagram, then the specific energy reduces. The energy level over the section of raised bed cannot increase so the local depth must decrease and the velocity must increase. The velocity head, therefore, increases and there must be a drop in water level over the raised section. If the bed is raised further, eventually the specific energy reaches the minimum value and the depth of flow over the raised section drops to the critical depth. Any further raising of the bed would reduce the energy below the required minimum value for that unit flow, and the only possible result is a reduction in flow over the raised sill. If the flow is to be maintained then there must be a rise in upstream water level to provide more energy to enable that flow to pass over the sill.

Critical depth of flow thus occurs when there is free discharge over a weir or gate. The flow utilizes minimum energy to pass the maximum discharge possible for the available energy head. The expressions *free* discharge, or *modular* discharge, are used to denote that the downstream conditions do not affect the flow. Clearly, if there were some control downstream, such as a gate, which

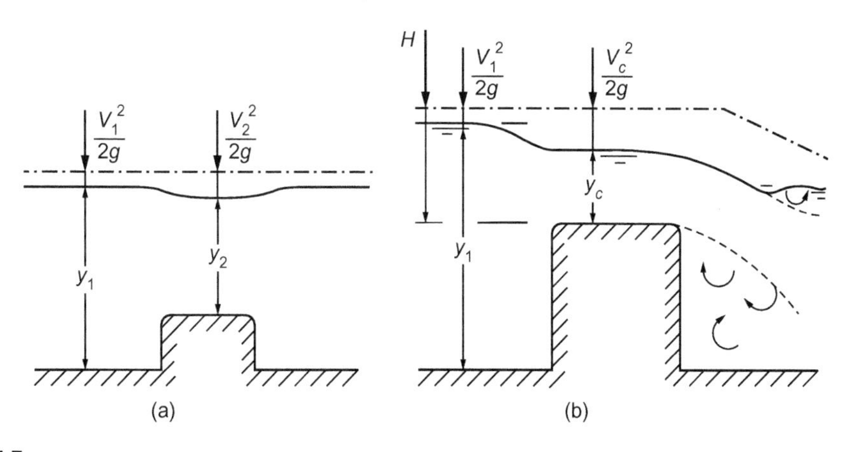

FIGURE 14.7

Flow over a raised sill: (a) low sill, (b) high sill.

was progressively closed then eventually the downstream level would back up to the extent that the free discharge over the weir would be drowned out; critical depth would not occur and the upstream water level would rise accordingly.

Critical depth conditions can also occur because of an increase in the unit discharge. If instead of raising the bed, the sides of the channel are brought in, squeezing the flow, the specific energy remains constant but the discharge per unit width must increase. Again, this can continue until the unit discharge reaches the value defined in Equation (14.14). If the channel is narrowed beyond this point then the specific energy must increase and the upstream water level must rise to provide this additional energy head. This effect is utilized in measurement flumes, which are discussed in Section 14.15. The entrance to a steep culvert is another example where critical depth may occur because the flow is squeezed through a narrower conduit. There are many instances where the flow passes through critical depth due to both a narrowing of the channel and the raising of the bed. The spillway of a dam is one such example. Another is illustrated in Figure 14.8 which illustrates the flow from a reservoir into a channel. The channel is narrower than the reservoir and clearly the bed is raised, so critical depth would be expected at the channel entrance, provided the channel is steep and that the downstream water level is not sufficient to drown out the entrance to the channel. The water surface would also be seen to drop as the flow accelerates into the channel.

The proviso above that the water level downstream of the entrance must not be so high as to drown out the entrance is important. This depends on the slope of the downstream channel. If the slope on the channel is too flat to maintain the flow, the water level rises and the entrance becomes drowned out with sub-critical flow throughout. At one particular slope the depth remains at critical depth and the control at the entrance remains. At steeper slopes than this *critical slope*, the flow accelerates away from the inlet with super-critical flow conditions.

The occurrence of critical depth in a system is described as a hydraulic control. This is because flow conditions in the super-critical reach downstream do not affect upstream conditions. At the point of critical depth the *rating curve*, that is, the relationship between depth (or level) and flow, is fixed and a function only of the local geometry. If such locations can be identified in a system they represent a starting point for assessing water levels upstream and downstream of the control.

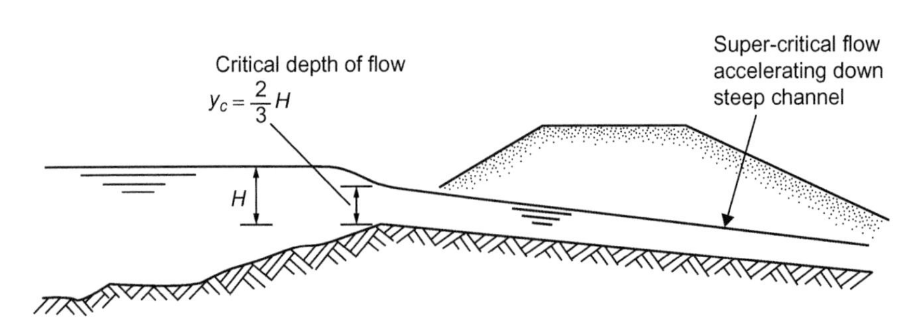

FIGURE 14.8

Critical depth of flow at a channel entrance with the condition that the slope is greater than the critical slope.

14.9 FROUDE NUMBERS

The state of flow in an open-surface conduit can be characterized by its Froude number. This is a non-dimensional grouping of the flow parameters, which can be shown to represent the relative magnitudes of inertial and gravitational forces acting on the fluid. The Froude number, F, is defined as:

$$F = \frac{V}{\sqrt{gy}} \tag{14.15}$$

where y is depth of flow, defined for non-rectangular sections as A/B where A is the flow area and B is the surface width. Its usefulness lies in the fact that at the minimum energy condition, the critical depth:

$$y_c = \frac{2H_c}{3}$$

as shown in Equation (14.14). Hence the kinetic energy of the critical velocity V_c equals the balance of H_S available, that is:

$$\frac{V_c^2}{2g} = \frac{H_c}{3} = \frac{y_c}{2}$$

Thus:

$$V_c = \sqrt{(gy_c)}$$

Therefore, as flow passes through critical conditions, its Froude number (Equation 14.15) has a value of 1.0. For sub-critical flow the depth is greater and the velocity lower, therefore the Froude number is always less than 1.0; for super-critical flow the opposite is true and the Froude number is always greater than 1.0. Calculation of the Froude number thus provides an immediate check on the type of flow and how near the flow conditions are to those at critical depth. The closer the Froude number is to 1.0 the more unstable the water surface becomes since small disturbances can cause flow to flip locally between two possible energy states (Section 14.7). This can lead to waves and surface disturbances. As a general rule it is preferable to design channels with Froude numbers outside the range of about 0.6–1.5.

It should be noted, however, that the control of super-critical flow is more difficult than sub-critical flow. Any changes to the alignment of a channel with super-critical flow result in the development of standing surface waves. Moreover, the high momentum of super-critical flow may create significant super-elevation effects at any bend. Whenever possible channels should be designed for sub-critical conditions. If super-critical conditions must exist, such as in a spillway chute, then the channel should be straight and, if any changes of direction are necessary, they should be made, if possible, in sub-critical flow sections.

14.10 HEADLOSSES IN CHANNELS

The concepts of headlosses in channels are exactly the same as those in pipes. There are both *friction* losses and local or *form* losses and the same equations can be applied. However, as noted in Section 14.7, water depth can vary so that the velocity is a function of both flow and depth.

Normal depth. In a long straight channel of constant cros ssection and bed gradient the flow reaches an equilibrium depth when the rate of loss of energy through the turbulence generated by the drag of the boundaries equals the rate at which potential energy is given up by the fall in elevation. When the flow reaches this condition, the water surface and the energy line are both parallel to the channel bed. This equilibrium depth of flow is known as the *normal depth* and applies to both sub- and super-critical flow. Strictly, the resistance equations for pipe flow apply to this condition of *normal depth* flow in channels but they need to be modified for the different cross-sectional parameters. Instead of the pipe diameter, the relevant dimension is the hydraulic mean depth, R, also commonly referred to as the *hydraulic radius*. This is defined as:

$$R = A/P$$

where A is the area of the cross section of the flow and P is the wetted perimeter, that is, the portion of the perimeter of the channel surface in contact with the flow.

Manning's formula. Although it is possible to use the full Colebrook–White equations for open-channel flow, by far the most widely used equation for calculation of normal depth in open-channel flow is Manning's equation (Equation 14.9) in metric units. In some countries, particularly in Europe, it is also known as Strickler's equation. It should be noted that this equation is not dimensionally balanced and hence n has dimensions of length to the power of one sixth. Some typical values of Manning's coefficient are given in Table 14.3. Despite the dimensional nature of the Manning's equation, it is almost universally used for open-channel headloss calculations and there is a great deal of guidance on suitable values for the coefficient. Its use in this context is strongly recommended. For more detailed information on channel roughness, see Ven te Chow (1959).

Table 14.3 Values of Manning's roughness coefficient n (in Equation 14.10)

Surface	Manning's coefficient n
Smooth metallic	0.012
Large welded steel pipes with coal-tar lining	0.011
Smooth concrete or small steel pipes	0.012
Riveted steel or flush-jointed brickwork	0.015–0.017
Rough concrete	0.017
Rubble (fairly regular)	0.020
Old rough or tuberculated pipes	0.025–0.035
Cut earth (gravelly bottom)[a]	0.025–0.030
Natural watercourse in earth	0.030–0.040
Natural watercourse in earth but with bank growths	0.050–0.070

Notes: [a]Values are for half-bankside depth of flow; at bankfull, the discharge may be 20% less.

Chezy formula. The equivalent equation to the Darcy–Weisbach equation is the Chezy equation, which is written as:

$$V = C(Ri)^{0.5} \tag{14.16}$$

where C is a coefficient related to the roughness of the channel and i is the hydraulic gradient $= H/L$ (the slope of the energy line). Like the Darcy equation this equation is dimensionally correct so that C has no units.

Comparing Equation (14.16) to the Darcy–Weisbach formula (Equation 14.3) and noting that for a pipe running full $R = \pi d^2/4\pi d = d/4$:

$$C = \sqrt{(8g/f)}$$

Therefore, it can be seen that, strictly, C varies, in a similar manner to the friction factor (f), as a function of relative roughness and Reynolds number. In practice the Reynolds numbers related to most open-channel conditions are such as to put the flow regime into the rough-turbulent region of the Moody diagram, where the variation of friction factor and hence of C is small. Thus the assumption that C is constant is reasonable in most cases.

The value of C can be derived from values of Manning's n because, for given values of i and V:

$$R^{2/3}/n = CR^{1/2},$$

i.e.

$$C = R^{1/6}/n$$

Local headlosses. As with pipeline fittings, headlosses occur in open channels at any change of channel geometry. The calculation of losses is made in a similar manner, that is, as a function of the velocity head. The complication with open-channel losses is that it is often not possible to use a single, standard velocity as in the case of a pipe of constant diameter. It is common, therefore, to express the headloss at a change of section in terms of the difference in velocity heads, thus:

$$\Delta H = K\left|\frac{V_1^2}{2g} - \frac{V_2^2}{2g}\right| \quad \text{where } |\ldots| \text{ denotes the positive value.}$$

where V_1 and V_2 are the upstream and downstream velocities, respectively and it is the positive value of the difference to which the empirical coefficient K is applied. Care must be taken in interpreting data on the coefficients, however, as this form of the expression is not universal. Some data may be presented in terms of a single reference velocity as is done for pipes. For example, losses at a channel bend of constant cross section can only relate to a single velocity. Table 14.4 provides some values of K.

Table 14.4 Local channel headloss coefficients (K)

Feature	Loss coefficient[a]	
Bends[b]		
$r/B > 3$ (B is surface width)	0.15	
$r/B < 3$ but >1	0.15–0.6	
90° single mitre	1.2	
45° single mitre	0.3	
Transitions	**Inlet**	**Outlet**
Square ended	0.5	1.0
Cylinder quadrant	0.4	0.6
Smooth tapered	0.1	0.2

Notes: [a]Loss coefficients for other elements can be taken as the same as for pipe elements provided the Froude Number of the flow is less than about 0.3.
[b]Based on average channel velocity.

14.11 HYDRAULIC JUMP

In the discussion of critical depth and its development over a weir crest or in a flume, it was assumed that the flow could accelerate smoothly from sub-critical conditions through critical depth to super-critical flow. Practical observation confirms this; accelerating flow tends to damp down any turbulence. The reverse is not true. To pass from super-critical to sub-critical flow is much more difficult without a significant energy loss. The flow is decelerating and expanding and almost always this leads to a region of instability with high turbulence and energy loss. The result is a *hydraulic jump* in which the water surface rises abruptly from the fast shallow flow to the deeper sub-critical flow downstream. Figure 14.9 illustrates a hydraulic jump downstream of a sluice gate.

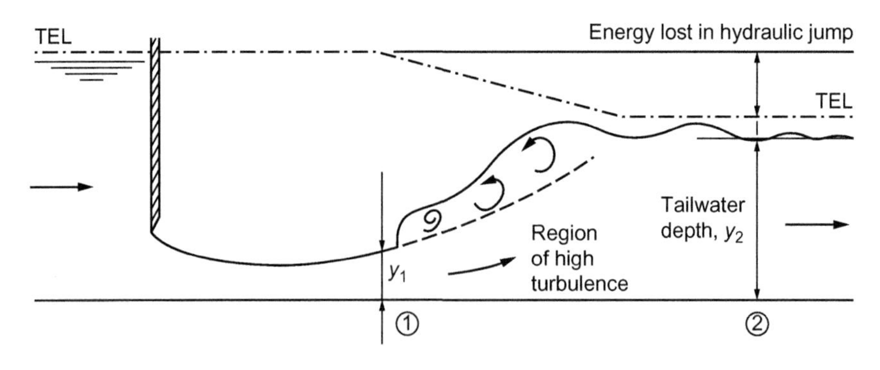

FIGURE 14.9

Hydraulic jump downstream of a gate.

A hydraulic jump is a good way of 'destroying' surplus energy (in reality, converting kinetic energy to heat) and is often deliberately introduced for that purpose, such as at the foot of a dam spillway or downstream of a control gate. It enables the potentially erosive power of the high velocity flow to be reduced in a controlled fashion before the flow is released into the river downstream. The region of high turbulence is contained locally within the structure.

Because there is a high energy loss across a hydraulic jump, it is not possible to use the energy equation to analyse the phenomenon. Instead, the theory of a hydraulic jump is based on the theory of momentum. With reference to Figure 14.9 it can be seen that the only external force on the flow is the friction on the floor of the channel. Within the jump itself this is relatively small and can be ignored, so that the assumption made is that the net pressure force acting on the flow, the difference in the downstream and upstream hydrostatic pressures, is equal to the net momentum loss of the flow. This is treated more fully in other hydraulic texts (Ven te Chow, 1959; Henderson, 1966) but it can be shown that for a jump to form, the downstream or tailwater depth y_2 must be equal to or greater than that given by:

$$y_2 = \left(\frac{y_1}{2}\right)\{(1+8F_1^2)^{0.5} - 1\}$$

where F_1 is the Froude number (Equation 14.15) of the upstream, super-critical flow of depth y_1. The value of y_2 given by this equation is known as the *sequent depth*.

If the tailwater level is less than the sequent depth the super-critical flow continues further downstream until it has lost enough energy or until there is sufficient tailwater depth for the jump to form. If, on the other hand, the tailwater depth is greater than the sequent depth the jump is forced upstream until the balance is re-established. Thus, for the jump downstream of the sluice gate shown in Figure 14.9, if there were another gate downstream controlling the tailwater level, then closing this second gate would increase the tailwater depth and force the hydraulic jump to move upstream. Ultimately it would drown out the first gate, which would then operate with a submerged, drowned discharge as discussed earlier.

14.12 NON-UNIFORM, GRADUALLY VARIED FLOW

So far only flow in a long straight channel of uniform gradient where the equilibrium of normal depth can be reached has been considered. In practice, long straight channels are something of a rarity, except for artificial irrigation channels. In many instances, for example, natural rivers, the channel shape changes or the slope varies. Even if the channel is straight and uniform, it may not be long enough for the flow to reach normal depth. Thus, although normal depth and uniform flow are useful concepts, in many cases this condition will not exist and neither the water surface nor the hydraulic gradient will be at a uniform slope.

One obvious example of conditions where the longitudinal profile of the water surface is curved, is the local drawdown of the flow as it speeds up to pass over a weir. This is a local effect and, as discussed in more detail in Section 14.13, where the simplifying assumptions underlying

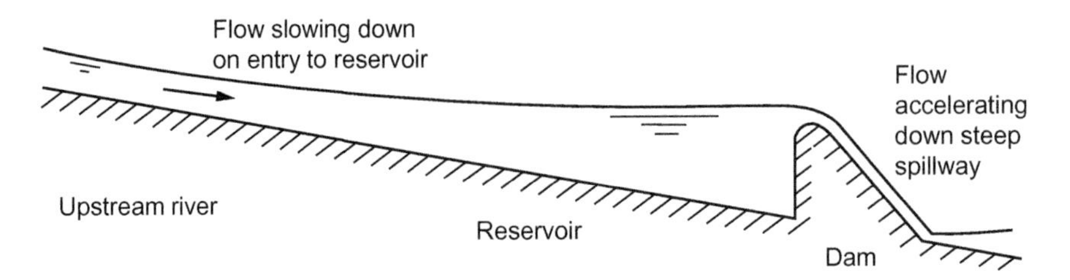

FIGURE 14.10

Examples of gradually varied flow profiles.

the theory no longer hold true. It is also the case that the depth of flow upstream of a weir is unlikely to be the normal depth in the channel, but moving further upstream the depth gradually approaches normal depth.

This can be more clearly visualized by considering the flow in a river entering a reservoir. The flow slows down and gets deeper, with the water surface gradient changing from a slope approximately parallel to the average river bed slope, to horizontal across the reservoir, as illustrated in Figure 14.10. On the other hand, downstream of the crest of the dam spillway, the flow accelerates away from the critical depth and becomes shallower and faster. Another good example of the curvature of the water surface is that of flow under a sluice gate into a horizontal channel downstream as illustrated in Figure 14.9. Energy must be lost through the resistance of the channel bed and the flow must slow down. This means that the depth must increase and the water surface curves upwards.

It is often necessary to calculate the water surface profile in these cases and for other situations, such as that of a river with varying cross section and with occasional obstructions such as bridges or weirs. In the following explanation, steady flow is assumed; the depth and velocity of flow may vary with position, but the total flow at any point remains constant. The introduction of changes to the flow with time adds another level of complexity to the problem; such conditions are then described as *dynamic* or *transient* and are almost always analysed using computer programs. They are beyond the scope of this book and reference should be made to any of the several open-channel flow text books available, for example, Ven te Chow (1959) and Henderson (1966).

Returning to the energy equation, consider the flow at two points; distance ΔL apart, in a non-uniform channel, as shown in Figure 14.11.

Between the two points, ΔL apart, the headloss, ΔH is given by:

$$\Delta H = H_1 + \Delta z - H_2 = \left(\frac{V_1^2}{2g} + y_1\right) + \Delta z - \left(\frac{V_2^2}{2g} + y_2\right)$$

Substituting Q/A for V:

$$\Delta H = \left(\frac{Q^2}{A_1^2 2g}\right) + y_1 + \Delta z - \left(\frac{Q^2}{A_2^2 2g}\right) - y_2$$

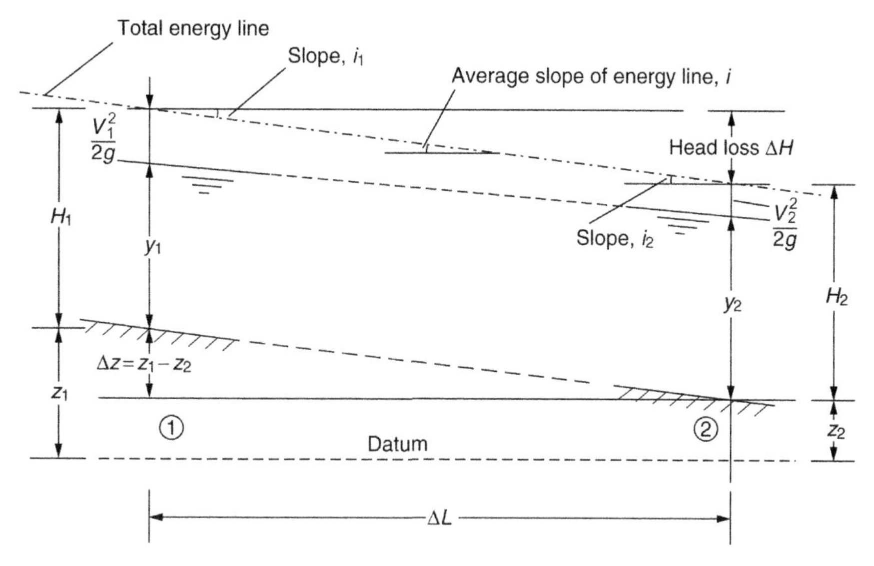

FIGURE 14.11

Energy loss in non-uniform flow.

If the length ΔL is reasonably short, the average slope, i, of the energy line can be taken as approximating to the mean of i_1 and i_2 so that:

$$\Delta H = \Delta L(i_1 + i_2)/2$$

where i_1 and i_2 are the slopes of the energy line at Sections 1 and 2.

There are thus two equations for ΔH. From the known geometry of the channel and starting from a known depth of flow y_1 at Section 1 for flow Q, the hydraulic gradient i_1 can be calculated from the Manning or Chezy formulae. Using $i_1 \Delta L$ as the loss of energy ΔH to Section 2, a first approximation of the depth of flow y_2 can be obtained (because $H_1 - \Delta H = Q^2/(A_2^2/2g) + y_2$). From this i_2 can be calculated, so that a revised estimate of the hydraulic energy loss $\Delta H' = \Delta L(i_1 + i_2)/2$ can be used to calculate a more accurate value of y_2. By further iterations, y_2 can be calculated to the desired degree of accuracy. Thus the computation can be carried out progressively for further points downstream giving the profile of the water level. For reasonable accuracy, the distance steps ΔL should be reasonably short, having regard to the change of slope of the hydraulic gradient.

Such calculations can be carried out by hand (though for non-rectangular channels they are very tedious), but are now normally undertaken with a spreadsheet or more probably with a computer program developed for the purpose. Many organizations have developed their own programs and there are a number commercially available, the most widely known probably being HEC 2 (now marketed as HEC2-RAS), developed originally by the Hydrological Engineering Centre in the USA.

There are three further points to make about this type of calculation.

1. The basic theory includes the inherent assumptions listed in Section 14.13 for weirs. Thus, the approach cannot be used accurately where there are rapid changes to the water surface, for example, close to a weir crest or through a flume.
2. The starting point for any calculation must be a known flow and depth. In many cases this is a hydraulic control, such as flow over a weir crest or through a gate where the water level can be directly calculated. Such a control defines the water profiles both upstream and downstream (Section 14.8). Thus, for the sub-critical flow, the calculation proceeds upstream from the control, whilst for super-critical flow the calculation can proceed in the downstream direction, that is, the profile in sub-critical conditions is affected by what happens downstream, and the super-critical flow profile is affected by what happens upstream.
3. In some cases there may be no hydraulic control that defines a specific depth at any point; for example, a long river reach with no control structures. In such a case it may be adequate to assume a normal depth control in the reach downstream of the length of interest (assuming sub-critical flow) and to check the sensitivity of the water levels to that assumption.

The most common calculation relates to finding the water profile of sub-critical flow approaching some hydraulic control. This type of calculation is known as a *backwater calculation* and the water-surface profile is known as a *backwater profile* as the calculations proceed upstream from the control point.

14.13 WEIRS, FLUMES AND GATES

Weirs and gates are used widely as hydraulic control structures. In the former case the weir is usually at a fixed level designed to control the upstream water level and or as flow measurement structures. Gates are movable devices which are used in several ways such as flow or level control. They are usually undershot – that is, the flow passes under the gate – but overshot gates are also used acting as a movable weir. Flumes tend to be used mainly for flow measurement. It should be noted that the term *flume* is also widely used for steep, artificial channels in the mining and other such industries. In this context the term is used to denote a narrowing of a channel to force the flow through critical depth.

Weirs

Equation (14.14) can be re-arranged in the form: $(2/3)H_S^3 = q^2/g$. Hence, flow per unit width $q = (2/3)\sqrt{(2g/3)}H^{1.5}$ and for a rectangular channel of width, b:

$$Q = (2/3)\sqrt{(2g/3)}bH^{1.5} \quad (14.17)$$

This is now in the form of a *weir* equation, which can be generalized as:

$$Q = C_d bH^{1.5} \quad (14.18)$$

where C_d is a *discharge coefficient.*

C_d in Equation (14.18) has dimensions as it involves $\sqrt{g}$. There are a number of other forms for this equation (Section 14.14), with the simplest involving a non-dimensional coefficient being:

$$Q = C'_d\sqrt{g}bH^{1.5} \tag{14.19}$$

In Equation (14.18) the discharge coefficient $C_d = (2/3)\sqrt{(2g/3)} = 1.705$ in metric units and, in Equation (14.19), $C'_d = (2/3)\sqrt{2/3} = 0.544$ in non-dimensional units. If the flow passes through critical depth over a weir crest then it might appear that C_d would always take that value. However, it is important to appreciate the inherent assumptions lying behind the theory that led to Equation (14.1) and all the subsequent equations derived from it:

- there is a uniform velocity distribution across the section so $\alpha = 1.0$ (Section 14.3);
- the streamlines are straight and parallel (i.e. there is no lateral pressure set up by curvature of the streamlines);
- the vertical pressure distribution is hydrostatic (i.e. the pressure is a linear function of depth);
- the effect of the longitudinal slope is negligible.

Provided these conditions are met, as is very nearly the case with a broad-crested weir such as illustrated in Figure 14.7(b), then the discharge coefficient C_d, is indeed about 1.705. However, many other crest profiles are used, ranging from simple plate *or sharp-edged* crests, to rounded tops of walls and triangular or ogee-profile crests. Their use depends on whether accurate flow measurement is required or whether they act merely as simple overflow hydraulic controls. Weir shapes used for flow measurement are discussed further in Section 14.14. The discharge coefficient can vary significantly from the basic broad-crested value of 1.705 and depends largely on the geometry of the crest, but it is also a function of the depth and velocity of the approach flow. The subject is wide-ranging and is beyond the scope of this book. As a rule of thumb, however, the discharge coefficient is likely to be greater than 1.705 if there is strong curvature to the flow, for example, over a half-round crest or an ogee crest. In the latter case, the profile of the crest is that of the underside of a free-falling jet of water, so, in theory, there should be little if any pressure on the solid surface. The pressure distribution through the depth of flow cannot, therefore, be hydrostatic and one way of considering the problem is that the back pressure on the flow over the crest is reduced, thus allowing an increased discharge and a corresponding increase in the discharge coefficient. Increases in C_d of 30–40% above the broad-crested weir value are possible.

Similarly, the discharge coefficient may be reduced if the weir crest is long (in the direction of flow), or very rough – as might be the case of flow over a grassed embankment. A value of 1.7 is however a good starting point for initial design or in the absence of more details of the weir shape.

It must be emphasized again that a weir only acts as a hydraulic control if it has free or 'modular' discharge; that is, the downstream water level is low enough to allow the flow to pass through critical depth. If the *tailwater* depth is high enough to affect the flow over the weir, the weir is said to be drowned. This condition is usually catered for in the weir equation by introducing a drowning factor f_d, which is a function of the height of the tailwater level above the crest. Referring to Figure 14.7(b) it would appear that, provided the tailwater level above the crest is not more than the critical depth, that is, two thirds of the upstream head above the crest, then critical depth flow will occur. Strictly, the tailwater level has a small effect even when it is at the level of the crest, particularly if lowered pressures can be generated below the *nappe* of the falling water (e.g. in an air pocket trapped behind the falling sheet of water), but this two thirds criterion is a useful rule of thumb. Although the crest shape does affect the drowning factor, in many instances where a weir is not being used for measurement it is reasonable to assume that the hydraulic control remains at the weir crest until the tailwater level rises to a depth greater than two thirds of the upstream head above the crest.

Flumes

The foregoing comments also apply to flumes. Figure 14.15 (Section 14.15) shows a typical flume with a narrow throat forcing the flow through critical depth. The weir equation can be applied to the throat and, provided the underlying assumptions as listed for weirs apply, the discharge coefficient will again be 1.705. Thus a flume, with parallel sides and level invert to the throat, has a discharge coefficient close to this base value, that is, in metric units:

$$Q = 1.705bH^{1.5}$$

If however the calculation of flow is related to the more easily measured water level and the water depth above the throat invert level h rather than the total energy head H, then:

$$Q = 1.705C_v bh^{1.5}$$

where C_v is a coefficient related to the approach velocity in the channel and b is the width of the throat of the flume. As a general guideline, a flume will also be drowned out when the downstream water level rises to a level greater than two thirds of the upstream head. In practice it is often possible with a well-designed exit transition from the throat to recover some of the velocity head. Thus it is possible to design a flume such that the downstream water level can rise above this critical level with an overall headloss less than $H/3$.

Gates

Water control gates, as distinct from other types of gates such as used in navigation locks, may be either undershot or overshot. The latter case is akin to the flow over a weir and only the former type is discussed here. There are many designs of gates within these broad categories including radial gates, rising sector gates and other designs for specific uses. The hydraulic principles are similar and are illustrated here for vertical-lift gates, also called *sluice gates* and *penstocks* which are by far the most common in water supply systems. This is another case of flow passing from sub- to super-critical flow, but with an undershot gate the flow does not pass through the critical depth at any meaningful location.

Figure 14.12(a) illustrates free discharge through a vertical penstock gate in the wall of a tank. The gate opening forms an orifice. The flow passing under the gate has a vertical contraction and, in the plane of the gate itself, there are vertical components of flow. Thus the minimum depth occurs downstream of the gate at a point referred to as the *vena contracta*, where the streamlines

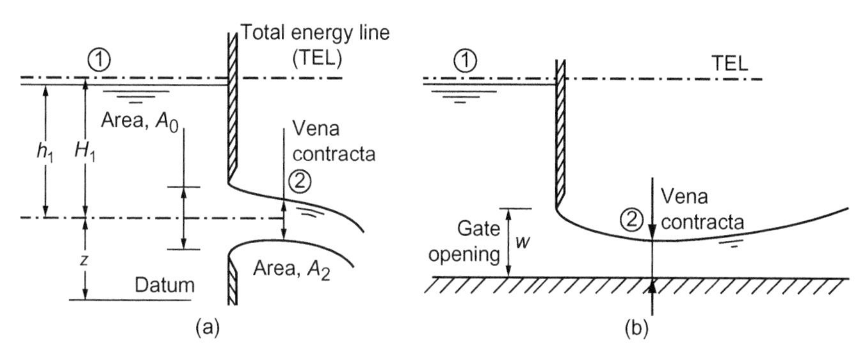

FIGURE 14.12

Orifice and undershot gate flow: (a) orifice flow, (b) undershot gate.

are parallel. Assuming negligible energy loss between the upstream section and the vena contracta, Bernoulli's equation can be applied along a streamline flow between points 1 and 2:

$$H_1 + z = V_1^2/2g + y_1 + z = V_2^2/2g + y_2 + z$$

where y_1 and y_2 are the upstream and downstream depths from the water surface to the centre of the opening. Assuming that the pressure throughout the free-falling jet at the vena contracta is atmospheric, then:

$$V_2^2/2g = H_1$$

and, since $Q = VA$:

$$Q = A_2\sqrt{(2gH_1)}$$

If the area of the gate opening is A_0 and the area of the vena contracta A_2 is $C_C A_0$, where C_C is a contraction coefficient, then:

$$Q = C_c A_0\sqrt{(2gH_1)}$$

In practice, as indicated for flumes, it is easier to measure the upstream water surface and depth and to relate the flow to the upstream depth using the more general equation:

$$Q = C_d A_0\sqrt{(2gy_1)}$$

where C_d is a discharge coefficient, which takes into account the approach velocity and the flow is related to the upstream depth rather than the energy head. Typically, for a sharp-edged opening, C_d is about 0.6. If the edges of the opening are more rounded, then the value approaches closer to 1.0.

For the case where the downstream water level is high enough to drown the opening and the jet is not freely discharging then the assumption of atmospheric pressure in the vena contracta is no longer valid and the downstream depth above the centre of the opening, y_2, becomes significant. Referring back to Bernoulli's equation above it can be shown that:

$$Q = C_d A_0 \sqrt{2g(y_1 - y_2)} = C_d A_0 \sqrt{(2g\Delta y)}$$

where Δy is the difference in water levels across the gate.

The same equations apply to the case of a sluice gate in a channel, as illustrated in Figure 14.12(b), but now the depth is the total water depth y to the channel invert rather than to the centre of the opening. For a full-width, freely discharging gate with sharp-edged lip, the discharge coefficient varies between about 0.5 for low depths of submergence ($y_1/w = 2.0$) up to about 0.58 for submergence ratios of 10 and more, where w is the height of the gate opening as shown in Figure 14.12(b). For a submerged gate opening the discharge coefficient is again about 0.6.

14.14 MEASUREMENT WEIRS

In Section 14.13 it is shown that, provided there is free, undrowned, discharge over the crest of a weir, the flow passes through critical depth and the upstream water level is uniquely controlled by the flow rate and the geometry of the crest. In theory, therefore, any weir can be used for measurement but, in practice, a limited number of standard crest shapes tend to be used, partly because experience has shown that these are the most accurate and practical structures for a particular application and partly because the discharge coefficients and the limiting conditions for their accuracy are well known. British and other international standards are available for such weirs, covering the standard geometries, measurement requirements and other features. Ackers (1978) provides a definitive discussion of the weirs and of flumes used for open-channel flow measurement. For non-standard shapes, either detailed model tests or site calibration must be carried out, but the resulting rating curve must have a weir-type relationship taking into account the overall geometry of the channel section at the weir.

For conditions which meet the underlying assumptions stated in Section 14.13, the upstream total head H is given by Equation (14.17), that is,

$$Q = (2/3)\sqrt{(2g/3)}bH^{1.5} = 1.705bH^{1.5}$$

with $g = 9.81$ m/s^2 in metric units, and the general equation is therefore:

$$Q = C_d bH^{1.5}$$

For *measurement weirs*, the general weir equation is often written as:

$$Q = C_d''(2/3)\sqrt{(2g)}bH^{1.5} \tag{14.20}$$

where, for equivalence to Equation (14.17), C_d'' has the value $1/\sqrt{3}$.

However, as noted in Section 14.13, there are other formulations of the weir equation. In particular, Equation (14.19) has a C'_d value for the broad-crested weir of 0.544, whereas Equation (14.20) gives $C''_d = 0.577$. The user must be careful to ensure the discharge coefficients relate to the correct formulation of the discharge equation. In practice C''_d varies from the value $1/\sqrt{3}$ according to the shape and type of weir crest.

Broad-Crested Weir

For the broad-crested weir, illustrated in Figure 14.7(b), the discharge coefficient C''_d is indeed about $1/\sqrt{3} = 0.577$. However, as indicated earlier it is easier to measure the actual water depth h above the weir crest, than the total energy head H (including the velocity of approach head), hence the equation is normally written as:

$$Q = C_v 0.577(2/3)\sqrt{(2g)}bh^{1.5} = C_v 1.705bh^{1.5}$$

where h is measured away from the local drawdown of the water surface as it passes over the crest and C_v is a *velocity coefficient* to account for the approach velocity head. If the approach velocity is small then C_v can be taken as 1.0 and for other cases the effect of approach velocity head can be calculated and included in C_v.

The generalized weir equation can be written as $Q = C_d C_v C_s C_p C_{bl}\sqrt{g}bh^{1.5}$ where C_d is a discharge coefficient, C_v is the velocity coefficient, both as defined before, C_s is a shape coefficient depending on the lateral shape of the crest (e.g. shallow 'Vee'), C_p is a coefficient to take into account the height of the weir and C_{bl} is a boundary layer coefficient to take into account the growth of the boundary layer across the length of the weir crest in the direction of flow. For accurate measurement, all these factors need to be taken into account and are discussed in detail in Ackers (1978).

Sharp-Crested or Thin-Plate Weirs

Sharp-crested weirs, usually formed from a metal plate, are used in a variety of situations. They are simple, and because the plate can be machined to a high degree of accuracy and can be arranged to be adjustable, they can be very accurate flow measurement devices. However, because the plate becomes heavy and unwieldy for large flows, they are used mainly for small and medium flows and not generally for river flow measurement. There are a number of standard shapes adopted, which are described by the shape cut out of the plate – rectangular of full channel width; rectangular with side contractions as illustrated in Figure 14.13; V-notch with a central angle, θ; or, less commonly, trapezoidal.

For a rectangular weir with no end contractions, that is, across the full width of the channel, the discharge coefficient, for use in Equation (14.19), can be taken from the empirically derived Rehbock formula:

$$C'_d = 0.602 + 0.083h/P$$

where h is the water depth over the weir crest, and P is the height of the weir crest above the floor of the channel.

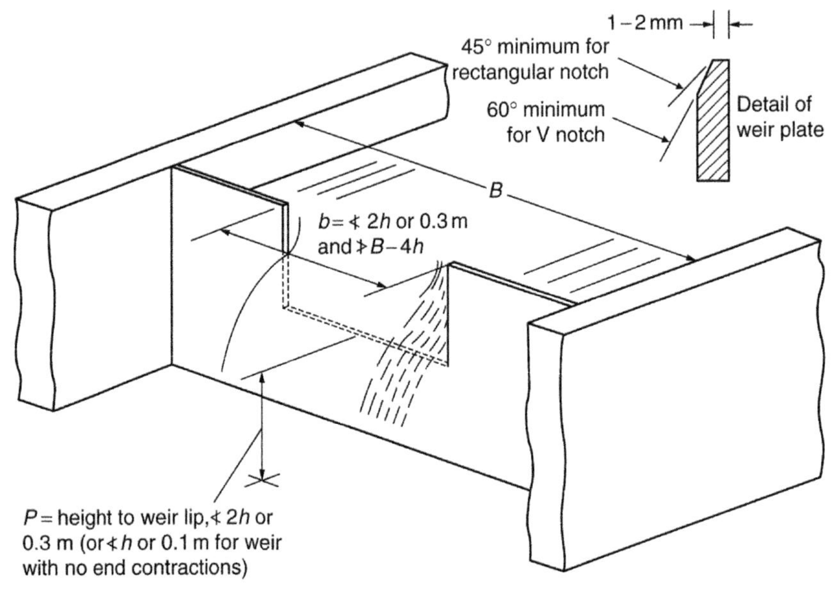

FIGURE 14.13

Rectangular weir with fully developed end contractions.

For a rectangular weir with fully developed end contractions as illustrated in Figure 14.13:

$$C'_d = 0.616(1 - 0.1h/P) \tag{14.21}$$

For a 90° V-notch weir, that is, with the central angle = 90°, C'_d is given by the following table:

h (m)	0.050	0.075	0.120	0.125	0.150	0.200	0.300
C'_d	0.608	0.598	0.592	0.588	0.586	0.585	0.585

If the base of the channel is less than $2.5h$ below the base of the 90° V-notch, or if the width of the approach channel is less than $5h$, where h is the depth of water measured above the vertex of the notch, a correction factor must be applied.

In all cases for thin-plate weirs there are a number of requirements for accurate measurements:

- the water depth should be measured at a distance between $3h$ and $4h$ upstream;
- if the waterway of the approach channel is less than 12 times the area of the waterway over the weir crest then a correction factor must be included to take into account the velocity of approach;
- the downstream water level must be below the level of the crest, that is, no submergence can be allowed; and
- most importantly, the *nappe* (or underside) of the falling water must be well aerated to ensure atmospheric pressure there. This means that, for a full width rectangular weir in a channel whose walls extend downstream, an air pipe must be provided to allow aeration of the pocket beneath the nappe.

A limitation of sharp-edged weirs is that they can be damaged by debris brought down by the flow, and the shape of the machined edge to the plate is important for accurate measurement.

Crump Weirs

Where new measurement weirs are constructed on rivers and streams in the UK the Crump weir is the most widely used, so-called after the engineer who developed it as a measuring device. The profile is triangular with an upstream slope of 1:2 and a downstream slope of 1:5 as shown in Figure 14.14.

There are several advantages to the Crump weir, not least of which is that the discharge coefficient is very nearly constant over a wide range of discharges. C_d for use in Equation (14.18) is almost exactly 2.0 (C''_d in Equation (14.20) thus equals 0.677). Other advantages are:

- the weir allows movement of sediment past the weir (deposition of silt against the vertical upstream face of most other types of weirs being a problem);
- the weir is easy to build accurately, particularly with a pre-formed bronze crest angle built into the concrete structure; and
- the high modular limit which the weir can tolerate before the discharge is significantly affected; the limit for the ratio of tailwater depth above the crest to the upstream depth being about 0.8.

Furthermore, the Crump weir can provide reasonably accurate flow measurement even if it is operating beyond the modular limit. The flow over a weir, once it is drowned, becomes a function of both the upstream and downstream water levels. For most weirs, this requires measuring the water level downstream in the area of high turbulence where the jet expands. This can be difficult and inaccurate, particularly as the critical difference between the tailwater and upstream depth is small. With a Crump profile weir, instead of the direct measurement of the tailwater level, the pressure in the separation pocket immediately downstream of the weir crest is measured (Fig. 14.14). The pressure in this pocket reflects the downstream tailwater level when the weir is drowned. It is a much more stable parameter than the turbulent water surface level downstream and can be measured more accurately. It thus enables the Crump weir to be used relatively accurately for measuring flow even when drowned and a rating curve can be developed for drowned flow using this parameter.

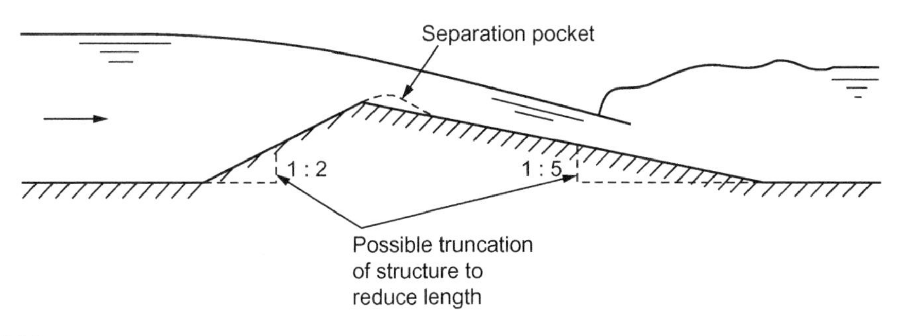

FIGURE 14.14

Crump weir profile.

The one disadvantage of a Crump weir is that it can be a fairly long structure with consequential extra costs but it has been extensively model tested and criteria have been developed for truncation of the shape, that is, neither the upstream nor downstream slopes need necessarily extend to the floor of the channel.

Crump weirs are often constructed with the crest sloping towards the centre so that the crest, in elevation, appears as a shallow Vee. This configuration has the advantage of concentrating low flows towards the centre of the weir and giving greater accuracy of measurement over a wide range of flows. The shape has been investigated thoroughly and the discharge coefficients have been calibrated.

14.15 MEASUREMENT FLUMES

The principles behind the use of a flume as a measurement device are identical to those for a weir. In essence a flume is a constriction in the channel such that the width is reduced, forcing the flow through critical depth (Section 14.13). Provided the requirements behind the simple theory, set out in Section 14.13 are met, Equation (14.17) applies to undrowned flow through a flume, that is,

$$Q = 1.705bH^{1.5}$$

Rectangular-Throated Flume

The flume equivalent to the broad-crested weir is the rectangular-throated flume as illustrated in Figure 14.15. A flume can also have a raised floor as well as a narrowed throat but since one of the advantages of a flume over weir is that it can more readily pass sediment, any raising of the floor probably needs to be limited and with as smooth an inlet transition as practical. Downstream of the throat the flow is super-critical and a hydraulic jump forms in the downstream channel. It is important that the tailwater level downstream of the jump does not drown out the critical depth in the flume and, if necessary, a fall must be introduced into the downstream channel to ensure that the water flowing away is at a low enough level.

For a rectangular-throated flume as in Figure 14.15, the length of the throat must be not less than 1.5 times the maximum total head upstream. It is also necessary that the surfaces of the flume are smooth, whether of concrete or constructed with a pre-formed steel or fibre-glass insert, that the divergence downstream must not exceed 1:6 and that the approach flow must be sub-critical. The upstream level measurement should be between $3h$ and $4h$ upstream of the flume inlet, where h is the upstream depth of water above the invert of the throat.

The discharge equation is usually modified to relate to the upstream depth of flow h in the channel above the invert of the throat of the flume:

$$Q = C_v 1.705bh^{1.5} \quad (\text{or } Q = C_d bh^{1.5})$$

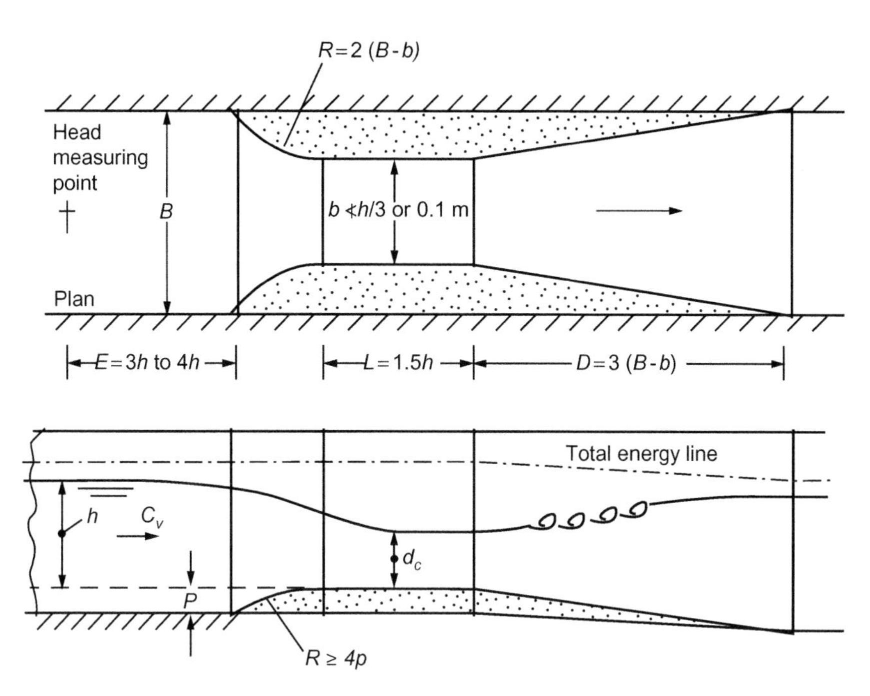

FIGURE 14.15

Rectangular-throated flume.

where C_v is a coefficient to account for the approach velocity. More generally the equation is written as:

$$Q = C_v C_s 1.705 b h^{1.5} \tag{14.22}$$

where C_s is a shape coefficient to take into the particular geometry of the flume. For the smooth-entry flume illustrated in Figure 14.15, the values for C_s depend on the ratios L/b and h/L as shown in Table 14.5, where L is length and b the width of the throat.

Table 14.5 Shape coefficient C_s for rectangular-throated flume

	Ratio of length L to width b of throat				
h/L	0.4–1.0	2.0	3.0	4.0	5.0
0.1	0.95	0.94	0.94	0.93	0.93
0.2	0.97	0.97	0.96	0.95	0.95
0.4	0.98	0.98	0.97	0.97	0.96
0.6	0.99	0.98	0.97	0.97	0.96

Other Standard Flumes

There are many other designs of flumes and the geometry determines the particular discharge coefficient C'_d for use in Equation (14.21), or the shape coefficient for use in Equation (14.22). In the USA, the Parshall flume is widely used but it has a relatively complicated geometry which varies depending on the range of flows to be measured and is rarely used in the UK. One disadvantage of the rectangular-throated flume is that it is relatively long and the cut-throat flume has been developed as a shorter alternative. This has tapering inlet and outlet sections which meet at an angle at the throat. The shape coefficient is reduced from that for a rectangular-throated flume.

For free or modular discharge for flumes and weirs, the tailwater level should be, as a general approximation, no higher than the level of the critical depth flow at the throat or crest. (For thin-plate weirs a much lower limit must be set.) Generally, therefore, at least one third of the upstream specific energy head must be lost at the structure. Since flumes force critical depth by increasing the unit discharge rather than reducing the specific energy as do weirs, a flume has a greater minimum critical depth in the throat than a weir in the same location. Thus a flume causes a greater minimum headloss than a weir. (For a flume the drop in water level is one third the depth of flow in the approach channel; for a weir it is one third the height of water above the weir sill, which is less. See Figures 14.7(b) and 14.15.) Hence, if headloss is a critical issue, then this must be a factor in the choice of measurement structure.

14.16 VENTURI AND ORIFICE FLOW METERS

These types of flow meter apply to flow in closed conduits and work on the principle demonstrated by the Bernoulli equation: if the velocity of flow is increased then the pressure must drop. In both Venturi and orifice meters the flow is passed through a constriction in the pipeline causing the velocity to increase. The two types of meter are shown in Figure 14.16 and the measurement principle is illustrated in Figure 14.17.

From the energy equation (Equation 14.2) and assuming the pipeline is level and that headlosses between sections A and B are negligible, the head difference between the two sections, A and B, is given by:

$$\Delta h = h_A - h_B = \frac{(V_B^2 - V_A^2)}{2g} \tag{14.23}$$

Since:

$$Q = \pi D^2/4V_A = \pi d^2/4V_B$$

where D and d and A and a are the pipeline and throat diameters and areas, respectively, Equation (14.23) can be re-written as:

$$Q = \frac{(\pi d^2\sqrt{2g\Delta h})}{\left(4\sqrt{1-(d/D)^4}\right)}$$

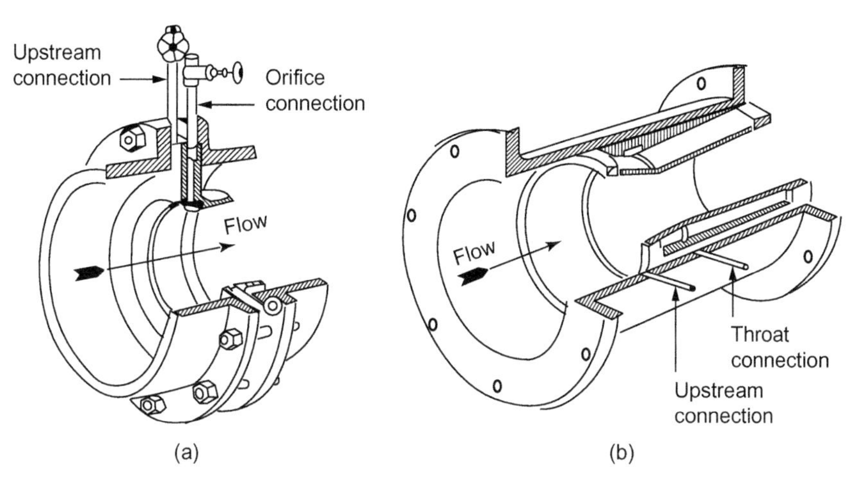

FIGURE 14.16

(a) Orifice meter, (b) Dall-type Venturi meter.

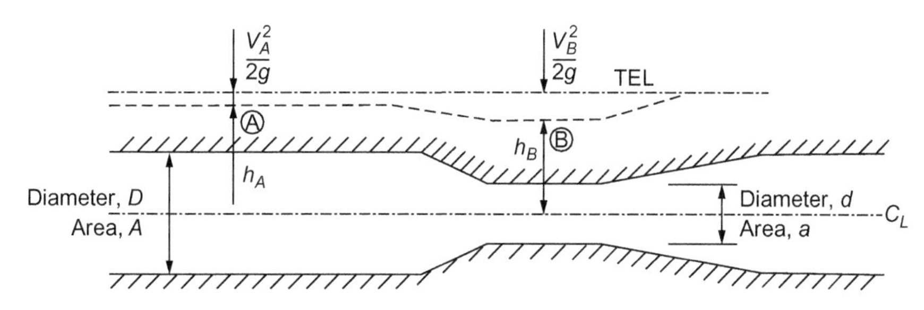

FIGURE 14.17

Venturi and orifice flow meter measurement principles.

or:

$$Q = ak\sqrt{\Delta h}$$

where:

$$k = \frac{\sqrt{(2g)}}{\sqrt{1-(d/D)^4}}$$

In practice there are headlosses between A and B and it is not necessarily the case that the streamlines at the constriction are parallel even for the venturi. Thus a further coefficient C must be introduced, so that:

$$Q = Cak\sqrt{\Delta h} \tag{14.24}$$

For a venturi meter C varies from about 0.95 to 0.99 depending on the exact geometry. Typically a well-designed meter with a parallel-sided throat has a value of about 0.98. The Dall tube illustrated in Figure 14.16 is a shortened version of the venturi meter and the value of C is less.

For an orifice meter Equation (14.24) equally applies. However, C varies much more, depending on the velocity and the ratio d/D. If the latter is in the typical range of 0.4–0.6 and for pipe diameters greater than 200 mm, the value of C will be in the range 0.60–0.61 for the usual velocities experienced in a pipeline.

The accuracy of measurement of both venturi and orifice meters depends on the lateral flow distribution through the device and can be severely affected by flow disturbances created by fittings in a pipe system. Detailed conditions for accurate measurement are laid down in BS EN ISO 5167-1:2003. These can generally be met by ensuring that for d/D ratios not exceeding 0.6, there are at least 20 diameters of straight pipe without a fitting upstream of the meter and 7 diameters of straight pipe downstream. The venturi is designed to minimize the headloss and this can be made very small with a well-designed expansion downstream of the throat providing good recovery of pressure head. The headloss through an orifice is substantially higher because of the sudden expansion of the diameter downstream. If headloss is an important consideration a venturi meter, such as the illustrated Dall tube, should be considered but venturis are more expensive and require greater space than an orifice meter.

14.17 OTHER FLOW METERS

A number of other types of flow meter, such as turbine, electromagnetic and ultrasonic are widely used in the water industry within distribution systems and as revenue meters. The principles on which they operate are not the same as those outlined in this chapter but they are described in Chapter 18.

14.18 COMPUTATIONAL FLUID DYNAMICS (CFD)

CFD is a computer modelling method which simulates fluid flow and heat transfer normally in three dimensions. Although initially developed for the aeronautical and nuclear industries it is now widely applied to almost any application involving fluid flow, such as wind turbines, ventilation in buildings and inkjet printers.

CFD has gained acceptance within the water industry as a powerful tool for investigating flow behaviour in cases that are difficult or unreliable to assess by standard hydraulic calculations. In addition, where process performance is determined by the flow field, CFD can enable performance to be predicted. For example, contact times or particle settlement can be simulated with CFD. In order to gain reliable output from CFD it is important to have a thorough knowledge and understanding of the principles of fluid flow, as well as the limitations of the CFD software in capturing the particular physical phenomena under investigation. No model is any better than the inputs provided and the modeller needs to be confident in identifying and rejecting spurious results.

CFD captures different physical phenomena to varying degrees of reliability, and the modeller is required to understand the limitations of the software in any given application.

A model can be relatively quick to set up using modern commercial CFD software. A mesh is defined that splits the fluid body into a large number of small elements. The software predicts flow patterns throughout the body by solving a series of simultaneous equations for conservation of mass, momentum and energy. Basic fluid flow, chemical reactions, heat transfer and multi-phase flow (liquids, gases and particles) can all be simulated. Computer processing speed and internal memory continue to rise sharply and it is increasingly common to solve models in parallel across multiple computers. This has led to a corresponding reduction in solver time, and opportunity for increasing mesh refinement and model accuracy. Simple models may be solved in less than an hour; more complex models still require many hours to solve.

CFD is used for a wide range of water engineering applications including:

- Headlosses for varying flow through complex structures and pipe details.
- Flow distribution through structures and over weirs (Plate 23(a)).
- Pump sumps to simulate approach conditions for different operating scenarios and to model improvements to the design such as installing baffles (Plate 23(c)).
- Simulated tracer tests for assessing contact time or mixing. Simulated tracer tests enable the modeller to predict the contact time and analyse options to improve hydraulic mixing and tank efficiency (Plates 21(b) and (c)). The analysis of existing tanks often demonstrates that installed baffles do not control the flow as intended. The technique can also be used to analyse the dispersion of contaminants through contact tanks and service reservoirs (Plate 21(a)).
- Reservoir turnover can be assessed by conducting simulated tracer tests and evaluating the coefficient of variation for the tracer concentration. The effect of temperature can also be simulated to assess stratification problems and to develop designs for the inlet to improve mixing within the tank (Plates 22(a)–(d)).
- Dynamic forces acting on hydraulic structures (Plate 23(b)).
- Velocity streamlines in pump inlets.

REFERENCES

Ackers, P., White, W. R., Perkins, J. A. and Harrison, A. J. M. (1978). *Weirs and Flumes for Flow Measurement*. John Wiley.

BS EN ISO 5167-1:2003. *Measurement of Fluid Flow by Means of Pressure Differential Devices Inserted in Circular Cross-Section Conduits Running Full. General Principles and Requirements*. British Standards (BSI).

Henderson, F. M. (1966). *Open Channel Flow*. Macmillan.

HR Wallingford (2006). *Tables for the Hydraulic Design of Pipes, Sewers and Channels*. 8th Edn. Thomas Telford.

Miller, D. S. (1990). *Internal Flow Systems*. 2nd Edn. BHRA.

Moody, L. F. (1944). Friction factors for pipes. *Transactions of the ASME* **66**, p. 672.

Ven te Chow (1959). *Open Channel Hydraulics*. McGraw-Hill.

CHAPTER

System Design and Analysis

15

15.1 INTRODUCTION

This chapter describes the principles and processes necessary when designing a single pipeline or network of pipes. Other chapters discuss related aspects:

- Demands and diurnal and seasonal variations in Chapter 1.
- Hydraulic principles in Chapter 14.
- Operational standards and asset maintenance planning in Chapter 16.
- Pipe materials and installation standards in Chapter 17.
- System valves in Chapter 18.
- Pumping in Chapter 19.
- Storage in Chapter 20.

Pipelines in a water distribution system can be divided into two functional categories:

Trunk mains convey water in bulk between facilities in the system, such as raw water intakes, treatment works, pumping stations and service reservoirs and the centres of demand. Their operational regime is a function of the demand characteristics of the area they supply, the availability of local storage and their transfer capacity. Consumers are not normally supplied directly from a trunk main.

Distribution pipes deliver water to the consumer from local storage and from connections off a trunk main through an integrated network. The pipes are sized to meet the hourly variation of consumers' demand and provide fire flows to any location in the network.

Within these two categories there are *critical mains*. A critical main is one where the consequence of failure poses an unacceptable risk of:

- interruption of supply to consumers;
- contamination of the water supply; or
- damage and disruption to third parties.

Most trunk mains and some key feeder pipes within distribution are designated as critical mains, the most critical being those where failure may affect the supply to sensitive consumers, consumers

Twort's Water Supply. DOI: http://dx.doi.org/10.1016/B978-0-08-100025-0.00015-6

with a special need for an uninterrupted supply, hospitals, large industrial and commercial consumers and communities with a single source of supply.

Where *resilience* is defined as *the ability of assets, networks and systems to anticipate absorb adapt to and/or rapidly recover from a disruptive event* (Cabinet Office, 2011), the operational management of all critical mains would be included in a utility's emergency resilience planning to ensure supplies can be maintained to customers in the event of a mains failure. Alternative methods of supplying an area affected by failure of a critical main need to be considered; these may include duplication of the main, an alternative feed into the area, tankering potable water into affected areas or, but expensive, standby local treatment facilities.

Trunk mains generally operate at relatively constant flow rates up to the service reservoir, the storage being used to dampen diurnal demand changes over the day. There are day-to-day and seasonal variations in demand and trunk mains must be designed to carry the maximum day demand, typically in the range 120–150% of the average daily demand (Section 1.12). Design may also need to take account of the possible use of the pipe for short-term emergency water transfers, for example: re-routing supplies following a burst main or to supply a major fire (Section 16.4); and for transfers between zones in order to maintain a balanced storage in service reservoirs on a weekly or seasonal cycle of demand.

Trunk mains supplying a zone without local storage should be designed to transfer the peak hour demand to the distribution network. However, since these pipes would normally convey water to large areas, the peaking factor to be applied tends to be less than that for individual distribution mains because the *diversity factor* effect applies, which is related to the size of the area served (Section 1.12).

The diurnal demand in a distribution system can vary by a factor of 2.5–3.0 or more of the annual average daily demand. A typical hourly variation of demand over 24 hours is shown in Figure 20.1. Urban, industrial and rural distribution systems exhibit different diurnal demand patterns between weekdays, weekends, public holidays and religious festival periods. They can also vary from one year to the next.

15.2 SYSTEM LAYOUTS

The geographic and physical characteristics of an area influence the layout of a system. An interconnected looped layout which enables water to flow in multiple paths to any part of a network provides maximum flexibility. Urban systems tend to comprise looped (or *reticulated*) networks, albeit that the network is subdivided into hydraulically discrete areas for leakage and demand management (Chapter 16). *Dendritic* (tree like) layouts are more common for trunk mains and local distribution in rural areas.

The most economic layout for a system is one that is gravity fed from a local service reservoir located as near as technically feasible to the service area it supplies, as shown in Plate 24(a)(i). Because the storage is used to meet the peak demands for water, the greater the distance from the service reservoir to the distribution network, the longer the pipeline designed for peak hourly flow rates will need to be (Plate 24(a)(ii)), and hence the more costly is the pipeline. Storage close to the centre of demand can better maintain supplies under emergency conditions and for firefighting, help reduce pressure fluctuations in the distribution system and aid economic development of the system.

If the service reservoir cannot be sited close to the area to be supplied the preferred layout is to have at least two major supply pipes from the reservoir, connected together at their extremities to form a ring main through the distribution network (Plate 24(a)(iii)). This *redundancy* enables the network to operate more efficiently under peak flow conditions, provides system resilience under emergency operational conditions and allows uninterrupted supplies to consumers while one of the pipes is being maintained.

Boosting is a pumping arrangement which augments the pressure or quantity of water delivered through a system. The term is sometimes wrongly used to mean simple pumping. Although the term is used in a wide range of applications, the three most important purposes with a distribution system are: to provide a fixed extra flow; to provide a fixed extra pressure; and to maintain a given pressure, irrespective of the flow.

One of the most frequent uses of a booster is to increase the pressure of water in a distribution system at times of high demand. During low demand the pressure may be adequate, but when demand is high system pressures may be too low. Instead of laying additional feeder mains into the distribution area, it may be more economical to boost the pressure at times of high flow. Figure 15.1 shows the hydraulic gradients that might apply before and after boosting.

A consequence of raising pressure is that flow rates increase. Prevailing low pressures may have restricted the flow taken by consumers leading to *suppressed demand*, so that historic flow records will not indicate the potential demand when the pressure is raised. Therefore, when sizing the booster pumps the potential increased demand should be estimated. The design may also include a margin for future increases in demand.

Much of the complication of controlling a booster pumping station can be avoided if ***balancing storage*** can be built into the system at an appropriate location and elevation. Balancing storage generally improves the resilience of the system and can be used to optimize operating costs. With balancing storage connected to a system, pumps can be operated by the simple use of level switches in the tank. When peak draw-off occurs, the water level in the storage falls thereby initiating a pump to start. With progressive lowering of the water level, further level switches may bring

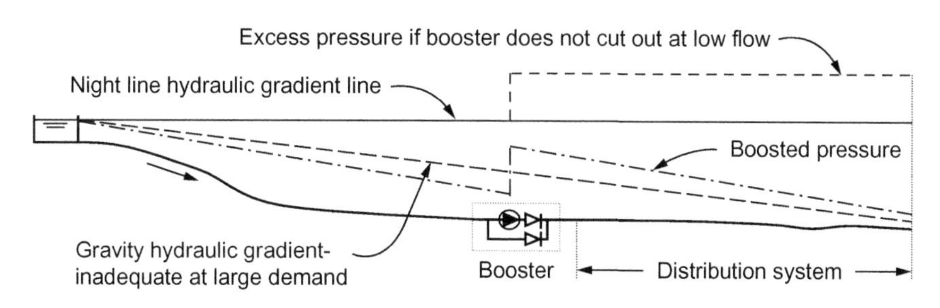

FIGURE 15.1

Effect of boosting pressure into a distribution system.

additional pumps into operation or increase the speed of the first pump. The benefits of this arrangement include:

- sudden large but short duration increases in draw-off can be handled by the balancing reservoir without causing the booster pumps to start;
- the maximum capacity required from the booster pumps is reduced;
- the head range against which the pumps have to work is reduced;
- pumping, once started, can continue until the balancing reservoir is refilled thereby increasing the load factor and efficiency of the pumps;
- control of the pumps is simple and positive and, with fixed-speed pumps, repeated stopping and starting of the pumps is much reduced. The output from variable speed pumps can be controlled to match demand, but at the cost of energy efficiency;
- balancing storage in an existing boosted system allows a range of options for extending the system for future network development.

The size of balancing tank or water tower required is often small because the periods of peak draw-off on a water distribution system are quite short. If a 5 Ml/d supply suffers from a peak draw-off 75% above normal for a period of 4 hours, the theoretical size of tank required to deliver an uninterrupted supply would be about 625 m^3, but a tank of one half or even one third of this size would greatly reduce the maximum duty required from the pumps.

A ***'rise-and-fall' main*** as illustrated in Plate 24(b)(i) can be adopted where the only possible location of a service reservoir is further away from the distribution system than is the source of water. If the rate of output from the source is equal to the average daily supply, water flows out of the service reservoir whenever the demand rate exceeds the average and flows into the reservoir when demand is less than the average. Alternatively, if the source output is made sufficiently large to fill the reservoir during part of a day, for example, during a day shift or to optimize the use of electricity tariffs by pumping only at night, when pumping ceases the distribution system is fed only from the service reservoir. Lower distribution pressure at times of low demand also helps to minimize night-time leakage. However, the design needs to ensure adequate turnover of water in the storage to prevent water quality problems arising, particularly during periods of low demand. Hydraulic models can be used to analyse this type of system using '*extended period simulation*' (quasi-dynamic) modelling over 24 hours (Section 15.15) to optimize the pumping and storage requirements and to ensure that the age of the water, particularly in the reservoir, does not become excessive and hence necessitate additional local disinfection facilities.

Elevated storage is often necessary on flat ground; Plate 24(b)(ii) shows a typical arrangement with a rise-and-fall main to an elevated storage tank. Alternatively, the tank pipework can be configured so that the pump delivers water to the top of the tank; with the distribution system being fed at all times by gravity. In either case a float-operated valve can prevent the tank from overflowing and water level sensors in it can control the number of pumps running or their output, to keep the supply roughly in step with demand. The advantage of the rise-and-fall main is that, during pumping, a higher pressure can be maintained in the distribution system than would be possible by gravity alone from the tank.

The same effect could be achieved without elevated storage by using pressure or flow measurements in the mains feeding the distribution system to control pump output. This is considerably more difficult to set up, to operate efficiently and to maintain and requires a thorough understanding of system demands and how they vary. Even a small tank is preferable because it permits simpler, water level, control of the pumps.

Elevated storage is more expensive to construct than ground storage, with costs escalating where the top water level rises above about 30 m high. Therefore it is seldom possible to provide the same amount of elevated storage as could be provided at ground level. Consequently, both ground-level and elevated storage may be provided for systems with flat topography, the ground-level storage providing the larger capacity. To ensure that the reserve supply in the ground-level tank can be used at all times, the transfer pumps filling the elevated storage usually have 50% standby capacity or more; an independent standby source of power, such as a diesel generator, may also be provided.

Many water companies prefer to pump directly into distribution. Although there are significant disadvantages with pumped distribution systems, as outlined above, the need to limit capital investment often outweighs the control advantages of elevated storage. It is true that control systems for direct pumping into a distribution system are becoming increasingly sophisticated, particularly with the increasing use of variable speed pumps, but they do require an appropriate level of technical maintenance which may not be available to water utilities with limited resources or in remote places where simple fail-safe systems are more reliable.

Historically, comparison of scheme options tended to focus on capital investment with limited attention to annual ongoing operational expenditure. However, with the rising cost of energy and with increasing awareness of environmental considerations, options are increasingly evaluated using whole-life cost–benefit analyses (WLCBAs; Section 2.10), taking into account current and projected future costs of energy and carbon costs.

15.3 PIPELINE AND SYSTEM PLANNING

New pipelines and distribution systems are required either to meet forecast increased demand in an existing supply area and from proposed developments adjacent to the area or to replace an existing pipe for structural, capacity or water quality reasons. The planning and design processes will size the pipe and determine its horizontal and vertical alignment within imposed physical constraints and operational conditions. However, a utility often has to plan a new pipeline or system not knowing precisely where the additional demand will be. Therefore, it is necessary to find a strategy that best covers a range of feasible medium and long-term development options while maintaining flexibility to accommodate changes to the anticipated development sequence.

The design process must include decisions as to which source should supply each area, which route is practical for new mains and how peak day demands are to be met when some sources have spare capacity and others have not. Oversized mains may be advisable in some areas to cater for possible additional future demand and ring mains may be needed to provide system resilience and to transfer water across the distribution system under different phases of development.

Design Considerations

Operational requirements and design standards should be determined to establish: minimum supply pressure requirements; fire risks, critical mains and system failure scenarios; a stable '*calm*' network by minimizing diurnal pressure variations; pressure management zones and areas (PMA) and district meter areas (DMA); consistent water quality by means of 'discrete' zones or areas, fed only from a single source.

The layout and capacity of the network should take account of demand and growth scenarios, including understanding and assessment of: current domestic and non-domestic consumption, characteristics of consumer categories and diurnal and seasonal profiles; existing population disposition and water related characteristics; the planning horizon and phasing opportunities; proposed and planned forecasts of growth in housing, industry and commerce by location and consumer type. Growth may occur both within the existing network and in planned new development areas adjacent to the existing system; where uncontrolled developments might occur.

In addition, the layout of an extension to a system should take account of: the capacity of existing system, assessed using hydraulic modelling techniques; alternative supply regimes and operating modes for both the existing system and the ultimate forecast demand and development; the optimum engineered or most physically practical alignments for strategic pipelines; use of ring mains or integrated networks of pipes to support supply flexibility; the resilience to support supply in drought and emergency conditions; and avoidance of the need to duplicate mains back to a source at a later date, unless:

- the current scheme includes provision for additional land, cross connections and the infrastructure for its future control, or
- there is uncertainty of the timing and/or location of the future development.

Aspects of the existing and future sources of supply need to be considered, including: peak outputs from existing sources and differences in their production costs; availability of potential sources versus seasonal and diurnal variation; possible locations and capacities for future sources; locations and capacities of existing strategic and local service reservoirs and pumping stations; and land that may be available for additional storage and booster pumps.

The location of additional storage depends on how each new area is supplied, where there are operational storage shortfalls and the topography of the surrounding area to allow the storage to be located at a sufficiently high level. However, if a ring main is developed, the location of storage is less critical because the integrated network should provide the hydraulic transfer capacity to manage the operational and development risks.

The design of a critical main must take account of how supplies are to be maintained if the main were to fail. Options can include duplicating the whole pipe or sections of it, strategic links into different parts of the system for supply diversion and additional valves and cross connections to allow local isolation and re-routing of supplies.

System Development Strategies

Three options for development strategy are:

1. *Use the existing network.* Provided there is sufficient hydraulic capacity, extend the existing networks into the new development areas, progressively creating PMAs and DMAs (Section 16.3) with the phased development. This would require minimal additional pipework, but does increase the number of consumers at risk of interruption to supply due to pipe failure.

2. *Create discrete supply areas for the new development.* Supply individual development areas directly from either a connection off, or an extension to, an existing trunk main or from a new main from an existing or new service reservoir located near to the development area.
3. *Lay a new trunk main through the whole of the proposed development area.* Create a ring main to improve hydraulic capacity, operational flexibility and reduce failure risk. However, this could incur both capital expenditure too early and the loss of financial contribution from developers of areas developed in the longer term.

Whatever strategy is adopted it would be beneficial to phase the pipe laying work to suit the building programme. However, this is unlikely to be achievable because the building programme is unlikely to suit logical order of network development.

The optimum layout will be the whole-life least-cost solution allowing for capital and operating costs including energy cost escalation, environmental and social costs and benefits. Alternatively, it will be a compromise between meeting known short-term demands and allowing for growth and hence development uncertainty in the later phased development areas. Greatest flexibility can be achieved where pipe laying can be delayed into those areas with the greatest uncertainty with respect to population and demand and location and phasing of new development.

Option Development and Evaluation

The design should achieve the optimum combination of pipe sizes and alignments to satisfy the various demand conditions. The first step is to identify route options using desk studies of maps, GIS (geographic information system), records of physical obstructions including the locations of other utilities and site visits to assess route viability and construction constraints. The studies should include reviewing archaeological, historic and environmental factors to minimize the disturbance of areas of special interest.

The analysis involves a logical assessment of all viable options for which there is either a clearly identified preferred solution that satisfies known current constraints or a compromise of the options that best fits with short- and medium-term uncertainties. The solution is invariably a compromise between financial, operational and third party requirements and constraints.

Options for both whole systems and individual trunk mains (to optimize pipe size and alignment) are compared using financial WLCBA. The selected scheme will exhibit a relatively high level of confidence for the early years where the basis of the development is founded on more reliable data. However, if the proposed distribution network (and other service assets) is fixed at the outset, the installed assets may not have the optimum capacity and layout for the ultimate final development. The WLCBA should therefore include a *Regret Cost* (Section 2.10). The suitability of parts of the system development to be executed later will be less certain; therefore, the development plan should include flexibility, and may even include identified options, to be able to manage divergence from the most likely scenario and to take advantage of the opportunities that may be presented.

Designing a new distribution network is relatively straightforward beyond the need to understand the building sequence of the development to be serviced by the new assets. The challenge is to ensure that domestic and non-domestic development can be supplied as soon as individual buildings become occupied, but without installing surplus assets that could become redundant if part of the development were to be reconfigured or cancelled. This is unlikely to be an issue for a small area being developed by a single developer, but is more likely to be relevant to large multiple phased developments extending over years.

Extending and enlarging an existing system to cater for a new development can be more problematic because the current capacity, forecast growth within the existing serviced area, existing system operational constraints, proposed network and source works replacement, enlargements and planned maintenance all need to be allowed for in addition to the new development area.

15.4 DISTRIBUTION SYSTEM CHARACTERISTICS

The performance requirements of a distribution system can be summarized as:

1. *Minimum pressures at peak hour demand* should be sufficient to serve the highest supply point in the network. Typically a mains pressure of not less than 15–20 m would be required to serve buildings up to three storeys high. Higher pressures may be necessary in some areas where there are significant numbers of dwellings exceeding three-storey height; but high-rise buildings are normally required to have their own boosted supply.
2. *Maximum pressures approaching static conditions during low demand periods*, typically at night when flow and velocities are low, should be as low as practicable to minimize leakage. For flat areas a maximum pressure during periods of low demand in the range 30–45 m is desirable. Pressure reducing valves can be installed to reduce high operating pressures but the valves must be regularly maintained if they are to operate reliably. Nevertheless, the pipework in the reduced pressure zone must be rated to resist the maximum possible static pressure if the valve fails or has to be bypassed for operational reasons. Higher static pressures may be unavoidable where undulating topography limits the opportunities for sub-dividing the network into discrete pressure zones, particularly where ground levels vary significantly over short distances.
3. *Surge pressures*. Positive pressures of at least 5 m head must be maintained at all points in the system particularly under the transient pressure conditions that may occur when a pump stops or pumping system shuts down on a power failure. The positive pressure will ensure that air valves cannot open allowing contaminated water to enter the pipe and will maintain the integrity of flexible joints designed to operate under positive pipe pressures. Generally it is advisable to have a surge analysis carried if there is any doubt and to install surge control equipment if the analysis indicates a problem.
4. *Fire demand* (Section 16.4). In the UK, fire demand requirements range from 8 l/s from a single hydrant in a one- or two-storey housing development to up to 75 l/s from one or more hydrants serving an industrial estate. Owners of properties requiring supplies for sprinklers or hydrants on their premises may need to enter special arrangements with the water supplier. In the USA higher fire flows are required.

5. *Consistent water quality* must be maintained throughout the system by establishing discrete hydraulic and source water quality areas. Wherever possible dead ends, long retention times, mixing of different waters within the distribution system, diurnal reversals of flow direction in mains and exceptionally high flow rates should all be avoided as they can cause water quality deterioration that may result in 'supplying water unfit for consumption'; a non-compliance of the water quality regulations in the UK.
6. *Spare flow capacity* must exist in the system sufficient to meet foreseeable rises in demand over the next few years and to provide resilience to minimize the number of consumers affected by interruptions to supply.

Lower supply pressures of about 10 m head are desirable where only standpipes are to be supplied since it allows use of low pressure, easily operable standpipe taps which give less maintenance problems and less wastage. However, standpipe-only systems are rare and it is not practicable to supply water over large areas at very low pressure. Further design considerations determined by the level of service standards provided by a water utility to its customers are discussed in Chapter 16.

Table 15.1 illustrates the proportions of pipes laid in different sizes in urban and sub-urban environments in regions of the world. The proportion of pipes over 450 mm size reflects the distance between sources or reservoirs and the centres of demand of the area served and hence the proportion of trunk mains in the system. The table indicates that pipes used for distribution are typically in the range 100–250 mm diameter and that these comprise the majority of pipes from which properties are served. Older systems often have a significant proportion of small mains, 80 mm and below. Smaller pipes can also be used as spur or rider mains feeding small groups of properties. Larger diameter critical distribution pipes are used to convey water to an area for further distribution. Critical mains vary in diameter according to the size of area served and some pipes may supply properties en route; for example, a critical main in a rural area may be as small as 100 or 150 mm diameter, whereas in most urban areas the minimum size of these feeder pipes may be 200 mm or larger.

Table 15.1 Representative percentages of mains by diameter

Diameter (mm) **Diameter (inches)**	**50–80** **2–3**	**100–150** **4–6**	**200–225** **8–9**	**250–350** **10–14**	**375–450** **15–18**	**500–600** **20–24**	**>600** **>24**
UK: city/urban	4	73	10	6	2	1	4
UK: urban/rural	26	50	8	6	4	3	3
SEA: city/urban	1	50	21	23	<1	3	2
SEA: urban/rural	8	63	8	3	14		4
USA: city/urban	1	42	24	16	4	4	9
USA: small urban		44	30		26		

Notes: SEA = South East Asia. *1.* 225 mm (9 in) and 375 mm (13 in) diameter pipes are no longer in production, but many were laid and are still operational. *2.* 175 mm (7 in) mains are also found in some older systems.

The cost and difficulty of making a service connection increases with the size of the pipe and, in general, service connections are not installed on 300 mm diameter pipes and larger. A 300 mm main is of such large capacity that it will normally be used to convey water in bulk to an area for further distribution. Ideally service connections should therefore be restricted to the smaller diameter pipes. In areas where there are large industrial consumers, or where the per capita consumption for domestic purposes is exceptionally high, some distribution pipework may be 400–450 mm diameter, however connections off larger mains are more typically created using pipe fittings rather that tapping the main.

15.5 DESIGNING TRUNK MAINS

There are few occasions when the design of a completely new distribution system is required, hence the majority of hydraulic studies and designs involve analysing the performance of an existing system with the purpose of improving its performance to meet demands from new developments and increased consumption from existing consumers.

The layout must form a sensible hierarchy, in which the larger mains feed the smaller ones. As well as carrying out a whole-life cost optimization as discussed in Section 15.3, the designer may need to take into account a water utility's preferred range of pipe diameters and materials, usually adopted to limit the quantity of spares that need to be held in store for emergencies. Care needs to be taken also to ensure the chosen network of preferred pipe diameters does not restrict potential growth or create unacceptable velocities, high or low. Pipe velocities at high demands should be limited to keep friction headlosses within acceptable limits. However, pipe sizes should also be chosen to achieve minimum velocities at least once in each diurnal cycle so that the age of the water in the pipes does not become excessive or that loose deposits do not build up locally to the extent that they could cause a dirty water incident if disturbed (Section 15.6).

Where a network is open and highly interconnected, there are opportunities to optimize the layout to maintain acceptable operating conditions under emergency flows while reducing retention times and minimizing pressure variations. However, the layout should also recognize that network operators need to manage the network efficiently by being able to monitor and control pressures and leakage and by operating stable hydraulic supply areas to support consistent water quality; the concept of a *calm network*.

The factors to be considered when designing trunk mains include:

- the optimum engineering or most physically practical route;
- proposed mode of operating the system; available resources versus seasonal and diurnal variations;
- location and capacity of sources and storage;
- provision of system flexibility without creating diurnal reversal of flow in mains;
- capital versus operating cost;
- confidence in demand projections.

The flow in a pipe is determined from the summation of demands that it feeds. The network demand is summed from its individual components working upstream from the extremities of the system towards the sources. The demand assessment process is similar for both trunk mains and distribution pipework except that the demands on trunk mains are the summations of the demands from the individual pipes they supply, not the individual customers.

15.6 DESIGNING DISTRIBUTION PIPEWORK

The design of new local distribution pipework is largely empirical. A pipe must be laid in every street along which there are properties requiring a supply. Pipes most frequently used for local distribution are 100 or 150 mm diameters (Section 15.4). However, distribution pipes tend to be oversized having been designed for fire flows, future growth, to provide operational supply flexibility and where a water company operates a policy to standardize pipe diameters. When replacing or rehabilitating pipes, there is therefore often an opportunity to reduce the size of the pipes while still achieving the required hydraulic capacity. In the Netherlands, following agreement of relevant stakeholders on reduced fire flow capacity (to 30 m^3/h), pipe design criteria were revised. Pipes are sized to achieve a daily 'self-cleansing' peak velocity of at least 0.4 m/s (van den Boomen, 1999) in order to prevent sediment accumulating. This approach produces new networks comprising of 40 and 63 mm diameter pipes feeding domestic consumers in local networks and dead ends where previously the minimum size pipe would have been 100 or 150 mm diameter to cater for the fire flows previously adopted.

The diameter of local distribution pipes depends primarily on the population density of the area to be served and how the pipe is supplied. Although guidelines can be developed for sizing new pipes when extending an existing network, the designer needs to understand the site specific conditions and performance of the system.

At any time, t, of the day, the demand from an area comprises:

$$\text{Average day domestic demand} \times D_{t1} + \text{Average day non-domestic demands} \times D_{t2} + \text{Allowance for leakage at time } t$$

where D_{t1} and D_{t2} are the diurnal factors applying to the domestic and non-domestic demands at time t and the leakage allowance varies with the system pressure at time t.

Peak hour demand occurs at different times for different types of demand in different parts of the system. Therefore, the demand characteristics need to be developed from field measurements of flow and pressure in the area. Alternatively a system demand profile can be derived from its constituent parts using records of typical demand, patterns of usage and recorded leakage, making assumptions only where better information is not available. A similar approach can be applied for deriving seasonal variations and to assess demands in trunk main systems.

Where dwelling numbers and occupancy ratios cannot be assessed readily from utility records and census data, figures for the average number of people per hectare or per kilometre of mains may be used, but with caution. Table 15.2 provides some guidance on population density and average numbers of connections per km of main for large utilities in regions of the world; the data include both domestic and trade connections. The number of people served by each pipe in a street is influenced both by the type and arrangement of the dwelling units and their occupancy ratio, and by the characteristics and constraints of the local water supply and distribution facilities. In densely populated low-income areas typically found in the developing world a large proportion of the population may be served by street standpipes. High-rise buildings result in high population densities, but each building may be fed by one or two large connections.

Sizing distribution mains should take account of the fire risk along the pipe alignment and hence the minimum fire demand associated with the risk. Fire flows are discussed further in Section 16.4. The full available head at a hydrant can be utilized to meet fire demands. A single fire hydrant flow of 2 m^3/min can be relatively easily delivered by a 150 mm main fed from both ends; the velocity in the main being 0.9 m/s with a rate of loss in the main of about 12 m/km. If only fed from one end, the velocity increases to 1.9 m/s and the headloss becomes 46 m/km which may be acceptable in urban areas if the lengths of pipe between cross connections are relatively short. Further away from the hydrant being used, the number of mains contributing to the flow increases so that the headloss in them is much less. The same flow in a 100 mm pipe fed from both ends would produce both high velocity (2.1 m/s) and headloss (100 m/km).

Table 15.2 Population served per kilometre of mains laid

		People	Connections
	People per household	Average number per km of main	
Densely populated low-income areas	7–9	3000–4000	300–550
Planned high occupancy dwellings	5–8	1000–2000	100–250
Residential urban areas:			
• Asia	4–7	400–800	70–100
• Europe and Australia	2–4	250–350	100–130
• North America	2–4	200–400	80–120
Medium to low density housing:			
• Asia	3–8	200	20–75
• Europe and Australia	2–5	100–200	50–80
• North America	3–5	20–200	5–65
Sub-urban areas with gardens	3–5	130–200	30–50
Populated rural areas with villages	1–4	110–160	10–50
Dormitory accommodation	n/a	750	n/a

15.7 HYDRAULIC DESIGN OF PIPELINES

Figure 15.2 illustrates the elements of a simple pumping system. It comprises a source tank or sump, a length of pipeline to the pumping station (the *suction main*), a pump and non-return valve (essential to prevent reverse flow when the pump is turned off) and a delivery pipeline discharging into a tank.

The *static lift* is the height, or equivalent head of water, between the water level in the source tank and that in the delivery tank or the level of the tank inlet if higher. This is the head across the pump (or, more correctly, the non-return valve) when there is no flow. The static lift does not depend on the level of the pump; only the difference between the level in the source tank or sump and the delivery level. In the case of an inline booster pumping station, for example, when pumping to a small demand area from an upstream network, the equivalent *sump level* will be the residual energy level from the upstream system; this level may vary depending on the demands and flows in that system. To deliver any flow the pump must overcome the static lift and the *pipe friction and other losses, ('fittings' or 'local' losses)*, in both the suction and delivery pipelines (Chapter 14). *Station losses*, the losses in the pipework within the pumping station itself, which would normally be the responsibility of the supplier of the pumps and station pipework, are assumed here to be included in losses in the suction and delivery pipelines.

In a pipeline of constant diameter and roughness the rate of loss of energy along the pipeline is constant. Thus, ignoring local losses, the *hydraulic grade* (or total energy) line can be represented by a straight line at constant gradient sloping in the downstream direction. This line represents the change in total energy level of the flow along the pipeline. Local or fittings losses occur wherever there is a change in geometry of the pipeline at bends, changes of diameter, valves and other fittings. Strictly they should be included as abrupt steps in this total energy line where they occur but they are often lumped in with the pipe friction losses and assumed to be spread over the length of the line. Typically in long relatively straight pipes local losses are likely to be less than 5% of the total loss. For initial design such a percentage allowance for local losses may be accurate enough but for detailed design each fitting or abrupt bend should, if practicable, be identified and the total losses more accurately assessed. Table 14.2 gives some guidance on the losses at pipeline fittings.

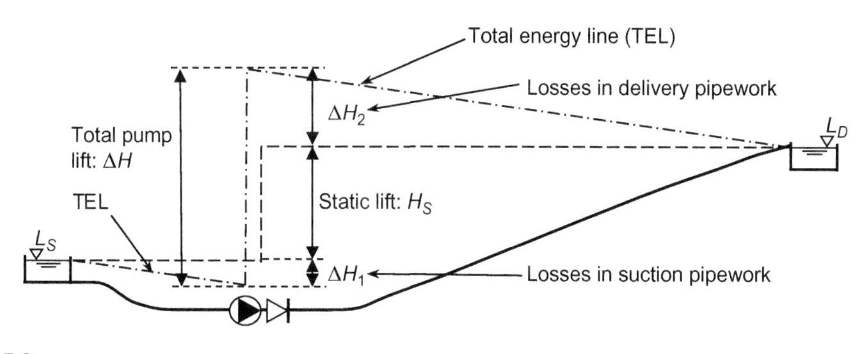

FIGURE 15.2

Definition of terms in pumping systems.

More extensive data is available (Miller, 1990). In systems where local losses dominate, for example in a short length of convoluted pipework within a treatment works, all fittings losses should be identified and the hydraulic grade line correctly drawn on a long section with the appropriate step for each local loss.

At the upstream end of the simple system illustrated in Figure 15.2 the level in the source tank or sump represents the energy level at that point. The hydraulic grade line illustrates the loss of energy along the pipeline and shows the grade line dropping below the static level on the suction side of the pump. Thus the maximum steady-state pressures in the suction pipework occur under static conditions when the pump is turned off. On the delivery side of the pump the static pressure is represented by the delivery level at the receiving tank. The hydraulic grade line slopes down to that final level and lies above the static pressure level. Maximum pressures in the delivery main thus occur when the pumps are working.

Energy is needed to be put into the system to overcome the static lift and energy losses. This is provided by the pump. The total *pump lift* (or *pump head*) is thus the sum of the static lift (H_S) and the headlosses in both the suction pipework and the delivery system, ΔH_1 and ΔH_2, respectively.

$$\text{Total pump lift (pump head)} = \Delta H = H_S + \Delta H_1 + \Delta H_2$$

The *pressure* at any point along the pipeline is the height of the hydraulic grade line above the pipe centre line. Strictly, the velocity head should be deducted from this value but in most pipeline systems this is small in comparison with the pressure and can be ignored. Only if the velocity is high or the pressure low would the velocity head component of the pressure become significant. It should be noted that the pressure is not necessarily greatest at the upstream end of a pipeline as it depends on the profile of the pipeline. Where it crosses a valley the highest pressure might occur at the lowest point of the profile. It is important to plot the profile of the main and to superimpose the hydraulic gradients to check the maximum steady-state pressures at all points in the system.

15.8 SYSTEM CURVES

A *system curve* is the relationship between the flow delivered into a system and the head required to deliver that flow. Figure 15.3 illustrates such a curve. It is made up of the static lift and the losses in the suction and delivery pipework, plotted as a function of the flow. As shown in Chapter 14, both the pipe friction and fittings headlosses are approximately proportional to the square of the velocity and hence, in a pipeline of known size, to the square of the flow rate. Thus the system curve is approximately parabolic.

For a design flow, Q_D, the required pump lift is H_D as illustrated. This is known as the *pump duty point*. However, a single system curve is rarely enough to define the range of the possible pump operation. The static lift varies with the levels in the source and discharge tanks. Furthermore the friction losses may change with time so it is necessary to consider the conditions when the pipeline is new and when it has been in service for many years. In addition, the fittings losses quoted

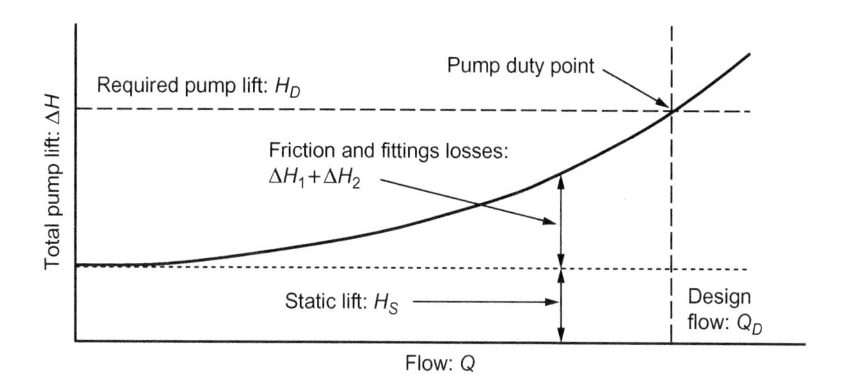

FIGURE 15.3

Simple system curve.

for design (Table 14.2) include a safety factor. If the fittings represent a significant proportion of the total losses then it may be necessary to consider the possible range of these losses with reference to more accurate, minimum values for these losses (Miller, 1990).

Thus, when considering a real system a range of system curves can be drawn. For the simple pipeline system considered in Figure 15.2, the envelope of possible pumping conditions can be drawn with two curves with the following parameters:

- maximum static lift (low sump level/high delivery tank level), maximum roughness and maximum fittings losses;
- minimum static lift (high sump level, low delivery tank level), minimum roughness and fittings losses.

Figure 15.4 shows this range of system curves with indicative pump characteristic curves added. For any flow, there is a range of pump lifts required: at the design flow, Q_D, the maximum pump lift under the most severe conditions is given by Point A. A fixed-speed pump selected for this duty would be delivering a flow greater than Q_D under all but the most severe conditions. Depending on the static lift and pipeline roughness, the pump would be operating along its characteristic curve somewhere between Points A and A′. Conversely, a pump selected for duty on Point B would in all but the most favourable conditions provide less than the design flow, operating between points B′ and B.

A more appropriate duty point might be at Point C. This represents the pump lift needed for the design flow at average static lift and maximum pipeline roughness. If the pump was sized for this duty point, it would operate between points C″ and C′. Over part of this range the pump would deliver less than the design flow and over part of the range more than the design flow. Provided the average flow rate is equal to or more than the design requirement then the pump should be adequately sized. By considering the average static lift and the possible long-term deterioration of the pipe roughness adequate performance should be ensured. However, if it is necessary to ensure the instantaneous design flow under all conditions then point A would be the appropriate pump duty.

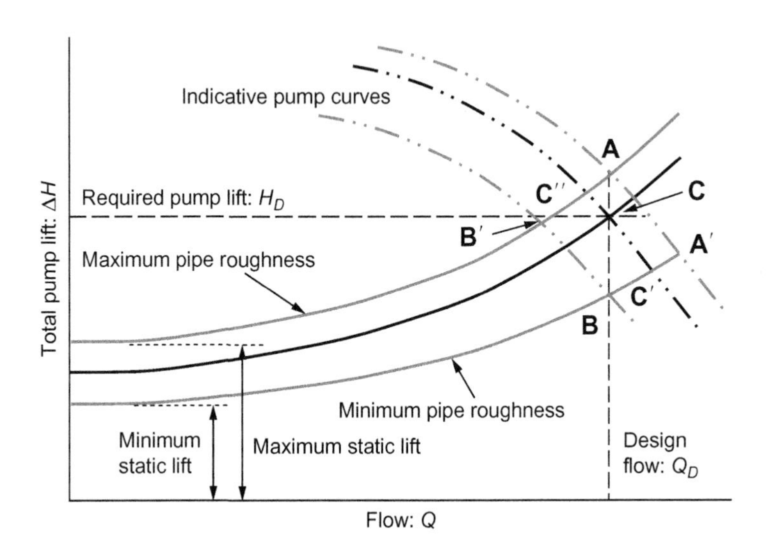

FIGURE 15.4

Maximum and minimum system curves and range of pump operation.

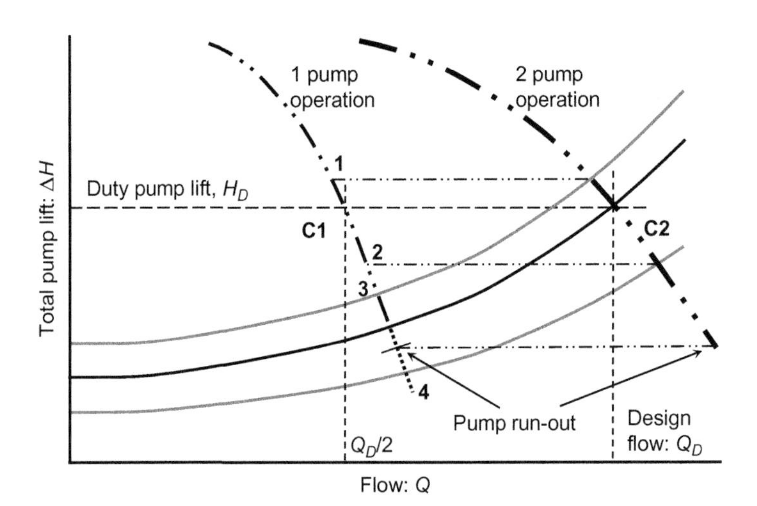

FIGURE 15.5

Multiple pump operation.

The need to consider the full range of possible pump operation is further illustrated in Figure 15.5. This shows the same system curves as in Figure 15.4 with a duty based on Point C but met by two pumps working in parallel. It is common for two or more pumps to be provided to meet a design duty. A further complication is pump replacement at the end of its design life; the pump characteristics required then are likely to be different to those required now.

In Figure 15.5 the duty point for operation of the station has been chosen as Point C2. The duty point for sizing the individual pumps is given by point C1: the same head requirement but with half the total flow through each of the parallel pumps. With two pumps operating, each pump may operate over the range between Points 1 and 2 but, with a single pump operating, the range of duties is between Points 3 and 4. It is essential to plot the range of system curves and proposed pump characteristics to check that the pumps can be allowed to operate over the whole range with respect to *Net Positive Suction Head* (NPSH) requirements (Section 19.4). As drawn in Figure 15.5, Point 4 is beyond the run-out point of the pump and this would be unacceptable. It would require either consideration of different pump units or the inclusion of throttling to raise the system losses when one pump is operating. Although this illustration is for two fixed-speed pumps, the same considerations apply to any multiple pump installation whether with fixed- or variable speed drives; although, in the latter case, reducing pump speed can allow it to operate at a more acceptable point on its characteristic. The full range of potential operating conditions must be checked against the pump characteristics.

A pump in a single pump installation operates at all times near to its duty point so it is logical to choose a pump with its best efficiency point close to that duty. In a multi-pump station there is a much wider range of operating conditions for any one pump. If, for example, the more common mode of operation is actually with only one pump running then it may be sensible to choose a pump with its best efficiency at such a duty. In the two-pump case illustrated in Figure 15.5 this might be a duty between Points 3 and 4.

15.9 LONGITUDINAL PROFILE

The starting point for any pipeline design is the longitudinal profile of the system (Section 17.5). It is essential that this is drawn out using topographic survey data or mapping – more commonly from the spreadsheet or hydraulic analysis program but if necessary by hand – with the hydraulic grade lines superimposed. Three example profiles of a pipeline with the same diameter, horizontal length and suction and delivery levels, L_S and L_D, are shown in Figure 15.6. Figure 15.6(a) illustrates the case where the hydraulic grade line is above the pipe profile and the static lift is the same for all flows up to the design discharge. The system curve is shown as the chain-dot line in Figure 15.7.

The profile for the system illustrated in Figure 15.6(b) however, has a summit on route, higher than the delivery point. At low flows this becomes the delivery point as 'seen' by the pumps. The static lift is higher but the effective length of the pipeline is reduced and the friction losses are smaller. Downstream of the summit, assuming that there is an air valve at that high point, the pipeline runs part full to the point at which the head is just sufficient to drive the flow through the full pipe. Note that a large-orifice air valve (Section 18.24) would almost certainly be located at the summit. Above some flow, when the pipe full losses downstream of the summit equal the available head ($L_H - L_D$) the pipeline runs full over its whole length and the system curve is the same as for Case (a). The full system curve for Case (b) is shown as the solid line in Figure 15.7.

The profile shown in Figure 15.6(c) has an even higher summit upstream of the delivery tank and even at maximum flow the controlling delivery point is at that summit with part-full flow

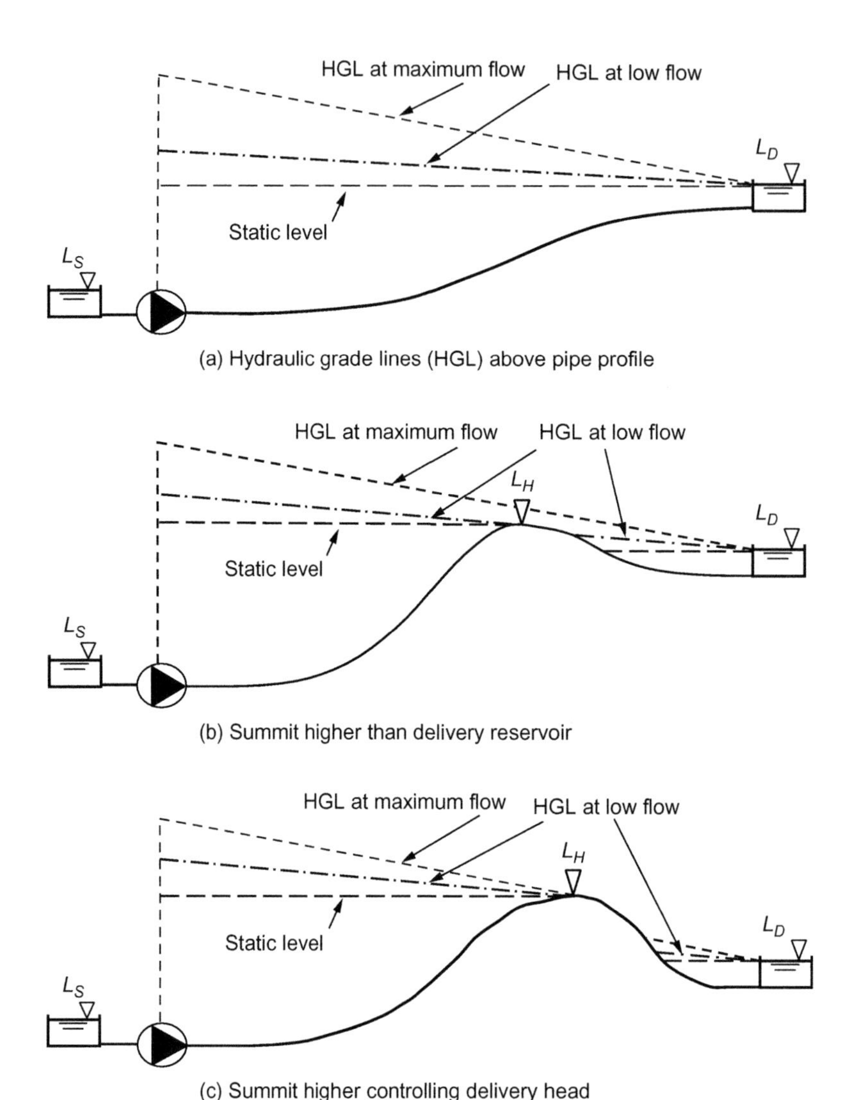

FIGURE 15.6

Effect of high point on pumped system.

downstream. The system curve must be based on the higher static lift and shorter pipeline and is shown as the chain-double dotted line in Figure 15.7. The pump requires a higher lift to meet the required duty.

The profile of the pipeline may thus affect the hydraulics and choice of pump significantly and it is essential that the long section and hydraulic grade lines are drawn. Similar issues can arise on a gravity system such as illustrated in Figure 15.8.

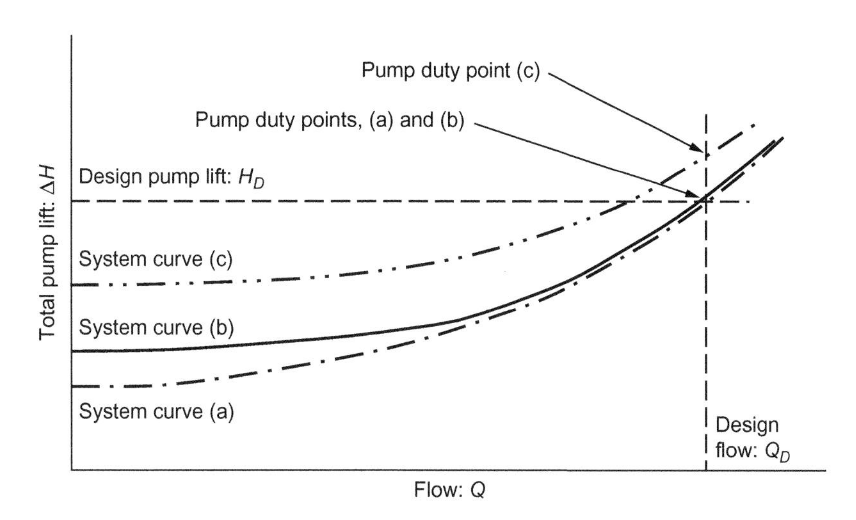

FIGURE 15.7

System curves for pipeline illustrated in Figure 15.6.

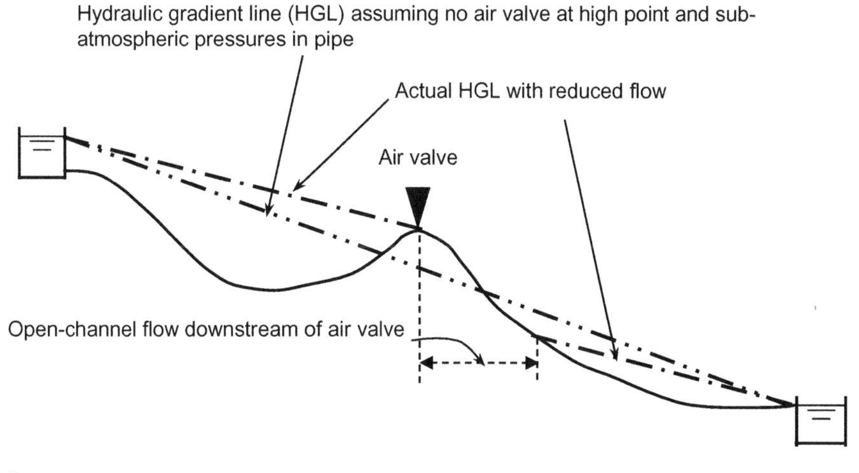

FIGURE 15.8

Effect of high point on gravity system.

In a raw water system the conditions illustrated in Figure 15.6(c) might be allowed but in a potable water system, part-full flow and open air valves are normally not acceptable as the latter could form entry points for contamination. Thus, even Figure 15.6(b) would be an unacceptable design as at low flows or when the pumps are turned off contamination could enter the system. If there is a risk of part-full flow in a potable water pipeline then means of maintaining pressurized flow and preventing air valve opening must be considered:

- re-routing the line to avoid the high point;
- including a pressure-sustaining valve at the discharge end of the line;
- providing a break tank at the summit with a control valve at the discharge operating in response to the water level in this tank; for a potable water system the ventilation would have to be designed to eliminate the risk of infiltration or airborne contamination (Section 20.9).

15.10 AIR IN PIPES

Air can be introduced into a pipeline during operational activities and as air dissolved in the water is released. Sources include air left after pipeline filling, poorly designed pipe and pump inlets with insufficient submergence that allow large amounts of air to be sucked in via vortices (Sections 19.13 and 20.26), air valves installed near the hydraulic profile that allow air to enter the pipe under certain conditions (Section 15.9) or locations where dissolved air comes out of solution, typically as internal pressure reduces along a pumped system.

The amount of air absorbed in water varies with water temperature and is directly proportional to pressure at a given temperature (Henry's Law). Water saturated with air at a given pressure and temperature releases some in order to reach equilibrium, if the pressure reduces or the temperature rises, so forming pockets. Table 15.3 shows the relationship between temperature and maximum air content in water at Standard Pressure (1013 mb). However, the amount of dissolved air cannot normally be greater than the amount introduced when the water was last exposed to air and may be less if the water has previously been subjected to higher temperature or lower pressure. In normal operating conditions the pressure in a pipeline is above atmospheric. Therefore, as a general rule, provided temperature remains constant, air does not come out of solution unless the pressure falls below atmospheric. There are three exceptions: where air is deliberately introduced at high pressure in a treatment process, for example, dissolved air flotation (Section 8.18); where surge vessels with air–water contact are used for surge suppression (Section 15.11); and where air trapped at joints or in the pipe lining gets absorbed as the pressure rises in test or operation. The first case is only relevant to pipework in a treatment works; the third may occur in any pipeline but should not represent large volumes of air. Likewise the effect of temperature rise is usually small and would only lead to air coming out of solution if pressure is near atmospheric.

Table 15.3 Temperature effect on maximum air content in water (at standard pressure)

Temperature °C	0	5	10	15	20	25	30	40	50
Maximum content (volume) %	2.78	2.53	2.53	2.12	1.95	1.80	1.66	1.43	1.25

A well-designed system should exclude all air on filling, prevent entry of air in operation and provide air release points. It is usual to maintain pressures at least 5 m above atmospheric at high points to ensure that air valves do not leak and, in potable water systems, under all operational conditions including surge and transient events, to eliminate the risk of contaminated water being drawn in valves particularly if the valve is located in a chamber below ground in areas of high water table. However, air must be allowed to enter a pipeline in a controlled way and at predetermined locations to enable the pipe to be emptied and to limit sub-atmospheric pressures and thereby prevent pipe buckling in the event of a sudden drop in pressure such as might occur on a pipe burst.

On pipeline filling air is exhausted at large-orifice air valves (Section 18.24) where the pipe is not yet full. Figure 15.9 illustrates where air can collect at high points and at changes of gradient in a pipeline in operation. Trapped air may get carried forward and accumulate at high points or may get absorbed into solution as the pressure rises. Air trapped at high points in a pipeline increases hydraulic losses with the accumulative effect of a series of high points along a pipeline profile seriously restricting flow. An air lock in domestic plumbing is an extreme example where water flow capacity is reduced to zero. Air expelled at the pipe discharge may also cause problems.

Air bubbles tend to rise to the soffit of the pipe where they may accumulate into larger pockets. Evidence from tests carried out at HR Wallingford (Escarameia, 2005a) confirms that air is transported along a pipeline mostly in the form of air pockets moving along the soffit of the pipe rather than as individual small bubbles. The air is moved forward in the direction of flow in an upward sloping pipeline but can also move forward along level pipes and along downward sloping pipes if the fluid velocity is high enough. However, if air is transported forwards in a pipe at shallow downward gradient it may not be transported forwards if the pipe gradient increases. Thus air can accumulate both at high points in a system and at 'knees' in downward sloping pipes. Air may move backwards against the flow in downward sloping pipelines operating at low velocity or at steep gradient, but there is a range of flow velocities below the *critical velocity* (V_C) where air bubbles and pockets may 'hover' with no clear movement in either direction. This '*hovering velocity*' is defined as the flow velocity at which air pockets neither move upstream or downstream. Even if the air does move in one or other direction it may be very slow, for example, when filling a 2 km length of main, although the flow from a downstream hydrant may soon be steady, it can be interrupted by occasional bursts of air for perhaps half an hour or more.

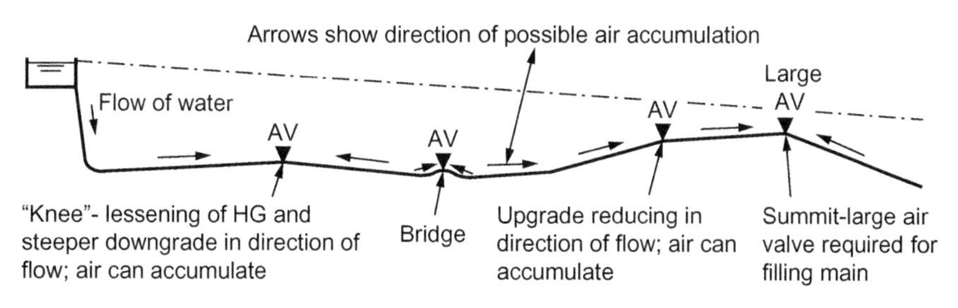

FIGURE 15.9

Locations of air accumulation on pipeline.

The HR Wallingford design manual for air problems in pipelines (Escarameia, 2005b) indicates that the critical velocity (V_C) needed to move air pockets forward in a downward sloping pipe is given by:

$$V_C/(gD)^{0.5} = a + 0.56(\sin\theta)^{0.5} \tag{15.1}$$

where a is a coefficient depending on the air pocket size represented by the parameter $n =$ volume of air pocket$/(\pi D^3/4)$; a ranging from 0.45 for $n < 0.06$–0.61 for n between 0.30 and 2.0; θ is the slope of the pipe from the horizontal and D is pipe internal diameter. Tests carried out to define the relationship in Equation (15.1) were performed in a 200 mm diameter pipe but guidance in the HR Wallingford manual suggests that the relationship is valid for pipe diameters up to about 1000 mm and slopes between 0 and 22.5 degrees. CFD modelling (Little, 2008) confirms this and suggests that the relationship generally remains valid for large diameter pipes although it is probable that a small bead of air remains in the soffit of larger pipes, particularly at undulations in pipe profile, even though pockets of air are moved forward. The tests at Wallingford show that in pipes laid horizontal or at near horizontal downward slopes the air pockets are long and thin and do not occupy a large proportion of the pipe, typically less than about 5%. This can be significant for behaviour of pipe bursts which may be more explosive than expected; may affect surge pressures under certain conditions and needs to be taken into account when designing submarine pipelines.

The HR Wallingford tests also showed that the hovering velocity was between about 0.85 and $0.95V_C$ with a recommended mean value of 0.9. Designers should try and avoid designing sections of pipeline where the mean velocity at the range of flows to be expected in normal operation is between say 0.85 and $1.0V_C$.

15.11 TRANSIENT PRESSURES: WATER HAMMER AND SURGE

In this context transient pressures are those generated in a pipeline or system by a sudden change in flow such as generated by the operation of a valve or, more commonly, the starting and stopping of pumps. The time scale of such events is likely to be measured in seconds, distinguishing them from the much more gradual flow changes that occur in a system as a result, for instance, of diurnal demand patterns.

There are two forms of transient pressures: the generation and transmission of pressure waves in a closed pipe system due to the compression and elasticity of the fluid, and mass oscillation where the whole water column moves between two open surfaces as illustrated by the U-tube experiment in school physics. The first type of event is potentially the more damaging but the designer needs to be aware of both possibilities. It must be stressed that the analysis of surge pressures and the design of remedial works is a specialist area of work and if it is considered that there could be a problem with surge then expert advice should be sought.

If a valve in a pipeline is suddenly closed the water immediately upstream is brought to an abrupt stop and is compressed by the momentum of the upstream water column. This results in a sudden large increase in pressure which propagates upstream as a positive pressure wave at the speed of sound in water (the noise made on reflection of this wave, particularly in older domestic plumbing systems, is known as water hammer). Similarly, downstream of the valve or downstream of a pump which has suddenly stopped, a rapid drop in pressure occurs as the momentum of the water column downstream moves it away; this is transmitted down the line as a negative wave, that is one in which the pressure drops.

Secondary pressure waves are generated as the initial wave passes any fitting or change to the pipeline, such as an enlargement or tee, but when a pressure wave reaches the closed end of a pipeline (such as a shut control valve at entry into a reservoir) it is reflected as a wave of the same type, that is, a positive wave is reflected as a positive wave, a negative wave as a negative one. Conversely, a pressure wave reaching an open end to a pipeline discharging to a reservoir or tank is reflected as a wave of the opposite type. Even in a relatively simple system, therefore, the pattern of secondary and reflected waves can become complicated. Wave amplitude is damped by friction and pipe wall elastic losses so it is often the passage of the initial pressure wave which gives rise to the most critical pressures. However, this is not always the case as secondary and reflected waves can interact positively and may cause air or check valves to shut suddenly, so generating further high pressures. Similarly, if the pressure falls low enough, vacuum cavities may form, which may also cause high shock pressures when they subsequently collapse on a rising pressure. The analysis of transient conditions can thus be very complicated and the results depend not only on the elements of the system, but also on the profile of the pipelines. Although methods have been developed in the past for hand or graphical computations, computer programs are now used almost universally to analyse transient conditions.

The main concern with transient pressures is that they should not be high enough to cause bursting of the pipes or fittings. To put the potential for damage into perspective, it is worth noting that the rise in pressure head, known as the *Joukowsky head* after the scientist who first developed the theory, on a sudden change of velocity ΔV in a pipeline, is given by $a.\Delta V/g$, where a is the velocity of wave propagation and g is the acceleration due to gravity. In a ductile iron pipeline the wave speed may be as much as 1200 m/s although it is normally between about 900 and 600 m/s as a result of small quantities of air present in the water. With this high value for the wave speed, the slamming of a valve in a pipeline operating at a velocity of 1.5 m/s could lead to a pressure rise of 180 m. Even with a wave speed of 600 m/s the pressure rise would be 90 m. Such surge pressures undoubtedly occur, but are often not recorded to the full extent by the ordinary Bourdon pressure gauge which is too 'sluggish' in operation to record the peak transient pressure.

Codes of practice for most pipe materials allow some transient overstressing above the allowable operating pressure (defined as the internal pressure, exclusive of surge, that a component can safely withstand in permanent service). However, other elements of the system including valves and jointing systems and the resistance of thrust blocks should be considered, particularly if an existing system is being uprated.

In potable water systems it is normal practice to avoid any negative pressures and consequent risk of contamination being drawn in through open air valves or through joints designed primarily to prevent leakage from high internal pressure. In addition, large diameter, thin-walled, pipes may collapse if negative pressures fall enough to induce buckling (Section 17.8) and certain plastic materials, particularly PVC-U, may suffer from fatigue failure if there are repeated excessive transient pressure fluctuations above a certain magnitude over the life of the system.

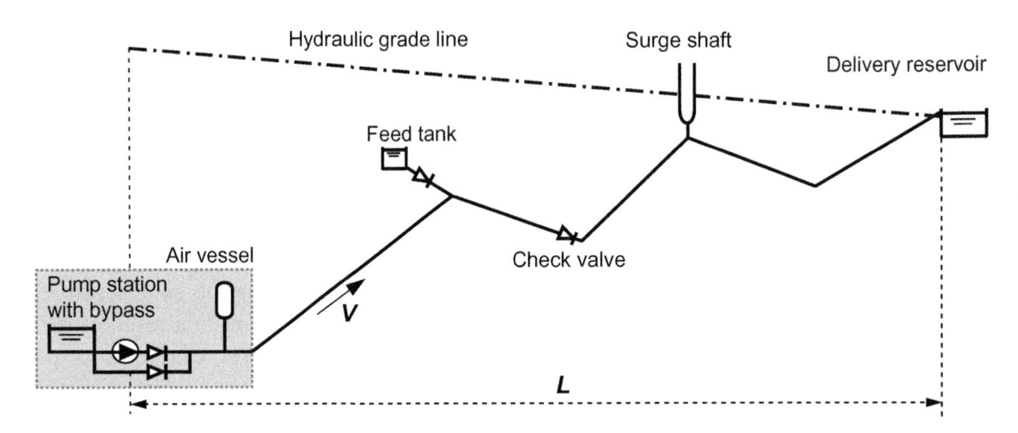

FIGURE 15.10

Pipeline profile with locations for installation of surge protection devices.

In the majority of systems the most likely causes of transient pressures are rapid valve closure and pump stoppage. Rapid valve closure can be controlled; it is always advisable to ensure that valves cannot be slammed shut and have a closure time long enough to limit the pressure rise to acceptable values. The minimum time of closure should, at least, be greater than the time it takes for the pressure wave to be reflected back to the valve from the far end of the line, $2L/a$, where L is the length of the pipeline and a is the wave speed as defined above. A longer closure time, or closure that allows the last 10–20% of the valve open area to be much slower than over the first 80%, may be required (Fig. 18.1). However, the most critical case for surge in a system is usually a power failure causing simultaneous stopping of all the pumps. Most modern low-inertia pumps stop producing forward flow of water in a few seconds when turned off, so they effectively stop almost instantaneously. Thus the problem of low pressures being transmitted through the system has become more severe. One solution is to add a fly wheel to increase the pump inertia but this goes against the trend for lighter and smaller pumps with reduced motor starting currents and may give rise to problems with bearings and drive shafts.

Once the potential problems are identified, it is necessary to consider their alleviation. It is rarely economic to increase the strength of a pipeline solely to cope with surge pressures so it is usually necessary to provide some other form of protection for the system. There are a number of protective measures that can be adopted as listed below (Thorley, 2004). An indication of suitable locations for installation of some of the devices mentioned is shown in Figure 15.10.

1. *Slower valve closure* by various mechanical means.
2. *Increased pump inertia*. Fly-wheels fitted to the pumps reduce the rate of deceleration of the pump and the corresponding rate of change of flow. In the right circumstances these can be the most effective and robust form of protection but there may be issues with modern pumps as noted above. If proposed, their use must be discussed with the pump manufacturers.

3. *Air vessels* (also referred to as '*surge vessels*') which comprise pressure vessels connected directly to the pipeline, part of their volume being occupied by compressed air. They are commonly used to feed water into the pipeline when the pumps stop but they also provide a cushion to absorb high pressures on the returning wave and on pump start up. Air is gradually absorbed into the water and compressor facilities are required to provide occasional topping up of air in the vessel. For this reason they are generally installed only at pumping stations. The absorbed air can find its way into the pipeline and come out of solution as pressure falls (Section 15.10).
4. *Accumulators* are similar to air vessels except that the air is separated from the water in the vessel by a flexible rubber membrane thus greatly reducing the loss of air by absorption. This eliminates the need for compressor facilities and allows the use of a gas such as nitrogen in place of air, topped up periodically from a portable cylinder.
5. *Surge shafts* (open-topped shafts connected directly to the line) can be constructed, if the topography permits, but they must extend above the hydraulic grade line. They should only be used on potable water systems where the ventilation facilities are designed to eliminate the risk of infiltration or airborne contamination, such as those fitted to potable water storage tanks. Such ventilation facilities must be designed to cope with the air flow appropriate to the maximum rate of change of volume in the shaft during a transient event.
6. *Feed tanks* operate by feeding water into the line to relieve low pressures. They can be located at high points below the hydraulic grade line as they are isolated by non-return valves which only allow flow into the pipeline. Again the issue of potential contamination must be addressed. There may also be an issue of water quality if the feed tank is not called upon to operate regularly and there is no turnover of the stored water.
7. *Air valves* of the large orifice type may be used to prevent low pressures in the line by opening to admit air when the pressure falls below atmospheric. Their use for this purpose is not generally permitted on potable water schemes because of the risk of contamination as mentioned above. There are potential drawbacks including the generation of high shock pressure on slamming, unless special non-slam valves are used.
8. *Pressure relief valves* can be set to open at a given pressure or operate in response to an initiating event thus limiting the maximum pressures at their location.
9. *Bypass pipework* can be fitted around the pumps to allow water to be drawn from the sump provided the pressure on the delivery side falls low enough. However, this may be insufficient to prevent other low pressure problems occurring down the line on a simple system, but they may be effective at booster stations where the pressure on the suction side of the pumps rises significantly on pump stoppage.
10. *Non-return check valves* can be used along a pipeline to reduce the effect of the returning positive pressure or water column, but they may give rise to adverse effects themselves so must be analysed carefully and used with care.

Pump delivery non-return valves need to be suitable for the system and its transient response, especially if a surge vessel is also provided because the flow in the connecting pipe to the air vessel may reverse very quickly. Ideally, the non-return valve should shut at the moment of flow reversal but if it reacts more slowly the reversed flow may slam the valve shut with the generation of a high shock pressure (Section 18.18). The dynamic response of the non-return valves should thus be matched to the transient characteristics of the pipeline system.

Surge protection system design must also take account of long-term performance; operations staff may be aware of the need to maintain surge equipment when the station is new but after 20 years, and after several changes of staff, the importance of components such as air valves may be forgotten. The approach should be for robust solutions in most cases and particularly for systems in remote locations.

15.12 CAVITATION

Cavitation occurs when the absolute pressure falls to the vapour pressure of the fluid. In water systems at normal temperatures this vapour pressure is close to absolute zero pressure (0.2 m absolute at 15°C but rising exponentially to atmospheric pressure at 100°C). With cavitation, in effect, the water tears apart forming cavities filled with water vapour. When the cavities move downstream to an area of higher pressure they may collapse or implode with the generation of shock pressures and very small high-velocity jets. The collapse of millions of vapour pockets against a surface can quickly erode even the hardest material and may destroy critical components but, if they collapse well clear of surfaces no damage should occur. The mechanism of surface damage is thought to be either local material (metal) fatigue under repeated intense pressure shocks or intense very high-velocity jets towards the surface as a result of the surface inhibiting pocket collapse on one side (Borden, 1998).

Cavitation may be bulk or local. It is a common risk where the velocity is high and pressure low but can be manifest at higher pressures if the velocity is high enough. Bulk cavitation occurs across the full flow cross section wherever pressure falls to below the vapour pressure of water for some reason. Local cavitation occurs where a jet of water is flowing fast past an abrupt or smooth irregularity and then separates from the boundary; pressure falls in the space left by the jet and cavitation pockets may be generated.

Cavitation may be a problem in a number of situations including:

- during a transient event as discussed Section 15.11;
- in a pump due to low pressure on the suction side;
- in a valve used for throttling or energy dissipation;
- at dam spillway or outlet works where the high head available from the stored water can lead to very high velocities.

In the case of a transient or surge event the low pressure occurs as a result of the passage of negative pressure waves, that is ones in which the pressure is falling, through the system. If the negative pressure is low enough and occurs at a defined 'knee' or high point in the system a vacuum cavity may form, separating the upstream and downstream water columns. After the separate columns come to rest they are accelerated back towards each other under high differential head and, with no cushion of air between them, may slam into each other generating very high shock pressures. This must be avoided and most analysts and designers of surge protection recommend that minimum pressures in any system are limited to no lower than about 0.5 bar absolute (i.e. −0.5 barg) even if the low pressure occurs for only a very short period and the risk of vacuum cavity formation is small. This applies even if the vacuum pressure is more widespread and there is no defined point for a large cavity to form preferentially. In a potable water system no sub-atmospheric pressures at all should be allowed (Section 15.10).

In a pump cavitation can occur if the velocities in the rotating impeller are high enough and the pressure on the suction side of the pump is too low; there are many examples of impeller erosion and damage from the collapse of such cavitation. The *absolute* pressure on the suction side of a pump is referred to as the NPSH (Section 19.4) and every pump has a minimum requirement (NPSHr) related to the flow it is producing. This is data supplied by the pump manufacturer. It is important that this requirement is met over the full range of pump operation (see Section 15.8 and Fig. 15.5). For most small typical installations with centrifugal pumps the NPSH requirement is likely to be of the order of 6–8 m. This is an *absolute* pressure requirement and represents a pressure range of 4–2 m below atmospheric – zero pressure absolute being about 10 m below atmospheric. Thus, provided the pump casing is below minimum water level in the sump the suction pressure should normally be adequate but if there is a long suction main the pressure on the pump suction must be carefully checked. If the NPSH available is insufficient then it may be necessary to lower the pump relative to the sump. In larger systems with high head pumps it is more common to provide low-lift booster pumps in series with the main pumps to increase the pressure on the suction side of the pumps. It is essential when choosing pumps to consider the NPSH requirements and to ensure that good design of the suction system enables those requirements to be met under all operating conditions.

In the same way that high velocities in pumps can lead to cavitation, the generation of high velocities in valves may also lead to low pressures and cavitation (Section 18.20).

15.13 SIZING A PUMPING MAIN

One issue that has not yet been discussed in relation to pipeline design is the sizing of the pipe. It is true to say that there is in every case an optimum size which minimizes capital and future costs. On major schemes it is desirable that such a 'whole-life cost' analysis (Section 2.10) is carried out but on most small schemes the uncertainties in the calculation make such an analysis less worthwhile and other factors, such as the availability of pipes, mean that the sizing is based most commonly on flow velocities.

The optimization calculation is based on balancing the capital cost, which increases with pipe size, against the future operating costs, over the life of the project, of which the major component, the cost of power, decreases with increasing pipe size due to reduced headloss and, therefore, reduced pumping head and power consumption. Figure 15.11 shows typical plots of the capital and operating costs, of which power costs are the major component, as a function of pipe diameter. Adding the two costs together gives the total net present cost curve which will always have a minimum value at some particular pipe size. That size is the optimum. It is, however, worth noting firstly that there are considerable uncertainties in future costs, particularly power costs and, secondly, the total cost curve is usually fairly flat in the region of the minimum. Judgement must therefore be made and it may be sensible to choose a pipe a little larger than the optimum.

Where the balance between present and future costs lies depends very much on the relative cost of power. Typically in the UK the optimum size of pipe for a pumping main gives a flow velocity

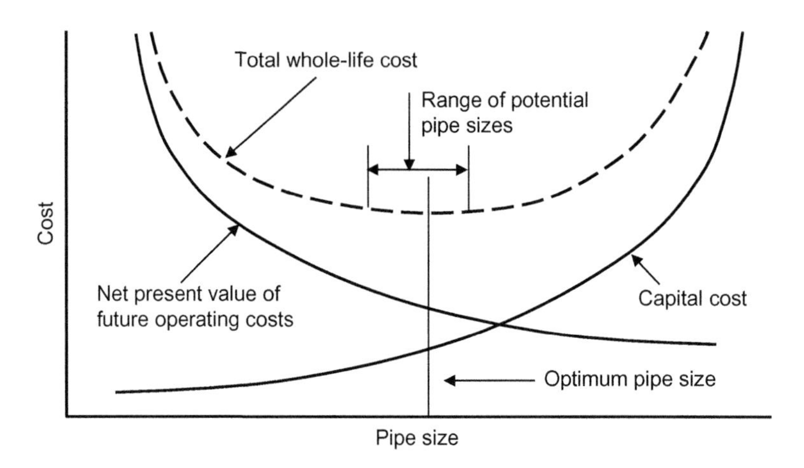

FIGURE 15.11

Pipeline size optimization.

between about 1.5 and 2.0 m/s whilst in the oil-rich countries of the Middle East where the cost of power is less the optimum pipe sizing leads to velocities nearer 3.0 m/s. In the absence of a detailed analysis smaller schemes in the UK are often sized on the basis of a velocity of about 1.75 m/s and experience suggests this is about right.

In practice, options for major schemes should be analysed in more detail and include alternative supply, demand and physical layout scenarios. The factors include:

- *change in annual energy requirements*: growth of flow over time resulting in increasing pumping costs until the pipe and pumps are operating at their capacities; pipes are seldom designed to operate at their maximum capacity from commissioning;
- *alternative demand scenarios*: phasing construction and capital expenditure, managing longer term demand forecasting risk;
- *change in Δh*: probable increase of pipe friction characteristics with time;
- *more accurate cost information*: pipe pressure class related costs, pumps, valves, electrical and surge protection equipment, structures, etc. including their annual maintenance costs;
- *pumping equipment and operating regimes*: variable, fixed speed, manual control or full automation;
- *energy costs*: the effect of future increases in the price of energy;
- *financing costs*: the scheme owner's financing and capital repayment terms.

If the calculation is set up using a spreadsheet, it can be extended to evaluate the different parameters and to determine the optimum year for possible phasing. However, the preferred solution may only emerge after considering also environmental and social costs or 'carbon footprint' (Section 13.5) differences between the options and after consultation with the client and specialists to ensure that there are no hidden engineering or operational problems, such as transient pressures or risks to water quality. The proposed pipeline route should be walked to see that it is practicable and some preliminary layout drawings may be necessary to ensure the costing is realistic.

15.14 DESIGN OF A GRAVITY MAIN

A gravity main can be designed using the principles described in the previous section. Velocities in gravity mains are generally significantly lower than in pumping mains since they are governed by the available head upstream and the required minimum service level pressure at the consumers tap under peak demand conditions; guidance velocities under seasonal peak demands being typically between 0.6 and 0.9 m/s. However, as discussed in Section 15.9 the longitudinal profile of a gravity pipe can be more critical.

1. The head available between source and delivery is fixed. Pipe diameter options are determined primarily by the head available to cater for the pipe friction and fittings losses. However, this may lead to excessive velocities if there is ample head and in practice velocities should be limited to no more than about 3 m/s; surge issues are more severe at higher velocities. It is also worth remembering that whereas it is possible to uprate a pumping scheme with larger pumps, options for uprating a gravity system involve either the replacement or duplication of the pipeline or provision of a booster pumping station. Therefore, consideration should be given to adoption of a larger pipe than required for the initial scheme.
2. Flow must be controlled from the downstream end. A throttling valve at the pipe exit raises the pressures in the pipeline as it is closed maintaining positive pressures, whilst a control valve at the exit from the upstream tank causes undesirable low pressures and potential vacuum in the pipeline downstream.
3. The pipe must be designed to withstand the maximum available pressure which occurs under static, no-flow conditions; the critical point is generally the lowest point in the pipeline.
4. The hydraulic gradient should be above the pipe profile for the range of flows and downstream control conditions expected (Section 15.9). Excess head should be dissipated to limit pressures if necessary or, preferably, used for energy generation using an inline turbine (Section 13.4).
5. The location of the downstream service reservoir should be as high as the pipeline hydraulic gradient allows; this avoids loss head, which is expensive to recreate, and can provide a margin of head on the distribution system to allow for unexpected growth in demand.
6. Where a break pressure tank is necessary to manage pressures, it should not be provided with a bypass to ensure that the high pressure cannot be transmitted to the downstream length of pipeline; instead the tank should have duplicate compartments to permit maintenance. An example of a break pressure tank is shown in Figure 15.12.

The hydraulic analysis and cost estimates may need to be repeated for a range of sizes of pipe and alternative schemes. Cost estimates tend to favour the smallest practicable pipe diameter. However, high flow velocity may reduce opportunities for increasing capacity in the future, in which case a larger diameter main may be preferred. Other options can be considered, such as relocating the treatment works closer to the service reservoir. However, the consequences for the performance of the raw water main and the treatment works would also need to be taken into account in the comparative analysis. Raw water pipes should be sized for the additional usage at the treatment works for process and cleaning water and to allow for higher pipe friction coefficients over the life of the pipe due to the quality of the raw water being transported.

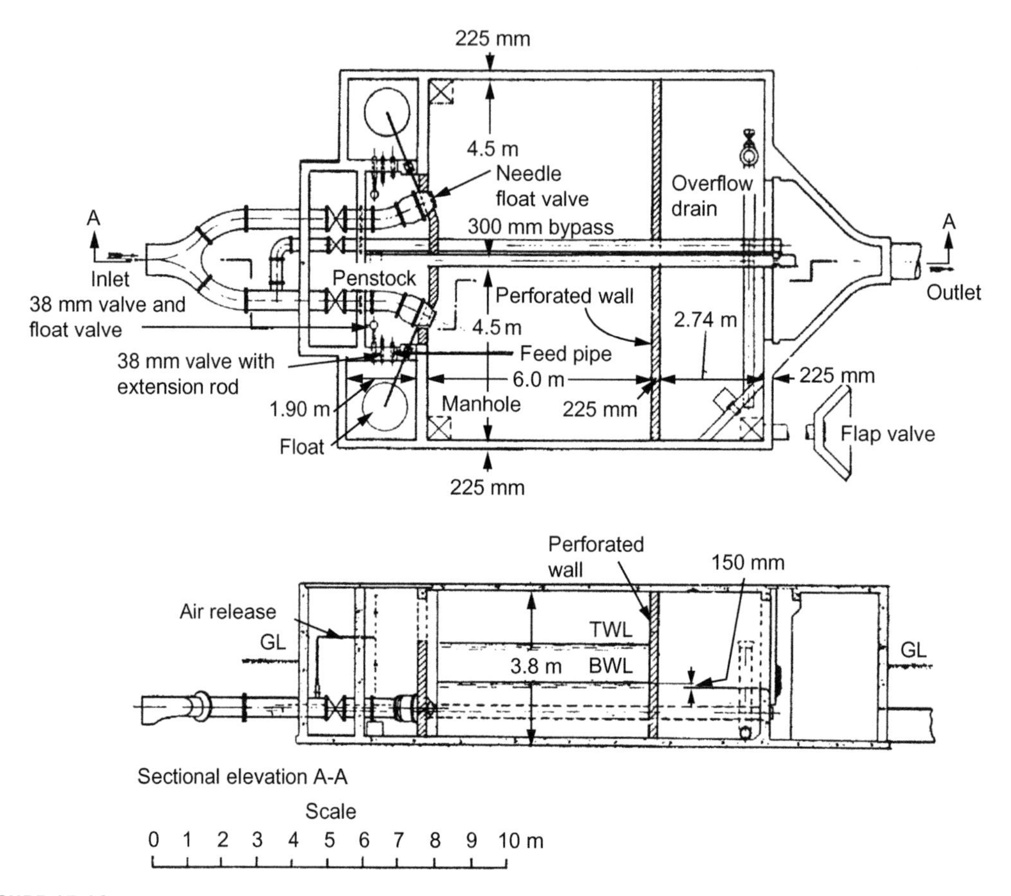

FIGURE 15.12

Break pressure tank.

15.15 PIPELINE DESIGN TECHNIQUES

Manual calculations can be used to analyse flows in simple pipe layouts and simple small networks. Hydraulic analysis software is appropriate for analysing more complex pipe arrangements and interconnected networks, for resolving operational management issues such as blending different sources in order to maintain water quality, for optimizing source and storage capacity, for testing security of supply and resilience, critical mains failure and fire demand scenarios, or where the design is integrated with base mapping (GIS) or computer aided drafting software (CAD).

Manual and Spreadsheet Calculations of Network Flows

Manual analyses can be a reliable basis for planning distribution improvements and extensions. When they take only a day or 2 to complete, manual and spreadsheet calculations are cheap and give the engineer an invaluable understanding of the way a system operates. The analysis can show which principal mains are overloaded, with high flow velocities and large friction losses, or have spare capacity; why areas of low pressure occur; and where new mains are required to meet additional demands or improve pressures.

Manual and spreadsheet calculations are particularly suited for analysing dendritic layouts. They can also be used for looped networks provided the network can be reduced to a 'skeleton layout' of the more significant and larger pipes. The total average daily demand is divided out over areas and allocated to 'nodes'. The simplest way of analysing the system is to divide the network into ring mains and 'tree branches'. Flows in the ring mains are analysed by assuming a division of flows at the input point and calculating the headloss along each branch of a loop. The assumed flow split is adjusted and headlosses are recalculated until a 'point of balance' is found in the ring main where friction losses in each leg are equal. Equilibrium can generally be reached within three or four iterations. This process can also be used to analyse multiple looped networks but may require more iterations to achieve a balanced solution. The analysis can be repeated using a peaking factor applied to the average daily nodal demands.

It is now common operational practice to measure distribution system pressures either continuously or periodically both for operational management and for level of service compliance monitoring. The field pressure records for average and peak hourly conditions can be used to draw pressure contour maps along a pipeline route or across a network. Provided inline valves are open, the contours can be used to estimate the headloss and hence friction coefficients in sections of pipe. The pressures can be used to check and, if necessary, adjust the friction coefficients used in the analysis. Where pressures records are not available, it is necessary to measure pressures and flows along the length of the pipe being analysed under a range of flow conditions.

A distribution system must have spare capacity to meet future increases of demand. However, the magnitude and locations of future demands are often not accurately known. Hence planning the development of a system is more often the choice of a strategy that best meets a range of viable future options than using precise calculation to satisfy some fixed theoretical future condition. Small mains of 200, 150 and 100 mm diameter are mostly ignored in a manual analysis, because they are normally designed according to local demand and fire flow conditions.

Analysing Existing Systems Using Modelling Software

Mathematical models can be used to analyse the hydraulic performance of existing trunk mains and distribution pipework, to design new networks and to assess system operational performance under a variety of supply and operating conditions. A detailed hydraulic analysis can define both the physical pipe performance and design parameters, control regimes for pumps, control valves and storage, establish and determine the behaviour of reservoirs, their inlets and outlets and the effects of a burst main or major fire demand. However, the modeller should always remember that the accuracy of a model is no better than the quality and availability of the data from which it has been developed. Confidence in the results requires a corresponding confidence in the input data, especially for estimated nodal demands.

Most modelling software uses similar data formats for physical system attributes and base demand information. Variations are usually associated with how network apparatus (pumps, valves, reservoirs) are represented mathematically. Each package has its strengths and weaknesses and the choice of software depends on client or user preference, the intended application, the availability and quality of source data, the type of output required, and whether the software is a component of the utility's integrated management information system (MIS). In practice, experienced modellers now regularly convert a model from one format to another. Software and hardware improvements have enabled larger models to be developed and to extend the models to analyse water quality parameters. It is now current practice to develop 'all mains models' that include all the pipework between sources and the service connections. Some software packages do impose physical size restrictions, but the overriding constraint is the ability of the analyst to assimilate and interpret the results from a large complex model.

With increasing need to achieve operational efficiency, maintain service levels and produce the most economic rehabilitation and reinforcement proposals, models need to be constructed and validated to more exacting criteria. A thorough systematic methodology must be followed to create both hydraulic and water quality models to ensure that they are to a consistent standard and are credible and so that they can be updated in the future. Where model construction procedures have been carefully developed from the outset, future updating becomes a relatively straightforward process.

Model data are derived from water utility corporate databases. It is increasingly common to find the hydraulic modelling software integrated with a utility's MIS. This can simplify model updating provided that GIS and related database errors identified and resolved through modelling are corrected before the next model update. Data abstracted from telemetry systems, such as from SCADA and levels of service permanent monitoring instruments, and from bespoke field tests designed to derive specific localized information are used for model calibration and operational control of the network. Telemetry data is now being used for 'real-time' modelling using current flows and pressures to analyse an event as it is occurring. Real-time modelling is a useful tool for improving the quality of short-term operational forward planning, for example: resource optimization, pump scheduling and managing emergencies. Database links can also be used to abstract demand data from customer information systems (CIS) for updating model demand.

The physical characteristics of the system are represented in the model by nodes and pipes (or 'elements'). The nodes, joined together by pipes, represent: pipe junctions, changes in pipe diameter and the locations of system attributes such as valves and demands. The node and pipe data sets contain geographic co-ordinates, ground levels, basic demand information, internal diameter and friction coefficients, pump curves, service reservoir geometry and valve performance characteristics.

Water demand is allocated to the node nearest to its draw-off point. Demands are distributed by the area which the nodes serve and by categories of demand including metered and unmetered domestic and non-domestic, industrial, commercial, institutional and leakage; demand data also includes patterns of usage. Where all consumers are metered, the demand is the measured consumption by user category. Demand is typically related to the postal address or postcode using a geographic reference system within the billing system. Where customers are not metered, demands are assessed based on unit consumption estimates applied to property counts or population estimates served by the node.

Each demand category exhibits its own diurnal demand profile; for example, industrial consumers may operate shift systems based on 24, 16 or 8 hour working days. Institution, commercial and office consumers have different usage patterns and some consumers, such as food processing industries, schools and colleges and sports facilities can display both diurnal and seasonal variations. The incidence of on-site consumer storage may also have a significant impact on the profile. The number of modelled demand types depends on the capacity of the software and the accuracy of the base data used to develop each demand type.

The analysis applies operational conditions to the network data such as diurnal demand patterns, times at which pumps start and stop or when valves are opened and closed. The analysis may be for a single 'snapshot' in time or a sequence of time steps, know as extended period (EPS) or quasi-dynamic simulation; each step representing a unique set of demand and operational conditions. An EPS analysis uses the initial set of demand profiles, reservoir levels and network operational conditions to calculate demands, pressures and flows in the network over the first time period to determine the operational status of automated pumps and control valves and the net reservoir inflows/outflows and thereby, using the reservoir geometry, changes in reservoir levels. The new reservoir levels together with the diurnal demands and operational changes become the starting values for the second time step. The analysis is repeated for each subsequent time step. The results can be displayed both graphically and in tabular form either for a single time step or a sequence of steps to illustrate the changing performance of the network and individual elements of the system over the period of the analysis.

The analysis is an iterative solution of a set of algorithms that simulate the hydraulic behaviour of the flow of water through the piped network, solving the equations to specified tolerances by successive approximations subject to the following rules:

- the algebraic sum of the flows entering and leaving a node must be zero;
- in any closed loop in the system the algebraic sum of the pressure losses must be zero;
- the combined inputs to the system must equal the total of the nodal demands.

For hydraulic design, simulation time steps are typically 60, 30 or 15-minute intervals over a 24-hour period. Operational management and predictive planning analyses can be used to simulate a 7-day period, seasonal variations or longer term planning and resource scenarios, for example, for a 25-year planning study.

Complex design and operational simulations should only be undertaken where the model has been calibrated either in steady state or preferably in extended period mode. Model calibration involves collecting and analysing contemporary flow and pressure measurements from the system being modelled over a period of time, typically 7 days, and comparing the measurements with the calculated flows and pressures derived from the model which has been set up to represent the field supply and demand conditions. As is true for the physical data used to develop the model, the quality of the data obtained from the field test has an impact on the credibility of the calibrated model. Field test activities include:

- measuring flows and pressures at temporary monitoring points;
- acquiring telemetry and compliance monitoring flow and pressure data, including preliminary data review to identify malfunctioning instruments;
- establishing pump head/discharge and efficiency characteristics;
- checking valve status to confirm that the system is operating as stated by the operations staff and as set up in the model;
- carrying out reservoir drop tests to quantify reservoir leakage;
- surveying all pressure monitoring points to a standard datum, Ordnance Datum for example in the UK;
- measuring internal diameter and velocity profiles at insertion probe flow locations;
- measuring metered consumer demands and establishing diurnal demand patterns;
- estimating pipe friction characteristics by internal visual inspection (if feasible);
- resolving anomalies identified during the field test and calibration processes.

Flows should be monitored at all significant meters including: source meters; internal system flow meters used to determine area or sub-area flows; inlet and outlet flow meters at storage facilities; and large usage non-domestic meters, typically the top 10 consumers by usage or those that represent 10–20% of the model demand. In the UK practically, all demand nodes average between 50 and 80 consumer connections per node but with a maximum of 100–200 per node. Pressures are generally measured at fire hydrants, the coverage being equivalent to between 20% and 35% of the demand nodes in the model. Pressures are also measured at control valves and pump suction and discharges. The quality of calibration improves, and the time required to complete the process reduces, with increased numbers of reference points. However, a field test can be time consuming to set up and expensive to run, especially in terms of the temporary equipment to measure flows and pressure. There is always a trade-off between maximizing data collection and minimizing equipment numbers. It is prudent to do preliminary field work to investigate and minimize the number of unknowns before a field test takes place and thereby avoid an abortive test. The preliminary investigations can identify equipment failures and lack of key flow meters and where equipment can and cannot be installed, enabling work needed to install equipment or repair network facilities, all of which need to be resolved before the test commences. A preliminary water balance should also be calculated to satisfy the modeller that the system and sub-system demands and diurnal profile appear reasonable and that there are no unrecorded inflows, exports or unmeasured large consumers.

Model calibration involves setting up the model to reflect the demands for a day during the field test that best represents a 'typical' day and fixing initial reservoir, pump and control valve settings. The model analysis results are compared with the measured flow and pressure data. Adjustments are made to pipe characteristics using engineering judgement until the two sets of data agree within specified tolerances. Where pipe roughness characteristics have to be adjusted beyond reasonable values for the type, age and condition of the pipe, the anomaly should be recorded, and data searches and field investigations initiated to attempt to explain the use of the anomalous value.

It is preferable to model such anomalies as a restriction at the suspected location of the anomaly. Where pipe flows or demands appear incorrect, demand data should never be arbitrarily reallocated to 'fit the field measurements'. All adjustments must be justified or identified as an unresolved anomaly to be investigated later or taken into account when using the model for predictive work. A model report should include the modelling techniques used, model characteristics, how to use the model, assumptions made and details of outstanding anomalies. The model can then be considered calibrated.

15.16 WATER QUALITY MODELLING

Water quality modelling software is now commonly available as a module of the hydraulic analysis package. Although there are differences in the modelling features and ways of presenting the results, they can generally analyse:

- age of water and thereby the retention time;
- the proportion of water from different sources, where there are multiple sources;
- concentrations of substances altered by mixing but not subject to growth or decay;
- decay or growth of substance concentrations which are time dependent.

The calculated ages and relative proportion of source waters at model nodes are used to correlate with the results from water quality sampling programmes, or from a sample taken following a consumer complaint. Source contribution analyses are used to track and predict the mix of water from different sources and the effects and extent of pollution incidents. Chlorine residual models, calibrated against field chlorine measurements, can be used in conjunction with chemical and bacteriological sampling data and statistical predictive modelling techniques to improve the efficiency of disinfection dosing and improve network retention times. Complex parameters, such as disinfection by-products, can also be modelled either directly or by modelling a surrogate parameter.

Water quality modelling must be based on a validated hydraulic model that accurately replicates velocities in the pipes and thereby the time of travel of water throughout the network; essentially a model is required that includes all pipes down to those in which the water quality parameters are being analysed. A '*skeletal*' model, a simplified hydraulic model that does not include some smaller and local distribution pipes, is generally not suitable because it overestimates velocities in the modelled pipes and thereby underestimates the time of travel and hence age of water at the modelled nodes. Age of water models are frequently included in model development specifications.

The key to all mathematical modelling is verification. The field instruments used to produce calibrated hydraulic models have developed to meet the need for increasingly accurate hydraulic models. However, equivalent field-deployable instruments for providing data for calibrating water quality models are still relatively expensive. They are therefore currently not available in sufficient numbers to achieve the coverage necessary for comprehensive calibration of all but relatively small models. There is no question that water quality modelling represents a potentially powerful tool for the future, but until field sampling equipment is sufficiently available, validation is constrained by the quality of data obtainable using existing equipment, sampling techniques and surrogate parameters.

15.17 UPDATING OF NETWORK MODELS

Most hydraulic analysis software can be integrated with the databases used by water utilities to manage their businesses. Although some software is designed to deal with particular construction issues, none has yet resolved the problem of repeatability or updating only where there are changes in the data sets. The question is therefore whether to build a completely new model each time a new generation of the model is required, or whether to update by exceptions; that is only revise sections of the model where there are known changes to the data. Either way, the decision must reflect the purpose for which the model is required. Possible changes requiring the model to be revised or rebuilt include:

- physical attributes: existing pipes and assets rehabilitated or replaced; new mains and extensions; valve status changes (hydraulic boundaries);
- base demand changes: changes in existing consumer usage and new consumers;
- operational changes: control rules, settings and profiles.

Small changes can be accommodated by revising an existing model, provided there are comprehensive records of the physical and operational changes since the model was developed. Changes in mains layouts or mains condition, for example, scraping and relining a pipe to improve its internal roughness or hydraulic boundary changes should be recorded on the GIS. Demands should be reviewed using billing records and telemetry flows to assess changes in average day demands, seasonal variations, reductions in leakage levels, increased per capita consumption or changes in consumption volumes recorded in metering records. Trade, industrial and institutional consumers are generally metered and any changes can be derived from billing records. Operational control rules are documented and can be validated from telemetry records.

Calibrated models should be validated periodically to confirm that they are still 'fit for purpose' within acceptable tolerances. Modellers have conflicting views on the frequency of re-validating a model, whether to undertake a full or partial model rebuild and whether to carry out a calibration field test. Unless there is strong evidence that the model is still performing within its specified tolerances of flows and pressures over the diurnal period, the model should be reviewed. Its performance should be validated against field measurements, using flow and pressure data at key locations derived from telemetry or a small field test. Where the model is assessed to be outside specified or acceptable tolerance, the decision is then whether to update the physical, demand and operational data in the existing model or to rebuild the model from scratch. In both cases the resulting model needs to be calibrated against a full field test.

15.18 SOFTWARE DEVELOPMENTS

Using information from mains record, customer billing, telemetry, levels of service and water quality data accessed through an integrated MIS, combined with on-line 'real-time' hydraulic analysis will provide future planners and operations engineers with a powerful suite of engineering analysis tools. Such tools are being developed by water utilities and software developers in different parts of the world, some as integrated suites of modules, others as stand-alone purpose made operational support tools.

The increasing use of GIS, CIS and associated system performance databases provides opportunities for building and updating hydraulic models electronically, thereby eliminating the costly manual process which has inhibited model revision in the past. Genetic algorithms add another dimension to model calibration, network optimization and asset management, all potentially deriving cost savings. Real-time modelling is expected to deliver demonstrable efficiency savings in the near future. However, whatever tools are available to the modeller, the quality of a model and credibility of the outputs are only as good as the data from which the model was developed and the technical ability and operational management understanding of the modeller. Where data standards are not maintained and critical business decision making relies on inexperienced modellers, the water utility is exposing its business to a risk of performance failure.

REFERENCES

Borden, G. and Friedmann, P. G. (1998). *Control Valves*. 3rd Edn. Instrument Society of America.

Cabinet Office (2011). *Keeping the Country Running: Natural Hazards and Infrastructure, A Guide to Improving the Resilience of Critical Infrastructure and Essential Services*. Cabinet Office, London.

Escarameia, M., Dabrowski, C., Gahan, C. and Lauchlan, C. (2005a). *Experimental and Numerical Studies on Movement of Air in Water Pipes*. HR Wallingford Ltd, Report SR661.

Escarameia, M., Burrows, R., Little, M. J. and Murray, S. (2005b). *Air Problems in Pipelines. A Design Manual*. HR Wallingford Ltd.

Little, M. J., Powell, J. C. and Clark, P. B. (2008). Air movement in water pipelines: some new developments. *10th International Conference on Pressure Surges*, BHR Group, May, Edinburgh.

Miller, D. S. (1990). *Internal Flow Systems*. 2nd Edn. BHr Group Ltd.

Thorley, A. R. D. (2004). *Guide to the Control and Suppression of Fluid Transients in Liquids in Closed Conduits*. 2nd Edn. Professional Engineering Publishing Ltd.

van den Boomen, M. and Vreeburg, J. H. G. (1999). Report No. SWE 99.011. Nieuwe Ontwerprichtlijen Yoor Distributientten. KIWA.

CHAPTER

Distribution Practice

16

16.1 CONDITION AND PERFORMANCE OF NETWORK ASSETS

The quantity and quality of water delivered to the customer are directly related to the physical condition of the network assets that transport the water between the source, treatment works or service reservoir and the customer's tap, and to how the assets are ultimately operated and maintained.

The pipelines contained within a system typically represent the largest capital asset that a utility possesses and the cost of maintaining these assets in good condition has become a major financial consideration for utilities. Historically investment tended to be focussed on the non-infrastructure assets, increasing capacity to meet consumer demand by developing resources and refurbishing and replacing treatment plants, and on building assets to deliver higher quality treated water. The infrastructure (network) assets were deemed to be lower priority, possibly due to the perception that underground assets exhibit long asset life, 60 to over 100 years, but also due to the difficulties of assessing asset condition; reactive maintenance was the norm.

However, globally large sections of networks, trunk and distribution, are reaching the end of their serviceable life. In addition, more robust regulatory conditions are being imposed requiring utilities to deliver defined levels of service to consumers including stricter water quality standards that oblige utilities to ensure that the condition of their systems does not cause the quality of the water to deteriorate before it reaches the consumer and that networks have an adequate level of resilience to ensure continuity of supply. Furthermore, with increased financial and environmental pressures on utilities to become more customer focussed and to improve operational and financial efficiency and reduce leakage and wastage, the emphasis has shifted to a better understanding and management of their underground assets.

Infrastructure assets are analysed to determine their current performance and thereby to forecast future performance, and to assess the risk and cost implications that the assets pose as their performance and condition deteriorates, be it structural condition, hydraulic capacity or water quality. The primary questions being addressed by the analysis are:

- What currently exists?
- What is its function?
- Is it performing the (perceived) function?
- What is its operational condition and serviceability?

Twort's Water Supply. DOI: http://dx.doi.org/10.1016/B978-0-08-100025-0.00016-8

Managing distribution assets is a major task due to the number of data sets used: pipes, service connections and ancillary equipment such as meters, control valves, pumps and storage facilities. However, a large quantity of data is collected and processed in a variety of ways by a utility in the course of managing its business and much of this data can be used in the analysis to assess performance and the risks associated with the failure of an asset. Table 16.1 illustrates the types of data and some of the processes and systems needed to manage underground assets. There is overlap in the systems and tools list because utilities tend to develop bespoke configured databases to meet their specific needs.

Table 16.1 Data and processes for management of underground systems

Examples of raw data held in company information systems	Systems, tools and processes used to collate and interpolate data
Critical Monitoring Point pressure data	Asset databases
Customer demand (billing information)	Burst records
Customer complaints	Business investment planning
Customer meter locations	Condition assessment and failure deterioration models
Customer surveys	District Meter Area monitoring
Demand surveys	Distribution Operation Maintenance Strategies (DOMS)
Facilities failures	Emergency plans
Flow and pressure measurements	Geographic Information Systems (GIS)
Inspectors' site reports	Health and Safety records
Mains and facilities record drawings	Hydraulic, surge and water quality modelling
Maintenance records	Investment management modelling
Measured leakage and other Non-Revenue Water uses and wastage	Job scheduling
Meter maintenance/audits	Leakage strategy, policy, modelling and management
Network maintenance reports	Maintenance schedules/programmes
Network performance by location	Management Information Systems (MIS)
Records of repairs and rehabilitation	Network control databases
Pipe age diameter, material, internal pipe diameter/class and condition	Operational Control Centres
Pipe failures by location and failure mode	Performance metrics
Pipe flushing sequences and programme	Pressure management
Pipe sampling (e.g. cut outs, inspections)	Regulatory reporting
Sediment composition, size and Specific Gravity	Risk registers
Staff training records	Risk and Value Management
Third party/shared databases and information	Supply and demand modelling
Valve registers (locations of shut valves, status change authorization/process)	Telemetry and metering
Valve maintenance records	Water Safety Plans
Water quality samples by location	Water quality databases

Analyses which extrapolate from historic condition and performance records to predict future asset performance can result in a programme to replace or refurbish an asset earlier than necessary and, thereby, an overestimate of the financial commitment.

The preferred integrated asset management approach takes account of the multiple factors impacting the asset. These include: the type, age, condition, physical and environmental characteristics, leakage and burst frequency of mains; performance aspects, which affect customer contacts such as flow, pressure and water quality; and, using statistical deterioration models, forecasts of the deterioration in service performance over the planning period. Deterioration models are used to predict failure rates, pipe bursts, rise in leakage and condition deterioration, for example, pipe wall thickness. The output is an assessment of the condition and potential remaining asset life of each asset and hence need for and timing of an intervention.

Deterioration modelling was used by utilities when developing business plans for the 2015–20 investment period in the UK, resulting in more robust evidence-based investment plans. However, for many utilities there is still a dearth of accurate and comprehensive performance and failure data relating to the physical and operational conditions of network assets. Consequently, asset investment plans for pipes and ancillary equipment are still likely to be a combination of both approaches until failure mode data and deterioration models become more readily available.

Advances in the quality of data and data processing speeds have allowed for semi-automated analyses to be performed, so providing a more structured, consistent and objective output and thereby allowing a better understanding of asset performance. The input and output data from hydraulic modelling software packages (Sections 15.15–15.18) together with the output from deterioration models can also be integrated with management information systems (MIS) and graphical information systems (GIS).

The system management analysis needs to take into account both the practical operational risk and consequences as well as the regulatory and reputational business risk consequences of a failure; these are assessed in terms of the financial, social and environmental consequences and the benefit of preventing the failure. The operational risk is often included, but cost–benefit assessments generally do not include for business risk. In practice, pragmatic engineering and operational judgement is required to ensure that model outputs are credible, but developments in data processing and improved quality of the raw data have reduced what was a labour-intensive process in the 1990s into an efficient and auditable semi-automated process today.

The output from the analysis is generally a schedule of interventions over the planning period to address the identified risks. Ideally the sequence of interventions would represent a programme of work but, in reality, financial constraints will inevitably require that the long list of interventions be reduced to an affordable shorter list; one which optimizes the conflicting drivers of maximizing the benefit to customers while minimizing the operational and business risk.

Statistical models are used to perform the optimization, for example, Monte Carlo simulations, generally with an *Asset Investment Manager* (AIM) tool. The cost and benefit of individual interventions are optimized to identify an optimum set of integrated interventions that deliver best value. For example, rather than implement three separate discreet schemes to reduce bursts in a District Meter Area (DMA), resolve low pressures and reduce water quality complaints problems, there may be a single alternative intervention that delivers the required outcomes but at a lower overall cost. Constraints are set within the optimizer such as: minimizing the risk of interruptions to supply; limits on capital expenditure; target numbers of customer complaints; that is, the required outcomes from the investment.

The output is an affordable ranked list of schemes to be included in the utility's Asset Investment Plan. The plan defines the short- and medium-term rehabilitation and replacement requirements of the assets in a system and the longer term infrastructure needs by broad categories of assets. Proposed investment is typically promoted through a *Business Investment Plan*, linked to future tariff structures. Historically the analysis would be repeated periodically to support revisions to the investment plan. However with improvements to the base data, analysis tools and computing power, utilities re-analyse their base data annually or biennially in order to update the scheme list to reflect changing asset performance, benefits, risk and regulatory requirements.

Individual interventions need to be specific and should include information which is not necessarily included in asset data sets, for example, operational issues and third party conflicts. Therefore, risk and benefit assessments need to be thorough, comprehensive, inclusive and specific. Generic risk and benefit factors may be appropriate for initial high level screening, but more detailed intervention-specific factors will be required for investment plan optimization.

For the 2015–20 planning period in the UK there has been a change in focus to 'the customer' in the regulatory arena. This has encouraged utilities to embrace the concepts of '*outcomes*', with clearly defined performance metrics, risk and uncertainty as opposed to the historical '*defined output*' approach where delivery of a specified scheme was deemed more important than any improved service to the customer or reduction in operational and business risk. In addition, there has been increasing recognition that operational interventions are equally valid options and that capital investment is not always required; this approach uses Totex (Total expenditure; the sum of Opex and Capex) in cost–benefit analyses, whereas previously comparisons would consider capital expenditure (Capex) and possible 'Opex impact' (change in operational expenditure resulting from the intervention) (Section 2.10).

Generally each analysis approach is bespoke to a utility's operational circumstances, business model, quality of data and data sets used. Intervention options tend to be common to all utilities but the reasons for and expectations of the outcome of an intervention are specific to a utility, water quality zone or a length of main in a network and to the internal and external performance targets of that utility.

16.2 SERVICE LEVELS

Service levels are the standards of performance which a water utility affords its customers. These standards can be targets for achievement set by a utility for itself, or outcome performance metrics set by an external authority or regulator, such as the Water Services Regulation Authority (WSRA), also known as Ofwat, for the privatized water companies in England and Wales. Service levels are often set by or agreed with international funding agencies such as the World Bank to define what a programme of rehabilitation or improvement should achieve. Records of how far such levels of service have been achieved can be used by the utility, customers and third parties to evaluate a utility's annual service delivery performance against intended outcomes and targets. A high level outcome could be the provision of a reliable (uninterrupted flow and pressure) and wholesome (safe and good to drink) water supply. The outcome would be assessed using performance metrics, such as the number of properties with low pressure, recorded interruptions to supply, the supply and demand balance, customer satisfaction via the number of complaints and water quality compliance.

Hydraulic performance defines the minimum pressure and flow domestic consumers should experience. In the UK, up to 2008 the reference level of service required by Ofwat when demand was not abnormal was a flow of 9 l/min at a pressure of 10 m head on the customer's side of the main stop tap at the property boundary. Since the adoption of the concept of outcomes, the UK Government Guaranteed Standards Scheme (GSS) Regulation 10 (HMSO, 2008) requires a utility to maintain a minimum pressure in the communication pipe of 7 m static head (0.7 bar). There is now no minimum flow or pressure requirement at the customer's tap. Variants of both requirements can be found worldwide.

Since pressure is difficult to measure at the customer's side of the main stop tap, a 'surrogate' pressure of typically 15 m in the main supplying the property may be used for design purposes. The use of a lower surrogate pressure is acceptable only where justified with supporting evidence. For two properties supplied through a common service pipe, the pressure reference level is the same for twice the flow. For common services feeding more than two properties, the increase in minimum mains pressure is linked to the size of the common service, number of properties supplied and the concept of '*loading units*' defined in BS EN 806-3 (2006). In practice utilities would plan for a constant water supply at a pressure which will reach the upper floors of existing dwelling units but not for buildings that use pumped systems, such as blocks of flats.

Continuity of supply is measured by the number, duration and circumstances relating to interruptions or deficiencies of supply. In the UK Ofwat uses a scoring system for assessing a water company's performance. Prior to 2013, the utility was required to report the number of properties affected by interruptions to supply of more than 3, 6, 12 and 24 hours by the four categories of:

- unplanned interruptions due to bursts, etc.;
- planned and warned interruptions due to planned/scheduled maintenance, new connections, etc.;
- unplanned interruptions caused by a third party, for example, another utility damaging a water pipe while excavating for their own service; and
- unplanned interruption due to overrun of a planned and warned interruption.

Following the development of the outcome approach, utilities in the UK have amended the way in which they measure and report interruptions to supply, particularly in the development of future performance targets. The measures vary across the industry, but essentially relate to the average number of hours or minutes across the whole utility customer base that customers experience interruptions. The above categories remain for reporting purposes, but utilities generally only specifically report incidents against interruptions greater than 12 hours. The GSS Regulations 8 and 9 require compensation payments to be made where a planned interruption is for more than 4 hours or where interrupted supplies are not restored within 48 hours of an emergency interruption to repair a leak or burst or 12 hours for any other emergency shut-off. The minimum payment for domestic customers is £20 for each event plus a further £10 for each 24-hour period the supply remains interrupted. Similar targets are appropriate where operation and maintenance of a public water supply system has been contracted out to a service provider company for a fixed term of years.

In lower income countries, particularly where utilities are publicly owned, similar target service levels may be the aim but achieving them could be challenging. For some a more realistic target is to achieve 4-hour supply morning and evening to connected consumers and with standpipe coverage for all other householders.

In addition to service levels for interruptions, there is increasing pressure on utilities to provide appropriate levels of resilience within their networks. Customers connected to a single source of water are at risk of loss of supply if the source fails or is forced to shut down. The utility is faced with providing emergency bottled water supplies until the piped supply is restored. Resilience measures address these issues by providing enhanced connectivity to ensure that customers are not adversely impacted by problems at specific treatment works or service reservoirs.

Water quality standards for water delivered to the consumer, together with sampling and reporting requirements, are set by legislation in most developed countries (Section 7.52). They apply universally to the quality of water delivered to consumers' taps. The quality of water leaving a treatment works may be acceptable but any decline of quality as it passes through the distribution system must be taken into account. All waters contain nutrients and long retention in mains can promote '*aftergrowth*', that is, increase of bacteria and small organisms in the mains. Taste, odours and suspended matter in the water can occur. Thus, the hydraulic design and operation of the distribution system needs to minimize these effects by ensuring good circulation of water, flow velocities and transients maintained within acceptable ranges, short retention times and cleanliness of the interior of mains. Sometimes it is found necessary to introduce additional disinfectant (usually chlorine) to the water at service reservoirs in order to inhibit aftergrowth in mains.

Whilst specific and measureable water quality issues can be addressed through additional treatment or catchment management, issues relating to taste and odour at customer taps remain slightly subjective. Customer complaints relating to taste, odour and discolouration are a measure of the acceptability of the water to the customer. There are occasions where the utility cannot improve one area of concern, without adversely impacting other areas. For example, where chlorine residual levels are at the lowest possible in the network, it is not possible to further address taste and odour issues without significant investment in additional treatment processes, which may not be either acceptable to customers or affordable.

16.3 DISTRIBUTION NETWORK MANAGEMENT

Organization

The general aspects of a water utility organization are described in Chapter 2. However the management of a distribution system involves both central and local activities. Utilities in the more industrialized countries increasingly are making use of telemetry, remote sensing and control and central or regional Operational Control Centres manned 24 hours a day to manage their networks. From the control centre remote sources, pumping stations and control valves can be monitored and controlled in order to maintain supplies to consumers under varying supply and demand conditions. For utilities with no central control facility, it is usual for local distribution managers to keep in regular contact with the operators running the sources, so that source outputs can be altered as necessary to meet demand variations. Regular reporting times are arranged; for example, at 09:00 hours to consider what source output and the water levels in service reservoirs are required according to the previous day's consumption. A second call at say 16:00 hours would agree further adjustments in the light of the current day's demand. Further communication would also be needed to respond to an emergency, for example, fire demand or a burst main.

Activities normally managed centrally include:

- Strategic and Asset Planning, Performance Monitoring and Regulation;
- Billing and Finance: meter reading, customer billing and income collection;
- 'Developer Services'; initiating and negotiating agreements for extension of mains and new services;
- Customer Services and Call Centre often attached to an Operational Control Centre; responding to customer queries and complaints:
- Water Quality and Central Laboratory: water quality sampling, testing and reporting;
- Engineering and Procurement: design, procurement, contract supervision and commissioning;
- Human Resources: training;
- Finance and Corporate Control: legal, budgetary, financial and governance control.

The activities managed locally, usually within a network operations team, comprise:

- Maintaining supplies to customers: zone engineer, supply engineer;
- Customer contact, investigation of customer complaints and network problems: district inspectors, network technicians, also known as 'turnkeys';
- Repair and maintenance of mains: direct labour gangs or contractors;
- Monitoring levels of service, flows and pressures: network technicians and analysts;
- Leak detection and repair, and waste reduction: leakage technicians, direct labour gangs and contractors;
- Inspection of plumbing systems and enforcement of water byelaws: Byelaws inspectors/technicians.

Managing Activities in the Network

The key operative at the local level is the '*District Inspector*' or '*Network Technician*'. The inspector liaises with the public and investigates customer complaints, monitors levels of service, inspects properties for waste and illegal consumption and keeps the area under surveillance for signs of visible leakage. The inspector can also be involved in leak and waste detection activities. Using their intimate local knowledge of the network, consumers and operational anomalies, often derived over many years working on the same part of the system, they are pivotal to the efficient and effective management of the network, diagnosing the causes of operational problems and directing remedial measures where necessary. The inspector would also be responsible for operating ('turning') valves which are likely to impact flow routings and hence water quality or pressure at the customers tap.

Local networks tend to be managed from a district or zone depot office. Customer contacts received centrally through a call centre are assigned electronically to the relevant district office and allocated to an inspector for action. The inspector will visit the location of the event, assess the scope of any necessary remedial intervention and initiate a *Work Order* on the organization's job scheduling system. Repairs are completed by direct labour repair gangs or a contractor. On completion of the work, the Work Order will be closed on the job scheduling system by the foreman of the gang executing the work, including listing water off and on times and materials used. If the utility retains direct labour gangs to carry out small scale work such as pipe repairs and maintenance, laying short lengths of main and installing service connections, valves and meters, the local depot will include stores, a plant yard and facilities to support the labour gang.

Recent communication technology developments have enabled some utilities to provide their inspectors with fully equipped vans using remote communication devices and computers for remote working. The inspector is sent details of customer contacts and network jobs directly from the control centre as the Work Order arises. The communications equipment can also be linked to corporate information systems so that the inspector can access the latest mains records and asset databases. This development allows inspectors to better manage their areas independently and to be able to respond more speedily to operational requirements.

The operation of key valves including boundary valves between adjacent hydraulic areas needs to be managed carefully to prevent unauthorized changes of valve setting that could affect the quality of the supply and the integrity of the flow monitoring. Therefore, before turning key supply valves, the inspector and zone engineer would complete a risk assessment in accordance with the utility's operational procedures and take appropriate mitigating actions to minimize the impact of the change on consumers.

Distribution Network Configurations

Water utilities in the UK are increasingly reorganizing their distribution networks into hydraulically discrete stable *(calm)* supply areas in order to control the network and manage performance service levels and leakage. The concept of a calm network is that the system is operated so that flows and pressures are maintained '*stable*' within operational ranges under '*normal*' demand conditions in order to avoid excessive velocities, pressure surges (transients) and flow reversals, all of which could cause deterioration in water quality or overstress the pipework.

A supply network is typically divided into hydraulically discrete supply areas termed '*zones*' or '*districts*', which are further subdivided into '*district meter areas*' (DMAs) and '*pressure management areas*' (PMAs). Preferably zones are configured to be supplied from a single source; but if two sources of different quality waters have to be used, the utility will attempt to blend the separate sources in a consistent ratio, typically in a service reservoir, to ensure a consistent water quality is delivered to the consumers.

DMAs, used for monitoring consumption and evidence of leakage, are configured to achieve stable flows under normal operational conditions (see also Section 16.2). PMAs are used to reduce and manage pressures in parts of a network that would otherwise be subjected to high or excessively variable diurnal pressures. PMAs can comprise whole zones, parts of a zone containing a number of DMAs or be a subdivision of a single DMA.

Network flow and pressure measurements provide the data to monitor and control each part of a system to ensure adequate supplies are maintained to consumers and to support leakage and waste detection activities (Sections 16.8 and 16.9). The data are derived from key monitoring points throughout the system including:

- flow meters at sources of supply, for district metering, pressure management control and on large consumer connections;
- pressure monitors for system management, valve and pump control and for level of service management and reporting.

The data can be transmitted by telemetry to the Operational Control Centre where it can be monitored, stored, processed and displayed graphically on schematics, as trend graphs and to trigger alarms. This real time decision-support data can also be used to identify deficiencies of supply, burst mains, fire demands, to ensure that the necessary staff are called out to deal with the emergency and to provide information to staff responsible for dealing with consumer complaints.

Intermittent Supply Conditions

An intermittent supply is the consequence of inadequate available resources or poor network asset condition. High leakage from a network may be the underlying causes of a supply/demand deficit and the consequent need to cut-off supplies for parts of the day to conserve resources and fill system storage. In low-income countries the situation may be further aggravated by insufficient financial resources to expand and maintain the network. Uncontrolled intermittent supply conditions should be avoided. They result in partial supplies to some consumers and little or no supply to those located at the extremities or higher parts of a network. The primary consequences are waterborne diseases due to infiltration and stagnation within the pipes.

Planned water rationing is implemented by operating distribution valves to apportion water to different parts of a network in rotation; for predefined periods or set quantities, or to the whole supply area for parts of a day; for example, twice a day, to provide water to consumers morning and evening. Under such circumstances other operational difficulties arise and maintaining water quality may be problematic. Valve operations need to be managed to minimize the risk of subjecting parts of the network to vacuum pressures which could cause groundwater infiltration and contamination of the network. This is a specific problem for branch mains off principal feed mains, where valves are operated to shut-off or re-open supplies, and where a service reservoir is allowed to drain down and empty during each supply cycle; in both cases resulting in the network emptying by draining to the lowest point. The design principles for DMAs are also applicable for designating areas to be used to implement planned water rationing.

However, consumers may themselves aggravate the situation by storing more water than they will use, draining off the surplus water and replacing the whole volume during each supply period. Most consumers would use the times of supply to fill storage containers for use during periods of non-supply and may leave their taps open causing the containers to overflow and waste water. Such behaviour represents wastage and inefficient use of valuable resources. The result is that more water has to be delivered than is actually required and that it has to be delivered in a limited period, under peak demand conditions, often through a network in poor condition which was originally designed for lower peak demands. The resultant higher delivery pressures further stress the network already in poor condition which might have been the cause of the problem in the first place.

If the quality of water is suspect, it is common for consumers with intermittent supplies to boil water for drinking; they may even be advised to do so by the supplier. Drinking water may also be available through private vendors though, unless the utility controls the vendors, the quality of water they supply can be very variable.

For systems subjected to severe supply constraints and water quality problems associated with infiltration and poor asset condition, it may be prudent to consider alternative methods of supply including tankering water to local storage connected directly to standpipes or to provide mobile storage at strategic locations in the supply area.

Distribution Network Extensions

The design and supervision of construction of network extensions and reinforcements, or the rehabilitation of mains (Section 16.14) is normally managed by the central engineering department of a water utility. However, in countries where local water utilities have limited technical resources, a separate regional organization may be responsible for all the design and construction of network extensions in its supply area. Where water companies come under regulatory control, as in the UK, a company's in-house design team may be required to be competitive against outside designers, and its direct labour section to be competitive against construction contractors.

16.4 FIREFIGHTING REQUIREMENTS

Under current UK legislation (FSRA, 2004) a water utility must install fire hydrants on mains where required by the fire authority. The fire authority must pay for the installation and maintenance of the hydrant. The only exception is that the utility is not required to install a fire hydrant on a trunk main. There are no requirements on the utility to provide any given pressure or flow at a hydrant, except that when the fire authority is dealing with a fire it can call upon the utility to provide a greater supply or pressure by shutting off water from mains or pipes in any area. The utility has, however, a duty to maintain a constant supply of water in its mains at a pressure sufficient to reach the topmost storey of every building in its area – but not to a greater height than that to which it will flow by gravitation through its water mains from the service reservoir or tank from which that supply is taken.

Table 16.2 summarizes the UK guidance on minimum flow requirements recommended for firefighting (Water UK, 2007). The minimum requirement is 0.5 m^3/min for two-storied housing, with up to 2.1 m^3/min for larger dwellings. This compares with the specified minimum capacity of fire hydrants of 2 m^3/min at a pressure of 1.7 bar at the inlet (BS 750:2012). When larger fires occur, suction hose lines are put out to additional hydrants further afield, thus spreading the water demand over more mains.

In the USA the NFPA1 Fire Code (NFPA, 2012), issued by the National Fire Protection Agency, stipulates minimum fire flow capacity of mains according to size of area and nature of property. These flow requirements are over and above domestic consumption and therefore dominate the design of distribution mains.

Fire appliances in common use have built-in pumps of capacities 2.3 and 4.5 m^3/min. Nozzle sizes are commonly 13 and 19 mm, discharging flows of 0.16 and 0.45 m^3/min, respectively. The largest practicable size for hand-held nozzles is 25 mm diameter with capacity up to 1.1 m^3/min. Table 16.3 summarizes hydrant spacing and flow fire requirements in Europe and the USA.

16.5 SERVICE PIPES

Service pipe connections from a main to a property are usually laid as shown in Figure 16.1. In the UK the length of service pipe from the main to the company stop tap (or consumer's meter if fitted) is termed the *'communication pipe'*; the balance of service pipe to the consumer's internal stop tap being the *'supply pipe'*. The communication pipe is maintained by the company and the supply pipe is the responsibility of the customer.

Table 16.2 UK guidelines on flow requirements for firefighting (Water UK, 2007)

Category	Description	'Minimum' requirements
Housing	Detached or semi-detached houses of not more than two floors	8 l/s through a single hydrant
	Multi-occupied housing with units of more than two floors	20–35 l/s through a single hydrant
Transportation	Lorry/coach parks – multi-storey car parks – service stations	25 l/s from any single hydrant on site or within 90 m
Industry	Recommended for industrial estates:	Site mains normally at least 150 mm diameter:
	• up to 1 ha	• 20 l/s
	• between 1 and 2 ha	• 35 l/s
	• between 2 and 3 ha	• 50 l/s
	• over 3 ha	• 75 l/s
Commercial	Shopping, offices, recreation and tourism development	20–75 l/s
Institutional	Village halls	15 l/s through any single hydrant on site or within 100 m of the complex
	Primary schools and single storey health centres	20 l/s through any single hydrant on site or within 70 m of the complex
	Secondary schools, colleges, large health and community	35 l/s through any single hydrant on site or within 70 m of the complex

Table 16.3 Fire hydrant and fire flow requirements

	England and Wales	USA[a]	Continental Europe
Hydrant characteristics	2 m^3/min at 1.7 bar	Up to 6.65 m^3/min	1.0–1.5 m^3/min minimum
Initial flow in close proximity to fire	1.4 m^3/min upwards	3.8 m^3/min for 1 hour	Generally 1.0 m^3/min
Subsequent rates of flow (as required) for a single fire	9 m^3/min or more, up to 25 m^3/min for major fires	Minimum flow 6–30 m^3/min related to fire area and construction type	3.6–6.0 m^3/min in high risk areas or more
Minimum residual pressure in mains	Preferably 0.7 bar but not less than zero	Not less than 1.4 bar at the required flow rate	Not stated
Possible quantity of water used in a fire	Maximum flow times 'several hours' say 6 hours	Minimum flow duration 2–4 hours	Maximum not stated; average incidents last less than 2 hours (France)
Spacing of hydrants	Generally 100–150 m but 30 m in high risk areas	Generally 60–100 m, maximum 90 m on dead ends and 200 m for looped systems	80–100 m in urban areas; 120–140 m in residential areas; wider in rural areas

Sources of information: Bernis, 1976; BS 750:2012; NFPA, 2012.

Note: [a]Reduced flow requirements of up to 75% are permitted where sprinklers are installed in buildings.

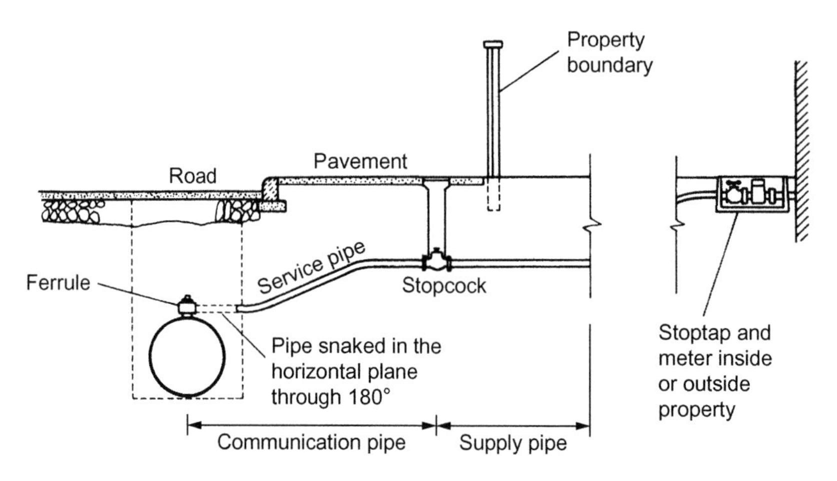

FIGURE 16.1

Typical service pipe connection.

Service pipe connections for houses are usually to a standard size according to the practice of the water utility, 15 or 20 mm nominal bore being the typical size for a one-family house or flat which is within 30 m of the distribution main. Although required to meet level of service standards for pressure (Section 16.2), in practice and subject to the physical characteristics and operating condition of the network, utilities typically try to maintain pressure in the mains between about 20 and 35 m. Higher operating pressures are maintained in the USA where there are buildings greater than three storeys and where a water utility is committed to provide higher mains pressures for contractual or firefighting reasons.

Installing a Service Pipe

Service connections to the distribution main are made with a fitting called a ferrule with diameter 6 mm less than the service pipe diameter, except in the case of a 13 mm service pipe diameter (the minimum size) which has a tapping of the same diameter as itself. There are usually two stopcocks on the service pipeline: one at the boundary to the consumer's property, which may be operated only by the water utility, and one just inside the consumer's property for his own operation. The ferrule is normally inserted into the main by means of an 'under-pressure' tapping machine as shown in Figure 16.2. The machine bores a hole into the main and taps a screw-thread in the hole. By rotating the head of the machine, the ferrule is brought into position over the hole and screwed into place. The ferrule has a plug in it which can be screwed down, thus cutting off the supply.

Service Pipe Materials

Materials used for service pipes include polyethylene, PVC-U, steel, copper, lead and lead alloys.

Polyethylene pipes. Polyethylene (PE) (Section 17.25) is now the preferred material used in the UK for service pipes and is being installed increasingly internationally, where locally

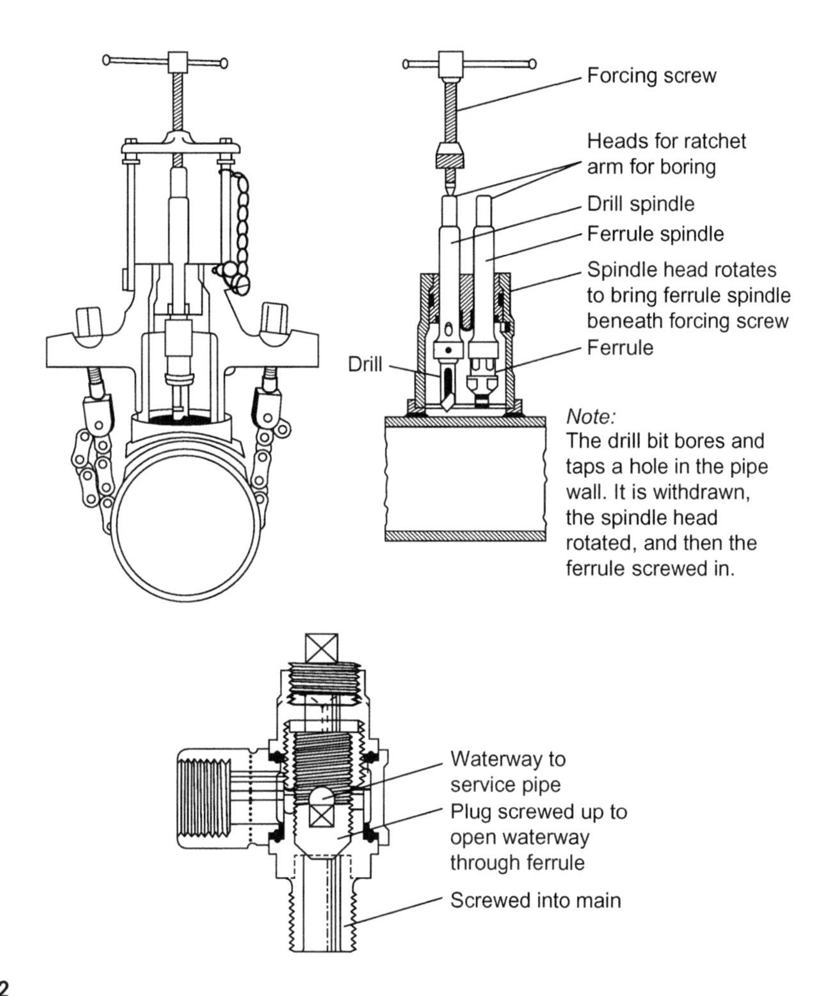

FIGURE 16.2

Under-pressure tapping in a service main showing the type of ferrule inserted.

manufactured or where material costs are comparable with other pipe materials. BS EN 12201:2011 specifies the requirements for the most commonly used sizes for cold potable water services according to pressure and with 20°C reference temperature. The standard refers to blue pigmented pipes typically used for below ground systems or where the pipes are protected from sunlight by enclosure in ducts or buildings, and to black pigmented pipes (with blue stripes to indicate potable water) generally laid above ground. Dimensions of polyethylene service pipes are given in Table 16.4.

PVC-U pipes. Unplasticized polyvinyl chloride (previously known as uPVC) piping is described in Section 17.18. It is used for service pipes in temperate climates and is increasingly being installed for cold water domestic plumbing pipework. It is not an approved material for hot water plumbing and is not wholly suitable for hot climates due to temperature derating requirements;

Table 16.4 Polyethylene and PVC-U for cold potable water up to 20°C and steel pipes

Nominal size/bore (mm/inch)	Polyethylene pipes to BS EN 12201:2011; Part 2: Tables 1 & 2 PE80 & PE100				PVC-U pipes to BS EN 1452-2:2009; Part 2: Tables B1 & B2				Steel pipes to BS EN 10255:2004		
	Mean OD		Wall thickness PN16 – PN20		Mean OD		Wall thickness		Outside diameter		Wall thickness (mm)
	Min. (mm)	Max. (mm)	Min. (mm)	Max. (mm)	Min. (mm)	Max. (mm)	Class PN12 (mm)	Class PN15 (mm)	Min. (mm)	Max. (mm)	
16	16.0	16.3	2.0	2.3	21.2	21.5	–	1.7	21.0	21.8	3.2
20	20.0	20.3	2.3	2.7	26.6	26.9	–	1.9	26.5	27.3	3.2
25	25.0	25.3	3.0	3.4	33.4	33.7	–	2.2	33.3	34.2	4.0
32	32.0	32.3	3.6	4.1	42.1	42.4	2.2	2.7	42.0	42.9	4.0
40	40.0	40.4	4.5	5.1	48.1	48.4	2.5	3.1	47.9	48.8	4.0
50	50.0	50.4	5.6	6.3	60.2	60.5	3.1	3.9	59.7	60.8	4.5
75	75.0	75.5	8.4	9.4	88.7	89.1	4.6	5.7	75.3	76.6	4.5

because the maximum working pressure reduces by about 2% per 1°C above 20°C. However, it does represent a low-cost pipe where locally manufactured. The pipes should be stored under cover to protect them from the ultraviolet rays of sunlight. The pipes are corrosion-resistant, light to handle and easy to joint using either solvent cement or compression joints. Solvent cement joints are not now favoured; if a solvent joint is broken it cannot be remade. PVC-U pipes are particularly susceptible to damage due to poor workmanship during laying and backfilling. Dimensions of PVC-U service pipes to BS 1452:2009 are given in Table 16.4.

Steel pipes are no longer used for service connections in the UK. However, they are still used in many countries because they are one of the cheapest forms of pipe material and can sustain high pressures. They may be supplied 'black' (untreated) or galvanized and with a range of internal and external protection systems. When laid unprotected in aggressive ground conditions, the life expectancy of steel can be as little as 2–5 years. Poor installation, particularly associated with the rigid screw-thread joints, can result in high leakage from an early stage in the life of the pipe. Many older properties have steel internal plumbing pipes. If steel pipes must be used for underground service connections, heavy grade steel tubes to BS EN 10255:2004 should be used, galvanized and, if laid in corrosive ground, wrapped with protective adhesive tape. For such pipe permissible working pressures are ample for the highest waterworks distribution pressures likely to be met in practice. Dimensions of steel service pipes are given in Table 16.4.

Copper pipes are widely used for plumbing and sometimes for service pipes where ground conditions are corrosive to iron. They are strong, durable, resistant to corrosion, easily jointed and capable of withstanding high internal pressures, but are expensive. Pipes and fittings are jointed either by compression or solder/capillary joints. However, lead has been found in water samples resulting from the solder and fluxes used to joint copper pipes in new buildings and when replacing existing plumbing fittings and there is concern about the contribution of soldered joints to the lead concentrations in water; particularly in the first draw after water has been standing in the service pipe overnight (Section 7.32).

Copper pipe can be used to replace sections of iron pipe provided 'dielectric union pipe fittings are used to connect the two dissimilar materials. The mechanical coupling incorporates a plastic spacer between the two materials to prevent metal to metal contact, consequential galvanic corrosion and ultimately, joint and pipe failures.

Lead pipes. Lead and lead alloy pipes and fittings are no longer installed but many older service pipes in the UK and overseas are of lead and still in use. An EC Directive required the maximum concentration value for lead in drinking water to be 10 μg/l from 2013 (Section 7.32). UK water utilities have been relying mainly on treatment solutions to achieve the standard. Where treatment solutions alone will not achieve the EC standard, utilities have needed to implement significant lead service pipe replacement programmes. In addition, if a consumer replaces his lead supply pipe with another pipe material, the utility is required to replace any lead piping used in the communication pipe. UK utilities have been replacing lead communication pipes systematically in parts of their networks when carrying out mains rehabilitation and service improvement works and are required to replace the lead communication pipe if regulatory water quality sampling results in a failure at the location of a random sample.

In the USA, the US EPA has set a maximum level of 15 μg/l for lead. If this is exceeded the water has to be treated to reduce its plumbosolvency and, where this treatment is not effective, lead pipe replacement must be undertaken.

Flow Requirements

The maximum demand rate imposed on a system from a house depends on the amount of storage, if any, provided on the premises. In most modern UK plumbing systems the only storage provided is that on the hot water system and in WC flushing cisterns. In such cases all cold water supplies to taps, showers, WC ball valves and the cold water feed tank to the hot water system are fed directly from the mains. In other cases, generally older properties, only the cold water tap in the kitchen is supplied directly off the mains; all other taps, WC cisterns and showers being fed via the cold water storage tank.

BS EN 806 Part 3:2006 recommends that domestic internal systems are designed to deliver a total supply not exceeding 2.0 m/s in header pipes, rising pipes, floor service pipes, and not exceeding 4.0 m/s in connection pipes to one fitting (dead legs). Simultaneous discharges are likely to cause reduced flows and could cause appliances to malfunction. If the pressure in the mains is high this effect may be less noticeable but when it is below about 30 m the impact is observable; for example, fully opening the kitchen tap tends to reduce the cold flow to washbasin and bath taps on the floor above. Conversely, the recent trend to install pump assisted (power) showers has significantly increased the potential peak demand rate from individual dwelling units. Therefore, the maximum rate of flow available to a property and to individual fittings and appliances will depend upon the type of plumbing arrangement within the property and the capacity of the existing service pipework. Table 16.5 summarizes the recommended design and minimum flow rates for a range of fittings. These flow rates should be used for sizing plumbing pipework only.

For 'high-rise' flats the cold water supply from the mains is usually boosted by a small pump to a roof tank which then feeds all the supplies to the flats below. The peak demand from the group of flats determines the amount of storage needed and the pump characteristics and controls.

Service Pipe Sizing

For sizing a service pipe the peak demand rate must be estimated and the minimum mains pressure must be known. The flow and headloss through the piping can be calculated using one of the formulae described in Sections 14.4 and 14.5. Table 14.1 suggests some friction coefficients that can be used in the Colebrook–White formula for new small diameter service pipes. For service pipes and fittings that have been in use for some years, friction coefficients may be much higher but are too variable to be able to quote reliable values, for example, the headloss in an old service pipe can range between 0.5 and 2.5 m/m of pipe at a flow of 9 l/min. Therefore, a 'realistic' friction value should be used bearing in mind that the k_s values in the Colebrook–White formula are meant to represent 'equivalent sand grain size' of the pipe's internal roughness and that allowance must also be made for the relatively high number of joints and fittings occurring on service pipes and internal plumbing. Table 16.5 suggests headlosses expected through typical fittings. Pressure losses in the tapping to the main, in stopcocks, and at ball valves and taps may represent the major pressure losses on the delivery line. It must be appreciated that no formula gives consistently identical results to those obtained even in laboratory tests and, in practice, discrepancies will be not less than $\pm 10\%$. Hence, for design purposes actual flows should be taken as less than the calculated value and headlosses as somewhat more.

BS EN 14154-1:2005 requires that the maximum headloss across water meters within 'Rated Operating Conditions' should not exceed 6.4 m; ranging from 1 m upwards depending on

Table 16.5 Design flow rates and headlosses through fittings (**For sizing pipes only**)

	Rate of flow (l/min)		Headloss (m) at flow rates		
Design flow rates	**Design**	**Minimum**	**7.5 l/min**	**15 l/min**	**25 l/min**
Domestic kitchen tap (DN 15)	12	6	0.7	1.7	4.9
Washing machines and dish washers	12	9			
Handbasin (pillar/mixer taps)	6	6			
Bath (DN 20)	24	18			
Shower head	12	9			
Ferrule 13 mm			0.3	0.7	2.4
Ferrule 19 mm			0.15	0.4	1.1
Float valve to tank – 6 mm orifice			1.7	4.7	–
Float valve to tank – 9 mm orifice			0.7	1.8	4.0
Service pipe fittings headlosses				**'K' factor in $V^2/2g$**	
Joints in 13 mm pipes				1.2	
Joints in 19 mm pipes				0.8	
Elbows				1.0	
Short radius bends				0.8	
Equal tees – flow to branch				1.5	
Screwdown stopcock – fully open				8.2	
Gate type stopcock				0.5	

Note: Losses for service pipe fittings are for average condition.

the Pressure Loss Class of the meter (Class ΔP 10 to Class ΔP 63); the class being chosen by the meter manufacturer. It is therefore best to consult manufacturers' literature for actual headlosses.

All supply meters need to be tested for accuracy every few years. Section 18.32 includes information on revenue meter accuracy and recommended meter calibration testing intervals.

16.6 WATERWORKS BYELAWS

Water utilities may have regulations or byelaws setting out a variety of requirements with respect to consumers' use of water and the materials and design of plumbing systems. In the UK all the water companies' individual byelaws were repealed and replaced by the *Water Supply (Water Fittings) Regulations 1999* (HMSO, 1999a). These Regulations, issued by the Secretary of

State for the Department for Environment, Food and Rural Affairs, follow requirements laid down by the EC. Although not as prescriptive in detail as some of the previous UK byelaws, they maintain many of the previous byelaw provisions, such as the requirement that *'no water fitting shall be installed, connected, arranged or used in such a manner which causes or is likely to cause waste, misuse, undue consumption or contamination of water supplied by a water undertaker'*. Additional provisions of the 1999 Regulations (in Section 5) include a requirement to notify the water undertaker of any proposal to install certain features likely to increase water demand including, inter alia: a bath of capacity over 230 litres; a pump or booster drawing more than 12 l/min; *'construction of a pond or swimming pool of capacity greater than 10 000 litres ...... designed to be replenished by automatic means and is to be filled with water supplied by the undertaker'*.

Schedule 2 to the Regulations gives details of requirements for plumbing fittings for cold and hot water supply systems. Among the principal provisions is that WC flushing cisterns installed after 1 January 2001 must not give a single flush exceeding 6 litres. Until then cisterns of 7.5 litres could be installed; but there is no time limit to the replacement of cisterns installed before 1 July 1999 by ones of smaller capacity. Both flushing cisterns and pressure flushing cisterns are permitted; but pressure flushing valves may only be installed in a building which is not a house. As siphonic flushing cisterns are no longer mandatory, flush cisterns equipped with flap valves are permitted (Table 1.4).

An overall requirement is that fittings and their installation must conform to *'an appropriate British Standard or some other national specification of an EEA [European Economic Area] State which provides an equivalent level of protection and performance'*. Further guidance is given in the *Water Industry Act, 1999* and the *Guidance Document to the Water Supply Regulations 1999* (HMSO, 1999b, c).

16.7 DISTRIBUTION SYSTEM MAINTENANCE

The distribution system comprises a complex set of interconnected pipes and fittings whose collective performance maintains the supply of water to the customer. Without ongoing maintenance the assets would deteriorate and ultimately fail. Replacing assets at predefined intervals or only on failure are the extreme approaches, which invariably do not satisfy stated outcomes. Therefore, the assets in a system need to be maintained, rehabilitated and replaced in a structured way that delivers a safe and reliable supply to the customer, maximizes the life of each asset and minimizes the cost. Striking a balance between these conflicting objectives is the responsibility of the asset planner supported by the frontline staff operating the network, the district inspectors and zone engineers.

The components of interventions used to manage and maintain a network are typically incorporated into a utility's Leakage Strategy which comprises:

- asset renewal to maintain and replace mains and service connections;
- pressure management to minimize supply pressures and leakage;
- district metering to monitor system flows and for water losses;
- targeting, detection and location of unreported leaks;
- minimizing repair times for visible and detected leaks.

The district inspector is the utility's interface with its customers on the doorstep. He is responsible for initiating risk assessments for and overseeing physical interventions on the network for planned and emergency maintenance activities and for dealing with consumer complaints. The zone engineer is responsible for the network, managing its performance and authorizing planned and unplanned interventions in order to minimize interruptions to supplies.

Customer complaints and reported operational anomalies are generally investigated by the district inspector. Pressure data can be retrieved from instrumentation at the nearest critical monitoring point (CMP), typically the highest supply point within the DMA or supply zone at which a permanent level of service pressure monitor is installed. Pressure measurements may also be taken at the customer's tap. Flow data is retrieved from bulk, district and, possibly, from revenue meters. A water sample might also be taken from the customer's tap for testing. Further investigations may then include analysing the system's performance using a hydraulic model. The objective is to determine whether the problem is an operational one, for which immediate remedial measures can be initiated, whether it is due to a gradual deterioration of the main requiring asset renewal that will need longer term planning and financing to resolve, or whether it is a problem with the customer's internal plumbing for which technical advice may be offered.

Interruptions to supply generally result either from equipment failure, for example, pump or electrical fault, a burst pipe or from planned maintenance of short duration. The impact of interruptions can be minimized either by repairing the failed asset or by rerouting supplies to restore the supply.

Flow and pressure complaints from customers may result from interruptions to supply or they may be caused by longer term network deterioration, in which case a more comprehensive analysis will be required to understand the problem and to develop the optimum rehabilitation or renewal solution. Dirty water complaints are often related to the disturbance of corrosion deposits from old cast iron mains or of manganese deposits caused by a change in the hydraulic conditions in the network. Dirty water problems can be rectified by mains cleaning or, if persistent, by mains renewal or relining. Some complaints may concern the presence of animals or insects due to larvae having infested the system (Section 7.78).

16.8 THE IMPORTANCE OF CONTROLLING WATER LOSSES

The public are generally unaware of the technical and operational complexities of delivering water into their premises until the supply is interrupted at which point any neutral or positive attitude towards the utility can rapidly deteriorate. National press and television coverage of an event often reinforces a sometimes inaccurate assessment of a utility's service and financial performance. Utility leakage and inefficiency tends to be the focus of attention of the public.

Water losses, a component of non-revenue water (NRW) (Fig. 1.1), are made up of apparent and real losses and unbilled authorized consumption. The two components of unbilled authorized consumption, unbilled metered and unmetered consumption, can be managed effectively either by installing permanent meters to measure consumption or through regular monitoring exercises to assess demand. Customer metering inaccuracies, usually the more significant component of apparent losses, can be minimized by regular monitoring of consumption, by inspection, recalibration and replacement of meters and by management of billing procedures to minimize recording errors. Unauthorized consumption, also a component of apparent losses, is more difficult to manage because of the difficulties in quantifying unauthorized consumption and locating illegal connections. Consequently these losses tend to be 'hidden' in the global water balance and per capita figures. Leakage from trunk and distribution mains and service connections, the real losses, generally represent the majority of NRW.

Consumer perception is that a distribution system should not leak, that any leakage represents an additional financial burden on the customers and demonstrates inefficient operation of the network by the utility. However, reducing water losses and wastage involves both the utility and the consumer. The utility needs to provide adequate resources to analyse and monitor performance, detect and repair leaks, identify and quantify apparent losses and to educate and encourage consumers to reduce leakage and wastage and practice water efficiency and conservation. The customer needs to become more aware of the value of water and how to reduce its profligate use both within the household and outside. Some leakage is unavoidable from a system.

Leaks are categorized as either '*reported*', visible leakage that is reported by the public or utility personnel, or '*detected*', identified during performance monitoring, field investigations or location surveys. Leaks continuously break out and therefore the quantity of water leaking from a system is the aggregate sum of the flow rate of each leak multiplied by the time it runs before repair. The hypothetical underlying rate of rise in leakage is termed 'the *total Natural Rate of Rise*' (NRRt) (UKWIR, 2011); defined as the growth in leakage which would occur if neither detected nor reported leaks were repaired.

Leakage is managed either by a passive or active control strategy. Passive control involves reacting to a reported leak or unexpected change in network performance, flow or pressure. Leaks are detected, located and repaired only when the utility has been made or becomes aware that there is a leak or a supply problem in the network. Where this approach is practiced, as in many developing countries or where water resources are plentiful, the undetected leaks and hence underlying system losses will gradually rise and the network assets will deteriorate.

Increasingly financial and environmental pressures are forcing water utilities to implement Active Leakage Control (ALC) strategies whose purpose is essentially to expend sufficient effort and resources to manage the NRR and thereby maintain leakage at, or reduce it to, a defined target. For each system, there is a '*baseline*' or '*Policy Minimum*' level of leakage: the lowest level of leakage that can be achieved through intensive active leakage control using conventional methods and reasonable effort (Ofwat, 2008). Policy minimum introduces the concept of the cost and benefit of the activity, recognizing that the cost of reducing leakage further increases disproportionately as leakage is reduced. Therefore, the policy minimum represents the lowest level of leakage that is economic to pursue, beyond which further reduction through ALC is not practicable. The corresponding underlying rise of leakage is called the '*detectable Natural Rate of Rise*' (NRRd). Therefore, NRRd is the difference between NRRt and the equivalent volume from reported leaks.

The Economic Level of Leakage (ELL) is when the cost of further leakage reduction becomes greater than the additional benefit gained; in other words: the level at which the marginal cost per m^3 of constructing and operating new water resources assets equals the operational and capital cost of leak detection and repair per m^3 of additional water saved. There are various methodologies for calculating the ELL, including ones which also take account of social and environmental costs and benefits, the *triple bottom line* (social, environmental and financial). Some components of these external factors are difficult to quantify in economic terms but they are fundamental to the principles of sustainable management of water resources. The calculation establishes the level of effort and thereby resources necessary for minimizing leakage for a defined set of conditions. In practice the output is the least cost set of interventions covering asset renewal, pressure management and ALC that delivers the minimum whole life cost (Ofwat, 2008; EA, 2012).

Regulated utilities manage leakage within targets set by or agreed with the regulator. For utilities above their ELL, annual targets may be set to achieve a 'glide path' to the ELL over a period of time. Once at the target level, the utility may agree to hold leakage at that level or offer to reduce leakage below the economic level for a variety of reasons; these may include managing their supply and demand water balance, recognizing the need for environmental sustainability or managing customers' and shareholders' perceptions of their performance. Utilities not subject to external regulation may also set themselves internal performance targets for similar reasons. Targets can be set either at the company level or, where the data are available and leakage levels justify a detailed calculation, for individual supply zones and DMAs. It is more common to calculate the ELL at company level and use the value to determine individual targets for DMAs.

Table 1.14 (Chapter 1) presents some typical 'background' leakage levels from UK water distribution networks; Table 1.13 shows rates of burst frequency for a range of materials in a number of countries worldwide.

The scale of the problem of detecting and locating leaks in the thousands of kilometres of pipework that make up a typical water distribution system should not be under-estimated. Modern leak detection equipment can be used to assist in locating leaks and flow monitoring can assist in identifying components of system losses, but they are only tools which aid leak detection. Reducing leakage is therefore inevitably a labour-intensive process which has to be pursued continuously.

16.9 ACTIVE LEAKAGE CONTROL

The objective of ALC is to shorten the duration of leaks and so reduce the quantity of water lost by minimizing:

- *awareness time*: the time between the start of the leak and the utility becoming aware of it;
- *location time*: the time to detect the leak and locate its position;
- *repair time*: including scheduling the work, obtaining permissions and carrying out enabling works where necessary.

District metering, ALC and minimizing repair times all contribute to reducing the time between development of a leak and its detection and repair. The process involves the sequence of:

Monitor – Target – Detect – Locate – Repair – Monitor

District metering provides the infrastructure for monitoring network performance and enabling the leakage technician to analyse and understand system and minimum night flows and consumption, and to identify those areas of the network where leakage is greatest. This process is also used to monitor flows at the end of the sequence to ensure that leakage has been reduced and assess the effectiveness of the overall process. Once located, a leak must be repaired speedily within the constraints of obtaining necessary permissions from third party organizations responsible for the public highways and open areas or private land owners through which the pipes are laid. Generally such permissions take longer to obtain than it would to organize the repair gang to execute the repair unless the burst or leak is sufficiently large to justify invoking emergency powers allowing the utility to commence the repair immediately. Visible leaks tend to be repaired as they are reported: the same day, or within a few days if the size of the leak is assessed to be small and there are more urgent repairs to be carried out. Leaks detected on mains and leaks and bursts on service pipes can take longer, typically 10–15 days for mains and 30–45 days for minor leaks on service pipes. Reporting repair times against internal targets can be used as a performance metric for assessing outcome performance.

Once leakage has been reduced in an area to a given target quantity – termed the *exit level* – ALC and repair activities in that area can be suspended. However, the area should continue to be monitored to detect as and when the supply into the area increases. The flow is analysed to assess whether any increase is a result of increased demand or leakage and, if the latter, whether it has risen to a level at which it is necessary to re-enter the area and to repeat the detection and repair process. This threshold level, termed the '*intervention level*', is determined by a combination of factors including: the ALC resources available within the organization; the unit cost of water in an area compared with the unit cost of reducing NRW in that area; other savings that can be made by reducing losses; and whether other areas should take priority. The aim of setting targets and thresholds is to maximize water loss savings by targeting the area or areas which will deliver the higher value or immediate returns. As the characteristics of each area are better understood, targets and thresholds can be reviewed and revised, taking into account the latest cost information on water production and manpower and the resources required to achieve the targets. Where a target has been exceeded or significant rehabilitation has taken place, it may be appropriate to reduce both the target and threshold to reflect the improved physical condition of the asset. Political and environmental considerations also influence intervention levels; a utility may need to show it has reduced leakage down to a target set by a regulating authority or to justify a proposal to develop a new source. Targets therefore need to be reassessed periodically to ensure that they remain relevant to the prevailing network and business conditions and strategies.

16.10 DISTRICT METERING

Targeting ALC activities relies on monitoring the system for evidence of leakage. The most effective method of monitoring is to divide the system into three levels of control:

- zones of 10 000–25 000 connections or perhaps more;
- DMAs of typically 500–3000 connections;
- waste districts of a few hundred connections.

The data derived from monitoring a zone is generally too coarse for detailed targeting and efficient deployment of ALC resources. However, zone monitoring can be considered as first stage monitoring for utilities that are starting to address leakage. Typically a zone would comprise a number of DMAs. Monitoring at the DMA level of network subdivision is the most common method. The data is at sufficient detail to be able to target ALC activities effectively to individual DMAs. Each DMA can comprise one or more waste districts. Waste districts are generally too small to be closed in permanently for efficient monitoring without introducing water quality and pressure performance problems but they are effective subdivisions for detecting and locating leaks within a DMA.

Ideally the design of these control areas should incorporate any existing system divisions and natural boundaries; physical or hydraulic. Permanent flow meters are installed on the feed mains to zones and DMAs. Preferably there should be a single feed main for each zone or DMA, although this is not always possible. However, DMAs fed by more than three meters are generally difficult to monitor and the flow analyses can be unreliable. For the waste districts, a bypass in a chamber is constructed around a valve on the main feed to the district, so that a waste meter can be temporarily inserted in the bypass when undertaking waste metering by step-testing. The boundary valves dividing zones and DMAs are normally kept closed unless, for operational reasons, some DMAs need to be combined. Pressure monitoring equipment is also installed at selected *'critical monitoring points'* in the system.

Historically, a venturi meter or a shorter form, the 'Dall tube' (Section 14.16), was installed in a large pipe feeding a zone; but these types of meters are no longer used because of their high cost. Instead electromagnetic and ultrasonic flow meters are now generally preferred. Inferential meters may continue to be used in smaller mains, equipped with electrical pulse flow recording equipment but they are being replaced progressively by electromagnetic meters (Section 18.28).

Zone and DMA flow and pressure measurement data can be transferred to the Operational Control Centre to be processed, stored and monitored using an integrated telemetry system, hard wired, by radio or GPS technology. The data is used to monitor daily demand and minimum night flows and is compared with historic data or calculated theoretical flows to identify differences and anomalous data. The data transfers and processing would typically be completed overnight, the output being a summary report which identifies areas where flows, pressures and processed data are outside predefined parameters, *'exclusion reporting'*, and indicating where further data analysis or investigation may be necessary. Alternatively, the data can be stored in on-site data loggers that are interrogated periodically, and their data transferred to the flow monitoring database for analysis. The data from on-site loggers is typically downloaded on a 28-day cycle, implying an inevitable delay before the information can be processed.

For many large water utilities in developing countries, setting up district metering or waste districts can be very expensive, because of the need to purchase and install new meters and boundary valves and to replace or repair many existing valves needed to create discrete hydraulic boundaries. Implementing district metering and flow monitoring has to be, therefore, a long-term plan starting with the zone metering of supplies, and progressively developing DMAs where zonal flows indicate losses are probably highest. For utilities only able to provide intermittent supplies, quite different methods for checking levels of consumption and losses have to be adopted, as described in Section 16.17.

16.11 PRESSURE MANAGEMENT

Reducing the operating pressure in a network has several benefits:

1. Research has shown that the quantity of water lost through a leak or burst is proportional to the operational pressure of the system. Therefore, reducing pressure, particularly during periods of low demand, will have a significant impact on leakage.
2. Lower and more consistent pressures can extend the asset life of the network assets by reducing the stress on pipes imposed by continuous high and transient pressures.
3. Reduced distribution pressures at times of low demand will also provide the consumer with a more consistent supply pressure and hence flow rate from fittings.
4. Consistent reduced pressure can also result in water conservation by consumers without them necessarily noticing. The flow rate from pressure dependent fittings supplied directly off the mains is reduced; examples are showers, taps and toilets that are equipped with flush valves rather than flushed through a cistern.

Pressure management is achieved by installing pressure-reducing valves (PRVs) on the inlet pipework supplying a DMA or PMA (Plate 25(a)). The size of a PMA is a function of the area to be pressure managed, the proposed change in pressure and the proposed type of valve. However, the PRVs require periodic maintenance which will increase network operating costs.

A PMA can comprise a group of DMAs, a single DMA or a part of a DMA. Hydraulic and transient models are useful tools to test a proposed design for pressure management and for developing control rules for areas supplied by multiple feeds where there is likelihood of '*hunting*' between the control valves.

Energy recovery turbines provide opportunities to generate electricity from the surplus energy that would otherwise be dissipated through a PRV. In practice energy recovery turbines are expensive, high maintenance items and therefore tend to be installed in distribution networks only where economically viable and where there are no environmental impact and regulatory restraints. See also Section 13.4.

16.12 WASTE METERING AND STEP-TESTING

Waste metering and step-testing is a method of localizing a leak by monitoring the flow to a small section of the distribution system. Sections of network are systematically shut down within the area in 'steps' in a predefined sequence, while monitoring the change in flow as each main is isolated. A flow reduction larger than expected from the properties that are connected to the main may indicate that the main that has just been isolated has a leak. The test results are interpreted and passed to the district inspector who arranges for further investigations to locate possible leaks. After all suspected leaks have been located and repaired the night test is repeated to confirm that the exercise has been effective.

The cost of installing and maintaining meters and valves for waste metering and step-testing tends to be high. A bypass including a 'waste meter' and isolating valve is installed on the feed main into the waste area (or district). During the test, all flow into the area passes through the meter with all other feeds shut-off. A DMA meter can also be used for step-testing provided it is sufficiently sensitive to measure the small changes of flow as each main is shut down.

Step-testing is only practicable on 24-hour supply systems; it is time consuming and generally has to be carried out at night when legitimate network flows are low. Computing and GPS technology developments have, however, enabled the inspector to monitor and analyse flows during the test remotely and thereby assess there and then whether the drop in flow relates reasonably to the number of domestic connections on the main and any legitimate non-domestic night-time demand or if there is a suspected leak.

Step-testing increases the risk of creating a water quality incident by disturbing sediments when closing and opening valves. For these reasons acoustic logger surveys are replacing step-testing for locating leaks.

16.13 LOCATING AND REPAIRING LEAKS

Many visible leaks are reported first by the public. However, more can be found by district inspectors looking for signs of leaks, particularly in areas with a history of pipe failures and leakage. Damp patches, trickles of running water and extra vegetation growth close to pipe alignments, valves or fire hydrants, or above ferrules or stop taps, may be indicative of leaks. The inspector is also able to detect signs of consumer wastage, such as overflows discharging outside properties. However, beneath metalled roads, or where mains are laid in freely draining ground, or adjacent to or below a watercourse, even large leaks may not show on the surface. Trunk main routes through open country should also be inspected periodically for indications of leakage.

Unreported leaks can be detected and located using ALC. The processes will initiate detection and location activities as the leakage flow rate increases beyond the trigger threshold. Leaks can then be located using one or more techniques such as 'sounding', leak noise correlation, acoustic logging, ground penetrating radar, in-pipe inspection, tracers and gas injection.

Locating Leaks by Sound

Monitoring the sound made by water leaking from a pipe is the basis of the majority of leak detection techniques and electronic equipment currently available. The traditional 'listening stick' is a light solid metal bar about 1.5 m long or the bar of a valve key. Purpose-made listening sticks can be equipped with an earpiece. One end of the stick is placed on an exposed part of the main or service pipe, such as a valve or hydrant spindle or stopcock, and the other is placed against the ear. The sound emitted by a leak, if audible, is 'a low drumming noise' or 'a continuous buzzing sound' and tends to be continuous without any change of audibility or quality. It stops abruptly when, and if, the water can be turned off. An experienced waste inspector using a listening stick can detect even a small leak at a distance of 10–15 m, if it is making a sound. By listening at another point of contact on the main, the inspector can judge by the difference in sound volume the probable location of the leak. Sounding is frequently carried out at night when background noise should be low. However, an experienced operator can still be successful during the day away from main roads and when there are lulls in traffic noise. Daytime sounding may also be preferred because, at night, parked cars can prevent access to valves and stopcocks.

Sounding all pipes in a system is not usually adopted because it is very labour intensive, relatively ineffective on non-metallic pipes and not efficient at identifying new or increased leakage as it occurs. In the UK where many service connections do not leak an inspector can sound up to 200 connections per shift. Overseas where pressures are lower and background conditions are less favourable, an experienced inspector should be able to sound 80–120 connections per day on average.

Ground microphones, first introduced in the 1960s, amplify the sound of a leak and are effective for leaks from non-ferrous pipes. It must be borne in mind, however, that not all leaks emit an audible sound, and the volume of sound emitted is mostly not related to the size of the leak.

Leak noise correlators have been available for about 40 years. They are electronic devices used to analyse the sound of a leak picked up by sensors in contact with the pipe. The software analyses the sounds from the two points of contact to estimate the positions of the leaks, using the sound time-delay between the sensors and pipe material, diameter and distance between sensors as input data. The technology continues to be developed and has improved such that portable digital correlators are effective for all pipe materials and easy to use by less experienced operators.

Acoustic loggers are used both to detect and locate leaks, either as permanently installed instruments at defined locations or temporarily deployed for a survey of an area. When permanently deployed, they monitor continuously for changes in noise characteristics, an indication of changes in network flows and hence possible leakage. The loggers are downloaded periodically either remotely or manually or specifically to investigate an identified anomaly. When temporarily deployed, sets of loggers, typically eight or 10 loggers per set, are installed in an area at predefined points. They are deployed for 24 hours and programmed to log during the minimum night flow period. A whole supply zone may be surveyed over two or more consecutive nights. Deployment may be for longer periods where the leakage team is attempting to reduce the baseline leakage level in the zone.

The data retrieved from both types of installation are analysed to filter out the background and 'normal' system noises and then combined with pipe diameter and material data and the length of pipe between logger locations, to correlate the location of possible leaks by triangulation. The same monitoring points are always used for regular repeat surveys within a system. System noise profiles can be developed and used to assess system changes and predict the need for intervention. A group of acoustic loggers can successfully detect and locate a number of leaks at different locations within the triangulation of the deployed loggers. The equipment suppliers suggest that the location accuracy provides sufficient detail for the utility to be able to instruct the repair contractor without the potential financial risk of digging a '*dry hole*'; excavating a pipe where there is no leak. However, where the utility is constrained by third parties when opening excavations in the public highway, located leaks should be confirmed on site using a leak noise correlator before instructing the repair contractor to proceed.

Some utilities are using acoustic loggers instead of setting up district metering, relying on the measurements from zonal metering to analyse flows. Although this does represent a short-term expedient where resources are constrained, the level of detail available from zone metering is too coarse to understand variations in subsystem flows and demands and hence to prioritize the critical leaks for location and repair. Therefore, in the longer term acoustic logging represents a key tool in leakage management but will not replace the detailed understanding of network operations and performance derived from monitoring flows in the smaller DMAs.

Other Technologies Deployed for Leak Location

Detecting and locating leaks in trunk mains and large critical mains can be particularly difficult. The pipes cannot be isolated for testing, are of large diameter, often constructed of different materials to those used in distribution networks, are laid across open terrain, or not readily accessible and include only limited numbers of valves and other fittings to which sensors can be attached. For these conditions, pipe integrity management systems and in-pipe acoustic technologies are appropriate.

Pipeline integrity management systems use continuously monitored flow and pressure data linked to hydraulic modelling software to provide on-line active leak detection on a pipeline. The technology, developed from the oil and gas industry, is suitable for application to lengths of trunk main where pipe failure could be damaging and would need immediate, automatic shutdown of a section of the main. The technology analyses flow and pressure measurements continuously to assess the performance of the main between the sensing points and compares the measured values against normal pipeline performance characteristics. Where the software detects abnormal measurements, the software calculates the location of a leak and initiates the closure of appropriate valves in order to isolate the fractured length of pipe. The equipment is expensive to install and maintain but cost effective in terms of the consequential damage that might otherwise occur if a large pipeline were to fail.

Tethered in-pipe sensor technology. Tethered pipe inspection tools comprise a sensor attached to a cable which is inserted into a pipe through a tapping point. The Sahara® sensor travels with the flow, assisted by a drag chute and controlled by the tension on the cable which is inserted through a 50 mm diameter minimum pipe tapping. The equipment is retrieved by rewinding the cable back up the pipe. The leak detection sensor is suitable for all material types and pipes of 150 mm diameter and larger. Readings from the sensor can be tracked on the surface and located to within 500 mm. The pipe wall assessment sensor, requiring a 100 mm diameter tapping, is used to assess the internal condition of metallic pipes of 400 mm diameter and larger.

The JD7™ 'tethered' technology involves installing the sensor through an in-line fire hydrant (Plates 25(c) and (d)). The sensor cable is inserted by hand up to 100 m either upstream or downstream of the insertion point. The cable is then withdrawn at a controlled speed by a drive mechanism while readings are taken at pre-set intervals. Different sensor heads can be deployed including camera, hydrophone for detecting leaks and ultrasonic scanner to produce a full diameter survey of the pipe wall thickness. The benefit of the technology is the accurate location of specific features and repeatability of the survey so that a 'library' of pipe condition can be developed over time to support assessments of asset failure risk and asset condition deterioration modelling.

Free swimming sensor technology. An acoustic sensor contained within a foam ball is inserted and retrieved through pipe tappings or other in-line network assets. The progress of the foam ball is tracked by detectors attached to fittings. Sensor data is recorded and on retrieval, downloaded and processed. Leaks can be located to within 3 m in pipes of 150 mm diameter and above including in larger diameter trunk mains. Sensors are also available to record flow rates along the pipe, the locations of valves, pipe joints and other fittings and the condition of the pipes.

Other techniques that have been developed include:

- robotic pipeline inspection vehicles suitable for inspecting larger diameter pipes;
- tracers and gas injection; gas injection is suitable for low-pressure systems and non-metallic and small diameter pipes;
- ground penetrating radar and thermal imaging to identify changes in soil moisture.

All these sophisticated tools tend to be expensive to deploy and usually require a specialist contractor and are therefore more appropriate for more complex surveys, such as for trunk mains and large pipes where the leaks are difficult to detect and locate.

Repairing Leaks

Confirmation of the existence of a leak is only obtained when the pipe is exposed and the leak located. However, pipe repairs can be expensive, especially if the leak is under a heavily trafficked road or road junction, and special planning of the repair operation may be necessary to minimize traffic disruption. The relevant road authorities have to be informed and all their requirements met. If the leak is known to be small and unlikely to increase in size rapidly, if the repair cost estimate is high or if the work would cause serious traffic disruption, the water utility may choose to delay repairing the pipe and accept the water loss until the leak can be repaired in a more cost effective and efficient way. One of the 'no-dig' rehabilitation techniques discussed in Sections 16.15 and 16.16 may offer a solution if internal access is feasible.

Where leak detection is outsourced to a detection contractor, the contract conditions should require the repair contractor to confirm the existence of detected leaks by excavation before the detection contractor is credited with finding the leak. Payment for detection services should be made against measured reduction in leakage flows, not the number of 'detected' leaks and theoretical leak rates. Repair contractors should be paid for all pipes excavated to repair a located leak, including 'dry holes'. The cost of excavating and refilling dry holes should be charged back to the detection contractor.

16.14 MAINS REHABILITATION

Asset renewal and pressure management (Sections 16.11 and 16.16) address the asset and operational conditions to reduce leakage but with the secondary benefits of reducing risk to water quality. Asset renewal is intended to target those pipes which are more vulnerable to bursts (visible leaks) or leaks (unreported leaks) or whose condition presents a risk to water quality and for which it would be more cost effective to replace the asset rather than rehabilitate it.

The asset planner is responsible for analysing a network's physical and operational performance, using analysis techniques, such as deterioration modelling to develop an asset maintenance and replacement plan. Increasingly performance and condition deterioration models are being used to identify pipes and network apparatus for intervention. Model output can be used in 'Asset Investment Manager' optimization tools to generate the 'Investment Plan' defining lengths of pipe to be rehabilitated or replaced. The models use physical, condition and performance data to predict failure or condition deterioration rates and need data on pipe material, wall thickness, class, age, jointing system, pressure, soil type and ground conditions, traffic loading and pipe failure and customer complaints history. Data on other factors such as lining and coating systems and condition may be included if available. The analysis compares the costs, benefits and risks of leakage interventions and a range of different options to develop and prioritize programmes of work for cleaning, rehabilitation or replacement of mains. Annual updates of the analysis can be run using the previous year's performance data but it is time consuming and may suit only small utilities.

Mains Rehabilitation and Cleaning

Rehabilitation techniques can be divided into short and longer term measures. Short-term actions include: rezoning supply areas to improve pressures and prevent interruption to supply; repairing visible leaks; active leakage control; and mains flushing to resolve dirty water problems. Valve maintenance should also be included since, for relatively small cost, air and gate valves can be repaired and leaking glands repacked. Longer term techniques include mains rehabilitation or replacement. Rehabilitation techniques can comprise non-aggressive and aggressive cleaning methods or use of non-structural and structural linings; while pipe replacement may utilize trenchless technology techniques.

Pipe Cleaning Methods

Mains cleaning techniques include flushing, air scouring, swabbing, ice pigging and scraping, of which the first four are non-aggressive. Flushing and air scouring are only effective for pipes up to 300 mm diameter.

Flushing is a well-established technique for small diameter mains where the mains pressure is adequate (Plate 26(a)). It is an effective technique for removing loose sediment and for flushing polluted water from a pipe. It is not appropriate where there is an underlying persistent problem with sediments and discoloration which can only be resolved by eliminating the source of the contaminants. For effective flushing, the velocity at the pipe invert must be sufficient to pick up material and hold it in suspension. This depends on the specific gravity of the material, pipe diameter and profile. In practice pipes up to about 100 mm diameter can be flushed through one or two hydrants supplied from both ends provided there is adequate pressure in the main. For larger diameter pipes, flushing generally needs to be through three or more adjacent hydrants, supplied from one direction only at a pressure of at least 4 bar. Achieving this may be difficult due to the system hydraulic conditions. Flushing is unlikely to be effective on mains over 300 mm diameter. When planning a flushing programme the following issues need to be taken into account:

- the longitudinal section of the pipe to be flushed including the pipe gradient and locations of washouts and hydrants;
- the characteristics of the material to be removed;
- the source of the contaminant as against its perceived location;
- the maximum achievable pipe velocity under realistic hydraulic conditions;
- the discharge capacity and locations of the hydrants or washouts in relation to the size of pipe to be flushed;
- acceptable methods of disposing of the flushing water.

Flushing a system must be carried out systematically from upstream to downstream, taking into account likely secondary water quality problems created by isolating lengths of pipe. Flushing is unlikely to be effective for a heavily tuberculated main.

Pressure jetting is a more aggressive technique for cleaning sediments and biofilm off the pipe wall. The technique can be used for short lengths of water pipes where access to the pipe can be gained and safe disposal of the waste water is viable but the technique is more applicable to unblocking and cleaning sewers.

Air scouring can be used to generate higher velocities during the flushing process without using as much water. The injection of filtered compressed air forces slugs of water along the pipe; these cause the disturbance of loose deposits on the pipe walls as the slugs form and collapse. The procedure is suitable for pipes up to 200 mm diameter and for lengths of up to 1000 m; its effectiveness relies on the skill of the operator in forming suitable air/water mixtures.

Foam swabs can be used to remove soft or loose material, such as organic debris, iron and manganese deposits, sand and stones. The process usually involves using a series of increasingly hard plastic swabs. The swab should have a diameter 25–75 mm larger than the bore of the pipe and a length of 1.5–2.0 times the bore. The softer swabs can pass through butterfly valves. They can be inserted into the main via a hydrant branch, but for harder or larger swabs the main must be opened. The optimum velocity of the swab is generally about 1 m/s. Its speed can be controlled by regulating the outflow at the discharge end, the speed of the swab being usually one half to three quarters of the water velocity because of flow past the swab. Swabbing can greatly improve the flow characteristics of a slimed main and is frequently used to clean out a newly laid main before being put into service.

Ice pigging, a patented technology developed in the UK between Bristol Water and Aqualogy, is a low risk method of removing sediment and biofilm from the inside of pipes using an ice–water slurry pumped into the pipe through a fire hydrant or similar suitable fitting. The slurry is moved along the pipe by the water pressure in the main to a downstream hydrant adjacent to a line valve. Water ahead of the slurry is discharged to waste in the normal way, but the slurry containing the pigging waste is collected in a tanker for safe disposal. The water–slurry interface is determined by monitoring the water temperature at the downstream hydrant. The pipe is then flushed in the conventional way prior to returning the section of pipe into service. The technique can be used to flush pipes up to 600 mm diameter and 3 km in length.

Scraping and relining is applicable to old tuberculated cast iron mains, which are structurally sound (Plate 25(b)). Pipe scraping methods include drag scraping, power boring and pressure scraping. Pigs or aggressive swabs incorporating grades of wire brush or studs may be used. For harder encrusted materials, several passes may be necessary and it is essential to have a good flow of water past the device to prevent debris accumulating in front of it. An electrical transmitter incorporated in the swab or pig assists in tracking its progress along the pipe. Scraping with pigs is effective in removing hard deposits but can also damage and remove linings. The latter may increase the polycyclic aromatic hydrocarbon levels in the water by exposing old coal tar linings (Section 7.42). Aggressive cleaning is usually followed by applying a secondary lining to the pipe. When considering the use of the technique, the asset planner should consider the volumes of material to be removed and replaced.

Long lengths of main can be cleaned aggressively in 'one go' if the run is fairly straight and no obstacles to the passage of the device exist. However, where line valve type constitutes an obstruction (e.g. a butterfly valve) or where tapers have been installed either side of valves so that smaller valves can be used, the length is limited to the distance between valve positions. Service connections off the main must be isolated and, after cleaning and relining the main, they must be cleaned by 'blowing back' with fresh water. The scraping and relining processes are likely to dislodge, damage or block ferrule connections, requiring them to be excavated and repaired or replaced, thereby increasing costs.

16.15 PIPE LINING METHODS

Cement mortar and epoxy resin linings applied to the cleaned internal surfaces of cast iron mains can improve hydraulic capacity, protect the pipe and reduce discoloration caused by corrosion. In situ linings can be applied to pipes from 75 mm diameter upwards. Cement mortar linings are typically 4–6 mm thick. They tend to increase the pH of the water; therefore, it is common practice to apply a seal coat particularly to small diameter pipes. For pipes up to 150 mm diameter, the reduction in pipe bore and relative roughness of the mortar surface may not provide adequate hydraulic capacity.

It is technically feasible to reline mains with epoxy resins up to 2000 mm diameter. The finished lining is 1–2 mm thick, has a projected life of up to 75 years and is very smooth providing a good hydraulic performance. The lining can be applied to most pipe materials, not only iron and steel mains. However, mixing the two part components must be carefully controlled and atmospheric conditions and cleanliness have to be carefully controlled to ensure the epoxy lining adheres properly to the pipe wall.

Structural lining methods are adopted where the structural integrity of the pipe has deteriorated. These lining methods involve inserting a flexible lining into the main by methods such as 'sliplining', '*Rolldown*', or '*Swagelining*'; the latter two being patented proprietary techniques. Rolldown involves inserting a polyethylene pipe, which has previously been reduced in diameter by up to 10% by squeezing in a 'rolldown' machine. When in position (Plate 26(b)), the pipe's 'elastic memory' is activated by internal pressure and it reverts to its former larger diameter to make a close-fit with the existing pipe wall. The process is suitable for relining pipes of 75–500 mm diameter. Sliplining and Swagelining can be used for pipes of up to 1200 mm or more and of length up to 1500 m. All these techniques reduce the internal bore of the host pipe but can significantly improve its hydraulic performance because of the smoothness of the lining. They are quicker to install and less disruptive than relaying the pipe in open cut, but require each service pipe to be reconnected.

In the '*Cured-in-place*' pipe lining method a new felt liner 'sock', impregnated with resin and reinforced with glass fibre for pressure pipe rehabilitation, is inserted into the pipe to be rehabilitated using water, steam or air pressure or is pulled into place. Lengths up to 500 m can be lined in one go. The resin is then cured by heating the water in the 'sock' by continuing to pass the steam through it or the resin is allowed to cure at ambient temperature. The inversion pressure is maintained until the resin has cured. Alternatively, inversion can be achieved using air pressure (air inversion method) and curing can be achieved using ultraviolet light or ambient resin curing systems. The method is more applicable to rehabilitating sewer pipes, although it is suitable for pipes with difficult access and complex shapes. However, the finished hydraulic surface is not as smooth as that of other methods.

16.16 PIPE REPLACEMENT

The main reasons for replacing a pipe are the risk of structural failure and inadequate hydraulic capacity. Causes of failure include: fracturing due to severe cold weather periods, prolonged dry spells causing ground movement, mining settlement, poor original workmanship when laid, age of pipes or their faulty manufacture, internal and external corrosion, inadequate cover for increased traffic loading, original class strength too low for current pressures, interference with pipe embedment or thrust restraint by other underground work, and cyclic stressing of pipes caused by operating conditions. Analysing the record of bursts on mains may reveal those which are particularly prone to failure. Where evidence indicates that a main has failed frequently during the last 5 years, typically more than two or three times per year per kilometre, it should probably be scheduled for replacement as part of an optimized integrated maintenance programme.

The traditional replacement method is to lay a new main adjacent to the pipe to be replaced. The service connections are transferred to the new main as a separate operation to minimize interruption of supply to consumers. As well as the traditional methods of trenching, newer techniques include narrow bucket open excavation, rockwheels, chain trenching machines and mole ploughing. Narrow trenching techniques (including chain excavators) for laying pipes up to 500 mm diameter are typically unsupported and avoid the need for an operative to enter the excavated trench and results in less handling of excavated material and reinstatement; but it can only be used for pipe materials that are sufficiently flexible for pipe lengths to be jointed before lowering into the trench. It is quicker and cheaper than traditional open cut excavation. However, it does increase the risk of disrupting other services and is therefore only suitable for uncongested locations, or where site possession is restricted or when rapid laying and reinstatement is required. Mole ploughing is only used for pulling small diameter pipes through soft ground.

The high cost of installing, replacing or renovating small diameter underground pipes and services by traditional methods, including surface reinstatement, has resulted in the increased use of trenchless construction methods. In addition to the relining methods discussed above, there is a range of 'no dig' or 'low dig' pipe replacement and tunnelling techniques, which include pipe bursting, directional drilling (Plate 26(c)), auger boring and microtunnelling. These techniques are typically used in urban areas, where the presence of many other underground services or traffic congestion makes trenching difficult and expensive. In pipe bursting a tapered bursting tool is winched through an existing pipe to break it up and displace the fractured material into the surrounding ground. A new PE liner pipe is pulled in behind the bursting tool. Pipes up to 300 mm diameter can be installed. However, extensive prior site preparation may be necessary to excavate or remove steel repair collars, pipe bends and fittings and concrete surrounds, and to disconnect mains and service connections to prevent their damage. Ground disturbance can occur and may be a problem.

Microtunnelling machines have been developed for tunnels up to 2 m diameter. Articulated shields, adjustable by means of remotely controlled jacks, allow the machines to be steered accurately along the required alignment which may be gently curved if required. Muck is removed by an auger or by fluid transport using bentonite or water. A permanent pipeliner is jacked into position behind the shield. When complete the water pipe is installed in the lined tunnel and the annular space filled by grout. Pipe jacking is a similar technique, usually used for pipes over 900 mm diameter. A tunnelling shield and string of tunnel lining is jacked into the ground from the drive shaft (Plate 26(d)) to a reception shaft. Lengths of several hundred metres up to about 2500 mm diameter can be achieved. The water pipe is installed inside the jacked 'tunnel'. Special ductile iron pipes without sockets but with in-wall joints have been manufactured for pipe jacking to avoid the use of a primary lining. Auger boring installs a sleeve, up to 1200 mm diameter, usually steel, between two pits. The pipe is installed as for microtunnelling.

Horizontal directional drilling can be used to install pipes under an obstruction, such as a road (Plate 26(c)), railway or river. The technique involves drilling a bore between two points either side of the obstruction, using a steerable drill string. When complete the hole is reamed out to the required size and the pipe is pulled through. Depending on the capacity of the pulling equipment, length of pull and size of pipe, pulls of up to 1500–2000 m and up to 1200 mm diameter of continuously welded steel pipes have been achieved. Care needs to be taken when pulling PE pipes through so that its permissible tensile stress is not exceeded. The radius of curvature of PE and steel pipe when installed should generally not exceed 50 times the pipe diameter.

16.17 IMPROVEMENT OF DISTRIBUTION SYSTEMS IN DISREPAIR

Worldwide, there are numerous public water utilities with distribution systems in need of rehabilitation and repair. Total NRW of 50–60% of the supply can be reported, made up of unknown proportions of distribution leakage, consumer wastage, and failure to meter or bill all consumers taking a supply (Chapter 1). It is not easy to improve the performance of a large system in a short time; resources invariably represent the major constraint and inevitably the utility has to adopt a strategy of progressive stages of improvement as finance becomes available. Ideally rehabilitation becomes self-financing as water losses are converted into revenue but, in practice rehabilitation may not keep up with the rate of deterioration and the asset life continues to reduce.

The difficulties of rehabilitation are compounded where supplies are intermittent due to scarcity of resources and large losses of water and revenue through leaks, wastage, illegal connections and unpaid supplies. Traditional methods of leak detection suitable for 24-hour supplies, such as waste metering, step-testing and acoustic logging, become impracticable because many consumers leave taps open, ready to discharge water into storage receptacles to cover periods of non-supply. Even the procedure of valving off lengths of main to put them under a pressure test for leakage after shutting down all service pipes, may be vitiated by lack of stop-valves on service pipes, by leaking ferrule connections or tappings to the main or by many illegal or unknown connections. Often a first attempt at a pressure test results in a burst main, particularly at joints, because the system has long been on intermittent supply at low pressures, or it may be difficult to gain any pressure because of leaking boundary valves or the existence of some unknown connection which has not been marked on the mains records.

A more productive approach can be to expose the soffit of lengths of pipe in selected areas to find the principal causes of loss and to investigate supply pipes and meters. Often it will be found that high losses are due to one or more of the following: badly made service pipe tappings on the main; illegal and unknown connections; illegal bypasses to customer meters; leaking service pipes not fitted with a stopcock; and service pipes continuously taking water because of waste on consumers' premises.

Exposing the soffit of pipes is time consuming and relatively expensive. However, it enables location and reduction of leakage directly and helps to find leaking joints. In many countries it is not as expensive as relaying a main since it is primarily labour-intensive work involving the use of less expensive repair materials and plant. Attempts to find all leaks on a pipe network known to be in poor condition using various methods of surface detection equipment and flow monitoring can be frustrating because of the repeated failure to achieve an acceptable result each time the system is re-tested. Repeated non-success after so much work reduces motivation to continue leak detection activities. The direct exposure approach produces positive results and, if pursued, can increase efficiency in finding leaks through experience and increases the amount of the system that is rehabilitated.

REFERENCE STANDARDS

British Standards (BSI)

BS 750:2012. *Specification for underground fire hydrants and surface box frames and covers.*

BS EN 806. *Specification for installations inside buildings conveying water for human consumption; 2:2005 Part 2 Design and 3:2006 Part 3 Pipe sizing – Simplified method.*

BS EN 1452-2:2009. *Plastics piping systems for water supply and for buried and above-ground drainage and sewerage under pressure – Unplasticized poly (vinyl chloride) (PVC-U), Part 2 Pipes.*

BS EN 10255:2004. *Non-alloy steel tubes suitable for welding and threading – Technical delivery conditions.*

BS EN 12201-2:2011 + A1:2013. *Plastic piping systems for water supply, and for drainage and sewerage under pressure – Polyethylene (PE), Part 2: Pipes.*

BS EN 14154:2005. *Water Meters; Part 1(+A2:2011) – General requirements, Part 2 (+A2:2011) – Installation and conditions of use, Part 3(+A1:2011) – Test methods and equipment.*

REFERENCES

Bernis, J. and Galan, F. (1976). Special Subject 7, Water requirements for firefighting. *Proc. IWSA Congress, J11–J17.*

EA (2012). *Benefits Assessment Guidance User Guide*. Environment Agency, UK.

FSRA (2004). *Fire and Rescue Services Act 2004*. HMSO.

HMSO (1999a). Water industry, England and Wales (1999). SI 1999/1148: *Water Supply (Water Fittings) Regulations 1999.* (Northern Ireland 2009 and Scotland Water Byelaws 2000). HMSO.

HMSO (1999b). *Water Industry Act (1999)*. HMSO.

HMSO (1999c). Water Regulations Advisory Scheme (1999). *Guidance Document for Water Supply (Water Fittings) Regulations 1999.* HMSO.

HMSO (2008). Statutory Instruments, 2008 No 594, Water Industry, England and Wales, *The Water Supply and Sewerage Services (Customer Service Standards) Regulations, 2008*. HMSO.

NFPA (2012). *NFPA1, Fire Code*. 2012 Edn. NFPA.

Ofwat (2008). PROC/01/0075. Leakage Methodology Review Project. *Providing Best Practice Guidance on the Inclusion of Externalities in the ELL.* Water Services Regulation Authority, UK.

UKWIR (2011). *Managing Leakage 2011*. UKWIR. UK.

Water UK (2007). *National Guidance Document on the Provision of Water for Fire Fighting*. 3rd Edn. Local Government Association and Water UK.

CHAPTER

Pipeline Design and Construction 17

17.1 INTRODUCTION

Hydraulic design, including loss calculation and diameter selection, surge analysis, longitudinal profile and air transport are discussed in Chapters 14 and 15. Valves and management of air are discussed in Chapter 18. This chapter deals with pipe material selection, structural design and construction.

17.2 PIPE DEVELOPMENT

Low-pressure earthenware pipes have been found at Knossos and in Mesopotamia. Copper is known to have been used for unburied water pipes from the same early period. Lead was used by the ancient Greeks to seal joints in earthenware pipes and the Romans used it for pipe. They also made pipe from clay and hollowed out tree trunks and such pipe persisted in use (being cheaper than lead) in the UK until the 17th century. Wood had a brief renaissance in North America at the beginning of the 20th century as a material for large diameter pressure pipelines for industrial use, but not for potable water. The method used preservative-treated Douglas fir staves bound together by threaded hoops. The method became defunct on the ready availability of large diameter steel pipe with reliable welding and coatings.

Iron pipe has been in use in Europe for over 500 years, cast initially in horizontal moulds and later in vertical moulds. Centrifugal casting was introduced in about 1920. All these iron pipes were joined by caulking yarn into the annular space at a spigot and socket joint and followed up by molten (run) lead but, after about 1950, flexible rubber joint rings were the norm. Ductile iron (DI) began to replace cast iron pipe in the UK a few years later.

Reinforced concrete pressure pipe was developed after the material's invention in the early 20th century while asbestos cement (AC) pipes were developed in Italy at about the same time. Prestressed concrete pipe was developed in the 1950s and has allowed concrete

Twort's Water Supply. DOI: http://dx.doi.org/10.1016/B978-0-08-100025-0.00017-X

to be used for higher pressures in very large diameters. Polymer concrete pipe was developed in the second half of the 20th century but its use is limited to low pressure and gravity applications. Plastic pipe materials were also introduced in the latter half of the 20th century and have since widened considerably in applicability with improved knowledge and quality of these materials.

Pipes found in water supply systems are generally of the following materials: cast or 'grey' iron; ductile iron (DI); steel; polyethylene (PE); PVC (polyvinyl chloride); GRP (glass reinforced plastic); prestressed concrete, cylinder or non-cylinder (PSC); reinforced concrete cylinder (RC) and asbestos cement (AC) which is no longer produced in the UK.

Other materials include galvanized iron, copper and lead which tend to be found in service pipes, plumbing, common connections and other small diameter pipes. Lead, extensively used in the past, is no longer installed because of the risk of plumbosolvency (Section 7.32), but many lead pipes remain in service for house connections and internal plumbing. Copper pipe can also give rise to plumbosolvency problems from lead solder used in joints. In the UK these materials now tend to be superseded by polyethylene; where lead in water is a problem, lead pipes are generally being replaced by polyethylene.

17.3 MATERIALS AND POTABLE WATER

Substances and products used in public water supply systems must not cause a water quality hazard. In the UK they are governed by:

- England – Regulation 31 of The Water Supply (Water Quality) Regulations 2000 (Statutory Instruments 2000 No. 3184);
- Wales – Regulation 31 of The Water Supply (Water Quality) Regulations 2000 (Welsh Statutory Instrument 2010 No. 994);
- Scotland – Regulation 27 of The Water Supply (Water Quality) (Scotland) Regulations 2001;
- Northern Ireland – Regulation 30 of The Water Supply (Water Quality) (Amendment) Regulations (Northern Ireland) 2009 (Statutory Rules for Northern Ireland 2009 No. 246).

Materials considered to comply on the water supply side are those published by Drinking Water Inspectorate (for England and Wales), and equivalent lists in Scotland and Northern Ireland.

In the US, products and materials are governed by CFR Part 141, National Primary Drinking Water Regulations. Minimum health effect requirements are established in NSF/ANSI 61.

17.4 ABBREVIATIONS AND DEFINITIONS

Table 17.1 lists the terms most widely used to define pipe characteristics or pressures applied to pipes or pipe systems (as in this chapter).

Table 17.1 Abbreviations for pipe and pipe system standards

Abbreviations used in standards and in this book:	
Pipes	
DN	nominal diameter in millimetres, e.g. DN 400
PN	nominal pressure in bar. For flanges it is the same as PFA – see below
Pressures (BS EN 805)	
PFA	allowable operating pressure, excluding surge
PMA	allowable maximum operating pressure, including surge
PEA	maximum allowable hydrostatic test pressure after installation
Pipe system pressures[a] (BS EN 805)	
DP	design pressure – maximum operating internal pressure of a system or zone fixed by the designer but excluding surge
MDP	as DP but including surge – designated MDPa when there is a fixed allowance for surge or MDPc where surge is calculated
STP	system test pressure as applied after installation

Note: [a]The pressure rating for a pipeline system may well be limited by fittings and particularly flange pressure ratings.

PIPELINE ALIGNMENT AND MATERIAL SELECTION

17.5 ALIGNMENT

The horizontal alignment of a pipeline should follow an available corridor giving minimum cost but adhering to constraints that may exist. Unless there is no alternative, laying a pipeline along a highway beneath the road surface should be avoided due to the cost and disruption to traffic. A route along the edges of agricultural or other open land is preferred to one crossing fields; however, if crossing is unavoidable pipeline features such as valve chambers and washouts should be sited near the boundaries. The alignment may make appropriate use of pipeline flexibility.

The vertical alignment of a pipeline should be arranged to minimize excavation while maintaining the required pipe cover and providing for forward and back gradients for drainage and air release purposes (Sections 15.10 and 17.34). Minimum cover is governed by frost and structural considerations (Section 17.8) or, at crossings, by the requirements of other authorities. Vertical bends or pipe flexibility should be used to effect changes in grade but care needs to be taken at hogging bends due to the upward thrust generated (Section 17.33) and potential for upheaval (BS PD 8010-1).

17.6 CHANGES IN DIRECTION

Changes in direction in rigid and semi-rigid pipelines have to make use of purpose-made fittings. Bends for such pipelines are available in DI and PVC and can be made in GRP, PE and RC (Sections 17.14, 17.25–17.26 and 17.29).

For flexible pipelines in materials such as PE and steel, use may be made of the natural flexibility of the pipe material but such bends have a large minimum radius (Sections 17.25 and 17.37).

17.7 SELECTION OF PIPE MATERIAL

The principal factors affecting the choice of pipe material are technical considerations, price, local experience and skills, ground conditions, preference and standardization. For pipelines it is usual to compare materials economically, including pipe purchase and laying costs and, where they differ, running costs due to hydraulic performance. Selection of pipe material for pipe systems should take account of requirements for joints which differ according to material; the pipeline should be considered as a mechanism limited only by the pipe structure, friction, soil stiffness and anchorages.

For the smaller pipe sizes in water distribution systems (DN 51 to 300), in the UK over 80% of new pipes are plastic (about 10% PE 100, 50% PE 80 and over 20% PVC-U and MOPVC) and over 15% DI. Plastic pipes offer cost advantages at small and medium diameters. They can be joined above ground to form continuous lengths which can be snaked into narrow trenches; this saves time and reduces social and environmental impact during construction. This advantage reduces with increasing diameter and pressure class and in hot climates where temperature derating is necessary.

PE can also be highly competitive with DI and is now being used in larger diameters up to 1600 mm, or larger for particular situations such as outfalls.

In the largest diameters DI pipe tends only to be used instead of steel if its price is competitive or where supply of skilled welders may be limited. The alternatives of prestressed concrete, concrete cylinder or GRP pipes tend to be used because of circumstances such as local preference and practice, price competitiveness, tied funding, in-country manufacture as opposed to importing, aggressive ground conditions or aggressive water to be conveyed, or (in the case of concrete pipes only) where a greater margin of safety is required against rough handling and backfilling. GRP pipes have the principal advantage that they are not attacked by ground conditions or by waters, such as desalinated water, which are severely aggressive to both iron and concrete.

Steel is predominantly used for large or high-pressure pipelines; it is the popular choice for pressures over about 25 bar. Steel pipes may also be useful in congested and urban areas where welded joints provide longitudinal strength and can avoid the need for thrust blocks (Little, 1986). However, the same situation complicates corrosion protection and the cathodic protection (CP) that may be needed. Steel pipes do not normally require rocker pipes at junctions with chambers and minor structures but may need them to accommodate major settlement where temporary excavation has to be backfilled at large structures. Anchorage points are nevertheless required at terminations and at connections to chambers where flexible couplings are to be installed at valves. Alternatively, the steel pipeline may be tied across flexible joints to provide longitudinal continuity of strength; the pipeline can then in principle be allowed to move longitudinally within the chamber and need not be anchored into the chamber walls. Leakage of groundwater into the chamber would have to be addressed but would not normally be an issue in dry conditions.

PIPELINE STRUCTURAL DESIGN

17.8 BEHAVIOUR OF BURIED PIPES

Buried pipelines are required to withstand internal hydrostatic pressure, external loads from soil, surcharge and traffic and are required to be safe against buckling (Young, 1983, 1986; Clarke, 1968). The soil load increases with depth and surcharge; traffic loads reduce with depth. Main road traffic load depends on the wheel load and configuration considered but typically accounts for less than 10% of the total load at 6 m depth. The minimum total external pipe load under main roads typically occurs at about 1.8 m depth. Loading from traffic differs between standards; for example, pressures determined at the crown of the pipe from traffic under normal roads according to BS EN 1295-1 and from HS-20 loading according to AWWA M45 differ significantly for pipe cover less than 5 m. At 1.0 m cover the pressure using BS EN 1295-1 is about three times higher and at 1.8 m cover the ratio is about 2.5 times reducing to about 1.2 times at 5 m cover (Little, 2004).

BS EN 1295-1 summarizes various European design methods and its UK National Annex A sets out procedures for structural design of different types of buried pipes. PD 8010-1 and -2 cover design of pipelines on land and under water. BS 9295 provides comprehensive guidance on the structural design of buried pipelines.

Pipes can be grouped as rigid, semi-rigid and flexible depending on their ring bending stiffness and response to external load. Structural stiffness is determined as S (kN/m^2) = EI/D^3 = 1000.E $(t/D)^3/12$, where E is the modulus of elasticity (N/mm^2), t is the mean pipe wall thickness (mm) and D is the mean diameter (mm) of the pipe (outside diameter (OD) less thickness). Structural stiffness varies widely depending on the material property and wall thickness to diameter ratio selected; typical values for DN 100 to 2000 pipe diameters range from 250 to 10 kN/m^2 for DI; 580 to 4 kN/m^2 for steel; 80 to 16 kN/m^2 for PVC-U; 17 to 8 kN/m^2 for MOPVC (Section 17.26); 10 to 5 kN/m^2 for GRP; and 80 to 4 kN/m^2 for PE. In comparison, reinforced concrete pipe stiffness is typically about 500–600 kN/m^2 for DN 800 and above, stiffer at smaller diameters.

Rigid pipes include concrete and AC. Rigid pipes have high structural stiffness and therefore deflect little; their load carrying capacity is derived from ring bending strength (as determined from crushing tests) but can be increased by bedding factors for various standardized bedding and surrounds. Rigid pipes tend to attract more load than more flexible pipes, particularly in wide trench situations.

Flexible and semi-rigid pipes have comparatively lower stiffness and deflect under load in inverse relation to pipe stiffness and overall soil modulus. DI pipe manufacture allows comparatively thin walled pipes; these may be classified as flexible or semi-rigid depending on wall thickness. Semi-rigid pipes derive their support partly from the soil and partly from pipe stiffness; their design takes into account ring bending stress. Permissible deflection limits for DI pipes are set to limit the ring bending stress in the pipe wall and for joint and lining performance.

Flexible pipes include thin walled steel (diameter to thickness ratio more than about 120) and plastics. As a general rule pipes classed as flexible are sized by OD while rigid and semi-rigid pipes are sized by internal diameter (ID); GRP pipes are sized by ID or OD depending on the manufacturing method (Section 17.27).

Flexible pipes derive their support primarily from passive soil resistance which develops as the pipe ovalizes under vertical load and deflects horizontally into the sidefill. The contribution of pipe stiffness is small and typically a flexible pipe is an order of magnitude weaker than the surrounding soil against compressive forces. (Note that unit structural stiffness of a flexible pipe is typically stated in kPa, while native soil and embedment stiffness is typically stated in MPa.) Overall soil modulus is a function of the pipe surround (embedment) material, compaction and depth and depends to some extent on the modulus of the native (trench wall) soil. If the trench wall material is weak the trench width may need to be increased or the embedment material and compaction improved or both. However, if the trench width is more than 4.3 times the pipe OD, further increase of trench width provides no benefit since the native soil modulus is then considered to have little effect.

The more flexible the pipe the greater the care needed in selection and control of embedment construction. Very flexible pipes are at risk of deflection to a 'squared' shape (as the pipe is deflected first inwards and upwards and then downward and outward as backfill is built up in layers). Pipes, particularly of plastic, of stiffness less than about 4 kN/m^2 are not recommended without great care in construction (Janson, 1995). This stiffness value is equivalent to a diameter/thickness (D/t) ratio for steel pipe of about 160, above which factory applied centrifugal cement mortar or spun concrete lining for steel pipes is not advisable. Despite case histories of much greater values, D/t for steel pipes should in no case be more than 200 (stiffness 2 kN/m^2) to retain sensible control during installation. The D/t ratio selected should be a matter of overall installed cost. Preferences and economics vary widely but some design and build contractors prefer stiffer pipe and pay less attention to backfill.

Characteristics of plastics pipes change with time; in particular, the modulus of elasticity (E) reduces very considerably with time. 'Short term' is defined as the one hour modulus. Long-term values are obtained by extrapolation of tests over various periods to obtain 50-year values. Long-term values must be taken into account in design; for PVC-U and PE, respectively, E values may typically be 50% and 25% of the short-term value, but depend on material formulation. 'Ultra short' term (10-second) values are significantly greater than the 1-hour values and are applied when calculating pressure surges.

Large diameter thin walled flexible pipes are particularly susceptible to buckling under external compressive loads due to soil pressure and sub-atmospheric pressure conditions. Detailed sectional design along a buried pipeline is often required to optimize wall thickness selection and special analysis may be needed for heavy loads from construction equipment.

17.9 LONGITUDINAL STRESSES IN BURIED PIPES

Longitudinal axial or bending stresses may arise where restrained pipes are subject to temperature change and, if there are no joints or where they are rigid, where differential settlement, lateral displacement or pressure forces induce bending. Longitudinal bending arising from uneven trench support or ground movement is not usually specifically considered except where pipes are designed to be supported at discrete intervals (Section 17.11) or at connections to structures and where differential settlement is expected. A short length of pipe installed with flexible joints (spigot and socket or flexible coupling) can be used to account for differential movement for example, the rocker pipe at exit from a structure. Pipe structural stiffness computed by the conventional formulae is valid for a continuous and infinitely long pipe; however, it drops significantly at discontinuous ends of the pipe. Design therefore needs to compensate for this loss of stiffness at pipe joints, at deeply buried rocker pipes for example.

Occasionally buried pipelines need to be designed for complex cases such as seismic conditions, permanent ground displacement, proximity to blast zones, etc. Equivalent stress (Von Mises) based design methods are used to address combined stresses generated in the pipe wall. The Von Mises equation for combined stress S_c at any point in the pipe wall is:

$$S_c = \sqrt{(S_h^2 - S_L S_h + S_L^2) + 3S_t^2},$$

where S_h and S_L are the hoop and longitudinal stresses, respectively and S_t is the torsional stress.

Comprehensive finite element analysis or CAE (Computer-Aided Engineering) based stress modelling becomes essential if pipe geometry is complex or if load combinations are difficult to interpret by simple stress calculations. The *Guidelines for the Design of Buried Steel Pipe* (ALA, 2001) is intended for pipe designed to one of the suite of ASME B31 codes. It provides guidance for calculating stresses from most of the loadings likely to be encountered and on the capacity of the pipe to accept them in combination.

Strain-based design approach is preferred in such complex load cases and soil–pipe stress modelling is often required to confirm pipe wall thickness selection. Buried pipe stress modelling requires detailed assessment of pipe–soil interaction; typical pipe–soil modelling approach is based on discrete non-linear springs. The maximum soil spring forces and associated relative displacements necessary to develop these forces are computed for modelling purposes. The majority of the commercially available CAE software applications used for buried pipe modelling are valid for pipes with D/t less than 100; for thin walled flexible pipelines where D/t ratio is in excess of 120, further checks for buckling stability are necessary.

17.10 BURIED FLEXIBLE PIPES

Flexible pipe design principles apply to steel and plastics but can also be applied to large diameter DI pipe under large loads. For steel pipes greater than about DN 750 (or D/t about 120–140, depending on conditions) the theoretical pipe thickness for normal water supply applications may be less than that required to limit deflection under backfill load. This also applies to high-pressure pipe using high-grade steels with relatively low wall thickness.

To save adding thickness to stiffen the pipe against backfill load, the pipe may be laid under 'controlled backfill' conditions on a thin layer of uncompacted sand or on a preformed circular invert (60° width) in sand or fine gravel and with selected and carefully compacted sidefill to achieve a required soil stiffness. At large diameters, temporary jacks may be inserted (in steel pipes) to maintain circularity or to pre-deflect the pipe upwards but care must be taken to spread jack loads to avoid damage to the lining. As the sidefill progresses the jacks may be removed depending on the degree of pre-deflection. Deflection is measured during embedment construction and, if it exceeds the permissible value, the backfill should be removed and replaced to obtain the necessary circularity. Measurements should be continued periodically after installation and checked at critical sections before the pipe is filled for testing.

Deflection limits vary according to pipe material. Limits quoted in AWWA M11 for steel pipe are as follows:

- mortar lined and coated — 2% of diameter
- mortar lined and flexible coated — 3% of diameter
- flexible lined and coated — 5% of diameter

If steel pipes are lined with cement mortar after installation or for flexible linings, deflections can be allowed to exceed the limit of 2% frequently quoted. However, if large deflections are to be permitted for steel pipes, analysis of buckling and ring bending should be carried out, allowing a design factor greater than 0.5 and taking into account the stress–strain characteristics of the material. Note that ASME B31.1 limits deflection to 5% for non-metallic piping while the water industry allows up to 6% for PE and PVC pipes.

Deflection of flexible pipes is usually calculated using the Spangler (Spangler, 1951) formula:

$$\frac{\Delta}{D} = \frac{k(P_e D_l + P_s)}{8S + 0.061E'},$$

where:

Δ = pipe deflection (assuming horizontal and vertical deflections are equal);
D = mean diameter of the pipe;
k = a constant, dependent on the angle, between contact points, over which the trench bed supports the pipe (typically 0.1 for 65°, 0.083 for 180°);
P_e = soil load per unit area (kN/m^2);
P_s = surcharge or traffic load per unit area (kN/m^2);
D_l = deflection lag factor, dependent on soil type and compaction;
S = diametrical ring bending stiffness = $1000E\,(t/D)^3/12$ (kN/m^2);
E = modulus of elasticity of the pipe wall (N/mm^2);
E' = soil stiffness (kN/m^2); and
0.061 is derived from assumed (parabolic) loading over a 100° lateral support angle.

For granular soils D_l is unity; long term, in clay, the value may be 3 or more. Where the cover is less than 2.5 m and where the pipeline will be under sustained pressure within a year of installation, long-term deflection may be reduced by a re-rounding factor:

$$D_R = 1 - (P_i/40)$$

where P_i is the internal pressure in bar.

Diurnal variations in pressure in distribution mains should be taken into account in deciding the re-rounding factor.

Embedment soil stiffness E_2' depends on the nature of the soil, the degree of compaction, the amount of overburden and degree of saturation. Values for granular soils are given in BS EN 1295-1 for different degrees of compaction. These values may be considered to be suitable for the least favourable situations – soil fully saturated and with little cover. Stiffness values quoted by AWWA and AASHTO tend to be rather higher and some sources show variation with soil cover and groundwater level (Little, 2004). Compaction is quoted either as percent Proctor (modified Proctor density (M_p) – which corresponds to the heavy compaction test to BS 1377 or ASTM D1557) or, for granular materials, to relative density to ASTM D6938. Soil stiffness can also be derived from laboratory tests. Native soil stiffness (Little, 2004) may be estimated from SPT or other test results with correction for depth or from undrained shear strength. AWWA M11 quotes the accuracy of predicted deflections for different degrees of soil compaction; it does not cover native soil modulus.

Ring bending stress is not addressed in thin wall steel pipe design but is included in design of PVC and PE (thermoplastics) pipe; bending strain is considered for design of GRP (thermosetting) pipe. Wall thickness for plastic pipe is usually expressed as a minimum but thickness tolerance is not covered. Bending stress is given by:

$$\sigma_{bs} = ED_f(\Delta/D)(t/D)$$

where E is the flexural modulus of elasticity of the pipe material and D_f is a strain factor, dependent on pipe and soil stiffness, and is given in BS EN 1295-1 and other references. For PVC and PE pipe the sum of bending stress and hoop stress is required to be less than the design value. The approach is similar for GRP but uses a criterion of strain. Designs for plastics pipes need to take into account both the initial short-term and the long-term characteristics.

17.11 ABOVE GROUND PIPELINES AND PIPING

Although it is preferable for pipelines to be buried for security and environmental reasons, laying a pipeline above ground may be necessary in some conditions, for example: where rock is close to the surface; to avoid contact with saline ground or in swamp, usually on piles.

Increasingly, water treatment and pumping facilities involve extensive and sometimes complex runs of piping which is not buried. Codes for the design of such piping include ASME B31.3 which is intended for process piping in petrochemical plants but covers all fluids involved including water. It deals with all the types of load discussed below and can be applied to most pipe materials. BS EN 13480 is an equivalent code of practice which covers above ground as well as buried installation of industrial grade piping.

Loadings on above ground pipelines include self weight, pipe contents, snow, wind, internal pressure (including end restraint and Poisson's ratio effects), support loads, thermal stresses and, in seismic areas, earthquake loads (ALA, 2002). Superimposed loads need not otherwise be taken into account unless the pipe is to carry an access walkway. Pipe supports may consist of plinths (concrete or steel) and saddles or brackets, fixed either to reinforced concrete structures or to a steel support framework. Pipe bridges can be designed as arch or suspension structures using bridge design methods.

In all cases piping should be treated as a structural mechanism. This applies particularly to pipes with flexible joints but also to pipes with flanged joints since such joints are not absolutely rigid.

Suitable lateral restraints must be included in the design, even in non-seismic areas; slight misalignments in otherwise straight runs can produce large forces through the multiplying effect of deflection, increased misalignment, increased force and so more deflection. Pipes with flexible joints such as spigot and socket DI pipe and GRP should be supported at least at every joint. For large diameter pipes with flexible joints additional supports may be needed to avoid excessive distortion and failure of the joint ring.

The ability of a pipe to support itself between saddle supports is determined by bending theory. However, buckling of the pipe at supports needs to be designed against. DI pipe is best supported just behind the socket which provides additional stiffness and anchors the flexible joint. For large diameter steel pipes with large spans additional stiffness is necessary at supports.

BS EN 545 requires that flanged joints of DI pipe (and therefore the pipe also) should be able to resist a bending moment as well as internal pressure without visible leakage for a given duration. For pipe with cast, screwed or welded flanges the moment is equivalent to that arising in a full pipe which is simply supported over a length of 12 m for DN ≥ 300 or 8 m for DN ≤ 250.

Above ground pipes are exposed to the elements. Greater provision for movements due to temperature and pressure changes has to be made. Use of special expansion bellows joints may be necessary adjacent to fixed points unless loops are incorporated.

AWWA M11 provides guidance for the design of ring girder stiffeners and of saddle supports for steel pipes as well as design guidance for the pipe bending condition, but this is difficult to interpret. Comprehensive guidance on saddle design of horizontal vessels is provided in the *Boiler and Pressure Vessel Code* (BPVC, 2015) and may be used for pipes. Depending on pipe geometry complexity, CAE stress modelling is often required to confirm support spacing and design.

Pipe supports need to provide adequate restraint, including lateral and end restraint, but must also allow movement where necessary. Except for very simple systems, with very few supports and restraints and where displacement strains are small, it is necessary to undertake a flexibility analysis; its object is to ensure that there is enough flexibility in the system so that the pipe, its supports and attached equipment are not overstressed. Analysis is complex but comprehensive CAE piping software is available and allows changes to the arrangement to be easily tested. Nevertheless, the analysis is best done after the pipe, support and joint scheme has been defined by an experienced piping engineer. This should minimize the need for repeat analyses after adjustments are made to support arrangements and provision for movement.

To protect against freezing, insulation may have to be provided in cold climates; smaller pipes might need to be trace heated as well. AWWA M11 provides guidance for pipes in freezing conditions.

17.12 BURIED PIPEWORK IN PLANTS

Buried pipework at a treatment or pumping facility is likely to be congested and may need to negotiate a plethora of existing pipes and other services. It may have to be laid in ground that has previously been disturbed, which is variable in stiffness and may be aggressive in places. Plant pipework may, at a later date, be subject to the effects of excavations and further pipe laying for plant upgrading and expansion. For these reasons it is sensible to provide buried plant pipework with an ample margin of strength and stiffness as well as provision for movement.

Buried plant pipework may be treated as a set of pipelines. Where joints are mostly flexible there is no need to undertake flexibility analysis but thrust blocks will be needed. Transfer of thrust to the ground by friction along the pipe should not be relied on where the ground conditions are variable or likely to be affected by future work. However, rigidly joined lengths which include tees and bends should be analysed to ensure that there is enough flexibility and strength in the length.

IRON PIPES

17.13 CAST OR 'GREY' IRON PIPES

Many cast iron pipes made towards the end of the 19th century are still in use; their walls were relatively thick and not always of uniform, 'Spun' grey iron pipes were formed by spinning in a mould and produced a denser iron with pipes of more uniform wall thickness; they comprise a large proportion of the distribution mains in many countries. Three classes of such pipes were available in the UK: B, C and D for working pressures of 60, 90 and 120 m, respectively; classes B and C were more widespread. Carbon is present in the iron matrix substantially in lamellar or flaky form; therefore, the pipes are brittle and relatively weak in tension and liable to fracture. The manufacture of grey iron pipes has been discontinued in most countries, except for the production of non-pressure drainage pipes.

17.14 DUCTILE IRON PIPES

DI pipes are normally cast by centrifugally spinning molten iron in high-quality steel moulds; fittings are cast in static moulds. The iron contains small quantities of magnesium to transform the lamellar form of carbon into a spheroidal form, thereby increasing tensile strength and ductility. DI pipes are available in sizes up to DN 2000 with socket and spigot ends suitable for forming push fit type joints, plain ends suitable for jointing with flexible couplings, or with flanged ends formed by welding on DI flanges. Pipes larger than DN 2000 are manufactured but with joints other than the push fit type. Standard lengths vary depending on the diameter of the pipe, the type of ends required and where the pipe is manufactured. In the UK the standardized length (overall length minus the depth of the socket) for spigot and socket ended pipe is typically 5.5 m for nominal sizes up to DN 800 and 8.15 m for DN 900 and larger; further information on other standardized lengths is provided in BS EN 545. In the US, standard lengths are 5.5 and 6.1 m. Plate 27(a) shows DI pipe being laid.

Until 2006 BS EN 545 required wall thickness, e in mm, for pipes and fittings to comply with the formula: $e = \mathrm{K}\ (0.5 + 0.001\ \mathrm{DN})$, where DN is in mm; K = 9 for socket/spigot, plain ended and welded flange straight pipe; K = 12 minimum for fittings without branches, for example, bends and tapers; K = 14 minimum for fittings with branches for example, tees. BS EN 545 sets out tolerances on thickness.

Table 17.2 PFA, PMA and PEA values for socket and spigot ductile iron pipe

Nominal diameter (DN) mm	Preferred class (Table 17) bar	PFA bar	PMA bar	PEA bar
40–300	40	40	48	53
350–600	30	30	36	41
700–2000	25	25	30	35

Source of information: BS EN 545.

However, BS EN 545 no longer includes the K class formula and the wall thickness is now defined according to pressure class (in the range 20–100), being the allowable operating pressure (PFA) in bars. The preferred pressure classes are provided in Table 17.2. Other 'non-preferred' pressure classes may be selected according to design requirements and availability. Caution should be used when selecting pipe according to pressure class, particularly if robustness is required (Section 17.12), as the wall thickness may be significantly less than that for an equivalent K9 pipe. BS EN 545 includes requirements for longitudinal moment resistance of pipes of pressure class 40 to 100 and DN up to 200. It also has requirements for moment resistance for flanged joints (Section 17.11). BS ISO 10803 also covers the design of DI pipe.

The allowable operating pressures excluding surge (PFA), and including surge (PMA), and the maximum allowable test pressure on site after installation (PEA), are shown in Table 17.2 for selected sizes for spigot and socket pipe of the preferred pressure classes.

The PFA and minimum wall thickness values in BS EN 545 take account of the 0.2% proof stress of the material and imply and use a safety factor of three on the required ultimate tensile strength of DI of 420 N/mm^2. PMA, the allowable maximum operating pressure including surge is approximately 20% more than the PFA. PEA, the maximum allowable hydrostatic test pressure after installation, is in general PMA plus 5 bar.

The nominal pressure rating (PN) of all components of the pipeline system should be considered when ordering pipes and fittings. For high pressures (PN typically 25 bar or more, but depending on the diameter and types of fittings), availability should be checked. Joints are required to be watertight at PEA and to durably withstand without leakage the PMA pressure under service conditions, including angular, radial and axial movement.

DI pipes are also widely used in the US, with pipe diameters ranging from 3 to 64 inch (76–1625 mm). AWWA M41 covers the design principles of DI pipe; AWWA C150 is the code of practice for thickness design of DI pipe and AWWA C151 is the relevant manufacturing standard for DI pipes in the US. AWWA M41 covers the standard pressure classes between Class 150 and 350 psi (10.3–24.1 bar) in 50 psi (3.4 bar) increments and according to pipe diameter but also includes requirements for special thickness classes. The minimum thicknesses quoted are somewhat larger than those for equivalent operating pressures in BS EN 545 and imply higher safety factors on the required tensile strength of 60 000 psi (413.7 N/mm^2). M41 includes pipe wall and restraint design information for buried and above ground situations, details of joints and fittings and installation and purchasing information.

BS EN 545 and AWWA M41 set out design methodologies and include tables of allowable depths of cover for defined pressure class, trench and surface loading conditions.

17.15 EXTERNAL COATINGS AND INTERNAL LININGS

Current European practice for external coating of DI pipe comprises various protection systems such as spray coating with zinc followed by a top coating of bitumen paint, and other more complex arrangements such as a zinc/aluminium (85/15%) alloy basecoat and epoxy topcoat. For aggressive soils polyethylene sleeving, either factory or site applied, is often adopted, plus imported backfill where appropriate. Minor damage and puncturing of the sleeving do not impair the efficiency of the protection, although any such tear should be patched with adhesive tape. The effectiveness of the wrapping, even though the sheeting is not watertight, is ascribed to insulation of the pipe from uneven soil contact (which can produce galvanic cells) due to uneven pipe bed and surround, particularly in the case of clay soils which are hard to compact near the bottom of the pipe. For highly aggressive soils the pipes can be wrapped with a heavy duty PVC backed bitumen adhesive tape, overlapped 25 mm or 55%. CP (Section 17.24) is not normally recommended but, if used, requires bonding across the joints to provide electrical continuity and tape wrap to reduce current and anode consumption. A points system, depending on soil resistivity, groundwater level and other characteristics, can be used to judge the potential severity of aggression in a given location. Where a joint, such as a mechanical joint (Section 17.16), includes corrodible components it should be protected. Typical systems include an inert putty fill and external tape wrap to isolate the components from the ground.

Earlier DI pipes were coated internally (and externally) with cold or hot applied bitumen or hot applied bitumen-based material sprayed or brushed onto the pipe metal. Use of coal-tar products was discontinued because coal-tar can give rise to polynuclear aromatic hydrocarbons in potable water (Section 7.41). Internal lining of DI pipes in use in the UK comprises spun mortar lining, using sulphate-resisting cement plus, for pipes DN 800 and smaller, an epoxy seal coat on top of the cement mortar. Details of cement mortar lining (CML) are given in Section 17.21.

17.16 JOINTS FOR IRON PIPES

Types of joints in use are shown in Figure 17.1.

The run lead joint for spigot and socket pipes has been superseded, but many mains still in use have been laid with this joint. Skilled workmanship is required to make the joint properly. The lead is heated to 400°C (at this point strong rainbow colours are revealed when the surface scum is drawn aside). A clip is placed around the pipe against the annular space of the socket and the lead is poured in one continuous pour through an opening left at the top of the clip. The operation is potentially dangerous and the socket space must be completely dry to avoid blowback of the lead by steam. The lead solidifies almost immediately after pouring and is then caulked up using a series of chisels. The joint is rigid and even slight movement tends to cause such joints to weep; however, countless mains have been satisfactorily laid with this type of joint in Europe, the USA and worldwide.

Flanged joints are covered by codes such as BS EN 1092-2 and ASME B16.42. Flanges are machined from heavy castings so that the face is perfectly flat. Bolt hole spacings, position and diameter are set by appropriate standards according to pressure. Flange gaskets should typically be manufactured from high-grade, non-biodegradable synthetic rubber to a thickness of 3 mm. Information on gaskets is provided in BS EN 1514 (flange gaskets), BS EN 681-1 (material

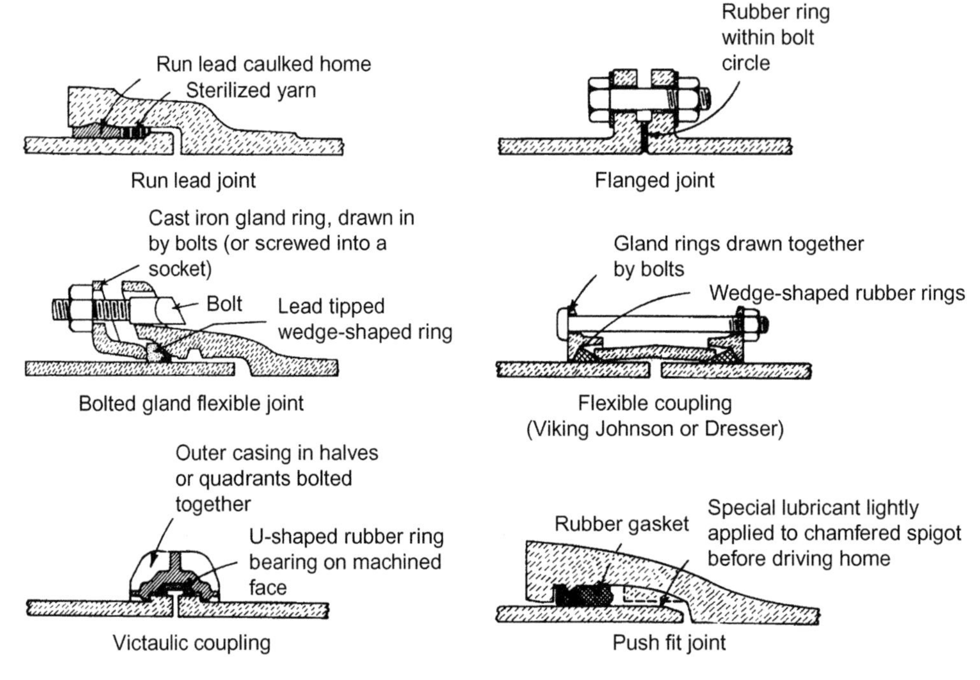

FIGURE 17.1

Joints for iron pipes.

requirements for elastomeric seals), AWWA M41 and from the (North American) Ductile Iron Pipe Research Association (www.dipra.org). Natural rubber gaskets are excluded (by AWWA C111/A21.11) and not available for DI pipe.

Bolted or screwed gland joints work by forcing a rubber ring such as EPDM into an annular space with a cast iron gland ring which is drawn by bolts or screwed into the socket but have not been used for water pipes for several years although versions of this joint incorporating restraint via a welded bead are offered by manufacturers.

Flexible couplings such as the Viking Johnson (or Dresser) couplings are used for connecting together lengths of plain ended pipe. This joint also uses rubber rings compressed into an annulus and can be used with many pipe materials including DI. This type of joint can accept some angular deflection (reducing with increasing pipe diameter). If specified without a central register the coupling can be moved along the pipe, allowing removal of a section. However, central registers are preferred where there is a risk that the coupling may move in service.

The Victaulic coupling is used in conjunction with shouldered ends of pipes, thus holding them together longitudinally. The joint can be unmade and remade without difficulty and is most often used for temporary pipelines laid above ground.

The push fit joint includes a specially shaped dual hardness synthetic rubber ring gasket fitted into the socket of a pipe before the spigot of the next pipe is pushed in. A little lubricant must be used on the inside of the gasket and on the outside of the pipe spigot before forcing it into the socket. Complete cleanliness of the socket, gasket and spigot is essential. These joints are flexible: BS EN 545 requires a minimum possible angular deflection of 3.5° up to DN 300, 2.5° for DN 350 to 600 and 1.5° for larger sizes. Allowable deflections for pipes produced in the UK are 5° up to DN 300 and 4° for larger sizes. AWWA M41 gives maximum joint deflection angles of 5° up to 12″ and 3° for larger sizes.

Push fit joints can also be restrained, for example, with anchor gaskets and other mechanical systems (such as weld bead and tie-bar). Allowable joint deflections are reduced for these systems but BS EN 545 requires that restrained joints shall be designed to be at least semi-flexible and that the allowable angular deflection shall not be less than half of the values given above for flexible joints. However, it should be noted that the specification of restrained joints may reduce the PFA values according to the specific joint arrangement, diameter and manufacturer. Further information on pipe restraint, anchorage and thrust blocks, is provided in Section 17.34.

STEEL PIPES

17.17 STEEL PIPE MANUFACTURE AND MATERIALS

BS 534 (for carbon steel pipes, joints and specials) is withdrawn but only partly replaced by BS EN 10224 (pipe ranging from 26.9 to 2743 mm OD using steel of yield strengths 235, 275 and 355 N/mm^2) and by BS EN 10311 for joints. BS EN 10312 covers stainless steel pipe. BS EN 10224 uses pipe consistent with BS EN 10216-1, 10217-1 and 10220, but pipe to BS EN ISO 3183 (API 5L) and other standards can also be used. CP 2010 Part 2 for design and construction of steel pipes on land remains current. Eurocode 3: BS EN 1993-4-3 applies to the design of steel pipelines which are not adequately covered by other European standards covering particular applications. This code requires consideration of five ultimate limit states, including fatigue, and three serviceability limit states including vibration. BS PD 8010-1 and -2 are intended primarily for oil and gas pipelines but apply to and provide useful design information for steel pipes for water conveyance.

BS EN 10224 covers four principal welding methods for manufacture: butt (BW) – OD up to 114.3 mm; electric (resistance) welded (EW) – OD up to 610 mm; seamless (S) – OD up to 711 mm and submerged arc welded (SAW) – OD 168.3–2743 mm. In ISO 3183 the designations EW and SAW are recognized but seamless pipe is designated SMLS and LW means laser weld.

Steel pipes are fabricated from steel plate bent to a circular form or they may be continuously produced from a coil of steel strip bent to a spiral and butt welded along the spiral seam. Joints between coil ends of spiral welded pipes are known as skelp end welds. Butt welded pipes are made from rolled strip with a longitudinal seam furnace butt welded by a continuous process. Lengths of pipe are usually in the range of 9–12 m dependent on manufacture, transport and project requirements.

Weld beads must be machined flush with the pipe surface at pipe ends to make them suitable for joint couplings. Spigot and socket ends, where shaped, are formed by die. Weld bead height needs to be limited for coating and lining. Electric (resistance) welding is carried out by passing an electric current (by induction or direct contact) across the edges which are joined under pressure, without filler metal. Heat treatment at least of the weld zone is usual in sizes larger than DN 200. EW pipes now tend to be welded using the high frequency induction process which avoids electrode contact with the pipe. Inspection typically includes chemical and mechanical material tests, ultrasonic inspection of plate and welds, radiography of welds and hydraulic pressure tests. The need for

Table 17.3 Steel grades to API 5L/ISO 3183

Grade		A	B	X42	X46	X52	X56	X60	X65	X70
Yield strength	psi	30 500	35 500	42 100	46 400	52 200	56 600	60 200	65 300	70 300
	N/mm^2	210	245	290	320	360	390	415	450	485

stress relieving after welding is dictated by the thickness of plate or weld throat according to code; ASME B31.4 requires stress relieving for effective weld throat more than 32 mm.

Steel grades in ISO 3183 and, as from 2008, the American Petroleum Institute standard API 5L are designated by grade and by yield stress in thousands of psi, as Table 17.3. Grades less than grade B would not normally be used. API 5L covers two quality categories, PSL1 and PSL2, the latter requiring much tighter control on steel composition (Carbon Equivalent) and higher fracture toughness. PSL1 is commercially available up to grade X70, whereas PSL2 can be sourced up to X120. Selection of steel quality category and grade is usually on overall cost and, where operating conditions warrant, it is often beneficial to use as high a grade as possible. However, pressures in water supply pipelines usually vary considerably over their length. In parts of the pipeline subject to lower pressures minimum pipe wall thickness criteria (for handling and laying conditions) will govern. In such situations the proportion of the pipeline with thickness governed by internal pressure reduces as steel grade increases. Use of different grades in the same pipeline may be considered but adds to the site quality control difficulties of managing varying wall thickness.

17.18 STEEL PIPE WALL THICKNESS

AWWA M11 gives a range of thicknesses and pressures and steels for diameters up to 4000 mm. Sizes in M11 are designated by OD below 30 inches (762 mm), otherwise by inside diameter.

Pipe wall thickness, t (mm) for internal pressure is determined by hoop stress, as follows:

$$t = \frac{PD}{2a\sigma e}$$

where P is the internal pressure (N/mm^2); D is the external diameter (mm); a is the hoop stress design or safety factor; σ is the minimum yield stress (N/mm^2); and e is the joint factor. The design factor, joint factor and definition of wall thickness depend on the design code. The joint factor is 1.0 for SAW pipes. Hoop stress design factors typically range from 0.4 to 0.8 and certain codes require the negative tolerance to be deducted from wall thickness. ASME codes B31.4 and B31.8 quote a basic hoop stress design factor of 0.72 and state that this includes for thickness tolerance. For typical water supply situations AWWA M11 and the WRc pipe selection manual (Trew, 1995) recommend a design factor of 0.5. This affords a degree of robustness which is useful in varied conditions. However, for higher pressure and long distance pipelines, a higher design factor may be acceptable. Whatever the design factor adopted, in sections of pipelines subject to lower pressures wall thickness is determined by handling and installation and the need to control diametric deflection.

Although AWWA M11 allows D/t up to 288 (240 if mortar lined) such pipe is difficult to handle and a lower limit is recommended (Sections 17.8 and 17.10). In the oil and gas industries a maximum D/t of 100 is often applied to reduce concerns about handling and backfill control and to speed laying; even at the cost of not using the benefits of higher strength steels.

Where necessary (Sections 17.9–17.11) the pipe should be analysed for ring bending, longitudinal bending, temperature changes, Poisson's ratio effects and combined (equivalent) stresses. Typical design factors recommended for estimation of longitudinal and combined equivalent stress are 0.80 and 0.90, respectively.

Pipelines operating at pressures over about 25 bar are frequent in the oil and gas industries and for hydropower but are less common in water supply. However, the mining industry is now requiring quite large supplies of water, sometimes to be carried over large distances and at pressures as high as 100 bar. For the purposes of this chapter high pressure is taken as any pressure over 25 bar. Note that this cut-off is not to be confused with the threshold in ASME B31.3 for high-pressure process piping – pressure class over 2500 (maximum operating pressure of over 430 bar).

At such pressures it is economically beneficial to adopt oil and gas practice (ASME B31.4), using higher strength steels and adopting hoop stress design factors higher than the 0.5 given in AWWA M11. However, use of higher design factors must be combined with the increased quality requirements for materials, pipe manufacture and for shop and field welding that are required in the relevant codes. A design factor of 0.72 is given for normal operation in several standards (ASME B31.4 and B31.8 and BS PD 8010-1); even 0.77 or 0.83 may be considered in some circumstances (PD 8010, BS EN 14161) subject to risk assessment. EN 1993-4-3 (Eurocode 3) recommends a load factor of 1.39 where certain conditions are met implying that hoop stresses dominate, and equates this to a design factor of 0.72.

Clearly fluids such as flammable gas and volatile hydrocarbons pose a serious risk if allowed to escape. Whilst there is no risk of explosion or fire due to water leaking from a pipeline, the energy and destructive power of water under a high pressure is great. Location can also be considered so that pipelines through unpopulated areas can be allowed to operate at higher hoop stress design factors than those through urban areas where the consequences of a high pressure burst would be more serious. In the selection of design factor, a risk-based approach should be used. BS PD 8010-1 considers location as well as fluid nature for design factor selection. The design factors in other codes reflect, to some extent, differences in risk.

17.19 FACTORY-MADE BENDS AND OTHER STEEL FITTINGS

Typically bends and fitting are made from sections of pipe. AWWA C208 and BS EN 10224 give dimensions for common fittings, for example, bends and branches. However, fittings can be made to any dimensions required. ASME B31.9 covers wrought fittings for butt welding. Since hoop continuity is compromised at mitre and lobster back bends and at openings for tees and nozzles the deficiency has to be made up in some way. This can be achieved by making use of any spare wall thickness that exists above that needed for normal hoop stress and, if necessary, by adding reinforcement in the form of welded saddles or girder stiffeners. AWWA M11 provides guidance on the design of such reinforcement. However, where the steel pipe design makes use of hoop stress design factors greater than 0.5 the relevant codes contain more onerous requirements for the design of bends and fittings. For hoop stress more than 20% of yield strength, ASME B31.4 does not permit mitre bends and requires the reinforcement, for welded branches with ratio of branch to main pipe diameter more than 50%, to be in the form of a complete sleeve around the main pipe.

Hot bends may be formed by the induction process (ASME B16.49 and BS EN 14870). In this, an intense magnetic field is set up by current passed though a ring of coils around the pipe so heating the pipe to working temperature. The pipe is slowly drawn through with the leading end held on a radius arm so forming a bend of the required radius as it cools.

17.20 EXTERNAL AND INTERNAL PROTECTION OF STEEL PIPE

Carbon steel pipes carrying water will corrode on the inside if unprotected. The nature, extent and rapidity of corrosion depend on the aggressivity of the water, temperature and presence of precursors (e.g. oxygen or sulphides). Corrosion may be localized in a form such as pitting or crevice corrosion and, for water transmission pipelines, is difficult to predict and control, even if the water is treated to provide carbonate stability and corrosion inhibition. For structural steel members where strength requirements predominate, it is common to add a corrosion allowance to thickness. However, the use of an allowance for corrosion of a steel pipeline is no longer recommended (AWWA M11). This is because the first consideration is leakage, for example, via pinholes, but also because reliable internal and external protection systems are available and are more economic than adding thickness sufficient to prevent leaks during pipeline life. For these reasons steel water pipelines are invariably lined and protected externally.

Corrosion protection may be carried out by the manufacturer at the place of manufacture or elsewhere or by a specialist applicator. Pipe out of roundness and straightness can affect the application of coatings and linings; both parameters must be specified to suit the process of application of the corrosion protection.

External Protection

Principal options for external coating to water pipes are bitumen sheathing; fusion-bonded epoxy (FBE); three layer polyethylene (3LPE) (Plate 27(b)) and liquid coatings (paints). Higher specification systems such as FBE and 3LPE require cleaning and preparation by grit blasting to white metal finish (SSPC-SP5/NACE No.1 or Sa3 to SIS 05 50 00). For other coatings near white metal finish is required (second quality to BE EN ISO 8501 (BS 7079), Sa 2.5 to SIS 05 59 00 or SSPC-SP10/NACE No.2). Surface preparation specifications also need to cover surface profile requirements.

Bitumen sheathing or wrapping to BS 534 consists of a hot applied bitumen with an inert filler, reinforced if required with a woven glass cloth, to a thickness of 3 mm for small diameter pipes rising to 6 mm for diameters exceeding 350 mm. Bitumen can biodegrade and needs protection from sunlight. As an alternative, coal tar has been used, particularly for underwater pipelines, but has lost favour in view of health and safety issues. Both bitumen and coal tar lose volatiles on exposure and then crack. Alternatives such as FBE and 3LPE for buried and exposed pipes are now preferred.

FBE has been in use since the early 1970s for external and internal coatings. The pipe surface is prepared by grit blasting followed by an optional phosphate or chromate pre-treatment. The FBE coating is applied (by either spraying or in a fluidized bed) as a powder mix of resin and hardener which is fused onto the pre-heated (to about 230°C) pipe surface and cured chemically irreversibly. Dry film thickness (DFT) typically is 300–400 μm for pipelines onshore. This may be increased to 500 μm for areas with high corrosion potential but thermal zoning can be an issue for high DFT. Offshore FBE alone is not sufficient protection. FBE coatings generally have better resistance to cathodic disbonding than brush or spray-applied coatings when used on cathodically protected pipelines. FBE is brittle and prone to impact damage during handling and installation; often specifications require use of additional external wrap of flexible PE mesh (e.g. Terram Rockshield™ or similar high impact resistance products). Protection at joints is completed by heat shrink sleeves, tape wrap, by compatible polyurethane or epoxy coatings. Codes of practice for FBE coating include AWWA C213, API RP 5L7 and 5L9 and BS EN ISO 21809-2.

3LPE systems have been in use since the early 1980s, and supersede the two layer system. EN ISO 21809-1 also covers 3LPE and 3LPP (polypropylene) systems although the latter are more suitable for higher operating temperatures and not generally recommended for use in water supply. The three layer system is more expensive than FBE but is considered cost effective in view of potential damage during shipping over long distances. The system comprises the following typical thicknesses in order of application:

fusion-bonded epoxy: 150–200 μm;
copolymer adhesive: 200–350 μm; and finally
polyethylene: 2.5 mm up to DN 750 and 3.0 mm for larger sizes.

The total thickness for the three layer system is thus between about 3 mm up to DN 750 and 3.5 mm for larger sizes depending on Class of duty. After surface preparation by grit blasting and phosphate or chromate pre-treatment, the pipe is pre-heated and application of all three components takes place while the pipe is rotated about its axis and moved forward through a specially designed booth. Timing is vital and curing takes place before the pipe reaches supporting rollers. Completion at joints is by heat shrink sleeves (favoured for larger pipe diameters) or by tape wrap.

Paint options include epoxy and polyurethane and may include a zinc-rich base layer. DFT typically is greater than that for FBE; the preferred application is by airless spray but, due to local constraints, brush application is often accepted with increased control on quality of application. Polyurethane coating (e.g. to AWWA C222) would typically be about 2 mm thick. Good preparation and protection is needed for fittings, which tend to be anodic and more prone to corrosion due to the additional welding and working. Small diameter steel pipes are galvanized. Other developments continue including blends of FBE, adhesive and PE and some manufacturers may also provide a thin additional coating of concrete for mechanical protection. Cement mortar external coating is covered by American standards.

The continuity of the applied protection (other than cement mortar) is checked with a 'holiday' detector. A scanning electrode in the form of a brush containing a high-voltage electrical charge is passed over the coating and lining; the voltage is set so as to produce a spark length of 10 mm or double the specified minimum thickness of the protective material whichever is the greater. Pinholes or breakages are disclosed by an electrical discharge to the steel of the pipe and this can be arranged to cause a buzzer to sound.

Internal Protection

Internal surfaces of steel pipes can be protected with concrete or cement mortar (Section 17.21), epoxy or polyurethane. Market forces and perceived quality issues currently favour epoxy. Epoxy coatings may be hot applied FBE or airless spray liquids and high solids systems. For FBE recommended DFT is typically about 300 μm and surface preparation is by grit blasting to a high standard. However, success of liquid systems is very sensitive to surface preparation and application conditions. These can be adequately controlled in the factory but with much more difficulty inside the pipe in the field.

Bitumen was a popular and successful option but is now little used. A hot applied bitumen with or without an inert filler is sprayed or brushed onto the pipe, after first priming it with a compatible priming coat; the usual practice was to apply a minimum thickness of 1.5 mm for diameters up to 300 and 6 mm for diameters exceeding 1000 mm.

17.21 MORTAR AND CONCRETE LININGS

As for DI, steel pipes can be lined with cement-sand mortar or spun concrete and these form some of the best and most cost effective types of interior protection. However, such cementitious linings are not appropriate for large diameter pipelines of high-grade steel operating at high hoop stress because the differential hoop strain is too great for the lining to accommodate.

The cement mortar or concrete lining generates a high pH environment which passivates the metal and prevents corrosion. CML may be either factory applied by centrifugal spinning or may be applied in situ by spraying (orange peel finish) followed, at large diameters and if required at smaller diameters, by trowelling to a smooth (washboard) finish. AWWA C205 and C602 cover shop applied and in situ mortar, respectively. BS EN 10298 covers mortar linings of steel pipe. Concrete lining has been chosen in preference to CML on some projects because it is thicker (typically 25 mm instead of 12 mm at DN 1400 and larger) and because the aggregate is larger, therefore reducing shrinkage and presenting a more robust and durable surface.

Aggressive waters, as indicated by the Langelier Index (Section 10.40), dissolve the cement and increase pH, causing water quality problems particularly for small diameter pipelines. However, the cement lining may be sealed with a thin (150–250 μm) layer of epoxy paint. The seal coat provides a barrier to solution of the cement mortar by aggressive low pH, low calcium waters but it is not particularly robust. Where pH cannot be raised, for example, by dosing with lime or, where dosing cannot be guaranteed due to materials supply or operational uncertainties, epoxy or other lining should be used.

CMLs and concrete linings are thicker than epoxy; the reduction in ID must be allowed for in hydraulic calculations. When new, the hydraulic roughness of mortar and concrete linings is much higher than that for FBE or other epoxy linings. However, after a while in service and with biofilm formation, the roughness of mortar, concrete and epoxy will be similar so that no distinction is needed between lining types when considering long-term roughness values. The trend away from cementitious linings and towards epoxy lining for new steel pipe is linked to market forces, and also to the approval process (Section 17.3) for material in contact with water used for the public supply. This applies to all constituent materials and requires tight control as is applied for DI pipe linings.

Bare metal pipe is spun at high speed on rollers and the mortar is poured as a slurry into the interior. The lining builds up by centrifugal force and when the required quantity has been poured the speed of rotation of the pipe is increased. The mortar compacts further and surplus water runs off as the pipe is tilted very slightly. The mortar mix comprises sand or fine crushed rock aggregate mixed with Portland or sulphate-resisting cement in the ratio of 2.5:1 or 3:1 by weight. The lining is usually so well compacted after spinning that the pipe can be immediately taken off the spinning bed and is then cured in a damp warm atmosphere for 21 days. The mortar lining thickness for steel is generally thicker than for DI pipes of the same diameter because of the flexibility of steel pipes compared with DI pipes. All fittings associated with mortar lined pipes are also mortar lined but this has to be trowelled on.

The discontinuity of the lining at joints in steel pipes should be filled if the pipe is of large enough diameter to give inside access. Steel pipes which are not large enough for this should either be jointed using flexible joints or be jointed by externally welded collar joints. With the latter the lining is brought flush with the pipe ends which are butted together; the heat input is not sufficient to damage the lining. The collar is coated before assembly; any water between pipe and collar is largely stagnant giving low risk of corrosion.

17.22 JOINTS FOR STEEL PIPES

The following types of joints are in common use: welded sleeve joints (Figs 17.2(a) and (b)); butt welded joints; flexible couplings (e.g. Viking Johnson type) (Fig. 17.1); flanged joints (Fig. 17.2(c)); and push fit joints.

The BS EN 10311 Type 1 taper sleeve joint (Fig. 17.2(a)) is common in the UK. It permits a small angular deviation (up to 1°) as long as the joint is arranged so that there is a penetration of at least four times the pipe wall thickness after deflection. The increased gap around the outside of the pipe requires the weld to be buttered: filler rod can be used. The internal joint gap should be fairly constant as the mating surfaces are brought together. The Type 1 parallel sleeve joint permits little angular deflection but fillet welding is easier. Pipes of DN 800 and larger should be welded internally and externally. The internal weld is the strength weld and the outer is a seal weld. The joint can then be tested by pressurizing the annular space between the two welds (usually with nitrogen at a pressure of 200 kPa); this permits the pipe joints to be tested before backfilling the trench. On pipes smaller than DN 700 the weld should be made on the outside only due to access considerations but completion and inspection of the internal protection remains an issue so that the alternative of using mechanical or other joints is preferred.

BS EN 10311 Type 2 collar joints comprise a short sleeve usually in two parts which is slipped around the pipe end and joined by fillet welding. It is used for jointing small diameter pipes where internal access is not possible and for jointing closing lengths on larger diameters. Where access is possible welds should be made on the inside as well and each side should be air tested.

The spherical joint (Fig. 17.2(b)) was developed to permit angular deflection without unduly increasing the size of the external weld and has been found practicable and more satisfactory than the short sleeve joint because of the smaller welds required. It is not recognized in BS EN 10311 but has been frequently and successfully used on large diameter pipes outside the UK. It allows

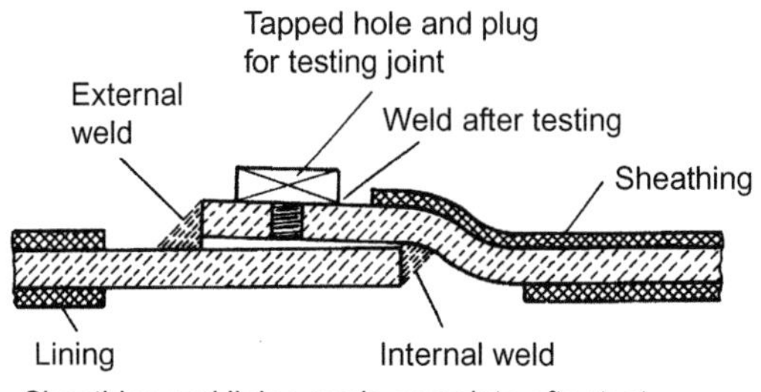

(a) Welded short sleeve joint for steel pipes

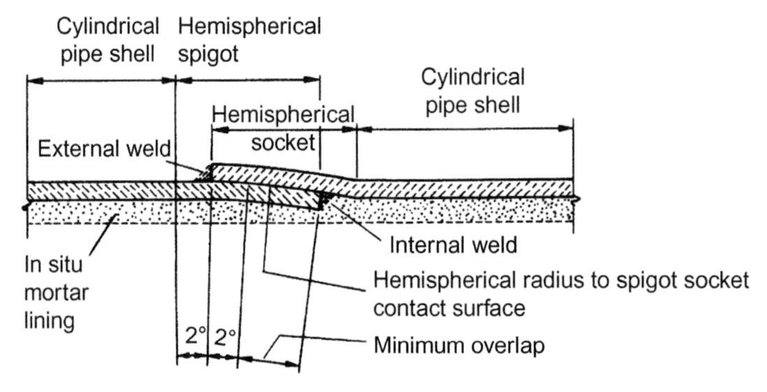

(b) Welded spherical sleeve joint for steel pipes

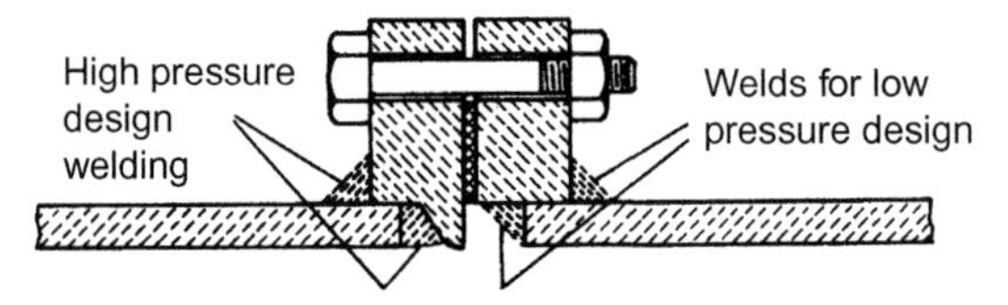

(c) Flanged joint for steel pipes

FIGURE 17.2

Joints for steel pipes.

larger deflections than the Type 1 joint, thus allowing gentle curves to be followed without use of manufactured bends. There must be an overlap of at least four times the pipe wall thickness.

For butt welded joints the ends of pipes are prepared by forming a 30° bevel on the full thickness of the pipe wall except for the inner 1.6 mm; the resulting V-groove between pipes is then filled with weld metal which is finished to stand proud of the external pipe surface. Butt welding is usually only adopted when the pipe wall thickness is substantial or when full longitudinal continuity of strength is required (such as in high pressure applications – Section 17.18), in underwater pipes or if pipes are to be snaked into the trench. Codes such as ASME B31.4 do not allow the use of any welded joint other than butt welds except where operating at low stress. Butt welds can be made by automatic welding machines designed for field use.

Flexible couplings permit modest angular deviation of pipes at joints. Longitudinal movement is also possible but is restricted to the amount allowed by shear movement within the rubber. Further movement will drag the rubber over the pipe surface eventually damaging the joint. Where temperature movements are not expected to be large these couplings may be used at every other pipe joint (24 m centres), with an anchored joint between, instead of expensive bellows joints.

Pipes for jointing with couplings or butt welds must be truly circular at the joint. Pipes that are intended to be cut for jointing must be ordered as 'true diameter throughout'. Out of roundness can cause difficulty in jointing when inserting a branch into an existing pipeline at a point where its diameter may not be true.

17.23 LINING AND COATING CONTINUITY AT FIELD JOINTS

Where the protection has been left off at pipe ends, for example, for welding, it will need to be completed after jointing. This can be done by entry into the pipe when its diameter exceeds about 800 mm or from outside when its diameter is smaller. In the latter case a former is inserted into the pipe and expanded against the pipe wall at the joint; protection material is injected through a hole in the coupling to fill the void at the collar. Alternatively a lining train can be used to apply a measured quantity of paint by roller or spray but surface preparation, cleaning and atmospheric conditions are critical and suitable facilities should be included in the lining train. Fully automated motorized internal lining modules capable of surface preparation, lining application, testing and inspection are available commercially for use with up to 60″ diameter pipes, but are limited with respect to pipe slope and bends.

The outside coating is completed by polyethylene heat shrink sleeves or by tape wrap with an overlap of up to 50%; an epoxy-based primer is often necessary to improve adhesion between the substrate and sleeve. For bitumen sheathed pipes the external protection is completed by 'flood coating' – putting a mould over the coupling, filling it with bitumen and allowing it to cool before removing the mould.

17.24 CATHODIC PROTECTION (CP)

CP is used for the secondary protection of the exterior of pipelines from corrosion, the primary protection being provided by the pipe external coating. Its use for the interior of pipes is not generally practicable. Clays, sulphate-bearing soils, moorland acid waters and saline ground waters are the principal causes of aggressive ground conditions. CP provides an additional measure of security to cover breakdown and deficiencies in the external coating.

In general a CP system requires:

- the pipeline to be coated externally;
- external coating to be resistant to cathodic disbondment – demonstrated by tests;
- electrical continuity along the pipeline;
- electrical isolation from other structures and earthing systems; these include transmission facilities such as pumping stations, storage and valve chambers.

The general principles of design of CP for buried structures such as pipelines are described in BS EN 12954. BS EN ISO 15589 covers design, installation and maintenance of CP systems for oil and gas pipelines but is equally applicable to steel water pipelines. Part 1 deals with on-land pipelines while Part 2 covers offshore pipelines.

CP uses electrolytic chemical action to protect pipes by providing the pipeline at intervals with sacrificial anodes or, alternatively, by impressing a (negative) voltage on the pipeline so that it becomes cathodic with respect to buried anodes in the medium of the wet soil which acts as a weak electrolyte. Corrosion takes place at the anode and not at the cathode. Sacrificial anodes are usually made of zinc, magnesium or aluminium. They are connected electrically to the pipe and cause a weak current to flow in the right direction since they are 'anodic' with respect to iron and carbon steel. They are buried a few metres away from the pipeline at 50–100 m intervals and are connected to it by insulated cables. If impressed current is used a d.c. negative voltage is applied to the pipeline and positive to the anodes which can be of silicon iron, titanium or mixed metal oxide; graphite is no longer used. This type of CP is used where the soil resistivity is high or where long stretches (typically more than 5 km) of pipe need protection since it can provide much higher potential driving forces than the sacrificial system. The lengths of pipe to be protected by a given anode must be electrically connected together. Where joints use rubber rings the pipes each side must be electrically connected by an insulated conductor welded to each.

CP should not be used without pipe coating to which it provides an additional defence. Where other services exist it should be used with care since effectiveness of the system may be reduced by draining of current to the other services and since corrosion of those services might be induced.

Where a steel pipeline passes parallel to and up to about 200 m away from overhead high voltage lines, stray electric potential may be induced in the pipeline. This requires the pipeline to be earthed, using either zinc anodes or through diodes which limit the current flow to one direction and only above a predetermined limit. A steel pipeline should be earthed during construction to prevent welding current from damaging coatings and to earth any induced currents. The length of pipe strung above ground should also be limited typically to about 400 m – and a strict safety regime implemented. Exact precautions need to be determined according to circumstances.

Use of sleeves for crossings should be avoided if possible because the casing may shield the carrier pipe from the impressed current. If a sleeve is unavoidable, the carrier pipe must be isolated from it and the annulus filled with a corrosion inhibiting material. Alternatively, the annulus may be filled with a material such as concrete or grout which allows current to reach the carrier pipe.

PLASTIC PIPES

17.25 POLYETHYLENE (PE) PIPES

PE is a thermoplastic and has been widely used for the production of pipes by an extrusion process for about 50 years. PE pipes are available in both solid wall and profile wall designs; solid wall pipes are primarily used for pressure services, whereas profile wall pipes are used for low pressure and gravity services.

Material classifications in current use are PE 80 and PE 100, where the number refers to the minimum required strength (MRS) in bar (e.g. 80 bar = 8 N/mm^2). The MRS is the pipe wall burst stress at 20°C at 50 years, extrapolated from shorter duration tests.

Materials terminology is confusing: current trend is to refer to grade (PE 80 or PE 100). PE 80 includes both medium-density polyethylene (MDPE) and high-density polyethylene (HDPE). MDPE was brought into service in the UK in the 1980s to replace low-density polyethylene (LDPE) and high-density polyethylene (HDPE). To take account of the risk of crack propagation (RCP) reported for gas pipes, PE 80 pipes DN 250 and larger were derated by about 20% and, above DN 315, were further derated for pipes containing more than 10% free air. This derating has been reviewed in the light of recent research; derating for fatigue is no longer considered necessary for modern PE 80 pipes that meet the stress crack requirements of relevant standards but is required for pipes that do not meet these high toughness requirements. PE 100 has been produced since 1990 and is also known as high-performance polyethylene (HPPE). Compared with PE 80 it has greater resistance to RCP for which it needs no derating. PE 100 costs more than PE 80 but its greater strength allows higher pressures or thinner walls and hence provides economy under certain conditions. PE 80 is less viscous than PE 100 and therefore allows larger manufacturing tolerances; PE 80 is the preferred material for profile wall pipes whereas PE 100 is used for manufacture of solid wall pressure pipes.

Maximum operating pressures are controlled by the sum of hoop stress and ring bending stress, which is limited to 0.8 times MRS (BS EN 12201-1). Maximum operating pressure (PMA) includes surge. Recommendations were that the range of surge pressure should be limited to half the maximum operating pressure. However, due to the low modulus of elasticity surge pressures in PE pipes are typically about 25% of those in steel for the same event. Research shows that PE pipes can sustain surge pressures at least twice their static pressure rating (the actual amount depending on the material and increasing with rate of strain).

Attention must be given to derating for operating temperatures above 20°C; BS EN 12201-1 gives derating coefficients up to 40°C. Long-term diametric deflection is allowed up to 6% of the diameter. However, this must be calculated taking into account creep in the soil and using long-term pipe stiffness – which may be one sixth of the initial stiffness.

Solid wall PE Pipe is produced in the UK up to 1600 mm OD and is available from Europe up to 2500 mm OD. Solid wall PE pipe is specified by nominal (minimum) OD, by standard dimensional ratio (SDR – the ratio of nominal OD to minimum wall thickness, typically 11, 17.6 and 26) and by material. Profile wall PE pipe is manufactured up to large sizes (3500 mm ID) but is not used for pressure applications.

Solid wall PE pipe in the UK is to be specified to BS EN 12201. AWWA C906 is the relevant manufacturing standard for solid wall PE pipes up to 63″ outer diameter in the US. This AWWA standard uses different designation for PE classes; ASTM F714 PE 4710 is the nearest equivalent classification to the PE 100 class used in the UK and Europe although it has a slightly better performance.

The advantages of PE pipes are that they are light and easy to handle, flexible but strong and resistant to cracking, do not corrode, are chemically resistant and can easily be cut to length. For HDPE (PE 100) a minimum radius of 60 times OD is recommended for a pipe under internal pressure or external load. This gives an immediate initial longitudinal bending stress in the pipe wall of about 4.8 N/mm^2 but, with creep, this decreases to about 2.2 N/mm^2 after 1 month and further thereafter. Where load on the pipe is insignificant, the minimum bend radius may be reduced. For solid wall pipes, SDR 11 and 17.6 pipes can be bent safely to radii 25 times OD, increasing to 35 times in cold weather and reducing to lower values in warm weather. For SDR 26 and 33 pipes these radii need to be 50% greater to minimize chances of buckling. Small diameter solid wall pipe can be supplied in coils and straight lengths can be joined above ground and snaked into narrow non-man-entry trenches.

Disadvantages are that the strength of PE pipe, defined as its ability to withstand hoop stress, decreases with time and reduces with increasing temperature. PE pipes are also liable to UV degradation if exposed overlong to sunlight. In the UK blue PE (for water supply) can be stored above ground for up to a year; for longer periods they should be covered. Under certain conditions they may be degraded or may be slightly permeable (and therefore give rise to taste problems) due to oils, organic solvents and by strong concentrations of halogens or acids. The *Handbook of PE Pipe* (PPI, 2007) provides information on salvation and permeability of PE to hydrocarbons. CP 312 Part 1 gives a list of chemicals and suitability with plastics. Where soil is polluted by hydrocarbons, typically in former gas works sites, fuel stations, scrap yards and other industrial areas, it may be enough to use imported backfill and impermeable membranes; alternatively other measures or other pipe materials may be needed.

The coefficient of linear expansion of PE 80 is about 1.5×10^{-4}, which is more than 10 times greater than for steel and is equivalent to about 9 mm for 10°C in a 6 m pipe length. PE pipes are not traceable underground with metal detectors and do not transmit leak noises like metal pipes. For distribution mains PE pipes are supplied in 6 or 12 m straight lengths although longer lengths may be supplied subject to transportation method; their flexibility makes the use of small angle bends largely unnecessary.

Solid wall PE pipes can be jointed mechanically with push fit joints (with longitudinal strength) but are more usually jointed by butt fusion or electrofusion thus forming a continuous string and do not need thrust blocks. Butt fusion requires a purpose-designed powered machine in which the prepared and cleaned pipe ends are clamped and held under longitudinal pressure against a thermostatically controlled heated plate (Plate 27(d)). After a given heating time the ends are pulled back, the plate is removed and the two melted ends of the pipe are brought together under longitudinal pressure to fuse and cool. Growth of the bead of melted material is observed during heating. After jointing the external bead is removed, numbered with the joint and twisted at several positions. Any split in the bead requires the joint to be cut out and made again; any further defect requires investigation and cleaning of the equipment and new trial joints before production is allowed to continue. Internal weld beads can be removed if required. Good quality assurance and quality control procedures are essential and should include tracking to monitor performance of individual and gang workmanship. Experienced and disciplined inspection is essential. Electrofusion joints are used at fittings and comprise a spigot and socket with in-built electrical heating coils which are energized to locally melt and fuse the fitting and pipe together. Equipment can be provided to read a bar code fixed to the fitting which contains information to control the weld. The equipment can then store weld data for downloading onto a computer database for traceability and quality assurance. Flow of melt material from indicator holes in the fitting gives visual evidence of weld completion.

Pressure testing must take into account the viscoelastic properties of the pipe material. Traditional pressure test techniques are not suitable due to creep and stress relaxation; pressure in a closed PE pipe falls with time even when the pipe is leak free. The accepted method is to fill the pipe with water, raise the pressure, measure the pressure decay with time, derive decay parameters using a 'three-point analysis' (more points are strongly suggested) and compare with a range of standard values according to pipe restraint. The pipe must be sensibly free of air and not above 30°C. Pressure should be released slowly. If a retest is required, a rest time, typically of about five times the previous test duration, must be allowed beforehand for the pipeline to recover from the previous conditions. WIS IGN 4-01-03 provides further guidelines on pressure testing of PE pipes. The test procedure in ASTM F2164 aims to compensate for creep of PE by testing for one hour at a pressure 1 bar lower than the initial expansion phase pressure held for 4 hours immediately before the test phase.

17.26 POLYVINYL CHLORIDE (PVC) PIPES

Four types of PVC pipes are available in the UK:

- PVC-U: unplasticized PVC;
- MOPVC: molecular orientated PVC;
- PVC-A: PVC with added impact modifiers – also known as PVC-M (modified PVC); and
- CPVC: chlorinated PVC.

BS EN 1452 is the manufacturing standard for PVC piping systems for water supply and replaced BS 3505 and the respective Water Industry Standards (WIS 4-31-06 and 4-31-07). Typical PVC-U pipe material has a relative density of 1.4 and a softening point of about 80°C. The coefficient of linear expansion is about 6×10^{-5} per °C which is more than five times the value for steel.

PVC pipes are classified by nominal OD and by pipe series (S) where: S = (SDR-1)/2. PVC-U pipes are available in the UK up to DN 630 and are supplied in 6 m lengths, with spigot and socket rubber ring push fit joints. In the US, AWWA C900 is the manufacturing standard for PVC-U pipes between 100 and 300 mm used for water supply distribution systems; AWWA C905 is the relevant standard for larger diameter PVC-U water transmission pipes, between 350 and 1200 mm diameter. Solvent joints have not been used since the mid 1980s and are no longer permitted for underground use (primarily due to problems arising from movement during bedding and backfilling) but are permitted for above ground use.

A number of failures of early PVC-U pipes led to lack of confidence in their use. The problems are now understood to result from local bending stress from point loads, typically uneven bedding. Material improvements have restored confidence but:

- pipe should be derated for temperatures above 20°C, derating coefficients are given in BS EN 1452-2 for up to 45°C;
- pipe with significant internal or external scratches should be rejected due to notch sensitivity;
- care is required with installation;
- care is required during tapping and saddles should not be over tightened; and
- derating is required for repeated cyclic (fatigue) loads (distinct from diurnal variations), particularly for pumping mains.

Wall thickness for PVC-U is based on an allowable design stress of 10 N/mm^2 for pipe OD up to 90 mm and 12.5 N/mm^2 for OD 110 mm and higher. These compare with the MRS specified in

BS EN 1452-2 of 25 N/mm^2 and the required minimum bursting stress of 35 N/mm^2 at 100 hours. The design stresses for MOPVC and typically for PVC-M are 22.5 and 18 N/mm^2, respectively. Full structural design would not normally be needed for DN 315 and smaller.

Maximum operating pressures are controlled by the sum of hoop stress and ring bending stress which should not exceed the design stress. Maximum operating pressure includes surge. Recommendations were that the range of surge pressure should be limited to half the maximum operating pressure, but because the wave speed in the pipe wall is lower, surge pressures in PVC pipes are less (typically 30% less) than those in steel for the same event. Research shows that PVC pipes can sustain surge pressures about twice their static rating (about 1.8 times for PVC-A and 2.2 times for MOPVC and PVC-U, the actual amount increasing with rate of strain). Care should be taken for pumping mains larger than DN 315 and fracture toughness tests should be specified as set out in BS EN 1452.

The main advantage offered by PVC is its resistance to corrosion; hence its use for chemical transfer lines in water treatment works. It is not suitable for use in ground contaminated, or likely to be contaminated, by detergents or solvents or from oil storage areas. It is light in weight, flexible and has easily made joints. Small size pipe can be joined above ground and 'snaked' into narrow non-man-entry trenches. Care should be taken when handling pipe at temperatures below freezing due to reduced impact resistance.

Fittings can be made in PVC-U or metal, usually DI. PVC-U pipes are degraded by ultraviolet light, the effect increasing with temperature so that the pipes must not be exposed to sunlight in hot climates for more than a day or two. In the UK, MOPVC, PVC-U and PVC-A can be stored above ground for up to a year; for longer periods if covered.

MOPVC is first formed at about half the diameter and double the final thickness and is then heated and expanded to the final diameter. The result increases the strength particularly in the hoop direction and allows use of reduced wall thicknesses (slightly more than half of those for PVC-U). It also increases significantly the resistance to failure from cyclic loading. MOPVC pipes are about 30% of the stiffness of PVC-U pipes due to their reduced thickness but taking into account their greater modulus.

PVC-A is an alloy of PVC-U, polyethylene (PE) and acrylics: wall thicknesses are about 80% of those for PVC-U.

Chlorinated PVC is post-chlorinated PVC which is inert to a wide range of aggressive chemicals. CPVC is formed by reacting additional chlorine with the PVC polymer to increase total chlorine content by mass from about 57% to about 67%. The additional chlorine content allows CPVC to withstand exposure to direct flame for longer and, with appropriate derating, the material is suitable for use at higher temperatures (up to 90°C). CPVC is widely used in chemical handling lines in water treatment plants and for firefighting pipework; ASTM D1784 is the standard specification for CPVC pipes.

17.27 GLASS REINFORCED PLASTICS (GRP) PIPES

GRP (known as Fibre Reinforced Plastic (FRP) in the USA) pipes are lightweight and corrosion resistant and are composite pipes made from polyester or vinyl ester or epoxy resin and glass fibre reinforcement, usually with a filler of silica sand. GRP pipes have been in use, primarily for trunk mains, since 1970. Current standards are AWWA C950 and BS EN 1796 (for pipes intended for use in water supply).

Filler inclusion allows increase in wall thickness at modest cost and thus provides additional stiffness, an advantage for laying. By adjusting the amount of glass fibre and the wall thickness, the pressure rating and stiffness can be varied independently, unlike other flexible pipes. GRP is thermosetting (sets irreversibly under heat) as opposed to thermoplastic (PE and PVC, which melt on heating). Pipes may comprise layers of different resins and different quantities and qualities of glass through the pipe wall: typically using resin rich and corrosion resistant qualities for the surface layers. Where different resins are used, the inner surface layer comprises a more flexible and chemically resistant resin, for example, bisphenol, and the structural and outer surface layers use a cheaper, stronger resin, for example, isophthalic or orthophthalic. Several formulations are used, different for each resin supplier and may be peculiar to each pipe manufacturer. The outer surface layer should be formulated to provide handling robustness and to resist environmental conditions.

Dual laminate thermoplastic lined GRP pipes are often used in chemical process industries. BS EN 1796 allows the ID of such pipe to be up to 3.5% smaller than the nominal size (DN). AWWA C950 requires pressure class derating for thermoplastic lined GRP pipes exposed to service temperatures in excess of 23°C. ASME RTP-1 Mandatory Appendix M-12 provides comprehensive guidelines on thermoplastic lining and dual laminate pipes.

Care is required to specify and ensure that the resins are suitable for encasement in concrete for example, vinyl ester or epoxy resin, if required; alternatively fittings in contact with concrete may be made of DI or steel. Materials and the process must be suitable for potable water. Like paints, thermosetting resins are slightly permeable. Consequently the whole pipe wall thickness is exposed to some extent to the liquid under pressure from one surface. Therefore, the types of resin and the glass (Types 'E' and 'C' are covered in BS EN 1796) should be selected to suit. Early problems with delamination have largely been overcome by improved materials and manufacturing processes but care has to be taken in selection of resins and in surface preparation of the glass fibres to ensure full wetting by the resin and resistance to corrosion.

Pipes can be made by the Hobas centrifugal casting process or may be filament wound on a mandrel. In the Hobas process metered weights of liquid resin plus filler and randomly orientated chopped glass strand are fed to the inside of a rotating mould. This produces a dense pipe wall with similar characteristics in the hoop and longitudinal direction, although fibre orientation can be adjusted if required to increase tensile capacity in a particular direction. The constant OD allows pipes to be cut and joined at any position.

Filament wound pipes are made in discrete lengths on a rotating mandrel. For reasons of production economy, this arguably allows more choice in design of the pipe wall but may require more glass fibre to retain (fluid) material as the mandrel rotates. Modern filament wound processes allow filler to be incorporated to produce economic pipe of higher stiffness. Pipes produced by the filament wound process typically have allowable hoop stress approximately twice that of allowable axial stress. Biaxial loaded pipes (i.e. pipes exposed to hoop stress and longitudinal stresses arising out of thrust restraints or bending) need complex laminate design and should be specified accordingly. The filament winding angle is a design parameter and typically varies between 55° and 73°; allowable stress values of the composite can be increased by increasing the winding angle.

The Drostholm process comprises a cantilevered mandrel wrapped by a continuous steel strip which advances helically around and along the mandrel, providing a continuous rotating surface before returning up the inside of the mandrel and commencing the helix again. Pipe materials are built up on this surface, being added sequentially along the pipe as the helix advances. Material is heated so that the pipe cures on the mandrel and can be cut into discrete lengths shortly after it leaves the mandrel. The Drostholm process is not used in the UK but pipe made by this process has been used on some major projects.

Manufacture of GRP pipe is a complicated process which must be carefully controlled. With improved materials and manufacturing control previous problems of delamination are now rare but deflections need to be controlled to avoid strain corrosion. Where such problems are suspected inspection using ultrasonic techniques should be adopted. Pipes must be laid under strictly controlled backfill conditions to prevent unacceptable distortion of the pipe wall or bending and must be designed for both short- and long-term conditions, typically using 50-year pipe characteristics extrapolated from shorter duration tests.

Pressure rating and stiffness can be tailored independently to suit project specific requirements; this benefits economy. The pipes are flexible and typically are designed to BS EN 1796 and installed to BS 8010 Section 2.5. The former standard states that pipes of stiffness less than 1000 N/m^2 are not intended for laying directly in the ground. However, pipes of stiffness less than 5000 N/m^2 require use of expensive embedment materials and very close control when placing and compacting the embedment (Plate 27(c)). Unless it is clear that such measures will be effective stiffness should not be less than 5000 N/m^2.

There is a significant difference between the stiffness measurement methods used in BS EN, ISO and the ASTM/AWWA standards. According to ISO stiffness is measured by dividing the force per unit length which causes 3% diametric ovalization of a specimen by the deflection; the stiffness is expressed in kPa. According to AWWA C950 the structural stiffness is measured as the ratio between the unit force required and vertical pipe deflection measured at 5% diametric ovalization ($Stiffness = F/dV$); the stiffness value in the AWWA standard is expressed in psi. The following table provides an approximate comparison between the stiffness class specifications used in the BS EN standards and AWWA.

Stiffness class (SN) as BS EN, ISO	Stiffness class as AWWA
SN2500 (2.5 kPa)	18 psi
SN5000 (5 kPa)	36 psi
SN10000 (10 Pa)	72 psi

Where GRP pipes are subject to cyclic loads the manufacturer must be informed of the detailed performance requirements. Thrust blocks at bends and fittings, on large diameter pipes particularly, need to be fully reinforced around the pipe because of the high strain in GRP under pressure. GRP pipes also need to be wrapped locally with elastic material at the exit from rigid structures; manufacturers can provide suitable details.

Lengths up to 18 m are produced (Plate 27(c)). BS EN 1796 sets out the initial minimum longitudinal tensile strengths for two cases: where the pipe (a) is, or (b) is not, exposed to longitudinal forces. Where such resistance is required (a) the minimum tensile strengths imply a short-term safety factor of about 2.8. It should be noted that the values set out in the BS for case (b) are higher in most cases than those obtained by application of case (a). Therefore, if case (a) applies the specifications should require the pipe to meet both requirements.

Joints generally are the flexible push fit spigot and socket or collar type. Alternatives are resin adhesive or flanges and screw threads. Angular joint deflection limits for push fit joints are at least 3° for DN 500 and less, 2° for DN 500 to 800, 1° for DN 900 to 1700 and 0.5° for DN 1800 and greater. Actual values at installation should not exceed half these values. Where required, rope or other locking strip can be threaded circumferentially into a preformed groove in push fit joints to lock and provide tensile strength across the joint, for anchorage and for installation under water. Polypropylene or similar non-metallic locking strips are provided by manufacturers; if metallic rings are used, adequate protection against corrosion should be provided.

GRP fittings such as bends and branches may be fabricated or moulded; the latter being made by building up mat, filament, resin and filler in a mould. Fabricated fittings are made from GRP pipe which is cut, glued together and laminated with glass fibre mat across the joint. Any bend angle can be supplied and complex specials may be made as a single unit. CAE software can be used for design and its output can be linked directly to automated fabrication machinery. Flanges can be fabricated in GRP and are provided with flat face finish; installation requires full-face gaskets to reduce secondary loads on flange laminate. An additional spacer is often necessary when a GRP flange is connected with a raised face valve or pipe flange. If any axial or bending loads are to be imposed on a GRP flange then these must be specified separately for flange laminate design. For smaller diameters (up to DN 600) DI and PVC-U fittings can be used.

GRP has a key advantage for carrying aggressive waters and or for laying in exceptionally aggressive ground conditions where both iron and concrete pipes would be severely attacked. A balance has to be drawn between resins with the best chemical resistance such as vinyl ester, which is comparatively brittle, and those with less chemical resistance but greater flexibility. Modern practice is to specify GRP pipe by performance and leave resin choice to the manufacturer, who has links to resin suppliers and can tailor resin to best suit all needs.

CONCRETE AND FIBRE CEMENT PIPES

17.28 PRESTRESSED CONCRETE PRESSURE PIPES

AWWA M9 provides guidance on design and use of concrete pressure pipe. Prestressed concrete pressure pipes are covered by BS EN 639 and BS EN 642 for cylinder and non-cylinder pipe respectively. The nominal size range is from DN 200 to 4000 and is the design ID. AWWA C301-92 for lined cylinder pipe and embedded cylinder pipe and C304-92 also cover large diameters. Examples of use of such pipe are typically DN 4000 up to 20 bar and DN 3600 up to 26 bar for the Great Man-Made River project in North Africa. Design is restricted in principle only by material properties and capability of the structural section. Pipes are made by tensioning high-tensile wire wound spirally around a cylindrical core of concrete, which is prestressed longitudinally, or of steel which has a thick spun concrete lining as shown in Figure 17.3. For large diameter prestressed concrete pipe the core is cast vertically around the inside and outside of the steel cylinder. Pipes with

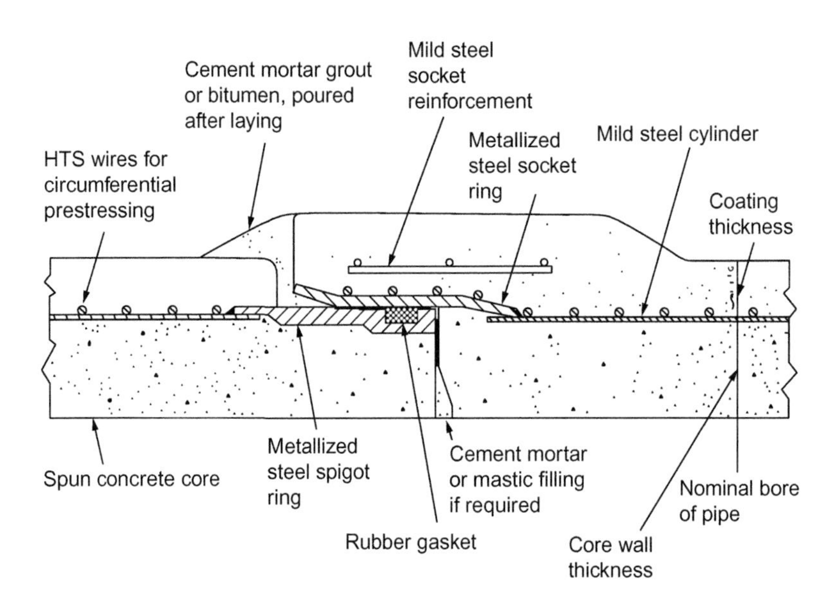

FIGURE 17.3

Joint in prestressed concrete cylinder pipe.

steel cylinders in the core are called 'prestressed concrete cylinder pipes'; those with longitudinal prestressing are termed 'non-cylinder'. When the wires have been wound onto the core, stressed and anchored, a relatively thin but dense cement rich mortar coating (20 mm minimum) is applied pneumatically or by machine at high velocity as the pipe is rotated externally. The resulting hard dense mortar coating provides mechanical and corrosion protection.

Prestressed concrete pressure pipes may offer a cost advantage over other pipes, particularly for large diameter. Prestressed pipes can be made to withstand higher pressures simply by increasing the number of turns of prestressing wire per unit length or by doubling up the layers of prestressing wire. Such pipes are proof against certain corrosive conditions, although they may need special protection if the groundwater or conveyed water is high in sulphates and chlorides or otherwise aggressive to concrete. CP (Section 17.24) may need to be considered but requires electrical continuity to the prestressing wires, the steel cylinder and joint rings at the ends of each pipe, with the pipes bonded together electrically after laying. Alternatively, or in addition, the pipes can be coated with urethane or epoxy. Coal tar enamel and or coal tar epoxy are not suitable as internal protection for pipes carrying water intended for human consumption. The pipes can also be made using sulphate-resisting cement. The pipes are used in the marine environment where buried and permanently submerged but attention is needed to prestressing steel quality and to combating aggressive conditions in the tidal and surf zone where the water level varies and where the oxygen supply is high.

The prestressed pipe steel joint rings provide a high-performance O-ring seal allowing joint rotations of up to 0.9° in any direction at each joint while maintaining water tightness. Deep joints are also available which can accommodate up to 1.4° of rotation at any joint. Full bevel and half bevel joints which allow several degrees of deflection can be built into standard sections of pipe, enabling prestressed concrete pipes to deal with severe changes of grade and alignment (up to about 5°) before having to use preformed bends. Joints can be welded if required and can then carry longitudinal force.

Concrete pipes are rigid and provide a degree of resistance against rough handling and poor backfilling techniques. However, in the larger diameters the pipes are very heavy and difficult to align in very soft ground unless they are placed on a prepared bed of granular material of adequate thickness to prevent uneven settlement. Connections can be made after the pipeline has been laid and while the pipe is in operation although to avoid shutdown, tees have to be incorporated in the line as it is laid.

Cracking of the mortar coating does not occur under normal operating conditions (up to the design operating pressures and external loads), but it may occur if the design operating conditions are accidentally exceeded through pipeline operational errors. However, the elastic behaviour of prestressed concrete pipe is such that the cracks close up after the abnormal conditions have been corrected. The mortar coating cannot be prestressed directly but does receive some compressive strain as the concrete core continues to shrink after completion of prestressing since the coating is applied immediately after the prestressing operation and is mechanically bonded to the prestressing wires. This feature together with a feature known as tensile softening allows the mortar coating to take much more strain than the standard calculated plain strain before cracking. If prestressed concrete pipes are to be tested for long durations (say 24 hours) at a pressure greater than the design operating pressure, then the pipe must be designed accordingly: that is, the sustained test pressure should be considered as the design pressure.

BS EN 639 and 642 contain requirements for factory pressure testing. US practice is to use cylinder pipe which has been pressure tested.

Joints for prestressed concrete pressure pipes are usually of the socket and spigot O-ring push fit type. The socket is normally mortared up afterwards and the joint then becomes rigid. Where ground conditions are known to cause differential settlement the joints may alternatively be filled with bitumen mastic which provides the necessary protection while allowing joint rotation to occur.

There has been concern about failures of prestressed concrete cylinder pipe; these can be catastrophic for large pipe. In 2008 AwwaRF carried out a survey and analysis (AwwaRF, 2008) of failures to PCCP pipe made to AWWA standards and its predecessors across the USA in the previous 65 years. Failures were found to have been attributed to a large number of different causes some of which were not peculiar to PCCP or even concrete pipe in general. However, two periods of manufacture coincided with higher failure rates – pre-1955 and 1972–78, both periods when the standards were less onerous for design and manufacture and allowed higher stresses. Questions used in the survey reflected risk factors acknowledged in the literature such as:

- use of Type IV prestressing wire;
- use of 8-gauge reinforcing wire not 6-gauge;
- use of cylinder thinner than 16-gauge;
- wire stress ratio more than 70% of yield;
- cementitious coating less than 22 mm thick;
- little or no review or inspection of the pipe design, manufacture or installation;
- pressure fluctuations in operation, for example due to surge;
- test pressure applied more that 20% greater than operating pressure.

17.29 REINFORCED CONCRETE CYLINDER PIPES

Reinforced concrete cylinder pipes (RCCPs) are similar to prestressed concrete pipes, except that there is no prestress and instead of using high-tensile steel wire for circumferential reinforcement, carbon steel rod is used. The reinforced cylinder is then covered internally and externally with concrete, the internal lining being spun on and the external coating being applied by impact and smoothed in layers. Size range is from 250 to 4000 mm ID with wall thicknesses from 78 to 320 mm, respectively. At larger sizes the pipe is cast vertically (wet mix) using a preformed spot welded reinforcement cage with either one or two layers of reinforcement according to design requirements. In the design of the pipe both the steel cylinder and the reinforcement are assumed to resist hoop tension stresses.

CCP pipes require less sophisticated manufacturing techniques than do prestressed concrete pipes. They can be used in the marine environment, albeit with care in the tidal and surf zone, and are widely used around the world in large sizes for power station cooling water systems and desalination plant inlets and outlets, as well as effluent outfalls and irrigation water transfer. Disadvantages are as mentioned for prestressed concrete pipes but, in addition, cracks in RCCPs caused by operational overload may not close up in the same way as for prestressed concrete. Some degree of shrinkage cracking is normal after manufacture and can be accepted – typically up to 1 mm in surfaces which will be continually wet after installation (internal surfaces and surfaces constantly below water level) as they will close after wetting and seal by the process of autogenous healing. Reinforcement size and spacing are designed to limit structural crack widths under load as in water retaining, maritime and other codes. For carrying axial forces, joints generally are welded and therefore are rigid requiring careful pipe bedding to limit differential movement. Rubber ring push fit joints and other joints can be made if required. Fittings are coated internally and externally with reinforced concrete.

BS EN 639 covers reinforced concrete pipes and BS EN 641 covers RCCPs. Effective lengths in the Europe are about 7.5 or 6 m for smaller sizes, reducing to 2 m at DN 4000 and are available up to 10 m from DN 400 to 1000. Up to DN 1000 in the UK the reinforcing rod is wound spirally under low tension on to a rotating steel cylinder, being welded to the cylinder at both ends. Reinforced concrete non-cylinder pipes are produced in the UK but are used for gravity flow and drainage.

AWWA M9 provides guidance on the design and use of concrete pressure pipe. AWWA C300 addresses RCCPs that are not prestressed or pretensioned. It covers pipe, which is treated as rigid, from DN 760 to DN 3600, typically with wall thicknesses between 89 and 305 mm (3.5–12 inches). This type of pipe has a relatively thick welded cylinder towards the inside and a layer of bar reinforcement towards the outside separated by concrete which renders the pipe rigid. AWWA C303 covers reinforced concrete cylinder pretensioned pipes (PtCCP). The pipe is reinforced with a steel cylinder which is helically wrapped with mild steel bar reinforcement under moderate tension, in sizes from DN 250 to DN 1520 (10–60 inches) inclusive and for working pressures up to 27 bar. The lining thickness is 13 mm minimum for DN 250 to DN 410 and 19 mm minimum for DN 460 to DN 1520. The cement mortar coating thickness is 19 mm over the reinforcing bar or 25 mm over the cylinder, whichever results in the greater thickness of coating. Compared with RCCPs, PtCCP pipes have thicker steel shells and thinner concrete and are semi-rigid. They are not used in the UK.

17.30 REINFORCED CONCRETE PRESSURE PIPES

Concrete pressure pipe without a steel cylinder (RCP) is made in large sizes (up to 4000 mm) for low-pressure applications. It is covered in AWWA C302 and by AWWA M9.

17.31 FIBRE AND ASBESTOS CEMENT (AC) PIPES

Although AC pipes were in use for over 50 years and continue to be produced in some countries, the danger to health posed by the handling of asbestos in pipe manufacture and on site, particularly due to the risks of inhaling asbestos dust during cutting, has resulted in cessation of production in many countries. There is however no evidence that their use for conveyance of drinking water presents a direct danger to health. UK standards warn against the hazards (asbestosis, lung cancer and mesothelioma) of breathing asbestos dust and refer to prohibitions on the supply of products containing amosite (brown asbestos) and crocidolite (blue asbestos). AC pipes made in the UK from 1982 are said to contain only chrysolite (white asbestos) but production ceased in 1986. BS EN 512 applies to two types: AT (Asbestos Technology), containing chrysolite asbestos and NT (Non-asbestos Technology) containing fibres other than asbestos. AWWA codes (C400, C403, C603 and M16) applicable to AC pipe have been withdrawn.

Fibre cement pipes are made of Portland cement and fibre mixed into a slurry, with or without finely divided silica, and deposited in layers on a cylindrical mandrel. Cement used for AC pipes in the USA may be Type I or II or slag or pozzolanic cement. When the required thickness has been built up, pipes are steam or water cured, cleaned, the ends turned down to an accurate diameter for some 150 mm, and then usually dipped into cold bitumen.

Fibre and AC pipes can often be produced from local resources more cheaply than other types of pipe, especially in countries which have to import steel or iron. They are also resistant to internal and external corrosion except in sulphated soils which attack the cement unless it is protected with bitumen. Fibre and AC pipes are brittle and need careful handling; their bedding and surround must contain no large stones and they should preferably not be laid beneath roads subject to vibration from heavy traffic unless surrounded by concrete. Ferrules tapped directly into a main without the use of a saddle are frequent causes of leakage; service pipe connections are made using a metal saddle clamped onto the pipe.

BS EN 512 gives standards for fibre cement pipes up to DN 2500. Pipes up to DN 1000 are classified according to nominal pressure up to 20 bar. Preferred classes are 4, 6, 10 and 16 bar but pipe classes greater than 20 bar can be supplied if required. The required nominal pressure rating is decided in relation to hydraulic and operating conditions and external load. Pipes exceeding DN 1000 are designed to suit particular requirements of the pipeline. The works test pressure (PT) is specified as double the nominal pressure up to DN 500 and 1.67 times the nominal pressure for larger sizes and is held for 30 seconds or, if the test pressure is increased by 10%, 10 seconds. Sizes for water supply are commonly DN 100 to 900 for working pressures of 75, 100 and 125 m. Nominal lengths are typically up to 5 m for DN 300 and smaller and up to 6 m for larger sizes.

Although some fittings are fibre cement, common practice is to use coated steel or standard DI fittings with spigots to match the class of pipe used. Fibre and AC pipes and fittings are all plain ended and hence are jointed using collars of asbestos cement or by use of flexible couplings. Great care is needed to ensure that O-ring joints have no twist in them when placed, otherwise the joint will leak. A leak under pressure at a joint may in time cut right through the pipe wall. A gap must be left between the ends of pipes to allow for deflection and, where appropriate, thermal movement.

17.32 POLYMER CONCRETE PIPES

Polymer concrete consists of aggregate, usually quartz sand and gravel, and a mineral filler in a polyester resin matrix which constitutes about 10% by weight. The material was developed in the 1960s but was not widely used until the 1980s when it was adopted in Germany and the USA for trenchless construction. Compressive strength is two to three times that of concrete used in pipes but the material is treated as brittle because the tensile strength is only about 10% of the compressive strength. In acidic environments both strength and pipe stiffness decrease with time.

Polymer concrete pipe is cast vertically in moulds. Joint rings are pre-fabricated in stainless steel or FRP. The standard for polymer concrete pipe is ASTM D6783 which covers gravity pipe which is unreinforced. In theory, the material can be reinforced with either steel or FRP bar. However, there is concern about corrosion of steel reinforcement in an aggressive environment if not embedded in the alkaline environment provided by cement concrete; resin, while having very low permeable, cracks in tension and has neutral pH. Polycrete pipe reinforced with FRP bar would offer no advantages over FRP pipe while being more complex to manufacture. As a result, polycrete pipe is used for gravity or low-pressure applications only.

RESISTANCE TO INTERNAL FORCES

17.33 FORCES DUE TO PRESSURE AND VELOCITY

Thrusts are developed on pipelines at changes in direction, tapers, junctions and closed ends (including valves) due to out of balance forces on surfaces exposed to the pipe fluid. Two types of thrust operate: that due to internal pressure and that due to velocity. The latter force is due to change in momentum and can usually be ignored in water supply pipe systems as velocities are low. The velocity force T on a bend is given by:

$$T = \frac{\pi}{2} D^2 \rho V^2 \sin\frac{\theta}{2}$$

where D is pipe internal diameter, ρ is fluid density, V is velocity and θ is the angle of the bend.

The pressure thrust is proportional to pressure; values for different fittings are given in Figure 17.4 which shows an assembly of the usual fittings. For pipework between flexible joints

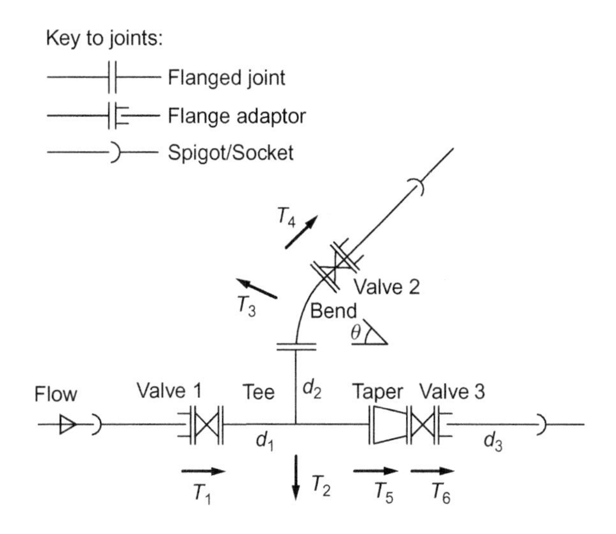

Case	A	B	C	D*	E	F	Thrust value
Valve shut	None	1	2	3	2 & 3	All**	
Thrust T_1		Yes				Yes (negative)	$\pi/4$ p d_1^2
Thrust T_2	Yes		Yes	Yes	Yes	Yes	$\pi/4$ p d
Thrust T_3	Yes		Yes	Yes	Yes	Yes	$\pi/2$ p d_2^2 sin $\theta/2$
Thrust T_4			Yes		Yes	Yes	$\pi/4$ p d_2^2
Thrust T_5	Yes		Yes	Yes	Yes	Yes	$\pi/4$ p $(d_1^2 - d_3^2)$
Thrust T_6				Yes	Yes	Yes	$\pi/4$ p d_3^2

d_1, d_2 and d_3 are internal diameters of joint socket or sleeve.
p=internal (or differential) pressure
*Net thrust is that from taper only (T_5) as T_2, T_3 and T_4 cancel each other out.
**Hypothetical situation equivalent to a pressure vessel—resultant of all thrusts is zero.

FIGURE 17.4

Thrusts at pipe fittings.

the relevant diameter is that at the centre of the joint ring. The loadings from internal pressure should be the thrust developed under operating and surge pressure or pipeline testing conditions, whichever gives the more critical result.

17.34 PIPE RESTRAINT AND THRUST BLOCKS

The designer must ensure that all thrusts on pipework can be taken safely particularly, but not only, where pipe joints are not able to transmit axial load. This is the case for flexibly jointed DI, GRP, PVC or concrete pipe. Inside reinforced concrete structures the various thrusts should be carried to the structure by blocks but the stability of the whole structure under all load combinations must be checked. For buried pipelines, thrust blocks (or walls) must be designed for the soil resistance available and must bear against ground undisturbed by the pipe laying process. The only exception is where the design uses parameters for fill which are suitable for the soil used, the compaction achieved and the level of quality control applied. Differential movement at thrust and anchor blocks needs attention: rocker pipes may be required.

Blocks or walls to resist horizontal thrust transfer it to the ground. The soil resistance that can be mobilized to resist thrust must be carefully analysed; specialist advice based on site investigation is needed to obtain appropriate geotechnical design parameters. Thrust block design in the UK is generally carried out in accordance with CIRIA Report 128 (CIRIA, 1994). Groundwater should be taken into account and a significant factor of safety should be applied to ultimate values of bottom friction and passive resistance to ensure that movements do not cause pipe joints to open. Report 128 recommends use of a resistance reduction factor, which includes a factor of safety, ranging from 2–3 for dense sand and gravel up to 4–5 for soft clay. Vertical thrusts have also to be allowed for, upthrust being resisted by tying the fitting down to a reinforced concrete block of sufficient weight or by mobilizing the weight of a sufficient length of buried pipe. Blocks or walls are usually suitable for horizontal thrusts at isolated fittings where there is room. However, thrust blocks designed following the CIRIA procedure may be large in weak soils, particularly where groundwater is high.

AWWA M11 quotes a factor of safety of 1.5 and recognizes that movement is needed to mobilize passive resistance. It lists a range of factors (from 0.2 to 1.0) to be applied according to natural soil type and laying conditions, a factor of 1.0 being given for compacted granular embedment in ground of zero or low plasticity. Allowances should be made for the presence of groundwater which is not discussed in M11.

Thrust resisting structures (and any concrete surround) must consider hoop strain in the pipe under pressure and the possible need for reinforcement in the concrete, and for elastic sleeving at entry and exits of the pipe from the concrete. For low-stiffness materials, such as GRP, hoop strain can be considerable and enough to split the concrete, rendering the block unserviceable, or to burst thin sections locally leaving sharp edges which can damage the pipe over time, particularly under cyclic pressure loading.

If space is restricted, for example, due to the presence of other services, or where thrusts have to be accommodated from several fittings, an alternative is to construct a reinforced concrete thrust slab beneath the pipework concerned and beneath adjacent services where appropriate. Thrusts are transferred by reinforced concrete upstands to the slab which should be designed as a thrust block. Otherwise, restrained joints may be used.

Restraint for joints is available in various forms, such as external ties, locking rings or toothed gaskets. A tied coupling is a flexible coupling across which threaded anchor bars are fixed between lugs or flanges welded to the pipe wall. Restrained flexible joints are available in GRP (FRP) as described in Section 17.27. Some DI pipe manufacturers offer various anchoring systems for their spigot and socket pipe. One type is the anchor gasket. This is similar to the standard push fit joint but the joint ring is armed with stainless steel teeth designed to bite into the outside of the pipe wall on tightening, thus providing capacity to carry tension. Allowance must be made for the longitudinal movement that is initially required for the joint teeth to bite. It is maintained that some such joints retain flexibility but users should be aware that repeated movement, due to cycles of expansion and contraction or angular displacement, is liable to cause the teeth to drag and consequently reduce locking capacity over time. If corrosion takes place it would further reduce anchor capacity. BS ISO 10804 covers testing of restrained joints and requires protection (of the joint system) against aggressive soils to be at least as good as that of the pipe. This may be achieved by application of works applied coatings or, on site, by used of corrosion inhibiting paste and tape wrap. It is suggested that anchor gasket joints be used only with supplementary corrosion protection, but not for larger pipes and higher pressures or where risks are higher and repairs would be awkward or expensive. Use of this type of joint to achieve resistance to thrust by friction along the pipe is not suitable if the ground is of variable quality or likely to be disturbed in the future.

AWWA M41 includes tables of resistance per unit length of pipe to movement and of the lengths of pipework with restrained joints that are required for combinations of pipe diameter, soil type and burial depth, all for an internal pressure of 100 psi (6.9 bar). The lengths take account of the bearing resistance and friction of the length of pipe. For pipe encased in polyethylene, M11 recommends use of a 0.7 modifying factor on friction resistance. The tables show that the necessary anchorage length can be large, particularly for poor soils, where the embedment is shallow and the water table is high. In granular soils the ratio of anchorage length to pipe diameter reduces with increasing diameter.

PIPELINE AND PIPE SYSTEM CONSTRUCTION

17.35 CHOICE OF PIPES

Pipe material selection is discussed in Section 17.7. Information on the use of pipes of different materials is set out in Sections 17.13–17.32. Although the design usually specifies the material to be used for construction, specifications for supply and lay work sometimes allow a choice of two or more materials. It can be advantageous to allow alternatives to be put forward by bidders since the equipment available to one bidder may be best suited to one material or one bidder may be able to secure a good price for supply of pipe of a particular material.

Given the option a bidder will tend to opt for the cheapest pipe material and the simplest and cheapest installation. These two aspirations are not usually compatible since the cheaper pipe materials require more expensive embedment materials and more effort and control in construction (Section 17.8).

17.36 PIPE LAYING AND INSTALLATION

Strict control needs to be exercised over the laying of pipes because of the high capital cost of pipelines and their swift deterioration if not properly laid. Factors key to successful installation of a pipeline are discussed below. Guidance for laying pipelines can be found in a number of the codes and other publications mentioned in this chapter. McAllister (2014) provides extensive guidance on practical aspects of construction of steel pipelines.

Typically a 'working width' for construction is defined in the contract documents to allow space for access, pipe stringing and separate stockpiles for topsoil and excavated spoil. In addition space is required for pipe storage and office and support facilities. The pipe 'easement' as finally granted relates to the pipe as laid and sets out rights for access and limits subsequent construction and other works. A 'wayleave' is an agreement similar in nature but less specific than an easement, granted by the landowner, permitting the pipeline owner to carry out the works.

Care should be taken to ensure that pipe supports on vehicles transporting pipes are adequate to prevent damage to the external coating (especially in hot climates where the coating may soften) and that flexible pipes, particularly those with concrete or cement mortar linings, are supported to maintain circularity and avoid damage to the lining. The stock dumps for pipes should be properly planned and pipes should only be stacked one above the other if they are properly provided with timber supports and packers. Care should be taken not to damage pipe ends: bevel protectors should be fitted to pipes to be joined by butt welding and flanges should be protected with plywood blanks. Pipes should be stacked in accordance with the manufacturer's recommendations; they should not be stacked in areas where long grass may grow and catch fire, damaging external protection. All pipes should be handled using purpose-made lifting slings of a wide fabric material, not chains, ropes or cables, so that the external coating is not damaged and so that the pipe does not slip. When pipes are delivered, and again just before they are lowered into the trench, they should be inspected for flaws. The Holiday detector should be passed over steel pipes; any coating or lining flaws detected should be made good. The interior of each pipe should be inspected as the pipe is lifted and any debris must be brushed out.

PE and PVC pipes can be joined on the surface and snaked into narrow non-man-entry trenches. Pipes are then usually installed on 100 mm granular bedding and backfilled to 100 mm above the pipe with free flowing granular material, above which the main backfill is placed. This allows trenches to be dug by chain excavator, rock-wheel or narrow bucket. If the pipe is laid directly on the soil then the trench bottom must give sensibly uniform support to the pipe: it must be stable, fine grained and free from flints and large stones or other material which may cause point loads. Additional excavation is required at pipe sockets. Bedding must be provided for support in soft ground and a minimum of 150 mm granular bedding must be provided in rock. The trench width for non-man-entry trenches should be at least 300 mm greater than the outside of the pipe. Trench sidefill must be brought up evenly either side of the pipe with soil selected and compacted to give the support required by the design. ASTM F1668 covers construction of plastic pipelines.

Steel pipes may also be snaked into the trench after jointing. Equipment for laying steel (and large PE) pipe with welded butt joints includes side-boom pipe-laying tractors working in series and supporting the pipe with slings, preferably of the roller type. With this system a pipeline may be snaked into its trench in a continuous process, while sections are welded together and coatings made good ahead of the laying. The laying tractor is provided with a counterweight and may have a facility to extend the counterweights. The boom angle is controlled by cable.

Steel, plastics and to a large extent DI, are flexible conduits which when buried rely on the pipe/soil-structure interaction for their load carrying capacity. Deflection under vertical load is limited by support obtained from the trench sidefill, which in turn transfers load to the trench sides. It is therefore essential that the pipes are bedded evenly and are surrounded in material which is well compacted and can transmit the lateral thrusts from the pipe to the trench sides and that the undisturbed ground does not become overstressed. Design and constructability considerations should determine whether the pipe can be laid on the trench bottom after trimming or whether a bedding must be used.

Successful construction requires care in preparation of the trench bottom, handling the pipe and in placing and compaction of embedment and backfill. It is vital to make an even bed for pipes, with joint holes previously excavated in the positions required. After trimming the trench bottom or compacting the bedding a depth of 50 mm of uncompacted soft sand may be left loose to form a uniform support for the bottom arc of the pipe. Alternatively the bedding must be shaped and compacted to match the curvature of the bottom of the pipe. Either way the work is required such that the specified deflections are not exceeded.

During backfilling, pipes must not be supported on timber or hard blocks; pegs should not be used in the trench bottom for setting out bedding levels. Similarly, all large stones that may damage the pipe or coating must be removed from the bed. Hard bands of rock and soft spots should be treated so that the pipe receives even support along the trench. The bedding, sidefill and initial backfill 300 mm above the pipe must be free from large stones (larger than say 15 mm) which could damage the pipe coating. Where the design requires the embedment and natural ground to support the pipe in the narrow section of fill above the pipe invert, compaction of the 'upper bedding' must be achieved. 'Pogo stick' type rammers may be needed for this purpose. Similarly, voids left by removal of trench supports must be filled by withdrawing trench supports as sidefill is raised. Compaction equipment must not come into contact with the pipe or be used within about 150 mm of the top of the pipe. The bedding not only gives support to the pipe but may also give protection against corrosion in aggressive soils.

Sidefilling to pipes should be placed in even layers either side of the pipe up to soffit level and, in addition, for non-rigid pipes the backfill must be carefully compacted to keep the pipe in a true cylindrical form. When the material from the trench is being excavated it should be inspected and instructions should be given for setting aside material that should not be used against the pipe. This material can be used in refilling the trench once the pipe has been properly covered with soft material.

Lengths of pipeline should be laid to even grades. Gradients should not be less than 1:500 which can be readily achieved and monitored during construction. Flatter gradients can be properly carried out only with much greater quality control to level and to backfilling (Plate 27(c)). Local backfalls caused by pipe level variations will produce minor ponding – which may not be important – and air traps, which can affect pressure tests. For distribution pipes which have service connections a flat grade matters less as air will be drawn off via the service connections but for pipelines, it is important to arrange even rises to air valves.

Trenches should be backfilled as soon as possible after laying and jointing and long lengths should not be left uncovered. This helps prevent flotation in the event of flood and helps limit thermal movement which can be particularly important for plastics pipes.

Considerable trouble is experienced when laying pipelines in urban areas where many other services, for example, gas, electricity, etc., have to be negotiated. Considerable efforts have been made in the last 20 years to record the locations of buried services and to store the information using GIS systems; however, records of these other services are seldom perfect or to the accuracy necessary to avoid all the problems that may be encountered. Close liaison with the utility operators is required and services should be located, before pipe trench excavation, by remote detection equipment from the ground surface, followed by opening of trial pits where appropriate. The trench should preferably be excavated well ahead of pipe laying, if the road authority permits this, so that the line and level can be adjusted in good time. Small angle deviations can be accommodated at pipe joints but preformed bends should be available to get around obstacles. Gusseted bends (otherwise termed mitred or lobster back bends) in steel pipes can be fabricated from straight pipe to suit any combination of vertical and horizontal angles, except for high-pressure pipelines for which induction or cold bends should be used.

If two pipelines are to be accommodated in one trench, a minimum spacing of 300 mm should be kept between the two lines. If a parallel pipeline is to be constructed at a future time enough space must be allowed in the pipeline reserve to construct a second trench without disturbing the original pipe or its embedment; consideration must be given to unbalanced thrusts if the original pipeline cannot be shut down for an extended period.

A principal requirement for satisfactory pipe laying is care in making the joints. Achieving cleanliness in a muddy trench is not easy; pipe layers should be provided with the facilities required, such as clean water and buckets, plenty of wiping rags, enough room to work and time to make the joint properly. Care taken in jointing can save weeks of extra work after failure of the pressure test.

Cover to pipes in fields should normally be not less than 0.9 m to provide some protection against physical (third party) damage, for frost protection and to limit the effect of seasonal ground movement. Depths of cover need to be increased in climates with very low and prolonged low temperatures. Frost cover can if necessary be reduced (e.g. to limit temporary environmental impact or to reduce excavation in rock) where flow in the pipe is continuous, even in areas liable to frost, but requires calculation of heat transfer. Cover below heavily trafficked roads (Sections 17.8 and 17.10) should not be less than 1.2 m. Cover may also be increased for flexible pipes which rely on support of soils that might otherwise be removed by excavation for other services – alternatively an easement must be created (and enforced) to prevent encroachment of other excavations on the pipeline soil support (which typically requires no soil movement within two pipe diameters each side of the pipe). Where shallow burial is necessary below roads, a reinforced concrete slab across and bearing on undisturbed ground each side of the trench can be used to transfer imposed load away from the pipe.

On steep gradients, the pipe trench may act as a drain, washing out granular embedment or allowing water to collect at low points and soften the embedment and natural soil. In such cases sections of less permeable embedment should be used to restrict passage of water. Cross walls should be used for stability where pipes are laid at gradients of 1 on 6 or steeper and may be needed on slopes between 1 on 6 and 1 on 12, depending on ground conditions. In weak soils particularly, backfill grading should be selected using normal filter rules (Section 5.9); in certain circumstances, it may be necessary to contain the pipe embedment within a membrane of filter fabric. Care should be taken during installation to ensure that joints are not put under undue strain. Pipelines with flexible joints at close spacings provide inherently more capacity for natural minor movement after installation; flexible joints at wide spacings are fewer but can increase local movement and shear forces. To allow for movement after laying, angular deflections at flexible joints on installation should be limited to 50% of the maxima allowable.

Flanges must be carefully aligned before the bolts are inserted and the flanges pulled together. The alignment must be almost as precise as that adopted for aligning motor couplings. To pull up misaligned flanges is likely to cause fracture of the pipe or flange. A fabric reinforced rubber gasket is inserted between the flanges. For metallic flanges the gasket is of such diameter that it lies inside the bolt circle but does not intrude into the pipe bore; for GRP flanges full-face gaskets should be used to avoid distortion of the flange. The faces of the flanges and the gasket must be perfectly clean before assembly and the bolts must be tightened up little by little in the sequence recommended by the pipe manufacturer so that an even pressure is maintained all round. No grease, paint, oil, dirt, grit or water should be permitted on the flange or rubber ring faces. The contact should be between clean dry metal and clean dry rubber. When making joint faces which are vertical, some difficulty may be experienced in keeping the rubber ring flat against the vertical flange face and, to counteract this, a little clear rubber solution may be used to tack the rubber ring on to the metal face. This is the only material whose use can be permitted in connection with rubber rings. If a greasy material such as a bitumen-based adhesive is used, tightening the flanges may cause the rubber ring to extrude into the pipeline and lead to leakage.

17.37 CHANGES IN DIRECTION (BENDS) IN THE FIELD

Gentle curves may be achieved in the field when laying flexible pipe such as PE (Section 17.25) and steel. In steel, a bend radius of 500 diameters produces a longitudinal wall stress of about 200 N/mm^2; 1000 diameters equates to a stress of about 100 N/mm^2. Therefore, the minimum acceptable radius of an elastic bend depends on the grade of steel used and the magnitude of other loads. ASME B31.4 allows a longitudinal stress factor of 0.8 subject to the combined stress (e.g. Von Mises – Section 17.9) factor not exceeding 0.9 (this limit is not reached unless there is appreciable torsion stress). The minimum bend radii for steel grades X42, X56 and X70 are therefore, 430, 320 and 260 diameters, respectively. It should be noted that appreciable force is needed to deflect steel pipes to these tighter radii and achieving them for large pipes may be impracticable or unsafe. Depending on the laying equipment used (Section 17.36), pipe laying tractors can exert appreciable horizontal force, up to about half the vertical lift. However, ASME B31.4 says that stresses induced in laying should be minimized and that pipe should fit into the trench without external force to hold it in place.

Otherwise, manufactured fittings need to be used for changes in direction with thrust restraint provided if flexible joints are used (Section 17.34). The exception is field cold bends in steel pipe. To achieve a complete bend a small segment of steel pipe is bent so that the extreme fibres yield. The process is repeated segment by segment until the required length is bent. Bends formed cold must be free from buckles and diameter must not be reduced by more than 2½%. To achieve this it may be necessary to use a mandrel. ASME B31.4 gives minimum radii of 18 diameters (*D*) up to 12″, ranging to 27*D* for 18″ and 30*D* for diameter of 20″ or more. Coatings are damaged in this process but can be made good on the outside of the pipe relatively easily. Reliable repair in the inside of the pipe is more difficult and limits the value of field cold bending for steel water pipelines.

17.38 TESTING OF PIPELINES

The test usually consists first of filling the pipeline with water and allowing it to stand and stabilize under working pressure. The test pressure is then applied slowly by pumping. BS EN 805 covers water supply systems, including raw and treated water pipelines, outside buildings and specifies the (site) system test pressure (STP). BS EN 805 requirements for STP may be interpreted as:

Surge pressures determined by analysis and included in maximum design pressure (MDPc):

- MDPc + 1 bar.

Surge pressure not determined by analysis but for which an allowance of not less than 2 bar is included in maximum design pressure (MDPa):
lower of:

- MDPa × 1.5 or
- MDPa + 5 bar.

The UK Water Industry (WIS) guide IGN 4-01-03 covers pressure testing of pressure pipes and fittings. It provides a review of the methods in BS EN 805 and, in addition, provides specific testing guidance for contractors (Appendix A5). Unsatisfactory leakage levels are considered a prime driver to identify pipes for replacement or renovation and the guide includes a standardized approach to assessing leakage rates in terms of litres/km/hr.

Procedures for testing plastic pipes, in particular polyethylene, need to allow for creep and reference should be made to the manufacturer. For steel pipelines AWWA M11 says that the test pressure must not cause stress exceeding 75% of yield.

BS EN 805 gives two options for test method, both with test duration of 1 hour:

a. Water loss method with measurement, either by (i) volume drawn off to replicate observed pressure drop in test period or (ii) volume pumped in to restore pressure.
b. Pressure loss method.

In both cases the acceptability criterion is independent of test pressure. The allowable pressure loss for method (b) is 0.2 bar for normal pipelines. For method (a) the acceptable water loss (or volume change) is calculated using a formula in which a change of pressure of 0.2 bar is inserted. The allowance factor used (1.2) results in the acceptable loss being 20% higher than the elastic volume change due to the pressure change. The formula includes terms for pipe wall thickness and mechanical properties. These should be taken from data provided by the manufacturer. If the pressure loss method is used a preliminary test (pressure drop test) should be carried out to check that the amount of air left in the pipeline is within acceptable limits. If the criterion is not met measures should be taken to eliminate the excess air and the drop test repeated. If the criterion cannot be met the water loss method should be used. The preparation for the main test is:

- after flushing and venting pipeline allow to relax for at least 1 hour;
- raise pressure continuously to STP as quickly as practicable;
- maintain STP (by pumping) for 30 minutes whilst observing for leaks;
- allow pipe to relax under pressure for 1 hour and measure remaining pressure;
- if pressure drop is less than 30% of STP proceed to main test.

Pressure and leak testing of pipelines to US standards are covered in AWWA manuals M9 for concrete, M11 for steel, M23 for PVC and M55 for PE. AWWA C600 covers testing of DI pipelines.

Pipe testing should proceed quickly after installation and results should be linked to payment for the pipeline. Initial tests should be made starting with short lengths for each pipelaying gang and pipe size and material. Test lengths can gradually be increased typically to 5 or 10 km or between section valves as a successful track record is developed. Test lengths depend on topography and availability of water.

Fluctuating test pressure results are likely to be caused by air locks in the pipe. To avoid air locks there must be suitable air valves on the pipeline (Section 18.24). Filling must proceed slowly, particularly on falling gradients, to vent air and avoid hydraulic jumps and entraining air. An equivalent velocity between about 0.2 and 0.5 m/s is sensible but can be varied to suit circumstances. Source of water and pumping arrangements should be designed to allow any entrained air to escape and to prevent air entrainment into the pipe.

Air must not be used for testing water mains. The test must be hydrostatic and take place between blank flanges, bolted or welded to pipe ends, or caps may be used if fully supported by anchor blocks. Where pipes have flexible joints the end pipe must be fully anchored. Testing should not take place between closed valves because if the valve is already inserted in the line it is not possible to detect any leakage past the valve and, if the valve is exposed at the end of a section of the line, it would be in the 'open end condition' and may leak unless it is designed for the 'closed end condition' (Section 18.6).

If a pipeline fails its test but has been backfilled, searching for leaks can be troublesome. It is best to leave the pipeline under pressure for a day or two so that external signs of the leakage can become visible, for example, a wet patch on the surface of the ground. The pipe may also be sounded for leakage using leak detection methods (Section 16.9). It is possible to use some kind of tracer element in the water but this is a skilled matter requiring a specialist. In practice, therefore, it is usual to expose the joints where leakage is suspected.

17.39 MAKING CONNECTIONS

If a socket and spigot tee has to be inserted into an existing pipeline the length cut out of the latter must be slightly greater than the overall length of the tee. The socket of the tee is pushed up to fit one end of the cut pipe and the resulting gap between the two spigots at the other end is joined by using a collar. If a double-socketed tee is used this must be inserted using a plain piece of pipe on one side, again joined by a collar to the pipeline.

An alternative is to use an under-pressure connection as shown in Figure 16.2. A split collar is clamped on to the main, the collar having a flanged branch on it to which is bolted a gate valve. A cutting machine is attached to the valve, the latter is then opened, and the cutter is moved forward through the valve and trepans a hole in the side of the pipe. The cutter is withdrawn with the trepanned piece of pipe wall and the valve is closed. The cutting machine is then removed and the branch connection can be made. Steel and iron pipes can be cut in situ using a rotating cutting tool which is clamped onto the main. A manually operated wheel cutter can be used on small diameter cast iron mains of 80 or 100 mm size. Oxyacetylene cutting of steel and iron pipes can be used but the cut is ragged and difficult to make exactly at right angles to the axis of the pipe. Cast and DI pipes can be cut above ground using a hammer and chisel. The pipe is placed on a timber baulk below the line of cut and is rolled back and forth as the chiselling proceeds: first to 'mark' the cutting line and then to deepen the chiselled groove. At a certain stage the pipe will come apart at the chiselling line.

17.40 UNDERWATER PIPELINES

Steel and PE are materials commonly used for pipelines laid below water for the crossings of rivers or estuaries. Both materials can be welded, thus providing longitudinal continuity of strength. Other materials can also be used including concrete, GRP, PVC and DI but require particular attention to tolerance on bed preparation, practicalities of level and position measurement underwater, plus the strength and deflection limits of flexible joints. Pipe laying is normally carried out by laybarge, reel barge, bottom pull or float and sink. The laybarge is a 'factory' for progressively adding pipes to a string whilst winching the barge along the pipeline route so that the string hanging in a catenary from the back of the barge is gradually lowered onto the river or sea bed or into a pre-dredged trench. The reel barge method is similar but is used for unreeling lengths of plastic tubing as the barge moves along the pipeline route. In the bottom pull method lengths of pipe, pre-fabricated onshore, are joined to form a pipe string which is progressively pulled into the water by a winch, mounted on a pontoon or on the far shore, until the crossing is complete. In the float and sink method lengths of pipe are made up into strings at a remote fabrication yard, the string is towed at or below the water surface to the crossing location where it is aligned into position and sunk by removing supporting buoyancy tanks or by filling with water.

Internal operating conditions are similar to those described for land pipelines but special attention needs to be given to external conditions as, once laid, access to a pipeline for remedial work is unlikely to be available. Attention has to be paid to stresses during laying, soil conditions, the effects of currents and waves (including soil liquefaction), sea bed topography and morphology, protection against corrosion and protection against damage by ships' anchors, dropped objects, fisheries (trawl boards) and other activities. Physical protection in shallow water typically requires burial with some cover below the lowest likely level of bed movement and depth of penetration of any ship's anchor. In deep water, where wave and current action is small, the pipeline may be laid without burial, subject to limitations of profile and hazards.

Because of their weight, pipelines are in general installed empty. Positive submerged weight ('negative buoyancy') is required to prevent the pipes from floating and for stability under the action of waves and currents. Pipes installed by float and sink would be towed in strings and manoeuvred into position using temporary buoyancy. Weight is added typically, for steel pipes, as a continuous concrete cladding; for polyethylene pipes concrete collars are used, spaced, shaped and sized to give the required underwater specific gravity and stability. Design considerations include buckling under external hydrostatic pressure, pull forces, longitudinal bending as the pipe is towed or as it conforms to the underwater profile, and ring bending under backfill loads.

The cladding thickness is typically about 10% of the pipe diameter. It is sometimes considered simply as a temporary weight coat but it does serve to protect the external corrosion coat and therefore can act as part of the permanent structure. The cladding can contribute significantly to bending stiffness particularly in ring bending; therefore differences at joints can raise stresses locally. Above about DN 1500 site applied cladding may be stiff and thick enough to be designed as a reinforced concrete structural ring. This can be used to support and reduce the thickness required to withstand buckling of the steel shell, typically by 40%, and carry external backfill loads. The combined structure can also be designed to withstand ships' anchors, thus saving considerably on the costs and environmental impacts of deep burial (Little, 1989). Backfill must be designed not to liquefy and to be stable under wave action. Rock armour can also be chosen and designed to save on burial costs. With care, considerable economies can therefore be achieved.

REFERENCE STANDARDS

British Standards (BSI)

BS 534 *Steel pipes, joints and specials for water and sewage.*

BS 1377 *Methods of test for soils for civil engineering purposes.*

BS 3505 *Unplasticized polyvinyl chloride (PVC-U) pressure pipes for cold potable water.*

BS 7079 *Preparation of steel substrates before application of paints and related products.*

BS 8010 *Code of Practice for Pipelines. Pipelines on land: design, construction and installation:*

Section 2.1 Ductile Iron.
Section 2.5 Glass reinforced thermosetting plastics.

BS 9295 *Guide to the structural design of buried pipelines.*

BS EN 512 *Fibre-cement products. Pressure pipes and joints.*

BS EN 545 *Ductile iron pipes, fittings, accessories and their joints for water pipelines. Requirements and test methods.*

BS EN 639 *Common requirements for concrete pressure pipes including joints and fittings.*

BS EN 641 *Reinforced concrete pressure pipes, cylinder type, including joints and fittings.*

BS EN 642 *Prestressed concrete pressure pipes, cylinder and non-cylinder, including joints and fittings.*

BS EN 681 *Elastomeric seals – Material requirements for pipe joints seals in water and drainage applications.*

BS EN 805 *Water supply – Requirements for systems and components outside buildings.*

BS EN 1092 *Flanges.*

BS EN 1295-1 *Structural design of buried pipelines under various conditions of loading – General Requirements.*

BS EN 1452 *Plastics piping systems for water supply. Unplasticized poly vinyl chloride (PVC-U).*

BS EN 1514 *Flanges and their joints – Dimensions of gaskets for PN-designated flanges.*

BS EN 1796 *Plastics piping systems for water supply with or without pressure. Glass reinforced thermosetting plastics (GRP) based on unsaturated polyester resin.*

BS EN 10216 *Seamless steel tubes for pressure purposes. Technical delivery conditions.*

BS EN 10217 *Welded steel tubes for pressure purposes. Technical delivery conditions.*

BS EN 10220 *Seamless and welded steel tubes. Dimensions and masses per unit length.*

BS EN 10224 *Non-alloy steel tubes and fittings for the conveyance of water and other aqueous liquids. Technical delivery conditions.*

BS EN 10298 *Steel tubes and fittings for on and offshore pipelines. Internal lining with cement mortar.*

BS EN 10311 *Joints for the connection of steel tubes and fittings for water and other aqueous liquids.*

BS EN 10312 *Welded stainless steel tubes for the conveyance of aqueous liquids including water for human consumption. Technical delivery conditions.*

BS EN 12201 *Plastics piping systems for water supply. Polyethylene (PE).*

BS EN 12954 *Cathodic protection of buried or immersed metallic structures. General principles and application for pipelines.*

BS EN 13480 *Metallic industrial piping.*

BS EN 14161 *Petroleum and natural gas industries – Pipeline transportation systems.*

BS EN 14870 *Induction bends, fittings and flanges for pipeline transportation systems.*

BS EN ISO 3183 *Petroleum and natural gas industries. Steel pipe for pipeline transportation systems.*

BS EN ISO 15589 *Petroleum, petrochemical and natural gas industries. Cathodic protection of pipeline transportation systems.*

BS EN ISO 21809 *Petroleum and natural gas industries. External coatings for buried and submerged pipelines used in pipeline transportations systems.*

BS ISO 10803 *Design method for Ductile Iron Pipes.*

BS ISO 10804 *Restrained joint systems for ductile iron pipelines.*

CP 312 *Code of practice for plastics pipework (thermoplastics material).*

CP 2010 *Code of Practice for Pipelines, Part 2 Design and construction of steel pipelines in land.*

Eurocodes

BS EN 1993-4-3 *Design of steel structures. Pipelines.*

Water UK Standards Board

IGN 4-01-03 *Pressure testing of pressure pipes and fittings for use by public water suppliers.*

IGN-4-31-06 *Blue unplasticised PVC pressure pipes, integral joints and post-formed bends for cold potable water.*

IGN-4-31-06 *Blue unplasticised PVC pressure fittings and assemblies for cold potable water.*

Draft European standards Published Documents

PD CEN/TR 1295-2 *Structural design of buried pipelines under various conditions of loading. Summary of nationally established methods of design.*

PD 8010-1, -2 *Code of practice for pipelines. Part 1 – Steel pipelines on land; Part 2 – Subsea pipelines.*

American Lifelines Alliance

ALA (2001) *Guidelines for the Design of Buried Steel Pipe.*

ALA (2002) *Seismic Design and Retrofit of Piping Systems.*

American Petroleum Institute

API 5L *Specification for line pipe. (Identical to ISO 3183).*

API RP 5L7 *Recommended Practices for Unprimed Internal Fusion Bonded Epoxy of Line Pipe.*

API RP 5L9 *External Fusion Bonded Epoxy Coating of Line Pipe.*

American Society of Mechanical Engineers

ASME B16.9 *Factory-made Wrought Buttwelding Fittings.*

ASME B16.42 *Ductile Iron Pipe Flanges and Flanged Fittings. Classes 150 and 300.*

ASME B16.49 *Factory-made Wrought Steel Induction Bends for Transportation and Distribution Systems.*

ASME B31.1 *Power piping.*

ASME B31.3 *Process Piping.*

ASME B31.4 *Pipeline Transportation Systems for Liquid Hydrocarbons and Other Liquids.*

ASME B31.8 *Gas Transmission and Distribution Piping Systems.*

ASME RTP-1 *Reinforced Thermoset Plastic Corrosion-Resistant Equipment.*

BPVC (2015) *Boiler and Pressure Vessel Code.*

ASTM International (previously American Society for Testing and Materials)

ASTM C296 *Asbestos Cement Pressure Pipe.*

ASTM D1557 *Laboratory Compaction Characteristics of Soil Using Modified Effort.*
ASTM D1784 *Rigid Poly (Vinyl Chloride) (PVC) Compounds and Chlorinated Poly (Vinyl Chloride) (CPVC) Compounds.*
ASTM D6783 *Polymer Concrete Pipe.*
ASTM D6938 *In-place Density and Water Content of Soil and Soil-aggregate by Nuclear Methods.*
ASTM F714 *Polyethylene (PE) Plastic Pipe (DR–PR) Based on Outside Diameter.*
ASTM F1668 *Construction Procedures for Buried Plastic Pipe.*
ASTM F2164 *Field Leak Testing of Polyethylene (PE) and Crosslinked Polyethylene (PEX).*
American Waterworks Association
AWWA C111 *Rubber-Gasket Joints for Ductile Iron Pressure Pipe and Fittings.*
AWWA C150 *Thickness Design of Ductile Iron Pipe.*
AWWA C151 *Ductile Iron Pipe. Centrifugally Cast.*
AWWA C205 *Cement-mortar protective lining and coating for steel water pipe – 4 in (100 mm) and larger – shop applied.*
AWWA C208 *Dimensions of Fabricated Steel Water Pipe Fittings.*
AWWA C213 *Fusion-Bonded Epoxy Coatings and Linings for Steel Water Pipe and Fittings.*
AWWA C222 *Polyurethane Coating for the Interior and Exterior Coating of Steel Water Pipe and Fittings.*
AWWA C300 *Reinforced Concrete Pressure Pipe, Steel Cylinder Type.*
AWWA C301 *Prestressed Concrete Pressure Pipe, Steel Cylinder Type.*
AWWA C302 *Reinforced Concrete Pressure Pipe, Noncylinder type.*
AWWA C303 *Concrete Pressure Pipe, Bar Wrapped, Steel Cylinder Type.*
AWWA C304 *Design of Prestressed Concrete Cylinder Pipe.*
AWWA C600 *Installation of Ductile-iron Mains and their Appurtenances.*
AWWA C602 *Cement-mortar lining of water pipelines in place-4in (100 mm) and larger.*
AWWA C900 *Polyvinyl Chloride (PVC) Pressure Pipe and Fabricated Fittings. 4 in Through 12 in.*
AWWA C905 *Polyvinyl Chloride (PVC) Pressure Pipe and Fabricated Fittings. 14 in Through 48 in.*
AWWA C906 *Polyethylene (PE) Pressure Pipe and Fittings, 4 in Through 63 in. for Water Distribution and Transmission.*
AWWA C950 *Fiberglass pressure pipe.*
AWWA M9 *Concrete Pressure Pipe.*
AWWA M11 *Steel Pipe – A Guide for Design and Installation.*
AWWA M23 *PVC Pipe – Design and Installation.*
AWWA M41 *Ductile-iron Pipe and Fittings.*
AWWA M45 *Fiberglass pipe design.*
AWWA M55 *PE Pipe – Design and Installation.*
Other codes
SSPC-SP5/NACE No.1 *Joint Surface Preparation Standard, White Metal Blast Cleaning.*
SSPC-SP10/NACE No.2 *Joint Surface Preparation Standard, Near-White Metal Blast Cleaning.*
SIS 05 59 00 Swedish Standard. *Pictorial Preparation Standards for Painting Steel Surfaces. Preparation of Steel Substrates before Application of Paints and Related Products.*

REFERENCES

AwwaRF (2008). *Failure of Prestressed Concrete Cylinder Pipe*. AwwaRF.

CIRIA (1994). *Design of Thrust Blocks for Buried Pressure Pipelines. Report 128*. CIRIA.

Clarke, N. W. B. (1968). *Buried Pipelines – A Manual for Structural Design and Installation*. Maclaren.

Janson, L.-E. (1995). *Plastics Pipes for Water Supply and Sewage Disposal*. Borealis, Stockholm.

Little, M. J. (1986). New pipelines on land and across Hong Kong Harbour. *JIWES, June*, pp. 271–286.

Little, M. J. (2004). Pipeline design – a UK perspective. *AWWA Conference, Florida*. USA. June.

Little, M. J. and Duxbury, J. A. (1989). Tolo channel submarine pipelines, Hong Kong. *Proc. ICE, Part 1, 1989, April*, pp. 395–412.

McAllister, E. W. (2014). *Pipeline Rules of Thumb, Handbook*. 8th Edn. Elsevier.

PPI (2007). *Handbook of Polyethylene Pipe*. 2nd Edn. Plastics Pipe Institute.

Spangler, M. G. (1951). *Soil Engineering*. International Textbook Company, Scranton.

Trew, J. E. (1995). *Pipe Materials Selection Manual. Water Supply*. WRc Publications.

Young, O. C. and O'Reilly, M. P. (1983). *A Guide to Design Loadings for Buried Rigid Pipes*. HMSO.

Young, O. C. and O'Reilly, M. P. (1986). *Simplified Tables of External Loads on Buried Pipelines*. HMSO.

CHAPTER

Valves and Meters 18

PART I VALVES

18.1 VALVE DEVELOPMENT

Valves have been in use on water pipe and lifting systems for over 2000 years. Bronze and brass valves dating from the Roman period have been found; these included plug cocks and flap valves. Leather flap valves may have been used in ancient Egypt in association with pumps and simple flaps have been used for several centuries on hand well pumps.

Valve technology took a leap forward at the end of the 18th century with the advent of steam and with improved metal working techniques. Plug cock, butterfly, flap, gate and pressure relief valves became widely used. By 1950 valves had become increasingly sophisticated and several new types of valve had been developed – diaphragm, ball and lubricated plug. Needle, sleeve, air release and hollow-jet valves were also invented in the 20th century and are now available in numerous variants.

18.2 VALVE FUNCTIONS

Valve type should be selected according to the required function. Valves for industrial process are classified in Europe as: isolating, regulating and control valves. The same classifications may be adopted for water supply valves. There are also other common functions such as air release, non-return, pressure relief and hydrant:

Isolating	Set either closed or fully open and normally not operated in flow conditions
Regulating	Set with any degree of opening to regulate flow and capable of periodic adjustment to opening
Control	Used with autonomous or external systems to respond to changes in flow or pressure conditions so as to achieve a set result which itself is capable of being reset
Non-return	Prevents reverse flow when downstream pressure is higher
Venting	To exhaust or admit air

Twort's Water Supply. DOI: http://dx.doi.org/10.1016/B978-0-08-100025-0.00018-1

It should be noted that the 'regulator' according to ANSI/ISA S75.05 is equivalent to the autonomous control valve – using the pressure of the fluid to actuate movement.

Operation to achieve the required function imposes conditions that the valve should be able to withstand. Such conditions may include load on components, vibration, wear, erosion and cavitation. Some of these conditions may, in certain circumstances, exacerbate corrosion.

18.3 ISOLATION

Isolating valves are usually required to be drop tight so that work can be carried out on parts of the system in the dry. Therefore, seals should be resilient. For safety reasons an isolating valve should be lockable so that it cannot be inadvertently opened. An isolating valve should not be left part closed unless designed for this condition, in which case it would be classed as a regulating valve. It is good practice to provide a means of checking valve status. Foam injected into the surface box shows if a valve might have been moved. More sophisticated is to use a 'WIZKEY' which is placed on the valve cap during operation and records, for later downloading to a database, the action taken.

18.4 REGULATION

Where necessary to reduce downstream head, a valve may be left part closed (throttled). The degree of closure would not be adjusted, except occasionally, therefore headloss is a function of flow. When flow stops the downstream head equals the upstream head. This decreases the utility of a regulating valve in a distribution system where flow varies considerably during the day. However, when used for regulation the valve must be capable of withstanding the expected conditions for long periods whilst throttling flow. Unless designed to be drop tight on closure such valves cannot be relied on to be drop tight and therefore cannot protect the system downstream from high static pressures.

18.5 CONTROL

When the setting of the valve has to be adjusted frequently to suit varying conditions this is usually arranged by some automatic means. This allows the valve to maintain a preset flow, pressure or water level. Control valves fulfil this function and must be capable of frequent movement under the range of conditions expected. The suitability of a valve type for control depends on its flow characteristic, its 'rangeability' (percentage of opening over which flow can be adequately controlled), mechanical play intrinsic to the design and, in some cases, resistance to cavitation. Information on the 'rangeability' of control valves is available from manufacturers but guidance is available in *Inherent Flow Characteristics and Rangeability of Control Valves* (ANSI/ISA S75.11, 1991). Control valves are not usually drop tight on closure and therefore cannot be relied on protect the system downstream from high static pressures.

Defined 'intrinsic' (i.e. due to valve only) flow characteristics include 'quick opening', 'linear' and 'equal percentage'. The first type produces a high flow rate at a relatively small spindle movement from closed and is not suitable for control valves, but is used for some on–off duties. A linear intrinsic flow characteristic is beneficial where system losses are small. The 'equal percentage' intrinsic characteristic is so called because any equal increment in opening produces an equal percentage increase in flow and shows as a straight line when plotted on semi-log paper. When combined with appreciable system losses it produces a near linear combined characteristic. Flow characteristics vary with the opening and are affected by body shape, any porting or special trim and by seating arrangements. For a given headloss, flow through a valve varies approximately with the square of the opening area. Figure 18.1 shows intrinsic flow characteristics through different types of valve ignoring the effect of the system in which they may be used. However, actual flow through a valve in a real situation depends on the overall system losses.

The sum of the headloss through the valve and the head lost in the system equals the head available. This may be fixed (e.g. between reservoirs) or variable as in a pumped system. Figure 18.2 illustrates the combined characteristics of DN 300 butterfly, ball and globe valves in pipelines of the same diameter having lengths zero, 10 m, 100 m, 1 km and 10 km and roughness k_s of 0.6 mm and with available head of 10 m. For good control, the combined characteristic should be as near linear as possible, particularly over the range in which the valve is expected to operate.

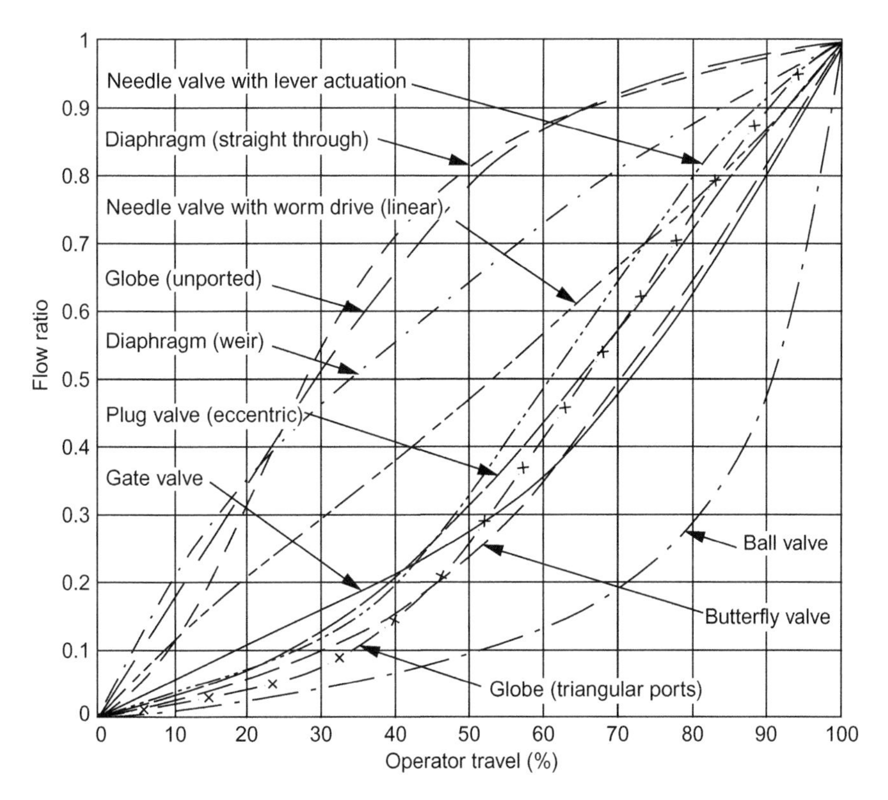

FIGURE 18.1

Valve intrinsic flow characteristics.

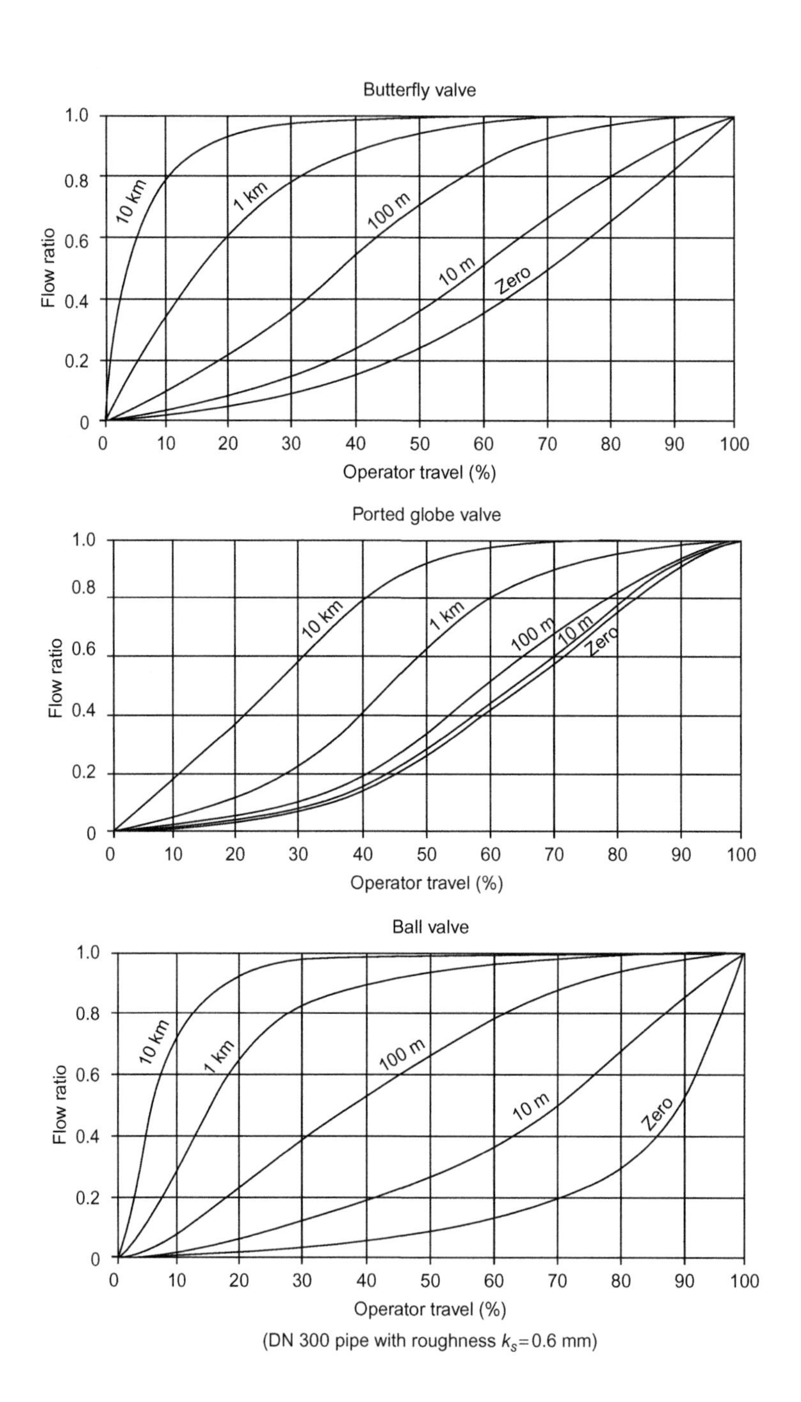

FIGURE 18.2

Effect of system length on valve flow characteristics.

The type of valve and any porting arrangement should be selected to achieve this in the system in which it is to be installed. The shape of ports (triangular) in the ported globe valve assumed for Figures 18.1 and 18.2 influences the shape of the characteristic. It is possible to produce almost any desired characteristic by means of shaped ports, holes of varying spacing or other special trim. Globe, sleeve and, to a lesser degree, needle valves are designed to accommodate such trim. Even more complex (and costly) porting arrangements can be applied to control valves of any type to reduce noise and vibration. Where the arrangements dissipate head in stages, the valve is less susceptible to cavitation. BS EN 60534 includes (in Parts 2-1 and 2-5) sizing equations for industrial valves, including those with multi-stage and multi-path designs, for incompressible fluids such as water and compressible fluids.

Automatic control requires the valve to respond to deviation from a set point, for example water level at a downstream discharge point. The signal initiating this response may be from a mechanical linkage (float valve), a hydraulic relay system (altitude valve) or a computer controller (programmable logic controller (PLC)) to which a measured value is transmitted from a level, pressure or flow transducer. Whatever method is employed the system designer needs to ensure that there is no instability sufficient to cause 'hunting'. Hunting is an oscillation of valve opening which causes increased valve and actuator wear and could, in extreme cases, lead to system failure. In a distribution system hunting is undesirable for the consumer. There should be sufficient delay and damping in the control arrangement to deal with slowness of response of the system.

Baumann (1998) discusses control characteristics and maintains that controllers need change of flow per unit change of valve opening to be constant (linear characteristic) within certain limits. However, he suggests that controllers can cope (without need for resetting) with changes of flow between 5% and 15% for each 10% change in valve opening. Just as important to successful holding of a set point is tightness in the actuator and operator mechanisms. In some types of valve such as the gate valve there is a degree of play between stem and nut. This, together with unfavourable flow characteristics of this type, renders the gate valve unsuitable for control. Any gear mechanism is a potential source of play (backlash) after wear and of flow induced vibration.

18.6 VALVE SELECTION AND SPECIFICATION

Valves should be selected to suit the function and duty and the materials used should be appropriate for the fluid being conveyed. For larger valves and those operating in difficult conditions selection is a specialist task. Helpful guidance is provided in the *Valve Selection Handbook* (Smith, 2004) and in *An Introductory Guide to Valve Selection* (Smith, 1995). For control valves the reader should refer to *Control Valves* (Borden, 1998) or *Control Valve Primer* (Baumann, 1998). The authors' suggestions for the use of the main types of valve employed in the water industry are given in Table 18.1.

Metal seated valves are designed to seal at a specified unbalanced pressure and may seal less well at higher or lower pressures. This is why the leakage test is carried out at a specified pressure. Valves with large bodies (particularly those of large diameter) should be specified as designed for the 'open end test' or the 'closed end test'. (Note that BS EN 12266 does not distinguish between these test conditions and implies open end.) A valve designed for drop tight closure under open end conditions may not close drop tight for closed end conditions due to body strain. Closed end siting

Table 18.1 Suggested valve utilization for water supply[a]

Valve type	Largest size (mm)	Typical valve use[b]	Waterway[c]	Cavitation resistance[d]	Headloss (fully open)	Manual closing speed[e]	Comments
Gate (sluice)	2400	Isolation (to DN 600), regulation at low head in smaller sizes	Clear	Poor	Very low	Very slow	Heavy, tall, simple and reliable
Knife gate		Isolation at low heads, cannot be buried	Clear	Very poor	Very low	Slow	Slim, tall – good for slurries
Butterfly	>4m	Isolation (above about DN 300), regulation and control	Obstructed	Poor	Low	Moderate	Light and compact
Globe[f]	900	Control in onerous conditions – PRV	Obstructed	High	High	Fast	Piston seal type
Globe[g]	900	Control under moderate head – PRV	Obstructed	Moderate	High	Fast	Diaphragm seal type
Ball	300	Isolation (friction significant under higher heads)	Clear[h]	Poor	Very low	Very fast	Bulky
Plug/cone	1800	Isolation, eccentric plug for larger sizes	Obstructed[i]	Poor	Low	Slow	Heavy
Diaphragm	350	Isolation, regulation and control	Obstructed[j]	–	Moderate	Slow	Slurry and chemicals
Pinch	300	Isolation and regulation	Obstructed	–	Moderate	Slow	Slurry and chemicals
Needle[k]	1800	Regulation and control in onerous conditions	Obstructed	Moderate	Low–moderate	Moderate	Bulky, expensive
Sleeve[l]	1600	Regulation and control in very onerous conditions	Obstructed	High	High	Moderate	Bulky, expensive

Notes: [a]Valve selection is for specialists. This table is intended as an approximate guide to the non-expert. Cost is an important factor but relative costs depend on valve size.
[b]Porting (by means of a toothed or perforated sleeve) can assist control characteristics for some valves where a sleeve can be fitted.
[c]Many valves do not present a clear opening for the transit of swabs. To keep costs low, a valve of smaller diameter than the pipeline may be used, but it obstructs swab passage.
[d]Cavitation resistance takes into account valve and flow geometry. Performance can be improved by arranging staged head dissipation or by using ported or other valve trim. Generally, good cavitation performance is obtained at the expense of high fully open headloss.
[e]Valve closing speed varies with valve size, operating pressure and is affected by any gearing provided. Unmotored closure times would vary from 1 to 90 minutes for 'very slow' valves such as gate valves (but these are normally actuated in larger sizes) to 1 to 5 seconds for 'very fast' quarter-turn ball valves.
[f]The characteristics are for a ported globe valve with flow inwards through the orifice.
[g]The characteristics are for an unported valve with flow outwards through the orifice.
[h]Ball valve openings are usually sized as the inlet pipe diameter.
[i]Plug valves usually have rectangular or trapezoidal orifices. Valves with circular water ways of the same diameter as the pipe are available up to about 200 mm diameter.
[j]In the weir type of diaphragm valve the waterway is obstructed.
[k]The needle valve indicated is one with no sleeve although they are designed to accommodate sleeves which can improve cavitation performance.
[l]The sleeve valve is of the angled or in-line type where flow is via circular orifices from the outside to the inside.

is defined as one which prevents valve expansion under load, as in the case of a double-flanged valve connected into rigidly held pipework. Open end siting allows the valve to expand, for instance where there is flange adaptor or other flexible coupling on or near one side, and is the usual situation.

Some valve types such as gate, plug and ball are heavy, bulky and therefore relatively expensive. It may be advantageous to use a valve 5/8 to 3/4 the diameter of the pipeline with carefully selected tapers each side to minimize headloss. This can reduce cost, weight, height and operating torque; however, it prevents transit of swabs for cleaning (Section 16.10).

Some valves such as gate and plug valves require high operating torque to overcome friction. A bypass around the valve may be fitted to allow pressure to be equalized before opening the main valve; it is also useful for pipeline filling. Bypasses need not be very large in diameter; a DN 100 bypass would normally be suitable for a DN 800 valve. However, a bypass is of little use where the pressure downstream cannot build up quickly, for example where the pipe network downstream is very extensive or where there are appreciable outflows that cannot be stopped. A permanently installed power driven actuator may be used to operate a valve, mounted either on the valve directly or on a headstock and coupled to the valve through an extension spindle. Alternatively, a portable actuator may be used if a suitable stub and gearbox are provided on the headstock.

18.7 GATE VALVES

Metal seated wedge-gate valves have not altered substantially over 100 years except that toroidal (O ring) stem seals were introduced for smaller valves. However, gate valves with a stuffing box at the top of the valve are still available and often preferred for larger sizes. Figure 18.3 shows a 300 mm diameter metal seated valve with a gland packing stuffing box; a large metal seated valve is illustrated in Plate 28(a). The machined sealing faces (seats) are usually made of gunmetal or phosphor-bronze and the seats are forced together by wedge action to produce a seal. Resilient seat gate valves (available in sizes up to DN 600) have a gate which is encapsulated in rubber (Plate 28(d)) and seals against a clear full bore typically without grooves in which dirt can collect and prevent full closure. A further development is the boltless design for the upper valve body: this facilitates mass production and, for valves up to and including DN 300, is cheaper to replace than refurbish. Parallel slide or 'knife gate' valves are used above ground for infrequent, low head isolating duties where space is restricted and are widely used with slurries and as block valves in power stations. Gate valves may be buried but are tall; pipelines of large diameter need to have more than the minimum cover where a gate valve is to be sited.

Gate valves have a low fully open headloss coefficient and good shut off; therefore, they are useful for isolation duties, particularly in pumped systems. However, they are not suitable for flow regulation or control due to poor flow characteristics, vibration and play in the stem. Gearing or special torque reduction devices, such as ball-bearing thrust collars (which can reduce torques by 50%), may be needed to overcome unseating forces and friction in large gate valves depending on the head. The disadvantage is that manual full opening or closing of a large gate valve requires considerable work and time, over an hour for a valve larger than DN 600. Regular operation of gate valves is advisable to keep the grooves clear and the stem and nut threads clean.

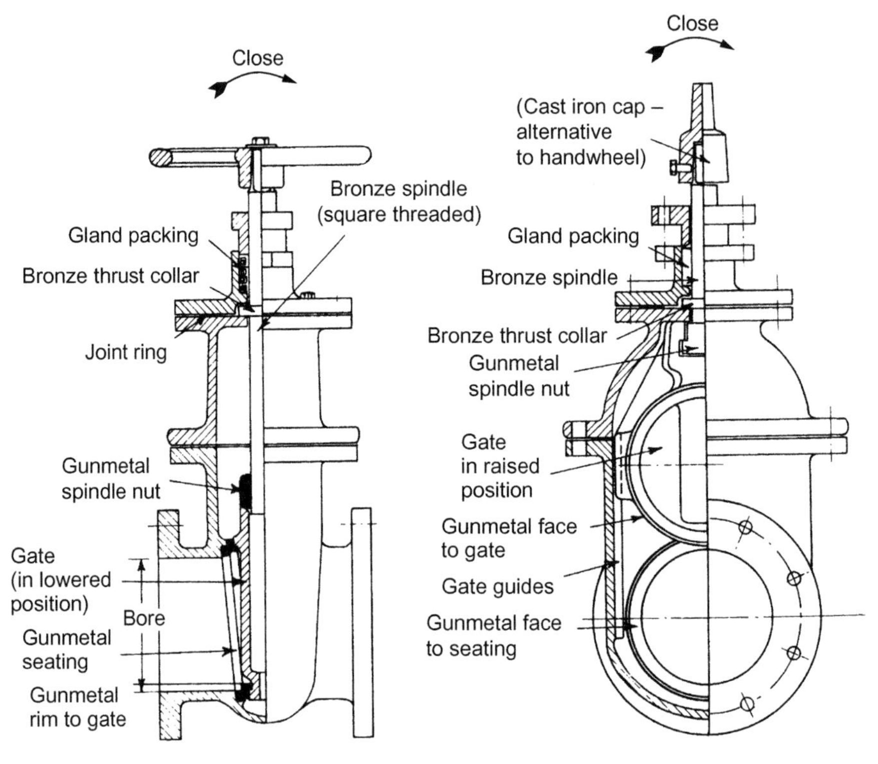

FIGURE 18.3

Metal seated gate valve (DN 300).

18.8 BUTTERFLY VALVES

Butterfly valves tend to be cheaper than gate valves because they require less material and less civil works. They are also easier to operate against unbalanced water pressures as the disc pivots about an axis on or near the pipe axis. Consequently butterfly valves are now commonly used in water distribution systems. Butterfly valves can be metal seated or resilient seated; in the latter case the seat is usually made of natural or synthetic rubber and is commonly fixed to the body of valves of smaller sizes or to the disc. Plate 28(b) shows a resilient seated butterfly valve.

Resilient seated valves can remain virtually watertight, even after prolonged use in silty water. Therefore, resilient seats are usually specified for isolating valves in distribution systems. Resilient seated valves may also be used for control purposes but, if operated at small openings, the seal may be damaged. Solid rubber is the material usually used for resilient seatings: inflatable seals have been used on very large valves but not always with success. Metal seated butterfly valves do not have tight shut-off characteristics and are mainly intended for flow control purposes where they need to be held in the partially open position.

Distribution network pipe systems are now designed to produce self-cleaning velocities at least once every 24 hours and should not need swabbing as part of normal operation. A transfer pipeline may need to be swabbed periodically. Butterfly valves on the line prevent the passage of foam swabs (except for very soft ones) but this does not usually pose a problem if the valves are spaced sufficiently far apart to allow the pipe to be cleaned in sections. Short lengths of pipe either side of the valve are made removable so that the cleaning apparatus can be inserted and removed.

Butterfly valves should normally be mounted with the spindle horizontal since this allows debris in the pipe invert to be swept clear as the valve is closed. Where the spindle is vertical solids can lodge under the disc at the spindle and cause damage to the seal. Disc position indicators are useful and strong disc stops integral with the body should be specified, so that the operator can feel with certainty when the disc is fully closed or fully open.

Butterfly valves have been made to very large diameters (10 m or more) operating under very high heads and at high water velocities (20 m/s or more) and have proved successful in use. However, when a butterfly valve is to be used for flow control purposes the maximum velocity of approach to the valve should be limited to 5 m/s. Resilient seated valves can be specified to have no visible leakage on seat test but the range of acceptable seat leakage rates for metal seated valves varies from about 0.004 to 0.04 l/h per 100 mm of nominal diameter (DN), at the specifier's choice. However, a low rate for a high pressure differential would be expensive to achieve and difficult to maintain with metal seats. For some control applications, an acceptable seat leakage rate of about 0.4 l/h per 100 mm DN may be appropriate.

If a valve may be required to remain in place closed on removal of the pipe on one side for a temporary operation, it must be flanged for bolting to a pipe flange on the other side. 'Wafer' butterfly valves whose bodies are sandwiched between pipe flanges do not achieve this. Use of such valves for isolation of air valves allows maintenance to be carried out on the air valve in situ with the pipeline in service but does not allow removal and replacement of the air valve under pressure. Since replacement of air valves is likely to be cheaper than in situ refurbishment, flanged isolating valves are preferred in such situations.

18.9 GLOBE VALVES

A globe valve consists of a circular orifice, usually with its axis at right angles to the pipe axis, against which a piston or disc obturator makes a seal. Movement of the obturator reveals a cylindrical opening which can be ported (provided with a serrated, slotted or perforated sleeve). The obturator is driven by a shaft, which can be operated by a device such as a spring. Spring actuation allows the globe valve to serve as a non-return valve. When configured for autonomous (self-acting) control, valve opening can be regulated automatically via a secondary hydraulic circuit acting on the shaft, using pilot valves and differential pressure across a diaphragm or piston. Such valves can serve duties such as pressure reducing, pressure sustaining, pressure relief and altitude (Section 20.8).

Globe valves are available in sizes up to DN 600 and are widely used for control.

Flow direction depends on the type of globe valve. Usually flow is outwards through the circular orifice of valves with diaphragms but inwards through the orifice of globe valves with secondary pistons. This difference is important where cavitating conditions are likely, since cavitating

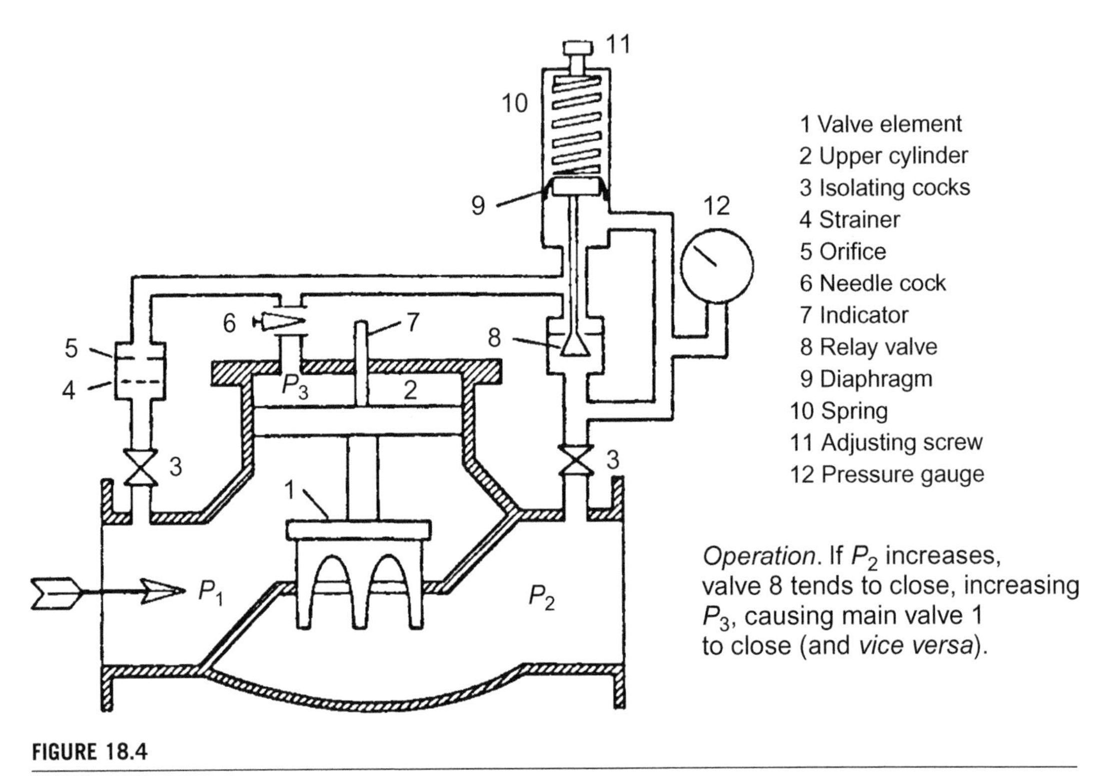

FIGURE 18.4

Globe valve with piston seal — with auxiliary circuit for PRV duty.

vapour pockets generated in a constriction should be prevented from collapsing against the valve body or other components where erosion damage could take place. This disadvantage of the outflow configuration in cavitating conditions can be overcome by dissipating some of the head just upstream of the circular orifice or by use of a sleeve with circular ports, but only at the expense of increased headloss when fully open. Plate 28(c) shows a cut-away illustration of a diaphragm sealed globe valve. Figure 18.4 shows an inflow type of globe valve with piston seal arranged as a pressure reducing valve (PRV).

18.10 SCREWDOWN VALVES

Screwdown valves are normally made only in small sizes but their operation is similar to that of the globe valve. The bib tap is a typical example. The body of the valve is cast so that the water must pass through an orifice which is normally arranged in the horizontal plane. A plug or diaphragm or, in the case of a bib tap or stopcock, a 'jumper' can then be forced down onto this orifice by a screwed handle, as shown in Figure 18.5. In small sizes, high pressures can be controlled, as in the case of the ordinary domestic tap. However, screwdown valves are not suitable for in-line flow regulation or isolation since they cause high headloss and their seatings need periodic renewal.

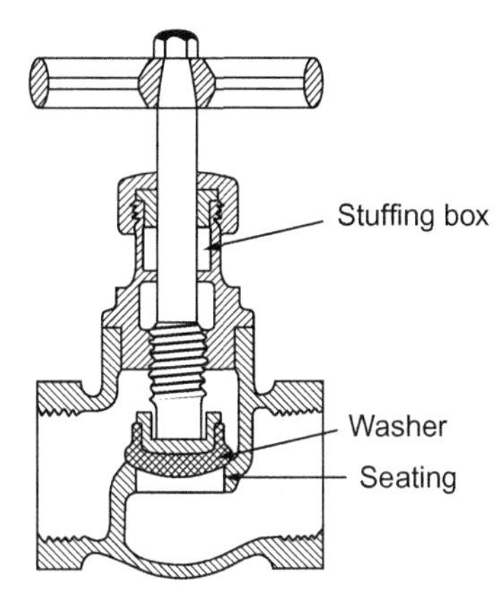

FIGURE 18.5

Screwdown valve (for service pipes).

18.11 BALL VALVES

Ball valves consist of a spherical obturator with a cylindrical hole, usually of the same diameter as the pipe, although it can be smaller. Operation is by rotation (1/4 turn) of a shaft mounted, often horizontally, with its axis at right angles to the cylindrical hole. Seals are usually resilient and can provide drop tight shut off. Ball valves are commonly used in small diameters (up to DN 300) although at least one manufacturer can make ball valves up to DN 1200. Ball valves are manufactured in one-piece, top entry, two-piece (Fig. 18.6) and three-piece bodies. A top entry body allows access to the ball and seats for maintenance without the need to remove the valve and is preferred for larger sizes.

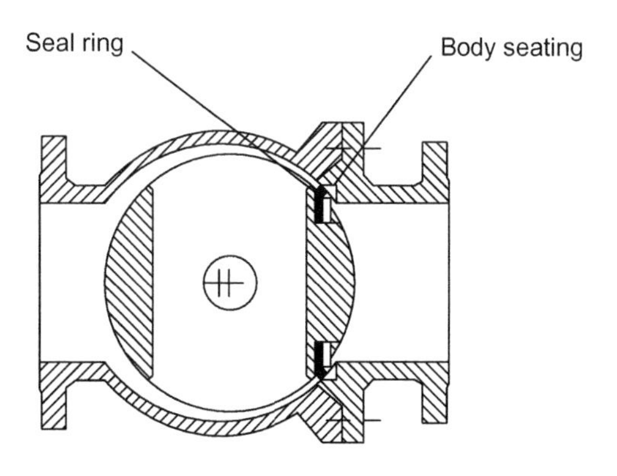

FIGURE 18.6

Ball valve (two-part body).

18.12 PLUG VALVES

The principle of the plug valve is similar to that of the ball valve in that closure is effected by a 1/4 turn of a spindle which, in this case, is usually mounted vertically to allow the weight of the plug to be taken on a bottom bearing. The opening is often rectangular rather than circular; it requires transitions upstream and downstream and prevents the passage of harder swabs. Plugs with circular openings are available but, except for the smallest valve sizes, the opening is smaller than the pipe. The plug may be a complete frustum of a cone or a sector only, in which case the obturator is called an eccentric plug or cam. The plug can be removed from the top of the valve body. Some designs allow the plug to be partly supported by upstream hydraulic pressure, thereby reducing wear. Eccentric plug valves are available up to DN 1800 and full cone plug valves can be made in similar sizes if required. Plate 29(a) shows a cut-away of an eccentric plug valve.

18.13 DIAPHRAGM VALVES

In a diaphragm valve the diaphragm is forced down onto a weir in the valve body (weir type – Fig. 18.7) or onto the invert (straight-through type – Fig. 18.8). The action is created by rotation of a threaded spindle of the rising type and is transmitted to the diaphragm through a shaped platen. Diaphragm movement for the weir type is less; consequently this type is preferred for higher pressure or partial vacuum conditions. As the fluid is separated from the moving parts the diaphragm valve is particularly useful for chemical liquids (if the body is lined) and those carrying solids. Diaphragm valves are made in sizes from about 6 mm to about DN 350.

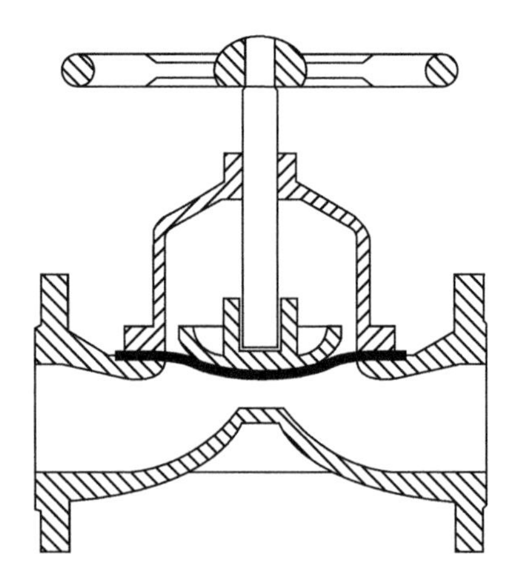

FIGURE 18.7

Diaphragm valve (weir type).

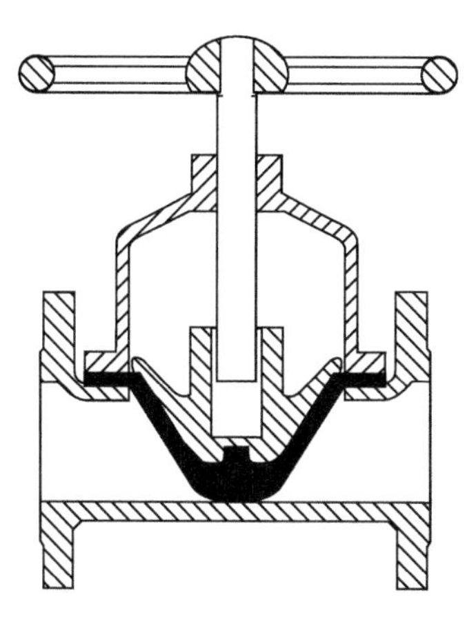

FIGURE 18.8

Diaphragm valve (straight type).

18.14 PINCH VALVES

A pinch valve consists of a diaphragm which is deflected by a cam or roller to reduce the gap between the diaphragm and the base of the valve opening. In some designs, an upper and a lower diaphragm are employed, both squeezed at the same time. Valves of this sort are particularly useful with very aggressive fluids if the liner/diaphragm is made of suitably resistant material.

18.15 NEEDLE VALVES

In a typical needle valve flow passes around the housing for a piston located centrally within the valve body. Flow can be gradually reduced by advancing the piston towards a circumferential seat in the downstream end of the valve body, often through a sleeve. The piston and central bulb may be streamlined to reduce losses with the valve fully open. Plate 29(b) shows an actuated needle valve. The full-open headloss of needle valves can be low whilst affording good control and performance in cavitation conditions is good – particularly when a perforated or slotted sleeve is employed.

For 'free discharge' high pressure duties such as occur during discharge at the base of a dam, the needle valve should be adapted with a jet disperser. This consists of a vaned insert located downstream of the valve; this causes the jet to break up and may impart a twist to the flow. If the jet has sufficient height and space to break up into falling water droplets and there is a pool to receive them, little protection of 'soft' river beds is needed.

18.16 SLEEVE VALVES

A sleeve valve consists of a fixed perforated sleeve where the openings are revealed by movement of a piston. In-line sleeve valves may be mounted with their axes on the pipe axis or at an angle. Flow in in-line sleeve valves is from the outside to the inside which provides a good basis for cavitation resistance. Sleeve valves are particularly suitable for control duties where high heads have to be dissipated. However, facilities should be provided for access to the sleeve for cleaning unless a self-cleaning mode is built into the design.

Where discharge of high head water into a tank is required a submerged discharge valve is often used. This is usually fitted with a sleeve and discharges radially outwards through the sleeve at the base of a concrete pit where excess energy is destroyed in turbulence.

18.17 HOLLOW-JET DISCHARGE VALVES

The hollow-jet discharge valve consists of a cone mounted in the waterway with its apex upstream on a shaft with a gearbox and drive spindle through the side of the body. The cone is moved downstream revealing an annular orifice. The jet of water issuing from the valve is diverging and is able to dissipate energy in a shorter distance than the needle valve. The original design was by Messrs Howell and Bunger in 1935 but similar valves are now made by several companies. The Howell–Bunger valve may be fitted with a hood to limit lateral spread of the jet and to cut down spray, although this partly negates the advantage of this design which can otherwise deal with heads up to 425 m and can be made in diameters up to 4.25 m.

18.18 NON-RETURN (OR CHECK) VALVES

Non-return or check valves are of six basic designs:

- a disc with a single hinge or off-centre pivot, usually closing against an inclined orifice;
- a disc similar to that of a butterfly valve but with an off-centre pivot and a counter weight (Plate 29(c));
- a spring loaded disc closing linearly against a circular orifice, either mounted in the horizontal plane as in a globe valve or in-line as in the needle or nozzle type;
- a split disc hinged in the middle like butterfly wings;
- falling ball type;
- conical diaphragm type.

With all these designs forward flow tends to move the obturator out of the way while the valves are designed to close as soon as possible after forward flow ceases. It is desirable for closure to be achieved without slamming, which generates surge pressures and may damage the valve or other parts of the system. Non-slam characteristics are achieved by either arranging for closure to be complete before reverse flow commences or by slowing closure down so that any reverse flow is reduced gradually. The latter method usually relies on air or hydraulic damping while the former may be assisted by external aids such as springs or weights. However, any assisted closure tends to force the obturator into the flow and therefore increases headloss. All non-return valves are thus a compromise between low headloss and speed of closure. It should be noted that damped closure can allow reverse flow and thus cause a pump upstream to turbine.

The most common type of non-return valve consists of a flat disc within the pipeline pivoted so that it is forced open when the flow of water is in one direction and forced shut against a seating

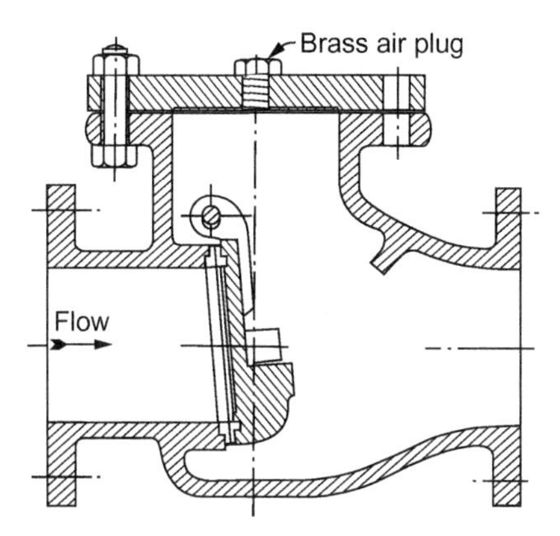

FIGURE 18.9

Non-return check valve.

when the flow tries to reverse (Fig. 18.9); it is often termed a 'swing check valve'. The seating is arranged slightly out of perpendicular when the valve is to be inserted into a horizontal pipe so that the flap closes by gravity when there is no flow. A more inclined seating reduces flap travel and speeds closure. Another way to reduce travel is to provide two or more smaller and lighter flaps in a single bulkhead but at the penalty of some increase in headloss. The globe type of non-return valve has a very short travel and with low spring pressures can close very quickly with effectively no reverse flow. However, it has high losses and is not usually employed on pumped systems due to wastage of energy. Spring loaded valves of the needle or nozzle type have lower losses (K of 0.75 for DN 300) than other types of non-return valve.

18.19 FLAP VALVES

Flap valves are used at drain or washout outfalls where a pipe discharges to a body of water with varying level. This is to prevent backflow at high water levels and consequent contamination. Flap valves can be circular or rectangular (flap gates) and are usually mounted against the outer face of a concrete wall, but can be flange mounted to the pipe. The materials used for gate and frame are usually cast iron but stainless steel is used for some large fabricated gates.

Factors to be considered in the selection (and design) of flap gates include:

- robustness and resistance to distortion under high (tidal) back pressure, particularly with debris trapped under the gate;
- ability to seal adequately after long use in difficult conditions; 'double-hung' (two hinges in series – see Fig. 18.10) flaps overcome some of the effects of wear and temperature change;
- avoidance of water and silt traps in the gate construction that can increase resistance to opening and aggravate corrosion;
- effects of waves and damage to seals and hinges; a stilling chamber should be considered.

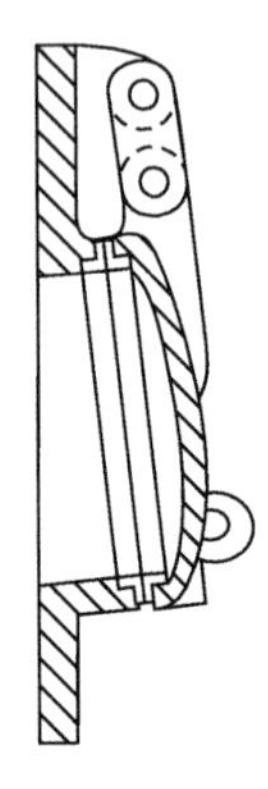

FIGURE 18.10

Flap valve.

18.20 CAVITATION AT VALVES

Cavitation is the generation of pockets (cavities) of water vapour and their subsequent collapse (Section 15.12). A part open valve presents an orifice which produces a high velocity low-pressure jet. The onset of cavitation is marked by a fall-off in discharge coefficient as flow is increased, as downstream pressure is decreased or as the orifice is made smaller since any of these factors tends to depress the pressure in the orifice. The fall-off in discharge coefficient occurs because vapour pockets start to occupy the orifice opening. This effect can be used to determine when incipient cavitation occurs and is identified as the point where the curve of headloss coefficient against cavitation number begins to rise markedly. At this level there is minimal noise, vibration or risk of damage. Another method of assessing the onset and severity of cavitation is by taking noise and vibration measurements in the valve. Three stages of increasing cavitation are described in *Control Valves* (Borden, 1998): incipient, full and supercavitation. In this context incipient cavitation is characterized by the irregular occurrence of cavitation instances at vortices. As cavitation increases, there is constant production and collapse of cavities and noise is steady and reaches a maximum level, after which (in supercavitation) noise decreases somewhat due to the dampening effect of the volume of vapour present and as the collapse area is pushed downstream.

At small openings the internal shape of the valve (according to type) has little effect on the headloss coefficient. Therefore, in such circumstances critical cavitation conditions are likely to present themselves at very similar loss coefficients irrespective of valve type. The shape of the opening has a minor effect on the onset of cavitation. However, both the nature of the opening and the shape of valve internals significantly affect the risk of damage once cavitation is occurring. Valves in which collapse of vapour pockets occurs away from the valve body and other components should not be damaged by cavitation. The distance a jet travels before head recovers sufficiently to collapse vapour pockets depends on jet size. Therefore, porting the opening to produce numerous jets can help keep the area of collapse away from critical components. An added advantage of numerous small jets is that noise and vibration are reduced.

Some materials commonly used in valve construction (cast iron and brass) are not particularly resistant to cavitation damage but some bronzes are more resistant. Various stainless steels, when hardened, particularly Duplex types, provide superior resistance (Borden, 1998). In very extreme cases titanium may have to be used.

18.21 VALVE OPERATING EQUIPMENT

Manual closing of a valve is usually by clockwise rotation of a hand wheel or tee key. However, for historic reasons some utilities may use valves with anti-clockwise closing in the whole or part of their systems. Clear labelling of the operating direction is essential. Where the closing direction is in doubt the valve can be 'sounded' to detect at which end of its travel there is noise of water rushing through a narrow opening. Hand wheels may be mounted directly on the valve. However, if this would put the operator below ground, where confined spaces procedures would be needed (to avoid danger of suffocation or noxious gases), a means of operation from above ground should be provided. This may require a headstock and, in many cases, an extension spindle running in brackets rigidly attached to the chamber walls. Spindles which are to be immersed in water, such as those for operating valves inside a reservoir, should be of manganese bronze or stainless steel. Although many old manually operated large gate valves without gearing are still in use, they have to be operated by large bar and ring key, sometimes by a team of operators. Valves for manual operation should now be specified to allow operation by one person with a force on each side of a hand wheel or tee key of not more than 200 N or, for unseating, 400 N.

If manual operation is not feasible (due to very long operating times) for a large geared valve for example, or where remote or automatic control is required, an actuator should be fitted. Actuators are hydraulic, electrohydraulic, pneumatic or electric. Pneumatic and hydraulic actuators are usually arranged to transform driving pressure into a linear force by a diaphragm (pneumatic) or piston or into a torque via a vane device for quarter-turn valves. The piston can be used with a cam, lever or pinion to convert linear into angular movement. Electric actuators apply rotational torque via a gearbox; further gearing may be used to obtain linear motion. The electrohydraulic actuator is local to the valve and uses an electric motor to drive a close-coupled hydraulic pump which then actuates movement as described above.

Electric actuators are usually provided with a hand wheel to allow manual operation in case of power failure and should be provided with local operation buttons. Where remote or automatic operation is needed signal cables are required. The sizing of actuators needs to be done with care since there should be enough torque to overcome resistance but not so much as to overstress the valve spindle or the obturator stops. For most types of valve, electric actuators can be position limited whilst for gate valves they need to be torque limited. It is usual to arrange pinned couplings which should shear at a torque above that required to operate the valve but below that which could cause damage.

Actuators for pipeline valves and other valves that are widely dispersed are usually electric. Pneumatic systems are used where a large number of, usually small, valves can be actuated off a central compressed air supply. Groups of valves may also be actuated off a single hydraulic power pack. Pneumatic actuators are cheapest but are more suited to on–off operation as positioning is not precise enough for control. They are very useful where fail-safe (open or closed) operation is needed as there is sufficient stored energy in the system to close or open a valve after power failure. Hydraulic systems are mainly used for very large installations – hydropower for example – where considerable actuator forces are required.

Guidance on actuators for valves can be found in the *Valve Section Handbook* (Smith, 2004).

18.22 VALVE CLOSURE SPEED

Valve closure is one of the causes of pressure transients (surge) in a pipe system. Closure decelerates the flow and causes a pressure build up according to the Joukowsky equation:

$$\Delta H = -c\Delta V/g$$

where ΔH is the change in head, c is the speed of a pressure wave, g is the acceleration of gravity and ΔV is the change in velocity (negative for a decrease).

Surge is discussed in more detail in Section 15.11. The magnitude of the pressure rise is roughly proportional to $(1/T_c)^{1.5}$ where T_c is the time of closure of the valve in seconds. Graphs published (by Thorley, 2004) indicate that, in a pipeline with initial flow velocity of 2 m/s, in order to limit the rise in head on valve closure to about 50 m, T_c needs to be about 150 times the length (L) in km of the pipeline upstream for gate valves. The equivalent figures for butterfly and ball valves are about $25L$ and $50L$, respectively. For an initial velocity of 1 m/s closure times would be about three quarters of the above values. Pipe diameter has little effect on these requirements; therefore it can be seen that there is greater risk of high pressures on closure of small valves since they can be closed more quickly. On the other hand, for pipelines of some materials there may be considerable spare capacity for higher pressures in smaller diameters. For actuated valves the closure speed should be specified taking into account the acceptable pressure rise. Where necessary surge analyses should be carried out to check that the selected closure speeds are appropriate, using the characteristics of the valve to be supplied (Section 15.11).

18.23 WASHOUTS

Despite the name, a washout is seldom used for scouring or 'washing out' a pipeline because its diameter is usually too small to create sufficient flow velocity in the main to wash out debris; its principal use is for emptying the pipeline or for the removal of stagnant or dirty water. Discharges may be subject to consent, as in the UK, but the relevant authority should always be consulted whether the discharge is to sewer, watercourse or other water body. If chlorinated water is involved its dechlorination may be required. In a major pipeline primary washouts may be installed to drain the majority of the length between section valves; secondary washouts of smaller diameter can then be used to empty undrained low points. Sizes, particularly of primary washouts, should be calculated according to the required drain-down time, which should typically not be longer than one working shift. Factors are the number of washouts, head available and limits on discharge, access and resources. Initial flows could be very high unless the washout valve is throttled but such throttling could cause cavitation unless the type of valve is selected accordingly (Table 18.1). The drain-down time is dominated by the low head available during the later stages. The washout diameters given below should allow the last 200 m length of a pipeline to be emptied in about 1 hour in typical situations:

Main pipeline diameter	Washout branch diameter
Up to 300 mm	80 mm
400–600 mm	100 mm
700–1000 mm	150 mm
1100–1400 mm	200 mm
1500–1800 mm	250 mm

In open country it is usual to install washouts at every low point with additional washouts being provided in each section where a main is subdivided into sections by stop valves. Each washout should discharge, wherever possible, by gravity to the nearest watercourse. The discharge should be to a concrete pit with overflow to the watercourse in order to prevent scour from the high velocity discharge. The washout branch on the main should be a 'level invert tee'. In flat country washouts should be spaced 2–5 km apart, depending on pipeline gradients and valving. Where it is not possible to get a free discharge to a watercourse the washout will have to discharge to a chamber from which the water can be pumped out to some other discharge point. In this case a means of prevention of backflow should be provided in addition to the washout valve. This could be a flap valve on the outlet to the chamber.

In distribution systems principal feeder mains are usually provided with washouts wherever this is convenient, regard being paid to the position of valves on the main and any branch connections and to the need to be able to empty any leg of the main in a reasonable time of 1 or 2 hours. On small mains, washouts are not normally provided because fire hydrants can be used to help empty the system. However, it would be usual to lay a specific washout to empty a part of the system where a convenient watercourse exists. Although pipe supply systems should be designed to avoid dead ends, spur mains may be necessary in some situations. Washouts should be placed at the end of every spur main; these usually comprise fire hydrants even though they may not be officially paid for by the fire authority and designated as such. They should be operated regularly to sweeten the water at the end of the main.

Care is essential in the design of washouts since, under high heads, the velocity of discharge can be very high and the consequent jet discharge can be destructive and dangerous. Manholes receiving the jet should be of substantial construction and valves should be lockable and slow opening for high heads. Two valves, one guard and one operating, may be installed where prolonged throttling at high head is required.

18.24 AIR VALVES

To fill a pipeline with water, there must be means for releasing air from it. Where hydrants are installed, for example in a distribution system, these can be opened to help release air, and to exhaust the first flush of water if required. However, unattended air release points should be provided with automatic valves which must close as soon as there is no more air present so that no water is lost. An air release valve contains a ball or other shape float which rises to close an orifice as the last air is excluded. The design must prevent the float from being sucked onto the orifice by

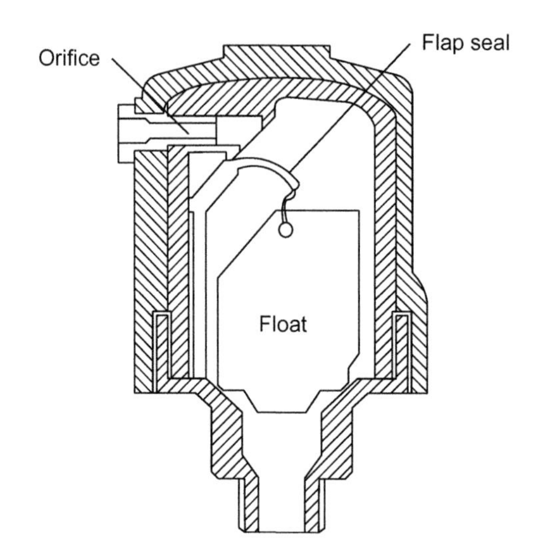

FIGURE 18.11

Small orifice flap type air valve. *Based on information from*: Aqua-Gas AVK.

high velocity air and must prevent oscillation of the float. Two sorts of air release valve are used, one being a 'large orifice' air valve, designed to release or admit large quantities of air at low differential pressure when a pipeline is being filled or emptied; this type does not open to release air under any appreciable pressure. The other is a 'small orifice' type (Fig. 18.11) which is designed for continuous operation, releasing small quantities of air under operating pressure as it collects at high points. Double orifice air valves (Fig. 18.12 and Plate 29(d)) combine one of each type in the same unit and are used at locations where both duties are required.

For raw water pipelines (as for sewage rising mains) it is important that air valve action is not obstructed by debris in the water. Air valves made for this purpose reduce the risk of float jamming or orifice obstruction. It is seldom necessary to put small orifice air valves on distribution mains since air is removed via the service pipe connections which should be soffit connected. Exceptions are where there is a sudden hump in the main, such as when it is laid over a bridge. For other pipelines, Lescovitch (1972) may be referred to but the authors advise the following:

1. Large orifice air valves are required for filling and emptying:
   - **a.** at high points;
   - **b.** at a steepening in gradient in a falling pipeline;
   - **c.** (possibly) at a flattening in gradient in a rising pipeline;
   - **d.** at about 2 km intervals on long lengths of pipeline (with no fire hydrants); and
   - **e.** between any intermediate line valves.
2. Small orifice valves are required at similar locations and additionally:
   - **f.** at high points relative to the slope of the hydraulic gradient;
   - **g.** on access manhole covers and other local 'humps' where air may collect;
   - **h.** at other locations on a pipeline supplied by pump if likely to introduce air:
     - downstream of the pumps,
     - downstream of any PRV if there is a substantial reduction of pressure, and
     - (possibly) at 0.5 km intervals on downward legs.

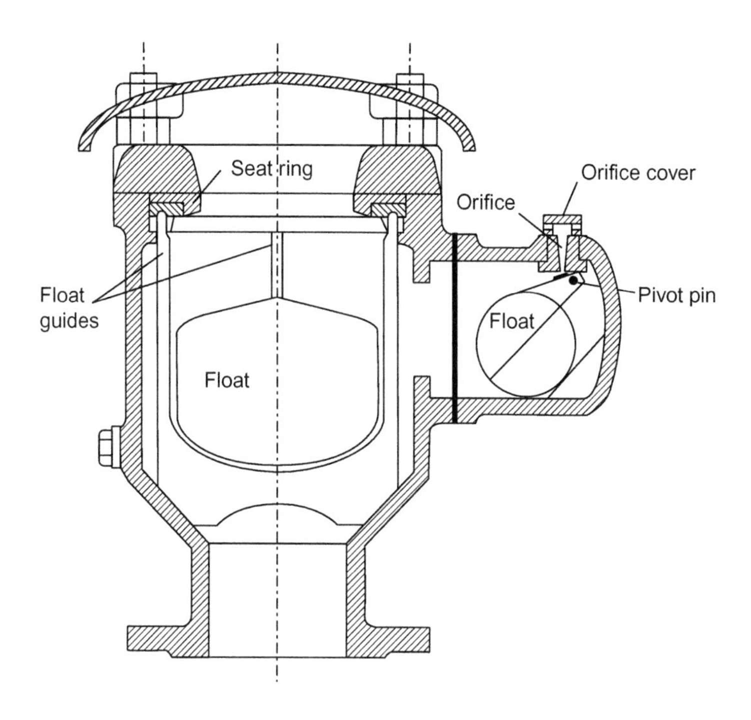

FIGURE 18.12

Double orifice air valve.

Large orifice air valves are usually sized to suit the maximum outflow expected during use of washouts, to cover the fastest likely rate of filling and in some cases to limit sub-atmospheric pressures in the event of a burst. However, fast filling is inadvisable and a planned filling rate equivalent to a pipeline flow velocity of about 0.3 m/s would be acceptable (AWWA M51). Where two or more large orifice air valves are available to exhaust the air the filling rate may be increased. The limitation is aimed at avoiding damage on closure of the air valve once all the air has been removed. It is suggested that air valves be sized for a differential pressure of 0.25 bar. The flow capacity of air valves should be taken from manufacturers' catalogues. With some designs the orifice size is smaller than the connection size. Table 18.2 shows quoted air valve capacities at 0.25 bar pressure differential; they vary considerably, therefore it is important to specify performance.

A flanged isolating valve (sluice valve, butterfly valve or stopcock) should be sited below each air valve, thus making it possible to remove the air valve for repair or replacement without shutting down the main. The restricting effect of valves and fittings between the pipeline and the air valve must be taken into account in assessing air valve capacity as installed. Although each pipeline should be treated specially, Table 18.3 provides a guide to the minimum air valve capacities for pipelines of selected diameters.

Small orifice air valves have an orifice size which should be related to the operating pressure in the pipeline: the higher the pressure the smaller the orifice. If it is assumed that 5% of the dissolved air in a pipeline may come out of solution (actual amounts should be much less in a well-designed system),

Table 18.2 Large orifice air valve capacities

Branch size (mm)	Range of quoted air flows at standard temperature and pressure for 0.25 bar differential (m^3/min)	
	Out	In
80	12–51	8–45
100	21–70	16–60
150	50–150	50–135
200	50–300	50–200

Table 18.3 Suggested minimum large orifice air valve capacities

Pipeline diameter (mm)	Outflow capacity (m^3/min)	Inflow capacity (m^3/min)
300	8	12
600	20	25
900	40	50
1200	70	80
1800	150	180

the total small orifice air valve capacity provided should equate to about 0.1% by volume of the water flow. A greater amount of small orifice air valve capacity at the upstream end of a pipeline than downstream should ensure early release of air. Capacities of small orifice air valves should be taken from manufacturers' catalogues, using the operating pressure at the air valve location.

On large pipes, air valves may be fitted to a blank flange on a DN 600 or larger tee to allow access to the pipeline. All air valves should be sited above the highest possible groundwater level that can occur in any pit; the pit should be free-draining with minimal maintenance; if this is not done, polluted water may enter via the air valve when the pipeline is emptied. Where the location of an air valve is not possible in the road (as cover at high points is usually a minimum), it may be necessary to connect the branch to an air valve in the verge, where air valve chambers can be raised above ground level if necessary.

On thin walled pipelines of steel or other flexible material anti-vacuum valves may be essential in order to admit air to limit sub-atmospheric pressures and prevent pipeline collapse on emptying or on a burst. If below atmospheric pressures are expected during a surge transient, this type of protection may be needed for large diameter or flexible pipes. Large orifice air valves can be used for this duty but special anti-vacuum valves may be necessary for very large air flows.

18.25 VALVE CHAMBERS

Large valves on trunk pipelines should be located in accessible chambers to facilitate maintenance. Smaller gate valves, for example on distribution systems, are usually buried. It is preferable to site butterfly valves in chambers so that the gearbox can be maintained, but for large butterfly valves some water companies put a chamber around the gearbox only.

A simple valve chamber for a gate valve is shown in Figure 18.13. The valve is anchored on the upstream side by an anchor flange in the chamber wall. Downstream of the valve a flange adaptor allows removal of the valve. The valve body is therefore free to expand downstream when the valve is closed under pressure and it must therefore be specified for the open end test conditions (Section 18.6). The pipe through the downstream wall is provided with a puddle flange to reduce ingress of groundwater. In order that the chamber can resist the thrust from the anchor flange, it should be designed in the same way as a thrust block, transferring the load to the soil through friction or cohesion and passive earth resistance (Section 17.34). Two flexible joints, with an intervening 'rocker pipe' two pipe diameters in length, should be provided on either side of the chamber to avoid damaging pipework if any differential settlement of the pipeline, relative to the chamber, should occur. The chamber cover can be made of precast reinforced concrete slabs, provided with sockets or other arrangements to receive lifting devices. Valve chambers should not be sited in roads if they can be located in the verge or other open ground instead but it is wise to ensure that the cover is strong enough to take the loading from heavy vehicles which may run-off roads.

Small valves, that is those DN 300 and smaller, are usually buried and a pipe (DN 80 or 100) is fitted as a sleeve for the spindle which is accessed through a surface box for tee key operation. Valves up to and including DN 600 may also be buried, depending on location and preference; the choice being on maintenance and the civil works required for giving access to the valve.

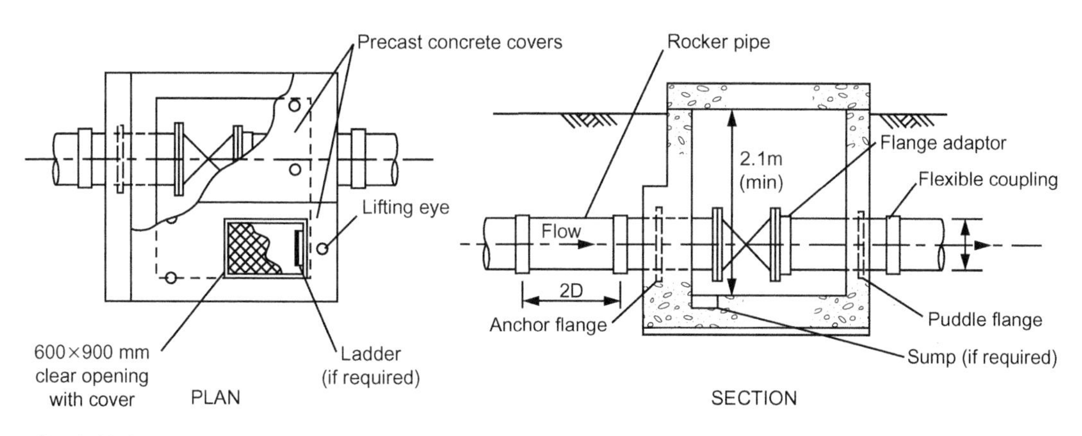

FIGURE 18.13

Valve chamber.

PART II MEASUREMENT OF FLOW AND CONSUMPTION

18.26 PURPOSES OF FLOW MEASUREMENT

There are three reasons for measuring flow:

- control (either manual or automatic) of a process;
- data acquisition for legal, record or operational purposes;
- billing to customers.

Automated control of a process using flow is based on instantaneous measurement by a flow meter instrument whose output is converted to an analogue (4–20 mA) or digital signal (Section 19.38). The signal is then transmitted to a PLC for automatic control of a valve, pump or other device or to a monitoring screen for manual intervention.

The data acquisition application is similar, but the signal is transmitted to a monitoring and control centre for recording or it may be recorded locally on a data logger. The use of chart recorders for this purpose has almost ceased. Typical uses include metering of pump output and of water supply zones or District Metering Areas which is essential for checking leakage and for monitoring consumption.

Billing of customers requires measurement of consumption at a water meter – a 'revenue' meter. For domestic and small commercial and industrial consumers the meter is usually of the mechanical volumetric type. These meters are increasingly provided with short-range transmitters to enable remote reading without having to enter premises. Revenue meters for larger consumers may use the same principles as those described below for flow meters.

18.27 TYPES OF FLOW METER

Flow meters are classed as volumetric or inferential, the latter term referring to meters that determine velocity from other variables such as pressure differences across a device such as an orifice plate. There is a large variety of flow measurement device, using numerous physical principles. Full discussion of the whole range of flow measurement device is out of the scope of this book but the reader will find a comprehensive reference in the *Flow Measurement Handbook* (Baker, 2000). Table 18.4 gives typical information on some of the flow meters usually encountered in the water industry.

The principles of orifice and venturi meters are discussed in Section 14.16. Two other kinds of inferential (or momentum) meter are the Dall tube and the V cone venturi. In both, flow accelerates through a constriction and leads to a pressure drop. The pressure difference is measured in the Dall tube and the V cone venturi as an indicator of velocity (and so flow) in the same way as for an orifice. The V cone venturi design is claimed to have a turn-down ratio of 25:1 and to be less affected by conditions upstream and downstream and can be fitted into shorter lengths of straight pipe than is recommended for other meter types. Further types of momentum meter are indicated in Table 18.5.

Table 18.4 Characteristics of meter types (data from manufacturers' catalogues)

Type	Typical pipe diameter (D) range (mm)	Typical velocity range (m/s)	Typical accuracy (% of full scale)	Typical repeatability (% of full scale)	Typical headloss coefficient (k)	Comment
Volumetric						
Electromagnetic	25–2500	0.5–10	0.25	0.1	0	
Ultrasonic[a]	Up to 7000	0.05–4	1	0.5	0	Sensitive to vibration and head alignment
Coriolis	15–80	0.5–5	1 (of mass)		15	May include density measurement with accuracy about 0.4%
Vortex	15–300	0.4–10	1	0.2	2	Not suitable for pulsating flow
Turbine	15–400	0.6–10	0.5	0.1	0.8	
Rotating vane	15–150	0.02–6	1		5.5	
Rotating piston	15–20	0.03–4.4	2		4	
Paddle wheel	15–1000	0.3–6	1	0.5	5.5	
Inferential					***Headloss as % of pressure difference***	
Venturi tube	50–1200		1.25		10–20	Length = 5D
Dall tube	150–2000		1.25		30	Length = 1.75D
V cone venturi	15–3000		0.5	0.1	Up to 75	
Orifice plate	25–200		3	1	Up to 90	

Note: [a]Time-of-flight ultrasonic meter.

Table 18.5 Further types of momentum meter

Type	Principle
Vortex meter	Detection by various means of the frequency of vortices shed by a 'bluff' body into the flow. Flow being proportional to frequency. Accuracy is claimed to be almost as good as for the ultrasonic type but there is some headloss due to the obstruction.
Swirl meter	Generation of a spiral flow stream in a throat by a vane array upstream and its detection via the pressure fluctuations caused at a tapping in the throat.
Fluidic meter	A jet of water is flipped from side to side of diverging conduits under the action of feedback from each side. The frequency of oscillations is proportional to velocity and is detected by pressure sensors.
Variable area meter	A 'float' sits in upward flow in a tube with increasing diameter and whose position is detected by optical transducer (principle used for the chlorinator).
Viscous meter	Measurement of the pressure difference across an array or labyrinth of fine tubes in the conduit.
Drag plate meter	The force or moment on a plate inserted into the flow is detected at its support in the wall.

18.28 VOLUMETRIC FLOW METERS

Electromagnetic (EM) flow meters. This type of meter (Plate 30(a)) is based on Faraday's Law of electromagnetic induction. The voltage induced in a conducting fluid as it moves through a magnetic field is measured between electrodes in the pipe wall and is proportional to the mean velocity of the fluid, the strength of the magnetic flux and the size of the conduit. Calibration produces a relationship between induced voltage and total flow through the conduit. The output is integrated across the pipe section and is not affected by small differences in velocity profile across the section. The magnetic field is generated by coils usually arranged as saddles around the pipe, powered by battery or, for large meters, from a mains electricity supply.

This type of meter is increasingly being used in the water industry owing to zero headloss and good accuracy and reliability. It does require fluid conductivity to be more than 5 μS/cm and air entrainment to be low but it can be used for a large variety of transported materials other than water. Deposits on the electrodes should be cleaned off periodically and any deposits on the pipe wall near the electrodes need to be rendered non-conducting. Both can be achieved by application of appropriate DC and AC voltages. EM meter accuracy can be affected by stray currents.

Calibration is best done 'wet' against a master meter since this is more reliable, although dry calibration (by measuring the magnetic field across the whole section) can be done (see also 'digital fingerprinting' – Section 18.29).

Ultrasonic flow meters make use of sound waves of frequency over 20 kHz. The transit time (time-of-flight) meter measures time differences due to the difference in velocity of sound waves transmitted upstream and downstream through the flow. The time difference is a function of average flow velocity along the beam. Doppler meters measure the change in frequency in sound waves reflected back from particles in the flow and, therefore, cannot be used for pure or treated water.

Ultrasonic flow meters of the transit time type are reasonably accurate if installed and operated correctly. The pipe wall must be of a hard material that transmits sound well. Deteriorated and porous pipe linings and deposits are likely to affect the accuracy of ultrasonic flow measurement and may cause drift. Accuracy is adversely affected by transducer misalignment and vibrations in the system. Doppler meters are generally less accurate but are used for dirty water and water with air bubbles. Ultrasonic meters do not intrude into the flow and cause no headloss but their accuracy is not as good as that of the magnetic meter. Clamp-on ultrasonic meters require no intrusion into a pipeline and, although their accuracy (about 3%) is not as good as for complete meters, are becoming increasingly used particularly for large conduits (over DN 500) and for temporary installations and check measurements. Clamp-on ultrasound meters are not suitable for concrete or GRP pipes or metal pipes with thick mortar or concrete linings.

The accuracy of single path ultrasonic flow meters is affected by uneven flow distribution across the section. In situations where this may be a problem, meters using multiple paths are recommended. Accuracy is also affected by presence of air bubbles.

Propeller or turbine meters use a freely rotating, bladed rotor positioned in the flow. The speed of rotation is measured, usually by a pick-up coil mounted in the housing to sense the passage of the rotor blades. Pulses are thus generated which can be counted over a known time. There is a small amount of resistance to rotation due to friction and the sensor. Therefore, the turbine does not rotate until velocity exceeds a threshold value. Above this level, the speed of rotation is proportional to the fluid velocity so that it is necessary to know the flow area to calculate flow. For large conduits it may be necessary to carry out velocity traverses across the section to establish the relationship between velocity at a particular location and the total flow rate. Provided the velocity profile is well known, the accuracy of the turbine meter for flow measurement is very good.

18.29 PERMANENT FLOW METER INSTALLATIONS

Source meters and meters on large mains have historically been of the Venturi or Dall tube type and a number of these may still be in use. Venturi and Dall tubes can have an accuracy of ±1.25%, but their performance is affected by upstream and downstream features such as bends and valves. If properly maintained they can continue to give satisfactory readings but the throats of such meters should be inspected from time to time to ensure that they are clean and free of slime or deposits, since they affect meter accuracy. For this purpose a hatch is usually provided over the throat of a venturi. The recording equipment, usually mechanical, also needs keeping in good order. To be certain that source outputs are known as accurately as possible, in situ volumetric testing of such meters is always advisable.

Most Venturi and Dall tube meters have been replaced with more modern types, such as the electromagnetic meter (Plate 30(a)), which is simpler to install, more accurate and which can be equipped with a full telemetry interface capability for district metering, customer billing, leakage control and treatment works applications. EM meters have an accuracy of ±1% but recent developments have enabled some manufacturers to quote ±0.25% accuracy for their meters. However, accuracy in the field is affected by pipe features. Plate 30(b) shows a pillar for housing district metering electronic equipment.

The flow meter manufacturer provides a specification and a flow calibration certificate from a recognized test laboratory. However, installation conditions usually differ markedly from the bench tests and invariably lead to differences in performance. Meters are frequently sited too close to fittings, valves or tees, or are affected by protruding gaskets and other factors such as vibration, flooding or large ambient temperature swings. To overcome the worst of installation effects manufacturers recommend that the length of straight pipe upstream and downstream of meters should be a certain multiple of pipe diameter. A commonly used figure for electromagnetic meters is 5D upstream and 3D downstream. For other types of meter and for precision installations a larger multiple and possibly flow straighteners may be needed. Extensive test data from various sources on the effects of nearby features on the accuracy of various types of meter is set out in the *Flow Measurement Handbook* (Baker, 2000). The data show that the effect of a pipe feature on meter accuracy depends on the nature of the feature (i.e. bend or part open valve), its distance from the meter, its orientation with respect to meter sensor orientation and on the type of meter.

In situ testing is often impractical and installed meters are likely to produce errors. Therefore, estimating leakage from the difference in flows measured at different locations can produce quite large errors. 'Digital fingerprinting' has been introduced to check for changes in calibration with time. This tests the electrical characteristics of a magnetic meter. Provided the electrical field remains constant and other key electrical parameters within the circuitry are also stable, it is possible to relate the electrical 'fingerprint' of a meter and transmitter back to a change in meter calibration. The meter is self-checking and the results can be sent back to the manufacturer annually for audit. In situ testing should be traceable and should be conducted by an experienced engineer.

18.30 TEMPORARY FLOW MEASUREMENT DEVICES

Insertion probe flow meters are installed for temporary measurement of flow for consumption surveys or for distribution networks analyses. These instruments are either the turbine or EM type, the latter becoming more common. Both are inserted into the pipe where flow measurement is required. The turbine type uses a small rotating vane at the end of a probe to record flow velocity. The vane is susceptible to damage, in which case the instrument has to be returned to the manufacturer for repair and recalibration. The turbine meter is inserted through a 40 mm diameter tapping in the pipe which has to be of at least 200 mm diameter. The EM probe (Plate 30(c)) uses an electromagnet at its end to apply a magnetic field to the water. Electrodes either side of the probe pick up the induced electromotive force (EMF) in the water which is proportional to the velocity past the electrodes. The tapping for an EM insertion probe is 20 mm diameter and can usually be installed in pipes of diameter 150 mm and greater. EM probes are made up to 1 m long; therefore, they cannot be used for pipes of diameter greater than 900 mm and are restricted to flow with velocity less than about 1.75–2.0 m/s due to the flexibility of the probe.

Insertion probes measure the velocity at the position of the measuring device. This can be at the pipe centreline or at defined points across the diameter. The measured velocity has to be converted to mean pipe velocity of flow by relating the measured value to the average velocity across the whole pipe. For this a velocity profile for the pipe is used, determined by using the same instrument to record velocities at set points across the diameter from crown to invert. The recorded measurements are corrected to take account of the disturbance caused by the instrument itself (increased local velocity). The disturbance coefficients are unique to each instrument and are provided by its manufacturer. For the conditions usually encountered the ratio of the mean velocity to the centreline velocity is 0.83 but can range from 0.7 to 1.0. Values differing widely from 0.83 should be viewed with caution and the cause investigated. However, satisfactory results should be obtained if the internal diameter is measured accurately and if the number of flow profile readings is sufficient – five for pipes of DN 150, nine for DN 300 and 13 for larger pipes. The measurements should be repeated at least three times to ensure the ratio is consistent and repeatable; the flow must be relatively consistent during each profile run. In practice poor field conditions often make precise measurement difficult so that several attempts may be necessary. Once the profile is established satisfactorily the instrument is set at the pipe centreline and the data logger is attached. However, pipes in poor condition produce a different velocity profile which changes with flow and can render very inaccurate measurement of flows using an insertion probe.

Making tappings and installing insertion probes pose a risk to water quality. Although such risks can be managed, ultrasonic strap-on flow meters are being used increasingly as an alternative and avoid the tedious exercise of velocity profiling. Versions of these meters can be installed on all sizes and materials of pipe used in distribution systems.

18.31 SUPPLY (REVENUE) METERS

Water supply meters in common use on consumer's service pipes are of two types: semi-positive and inferential. The semi-positive meter, typically sized in the range 15–40 mm, is almost universally used in the UK for metering domestic and small trade supplies. The most usual of the semi-positive meters in the UK is the rotary piston (or rotary cylinder) meter which has an eccentrically pivoted, light weight, freely moving cylinder which is pushed around by the water inside the cylindrical body, opening and closing inlet and outlet ports as it turns (Plate 30 (d)). This movement operates a counter mechanism which summates the total flow. Another type uses the nutating disc which wobbles around in a circle as water is drawn through the meter and endeavours to pass above or below the disc. This meter does not measure low flows to the same accuracy as the rotary piston meter but is widely used in the USA where domestic flows are generally higher. Other types of semi-positive meter include the single or multiple orifice vane meters and the paddle wheel type (Fig. 18.14). All semi-positive meters should incorporate a strainer upstream as the meter is only suitable for water free from grit or other suspended matter.

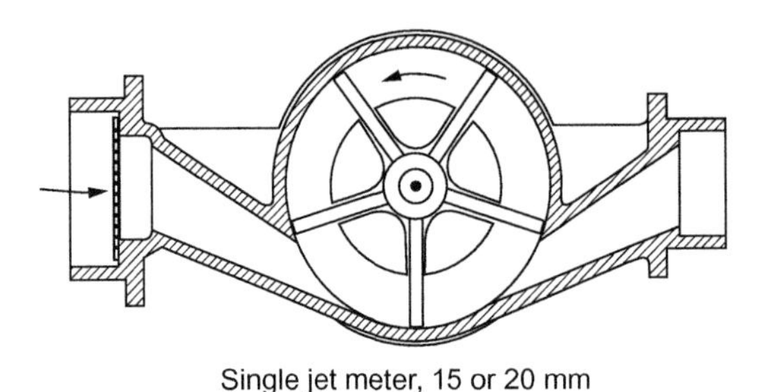

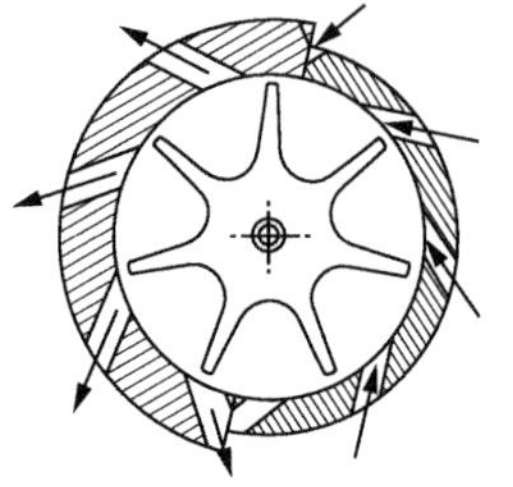

FIGURE 18.14

Vane meters.

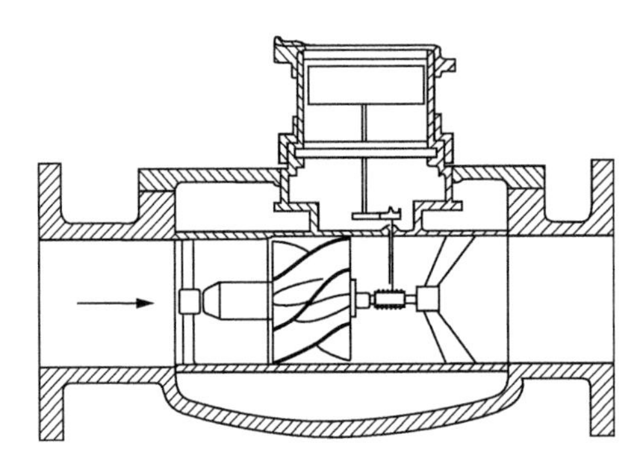

FIGURE 18.15

Turbine meter (Woltman type).

The inferential meter has a bladed turbine which is turned by the flow. The quantity of water is 'inferred' by counting the revolutions of the turbine (Fig. 18.15) and must be calibrated at the maker's factory. It is primarily used on industrial supplies, being suitable for large flows. It can summate the total flow either through counter gearing or each revolution of the propeller may initiate an electrical pulse which an electrical logger can summate for the total flow and also record flow rates over short time intervals. This data can be locally recorded or transmitted elsewhere by telemetry. For the measurement of widely fluctuating flows, beyond the range of any single meter, past practice has been to use two meters of different sizes in parallel; an automatic device ensures that the smaller flows pass through the small meter only. However, such combination meters were expensive to install and have now been superseded by the electromagnetic meter (Section 18.28) which gives good accuracy over a wide flow range and is cheaper to install than the mechanical volumetric type.

All supply meters need to be regularly tested for accuracy every few years. A typical small meter-testing bench is shown in Figure 18.16. Large industrial supply meters may be tested in situ using a turbine or electromagnetic flow probe as described in Section 18.30.

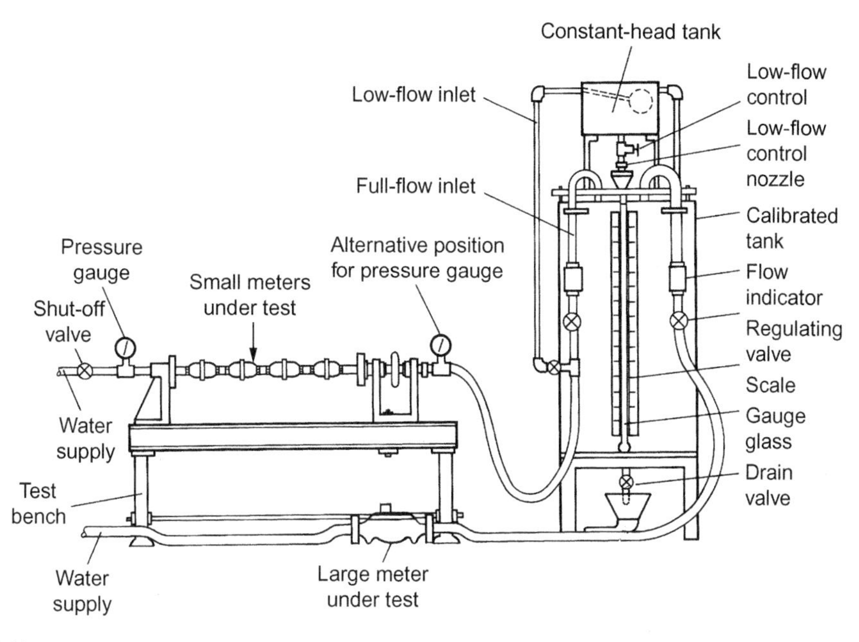

FIGURE 18.16

Small meter-testing bench.

18.32 THE ACCURACY OF WATER METERS

BS EN 14154-1 sets out the criteria for accuracy of water meters in the UK. The accuracy classes defined in the superseded BS 5728-1 have no equivalent in the new standard. Instead manufacturers are expected to offer meters which suit typical applications, stating the values of Q_1 (minimum flow), Q_2 (transitional flow), Q_3 (permanent flow) and Q_4 (overload flow). BS EN 14154-1 requires the measurement error not to exceed 5% in the lower flow range Q_1 to Q_2 and 2% in the upper flow range Q_2 to Q_4. There is no requirement for the minimum flow at which the meter must start to register but some manufacturers state approximate values. The smaller meters used for domestic supply are usually 'semi-positive meters' in which an eccentrically pivoted plastic cylinder is caused to rotate by the through-flow of water. Larger meters, predominantly used on trade supplies, are usually 'inferential' meters of the rotating vane type. Domestic supply meters would normally have a maximum capacity of between 2 and 5 m^3/h. These values may be compared to the design flow rates shown in Table 16.5 for different domestic fittings.

Small water meters typically over measure in the upper flow range and under measure in the lower flow range; indeed they cease to turn at flows below about 3 l/h, even when new. This cut-off should be compared with common situations in domestic premises: a relatively fast dripping tap (4 drips/s) would waste 3–4 l/h, and the thinnest continuous stream about 6 l/h. In a direct system all cold taps and WC cisterns are fed direct from the mains. In indirect systems only the cold water drinking taps in the kitchen and bathroom are fed direct from mains, the rest are fed from a float-valved roof storage tank. As a ball float valve approaches closure flow decreases and for a finite period before full closure flow is less than that which the meter can measure. The WRc found that low flows caused by near-closed float valves to WC cisterns and storage tanks were seriously under-recorded, resulting in a mean under-registration of 2.5% for 'direct' supply systems and 6% for 'indirect' systems (Welton, 1984). The accuracy of all meters also deteriorates with age. Most tests show that the great majority of domestic meters under-record total consumption.

In the National Metering Trials in England 1989–1992 (Section 1.10) it was found that, of 200 meters withdrawn annually for testing, approximately 20% had failed, most by under-recording; about one sixth of them due to blockages from particles in the flow (Hall, 1992). Estimates of the under-recording of revenue meters reported by undertakings in England and Wales for 2009–10 were 4.1% (range 2.9–7.3%) for household revenue meters and 4.8% (range 3.0–8.1%) for non-household revenue meters. The two companies having the largest number of domestic meters installed, namely Anglian Water and South West Water, estimated that their household meters under-registered by 3.0% and 6.1%, respectively and their trade meters by 3.1% and 3.3%, respectively (Ofwat, 2010). Generally 3% under-recording would be considered an average for semi-positive meters and 5% under-recording for inferential meters; but if meters are over 10 years old and have not been regularly removed for testing and refurbishment, substantially greater under-recording must be suspected.

Accuracies considerably better than 2% may be claimed by some manufacturers for the upper flow range, particularly for larger inferential (vane) meters. However, such meters have to be carefully sited because an adjacent upstream bend or tee can seriously affect their accuracy. Because of this WRc thinks that their under-recording in practice is greater than that of domestic supply meters. The 'multi-jet' meter (dividing the flow into several streams) was developed to improve the accuracy of such meters.

18.33 FUTURE TRENDS IN METERING

Automatic meter reading (AMR) technology has been installed by a number of utilities worldwide. AMR comprises automatic remote reading of consumer meters and transmitting the data to the utility's billing database. The readings are made automatically by the meter reader when walking or driving past the premises, using devices, handheld or attached to the vehicle. The data can be downloaded from the device by wire connection, GPS radio telemetry or power line carrier transmission. Provided the meter location does not require access to the property, AMR offers potentially significant cost and manpower savings. However, application is currently constrained by the significant investment required.

Smart metering is an extension of AMR whereby more sophisticated meters can be read automatically to help the utility to operate and manage the distribution network. The systems are able to measure, analyse and store consumptive data throughout the day or at preset times and thus enable both the utility to understand patterns of demand and set variable tariffs according to diurnal and seasonal consumption and to provide the customer with real-time consumption data. Development is at present constrained by issues of cost and the accuracy of meters at different flow rates.

REFERENCES

ANSI/ISA S75.05 (1983). *Control Valve Terminology*. Research Triangle Park, NC; American National Standards Institution (ANSI), Washington; Instrument Society of America (ISA), North Carolina, USA.

ANSI/ISA S75.11 (1991). *Inherent Flow Characteristics and Rangeability of Control Valves*. ISA, Research Triangle Park, NC, USA.

AWWA M51 (2001). *Air-Release, Air/Vacuum and Combination Air Valves*. AWWA, USA.

Baker, R. C. (2000). *Flow Measurement Handbook*. Cambridge University Press.

Baumann, H. (1998). *Control Valve Primer*. Instrument Society of America.

Borden, G. and Friedmann, P. G. (1998). *Control Valves*. 3rd Edn. Instrument Society of America.

BS EN 12266. *Industrial Valves. Testing of Valves*. BSI.

BS EN 14154:2005. *Water Meters. Part 1 (+ A2:2011) – General Requirements; Part 2 (+ A2:2011) – Installation and Conditions of Use; Part 3 (+ A1:2011) – Test Methods and Equipment*. BSI.

BS EN 60534. *Industrial Process Control Valves – Flow Capacity*. BSI.

Hall, M. (1992). Technological developments in metering. IWEM Symposium 'Paying for Water', January 1992. *Journal IWEM*, August, p. 517.

Lescovitch, J. E. (1972). Locating and sizing air release valves. *Journal AWWA*, Water Technology/Distribution, July.

Ofwat (2010). *Service and Delivery – Performance of the Water Companies of England and Wales, 2009–2010 Report*. Ofwat.

Smith, E. and Vivian, B. E. (1995). *An Introductory Guide to Valve Selection*. Mechanical Engineering Publications Limited, London.

Smith, P. and Zappe, R. W. (2004). *Valve Selection Handbook*. 5th Edn. Elsevier.

Thorley, A. R. D. (2004). *Fluid Transients in Pipeline Systems*. 2nd Edn. Professional Engineering Publishing Ltd.

Welton, R. J. and Goodwin, S. J. (1984). *The Accuracy of Small Revenue Meters*. Report TR 221. WRc.

CHAPTER

Pumping, Electrical Plant, Control and Instrumentation

19

PART I PUMPS

19.1 PUMPING PLANT

Using machines to lift water is a very ancient art, developed to satisfy the most basic of human needs, water for domestic use and irrigation of crops. Pumping machinery has developed a long way from the hand, foot or animal-driven shadufs and water-wheel pumps of the ancient world but most modern pumps strongly resemble those of 100 years or so ago, although with many improvements. Examples are better materials, improved bearings, protective coatings, better designed and finished hydraulic passages and better methods of drive and control. These have contributed to great improvements in performance and reliability.

The laws of physics dictate the minimum power required to lift a given mass of water through a given height in a given time (1 m^3/s lifted 1 m requires 9.81 kW at 100% efficiency). High pump efficiency is of great importance since most water supply pumps operate for long continuous periods. Pumping costs dominate the running costs of most water supply systems and energy dominates the whole-life cost of owning pumps in nearly continuous use (Section 10.2).

19.2 CENTRIFUGAL PUMPS

Pumps which operate by rotary action are called rotodynamic pumps and the centrifugal pump is the first type to be considered. Other types of pump still have their uses, particularly for pumping chemicals, slurries and sludge (Sections 12.3 and 13.10), but the centrifugal pump is the most commonly used for pumping water because of the wide range of duties possible and the comparatively high efficiency and low cost.

Centrifugal pumps are available in many different arrangements, as single- or multi-stage units, and mounted vertically (Plate 31(a)) or horizontally (Plate 31(c)) to suit particular needs. The centrifugal pump comprises an impeller which is rotated at high speed in a casing. The impeller

Twort's Water Supply. DOI: http://dx.doi.org/10.1016/B978-0-08-100025-0.00019-3

usually consists of two discs with a number of shaped blades between them; it is manufactured as a single casting and different materials are used for different applications. For fresh water bronze is a satisfactory material for resistance to corrosion, abrasion and cavitation damage, combined with ease of casting, good machining properties and moderate cost. For sea water (or brine) super duplex stainless steel is necessary (for impeller and casing).

One of the discs is fixed to the shaft of the pump and the other has a central hole in it, making an annular space around the shaft – the 'eye'. When the impeller rotates, water is drawn in through the eye and is thrown off the edge of the impeller with high kinetic energy. In the diffuser chamber around the impeller, part of the kinetic energy is converted to pressure, part to forward movement of the water through the connected system and part is lost in turbulence and friction. Plate 31(b) shows a cut-away view of a double suction centrifugal pump. The efficient conversion to useful pressure rise and forward movement of the water is done by careful design of the impeller and diffuser chamber.

With good design the maximum efficiency of the centrifugal pump can exceed 80%, including the energy lost in bearing friction as well as the hydraulic losses within the pump. Efficiency depends on several factors including the size of the pump; larger sizes are generally more efficient than small. The efficiency of a centrifugal pump must vary at different flow rates. Maximum efficiencies of 80–90% are possible with special designs and large machines but these are obtained partly by fine clearances between the moving and static parts of the pump so that efficiency may fall sharply with wear. In practice, all centrifugal pump efficiencies are likely to fall with time; a figure of 1% per annum is often used as a basis for planning but the actual rate depends on hours run and operating conditions, solids in suspension and cavitation. The operating efficiency is a useful indicator of need for refurbishment or replacement, since the efficiency inversely affects the running cost.

19.3 TYPES OF CENTRIFUGAL PUMP

All centrifugal pumps work on the principle set out above but their construction varies greatly according to the duty required. For example multi-stage pumps (Plate 31(d)), used to generate high heads, consist of several impellers and diffuser chambers arranged in series, the impellers being fixed to one shaft. The water from one diffuser chamber is led to eye of the impeller of the next stage so that the pressure developed increases stage by stage. For general waterworks duties the maximum pressure normally developed by one impeller may be between 80 and 100 m. Higher head (H) can be produced by higher speeds of rotation and larger impellers (H is proportional to D^2 for impellers of different diameter (D) in the same body), although this increases the cost and is limited by available suction conditions.

If it is known that increased head will be required in the future a multi-stage pump with a dummy first stage can be specified. This is simply a diffuser chamber without an impeller, allowing the later addition of an impeller to develop more pressure. The efficiency of a pump is not much altered by the 'dummy stage' but making provision for it can prolong its useful life. The driving motor must of course have enough power to drive the pump when the impeller is added, or a new motor will be needed.

A 'split casing' (Plate 32(a)) is the preferred centrifugal pump arrangement in which the upper half of the casing, which is easy to remove, gives access to the impellers and diffuser chambers for inspection and any needed maintenance without disconnecting the pipework or the drive.

Axial thrust arises in a centrifugal pump due to pressure imbalance on the two plates. Several ways can be used to balance this thrust, which would otherwise quickly cause wear on the pump and shorten its life. Small pumps can absorb the end thrust by the use of thrust bearings. For larger pumps a double-entry, back-to-back, impeller design may be used, with the water entering the impeller from both sides so balancing the end thrusts. A multi-stage pump does not normally have double-entry for each impeller, although special designs have been successful for very high heads in which there are double-entry impellers or balanced single-entry impellers on the same shaft. Another common device for overcoming end thrust on smaller multi-stage centrifugal pumps is to incorporate a balancing disc on the shaft; high-pressure water being led to one side of it so that most of the end thrust is taken by the disc.

Vertical spindle (or vertical turbine or line shaft) pumps (Fig. 19.1 and Plate 32(b)) are often used for pumping water from a well and for intakes. The driving motor is at the surface, mounted above flooding level, but the pump is immersed in the water. With enclosed shafting, the spindle rotates within a protective tube or sleeve, perhaps 75–125 mm diameter, and held centrally in the riser pipe by 'spider' bearings. The pumped water is delivered to the surface through the annular space between the sleeving and the riser. A typical arrangement would be a 250 mm diameter riser pipe in 3 m lengths bolted together with flanged joints, the sleeve tube being perhaps 100 mm diameter, with bearings for the spindle at every joint in the riser pipe. These bearings are nearly always water lubricated (oil or grease lubrication risks contamination of the pumped flow), the water being taken from the delivery pipe via filters and fed through the sleeving to the bearings. The alternative arrangement of open 'line shafting' exposes the shaft and its bearings to the pumped fluid and is not suitable for water containing solids. The whole weight of spindle and pump impellers and the hydraulic thrust generated, is taken by a Michell thrust bearing at the top of the shafting, just below the coupling to the motor. Pumps of this type are very reliable, being robust and suitable for continuous heavy duty. However, they are expensive and take time and skill to dismantle or erect when repairs are necessary. Their capital cost may be double that for a horizontal spindle pump and they are now much less common with the increased use of cheaper submersible pumps.

Submersible Pumps

Submersible pumps should strictly be termed 'submersible motor' pumps or 'submersible pumpsets'. The motor design (Plate 32(c)) is the main difference from more conventional designs. The pump, driven by a submersible motor, is very similar to a pump driven by a vertical spindle 'dry' motor, although some differences are given below. Submersible pumps gained in popularity because they usually result in a cheaper installation than one using dry motors. Motor reliability and its location out of the sight and hearing of any attendant were issues but these have been largely overcome by improvements in the motor design, particularly in the insulation and in the instrumentation used for monitoring pump performance. Properly chosen submersible pumps have proved reliable in service over many years; submersible designs are now available from specialist manufacturers for a very wide range of duties.

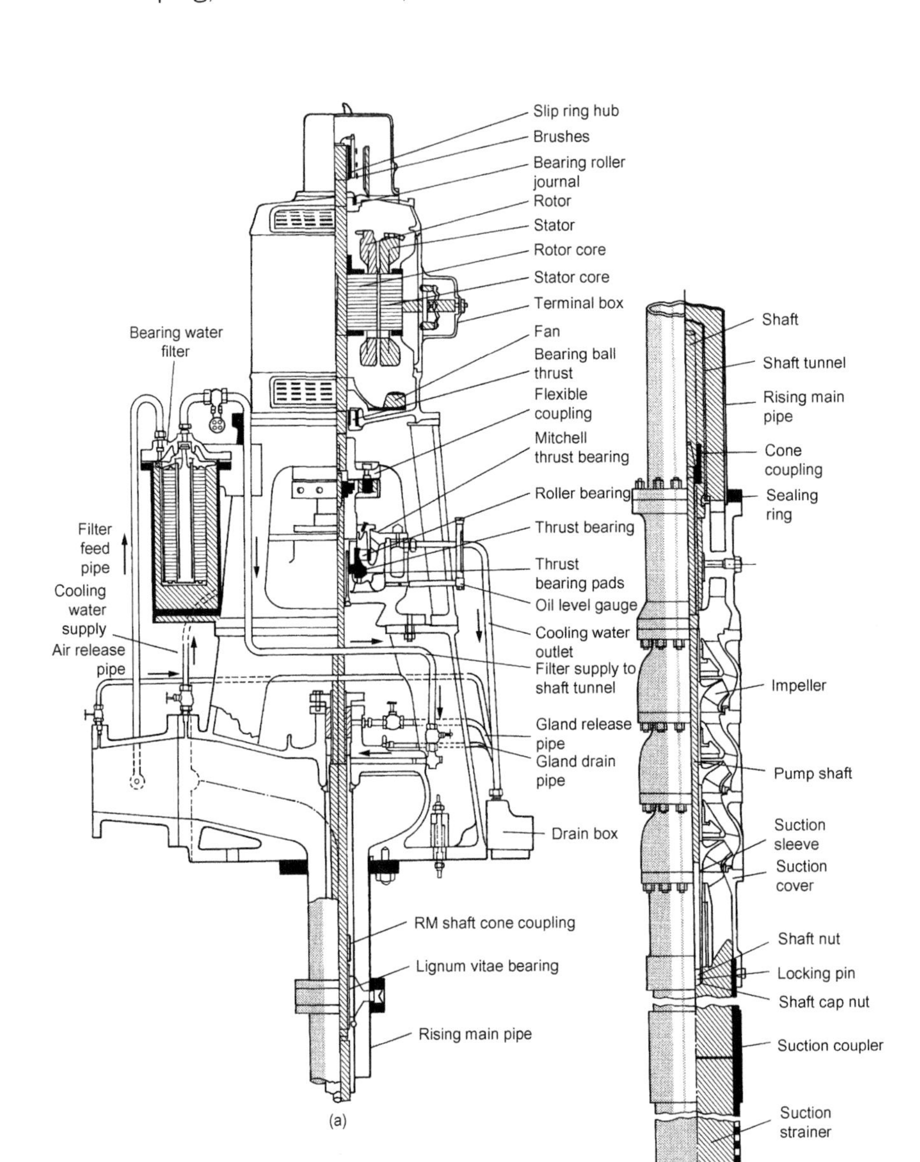

FIGURE 19.1

Vertical spindle (line shaft) pump and motor for borehole.

Many submersible pumps in water supply are installed in drilled boreholes. The high cost of drilling is affected by the borehole diameter; therefore the diameter of the submersible pump is of great importance. Designers must produce pumps and motors of small diameter. Mixed flow pumps produce more flow at a given casing diameter than radial flow pumps and are suitable for borehole pumps. However, they produce less head and more pump stages are needed. This results in pumps longer and narrower than more conventional designs. For the same reasons, submersible motors are longer than equivalent dry motors. They are nearly always two-pole designs (Section 19.21) to develop more power from a given size motor and to run at the highest available speed to maximize pump output, hence reducing overall cost. Naturally the mechanical design of the pump, especially its bearings, must be appropriate for the chosen speed. The disadvantages of a higher speed are increased wear, particularly if the pumped water contains abrasive solids, and reduced suction capability, so that deeper submergence may be required.

In a typical borehole installation, the pump is directly coupled to the submersible motor, which is underneath, and power is supplied to the motor through waterproof cables clipped to the outside of the riser pipe. The water inlet is between pump and motor with the outlet from the final pump stage leaving axially. The motor is normally a water-cooled fixed-speed caged induction motor, specially designed for underwater running. However, where varying output is needed, a variable frequency power supply may be used (Section 19.24), although at significant extra cost. If there is any risk that inflow to the borehole could be predominantly from a higher level than the pump inlet, or if the pump is installed in a large body of water so that the pumped flow does not pass over the motor, a motor shroud should be used to ensure a cooling flow passes over the motor.

Whilst all submersible motor windings will meet the required temperature class there can be an issue with the insulation breaking down due to the voltage withstand level that the motor can accept (Section 19.24).

Submersible pumps are relatively quick and easy to install. The rising main is free of the spindle and sleeving needed with the vertical spindle pump; a large thrust bearing to support the heavy rotating parts is not required. Submersible pumps need not be installed truly vertically, which may be a big advantage in very deep wells. They are even sometimes used horizontally as booster pumps in distribution mains (Plate 32(d)). Submersible-pump reliability in non-corrosive waters has been proven over the years; even in corrosive waters they can be withdrawn for attention or replacement more easily than pumps of the vertical spindle design. Some modern borehole installations are now designed without any surface housings, although provision still needs to be made for access for a mobile crane or sheerlegs for withdrawal of the pumpset and its riser pipe. This simplification can make substantial cost savings.

Submersible pumpsets may be less efficient than the vertical spindle design, partly because of the special design of the motor but also because of the higher number of stages needed to achieve a given duty. This can be important if the pumping duty is wrongly estimated, because of the pronounced peak in the efficiency curve with the multi-stage unit. However, submersibles gain by avoiding the transmission shaft losses of the vertical spindle design.

19.4 CHARACTERISTICS OF CENTRIFUGAL PUMPS

Characteristic curves for a typical true radial flow centrifugal pump are shown in Figure 19.2. The head/flow curve is relatively flat up to the Best Efficiency Point (BEP) and the power at zero flow is only about 40% of that required at the design duty. This allows the centrifugal pump to be started against a closed valve; they are also shut down with the valve closed first. The pump is unaffected provided the valve is opened (or the pump stopped) before the pump becomes overheated. The pressure rise in the delivery system on starting can be managed by controlling the rate of opening of the valve. The reduced power required for starting reduces the starting current when an electric motor is used to drive the pump.

The maximum head generated by the pump, for a given speed, is usually not greatly in excess of the design duty head. However, in the example shown, the head–output curve is unsatisfactory since for heads higher than the duty head there are two possible outputs. The pump is therefore unstable which could cause trouble if operated in parallel with another pump. The head–output curve should preferably fall continuously from the 'shut valve' head as flow increases. Specifications often require the curve to be steep over the expected operating range so that pump output will not vary much if the head alters somewhat during operation. However, the maximum head at zero flow (the 'shut valve' head) should not be excessive since this governs the design pressure for the connected system. The efficiency curve should be reasonably flat about the design duty point so that there is no great reduction in efficiency if the actual head is slightly different from the expected duty, which is often difficult to estimate exactly at the design stage and may alter in operation.

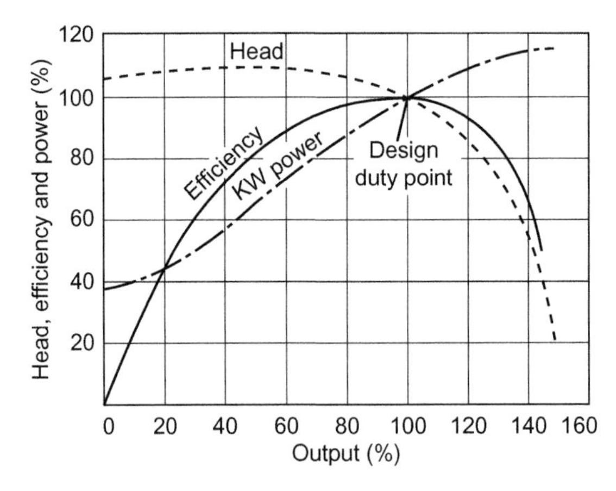

Note: *The peak in the flow characteristic that could cause instability.*

FIGURE 19.2

Characteristic curves for a radial flow centrifugal pump at constant speed.

Affinity Laws

When the rotational speed (N) of a centrifugal pump is changed, there may be little change in efficiency, depending on the amount of the change, but the output (Q), head developed (H) and power required (P) are altered according to the following relationships, known as the 'Affinity Laws':

$$Q_1/Q_2 = N_1/N_2$$

$$H_1/H_2 = (N_1)^2/(N_2)^2$$

$$P_1/P_2 = (N_1)^3/(N_2)^3$$

The theoretical effect of changing the speed of the pump is shown in Figure 19.3.

Gas (air or water vapour), even 1%, in a water pump reduces its efficiency; 5% can cause choking (Rayner, 1995). Dissolved air will come out of solution if the pressure is lowered below that at which the water is saturated with air. This is normally at some pressure below atmospheric depending on previous exposure to air (Section 15.10). Therefore, there should normally be no loss of efficiency due to air coming out of solution if the pump inlet pressure is above atmospheric. However, obtaining such conditions is not always feasible. Whether air is present or not water vaporizes at a pressure which depends on temperature. This pressure is lower than that at which air may come out of solution so that, as pressure drops in the pump, air

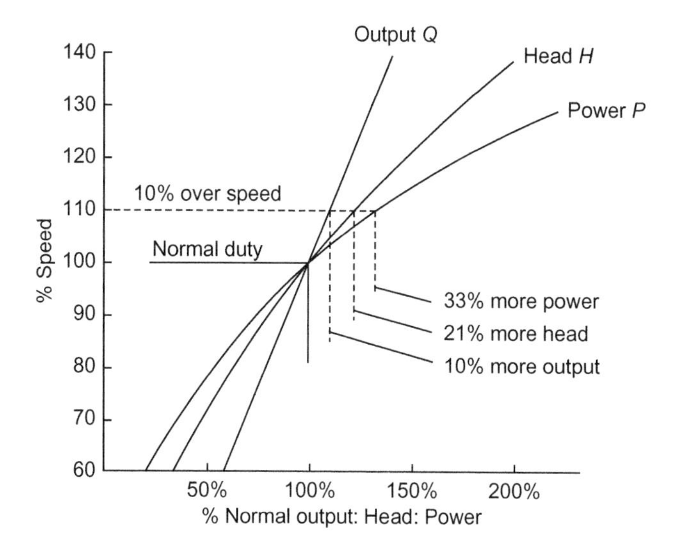

FIGURE 19.3

Effect of changing speed of a centrifugal pump.

bubbles form first and then the water starts to vaporize. On subsequent rise in pressure water vapour pockets collapse instantaneously causing noise and erosion if adjacent to a solid surface. However, air does not go back into solution as quickly and bubbles remain, albeit compressed, as pressure rises. Remaining bubbles form a cushion which can reduce cavitation damage, depending on flow patterns in the pump, but air does cause a hissing noise. A hard crackling noise heard from a pump, almost as if there is gravel inside it, is an indication of cavitation. If such a sound does not disappear shortly after starting, the cause should be investigated and eradicated before damage results.

Net Positive Suction Head

Each pump has a minimum 'net positive suction head requirement' (NPSHR), which varies with flow. This is the head that causes water to flow into the eye of the impeller and is the minimum suction pressure required to prevent cavitation. Its value varies with the speed and capacity of the pump and will normally be given by the manufacturer, based on the results of tests. By convention NPSHR is the NPSH at which the pump head reduction (due to cavitation) is 3%. NPSHR curves rise as pump flow increases with lower head. These are the critical conditions for pump cavitation since the NPSH available (NPSHA) is reduced due to increased suction losses at higher flows. NPSHA is given by:

$$NPSHA = Z + \frac{(P_a - P_{vp})}{\gamma} - h_f$$

where Z is the difference between the pump impeller eye level and the suction water level, P_a is the absolute atmospheric pressure, P_{vp} is the absolute vapour pressure of the liquid at the pumping temperature, γ is the specific weight of liquid at pumping temperature, h_f is the head lost in the suction pipework. Values of atmospheric pressure (P_{vp}) below which water vaporizes at different temperatures are indicated in Section 15.12. Atmospheric pressure reduces with altitude and, to a smaller extent, the weather; the figures in Table 19.1 take account of the average temperature lapse rate from a base of 101 kPa (1013 mb) and 15°C at sea level.

The value of NPSHA must always be greater than NPSHR and a margin of perhaps 1 m is usually specified to cover any tendency for cavitation at slightly higher NPSH values, although some sources recommend a margin of 1.5 m (Sanks, 1998). This allows for any minor differences between calculated and actual figures, as well as changes with time.

Table 19.1 Atmospheric pressure at altitude

Altitude (m)	−300	Sea level	200	500	1000	2000	3000	4000
Pressure (kPa)	105	101.3	98.9	95.5	89.9	79.5	70.1	61.6

Specific Speed

To classify geometrically similar pumps, the numerical quantity 'specific speed' has been adopted. Specific speed (N_s) is the speed required to deliver unit flow against unit head; it varies in accordance with the system of units used.

$$N_s = \frac{NQ^{1/2}}{H^{3/4}}$$

where N is pump impeller speed in rpm; Q is output at maximum efficiency (m^3/s); H is delivery head at maximum efficiency (m). Specific speeds (in metric units) fall approximately into the following categories:

Type of pump	Specific speed
Radial flow	10–90
Mixed flow	40–160
Axial flow	150–420

19.5 AXIAL FLOW AND MIXED FLOW PUMPS

Axial flow pumps are of the propeller type, in which the rotation of the impeller forces the water forward axially, and do not strictly qualify as centrifugal pumps. Mixed flow pumps act partly by centrifugal action and partly by propeller action, the blades of the impeller being given some degree of 'twist'. However, in practical terms there are no precise dividing lines between radial flow (centrifugal), mixed flow and axial flow pumps. In general, axial and mixed flow pumps are primarily suited for pumping large quantities of water against low heads, while centrifugal pumps are best for pumping moderate outputs against high heads. Axial flow pumps have poor suction capability and must be submerged for starting. They are most often used for land drainage or irrigation or for transferring large quantities of water from a river to some nearby ground-level storage. The pumps shown in Plates 32(b) and (c) are mixed flow.

Characteristic curves for typical mixed flow and axial flow pumps are given in Figure 19.4. The starting power required by the mixed flow pump shown is about the same as the duty power, but for the axial flow pump the starting power is substantially greater than the duty power. Axial flow pumps are therefore not started against a closed valve, which would overload a motor correctly sized for the expected duty. They are either started against an open valve to minimize the starting power and current required, or installed in systems specially designed to ensure no delivery valve is needed, for example with a siphonic delivery.

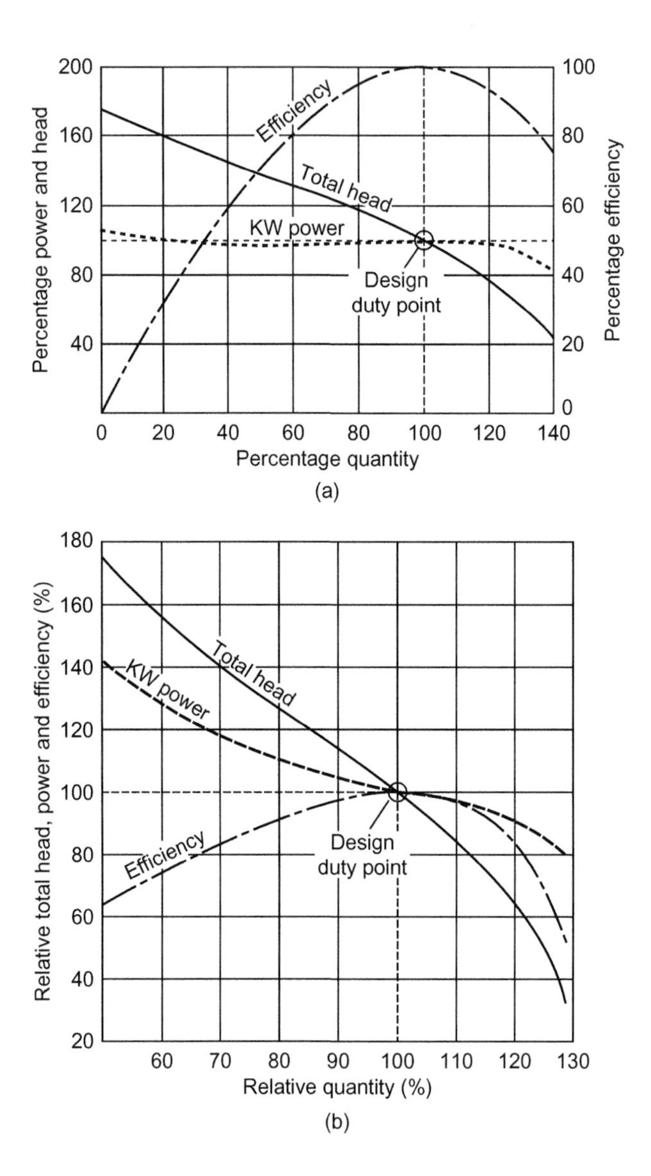

FIGURE 19.4

Characteristic curves for (a) a mixed flow pump and (b) for an axial flow pump.

19.6 RECIPROCATING PUMPS

Most pumps used for water supply nowadays use a rotating impeller, but the reciprocating pump still has its uses. The ram pump is the most common form of reciprocating pump; it consists of a piston reciprocating within a cylinder provided with water inlet and outlet valves. Water is drawn in by one stroke of the piston and expelled by the next. With all ram pumps the output fluctuates cyclically, but if three or more cylinders are used, a reasonably constant flow is maintained. The type of pump with three cylinders is termed a 'triple-throw ram pump', the pistons being connected to the operating crank 120° apart.

Ram pumps were common for high head duties in water supply before the widespread introduction of centrifugal pumps. Their slow speed suited the reciprocating steam engines which usually drove them. Driven nowadays by an electric motor, the triple-throw ram pump may sometimes be used where exceptionally high heads must be produced, since no special design is necessary other than making the working parts strong enough to sustain the pressure. Efficiencies of more than 90% can be obtained, falling off with piston and cylinder wear. However, since very high head centrifugal pump designs are now available, the use of new ram pumps for high head duties in water supply has almost completely ended.

Small ram pumps, usually driven by variable speed electric motors, are however still in widespread use for the injection of the chemical solutions needed for dosing in water treatment. Their property of delivering a constant volume of liquid per stroke makes them well suited to the metering duty required. With appropriate design of the pump and its driving system, both the pump speed and its stroke are varied to maintain any desired rate of chemical injection per unit volume of water.

The bucket pump is another form of the ram pump, arranged to operate vertically for drawing water from a well or borehole. Instead of a piston, a 'bucket' is given a vertical reciprocating motion in the rising main. The bucket incorporates a valve which opens as the bucket descends below water level. When the bucket rises, the valve closes and water is lifted. An alternative arrangement makes the descent of the bucket force water up the rising main through a series of valves. This type of pump is known as a 'lift and force' pump. The village pump is a single-throw bucket pump, operated manually; it illustrates one of the chief advantages of the bucket pump apart from its low cost: it is always ready for use and able to function even if erratically operated. Modern versions of these machines are still being installed for village water supplies in remote regions of the world. Small single-throw bucket pumps are also still used, often wind powered, to provide water supplies for remote farmland.

The hydrostat or *hydraulic ram* pump is another example of reciprocating pump still in occasional use. A large volume of water flowing in one pipe is used to drive a ram which is connected to a smaller ram pump, which pumps part of the water to a higher elevation through a branch pipe. The flow which continues down the main pipeline suffers a loss of head. The proportion of flow pumped depends on the relative pressures. Hydraulic rams are not convenient in public water supply systems because any increase of flow on either the high or low pressure side quickly invalidates the set-up. However, they do find use in special situations, such as when a small volume of water is to be lifted in a remote location without an accessible power supply.

19.7 SELECTION OF PUMPS FOR WATER SUPPLY

The horizontal centrifugal pump is suitable for nearly all waterworks duties, except handling very large volumes of water against low heads and pumping from wells and boreholes. The main advantages of the horizontal pump are that it is relatively low in first cost and it can be arranged to provide easy access for maintenance. A great variety of designs are available to meet required pumping conditions but the most common arrangement is to use the horizontal, split-casing double-suction design (Plate 31(b)) which has been developed over many years of water supply duties. For a single unit the output can range from about 50 Ml/d at 60 m head to 10 Ml/d at 200 m head. The most common waterworks duty is from 10 to 25 Ml/d per unit at 30–120 m head, and in this range the horizontal pump is cheapest.

The pump should have stable characteristics and should be 'non-overloading': the power absorbed should not increase much if the delivery head drops. This is not always possible to arrange and protection against overload must be provided for the motor. Low head could result if the delivery main were to burst near to the pumping station or when pumping into an empty rising main. The efficiency curve should also indicate no severe fall in efficiency for moderate variations of flow and head about the duty point. When specifying a pump, the manufacturer must be told of the complete range of duties the pump is intended to meet, including whether series or parallel operation is required (Section 19.10). To specify the duty point only could lead to large efficiency loss when actual running conditions vary from that specified. In order to minimize energy consumption, the design duty point(s) should be determined so that, within the range of duties at which the pump will operate most of the time, efficiency will be as near to the best as possible (BEP). However, the selected point(s) may need to be adjusted to avoid the pump from operating outside its acceptable range, for NPSH or other reasons.

For wells and boreholes, the choice lies between vertical spindle pumps (Fig. 19.1 and Plate 32(b)) and submersible motor pumps (Plate 32(c)). The vertical spindle pump may be regarded as a 'heavy duty' pump and can be driven by a synchronous electric motor (Section 19.21) which enables the power factor to be brought to unity. The consequent cost of power is reduced, although the motor is more expensive initially and costs more to maintain. Variation in speed is sometimes necessary when pumping from a well or borehole because the output from the well may have to be kept in step with the demand, irrespective of seasonal fluctuation in water level in the well. A thorough appraisal of all the possible operating conditions must be made before choosing the right pump and drive for the duty.

The vertical spindle motor driving a mixed flow pump (Fig. 19.1) is suitable for pumping large quantities of water against low heads; the pump is immersed in the water and the motor sited above the highest flooding level. An alternative arrangement for pumping water from a tank or main is to site the cheaper horizontal centrifugal pump in a dry well, with its centreline below the bottom water level in the tank so that the pump is always primed. Centrifugal pumps should preferably not be sited so that they have to produce any suction lift, because their efficiency drops and cavitation may result (Section 19.4). However, if essential and depending on the design and NPSHR curve, up to 4 m lift can be managed, with 5.5 m the maximum possible.

For high lift pumping the most common choice is the fixed-speed, horizontal multi-stage centrifugal pump (Plate 31(d)). Submersible pumps are also used for this duty, either for pumping direct from a well to a high level tank or for inserting as boosters in a pipeline in a pit below ground level, thus avoiding construction of a building to house the normal arrangement of horizontal pump and motor.

19.8 PUMP BODY AND IMPELLER SELECTION

Pumps for typical water supply duties are offered by manufacturers in different body sizes, each with a range of impellers. This allows a wide range of duties to be covered by a relatively small number of different pump bodies and gives the operator some flexibility to change the impeller later when demand grows, provided the maximum size of impeller has not already been installed. Selecting a pump with this capability is prudent.

19.9 STANDBY PUMPING PLANT

In water supply pumping stations, a single pump is seldom relied on for the full output; adequate standby is essential to ensure continuity of supply. If the full duty can be handled by a single pump then a duplicate of equal capacity should be installed; this is a common arrangement when the unit sizes are small. However, if the pumps are each sized for 50% of total required output, the installation of three similar pumps ensures 100% output on breakdown of any one machine, 50% in the much more unlikely event of two pumps failing at the same time. This reduces the cost of the standby plant. For large installations with many pumps of the same size a single standby unit should suffice.

Provision of standby requires additional space in the structure as well as associated piping and electrical plant. Where the supply system allows, for example where there is adequate storage or an alternative source, it can be cheaper to provide 'boxed spares'. These are pumps and motors kept available for replacement within a short time.

19.10 BOOSTING

Booster pumping augments the pressure or quantity of water delivered through an existing system. The use and control of boosting are discussed in Section 15.2. The three most important boosting arrangements are:

- addition of a fixed extra flow to an existing supply;
- addition of a fixed extra pressure to an existing supply;
- maintenance of a given pressure, irrespective of the flow.

Addition of Fixed Extra Flow or Pressure

To increase the flow rate, two similar pumps can be connected in parallel; to increase the pressure two similar pumps can be connected in series. For lesser increases the pump added in parallel or series is of a smaller rating. In both cases the characteristics of the system into which the pumps are to deliver must be considered. Figure 19.5 shows the characteristic flow–pressure curve for a pump A. Combined characteristic curve (A + A) is for series working of two such pumps and combined characteristic curve (A‖A) is for parallel operation. The head–flow relationship for the system into which the pumps are delivering is shown as system curve S. The points of intersection of

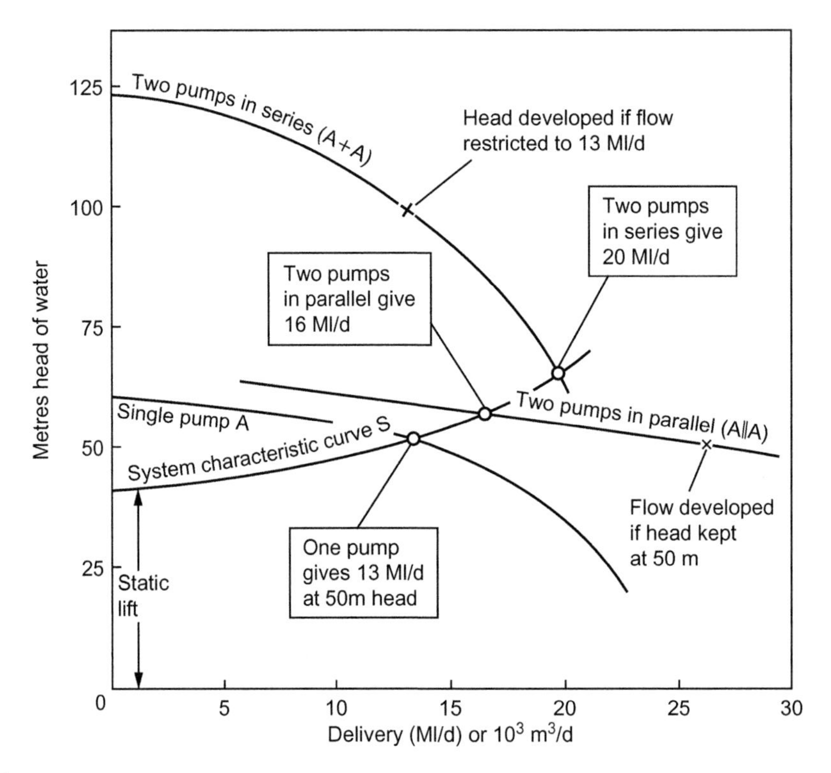

FIGURE 19.5

Output of two pumps connected in series or parallel.

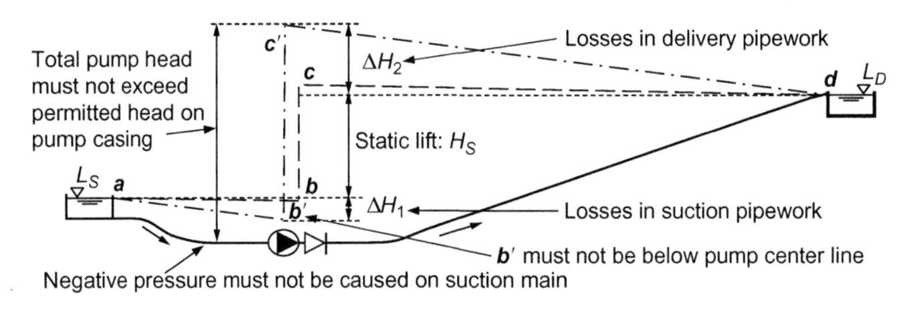

FIGURE 19.6

Effect of boosting in a rising main to a reservoir.

the curves (A + A) and (A$\|$A) with the system curve S indicate the resulting output of the two pumps A operating in series and parallel respectively.

A hydraulic gradient must be drawn for the system in which the pumps are to work to confirm that the siting of the added pump is correct. Referring to Figure 19.6, if pump A initially draws water from reservoir R_1 and pumps it to reservoir R_2 the hydraulic gradient will be the line ***abcd***.

If the pumped flow is increased the hydraulic gradient must change to some line ***ab′c′d*** where ***b′*** is lower in elevation than ***b***, and ***c′*** is higher in elevation than ***c***. The difference ***cb*** is the pump lift from pump A alone and the difference ***c′b′*** is the new lift developed by pumps (A + A). The following conditions must be satisfied for series operation; similar conditions apply to parallel operation:

1. The level of point ***b′*** must satisfy NPSH requirements (Section 19.4) and any requirements for self priming. When pumping treated water there should be no negative pressure in the suction pipeline to avoid risk of infiltration of groundwater. If necessary the suction pipeline could be duplicated to reduce losses and raise point ***b′***.
2. Point ***c′*** must not imply a pressure beyond the safe rated working pressure of the pump bodies, valves, fittings and pipeline to reservoir R_2. For series operation the second pump could be inserted sufficiently far along the delivery pipeline to avoid component overpressurization. This would avoid the need to replace the delivery pipeline.
3. Pumps operating in series will experience the same flow; those in parallel will experience the same head. The operating point of each pump must be checked with respect to efficiency, power and NPSHR. If not satisfactory, replacement of the impeller of pump A may be necessary. Alternatively pump A could be replaced by one capable of the whole duty.

Maintenance of a Given Pressure

The operation of this kind of booster is discussed in Section 15.2. The booster pump is usually arranged to start automatically when the pressure downstream of the pump reaches a certain low value. Safeguards should be included in the pump controls to prevent hunting and ensure correct interpretation of events such as a burst main near the pumping station. The duties required from a booster pump of this kind are usually too wide for a fixed speed machine and a variable speed drive (VSD) may be needed. Alternatively a range of pumps could be provided so that these can be brought into use one by one. In both cases, pressure surges must be controlled (Section 15.11).

19.11 INCREASING PUMPING STATION OUTPUT

If an increase in pumping station output is needed there are several ways of doing this without the substantial cost of complete replacement of the pumps and making large alterations to the building:

Running some of the standby plant. This method is simplest, since no modifications are needed. However, it may not provide all the required extra flow and the loss of standby may be unacceptable except for very short periods.

Replacing the pumpsets with units of larger capacity. Usually building design allows installing somewhat larger machines with relatively minor modifications, but the extent of equipment replacement needs careful consideration. The pump inlet and outlet pipework and valves may also need replacement and suction conditions may need to be enhanced to ensure adequate NPSH is provided for the increased flow. The new pumps will use more power and may need new motors and starters; the complete electrical system will need careful review. New higher power pumps and motors may weigh more; therefore, the capacity of any station crane and its supports will need checking. Similarly, structural support for the new machines when installed will need to be checked.

Replacing only the pump impellers. Replacing only the pump impellers to increase the pump output should be possible if the maximum size impeller for the casing has not already been installed (Section 19.8). Impeller replacement is cheaper than pump replacement but power and NPSH requirements still need consideration. The manufacturer must be consulted to ensure that larger impellers can be fitted and that other pump components are capable of the new duty.

Replacing the motors to drive the existing pumps at a higher speed. Very occasionally this may be possible but careful checks in collaboration with the manufacturer are essential. As well as considerations of power, weight and NPSH, other components of the system should be checked, including the driving arrangements, which may need replacement for higher speed operation, particularly if long vertical shafts are involved.

19.12 STATION ARRANGEMENT AND PLANT LAYOUT

Wet Wells

A wet well, a chamber where flow accumulates for onward pumping, is necessary:

- when inflows are intermittent or small and require only intermittent pumping, for water treatment waste and drainage flows for example, or
- when it is required for control, for example between treatment processes; however, breaking the pressure should be avoided where there is excess head which could otherwise be utilized.

Options for pumping from wet wells are to use:

- immersible pumps such as vertical turbine or submersible pumps in the wet well, or
- centrifugal pumps in an adjacent dry well (Fig. 19.7).

The second alternative often requires a large structure and is more expensive, particularly for deep wells with several pumps. However, centrifugal pumps are more efficient than immersible pumps and whole-life costs may be less for a dry well installation if it is shallow. If the fluid is unscreened raw water it may be necessary to use a mixed flow pump such as a vertical turbine unit. Such considerations will dictate the need for a wet well. If a wet well is adopted, design for suitable hydraulic suction conditions is essential (Section 19.13).

Station Arrangement

Once decisions have been made on the use of a wet well and on the type, capacity and number of pumps, the layout of the station can be developed. Station arrangement is influenced by such things as land availability, station duty, required standards and the preferences of the engineers

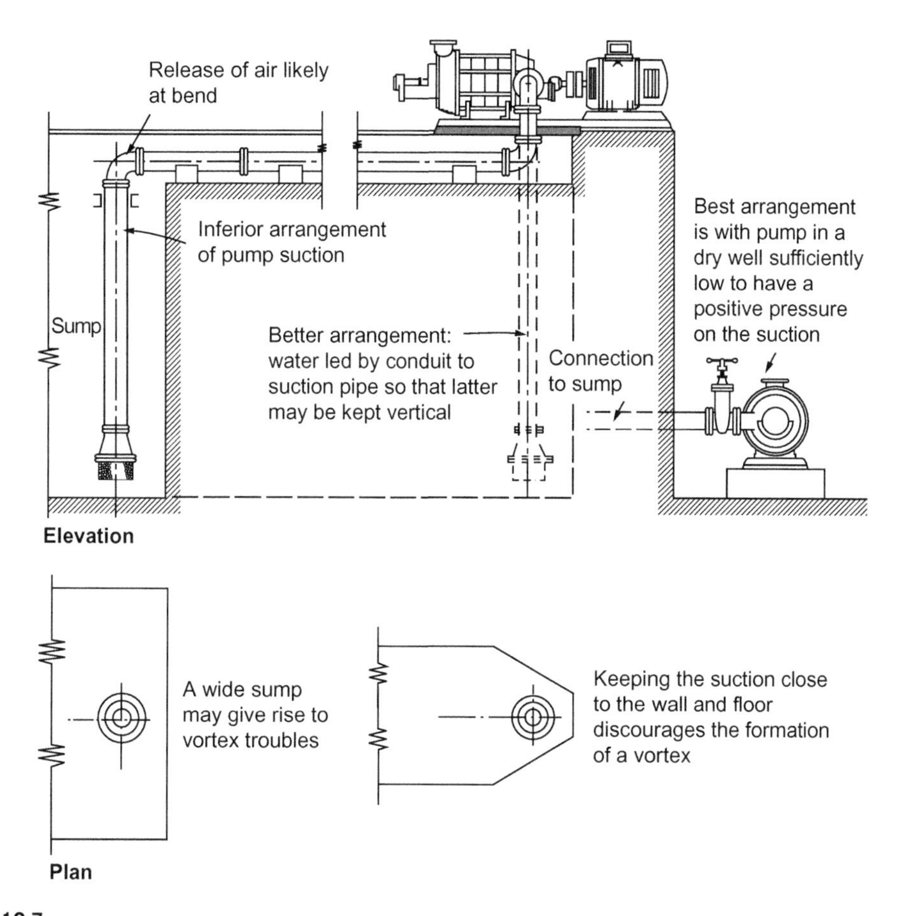

FIGURE 19.7

Pump suction arrangements.

responsible. Cost is always a consideration but compromising by saving initial cost at the expense of future benefits from reduced running costs or inaccessibility for maintenance should be avoided. Other considerations include:

1. Adequate space needs to be allowed for safe working and access.
2. Facilities for lifting and loading must be provided. A travelling crane is essential for all but the smallest pumping units; power operation can usually be justified. The crane should be arranged to serve all the heavy plant and to cover the loading area, which should be big enough for access by a vehicle suitable for carrying the largest item. For borehole installations a monorail hoist or a mobile crane should be considered. For large pumps there should be space for stripping down a unit for repair; the loading bay may be used.

3. High-voltage (HV) switchgear and transformers are usually needed for all but the smallest pumping stations. HV switchgear must be sited in a locked room separate from the main rooms, with access restricted to authorized personnel. Transformers are usually mounted outside the building although for resin insulated units in cold climates, use can be made of the heat they generate to provide some background heating to the station if they are inside. Main switchgear, both high and low voltage, and control panels, must have adequate space both in front and behind them. About 1 m should be regarded as the minimum at the back, and 2 m in the front. All switchgear should as far as possible be in line so that the main busbars are kept short; cable routes from switchgear to motors should also be kept as short as practicable.
4. All cabling and pipework should be in conduits of ample size and runs must be carefully laid out allowing for minimum bending radii for cables (Section 19.32). Pipework T-junctions or 90° bends in the line of flow should be avoided. Radiused junctions and large radius bends should be used to reduce station hydraulic losses. All moving machinery must be securely bolted to adequate foundations; a vibration analysis may be needed to ensure that the frequency of machinery vibration is not close to the natural frequency, or its harmonics, of the supporting structure.
5. If air vessels for surge protection are needed, these can be placed outside the pumping station if the climate permits, or inside in an unheated part of the station. Outdoor vessels may affect the appearance of the station and in severe climates need frost protection. This can be done with insulation (which may only delay the onset of freezing) or by arranging a trickle flow of warmer water from the supply main to keep the temperature above freezing. In extreme cases trace heating and insulation may be needed. If the vessels are inside, condensation may occur on their surfaces and cause staining of floors and rusting of the vessels.
6. Facilities for chlorination may be needed, but dosing points and chlorination equipment must be in a separate room or building with restricted access (Section 12.13).
7. Some facilities are usually provided for the station attendants even if the site is not permanently manned; a small workshop may be justifiable as well as a mess room containing a sink and means for cooking light meals. Sanitation of the highest standard is needed, together with hot water and washing facilities.

19.13 PUMP SUCTION DESIGN

Pumps must be provided with sufficient suction pressure so that they self-prime and so that NPSHA is well above the NPSHR (Section 19.4) over the expected operating range. 'Self-priming' pumps can be specified if necessary or the pump should be lowered (Fig. 19.7). The suction arrangement should minimize pressure losses, prevent air being released and avoid vortexing, which can be reduced by fitting flow-straightening vanes at the inlet. *The Hydraulic Design of Pump Sumps and Intakes* (Prossor, 1977) provides guidance on wet well pump arrangements but it may be necessary to check flow patterns by modelling (Section 14.18).

The reader may also like to refer to *American National Standard for Rotodynamic Pumps for Pump Intake Design* (HI 9.8, 2012).

The diameter of pump suction pipes is usually larger than the delivery pipe diameter to reduce headlosses. Suction pipes for raw water may need to be equipped with a strainer which should have an effective area of opening at least double that of the suction pipe. Strainers should be kept in good condition. Foot valves, that is non-return valves, fitted at the inlet to vertical suction pipes, are a potential source of trouble and should be avoided. If installed they must be of high quality and be well maintained or they may tend to stick open. They are sometimes installed for keeping a pump primed when it is idle but often they are not effective if the pump is stopped over a long period; other priming methods are better. Foot valves are also sometimes used to prevent reverse rotation on stoppage of a vertical spindle pump; a high speed of reverse rotation can be caused by the falling column of water. Restarting a pump when it is rotating in reverse may cause shaft breakage which must be prevented. Instead of a foot valve, a pump may be specified as suitable for being 'turbined' under reverse flow; a time delay switch is then incorporated in the switchgear to prevent restarting before the reverse rotation has stopped.

19.14 THERMODYNAMIC PUMP PERFORMANCE MONITORING

As a pump gets older, its performance inevitably declines, until its upgrading by servicing or replacement is needed to maintain high efficiency and minimize operating costs. Regular monitoring of individual pump performance is needed to do this and a record must be kept of pump output compared with energy consumed. The record needs to be regularly updated, perhaps annually, but most water supply pumping stations are not equipped to monitor individual pumpsets although the total station power consumption and flow delivered are usually logged. The difficulty lies in measuring the individual pump flow.

Thermodynamic pump performance monitoring provides one possible solution, which has been developed and improved so that good accuracy of measurement is now possible. Most of the pump energy losses, which lead to reduced efficiency, result in heating the pumped water by a small amount. Thermodynamic monitoring relies on accurate sensors to detect the resulting small temperature rise across a pump, which is proportional to the wasted energy. The great advantage of this method is that there is no need for an individual flow meter to measure the pumped flow directly to determine the pump efficiency.

These devices are not yet suitable for every installation. They work best with higher head machines because the resulting temperature rise is then higher for the same pump efficiency and power. They are not always very effective when used with borehole pumps because of difficulty in measuring the inlet water temperature. Development work is continuing however, and the performance of these devices continues to improve.

In a typical installation, temperature probes and pressure transducers are mounted on the pump pipework (Yates, 1989) to detect the inlet and outlet conditions (Fig. 19.8). A microprocessor, usually mounted in portable equipment, is used to analyse the data and to display efficiency, flow, head and power consumption. Thermodynamic pump testing is accepted in *Pump Test Standard* BS EN ISO 5198 as a precision-class test.

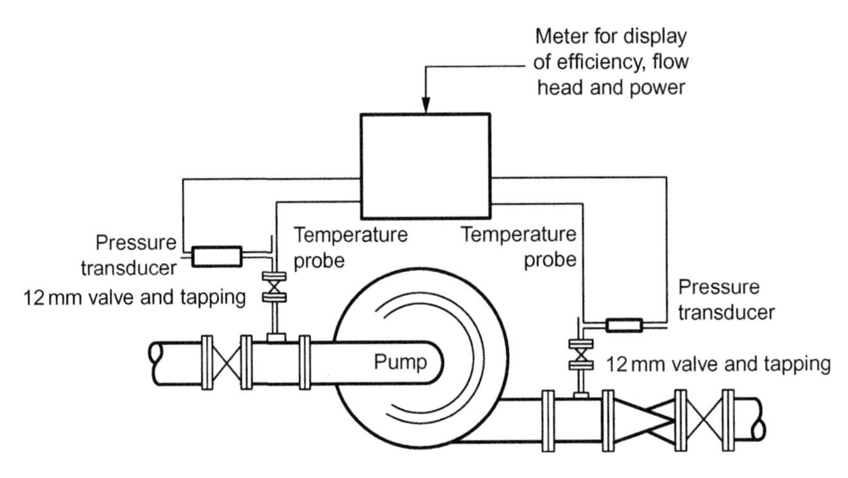

FIGURE 19.8

Pump performance monitoring.

19.15 CAVITATION DAMAGE

Cavitation occurs when the absolute pressure in the pumped water falls below its vapour pressure (Section 15.12). Pockets of water vapour are then released in the impeller. On subsequently entering a higher pressure zone, against the blade of the impeller or elsewhere, they collapse. A stream of vapour pockets continuously collapsing on the same area quickly erodes the blade material. Cavitation damage can be avoided by good design of the pump and suction pipework and by maintaining sufficient NPSH (Section 19.4) and, if possible, a positive pressure on the suction side, by siting the pump below or as near as possible to the suction-side water level.

19.16 CORROSION PROTECTION

Corrosion protection can be provided for most pumping plant by the correct choice of pump materials to suit the water. Care should be taken to avoid use of metals with markedly different electropotentials. For sea water applications, for desalination plants for example, special stainless steels with Pitting Resistance Equivalent Number (PREN) (see Section 10.38) greater than 40 provide adequate resistance to crevice and erosion corrosion and to pitting.

Ceramic coatings have been developed for pump internal parts. One commonly applied material is a glass-filled resin-based coating which has been successfully used for application to the internal surfaces of pipework and valves, as well as pump bowls and impellers, to restore performance following corrosion or erosion. They are also sometimes applied to new pumps to prevent damage and improve efficiency by providing smoother surfaces. Metal surfaces require careful surface preparation to obtain the full benefits of the coating material. Since the coatings have significant thickness, care is needed to ensure that pump passages, particularly at the pump inlet, are not reduced in area so much that performance is impaired. Clearly this is more significant with smaller pumps and, for this reason, combined with the cost, coatings are not justified on pump sizes of less than about 300 mm branch diameter.

19.17 TRANSIENT PRESSURES: WATER HAMMER AND SURGE

The key causes, features and consequences of surge are discussed in Section 15.11. The most critical case is usually a power failure causing simultaneous stopping of all the pumps in a station. Most modern low inertia pumps stop producing forward flow of water in a few seconds when the power fails. Measures to control the magnitude of pressure transients are set out in Section 15.11 and are described in Thorley (2004). Those that would need to be accommodated at the pumping station are listed below:

1. Reduce delivery valve closure speed.
2. Increase pump inertia by fitting fly-wheels to the pumps.
3. Have surge vessels, accumulators or surge shafts connected to the delivery pipeline just downstream of the pumping station.
4. Incorporate bypass pipework around the pumps.

With the emphasis of pump manufacturers on producing lighter pumps in order to reduce motor starting torques and currents, fly-wheels and added inertia are out of fashion but, in the right circumstances, they are the most reliable and effective form of protection. More commonly on pumping systems a surge vessel or accumulator is used.

Pump delivery non-return valves need particular attention to ensure they are suitable for the system and its transient response, especially if a surge vessel is also provided, because the flow in the connecting pipe to the air vessel may reverse very quickly. Ideally the non-return valve should shut at the moment of flow reversal but if it reacts more slowly the reversed flow may slam the valve shut with the generation of a high shock pressure. The dynamic response of the non-return valves should thus be matched to the transient characteristics of the pipeline system.

19.18 EFFICIENCIES AND FUEL CONSUMPTIONS

Quoting efficiencies and fuel consumptions is difficult because they vary so widely; however, the data in Tables 19.2 and 19.3 are intended to give a guide to the values normally to be expected. There are wide variations according to the power rating and type of pump and motor.

Table 19.2 Efficiencies

Pumps:	
Horizontal centrifugal	Medium size 80–82%, perhaps 85% large size. Even higher with special construction but at higher price.
Vertical spindle shaft driven	Tending towards about 3% less than the horizontal centrifugal.
Submersible	75–81% and can be lower, to about 70% for small sizes. Generally about 3% less again than the vertical spindle pump; the reason being that the pump is restricted in diameter.
Electric motors:	
For horizontal pumps	93–95%, for fixed speed AC induction.
For vertical pumps	90–94%, for fixed speed AC induction.
For submersibles	85–89%. Less than the above because of the restrictions imposed on the design.
Variable speed	About 3–5% less than with a caged AC motor.

Table 19.3 Overall fuel consumptions

Electrically driven pumps	About 1.0 kW for every 0.75 kW of water power output; this implies an overall efficiency of about 75% which would be usual. Up to 1.3 kW per 0.75 kW water power output or higher for small pumps or variable speed pumps.
Diesel engines	0.21 kg of diesel fuel oil consumed per kWh of engine power exerted would be considered good, 0.28 kg per kWh being not unusually high. For lubricating oil add 5% to fuel oil cost.

19.19 PUMP DRIVES

Pumps may be driven by almost any prime mover of suitable power and speed. Most water supply pumps are nowadays driven by electric motors but there are still some driven by other means, usually a diesel engine. Where large powers are needed, there are some water supply pumps powered by steam turbine. However, the widespread availability of reliable and economically priced electricity supplies and the comparative cheapness, wide choice of designs and sizes and ease of operation and control of electric motors make them clear favourites in most circumstances.

PART II ELECTRICAL PLANT

19.20 ELECTRIC MOTORS FOR PUMP DRIVES

The three-phase alternating current (AC) motor is the most common type of electric motor used for driving pumps. BS EN 60034-1 covers rating and performance of electrical machines. Motor efficiencies for the general purpose motor are defined in BS EN 60034-30:2010 which provides for the global harmonization of energy efficiency classes for electric motors.

The older standard (60034-30:2009) defines three classes IE1, IE2 and IE3 for use on fixed and variable speed drives and covers single speed, three phase, 50 and 60 Hz cage induction motors up to 1000 V 3ph between 0.75 and 375 kW.

The latest standard (60034-30-1:2014) defines four classes IE1, IE2, IE3 and IE4 for use on fixed speed drives and covers single speed, three phase, 50 and 60 Hz motors up to 1000 V 3ph between 0.12 and 1000 kW. It now covers all technical constructions rather than just cage induction motors but does not include VSD driven systems. When 60034-30-2 is published, 60034-30:2009 will be totally withdrawn since 60034-30-2 will cover motors designed for variable voltage and frequency supply.

EU Regulation No 640/2009 further restricts new motor installations in the EU to classes IE2 and IE3 or greater. All standard motors between 7.5 and 375 kW shall at least meet the IE3 efficiency level or meet the IE2 efficiency level and be equipped with a VSD.

Alternating current motors can be classified as follows:

- Induction motors, either (1) 'cage' or 'caged' (formerly known as 'squirrel cage' motors) or (2) 'wound rotor' motors (sometimes called 'slip ring' motors) – see below.
- Synchronous motors.
- Commutator motors.

There are new styles and types of motor being developed to further improve the motor efficiency. These new styles are becoming more apparent as BS EN 60034-30 is updated to include IE4 (Super-Premium Efficiency). These new types of motor include:

- Switched Reluctance.
- Enhanced squirrel cage motors (approximately 50% more copper than existing motors), very useful for retrofits but can be more expensive than other IE4 motor types.
- Permanent Magnet.
- Hybrid squirrel cage/permanent magnet machines.

Induction motors of the caged type are the most widely used because of their simplicity, robustness, reliability and relatively low cost. They are inherently fixed speed machines when connected to conventional fixed frequency supplies. This is a handicap for centrifugal pumps, as the pumps themselves are capable of a wide range of duties without modification, if the speed can be varied. Since many water supply systems require pumping installations of output varying at different times, fixed speed pumps are often a disadvantage. VSDs have been developed to a point where their historical disadvantages of reduced overall efficiency, limited speed range, limited power capability and high capital cost have largely been overcome. The latest generations of drive have high overall efficiency levels and flexibility although the cooling requirements can still be an issue, especially for large drives and in hot countries. According to the findings of the IEA 7 July 2006 Motor Workshop (BS EN 60034-30), electric motors with improved efficiency in combination with frequency converters can save about 7% of total worldwide electrical energy consumption.

There is a tendency for motors to be oversized: the hydraulics engineer identifies a worst case hydraulic design; the pump manufacturer designs the pump with a specific minimum rotor power, often slightly greater than the value required to ensure that the pump will run successfully; he then selects the next size of standard motor for their system which is, as a result, usually larger than needed; and the motor manufacturer practice is for the motor supplied to exceed the quoted 100% motor power to ensure satisfactory operation. The result can mean that the motor will run at 70–80% of its power rating with a reduction (albeit small) in efficiency. One means of reducing these inefficiencies is to reduce the motor voltage to reduce the losses in the motor – and this is achieved by the current generation of VSDs.

The VSD and cage induction motor have now largely replaced the synchronous and wound rotor motors within the water industry due to lower capital cost, greater efficiency and lower maintenance requirements.

The *synchronous motor* costs more because of the need for more complex control equipment. A direct current (DC) supply is also required (supplied by an 'exciter' driven by the motor itself) for the rotor. This results in the rotor turning at the same speed as the rotating magnetic field created by the stator. The synchronous motor (which has to be started up by short-circuiting the rotor windings in such a way that it acts as an induction motor) is normally used for applications requiring constant speed operation under varying load conditions, or where its ability to provide power factor correction or improve the voltage regulation of the supply system is justified by the extra cost. If used, it is usually only for pumping large steady outputs.

The *switched reluctance motor* is an old concept, a form of stepper motor designed for continuous running. The rotor is constructed from a series of poles of soft magnetic material. As for any magnetically conductive material, the rotor pole will try and orient along the magnetic field, keeping the magnetic field as short as possible. By keeping the magnetic field rotating just ahead of the pole of the rotor, the rotor is pulled round. The reason for it not being commercially developed was the need for a variable frequency source and a means of controlling it. The availability of high power inverters has overcome the first issue and the powerful control algorithms now available means this option is now available at a similar price to a VSD with induction motor.

The *permanent magnet motor* is a type of synchronous machine which has dispensed with the rotor winding, DC supply, etc. This is replaced by a rare earth permanent magnet which provides the required magnetic field. These are ideal for small motors but are very costly for large motors and do not cope with overheating. Supplies of the required rare earths are also restricted making them more expensive.

Hybrid machines are a mixture of the above types, trying to develop the increased efficiency through the inclusion of a permanent magnet but keeping the overall cost down through the inclusion of the squirrel cage winding.

Because by far the most commonly used motors for driving pumps are induction motors of the caged or wound rotor type; only these are described in more detail below.

19.21 THE INDUCTION MOTOR

The induction motor comprises a stator which incorporates a distributed winding, connected to the three-phase electrical supply, and a laminated steel rotor which in its simplest form has embedded large section bars which are short-circuited at each end. The motor thus consists of two electrically separate windings which are linked by a magnetic field forming a transformer, with an air gap magnetic circuit. The three-phase current in the stator winding produces a smoothly rotating magnetic field whose rotational speed N (synchronous speed in rpm) is given by the equation:

$$N = \textit{supply frequency}\ (\text{Hz}) \times 120/\textit{number of poles}$$

Thus for a supply frequency of 50 Hz a two-pole motor has a synchronous speed of 3000 rpm, a four-pole motor 1500 rpm, and a six-pole motor 1000 rpm and so on.

The rotating magnetic field cuts the rotor bars and induces a current in them. The rotor current produces a magnetic field which interacts with the stator field, producing an accelerating torque. The motor will run-up to a speed at which the developed torque is equal to the torque required by the driven load, including that required to overcome friction and windage losses.

In practice, the rotor cannot reach synchronous speed with the magnetic field when loaded, because under such a condition no current would be induced in the rotor conductors and hence no magnetic flux and torque would be produced. This explains why, for example, a loaded four-pole motor may actually run at say 1440 rpm. The difference between the actual speed of rotation and synchronous speed with the magnetic field is termed the slip. The slip increases with the load and is normally expressed as a percentage of the synchronous speed. The slip at full load typically varies from about 6% for small motors to 2% for large motors. The starting torque and speed characteristics of the cage induction motor of basic standard design are shown in Figure 19.9.

The *cage induction motor* has a rotor core which is made up of laminations and conductors of aluminium, copper or copper-alloy non-insulated bars in semi-enclosed slots, the bars being short-circuited at each end by rings. For the smaller motors, the rotor bars and end rings are often cast. The advantages are simple construction, low cost and low maintenance. The limitations of the cage induction motor of basic standard design are low breakaway torque and high starting current, the former typically ranging from 0.5 to 2.0 times and the latter three to seven times rated values, depending on motor rating and number of poles. These limitations can be improved by the use of motors with multi-cage rotors. The multi-cage rotor, in its simplest form, comprises a low reactance and high resistance outer or starting cage, whose influence predominates during the starting period and results in increased torque and reduced current, and an inner cage of lower resistance which is dominant under running conditions.

The *wound rotor induction motor* design incorporates a three phase, star configuration rotor winding connected to slip rings, to which external resistance is connected for starting. This type of motor is used when high starting torque, reduced starting current and controlled acceleration characteristics are required. The magnitude of starting torque and current, and acceleration period are determined by the value of the external resistance. Under fixed speed running conditions the slip rings are shorted-out whilst for variable speed running conditions the external resistance is maintained but the system efficiency soon decays. Major disadvantages of the wound rotor motor are higher capital cost and regular slip ring and brush gear maintenance and lower overall efficiency compared to a squirrel cage motor and VSD solution. The only major benefit of the wound rotor solution is a very low starting current (something which can also be achieved with the latest VSD drives).

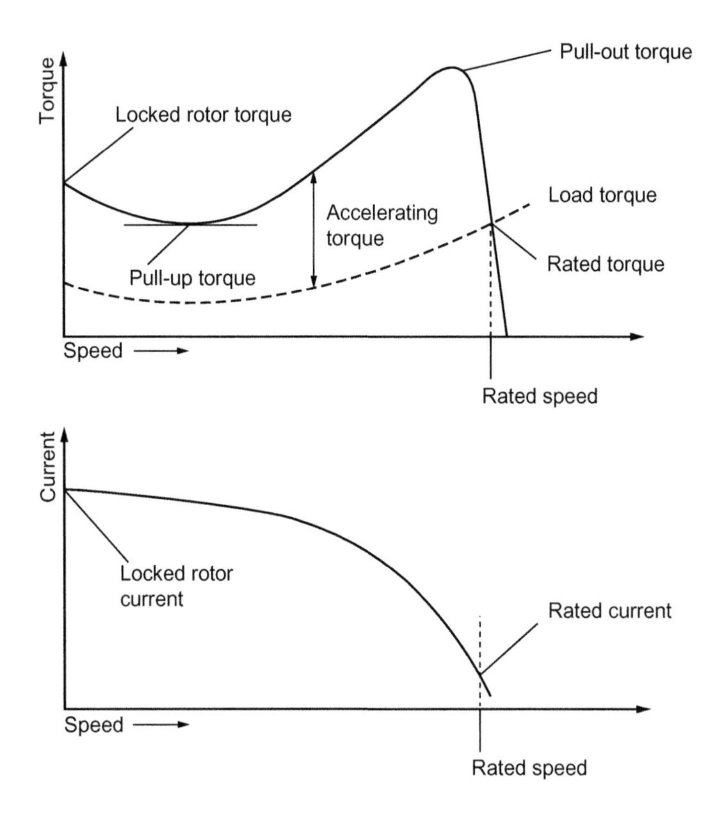

FIGURE 19.9

Torque and speed characteristics of the standard cage induction motor.

Rated Output, Starting Torque and Start Frequency

The *rated output* of an induction motor is defined for the designated duty when the ambient temperature of the coolant air does not exceed 40°C or, for a water-cooled motor, when the temperature of the water entering the heat exchanger does not exceed 25°C, both at a height above sea level not exceeding 1000 m. The supply voltage is allowed to deviate between 95% and 105% of the rated voltage of the motor without affecting the rated output. VSD operation of the motor actually varies the motor voltage to improve the efficiency of the drive system.

When the motor is operated in conditions different from the reference values, motor rated output will be affected. This must be considered at the time of motor selection. The normal requirement is for the motor to operate continuously at voltages not differing from the rated value by more than 5%. When the motor is required to operate under varying and cyclic load conditions, which may include periods of either no-load or standstill, reference must be made to the manufacturer giving details of the load inertia and the load/time duty sequence.

Starting torque. The standard cage induction motor is designed to produce a starting torque over the range between standstill and that at which pull-out (minimum) torque occurs, which is not less than 1.3 times a torque characteristic which varies as the product of the square of the unit speed and the rated torque. This is representative of the run-up characteristic of a typical centrifugal pump. The factor of 1.3 is chosen to allow for a voltage of 90% rated value at the motor terminals during the starting period. The load and motor torque characteristic during the starting period must be considered at the time of motor selection. This is especially important where the motor is required to start with a voltage drop greater than 10%, typically for direct-on-line starting control.

Frequency of starting. The standard cage induction motor is designed to allow two starts in succession (running to rest between starts) from cold or one start from hot after running at rated conditions. Further starting is permissible only if the motor temperature does not exceed the steady state temperature at rated load. Because the number of starts directly affects motor service life, they should be kept to a minimum. More onerous starting requirements must be considered at the time of motor selection. For the wound rotor induction motor, the starting frequency is generally dictated by the short-time rating of the starting resistor. When using the standard cage induction motor with an electronic VSD these limitations are reduced since the starting conditions are less onerous. The mechanical stress on the motor and the windings is also reduced.

19.22 INDUCTION MOTOR STARTING METHODS

The starting current drawn by an AC induction motor depends on its type, rating, voltage and starting method. When starting a motor the resulting current causes a voltage dip which has to be kept within defined limits, stipulated by the electricity supply company, to avoid affecting other consumers and the mal-operation of connected equipment. A detailed electrical system analysis is required in special cases to assess whether unwanted effects are likely. However, experience shows that general guidelines can be applied based on permitted transient voltage dip at the point of common coupling (i.e. the busbar from which other consumers are supplied) and the frequency of starting. Typical permitted voltage dip limits at the point of common coupling are 8% for motors started infrequently, say at intervals longer than 2 hours, and 4% for motors started frequently. Where large motors are to be installed it is important to establish the transient voltage dip and starting frequency criteria at the design stage to decide the method of starting. In addition, the ability of the motor to start and accelerate if the voltage drop at the motor terminals is greater than 10% and the effect of motor starting on other parts of the consumer's installation must be assessed. These aspects can be significant if the electricity supply is taken from a low fault level rural distribution system or from small capacity transformers.

The starting method depends on motor starting torque and current requirements and, in some applications, the need to control acceleration. Starting methods for the cage induction motor can be classified as:

- full voltage, or direct-on-line;
- reduced voltage (star-delta, autotransformer and electronic soft start);
- electronic VSD;
- rotor resistance starting (for wound rotor motors only).

Direct-on-line starting is the cheapest, simplest and most reliable method, and is therefore the most widely used. However, direct-on-line starting causes a high starting current and may not be suitable if the electricity supply company requires a reduced starting current (or limited voltage dip) if shock-free controlled starting is required. With direct-on-line starting, the applied voltage, starting torque and current are 100% rated values and rapid run-up at maximum available torque is achieved. An advantage is that direct-on-line starting simplifies motor construction, as only one end of each phase winding need be brought out to terminals. For low voltage (LV) motors, the size that can be started by the direct-on-line method is often arbitrarily limited by the station designer to the smaller ratings, typically up to 15 kW. However, unless the maximum size of motor is stipulated by the electricity supply company, or is limited by supply transformer transient loading considerations, there is no restriction on the size of motor which can be started direct-on-line, provided transient voltage dip and the mechanical impact loading criteria are met. These design limitations apply equally to HV motors.

Star-delta starting is the most usual method used when reduced voltage starting is needed and involves connecting the stator winding of the motor initially in 'star' until an optimum speed is achieved, when it is switched to 'delta'. For a three-phase AC supply in star connection the stator windings are connected phase to neutral and in delta connection the windings are connected phase to phase. When connected in star, the voltage across each phase winding is 58% ($1/\sqrt{3}$) of the supply voltage and the starting current and torque are reduced to 33.3% of the full voltage values.

This starting method is relatively simple and inexpensive but its use is limited to low inertia drives because of the reduced starting torque available from the motor. The disadvantage of this starter type is the possible high transient torque and current which could occur when switching from star to delta. The number of starts per hour is not normally restricted by the starter, although consideration has to be given to the characteristics of the main circuit, short circuit and thermal overload protection. To eliminate or reduce the high transient torque and current when switching from star to delta it is appropriate to use the closed transient star-delta (Wauchope) type starter in which resistance is inserted when changing over from star to delta to provide a no-break transition. This closed transition type starter provides three steps of acceleration against the two steps provided by the standard star-delta starter. Because of the additional resistors and control equipment, the closed transition starter is more expensive than the standard starter. The number of starts per hour is limited by the short-time thermal rating of the transition resistors.

Autotransformer starting provides more flexibility than star-delta, because the applied voltage and hence the starting torque and current can be varied by changing the transformer voltage tappings. The motor starting torque and current are a function of the square of the transformer tapping with respect to rated voltage. For example, at 50% voltage tapping, the torque and current are 25% of the full voltage values, and at 80% voltage tapping, the torque and current are 64% of the full voltage values. The ability to adjust the voltage of the autotransformer provides a convenient means for closely matching the starting torque to the driven load and for reducing the starting current. For this reason, the autotransformer method is suitable for starting larger motors than the star-delta method but starting is hard to achieve on loads with high inertia at zero speed.

The disadvantages of the autotransformer starter are cost (it is considerably more complex than the star-delta starter) and the possibility of high transient torque and current when switching from reduced to full supply voltage. To eliminate or reduce the high transient torque and current when switching from reduced to full voltage, a closed transition configuration can be used. Its operating principle is similar to that of the closed transition star-delta starter. It involves the use of a more costly transformer and additional control equipment. The number of starts per hour is limited by the short-time thermal rating of the autotransformer.

Electronic soft starting. The disadvantages of the above electromechanical starting methods can be partly mitigated by electronic soft starting. With this method, the motor supply voltage is gradually increased linearly up to rated value, providing smooth acceleration, controlled and reduced starting current (typically between 200% and 300% of full load value) and controlled motor torque. At full speed, the electronic controls would normally be bypassed and the motor connected directly to the mains supply. Electronic soft starters are more complex and expensive than the electromechanical methods discussed above but are being continuously improved, both in ability to provide reduced starting current and for reliability and durability. These are usually short-time rated only and cannot be used to start loads with a long ramp-up time. In these cases a full electronic VSD is used but it can then the bypassed for fixed speed control.

Electronic VSDs. With this method the frequency and voltage applied to the motor is ramped up to provide rated torque from zero speed (for some drives) up to rated speed. This is used in instances where full variable speed control is required or where the starting time is too long for the simpler Electronic Soft Start. Where the load requires a simple fixed speed drive this VSD can be bypassed when running to allow a single VSD to start multiple drives. Note that if the ramp-up time is extended, or periods of slow speed running are required, consideration should be given to a separate motor driven cooling fan rather than the integral fan normally used.

Rotor resistance starting of wound rotor induction motors is used where the applied load has large inertia and requires a high starting torque, where the starting current needs to be limited by supply system voltage drop limits or if controlled acceleration is required. With the connection of external resistance into the rotor circuit via the slip rings, high starting torque and low starting current can be obtained. Typically, for full load torque at standstill, the starting current is in the order of 125% of the full load value. By selection of resistance, starting torques of 200–250% can be attained with corresponding currents of 250–300%. The resistance is either the multi-stage metallic non-inductive grid or the liquid type, the latter providing smooth control of acceleration. The number of starts per hour is limited by the short-time thermal rating of the rotor resistance. Rotor resistance speed control can also be utilized with the same motor but it has the disadvantage that a 5% reduction in speed results in a 5% reduction in overall efficiency unless a complex energy recovery system (VSD) is installed on the rotor circuit.

19.23 INDUCTION MOTOR PROTECTION

The function of the protection is to initiate disconnection of the motor from the supply to prevent or limit damage caused by overheating due to abnormal load, failure of winding insulation or some other fault. For LV motors, the protection provided needs to be chosen by considering factors such as motor cost and the characteristics and importance of the drive. The degree of protection provided can range from the thermal overload relay to a motor protection relay providing high-set over current, overload, earth fault, negative phase sequence (unbalance) and stall protection. A comprehensive motor protection relay would not normally be used for motors rated below 50 kW. HV motors, irrespective of rating, are provided with high-set over current, overload, earth fault, negative phase sequence (unbalance) and stall protection. For the larger motors, typically 1000 kW and above, high speed differential protection is often provided to minimize damage to the stator core in the event of a stator winding fault. For motors where a reversal in the direction of rotation could cause damage to the driven load, phase reversal protection should be provided.

Additional motor winding and bearing protection can be provided by thermistors, thermocouples and resistance elements. These methods give protection against faulty conditions which are not reflected in the line current of the motor. For applications requiring temperature indication, in addition to alarm and motor tripping initiation, the thermocouple or resistance element is used. Thermocouples can be located in stator slots, stator end windings, cooling air circuits and bearings. The location of resistance elements is the same as for the thermocouple, except for the stator end winding. The degree of protection provided by the three devices is good and response to temperature change is fast.

19.24 SPEED CONTROL OF INDUCTION MOTORS

The standard induction motor is a constant speed machine, the speed being determined by the number of poles and the power supply frequency (Section 19.21). When speed control is needed, no single solution is of universal application; costs and benefits must be assessed for each case. The economic case for a solution has to consider the capital cost of the drive solution and the operational costs.

Whilst provision of a means of pump speed control such as VSD increases capital cost, this may be more than compensated by reduction in operating costs if it avoids the need for throttling of excess head, a waste of energy, and by avoiding repeated pump starts which produce losses in the starter.

The two most commonly used methods of speed control are based on changing the number of poles or changing the supply frequency, although limited speed variation can be obtained by using the wound rotor type induction motor and varying the rotor resistance. However, the use of variable rotor resistance has now largely been abandoned in the water industry because of its inefficiency and the development of slip energy recovery systems. The most commonly used methods of speed control are summarized below.

Pole changing motor. The simplest form of pole changing motor has a single-tapped winding, also known as a Dalhander winding, which provides two speeds in a ratio of 2:1, for example 3000:1500 rpm for a motor connected to a 50 Hz supply. A variant on the single-tapped winding is a motor with two separate windings, which can provide any combination of two speeds. By combining the tapped and two winding arrangements, up to four speeds can be obtained, for example 3000, 1500, 1000, 500 rpm at a supply frequency of 50 Hz.

Pole amplitude modulated motors. The pole amplitude modulated (PAM) motor is a development of the single-tapped 2:1 speed ratio winding; speed change is achieved by reversing one half of each phase which changes the magnetic flux distribution and produces a resultant field of different polarity. Various combinations of speed ratios can be obtained with this motor design. Other than speed ratio, the advantages of PAM motor design over the pole changing type are better utilization of active materials resulting in smaller physical size for a specific speed ratio and rated power and improved efficiency and power factor.

Variable speed drive (VSD). This is also known as variable frequency drive (VFD) or adjustable frequency drive. The current version of this drive uses a standard cage induction motor with variation of the supply frequency and voltage from the VSD. The frequency is varied to change the motor speed and the voltage is varied to reduce the iron losses within the motor; if a motor is only running at 50% load then the motor does not need line voltage on the terminals. This then reduces the iron losses within the motor. VFD controller itself is of the static design, the basic components of which are a rectifier connected to the AC supply, an inverter to provide a variable frequency/voltage supply to the motor and a DC link between the rectifier and the inverter. The motor speed can be regulated typically over the range 10–200% motor rated speed, although mechanical design considerations make speeds of greater than 100% unusual. The full load efficiency of a VSD is typically about 97% but falls off (to about 95% at 25% load) at lower loads (DoE, 2012). The best VSD efficiency available is about 98%. The standard rectifier produces significant harmonic currents, which reflect into the electricity supply system and can cause interference with other consumers or be detrimental to connected plant such as capacitors, generators and motors. It also reduces the effective supply cable capacity since the supply cables have to carry significant harmonic currents which result in increased cable heating and increased volt drop in the supply but which provide no useful power. This means significantly larger supply cables have to be installed for the same load.

Electricity supply companies lay down guidelines for the permitted magnitude of individual harmonic currents and harmonic voltage distortion. Methods available for reducing harmonic currents are to:

1. increase the rectifier from 6-pulse to 12- or 24-pulse, depending on the power rating of the drive;
2. install active or passive harmonic filters (Section 19.25);
3. use phase shifting supply transformers; or
4. use a low harmonic drive which has an active front end which draws a near sinusoidal waveform from the supply rather than the square wave of the normal 6-pulse drive.

Increasing rectifier pulse number or installing harmonic filters considerably increases the cost of the drive. When selecting an induction motor for use with a VFD controller, the following factors need to be considered:

- increased losses and hence heating in the magnetic circuit caused by the harmonics in the inverter output waveform;
- power and torque requirements throughout the speed operating range;
- possibly impaired cooling at low speed operation (an important consideration for constant torque loads) since motor cooling fans are normally driven by the motor shaft and so produce less cooling at lower motor speeds; separately powered fans can be used to avoid this difficulty;
- increased voltage withstand levels for the winding insulation (this should not be confused with the winding temperature class); whilst a simple fixed speed cage induction motor may only see 415 V between phases, modern VSD drives can provide short-term spikes of 1200–1400 V; the winding insulation of low-cost cage induction motors can withstand only 800 V but use of a VSD requires a motor voltage withstand level of 1500 V phase to phase.

The voltage withstand level of a typical borehole pump is only 500 V peak to peak. One solution is to put a sine filter on the inverter output; this only allows the 50 Hz sinusoidal waveform through but it doubles the size and cost of the unit and also reduces the system efficiency (DS/CLC/TS 60034-17). Borehole pumps with higher voltage withstand levels and now becoming available.

Slip energy recovery. The slip energy recovery VSD, also known as a Kramer drive, uses a wound rotor type induction motor and operates by recovering rotor energy and feeding it back into the supply so it has a higher efficiency than a standard wound rotor type motor. The mechanism for doing this is to convert the slip ring frequency power to DC and, for the static type drive, return the power to the mains supply via an inverter. An alternative to the inverter would be a DC motor driven asynchronous generator. A direct connection between the rotor and the mains is not possible because both the rotor voltage and frequency vary with motor speed. Harmonic currents are produced by the inverter and the considerations above for the VSD harmonics apply. This type of drive has the increased cost and maintenance of the slip rings/wound rotor and the losses associated with a VSD from the slip recovery inverter as well as the costs of the wound rotor motor itself. It has largely been replaced by VSD drives and standard cage induction motors.

19.25 HARMONIC FILTERS

There are two types of harmonic filter:

1. Passive, where a series of inductors and capacitors try to filter the high frequency components, preventing them reaching the mains. This has the disadvantage that it is also affected by the external harmonics on the mains and can overheat. It is impossible to prevent the external harmonics from being absorbed by the filter and it is very hard to size such filters because the effect of the external harmonics cannot easily be identified and designed for.
2. Active, where a semiconductor based active front end (similar to a VSD active front end) detects and measures the harmonic frequencies generated by the drives and injects equal amounts of harmonic current in anti-phase to the generated harmonics. These cancel out resulting in an acceptable level of harmonics. One of the greatest benefits of these units is that they are controlled and current limited and will not exceed their thermal limits. This means that if the harmonics should increase the unit will not overheat, although it might not remove all of the primary harmonics. Since the unit is also microprocessor controlled, alarms can be automatically generated if this should occur. Some active units provide a broad adjustment over four or five harmonic frequencies whilst others allow individual frequencies to be targeted.

19.26 EFFECT OF ELECTRICITY TARIFFS

The charges made by electricity supply companies are designed to favour the consumer who takes a steady supply of electricity at high power factor. The consumer who wants large power intermittently is penalized. Tariffs often comprise two, three or even four separate charges as listed below.

If the supply is required for a new installation in a remote area, the supply company will usually also ask for a connection charge to recover the cost of providing the supply.

1. A 'maximum demand charge' per kW or kVA of maximum demand in that month above a given figure (and sometimes, more onerously, per kW or kVA of maximum demand for the last 12 months). The charge may differentiate between summer and winter maximum demand or may be varied month by month.
2. A 'unit charge', which is a monthly charge for the number of units of electricity consumed. The charge per unit can be substantially higher for daytime consumption. It is often on a sliding scale allowing reductions when large numbers of units are consumed.
3. A 'fuel clause', which increases the unit price of electricity according to the increase of basic fuel cost.
4. A 'power factor clause', which increases the maximum demand charge if the power factor drops below a certain figure (usually 0.90).
5. Other tariffs can also include time of day costs which can make certain times of day much more expensive (early morning and early evening are typical) whilst overnight is at a significantly lower cost.

The effect of the maximum demand charge (especially when based on 12 months) can be onerous. The larger the gap between the maximum demand and the average demand, the more penal the maximum demand charge becomes and the more costly the charges per unit. Where such charges apply it is very important for pumping stations to be run at a high 'load factor', that is the average power consumed should be as close as possible to the maximum power required at any time during a month (or year). This means that pumps should be run for long periods rather than short; pumping for short intermittent periods at high outputs is expensive. Similarly, the occasional running of pumps at full load when they normally run only at part load, or bringing in extra pumps for short periods can also be expensive. Even the testing of pumps when first installed, perhaps running several together on test, can bring a heavy extra cost for a run as brief as 30 minutes.

Choosing the right duty for the pumps, the right hours of working and the right amount of standby for any pumping station can all help reduce costs. If there is adequate water storage in the supply system it may be feasible to design the installation to maximize pumping at periods of low tariff and avoid pump operation at the peak tariff periods. Alternatively, storage may be used for levelling out pumping rates thus reducing friction losses and the maximum power demand, permitting steady running at a high load factor and minimizing electricity charges. With the competition now present in the electricity market, water companies should be in a good position to negotiate favourable rates from the electricity supplier; unlike many users, water companies are usually large electricity consumers with a steady load for most of the year.

19.27 ELECTRICAL POWER SUPPLIES

The preferred source of electricity supply for a pumping station or other water supply facility is from the local electricity supply company. If the maximum demand is not high (typically below 250 kW), supply can be arranged from the nearest existing substation, at a low voltage. However, for larger maximum demands, dedicated transmission lines and substations may need to be built. For sites with large maximum demand higher voltage supplies will be required. Early negotiation between the user and the electricity provider is essential for any capital project. Information needed for these discussions includes the load and the nature of the load, types of motor starting proposed and intended use of any variable frequency devices which could induce harmonics in the electrical distribution system.

Public electricity supplies are to the national standard with respect to frequency and voltages available. Voltages and frequencies of electricity supplies vary considerably around the world and even from city to city in some countries. *Technical Manual 5-688* provides a comprehensive listing of supply voltages and frequencies (USACE, 1999).

The requirements for the electricity supply include the load and the required voltage which depends on the voltage suitable for large units such as motors. It also depends on the distance from the facility to the point where the new supply is to be taken. Except for the very smallest installations supplies will normally be three phase. For a very large load of strategic importance, where power supply failure is an unacceptable risk, supplies from two independently fed substations may be advisable.

Increasing the operating voltage of motors can reduce the cost of cabling and lowers operating costs due to reduced cable and winding losses. However, higher voltage may increase motor and starter costs so that there is no simple relation between motor size and optimum voltage.

For installations where all electrical components operate at the same time the connected load may be used for sizing the power supply. However, this rarely applies to water supply facilities where standby units such as pumps are available in case of failure or need for maintenance. In addition, many loads are not usually expected to operate simultaneously and diversity factors should be applied to allow for this. The generation of these diversity factors is process specific. For a rapid gravity filter with separate backwash and air scour the maximum load experienced by the electrical supply will be the larger of these whilst for a combined air scour/backwash the load would be a contribution from both loads.

The electrical load should be determined using a comprehensive list of all drives and other loads. Pump motor loads will tend to dominate but on-site ozone and hypochlorite generation will also add significant load. All other plant should be covered and allowances should be made for building services, including heating, ventilation and air conditioning as well as lighting and small power. National codes (e.g. BS 7671) give recommendations on demand factor (ratio of maximum coincidental load to sum of all connected loads) for building services, etc.

The electricity supply company will require the supply to be metered and a disconnection switch provided. The location of these facilities is often at the site boundary but, in any case, has to be secure while being accessible to the supplier who will notify such requirements.

Electrical loads, such as motors, which comprise windings have inductance. The consequence is a power factor (kW/kVA) which is less than unity. Electricity suppliers often impose a penalty on low power factor loads. Therefore, it is common to correct low power factors by introducing capacitance of a value sufficient to raise the power factor enough to avoid the penalty. The capacitors may be installed for individual motors or as a group connected to the user's main switchboard busbars. Running a site at a low power factor also has additional costs since the power cables carry more current than required resulting in excessive cable heating and volt drop issues across the site.

19.28 STANDBY AND SITE POWER GENERATION

Although in many places the reliability of electrical power supplies has greatly improved, so that no consideration need be given to outages of any long duration, there are still installations where power failure must be taken into account. A typical example is where severe weather can affect transmission lines to remote pumping stations. Some form of standby generating plant will then be needed. Except for very large unit sizes (over about 5 MW), for which gas turbine generators are usually used, diesel-driven generators are preferred. Where a single large water source works provides a major part of the water supply, there may be no escaping the costly provision of a fixed, dedicated standby generating station, possibly serving high- and low-lift pumps and a water treatment works as well. The generating capacity needs to be sufficient for all the plant which must be kept operational during the outage and the size of the largest motor to be started. The plant to be included may be affected by the duration of the outage to be covered.

If provision of a large generating plant is essential, substantial capital investment will be needed. It could stand idle for most of its life but, alternatively, it could be operated in a number of ways to recover some of this cost, for example:

1. continuously to cover a base load, or
2. for peak lopping to minimize electricity usage during periods of high unit cost; this helps avoid the financial penalties associated with some tariffs.

Leaving generating installations unused leaves them prone to deterioration and the operating staff unused to running the plant. It is preferable to exercise the standby generating plant on-load; running a large diesel engine on no-load can result in the engine coking up and requiring expensive maintenance. If neither operating mode (1) nor (2) is adopted the standby generator should be test run at more than 50% electrical load. This can be achieved either by running part of the works on the generators or by installing dedicated load banks for test purposes.

An alternative, which may sometimes be possible when a number of source works contribute to the total water supply, is to set up a pool of mobile generating sets, which can be quickly transported to any site where a power outage has occurred, if the same outage is not expected to affect all works simultaneously. This reduces the amount of standby generating capacity needed and makes for more cost effective use of the generators; however, transporting large generators can be awkward and expensive. The solution may not always be effective since severe storm or other conditions causing the original electrical outage might affect more than one installation and could also make roads impassable. Simple and speedy arrangements for receiving and connecting the generators, and for fuelling them, must be made at each point of use and more than one size of generator may be needed. Arrangements must also be made for safe storage and maintenance of the generators to ensure they are always ready for emergencies.

If pumping stations and treatment works have to be built where there is no public electricity supply available within an economic distance, power may have to be provided by a diesel or gas turbine engine on site. If more than one pump has to be operated, the diesel engine would drive a generator to make electrical power available to all pumps and provide power for other uses such as lighting and instrumentation. In rare cases of small isolated pumping stations with only one pump, a small diesel engine can be direct coupled to the pump. Normally, however, the choice is between building a power station to supply the works, which will have to include some standby capacity, or negotiating with the electricity supply authority to make new supply arrangements. The economics of these situations can be complex, being dependent on the charge the electricity supplier makes for bringing a power line to site, the terms of the supply and the need to take into account likely future developments in both power and water needs. The capital, operating and maintenance costs of on-site generating plant and of providing adequate fuel storage tend to be high, so that such a set-up is rarely economic in developed countries where high-voltage electrical networks make supplies widely available.

19.29 TRANSFORMERS

Transformers are alternating current devices which, in water supply installations, are normally used for two purposes: voltage change and isolation of instrumentation. Voltage or power transformers have primary and secondary windings around an iron core in which a magnetic field is set up when alternating voltage is applied to the primary winding.

Winding insulation is either 'liquid' or 'dry'. Liquid insulation consists of oil of low flammability; dry insulation is a resin. Oil used in 'liquid' transformers is a potential source of fire so such transformers are always located in enclosures separated from other building spaces (Plate 33(a)). In case of leaks of oil, oil insulated transformers are installed on plinths surrounded by open gravel in a concrete bund so that any spillage is contained. Traditionally this was a mineral oil but modern oil filled transformers can use an environmentally friendly oil (Midel) which has much greater biodegradability. Resin insulated transformers may be located within buildings subject to any noise and cooling constraints. Smaller transformers below 200 kVA on incoming supplies are often pole mounted at the supply connection.

Power transformers are normally provided with off-load tap changers to cope with long-term voltage changes. Such taps are usually 2.5–5% above and below normal supply voltage. Alternatively, for certain supplies which have fluctuating voltages, a more expensive on-load automatic tap changer could be considered to maintain a more stable secondary supply voltage.

Transformer impedance for transformers above 500 kVA is usually in the range 5–8%. Impedance has differing effects, for example cost, short circuit currents, switchgear ratings and motor starting voltage drop, and needs careful consideration when designing and selecting components for the electrical distribution system.

19.30 HV AND LV SWITCHBOARDS

Switchboards are split into two voltage ratings, LV (Low Voltage – less than 1000 V AC) and HV (High Voltage – above 1000 V AC). The voltages between 1000 and 33 000 V can often be described as MV (Medium Voltage) but this has no recognized meaning in electrical engineering.

Switchboards are enclosed panels (usually of coated steel) arranged in sections with an interconnecting busbar which usually runs at the top of the switchboard in a segregated compartment. Panels may be arranged for top or bottom entry for site cabling. For the latter a cable trench or basement area is necessary and for the former high level cable trays or ladder would run to the panel. The switchboards are normally provided with segregated sections which comprise circuit breakers, fuses, switches, indicators and meters.

HV switchboards should be located in a room separate from LV and other panels, as is usually required by regulations, since different safety procedures and permits usually apply to HV switchgear.

It is sensible for main switchboards to be arranged with a bus-coupler in order for each incoming section to be isolated individually. This permits maintenance on one side of the switchboard without de-energizing the complete panel. This principle would normally be adopted for improving the security of supply; it includes the provision of two incoming transformers.

HV switchboards are also subject to significant blast relief ventilation requirements. If an HV internal arc fault should occur there will be an overpressure generated in the panel during the isolation of the fault and this is often directed out of the back or top of the panel. As a result these areas are restricted access to prevent injury to personnel.

19.31 MOTOR CONTROL CENTRES

A motor control centre (MCC) (Plate 33(b)) is an electrical panel similar to a switchboard but including motor starters and controls. It may also contain Programmable Logic Controllers (PLCs) and Human–Machine Interfaces (HMIs) (Section 19.41) and a number of analogue or digital indicators and lights. MCC panel layout should be logical and should be designed for either front or rear entry.

The MCC is usually divided into vertical sections which have separate compartments for different functional devices. MCCs are of different assembly type depending on the segregation between items (BS EN 61439-2):

- Form 1 has no segregation between components in the MCC;
- Form 2 segregates the busbars from the functional units but there is no segregation between functional units;
- Form 3 segregates the busbars from the functional units and segregates the functional units; there is no segregation between terminals for external conductors;
- Form 4 segregates the busbars from the functional units and segregates the functional units; this also provides segregation between terminals for external conductors.

The designer should decide on the degree of segregation taking into account required maintenance activities; personnel should not be required to access a live MCC compartment. If the MCC is Form 1 or 2 then the whole MCC would have to be isolated before any maintenance activities can occur. This may be acceptable for a single process unit (centrifuge or screen, etc.) but, where the MCC feeds multiple process units, a Form 3 or Form 4 MCC should be specified. Form 4 is usually required in the UK water industry.

Traditional MCCs with individual starter door furniture such as indicating lamps, pushbuttons and ammeters can take up a large amount of space in the switchroom. An alternative is the intelligent MCC which minimizes the door furniture and therefore reduces the space requirement by providing a HMI to allow the operator to start and stop pumps and motors from a touch screen rather than pushbuttons on individual starters.

Intelligent MCCs can reduce the overall space requirements since the number of components is significantly reduced but these usually include a network solution between starters (Profibus or DeviceNet are often used). Whilst two or more networks can be included, failure of a single network can result in multiple drives failing unless care is taken with the design and construction of the unit.

Some Intelligent MCCs include full local intelligence such that network loss allows the drive to continue operating using its 'local' intelligence whilst other designs rely on the Central PLC. Obviously the full local version is a more resilient solution but is only economically available with certain network solutions.

19.32 ELECTRICAL CABLING

Power cables comprise a number of conducting cores of copper or aluminium, insulation and, for some situations which require mechanical protection, armouring. Insulation is usually XLPE (cross-linked polyethylene), LSF (low smoke and fume) – for occupied buildings or EPR (ethylene-propylene). Armouring, usually specified to provide physical protection against subsequent damage (digging, etc.) and from rodent attack, is usually steel wire (SWA) or steel braid for three-phase cables, or aluminium for single core cables; it is itself sheathed to prevent corrosion.

Cables serving the motors of large pumps or other large loads have a large cross-sectional diameter and require a large bend radius to be considered for their installation. This affects the space needed in cable trenches under panels and at other changes in direction such as drawpits. All cables have restrictions on bending radius and HV cables are worse than LV cables in this respect. Some typical cable types and minimum bending radii are listed in Table 19.4.

Table 19.4 Typical minimum cable bend radii

Cable type	Cable size	Voltage rating	Minimum bending radius
XLPE/PVC (unarmoured)	$1 \times 120\ mm^2$	600/1000 V	90 mm
XLPE/PVC (unarmoured)	$1 \times 400\ mm^2$	600/1000 V	210 mm
XLPE/SWA/PVC	$1 \times 120\ mm^2$	600/1000 V	150 mm
XLPE/SWA/PVC	$1 \times 120\ mm^2$	6350/11 000 V	400 mm
XLPE/SWA/PVC	$1 \times 400\ mm^2$	600/1000 V	250 mm
XLPE/SWA/PVC	$1 \times 400\ mm^2$	6350/11 000 V	550 mm
XLPE/SWA/PVC	$3 \times 120\ mm^2$	600/1000 V	330 mm
XLPE/SWA/PVC	$3 \times 120\ mm^2$	6350/11 000 V	780 mm
XLPE/SWA/PVC	$3 \times 400\ mm^2$	600/1000 V	540 mm
XLPE/SWA/PVC	$3 \times 400\ mm^2$	6350/11 000 V	1100 mm

19.33 HEATING AND VENTILATION

Electric motors have losses which generate heat and noise. Very large motors are usually water cooled. Other electrical plant such as transformers and VFDs also give off heat and are usually air cooled although water-cooled VFDs are now available as standard items. Such heat needs to be removed and this is normally done by ventilating the room housing the plant. The number of air changes per hour needed should be calculated from the heat output and permissible temperature rise above ambient summer conditions, taking into account solar gain and other effects.

Condensation is potentially harmful to electrical plant. This can arise in any season when the temperature of equipment such as motors and panels falls below the dew point of the adjacent air. Electric motors are usually fitted with built-in anti-condensation heaters, which are energized automatically when the motor is stopped. These are very effective and take little power: perhaps 0.25 kW for motors under 50 kW and up to 1.5 kW for very large motors. Switchboards and MCCs are also equipped with anti-condensation heaters for the same reason.

PART III CONTROL AND INSTRUMENTATION (C&I)

19.34 INTRODUCTION

The control and instrumentation (C&I) of pumping plant, distribution network flows and water treatment plant has changed considerably with the introduction of stricter requirements for service levels, maintenance of water quality and the development of many specialized and advanced water treatment processes. These require the processing of thousands of signals and large amounts of data; a typical modern water treatment works control system includes ten to fifteen thousand signals to be processed. The automation of such plant is becoming essential due to this complexity and the amount of data to be handled. Automatic control systems are also seen as a way to reduce labour operating and supervisory costs, whilst maintaining product quality.

The use of computers in control applications is fundamental, particularly in the control of water treatment processes and pumping systems. Making the best use of electricity supply tariffs, concentrating pumping at times of low-cost electricity and avoiding incurring high maximum demand charges (Section 19.26) involves calculations which can be complex. Such computerized control systems can also be readily designed to take account of a wide variety of conditions. Examples include: preventing the over-frequent starting of electric motors; sharing the hours run equally between a number of pumps; coordinating the operation of different pumping stations serving the same supply area; and regulating the operation of pumping stations arranged in series along a long transfer main.

For detailed information on process measurement and control and on networks the reader is referred to the *Instrument Engineers' Handbook*, Volumes 1, 2 and 3 (Liptak, 2003, 2005, 2011).

19.35 CONTROL

Control Level

Four levels of control are practiced in the water industry for control of plant generally: local and remote manual, and local and remote automatic.

Manual control is the most basic form of controlling plant and relies primarily on the operator to oversee and to react to any condition that requires alteration of plant operation. Although manual control may be impracticable for more complex plant, it must still be available wherever possible as a fall-back method for emergencies and when other systems are unsuitable or fail. Manual control may be classed as Local or Remote Manual control.

Local Manual refers to control at the equipment, MCC, or other local control panel and are often located adjacent to the equipment being controlled. For motorized plant with starting equipment in a MCC, the manual control facilities are typically located on the starter section of the MCC. For local control provision is usually provided for packaged equipment. Local manual control also serves to assist with equipment maintenance.

Remote Manual refers to equipment which is selected to receive control from a remote location and is provided with full control capability from that remote location. The remote locations are typically from a remote control room via computer displays and PLCs (programmable logic controllers), DCUs (digital control units), etc. Remote Manual control is typically provided as an option to Remote Automatic control.

Automatic control aids the operator to cope with treatment processes which are too complex for control without constant assistance. Examples are ozone production and the start-up of plant processes applying numerous chemicals. Some operations, such as filter flow control or washing, which were traditionally manual, are now frequently automated. Automatic facilities tend to be located centrally, but can be remote, to the equipment being controlled. Automatic control can therefore be classified as Local Automatic control and Remote Automatic control.

Local Automatic control is traditionally provided with local panel loop controllers that maintain control set points for various processes such as flow, level, etc. Local automatic control continues to be utilized for smaller processes that are intended to operate as stand-alone processes such as chemical feed flows and level control. Local Automatic control is progressively being consolidated into computer based PLC or remote terminal units (RTUs) with local operator terminal interface devices.

Remote Automatic control is utilized with complex processes with the minimum of operator intervention. This type of control must ensure a high quality, safe product and that corrective action is taken on equipment failure. Operator intervention is usually limited to entering key set point parameters related to output quality and quantity. Plant operators are usually provided with facilities for automatic control from a central monitoring point such as a remote control room. Although the control logic performing the automatic functionality may be located within controllers distributed around the plant, the control room facilities are linked with the plant by means of digital data communication.

Remote automatic controls typically provide multiple modes of automatic operation depending on the process controlled. For example, filter flow control in plants with multiple filters typically provide individual filter flow rate control for each filter and a filter inlet flume level cascaded filter flow rate control as two separate remote automatic control mode selections for each filter.

Automatic Control of Chemical Dosing

Control of chemical dosing makes use of a range of instruments, some described in Sections 12.33–12.35 and 19.39, and usually employs some degree of automatic control. Various forms of automatic control of dosing are described in Sections 12.26–12.31.

19.36 AUTOMATION

Automation of water treatment plant involves the control system opening and closing valves and starting and stopping equipment in predefined sequences to complete specific tasks or to provide the desired process plant output. To achieve these results the automation system relies on signals from correctly selected and placed instruments, devices such as actuators and motor control circuits and reliable control logic. The degree of automation to be used is fundamental to developing an automation system.

A *functional design specification* (FDS) must be prepared in detail setting out all the functional requirements for each item of equipment and the controls to be applied. The FDS is critical to achievement of a satisfactory monitoring and control system that meets the requirements. In the information technology industry the following terms are in common usage:

Hardware (Section 19.41) means the device which controls and monitors the operation of an item of plant.

Software is the programming logic or set of instructions which designates the tasks to be carried out by a group of hardware. These tasks can range from starting a pump to calculating, from monitored process variables, the appropriate hardware actions to be taken.

Firmware means a permanent form of software (embedded in hardware) which is normally not available for alteration by the operator but may have features permitting selection from a standard menu.

'Application' or 'bespoke' software is software which is custom designed for a specific application. Most control systems involve the use of some bespoke software.

Where two items of instrumentation equipment are required to operate as a system the interface between them must ensure proper data communication. When selecting components for a system, the designer must ensure that they will operate satisfactorily together.

19.37 CONTROL SYSTEM DEFINITION

Control requirements should be defined in an FDS as fully and as early in the project as possible in order to:

1. provide a clear understanding of the hydraulic, mechanical, electrical and control requirements before the design is started;
2. reduce the number of changes made during the design and construction;
3. produce better control requirements which are less onerous on mechanical equipment and which lead to smoother, more stable plant operation.

An example of less onerous control is a variable speed pump drawing from a sump subject to rapid changes in inflow. Two options for this are:

1. Pump sump level maintained at a preset point for all sump inflows.
2. Pump speed (and so flow) proportional to sump level with a low speed for a low level.

The first option requires the control loop to be tightly controlled and results in a fast change in pump speed, fast flow changes and increased wear on the plant. It requires a tightly set PID (full three-term proportional, integral and differential) control loop.

The second option requires only a proportional control loop which is easily set and maintained. It makes use of the storage available in the sump and allows the system greater time to respond. The control loop can be slower acting, resulting in fewer process disturbances downstream, less wear on the pumps and fewer process trips.

19.38 INSTRUMENTATION

Only a few process parameters make up the majority of those measured in the water industry. Of these parameters, flow, level, pressure and water quality are the most common. Common types of flow, level and pressure instruments are described in more detail below. Parameters such as chlorine residual, colour, turbidity and pH are among the most frequently used water quality measurements. Permanently installed on-line instruments can be selected to measure these parameters and provide electrical signals for use within a control and monitoring system.

Measurement of process parameters can be digital or analogue. Digital measurement is used for reporting the status of plant, for example whether it is running or stopped. Analogue measurement is used to represent variable values, such as chlorine residual or pump output. This data is either collected via traditional hard-wiring or over digital communication buses.

There are several standard analogue interface standards used for hard-wired signals. The water industry uses the (0) 4–20 mA current loop as the standard analogue interfaces. The 4 mA (or 0 mA) value usually represents the lowest end of the value range and the 20 mA the highest. When the 4 mA value is selected it is referred to as ‘live zero’, so called that an actual zero milliamp value can be attributed to a problem such as a broken connection or defective transmitter.

Whilst various networked communication standards have been developed (generically these are termed FieldBuses) there is a large installed base of 4–20 mA instrumentation and some users have decided not to change. Whilst there are many benefits to the new networked instrumentation standards there is also a significant change in maintenance procedures.

Four of these protocol standards are HART (highway addressable remote transducer), Foundation Fieldbus, Profibus and ‘ModBus’ (see Section 19.40 for networks). Although instruments with these capabilities have been available for some time, only in the last 15 years has the water industry begun to use them extensively. The benefit of these standards include:

1. Multiple instruments on one loop, reducing installation costs.
2. Significant additional instrument health and status information.
3. Remote configuration of instruments.

4. Reduction in ranging errors. A 4–20 mA instrument has to convert the measured value into a 4–20 mA analogue signal whilst the PLC and SCADA has to convert this back to display the actual reading. These protocols allow the actual reading to be transmitted reducing the possibilities for signal conversion errors.
5. Increased accuracy in flow totalization performed at the source flow transmitter.

The operation and characteristics of flow meters are described in Sections 14.16, 14.17, 18.27 and 18.28. Methods commonly used to measure other key parameters in the water industry are described below.

Level

Ultrasonic level measurement is a non-contact method used for measuring the distance between a surface and the instrument sensing element. Ultrasonic waves are transmitted from and received at the sensing element; the time taken is used to calculate the distance to the surface. Once the sensing element is installed over the surface the instrument is programmed with datum information to allow the process surface level, in a tank or channel, to be measured in useful terms. When programmed with tank or container dimensional information, the ultrasonic level instrument can also provide volumetric readings. Precautions must be taken when locating the sensing element to avoid reflections off fixed surfaces that may confuse the instrument and make the reading unreliable. A drawback is that floating debris and foam can adversely affect accuracy. Using ultrasonic level for measurements with rapidly varying temperature in the measurement area can result in unstable operation.

Radar level measurement is another non-contact method used for measuring the distance between a surface and the instrument sensing element. It has the benefit that changes in the speed of sound in air due to temperature variations or to differing gas concentrations (ozone) don't affect the reading. Precautions do have to be taken to prevent large amounts of condensation building up on the face of the sensing element but these are relatively easily achieved. Different wavelengths can be used for different materials due to the different reflective properties of water, sludge, sludge cake, etc. This can allow the unit to see the actual liquid level through foam, something the ultrasonic unit cannot do.

Laser level measurement is another non-contact method used for measuring up to at least 100 m in deep tunnel or highly crowded structures requiring level detection. The level transmitter emits a pulsed laser light beam that reflects off of the measured surface and is detected by a sensor on the transmitter. The measured time between transmission and reception is converted and produces the level output. An advantage of laser level transmitters is that the minimal clearance required from obstructions for the beam to be effective is very small – 40 mm or less. Laser level transmitters are adversely affected by highly dusty environments which can coat instrument optics and attenuate the laser beam.

Hydrostatic pressure instruments use the static head of the process fluid to deflect a diaphragm in a sensing element which is placed at the bottom of the fluid being measured. The deflection is monitored and, based on the specific gravity of the fluid, a signal is produced in proportion to the depth of the fluid above the element. This type of instrument is unaffected by floating debris and surface foam but may require frequent cleaning in some applications.

Differential pressure level instruments are frequently used to measure fluid surface level in pressurized vessels where hydrostatic measurement would be useless. The differential pressure instrument has a high-pressure port and a low-pressure port. For this application the high-pressure port is connected to the bottom of the vessel and the low-pressure port is connected to the top of the vessel. The instrument calculates the level of the interface between the fluid in the vessel and the space above the fluid by finding the difference between the two pressures. This type of instrument can also operate in open air tanks by simply leaving the low-pressure port open to atmosphere.

Conductivity level probes are commonly used to detect discrete surface level points. Conductivity is established between a reference (earth) probe and the probe set to the desired switching level when the process fluid wets both probes. The conductivity is lost when the fluid level drops below the switching level probe. The signal produced can be used for discrete control purposes.

Float operated level switches are simple devices that operate by tilting a floating bulb with an internal gravity operated switch. The float switch is connected to a flexible cable that acts as the tether for the float and for transmitting the signal to the control system. These switches are suitable when precise level switching is not required but are reliable and easy to maintain.

Pressure

Pressure measurement using diaphragm deflection involves the use of variable electronic resistors fixed to a flexible surface (diaphragm). As the pressure deflects the diaphragm an electronic circuit detects minute changes in a group of resistors, known as a 'Wheatstone Bridge', and produces a signal proportional to the pressure causing the deflection.

Most pressure gauges in the water industry are Bourdon tube type. The principle utilizes a curved tube that straightens as pressure is applied internally. The pressure gauge display pointer is fastened mechanically to the end of the Bourdon tube. Pressure gauges can be fitted with switches to provide pre-set alarms to the control system but cannot provide an analogue electrical signal.

Pressure switches in the water industry are Bourdon tube type and piston spring type switches. The Bourdon tube type operates as described above but provides a contact closure when a pressure set point is achieved. The principle of the piston spring type utilizes a diaphragm displacement against a mechanical spring when exceeded by the pressure set point provides a contact closure.

Water Quality

Water quality instruments commonly used in the water industry are:

- chlorine residual (Section 12.35), for example potentiostatic, colorimetric (DPD), amperometric and gas stripping technologies;
- colour for example light diffusion;
- turbidity (Section 12.36), for example light scatter or particle counter;
- pH (Section 12.34) and ORP (oxidation reduction potential).

The technologies associated with these water quality instruments are discussed in Chapter 12. Selecting the correct type of instrument for a given application is critical to the reliability and cost effectiveness of the application and should be done by an instrumentation expert taking into account the operating conditions, environment, maintenance, consumable components and overall cost, drawing on experience and research of products.

19.39 SYSTEMS

A 'system' may involve only one measurement, a controlled device and a controller to maintain a set point of a single parameter; more commonly it will handle data from many monitoring points and propagate many plant control commands. For any given treatment process, the combination of these items defines the control system. The control logic currently used for automation is usually installed in programmable devices which evolved from the electronics and information technology industry. However, modern control systems still use traditional hard-wired monitoring methods, primarily as a backup, fail-safe, system for a critical process or piece of plant.

Safety shut down systems prevent damage and danger to plant or personnel and ensure there is no failure to maintain output quality. The criteria used and circumstances for this vary depending on works location but can include:

- low chlorine residual
- high turbidity
- low pH
- adverse colour
- overflow
- chlorine or other chemical leak or spill.

Alarms notify the need for operator attention when equipment, plant, process, personnel or output quality and quantity may be at risk.

Alerts draw operator attention to any state that may develop into an alarm if left unattended, for example tank water level high or process flow low.

Events inform the operator of a confirmed 'normal' operating occurrence that may or may not require attention, for example a pumping sequence completed or a valve opening.

Hard-wired systems. Traditional hard-wired systems utilize direct wiring between electromechanical relays (Plate 33(c)) to perform control logic operations and are directly wired to the controlled equipment. These systems do not rely on programming software and are therefore rigid in terms of functionality and often very difficult to modify. However, well designed hard-wired systems have a reputation for being very reliable. For these reasons independent hard-wired systems often supplement programmable systems, primarily for simple backup controls and safety critical applications such as alarm and emergency shutdown systems.

Programmable control. There are many PLC devices used in control systems. Most electronic equipment now has some form of programmable functionality. Any device that can be programmed to carry out predefined sequences to provide control outputs could be called a programmable controller. A program that performs specific logic or tasks is referred to as a software 'algorithm'. The industry workhorse is the PLC (Section 19.41). When networked with others, a system of PLCs can be created that is capable of controlling even the largest and most complex water treatment and distribution works.

Distributed control. A 'distributed control system' (DCS) can also be found in many works (Fig. 19.10). In it the control logic is distributed in outstations around the plant being controlled. Until it was practicable to locate processors at outstations the entire computing power was held centrally. Such centrally based systems are now mostly obsolete. An up-to-date DCS comprises a number of discrete process control units, for example PLCs, outstations or RTUs, linked by a

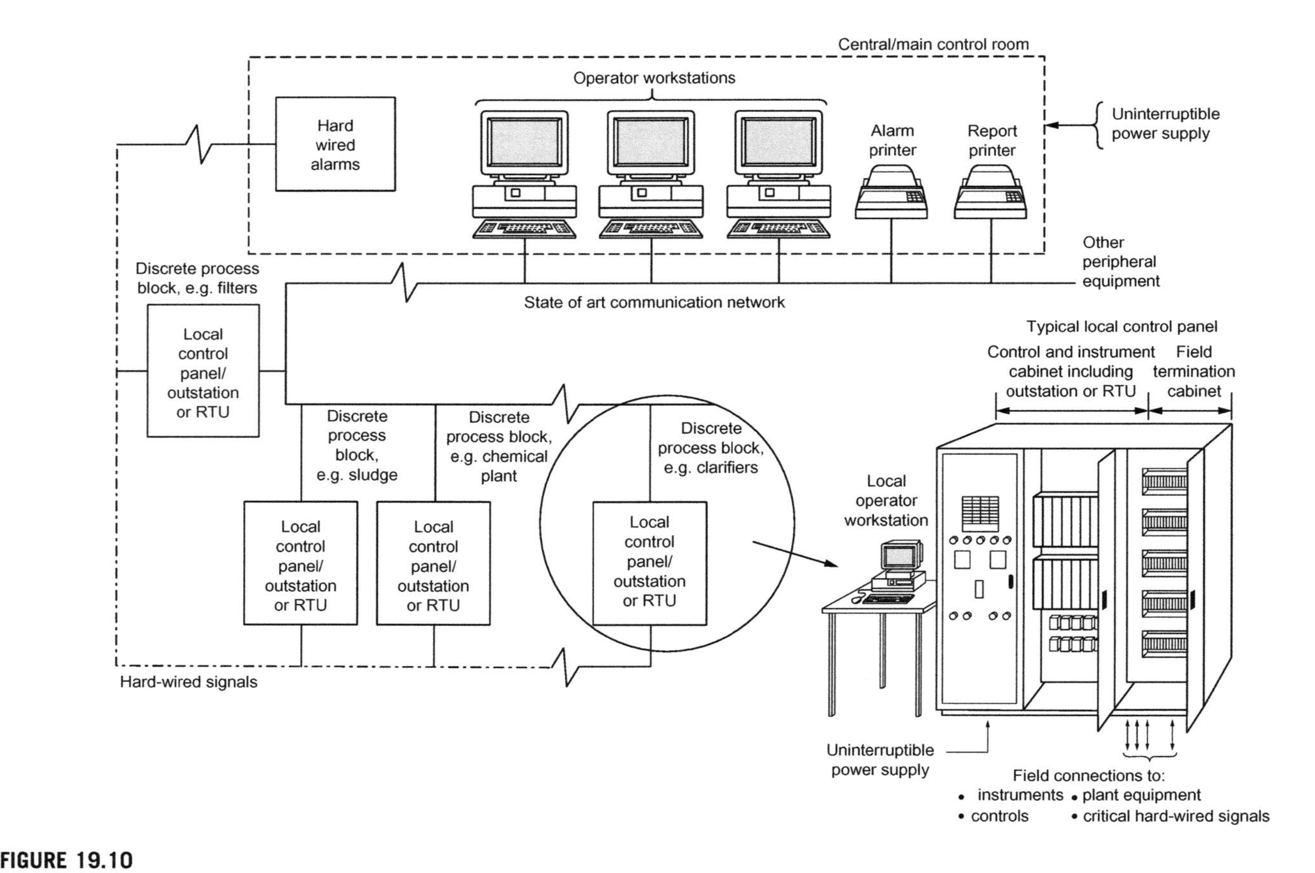

FIGURE 19.10

Typical distributed control system (DCS) linking a number of discrete process block control units.

standard communication network (Fig. 19.10). The automatic control logic for each process area is contained in the control units and can therefore function independently of the overall monitoring and control system. A DCS links discrete process blocks to allow:

- transfer of data round the system, including changes to control;
- changing parameters at RTUs, such as adjustment of set points or duty plant selection;
- monitoring of individual RTUs;
- display and recording of data, events and alarms;
- safe reaction to changes or failure of individual RTUs which could affect product criteria, process performance or plant safety.

The distinction between PLC and DCS is blurred with many PLC units being used within DCS systems and some DCS systems acting more like traditional PLC systems. DCS systems historically were utilized for extensive analog related control and computational controls in process oriented plants (chemical, refining and others). PLC systems historically were developed primarily as a manufacturing replacement to relay logic control. Early PLCs were utilized for discrete repeating functions typically utilized in industries such as automotive manufacturing.

As the development of both systems has progressed over the decades, PLCs have developed extensive analog calculation and control software capabilities. The DCS systems have also progressed to support more digital sequential controls similar to PLCs.

SCADA (supervisory control and data acquisition) refers to virtually any data acquisition system, but usually one which exercises monitoring and supervisory control of a number of sites from a control centre. Such systems are widely used in the water industry so that a 24 hour manned control centre can react to any problems arising at sources or throughout a water production facility or a water distribution system.

Telemetry is the system for transmitting data from one location to another. The distance between the two locations can vary from a few metres to thousands of kilometres; with only the limitations of the transmission medium applying (e.g. hard-wired data communication networks, telephone networks, microwave systems or radio systems, fibre-optic transmission, etc.). Telemetry systems typically used by most monitoring and control systems include DCS and SCADA systems (Plate 33(d)).

Web-based technology has changed the way in which control system data is transmitted around the world. The introduction of web-based global Internet technology is likely to change the way in which control systems are used in the water industry. Already, there are web-based applications that provide dashboard remote monitoring and control facilities and allow plant to be interrogated, controlled and modified from any enterprise connected computer that has web access and the appropriate software. For monitoring purposes this technology offers limitless benefits but, for control of critical plant, use of the Internet raises concerns about security.

19.40 COMMUNICATIONS

Networks generally. Communications networks have been used to transmit instructions and data for process monitoring and control for many years. Process control networks have used various technologies and topologies, as have business and computer networks in less critical situations. Process control networks require robustness, determinacy and compatibility.

Robustness is a measure of the reliability of the network to perform its function throughout the life of the installation. The designer of a network must evaluate the need for redundancy of network components and cabling, as well as error testing and error correction facilities, to provide a network that meets the needs of the system it serves. *Determinacy* is a specific guarantee that messages enter the network and reach their destination within known times. Non-deterministic networks cannot guarantee message delivery in a specific time but recent technologies have allowed some non-deterministic networks (such as Ethernet) to function satisfactorily in the process control industry. *Compatibility* describes the ability of the network to communicate between equipment from various manufacturers or suppliers without protocol conflicts.

The obvious advantage of using a network rather than traditional cabling is that all the data is passed along a single cable therefore reducing significantly the number of cables to be installed.

Ethernet. CSMA/CD (carrier sense multiple access/collision detection) performance has evolved to a very high standard since its introduction in 1973. Although it is non-deterministic, CSMA/CD allows every device on a network to check if any other device is transmitting before attempting to transmit and, if multiple devices transmit simultaneously, the collision detection discards corrupted messages (packets) and instructs the devices to resend their messages after random wait times. The network is non-deterministic because there is no guarantee that a message will reach its destination is a specific time. However, industrialized Ethernet networks have become so reliable that they are the standard in the industry for linking site-wide process control systems. The Ethernet design specification must be followed regarding connectivity and cable lengths, for example using a maximum of 100 m of copper cable from a network switch.

Ethernet networks can form part of local area networks (LAN) or wide area networks (WAN). Although there are no specific definitions, a site-wide SCADA system for a process plant would generally be referred to as a LAN and a regional network linking several sites would be considered to be a WAN.

Whilst Ethernet is becoming more prevalent it is unlikely to totally replace other network protocols in the water industry. One of the disadvantages of Ethernet is that it is not a multi-drop system so each separate device needs its own individual cable back to the local communications switch device. Whilst this may not be an issue in a factory or small building it can be a major issue with large banks of filters or tanks. The maximum cable length from the Ethernet switch is also more limited than other techniques (usually 100 m). Ethernet communication speeds are typically 10–100 Mbps (Megabits per second) and some networks currently operate at 1000 Mbps.

A ***bus network*** is an arrangement in a LAN where a single multi-conductor copper cable links multiple devices. The cable is the 'bus' to which 'nodes' are connected. Each node generally corresponds to a specific item of equipment. Bus networks are simple and reliable and reasonable fail-safe. If one node device fails the bus continues to operate with the remaining functional devices. Only if the bus cable itself is broken would there be serious communication problems in the network. Bus networks provide a simple means of expansion because generally they allow nodes to be added fairly easily.

The limitations of bus networks are primarily the physical properties of the bus cable itself. As cable length increases, the losses affect the reliability of the data being transmitted. Therefore the topology of the bus network needs good design. Other network topologies such as 'ring' or 'star' may sometimes provide better flexibility and may be cheaper.

Table 19.5 Comparison of Profibus DP and Profibus PA

	Profibus DP	Profibus PA
Communication signal	Voltage based, using RS-485 standard	Current loop technology
Cables	Copper or fibre-optic	Copper
Intrinsically safe solutions	Not available	Offered
Maximum network length	100–1200 m depending on network speed	1900 m
Network configuration	Point to point unless repeaters are included	Star
Means of segment segregation	Use of repeaters	Use of multiple Profibus PA networks

Where a point to point link is required a fibre-optic link can be used to extend the distance achievable and provide lightning protection, but this is unsuitable for multiple connections without additional switching devices.

Profibus is one bus technology that the industry has generally adopted. Profibus is a high speed digital communication system which utilizes a single cable (bus) to link devices. Many manufacturers of electrical, electromechanical and instrumentation equipment now provide Profibus compatible products. It is common to link a number of related plant items by a Profibus network while then linking this plant area to other plant areas by a different network such as Ethernet. As with Ethernet networks it is important for the design of a Profibus network to comply with Profibus specifications.

There are three types of Profibus namely Distributed Protocol (DP), Process Automation (PA) and ProfiNET. There are two parts to the Profibus specification; the language used which is common across the three types of Profibus (called the application layer communications) and the physical media which is different in each case. ProfiNET is not a unique bus technology since it runs over Ethernet but it allows Profibus communications across Ethernet networks.

The choice of type of Profibus network depends on a review of all functional requirements for the network. Profibus DP and Profibus PA are compared in Table 19.5. Both communicate over two-wire twisted pair cables but there are significant differences.

Within the UK WIMES 3.02 (2014) suggests a Profibus DP speed of 1.5 Mbit/s (which limits the segment cable length to 200 m) but most works can operate successfully with network speeds of 500 Kbit/s (500 m cable length) or even 187.5 Kbit/s (1000 m cable length). The lower speed networks are better able to withstand noise and interference but data takes slightly longer (probably an extra 0.5 seconds) to travel across the network. This is not seen as an issue on a water treatment works where instrument sample times can be many orders greater than this.

Whilst the network cabling is relatively simple (twisted pair in both cases) it is worth investing in a diagnostic test set which will check the network health in real time. Various versions are available: Profitrace2 from Procentec and the Fieldbus Diagnostic Module from Pepperl & Fuchs are two units.

It is important to refresh SCADA screens quickly from the data held on the local PC, otherwise operators will fail to use the system if the screen doesn't refresh quickly when changing from one screen to another. However, having the field data take 1–2 seconds to travel from the field instrument to the PC is not an issue (a Profibus DP network speed of only 93.75 Kbit/s will easily achieve this). This 'delay' will not affect most water treatment processes and can reduce the installation cost and make the network more resilient to noise and interference.

Wireless data communication networks have become very popular for business and domestic use and for some process monitoring applications but there remain many safety and security concerns with its use for control of plant. For these reasons, wireless networks should be used for control of plant only after careful consideration and where directly connected systems are impractical or impossible to install.

Wireless technologies such as low power radio and microwave have been used for control of some equipment but should always be used with caution as risks associated with plant protection and personnel safety must always be considered. Typically, these technologies are used over relatively small areas and primarily for transmitting status monitoring signals.

Remote Configuration. The latest implementations of Profibus and Ethernet allow the engineer to configure all of the devices on the network from a central location. The two main packages use either FDT/DTM (Field Device Type/Device Type Manager) or EDDL (Extended Device Description Language) to communicate with the instruments.

19.41 HARDWARE

The essential components of ICA system hardware are described briefly below. MCCs are described in Section 19.31; these, particularly intelligent MCCs, may house complex ICA equipment or it may be located in dedicated control panels.

PLCs (Programmable Logic Controllers) are computers with components similar to those of the desktop PC, but configured for multiple inputs and outputs and capable of withstanding a wider range of environmental conditions such as temperature, dust and vibration. *Programmable Logic Controllers* (Bolton, 2006) provides a useful guide to PLCs. Plates 33(c) and (d) show small PLCs in ICA sections of an MCC.

HMIs (Human–Machine Interfaces) provide an interface between an operator and the plant and allow the operator to monitor, control, diagnose and manage the plant. In the traditional MCC, key plant parameters are displayed on analogue or digital instruments mounted on the front, with plant control by push button and switch. The modern HMI makes use of VDUs and keyboards or touch screens and allows MCC size to be reduced. HMIs can be as diverse as local touch screen type devices (OIT – operator interface terminals) or plant wide access and monitoring (personal computer based SCADA software). Portable handheld digital devices (such as tablets, pads, etc.) are becoming more available with wireless data communications systems.

Control rooms are dedicated areas where process control functions are gathered to allow plant to be monitored and controlled by the minimum number of operators. In the traditional control room there may be a mimic diagram consisting of a large fixed display of the system, with live plant data indicated at appropriate points, and a control console with indicators and push button controls. The modern arrangement consists of a more simple control desk arranged with VDUs and other HMI elements. The VDUs are likely to be large flat screen LED/LCD monitors, arranged to display schematics of sections of the plant, which can be selected by the operator. The operator will also be able to call up information and trends from memory and will be able to produce a variety of reports of plant operation and failures.

The HSE publishes guidance on certain industrial plant and includes on its web site a Technical Measures Document on *Control room design* – www.hse.gov.uk. Relevant reference standards include BS EN ISO 11064, BS EN 894, BS EN 60073 and NUREG-0700.

19.42 ANCILLARY EQUIPMENT

Uninterruptible power supplies are often used in support of water industry monitoring and control systems to maintain the control system operation for a predefined period during a utility power supply failure. Plant behaviour under power failure conditions must always be reviewed; each piece of equipment must be controlled into a fail-safe condition when power supply fails. Ancillary plant such as mechanical or pneumatic backup may be necessary to fulfil these requirements. In addition, plant start up following a power failure is another key consideration; and automatic start up or manual start up must be reviewed accordingly and specified clearly.

19.43 AUTOMATION AND OPERATION AND MAINTENANCE

Operators of automated treatment facilities need an understanding of electronic control and software programs, as well as a full knowledge of the treatment processes. Maintenance of automatic control systems is a specialist field: this aspect of modern system installation is often overlooked. The system buyer must fully appreciate the maintenance commitment that is involved. The shift from traditional plant, with a high level of operator input, to plant which has complex software and electronics control has not significantly reduced overall operational and maintenance costs of water supply largely because maintenance costs associated with C&I systems for a treatment works tend to offset the savings due to reduction in plant operational personnel needed. Careful planning of shifts can maximize these cost reductions.

A more significant driver for automated treatment facilities is the wide variety of specialist processes now involved in water treatment; these require more complex systems of real-time monitoring and control than would be possible manually to maintain highly regulated water plant output.

REFERENCES

Bolton, W. (2006). *Programmable Logic Controllers.* Elsevier Science.

BS 7671. *Requirements for Electrical Installations.* BSI.

BS EN 894. *Safety of Machinery. Ergonomics for the Design of Displays and Control Actuators.* BSI.

BS EN 60034-1. *Rotating Electrical Machines. Rating and Performance.* BSI.

BS EN 60034-30-1:2014. *Rotating Electrical Machines. Part 30-1. Efficiency Classes of Line Operated AC Motors.* BSI.

BS EN 60073. *Basic and Safety Principles for Man-Machine Interface, Marking and Identification – Coding Principles for Indicators and Actuators*. BSI.

BS EN 61439. *Low Voltage Switchgear and Controlgear Assemblies*. BSI.

BS EN ISO 5198. *Centrifugal, Mixed Flow and Axial Pumps: Code for Hydraulic Performance Tests*. BSI.

BS EN ISO 11064. *Ergonomic Design of Control Centres*. BSI.

DOE (2012). *Adjustable Speed Drives, Part 1 – Load Efficiency. Energy Tips: Motor Systems*. Advanced Manufacturing Office. Energy Efficiency and Renewable Energy. US Department of Energy.

DS/CLC/TS 60034-17. *Rotating Electrical Machines. Part 17. Cage Induction Motors When Fed From Converters – Application Guide*. Dansk Standard.

EU Regulation No 640/2009. Implementing Directive 2005/32/EC of the European Parliament and of the Council with regard to ecodesign requirements for electric motors.

HI 9.8 (2012). *American National Standard for Rotodynamic Pumps for Pump Intake Design*. ANSI/HI 9.8.

Liptak, B. (2003). *Instrument Engineers' Handbook. Process Measurement and Analysis*, Vol. 1. 4th Edn. CRC Press.

Liptak, B. (2005). *Instrument Engineers' Handbook. Process Control and Optimization*, Vol. 2. 4th Edn. CRC Press.

Liptak, B. (2011). *Instrument Engineers' Handbook. Process Software and Digital Networks*, Vol. 3. 4th Edn. CRC Press.

NUREG-0700. *Human-System Interface Design Review Guidelines*. Rev. 1. US Nuclear Regulatory Commission. Washington.

Prossor, M. J. (1977). *The Hydraulic Design of Pump Sumps and Intakes*. CIRIA and BHRA Report. July.

Rayner, R. (1995). *Pump Users Handbook*. 4th Edn. Elsevier Science.

Sanks, R. L. (1998). *Pumping Station Design*. 2nd Edn. Butterworth-Heinemann.

Thorley, A. R. D. (2004). *Fluid Transients in Pipeline Systems*. 2nd Edn. Professional Engineering Publications Ltd.

USACE (1999). *Technical Manual 5-688. Foreign Voltage and Frequencies Guide*. Publication of the Headquarters. USACE.

WIMES (2014). *Water Industry Mechanical and Electrical Specifications 3.02*. Issue 4 April 2014. Pump Centre, ESR Technology.

Yates, M. A. (1989). A meter for pump efficiency measurement. *World Pumps*, **January**.

CHAPTER

Treated Water Storage

20

20.1 FUNCTIONS OF TREATED WATER STORAGE

Treated water storage became important with the expansion of piped water supply systems in the 19th century. Its subsequent development has been driven by the need to protect water supplies from contamination and deterioration and by increasingly sophisticated operation of supply systems. Treated water storage may be provided at the treatment works or further downstream; it can be located at ground level or in elevated tanks.

Treated water storage is used to balance relatively consistent or stepped changes in source output with the variable and less predictable demand of consumers. The storage covers diurnal demand variations and may additionally, during peak seasonal demand periods, provide a balance over a week. Treated water storage is also used to maintain supplies during failure of a source works or critical pipeline or to meet fire or other emergency demands. Such storage is located strategically to ensure resilience within the trunk main network and to support local demand centres.

The need for treated water storage depends on the facilities supplying water to a distribution zone and on variation in demand in the zone. It is seldom possible or economic for a water treatment works to provide a fluctuating output in step with demand; treatment processes need to be run 24 hours a day with only infrequent, carefully controlled changes of output. For maximum efficiency and to avoid risk of cavitation, pumps should be operated near their design duty point; electricity tariffs may influence pump running times (Section 19.26). It is not usually economic for a long supply pipeline to have a capacity large enough to meet the peak demand of a few hours duration. Introducing water storage to the system can reduce whole-life costs and overcome such technical difficulties. The various storage, resilience and siting options should be evaluated technically and in terms of whole-life costs: financial, environmental and social (Sections 2.9–2.12).

The minimum water level feeding distribution by gravity should generally be just high enough to maintain the required minimum pressures in the distribution system at peak flows during the planned life of the scheme. A balance has to be struck between having an elevation just high enough to maintain the required pressures and having some reserve elevation to meet future forecast needs at the expense of increasing pumping costs.

Twort's Water Supply. DOI: http://dx.doi.org/10.1016/B978-0-08-100025-0.00020-X

20.2 STORAGE CAPACITY REQUIRED

Minimum Storage to Even Out Hourly Demand

A typical graph of the hourly variation of demand for a UK town of about 75 000 population is shown in Figure 20.1(a). The demand reaches a peak typically between 07:00 and 09:00 hours and remains high until after midday. It slackens off in the afternoon but rises to a second peak in the evening when people wash and prepare an evening meal. In summer the evening peak may be higher and more prolonged due to garden watering. The demand pattern is usually different at weekends and on holidays with peaks occurring later and lasting longer. Daily demand profiles in tropical countries, or those where there are different industrial, work or social regimes, should be assessed. Garden watering can form a large part of peak demand and may require special attention.

The average demand for the day from Figure 20.1(a) is assumed to represent the input to the reservoir; Figure 20.1(b) shows the consequent net outflow starting at 07:00 when demand rises above input. A total of 3750 m^3 of stored water is used for an average demand of 960 m^3/h. This is almost 4 hours' storage and is representative for towns of this size in temperate climates. The number of hours of storage needed for balancing or equalizing demand through the day is more for smaller systems and less for larger systems, particularly those with significant 24 hour industrial demand.

In practice a greater storage volume will be required because:

1. daily peak demands and diurnal profile will vary from day to day and with the season;
2. a full reservoir cannot be guaranteed before the start of the peak demand period; and
3. provision must be made for contingencies (see below) and some storage volume must be retained in the bottom of the reservoir for settling sediments.

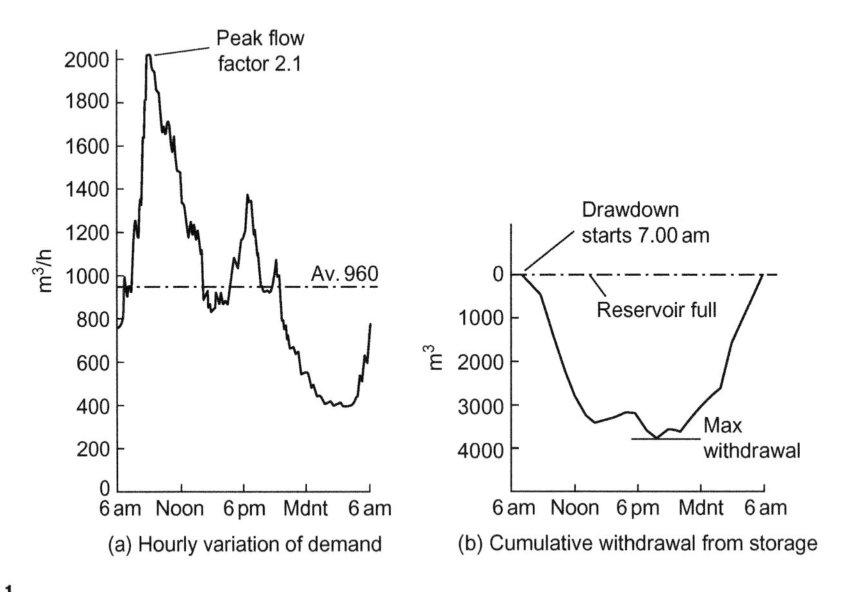

FIGURE 20.1

Typical variation in demand and use of storage for 23 Ml/d supply.

As a guide, for a town with a population of about 75 000, adequate flexibility is provided if the capacity of the reservoir is increased by about 50%, that is to about 6 hours' supply or 25% of the average daily demand. This is not, however, sufficient to safeguard the continuance of the supply against all contingencies and applies to locations where flow from source works is continuous.

Contingency Storage

Some contingency storage should be provided to cover intermittent source operation, breakdowns at sources, loss of supplies after major bursts, and consumption for major fires. Pollution incidents may cause a source to be shut down until the hazard has passed (Section 8.1). The storage required for these contingencies depends on the ability to change source output, the availability of alternative sources, the layout of the pipe supply network; and what fire regulations or safety precautions may require. In some countries storage stipulated for firefighting is sacrosanct and has to be segregated with its own outlet. If combined with storage for other purposes in the same structure, the fire storage needs to be arranged to fill first.

Risk assessments should be carried out for critical sources and pipelines; the likelihood of a failure event, its consequences and any possible mitigation should be determined. The results should be covered in emergency action plans. The loss of water following a major burst should also be considered and how such loss would be regained. A major fire can use 5000–15 000 m^3 of water, but it should be possible to increase the output from sources accordingly. Allowance should be made for the event to occur when the reservoir is already low.

Shut downs for planned maintenance and due to power supply cuts must be allowed for, particularly for sources such as borehole pumps or booster stations with a single source of electricity supply and no standby generator. Such outages may last a few hours.

Pipeline bursts cause supply to stop for rather longer – the following figures are indicative only of the least time that may be needed; if things go wrong, double the time may elapse before supply can be restored.

Burst reported	1/2 hour (say)
Mobilizing repair gang and closing valves	2 hours
Repairing pipeline:	
– up to 600 mm diameter	6–8 hours
– over 600 mm	8–12 hours
Refilling and disinfecting	2–4 hours

Taking into account the need for diurnal, seasonal and strategic contingency storage for routine and emergency conditions, network resilience and source works operational constraints, the overall desirable storage within a system is 18–24 hours – subject to risk assessment, levels of service commitments and availability of land. This is a useful target but larger amounts of strategic storage increase the risk of water quality deterioration unless the storage is carefully managed to ensure regular turnover. The location of the storage should be selected so that supplies can be re-routed to areas cut-off by any burst pipe or by maintenance work. It may be necessary to employ a water balance/strategic transfer model, especially where a reservoir forms one component of a complex integrated system. Each water authority should decide on suitable peak hour and peak week demand factors; on seasonal demand variations, for example in holiday areas where visitors can cause local demands to increase by as much as 50%; on allowance for high seasonal garden watering; and on policy for use of cheap-rate off-peak electricity supplies for pumping to minimize pumping costs (Section 19.26).

Volume not Available for Balancing or Non-Fire Contingencies

Some parts of the gross volume of a service reservoir are not available for normal use. These include:

- dead storage, below bottom water level (BWL; Section 20.14) – storage which cannot be discharged at all, or at the design flow or with the required pressure;
- volume between top water level (TWL) and overflow level (Section 20.14);
- firefighting storage;
- control margins – the volumes (usually small) between the limits of pump, inlet or outlet operating bands and BWL or TWL.

20.3 GROUND OR ELEVATED STORAGE

If storage is required there is usually some flexibility in selection of its location. It may be possible to site it a little way from the ideal point of connection to the system. This could allow it to be located on a hill in order to maintain the pressure required in the distribution system. However, where there is no convenient high ground, some other solution is necessary. An alternative is to provide elevated storage at a water tower, but local objections may prevent granting of necessary consents. If a water tower is unacceptable in such circumstances, the only alternative is to provide ground-level storage and boost the water into local supply. Except for environmental acceptability, the choice of ground or elevated storage is a matter of economics and operational factors. Elevated storage is more expensive to construct and maintain but might allow shorter connection pipelines. There is a practical limit to the size of elevated storage so that, in theory, the choice is between a large ground-level service reservoir serving a large area and several elevated water towers each serving smaller areas. The choice is then influenced by the configuration of the distribution system (Section 15.2).

20.4 STATUTORY CONSENTS AND REQUIREMENTS

Reservoirs and associated structures require the consent of the national and local Planning Authority. Regulations vary from country to country but in the UK reference has to be made, under the Town and Country Planning Act 1990, for consent. This action normally attracts comment from various statutory bodies, local organizations, environmental groups and other stakeholders depending on the sensitivity of the location. Early consultation with these bodies is necessary to avoid delay or refusal of the planning application. Section 5.27 sets out the relevant legislation in the UK and the types of body that may need to be consulted.

Where the proposed works may affect the apparatus of other statutory undertakers, they must be consulted, and Notices may need to be served (in the UK – under the New Roads and Street Works Act 1991) for any street works. Where public roads are affected, the Highway Authority should be consulted at an early stage. Discharge consent from the Environment Agency will be required for overflow and drain down discharge into a sewer, soakaway or watercourse. A range of non-statutory consultation may also be required (e.g. The Countryside Agency for landscape, and English Heritage for archaeology).

Under the UK Reservoirs Act 1975 a service reservoir capable of storing more than 25 000 m^3 above adjacent natural ground level is subject to having its design and construction supervised by a Construction Engineer and being inspected at least every 10 years by an Inspecting Engineer, both being qualified civil engineers appointed by the reservoir owner from a panel of engineers maintained under the Act. In 2010 in England and 2011 in Scotland the arrangements were changed (Section 5.23). There is no longer a fixed volume threshold; large raised reservoirs are designated on a risk basis as 'high risk' or 'not high risk' (England), or 'high risk', 'medium risk' or 'low risk' in Scotland. Non-compliance now constitutes a criminal offence.

20.5 WATER QUALITY CONSIDERATIONS

There is increasing awareness of the influence of service reservoirs on water quality. Ingress of pollutants is a particular concern and the reservoir structure, particularly the roof, must be sufficiently impermeable (Section 20.15). High storage times can be detrimental to water quality by allowing decay of disinfectant residuals and growth of disinfectant by-products such as trihalomethanes (THMs) (Section 7.24). Of particular concern is the potential for stagnant regions to form. Reservoirs with common inlet and outlet pipework are particularly susceptible to water quality deterioration since demand tends to be met direct from the reservoir inlet; this leads to low turnover of reservoir contents and high stored water age. To maintain good water quality, it is important to ensure that there is good turnover of water throughout the reservoir. This can be achieved by:

1. installing baffles to promote plug flow and prevent dead spots; or
2. positioning inlets and outlets to create good mixing and prevent stratification; and
3. using the operational regime to generate diurnal fluctuations in water level.

Perfect mixing or perfect plug flow in a tank is not achievable and even reasonable mixing or plug flow can be surprisingly difficult to engineer. The tank characteristics required for plug flow are very different to the characteristics required for good mixing. It is therefore best to decide which type of flow the basic tank design is most likely to promote and then modify the design to optimize those flow characteristics, using suitable modelling. This is now best achieved with Computational Fluid Dynamics (CFD) (Section 14.18). Useful guidance is given in the AwwaRF report *Water Quality Modeling of Distribution System Storage Facilities* (Grayman, 2000).

20.6 SAMPLING AND WATER TESTING

Monitoring of water quality in service reservoirs and water towers is required in the UK under the Water Supply (Water Quality) Regulations 2000 but, irrespective of any statutory requirement, routine monitoring of water quality is good practice to ensure that water in a service reservoir has not become polluted. At sites with more than one service reservoir or compartment, each unit should be sampled separately unless they are sufficiently inter-connected that a sample at the combined outlet is representative of both storage units. The sampling arrangements should always ensure that the water sampled is from the body of the reservoir or that leaving the reservoir. If there is no suitable location for a sample tap at the reservoir site, then a tap should be provided on the outlet main at the nearest possible point to the reservoir.

Tanks with combined inlet and outlet pipework should be sampled at different depths. For large tanks with poor mixing or geometry not conducive to plug flow it may be necessary to consider additional sampling, for example in the reservoir centre, near the roof or from areas where water could become stagnant. Dip sampling is not recommended and should only be used as an emergency measure. Some water companies have particular sampling requirements and may make it necessary to provide several sample points from each compartment.

Each sample pipe should be brought to a location where the sample can discharge, preferably by gravity, but using a pump if necessary, into a common sink which is not located in a basement considered as a confined space. All materials used must be suitable for contact with drinking water (Section 17.3) and should have no effect on the bacteriological or chemical quality of the sample. Stainless steel pipes and valves are preferred. The sample flow rate should be enough to flush the whole length of pipe at least three times in 5 minutes.

20.7 INSTRUMENTATION

The following instruments and equipment are normally required (see Section 20.14 for water level terminology and Section 19.38 for instrumentation and control).

1. Water level measuring equipment, preferably ultrasonic, in each compartment. If the location is subject to waves or rapid level fluctuation a stilling tube should be used.
2. Equipment for displaying the level measurement in digital or analogue form in the valve house or other secure location.
3. Level electrodes or float switches at TWL (high) and BWL (low) for normal on/off pump control and alarm purposes. Additional switches at intermediate levels may be required for some pump control schemes.
4. Level electrodes or float switches at OWL (extra high) and just below BWL (extra low) for alarm purposes and emergency control of pumps feeding to or from the reservoir.
5. A panel for display of level and alarm indications and for housing power supplies and a telemetry outstation if required.
6. A telemetry link to the pumping station or control centre to transmit water levels, flow rates through the inlet and outlet pipes, valve positions, power supply failure, telemetry failure and to carry signals for an intruder alarm and for remote control of valves, where appropriate.

Where reservoir foundation or structural movements are thought to be a risk, facilities for monitoring should be provided. These may include strain gauges and inclinometers as well as physical reference points at selected external locations.

20.8 OVERFLOW AND DRAIN DOWN DESIGN

Inflow controls are normally provided to prevent water level rising above a set normal maximum or top water level. The control may be via telemetry on source pumps or inlet valves or may use autonomous mechanical or hydraulic devices such as float or altitude valves (Section 18.9). However, failure of such controls is possible so that some fail-safe means of restricting water level has to be provided by means of an overflow. Reservoir roofs are not normally designed for uplift and flow out of vents or access openings is undesirable. Both could lead to structural damage or even reservoir failure.

The capacity of the overflow should be not less than the maximum likely inflow. For small reservoirs with small inlet pipework in a simple system it may be sufficient to set design overflow rate at the maximum inflow that the source could produce. However, provision of such overflow capacity and disposal in an urban area may be very onerous for a large reservoir. If the system supplying the reservoir is complex a risk analysis should be carried out with the aim of making reasonable provision for overflow without necessarily covering combined risks of very low probability, particularly where overwhelming of the overflow would not lead to loss of life or other serious consequences.

When it is necessary to drain a service reservoir down, as much of the contents as possible should be released to supply via the usual outlet arrangements. However, separate facilities should be provided to allow the rest of the contents to be drained to waste. These may have to be used for emptying the complete contents if contamination prevents use of the water in supply. Drain down times of 8–12 hours for reservoir compartments under 2000 m^3, and between 24 hours and 3 days for larger compartments, may be appropriate.

The overflow and drain down systems must not be a route for contamination of reservoir contents. The 'Ten States Standards' (Health Research, 2012) which are widely used in the USA require that overflows are not connected directly to sewers or storm drains but should discharge freely between 300 and 600 mm above ground or the top of a drain inlet. This allows overflows to be visible for corrective action and helps prevent vermin entry. Practice in the UK is to use a water trap; discharge to piped drainage is often permitted. A water trap is likely to become a breeding ground for mosquitoes so this solution is not appropriate in warmer climates. It should be noted that rats and mice can climb vertical surfaces of many finishes, can jump up to 900 mm (450 mm for mice) and can swim under small water traps as used in plumbing.

The 'Ten States Standards' also require the overflow pipe to be equipped with an internal mesh screen (Section 20.15) and suggest use of a duckbill valve at the exit. Mesh and non-return valves restrict flow and their use should be subject to hydraulic validation. Where overflow volume and conduit size are large a combination of mesh screens and flaps should be considered.

20.9 VENTILATION

Ventilation of reservoir compartments is needed to maintain a fresh supply of air above the water surface, for temperature control of that air and to admit or release air displaced by varying water levels in the compartment. The capacity of the ventilation system should be subject to risk analysis. It should be sufficient for the fastest rate of fall (or rise) of water level that is likely. Drain down must be allowed for but pipe burst may also need to be taken into account to avoid roof collapse.

The cross-sectional area of ventilation ducts or openings should be based on a suitable air speed (say 15 m/s). The air vents need to be insect and vandal proof and allowance should be made for the reduction in effective air passage area caused by insect screens (Section 20.15). In many cases, traditional mushroom-type roof ventilators have been found unsatisfactory in long-term service and to be a potential source of pollution. 'Vented' access covers have sometimes been used for small reservoirs but are also vulnerable to pollution. Alternatives include piped systems above the reservoir roof, leading to one or more ventilation chambers or ventilation ducts.

At unmanned sites or at very sensitive installations and where serious malicious intrusion is a risk, ventilation (and access) facilities should be designed against introduction of chemicals and explosives. The way this is done should be agreed with the reservoir owner but should not be made public (Section 20.15).

20.10 LOADINGS ON WATER STORAGE RESERVOIRS AND TANKS

Water storage structures must be designed to resist all loads which can arise. These include: water and structure self-weight, wind, water pressure, earth pressure, negative and positive air pressure (affected by vent capacity), thermal loads and dead, snow and live loads on the roof, including allowance for dynamic effects. The loadings to be applied in structural design are dealt within the codes discussed in Sections 20.11–20.13 and 20.20–20.23 for various types of water storage structure.

In seismic areas the hydrodynamic pressures due to water need to be taken into account as well as forces due to inertia of the structure and the modifying effect on any earth pressures on its outside (ACI 350.3). The hydrodynamic pressures are considered to have an impulsive component and a convective component; both are taken to act negatively on the inside of one wall and positively on the inside of the other. The impulsive component is due to the inertia of the water; the convective component is due to the lower period sloshing of the water excited by the earthquake. Apart from the pressures on the walls, sloshing needs to be taken into account in the freeboard and, if the wave reaches the roof, in the loading on the roof. The sloshing effect can be very pronounced, particularly at the corners of rectangular tanks.

20.11 WATER RETAINING CONCRETE DESIGN

British practice since 1987 has been to follow the procedures set out in BS 8007 for the design of liquid retaining structures. However, by 2010 in the UK, with the withdrawal of BS 8110 and BS 8007, design of concrete structures has to follow BS EN 1992-3, Eurocode 2. As with BS 8007 in relation to BS 8110, BS EN 1992-3 is based on limit state philosophy as used for the design of reinforced concrete structures to BS EN 1992-1-1. BS EN 1992-3 defines four classes of water tightness (0, 1, 2 and 3) for checking cracking serviceability. The first applies where some degree of leakage is acceptable or irrelevant. Tightness Class 1 is the case usually applying to water retaining structures where small leakage leading to damp patches or staining is acceptable. Tightness Classes 2 and 3 require appearance not to be impaired by leakage or, in the latter case, there to be no leakage. For both Classes 2 and 3 measures are needed to ensure that part of the concrete section remains in compression at all times or that a supplementary barrier such as a liner is applied. For Tightness Class 1 the width of any crack that is expected to pass right through the section is to be limited to:

$$w_{k1} = 0.875/(h_D/h) + 0.025$$

(with limiting values of 0.2 and 0.05 mm), where h_D is the hydrostatic pressure and h is the concrete section thickness.

Such cracks can be expected to heal in time subject to certain provisos. Where Tightness Class 2 applies (perhaps for exposed surfaces of tank walls) the minimum thickness of the concrete which is to be always in compression is the lesser of 50 mm and $0.2h$. The procedure involves the determination of crack widths and the reinforcement needed. Checks are made for other serviceability limit states as well as ultimate limit states. BS EN 1992-3 limits steel tensile stresses and bar spacing for different design crack widths for sections under axial tension.

Under BS 8007 water retaining structural concrete should have at least 325 kg/m^3 of cement with a maximum water/cement ratio of 0.55, reduced to 0.50 when pulverized fuel-ash or GGBS (ground granulated blast-furnace slag) forms part or all of the cementitious material. It also requires cover to reinforcement to be not less than 40 mm. There are no equivalent requirements in BS EN 1992-3 but the durability requirements of BS EN 206-1 and BS 8500, and any general requirements of BS EN 1992-1-1, should be followed in respect of exposure to water as well as aggressive ground, liquids or environments. Additional protective measures (APMs) may be required for protection of concrete; examples are: (1) application of an external waterproof membrane – usually a bitumastic material on a carrier film, protected either by a fibre board at least 10 mm thick or by concrete blockwork or mass concrete; and (2) increased concrete cover. It should be stressed that good quality control is essential for achieving satisfactory water retaining concrete.

In the USA and many other countries, water retaining concrete design is to ACI 350 using the load factor method. The requirements in ACI 350 are aimed at achieving a design life of 50–60 years. Improved durability requirements, for example tighter control of crack widths and concrete quality, may be needed if a longer life is required. Crack widths are not limited to particular values but are taken into account in the determination of maximum allowable stress in tensile reinforcement, a lower value of which is used for severe exposure. ACI 201.2 R provides guidance on durability of concrete. The parameters selected for concrete durability, including any APMs, may be tested via a 'life-cycle analysis' for which software is readily available. ACI 350 covers unreinforced, reinforced, prestressed and precast concrete as well as wire wound prestressed circular tanks.

In countries where neither European nor US standards are used, relevant local codes of practice should be followed. In the USA and elsewhere, it is a legal requirement that structural designs are done by an engineer registered in the state concerned, or signed-off by one. It is prudent to take account of local constructional abilities and the quality of materials available and amend factors of safety if necessary. Design practices may need to be modified to take account of local prices so that an economic construction results.

20.12 PRESTRESSED CONCRETE CIRCULAR TANK DESIGN

Design of prestressed/post-tensioned concrete tanks is covered in AWWA D110 and D115 and in ACI 350, ACI 372 and ACI 373. The last has been withdrawn but has not been replaced and is still referred to in the AWWA standard.

It should be remembered that, as with any post-tensioning, tendons stressed earlier in the process lose tension as other cables are stressed later. This has to be allowed for in the design, along with concrete creep, temperature and moisture content effects, either by providing excess tension at the outset or by returning to apply additional load. Friction is also a serious issue and leads to load at the jack exceeding load in the middle of the tendon by a significant margin. Distortion of the cylindrical shell can arise unless the tendon arrangement and order of stressing are carefully planned.

20.13 WELDED STEEL PLATE DESIGN

Large steel plate tanks are usually cylindrical with their axes vertical. Steel tank design in the UK is covered in BS EN 14015 and must now follow BS EN 1993-4-2, Eurocode 3. Under this Eurocode steel tanks for water come under Consequence Class 1 for which membrane theory may be used for determining principle stresses and simplified expressions may be used to determine local bending effects. Loadings are to be as defined in BS EN 1990, Eurocode 0. The shell is to be checked for plastic limit, cyclic plasticity, buckling and fatigue. Serviceability checks are to be made for deformations, deflections and vibrations. Minimum carbon steel plate thickness for tank bottoms excluding corrosion allowance is 5 mm for butt welded plates and 6 mm for lap welded plates. Reinforcement at openings is calculated by the area replacement method.

Otherwise, design of steel tanks for storing water can be to AWWA D100 which, for certain details, makes reference to API standard 650 (oil tanks). Both these codes include basic and refined design procedures. The basic procedures use conservative allowable stresses (the same in both codes) and are based on simplified design rules. With these, the steel plate selected is usually the cheapest that satisfies the rules for the intended service although a wide range of steel grades are permitted. The refined design procedures recognize the benefits of higher grade steels, an advantage for higher loaded members such as walls. The steel grade must be weldable and suitable for the stress and temperature ranges expected. Toughness needs to be taken into account for higher strength steels; it reduces with increased thickness but is improved for fine-grained steels and for those with higher manganese content.

One difference between the AWWA and API codes is that the latter allows the excess thickness at the top of one wall plate to be taken into account in determining the thickness of the plate above it.

D100 requires the plate thickness to be based on the stress in the highest loaded extremity. Loads to be allowed for in the tank walls include those from the contained liquid and either wind or seismic effects. Shape factors are given in the standards and provide a convenient means of determining wind pressure. Such lateral loads may induce buckling in tall shell cylinders if the roof provides insufficient bracing. The factor of safety against buckling should be calculated and wind girder stiffeners included if necessary (Rajagopalan, 1990). AWWA D100 contains factors for determining increases in plate thickness for different tank heights and seismic accelerations. Sloshing of tank contents may need to be considered separately since it can occur in earthquakes, depending on the tank size and the frequency of seismic oscillations (Section 20.10). Distortion may result but failure is rare. When a tall tank is empty wind loads can cause floor lift near the shell on the windward side; anchors may have to be provided to prevent this.

Under AWWA D100 roof loads should include dead weight, snow or other live loads, wind and vacuum or internal pressure commensurate with the capacity of the ventilation system. Buckling safety factors should be determined for spherical or ellipsoidal roofs. Minimum plate thicknesses given in the codes are 6.35 mm (1/4″) for floor and 4.76 mm (3/16″) for roof. Care should be taken with penetrations of the shell, floor or roof where additional stiffening or reinforcement may be needed. Guidance on these and on other details is given in both AWWA D100 and API 650. Care should also be taken at the joints between wall shell and both roof and floor. Five designs of the latter are covered:

1. Tank founded on 75 mm of oiled sand with the shell located on and bedded in grout on a reinforced concrete ring beam.
2. Tank founded on a reinforced concrete slab covered by 25 mm of oiled sand or 13 mm of joint filler.
3. Tank founded on 150 mm of oiled sand with the shell located inside a reinforced concrete ring beam leaving a gap of at least 19 mm.
4. Tank founded on a platform of graded stone or gravel with side slopes of 1 in 1.5 and surrounded by a level berm of at least 1 m width.
5. As for Type 4 but with a steel retaining ring extending into the gravel platform.

Rigorous corrosion protection is required including: meticulous surface preparation; use of galvanized plates; zinc rich priming; and paint systems such as epoxy resin, glass flake resin and elastomeric coatings (Section 20.22). Access has to be provided, except under the floor, for the inspection and maintenance of coatings in all areas.

SERVICE RESERVOIRS

20.14 RESERVOIR SHAPE AND DEPTH

The most cost-effective shape of a reservoir is circular in plan but the area of land required is greater. Except where a storage facility comprises several storage tanks, service reservoirs are generally built with at least two compartments so that one can be drained for maintenance. Reservoirs which are circular in plan are less suitable for subdivision. Nevertheless circular tanks permit the use of prestressed concrete or steel, which may offer cost advantages. For a two-compartment rectangular reservoir the most economic plan shape is usually obtained when its length (measured perpendicular to the division wall) is 1.5 times its breadth. These proportions may need alteration in the light of the shape and slope of the site, the cut and fill balance, pipework configuration for circulation, and any future extension likely or amenity requirements. If significant or abnormal soil settlements are expected, there may be advantages in providing two adjacent structurally independent single-compartment reservoirs instead of one two-compartment reservoir.

There is an economic depth for any service reservoir of a given storage capacity. The greater the depth the less length of wall and area of roof and floor that are needed, though the unit cost of the wall increases with increased water depth. There can, however, be other constraints on the depth such as foundations, the character of the available site or the desirable range of distribution pressures. Depths most usually used for rectangular concrete reservoirs are:

Size (m^3)	Depth of water (m)
Up to 3500	2.5–3.5
3500–15 000	3.5–5.0
Over 15 000	5.0–7.0

The following parameters are key to the design:

- TWL, usually the level at which the supply into the reservoir is to be shut off;
- the overflow weir level (OWL), giving a small margin above TWL;
- the MWL needed to discharge the maximum possible inflow over the overflow weir;
- BWL, being the lowest level to which the water should be allowed to fall for the purposes of supply;
- lowest roof soffit level – allowing for roof slope.

A freeboard between MWL and the roof soffit is required for ventilation and should not be less than 150 mm above MWL or less than 300 mm above TWL. Settlement of precipitated or suspended solids may occur in reservoir compartments. To prevent turbid water being drawn into supply, BWL should be not less than 150 mm above the highest level of the floor. It may need to be higher, depending on the outlet arrangements (Section 20.26).

20.15 COVERING AND PROTECTING RESERVOIRS

Reservoirs for treated ('finished' in the US) water must be protected from contamination, vandalism and criminal or terrorist acts. Contamination can enter via leaks in the roof or other parts or carried in by vermin, birds or insects (van Lieverloo, 2006).

In temperate climates flat-roofed concrete reservoirs are usually covered over with earth and grass for appearance and temperature insulation. This involves maintenance and grass cutting, but an uncovered reservoir may result in amenity objections. The earth cover to the roof should comprise grassed topsoil 150 mm thick, over a fabric filter membrane laid over 100 mm of single size

20 mm round gravel forming a drainage layer. The drainage layer is laid over a waterproof membrane (Section 20.24). A 150 mm thick gravel layer on the waterproof membrane, with no topsoil, can be used in arid countries if there are no amenity objections; this provides thermal insulation which is essential in hot climates. The earth banks against the external reservoir walls must be designed to stable slopes and, for ease of grass cutting, should not be steeper than 1 on 2.5. Topsoil cover to banks should be not less than 150 mm vertical thickness.

Reliable options for the waterproof membrane include:

- unbonded high density polyethylene sheets, at least 2 mm thick, bonded together on site by fusion welding, laid without protection board;
- self-adhesive, two or three ply, cross-laminated carrier film, laid with lapped joints on a thoroughly prepared and cleaned surface; laid with protection board if the gravel layer, or the placement of it, risk puncture.

To ensure waterproofing, all openings for access, ventilation and instruments must be contained within upstands raised at least 200 mm above the finished level of the earth or gravel covering, with the membrane protection taken up and sealed into a groove near the top of the covering. At free edges the membrane should be taken over the edge of the roof and bonded to the top of the wall at least 200 mm below the underside of the roof slab.

All openings not sealed with water at all times should be covered by strong, rot-proof insect mesh. Practice in the USA is to use number 24 mesh; woven stainless steel mesh of this size with 0.345 mm wire has square openings of about 0.7 mm and a net open area of 45%. Another source (Ainsworth, 2004) recommends a mesh opening size of 0.5 mm, equivalent to number 34 mesh woven with 0.25 mm stainless steel wire. The aim is to prevent access by all insects.

Special attention has to be paid to ensure service reservoirs are secure against vandalism, acts of terrorism and theft. Guidance is given in the UK water industry *Code of practice for the security of service reservoirs, 1997* (this is a restricted document, accessible through managers responsible for security in each of the water companies). An impact and vulnerability assessment should be undertaken to determine the level of risk and hence the security measures necessary. Secure perimeter fencing can minimize ordinary vandalism but is not sufficient protection by itself. Access openings to the reservoir can be screwed and locked down; but, if possible, concealment is better, their location being known. Roof air vents are a problem because they are a potential source of pollution and access (Section 20.9). Valve and instrument houses and their doors should be of strong construction and should have no windows. Sampling points are necessary on both inlet and outlet mains and they too should be protected. All reservoirs should be visited frequently to ensure that none of the protective measures have been tampered with.

20.16 SERVICE RESERVOIR STRUCTURES

The earliest service reservoirs were built in masonry (usually brick) on a concrete or masonry base. Roofs were commonly vaulted and supported on masonry columns. Many such reservoirs are still in use in the UK – some, such as Honor Oak in South London, being very large. Smaller brick reservoirs were often mortar lined to assist water tightness. Masonry is a flexible material that can accommodate movement but the cracking that may result and the gradual erosion of mortar lead to ongoing maintenance problems so that old masonry reservoirs may need to be replaced (in reinforced concrete).

With the advent of reinforced concrete in the first half of the 20th century, this material became very widely used for service reservoirs. Until about 1980 reservoirs were jointed but after that monolithic construction became common because:

1. Joints require considerable care in construction and frequently are the cause of poor concrete and leaks.
2. Monolithic construction usually requires less concrete and reinforcement.
3. There is a better understanding (with modern codes) of the shrinkage and stress cracking of concrete.
4. There is a better understanding of soil–structure interaction and a better ability to model it to establish whether monolithic construction can take the movements.
5. Piling has become cheaper and has enabled monolithic construction on poor ground to be feasible.
6. Site practice can now achieve satisfactory concrete with high wall pours and large distances between construction joints.

The materials adopted for reservoirs today depend on their availability and unit cost, local skills, client preferences and on the topography and geology of the site. The materials that should be considered are the following (starting with those suitable for large reservoirs and ending with those for small tanks): concrete (reinforced (RC) or prestressed), steel, glass reinforced plastic (GRP) and polypropylene.

20.17 RECTANGULAR JOINTED CONCRETE RESERVOIRS

Jointed concrete reservoirs normally have joints: between lengths of wall; between the top of walls and the roof; between floor panels and their junction with wall bases and columns; and in roofs dependent on their area. The floor and roof are usually parallel so that walls and columns are of constant height.

Walls are usually reinforced concrete free-standing and cantilevered from a substantial base and stable against sliding and overturning (Fig. 20.2) under soil or water loadings. An unreinforced mass concrete wall may be used if reinforcement is locally difficult to obtain for some reason. A sliding joint is normally provided between the top of the wall and the roof to prevent transfer of load due to roof thermal movements. Vertical joints with waterstops and sealing grooves are provided in the wall at spacings of about 12.5 m for contraction joints and not exceeding 30 m for expansion joints.

Columns of reinforced concrete are normally arranged on a rectangular (usually square) grid pattern. A column spacing of 5 m results in a flat-slab roof of economic thickness without the need for dropped panels. The side dimension or diameter should be not less than 300 or 350 mm, respectively and not less than one twentieth of the height from reservoir floor to bottom of column head.

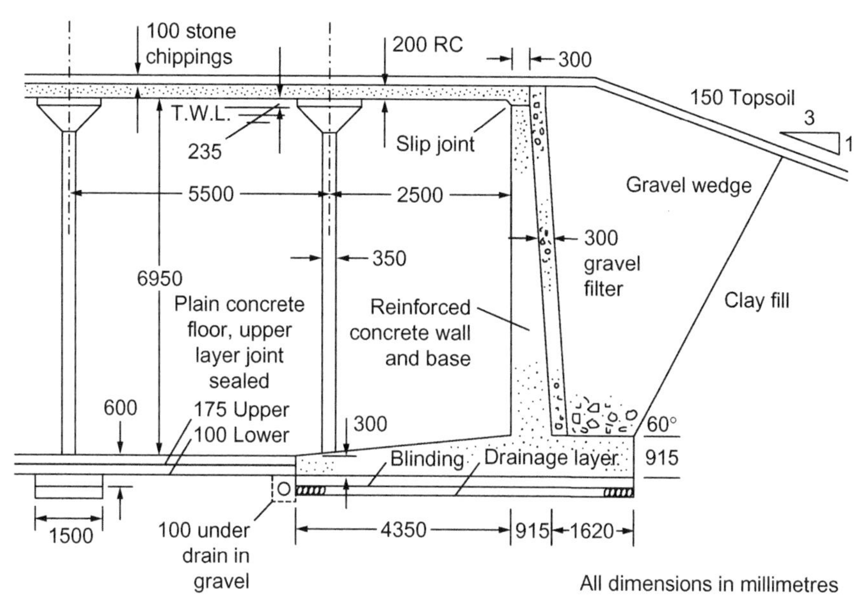

FIGURE 20.2

Section of jointed reinforced concrete service reservoir with two-layer floor slab.

The ***floor*** is cast as a single or two-layer slab in square panels having a side length equal to the column spacing. The single layer slab, typically 175 mm thick, is suitable for founding on a firm, non-compressible material. It is laid on a membrane of low frictional resistance; it may be unnecessary to provide reinforcement if the subsoil is firm and of uniform bearing capacity. The two-layer slab has an upper layer, typically 175 mm thick, over a lower layer, typically 100–125 mm thick. A membrane between the layers permits sliding of the upper layer. This design is suitable for a clay subsoil. Usually only the top layer is reinforced, the reinforcement being discontinuous through the contraction joints. With these types of jointed floors, uplift pressures must be prevented by provision of an under drainage system which has a free discharge to a lower level.

Joints separating floor slabs should be of the 'complete contraction' type, incorporating a joint sealant at the water face (Fig. 20.3). Externally placed waterstops are generally used on the underside of the base slab since these allow better compaction of the concrete at the waterstop than with the centrally placed type. In a two-layer floor, the joints in each layer should be staggered to avoid vertical alignment. Where possible, the upper floor slab should not be cast until the reservoir, including the roof, is substantially complete. This helps to avoid excessive shrinkage, temperature movements and joint damage and fouling before the joint sealant is applied.

The ***roof*** is a reinforced concrete slab of uniform thickness, minimum 200 mm, and is monolithic with the column heads. This is acceptable because the columns are flexible enough to permit roof expansion and contraction. If expansion joints are needed (depending on exposure and insulation) waterstops must be of the centre bulb type; such waterstops must be provided at any other joints such as construction joints. The roof design must allow for the impact loading of construction plant placing gravel and soil on the roof and for any other live loading that may occur.

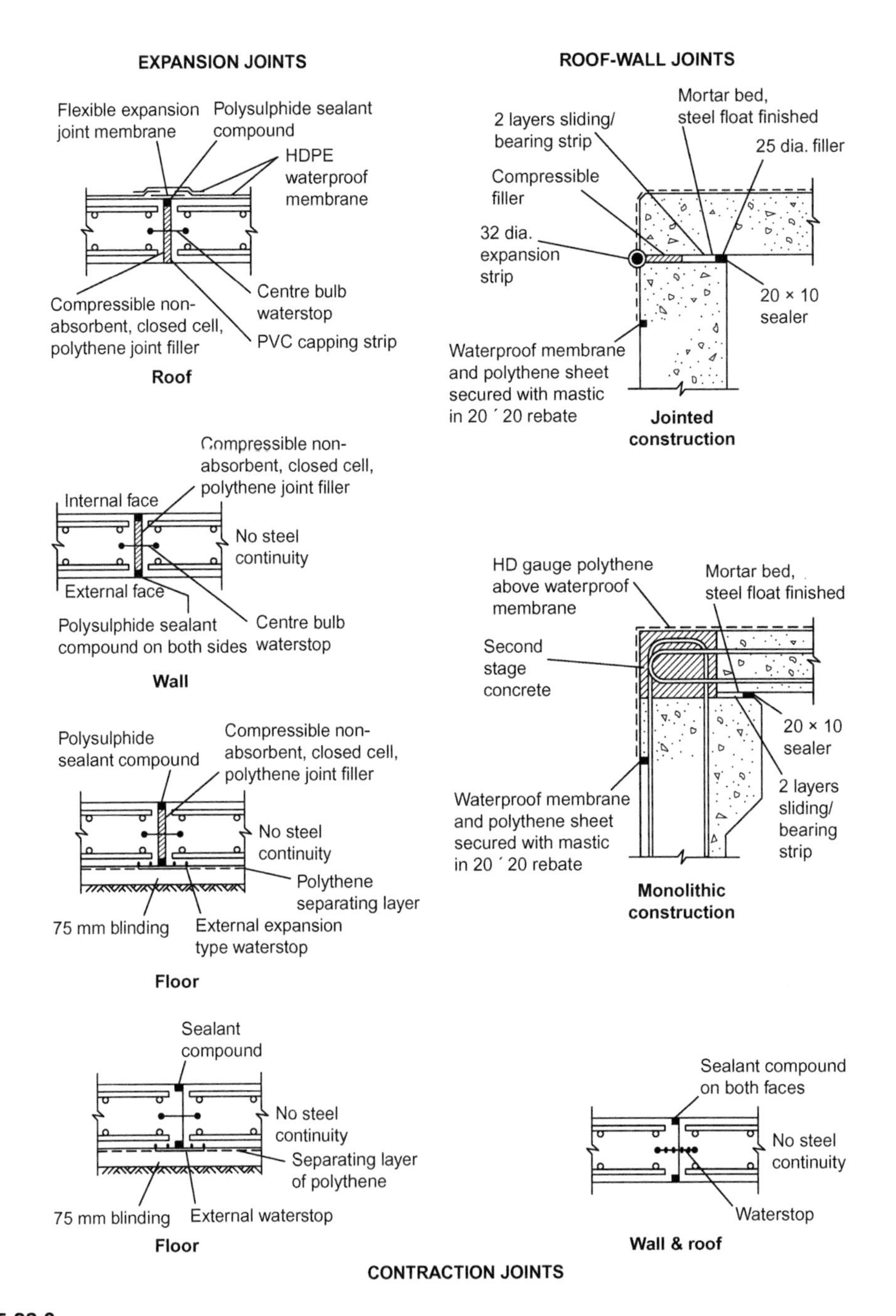

FIGURE 20.3

Typical joints for reinforced concrete reservoirs (and other water retaining structures).

20.18 MONOLITHIC CONCRETE RESERVOIRS

A monolithic concrete reservoir has reinforced concrete walls, floors, columns and roof, in which there are few (if any) permanent movement joints. In some cases the walls and floor are monolithic but there are sliding joints between roof and the top of walls. This type of design has been found to be structurally economical in most situations where the underlying ground (after improvement if necessary) can support the load without risk of appreciable differential settlement. The reservoirs are normally rectangular in plan (Fig. 20.4) but circular and other shapes are feasible.

External walls are usually vertical or near vertical on the inner face but battered on the outer face to give the tapered section appropriate to the form of loading (Fig. 20.5). Depending on the height of the walls and the length of the roof slab, monolithic connections with floor and roof slabs can result in lower bending moments and shear forces (especially in the vertical plane) than is the case with jointed structures. The roof is normally constructed in two stages; the second stage at the wall interface being cast after the initial thermal shrinkage has taken place in the roof slab. Within the walls, joints are usually restricted to partial contraction joints (discontinuities in the concrete with 50% of the main horizontal reinforcement passing through) with a sealing groove on each face. The maximum spacing of partial contraction joints should be 7.5 m to avoid unacceptable cracking. For operational reasons, the division wall is usually full height and can therefore assist in supporting the roof. The columns are arranged on a square grid, the span to external walls being typically reduced to three quarters of the normal spacing.

An economical form of floor is a reinforced concrete slab of uniform thickness except at the perimeter, where it should be thickened to cater for moments transferred from the walls or resulting from differential vertical movements as between perimeter and centre (Fig. 20.5). Local thickening of the floor below columns should be avoided as it can be awkward and costly to construct; instead additional reinforcement under the columns can be used to increase the shear strength. Local thickening is usually required at drainage channels and sumps where these are included. Joints are normally restricted to construction joints. Plate 34(a) shows an internal view of a monolithic RC reservoir nearing completion.

20.19 CIRCULAR REINFORCED CONCRETE RESERVOIRS

The circular reservoir makes the most efficient use of materials since it needs a minimum wall length for a given plan area. Part of the water load on the walls is taken in tension by hoop reinforcement. As the reservoir size increases crack control becomes more difficult but this design has been used extensively for smaller reservoirs, both buried and unburied. However, the curved formwork required for the walls and the double curvature formwork usually needed for the roof are expensive and tend to outweigh the savings in concrete materials by adopting the circular shape.

Inlet and outlet pipework is usually arranged through the base with access and ventilation through the roof. If a division wall is required it can be in the form of a concentric inner wall with an internal radius about 70% of that of the outer wall. If a diagonal division wall is used, thickening is needed at the joints with the circular wall to cope with the horizontal moments; this reduces the simplicity of the design. Circular reinforced concrete reservoirs are of monolithic construction

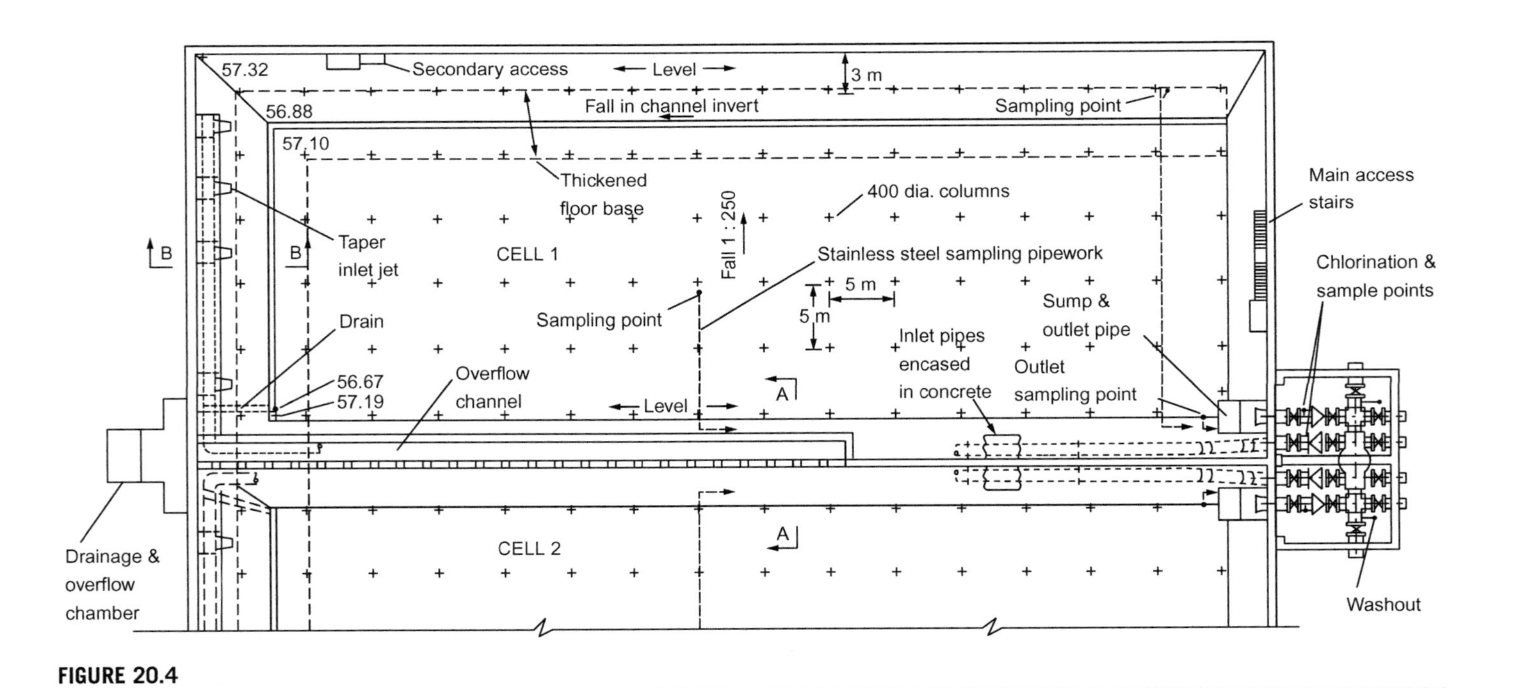

FIGURE 20.4

Plan of monolithic reinforced concrete service reservoir.

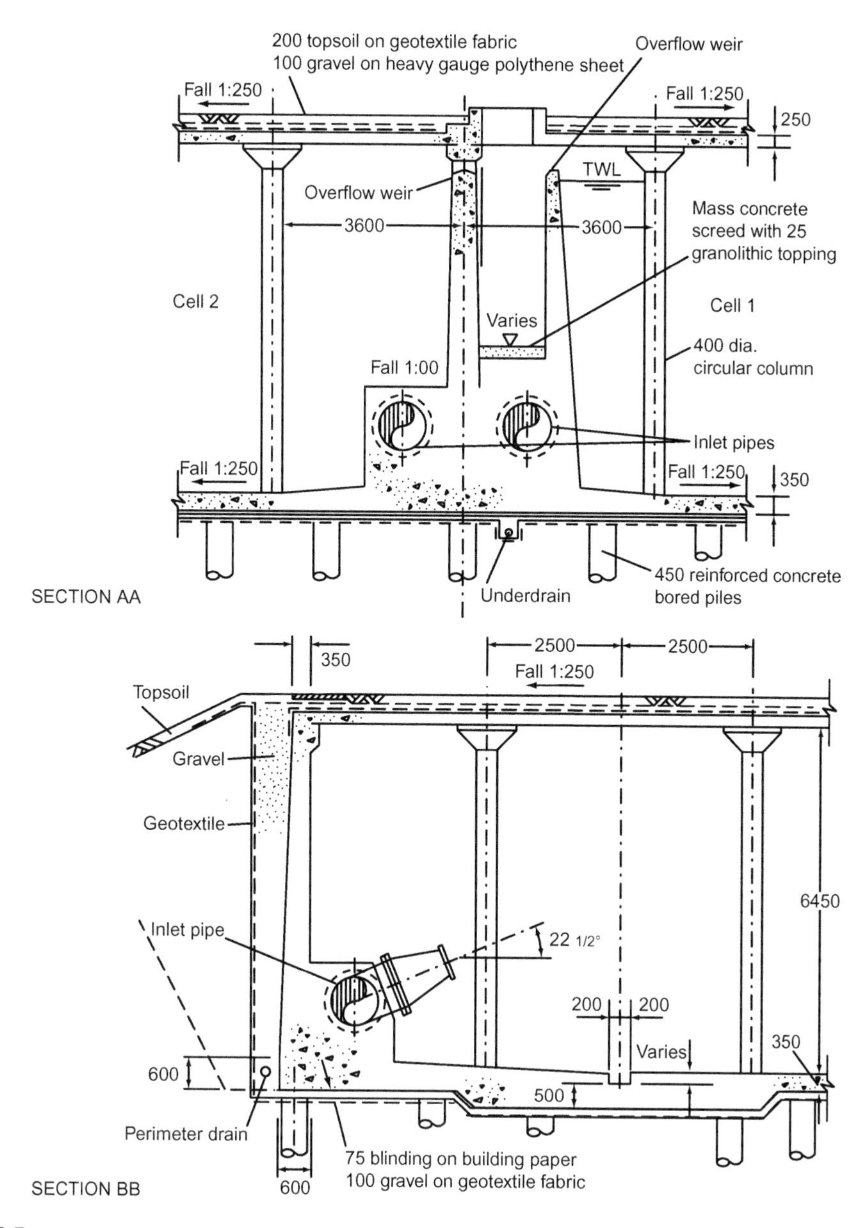

FIGURE 20.5

Sections of monolithic reinforced concrete service reservoir.

with stiffening beams at the top, and bottom if necessary, of the wall. Roofs may be of the self-supporting thin shell type or flat with columns. A full hemispherical roof imposes no radial load on the wall but this shape is undesirable for aesthetic and planning reasons and is difficult to construct. A satisfactory span to height ratio is about 8:1 but requires thrust to be taken at the top of the wall by a ring beam.

20.20 CAST IN-SITU PRESTRESSED CONCRETE RESERVOIRS

For larger circular reservoirs hoop tension in the walls needs to be resisted by stressed tendons to eliminate cracks in the concrete. The compression in the concrete is usually arranged to match the tension caused by the maximum internal water load plus a margin to ensure that the concrete is always in compression. Two methods of wall circumferential prestressing have been used, both strictly speaking being post-tensioning. With the first, tendons inserted in ducts cast into the concrete are later stressed (AWWA D115). The second is where wires or strands are wrapped around the outside of the wall and later covered with sprayed concrete (AWWA D110). A particular case of this involves winding the cables under tension. This allows the cable spacing to be varied up the wall to provide exactly the required load profile. Otherwise, the tendons are arranged in groups and provide a stepped load profile which sets up secondary vertical bending moments in the wall. Wires and strands should be galvanized or fusion bonded epoxy coated.

Except in seismic areas, were it not for the need to ensure a good seal at the base of the wall, a frictionless sliding joint would be ideal since it induces no shear or bending in the wall base. However, a completely frictionless joint cannot be achieved so that some horizontal load transfer takes place, even with a sliding joint. To deal with this, additional prestress has to be applied at the base of the wall to compensate and thereby achieve the correct wall profile. A more usual joint is the pinned joint. This may take various forms, some involving 'pinning' after cable stressing. Sealing is achieved by use of a waterstop and sealants. Where the joint is monolithic the resulting shears and moments have to be allowed for. For pinned or fixed walls over about 8 m high the walls are usually prestressed in the vertical direction. Otherwise vertical reinforcement is sufficient. For large reservoirs, where radial loads on the base slab would otherwise be difficult to resist, the slab may be radially prestressed, with the cables anchored through the base of the wall in some cases. Under seismic loading the walls of a circular tank tend to lift; this effect may need to be counteracted by the use of 'seismic cables' (AWWA D115) between the footing and wall.

Roof construction may be either self-supporting spherical or ellipsoidal shells or flat slabs supported on columns as described in Section 20.17. Precast concrete profiled planks have been used as a permanent shutter for an in situ concrete roof. This reduces the amount of internal falsework needed. In theory, a shell roof need only be lightly reinforced to carry the load. However, thermal effects (usually not uniform) result in movement and cracking. The working of such cracks and their penetration by foreign matter over time can cause gradual increase in load on the wall ring beam and may result in settlement of the roof profile. Such settlement can further increase the load on the top of the wall and lead to failure.

While much used in the second half of the 20th century, prestressing of water tanks is now less common. One reason is that the thinner concrete sections and the use of prestressing require a high degree of skill and control in construction. Another reason is the ongoing inspection requirement, particularly with the 'wire wound and shotcreted' form of construction, as this is subject to concrete spalling and wire corrosion. A significant factor is the increased attention to health and safety in the built environment. High locked-in stresses in prestressed concrete represent a considerable hazard during demolition since the energy released can be damaging and unpredictable. This requires controlled demolition in stages and with a number of additional precautions.

Although published in 1961, Creasy's book *Prestressed Concrete Cylindrical Tanks* (Creasy, 1961) remains a valuable reference to those coming across this form of construction.

20.21 PRECAST CONCRETE PANEL RESERVOIRS

Precast concrete has been used for many years in the construction of tanks and containers required for waste and water treatment and other industries. With improved design, joint detailing and with the benefits of reduced time on site and cost, use of precast concrete in treated water reservoir construction has become widely accepted. Apart from roof construction using precast concrete beams and slabs, precast concrete is now increasingly used in panels for walls. These can be simply reinforced but, for the higher walls needed for storage reservoirs panels, they are pre-tensioned in order to minimize weight and cost. Panels may be used for small rectangular tanks where supported top and bottom or with internal braces; however, to ensure watertight joints between the panels either the joint section is cast in-situ or horizontal post-tensioning is used to keep the joints in compression.

Precast concrete panel walls are most beneficial for circular tanks where they can be used in conjunction with post-tensioning, designed to withstand all of the internal loads on the walls. Post-tension stressing may be applied as for in-situ cast circular post-tensioned tanks (Section 20.20) by wires wound around the assembly after erection and then covered in sprayed concrete. However, precast panels can incorporate ducts for tendons (AWWA D115), with closer duct spacing in the lower portions. Ducts may be diverted around wall penetrations but large penetrations are best avoided due to the weakness created and because of the friction increase when tensioning tendons. There should be at least two tendon stressing locations around the perimeter where the panel is arranged for tendon exit and anchorage. Tendons may be bonded or unbonded to the duct. Unbonded tendons must be sheathed with high density polyethylene or polypropylene with continuity arranged at anchorage points. Damage to sheathing must be avoided. Non-metallic ducts and tight duct position tolerances assist. Corrosion resistance can also be provided by epoxy coating the tendons and grouting them after completion.

Base of wall details and roof construction considerations are similar to those for in-situ post-tensioned circular tanks.

Demolition of tanks post-tensioned with unbonded tendons is hazardous and requires the exposure of each anchorage and the systematic release of tension in each tendon by use of jacks.

20.22 STEEL PLATE RESERVOIRS

Circular steel above ground reservoirs have been used for water storage since before World War II. Steel reservoirs with capacity up to 100 000 m^3 are now in use in many countries, particularly in North America and the Middle East (Plate 34(b)). The design is of all welded steel plate (Fig. 20.6) and is very similar to that used for oil storage but greater attention is paid to coatings. The floor of a circular steel reservoir is made up of rectangular plates of thickness sufficient to take any radial tension and provide some contingency for corrosion, where required by design standards. The plates are either butt-welded together or lapped and fillet welded on one side only. Walls are made of butt welded rectangular plates but are thicker. Thickness is matched to circumferential tension according to the position in the wall. The wall base is fillet welded to a ring plate at the edge of the floor. This is thick and wide enough to prevent rotation and lifting of the floor under water load. Roofs are either conical or of low rise spherical shape, sometimes with a tighter curvature at the perimeter (the 'torispherical' shape). Purlins and light truss supports on columns may be used or the roof may be self-supporting.

The tank base is usually founded on a concrete slab for small tanks or on a bed of sand – oiled sand has been used in arid locations – or fine gravel. Where foundation loadings require it the granular bed should be retained by a reinforced concrete ring beam placed centrally under the wall shell. This prevents local shear failure of the foundation. In common with all ground tanks, foundation settlements should be evaluated. These may be more in the centre than at the perimeter. Differential settlements around the perimeter tend to cause the shell to cant or twist and the walls to get out of vertical. Jacking distorted tanks back into shape has been successful where movement was excessive but is best avoided.

Overflow and inlet and outlet pipework is usually arranged to penetrate the wall 'shell' via circular nozzles around which the shell plate is reinforced. Washout and drain down pipework usually exit via the ring plate into a sump below the wall. However, thermal movements of the tank

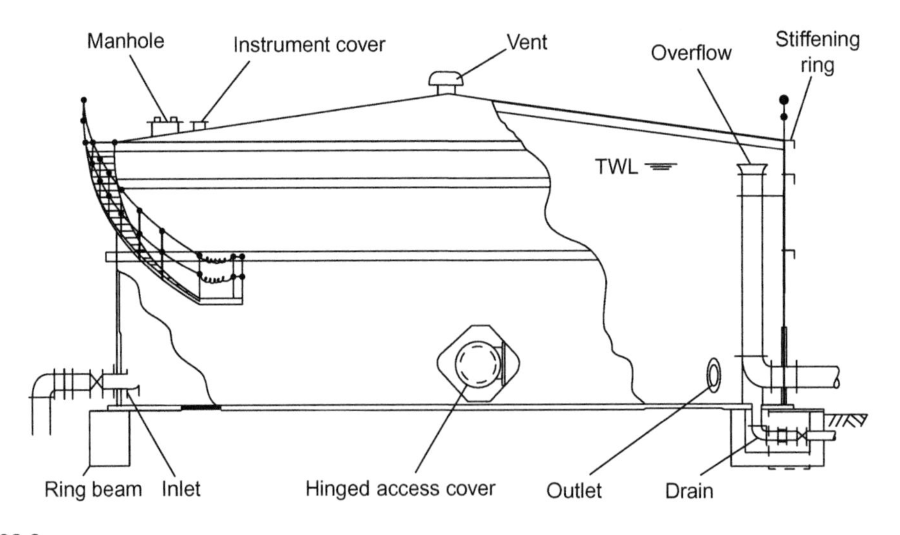

FIGURE 20.6

Welded steel ground tank.

perimeter may be large and may need to be accommodated at pipework by special expansion and movement joints. Access to the tank is via flanged manholes in the shell. Access to the roof is often provided by external ladders or stairs. In both hot and temperate climates the water temperature is increased by action of the sun on the steel surfaces and can lead to bacteriological issues.

The light construction cannot resist uplift forces when the tank is empty; therefore the design is not suitable for sites subject to flooding or high ground water. The tank walls should be left exposed, for inspection of coatings, with a level space 1.2 m wide all round. The finished ground should be battered back from the level strip if the tank base is below original ground level and suitable drainage should be provided.

Coatings for steel plate tanks are covered in AWWA D102. A shop primer is used to protect surfaces until field coatings can be applied after tank construction. A pre-construction priming coat may also be applied. Such primers should be compatible with the coating system and should be either formulated for welding or reduced or removed in the weld area. All paints are permeable to some, usually minor, extent. Permeability reduces with thickness but it important to ensure that the coating adheres well to the steel substrate under all conditions expected. Surface preparation of tank interior surfaces must be to a high standard, as for steel pipes (Section 17.20). Ambient conditions for coating application need to be within acceptable limits for the system; this may require climate control inside the tank and covering outside. Plates used for the floor may be primed on the underside but this can be affected by joint welding leaving the joint area more at risk. If corrosion of the underside of the floor is a concern cathodic protection should be considered (BS EN 16299).

Interior coatings for steel plate reservoirs must be suitable for use with drinking water (Section 17.3) and should have a zero or low volatile organic carbon content and high (near 100%) solids content. Interior coatings are usually epoxy or polyurethane two-component liquid systems with total dry film thickness varying from 200 μm (8 mil) to 640 μm (25 mil) depending on the system. Inaccessible surfaces should be minimized but some are unavoidable. These may be sealed in some way or left untreated. Sacrificial anode or impressed current cathodic protection may be used in conjunction with coating to protect wetted surfaces (AWWA D104 and D106).

Exterior coating may be alkyd systems with a rust inhibitive primer, or acrylic, polyurethane or epoxy systems each with zinc rich primers. Alkyd systems are not suitable on surfaces subject to condensation. Total dry film thickness is at least 150 μm (6 mil) or 100 μm (4 mil) for an acrylic system without a second primer coat. The coating system should be decided on the basis of life to first maintenance for the exposure conditions expected.

The glass coated steel plate tank has become more popular in recent years, particularly for the storage of some chemicals and for smaller water tanks. The plates are pre-cut, drilled and then coated with a bonded glass coating. The plates are bolted together on site to a pattern with a sealer applied at the contact. The tanks appear to perform well; any leaks at joints are visible and can be resealed. However, the coating is delicate and is difficult to make good if damaged.

20.23 PANEL OR SECTIONAL TANKS

Panel or sectional tanks were developed for military purposes to allow mass production, adaptation to many capacities and configurations and for easy transport and assembly. Originally, sections were of pressed steel with flanges drilled for bolting together through a gasket. Panel sizes were then 4 foot square and this size is still available (1.22 m) but 1.00 m square and rectangular panels are now produced. Coatings of steel panels are now usually epoxy or borosilicate glass and are covered in AWWA D103.

GRP panels are increasingly used but their flanges are not as robust as those of steel and need to be treated with care. Other plastics have been tried but are not favoured due to temperature distortion.

The use of internal stiffening braces theoretically allows any capacity to be achieved, albeit limited to a depth of about 6 m for steel (or 4 m for GRP). Sectional ground tanks are raised off the ground on dwarf walls at spacings to suit the panel size. These allow access for assembly and maintenance. Panels can be made with flanged nozzles for pipe entries and instrumentation.

20.24 DRAINAGE AND WATERPROOFING CONCRETE SERVICE RESERVOIRS

Where the reservoir external wall is designed as a free-standing cantilever or is of mass concrete, the backfill against the wall should comprise a vertical drainage layer of gravel about 300 mm thick, extending the full height of the wall and continuing down over the wall heel to link with a drainage system. Where the wall is monolithic with the reservoir floor it is still advantageous to retain the vertical gravel layer to control the ground water level and also to transmit water draining from the roof (Fig. 20.5). The wall and roof drainage systems must be kept separate from any underfloor drainage system. Wall drainage pipework should discharge into an observation chamber to help locate any leaks.

Ingress of water and pollution through the roof must be prevented by a positive waterproof membrane. For new reservoirs an adhesive membrane such as Bituthene DW is recommended, protected by heavy duty polyethylene sheet under the gravel drainage layer. The roof gradient should be no flatter than 1:250 for drainage; the floor slope should be made parallel to it so as to maintain constant wall and column heights. The simplest way to achieve this is to provide the slope in one direction only. Where there is no vertical wall drainage layer and where support from the embankment is essential for wall stability, a low peripheral kerb provided along the lowest edge of the roof can act as a collector for piping the water away. Inside the reservoir, the floor should have a shallow collecting channel leading to a drainage sump to aid cleaning of the floor of the reservoir.

The underfloor drainage system is usually laid to a rectangular pattern, normally comprising porous pipes surrounded in gravel in a trench below the floor. The layout should make it possible to observe drainage or leakage flow from separate areas of the floor. The porous pipes from each area are continued in ductile iron piping laid in concrete below the wall and the embankment and discharge individually to collector manholes, from which there is a free outfall pipe to some lower point. With jointed reservoir construction there must be no possibility of the drain outfall being submerged. If, on the other hand, conditions are such that uplift below the floor is unavoidable, then monolithic construction must be adopted, with the unit area weight of the floor, columns and roof being made greater than the design uplift pressure by an appropriate margin.

20.25 ACCESS TO SERVICE RESERVOIRS

Access to each reservoir compartment is needed for personnel, plant and materials. Access openings are usually sized to allow entry by a person wearing breathing apparatus. Access openings for plant and materials should be larger. Upstands should be provided around each opening to prevent surface water entering the reservoir. Covers to all openings must be robust but they do not normally need to be designed to support heavy loadings. They must be secure to prevent unauthorized access and must not allow rainwater to enter the reservoir. Lift-off covers risk introduction of mud and debris into the reservoir; therefore hinged covers are preferred but they must have an effective system for holding them in the open position when the access is in use.

For personnel entry into the reservoir the preferred arrangement is an inclined ladder leading to a platform about 2.5 m below the roof and a stairway leading from the platform to the floor. Where a stairway height exceeds 3 m, an intermediate landing is required. Reinforced concrete construction is recommended for platforms and stairways as this needs less long-term maintenance. The platforms can either be supported on columns or, in some cases, cantilevered from the walls. Alternatively the platforms and stairways can be fabricated in galvanized steel or anodized aluminium alloy. The same material should be used for the ladder. Typically two separate human accesses should be provided into each compartment, near opposite corners to assist ventilation of the compartment when work is in progress and to provide an escape route in an emergency.

Access for plant and materials has to be unobstructed to allow items to be lowered vertically to the compartment floor. The clear opening needed for small plant and materials for normal maintenance should be not less than 1.5 m $\times$ 1.0 m to allow a wheelbarrow to be lowered. Consideration should be given to the provision of removable handrailing around such openings, or of sockets into which it could be fitted. For reservoir compartments exceeding about 10 000 m^3 a second and larger access for plant and materials should be considered if larger mechanical equipment might be needed for cleaning or major repairs. It is important to ensure that unauthorized vehicles cannot reach the roof or be used outside any specially strengthened areas of the roof.

20.26 SERVICE RESERVOIR PIPEWORK

Reservoir pipework normally comprises: inlet(s), outlet(s), overflow, drawdown, reservoir bypass and drainage pipes. The outlet may comprise a suction main to a site pumping station. An 80 mm diameter valved pipe through the division wall of a two-compartment reservoir should be provided so that water is available for hosing down a compartment when taken out of service for cleaning. Unless separate connecting pipes are used for flow in each direction, the control valve on the connection must be operable from outside the reservoir. Flexible joints should be incorporated between embedded or rigid pipes and external pipelines to accommodate differential settlement (Section 18.25).

Inlet and outlet pipes should bifurcate to serve each reservoir compartment equally. The inlet pipe can discharge at TWL or near BWL. One of the disadvantages of the latter is that, in the event of a burst on the incoming main, the reservoir contents will be lost unless a suitable non-return valve is provided. On the other hand, if the incoming supply is pumped, a high-level entry will forfeit the energy savings potentially available when the reservoir is operating below TWL.

Inlet Pipework

The inlet piping arrangement needs to either achieve complete mixing of the inflow with the stored water or produce plug flow and thus avoid build-up of stagnant water areas. This involves suitable siting of the inlet and outlet pipes and, if necessary, the use of baffle walls. For the design of large reservoirs, and where there are water quality problems, it is becoming more common to use 3D (CFD) modelling techniques (Section 14.18) to optimize inlet and outlet arrangements and baffle wall (if any) placement. Options for encouraging circulation comprise: placing the inlet and outlet at opposite ends of the compartment (Fig. 20.4); distributing the incoming flow as evenly as possible along an end wall by the use of a long inlet weir; using a tapered diffuser pipe with several openings; or delivering to a semi-circular terminal box with slotted outlets.

Outlet Pipework

The most common (and simplest) outlet system uses only one draw-off point per compartment, but this is likely to leave some potentially stagnant areas in one (or both) corners at the outlet end of the compartment. To avoid this, the outlet may draw water from a number of points along an end wall if flow distribution is used as the sole means of avoiding stagnation.

If the inlet and outlet pipelines are to terminate at opposite sides of the reservoir compartment, it may be appropriate to have separate inlet and outlet valve houses. However, for economy and ease of operation, a single valve house containing controls for both inlet and outlet is usually preferable. With this arrangement, one pipeline (usually the inlet) is normally laid within the reservoir compartment to feed water to the far end of it. This pipeline should be placed alongside the wall and encased in concrete to avoid 'dead' spaces and to inhibit external corrosion of the pipes. If the reservoir is of the jointed design, the internal pipework (and its surround) must have flexible joints corresponding with the joints in the structure.

The outlet pipe can be laid horizontally, either through the reservoir compartment wall, or under the floor with a 90° vertical bend. It is usual to provide an entry bellmouth, to reduce hydraulic losses. The outlet bellmouth must be sufficiently submerged at BWL to prevent the entrainment of air into the flow, particularly where the flow will be pumped. For a bellmouth in the horizontal plane (i.e. vertical axis), a safe rule for minimum submergence of the bellmouth lip is:

$$S/D = 1.0 + 2.3F$$

where S = submergence below BWL; D = bellmouth diameter at lip; and F is the Froude number $V_D/(gD)^{0.5}$, where V_D = the average flow velocity through the bellmouth opening. With a bellmouth in the vertical plane (horizontal axis) the same equation may be used, but S is measured from the bellmouth axis. The reader is referred to Knauss (1987) and ANSI/HI 9.8 for further guidance for outlet pipework which supplies pumps.

For a gravity supply outlet, the submergence requirement can be somewhat relaxed depending on the acceptability of air entry into the pipeline, but should not be less than D. The required submergence may create an uneconomic depth of 'dead' water unless the reservoir outlet is lowered by means of a sump in the floor. The sump should be generously sized to avoid undesirable hydraulic turbulence. The bottom of the sump collects floor deposits and should be not less than 300 mm below the bellmouth lip. Safety features for maintenance personnel need to be considered in the detailed design of a sump. The outlet sump can also serve as the drain sump.

Where a service reservoir has a common inlet/outlet main, circulation inside the reservoir can be achieved by dividing the common main into inlet and outlet pipes before these pipes bifurcate to each compartment. If a low-level entry design is used, both inlet and outlet must be fitted with non-return valves. With the high-level type entry, a non-return valve is required on the outlet pipes only. Common inlet/outlet arrangements are not preferred due to the risk to water quality from stratification and stagnation (Section 20.5).

Overflow and Drain Down Arrangements

Adequate overflow arrangements must be provided in case of an inflow control malfunction. Each compartment should be provided with an overflow capacity equal to the maximum likely inflow possible into that compartment with the other in or out of service (Section 20.8). The simple provision of a vertical pipe with bellmouth attached as an overflow has limited capacity and a horizontal weir is usually required. A convenient arrangement in a two-compartment reservoir is a weir box in the central division wall with weir entries from each compartment. The weir box often discharges to a pipe laid through the valve house, which can also receive the washout pipework connections.

The combined overflow/washout system should preferably discharge into an open watercourse or, failing that, to a sewer; both of which must be of adequate capacity. It may be necessary to consult with the land drainage authority or sewerage agency concerned for any permissions needed. A break pit must be provided before final discharge to allow levels to be monitored and dechlorination to be carried out if necessary. Drainage pipes should be connected to a drainage sump in each compartment and sized to allow emptying in an acceptable time (Section 20.8).

A reservoir bypass (between inlet and outlet pipes) is necessary in the case of a single-compartment reservoir, or where the whole reservoir may need to be taken out of service.

Valves

Stop valves (gate or butterfly) must be provided on inlets, outlets, drain down pipes and the reservoir bypass but must not be provided on the overflow or on any wall or underfloor drainage systems. Gate valves become impracticable for normal reservoir use above about 600 mm diameter, when resilient-seated butterfly valves should be provided (Section 18.8). The valve size can be less than that of the pipeline, though the saving in cost of the valve is at least partly offset by the need for tapers and the increased space occupied by the pipework in a valve house. If a smaller size of valve is selected, a check should be made that the maximum velocity through the valve does not exceed that recommended by the valve manufacturer.

Autonomous over-velocity valves, designed to close automatically when the water velocity in the pipeline exceeds a predetermined rate, have fallen out of general favour because of their high cost and infrequent use. They may still be appropriate in special circumstances, for example where a large reservoir provides the major supply to a distribution area, or where the loss of water from a failed outlet main would be severe because of high head. The possible need for such valves should therefore be reviewed in reservoir planning and electrically operated butterfly valves should be considered as an alternative.

Wherever they are located, all butterfly valves and special control valves should preferably be installed in chambers or houses so that they are accessible for maintenance. Important gate valves (such as the isolating valves on any pipes connecting into the reservoir) should also be placed in chambers but others can be buried.

Isolating valves on pipes leading into or out of the reservoir should be bolted to flanged pipes cast into the reservoir wall. Otherwise any differential movement between reservoir and valve could cause a joint to fail and release of the entire reservoir contents. The same principles apply to outlet or drain pipework built into the reservoir floor.

20.27 VALVE HOUSES FOR SERVICE RESERVOIRS

It is often convenient for all reservoir control valves to be concentrated in one chamber or valve house. For security of supply, the valve house should be as close as possible to the reservoir and is usually part of the reservoir structure. Access to the valve house may be by top entry through the roof or side entry through a wall. Top entry may result in the whole of the interior of the house being classified as a 'confined space', with consequent safety constraints on entry. These could give rise to unacceptable delays in gaining access if, for example, it becomes necessary to isolate the reservoir in an emergency. Side-entry valve houses are therefore preferred but may give rise to unacceptable visual impact in environmentally sensitive areas.

The pipework within the valve house must be arranged so that it is possible to install, maintain or remove any valve without great difficulty. If a valve is too heavy to be manhandled, it is important that there is clearance for a straight vertical lift out of the building (if top entry is provided) or to a position where it can be transferred to a trolley or road vehicle (if side entry is provided). Fixed monorail hoists, lifting eyes, davits or portable hoists may need to be provided.

In addition to pipework, valve houses are often used to accommodate sampling pumps and pipework, level recording and indicating equipment, telemetry and site monitoring equipment for flowmeters on inlet and outlet pipework, ventilation and dewatering equipment. Provision must also be made for dealing with any water resulting from spillages during maintenance work or leakage from pipework components. As a minimum, this should comprise a sump into which the suction hose of a portable pump can be inserted.

20.28 BAFFLES IN SERVICE RESERVOIRS

The purpose of baffle walls or curtains is to achieve plug flow through the reservoir if this is required instead of good mixing. This is achieved by directing flow from inlet to outlet by a circuitous route. The optimum arrangement of baffles depends on the shape of the compartment and the most convenient positions for the inlet and outlet. For construction convenience, baffles are normally installed between the columns supporting the reservoir roof.

Solid baffle walls are normally made of reinforced concrete, brickwork or blockwork. Lightweight, hollow blockwork should be avoided because the free chlorine normally present in a reservoir air-space has been known to cause deterioration of some blocks of this type. Plastic or rubber curtains are sometimes used. Where appropriate, openings must be provided along the bottom of the baffle walls to allow all areas of the reservoir floor to drain and to facilitate cleaning and maintenance. Openings should also be provided at the top of the walls to assist with ventilation. Joints in solid baffle walls must be provided at all points where they bridge movement joints in the floor or roof. Additional joints may be needed to allow for thermal movements or for the flexing of the reservoir walls as water levels rise and fall.

Baffle curtains may be less robust but are easier to install. The material should be fibre reinforced to reduce elasticity and should be resistant to chlorinated water and approved for use with potable water. The edges of each curtain are formed by a seam in which holes are cut out at intervals for attachment points and through which a stainless steel rod is inserted. The recommended method for supporting the top edge of the curtain is to build a dove-tail galvanized or stainless steel channel into the roof soffit for the subsequent attachment of hangers although, for columns, lashing or strapping may be acceptable. The bottom edge must also be firmly anchored, preferably by fixing it to stainless steel or GRP angles bolted to the floor. The use of precast concrete blocks which are not fixed to the floor is not recommended. Curtain alignment should be chosen so that they are not subject to high velocity or turbulent flow such as may occur near an inlet. At the inlet a dwarf concrete wall, bonded to the floor, should be provided below the curtain to protect it when the reservoir is filling. This wall should be about 500 mm high and the bottom of the curtain should be anchored to the top of the wall.

WATER TOWERS

20.29 USE OF WATER TOWERS

Water towers are used as a local source of water at times of peak demand where it would not be economical to increase the size of the supply pipeline and add a booster pump installation. In undulating terrain ground-level storage can provide the pressure needed but in areas of flat topography the storage must be elevated. Many shapes and design features are possible but the designer should aim to produce a structure that meets the requirements of both water supply and planning authorities, bearing in mind that it will become a landmark in the community which it serves. Ancillary equipment including pipework, valves, ladders, instrumentation and booster pumps, if required, can all be hidden in the cylindrical shaft.

The optimum depth/diameter ratios should be determined taking into account the most efficient shape and the needs of the distribution system. It is usually advisable to avoid large pressure fluctuations in distribution that may be caused by drawdown or filling in excessively deep tanks.

Water towers need to be leak tested as described for reservoirs (Section 20.34).

20.30 CONCRETE WATER TOWERS

Concrete water towers are built with capacities up to about 5000 m^3. They are usually circular in plan although rectangular concrete towers have been built (Plate 34(c)). The diameter of circular

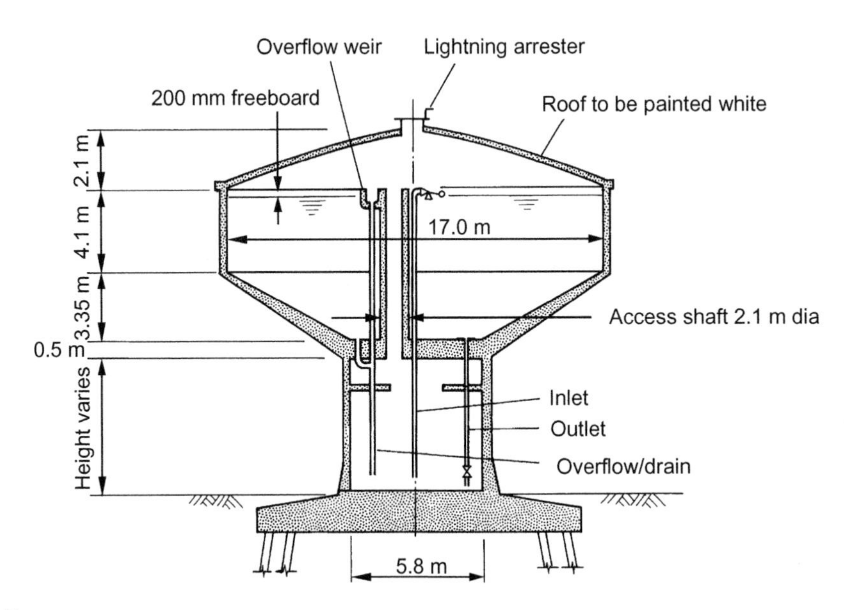

FIGURE 20.7

Reinforced concrete water tower.

water towers is not usually sufficient to warrant the use of prestressing since cracks can be controlled by applying normal water retaining concrete criteria. Concrete water towers allow some scope for architectural statement so that the result can be regarded as a visual asset. Typical dimensions adopted for the reinforced concrete design shown in Figure 20.7 are:

Size (m^3)	Depth of water (m)	Internal diameter (m)
1200	7.5	17.0
2000	9.1	19.4
3000	10.2	22.6

Rectangular water towers are designed as small monolithic service reservoirs with the floor slab supported on some form of open column and beam framework or on a hollow vertical shaft, itself founded on a base slab, piled if necessary. Wind and seismic loads should be taken into account in the design of tank, supports and foundations. Circular concrete water towers allow more scope for different styling from a simple cylinder with a flat base to a sophisticated form such as the hyperbolic-paraboloid of the 39 m high Sillogue tower near Dublin airport built in 2006. In this case the vase shape resembles an inverted version of the nearby control tower. The Intze type water tower (Rajagopalan, 1990) is designed so that bending moments are as near zero as possible at all sections (Fig. 20.8). The radial thrusts from the outer conical section of base on the supporting ring balance those from the spherical centre section. Roofs may be flat for small tanks, conical or, for larger tanks, spherical as described in Section 20.19.

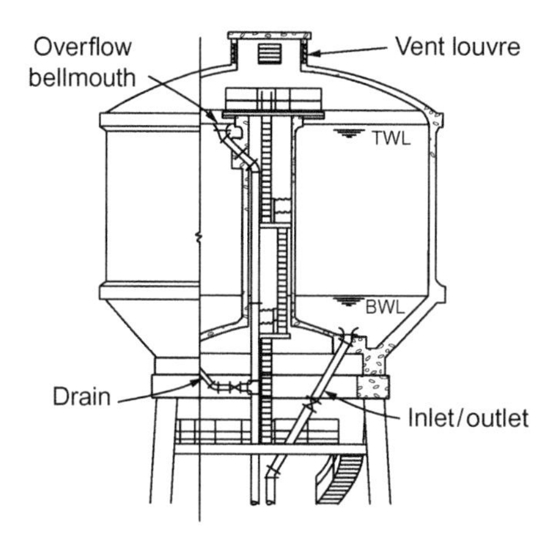

FIGURE 20.8

Reinforced concrete water tower (Intze type).

20.31 WELDED STEEL WATER TOWERS

Relatively small welded tanks have been used for over 100 years for industry and rail transport. These were usually small radius cylinders supported on a framework of steel columns with braces or ties. Welded steel water towers of capacities up to 15 000 m^3 are now available and have been widely used all over the world, particularly in North America, the Middle East and the Far East. These are now constructed of butt welded steel plate in several configurations:

- spheroids or ellipsoids on tubular columns belled out at the base;
- cylindrical or spherical shapes with conical bases and supported on wide steel columns which help resist seismic loads and provide space for plant rooms or offices or on a reinforced concrete frame (Plate 34(d)).

Whilst the forms available for welded steel water towers do not offer much scope for architectural treatment, the coatings provide an opportunity for decoration and can be attractive.

20.32 SEGMENTAL PLATE TANKS

The type of steel or GRP panel construction described in Section 20.23 can also be used for elevated storage. However, it is unlikely that segmental plate tanks would be used for anything other than industrial or emergency water storage since their poor visual appearance is exaggerated by height. Where they are used, the bases are placed on a series of beams which are supported on a framework of braced columns.

20.33 PIPEWORK AND ACCESS FOR WATER TOWERS

Pipework and access facilities below water towers are usually concealed within, or obscured by, the tank supports. A dry access shaft in the centre of the tank allows access to equipment above water level such as water level instruments and inlet float valves and may permit access out onto the roof. Old designs used to provide facilities for external access including circular walkways and revolving ladders. However, such facilities themselves need maintenance and cannot easily meet modern safety standards. This means that external coatings for steel tanks must be of the highest quality to minimize time to first maintenance.

TESTING SERVICE RESERVOIRS AND TANKS

20.34 TESTING PROCEDURE

Service reservoirs (ground and elevated) should be tested for water tightness before being put into service. The test should be carried out before placing any backfill or embankments against the outside walls unless the wall design relies on the embankment to resist hydraulic forces. The roof should be complete including any second-stage concrete.

Each reservoir compartment should be tested separately, with the other compartment empty. The compartment should be filled with treated water, to a test level about 75 mm below the overflow sill, at a uniform rate not exceeding 2 m vertical rise in water level per 24 hours. In concrete tanks, the water should then be left to stand for at least 7 days to allow for absorption into the concrete. Longer periods (up to 21 days) may be required by some specifications. The water level should then be measured and recorded using a hook gauge with vernier control, or by other approved means of no less accuracy, and the water allowed to stand under test for 7 days. For elevated tanks the acceptance criteria can be no visible leaks. For tanks and reservoirs where the outside and underside is not accessible, at least once each day during this period, the water level should be measured and recorded. During the 7-day test period, the effects of evaporation from the water surface can be reduced by closing all air vents and access openings (except for one vent left open for pressure balance).

Any flows in the underdrain and wall drain systems should be measured and recorded throughout the test, from a time at least 24 hours before beginning to fill, until 24 hours after emptying or on completion of a final water level measurement. Taking such measurements in chambers on the drain systems normally requires safety precautions appropriate to confined spaces. The outfalls of all pipes connected to the reservoir should be inspected during the test to ensure that all isolating valves are shut tight. Any significant leakage through them should be measured. In some circumstances it may also be necessary to keep records of evaporation losses from the water surface.

The test may be deemed successful if the drop of water level over the 7-day test period does not exceed the lesser of $1/500 \times$ average water depth or 10 mm, after deducting any measured leakage through valves and making allowance for any evaporation or condensation. If the test fails, any increase in underdrain or wall drain flow during the test period should be investigated to identify, if possible, the part of the reservoir that leaked. The test compartment should then be emptied and closely inspected for faults likely to cause the leakage. Investigating reservoir leakage can be troublesome and time consuming. The interior of the reservoir, especially any joints, should be closely inspected before filling with water.

20.35 SEARCHING FOR LEAKS

Location of leaks in elevated storage tanks is straightforward since the leak should be visible. It is not easy to make a large reservoir fully watertight and the following notes may help track down the point of leakage.

1. Flows from underdrains should be examined. If they serve known areas of floor and are not joined to one common outlet point they may help narrow the search.
2. The inlet and outlet valves must be observed to ensure that they are not passing water into or out of the reservoir. The only secure way of knowing this is by withdrawing flange adaptors from the valve or by removing or temporarily leaving out a section of pipe next to the valve. Any outflow should be measured.
3. The rate of leakage at full depth, half depth, and with about 0.6 m of water in the reservoir should be measured. It is not likely that any revealing mathematical relationship between rate of leakage and depth of water in the reservoir will be established but leakage is usually less when the depth of water is less. No leakage at all below a certain level points to a leak at higher sections of wall.
4. After attempts to narrow the search for leaks the reservoir must be emptied and subjected to the most careful internal inspection with good lights, adequate ladders, plenty of time and a consistent pattern of examination. It is very easy to miss a faint crack in the wall or floor. Walls (particularly the joints next to the corners) should receive special attention for it is here that there is most likelihood of movement having occurred. After emptying, reservoir walls should be kept under observation when drying off since temporarily visible damp patches may be evidence of water held in underlying cracks or poor concrete. If first inspection does not reveal the causes of the leaks, it is worth repeating the inspection with even more care.
5. The floor joints should be inspected. Jointing material should be examined to see if it has sunk, has holes in it, or has come away from or failed to bond to the concrete of the sealing grooves. The majority of leakages arise from defects of this kind. Wall joints should similarly be examined.
6. If failure still results, about 0.6 m of water should be put into the reservoir and be left to stand until the water is quite still. Then crystals of potassium permanganate may be dropped into the reservoir, widely spaced, and left for a considerable time. Observed from a pre-arranged walkway (so as not to disturb the water) with a good light, streaks of colour may be noticed from the permanganate crystals showing some definite flow towards a point of leakage.
7. As an alternative to method (6), about 150 mm of water can be put into the reservoir, a hole or several holes bored through the floor, and compressed air can be introduced under the floor. In certain conditions of floor foundation air bubbling upwards through the water may indicate where faulty floor joints occur.
8. If, despite all these attempts, the cause of leakage is still unaccounted for then more drastic measures may have to be undertaken, such as digging pits in the bank to inspect the rear of the wall

joints, placing further sealing strips (such as glass fibre embedded in bitumen) over joints, or even rendering wall face areas. Sealing of leakage through a reservoir floor has been achieved by gravity grouting. About 450 mm of thin grout mix is put into the reservoir and the cement is kept in suspension by continually sweeping the floor and disturbing the water with squeegees for two successive days. Thus grout passes into the unknown paths of leakage and the cement sets. It should not be necessary, however, to adopt these measures unless poor construction has taken place.

20.36 CLEANING AND DISINFECTION

After successful testing the reservoir must be cleaned and disinfected. In the US, water storage facilities are disinfected to AWWA C652. Guidance for cleaning and disinfecting is given in Section 11.29 but individual water utilities may have their own procedures.

Cleaning of service reservoirs is recommended every 3 years or so. Any sediment or deposits on surfaces should be removed and the reservoir disinfected before refilling.

20.37 INSPECTION AND REPAIR OF SERVICE RESERVOIRS

Many service reservoirs over 100 years old are still in use. The earliest are brick, some with vaulted roofs, and may be very large. There are also examples of reservoirs with gravity concrete walls. In some cases work may have been done to replace or repair the roofs to make them watertight but structural cracking due to settlement and leaking vaulted roofs are increasingly leading to replacement with new structures.

Some of the earliest prestressed concrete tanks are now well over 50 years old and nearing the end of their lives. Many of these have thin shell roofs. Corrosion of reinforcement and prestressing wire are issues with this sort of construction which is difficult to repair once the damage has gone too far.

Aside from statutory requirements, owners need to inspect older reservoirs regularly and to plan for their repair or replacement if necessary. CIRIA Report 138 (CIRIA, 1995) provides guidance for the repair of underground service reservoirs.

REFERENCE STANDARDS

ACI 201.2 R *Guide to Durable Concrete*. ACI.

ACI 350 *Code Requirements for Environmental Engineering Concrete Structures*. ACI.

ACI 350.3 *Seismic Design of Liquid Containing Structures*. ACI.

ACI 372 R *Circular Wire- and Strand-Wrapped Prestressed Concrete Structures*. ACI.

ACI 373 R *Circular Prestressed Concrete Structures With Circumferential Tendons*. ACI.

ANSI/API STD 650 *Welded steel tanks for oil storage*. ANSI.

ANSI/HI 9.8 *American National Standard for Rotodynamic Pumps for Pump Intake Design*. Hydraulic Institute.

AWWA C652 *Disinfection of Water Storage Facilities*. AWWA.

AWWA D100 *Steel tanks for water storage*. AWWA.

AWWA D102 *Coating Steel Water Storage tanks*. AWWA.

AWWA D103 *Factory-Coated Bolted Carbon Steel Tanks for Water Storage*. AWWA.

AWWA D104 *Automatically Controlled Impressed-Current Cathodic Protection for the Interior Submerged Surfaces of Steel Water Storage tanks*. AWWA.

AWWA D106 *Sacrificial Anode Cathodic Protection Systems for the Interior Submerged Surfaces of Steel Water Storage tanks*. AWWA.

AWWA D110 *Wire- and Strand-Wound Circular, Prestressed Concrete Water Tanks*. AWWA.

AWWA D115 *Tendon Prestressed Concrete Water Tanks*. AWWA.

BS 8007 *Code of Practice for Design of Structures for Retaining Aqueous Liquids*. BSI.

BS 8110-1 *Structural Use of Concrete. Code of Practice for Design and Construction*. BSI.

BS 8500-1 *Concrete. Method of specifying and guidance for the specifier*. BSI.

BS EN 206-1 *Concrete. Specification, performance, production and conformity*. BSI.

BS EN 1990 Eurocode 0. *Basis of structural design*. BSI.

BS EN 1992-1-1 Eurocode 2. *Design of concrete structures. General rules and rules for buildings*. BSI.

BS EN 1992-3 Eurocode 2. *Design of concrete structures. Liquid retaining and containing structures*. BSI.

BS EN 1993-4-2 Eurocode 3. *Design of steel structures. Tanks*. BSI.

BS EN 14015 *Site built, vertical, cylindrical, flat-bottomed, above ground, welded steel tanks for the storage of liquids*. BSI.

BS EN 16299 *Cathodic protection of external surfaces of above ground storage tank bases in contact with soil or foundations*. BSI.

REFERENCES

Ainsworth, R. (2004). *Safe Piped Water: Managing Microbial Water Quality in Piped Distribution Systems*. IWA Publishing.

CIRIA (1995). *Underground Service Reservoirs. Waterproofing and Repair Manual*. CIRIA.

Creasy, L. R. (1961). *Prestressed Concrete Cylindrical Tanks*. John Wiley & Sons.

Grayman, W. M. et al. *Water Quality Modeling of Distribution System Storage Facilities*. AwwaRF and AWWA, Denver, CO.

Health Research (2012). *Recommended Standards for Water Works*. Health Research Inc. for Great Lakes – Upper Mississippi River Board of State and Provincial Public Health and Environmental Managers.

Knauss, J. (1987). *Swirling Flow Problems at Intakes*. Balkema.

Rajagopalan, K. (1990). *Storage Structures*. Balkema.

van Lieverloo, J. H. M., Blockker, E. J. M., Medema, G. et al. (2006). *Contamination During Distribution*. MicroRisk Project, EVK1-CT-2002-00123. European Commission.

Conversion Factors

	Metric to US	US to Metric
Length:	1 mm = 39.37 thou (in/1000)	1 in = 25.40 mm
	1 m = 39.37 inches (in)	1 foot (ft) = 304.80 mm
	1 m = 3.2808 ft = 1.0936 yd	1 yard (yd) = 0.9144 m
	1 km = 0.61237 miles	1 mile = 1.60934 km
Area:	1 m^2 = 1.196 yd^2 = 10.764 ft^2	1 ft^2 = 0.0929 m^2
	1 ha = 2.471 acres	1 yd^2 = 0.8361 m^2
	1 km^2 = 0.3861 sq. miles	1 acre = 0.4047 ha
	1 km^2 = 100 ha	1 sq. mile = 2.590 km^2
	1 ha = 10 000 m^2	*1 acre = 4840 yd^2*
Volume:	1 ml = 0.06102 in^3	1 in^3 = 16.387 ml
	1 litre (l) = 61.02 in^3	1 ft^3 = 28.32 l
	1 litre = 0.2642 gal (US)	1 quart (US) = 0.9463 l
	1 m^3 = 35.315 ft^3	1 yd^3 = 0.76456 m^3
	1 m^3 = 1.3080 yd^3	1 gal (US) = 3.785 l
	1 Ml = 0.2642 million gal (US)	*1 gal (US) = 0.8326 gallon (UK)*
	1 litre = 0.220 gallon (UK)	*1 gallon (UK) = 4.546 l*
Mass:	1 g = 0.03527 ounces (oz)	1 lb = 0.4536 kg
	1 kg = 2.2046 pounds (lb)	1 cwt = 50.802 kg
	1 tonne = 19.684 hundredweight (cwt)	1 ton = 1.01605 tonne (t)
	1 t = 0.9842 tons	
Mass unit area:	1 kg/m^2 = 0.2048 lb/ft^2	1 lb/ft^2 = 4.882 kg/m^2
	1 g/mm^2 = 22.76 oz/in^2	1 oz/ft^2 = 305.2 g/m^2
Force:	1 Newton (N) = 0.2248 lb(f)	1 lb(f) = 4.4482 N
	1 kN = 0.10036 ton(f)	1 ton(f) = 9.964 kN
	9.81 N = 1 kg(f) at sea level	
Pressure:	1 N/mm^2 = 145.038 lb(f)/in^2 (psi)	1 lb(f)/in^2 = 6895 N/m^2
	1 N/mm^2 = 9.324 ton(f)/ft^2	1 ton(f)/ft^2 = 10.937 t(f)/m^2
	1 kN/m^2 = 20.885 lb(f)/ft^2	1 ton(f)/ft^2 = 107.25 kN/m^2
	1 m head of water = 1.422 lb(f)/in^2	1 lb(f)/ft^2 = 47.88 N/m^2
	1 N/m^2 = 1 pascal (pa)	
	0.1 N/mm^2 = 1 bar (≡ pressure from 10.20 m of water at sea level)	
Density:	1 kg/m^3 = 0.06243 lb/ft^3	1 lb/ft^3 = 16.018 kg/m^3
	1 t/m^3 = 0.7525 ton/yd^3	1 ton/yd^3 = 1.329 t/m^3
Unit weight:	1 kN/m^3 = 6.3658 lb(f)/ft^3	1 lb(f)/ft^3 = 0.15709 kN/m^3

(Continued)

Conversion Factors *(Continued)*

	Metric to US	US to Metric
Velocity:	1 m/s = 3.2808 ft/s = 2.237 mph 1 km/h = 0.6214 mph	1 ft/s = 0.3048 m/s 1 mph = 0.447 m/s = 1.609 km/h
Acceleration:	1 m/s^2 = 3.2808 ft/s^2	1 ft/s^2 = 0.3048 m/s^2
Flow:	1 m^3/s (cumec) = 35.315 ft^3/s (cfs) 1 m^3/h = 4.403 gal (US)/min (gpm) 1 l/s = 15.851 gpm (US) Ml/d = 0.2642 million gal (US)/d (mgd)	1 ft^3/s (cusec) = 0.0283 m^3/s 1 gpm (US) = 0.2271 m^3/h 1 gpm (US) = 0.0631 l/s 1 mgd (US) = 3.785 Ml/d *1 mgd (US) = 0.8326 mgd (UK)*
Power:	1 kW = 1.3410 hp 1 W = 0.73756 ft.lb(f)/s *1 J/s = 1 watt (W)*	1 hp = 0.7457 kW 1 ft.lb(f)/s = 1.3558 W
Energy:	1 Joule (J) = 0.7376 ft.lb(f) *1 kJ = 0.9478 Btu* *1 Wh = 3.412 Btu*	1 ft.lb(f) = 1.3558 J *1 Btu = 1055 J*
Filtration rate:	1 m^3/h.m^2 = 0.4091 gal (US)/ft^2.min	1 gal (US)/ft^2.min = 2.444 m^3/h.m^2
Membrane flux:	1 l/m^2.h = 0.5891 gal (US)/ft^2.d (lmh) (gfd)	1 gal (US)/ft^2.d = 1.697 l/m^2.h (gfd) (lmh)
Hydrological:	1 mm on 1 km^2 = 0.8107 acre-ft 1 l/s per km^2 = 0.09146 ft^3/s per 1000 acres	1 acre-foot = 1.2335 mm.km^2 1 acre-foot = 1233.5 m^3 1 ft^3/s per 1000 acres = 6.997 l/s per km^2
Permeability:	10^{-6} m/s = 103.5 ft/yr 1 lugeon = 1 l/m of test length/min at 10 bar injection pressure = 0.01076 ft^3/ft/min at 142 psi *1 lugeon is approximately equivalent to permeability of 1 × 10^{-7} m/s*	1 ft/yr = 9.66 × 10^{-9} m/s
Dynamic viscosity (μ):	1 N.s/m^2 = 0.02089 lb(f).s/ft^2	1 lb(f) s/ft^2 = 47.88 N.s/m^2
Kinematic viscosity (υ):	1 m^2/s = 10.764 ft^2/s *1 m^2/s = 1 × 10^6 centiStokes (cSt)*	1 ft^2/s = 0.0929 m^2/s
Temperature:	t°C = 5/9(t°F − 32)	t°F = 9/5 t°C + 32

Notes:
1. US masses (oz, lb) are avdp (avoirdupois).
2. US ton and cwt are 'long' (2240 and 112 lb, respectively). 1 'short' US ton = 2000 lb.

Index

Note: Page numbers followed by "*f*" and "*t*" refer to figures and tables, respectively.

A

B

C

D

E

F

G

H

I

N

O

P

Q

R

S

X

Y

Z

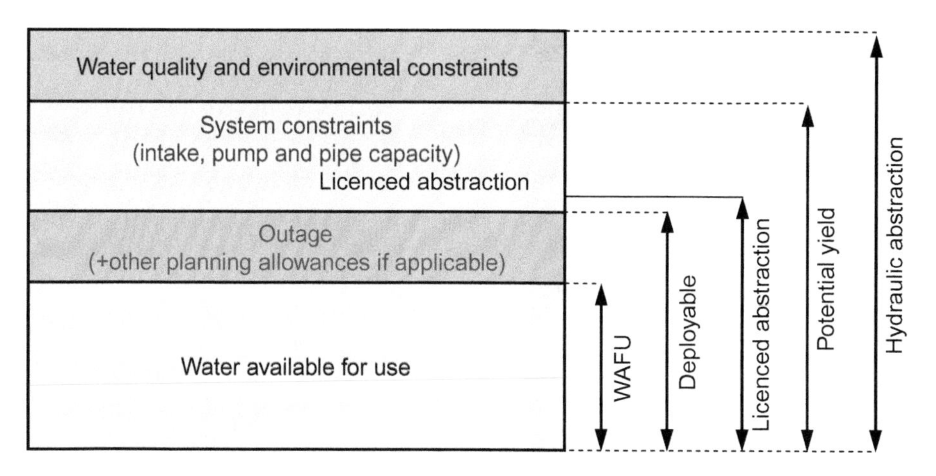

(a) Relationship between yield-related definitions and terms. *Based on information from: Water Resources & Supply: Agenda for Action* (DoE, 1996).

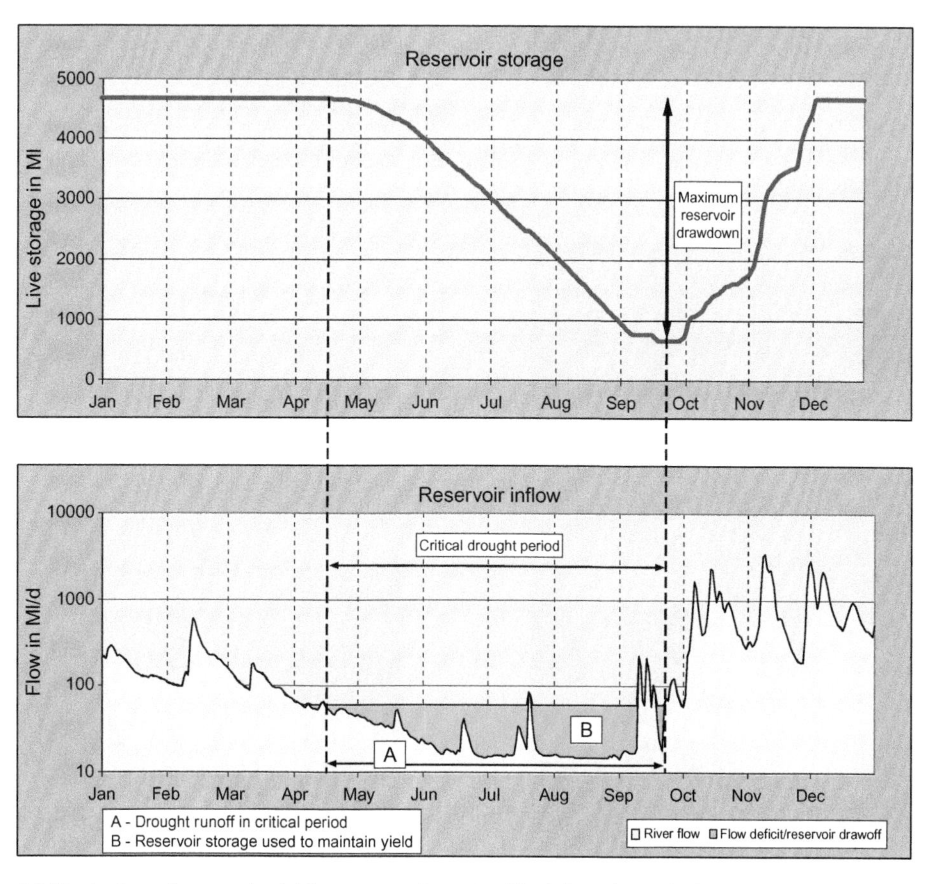

(b) Illustration of reservoir yield components and critical drought period.

PLATE 1

(a) Automatic recording gauging station with stilling well and cableway, Philippines. (Courtesy of John Hall).

(b) Crump weir and recording station on a stream in the UK. (Courtesy of John Hall).

(c) Flow gauging with electromagnetic flow meter on a small stream in the UK. (Courtesy of John Hall).

(d) Automatic rain gauge with telemetry link, Philippines. (Courtesy of John Hall).

PLATE 2

(a) Jari 42 m high earth dam, Mangla, Pakistan. (Engineer: Binnie & Partners).

(b) Llyn Brianne 91 m high rockfill dam, Wales; side spillway: capacity 850 m^3/s. (Courtesy of Aled Hughes) (Engineer: Binnie & Partners).

PLATE 3

(a) Olivenhain dam under construction, USA. RCC being placed at a rate of about 750 m^3/h. (Courtesy of Dr M.R.H. Dunstan).

(b) Wadi Dayqah 75 m high curved concrete gravity dam, Oman. (Owner's Engineer and Supervisor: WDDJV – Lead Consultant: Black & Veatch).

(c) Nant-y-Moch 53 m high buttress dam, Wales. (Courtesy of John Sawyer).

PLATE 4

(a) Chira River closure and diversion for Poechos Dam, Peru. (Courtesy of R. L. Brown).

(b) Mudhiq concrete arch dam, 73 m high, Saudi Arabia. (Engineer: Binnie & Partners).

(c) Waterfall masonry faced concrete gravity dam, 55 m high, Tai Lam Chung, Hong Kong. (Engineer: Binnie & Partners).

PLATE 5

(a) Lower Lliw dam side spillway, Wales. (Courtesy of Welsh Water).

(b) Taf Fechan dam draw-off tower and bellmouth spillway, Wales. (Engineer: Sir Alex Binnie, Son & Deacon).

PLATE 6

(a) Main spillway of 50 m high earth dam at Poechos, Peru – capacity 7200 m^3/s.

(b) Main spillway of 138 m high earth dam at Mangla, Pakistan – capacity 25 000 m^3/s. (Engineer: Binnie & Partners) (Courtesy of O.J. Berthelsen).

PLATE 7

(a) 14 000 m^3/s capacity fuse plug emergency spillway at Poechos Dam, Peru.

(b) Concrete tipping gates at dam in the UK. (Courtesy of John Ackers).

(c) Reinforced concrete labyrinth fusegates at Terminus Dam, California. Capacity 7545 m^3/s after tipping of all gates. (Courtesy of Hydroplus, Inc., VA).

PLATE 8

(a) Coarse screen with cable-operated cleaning arrangement. (Information provided therein courtesy of Ovivo, Inc.).

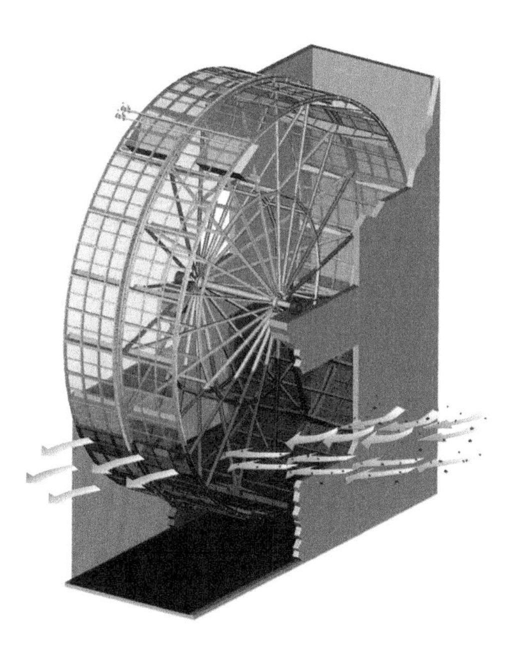

(b) Cut-away graphic of band screen. (Information provided therein courtesy of Ovivo, Inc.).

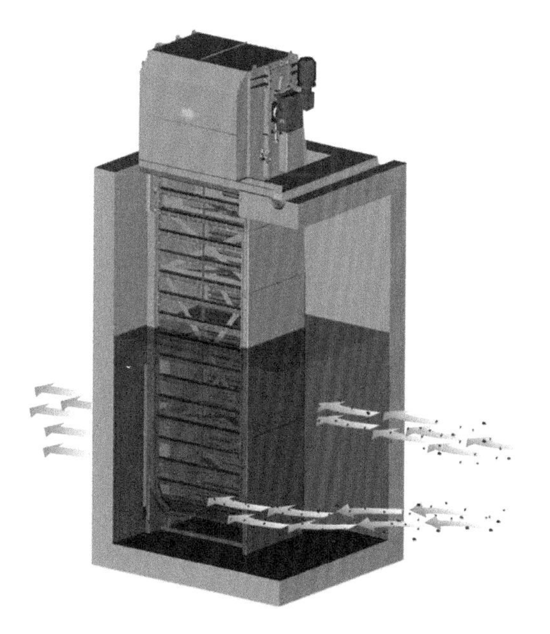

(c) Cut-away graphic of double-entry drum screen. (Information provided therein courtesy of Ovivo, Inc.).

PLATE 9

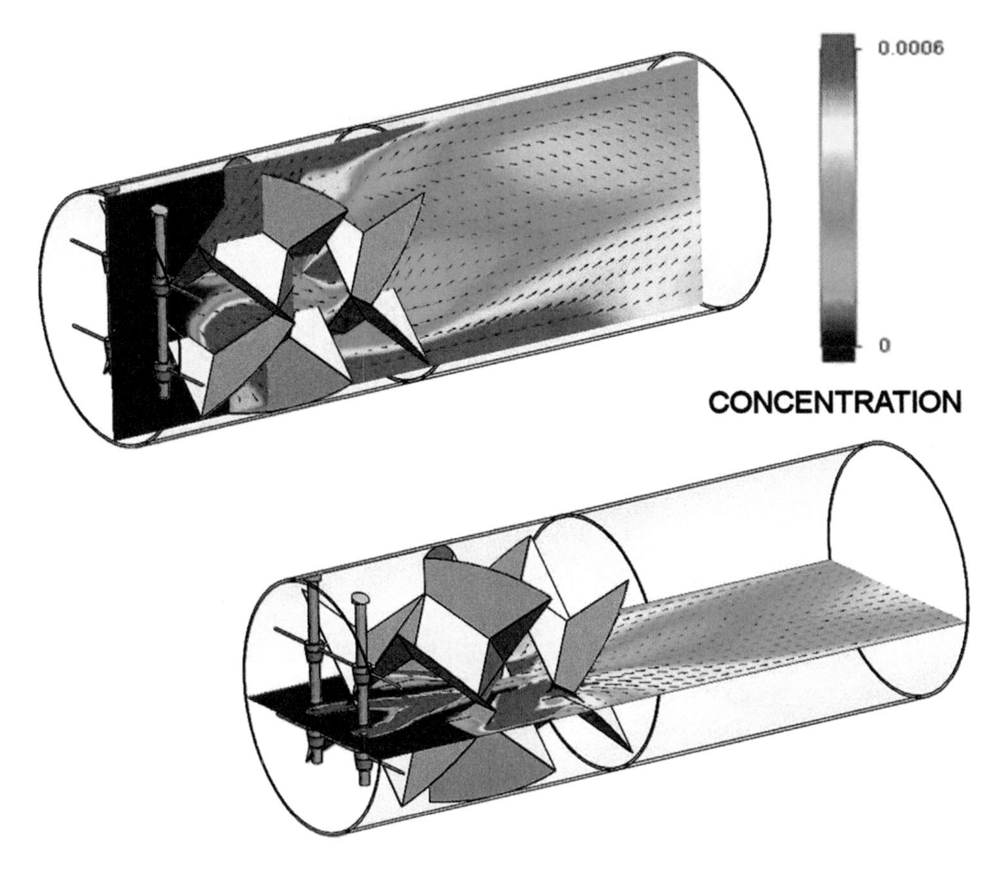

(a) CFD simulation of chemical dosing sparge and static mixer – concentration shown in vertical and horizontal planes. (Manufacturer: Statiflo International).

(b) Pipeline static mixer. (Manufacturer: Statiflo International).

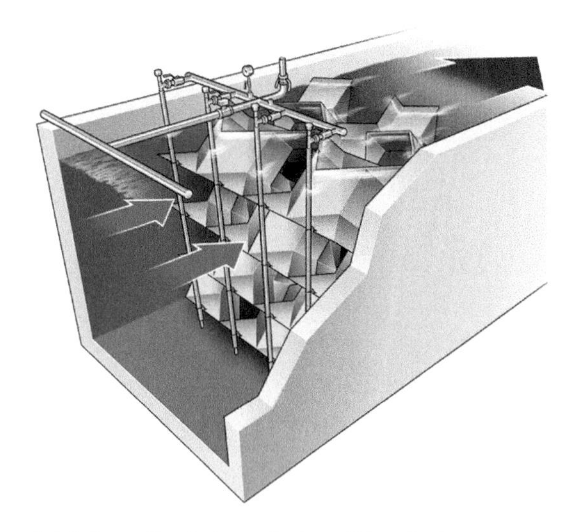

(c) Schematic of channel mixer. (Manufacturer: Statiflo International).

PLATE 10

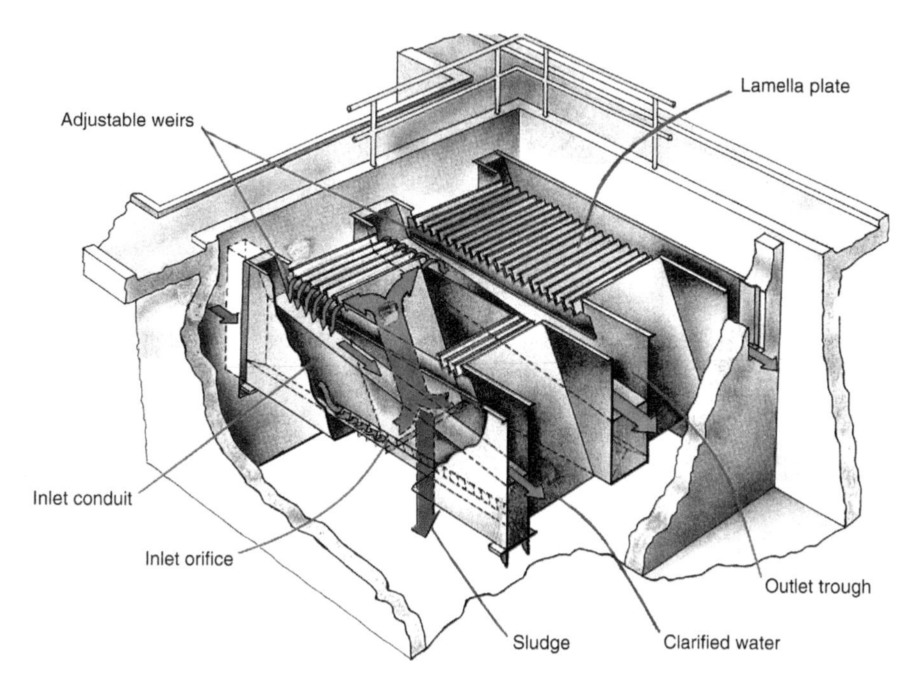

(a) Schematic of lamella sedimentation clarifier. (Purac, Sweden AB).

(b) Lamella clarifier, capacity 3155 m^3/d at the 472 000 m^3/d Water District #1 of Johnson County, Kansas, pre-sedimentation facility. (Engineer: Black & Veatch; manufacturer: Parkson Corporation).

PLATE 11

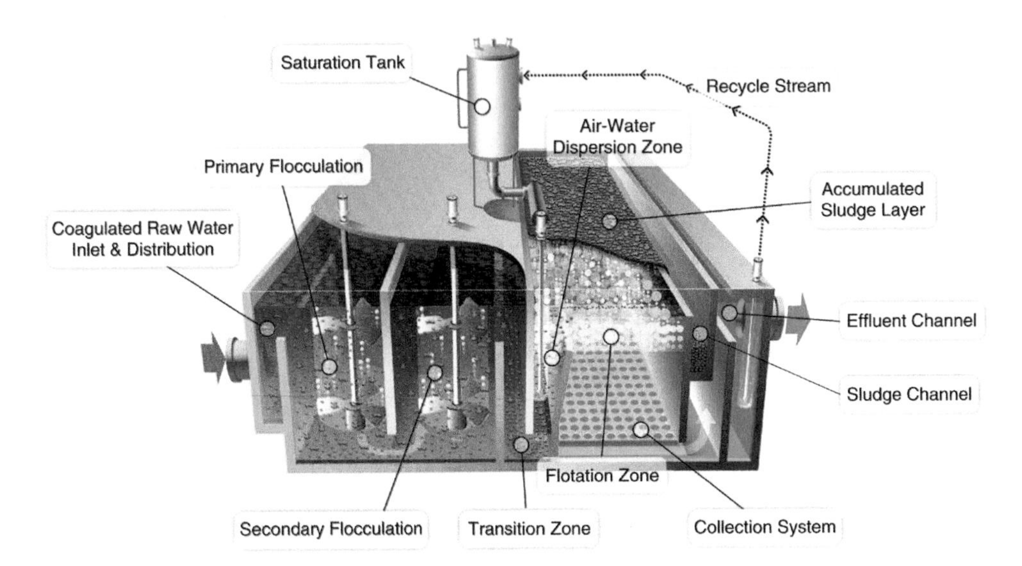

(a) Schematic of AquaDAF® flotation clarifier. (Courtesy of Infilco Degremont Inc., USA).

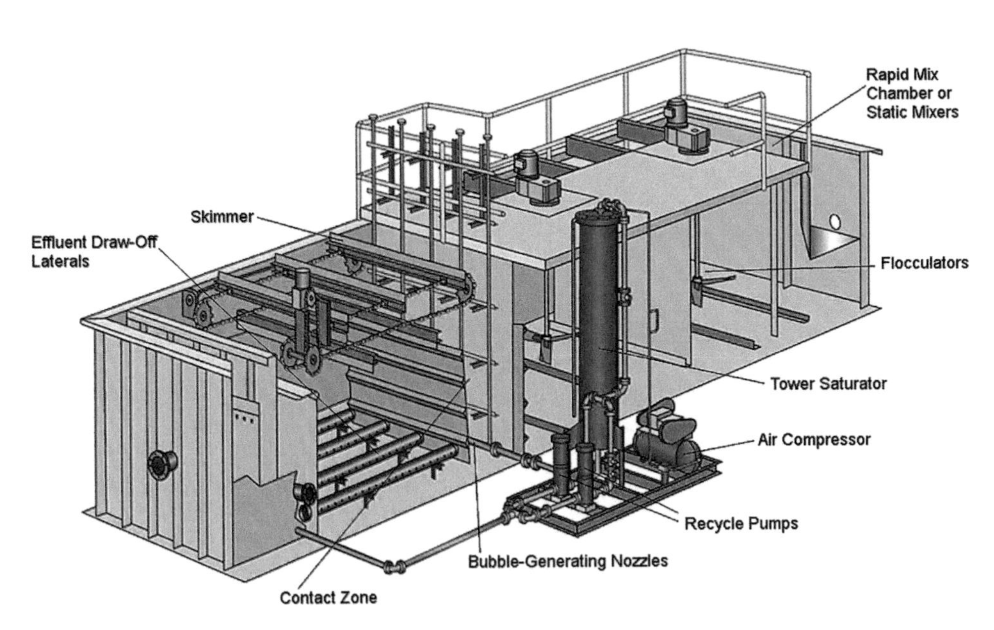

(b) Schematic of Clari-DAF® dissolved air flotation system. (Courtesy of Xylem Inc., USA).

PLATE 12

(a) Cross section of UF hollow-fibre membrane; internal diameter 0.8 mm and external diameter 1.2 mm. (Courtesy of Pentair, The Netherlands).

(b) Membrane module containing more than 12 000 fibres with total area about 40 m^2. (Courtesy of Pentair, The Netherlands).

(c) XIGA UF membrane system at a 273 Ml/d waterworks, comprising 10 duty and one standby trains, each of area 15 360 m^2 in 96 pressure vessels with a design flux of 87 lmh. (Courtesy of Pentair, The Netherlands).

PLATE 13

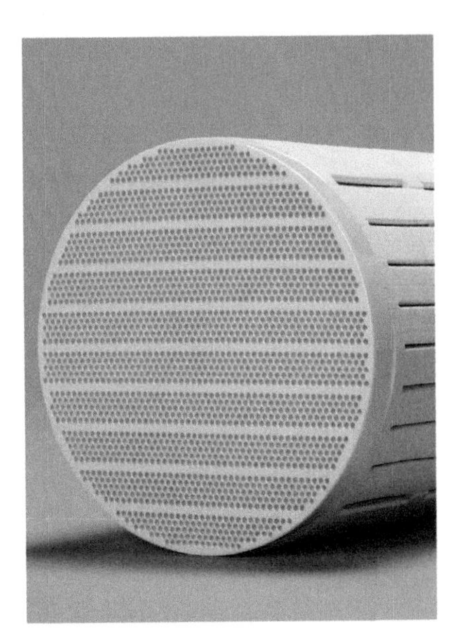

(a) Cross section of a ceramic membrane; diameter 0.18 mm, length 1.5 m and filtration area 25 m^2. Pore size 0.1 micron. Each module has 2000 feed channels of diameter 2.5 mm. (Courtesy of Metawater, The Netherlands).

(b) Drawing of CeraMac® filtration vessel with 192 membrane elements and dedicated backwash vessel on the right. (Courtesy of PWN Technologies, The Netherlands).

(c) 120 Ml/d Andijk III Water Treatment facility in the Netherlands using SIX® and CeraMac® technologies. (Courtesy of PWN Technologies, The Netherlands).

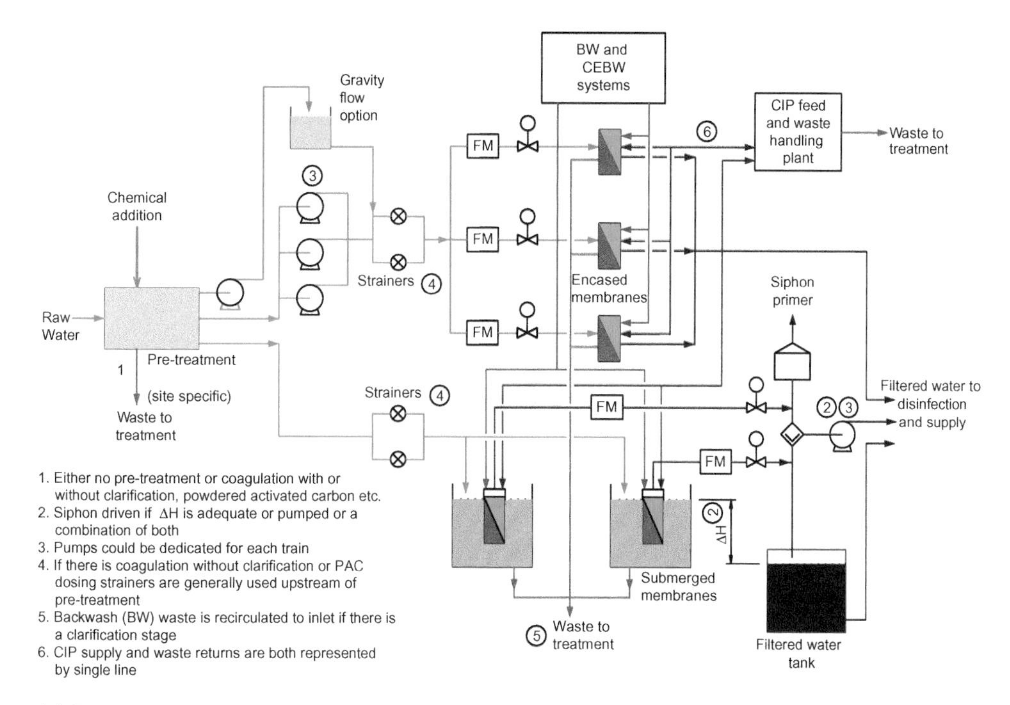

(a) Flow schematic of MF/UF membrane plant – encased and submerged types.

(b) GE's ZeeWeed* 500D immersed membrane cassette being lowered into a tank.

(a) Membrane filter press plant of side bar design at a 273 Ml/d waterworks comprising 2×3.8 t/d, 78 chambers formed from 1.5×2.0 m polypropylene plates with automatic cake dislodging system mounted on top. (Manufacturer: Andritz, Singapore).

(b) A three-step cascade aerator (each fall 0.370 m) for 1.35 m^3/s flow rate at a waterworks in India. (EPC Contractor: VATech Wabag, Chennai, India).

PLATE 16

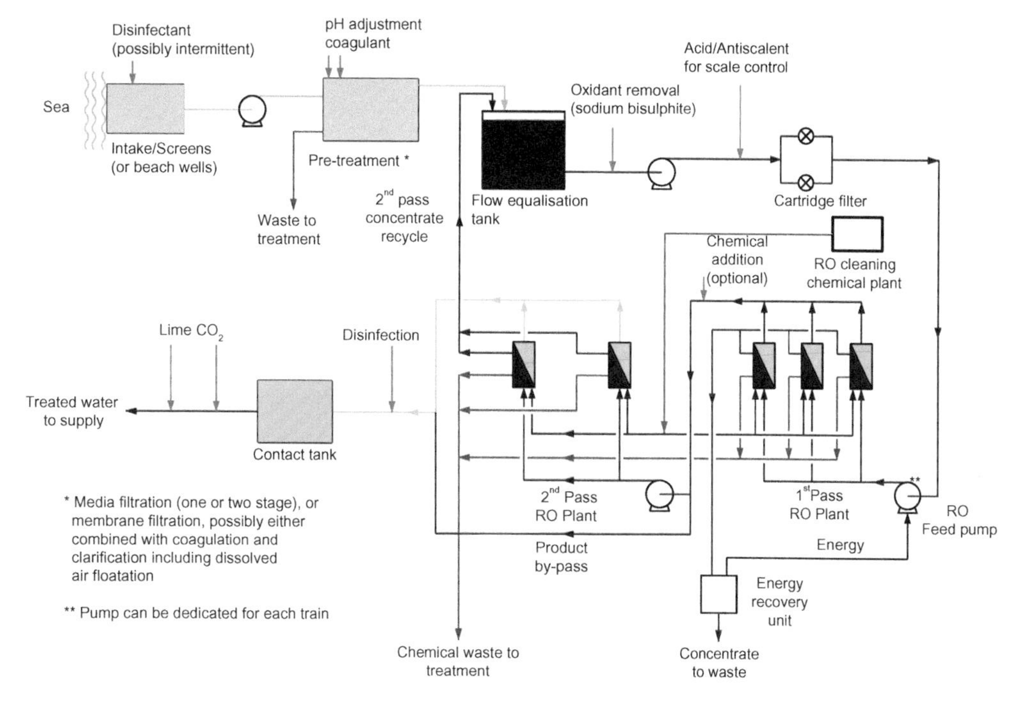

(a) Flow schematic of reverse osmosis desalination plant for sea water.

(b) RO membrane skids (12 No. 360 m^3/h at 70 bar). 100 Ml/d SWRO plant at Nemmeli, Chennai, India. (EPC Contractor: VATech Wabag, Chennai, India).

(c) High pressure feed pump (12 + 3 No. 360 m^3/h at 70 bar), 100 Ml/d SWRO plant at Nemmeli, Chennai, India. (EPC Contractor: VATech Wabag, Chennai, India).

PLATE 17

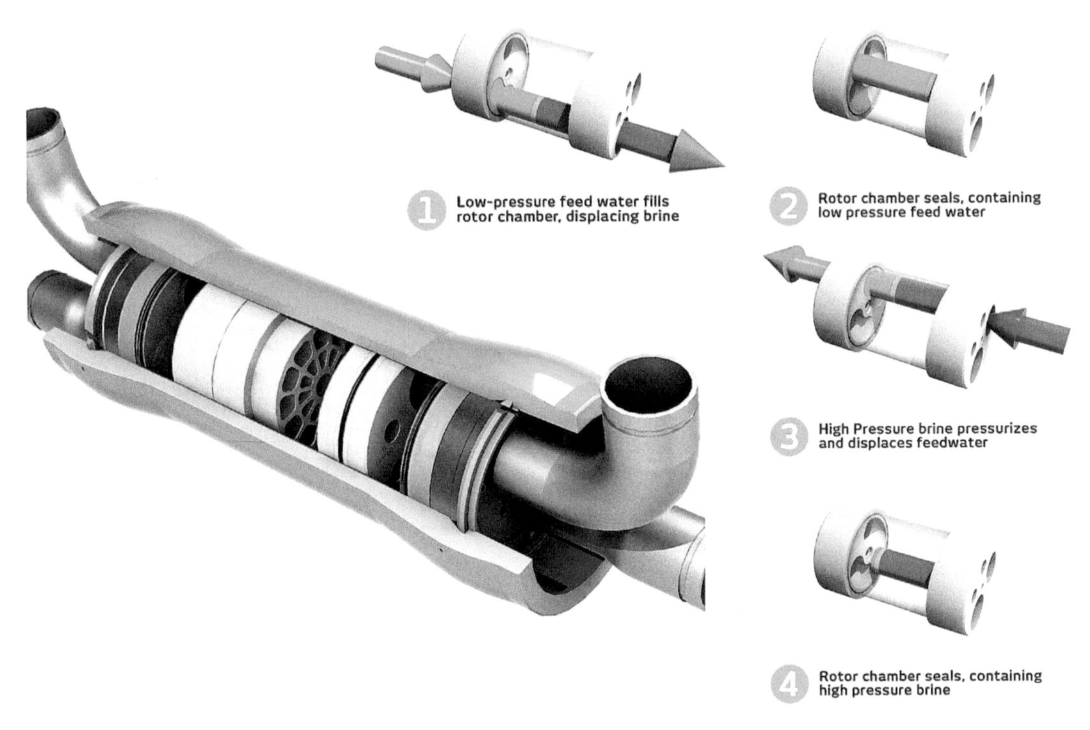

(a) Pressure exchanger (PX) energy recovery device – cut-away graphic and step diagram showing operation. (Courtesy of Energy Recovery, USA).

(b) and (c) Turbine or turbocharger energy recovery devices (Courtesy of Fluid Equipment Development Company LLC): (b) High pressure direct coupled (HPB™) unit for SWRO; (c) Low pressure pump and turbine (LPH) unit for brackish water, Orange County, California (Engineer: Black & Veatch).

PLATE 18

(a) Chlorine drum store with evaporator in the background. (Manufacturer: Portacel).

(b) Automatic drum shutdown system. (ChlorGuard).

(c) 4 × 680 kg/d ClorTec® on-site hypochlorite generation system, Pecos Road WTP, Chandler, Arizona. (Engineer: Black & Veatch; Manufacturer: Severn Trent Water Purification Inc.).

(a) Ozone generator with dielectrics. (Courtesy of Ozonia, USA).

(b) LPHO UV reactors (2 duty + 1 standby) disinfecting 227 Ml/d at the Lloyd W. Michael Water Treatment Plant (Cucamonga Valley Water District, Rancho Cucamonga, CA). (Engineer: Black & Veatch; Manufacturer: Trojan Technologies).

PLATE 20

(a)

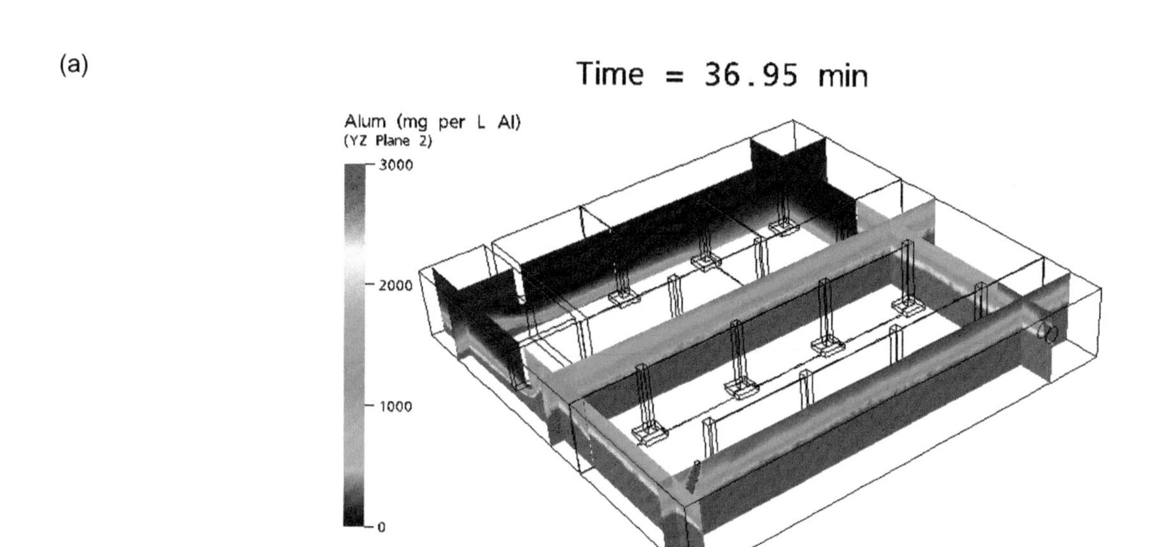

0.00 0.13 0.25 0.38 0.50

Velocity (Streamline 1)
(ms^ – 1)

(b)

0.00 0.30 0.60 0.90 1.20 1.50 1.80

Velocity [m s^-1]

Filename: Chestnut_PORT_04_01_001.res

(c)

CFD modelling (Black & Veatch):

(a) Contact tank pollution concentration.

(b) Chlorine contact tank flow – mixing and streamlines.

(c) Assessment of ozone contact efficiency for a post-ozone reaction tank with up and over baffles.

PLATE 21

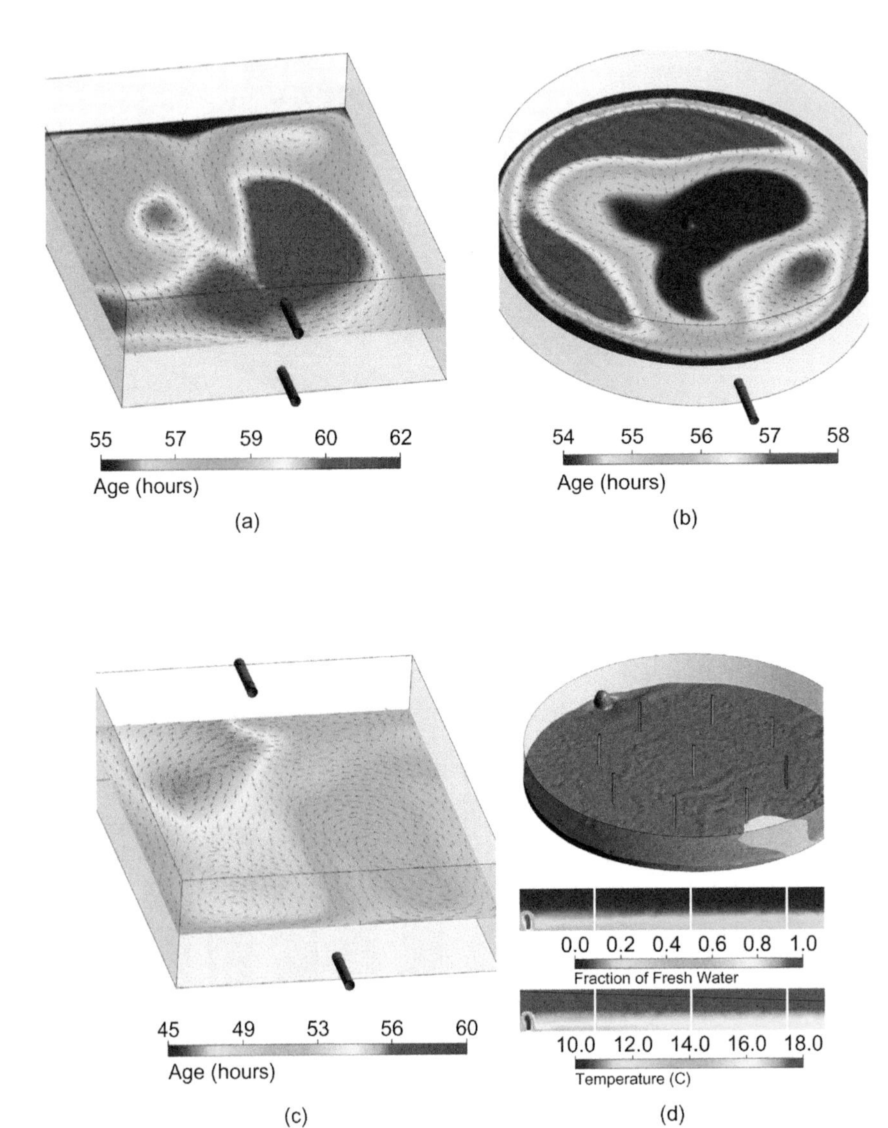

CFD modelling (Black & Veatch):

(a) Rectangular tank with high level inlet and low level outlet at same location – age of water at mid depth.

(b) Circular tank with high level central inlet and low level outlet – age of water at mid depth.

(c) Rectangular tank with high level inlet and low level outlet on opposite sides – age of water at mid depth.

(d) Circular tank with low level inlet and outlet – isotherm at tank base and cross sections of fresh water fraction and temperature.

PLATE 22

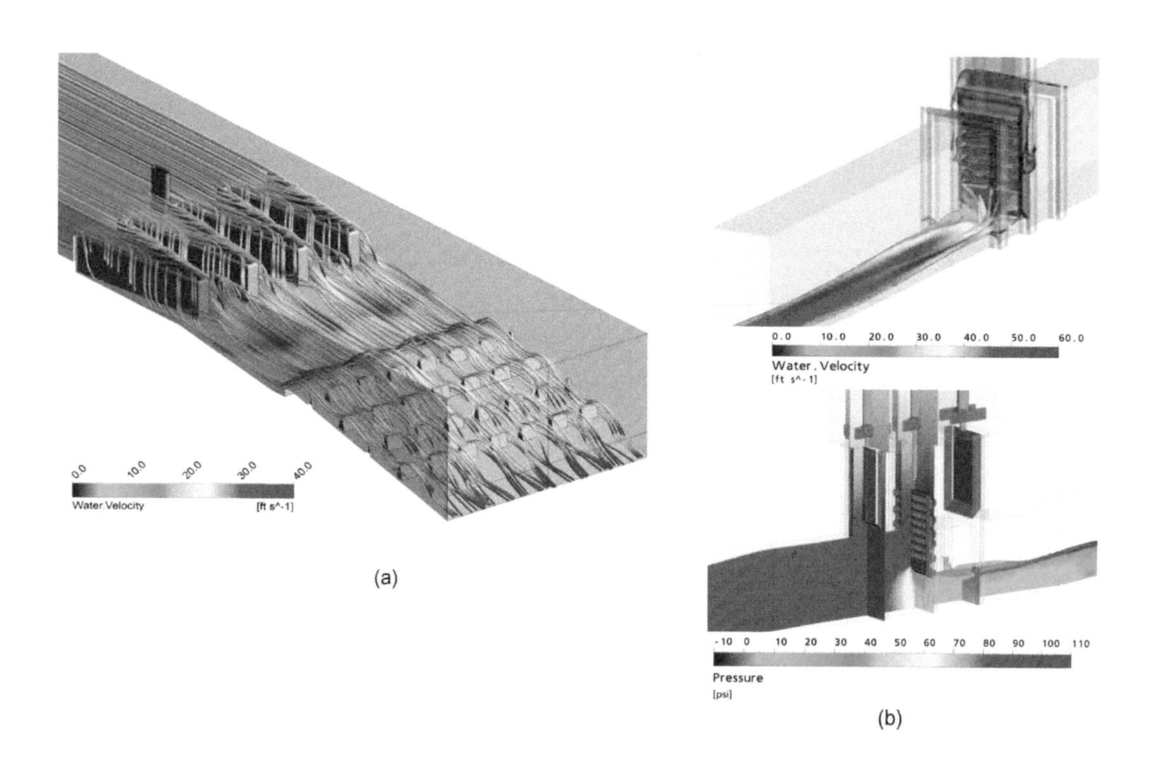

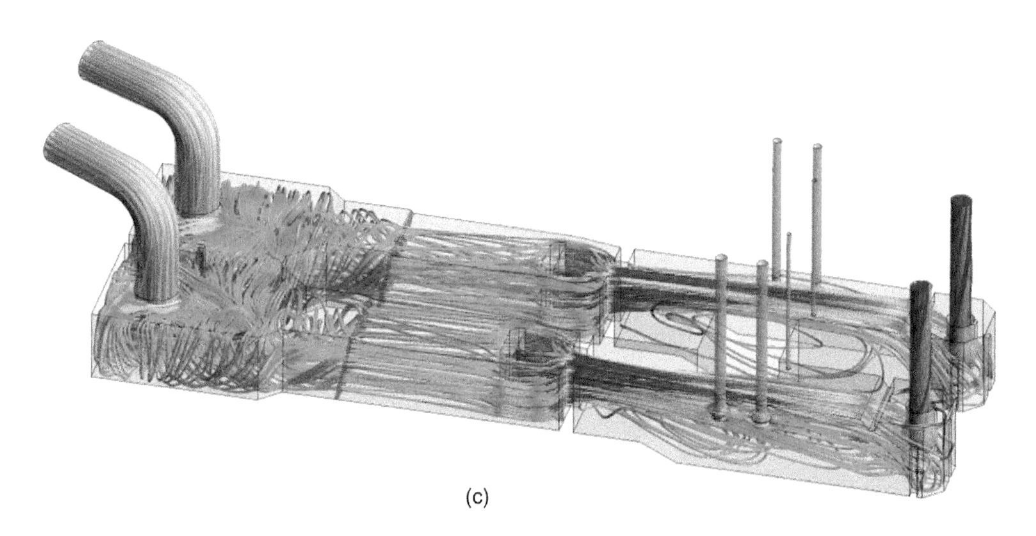

CFD modelling (Black & Veatch):
(a) Labyrinth weir with baffle blocks: velocity streamlines.
(b) Dynamic forces acting on hydraulic structure.
(c) Velocity streamlines in pump inlet structure.

PLATE 23

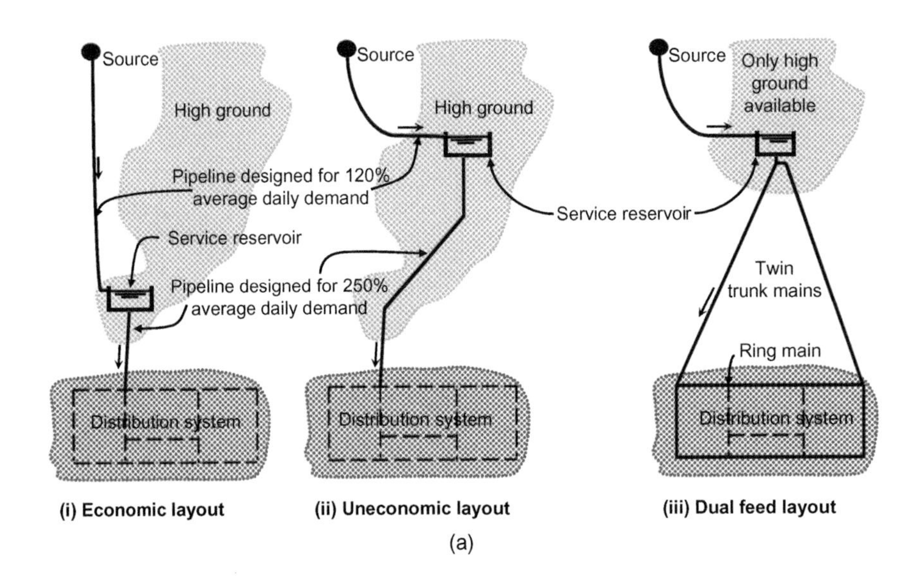

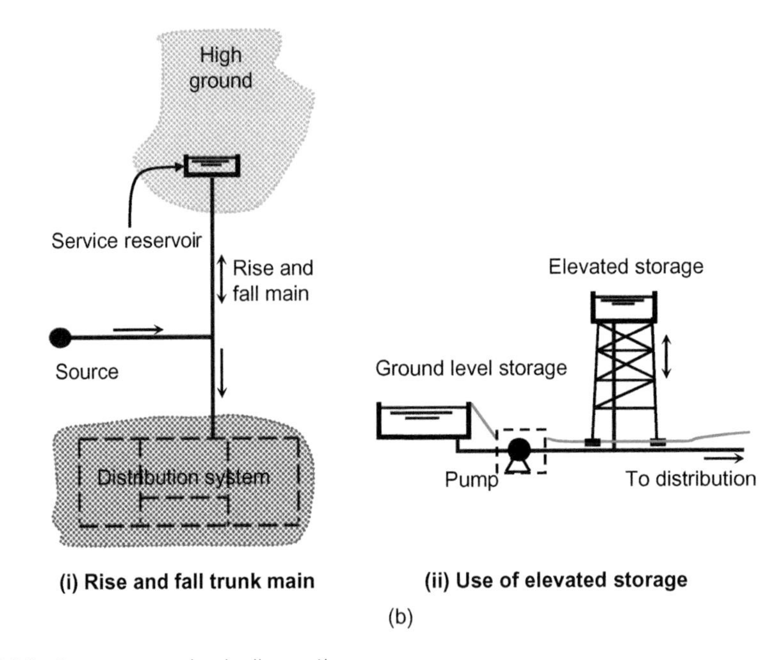

Distribution area supply pipeline options:
(a) Single trunk main and ring main feeders to distribution system.
(b) Rise and fall trunk main and elevated storage.

PLATE 24

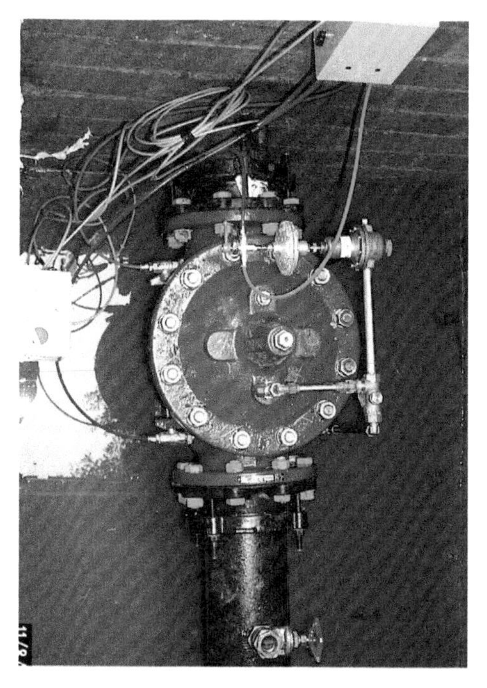

(a) Globe pressure reducing valve in chamber. (Black & Veatch).

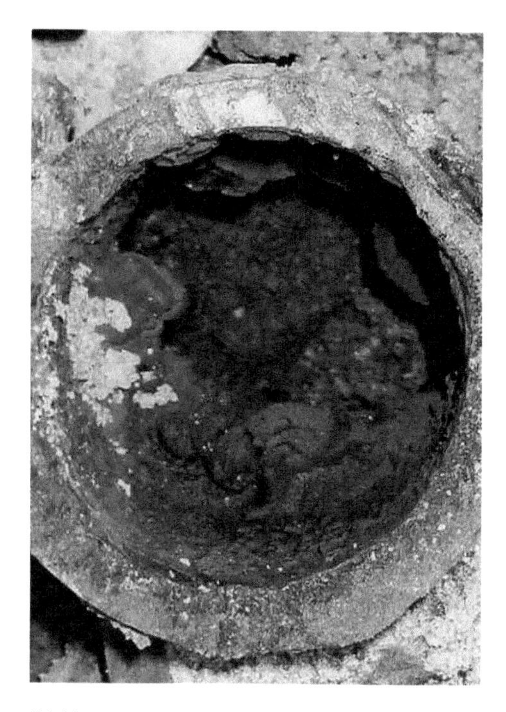

(b) Heavily tuberculated cast iron pipe. What is the value of pipe roughness?

(c) JD7™ sensor head.

(d) JD7™ sensor after insertion into pipeline.

PLATE 25

(a) Flushing a 100 mm diameter cast iron main. Debris collects in net for grading and analysis. Flow meter records the quantity of water used. (Black & Veatch).

(b) Exit pit – 757 mm OD PE still in tension and in its 'swaged' reduced diameter. (Black & Veatch).

(c) Directional drilling for a 350 mm OD pumping main. (Black & Veatch).

(d) Pipe jack under railway line for sleeve for 300 mm diameter trunk main. (Black & Veatch).

PLATE 26

(a) Twin 1000 mm diameter ductile iron pipe laying in Middle East. (Black & Veatch).

(b) 1524 mm diameter steel pipeline awaiting backfilling in Middle East. (Binnie Black & Veatch).

(c) 900 mm diameter GRP pipeline installation in Middle East – note attention to sidefill compaction. (Binnie Black & Veatch).

(d) Automatic butt fusion jointing of PE pipe. (Courtesy of The Fusion Group, Chesterfield, UK).

PLATE 27

(a) Metal seated gate valve. (Glenfield Valves Ltd, Kilmarnock, Scotland).

(b) Resilient seated DN 150 butterfly valve.(by courtesy of ERHARD GmbH & Co. KG (now part of TALIS Management Holding GmbH)).

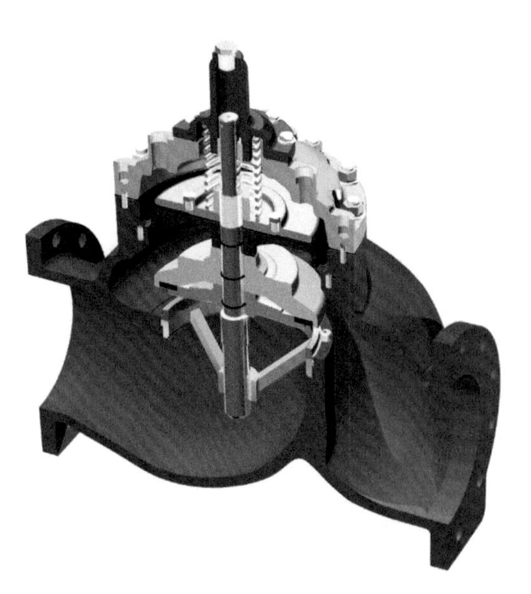

(c) Diaphragm sealed globe control valve – flow left to right. (Singer Valve Inc., British Columbia, Canada).

(d) Resilient seated (Multamed) gate valve. (by courtesy of ERHARD GmbH & Co. KG (now part of TALIS Management Holding GmbH)).

PLATE 28

(a) Eccentric plug valve (Cam-centric) (Val-Matic Valve & Mfg Corp., Illinois, USA).

(b) Larner Johnson (needle) valve – inlet 450 mm, throat 229 mm. (Blackhall Engineering Ltd., UK).

(c) Tilting disc check valve.(by courtesy of ERHARD GmbH & Co. KG (now part of TALIS Management Holding GmbH)).

(d) Double orifice 80 mm air valve (HiVent). (GA Valves Sales Ltd).

PLATE 29

(a) Electromagnetic district flowmeter installation. (Black & Veatch).

(b) District metering roadside cabinet. Control unit, data logger and pressure gauge. Telemetry and modem can also be installed. (Black & Veatch).

(c) Electromagnetic flow meter insertion probe with pressure transducer and data logger. (Black & Veatch).

(d) Semi-positive rotary piston meter (V100) for 15–40 mm supply pipes. (Manufacturer: Elster Metering Ltd, Luton, UK).

PLATE 30

(a) Vertically mounted close coupled centrifugal pump – 20 500 m^3/h @ 21 m lift. (Sulzer Pumps (UK) Ltd).

(b) Cut-away diagram of single stage double suction split case centrifugal pump. (Courtesy of Clyde Union Pumps, Glasgow (now part of SPX Corporation)).

(c) Horizontally mounted double suction centrifugal pumps – 1650 m^3/h @ 53 m lift. (Sulzer Pumps (UK) Ltd).

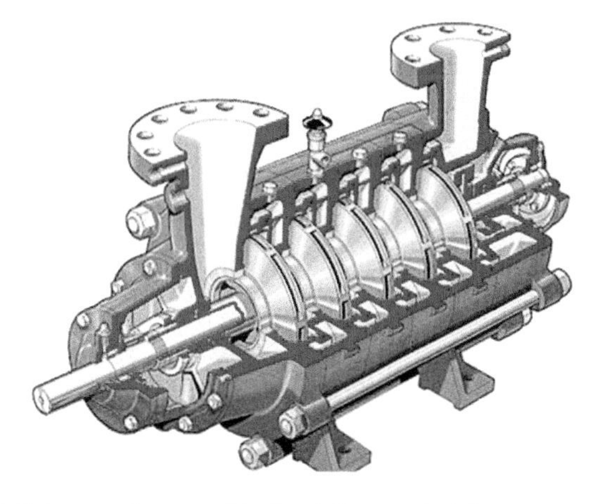

(d) Cut-away diagram of horizontally mounted five stage centrifugal pump. (Courtesy of Clyde Union Pumps, Glasgow (now part of SPX Corporation)).

PLATE 31

(a) Vertically mounted split casing centrifugal pumps – 900 m^3/h @ 119 m lift. (Sulzer Pumps (UK) Ltd).

(b) Two stage line shaft wet well mixed flow pumps – up to 40 000 m^3/h @ up to 100 m lift. (Courtesy of Clyde Union Pumps, Glasgow (now part of SPX Corporation)).

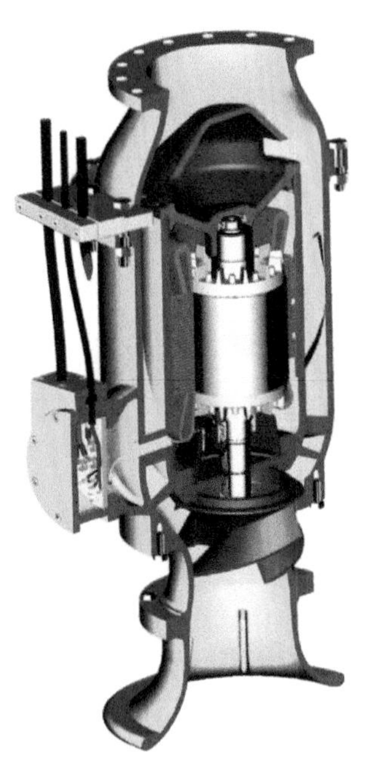

(c) Cut-away diagram of mixed flow submersible motor intake pump. (Bedford Pumps Ltd., UK).

(d) Submersible motor in-line booster pump – 1500 m^3/h @ 10 m lift. (Bedford Pumps Ltd., UK)

PLATE 32

(a) Oil insulated transformer with gravel filled sump.

(b) MCC – sections from left to right: incomer and spare; 4 × plant motor starters; cableway; 2 × pump starters; 2 × pump starters.

(c) ICA section of MCC with conventional hard-wired relay logic and PLC.

(d) ICA section of MCC with small PLCs and telemetry equipment.

PLATE 33

(a) Rectangular reinforced concrete monolithic service reservoir under construction in the UK. (Engineer: Binnie & Partners).

(b) 45 000 m^3 circular steel plate ground storage tanks under construction in the Middle East. (Engineer: Binnie & Partners).

(c) 2270 m^3 reinforced concrete water tower, Malaysia. (Engineer: Binnie & Partners in association with SMHB).

(d) 700 m^3 welded steel water tank on reinforced concrete tower, Malaysia. (Engineer: Binnie & Partners in association with SMHB).

PLATE 34

Printed and bound by CPI Group (UK) Ltd, Croydon, CR0 4YY
08/05/2025
01864857-0001